GEOSTATISTICS TRÓIA '92

Volume 1

Quantitative Geology and Geostatistics

VOLUME 5

Series Editor:

F. M. GRADSTEIN
Atlantic Geoscience Centre,
Bedford Institute of Oceanography,
Dartmouth, N.S., Canada

GEOSTATISTICS
TRÓIA '92

Volume 1

Edited by

AMILCAR SOARES
Centro de Valarizacao de Recursos Minerais,
Instituto Superior Técnico,
Lisbon, Portugal

Kluwer Academic Publishers
Dordrecht / Boston / London

Library of Congress Cataloging-in-Publication Data

Geostatistics Tróia '92 / edited by Amilcar Soares.
p. cm. -- (Quantitative geology and geostatistics ; v. 5)
"[Papers from] the Fourth Congress Troia '92"--Foreword.
Includes index.
ISBN 0-7923-2157-X (set : acid-free). -- ISBN 0-7923-2155-3 (v. 1 : acid-free). -- ISBN 0-7923-2156-1 (v. 2 : acid-free)
1. Geology--Statistical methods--Congresses. 2. Petroleum--Geology--Statistical methods--Congresses. I. Soares, Amilcar. II. Series.
QE33.2.S82G46 1993
550'.72--dc20 92-46924

ISBN 0-7923-2155-3 (vol. 1)
ISBN 0-7923-2156-1 (vol. 2)
ISBN 0-7923-2157-X (set)

Published by Kluwer Academic Publishers,
P.O. Box 17, 3300 AA Dordrecht, The Netherlands.

Kluwer Academic Publishers incorporates
the publishing programmes of
D. Reidel, Martinus Nijhoff, Dr W. Junk and MTP Press.

Sold and distributed in the U.S.A. and Canada
by Kluwer Academic Publishers,
101 Philip Drive, Norwell, MA 02061, U.S.A.

In all other countries, sold and distributed
by Kluwer Academic Publishers Group,
P.O. Box 322, 3300 AH Dordrecht, The Netherlands.

Printed on acid-free paper

TABLE OF CONTENTS

PETROLEUM

OTHER APPLICATIONS

ENVIRONMENT

HYDROLOGY

SOIL SCIENCES

NATURAL RESOURCES

HUMAN SCIENCES

MINING

FOREWORD

From Frascati to Troia

In the Seventeen years between the first International Geostatistics Institute in Frascati and the Fourth Congress Troia'92, the geostatistical community has grown in number and diversity and has reached recognition yet holding on its trademark. In recent years it has emerged from its original field of application in the mining industry to include applications in petroleum, oceanography, soil, entomological and environmental sciences. The "others" category of past congresses has grown into the most important, and had to be subdivided into many new sub classifications.

Troia'92 has been a most fruitful forum where more than 200 participants from 28 different countries challenged each other's understanding of a still fast moving theory and learnt from unexpected applications from remote fields. Troia'92 will be remembered as the Congress of consolidation when Geostatistics reached its place within the dynamics of the roaring scientific growth of this century end.

The Organizing Committee.

Margaret Armstrong
André Journel
Peter Dowd
Amilcar Soares (General Secretary)

Acknowledgements

TROIA'92 was organized by CVRM - Instituto Superior Técnico (Lisbon).

The Organizing Committee would like to thank the members of the Scientific Committee who had selected the abstracts and reviewed the papers. The quality of the final version of the articles is due to their conscientious work.

The Congress was made possible by the financial aid of the following institutions and companies which have sponsored the technical sessions:

- Commission of the European Communities - DGXII
- SCRF - Stanford Center for Reservoir Forecasting
- SOMINCOR
- SPE - Sociedade Portuguesa de Empreendimentos
- RTZ Corporation PLC
- COGEMA
- EDM - Empresa de Desenvolvimento Mineiro
- AFAIRS - Associação Universidade - Empresa para Formação no Âmbito da Industria e Recursos do Sub-solo.
- INIC - Instituto Nacional de Investigação Científica
- DGGM - Direcção Geral de Geologia e Minas
- FLA - Fundação Luso Americana

The Organizing Committee is deeply obliged to these institutions and companies.

The extraordinary success of the Congress would not have been possible without the dedication of the students of the CVRM - Instituto Superior Técnico.

LIST OF PARTICIPANTS

Abrahamsen, Peter, NORWEGIAN COMPUTING CENTER
Postbox 114 BLINDERN 0314 OSLO 3, NORWAY

Ajlan, Adel Hamoud, NIL
P.O. Box 345 Jeddah 21191, SAUDI ARABIA

Alabert, François, Societe Nationale ELF AQUITAINE SNEA(P), CSTJF
Avenue Larribau 64018 PAU CEDEX, FRANCE

Albuquerque, Alexandre, SPE, Sociedade Portuguesa de Empreendimentos, SA
Rua dos Fanqueiros, 12 - 2°, 1100 LISBOA, PORTUGAL

Albuquerque, Maria Teresa, INSTITUTO SUPERIOR TÉCNICO - Centro de Valorização de Recursos Minerais
Av. Rovisco Pais, 1096 LISBOA, PORTUGAL

Alfaro, Marco, CODELCO-CHILE
Division Chuquicamata Planification Mina CHUCHICAMATA, CHILE

Allard, Dennis, ÉCOLE DES MINES DE PARIS - Centre de Geostatistique
35, rue Saint-Honore 77305 Fontainebleau, FRANCE

Allison, Henry, CSIRO - DIVISION OF WATER RESOURCES
Private Bag PO Wembley WA 6014, AUSTRALIA

Almeida, Alberto, APPLIED EARTH SCIENCES DEPT. Stanford University
Stanford, CA 94305-2225, U.S.A.

Almeida, José António, INSTITUTO SUPERIOR TÉCNICO - Centro de Valorização de Recursos Minerais
Av. Rovisco Pais, 1096 LISBOA, PORTUGAL

Araktingi, Udo, CHEVRON OIL FIEL RESEARCH COMPANY A CHEVRON CORPORATION SUBSIDIARY
1300 Beach Boulevard P.O.Box 446 La Habra, CA 90633-0448, U.S.A.

Armstrong, Margaret, ECOLE NATIONAL SUPERIEURE DES MINES DE PARIS - Centre de Geostatistique
35, Rue Saint-Honore 77305 FONTAINEBLEAU, FRANCE

Attas, Abdallah Mohsen Al, NIL
P.O. Box 345 Jeddah 21191, SAUDI ARABIA

Baafi, Ernest, UNIVERSITY OF WOLLONGGONG
NSW 2500 Wollongong, AUSTRÁLIA

Barata, Teresa, INSTITUTO SUPERIOR TÉCNICO - Centro de Valorização de Recursos Minerais
Av. Rovisco Pais, 1096 LISBOA, PORTUGAL

Barua, Sukhendu L., ISLAND CREEK CORPORATION
250 West Main Street, P.O.Box 11430 LEXINGTON, KY 40575 - 1430, U.S.A.

Beckner, Bret, MOBIL RED
P.O. Box 819047, Dallas, U.S.A

Beier, Richard, CONOCO INC.,
PO BOX 1267, PONCACITY, OK 74603, U.S.A.

Beucher, Helene, ÉCOLE NATIONALE SUPERIEURE DES MINES DE PARIS Centre de Geostatistique
Ecole des Mines 35 rue Saint-Honore 77305 Fontainebleau, FRANCE

Beurden, Arthur Van, R.I.V.M.
Ant. van Leeuwenhoeklaan 9 P.O.Box 1 3720 BA Bilthoven (NL), THE NETHERLANDS

Bezerra, José Carlos Leal, ODEBRECHT
Alameda da Espatódias, 915 Pituba, Salvador-Bahia, BRAZIL

Bierkens, M.F.P., UNIVERSITY OF UTRECHT GEOGRAPHICAL INSTITUTE
P.O.Box 80.115, 3508 TC Utrecht, THE NETHERLANDS

Blaisé, Jean Réné, SOCIETE DES MINES DU BOURNEIX
BP 36 87500 St Yriex la Perche, FRANCE

Blanchin, Raymond, BRGM
BP 6009 45060 Orleans cedex 2, FRANCE

Bonaccorsi, Giusepi, ANG-ONG
Via XX Setembre, 75 Tredozio SO-zip 47019 - ITALY

Bongarçon, Dominique F., INCO EXPLORATION AND TECHNICAL SERVICES, INC.
1630 Paris ST #209 Sudbury, Ont. P3E 3CI, CANADA

Bortoli, Louis, APPLIED EARTH SCIENCES DEPT. Stanford University
Stanford, CA 94305-2225, U.S.A.

Bourgault, Gilles, ECOLE POLYTECHNIQUE MONTREAL
Dept. Genie Mineral, 5 etage C.P. 6079, CANADA H3C 3A7

Braga, Luis Paulo V., UFRJ
C.P. 2386, 20 001 - RIO DE JANEIRO, BRASIL

Braud, Isabelle, UNIVERSITE SCIENTIFIQUE TECHNOLOGIQUE ET MEDICALE DE GRENOBLE
Institut de Mecanique de Grenoble B P 53 X, 38041 Grenoble, FRANCE

Brito, Graça, INSTITUTO SUPERIOR TÉCNICO - Centro de Valorização de Recursos Minerais
Av. Rovisco Pais, 1096 LISBOA, PORTUGAL

Bruno, Roberto, UNIVERSITA DEGLI STUDI DI ROMA "LA SAPIENZA"
Dipartimento di Ingegneria Chimica, dei Materiali, delle Materie Prime Metall.
Via Eudossiana, 18 - 00184 ROMA, ITALY

Buxton, Bruce, BATTELLE
505 King Avenue Columbus, Ohio 43201-2693, U.S.A.

Camisani, F.A., ANGLOVAAL LIMITED
Anglovaal House 56, Main Street JOHANNESBURG 2001, SOUTH AFRICA

Carolino, Alexandra, PEGMATÍTICA
Soc. Mineira de Pegmatites Lda., Mesquitela, 3530 Mangualde, PORTUGAL

Carolino, B., PEGMATÍTICA
Soc. Mineira de Pegmatites Lda., Mesquitela, 3530 Mangualde, PORTUGAL

Carolino, José, PEGMATÍTICA
Soc. Mineira de Pegmatites Lda., Mesquitela, 3530 Mangualde, PORTUGAL

Carr, James R., UNIVERSITY OF NEVADA, Reno Department of Geological Sciences/168
M S 168 Reno, NV 89557-0138, U.S.A.

Carrasco, Pedro, CODELCO
Diego Portoles 510, Salvador, CHILE

Carrera, Jesus, DIT-UPC Dep. Ing. Terreno
UPC, Campus Nord, Mod D-2 08034, BARCELONA, SPAIN

Carter, L., ANGLOVAAL Ltd.
P.O.Box 62379 Marshalltown 2107, SOUTH AFRICA

Carvalho, Justino da Encarnação, PARTEX-CPS
Av. 5 de Outubro, 260, 1000 LISBOA, PORTUGAL

Castillo, Carlos L., INTEVEP, S.A
Sector el Tambor, Santa Rosa los tegues Edu Miranda, VENEZUELA

Caumartin, COGEMA
BP 4 78141 Velizy Villa Coublay, FRANCE

Chambel, Luís,SOCIEDADE PORTUGUESA DE EMPREENDIMENTOS
Rua dos Fanqueiros, 12, 5º 1100 lisboa - PORTUGAL

Chica-Olmo, Mario, UNIVERSIDAD DE GRANADA - Dept. Geodinamica
Av. da Fuentenueva, S/N 18071 GRANADA, SPAIN

Chilès, Jean-Paul, BRGM
BP 6009, 45060 Orleans Cedex 2, FRANCE

Clarici, Elena, IMPERIAL COLLEGE OF SCIENCE, Technology & Medicine
Royal School of Mines Prince Consort Road, London SW7 2BP, ENGLAND

Clark, Isobel, GEOSTOKOS LIMITED
Alloa Business Centre Whins Road Alloa FK10 3SA, Central Scotland

Coleou, Thierry, PETROSYSTEMS/CGG
2500 Wilcrest, suite 200, Houston, TEXAS, 77042, U.S.A.

Cortez, Leopoldo, INSTITUTO SUPERIOR TÉCNICO - Centro de Valorização de Recursos Minerais
Av. Rovisco Pais, 1096 LISBOA, PORTUGAL

Cressie, Noel A.C., IOWA STATE UNIVERSITY - Departments of Statistics
102E Snedecor Hall Ames IA 50011, U.S.A.

Dagbert, Michel, GEOSTAT SYSTEMS INTERNATIONAL INC.
800 Chomedey Blvd -Tower A -Suite 240 Laval, QUEBEC CANADA H7Y3YA, CANADA

Daly Colin, B.P RESEARCH CENTER
Chertsey Road Sunbury-on-thames, Middx TW 167 LN, ENGLAND

Daly, Mike, ESKOM FUEL AND WATER DIVISION
SOUTH AFRICA

Demange, Claude, COGEMA
BP 4 78141 Velizy Villa Coublay, FRANCE

Deutsche, Clayton, Exxon Production Research Company
P.O. Box 2189, Houston, Tx 77252-2189, U.S.A.

Dimitrakopoulos, Roussos, McGILL UNIVERSITY Dept. of Mining and Metallurgical Eng.
3480 University Street Montreal, P.Q., CANADA H3A 2A7

Donato, Matteo, COMMISSION OF EUROPEAN COMMUNITIES
Head of the Row Materials Programme DG XII/C-5 200 Rue dela Loi B-1049, Brussels, BELGIUM

Dosi, Paola, DIGITER, SRL
U. Roccadipapa, 00184 ROMA, ITÁLIA

Dowd, Peter, THE UNIVERSITY OF LEEDS - Dept. of Mining and Mineral Engineering
Leeds LS29JT, ENGLAND

Dubrule, Olivier,SOCIETÉ NATIONALE ELF-AQUITAINE
CSTJF, 64018 Pan Cedex, FRANCE

Durão, Fernando, INSTITUTO SUPERIOR TÉCNICO- Centro de Valorização de Recursos Minerais
Av. Rovisco Pais, 1096 LISBOA, PORTUGAL

Englund, Evan J., U.S.EPA
P.O.Box 93478 Las Vegas NV 89193-3478, U.S.A.

Felix, Manfred, GEOLOGICAL SURVEY OF SAXONY; GERMANY
Halsbrucker Str. 31 a 0 - 9200 Freiberg, GERMANY

Ferra, Pedro, INSTITUTO SUPERIOR TÉCNICO - Centro de Valorização de Recursos Minerais
Av. Rovisco Pais, 1096 LISBOA, PORTUGAL

Ferreira, Alfredo Virgílio, SOMINCOR
Apartado 12, 7780 Castro Verde, PORTUGAL

Fouquet, Chantal de, ÉCOLE NATIONALE SUPERIEURE DES MINES DE PARIS
Centre de Geostatistique
35, Rue Saint-Honoré, 77305 Fontainebleau, FRANCE

Freulon, Xavier, ÉCOLE NATIONALE SUPERIEURE DES MINES DE PARIS
Centre de Geostatistique
35, Rue Saint-Honoré, 77305 Fontainebleau, FRANCE

Froidevaux, Roland, FSS INTERNATIONAL
10 Chemin de Drize 1256 Troinex, SWITZERLAND

Galli, Alain, ÉCOLE NATIONALE SUPERIEURE DES MINES DE PARIS Ecole - Centre de Geostatistique
35, Rue Saint-Honoré, 77305 Fontainebleu, FRANCE

Garcia, Jorge Delgado, UNIVERSIDAD DE GRANADA - Dpto de Geodinamica
Av da. Fuentenueva s/n 18071 - GRANADA, SPAIN

Garcia, Michel, ÉCOLE NATIONAL SUPÉRIEURE DES MINES DE PARIS - Centre de Geostatistique
35, Rue Saint-Honoré, 77305 Fontainebleau, FRANCE

Gedler, Gustavo, INTEVEP, S.A
Vr. Santa Rosa, Sector el Tambor los Teques, VENEZUELA

Geiger, Janos, "MOL"RT-NKFV-SLKFI Nagyalfold Oil and Gas Production Co., P.O.Box 86 5000 Szolnok, HUNGARY

Georgsen, Frode, NORWEGIAN COMPUTING CENTER/SAND
Postbox 114 Blinden N-0314 OSLO 3, NORWAY

Gois, L., INSTITUTO SUPERIOR TÉCNICO - Centro de Valorização de Recursos Minerais
Av. Rovisco Pais, 1096 LISBOA, PORTUGAL

Gomez-Hernandes, Jaime, UNIVERSIDAD POLITECNICA DE VALENCIA Dep. Hidraulica y Medio Ambiente
46071 Valencia, SPAIN

Gonzaga, José Luís, INSTITUTO SUPERIOR TÉCNICO - Centro de Valorização de Recursos Minerais
Av. Rovisco Pais, 1096 LISBOA, PORTUGAL

Goovaerts, Pierre, UNIVERSITE CATOLIQUE DE LOUVAIN Faculte des Sciences Agronomiques
Place Croix du Sud, 2 Bte 16 Louvain-la-Neuve, BELGIUM

Gossa, Hedhili, SOCIETE ITALO TUNISIENNE D'EXPLOTATION PETROLIERE
Centre Urbain Nord B.P. 424 1080 Tunis Cedex , Tunis, TUNISIE

Goulard, Michel, INRA LABORATOIRE DE BIOMETRIE
INRA BP 91 84140 MONTFAVET, FRANCE

Gruyter, Joop J. de, WIMAND STARING CENTRE
Postbus 125 NL-6700 AC Wageningen, THE NETHERLANDS

Guardiano, Felipe, STANFORD UNIVERSITY, Dept. of Applied Earth Sciences
Stanford, CA 94305-2225, U.S.A.

Guimarães, Carlos, INSTITUTO SUPERIOR TÉCNICO - Centro de Valorização de Recursos Minerais
Av. Rovisco Pais, 1096 LISBOA, PORTUGAL

Guterres, Ana, INSTITUTO SUPERIOR TÉCNICO - Centro de Valorização de Recursos Minerais
Av. Rovisco Pais, 1096 LISBOA, PORTUGAL

Grzebyk, Michel, ÉCOLE NATIONAL SUPÉRIEURE DES MINES DE PARIS - Centre de Geostatistique
35, rue Saint-Honore, 77300 Fontainebleau, FRANCE

Haas, André, SOCIETE NATIONALE ELF AQUITAINE
SNEA(P), CSTJF Avenue Larribau 64018 PAU Cedex, FRANCE

Halleux, Isabelle, ISSP - INSTITUT SCIENTIFIQUE DE SERVICE PUBLIC
Rue du Chera, 200 B-4000, Liege 1, BELGIUM

Hektoen, A. L.- NORWEGIAN COMPUTING CENTER
P.B.Box 114 - Blindern 0314 Oslo, NORWAY

Heuvelink, G.B.M., UNIV. OF UTRECHT Geographical Institute
P.O.Box 80115, 3508 TC Utrecht, THE NETHERLAND

Hewett, Thomas A., STANFORD UNIVERSITY, Dept. of Applied Earth Sciences
Stanford California 94305-2220, U.S.A.

Hohn, Michael, WV GEOLOGICAL & ECONOMIC SURVEY
P.O.Box 879 Morgantown, WV 26507-0879, U.S.A.

Hollund, Knut Utne, NORSK HYDRO A.S.
P.O.Box 200 N-1321 STABEKK, NORWAY

Hu, Linying Y., ÉCOLE NATIONAL SUPÉRIEURE DES MINES DE PARIS - Centre de Geostatistique
35, rue St.Honore, 77305 Fontainebleau, FRANCE

Hudson, Gordon, MACAULAYLAND USE RESEARCH/NST.
Craigiebuckler Aberdeen AB9 2QJ, SCOTLAND

Jepsen, Jens Bach, STATOIL
Box 300, Forus N-4001 Stavanger, NORWAY

Jeulin, Dominique, ÉCOLE NATIONAL SUPÉRIEURE DES MINES DE PARIS - Centre de Geostatistique
35, rue St. Honore, 77305, Fontainebleau, FRANCE

Jimenez-Espinosa, Rosario, UNIVERSIDAD DE GRANADA - Dept.Geodinamica - Facultad de Ciencias
c/ Fuentenueva s/n 18071 Granada, SPAIN

Journel, André, STANFORD UNIVERSITY, Dept. of Applied Earth Sciences
Stanford, CA 94305-2225, U.S.A.

Kalkomey, Cynthia T., MOBIL RESEARCH & DEVELOPMENT
P.O.Box 819047 Dallas, TX 75381, U.S.A.

Kleingeld, W.J., ANGLO AMERICAN CORP.
P.O.Box A 1848, Craighall 2024, SOUTH AFRICA

Krige, Daniel - UNIV. OF THE WITWATERSRAND
P.O.Box 121, Florida Hills, 1716, SOUTH AFRICA

Kupfersberger, Hans, INSTITUT FUR HYDRAULIK GEWASSERKUNDE & WASSERWIRTSCHAFT
Technische Universitat Wien Karlplatz 13/223 A-1040 Wien, AUSTRIA

Laine, Eeva Liisa, HELSINKI UNIVERSITY OF TECHNOLOGY LABORATORY of Eng. Geology
Vuorimiehentie 2A 02150 ESPOO, FINLANDE

Lajos, Kovacs, HUNGARIAN GEOLOGICAL SURVEY
Stefania ut 14 Budapest H - 1143, HUNGARY

Langlais, Valerie, BP EXPLORATION INC.
5151 San Felipe P.O.Box 4587 Houston, TEXAS 77210 , U.S.A.

Lantuéjoul, Christian, ÉCOLE NATIONAL SUPÉRIEURE DES MINES DE PARIS - Centre de Geostatistique
35, Rue Saint-Honore, 77305 Fontainebleau, FRANCE

Lemmer, Carina
15 Chiselhurst Drive ROSSMORE 2092 Johannesburg, SOUTH AFRICA

Liao, Hong Tao, PETROSYSTEMS
1 rue Leon Migaux 91341 Massy Cedex, FRANCE

Lokosha, Osango, UNIVERSITE DE LAUSANNE - Institut de Mineralogie et Petrographi
UNIL BFSM2 1015 Lausanne, SUISSE

Luís, João, INSTITUTO SUPERIOR TÉCNICO - Centro de Valorização de Recursos Minerais
Av. Rovisco Pais, 1096 LISBOA, PORTUGAL

MacCarthy, Jane, BHP Tower - Bourke Place
600 BOURKE STREET, MELBOURNE, VICTORIA 3000, G.P.O.Box 86A, MELBOURNE, AUSTRALIA 3001

Magri, Ed, NCL Ingenierie y Construction S.A., Ricardo Matte Perez 479-485
Providência - Santiago - CHILE

Maignan, Michel, UNIVERSITE DE LAUSANNE
Mineralogie BFSM2 1015 Lausanne-Dorigny, SUISSE

Mandallaz, Daniel, ETH- Eidgenossische Technische Mochschule Zurich
ETH-Zentrum, HG E 12 CH-8092 Zurich, SUISSE

Marbeau, Jean Paul, PETROSYSTEMS/CGG
1, rue Leon Migaux 91302 MASSY Cedex, FRANCE

Marcotte, Denis, ECOLE POLYTECHNIQUE
Case Postale 6079 Sucursale A Montreal, Quebec H3C G47, CANADA

Meer, Freek van der, INTERNATIONAL INSTITUT FOR AEROSPACE SURVEY AND EARTH SCIENCES
350 Boulevard 1945 P.O.Box 6, 7500 AA Enschede, THE NETHERLANDS

Monestiez, Pascal, LABORATOIRE DE BIOMÉTRIE, INRA
B.P.91 84140 MONTFAVET, FRANCE

Mucha, Jacek, ACADEMY OF MINING AND METALLURGY
IHiGI AGH Al. Mickiewicza 30 30-059 KRAKOW, POLAND

Muge, Fernando, INSTITUTO SUPERIOR TÉCNICO - Centro de Valorização de Recursos Minerais
Av. Rovisco Pais, 1096 LISBOA, PORTUGAL

Munthe, Kristin, NORWEGIAN COMPUTING CENTER
P.O.Box 114-Blindern 0314 Oslo, NORUAY

Murillo, Alvaro, Calle 129 BIS, nº SA-26/56 Ap. 315
Santa Fe de Bogota, COLOMBIA S.A.

Murray, Christopher, STANFORD UNIVERSITY, Dept. of Applied Earth Sciences
Stanford, CA 94305-2225, U.S.A.

Myers, Donald, UNIVERSITY OF ARIZONA - Department of Mathematics
Tucson AZ 85721, U.S.A.

Nulsen, Marcus, WESTERN ATLAS INTERNATIONAL
455, London Road Isleworth, Middx, ENGLAND

Okx., J.P., TAWN INFRA CONSULT B.V.
P.O.Box 479 7400 AL DEVENTER, THE NETHERLANDS

Oliver, A., THE UNIVERSITY OF BIRMINGHAM School of Geography
Edgbaston BIRMINGHAM, B152TT U.K., ENGLAND

Omre, Henning, NORWEGIAN COMPUTING CENTER
Postbox 114 BLINDEN N-314 Oslo 3, NORWAY

Oosterveld, M.M., ANGLO AMERICAN CORP. OF SA LTD
P.O.Box 41848 2024 CRAIGHALL, SOUTH AFRICA

Pardo-Iguzquiza, Eulogio, UNIVERSIDAD DE GRANADA, FTAD.Ciencias - Dpto.Geodinamica
Av da. Fuentenueva, 18071 GRANADA, ESPANA

Parker, Harry M., MINERAL RESOURCES DEVELOPMENT
Suite 302 1710 South Amphlett Blud. San Mateo, California, 94546 U.S.A.

Patinha, Pedro, INSTITUTO SUPERIOR TÉCNICO - Centro de Valorização de Recursos Minerais
Av. Rovisco Pais, 1096 LISBOA, PORTUGAL

Pebesma, E.J., UNIVERSITY OF UTRECHT Geographical Institute
P.O.Box 80115, 3508 TC UTRECHT, THE NETHERLANDS

Pereira, H.G., INSTITUTO SUPERIOR TÉCNICO - Centro de Valorização de Recursos Minerais
Av. Rovisco Pais, 1096 LISBOA, PORTUGAL

Pereira, Maria João, INSTITUTO SUPERIOR TÉCNICO - Centro de Valorização de Recursos Minerais
Av. Rovisco Pais, 1096 LISBOA, PORTUGAL

Pilz, Jurgen, BERGAKADEMIC FREIBERG
Fachbereich Geowissenschaften G.-Zeuner-Str. 12 D - 9200 Freiberg (Sachs), GERMANY

Pina, Pedro, INSTITUTO SUPERIOR TÉCNICO - Centro de Valorização de Recursos Minerais
Av. Rovisco Pais, 1096 LISBOA, PORTUGAL

Posa, Donato, Dipartminento de Mathematica, IRMA
Via G.Fortunato Campus Universitario 70125 Bari, ITALY

Quental, Luís Ramos, PARTEX
Avenida 5 de Outubro, PORTUGAL

Raspa, Giuseppe, UNIVERSITA dI ROMA "LA SAPIENZA"
Dipartimento di Engegneria Chimica de Materiali, delle Materie Prime e Metallurgia
Via Eudossiana, 18, 00184 Roma, ITALY

Ravenne, GEOVARIANCES
Avenue Franklin Roosevelt 38, 77210 Avon, FRANCE

Ravenscroft, Peter J., RTZ TECHNICAL SERVICES LIMITED
P.O.Box 50, CASTLEMEAD LOWER CASTLE STREET BRISTOL BS99 7YR, UNITED KINGDOM

Remacre, Armando Zaupa, UNIVERSIDADE ESTADUAL DE CAMPINAS
C.P. 6152 - Instituto de Geociências - CAMPINAS, BRASIL

Renard, Didier, ÉCOLE NATIONAL SUPÉRIEURE DES MINES DE PARIS - Centre de Geostatistique
35, rue Saint-Honore, 77305 FONTAINEBLEAU, FRANCE

Ribeiro, Jorge, INSTITUTO SUPERIOR TÉCNICO - Centro de Valorização de Recursos Minerais
Av. Rovisco Pais, 1096 LISBOA, PORTUGAL

Ribeiro, Luís Tavares, INSTITUTO SUPERIOR TÉCNICO - Centro de Valorização de Recursos Minerais
Av. Rovisco Pais, 1096 LISBOA, PORTUGAL

Roditis, Yani, GOLD FIELDS MINING CO.
1687 Cole Blvd. Golden, CO, 80401, U.S.A. (3 books)

Rodrigues, João José, SOMINCOR - Sociedade Mineira de Neves Corvo, SARL
Apartado 12, 7780 CASTRO VERDE, PORTUGAL

Rogado, José Quintino, INSTITUTO SUPERIOR TÉCNICO - Centro de Valorização de Recursos Minerais
Av. Rovisco Pais, 1096 LISBOA, PORTUGAL

Rohrlich, Vera, TECHNION Geotechnology
Technion City 32000 Haifa, ISRAEL

Rossi, Mario, MINERAL RESOURCES DEVELOPMENT
1710 So. Amphlett Blvd. Suite 302 San Mateo, CA, 94402, U.S.A.

Rossini, Claudia, AGIP S.P.A.
Via Fabiani n 1 S.Donato M. Milano , ITALY

Roth, Chris, BRGM
BP 6009, 45060 Orleans Cedex 2, FRANCE

Rovellini, Marco, AGIP S.P.A.
Via Fabiani n°1 S.Donato M. Milano, ITALY

Salguero-Perez, Antonio, UNIVERSIDAD DE GRANADA/IAGM
Dpto. Geodinamica / AUDA. Fuentenueva S/N, 18071 Granada, SPAIN

Samper, Javier, UNIVERSITAT POLITECNICA DE CATALUNYA
Departamento de Ingenieria del Terreno y Cartografica c/Gran Capitan S/N
Modulo 2 08034 BARCELONA, SPAIN

Sanchez-Vila, Xavier, UNIVERSITAT POLITECNICA DE CATALUNYA
Gran Capitan, s/n Modulo D-2 Barcelona 08034, SPAIN

Schad, Hermann, INSTITUT FUR WASSERBAU
University of Stuttgard, GERMANY

Schaffrin, Burkhard, OHIO STATA UNIVERSITY Depart.of Geodetic Science Surveying
The Ohio State University 1958 Neil Ave., Columbus/Ohio 43210, 43219, U.S.A.

Schafmeister, Maria-Th., FREE UNIVERSITY BERLIN Institute for Applied Geology
Wichernstr. 16 D - 1000 BERLIN 33 F.R.G., GERMANY

Schamber, Jennifer, MICHIGAN TECHNOLOGICAL UNIVERSITY
1400 Townseno Dr. Houghton MI 49931, U.S.A.

Schofield, Neil, FSS INTERNATIONAL INC.
P.O.Box 657 Epping NSW 2121, AUSTRALIA

Seguret, Serge, ÉCOLE NATIONAL SUPÉRIEURE DES MINES DE PARIS - Centre de Geostatistique
35, Rue St. Honore, 77305 Fontainbleau, FRANCE

Silva, A. Costa e, PARTEX, cps
Av. 5 de Outubro, 160, 1000 LISBOA, PORTUGAL

Silva, Emídio, University of Roma
Via Eudossiana, 18-00184 ROMA, ITÁLIA

Silva, Fernando, PARTEX
Av. 5 de Outubro, Lisboa, PORTUGAL

Silva, Francisco José, SOMINCOR - Sociedade Mineira de Neves Corvo, SARL
Apartado 12, 7780 CASTRO VERDE, PORTUGAL

Simard, Yvan, DEPT.FISH.& OCEANS INSTITUT MAURICE-LAMONTAGNE
C.P. 1000 MONT-JOLI QUÉBEC G5H 3Z4, CANADA

Soares, Amílcar O., INSTITUTO SUPERIOR TÉCNICO - Centro de Valorização de Recursos Minerais
Av. Rovisco Pais, 1096 LISBOA, PORTUGAL

Sonnet, Philippe, UNIVERSITE CATHOLIQUE DE LOUVAIN
Laboratoire de Geologie et Mineralogie 3, Place L. Pasteur 1348 Louvain-a-Neuve, BELGIUM

Soulié, Michel, UNIVERSITE DE MONTREAL Ecole Polytechnique
Section Geotechnique CP 6079, Succ A. Montreal (Qc) H3C 3A7, CANADA

Sousa, António Jorge de, INSTITUTO SUPERIOR TÉCNICO - Centro de Valorização de Recursos Minerais
Av. Rovisco Pais, 1096 LISBOA, PORTUGAL

Srivastava, Mohan, FSS INTERNATIONAL
#24 The Creekhouse 1551 Johnston Street Vancouver, B.C., CANADA V6H 3R9

Steffens, Francois E., UNIVERSITY OF SOUTH AFRICA Statistics Dept.
P.O.Box 392 0001 PRETORIA, SOUTH AFRICA

Stein, Alfred, AGRICULTURAL UNIVERSITY Dept. Soil Science and Geology
P.O.Box 37 6700 AA Wageningen, THE NETHERLANDS

Suro-Perez, Vinicio,PEMEX
Av.Marina Nacional 320, Torre Ejecutiva, Piso 32 11311 México DF, MÉXICO

Taheri, Mojtaba, INTEVEP, S.A.
P.O.Box 76343 Caracas 1070-A, VENEZUELA

Tavchandjian, Oliver, Université of Quebeque in Chicoutimi
555 Boul. Université, Chicoutimi, 67H-3R5, QUEBEC, CANADA

Tercan, A.E., UNIVERSITY OF LEEDS Mining and Mineral Engineering
Leeds LS2 9JT, ENGLAND

Thadani, Suresh G., STOCHASTIC SYSTEMS, INC.
1524 Mockingbird Plano, Texas 75075, U.S.A.

Tjelmeland, Hakon, NORWEGIAN COMPUTING CENTER
Postbox 114 Blindern 0314 OSLO, NORWAY

Tran, Thomas, SHELL DEVELOPMENT CO.
P.O.BOX 481 HOUSTON, TX 77001, U.S.A

Travassos, José Filipe, INSTITUTO SUPERIOR TÉCNICO - Centro de Valorização de Recursos Minerais
Av. Rovisco Pais 1096 LISBOA, PORTUGAL

Tyler, Kelly, Dept. for Res. Tech. STATOIL
PO Box 300 N-4001 Stavanger, NORUAY

Vairinho, Margarida, MINISTÉRIO DA INDÚSTRIA E ENERGIA
Rua Diogo Couto nº1 2º, Lisboa, PORTUGAL

Valente, Jorge, CONSULTOR
R. Martin Francisco, nº331, sp. 402, 30-430, BELO HORIZONTE, BRASIL

Varsanyi, Iren, INSTITUTE OF MINERALOGY
H - 6701 Szeged Pf 651, HUNGARY

Verly, Georges, B.P RESEARCH CENTER
Chertsey Road Sunbury-on-thames Middlesex TW 167 LN, ENGLAND

Verrezen, Lucien, GENMIN
P.O.Box 1404 Roosevelt Park Johannesburg 2129, SOUTH AFRICA

Vieler, Jonathan, RIO TINTO MANAGEMENT SERVICES
P.O.Box 784467 SANDTON 2146, SOUTH AFRICA

Vicente, Yolanda, SOCIEDADE PORTUGUESA DE EMPREENDIMENTOS
Rua dos Fanqueiros n°12 5°, 1100 Lisboa, PORTUGAL

Wartenberg, Daniel, DEPARTMENT OF ENVIRONMENTAL AND COMMUNITY MEDICINE
UMDNJ-Robert Wood Johson Medical School Piscataway NJ 08854, U.S.A.

Weber, Dennis D., UNIVERSITY OF NEVADA Environmental Research Center
4505 Maryland Parkway LAS VEGAS, NV 89154, U.S.A.

Webster, Richard, ETH Zurich Institut fur terrestrische Okologie
Grabenstrasse 3 8952 Schlieren, SWITZERLAND

Werovszkyné, Veronika Pipic, "MOL"RT-NKFV Nagyalfold Oil and Gas Production Co., P.O.Box 86 5000 Szolnok, HUNGARY

Yfantis, E.A., UNIVERSITY OF NEVADA Computer Science Department
LAS VEGAS, NV 89154, U.S.A.

Zeissler, Karl-Otto, GEOLOGICAL SURVEY OF SAXONY
Halsbrucker Str. 31 a 0 - 9200 Freiberg, GERMANY

Zepic, Franc, INSTITUT ZA GEOLOGIJO, Geotehniko in Geofiziko
Geoloski Zavod Ljubljana Dimiceva 14 61000 LOJUBLJANA SLOVENIA, YOUGOSLAVIE

Formatting and Integrating Soft Data: Stochastic Imaging via the Markov-Bayes Algorithm

Hua Zhu* and André G. Journel
Stanford University, CA 94305, U.S.A.

Abstract

Under a Bayesian framework both hard and soft data can be encoded as local prior (also called pre-posterior) probability distributions. These local prior distributions are then updated into posterior distributions using the nearby hard and soft data. The posterior distributions provide models of uncertainty prevailing at unsampled locations.

The Markov-Bayes algorithm for such updating can be seen as a cokriging capitalizing on spatial correlations between prior distribution values. A Markov-type assumption stating that hard data prevail over collocated soft data, allows determination of the hard-soft data coregionalization model through some simple calibration parameters. This alleviates the burden of structural modeling, the main obstacle in the practice of cokriging.

Stochastic imaging of the phenomenon can be performed by drawing sequentially and repetitively from these posterior probability distributions. Joint spatial uncertainty about the phenomenon can then be assessed through those alternative images.

A case study using a topographic elevation derived-data set (the Walker Lake data set) is included. The case study shows that the Markov-Bayes algorithm succeeds in reproducing the spatial distribution and spatial structure of the attribute under consideration. The influence of soft data is seen to be substantial, especially when hard data is sparsely sampled with respect to soft data and when there is a good calibration correlation between hard and soft data.

1 Introduction

In earth science, precise measurements (hard data), such as given by core samples, are often scarce. These precise measurements, however, are often supplemented by a large amount of indirect measurements (soft data). For example, in a petroleum reservoir, hard data on porosity from wells are usually sparse, but geological interpretation, geophysical data, seismic profiles, sample data from a similar geological environment, or expert experience can provide additional information about the spatial distribution of porosity. These soft data are typically more abundant and usually much less expensive and more accessible than hard well data. Therefore, integration of soft data in petrophysical modeling is important, and could be particularly critical whenever there is a severe lack of well data.

*presently with Shell Development Co., Houston, U.S.A.

To integrate soft data, calibration is essential. Typically, only the regression (most often linear) statistics are extracted from the calibration, providing a regression-type relationship between hard and soft data variables. There are cases, where there is much more information to be collected than a mere regression. For example, shown in Figure 1 is a scatter plot of two variables $Z-V$, where Z is the variable of primary interest, and V is the soft data variable. On this scatter plot, not only the regression of Z on V is non-linear but the corresponding residual variance is seen to be strongly heteroscedastic, i.e. dependent on the V value: low V-values appear to be much better correlated to Z values than high V-values. If the estimate of Z derived from V is to be informed with a measure of accuracy, then the observed dependence of the residual variance on the V-values must be modeled and exported together with the regression curve. The entire conditional distribution of Z given classes of V needs to be extracted and exported. These conditional distributions may be widely different for different classes of V values, both in terms of variance and distributional shape (see Figure 1).

At any spatial location x_α where the attribute Z is not sampled but a V-value is available, the conditional distribution is extracted from the calibration scatter plot:

$$Prob\{Z(x_\alpha) \leq z | v(x_\alpha) \in [v_j, v_{j+1}]\} \tag{1}$$

This distribution represents the "prior" information available about the unsampled value $Z(x_\alpha)$. This prior information (prior distribution) will have to be updated by any other neighboring Z and/or V information into the "posterior" distribution

$$Prob\{Z(x_\alpha) \leq z | \text{ all information}\} \tag{2}$$

representing the state of knowledge about $Z(x_\alpha)$.

The objective of this paper is to provide a model of such posterior distribution taking into account all information. First, a Bayesian coding of information is presented; then the Markov-Bayes model for updating prior distributions is illustrated. Finally, a case study applying the Markov-Bayes model is included.

2 A Bayesian coding of information

Under a Bayesian framework, all data (hard or soft) can be put under the format of local prior distributions. Consider the stationary random function (RF) $Z(x)$ representing the primary variable of interest, where x represents the vector of spatial coordinates. Any prior local information about the value $z(x_\alpha)$ at a particular location x_α is modeled by a "local prior" or "pre-posterior" distribution of the random variable $Z(x_\alpha)$ at that location. Denote this local prior distribution by the variable $y(x_\alpha; z)$:

$$y(x_\alpha; z) = Prob\{Z(x_\alpha) \leq z | \text{ local information at } x_\alpha\} \tag{3}$$

$y(x_\alpha; z)$ is a scalar probability valued in $[0, 1]$; it represents the local information available at x_α prior to any correction or updating based on neighboring information, hence the qualifier "local prior " or "pre-posterior" used in Bayesian jargon (Berger, 1985)

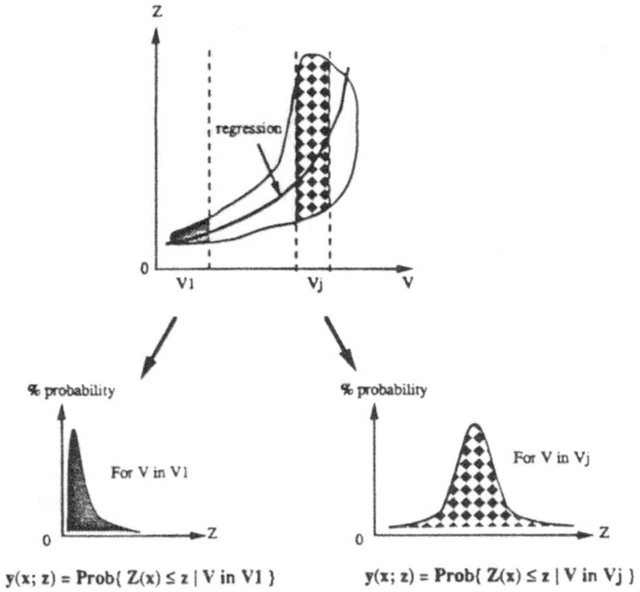

Figure 1: Calibration scattergram of Z vs. V

A hard datum $z(x_1)$ is but a particular local prior cdf (cumulative distribution function) $y(x_1; z)$ with zero-variance, i.e. zero-uncertainty. The corresponding cdf is a unit-step function of the threshold z:

$$\text{Hard data local prior: } y(x_1; z) = \begin{cases} 0, & \text{for all } z < z(x_1) \\ 1, & \text{for all } z \geq z(x_1) \end{cases} \tag{4}$$

A constraint interval $z(x_2) \in (a_2, b_2]$ is coded as an incomplete unit-step function where the step location $z(x_2)$ is unknown in the scalar interval $(a_2, b_2]$, i.e.

$$y(x_2; z) = \begin{cases} 0, & \text{for all } z < a_2 \\ \text{undefined for } z \in [a_2, b_2] \\ 1, & \text{for all } z > b_2. \end{cases} \tag{5}$$

Local indirect information, originating from a secondary variable datum $v(x_3)$ may be calibrated into a local prior cdf of type (1) (Journel and Zhu, 1990):

$$y(x_3; z) = Prob\{Z(x_3) \leq z | v(x_3)\} \quad \in [0, 1] \tag{6}$$

Similarly, geological interpretation may indicate that at location x_3 a particular facies or rock type r prevails. Ancillary information may provide the probability distribution of the attribute Z within that rock type r, say: $F(z|r)$. The local prior cdf at x_3 is then identified to that distribution:

$$y(x_3; z) = Prob\{Z(x_3) \leq z \mid r \text{ at } x_3\} = F(z|r) \tag{7}$$

Another example of soft information would correspond to an expert guess about the possible value $z(x_4)$ at location x_4. That guess may result into a triangular probability density function (pdf) with most probable value at $c_4 \in (a_4, b_4]$ and zero-probability for any value $z(x_4)$ outside the interval $(a_4, b_4]$.

Where there is no local prior information, the pre-posterior cdf would reduce to the common global cdf $F(z)$ as obtained from the prior decision of stationarity.

The local prior cdf's $y(x;z)$ take the form of vectors of probability value, one vector for each location x with up to K members $y(x;z_k), k = 1, \ldots, K$, if K threshold values z_k are retained to discretize the z-interval of variability. Prior cdf data $y(x;z_k)$ are, thus

- either hard bits, valued 0 or 1, for hard data;
- or, fuzzy bits, valued between 0 and 1, for soft data;
- or, missing bits, as in the case of constraint intervals;

3 A Bayesian updating of local prior distributions: The Markov-Bayes formalism

To provide a model of uncertainty at location x, the local prior cdf $y(x;z)$ must be updated using all neighboring information, that is the other prior cdf's $y(x_\alpha;z)$ available at neighboring locations. This updating would result in a usually different cdf, called the posterior distribution:

$$Prob\{Z(x) \leq z | \text{ All information available}\} \tag{8}$$

The local prior distributions $y(x;z)$ can be interpreted as realizations of a random function $Y(x;z)$. Then, regression algorithms can be used to update a y-value from other neighboring y-values:

$$Prob^*\{Z(x) \leq z|(n)\} - F(z) = \sum_{\alpha=1}^{n_1} \lambda_\alpha [i(x_\alpha : z) - F(z)] + \sum_{\beta=1}^{n_2} \nu_\beta [y(x'_\beta; z) - E\{Y(z)\}] \tag{9}$$

where $F(z)$ is the stationary cdf (global prior), $E\{Y(z)\}$ is the stationary expectation of the soft variable, $i(x_\alpha;z)$ are the prior distributions (hard bits) generated by the hard data at n_1 locations x_α, and $y(x'_\beta;z)$ are the priors generated by soft information at n_2 other locations x'_β. The short notation "$|(n)$" means "conditional to all information available at location x and its neighborhood". The weights λ_α and ν_β are given by a generalized linear regression system.

$$\begin{bmatrix} C_I(x_1 - x_1) & \cdots & C_I(x_1 - x_{n_1}) & C_{IY}(x_1 - x'_1) & \cdots & C_{IY}(x_1 - x'_{n_2}) \\ \vdots & \vdots & \vdots & \vdots & \vdots & \vdots \\ C_I(x_{n_1} - x_1) & \cdots & C_I(x_{n_1} - x_{n_1}) & C_{IY}(x_{n_1} - x'_1) & \cdots & C_{IY}(x_{n_1} - x'_{n_2}) \\ C_{YI}(x'_1 - x_1) & \cdots & C_{YI}(x'_1 - x_{n_1}) & C_Y(x'_1 - x'_1) & \cdots & C_Y(x'_1 - x'_{n_2}) \\ \vdots & \vdots & \vdots & \vdots & \vdots & \vdots \\ C_{YI}(x'_{n_2} - x_1) & \cdots & C_{YI}(x'_{n_2} - x_{n_1}) & C_Y(x'_{n_2} - x'_1) & \cdots & C_Y(x'_{n_2} - x'_{n_2}) \end{bmatrix} \cdot \begin{bmatrix} \lambda_1 \\ \vdots \\ \lambda_{n_1} \\ \nu_1 \\ \vdots \\ \nu_{n_2} \end{bmatrix} = \begin{bmatrix} C_I(x - x_1) \\ \vdots \\ C_I(x - x_{n_1}) \\ C_{IY}(x - x'_1) \\ \vdots \\ C_{IY}(x - x'_{n_2}) \end{bmatrix} \tag{10}$$

For each cutoff z, the regression (9) requires three covariance models, the auto-covariances of the two RF models $I(x;z)$ and $Y(x;z)$ and their cross-covariance, i.e. C_I, C_Y and C_{IY} in (10). The inference of so many covariance models is impractical. Fortunately, under a Markov-type assumption, considerable simplification can be achieved and a coregionalization model of hard-soft data variable can be conveniently established.

The Markov approximation states that "hard data always prevail over (screen) any collocated soft data". This amounts to hypothesizing the following relations:

$$\begin{cases} P\{I(x+h;z)|I(x;z)=i,\ \ Y(x;z)=y\} = P\{I(x+h;z)|I(x;z)=i\} \\ P\{Y(x;z)=y|I(x;z)=i, Y(x+h;z)=y'\} = P\{Y(x;z)=y|I(x;z)=i\} \\ P\{Y(x;z)=y|I(x;z)=i, I(x+h;z)=j\} = P\{Y(x;z)=y|I(x;z)=i\} \\ \qquad \text{for } i,j=0,1, \quad \forall y \in [0,1] \quad \text{and} \quad \forall h \end{cases} \tag{11}$$

Consider the calibration parameters:

$$\begin{aligned} m^{(1)}(z) &= E\{Y(x;z)|I(x;z)=1\} \in [0,1] \\ m^{(0)}(z) &= E\{Y(x;z)|I(x;z)=0\} \in [0,1] \\ B(z) &= m^{(1)}(z) - m^{(0)}(z) \in [0,1] \\ \sigma^{2(1)}(z) &= Var\{Y(x;z)|I(x;z)=1\} \\ \sigma^{2(0)}(z) &= Var\{Y(x;z)|I(x;z)=0\} \end{aligned}$$

which can be evaluated from a calibration scatter plot. The following relations can be established (Zhu, 1991):

- The mean of $Y(x;z)$ is, from its definition:

$$E\{Y(x;z)\} = E[Prob\{Z(x) \leq z|v\}] = F(z) = E\{I(x;z)\} \tag{12}$$

- The variance of $Y(x;z)$ is:

$$Var\{Y(x;z)\} = E\{Y^2(x;z)\} - [E\{Y(x;z)\}]^2 \tag{13}$$

  Interpreting $Y(x;z)$ as a function of $V(x)$, as in

$$y(x;z) = P\{Z(x) \leq z|v(x)\} = q\{v(x);z\},$$

  it follows that:

$$\begin{aligned} E\{Y^2(x;z)\} &= \int_{-\infty}^{\infty} q^2\{v(x);z\} \cdot f_V(v)dv \\ &= \int_{-\infty}^{\infty} q\{v(x);z\} \cdot P\{Z(x) \leq z, V(x)=v\}dv \\ &= \int_{-\infty}^{\infty} q\{v(x);z\} \cdot P\{V(x)=v|Z(x) \leq z\} \cdot F(z)dv \\ &= F(z) \cdot m^{(1)}(z) \end{aligned} \tag{14}$$

  finally:

$$\begin{aligned} Var\{Y(x;z)\} &= F(z)m^{(1)}(z) - F^2(z) = F(z)[1-F(z)] \cdot B(z) \\ &= B(z) \cdot Var\{I(x;z)\} \end{aligned} \tag{15}$$

- Cross-covariance between hard and soft data $C_{IY}(h;z)$:

$$C_{IY}(h;z) = E\{I(x+h;z)\cdot Y(x;z)\} - E\{I(x+h;z)\}\cdot E\{Y(x;z)\} \quad (16)$$

developing the joint probability as:

$$P\{I(x+h;z)=1, Y(x;z)=y\} = P\{I(x+h;z)=1, Y(x;z)=y|I(x;z)=1\}\cdot F(z)$$
$$+P\{I(x+h;z)=1, Y(x;z)=y|I(x;z)=0\}\cdot[1-F(z)]$$

and accounting for the Markov-approximation (11):

$$E\{I(x+h;z)\cdot Y(x;z)\} = K_I(h;z)[m^{(1)}(z) - m^{(0)}(z)] + F(z)\cdot m^{(0)}(z)$$

where $K_I(h;z) = E\{I(x;z)I(x+h;z)\} = C_I(h;z) + F^2(z)$. Then:

$$C_{IY}(h;z) = B(z)\cdot[K_I(h;z) - F^2(z)] = B(z)\cdot C_I(h;z) \quad (17)$$

- Auto-covariance of the soft data $C_Y(h;z)$:

Similar to the development of the cross-covariance of the hard-soft data, the auto-covariance of the soft data $C_Y(h;z)$ can be derived by first developing the joint probability of $[Y(x;z), Y(x+h;z)]$. Accounting for the Markov approximations, it can then be established

$$C_Y(h;z) = Cov\{Y(x;z), Y(x+h;z)\} = B^2(z)\cdot C_I(h;z), \quad \text{for } h>0. \quad (18)$$

Under the Markov approximation, the auto-covariance of the soft data $C_Y(h;z)$ and the cross-covariance between the hard and soft data $C_{IY}(h;z)$ can be deduced directly from the corresponding indicator covariance $C_I(h;z)$. However, notice that there is a discontinuity at the origin of the auto-covariance of the soft data, since

$$\lim_{h\to 0^+} C_Y(h;z) = B^2(z)C_I(0) < B(z)C_I(0) = C_Y(0)$$

The Markov-Bayes model of hard-soft coregionalization is positive semi-definite, similar to the linear model of coregionalization. Using the Markov-Bayes model, the inference required by the cokriging estimate (9) reduces to that of the indicator covariances $C_I(h;z)$, equivalent to that required by the multiple indicator kriging.

4 Updating of the soft data: An illustration

Consider the posterior distribution at location x where there is a soft datum $y(x)$, and at a distance vector h away, there exists a hard datum $i(x+h)$. The posterior distribution at x is, from system (9):

$$Prob^*\{Z(x)\le z|(n)\} - F(z) = \lambda_1[i(x+h;z) - F(z)] + \nu_1[y(x;z) - E\{Y(z)\}] \quad (19)$$

where λ_1 and ν_1 are determined by system (10) which is written

$$\begin{bmatrix} C_I(0) & C_{IY}(h) \\ C_{IY}(h) & C_Y(0) \end{bmatrix} \cdot \begin{bmatrix} \lambda_1 \\ \nu_1 \end{bmatrix} = \begin{bmatrix} C_I(h) \\ C_{IY}(0) \end{bmatrix} \tag{20}$$

Then:

$$\lambda_1 = \frac{C_Y(0)C_I(h) - C_{IY}(0)C_{IY}(h)}{C_Y(0)C_I(0) - C_{IY}^2(h)} = \frac{[1-B]C_I(0)C_I(h)}{C_I^2(0) - BC_I^2(h)} = \frac{[1-B]\rho_I(h)}{1-B\rho_I^2(h)} \tag{21}$$

and

$$\nu_1 = \frac{C_I(0) - \lambda_1 C_I(h)}{C_I(0)} = 1 - \lambda_1 \rho_I(h) = \frac{1-\rho_I^2(h)}{1-B\rho_I(h)} \tag{22}$$

From the results above, it is clear that the posterior distribution $Prob^*\{Z(x) \leq z|(n)\}$ is different from the prior soft distribution $y(x;z)$. In other words, the local prior distribution $y(x;z)$ is updated by the hard datum $i(x+h;z)$. Notice when $B=1$, i.e. the soft datum is equivalent to a hard datum, hence $\lambda_1 = 0$ and $\nu_1 = 1$. Any hard datum is honored exactly by the Markov-Bayes algorithm.

5 The Walker Lake Case Study

The exhaustive Walker Lake data base (Isaaks and Srivastava, 1989) is a digital terrain related data set consisting of 78000 data locations on a regular grid of 260×300 nodes, informed for two variables denoted as Z and V. The variable Z represents the smoothness of the topography within a spatial area and V represents the average elevation of the same area. In this case study, Z is considered as the primary variable of interest, and V is the secondary variable (a source of soft data).

- The calibration data set:

  A calibration data set of 224 data is sampled from the exhaustive data base for both variables. Ten cutoffs are retained for both the hard data variable Z and the soft data variable V, corresponding to, roughly, the deciles of the calibration sample histogram of each variable. The prior local cdf's,

  $$y(x;z_c|V) = Prob\{Z(x) \leq z_c|V(x) \in \text{class } v_j\}$$

  are derived directly from the calibration scattergram and are summarized in Table 1. The calibration parameters $B(z_k)$ for each of the nine threshold values $z_k, k = 1,\ldots,9$, are presented in Table 2.

  Also from the calibration data, the indicator variograms are calculated and modeled. Those indicator variograms, together with the calibration parameters provide all the structural information needed for the determination of posterior cdf's using the Markov-Bayes procedure.

Table 1: Conditional distributions of $Z(x)$ given that $V(x)$ falls into one of ten classes:

		z_k									
		0.38	3.83	15.00	31.16	61.10	120.77	220.55	459.39	834.76	z_{max}
	8.67	0.773	1.000	1.000	1.000	1.000	1.000	1.000	1.000	1.000	1.000
	44.71	0.182	0.591	0.818	0.909	0.955	1.000	1.000	1.000	1.000	1.000
	116.15	0.044	0.391	0.609	0.783	0.913	0.957	1.000	1.000	1.000	1.000
	172.97	0.000	0.000	0.273	0.546	0.682	0.864	0.955	1.000	1.000	1.000
v_j	224.46	0.000	0.000	0.182	0.500	0.727	0.818	0.909	1.000	1.000	1.000
	297.53	0.000	0.000	0.091	0.091	0.364	0.591	0.727	0.864	0.909	1.000
	399.33	0.000	0.000	0.046	0.136	0.273	0.409	0.682	0.864	0.955	1.000
	520.21	0.000	0.000	0.000	0.000	0.087	0.348	0.522	0.739	0.957	1.000
	692.64	0.000	0.000	0.000	0.000	0.000	0.000	0.191	0.524	0.810	1.000
	v_{max}	0.000	0.000	0.000	0.000	0.000	0.000	0.000	0.044	0.391	1.000

Table 2: Calibration parameters $B(z_k)$

k	1	2	3	4	5	6	7	8	9
Cutoff z_k	0.38	3.83	15.00	31.16	61.10	120.77	220.55	459.39	834.76
$B(z_k)$	0.593	0.695	0.590	0.604	0.582	0.575	0.553	0.553	0.374

- The conditioning data set:

  Out of the 78,000 Z-V reference data, only 30 Z-sample values located on a regular grid are retained as conditioning hard data. In addition, a larger sample of 585 V-data on a pseudo-regular grid is retained to provide soft information about the primary Z-attribute.

Stochastic imaging via the Markov-Bayes algorithm:

Using the 30 hard data and the 585 soft data as conditioning data set, the entire Walker Lake area, comprising $260 \times 300 = 78000$ grid nodes, is simulated applying the Markov-Bayes procedure for determination of the local posterior distributions.

Figure 2 gives two realizations of the simulated class values for the primary variable Z using both hard and soft data. Comparing these images to the exhaustive data base shown in Figure 3(a), it appears that the major features of the Walker Lake are reproduced. The continuous lake area of low values and the mountain range of high values are well reproduced on both simulations. However, the two images do differ locally; this difference provides a measure of spatial uncertainty.

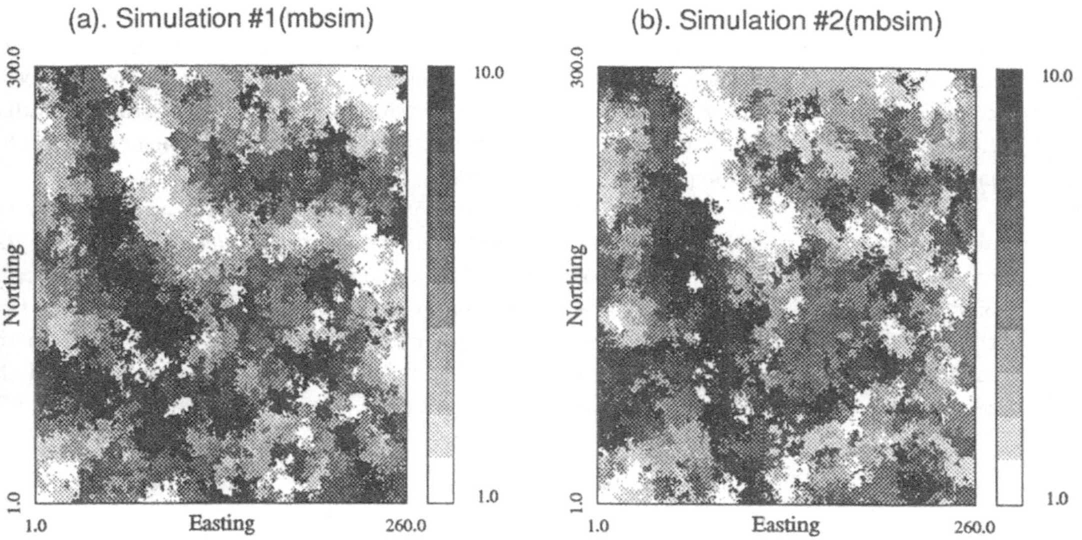

Figure 2. Two Markov-Bayes simulated realizations

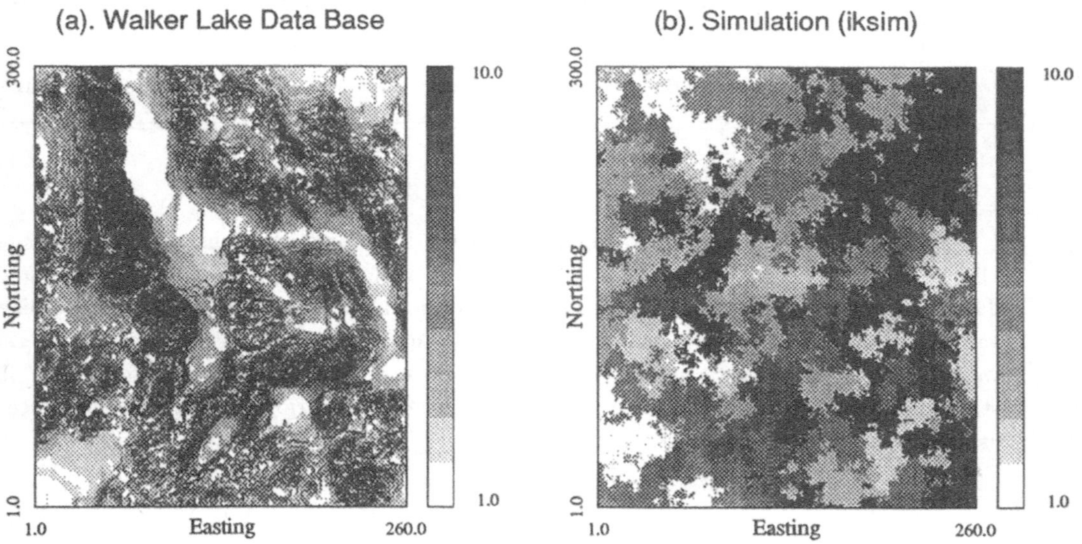

Figure 3. (a) The exhaustive Walker Lake data, (b) a simulation using only the 30 hard data

To evaluate the influence of the soft data in the mapping of the primary variable Z, Figure 3(b) shows a simulation realization using only the 30 hard data. This simulation is equivalent to multiple indicator simulation (Journel, 1989) using only these hard data. With only 30 conditioning hard data, the simulation image fails to reflect the major features of the Walker Lake area. This indicates that in presence of sparse hard data, using soft data may considerably improve the characterization of the primary variable.

Table 3 gives the proportions of values falling in each of the 10 classes as defined by the nine cutoffs z_k retained, for the two Markov-Bayes simulation realizations (MBSIM), the 224 calibration data, the original 30 conditioning hard data, the multiple indicator simulation using only the 30 hard data (IKSIM), and the true exhaustive Walker Lake data base. It is seen that the MBSIM realizations reproduce correctly the proportion of each class as defined from the calibration data set, whereas the IKSIM realization reproduces the class proportions of the conditioning 30 hard data. This is as one would expect.

Table 3: Proportion of each class

Class k		1	2	3	4	5	6	7	8	9	10
Upper bound z_k		0.38	3.83	15.00	31.16	61.10	120.77	220.55	459.39	834.76	z_{max}
	N					Proportion in each class					
Cond. data	30	0.1000	0.0667	0.1000	0.1333	0.1000	0.1333	0.0667	0.0667	0.0667	0.1667
IKSIM(12)	78000	0.0828	0.0772	0.0786	0.1228	0.0938	0.1231	0.0679	0.0650	0.1019	0.1869
Calib. data	224	0.0982	0.0982	0.1027	0.0982	0.1027	0.0982	0.0982	0.1027	0.0982	0.1027
MBSIM(12)	78000	0.1118	0.1296	0.0900	0.0928	0.0918	0.0765	0.0755	0.1289	0.1112	0.0919
MBSIM(1234)	78000	0.1237	0.1174	0.0957	0.0969	0.0967	0.0869	0.0652	0.1191	0.1073	0.0911
True	78000	0.1200	0.0913	0.1144	0.0933	0.0908	0.0973	0.0876	0.1185	0.0951	0.0917

To check the reproduction of the indicator spatial structures, the indicator variograms, at all cutoffs, for the realization of Figure 2(a) are compared to the model indicator variograms. This comparison is presented in the form of $\gamma-\gamma$ plot in Figure 4, in which the two standardized variograms are plotted against each other for all lag distances up to 20 in north-south and east-west directions. A perfect reproduction of the model variogram would correspond to the 45^0 straight line at all cutoffs. It appears that the model indicator variograms are well reproduced, especially at the smaller lags.

6 Conclusions

Hard and soft data, once encoded as local prior distributions, can be readily incorporated into the modeling of spatial distributions. This makes it possible to merge data from different sources and different qualities prior to their common processing. The Markov-Bayes procedure

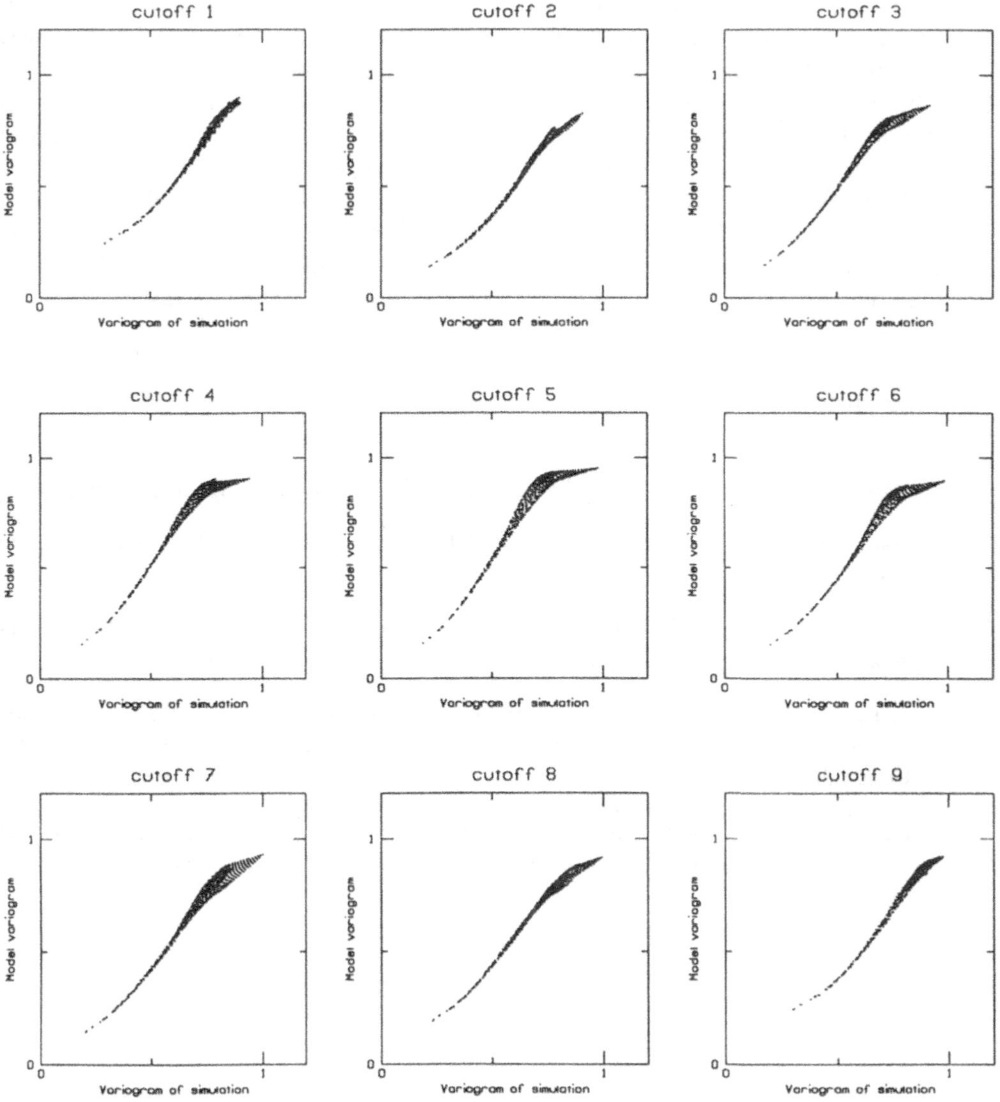

Figure 4. Comparison of the standardized experimental indicator variograms with the input models, for the Markov-Bayes simulated realization #1 shown in Figure 2(a).

provides a simple, easy way to derive local posterior distribution models, hence, stochastic images of the spatial phenomenon under study. It distinguishes hard data from soft data by considering different models of spatial continuity. This allows a full updating of local prior information as provided by soft data into a posterior probability distribution accounting for nearby hard and soft information. The model inference called by the Markov-Bayes procedure is equivalent to that involved in a multiple indicator kriging. The case study presented shows that the Markov-Bayes procedure succeeds in reproducing the spatial distribution and spatial structure of the primary variable. The influence of soft data in the characterization of the primary variable is seen to be quite substantial. This is especially true when hard data are sparse, and there is a good correlation between the hard and the soft variables.

References

1. Berger, J.O. (1985) Statistical Theory and Bayesian Analysis, Springer Verlag, 617 p.

2. Isaaks, E.H. and Srivastava, R.M. (1989) Introduction to Applied Geostatistics, Oxford Press, 561 p.

3. Journel, A.G. (1989) Fundamentals of Geostatistics in Five Lessons, Short course in geology, v.8, American Geophysical Union Press, Washington D.C., 40 p.

4. Journel, A.G. and Zhu, H. (1990) "Integrating soft seismic data: Markov-Bayes updating, an alternative to cokriging and traditional regression", SCRF (Stanford Center for Reservoir Forecasting), Report 3.

5. Zhu, H. (1991) Modeling Mixture of Spatial Distributions with Integration of Soft Data, Ph.D Thesis, Department of Applied Earth Sciences, Stanford University, p.310

COMPARISON OF DISJUNCTIVE KRIGING TO GENERALIZED PROBABILITY KRIGING IN APPLICATION TO THE ESTIMATION OF SIMULATED AND REAL DATA

James R. Carr[1] and Nai-hsien Mao[2]

[1] Department of Geological Sciences, University of Nevada, Reno Nevada, 89557 U.S.A.

[2] Lawrence Livermore National Laboratory, P.O. Box 808, L-208, University of California, Livermore, California 94551 U.S.A.

Abstract. Disjunctive kriging has been compared previously to multigaussian kriging and indicator cokriging for estimation of cumulative distribution functions; it has yet to be compared extensively to probability kriging. Herein, disjunctive kriging and generalized probability kriging are applied to one real and one simulated data set and compared for estimation of the cumulative distribution functions. Generalized probability kriging is an extension, based on generalized cokriging theory, of simple probability kriging for the modeling of conditional distributions. The disjunctive kriging and the generalized probability kriging give similar results for simulated data of normal distribution, but differ considerably for real data with non-normal distribution.

1. INTRODUCTION

Disjunctive kriging (c.f., Matheron, 1976a,b; Armstrong and Matheron, 1986 a,b) and probability kriging (c.f., Isaaks, 1984; Sullivan, 1985) are interpolation methods capable of returning estimates of the cumulative distribution functions (cdfs); another such method is multigaussian kriging (Verly, 1983). Probability kriging and disjunctive kriging have yet to be extensively compared. Disjunctive kriging has been directly compared to multigaussian kriging (Verly, 1983) in which multigaussian kriging was found to yield more consistent results over all cutoffs. Probability kriging has been directly compared to multigaussian kriging (Verly and Sullivan, 1985) wherein the two methods were found to perform equally well. Other related studies are that of Carr and Deng (1987), in which indicator cokriging and disjunctive kriging are directly compared for the estimation of the cdfs (this study showed that the indicator approximation of the cdfs is similar to the disjunctive kriging of the cdfs), and Hu (1989), which presents analyses comparing disjunctive kriging (two different models) to indicator kriging.

Herein, generalized probability kriging theory is presented based on generalized cokriging theory (c.f., Myers, 1982) to extend simple probability kriging to the joint estimation of the indicator and uniform transforms. This method is compared to disjunctive kriging in application to one real and one simulated data set to identify similarities and differences between these models.

2. THEORY OF GENERALIZED PROBABILITY KRIGING

Equations for simple probability kriging are found in Isaaks (1984) and Sullivan (1985). These equations are simply extended for generalized probability kriging by using the theory of generalized cokriging (c.f., Myers, 1982). This extension is described in detail in Carr and Mao (1992).

With probability kriging, simple or generalized, the cdfs are approximated using indicator functions developed for specific cutoffs, Z_k. Often, and throughout the remainder of this discussion, the cutoffs, Z_k, are chosen at the quantiles of a set of data. Once the cutoffs are chosen, the indicator function is assigned as:

$$i_k(x) = 0, \text{ if } Z(x) > Z_k; \quad i_k(x) = 1, \text{ otherwise.} \tag{1}$$

Additionally, the uniform transform is developed as (e.g., Isaaks, 1984):

$$u(x_i) = \sum_{i=1}^{R_i} w_i \tag{2}$$

where R_i is the rank order of datum, $Z(x_i)$, and w_i are declustering weights (c.f., Journel, 1983; Deutsch, 1989).

Once the indicator and uniform transforms are developed, generalized probability kriging proceeds using N nearest neighboring sample locations as:

$$Z_c^*(x_0) = \sum_{i=1}^{N} \Gamma_i \overline{Z_c}(x_i) \tag{3}$$

where

$$\overline{Z_c^*}(x_0) = (i_c^*(x_0),\ u^*(x_0));\ \overline{Z_c}(x_i) = (i_c(x_i),\ u(x_i));\ \sum_{i=1}^{N} \Gamma_i = I.$$

From the estimates of the indicator function over all cutoffs, an estimate of the expected value of Z is developed as (c.f., Isaaks, 1984):

$$E^*[Z(x_0)] = \sum_{j=2}^{K} g_j [i_j^*(x_0) - i_{j-1}^*(x_0)] \quad (4)$$

where K is the number of cutoffs and:

$$g_j = \sum_{i=1}^{NK} w_i Z(x_i) \; / \sum_{i=1}^{NK} w_i$$

where $Z(x_i)$ is an element of $[Z(x_{j-1}), Z(x_j)]$ and NK is the number of values, Z(x), in this interval.

2a. *Cross-covariance models based on paired-sum semi-variograms.*

A distinction is drawn between generalized probability kriging (as implemented herein) and simple probability kriging (c.f., Isaaks, 1984; Sullivan, 1985). Herein for generalized probability kriging, cross-covariance models between an indicator function, i, and the uniform transform, u, are developed using covariances based on semi-variograms of the paired-sum, i + u, i alone, and u alone such that:

$$CCOV_{iu}(h) = 0.5(COV_{i+u}(h) - COV_i(h) - COV_u(h))$$

There are several advantages to modeling cross-covariance in this manner over modeling cross-covariance directly, such as has been used with simple probability kriging (see, for example, Isaaks (1984)). First, provided that COV_{i+u}, COV_i, and COV_u are modeled using positive, semi-definite functions, the model for $CCOV_{iu}$ is positive, semi-definite because a linear combination of positive, semi-definite models is also positive, semi-definite. Second, the three covariance parameters used to compute $CCOV_{iu}$ are developed from standard semi-variogram models, such as spherical; this is convenient.

2b. *Software Used in This Study for Generalized Probability Kriging*

COKRIG (Carr, Myers and Glass, 1985) is modified to perform generalized probability kriging. This modification involves: 1) adding cell declustering using a modified version of the computer algorithm given in Deutsch (1989); 2) adding automatic indicator function assignment based on specified cutoffs; 3) adding the automatic development of the uniform transform; 4) adding a loop over the number of cutoffs to solve Eq. (3) for each cutoff; and 5) adding the capability to solve Eqs. (4) and (5).

3. THEORY OF DISJUNCTIVE KRIGING

This section is basically that presented in Carr, Deng, and Glass (1986) based on references cited therein. When applying disjunctive kriging to the estimation of raw data, Z, the first step is the transformation of Z to normally distributed data, Y, having a mean of zero and unit variance; Hermite orthogonal polynomials are often used for this purpose, such that:

$$Z(x) = \phi(Y(x)) = \sum_{N=0}^{K-1} C_N \chi_N(Y(x)) \tag{5}$$

where K is the order of the Hermite polynomial expansion, the C_N are coefficients of this expansion, and χ is the Hermite polynomial. Once this transformation is complete, the conditional probability that $Z(x_0)$ is greater than a cutoff value, Z_k, is given as:

$$P[Z(x_0) \geq Z_k] = 1 - G(Y_0) - g(Y_0) \sum_{N=1}^{K} \frac{\chi_N - 1}{N!} [\sum_{\beta=1}^{M} \lambda_N^\beta \chi_N(Y_\beta)] \tag{6}$$

where G and g are, respectively, the distribution and density of Y. In this expression, M is the number of nearest neighboring data locations used to obtain an estimate, and the weights, λ, are obtained by solving the following expression:

$$\sum_{j=1}^{M} \rho_{ij}^N \lambda_N^j = \rho_{0i}^N, \quad N = 1, 2, \ldots, K; \quad i = 1, 2, \ldots, M \tag{7}$$

in which, using the semi-variogram for Y:

$$\rho(h) = 1 - \frac{\gamma_Y(h)}{\sigma_Y^2} \tag{8}$$

where the semi-variogram values for Y are normalized by the variance of Y because, once the Hermitian transform is applied to Z to obtain Y, the variance of Y is approximately equal to one, but rarely exactly equal to one.

Finally, an estimate, $Z^*(x_0)$, is computed in the disjunctive kriging as:

$$Z^*(x_0) = \sum_{N=0}^{K-1} C_N \sum_{\beta=1}^{M} \lambda_N^\beta \chi_N(Y_\beta) \tag{9}$$

In disjunctive kriging, the cumulative distribution function is represented by a single covariance function for the transform data, Y. This distinguishes disjunctive kriging from simple and generalized probability kriging as previously described.

3a. *Software Used in This Study for Disjunctive Kriging*

HNORM and DKNEW (Kim, Myers and Knudsen, 1977) are used to perform, respectively, the Hermitian transformation of Z to Y and the disjunctive kriging. HNORM is modified to replace its original sorting algorithm with a heap sorting algorithm from Press and others (1986); this change is made to allow HNORM to sort x and y coordinates along with data values, to facilitate the creation of an output file for use in the disjunctive kriging.

4. COMPARISON OF GENERALIZED PROBABILITY KRIGING AND DISJUNCTIVE KRIGING FOR THE ESTIMATION OF REAL DATA

A real data set given in Isaaks (1984) is chosen to evaluate the two models. These data are lead values measured adjacent to the Trinity River in Dallas, Texas (U.S.A.). Statistical attributes of these data are as follows: mean = 428.67; median = 158.70; variance = 994,000.00, and the coefficient of skew = 6.54; these data have a non-normal distribution. Please note that one very unusual value, 10,400.0, occurs in this data set and was left in the data set for the following analysis; no data filtering was implemented for the subsequent comparison.

The following steps are implemented:

1) in preparation for disjunctive kriging, the data, Z, are transformed to normal data, Y, using a Hermitian model of order 13; the quality of this model is shown in Table 1. The Hermitian model under-represents the data variance in this case because of the one very unusual value, 10,400 ppm, present in the data set;

2) for representation/estimation of the cdfs, ten cutoffs are chosen at the quantiles and are listed in Table 2;

3) to enable the application of generalized probability kriging, indicator functions are formed for each cutoff value, and the uniform transform is developed;

4) semi-variogram analysis is performed on the transformed data, Y, the indicator functions, and the uniform transform; Figure 1 displays the results;

5) estimates are obtained of the cdfs.

Table 1. Properties of the Hermitian Model, Isaaks (1984) Data.

	original data, Z	retransformed data, Z	%difference
mean	428.67	403.65	-5.8
var.	994,000	588,654	-40.8

13 coef: .403645E+03 .528328E+03 .333096E+03 .107592E+03 .833933E+01 -.5859832E+01 -.168172E+01 -.444478E-01 -.737359E-01 -.73167E-01 -.822915E-02 .863971E-02 .321113E-02

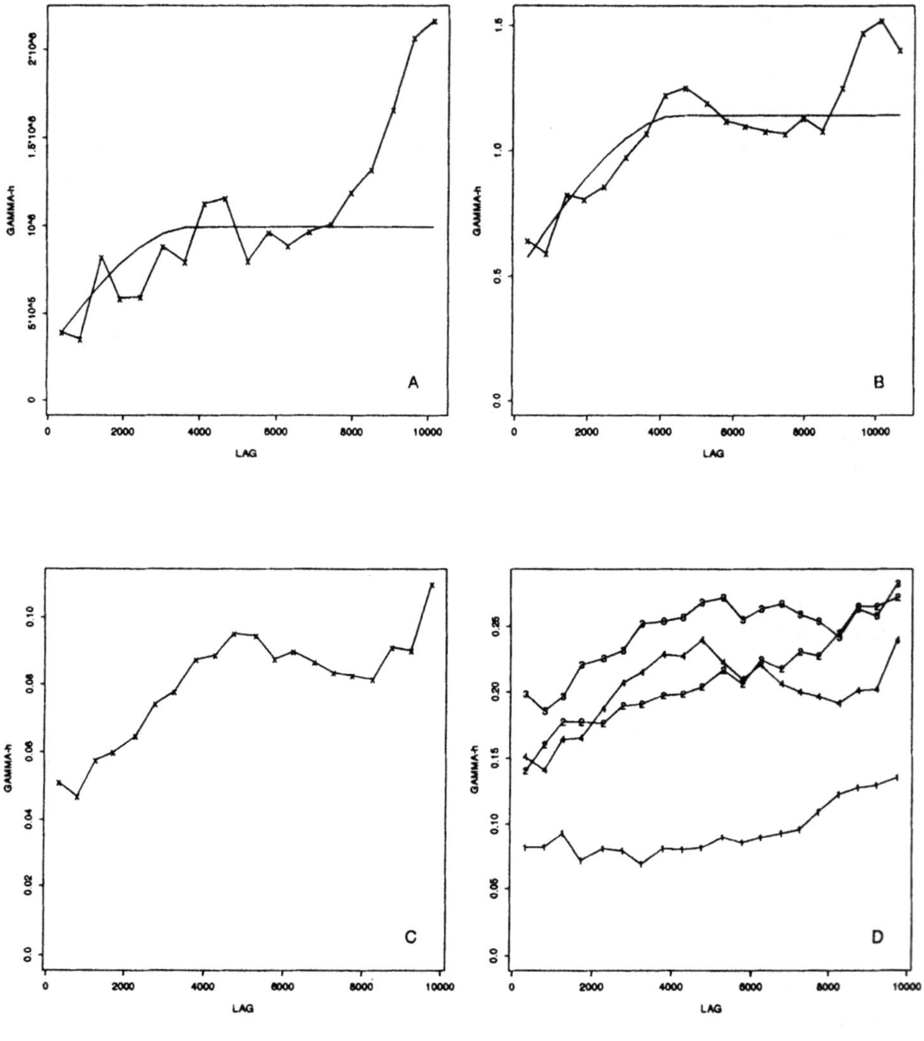

Figure 1. Semi-variogram analysis of Isaaks (1984) data. A: raw data; B: after Hermitian transform; C: uniform transform; D: four selected indicator transforms for cutoffs: 1=71.0; 2=121.0; 3=202.0; 4=422.0.

Table 2. Kriging attributes and cutoffs for estimating cdfs.

Grid: 10 rows, each having 10 columns;
max. y = 10,000 ft; min. x = 0.0
y dimension between rows = x dimension between columns = 1000
10 quantiles: 1=49.0; 2=71.0; 3=91.8; 4=121.0; 5=159.0
6=202.0; 7=273.0; 8=422.0; 9=771.0; 10=10,400.0

4.1. *Results*

A ten by ten (10 x 10) grid is chosen as described in Table 2; this grid is used for both generalized probability kriging and disjunctive kriging; the selection of this grid is arbitrary and simply serves as a means by which to compare the two estimation models.

Selected estimates of the cdfs are shown in Figure 2 [chosen to represent situations in which estimates from the two models differ (Figure 2a) and situations in which estimates are similar (Figure 2b)]. In the situation where there is a discrepancy between the cdfs (Figure 2a) for the two models, estimates of Z yielded by each estimation model (using Eqs. (4) and (9) respectively) are 931.5 (pk) and 526.5 (dk). The average estimate of the uniform transform for this location (i.e., the average of all estimates from Eq. (3) over all cutoffs) is 0.48, indicating a median data magnitude at this location approximately equal to the 5th quantile (159.0), a quantity much less than what is yielded by Eqs (4) and (9). At this estimation location, there is a sizeable difference in estimation results from the two models.

In the other example, Figure 2b, estimates of the cdfs are similar; moreover, the estimates from Eqs. (4) and (9) are 136.3 (pk) and 186.7 (dk). At this estimation location, the two estimation models yield almost identical results. Further, the average estimate of the uniform transform is 0.40, indicating a quantity equal to the 4th cutoff (121.0), a quantity more similar to the estimates obtained from Eqs. (4) and (9) than what was obtained in the example shown in Figure 2a. These two examples show that the performance of probability kriging and disjunctive kriging can vary widely, depending on the estimation location and the distribution of the data. In this case, results from probability kriging are perhaps preferred because this method is a nonparametric one and is therefore less sensitive to the original data distribution in comparison to disjunctive kriging. It may be worthwhile to apply both methods, and multigaussian kriging as well, to such data to examine the variability in results and use some other information, perhaps subjective inference, to determine which result is preferred.

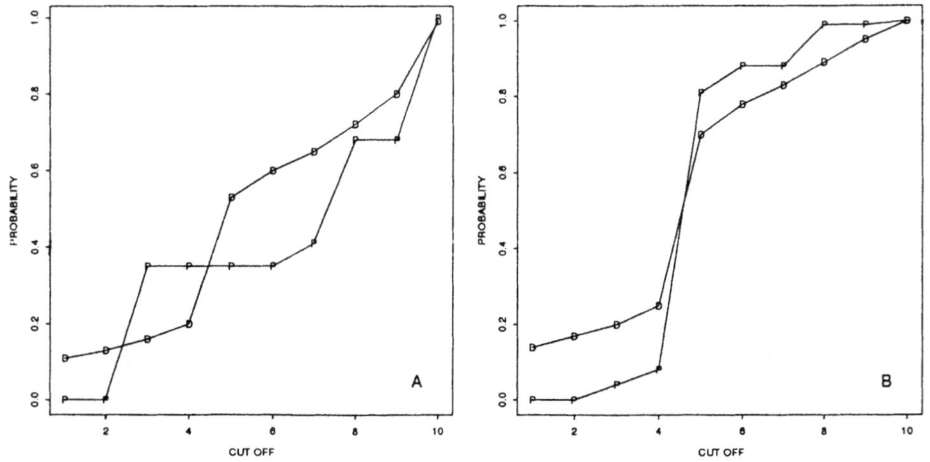

Figure 2. Selected estimates of cumulative distribution functions for Isaaks (1984) data. 2a: result for grid location: row = 1; col = 9; estimates of Z (from Eqs. (4) and (9)): 931.5 (pk) and 536.5 (dk); the average value of the estimates (from Eq. (3)) of the uniform transform is 0.48, almost equal to the fifth quantile: 159.0; 2b: result for grid location: row = 2, col = 2; estimates of Z: 136.3 (pk) and 186.7 (dk). The average value of the estimates of the uniform transform is 0.4, equal to the 4th quantile, 121.0.

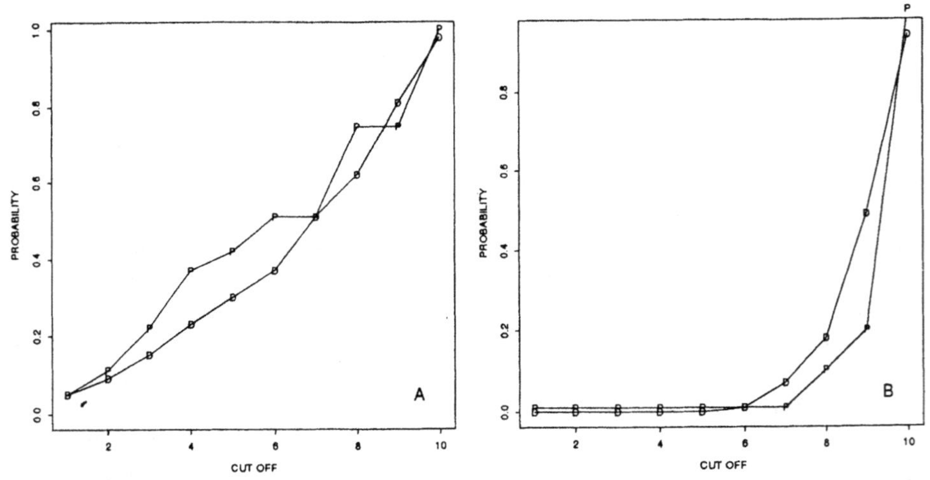

Figure 3. Selected estimates of cumulative distribution functions for simulated data. 3a: result for grid location: row = 1; col = 1; estimates of Z (from Eqs. (4) and (9)): 28.3 (pk) and 29.7 (dk); the average value of the estimates of the uniform transform (from Eq. (3)) is 0.63, falling between the 6th and 7th quantiles (27.2 and 30.6, respectively); 3b: result for grid location: row = 3, col = 5; estimates of Z: 42.4 (pk) and 42.2 (dk); average value of the estimates of the uniform transform is 0.9, equal to the 9th quantile, 39.3.

5. COMPARISON OF GENERALIZED PROBABILITY AND DISJUNCTIVE KRIGING FOR THE ESTIMATION OF SIMULATED DATA

Because the previous application uses a data set associated with a non-normal distribution, an application to normally distributed data, specifically data for which the Intrinsic Hypothesis holds, is sought. Nonconditional simulation is used to develop such a data set; COSIM (Carr and Myers, 1985) is used for this purpose relying on the attributes listed in Table 3. A 50 x 50 grid is used to develop the simulation; a data set is developed from these 2500 simulated values by using a random number generator to select 250 of these grid locations. Statistical attributes of these 250 simulated values are as follows: mean = 25.0; median = 25.1; variance = 117.4; coefficient of skew = 0.12. A 10 x 10 kriging grid is designed, as described in Table 4; moreover, ten cutoffs, chosen at the quantiles, are used and are also listed in Table 4. The same five (5) data preparation steps described above in Section 4 for the real data are followed here for the simulated data. Table 5 reviews the quality of the Hermitian transform.

5a. Results

Selected estimates of the cdfs are shown in Figure 3; semi-variogram analysis is shown in Figure 4. In these two cases [chosen to be representative of results obtained at the 100 estimation locations], estimates of the cdfs from disjunctive kriging and generalized probability kriging are very similar. The indicator function approximation to the cdfs is seen to be very good for these normally distributed data as judged against disjunctive kriging. Estimates of Z from Eqs. (4) and (9) are also very similar. Moreover, the average of the estimates of the uniform transforms for both Figures 3a and 3b are approximately equal to the quantiles matching, in numerical magnitude, the estimates obtained from Eqs. (4) and (9). For these simulated, normally distributed data, then, estimation results from both models are very similar; moreover, the averages of the estimates of the uniform transforms indicate quantile intervals approximately matching estimates obtained from Eqs. (4) and (9).

Table 3. Parameters for Non-Conditional Simulation Using COSIM.

Grid:	50 rows, each having 50 columns; maximum y = 500.0; minimum x = 0.0
	y dimension between rows = x dimension between columns = 10.0
	Beginning random number seed = 711111;
	Linear transformation applied: mean = 25.0, variance = 117.0

Table 4. Kriging Attributes and Cutoffs For Simulated Data.

Grid:	10 rows, each having 10 columns; maximum y = 330.0; minimum x = 0.0
	y dimension between rows = x dimension between columns = 10.0
	10 Cutoffs: 1=11.1; 2=15.4; 3=19.1; 4=22.4; 5=25.1
	6=27.2; 7=30.6; 8=33.3; 9=39.3; 10=54.7

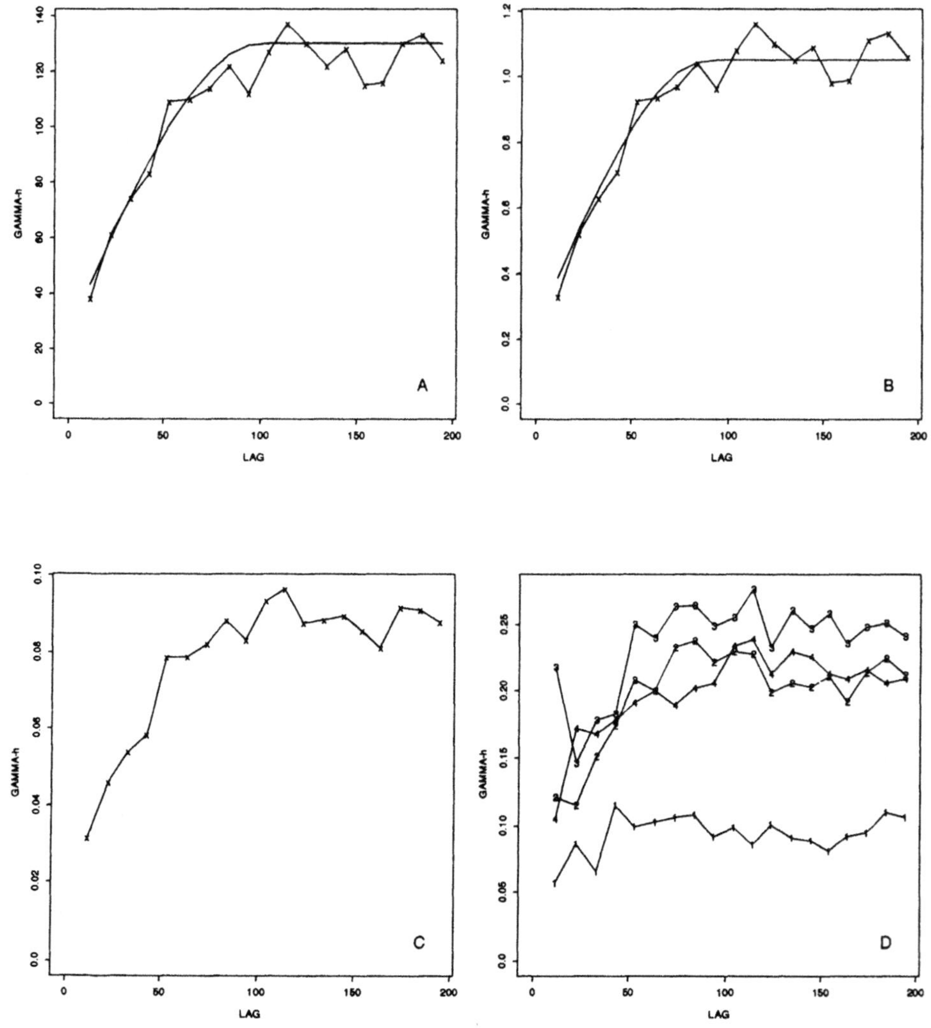

Figure 4. Semi-variogram analysis of simulated data. A: raw data; B: after Hermitian transform; C: uniform transform; D: four selected indicator transforms for cutoffs: 1 = 15.4; 2 = 22.4; 3 = 27.2; 4 = 33.3.

Table 5. Hermitian Model for Transformation of Simulated Data.

	original data Z	retransformed data, Z	%difference
mean	25.0	24.8	-0.9
var.	117.4	116.2	-1.0
10 coefficients:	.2477541E+02	.1075969E+02	-.353437E-02
-.5416419E-01	.3228697E-01	-.4389054E-01	-.954316E-02
-.8148294E-03	.4362029E-03	.2832388E-03	

6. CONCLUSION

Applied to highly non-normal data sets, these models are found to vary widely in estimation performance. Perhaps because probability kriging (simple and generalized) is a non-parametric estimation model, thus requiring no assumptions for the data distribution, this model is preferred for such data. Applied to normally distributed data, both estimation models yield very similar, almost identical results; of course, multigaussian kriging is an excellent model for such data as well. For such data distributions, because there are fewer semi-variograms (and no cross-variograms) to model, disjunctive kriging is perhaps preferred. These are subjective statements and are intended only as very general guides.

ACKNOWLEDGMENT

We thank Paul Kasameyer for reviewing the manuscript. Work performed under the auspices of the U.S. Department of Energy by the Lawrence Livermore National Laboratory under Contract W-7405-Eng-48.

REFERENCES

Armstrong, M. and Matheron, G. (1986a) "Disjunctive kriging revisited: part I", Mathematical Geology 18, 711-728.

Armstrong, M. and Matheron, G. (1986b) "Disjunctive kriging revisited: part II", Mathematical Geology 18, 729-742.

Carr, J.R. and Deng, E.D. (1987) "Comparison of two techniques for applying disjunctive kriging: the Gaussian anamorphosis model versus the direct statistical inference of the bivariate distributions", Mathematical Geology 19, 57-68.

Carr, J.R. and Mao, N. (1992) "Generalized probability kriging for estimation of the indicator and uniform transforms", Mathematical Geology (in review).

Carr, J.R. and Myers, D.E. (1985) "COSIM: A FORTRAN-IV program for coconditional simulation", Computers and Geosciences 11, 675-705.

Carr, J.R., Deng, E.D., and Glass, C.E. (1986) "An application of disjunctive kriging for earthquake ground motion estimation", Mathematical Geology 18, 197-214.

Carr, J.R., Myers, D.E., and Glass, C.E. (1985) "Cokriging-A computer program", Computers and Geosciences 11, 111-128.

Deutsch, C. (1989) "DECLUS: a FORTRAN 77 program for determining optimum spatial declustering weights", Computers and Geosciences 15, 325-332.

Hu, L.Y. (1989) "Comparing gamma isofactorial disjunctive kriging and indicator kriging for estimating local spatial distributions", in M. Armstrong (ed.), Geostatistics, Vol. 1, Kluwer Academic Publishers, Dordrecht, pp. 335-346.

Isaaks, E.H. (1984) "Risk qualified mappings for hazardous waste sites: a case study in distribution free geostatistics", unpublished Master of Science thesis, Department of Applied Earth Science, Stanford University, Stanford, California (U.S.A.), 85p.

Journel, A.G. (1983) "Nonparametric estimation of spatial distributions", Journal for the International Association for Mathematical Geology 15, 445-468.

Kim, Y.C., Myers, D.E., and Knudsen, H.P. (1977) "Advanced geostatistics in ore reserve estimation and mine planning (practitioner's guide)", prepared for the United States of America Department of Energy, Grand Junction, Colorado, report no. GJBX-65(77), 154p.

Matheron, G. (1976a) "A simple substitute for conditional expectation: the disjunctive kriging", in M. Guarascio, M. David, and C.J. Huijbregts (eds.), Advanced Geostatistics in the Mining Industry, D. Reidel Publishing Co., Dordrecht, pp. 221-236.

Matheron, G. (1976b) "Forecasting block grade distributions: the transfer functions", in M. Guarascio, M. David, and C.J. Huijbregts (eds.), Advanced Geostatistics in the Mining Industry, D. Reidel Publishing Co., Dordrecht, pp. 237-251.

Myers, D.E. (1982) "Matrix formulation of co-kriging", Journal of the International Association for Mathematical Geology 14, 249-257.

Press, W.H., Flannery, B.P., Teukolsky, S.A., and Vetterling, W.T. (1986) Numerical Recipes: The Art of Scientific Computing, Cambridge University Press, New York, 818p.

Sullivan, J. (1985) "Non-parametric estimation of spatial distributions", unpublished doctoral dissertation, Department of Applied Earth Sciences, Stanford University, Stanford, California (U.S.A.), 367p.

Verly, G. (1983) "The multigaussian approach and its application to the estimation of local reserves", Jour. International Association for Mathematical Geology 15, 259-286.

Verly, G. and Sullivan, J. (1985) "Multigaussian and probability krigings-application to the Jerrit Canyon deposit", Mining Engineering 37, 568-574.

AGGREGATION IN GEOSTATISTICAL PROBLEMS

NOEL CRESSIE
Iowa State University
Department of Statistics
Ames, IA 50011-1210
USA

Abstract: A random process $Z(\cdot)$ defined on point support has quite different characteristics to those of aggregations of $Z(\cdot)$. For example, $Z(\underset{\sim}{s})$ has larger variance than $Z(B) \equiv \int_B Z(\underset{\sim}{u})d\underset{\sim}{u}/\int_B d\underset{\sim}{u}$, where $\underset{\sim}{s}$ is a point chosen at random within the block B; B is often referred to as the support of Z(B). Suppose that a resource $Z(\cdot)$ is sampled, yielding data $\underset{\sim}{Z} \equiv (Z(\underset{\sim}{s}_1),...,Z(\underset{\sim}{s}_n))'$. However, the resource is extracted in blocks $B_1,...,B_N$. Let B denote a generic block and suppose that it is desired to predict g(Z(B)) based on the data $\underset{\sim}{Z}$. The conditional expectation, $E\{g(Z(B))|\underset{\sim}{Z}\}$, minimizes the mean-squared prediction error, but it is impossible to estimate it without making over-utopian parametric assumptions. Current approaches to the problem require knowledge of "block-to-block" and "block-to-sample" parameters that cannot be estimated from the data. This paper proposes alternatively to make the kriging predictor more variable; the result is an optimal predictor for g(Z(B)) that is unbiased for a Gaussian process and approximately unbiased for a non-Gaussian process and sufficiently smooth g.

1. Introduction

Let $\{Z(\underset{\sim}{s}): \underset{\sim}{s} \in D\}$ denote a random field (or random process) defined on the d-dimensional domain $D(d = 1,2,...)$. When convenient, also denote it as $Z(\cdot)$. A finite number of observations

$$\underset{\sim}{Z} \equiv (Z(\underset{\sim}{s}_1),...,Z(\underset{\sim}{s}_n))' , \tag{1.1}$$

of $Z(\cdot)$, are taken at locations $\{\underset{\sim}{s}_1,...,\underset{\sim}{s}_n\} \subset D$. The data $\underset{\sim}{Z}$ are used to predict some known functional $h(Z(\cdot))$ of the random field. For point prediction,

$$h(Z(\cdot)) = Z(\underset{\sim}{s}_0) , \tag{1.2}$$

and for block prediction,

$$h(Z(\cdot)) = Z(B) \equiv \int_B Z(\underset{\sim}{u})d\underset{\sim}{u}/|B| , \tag{1.3}$$

where |B| is the volume of the block B. Throughout this article, h is real-valued; some discussion of vector-valued h is given in Cressie (1991, sec. 3.4.5).

To simplify the initial presentation, assume that $Z(\underset{\sim}{s}_0)$ is to be predicted. Denote $L(Z(\underset{\sim}{s}_0), p(\underset{\sim}{Z}; \underset{\sim}{s}_0))$ as the loss incurred when $Z(\underset{\sim}{s}_0)$ is predicted with $p(\underset{\sim}{Z}; \underset{\sim}{s}_0)$. An optimal predictor p^o is one that minimizes the so-called Bayes risk $E\{L(Z(\underset{\sim}{s}_0), p(\underset{\sim}{Z}; \underset{\sim}{s}_0))\}$ with respect to p. From Bayesian decision theory (e.g., DeGroot, 1970), p^o minimizes the *posterior expected loss*,

$$E\{L(Z(\underset{\sim}{s}_0), p(\underset{\sim}{Z}; \underset{\sim}{s}_0))|\underset{\sim}{Z}\} \tag{1.4}$$

with respect to p. Thus, an optimal predictor of unobserved parts of $Z(\cdot)$ is obtained conditionally on the observed part $\underset{\sim}{Z}$.

In many prediction problems, *squared-error loss*,

$$L(Z(\underset{\sim}{s}_0), p(\underset{\sim}{Z}; \underset{\sim}{s}_0)) = (Z(\underset{\sim}{s}_0) - p(\underset{\sim}{Z}; \underset{\sim}{s}_0))^2 , \tag{1.5}$$

is used. In this case, the Bayes risk is simply the mean-squared prediction error (mspe) of p, namely,

$$\sigma^2(\underset{\sim}{s}_0; p) \equiv E(Z(\underset{\sim}{s}_0) - p(\underset{\sim}{Z}; \underset{\sim}{s}_0))^2 . \tag{1.6}$$

Now, the posterior expected loss is

$$E\{(Z(\underset{\sim}{s}_0) - p(\underset{\sim}{Z}; \underset{\sim}{s}_0))^2|\underset{\sim}{Z}\} = var(Z(\underset{\sim}{s}_0)|\underset{\sim}{Z}) + \{E(Z(\underset{\sim}{s}_0)|\underset{\sim}{Z}) - p(\underset{\sim}{Z}; \underset{\sim}{s}_0)\}^2 , \tag{1.7}$$

and hence (1.4) is minimized by the optimal predictor

$$p^o(\underset{\sim}{Z}; \underset{\sim}{s}_0) = E(Z(\underset{\sim}{s}_0)|\underset{\sim}{Z}) . \tag{1.8}$$

For squared-error loss, the minimized posterior expected loss is, from (1.7), $var(Z(\underset{\sim}{s}_0)|\underset{\sim}{Z})$ and the minimized mspe is

$$\sigma^2(\underset{\sim}{s}_0; p^o) = var(Z(\underset{\sim}{s}_0)) - var(p^o(\underset{\sim}{Z}; \underset{\sim}{s}_0)) .$$

It should be noted that, as the variance of the optimal predictor increases, $\sigma^2(\underset{\sim}{s}_0; p^o)$ decreases. Intuitively, when $\underset{\sim}{s}_0$ is in a region where there are few data, the optimal predictor has to do a lot of smoothing, which results in small variability and hence large mspe.

Regardless of choice of L, it is clear from (1.4) that the optimal predictor *always* depends on the posterior distribution of $Z(\underset{\sim}{s}_0)|\underset{\sim}{Z}$. Consequently, much of the recent literature in geostatistics has discussed the estimation of

$$F_{\underset{\sim}{s}_0}(z|\underset{\sim}{Z}) \equiv Pr(Z(\underset{\sim}{s}_0) \le z|\underset{\sim}{Z}) ; \quad -\infty < z < \infty . \tag{1.9}$$

Having estimated (1.9), any predictor of the form $E\{g(Z(\underset{\sim}{s}_0))|\underset{\sim}{Z}\}$ can in principle be estimated from

$$E\{g(Z(\underset{\sim}{s}_0))|\underset{\sim}{Z}\} = \int g(z)F_{\underset{\sim}{s}_0}(dz|\underset{\sim}{Z}) . \tag{1.10}$$

By a similar argument to the one yielding (1.8), the predictor (1.10) is seen to be an optimal predictor of $g(Z(\underset{\sim}{s}_0))$, assuming squared-error loss.

Notice that (1.9) is itself an optimal predictor: Suppose that it is desired to predict whether or not $Z(\underset{\sim}{s}_0)$ is greater than some specified threshold z_0; i.e., predict $I(Z(\underset{\sim}{s}_0) > z_0)$, where the indicator function is defined by

$$I(A) = 1 \text{ if A is true; } = 0 \text{ otherwise} . \tag{1.11}$$

Such a prediction problem might arise in the case where compliance to environmental regulations is being monitored. Then, assuming squared-error loss, the optimal predictor is

$$E\{I(Z(\underset{\sim}{s}_0) > z_0)|\underset{\sim}{Z}\} = \Pr(Z(\underset{\sim}{s}_0) > z_0|\underset{\sim}{Z}) = 1 - F_{\underset{\sim}{s}_0}(z_0|\underset{\sim}{Z}) .$$

Suppose one wishes to predict a block average $Z(B)$. All the expressions derived above remain true upon substituting $Z(B)$ for $Z(\underset{\sim}{s}_0)$; interest now centers on

$$F_B(z|\underset{\sim}{Z}) \equiv \Pr(Z(B) \le z|\underset{\sim}{Z}); \quad -\infty < z < \infty . \tag{1.12}$$

Section 2 considers linear prediction of $Z(B)$. The new predictor proposed for nonlinear functions of $Z(B)$ is established in Section 3. Section 4 assesses the performance of the proposed predictors by simulation. This is followed by a discussion and some recommendations.

2. Linear prediction of Z(B)

It is difficult in general to obtain estimates of quantities like (1.9) and (1.12), since they involve the (n+1)-dimensional distribution of $(Z(\underset{\sim}{s}_0),Z(\underset{\sim}{s}_1),...,Z(\underset{\sim}{s}_n))'$ or of $(Z(B),Z(\underset{\sim}{s}_1),...,Z(\underset{\sim}{s}_n))'$. Alternative approaches typically restrict in some way the class of predictors from which the optimal one will be sought.

Kriging (ordinary, universal, IRF-k) predicts $Z(\underset{\sim}{s}_0)$ or $Z(B)$ with a linear combination, $\underset{\sim}{\lambda}'\underset{\sim}{Z}$, of the data $\underset{\sim}{Z}$, that is optimal in the class of homogeneously linear unbiased predictors (e.g., Matheron, 1971; Cressie, 1991, Ch. 3). Indicator kriging predicts $I(Z(\underset{\sim}{s}_0) \le z)$ with a linear combination of $\{I(Z(\underset{\sim}{s}_i) \le z)\colon i = 1,...,n\}$ that is optimal in the class of homogeneously linear unbiased predictors. But, instead of the resulting predictor $\sum_{i=1}^{n} \lambda_i I(Z(\underset{\sim}{s}_i) \le z)$ being used to infer a value for $I(Z(\underset{\sim}{s}_0) \le z)$, the proponents of indicator kriging (e.g., Journel, 1983) have proposed it as an estimator of $\Pr(Z(\underset{\sim}{s}_0) \le z|\underset{\sim}{Z})$. Is this sensible?

Recall the relationship (1.7). It is not difficult to show that, under the present circumstances, the analogous expression is,

$$\begin{aligned} E\{(I(Z(\underset{\sim}{s}_0) \le z) - p(\underset{\sim}{Z};\, \underset{\sim}{s}_0))^2|\underset{\sim}{Z}\} &= \Pr(Z(\underset{\sim}{s}_0) \le z|\underset{\sim}{Z})\{1 - \Pr(Z(\underset{\sim}{s}_0) \le z|\underset{\sim}{Z})\} \\ &\quad + \{\Pr(Z(\underset{\sim}{s}_0) \le z|\underset{\sim}{Z}) - p(\underset{\sim}{Z};\, \underset{\sim}{s}_0)\}^2 . \end{aligned} \tag{2.1}$$

Since the first term on the right-hand side of (2.1) does not depend on p, it is clear that the mspe is decreased whenever

$$E\{\Pr(Z(\underset{\sim}{s}_0) \le z|\underset{\sim}{Z}) - p(\underset{\sim}{Z};\, \underset{\sim}{s}_0)\}^2 \tag{2.2}$$

is decreased, and *vice versa.* Thus, a good predictor of $I(Z(s_0) \leq z)$ yields a good estimator of $\Pr(Z(s_0) \leq z|Z)$.

Clearly, the information in $\{I(Z(s_i) \leq z)$: $i = 1,...,n\}$ is inferior to the information in $\{Z(s_i)$: $i = 1,...,n\}$. Now enrich the indicator information by adding more thresholds to obtain

$$\{I(Z(s_i) \leq z_j): \; j = 1,...,m_i; \; i = 1,...,n\} , \tag{2.3}$$

where one of $z_1,...,z_{m_i}$ is z; $i = 1,...,n$. This yields an optimal (minimum mspe) predictor (unbiased, linear in indicator functions):

$$\sum_{i=1}^{n} \sum_{j=1}^{m_i} \lambda_{ji} I(Z(s_i) \leq z_j) , \tag{2.4}$$

which is *always* superior to $\sum_{i=1}^{n} \lambda_i I(Z(s_i) \leq z)$ since $\{I(Z(s_i) \leq z)$: $i = 1,...,n\}$ is contained in (2.3). Journel (1983) claims that there is not enough improvement in the indicator cokriging predictor (2.4) to make it worthwhile but a comparison in Lajaunie (1990) suggests otherwise. From (2.1), the improvement in using (2.4) to estimate $\Pr(Z(s_0) \leq z|Z)$, rather than indicator kriging, can be measured by taking the difference of their respective mean-squared prediction errors.

Once one realizes that any measurable function of $Z(s_i)$ can be approximated by a sequence of step functions (e.g., Loeve, 1977, Sec. 5.3), which are in turn linear combinations of indicator functions of the type seen in (2.3), it is clear that the predictor (2.4) is optimal (up to discretization) in the class of predictors

$$p(Z; s_0) = \sum_{i=1}^{n} f_i(Z(s_i)) , \tag{2.5}$$

where $\{f_i(\cdot)\}$ are measurable square-integrable functions. But (2.5) is precisely the form of the *disjunctive kriging* predictor proposed by Matheron (1976).

To make progress mathematically, disjunctive kriging goes on to make parametric assumptions about bivariate distributions of the random process. To make progress computationally, indicator kriging has purged the approximation family of many functions. It is argued here that indicator *co*kriging given by (2.4) provides a compromise between these two extremes. Further, the result (2.1) allows one to assess the degree of improvement (for estimation and, obviously, prediction) by comparing mean-squared prediction errors for all three.

When one is interested in predicting functions of the block average, say $g(Z(B))$, the optimal predictor is $E\{g(Z(B))|Z\}$, which means that approximations to $F_B(z|Z)$ defined by (1.12) are sought. The same type of arguments as above apply, so that one can make a choice between indicator kriging, indicator cokriging, or disjunctive kriging of $I(Z(B) \leq z)$. However, to implement these methods in actual studies, knowledge of "co-dependence" between $Z(B)$ and $Z(s_i)$ is needed. That is, to construct the (co)kriging equations, one needs to know $\Pr(Z(B) \leq z, Z(s_i) \leq z_j)$; $1 \leq j \leq m_i$, $1 \leq i \leq n$ (as well as $\Pr(Z(B) \leq z)$). I have yet to see any satisfactory models or estimation techniques for these quantities. (Notice that, in the special case of predicting Z(B) from a linear combination of $\underset{\sim}{Z}$, one has to know the co-dependencies, $\mathrm{cov}(Z(B), Z(\underset{\sim}{s}_i))$, which are easily calculated as $\int_B \mathrm{cov}(Z(\underset{\sim}{u}),Z(\underset{\sim}{s}_i))d\underset{\sim}{u}/|B|$.) Only in the case where $h(Z(\cdot))$ is *linear* in $\{Z(\underset{\sim}{s}): \underset{\sim}{s} \in D\}$ can one write $E(h(\{Z(\underset{\sim}{s}): \underset{\sim}{s} \in D\})|\underset{\sim}{Z}) = h(\{E(Z(\underset{\sim}{s})|\underset{\sim}{Z}): \underset{\sim}{s} \in D\})$ and hence can the "block-to-sample" co-dependencies be expressed in terms of the "point-to-sample" co-dependencies. Since most models usually only give the latter but *not* the former, it is rare that indicator and disjunctive kriging will allow predictions on quantities defined in terms of *block* support.

3. Prediction of nonlinear functionals

In Sections 1 and 2, several examples of functionals have been considered:

$$h(Z(\cdot)) = Z(\underset{\sim}{s}_0) \tag{3.1}$$

$$h(Z(\cdot)) = Z(B) \tag{3.2}$$

$$h(Z(\cdot)) = I(Z(\underset{\sim}{s}_0) > z_0) \tag{3.3}$$

$$h(Z(\cdot)) = I(Z(B) > z_0) \;. \tag{3.4}$$

An important example in mining is

$$h(Z(\cdot)) = Z(B)I(Z(B) > z_0) \;, \tag{3.5}$$

which corresponds to predicting the *quantity of metal* above the cutoff grade z_0 in the mining block B. Using squared-error loss given by (1.5), the optimal predictor of (3.5) is $E\{Z(B)I(Z(B) > z_0)|\underset{\sim}{Z}\}$.

Henceforth, consider the (typically nonlinear) functional

$$h(Z(\cdot)) = g(Z(B)) \;. \tag{3.6}$$

My goal in this section is to predict $g(Z(B))$ using a predictor of the form $g(\underset{\sim}{a}'\underset{\sim}{Z})$.

Assume initially that

$$Z(\underset{\sim}{s}) = \mu + \delta(\underset{\sim}{s}) \;;\; \underset{\sim}{s} \in D \;, \tag{3.7}$$

where the (constant) mean μ is unknown and $\mathrm{cov}(\delta(\underset{\sim}{s}),\delta(\underset{\sim}{u})) \equiv C(\underset{\sim}{s},\underset{\sim}{u})$ is known for all $\underset{\sim}{s},\underset{\sim}{u} \in D$. Since the ordinary kriging predictor $\underset{\sim}{\lambda}'\underset{\sim}{Z}$ of Z(B) is a linear unbiased predictor that minimizes mean-squared prediction error, one might guess that $g(\underset{\sim}{\lambda}'\underset{\sim}{Z})$ is an appropriate predictor of $g(Z(B))$. However, to predict nonlinear functions of Z(B), $\underset{\sim}{\lambda}'\underset{\sim}{Z}$ is *too smooth*; see Section 4.

Suppose for the moment that $Z(\cdot)$ is a Gaussian process. Then Z(B) is Gaussian with mean μ, and so is $\underset{\sim}{a}'\underset{\sim}{Z}$. Therefore, if $\mathrm{var}(\underset{\sim}{a}'\underset{\sim}{Z}) = \mathrm{var}(Z(B))$, then $Z(B) = \underset{\sim}{a}'\underset{\sim}{Z}$ in distribution. As a consequence, $g(\underset{\sim}{a}'\underset{\sim}{Z})$ is an *unbiased* predictor of $g(Z(B))$ for *any* measurable function g. An optimal choice of $\underset{\sim}{a}$ is obtained by minimizing the mspe $E\{Z(B) - \underset{\sim}{a}'\underset{\sim}{Z}\}^2$, subject to the unbiasedness constraint

$$E(\underset{\sim}{a}'\underset{\sim}{Z}) = E(Z(B)) \tag{3.8}$$

and the variance constraint

$$\text{var}(\underset{\sim}{a}'\underset{\sim}{Z}) = C(B,B) \equiv \int_B \int_B C(\underset{\sim}{s},\underset{\sim}{u})d\underset{\sim}{s}d\underset{\sim}{u}/|B|^2 = \text{var}(Z(B)) \; . \tag{3.9}$$

My suggestion in this article is to use $g(\underset{\sim}{a}'\underset{\sim}{Z})$ to predict $g(Z(B))$ even when $Z(\cdot)$ is non Gaussian. For sufficiently smooth functions $g(\cdot)$ that possess at least two continuous derivatives, the δ-method (e.g., Bishop, Fienberg, and Holland, 1975, p. 486) yields

$$\begin{aligned} E(g(\underset{\sim}{a}'\underset{\sim}{Z})) &\simeq g(E(\underset{\sim}{a}'\underset{\sim}{Z})) + g''(E(\underset{\sim}{a}'\underset{\sim}{Z}))\text{var}(\underset{\sim}{a}'\underset{\sim}{Z})/2 \\ &= g(E(Z(B))) + g''(E(Z(B)))\text{var}(Z(B))/2 \simeq E(g(Z(B))) \; , \end{aligned}$$

regardless of whether $Z(\cdot)$ is Gaussian.

Notice that, although some optimization has been carried out, $g(\underset{\sim}{a}'\underset{\sim}{Z})$ is in general suboptimal to the predictor $E(g(Z(B))|\underset{\sim}{Z})$, assuming squared error loss. However, the advantage of $g(\underset{\sim}{a}'\underset{\sim}{Z})$ is that it depends only on $C(\cdot,\cdot)$ and the range of values it can take matches exactly that of $g(Z(B))$. Matheron (1985) and Lantuejoul (1988) caution us against basing change-of-support calculations only on $C(\cdot,\cdot)$. However, in their demonstrations, they are concerned with inferring the distribution of $Z(B)$, whereas my concern is with the *conditional* distribution of $Z(B)$ given $\underset{\sim}{Z}$. This conditioning on the data will undoubtedly mitigate the circumstances described by Matheron and Lantuejoul.

From (3.8) and (3.9), the appropriate objective function to minimize is

$$E\{\underset{\sim}{a}'\underset{\sim}{Z} - Z(B)\}^2 + 2m_1(\underset{\sim}{a}'\underset{\sim}{1} - 1) + m_2(\underset{\sim}{a}'\Sigma\underset{\sim}{a} - C(B,B)) \; , \tag{3.10}$$

where $\Sigma = \text{var}(\underset{\sim}{Z})$ and m_1,m_2 are Lagrange multipliers. The expression (3.10) is quadratic in elements of $\underset{\sim}{a}$; upon differentiating it with respect to $\underset{\sim}{a}$, m_1, and m_2 and setting the results equal to zero, one obtains

$$\underset{\sim}{a}' = \{(1/m_2)\underset{\sim}{c}'\Sigma^{-1}\} - \{(m_1/m_2)\underset{\sim}{1}'\Sigma^{-1}\} \tag{3.11}$$

$$m_1 = \{-m_2/(\underset{\sim}{1}'\Sigma^{-1}\underset{\sim}{1})\} + \{(\underset{\sim}{c}'\Sigma^{-1}\underset{\sim}{1})/(\underset{\sim}{1}'\Sigma^{-1}\underset{\sim}{1})\} \tag{3.12}$$

$$m_2 = \{(\underset{\sim}{c}'\Sigma^{-1}\underset{\sim}{c})(\underset{\sim}{1}'\Sigma^{-1}\underset{\sim}{1}) - (\underset{\sim}{c}'\Sigma^{-1}\underset{\sim}{1})^2\}^{1/2}/\{(\underset{\sim}{1}'\Sigma^{-1}\underset{\sim}{1})C(B,B) - 1\}^{1/2} \; , \tag{3.13}$$

where $\underset{\sim}{c} \equiv (C(B,\underset{\sim}{s}_1),\ldots,C(B,\underset{\sim}{s}_n))'$ and $C(B,\underset{\sim}{s}) \equiv \int_B C(\underset{\sim}{u},\underset{\sim}{s})d\underset{\sim}{u}/|B|$. Now the numerator of (3.13) is well defined by the Cauchy-Schwarz inequality and the denominator is well defined if

$$\text{var}(Z(B)) > (\underset{\sim}{1}'\Sigma^{-1}\underset{\sim}{1})^{-1} = \text{var}(\hat{\mu}_{g\ell s}) \; , \tag{3.14}$$

where $\hat{\mu}_{g\ell s} \equiv (\underset{\sim}{1}'\Sigma^{-1}\underset{\sim}{Z})/(\underset{\sim}{1}'\Sigma^{-1}\underset{\sim}{1})$ is the generalized least squares estimator of μ. Intuitively speaking, the condition (3.14) says that it is not possible to predict with the same variation as $Z(B)$ if $\text{var}(Z(B))$ becomes too small. (This is more likely to happen as the block B increases in volume.)

Now suppose that

$$\mu(\underset{\sim}{s}) \equiv E(Z(\underset{\sim}{s})) = \underset{\sim}{x}(\underset{\sim}{s})'\underset{\sim}{\beta} \ , \tag{3.15}$$

where $\underset{\sim}{\beta}$ is a $q \times 1$ vector. Hence, $E(\underset{\sim}{Z}) = X\underset{\sim}{\beta}$, where $X \equiv (\underset{\sim}{x}(\underset{\sim}{s}_1),\ldots,\underset{\sim}{x}(\underset{\sim}{s}_n))'$. The objective function analogous to (3.10) is easily seen to be

$$E\{\underset{\sim}{a}'\underset{\sim}{Z} - Z(B)\}^2 + 2(\underset{\sim}{a}'X - \underset{\sim}{x}(B)')\underset{\sim}{m}_1 + m_2(\underset{\sim}{a}'\Sigma\underset{\sim}{a} - C(B,B)) \ , \tag{3.16}$$

where $\underset{\sim}{x}(B) \equiv \int_B \underset{\sim}{x}(\underset{\sim}{u})d\underset{\sim}{u}/|B|$. The expression (3.16) is to be minimized with respect to $\underset{\sim}{a}$ ($n \times 1$ vector), $\underset{\sim}{m}_1$ ($q \times 1$ vector), and m_2 (scalar). Upon differentiating and setting the results equal to zero, one obtains

$$\underset{\sim}{a}' = \{(1/m_2)\underset{\sim}{c}'\Sigma^{-1}\} - \{(\underset{\sim}{m}_1'/m_2)X'\Sigma^{-1}\} \tag{3.17}$$

$$\underset{\sim}{m}_1' = \{-m_2\underset{\sim}{x}(B)'(X'\Sigma^{-1}X)^{-1}\} + \{(\underset{\sim}{c}'\Sigma^{-1}X)(X'\Sigma^{-1}X)^{-1}\} \tag{3.18}$$

$$m_2 = \{\underset{\sim}{c}'\Sigma^{-1}\underset{\sim}{c} - \underset{\sim}{c}'\Sigma^{-1}X(X'\Sigma^{-1}X)^{-1}X'\Sigma^{-1}\underset{\sim}{c}\}^{1/2} / \{C(B,B) - \underset{\sim}{x}(B)'(X'\Sigma^{-1}X)^{-1}\underset{\sim}{x}(B)\}^{1/2} \ . \tag{3.19}$$

As before, the numerator of (3.19) is well defined since, if $\underset{\sim}{b} \equiv \Sigma^{-1/2}\underset{\sim}{c}$ and $\underset{\sim}{W} \equiv \Sigma^{-1/2}X$, it is expressible as $\underset{\sim}{b}'(I - W(W'W)^{-1}W')\underset{\sim}{b}$ and the projection matrix $I - W(W'W)^{-1}W'$ has only nonnegative eigenvalues (equal to 1 or 0). The denominator of (3.19) is well defined if

$$\mathrm{var}(Z(B)) > \underset{\sim}{x}(B)'(X'\Sigma^{-1}X)^{-1}\underset{\sim}{x}(B) = \mathrm{var}(\underset{\sim}{x}(B)'\hat{\underset{\sim}{\beta}}_{gls}) \ , \tag{3.20}$$

where $\hat{\underset{\sim}{\beta}}_{gls} \equiv (X'\Sigma^{-1}X)X'\Sigma^{-1}\underset{\sim}{Z}$ is the generalized least squares estimator of $\underset{\sim}{\beta}$. Clearly, (3.17), (3.18), and (3.19) reduce to (3.11), (3.12), and (3.13) in the special case where $\underset{\sim}{x}(\underset{\sim}{s}) = 1$ and $\underset{\sim}{\beta} = \mu$.

Define the *constrained kriging predictor* of $Z(B)$ to be

$$p^{@}(\underset{\sim}{Z};\, B) \equiv \underset{\sim}{a}'\underset{\sim}{Z} \ , \tag{3.21}$$

where $\underset{\sim}{a}$ is given by (3.11) or (3.17); the constraints are $E(p^{@}(\underset{\sim}{Z};\, B)) = E(Z(B))$ and $\mathrm{var}(p^{@}(\underset{\sim}{Z};\, B)) = \mathrm{var}(Z(B))$. Consider now some properties of $g(p^{@}(\underset{\sim}{Z};\, B))$ as a predictor of $g(Z(B))$.

By construction, $p^{@}(\underset{\sim}{Z};\, B)$ is *unbiased* for $Z(B)$ and its optimal mspe is

$$E\{Z(B) - p^{@}(\underset{\sim}{Z};\, B)\}^2 = 2C(B,B) - 2\underset{\sim}{a}'\underset{\sim}{c} \ . \tag{3.22}$$

Further, for smooth $g(\cdot)$, the δ-method yields an approximate unbiasedness result, $E\{g(p^{@}(\underset{\sim}{Z};\, B))\} \simeq E\{g(Z(B))\}$, and approximate mspe

$$E\{g(Z(B)) - g(p^{@}(\underset{\sim}{Z};\, B))\}^2 \simeq \{g'(E(Z(B)))\}^2\{2C(B,B) - 2\underset{\sim}{a}'\underset{\sim}{c}\} \ . \tag{3.23}$$

Since the rhs of (3.23) has been optimized, so too has the lhs (approximately).

Although discussion so far has concentrated on prediction of g(Z(B)) for one B, the more important problem is predicting functionals of *many* blocks. In mining, these blocks represent the units that can be mined selectively and sent to the mill. Clearly, a mine will be more profitable if the high grades are concentrated in easily extractable units. If instead those same high grades are scattered throughout the deposit in pockets that are expensive to access, the mine may be unprofitable, even though the average grade of the deposit is unchanged.

Think of the region D as being made up of N disjoint selectable mining units:

$$D = \bigcup\{B_i\colon\ i = 1,\dots,N\}\ , \tag{3.24}$$

where $B_1,\dots,B_N$ are disjoint. For resource-evaluation purposes, several quantities are of interest. Consider the (weighted) proportion of regions whose grade is above the cutoff z_0:

$$h(Z(\cdot)) = \sum_{i=1}^{N} |B_i| I(Z(B_i) > z_0)/|D|\ . \tag{3.25}$$

Also consider the total quantity of metal above the cutoff z_0:

$$h(Z(\cdot)) = \sum_{i=1}^{N} |B_i| Z(B_i) I(Z(B_i) > z_0)\ . \tag{3.26}$$

The optimal predictor (for squared-error loss) is, in the case of (3.25),

$$\sum_{i=1}^{N} |B_i| E\{I(Z(B_i) > z_0)|\underset{\sim}{Z}\}/|D| \tag{3.27}$$

and is, in the case of (3.26),

$$\sum_{i=1}^{N} |B_i| E\{Z(B_i) I(Z(B_i) > z_0)|\underset{\sim}{Z}\}\ . \tag{3.28}$$

Each of the summands in (3.27) and (3.28) can be expressed as $E\{g(Z(B))|\underset{\sim}{Z}\}$, which can be replaced by $g(p^{@}(\underset{\sim}{Z};\ B))$, where $p^{@}$ is given by (3.21). Thus, to predict (3.25), I propose

$$\sum_{i=1}^{N} |B_i| I(p^{@}(\underset{\sim}{Z};\ B_i) > z_0)/|D|\ , \tag{3.29}$$

and to predict (3.26), I propose

$$\sum_{i=1}^{N} |B_i| p^{@}(\underset{\sim}{Z};\ B_i) I(p^{@}(\underset{\sim}{Z};\ B_i) > z_0)\ . \tag{3.30}$$

A third quantity of interest in resource evaluation is the average grade in those blocks above the cutoff z_0:

$$h(Z(\cdot)) = \sum_{i=1}^{N} |B_i| Z(B_i) I(Z(B_i) > z_0) \Big/ \sum_{i=1}^{N} |B_i| I(Z(B_i) > z_0) . \qquad (3.31)$$

Because (3.31) is a ratio, it is, unlike (3.25) and (3.26), *nonlinear* in $\{g(Z(B_i)): i = 1,...,N\}$. Some have incautiously suggested predicting (3.31) with

$$\sum_{i=1}^{N} |B_i| E\{Z(B_i) I(Z(B_i) > z_0) | \underset{\sim}{Z}\} \Big/ \sum_{i=1}^{N} |B_i| E\{I(Z(B_i) > z_0) | \underset{\sim}{Z}\} ,$$

but the nonlinearity will cause it to be *biased*. To use the approach taken in this section, one would have to develop multivariate linear predictors $A\underset{\sim}{Z}$ for $\underset{\sim}{Z}(\underset{\sim}{B}) \equiv (Z(B_1),...,Z(B_N))'$ such that $E(A\underset{\sim}{Z}) = E(\underset{\sim}{Z}(\underset{\sim}{B}))$ and $\text{var}(A\underset{\sim}{Z}) = \text{var}(\underset{\sim}{Z}(\underset{\sim}{B}))$.

4. Examples and discussion

It was established in Section 3 that the predictor $g(p^{@}(\underset{\sim}{Z}; B))$ of $g(Z(B))$ will be unbiased if either

(i) $Z(\cdot)$ is Gaussian, or (ii) $Z(\cdot)$ is non Gaussian and $g(\cdot)$ is smooth.

In this section, I shall investigate the practical worth of (i) and (ii) as well as the case where $Z(\cdot)$ is *not* Gaussian and $g(\cdot)$ is *not* smooth.

A simple simulation experiment in $\mathbb{R}^2$ was conducted. Data $Z(\underset{\sim}{s}_1),...,Z(\underset{\sim}{s}_6)$ were generated at the locations $\underset{\sim}{s}_1 = (0,0)$, $\underset{\sim}{s}_2 = (8,0)$, $\underset{\sim}{s}_3 = (16,0)$, $\underset{\sim}{s}_4 = (16,16)$, $\underset{\sim}{s}_5 = (0,16)$, and $\underset{\sim}{s}_6 = (0,8)$, and were used to predict functions of

$$Z(B_j) \equiv n_j^{-2} \sum_{\ell=1}^{n_j} \sum_{m=1}^{n_j} Z(9-\ell, 9-m); \quad j = 1,2,3 , \qquad (4.1)$$

where $n_1 = 2$, $n_2 = 3$, $n_3 = 5$. The functions chosen were g_k; $k = 1,2,3,4$:

$$g_1(z) = z; \quad g_2(z) = z^2; \quad g_3(z) = I(z > z_0); \quad g_4(z) = zI(z > z_0) , \qquad (4.2)$$

where z_0 is defined to be the 90-th percentile of a $N(E(Z(B)),\text{var}(Z(B)))$ distribution. The predictors of $g_k(Z(B_j))$ can then be compared to the simulated (true) value generated at the same time as the data $Z(\underset{\sim}{s}_1),...,Z(\underset{\sim}{s}_6)$.

Three random fields,

$$\{Z_i(\underset{\sim}{s}): \underset{\sim}{s} \in \mathbb{R}^2\}; \quad i = 1,2,3 , \qquad (4.3)$$

were used to generate the data and (4.1); each one derives in one way or another from a Gaussian random field with mean zero and "spherical" covariance function

$$S_h(\underset{\sim}{s},\underset{\sim}{u}) \equiv 1 - \{(3/2)(\|\underset{\sim}{s}-\underset{\sim}{u}\|/a_h) - (1/2)(\|\underset{\sim}{s}-\underset{\sim}{u}\|/a_h)^3\}; \quad \underset{\sim}{s},\underset{\sim}{u} \in \mathbb{R}^2,\ h = 1,2,3 ,$$

where the degree of spatial dependence is given by

$$a_1 = 15 \text{ (weak dependence)}, \ a_2 = 30 \text{ (moderate)}, \ a_3 = 150 \text{ (strong)} . \qquad (4.4)$$

The first random field $Z_1(\cdot)$ is *Gaussian* with mean zero and covariance function

$C_1(\underset{\sim}{s},\underset{\sim}{u})$ given by the spherical covariances above. The second random field is $Z_2(\cdot) \equiv k_2 \exp\{Z_1(\cdot)\}$, the so-called *log Gaussian* random field. The constant k_2 is chosen so that $\mathrm{var}(Z_2(\underset{\sim}{s})) \equiv 1$. Its mean and covariance function $C_2(\underset{\sim}{s},\underset{\sim}{u})$ is easily calculated from C_1 using formulas such as those given by Cressie (1991, p. 135). Marginal distributions are lognormal and hence highly skewed.

The third random field is a *mixed Gaussian:*

$$Z_3(\cdot) \equiv k_3\{I(\cdot)U(\cdot) + (1 - I(\cdot))V(\cdot)\} ,$$

where $I(\cdot)$ is a 0–1 random field such that any $I(\underset{\sim}{s}),I(\underset{\sim}{u})$; $\underset{\sim}{s} \neq \underset{\sim}{u}$, are independent Bernoulli(3/4) random variables; $U(\cdot)$ is a Gaussian random field with mean (−3/8) and covariance function C_1; $V(\cdot)$ is a Gaussian random field with mean (9/8) and covariance function $(1/3)^2 C_1$; $I(\cdot)$, $U(\cdot)$, and $V(\cdot)$ are mutually independent random fields; and finally the constant k_3 is chosen so that $\mathrm{var}(Z_3(\underset{\sim}{s})) \equiv 1$. The mean of $Z_3(\cdot)$ is zero and straightforward calculations yield the covariance function $C_3(\underset{\sim}{s},\underset{\sim}{u})$. Marginal distributions are bimodal and skewed.

The simulation experiment has a factorial structure: Strength-of-spatial-dependence factor (3 levels; see (4.4)), Random-field factor (3 levels; see (4.3)), Size-of-block factor (3 levels; see (4.1)), and Function-of-block factor (4 levels; see (4.2)). For each of the 3×3 levels of the first two factors, 5,000 realizations of $\{Z_i(x,y):\ x = 0,1,\ldots,16,\ y = 0,1,\ldots,16\}$ were obtained. From the data $\underset{\sim}{Z} \equiv (Z(\underset{\sim}{s}_1),\ldots,Z(\underset{\sim}{s}_6))'$, the constrained kriging predictor $p^@(\underset{\sim}{Z};\ B) = \underset{\sim}{a}'\underset{\sim}{Z}$, given by (3.21), was computed, as was the ordinary kriging predictor $\hat{p}(\underset{\sim}{Z};\ B) \equiv \underset{\sim}{\lambda}'\underset{\sim}{Z}$, where $\underset{\sim}{\lambda}'$ is given, e.g., by Cressie (1991, p. 122). The simulation study compares the performances of $g(\underset{\sim}{a}'\underset{\sim}{Z})$ and $g(\underset{\sim}{\lambda}'\underset{\sim}{Z})$ as predictors of $g(Z(B))$. Their biases are given in Table 1 (see below).

The results of the simulation study show that, in terms of prediction bias of $g(Z(B))$, $g(\underset{\sim}{a}'\underset{\sim}{Z})$ is generally an improvement over $g(\underset{\sim}{\lambda}'\underset{\sim}{Z})$, sometimes by an order or two of magnitude. When small bias is of paramount importance, such as in resource evaluation, the predictor $g(\underset{\sim}{a}'\underset{\sim}{Z})$ based on constrained kriging is to be preferred. However, its use is not without a price; mspe's of $g(\underset{\sim}{a}'\underset{\sim}{Z})$ were generally found to be between 20% and 50% larger than those of $g(\underset{\sim}{\lambda}'\underset{\sim}{Z})$.

In conclusion, nonlinear prediction of aggregations of a random process can be achieved simply and accurately via the constrained kriging predictor $p^@(\underset{\sim}{Z};\ B) = \underset{\sim}{a}'\underset{\sim}{Z}$. The coefficients $\underset{\sim}{a}$ depend only on the covariance function $C(\cdot,\cdot)$ and the support B.

Acknowledgments

Funding for this research came from the National Science Foundation, the National Security Agency, and the Office of Technology Development, Office of Environmental Restoration and Waste Management, U.S. Department of Energy.

References

Bishop, Y.M.M., Fienberg, S. E., and Holland, P. W. (1975). *Discrete MultivariateAnalysis: Theory and Practice.* The MIT Press, Cambridge.

Cressie, N. (1991). *Statistics for Spatial Data.* Wiley, New York.

DeGroot, M. H. (1970). *Optimal Statistical Decisions.* McGraw-Hill, New York.

Journel, A. G. (1983). Nonparametric estimation of spatial distributions. *Journal of the International Association for Mathematical Geology*, **15**, 445-468.

Journel, A. G. (1984). mAD and conditional quantile estimators. In *Geostatistics for Natural Resources Characterization, Part 1*, G. Verly, M. David, A. Journel, and A. Marechal, eds. Reidel, Dordrecht, 261-270.

Lajaunie, C. (1990). Comparing some approximate methods for building local confidence intervals for predicting regionalized variables. *Mathematical Geology*, **22**, 123-144.

Lantuejoul, C. (1988). On the importance of choosing a change of support model for global reserves. *Mathematical Geology*, **20**, 1001-1019.

Loeve, M. (1977). *Probability Theory I, 4th edn.* Springer, New York.

Matheron, G. (1971). The Theory of Regionalized Variables and Its Applications. *Cahiers du Centre de Morphologie Mathematique*, **No. 5**. Fontainebleau, France.

Matheron, G. (1976). A simple substitute for conditional expectation: The disjunctive kriging. In *Advanced Geostatistics in the Mining Industry*, M. Guarascio, M. David, and C. Huijbregts, eds. Reidel, Dordrecht, 221-236.

Matheron, G. (1985). Change of support for diffusion-type random functions. *Journal of the International Association for Mathematical Geology*, **17**, 137-165.

Table 1. Entries (in units of 10^{-3}) show bias = $E\{g_k(\underset{\sim}{a}'\underset{\sim}{Z}) - g_k(Z(B_j))\}$; $E\{g_k(\underset{\sim}{\lambda}'\underset{\sim}{Z}) - E(g_k(Z(B_j))\}$, estimated from 5,000 simulations. Underlined entries denote bias significantly different ($\alpha = .01$) from zero.

	$a_1 = 15$			$a_2 = 30$			$a_3 = 150$		
	B_1	B_2	B_3	B_1	B_2	B_3	B_1	B_2	B_3
					Gaussian				
g_1	22; 13	17; 17	17; 18	18; 14	12; 14	12; 15	8; 5	5; 6	5; 6
g_2	1; $\underline{-575}$	−10; $\underline{-507}$	27; $\underline{-327}$	11; $\underline{-417}$	5; $\underline{-384}$	23; $\underline{-277}$	4; $\underline{-95}$	6; $\underline{-85}$	10; $\underline{-61}$
g_3	2; $\underline{-79}$	−5; $\underline{-75}$	7; $\underline{-51}$	0; $\underline{-60}$	0; $\underline{-49}$	7; $\underline{-36}$	2; $\underline{-13}$	2; −5	−3; $\underline{-13}$
g_4	5; $\underline{-135}$	−7; $\underline{-128}$	12; $\underline{-84}$	1; $\underline{-106}$	2; $\underline{-90}$	14; $\underline{-65}$	2; $\underline{-25}$	3; $\underline{-13}$	−1; $\underline{-20}$
					Log Gaussian				
g_1	16; 9	7; 4	21; 22	16; 13	11; 9	19; $\underline{21}$	7; 5	6; 6	10; $\underline{11}$
g_2	60; $\underline{-549}$	79; $\underline{-481}$	75; $\underline{-301}$	108; $\underline{-358}$	127; $\underline{-336}$	64; $\underline{-230}$	94; −47	53; −75	51; −38
g_3	9; $\underline{-39}$	1; $\underline{-39}$	9; $\underline{-24}$	4; $\underline{-26}$	3; $\underline{-19}$	$\underline{11}$; $\underline{-10}$	−1; −4	1; 0	3; −2
g_4	18; $\underline{-144}$	−1; $\underline{-140}$	23; $\underline{-88}$	16; $\underline{-97}$	15; $\underline{-80}$	$\underline{33}$; $\underline{-47}$	2; $\underline{-19}$	6; −9	13; −6
					Mixed Gaussian				
g_1	5; 12	16; 23	−12; −15	7; 13	8; 14	−10; −13	4; 7	8; 10	*
g_2	$\underline{-86}$; $\underline{-423}$	$\underline{-99}$; $\underline{-329}$	$\underline{-55}$; $\underline{-203}$	$\underline{-79}$; $\underline{-332}$	$\underline{-76}$; $\underline{-262}$	$\underline{-40}$; $\underline{-163}$	$\underline{-39}$; $\underline{-105}$	$\underline{-30}$; $\underline{-52}$	*
g_3	−1; $\underline{-92}$	−11; $\underline{-87}$	$\underline{-19}$; $\underline{-73}$	−1; $\underline{-69}$	−12; $\underline{-56}$	$\underline{-17}$; $\underline{-48}$	10; $\underline{-11}$	−1; −4	*
g_4	−9; $\underline{-122}$	$\underline{-22}$; $\underline{-111}$	$\underline{-29}$; $\underline{-87}$	−6; $\underline{-94}$	$\underline{-24}$; $\underline{-81}$	$\underline{-27}$; $\underline{-65}$	10; $\underline{-18}$	−7; $\underline{-11}$	*

SUBSTITUTION RANDOM FUNCTIONS

Christian LANTUÉJOUL
Centre de Géostatistique
Ecole des Mines
35 rue Saint-Honoré
77305 Fontainebleau
France

ABSTRACT. The construction of the new family of random functions presented here was inspired by cartographic techniques. In exactly the same way that a map is a combination of spatial information and an appropriate representation, a substitution random function is a combination of a spatial random function and a coding process. Explicit calculations can be derived on substitution random functions. Moreover, non conditional and conditional simulations can be carried out to produce outcomes that respect some prescribed morphology.

1 Introduction

In order to introduce substitution random functions (s.r.f.), it is convenient to resort to an analogy coming from cartography. A map is commonly defined as a conventional representation of a spatial phenomenon. Therefore a map combines two distinct ingredients: firstly the spatial phenomenon itself, secondly its representation. Although it seems of minor interest, the second ingredient should not be neglected since it makes the map legible. It usually relies on a coding of the spatial information. In a more formal setting, a map can be defined as $X[T(x)]$ where $T(x)$ denotes the spatial phenomenon under study and $X(t)$ stands for the coding process.
Substitution random functions are built exactly in the same way as maps. $Y(x)$ is said to be a substitution random function if it can be written as $Y(x) = X[T(x)]$ where $T(x)$ is a random function (from now on, called the *directing function*), and $X(t)$ is a stochastic process (called the *coding process*). Various models for $X(t)$ and $T(x)$ are available which can be combined in such a way that $Y(x)$ has outcomes with some prescribed morphology (cf. sections 2 and 3). In the case where $X(t)$ and $T(x)$ are independent, explicit calculations can be derived, namely the bivariate distribution of $Y(x)$ (cf. section 4). Finally, algorithms for the non conditional and for the conditional simulation of $Y(x)$ will be presented (cf. section 5).

2 A typical example

In this example, we adopt for the directing function T the construction by Chentsov (1957). Consider a network of Poisson lines in $\mathbb{R}^2$. Each line d of the network separates $\mathbb{R}^2$ into two half-planes. All the points on one half-plane are given a random weight U_d, and all the points on the other $-U_d$. We thus have defined a random function on $\mathbb{R}^2$, denoted by V_d, and which takes only two values U_d or $-U_d$ (the points of the line d need not be valued since they constitute a negligible set in $\mathbb{R}^2$). We proceed the same way independently for all the lines of the network. Then we put

$$T(x) = \sum_d [V_d(x) - V_d(o)] \tag{1}$$

Since $V_d(x) = V_d(o)$ when the two points o and x belong to the same half-plane limited by the line d, formula (1) involves only a finite summation over the lines hitting (o, x). A standard result is that T has stationary increments, but is not stationary ($T(o) = 0$).

As for the coding process X, we consider a two-state continuous Markov chain (Feller, 1971) independent of T. If $a_0 \exp(-a_0 t)$ and $a_1 \exp(-a_1 t)$ are the exponential sojourn time distributions at the states 0 and 1, the process X has the stationary univariate distribution

$$p_0 = \frac{a_1}{a_0 + a_1} \qquad p_1 = \frac{a_0}{a_0 + a_1} \tag{2}$$

and its bivariate distribution is given by

$$p_{ij}(t) = p_i p_j + (-1)^{|i-j|} p_0 p_1 e^{-a|t|} \qquad 0 \le i, j \le 1 \tag{3}$$

with $a = a_0 + a_1$.

By randomizing with respect to $T(x)$ and $T(y)$, we immediately derive a formula for the bivariate distribution of $X[T]$

$$P\{X[T(x)] = i, X[T(y)] = j\} = p_i p_j + (-1)^{|i-j|} p_0 p_1 E\{e^{-a|T(x)-T(y)|}\} \qquad 0 \le i, j \le 1 \tag{4}$$

The calculation of the Laplace transform of $|T(x) - T(y)|$, shown in Appendix, gives

$$E\{e^{-a|T(x) - T(y)|}\} = \frac{1}{\pi} \int_{-\infty}^{+\infty} \frac{e^{2\mu|x-y|[\Phi(2au) - 1]}}{u^2 + 1} du \tag{5}$$

where μ is the Poisson intensity of the network and Φ is the Fourier transform of the weights assigned to the half-planes. Since the bivariate distribution is a function of the increment $x - y$, $X[T]$ is second order stationary. Its univariate distribution is p_0 and p_1, as can be seen by setting $x = y$ in equation (4). Note also that the Poisson intensity μ appears as a multiplicative factor of $|x - y|$. Therefore μ is a scale factor.

In order to bring out the importance of the weights to the morphology of the outcomes, two different examples are going to be considered:

i) the weights have a stable distribution

$$\Phi(u) = e^{-|u|^{\alpha}}$$

with $\alpha = 0.47$. The outcome depicted on Figure 1 gives a strong impression of spatial homogeneity. What we see is a texture.

Figure 1: Example of an outcome of a s.r.f. The strong spatial homogeneity suggests a texture.

ii) The weights take only the two values $+1$ and -1

$$\Phi(u) = \cos u$$

Here, in contrast, we no longer have the impression of spatial homogeneity (see Figure 2). The outcome has been displayed at 6 different scales, $1, 2, 4, 8, 16$ and 32, and at each scale we can observe the emergence of new structures.

What makes the two outcomes look so different? In the first example, the weight distribution is heavy-tailed. As a consequence, two close points x and y can have very different directing values $T(x)$ and $T(y)$, which in turn implies that $X[T(x)]$ and $X[T(y)]$ are practically uncorrelated. In the second example, the weights are never large, and the presence of channels on Figure 2 indicates that two points that are far apart can have a non-negligible correlation. Therefore, the two outcomes are derived from two random functions having their correlation functions with totally different rates of decrease.
Using equations (4) and (5), the correlation function of a s.r.f. is

$$\rho(h) = E\{e^{-a|T(x) - T(x+h)|}\} = \frac{1}{\pi}\int_{-\infty}^{+\infty} \frac{e^{2\mu|h|[\Phi(2au) - 1]}}{u^2 + 1}\, du \tag{6}$$

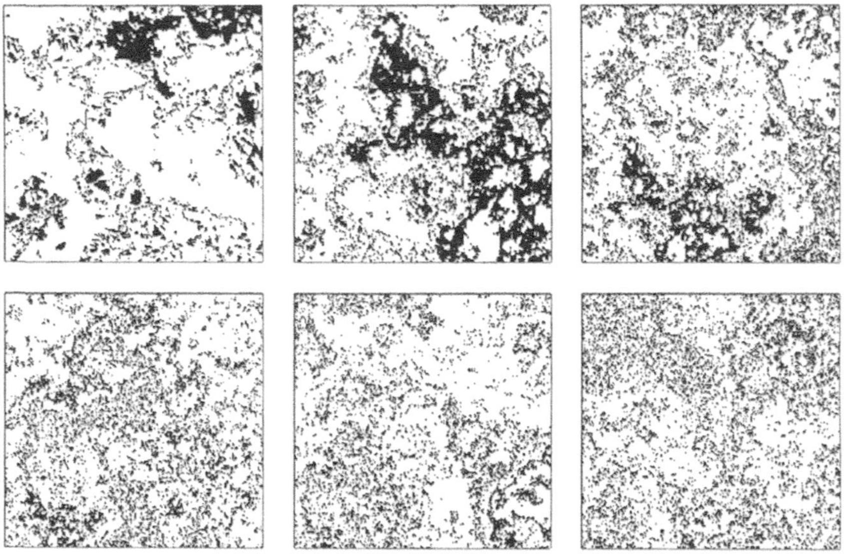

Figure 2: An outcome of another s.r.f. at 6 different scales. On each scale new structures are encountered, which bars any impression of spatial homogeneity.

In order to quantify its rate of decrease, it is convenient to introduce its integral range (Yaglom, 1986; Matheron, 1988; Lantuéjoul, 1991)

$$A = \int_{\mathbb{R}^2} \rho(h)dh \tag{7}$$

An integral range may be finite or infinite. It is finite on the first example (in the case where the weights follow a stable distribution the integral range is finite if $\alpha < 0.5$), and infinite on the second one.

Remark 1: The broken aspect of the two outcomes may suggest a fractal interpretation. As a matter of fact, these two outcomes are not fractal for their specific perimeter is finite

$$L_A = 2\mu p_0 p_1 \int_{-\infty}^{+\infty} \frac{1 - \Phi(2au)}{u^2 + 1} du < +\infty \tag{8}$$

Nor are they self-similar since they are derived from stationary random functions.

Remark 2: In the second example where $\Phi(u) = \cos u$, it is possible to derive an "explicit" formula for the correlation function

$$\rho(h) = e^{-2\mu|h|} \sum_{k=-\infty}^{+\infty} I_k(2\mu|h|)e^{-2a|k|} \tag{9}$$

where the I_k stand for the classical modified Bessel functions of the second kind (Abramovitz and Stegun, 1965).

3 On the morphology of the outcomes

Is it possible to compel the outcomes of a s.r.f. to respect some prespecified morphology? This question is important in many fields of application, such as the mining or oil industries, where structural information has been made available by the geologists and should be taken into account in a simulation exercise. Unfortunately, it is not possible to provide it a general answer, since the word morphology has a generic meaning which encompasses various metric, topologic, and geometric notions. In what follows, we merely give three illustrative examples.

3.1 Anisotropy

Anisotropies can be generated using Chentsov's construction by considering Poisson lines with a non-uniform orientation. If F is the distribution function of the orientations, then

$$\rho(h) = \frac{1}{\pi}\int_{-\infty}^{+\infty} \frac{e^{2\bar{\mu}(h)\,[\Phi(2au)-1]}}{u^2+1}du \tag{10}$$

with

$$2\bar{\mu}(h) = \int_0^{\pi} |< h,\theta >|\, dF(\theta) \tag{11}$$

This yields geometric, functional, or pure zonal anisotropies. Another possibility suggested by Prof. Matheron consists in using isotropic Poisson lines and in assigning the weights to the half-planes in a non-isotropic way. More precisely, let ω be an unit vector. Given any line d, we assign a positive weight to the half-plane that contains all the vectors in the direction ω, starting out from the line. This gives Figure $3-1$. Since the directing function T is monotonic increasing along the direction ω, the s.r.f. is markovian along that direction, and not markovian along the other ones. It is possible to complicate the model by introducing a disorientation factor. The weight assigned to the "supposedly positive" half-plane is positive with probability $1-r$, and has an arbitrary sign with the complementary probability r. Figures $3-2,\cdots,3-6$ depict several outcomes obtained using ± 1 weights and disorientation factors ranging from 0.2 to 0.8.

Let ϕ the Laplace transform of the positive weights. We put

$$\Phi(u) = \frac{\phi(iu)+\phi(-iu)}{2} \qquad \Psi(u) = \frac{\phi(iu)-\phi(-iu)}{2i} \tag{12}$$

to express the correlation function of the s.r.f.

$$\rho(h) = \frac{1}{\pi}\int_{-\infty}^{+\infty} \frac{e^{2\mu|h|[\Phi(2au)-1]}\cos[2\mu(1-r) < h,\omega > \Psi(2au)]}{u^2+1}\, du \tag{13}$$

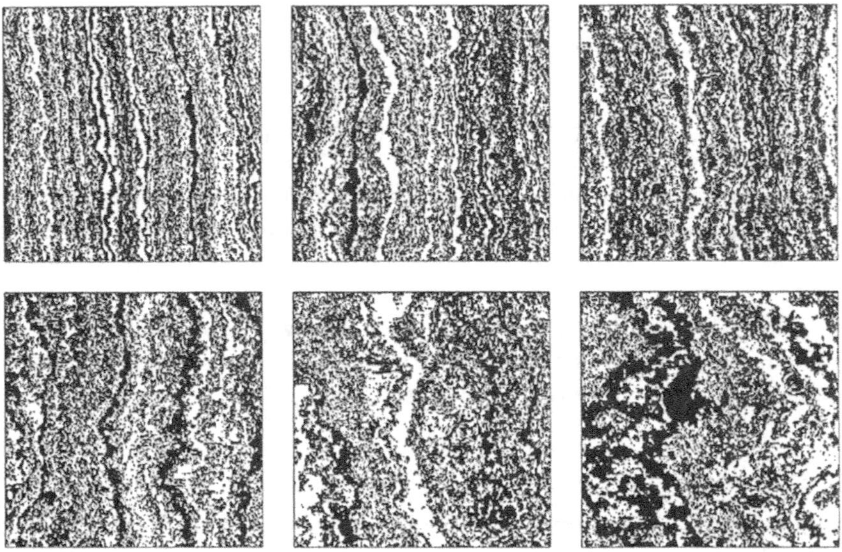

Figure 3: Some outcomes of anisotropic s.r.f.'s The direction of maximal variability is horizontal. The weights are ± 1. From top to bottom and left to right, The disorientation factors are equal to $0., 0.2, 0.4, 0.6, 0.7, 0.8$.

In this formula, the vector h appears in two places, firstly via its modulus $|h|$, and secondly via the scalar product $< h, \omega >$ accounting for the relative direction of h with respect to the direction of maximal variability ω. This gives rise to a functional anisotropy. One possible way to quantify this anisotropy is to assess the directional integral ranges

$$A(\alpha) = \int_{-\infty}^{+\infty} \rho(t, \alpha) \; dt \tag{14}$$

(see Figure 4). The integral range is finite except in the direction orthogonal to ω.

Remark 3: As it could be expected, disorientation only acts as a multiplicative factor of the scalar product in formula (13). In the particular case $r = 1$ (total disorientation), this formula simplifies to (6).

Remark 4: If the positive weights are set to 1, then the correlation function admits an expansion in terms of Bessel functions

$$\rho(h) = e^{-2\mu|h|} \sum_{k,l=-\infty}^{+\infty} I_k(2\mu|h|) \; J_l(2\mu(1-r) < h, \omega >) e^{-2a|k+l|} \tag{15}$$

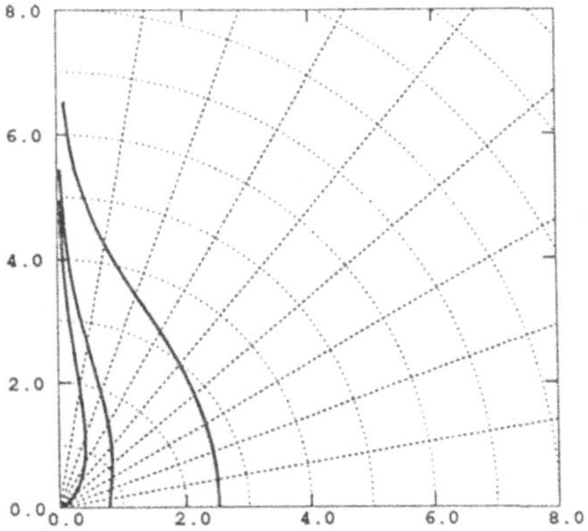

Figure 4: Plotting the logarithm of the integral range versus the direction. The disorientation factors are 0, 0.4 and 0.8.

3.2 Dilution

We have just seen how to build outcomes with some lineations in one or several directions. In contrast to this, the aim of this subsection is to design outcomes with locally radial structures. Let $\mathcal{P}$ be a planar Poisson point process. The distance function $d(x, \mathcal{P})$ to $\mathcal{P}$ is stationary. For the directing function we take

$$T(x) = d(x, \mathcal{P}) + T_0(x) \tag{16}$$

where $T_0(x)$ is a random function with stationary, isotropic increments. A typical example is given Figure 5.

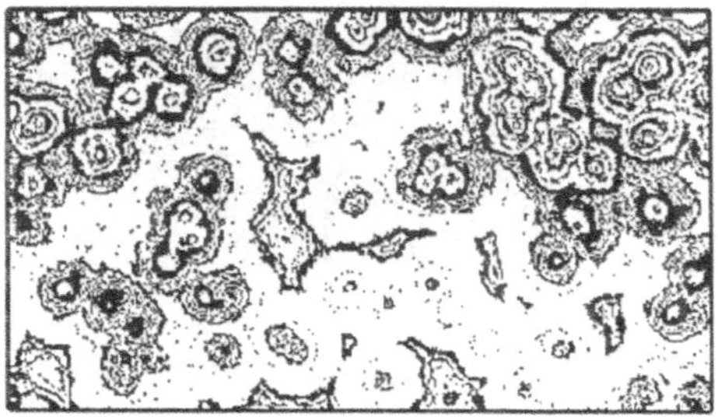

Figure 5: Dilution from a set of points.

In this example, $T_0(x)$ has been been obtained using Chentsov's construction. The coding process is a two-state continuous Markov chain with $p_1 = 0.7$ and $p_0 = 0.3$.

Remark 5: The same procedure can be applied by replacing the Poisson point process $\mathcal{P}$ by any stationary random set.

3.3 Contiguity

Suppose that we want to model some spatial phenomenon made up of several facies. The facies cannot be independent since they partition the whole space. There exists therefore some dependence relationships between the different facies. These relationships are particularly apparent if they concern the neighbourhoods of the facies. For instance, two different facies may not be in contact, or one facies must be surrounded by another one. Is it possible to set up a model for the spatial phenomenon, that fulfills these neighbourhood relationships?

A graph is quite convenient to describe the neighbourhood relationships between the different facies. To each facies i, a vertex, also denoted by i, is associated. Two different vertices i and j are connected by an edge if the corresponding facies are in contact (see Figure 6).

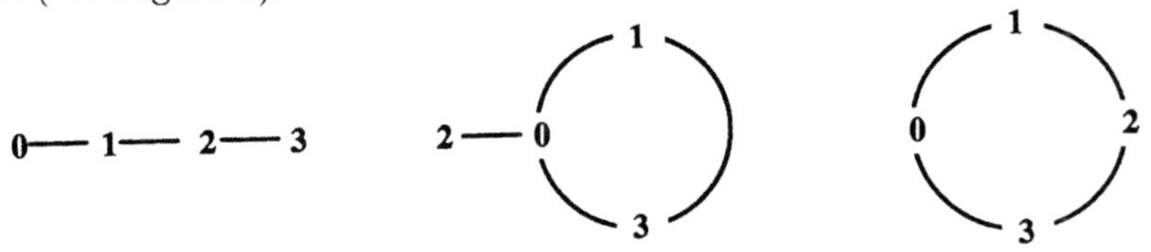

Figure 6: A graph is introduced to display the neighbourhood relationships between the different facies. Left, facies 0 is surrounded by facies 1. Middle, facies 0 surrounds facies 2, but has also contacts with facies 1 and 3. Right, facies 0 has no contact with facies 2.

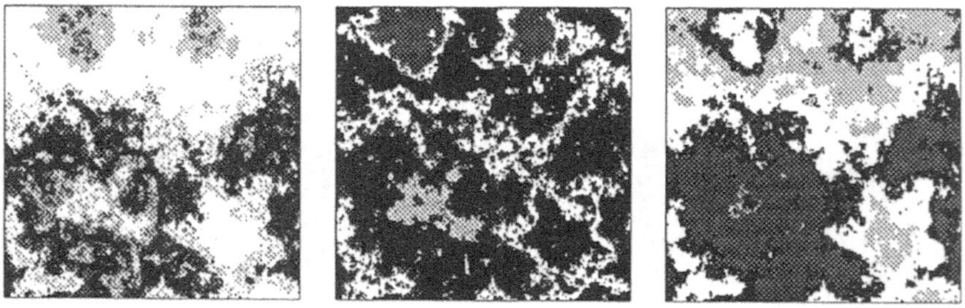

Figure 7: The neighbourhood relationships of these three outcomes are given by the three graphs of Figure 6. Facies 0 is black.

If judiciously chosen, the coding process makes it possible for the s.r.f. to respect the neighbourhood relationships. We take for the coding process a stationary, continuous process with a finite number of states that is equal to the number of facies. A direct transition from state i to state j can take place if and only if the two facies i and j are in contact. On Figure 7, we can see three outcomes corresponding to the three graphs of Figure 6. The three outcomes have been obtained using Markov coding processes.

Remark 6: To ensure that the contacts between the different facies of the s.r.f. are given by the direct transitions of the coding process, the directing function must have continuous outcomes.

4 Some properties of s.r.f.

In the previous sections, we have considered directing functions with stationary increments and stationary coding processes. Although such choices are in no way compulsory, they are agreeable for they convey some nice properties to the s.r.f.'s. If T and X are independent, then $X[T]$ is second order stationary, and moreover $X[T]$ and X have the same univariate distribution. To see that, let us denote by $\Phi_X(u)$ the univariate Fourier transform of X, and by $dF_x(t)$ the distribution of $T(x)$. The Fourier transform of $X[T(x)]$ can be written

$$E\left\{e^{iuX[T(x)]}\right\} = \int_{-\infty}^{+\infty} E\left\{e^{iuX(t)}\right\} dF_x(t) = \int_{-\infty}^{+\infty} \Phi_X(u) dF_x(t) = \Phi_X(u) \tag{17}$$

If $C(h)$ stands for the covariance function of X, we have

$$\text{Cov}\left\{X[T(x)], X[T(y)]\right\} = E\left\{C\left(|T(x) - T(y)|\right)\right\} \tag{18}$$

and since T has stationary increments

$$\text{Cov}\left\{X[T(x)], X[T(y)]\right\} = E\left\{C\left(|T(x-y) - T(o)|\right)\right\} \tag{19}$$

So the covariance between $X[T(x)]$ and $X[T(y)]$ depends only upon $x - y$.

Another interesting feature is isofactorial permanence. If the bivariate distribution of X admits an isofactorial expansion

$$E\left\{e^{iuX(s) + ivX(t)}\right\} = \sum_{n=0}^{+\infty} \rho_n(|s-t|)\hat{\chi}_n(u)\hat{\chi}_n(v) \tag{20}$$

where the $\hat{\chi}_n$'s denote the Fourier tranform of the factors, then we have

$$E\left\{e^{iuX[T(x)] + ivX[T(y)]}\right\} = \sum_{n=0}^{+\infty} E\left\{\rho_n(|T(x) - T(y)|)\right\} \hat{\chi}_n(u)\hat{\chi}_n(v) \tag{21}$$

and since the distribution of $T(x) - T(y)$ depends only on $x - y$, we can write

$$E\left\{e^{iuX[T(x)] + ivX[T(y)]}\right\} = \sum_{n=0}^{+\infty} \sigma_n(|x-y|)\hat{\chi}_n(u)\hat{\chi}_n(v) \tag{22}$$

which is another isofactorial expansion with the same factors (Matheron, 1989).

Remark 7: If T is stationary, it can be shown that $X[T]$ is second order stationary, but not ergodic. The assumption that T is not stationary but has stationary increments guarantees that $X[T]$ is second order stationary and ergodic.

5 Simulations

The previous sections of this paper have been illustrated with non conditional simulations. Carrying out such simulations is rather straightforward provided that one can simulate the directing function and the coding process. The algorithm for the non conditional simulation of a s.r.f. $X[T]$ in a field D is given below:

i) simulation of the directing function T in the field D.
ii) simulation of the coding process X at the values taken by the simulated directing function.
iii) assignment of the values.

To facilitate ii), it is recommended to sort the values taken by the simulated directing function into increasing order.

The conditional simulation of a s.r.f. is more tricky. In order to simulate $X[T]$ in a field D under the condition that $X[T(x_\alpha)] = z_\alpha$ for $\alpha = 1, ..., n$, the following algorithm can be used:

i) non conditional simulation of the coding process X.
ii) simulation of the $T(x_\alpha)$ under the constraint that $X[T(x_\alpha)] = z_\alpha$. Let t_α the generated values.
iii) conditional simulation of T in the field D given $T(x_\alpha) = t_\alpha$.
iv) conditional simulation of X at the values taken by the directing function given $X(t_\alpha) = z_\alpha$.
v) assignment of the values.

Figure 8: Example of a conditional simulation of a s.r.f. Left, a non conditional simulation. Middle, the conditioning points. They have been assigned the values of the non conditional simulation. Right, the conditional simulation.

This algorithm requires conditional simulations of both the directing function and the coding process (steps iii) and iv)). But the most difficult step is certainly ii). To carry out a simulation under constraints, one can resort to an algorithm inspired by Metropolis (1953), but other algorithms are available, namely in a multigaussian

context (Langlais (1990), Freulon (1992)). A simulation under constraint is certainly not possible in all cases. Some limitations must be placed on the coding process that prescribes the constraints. For instance, a coding process with fractal trajectories (e.g. continuous diffusion process) should be excluded.
Figure 8 gives an example of a conditional simulation of a two state s.r.f. The values taken by the conditioning points (middle) come from a non conditional simulation (left). Although conditioning aims at respecting local features, it can be noted that regional structures are rather well preserved at the vicinity of the conditioning points.

References

Abramowitz M. and Stegun I. (1965) Handbook of Mathematical Functions, Dover Pub., New York.
Chentsov N. N. (1957) "Lévy brownian motion for several parameters and generalized white noise", Th. of Prob. and its Applic., pp. 265,266.
Feller W. (1971) An introduction to Probability Theory and its Applications, Vol. 2, Wiley, New-York.
Freulon X. (1992) "Conditionnement du modèle gaussien par des inégalités ou des randomisées", Geostatistics Doctoral Thesis, School of Mines of Paris.
Freulon X. and de Fouquet C. (1992) "Conditioning a Gaussian model with inequalities", in these Proceedings.
Langlais V. (1990) "Estimation sous contraintes d'inégalités", Geostatistics Doctoral Thesis, School of Mines of Paris.
Lantuéjoul Ch. (1991) "Ergodicity and Integral Range", Journal of Microscopy, Vol. 161-3, pp. 387-404.
Matheron G. (1988) Estimating and Choosing, Springer-Verlag, Berlin.
Matheron G. (1989) "Two classes of isofactorial models", in Armstrong M. (ed.) "Geostatistics", Kluwer Academic Publishers, Dordrecht, pp. 309-322.
Metropolis N. and al. (1953) "Equations of State Calculations by fast Computing Machines", The Journal Of Chemical Physics, Vol 21-6, pp. 1087-1092.
Yaglom A. M. (1986) Correlation Theory of Stationary and related Random Functions, Springer Series in Statistics, New York.

Appendix: calculation of the correlation function of $X[T]$.

According to equation (6), the Laplace transform of $|T(x)-T(y)|$ has to be calculated. Let U be a Cauchy variable of density

$$f(u) = \frac{1}{\pi}\frac{1}{u^2+1}$$

It is well known that its Fourier transform is

$$E\left\{e^{itU}\right\} = e^{-|t|}$$

Then, If U is independent of T, we have

$$E\left\{e^{-a|T(x)-T(y)|}\right\} = E\left\{e^{iaU\,(T(x)-T(y))}\right\}$$

and, by deconditioning on U, we get

$$E\left\{e^{-a|T(x)-T(y)|}\right\} = \frac{1}{\pi}\int_{-\infty}^{+\infty} \frac{E\left\{e^{iau\,(T(x)-T(y))}\right\}}{u^2+1} du$$

Thus we have to calculate the Fourier transform of $T(x)-T(y)$. Since equation (1) implies

$$T(x)-T(y) = \sum_{d} [V_d(x) - V_d(y)] = 2 \sum_{d\uparrow[x,y]} V_d(x)$$

(the notation $d \uparrow [x,y]$ means that the line d crosses the segment $[x,y]$) and since the number of Poisson lines crossing the segment $[x,y]$ is Poisson distributed with mean of $2\mu|x-y|$, we only have to calculate the Fourier transform of the sum of a Poisson number of random weights. This classical calculation is achieved by randomizing upon the Poisson variable. We obtain

$$E\left\{e^{iau\,(T(x)-T(y))}\right\} = e^{2\mu|x-y|\,[\Phi(2au)-1]}$$

and finally

$$E\left\{e^{-a|T(x)-T(y)|}\right\} = \frac{1}{\pi}\int_{-\infty}^{+\infty} \frac{e^{2\mu|x-y|[\Phi(2au)-1]}}{u^2+1} du$$

TOWARDS A DIRECT STRUCTURAL ANALYSIS OF AN IRF-k

Roberto BRUNO, Giuseppe RASPA
Department of I.C.M.M.P.M.
Univ. of Rome "La Sapienza
Via Eudossiana, 18
00184 Rome, Italy

ABSTRACT. The automatic structure recognition of an Intrinsic Random Function of order k (IRF-k) is the solution that is available today and fully implemented for modelling the Generalised Covariance (GC). However, in many cases, still working in the frame of the IRF-k, it could be interesting to be able to infer the GC directly. This paper proposes a way for the direct modelling of an IRF-k based on the identification of particular authorised measures. These measures are such that the variances of the consequent experimental generalised increments, in accordance with the usual elementary polynomial models of GC, could be expressed as functions of only one parameter. Hence, it is possible to try to model the GC practically like the stationary case where the experimental variogram is plotted versus the distance h. Any such particular measure can be identified by solving one non-linear system. This paper presents the formalism of the methodology, then provides some comparisons for different cases of GC modelling, by automatic recognition and by the proposed direct modelling.

1. INTRODUCTION

In geostatistics practice many problems must be addressed by working in the framework of non-stationary random functions (RF), and the IRF-k approach, where the structural analysis that is proposed and implemented is essentially an automatic procedure, is always attractive (Renard, 89).

In fact, automatic recognition of the generalised covariance (GC) model has always left a sense of dissatisfaction in people like "stationary geostatisticians", who are accustomed to adjusting the variograms manually. The latter could be considered members of the "school of thought" that prefers "human" control over the selection of the models. This does not mean that the controls of the solutions proposed and implemented by the automatic structural analysis programmes of an IRF-k are not adequate, only that this is still automatic control.

Interactive modelling has been proposed for cases where the data are arranged according to a regular grid. However, the "generalised variogram" or the "variogram and covariance of increments" introduced by Matheron (1971) and studied by Chiles (1979) and Chauvet (1981, 1989) have apparently not been widely applied; this is understandable because the data that are available are rarely arranged on a regular grid.

The problem of how a direct inference of the GC can be obtained therefore remains. This paper proposes a technique for the non-automatic modelling of the GC that starts by identifying authorised measures with a special property: namely the ability to generate the variances of the increments (or equivalent generalised variogram) as a function of a single parameter which is selected on the basis of the elementary models of the GC.

2. PRINCIPLES OF THE METHOD

Given an IRF-k, the central formula of the variance of an increment of order-k is:

$$E[Z(\lambda)^2] = \Sigma_\alpha \Sigma_\beta \lambda_\alpha \lambda_\beta K(h_{\alpha\beta}) \quad (1)$$

where, as usual, K(h) represents the GC and λ_α, λ_β the weights of the linear combination, or measure, authorised for order k (ALC-k) attributed to the n data: $\{\lambda_\alpha \ \alpha = 1,n\}$.

It is noteworthy that this relation is valid for any ALC-k. Thus, many different increments can be defined for a same set of data $\{Z_\alpha \ \alpha = 1,n\}$, and all of which will be authorised linear combinations for the same data but using different sets of weights $\{\lambda_\alpha\}$. Different variances are obtained for each authorised measure because the increments are of course different. Each variance can be expressed using the quadratic form (1), a function of the variables λ_α and whose coefficients matrix, $K(h_{\alpha\beta})$, is fixed for a fixed set of data, being unequivocably defined by the model of the GC and by the arrangement of the available data.

Let us now consider the case of interactive modelling of a variogram under stationary conditions: this is essentially based on an analysis of the experimental estimates of the second moment of an increment of order 0 examined as a function of one parameter: distance h between two points. In practice, having selected a distance class {h'}, the average squared difference of the RF values is constructed for the pairs where the distance h falls in the class $\{h'\}$. The modelling consists in fitting a model function $\gamma(h)$, whose independent variable is precisely the parameter h, to the experimental values.

Estimates of the second moment of increment of order k, namely the variance of the increments, can also be constructed under non-stationary conditions. Continuing the comparison between intrinsic conditions of orders 0 (I-0) and k (I-k), we see that:

- in both cases, the variance of the increments is dependent on the model of spatial self-correlation $[\gamma(h), K(h)]$; this model is the goal of this phase of the structural analysis;
- the configuration depends on a single parameter in the case of I-0 (h), but on a matrix of parameters in the case of I-k: assuming that n points are arranged at random, with $n \geq k+2$, this matrix of dimension (nxn) is given by the set of the distances $\{h_{ij}\}$ which are generally different;
- the authorised measure is fixed in the case of I-0 ($\lambda_1 = 1$, $\lambda_2 = -1$), but variable in the case of I-k, even for the same configuration of the data.

Direct modelling of the GC on the basis of a parametric representation of the variance of the increments of order-k is therefore not immediate. But it is worth recalling that parametrisations of variances expressed in complex forms have already been made under stationary conditions. For example, in Journel and Huijbregts (1978) it is possible to find some parametrisations of variances for a given geometrical figure (rectangle or parallelipiped), based on a fixed configuration of the data (the four vertices, the central section) as a function of a geometrical parameter (the side of the square, the ratio between the sides). These variance parametrisations vary, of course, following the model of the covariance/variogram function.

The idea is, therefore, to find a parametrisation [p] of the generalised increment $\{\lambda_\alpha Z_\alpha\}$ such that its variance $E[Z(\lambda)^2] = \Sigma_\alpha \Sigma_\beta \lambda_\alpha \lambda_\beta K(h_{\alpha\beta}) = K(\lambda_\alpha, h_{\alpha\beta})$ can be expressed as K(p), thereby permitting the modelling of the GC but without imposing conditions with regard to the location of the points $\{x_\alpha\}$ considered by the measure.

The first observation is that the conditions on the weights $\{\lambda_\alpha\}$, which define an authorised measure of order-k, must be maintained:

$$\Sigma_\alpha \lambda_\alpha x_\alpha^l = 0 \qquad l = 0,k$$

These are in fact the (necessary) conditions that guarantee the possibility of expressing the variance of the increments of an IRF-k as linear combinations of the GC (1).

Second, it is seen that, chosen the arbitrary class of polynomial GC's models, in practice a combination of four elementary models is used:

$$K(h) = b_0\,\delta(h) + b_1\,h + b_3\,h^3 + b_s\,h^2\log(h) = \Sigma_m\, b_m\, f_m(h) \qquad (2)$$

where:

- $\delta(h)$ is the Dirac measure ($\delta(0) = 1$, $\delta(h) = 0$ for $h > 0$);
- the coefficients $\{b_m\}$ must respect certain conditions as a function of the dimensions of the considered space;
- h, as usually, represents the distance between two points.

Therefore, the variance of the increments as a function of the elementary terms of the GC is expressed as follows:

$$E[Z(\lambda)^2] = K(\lambda) = \Sigma_\alpha \Sigma_\beta \lambda_\alpha \lambda_\beta \Sigma_m\, b_m\, f_m(h_{\alpha\beta}) \qquad (3)$$

When modelling the GC, the $\{\lambda_\alpha\}$ are known while the $\{b_m\}$ coefficients are unknown.

Let us therefore assume that the variance (3) can be expressed as a function of one parameter p by means of a linear combination of the known functions $g_m()$ and of the unknown coefficients b_m which we want to infer:

$$E[Z(\lambda)^2] = K(p) = \Sigma_m\, b_m\, g_m(p) \qquad (4)$$

Recalling that the variogram coincides with the variance of an ALC-k, multiplied by a constant, expressed as a function of the parameter h, it is an attractive idea to consider K(p) to be an "equivalent generalised variogram".

The equality of the two expressions of the variance of the increments provides us with (cfr. 3 and 4):

$$\Sigma_m\, b_m\, g_m(p) = \Sigma_\alpha \Sigma_\beta \lambda_\alpha \lambda_\beta \Sigma_m\, b_m\, f_m(h_{\alpha\beta}) \qquad (5)$$

And for the relation (5) to be true regardless of the value of the coefficient $\{b_m\}$, it is sufficient that the following be true for each m:

$$g_m(p) = \Sigma_\alpha \Sigma_\beta \lambda_\alpha \lambda_\beta f_m(h_{\alpha\beta}) \qquad (6)$$

For (6) to be easy to solve, the functions $g_m(p)$ should be simple functions of p, or at least easily identifiable. It therefore seems natural to start by ensuring that the $g_m()$ coincide with the $f_m()$, which are the elementary polynomial models (2).

However, a closer analysis of this proposal imposes several modifications and suggestions.

With regard to the nugget effect ($m=0$), the identity $g_0()=f_0()$ will give the relation $\delta'(p) = \Sigma_\alpha \Sigma_\beta \lambda_\alpha \lambda_\beta \delta(h_{\alpha\beta}) = \Sigma_\alpha \lambda_\alpha^2$ which cannot be solved because, unlike the Dirac measure of the parameter p, the quadratic form of the nugget terms of distances will always result in a strictly positive value, variable with changes in the measures. Let us therefore redefine the $g_0()$ as the complement of the density of Dirac $g_0(p) = 1 - \delta(p) \Rightarrow g_0() = 1 - f_0()$. In this case it is possible solving the relation obtained: $g_0(p>0) = 1 = \Sigma_\alpha \lambda_\alpha^2$.

Observing the relation (6) with the linear term ($m=1$), in case of identity $g_1()=f_1()$ we obtain: $p = \Sigma_\alpha \Sigma_\beta \lambda_\alpha \lambda_\beta h_{\alpha\beta}$, which means that the value of the parameter p should always be negative. In fact, as $b_1 p = b_1 \Sigma_\alpha \Sigma_\beta \lambda_\alpha \lambda_\beta h_{\alpha\beta}$ is the variance of an increment, it is always a positive quantity and always $b_1 < 0$. It is, of course, necessary to ensure that the parameter p remains positive, if for no other reason than that it appears in the spline term of the GC as an argument of a logarithm (cfr. 7). It is thus necessary to redefine $g_1() = -f_1()$.

After taking account of the choices made and maintaining $g_s()=f_s()$ and $g_3()=f_3()$, we obtain the system (7). Observing the third equation of this system, it is seen that the term to the left of the equal sign can be negative for $p<1$, whereas the term to the right is almost always positive. The simplest solution would seem to be to focus on controlling the values of parameter p to be used.

Considering all the above, we propose to solve the following system:

$$\begin{aligned} g_0(p) &= 1 - \delta(p) = \Sigma_\alpha \Sigma_\beta \lambda_\alpha \lambda_\beta \delta(h_{\alpha\beta}) = \Sigma_\alpha \lambda_\alpha^2 \\ g_1(p) &= -p = \Sigma_\alpha \Sigma_\beta \lambda_\alpha \lambda_\beta h_{\alpha\beta} \\ g_s(p) &= p^2 \log(p) = \Sigma_\alpha \Sigma_\beta \lambda_\alpha \lambda_\beta h_{\alpha\beta}^2 \log(h_{\alpha\beta}) \\ g_3(p) &= p^3 = \Sigma_\alpha \Sigma_\beta \lambda_\alpha \lambda_\beta h_{\alpha\beta}^3 \end{aligned} \qquad (7)$$

With the normal method of identifying the authorised measures $\{\lambda_\alpha\}$, it is practically impossible to obtain a parameter p that can automatically satisfy all the conditions of the system (7).

Conversely, it is possible to consider constructing "ad hoc" authorised measures {1}. In fact, if the conditions that transform {1} into authorised measures and the conditions expressed by the system (7) are considered, the following sequence of non-linear equations is obtained:

$$\begin{aligned} &\Sigma_\alpha \lambda_\alpha x_\alpha^l = 0 && l=0,k \\ &\Sigma_\alpha \Sigma_\beta \lambda_\alpha \lambda_\beta f_m(h_{\alpha\beta}) - g_m(p) = 0 && m=0,1,3,s \end{aligned} \qquad (8)$$

3. IMPLEMENTATION OF THE ALGORITHM

First, several roots are generally admissible because system (8) is non-linear in $\{\lambda_\alpha\}$. Further, two different strategies can be proposed for searching for solutions:

- solution with an unknown parameter: in two dimensions, the system (8) for an IRF-2 comprises ten equations: it therefore is appropriate to consider increments of at least nine points since the tenth unknown is the parameter p;
- solution with a known parameter: when the value of parameter p is imposed, the system (8) in two dimensions - again for an IRF-2 - calls for ten unknowns; 10 is therefore the minimum number of increments to be considered.

The approach with the known parameter was adopted to try to solve this system. Although it cannot be said that selecting a value for parameter p is easy, also because there is a significant risk that the system may not be solvable, careful selection of the values of p (well-matched values) will facilitate the modelling phase described later. Further, the homogeneity in terms of size and number of the unknowns will enhance the robustness of the algorithms developed for the solution.

In practice, the procedure begins by reducing system (8) to a system of four quadratic forms with four unknowns by using the linear equations to solve for some of the λ_αs in terms of the remaining four. The method of successive substitutions is then applied to search for the solution by adopting different values for the parameter p.

The algorithm implemented proceeds in the following phases, by trial and error:

1) automatic definition, using a fixed criterion, of the interval of the values for the parameter p, of the corresponding sampling increment Δp and of the neighbourhood wherein the solution is to be sought;
2) search for the solution for all the values assigned to p: if the solution exists, it will generally be found;
3) if one solution is found, reduction of the sampling increment, e.g. $\Delta p' = \Delta p/10$, then detection and memorisation of all the possible solutions for the successive values of p.

The algorithm seems robust and provides valid results provided that a maximum of two quadratic forms are considered. This does not appear to be a particularly heavy limitation at the current stage of the study, because the modelling of the GC rarely involves more than two elementary structures. On the other hand, the programme, which is implemented in Turbo-Pascal, permits the interactive selection of any two of the four elementary structures considered before proceeding with the verification of the different solutions.

It is useful to carry out a preliminary structural analysis in the traditional automatic manner, both to identify the order k of the regionalisation and to obtain an initial idea of the probable GC model. This information makes it possible to run the programme for direct modelling of the GC interactively, by stating the number of constraints for generating the authorised measures for order-k and by orienting the search for the two elementary models to be considered. However, even an exhaustive search would not involve more than six runs.

The whole set of available data is sampled iteratively by means of a "moving neighbourhood" type of technique, in order to obtain increments that describe the whole field. Neighbourhood dimensions can be modified. At any rate, for each configuration that is found, several solutions will be generated as a function of the different values imposed for the parameter p and this will make it possible to generate many increments.

4. MODELLING OF THE GENERALISED COVARIANCE

To solve system (8) for different increments means that an authorised measure $\{\lambda\}$ must be identified for each increment and, generally, an associated value for the parameter p.

It is then possible to plot, on a plane $K(\lambda)/p$, the experimental values of the variance of the increments $K^*(\lambda) = \Sigma_\alpha \Sigma_\beta \lambda_\alpha \lambda_\beta Z_\alpha Z_\beta$ calculated using the measures obtained by solving system (8) as a function of the parameter p associated with the measures themselves.

The modelling is the immediate result of the estimation of the coefficients b_m^* that generate, with respect to the experimental points, the best fit for the function:

$$K(p) = \Sigma_m b_m g_m(p) =$$
$$= b_0 [1 - \delta(p)] + b_1 (-p) + b_3 p^3 + b_s p^2 \log(p) \quad (9)$$

This is just the procedure for stationary modelling of a variogram with a nested structure.

If the coefficients b_m^* describe the true b_m correctly, we know from (4) that we are actually estimating the variance of the increment associated with each parameter p and hence the coefficients of the model of the GC of the IRF-k. And we know, in fact, that $E[K^*(\lambda)] = K(\lambda)$ [Matheron, 1971].

We know that the coefficients of the GC cannot have just any value, because the elementary models of the GC considered must express functions defined as positive. In particular, in the plane (in R^2):

$$b_0 \geq 0, \quad b_1 \leq 0, \quad b_s \geq -3/2\,(|b_1|\,b_3)^{1/2}, \quad b_3 \geq 0$$

All the above conditions are easily verified. In fact, since two elementary models are examined each time, it suffices to impose the condition that the spline coefficient, b_s, must be positive since it is assumed that either b_1 or b_3 is zero.

As we cannot know, a priori, the pair of elementary covariances (f_m) to which the IRF-k will be associated, a few attempts may be necessary. Often it is sufficient the visual observation of the form of the $K^*(p)$ obtained with any two models to detect the congruence. The most difficult situations can be addressed by attempting to carry out the complete modelling, then assigning numerical values to the two b_m coefficients considered at the time and verifying the fitting.

An essential condition for direct modelling is the availability of experimental data that allow us to define well-matched values for the parameter p. It is clear that measures clustered around a narrow interval of values for p will make modelling difficult. As in the case of automatic recognition, it is necessary to have well-matched increments of order-k.

5. CASE STUDY

To test the operational characteristics of the proposed method, several IRF-k in R^2 were simulated for different orders-k, different models for GC and using a square grid with 2,500 points. A set of 100 data items (fig. 1) was selected at random from the simulation and treated using both the classical method of automatic structural analysis [Renard, 1989] by means of the FAIKPACK software package

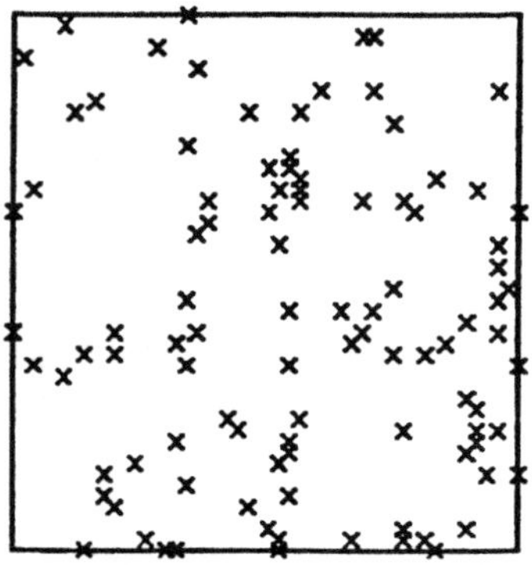

Fig.1 - Sample data location selected from simulations of a 50x50 sqare grid

STRUCT. AN. 2	FAIKPACK Universita' di Roma "La Sapienza"	Release 1.2

ORDER-K IDENTIFICATION

A)

order-k	average rank	mean sq. err.
1	1.53448300	.50704870E-02

AUTHORISED COVARIANCE MODELS COEFFICIENTS

N.	B0	B1	B3	Bs	Jack-knife
1	.000000	.000000	.000000	.689392E-08	1.0771
2	.000000	.000000	.46036E-11	.000000	1.1127
3	.000000	-.774304E-05	.000000	.000000	1.1804
7	.949938E-05	-.771731E-05	.000000	.000000	1.1812

B)

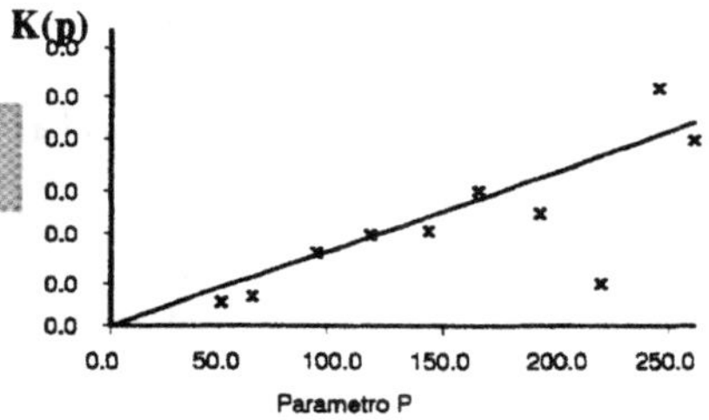

Fig.2 - Simulation of a FAI-1 with $K(h) = -1.4\times10^{-5}$ h:
A) Automatic structure inference ($b_1 = -0.77\times10^{-5}$);
B) Interactive structure inference ($b_1 = -1.3\times10^{-5}$).

[Bruno and Raspa, 1991] under PC-MS/DOS (and VAX-VMS), and the interactive structural analysis procedure described in this paper.

Figures 2a and 2b show the results of an example of comparison between the automatic (a) and direct (b) structural analyses. An IRF was simulated with an order $k=1$ and a linear GC with $b_1=-1.4\times10^{-5}$. The correct order-k was detected experimentally in the moving neighbourhood with the traditional techniques. The application of the automatic technique in a moving neighbourhood was unable to obtain a clear interpretation of the GC model. In fact, the linear model is relegated preferentially to the third order on the basis of the "jack-knife" criterion. The linear nature of $K(\lambda)$ was brought out more clearly by the direct analysis of the GC and selecting the linear and spline models as the pair of elementary models. Further, the estimate of the coefficient seems to be more robust.

Figures 3a and 3b show the results of the comparison of the structural analysis on an other example. An IRF-2 was simulated with a GC that is the sum of a nugget term and a linear term ($b_0=1/12$, $b_1=-1.4\times10^{-5}$). In this case as well, it is not easy to identify the true GC model using the automatic method, as corroborated by the

FAIKPACK

STRUCT. AN. 2 | Universita' di Roma "La Sapienza" | Release 1.2

A)

ORDER-K IDENTIFICATION

order-k	average rank	mean sq. err.
2	1.23706900	.2578379000

AUTHORISED COVARIANCE MODELS COEFFICIENTS

N.	B0	B1	B3	Bs	Jack-knife
4	.137072	.000000	.000000	.000000	.66523
2	.000000	.000000	.105987E-08	.000000	.45680
3	.000000	-.637797E-03	.000000	.000000	.39798
1	.000000	.000000	.000000	.842691E-06	.38366

B)

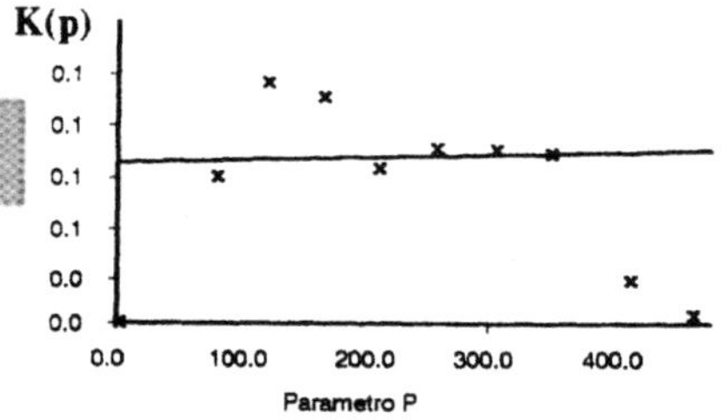

Fig.3 - Simulation of a FAI-2 with $K(h) = 0.083 - 1.4 \times 10^{-5}\ h$:
A) Automatic structure inference ($b_0 = 0.137$);
B) Interactive structure inference ($b0 = 0.08$, $b_1 = -1.3 \times 10^{-5}$).

unsatisfactory values of the "jack-knife". Direct analysis seems to be even more effective in this difficult case.

Similar results have been observed on the several IRF simulated and only in few cases the interactive inference has been very difficult.

6. CONCLUSIONS

Direct structural analysis of a generalised covariance appeals to people working with IRF-k. The proposed solution of searching for appropriate measures that permit the parametrisation of the variance of increments seems to be a potentially valid route that has provided encouraging results. However, full testing of the method will require targeted research aimed at establishing the theoretical reliability of the results, at making the non-linear algorithms more general and effective, and at identifying stable and improvable procedures, e.g. for selecting the configurations of the data. The authors feel that the prospects for further research on direct analysis of the GC are promising.

REFERENCES

1) Renard, D. (1989), "Automatic structure recognition", in Geostatistics, Vol. II, Kluwer Academic Publisher, Dordrecht, pp. 579-590.

2) Matheron, G. (1971), "La theorie des fonctions aleatoires intrinseques generalisees" (The theory of generalised intrinsic random functions), Centre de Geostatistique, ENSMP, Fontainebleau, pp. 64.

3) Chiles, J.P. (1979), "Le variogramme generalisee" (The generalised variogram), Centre de Geostatistique, ENSMP, Fontainebleau, N. 612, pp. 22.

4) Chauvet, P. (1981), "Remarques sur le variogramme generalise" (Remarks on the generalised variogram), Centre de Geostaistique, ENSMP, Fontainebleau, N. 719, pp. 27.

5) Chauvet, P. (1989), "Quelques aspects de l'analyse structurale des FAI-k a l dimension" (Some aspects of the structural analysis of the IRF-k with 1 dimension), in Geostatistics, Vol. I, Kluwer Academic Publisher, Dordrecht, pp. 139-150.

6) Journel, A.G. and Huijbregts, C.J. (178), Mining Geostatistics, Academic Press, London, pp. 108-144.

7) Bruno, R. and Raspa, G. (1991), FAIPACK. Vers. 1.2.Manuale Vol. I and II, Dip. ICMMPM, Univ. of Rome, "La Sapienza", Rome.

FILTERING NON-LINEAR FUNCTIONS

Colin DALY
BP Research
Chertsey Road
Sunbury-on-Thames
Middlesex TW16 7LN
England

ABSTRACT.Two methods of estimating non-linear functions of a signal which has been observed in the presence of contaminating noise are considered based on disjunctive kriging and conditional expectation. For both methods a reliable estimate of the probability density function of the underlying variable must be obtained first.

1. Introduction

In some branches of Physics and Biology, such as microprobe imaging of chemical concentrations in steel samples, we obtain images which are subject to a destructive noise term at each pixel. To answer questions about the underlying, or true, chemical concentration map it is usually necessary to filter the image. For example we might want to know about the spatial distribution of the values with a 'high' chemical concentration (e.g. the highest 5% for the underlying map). The simplest approach to this is to apply a filter directly to the image to estimate the underlying map and thereafter calculate the percentage of high values and their distribution.

Unfortunately most filtering techniques do not even preserve the underlying histogram so that they certainly can not reproduce the spatial distribution (for example the linear kriging filter, which we recall briefly below in §2 tends to over smooth the data, underestimating the high values). One way around this is to try and estimate the underlying distribution directly. This will give an indication of the percentage of high or low values but is insufficient to give an indication of their spatial distribution, a problem which is important for some applications. We can get a partial answer to this problem by estimating histograms locally, so that we estimate the probability of having a high value at a particular point.

If our noisy data is modeled by a random function $Z(x)$ and the underlying data is modeled by $Y(x)$ then our local histogram in a neighborhood v of a point x may be represented as $P[Y(x) < y | x \in v]$ which is equal to the mean of the indicator $\frac{1}{v}\int_v \mathcal{I}_{Y(x)<y} dx$ so that we can estimate this term provided we have an estimate of $\mathcal{I}_{Y(x)<y}$. Throughout this text we will assume that the noise terms at different points are independent,

that is, the random variables $(Z(x)|Y(x))$ and $(Z(y)|Y(y))$ are independent for $x \neq y$.

The local histogram may thus be estimated in terms of a non linear function of $Y(x)$. More generally we may be interested in an arbitrary function of $Y(x)$ so that the problem is; Given $Z(x)$ can we estimate a non-linear function of $Y(x)$, say $g(Y(x))$? To obtain a solution to this problem we need to know the global histogram $f^Y(y)$. Unfortunately there is no general method to obtain f^Y from f^Z, but in certain important cases we may obtain a reasonable estimate. We will discuss this in §3. In §4 we consider the non-linear filters themselves and apply them in §5 to the problem of Poisson noise. But before that, let us recall briefly the linear kriging filter.

2. Linear Kriging

It has been known for a long time that linear kriging may be used as a filter (Matheron, 1970). Moreover this filter is equivalent to the classical Wiener filter (Wiener, 1949) in the stationary case. It was applied in the multivariate case in Daly, Lajaunie and Jeulin,1989, and several further developments are considered in Daly, 1991. We apply it to obtain an estimate of $Y(x)$ from neighboring noisy values $\{Z(x_i)\}_{i=1}^n$. Note that we only estimate the underlying map itself and are not trying to estimate any non-linear function of this map.

A linear estimate is of the form

$$Y^*(x) = \sum_{i=1}^{n} \lambda_i Z(x_i) + \lambda_0$$

For the kriging filter we assume that $Y(x)$ and $Z(x)$ are both stationary random functions and that the $\{\lambda_i\}$ are chosen so that the variance of error $\mathrm{Var}[Y(x) - Y^*(x)]$ is a minimum. Following the usual development of kriging we can show that the $\{\lambda_i\}$ are solutions to

$$\sum_{i=1}^{n} \lambda_i \mathrm{Cov}(Z(x_i), Z(x_j)) = \mathrm{Cov}(Y(x), Z(x_j))$$

and $\lambda_0 = (1 - \sum_{i=1}^n \lambda_i)\mathrm{E}[Z]$. The term $\mathrm{Cov}(Z(x_i), Z(x_j))$ may be calculated from the experimental data, while the term $\mathrm{Cov}(Y(x), Z(x_j))$ is not available solely on the basis of the data set unless we know some more additional information about the relationship between $Y(x)$ and $Z(x)$ (eg the form of the noise). Alternatively, if we have two acquisitions $Z^1(x)$ and $Z^2(x)$, which is often possible in the case of electron microscopy, then under general conditions (Daly, 1991), we have

$$\mathrm{Cov}(Z^1(x), Z^2(y)) = \mathrm{Cov}(Y(x), Z^1(y)) \quad \forall x, y$$

As a consequence the linear kriging filter is fairly straightforward to put into practice provided $\mathrm{Cov}(Y(x), Z(x_j))$ is known. For images of diffuse type linear kriging provides a good robust estimation of the underlying map $Y(x)$.

Finally we can see that in this case linear kriging is equivalent to the Wiener filter. This is because both are 'the' best linear estimate in the sense of giving a minimum variance estimate, so that they must be identical. The linear kriging filter has perhaps an advantage in that we may use the usual extensions of kriging to local stationarity which allow us to increase the robustness of the estimate.

3. Estimation of the underlying distribution.

Let us assume as before that $Z(x)$ is the random function which models the noisy data and that $Y(x)$ models the underlying map. Let the respective probability laws be $F^Z(x)$ and $F^Y(x)$. In many cases of interest the noisy variable is a randomization of Y. That is, there is another random variable X_θ of parameter θ such that $Z = X_Y$. Thus Z is the variable X with random parameter Y.
Examples: 1)The most commonly applied hypothesis for noise is that it is additive and independent of the signal.

$$Z(x) = Y(x) + \epsilon(x)$$

with $\epsilon(x)$ the noise term independent of $Y(x)$. In this case $Z = X_Y$ with the variable X_Y defined as $X_Y = Y + \epsilon$.

2) Multiplicative noise. Here $Z = X_Y$ with $X_Y = Y \times \epsilon$

3) Poisson noise. This is the type of noise appropriate for many of the electron microscopy problems. Here we have $Z = X_Y$ with $X_Y = \mathcal{P}(Y)$ a Poisson random variable with (random) parameter Y. As we can see, Z is a discrete random variable.

We may obtain the probability law F^Z from the law of F^Y provided we know the law of the noise. So let us assume that $F(x,\theta)$ is the probability distribution of X when the parameter takes the value θ. The law of total probability gives us

$$F^Z(x) = \int F(x,\theta)dF^Y(\theta) \tag{1}$$

This is the equation which links the law of Z to the law of Y. In practice of course we want to estimate F^Y from F^Z, or at least from the estimate of F^Z that we can obtain from the experimental data, consequentely we must invert this equation. Unfortunately it is not always invertible. Moreover even when it is invertible it may be the case that the solution is unstable. Let us see this with the help of a couple of examples.
Example 1) Let us consider again the case of additive noise. $Z = Y + \epsilon$ with $F(.)$ the law for the noise. Then we have $F^Z(x) = \int F(x-y)dF^Y(y)$ and if all the distributions admit density functions

$$f^Z(x) = \int f(x-y)f^Y(y)dy$$

Taking Fourier transforms,

$$\hat{f}^Z = \hat{f}\hat{f}^Y \quad \text{so that} \quad \hat{f}^Y = \frac{\hat{f}^Z}{\hat{f}} \quad \text{if } \hat{f} \neq 0$$

and we have $f^Y = \mathcal{F}^{-1}\left(\frac{\hat{f}^Z}{\hat{f}}\right)$, where $\mathcal{F}^{-1}$ means the inverse Fourier transform. So provided $\hat{f}$ does not vanish f^Y is determined from f^Z. On the other hand if $\hat{f}$ vanishes, as for example if it vanishes outside an interval, then we have a case where f^Y may not be determined from f^Z.

Even if the inverse does exist it will be very unstable. For example in practice we will not know f^Z exactly but only up to some perturbation $f^W = f^Z + \delta$. In this case our estimation of f^Y, call it $f^{Y'}$ say, will be

$$f^{Y'} = \mathcal{F}^{-1}\left(\frac{\hat{f}^W}{\hat{f}}\right) = f + \mathcal{F}^{-1}\left(\frac{\hat{\delta}}{\hat{f}}\right)$$

Now if $\hat{\delta}$ is a white noise, so that $\hat{\delta}$ never vanishes, the estimate will not be defined. Of course in practice we would truncate the spectrum and so will always obtain a 'solution'. This will not however be stable.

2) For the Poissonization problem we have $Z = \mathcal{P}(Y)$ so if $\{p(n)\}$ is the probability function for Z and $f(x)$ is the density for Y we have

$$p(n) = \int e^{-x}\frac{x^n}{n!}f(x)dx$$

Taking the generating function of Z we get

$$G(s) = \mathrm{E}[s^Z] = \mathrm{E}\left[\mathrm{E}[s^Z|Y]\right] = \mathrm{E}[e^{(s-1)Y}] = L_Y(1-s) \quad 0 \leq s \leq 1$$

where L_Y is the Laplace transform of Y. Thus if we know $\{p(n)\}$ exactly we know the Laplace transform on $[0,1]$ and it can be shown (Daly, 1991) that this is sufficient to specify the density of Y completely. However it can also be shown that this problem, like the previous example, is unstable with respect to small variations of $\{p(n)\}$.

So we have seen that sometimes even with perfect knowledge of f^Z we can not specify a unique solution f^Y. In this case we can usually find several well behaved solutions for f^Y which give the same f^Z. We can only decide among these solutions by making other non trivial assumptions about the nature of f^Y. We will not consider such problems here. In the other case where a unique inverse exists, but is unstable with respect to small variations in the observed data, the instability of the numerical solutions tends to manifest itself in the form of pathological, or physically unreasonable looking solutions. Problems of this type are known in the literature as ill-posed problems. For ill-posed problems there are ways of obtaining more reasonable solutions. These methods have been studied in great generality by Tikhonov, 1976 and others. We will consider an application of the most well known of these techniques, regularization, to the estimation of the underlying density. For a more detailed description of the contents of this paragraph see Daly, 1991.

The examples we have considered up to now are linear ill-posed problems. For example in the case of additive independent noise we have

$$Af^Y = f^Z = \int f(x-y)f^Y(y)dy$$

and we see that A is a linear operator on an L^2 space of functions.

In practice we can try to solve these equations by discretizing the functions, so that the operator A becomes a matrix and the problem becomes

$$f_i^Z = A_i^j f_j^Y$$

Thus we have $f^Y = A^{-1}f^Z$. However in practice f^Z is not known exactly. All we have is a perturbation f^W of f^Z due to sampling. We might try to estimate f^Y using least squares, so we look for f^Y by

$$f^{Y^*} = \min_g (Ag - f^W)^2$$

or more generally if we want to weight the least squares by a positive definite matrix

$$f^{Y^*} = \min_g (Ag - f^W)^t B(Ag - f^W) \tag{2}$$

Unfortunately because of ill-posedness the least squares solution may turn out to be very unstable. To overcome this Tikhonov proposes a regularization method whereby we choose a continuous positive functional $\Omega(f)$ and a parameter, known as the regularization parameter λ and instead of minimizing in equation (2) we obtain our estimate by

$$f^{Y^*} = \min_g \left[(Ag - f^W)^t B(Ag - f^W) + \lambda\Omega(g)\right] \tag{3}$$

The rational behind this is as follows. There are many 'solutions' which approximately minimize (2) (and note that the true f^Y, though giving a value close to the minimum value of the least squares, will not be the actual minimum), so to distinguish amongst these solutions we demand that our chosen solution be small with respect to the functional $\Omega(f)$. This often gets rid of the more pathological solutions which the least squares can give (see figures 5,6 and 7 for example). The functional $\Omega(f)$ is often chosen in the literature to be either the entropy $\sum g(i)\log g(i)$ or to be a positive quadratic form (in the continuous case we might take a spline $\int |g''(x)|^2 dx$). We will only consider the positive quadratic form case here. As before we assume a discretization has taken place so that Ω is represented as a matrix $\Omega(f) = f^t\Omega f$. The regularization parameter λ determines the influence of Ω; $\lambda = 0$ reduces to the pathological least squares and $\lambda \to \infty$ gives a solution which depends only on Ω and is independent of the data. We can see then that the choices of Ω and above all of λ are of paramount importance. The regularized solution is now found simply by solving

$$f^{Y^*} = \min_g \left[(Ag - f^W)^t B(Ag - f^W) + \lambda g^t \Omega g\right] \tag{4}$$

To try to shed some light on the meaning of B, Ω and λ let us consider a different way of estimating f^Y.

Up to now we have thought of the equation

$$f^Z = Af^Y$$

in terms of probability density functions of the noisy and underlying image and the relationship between them. More generally let us now consider a random function $R(x)$ being distorted by an operator A, so that $S' = AR(x)$ (for example a convolution is given by $AR(x) = \int K(x-y)R(y)dy$ for some kernal function $K(x-y)$ (Jeulin and Renard, 1991)). Now we do not observe $S'(x)$ exactly but only with noise. We denote the perturbed data by $S(x)$,

$$S(x) = AR(x) + \epsilon(x)$$

We may now try to estimate $R(x)$ from the observed $S(x)$ by a kriging procedure. Let us take $\epsilon(x)$ to be a noise random function with a strictly positive definite covariance function given by $\mathrm{Cov}(\epsilon(x), \epsilon(y)) = \Sigma(x,y)$ and take $R(x)$ to be an intrinsic random function of order k with generalized covariance $K(x,y)$ and with 'drift' functions $\{h^l(x)\}_{l=1}^k$. Then in the usual way we can consider estimates of the form

$$R^*(x) = \sum \lambda^\alpha S(x_\alpha)$$

such that $\mathrm{Var}[R(x) - R^*(x)]$ is a minimum. The weights $\{\lambda^\alpha\}$ are given by kriging equations (here shown in vector form)

$$\lambda^t[AKA^t + \Sigma] + C^t(Ah^l) = KA^t$$

$$\lambda^t A h^{l^t} = h^l$$

where $C = \{C_l\}_{l=1}^{k}$ is a vector of Lagrange parameters Note that if we apply this formalism to the case of probability density functions, (ie. if we model $f^Y(x)$ as the random function $R(x)$ and $f^Z(x)$ as $S(x)$) this kriging interpretation means that we are modeling the density functions themselves as a random function! This might seem curious at first, but on reflection there is no reason why we might not model them, at least approximately, as a differentiable non-stationary random function.

The link between the kriging problem and the regularization problem is given in the following theorem. Let $\Omega' = \lambda\Omega$.

THEOREM: If we take $\Sigma^{-1} = B$, Ω' such that $\Omega' K \Omega' = \Omega'$ (Ω' is the generalized inverse of K) and the drift functions h^l to be a basis for the subspace $\{f; \Omega(f) = 0\}$ then the regularization and the kriging problems have the same solution.

This theorem is proved in Daly, 1991, in the general case where A is an operator between Hilbert spaces of functions H_1 and H_2 with $f^Y \in H_1$ and $f^Z \in H_2$. The proof draws on the demonstration of the equivalence between splines and kriging (Matheron, 1980) but is fairly long and is complicated by the awkwardness of the notation involved so that we will just assume the result here.

The interest of this equivalence theorem is that it gives a way of choosing the regularization function Ω and the noise quadratic form B, at least heuristically. They are none other than the inverses of the covariance matrices of the signal and the noise respectively. In fact if there are sufficient data, it may be possible to develop an inference technique for the regularization function if f^Y is modeled as an IRF-k in the same sort of way that generalised covariances are usually infered (Daly 1991).

While the inference of the regularization function is of interest, it is even more important to obtain an estimate of the regularization parameter, which due to the equivalence theorem may be thought of as a sort of signal to noise ratio term. In the regularization literature, this is usually found by a cross validation technique (O'Sullivan 1986). A modification of this method which appears to be more stable is available when we have two acquisitions. This is not really a constraint for the histogram problem, we simply take every second point of the noisy image to form two noisy histograms. We can then obtain the regularization solution by minimizing equation 4 for several values of λ for both of the acquisitions. This will give us two functions if we calculate the square difference between each solution and the other noisy histogram.

$$CV_1(\lambda) = \sum_i \left(Ag_1(i) - f_2^Z(i)\right)^2 \text{ and } CV_2(\lambda) = \sum_i \left(Ag_2(i) - f_1^Z(i)\right)^2$$

We then choose as regularization parameter the average of the minima of these two functions.

4. Non-linear filtering

Once we have found the underlying stationary law f^Y we may now try to estimate non-linear functions of the underlying data, say $g(Y(x))$. We could do this either by producing a full multivariate model of the underlying image and of the randomization law and then by finding the conditional expectation within this model (usually by some Monte-Carlo type technique, Besag 1991, Freulon 1992) or alternatively we can use a technique based on disjunctive kriging, Matheron, 1973, being modified as a filter, Daly, 1991. Both of these methods have the disadvantage of being computer intensive. However approximations to both exist based on making a fairly strong assumption about the randomization law which allows us to use an isofactorial approach which significantly increases computer efficiency. We will firstly discuss the disjunctive kriging approach (DK) and then, briefly, the conditional expectation method (EC).

4.1 DISJUNCTIVE FILTERING

With a disjunctive filter we try to estimate $g(Y(x))$ using only information on the bivariate law of $(Y(x), Y(y))$ and of course a knowledge of the randomization law $(Z(x)|Y(x))$. So we look for estimates of the form $g(Y(x)) = \sum_{i=1}^{n} f_i(Z(x_i))$ with the functions f_i chosen so that our estimate is best in an L^2 sense. Let us use the notation $Z_i = Z(x_i)$. It can be shown that for the filtering problem the functions f_i are the unique solutions to the equations

$$E[g(Y(x))|Z_k] = \sum_i E[f_i(Z_i)|Z_k] \quad \forall k$$

or if $P^{Z_i|Z_k}$ is the conditional law of Z_i given Z_k and $P^{Y|Z_k}$ the conditional law of Y given Z_k then these equations may be written,

$$\int g(y)P^{Y|Z_k}(dy) = \sum_{i=1}^{n} \int f_i(y)P^{Z_i|Z_k}(dy) \quad \forall k \tag{5}$$

We could try to solve these equations directly by discretization. All we need to know is the bivariate law of (Y_i, Y_j) as the two conditional laws used in (5) may be found from these by randomization. In principle we can try to estimate these bivariate laws directly by the methods used in §3, but this appears to be computer intensive. Firstly we would have to solve for $(Y(x), Y(y))$ for several separation vectors $(x - y)$ and then find a random function with bivariate law compatible with these experimentally obtained bivariate laws. This procedure has not yet been carried out in practice although it seems to be the only sure way of obtaining an estimate which respects both the spatial distribution of the data and the randomization law and which does not make supplementary untestable hypothesis about the nature of $Y(x)$.

For application studies we need to have an estimate which is quite efficient in terms of computer time used. We can trade some of theoretical properties of the above true disjunctive filter for computer efficiency by using an isofactorial model as an approximation to the true disjunctive kriging solution. There are many types of isofactorial model (Matheron, 1973,1984) and many of them may be used for disjunctive filtering (Daly 1991). All we need is an isofactorial change of support model. However here we will only consider briefly, and without proofs, the most commonly used Gaussian model. We suppose that the probability density for the underlying variable has been estimated and is f^Y. Now transform $Y(x)$ to a Gaussian random $\tilde{Y}(x)$, such that $Y = \phi\left(\tilde{Y}(x)\right)$. We assume that after transformation the random function $\tilde{Y}(x)$ has bigaussian laws. We may calculate $\rho(h)$, the correlation function for $\tilde{Y}(x)$, and it can be shown that the bivariate law for $(\tilde{Y}(x), \tilde{Y}(y))$ takes the form

$$g^{\tilde{Y}}(x, y) = \sum_{n=0}^{\infty} \rho^n(x - y)\eta_n(x)\eta_n(y)g(x)g(y)$$

where $g(x) = \frac{1}{\sqrt{2\pi}} e^{-x^2/2}$ and $\eta_n(x)$ is the normalized Hermite polynomial of order n. Let us define

$$\tilde{Z}(x) = r\tilde{Y}(x) + \sqrt{1-r^2}\epsilon(x) \tag{6}$$

with $\epsilon(x)$ a random function representing the noise which is independent of $\tilde{Y}(x)$ and $0 \le r \le 1$. Then $\left(\tilde{Y}(x), \tilde{Z}(y)\right)$ is bigaussian with correlation function $r\rho(x-y)$ and $\left(\tilde{Z}(x), \tilde{Z}(x)\right)$ is also bigaussian with correlation function

$$\rho_{\tilde{Z}(h)} = \begin{cases} r^2\rho(h), & \text{if } h > 0; \\ 1, & \text{if not.} \end{cases}$$

Finally transform the probability law for our experimental noisy data which we call f^Z to a Gaussian law $f^{\tilde{Z}}$ by $Z(x) = \psi\left(\tilde{Z}(x)\right)$. We identify this transformed variable with the randomized variable defined by (6). Since the covariance function for $\tilde{Y}$ is known, only the randomization parameter r remains to be determined. This parameter r may be determined by solving in

$$\text{Var}[Y] = \sum_{n=1}^{\infty} r^n \psi_n \phi_n \tag{7}$$

where ψ_n and ϕ_n are the coefficients of the developement of the functions ϕ and ψ in terms of Hermite polynomials. Notice that the model assures us that f^Y and f^Z will be correctly represented but it does so in a fixed manner (with the randomization defined by (6)) so that it will not exactly fit other types of randomization (such as Poissonization for example). How good the approximation will be is problem specific, but applications show that the approximation to the Poisson, is sufficiently good to make the isofactorial DK filter a useful and effecient way of estimating non-linear functions.

If we wish to estimate some function $g(Y(x)) = (g \circ \phi)(\tilde{Y}(x)) \equiv h(\tilde{Y}(x))$, we do so by developing $h(x)$ in terms of Hermite polynomials $h(x) = \sum h_n \eta_n(x)$. The DK estimate is then $g^*(Y(x)) = h^*(\tilde{Y}(x)) = \sum_{n=0}^{\infty} h_n \eta_n^*(x)$ where $\eta_n^*(x) = \sum_{\alpha=1}^{n} \lambda_\alpha^n \eta_n(\tilde{Z}(x_\alpha))$ and the weights λ_α^n are solutions to the linear kriging equations

$$\sum_{\alpha=1}^{n} \lambda_\alpha^n \rho_{\tilde{Z}}^n (x_\alpha - x_\beta) = r^n \rho^n (x - x_\beta) \quad \beta = 1, \ldots, n$$

Theoretically we would have to solve these equations for each value of $n = 1, \ldots, \infty$, but in practice convergence is usually quick and we would rarely have to solve for more than the first 8 or 9 terms.

4.2 CONDITIONAL EXPECTATION

Besag, 1991, adopts a Bayesian approach to the filtering problem. This method is developed further by Freulon (1992). As in §4.1 let $\tilde{Y}(x)$ be the Gaussian transformed version of the underlying random function. We may obtain the conditional expectation of any function $h(Y(x))$ by making several simulations of $\tilde{Y}(x)$ conditional on the observed but noisy data $Z(x)$ and then averaging h on these simulations. So firstly we must simulate $\tilde{Y}(x)$ conditional on $Z(x)$. Let I be the set of sites at which we have observed the noisy data. The conditional law of $\tilde{Y}(x)$, $f^{\tilde{Y}|Z}(x)$ may be written, using Bayes' law as

$$f^{\tilde{Y}|Z}(x) = \frac{f^{\tilde{Y}}(x) \prod_{\alpha \in I} f^{\tilde{Z}|\tilde{Y}}}{\int f^{\tilde{Y}}(u) du \left[\prod_{\alpha \in I} f^{\tilde{Z}|\tilde{Y}}\right]}$$

In the general case it is very difficult to calculate the conditional expectation directly due to the normalization term. Instead we sample realizations of $\tilde{Y}(x)$ from this distribution using one of the Monte-Carlo simulation techniques which have become very popular (Ripley 1987, see Freulon, 1992, for a detailed description). The conditional expectation is then found by taking the average of $(g \circ \phi)(\tilde{Y}(x))$ over the realizations.

As with the disjunctive kriging, we could approximate this conditional expectation by making the hypothesis that $Z = rY + \sqrt{1-r^2}\epsilon$ and allowing this to determine the randomization. We then place ourselves in a full multigaussian context and may find the conditional expectation directly (that is without using simulations). This dramatically improves the computation time but we find that the approximation is not as robust (Daly, 1991) as the corresponding one for disjunctive kriging and in general gives worse results.

5. Applications to the Poissonization problem

In microprobe images of solidification studies of steels, the images are contaminated by a noise which we model as Poisson. Estimation of the local law is useful to determine the degree of segregation locally. In Daly, Jeulin and Benoit (1992) it was shown that a disjunctive kriging approach with an isofactorial approximation provides good estimates of the local law for real case studies. Here we show the steps necessary to apply this model, and apply them to a synthetic image. The reason we apply the disjunctive filter with the isofactorial approximation is that it is numerically more efficient than the other techniques and provides good results, so making it attractive for routine applications.

In figure 1, we show the real or underlying image $Y(x)$. By randomizing this with a Poisson noise we end up with figure 2, the noisy (or observed) image $Z(x)$. In figure 3 we show the histograms of the two images on the same scale. In figure 4, we show the true local law $\frac{1}{v}\int \mathcal{I}_{Y(x)>y}dx$ where v is an 11×11 neighborhood and y, the cutoff is chosen to be 250. We want to estimate the local law in this section.

Before we start the non linear filtering we must first estimate the underlying histogram from the observed one.

In Poissonization the randomization procedure changes the law according to

$$G(n) = \int e^{-x}\frac{x^n}{n!}f^Y(x)dx$$

Suppose that the underlying law has been discretized and takes the values $\{f_i\}_{i=1}^N$ at the nodes $\{x_i\}_{i=1}^N$. Then we have $G(n) = \sum_i P_{in}f_i$ where P_{in} is the matrix with entries $P_{in} = \{e^{-x_i}x_i^n/n!\}$. In practice we do not observe the true Poissonised histogram $G(n)$ but only a sample histogram $g(n)$ so that a direct inversion of the equation $G = Pf$ is not sensible. We might try to apply a least squares method as mentioned in §3. However this turns out to be very unstable. For example in figure 5 we see a true and Poissonised histogram together. A least squares estimate of the underlying law is shown in figure 6. Clearly this is not a very good estimate of the underlying law. To stabilize the least squares method we use the regularization method

$$\min_{\hat{f}_i} Q = \sum_{n,m}(g(n) - \hat{G}(n))B_{nm}(g(m) - \hat{G}(m)) + \lambda \sum_{i,j} \hat{f}_i \Omega_{ij} \hat{f}_j$$

where $\hat{G}(n) = \sum_i P_{in}\hat{f}_i$, B_{nm} is some weighting matrix (usually we take $B_{nm} = \delta_{nm}$ or $B_{nm} = \delta_{nm}/g(n)$) and Ω_{ij} is the inverse of a covariance matrix which we suppose to come from an appropriate model for the covariance function of the underlying histogram treated as a random function. In this case as we have $(256)^2$ data points making up the underlying law, we expect it to be very smooth and we use a cubic covariance function model for the underlying law. The parameter λ is found by cross validation. The resulting regularization solution is shown in figure 7.

Applying this technique to our histograms for the images of figures 1 and 2 we obtain the regularization solution shown in figure 8. We now find a transformation of this law to Gaussian in terms of Hermite polynomials and do likewise for the noisy law shown in figure 4. It is easy to show that for Poisson noise we have $\mathrm{Var}Z = \mathrm{Var}Y + m_Y$ with $m_Y = m_Z$ being the mean of Y or Z, so that we may obtain an estimate of $\mathrm{Var}Y$ from the experimental data. Using equation 6 we may then estimate the correlation parameter r. We then estimate $\eta_n^*(x)$ by solving the DK equations. Then by expressing $\mathcal{I}_{Y(x)>y}$ in terms of Hermite polynomials we get

$$\mathcal{I}^*_{Y(x)>y} = \sum_{n=0}^{\infty} h_n \eta_n^*(x)$$

with $\{h_n\}$ the parameters of the Hermite polynomial development. Finally we get the local histogram $1/v \int \mathcal{I}_{Y(x)>y} dx$ simply by averaging in the volume v. The final result is shown in figure 9 and compares well with figure 3.

6. Conclusions

To obtain an estimate of an underlying image $Y(x)$ from an observed noisy image $Z(x)$ is a relatively straightforward matter in many cases of interest. Many types of estimation give comparable results. One filter with several useful properties is the linear kriging filter. However an estimate of $Y(x)$ is often not of too much interest in itself and we are often concerned by more complicated properties of $Y(x)$ such as the spatial distribution of high and low values. To obtain estimates of these functions is not as simple as linear kriging and necessitates an estimate of the underlying law. As a first step we have seen that the regularization technique gives us a reasonably good estimate of this law in many cases. The version of regularization described here admits an interpretation in terms of kriging which allows us to choose the regularization parameters in a meaningful way. Indeed for certain problems it may even be possible to infer the parameters directly from the data.

Once this underlying law is known we may proceed to the non-linear filtering itself. We have described two methods, the disjunctive kriging and conditional expectation. Furthermore both of these admit approximations but at the cost of modifying the randomization law. In this paper only the isofactorial DK method has been applied. From Daly, 1991 it would appear that the approximate DK gives better results than the approximate EC, so we may conclude that it is more robust. For routine applications the approximate (isofactorial) DK method seems to provide an unbiased and numerically efficient method of estimating non-linear functions. It does not appear to be quite as good as the full EC method if we know the randomizing law exactly (Freulon, 1992), but the differences do not appear very large and the DK is significantly quicker. Moreover in many applications the randomization law will be only approximately known, and, as we have already noted, the EC is less robust than DK to an approximation of the randomization law. If accuracy is of importance, the randomization law is known, and computing time not a great problem the

EC method is probably better, however it would be interesting to test the full DK solution in such a case. It may well turn out to be better than the full EC because of robustness with respect to the spatial law. Finally the full EC method is obligatory if we wish to estimate complicated non-linear functions of more than one point which depend explicitly on the multivariate laws. It is worth remembering in this case that we would be using information that is in no way inferred from the data, so that a careful criticism of the results from the standpoint of a physical model should be made. A by-product of the EC method, namely simulations of the underlying signal conditional upon the noisy data may also be of interest in certain studies.

7. References

Besag, J., York, J., Mollié, A. (1991),'Bayesian image restoration with two applications in spatial statistics', *Ann. Inst. Statist. Math.*, Vol 43, No1, 1-59.

Daly, C., Lajaunie, Ch., Jeulin, D., (1989) 'Application of multivariate kriging to the processing of noisy images', Proc. 3rd Intern. Geostatistics Congress, Sept. 5-9, 1988, Avignon, *Geostatistics*, ed. by M. Armstrong, Vol. 2, Kluwer Academic Publ., pp. 749-760

Daly, C., (1991) *Applications de la Géostatistique a quelques problèmes de filtrage.*, Thèse de Docteur en Géostatistique. ENSMP, Fontainebleau.

Daly, C., Jeulin, D. and Benoit, D., (1992) 'Nonlinear statistical filtering and applications to segregation in steels from microprobe images.', Proc. 10th Pfefferkorn conference, to appear in Scanning Microscopy International.

Freulon, X., (1992) *Conditionment du modèle Gaussienne par des inégalités ou des randomisées.*, Thèse de Docteur en Géostatistique, ENSMP, Fontainebleau.

Jeulin, D. and Renard, D. (1991) 'Practical Limitations of deconvolution kriging.', in *Compte-Rendu des journées de Géostatistique.*, Fasc. 1, Cahiers de Géostatistique. ENSMP. pp 173-203.

Matheron, G., (1970) *La théorie des variables régionalisées et ses applications.*, Fasc. 5, Cahiers du centre de Morphologie Mathématique, ENSMP, Fontainebleau.

Matheron, G., (1973) Le Krigeage Disjonctif, Centre de Géostatistique, ENSMP, Fontainebleau.

Matheron, G., (1981) 'Splines and Kriging, their formal equivalence.', in *Down to Earth Statistics. Solutions looking for Geological Problems*, Syracuse University Geology Contributions.

Matheron, G., (1984) 'Isofactorial models and change of support', in Proc. 2nd NATO ASI, *Geostatistics for Natural Resources Characterisation.'*, G. Verly et al. eds., Reidel Publ. Co., Dordrecht, Holland Part 1, pp 449-467.

O'Sullivan, F., (1986), 'Ill posed inverse problems',*Statistical Science.* Vol 1, no. 4, pp 502-519.

Ripley, B., (1987), *Stochastic Simulation.*, John Wiley, N.Y.

Tikhonov. A and Arsénine V., (1976) *Méthodes de résolution de problèmes mal posés.* Ed. Mir, Moscow.

Wiener. N., (1949) *Extrapolation, interpolation and smoothing of stationary time series.* MIT Technology Press and John Wiley, N.Y.

Figure 1 Underlying image.

Figure 2 Noisy image.

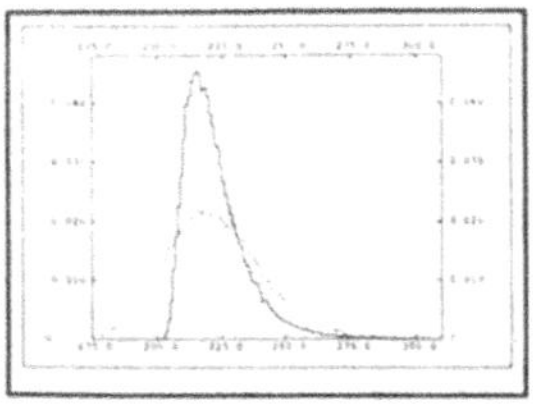

Figure 3 Histograms of underlying (continuous line) and noisy images.

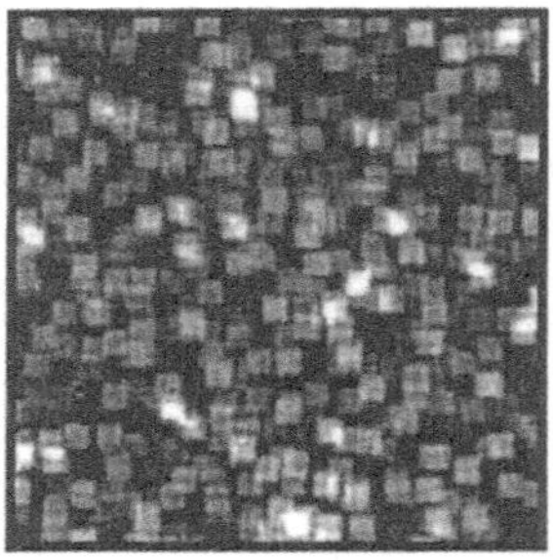

Figure 4 True local law $1/v \int_v \mathcal{I}_{Y(x)>y} dy$

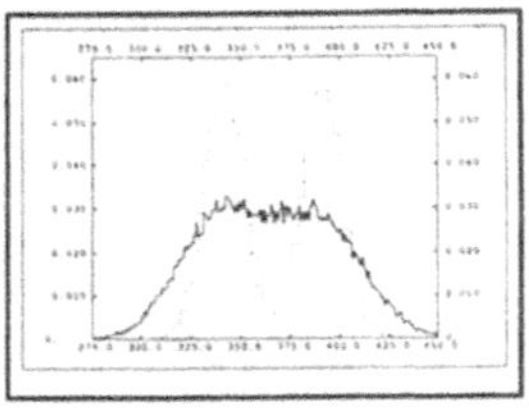

Figure 5 Underlying and noisy histograms of test case.

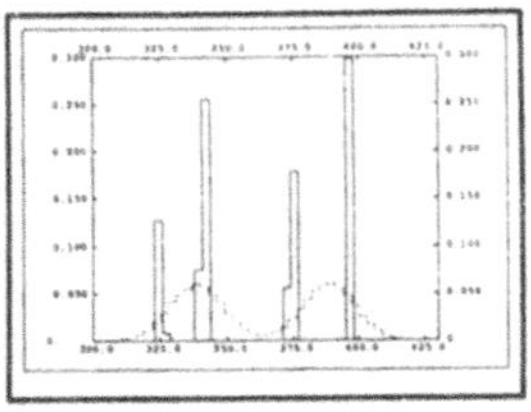

Figure 6 Least squares estimate of test case.

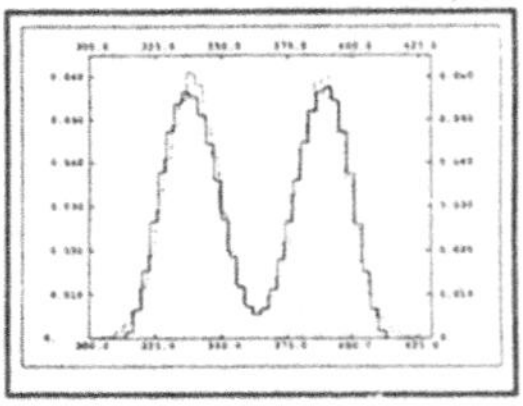

Figure 7 Regularization estimate of test case.

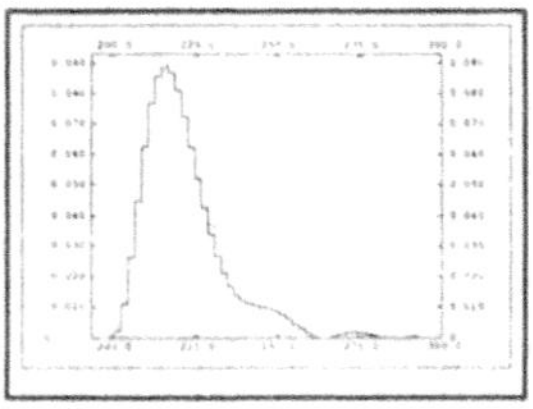

Figure 8 Regularization estimate for image histogram.

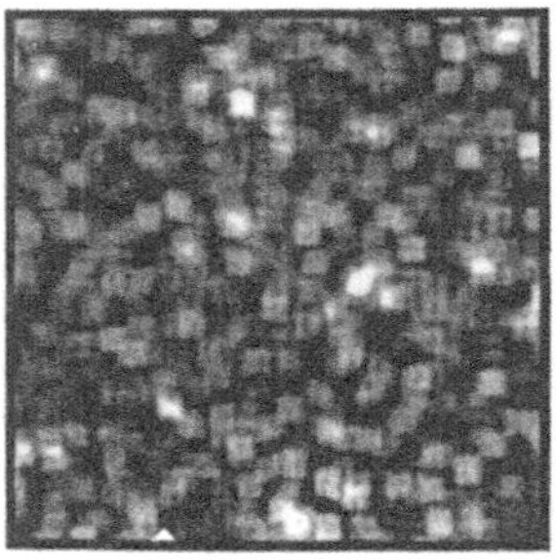

Figure 9 Disjunctive kriging estimate of local law.

PROBABILITY FIELD SIMULATION

Roland FROIDEVAUX
FSS International
10, chemin de Drize
1256 Troinex, Switzerland

Abstract

A new approach for conditional simulation is proposed. The key idea of this method is to dissociate the task of estimating the local probability distribution functions from the task of producing equi-probable images. Probability Field Simulation starts with the premise that the local conditional distributions are known. The conditional simulations are then obtained by drawing realizations from these cdfs. The probability values used to draw from the local cdfs constitute a probability field, and are viewed as outcomes of a random function characterized by a uniform distribution and a given covariance function. Probability Field Simulation, thus, consists of the following steps: inferring the univariate and bivariate characteristics of the probability field, generating a non-conditional simulation of the probability field and , finally, using the simulated probability values to draw from the local cdfs. Guidelines on how to infer the distribution characteristics of the probability field are discussed and an example of application is presented.

Introduction

Among all the geostatistical tools available, stochastic simulation is certainly the technique which has the greatest appeal for the oil industry and which has the greatest potential for further development (Haldorsen & Damsleth, 1990). Currently, simulation is being used to address issues such as modelling reservoir heterogeneities (Journel & Alabert, 1988), risk analysis for reservoir management (Srivastava, 1990) and volumetric uncertainty characterization (Ruffo & Scola, 1992).

Because of the rising interest in stochastic simulation, many new algorithms and techniques have been developed over the past few years: sequential indicator simulation (Gómez-Hernández & Srivastava, 1990), truncated Gaussian simulation (Matheron & als. 1988), sequential Gaussian simulation (Isaaks, 1990), fractal simulation (Hewett & Behrens, 1988), Boolean simulation (Damsleth & als., 1990).

What then is the rationale for introducing yet another simulation technique?

Probability Field Simulation should be viewed as an engineering solution to a class of problems which cannot be dealt with easily using existing techniques. Its two major objectives are·

- To provide a tool which dissociates the task of estimating the local conditional probability distribution function of the simulated attribute from the task of producing equi-probable images of that attribute, and
- To provide a simulation tool which allows an easy integration of very sparse hard information with exhaustive soft information.

The need for such a tool arises frequently in the context of volumetric uncertainty characterization, where the single most important parameter, depth to the top of the reservoir, is derived from 2D or 3D seismic data and is obtained by multiplying the seismic reflection time by the estimated velocity. Two ways travel times are typically regularly gridded and very abundant; it is not uncommon to have hundreds of thousands of such measurements. Direct measurements of velocity, on the other hand, are typically irregularly located and very sparse; in many reservoirs there are only a few wells in which "hard" velocity measurements are available. It is clear that different approaches have to be used to infer the local probability distribution functions: for the two-ways travel times they will be provided directly by the seismic interpreter who can assess experimentally the margin of error at each grid node; for the velocities, on the other hand, they will be inferred using a stochastic estimation procedure.

Probability Field Simulation provides an ideal framework for addressing this type of problem and, although there is little literature available on this technique, it is currently used, in one form or another, by several oil companies (Alabert; Dubrule; Hewett; personal communications).

1. The Basis of Probability Field Simulation

Consider a grid at the nodes of which an attribute Z must be conditionally simulated. At grid node x the range of possible outcomes for the unknown true value is characterized by a local conditional cumulative distribution function (ccdf) given by:

$$F(x,z) = Prob\{Z(x) \leq z | N(x)\}; \qquad N(x) \subset \{z_1, \dots, z_{Nt}\}$$

where:

$z_1, \dots, z_{Nt}$ are the data available over the entire grid

and :

$N(x)$ represents the local conditioning information.

Once an equi-probable image of the attribute has been generated, the simulated attribute value z_s at location **x** corresponds to the p(x) quantile of the local ccdf, i.e. $z_s(x)$ is such that:

$$F(x, z_s(x)) = p(x)$$

and the local probabilities p(x) can be interpreted as the outcomes of a random function P(x).

Given this general framework, probability field simulation makes the two following constitutive assumptions:

1) The local pdfs are available, and

2) The univariate and bivariate characteristics of P(x) can be inferred.

If these two assumptions are verified, then conditional simulation using the Probability Field Simulation approach is straightforward:

First a non conditional simulation of the probability field P(x), reproducing the specified a-priori distribution and covariance function , is generated. Then, at each grid node, a realization $z_S(x)$ is drawn from the local ccdf F(x,z) using the local value of the probability field p(x) i.e:

$$z_S(x) = F^{-1}(x, p(x))$$

These two steps are repeated for each equi-probable image to be generated.

There are therefore two critical issues for probability field simulation: the building of realistic local probability distribution functions and the proper inference of the probability field univariate and bivariate distribution.

2. Building the Local Ccdfs

Contrary to classical geostatistical simulation algorithms, Probability Field Simulation completely dissociates the task of building the local ccdfs from the simulation process itself: local distribution functions are input parameters. Although Probability Field Simulation does not control the way in which these local ccdfs are built, the relevance of its results depends largely on the relevance of these local ccdfs.

In practice the local cdfs can have two distinct origins, depending on the nature of available information.

Sparse hard information. If irregularly spaced hard information is available, then the local cdfs will probably be interpolated using one of the many kriging techniques available, and considering either the primary variable alone or also taking into account some co-variable. This would be the preferred way of building local cdfs for petrophysical attributes which are measured only at well locations .

Exhaustive soft information. If no hard information is available, but, instead, exhaustive (i.e. available at each grid node) soft or indirect information exists, then probabilistic interpolation techniques are generally not called for, and the local cdfs are built using expert judgement and/or experimentation.

Consider, for instance, the situation in which the depth to the top of a structure is derived from a dense 3D seismic survey. Using different reasonable assumptions on various parameters (such as velocity field) the seismic interpreter can produce a series of equi-probable interpretations of the depth. Therefore, at each grid node, a range of possible values for the attribute of interest is determined. If enough plausible scenarios are available, then one could use them to discretize directly the local ccdf. If, on the other hand, there are only a few, they can be used to provide the bounds of the local ccdfs, and the interpreter would have to supply some indications on the anticipated shape of the local distributions between these two bounds.

One of the distinctive advantage of Probability Field Simulation is that it provides a framework which allows the easy integration of these two types of information.

3. Inferring the Probability Field Parameters

The second critical issue is the correct inference of the univariate and bivariate characteristics of the distribution function P(x), i.e. the shape of its probability distribution function and its spatial continuity characteristics.

From an intuitive point of view, it seems that the following assumptions can be made:

- Since P(x) represents a probability field, it follows a uniform distribution.
- There is a "permanence of characteristics" between the uniform transform U(Z(x)) and P(x):

$$C_P(h) \approx C_U(h)$$

with :

$$U(x) = F(Z(x))$$

The important features of spatial continuity, such as direction and range of maximum continuity, anisotropy ratio and relative nugget effect, are similar for the uniform transform of the attribute Z(x) and for its probability field P(x).

The obvious question at this stage is whether these assumptions on the univariate and bivariate characteristics of P(x) are reasonable and can be used in practice.

Since experimental local probability values are never available, the validity of these hypotheses is checked through the following experiment:

- The local ccdfs of the attribute to be simulated are empirically defined at the nodes of a grid and assumed to be all identical:

$$F(x, z) = F(z), \qquad \forall x$$

with F(z) being an a-priori empirical distribution function.

- Sequential Gaussian Simulation is then used to generate a non-conditional realization of Z(x), honoring the a-priori distribution F(z) and a specified covariance function $C_Z(h)$.

- The local values of the probability field $p_S(x)$ are then given by:

$$p_S(x_i) = F(x_i, z_S(x_i))$$

with $z_S(x_i)$ being the simulated attribute value at node x_i

A grid of size 50 x 50 units was considered, and the a-priori distribution of the attribute Z(x) is given on Figure 4A Four non-conditional images of Z(x) were generated, each honoring the same a-priori distribution, but each assuming a different covariance model, ranging from a pure nugget effect to an isotropic spherical model with a correlation distance of half the grid size:

case A : $C(h) = 1.0$;
case B : $C(h) = 1.0 - Sph_5(h)$;
case C : $C(h) = 1.0 - Sph_{10}(h)$;
case D: $C(h) = 1.0 - Sph_{15}(h)$;

These four simulations resulted in four different sets of probability field values which were submitted to standard univariate statistical analysis and variography. The histograms of these probability field are shown on Figure 1, their summary statistics are listed on Table 1 and the experimental variograms are presented on Figure 2.

Table 1 : Summary statistics for the simulated probability fields

	Case A	Case B	Case C	Case D
Average	0.502	0.499	0.488	0.486
Variance	0.084	0.084	0.085	0.087
1st quartile	0.252	0.248	0.231	0.223
Median	0.500	0.496	0.500	0.437
3rd quartile	0.752	0.747	0.750	0.757

These preliminary results seem to confirm, to a large extent, the subjective assumptions set forth above. In particular:

- The experimental variograms of Z(x) and P(x) share similar features: same relative nugget effect, same range of influence. The shape however differs: whereas the covariance model for the attribute is spherical, the resulting probability field variograms are clearly Gaussian. In all cases an acceptable fit was obtained by using a Gaussian model with the range of influence of the a-priori spherical model.

- The histograms confirm that it is reasonable to assume a uniform distribution for P(x). It is worth noticing, however, that in Case D there are some indications that P(x) may depart from a strict uniform distribution. This is probably due to the fact that the range of correlation becomes large with regards to the size of the grid. For very large range of influence (zonal anisotropy) and small grids, the uniform distribution hypothesis may be questionable.

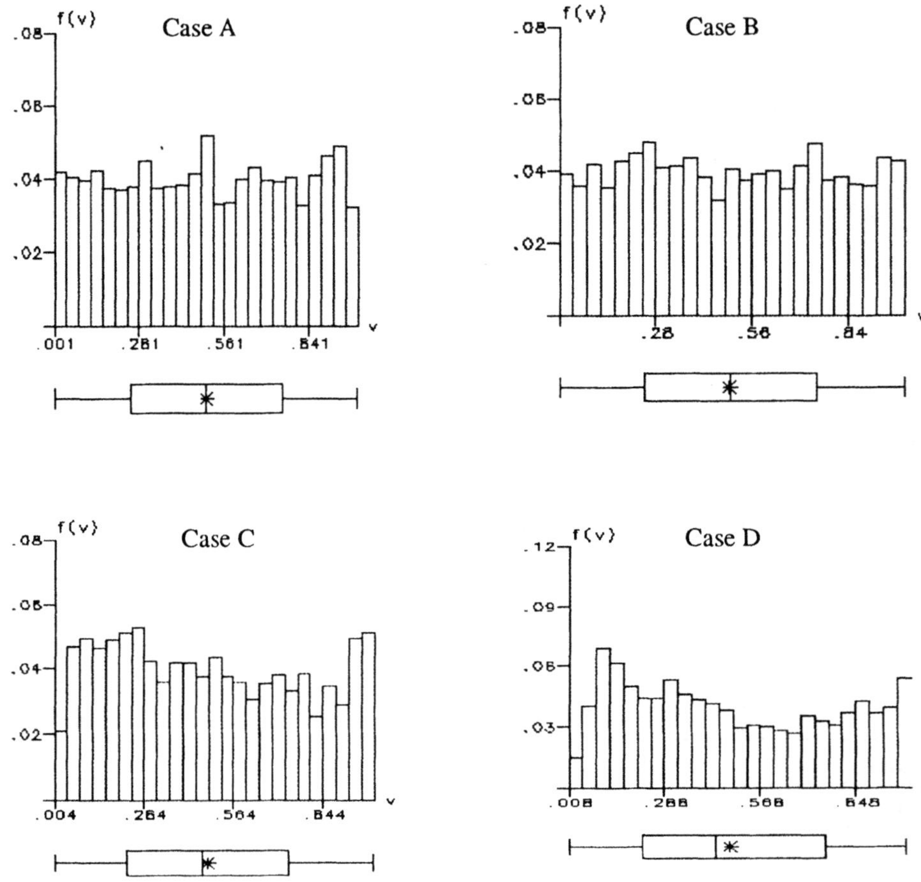

FIGURE 1 : Histograms of experimental probability fields

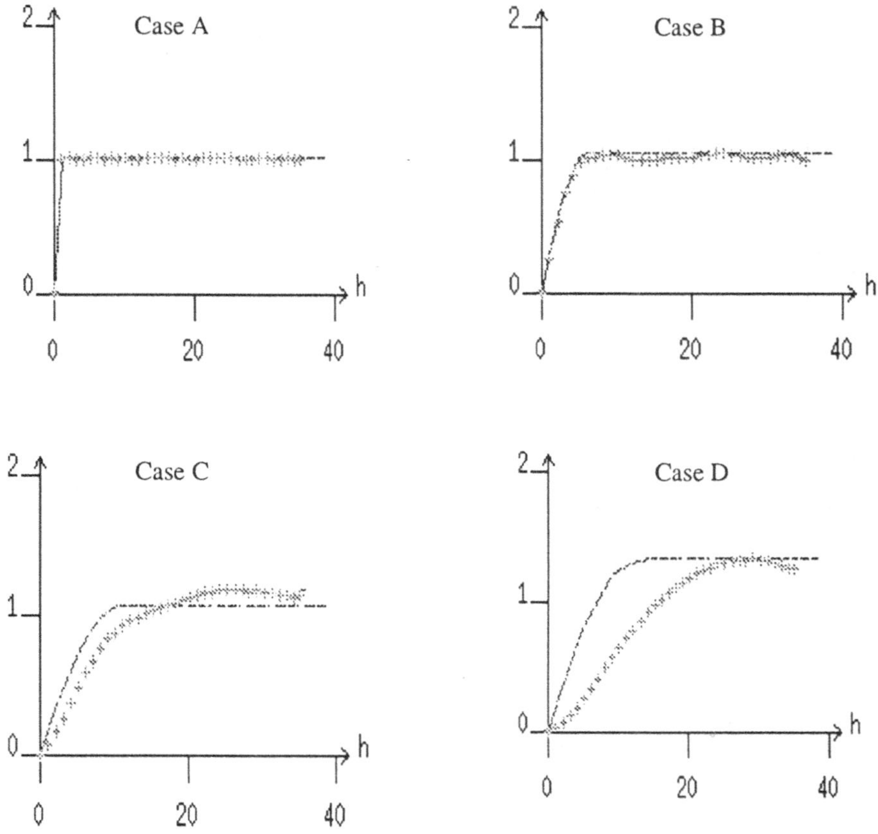

Figure 2 : Variogram models for the attribute (dashed line) versus experimental variograms of the probability fields

4. The Practice of Probability Field Simulation

Implementing Probability Field Simulation, thus, involves the following steps:

1.- Define a grid of suitable size covering the area of interest.

2.- Obtain, for each grid node, the local cdfs F(x,z) of the attribute to be simulated. These cdfs may either be the result of a stochastic estimation process or be defined empirically based on expert judgement.

3.- Calculate and model the variogram of the uniform transform of the attribute to be simulated. Assume that the covariance of U(Z(x)) and P(x) are similar:

$$C_U(h) \approx C_P(h)$$

4.- Assume that the probability field P(x) follows a uniform distribution.

5.- Generate, at each grid node, a non-conditional simulation of P(x) honoring the a-priori uniform distribution and the covariance $C_P(h)$. Any simulation technique which honors an histogram and a covariance (or variogram) will do. The choice should be based, foremost, on speed and efficiency criteria.

6.- A each grid node draw a realization $z_S(x)$ from the local cdf F(x,z) using the probability field value p(x):

$$z_S(x) = F^{-1}(x, p(x))$$

7.- Repeat step 5 and 6 for each equi-probable image to be generated.

5. Example of Application

Probability Field Simulation was run to simulate the thickness of a reservoir in the North Sea. The available data consisted of 89 reservoir thicknesses measured at well locations. Experimental variography indicated that the direction of maximum continuity was approximately N45W, with an anisotropy ratio of 0.6. An appropriate covariance model for the uniform transforms of the thickness was found to be :

$$C_U(h) = 0.25 \cdot (1.0 - Exp_{5000}(h)) \,; \qquad \text{in the direction N45W}$$

Generating a conditional, equi-probable image of the reservoir thickness via Probability Field Simulation involved the following steps:

- Multigaussian kriging (Verly, 1983) is used to estimate the local cdfs of the reservoir thickness at each node of a 48 x 32 grid, with a node spacing of 250 metres.

- The covariance of the probability field is assumed to be identical to the covariance of the uniform transform:

$$C_P(h) \approx C_U(h)$$

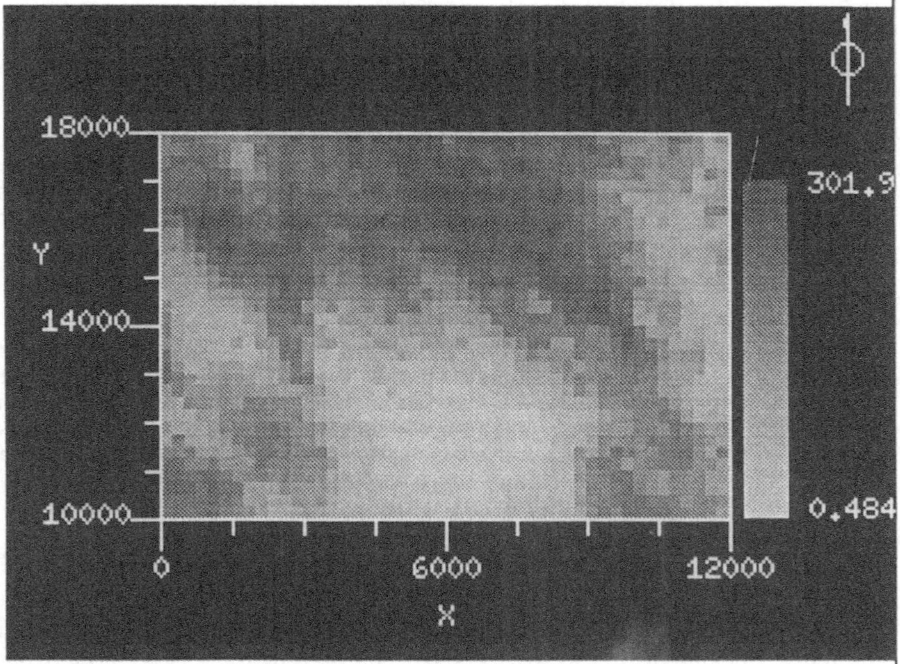

Figure 3 : Simulated reservoir thickness

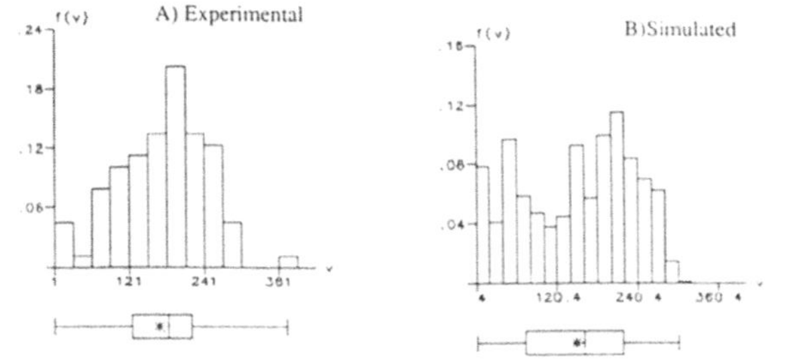

A) Experimental		B) Simulated	
Mean	171.9	Mean	149.9
Variance	5044.1	Variance	6811.5
Minimum	1.1	Minimum	0.5
1st quartile	126.1	1st quartile	72.01
Median	183.2	Median	160.8
3rd quartile	220.8	3rd quartile	217.6
Maximum	374.7	Maximum	302.0

Figure 4 : Univariate statistics of experimental and simulated thicknesses

- Sequential Gaussian Simulation is then used to generate a non-conditional simulation of $P(x)$ honoring the a-priori uniform distribution and the covariance $C_P(h)$.

- The conditional simulation of thickness is then obtained by drawing, at each grid node, a realization from the local cdf using the probability field value $p(x)$.

Figure 3 shows the simulated thickness map, Figure 4 shows the univariate statistics of the experimental and simulated thicknesses and Figure 5 compares the variograms of the simulated values with the original variogram model. As can be seen, there is a fair agreement between the univariate and bivariate characteristics of the simulated thicknesses and the a-priori models.

These results call, however, for the following comments:

- The histogram of the experimental thicknesses shows the existence of two modes one with an average thickness below 60 metres and the other one with an average of approximately 200 metres. This bi-modality is present again in the simulated thicknesses. This time, however, the first mode is more strongly represented. This stems from a simple clustering effect: the vast majority of the wells are located on the thick part of the reservoir, and there are large portions of the grid area, where the reservoir is indeed thin if it exists at all, which are not penetrated by any well

- The variograms of the simulated thicknesses show that the important features of the thickness spatial continuity are correctly reproduced: same relative nugget effect, same anisotropy ratio and approximately same correlation distances. The shape, however, is different: for the experimental thickness it is exponential, and for the simulated thicknesses it appears gaussian.The reasons for this apparent divergence in shape is not yet clear. It should be remembered, however, that this variogram was calculated on one realization only and does not represent the final ensemble average variogram.

6. Conclusions

Probability Field Simulation provides a promising alternative to existing geostatistical simulation techniques. Its most attractive features are:

- It provides an ideal framework for integrating sparse hard information with exhaustive soft information

- It makes no assumption on the way in which local cdfs are to be built and, therefore can accommodate cdfs which are inferred stochastically with cdfs which are built empirically.

- It is fast, has minimal hardware requirement (a 386 machine will do) and is well adapted to fast paced working environments.

As it stands now, however, Probability Field Simulation still represents an engineering solution to specific problems, and more work needs to be done to put the inference of the probabilty field covariance on a firmer theoretical ground.

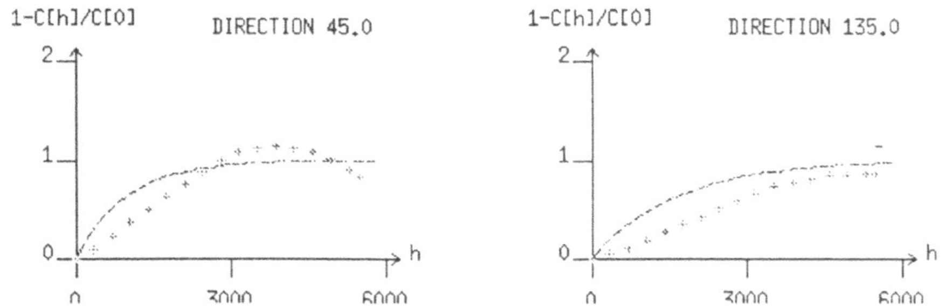

Figure 5: Experimental variograms of simulated values (+) versus variogram models (dashed line)

7. References

Damsleth E., Tzolesen C.B, Omre K.H. and Haldorsen, H.H., 1990, "A two-stage stochastic model applied to a North Sea reservoir", *SPE Paper No. 20605.*

Gómez-Hernández J.J and Srivastava R.M., 1990, "ISIM3D: An ANSI-C 3 dimensional multiple indicator simulation program", *Computers and Geosciences,* v.16, n.6, pp 395-440.

Haldorsen H.H. and Damsleth E., 1990, "Stochastic Modeling", *Journal of PetroleumTechnology,* April 1990, pp. 404-412.

Hewett T. and Behrens R.A. (1988) " Conditional simulation of reservoir heterogeneity with fractals", *SPE Paper No. 18326.*

Isaaks E.H., 1990, "The Application of Monte Carlo Methods to the Analysis of Spatially Correlated Data", PhD thesis, *Stanford University,* 213p.

Journel A.G. and Alabert F.G., 1988, "Focusing on Spatial Connectivity of Extreme-Valued Attributes: Stochastic Indicator Models of Reservoir Heterogeneities", *SPE Paper No 18324.*

Matheron G., Beucher H, Fouquet Ch. de, Galli A and Ravenne Ch., 1988,"Simulations Conditionnelle à Trois Faciès dans une Falaise de la Formation du Brent", *Sciences de la Terre,* No. 28, pp. 213-249.

Ruffo P. and Scola V., 1992, "An Application of Geostatistical Simulation to Probability Estimation of Prospects Geometry", *Mediterranean Petroleum Conference,* Tripoli, 10p.

Srivastava R.M., 1990, "An Application of Geostatistical Methods for Risk Analysis in Reservoir Management", *SPE Paper No 20608.*

Verly G., 1983, "The Multigaussian Approach and its Applications to the Estimation of Local Reserves", *Mathematical Geology,* vol. 15, No 2, pp. 259-286.

Joint Sequential Simulation of MultiGaussian Fields

J. Jaime Gómez-Hernández, Stanford University, Department of Applied Earth Sciences. Current address: Departamento de Hidráulica, Universidad Politécnica, 46071 Valencia, Spain.
André G. Journel, Stanford University, Department of Applied Earth Sciences, Stanford, CA 94305-2225, USA.

Abstract

The sequential simulation algorithm can be used for the generation of conditional realizations from either a multiGaussian random function or any non-Gaussian random function as long as its conditional distributions can be derived. The multivariate probability density function (pdf) that fully describes a random function can be written as the product of a set of univariate conditional pdfs. Drawing realizations from the multivariate pdf amounts to drawing sequentially from that series of univariate conditional pdfs. Similarly, the joint multivariate pdf of several random functions can be written as the product of a series of univariate conditional pdfs. The key step consists of the derivation of the conditional pdfs. In the case of a multiGaussian fields, these univariate conditional pdfs are known to be Gaussian with mean and variance given by the solution of a set of normal equations also known as simple cokriging equations. Sequential simulation is preferred to other techniques, such as turning bands, because of its ease of use and extreme flexibility.

Introduction

In many mapping applications reproduction of the spatial dependence of several variables is critical. For example, reservoir engineering and groundwater applications calls for joint modeling of the spatial distribution of porosity and permeability. Likewise, a realistic model for a porphyry copper-gold deposit must account for the cross-correlation of copper and gold grades.

The sequential simulation algorithm (Johnson, 1987; Journel, 1989) can be generalized for the joint simulation of multiple variables. That generalization is straightforward, the only additional problem is the inference and modeling of the cross-covariance matrix. The sequential simulation algorithm is chosen for its extreme flexibility. Its features are: (1) a straightforward theoretical basis, (2) it is not limited to the generation of Gaussian fields, (3) conditioning to local data does not require any additional step, (4) it imposes no limitation on the type of covariance or anisotropy, (5) it can be applied on any grid regular or not.

A simple example

A simple example will serve to present the key ideas behind sequential simulation. Consider three binary random variables $\{Z_1, Z_2, Z_3\}$. Any realization of this set of random variables consists of one of 8 possible triplets of 1's and 0's. The probabilities of these eight realizations are given by the trivariate probability density function (pdf): $f(z_1, z_2, z_3)$ enumerated as follows:

$$
\begin{array}{llll}
f(0,0,0) = 0.10 & f(0,0,1) = 0.15 & f(0,1,0) = 0.20 & f(0,1,1) = 0.20 \\
f(1,0,0) = 0.10 & f(1,0,1) = 0.10 & f(1,1,0) = 0.10 & f(1,1,1) = 0.05
\end{array}
$$

One can easily construct a simple algorithm to draw random sequences of triplets which would reproduce these probabilities. For instance, label each of 100 balls with one of the 8 possible realizations according to the relative proportions given above, that is, 10 balls have label (0,0,0), 20 balls, label (0,1,1), etc. Then draw, with replacement, any number of these balls at random. This algorithm generates equally likely random triplets according to the trivariate pdf.
An alternative approach consists in drawing sequentially from three univariate conditional pdfs obtained from the trivariate pdf by recursive application of the conditional probability axiom. For the trivariate case, as many as six such sequences are possible, three of which are given below:

$$
\begin{aligned}
f(z_1, z_2, z_3) &= f(z_1|z_2, z_3) \cdot f(z_2|z_3) \cdot f(z_3) \quad \text{or} \\
f(z_1, z_2, z_3) &= f(z_2|z_1, z_3) \cdot f(z_1|z_3) \cdot f(z_3) \quad \text{or} \\
f(z_1, z_2, z_3) &= f(z_3|z_2, z_1) \cdot f(z_2|z_1) \cdot f(z_1)
\end{aligned} \tag{1}
$$

Drawing a triplet at random from $f(z_1, z_2, z_3)$ is equivalent to drawing successively each component of the triplet from the univariate pdfs of any one of the previous sequences. The only requirement is the ability to determine the univariate conditional pdfs involved. In this example, given the trivariate pdf, all conditional pdfs are easily determined. For instance, the conditional probability of Z_1 given that both Z_2 and Z_3 are equal to zero are

$$
\begin{aligned}
f_{z_1|z_2,z_3}(0|0,0) &= \frac{f(0,0,0)}{f(0,0,0)+f(1,0,0)} = \frac{0.1}{0.1+0.1} = 0.5 \\
f_{z_1|z_2,z_3}(1|0,0) &= \frac{f(1,0,0)}{f(0,0,0)+f(1,0,0)} = \frac{0.1}{0.1+0.1} = 0.5
\end{aligned}
$$

Consider now that the first decomposition (1) is chosen. Drawing a triplet $\{z_1, z_2, z_3\}$ using the sequential simulation algorithm consists of the following steps:

1. Compute $f(z_3)$. In this case, $f_{z_3}(0) = 0.5$ and $f_{z_3}(1) = 0.5$. Draw at random a value z_3 from this distribution, e.g., 0.

2. Compute the conditional distribution of Z_2 given the previously simulated value, $f_{z_2|z_3}(z_2|0)$. This conditional distribution is $f_{z_2|z_3}(0|0) = 0.40$ and $f_{z_2|z_3}(1|0) = 0.60$. Draw at random a value z_2 from this distribution, e.g., 1.

3. Last, compute the conditional distribution of Z_1 given the two previously simulated values for Z_2 and Z_3. This conditional distribution is $f_{z_1|z_2,z_3}(0|1,0) = 0.67$ and $f_{z_1|z_2,z_3}(1|1,0) = 0.33$. From this distribution a value z_1 is drawn, e.g., 1.

The triplet drawn is $(1, 1, 0)$. To draw a second triplet repeat steps 1 through 3 retaining the same first sequence (1) or any other of the 6 possible.
Notice that, in general, the conditional pdfs cannot be computed before the conditioning values are known. For example, the value for $f(z_2|z_3)$ cannot be determined until the value z_3 is known, nor can $f(z_1|z_2, z_3)$ be determined until the values z_2 and z_3 are known.
All the previous discussion refers to drawing an unconditional realization (triplet). If the triplet is to be conditioned, say, to $Z_3 = 1$, the algorithm remains the same simply replacing step 1, i.e., the drawing from $f(z_3)$, by setting directly the value z_3 equal to 1. For conditioning, it suffices that the sequence of type (1) be such that the conditioning data locations are visited first and assigned the conditioning data values.

Sequential simulation for multiple variates

The algorithm described above can be generalized to handle multiple variables representing several attributes over an arbitrarily large number of locations. Consider the joint distribution of a total of N random variables with N possibly very large. The N random variables comprise K different attributes at the N' nodes of a grid, with $N = KN'$. The generic notation Z_i, with $i \in \{1, \ldots, N\}$ is used to denote any such random variable no matter its location or its attribute-type. The specific notation Z_{jk} with $j \in \{1, \ldots, N'\}$ and $k \in \{1, \ldots, K\}$ is used to denote a random variable at location j and corresponding to attribute k. The generic notation Z_i, with $i \in \{1, \ldots, N\}$, is used when neither the attribute nor the location is relevant to the discussion.
Next, consider the conditioning of these N random variables by a set of N_0 data of any type symbolized by the notation $|(N_0)$. The corresponding N-variate conditional pdf can be decomposed using the conditional probability axiom in the product of N univariate conditional pdfs:

$$\begin{aligned} f(z_1, \ldots, z_N)|(N_0)) &= f(z_N|(N-1) \cup (N_0)) \cdot f(z_{N-1}|(N-2) \cup (N_0)) \cdot \\ &\quad \cdots f(z_2|z_1 \cup (N_0)) \cdot f(z_1|(N_0)) \end{aligned}$$

where, e.g., $f(z_2|z_1 \cup (N_0))$ denotes the conditional pdf of z_2 given the realization z_1 and given the set of N_0 original data values.
This decomposition allows drawing realizations of the N random variables Z_i conditioned by the N_0 original data. This is done by drawing recursively from the N univariate conditional pdfs of the decomposition.
The procedure is completely general and allows conditioning to prior (N_0) from the onset. The algorithm is independent of the order used to visit the N random variables. The only requirement for its implementation is the ability to compute at each step the model conditional pdf given both the original data and the values taken by all previously simulated random variables.

Determination of the conditional pdfs

The conditional pdfs in (2) can be obtained straightforwardly if a joint multivariate normal distribution can be assumed. The reader is referred to Alabert (1987) and Journel (1989) for a discussion of non-Gaussian cases, and to Gómez-Hernández and Srivastava (1990) for the corresponding C code.

Assume that the N random variables have joint Gaussian multivariate distribution. This distribution is fully characterized by the expected values $E\{Z_i\}$ and variances $\text{Var}\{Z_i\}$ of all random variables and by the covariance matrix $\text{Cov}\{Z_i, Z_j\}, i,j = 1,\ldots,N$. In practice, it is also assumed that all random variables related to the same attribute but at different locations have the same mean and variance (stationarity):

$$\begin{aligned} E\{Z_{jk}\} &= m_k \quad \forall j = 1,\ldots,N' \\ \text{Var}\{Z_{jk}\} &= \sigma_k^2 \quad \forall j = 1,\ldots,N' \end{aligned}$$

The covariances and cross-covariances are also assumed stationary.

$$\text{Cov}\{Z_{ik}, Z_{i'k'}\} = \text{Cov}\{Z_{jk}, Z_{j'k'}\} = C_{kk'}(\mathbf{x}_i - \mathbf{x}_{i'})$$

for all pairs $(i,i'),(j,j')$ so that $\mathbf{x}_i - \mathbf{x}_{i'} = \mathbf{x}_j - \mathbf{x}_{j'}$, where $\mathbf{x}_i - \mathbf{x}_{i'}, \mathbf{x}_j - \mathbf{x}_{j'}$ represent the separation vectors between the random variables. Any of the covariances could be the sum of any of number of nested structures with any type of anisotropy.

Next, consider a set of M conditioning data comprising values from any or all K attributes not necessarily at the same location $\{z_{i_1 1}, i_1 = 1,\ldots,n_1\} \cup \{z_{i_2 2}, i_2 = 1,\ldots,n_2\} \cdots \cup \cdots \{z_{i_K K}, i_K = 1,\ldots,n_K\}$, and $M = n_1 + n_2 + \cdots + n_K$.

It can be shown (Anderson, 1984) that the conditional density function of Z_{jk} given the set of M conditioning data is Gaussian with mean and variance given by the solution of a set of normal equations also known as simple cokriging system. The mean has the expression:

$$\begin{aligned} E\{Z_{jk}|(M)\} &= m_k + \sum_{i_1=1}^{n_1} \lambda_{i_1}(z_{i_1 1} - m_1) + \sum_{i_2=1}^{n_2} \lambda_{i_2}(z_{i_2 2} - m_2) \\ &\quad + \cdots + \sum_{i_k=1}^{n_k} \lambda_{i_k}(z_{i_k k} - m_k) + \cdots + \sum_{i_K=1}^{n_K} \lambda_{i_K}(z_{i_K K} - m_K) \end{aligned}$$

and the variance has the expression:

$$\begin{aligned} \text{Var}\{Z_{jk}|(M)\} &= \sigma_k^2 - \sum_{i_1=1}^{n_1} \lambda_{i_1} C_{k1}\{\mathbf{x}_j - \mathbf{x}_{i_1}\} - \sum_{i_2=1}^{n_2} \lambda_{i_2} C_{k2}\{\mathbf{x}_j - \mathbf{x}_{i_2}\} \\ &\quad - \cdots - \sum_{i_k=1}^{n_k} \lambda_{i_k} C_{kk}\{\mathbf{x}_j - \mathbf{x}_{i_k}\} - \cdots \\ &\quad \sum^{n_K} \lambda_{i_K} C_{kK}\{\mathbf{x}_j - \mathbf{x}_{i_K}\} \end{aligned}$$

where the coefficients $\{\lambda_{i_1}, i_1 = 1, \ldots, n_1\} \cup \cdots \cup \{\lambda_{i_K}, i_K = 1, \ldots, n_K\}$ are the solution of the following set of simple cokriging equations:

$$\begin{aligned}
\sum_{i_1=1}^{n_1} \lambda_{i_1} C_{11}(\mathbf{x}_{i_1} - \mathbf{x}_{i'_1}) + \cdots + \sum_{i_K=1}^{n_K} \lambda_{i_K} C_{K1}(\mathbf{x}_{i_K} - \mathbf{x}_{i'_1}) &= C_{k1}(\mathbf{x}_j - \mathbf{x}_{i'_1}) \\
&\quad \forall i'_1 = 1, \ldots, n_1 \\
\sum_{i_1=1}^{n_1} \lambda_{i_1} C_{12}(\mathbf{x}_{i_1} - \mathbf{x}_{i'_2}) + \cdots + \sum_{i_K=1}^{n_K} \lambda_{i_K} C_{K2}(\mathbf{x}_{i_K} - \mathbf{x}_{i'_2}) &= C_{k2}(\mathbf{x}_j - \mathbf{x}_{i'_2}) \\
&\quad \forall i'_2 = 1, \ldots, n_2 \\
\vdots \qquad \vdots \qquad \vdots \qquad &\vdots \qquad \vdots \\
\sum_{i_1=1}^{n_1} \lambda_{i_1} C_{1K}(\mathbf{x}_{i_1} - \mathbf{x}_{i'_K}) + \cdots + \sum_{i_K=1}^{n_K} \lambda_{i_K} C_{KK}(\mathbf{x}_{i_K} - \mathbf{x}_{i'_K}) &= C_{kK}(\mathbf{x}_j - \mathbf{x}_{i'_K}) \\
&\quad \forall i'_K = 1, \ldots, n_K
\end{aligned}$$

Implementation

The Markov approximation

The sequential simulation algorithm is completely general and suffers no exception. Expressions for all required conditional probabilities can be derived by simple kriging or cokriging in the case of jointly multiGaùssian random variables. However, the number of conditioning data thus of normal equations increases along the sequence with the number of previously simulated random variables, and can become quickly unwieldy. At the Nth step, the normal probability of Z_N given $(N - 1 + N_0)$ conditioning values has to be estimated. If N is larger than a few hundred, the solution of such large systems becomes extremely CPU-intensive not mentioning poor numerical accuracy. Approximations are therefore required.

An approximation of the conditional probabilities can be obtained by retaining only those conditioning values which are most consequential. For example, one could approximate $f(z_N|(N-1))$ by $f(z_N|(n))$ with $(n) \subset (N-1), n \ll N-1$. Of all possible subsets of size n we should retain the one that allows the best approximation of the conditional pdf $f(z_N|(N-1))$. In most cases, the subset of size n that provides such best approximation is constituted by the conditioning values which are "closest" to Z_N; "closest" being defined in terms of some correlation/cross-correlation function. Retaining the "closest" values amounts to assume a Markovian behavior in the sense that the closest data screens out the influence of further away data.

In practice, an ellipsoidal search neighborhood homothetic to the anisotropy of covariance of the random variable being simulated is defined and only the closest random variables within sectors of this ellipsoid are retained.

The use of a search neighborhood implies that only covariance and cross-covariance values for lags up to the dimensions of the search neighborhood can be reproduced in the final set of realizations.

Visiting sequence

The algorithm is, in principle, independent of the sequence used in the decomposition (1). However, a random sequence is recommended to avoid spreading random artifacts over too close (or too correlated) neighboring nodes. This is particularly true when using a limited neighborhood for selecting the conditioning data.
Other considerations can be used to define the visiting sequence. For example, if one specific attribute is particularly critical and the remaining attributes secondary, that should be visited (simulated) first. Also, in such a case, one may use different search neighborhoods for the simulation of each of the attributes: the principal attribute could be simulated using a large search neighborhood whereas the secondary ones could use a smaller search neighborhood. At the limit, the search neighborhood for the secondary variables could be limited to include only the collocated with the principal random variable.

Multiple-steps simulation

To allow the reproduction of large scale correlation distances without having to use for all nodes a very large search neighborhood, a two- or multiple-steps approach is suggested. In the two-steps approach, the visiting sequence is divided in two paths. The first path visits the random variables located over a coarse grid, the spacing of which is a multiple integer of the spacing of the final fine grid. The second path visits the remaining nodes located on the fine grid. During the simulation of the nodes on the first path, a search neighborhood large in size but small in the number of data retained is used. During the simulation of the random variables on the fine grid, the search neighborhood is reduced in size. The simulation over the coarse grid using the large search neighborhood allows reproduction of the large scale correlation structure; whereas the simulation over the small grid using the smaller search neighborhood will reproduce the small scale correlation.

Example

The public domain ANSI-C code GCOSIM3D has been developed for the implementation of the sequential simulation algorithm applied to jointly multiGaussian random fields. The algorithm is demonstrated in two dimensions for the generation of block permeabilities conditioned to point values. This exercise is particularly important in petroleum engineering and hydrogeology where the measurements are taken at a scale much smaller than the size of the blocks later used in the numerical flow simulators. The three variables considered are the point permeabilities K and the two components of the block permeabilities K_{xx} and K_{yy}. The logarithms of these three values are modeled by a multivariate Gaussian distribution with the following parameters: expected values, variances and correlograms of $\ln K$, $\ln K_{xx}$ and $\ln K_{yy}$ as well as the cross-variances and cross-correlograms between $\ln K$-$\ln K_{xx}$, $\ln K$-$\ln K_{yy}$ and $\ln K_{xx}$-$\ln K_{yy}$. All these parameters are displayed in Figures 1 and 2. The inference of these parameters is not straightforward since data on block values are not usually available. The reader interested in how such inference can be done is referred to Gómez-Hernández (1991, 1992).

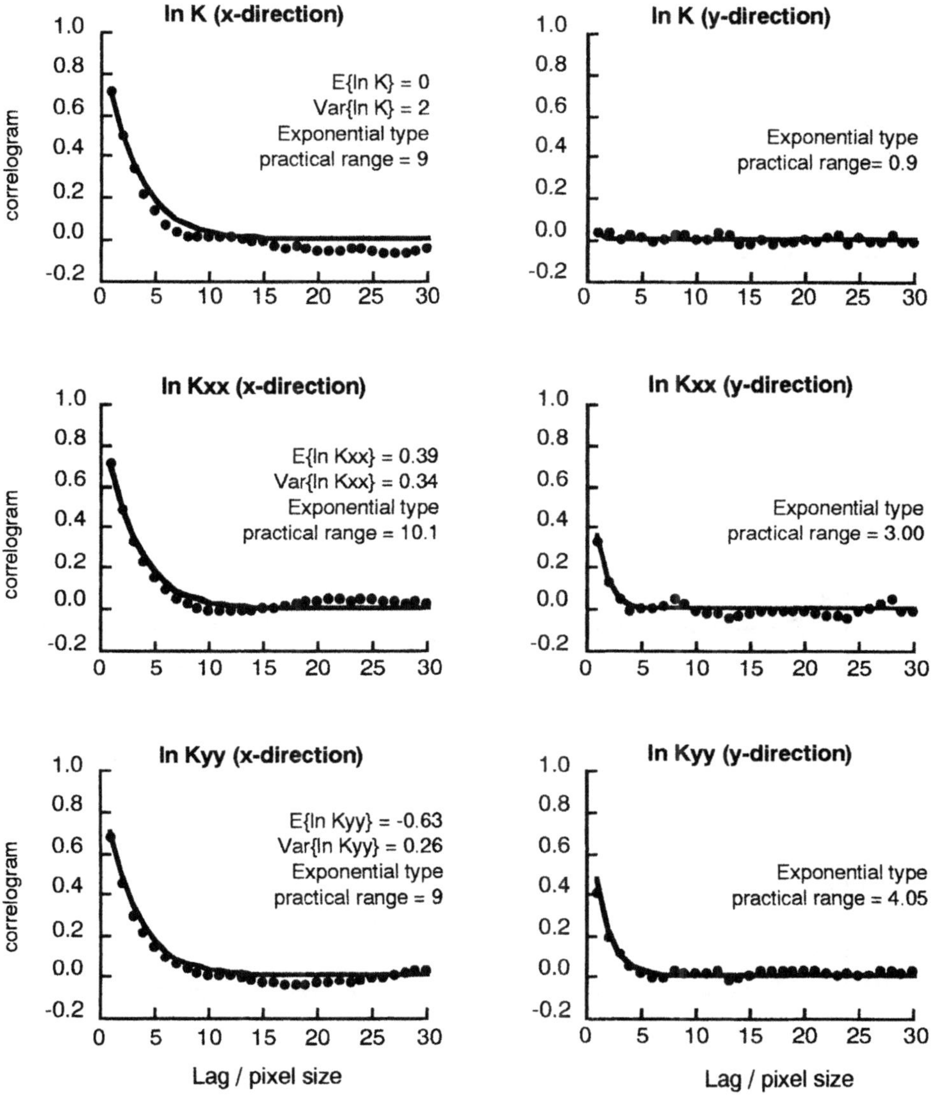

Figure 1: Correlation functions. The correlation models input to the sequential simulation algorithm (solid lines), and the correlation functions inferred from the generated maps (black dots) are given. The agreement is very good.

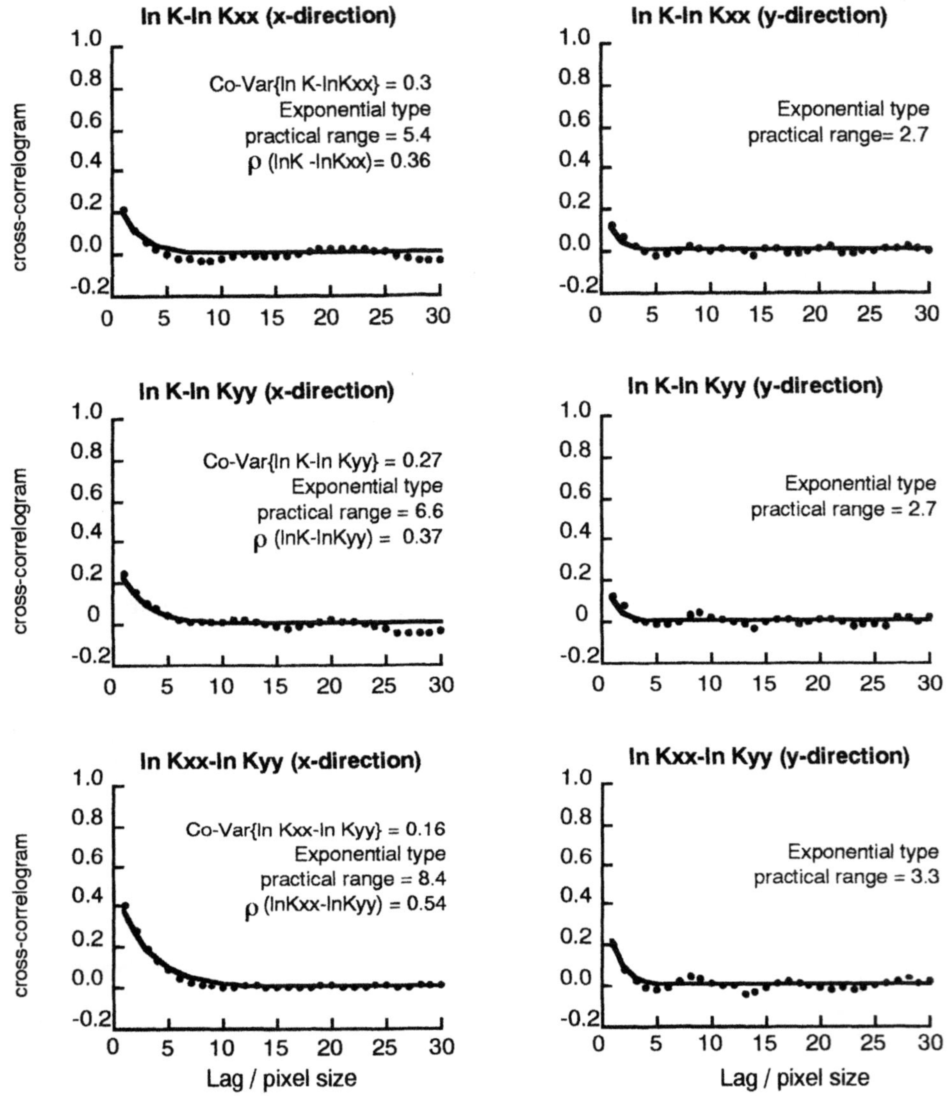

Figure 2: Cross-correlation functions. The cross-correlation models input to the sequential simulation algorithm (solid lines), and the cross-correlation inferred from the generated maps (black dots) are given. The agreement is very good. The coefficient of correlation is also shown.

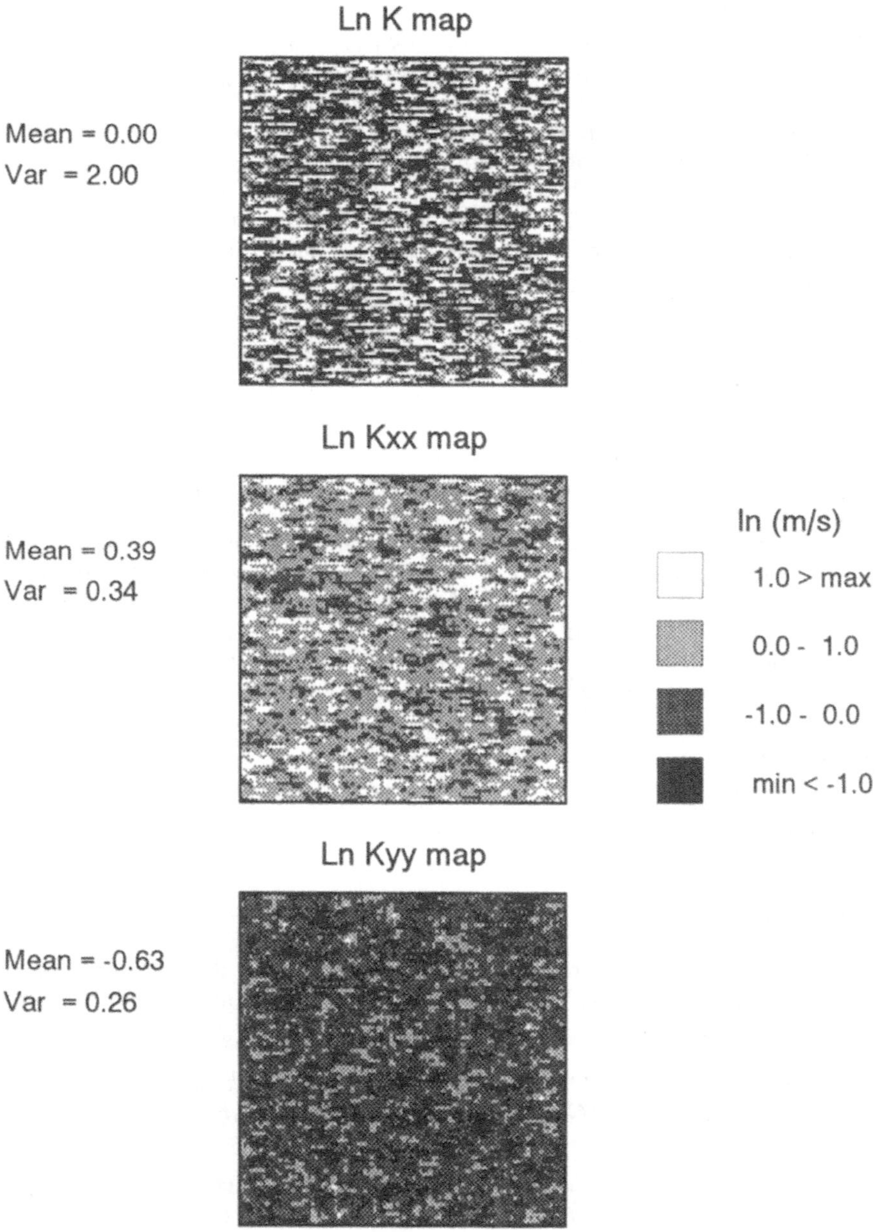

Figure 3: One realization of the three variables. A joint realization of the three variables conditioned to the $\ln K$ data along the left and right edges were generated using the correlation and cross-correlation functions shown in Figures 1 and 2.

A 100 by 100 pixels, two-dimensional vertical cross-section, is generated. Conditioning data on $\ln K$ is available along two vertical wells at both extremities of the cross-section. Grayscale maps of $\ln K$, $\ln K_{xx}$ and $\ln K_{yy}$ corresponding to one simulated realization are displayed in Figure 3. The values of the mean and variance of the realizations are shown to reproduce the input model values. Likewise, the correlograms and cross-correlograms of the realizations are displayed as black dots in Figures 1 and 2 and show a very good agreement with the input correlograms and cross-correlograms. The size of the search neighborhood used in the simulation was 15 pixels in the direction of maximum continuity and 5 pixels in the direction of minimum continuity.

Conclusion

The sequential simulation algorithm has several advantages over other simulation techniques of which the two most important are: (1) it is completely general and (2) it is conditional by construction.
The sequential simulation algorithm is ideally suited for the joint generation of multi-Gaussian fields with excellent reproduction of the model means, covariances and cross-covariances. The algorithm can handle any type of cross-covariance matrix, no matter how complex its analytical expression. Also, the algorithm remains efficient for the generation of very large fields. Its major drawback is the need to progressively limit the set of conditioning data to those within a search neighborhood. This approximation entails reproduction of correlation lengths limited to the dimensions of the search neighborhood.

Acknowledgments

This research was funded by the Stanford Center for Reservoir Forecasting. Part of the code development was funded by Empresa Nacional de Residuos S.A.

References

Alabert, F. G. (1987). Stochastic imaging of spatial distributions using hard and soft data. Master's thesis, Stanford University, CA, Branner Earth Sciences Library.

Anderson, T. W. (1984). *Multivariate statistical analysis*. Wiley, New York.

Gómez-Hernández, J. J. (1991). *A Stochastic Approach to the Simulation of Block Conductivity Values Conditioned Upon Data Measured at a Smaller Scale*. PhD thesis, Stanford University.

Gómez-Hernández, J. J. (1992). Regularization of hydraulic conductivities: A numerical approach. In *Proc. of the 4th Int. Geostat. Congress*.

Gómez-Hernández, J. J. and Srivastava, R. M. (1990). ISIM3D: An ANSI-C three dimensional multiple indicator conditional simulation program. *Computer and Geosciences*, 16(4):395–440.

Journel, A. G. (1989). *Fundamentals of Geostatistics in Five Lessons*, volume 8 of *Short Courses in Geology*. AGU, Washington D.C.

GEOSTATISTICAL MODELLING OF A SINGLE FRACTURE

Jean-Paul Chilès and Sylvie Gentier

BRGM, BP 6009, 45060 Orléans Cedex 2, France

ABSTRACT. The mechanical and hydrological behaviour of fractured media depends on the morphology of single fractures. Geostatistics enables modelling of the topography of both fracture surfaces and the aperture. The small-scale variations of a fracture surface depend on the angularity of individual grain facets and are modelled by means of a combined variography of the elevation and its first derivatives. The topography of the two fracture sides and the void height are studied simultaneously and consistently with specific tools, such as cross-covariance and the translatogram.

INTRODUCTION

The mechanical and hydraulic behaviour of a rock mass depends on the stress imposed on it, and on its intrinsic characteristics, i.e. physical and mechanical properties (velocity, elasticity, compression strength), petrographical properties (mineralogy, texture, type of grain joints, alteration), and structural properties. The structural characteristics include the spatial distribution of the various joint sets, as well as the morphological properties of individual joints, both of which are important. Modelling of joint sets describes the number, relative location, extension, orientation, and termination of the fractures, which are essential characteristics for analysis of the connectivity of the network, and for the simulation of the mechanical and hydraulic behaviour of the fractured rock mass. Individual morphological properties include the roughness of the fracture surfaces and the spatial variations of the aperture ; these define how flow and transport occur within the fractures, and determine the fracture's ability to resist a shear and modify the fluid flow as a function of the stress. The 3D modelling of fracture networks from actual data, using probabilistic and geostatistical tools, has been presented by Chilès (1989a, b), with reference to other authors ; at this level, the fractures, joints or faults are simplified and considered as planar. This article is concerned with the actual shape of fractures, i.e. the modelling of a single fracture.

TYPES OF DATA

A fracture appears as a set of two surfaces, separated by a void space that can or cannot be filled. We shall consider here natural unfilled fractures. According to the size of the fracture sample, several types of data can be recorded :

- The precise shape of a fracture can be studied at the laboratory scale on a core that is perpendicular to the average fracture plane. Profiles are sampled along both fracture surfaces with a laboratory profilometer (Gentier, 1986 ; Gentier and Riss, 1990). As the two surfaces are sampled successively, a careful measuring method is used for the recording of points at the same location on the two surfaces. The void height is obtained as the difference between the two surface heights.
- The void space can also be obtained by means of a soft translucent cast (Billaux and Gentier, 1990). Its grey-tone image can be transformed into a void-height image by reference to a calibrated standard wedge (Billaux, 1990). This method offers the advantage of a complete description of the aperture, but gives no information concerning the shape of the fracture surfaces. The translucent-cast data are sufficient for flow studies, but for transport and mechanical studies it is also necessary to consider the fracture surface profiles. Note that it is possible to get a cast of each fracture part, but that the transformation of their grey-tone images into surface-height values will be inaccurate, due to the thickness of such casts.
- At a greater scale (up to several metres), a fracture can now be sampled on site by using a terrain profilometer. This enables sampling of a single side of the fracture (the other one is not available, or cannot be spatially related to the first one).

MODELLING TECHNIQUES FOR A SINGLE FRACTURE

Here, we are interested in the analysis of fracture surfaces and void height. We consider fractures whose surfaces present no overhang, so that the surface heights and void height can be considered as regionalized variables, i.e. functions of a point x in the average fracture plane. The spatial features of the fracture can be studied by means of a geostatistical approach or a spectral approach. Geostatistical tools for the study of rough surfaces have been in use since several decades (e.g., Sayles and Thomas, 1977, with references to previous work since 1945). Applications to natural rock surfaces are described for instance by Brown and Scholz (1985) (spectral methods) and by Gentier (1986) (geostatistical and spectral methods).

The widely used fractal approach is a by-product of geostatistical or spectral analysis (e.g., Mandelbrot et al., 1984, in the study of a metal-fracture surface). Here, the Hausdorff dimension D is related to the behaviour of the variogram at the origin ; for a behaviour in $|h|^{\alpha}$, the Hausdorff dimension is $3-\alpha/2$ if the topography is of diffusion type, $3-\alpha$ if it is of mosaic type (the appropriate tool is then the variogram of order 1, computed with the absolute values of the increments rather than with square increments, which behaves like $|h|^H$ with H=3-D in both cases) (Matheron, 1989). The geostatistical and spectral approaches are theoretically equivalent, but the geostatistical one facilitates analysis of the relationships among the three variables (the two fracture surfaces, and the corresponding aperture). Only this approach will be used here. A limited spectral approach of this relationship can be found in Durham and Bonner (1992).

The overall objective of a geostatistical approach is two-tiered. First, it characterizes the spatial variations of the topography of the fracture surfaces and those of the aperture ; second, these characteristics are taken into account for the kriging of the fracture, or for its simulation at a larger scale. This numerical model can then be used for simulation of the mechanical behaviour of the fracture under stress, or of flow and transport within the fracture.

The standard geostatistical tools (variogram, kriging, simulation) are well known and need no introduction. Hereafter, we examine two points that are specific to the analysis of fracture surfaces :

- Small scale variations in the fracture surface are very important, for they are due to the presence of asperities on the fracture surface and condition the behaviour of the fracture under shear. The asperities commonly consist of planar facets, and are better modelled by considering not only the variogram of the topography, but also those of the first derivatives of the topography.
- The aperture is the difference between the two surface heights ; these are also related to the type of stress that created the fracture (tension or compression, with or without a lateral component) and in places also to the subsequent hydrological and/or metamorphic process (dilution, filling). Hence, the three variables must be modelled simultaneously and consistently. Specific tools, such as cross-covariance and the translatogram, are particularly suited for this study.

MODELLING OF A SINGLE FRACTURE SURFACE

Consider a core sample from one side of a fracture. The topography of the fracture surface can be considered as (a realization of) a random function Z(x) of a point x in a reference plane (e.g., the average fracture plane, or a plane perpendicular to the core axis). The geostatistical analysis of Z(x) presents two types of problems :

i) the fracture surface is not necessarily subparallel to the reference plane, so that Z(x) may be affected by a trend or drift ;
ii) at very small distances, because of the planarity of the grain facets, the surface is very regular, which makes the variogram analysis inaccurate because of possible strong fluctuations.

Two approaches can be used to overcome these problems :

i) filter the drift by using a variogram of residuals or a generalized variogram ;
ii) characterize the small-scale variations by studying the variogram of the first derivatives.

The use of a variogram of residuals is well known and will not be presented. However, it should be remembered that such variograms are biased for medium and long distances (Chilès, 1977).

Filtering the drift : the generalized variogram

The usual variogram filters any constant component by considering increments Z(x+h)-Z(x) : Z(x) and Z(x)+m have the same variogram. By considering increments of order k+1 instead of increments of order 1, a generalized variogram of order k is obtained that filters any polynomial component of degree k (the usual variogram is the generalized one of order 0). The generalized variogram can be used for stationary or intrinsic random functions, and also for intrinsic random functions of order k. For the theory and practice of generalized variograms, see Chilès (1979).

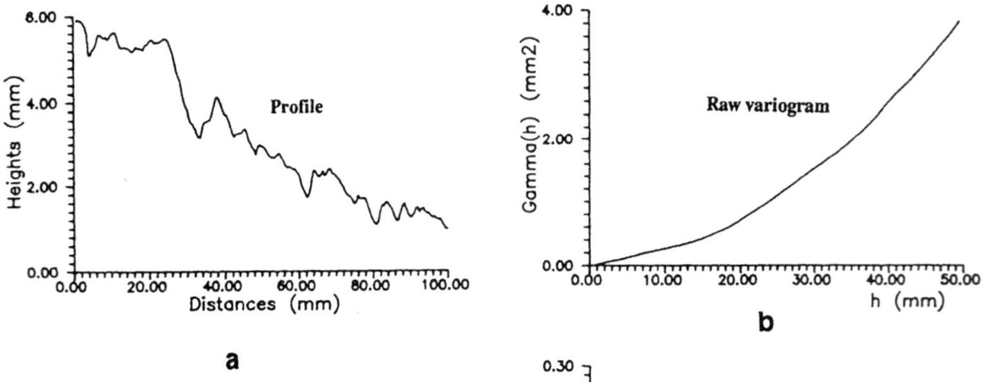

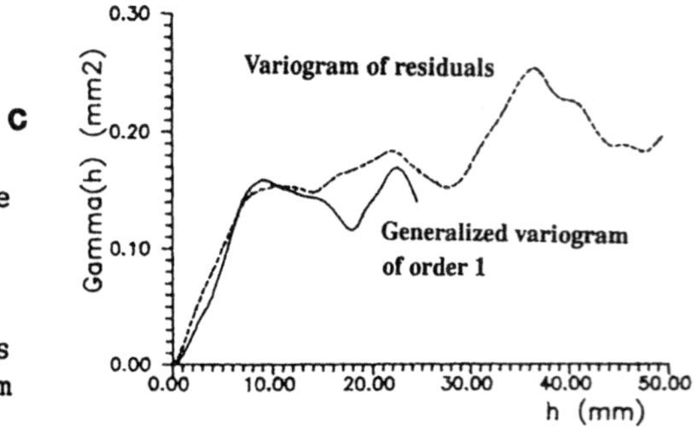

Fig. 1. Variography of a fracture profile :
a) profile ;
b) raw variogram ;
c) variogram of residuals and generalized variogram of order 1.

Gentier (1986) and Gentier and Riss (1990) describe applications to a natural rock fracture. Figure 1 presents a fracture profile with a clear drift, its raw variogram, its variogram of residuals, and its generalized variogram of order 1. The raw variogram is of course mainly parabolic and thus unusable except over very short distances, whereas the other two variograms are quite coherent ; according to the theory, they have the same range and sill, and the slope at the origin for the generalized variogram is 2/3 of the slope of the variogram of residuals.

Modelling small-scale variations : the variogram of the first derivative

When Z(x) is very regular over short distances, as is the case for most natural fractures, the behaviour of its variogram g(h) in the vicinity of the origin is in $|h|^{\alpha}$ with $2<\alpha<4$ (the regular h^2 and h^4 terms shall not be taken into account, and a regularity higher than that of h^4 has never been observed). The determination of g(h) for short distances is difficult, because strong fluctuations are possible : Z(x) is not micro-ergodic as soon as $\alpha>2$ (Matheron, 1965, Chap. XIII). But the variogram of a first derivative of Z has a behaviour in $|h|^{\alpha-2}$ and can be determined with good precision since $0<\alpha-2<2$, if data are available.

We have no data concerning the first derivatives, but as the profiles are sampled at regular intervals, it is easy to consider a discrete form of the first derivative. If x_o is the first point of a profile and u the vector representing the oriented distance from one point to the next, the discrete first derivative at x is :

$$Z'_u(x) = [Z(x)-Z(x-u)]/|u|$$

Its variogram $g_u(h)$ derives from that of Z(x) by :

$$g_u(h) = [2g(u)+2g(h)-g(h-u)-g(h+u)]/u^2$$

Good variogram modelling is obtained when g(h) is chosen so as to produce a good fit of the experimental variograms of Z and Z'_u. Several variograms of Z'_u can be considered, if the profiles have been sampled along several directions and g(h) is not isotropic. This kind of modelling was described by Chauvet et al. (1976) for meteorological fields : in the framework of the geostrophical approximation, wind is the gradient of pressure (or of geopotential) up to a multiplicative factor and a $\pi/2$ rotation. The difference is that direct measurements of the gradient are available in this case.

Gentier et al. (1991) presented the analysis of a natural fracture in the granite of Guéret (France), which was sampled in two directions corresponding to the apparent minimum and maximum variability. The variogram g(h) (Fig. 2 for the direction of maximum variability) can be modelled by a single cubic model, but in this case the variogram of the discrete first derivative is not correctly reproduced, leading to an underestimation of the possible variations. A good fit of the two variograms is obtained by defining g(h) with two nested cubic models.

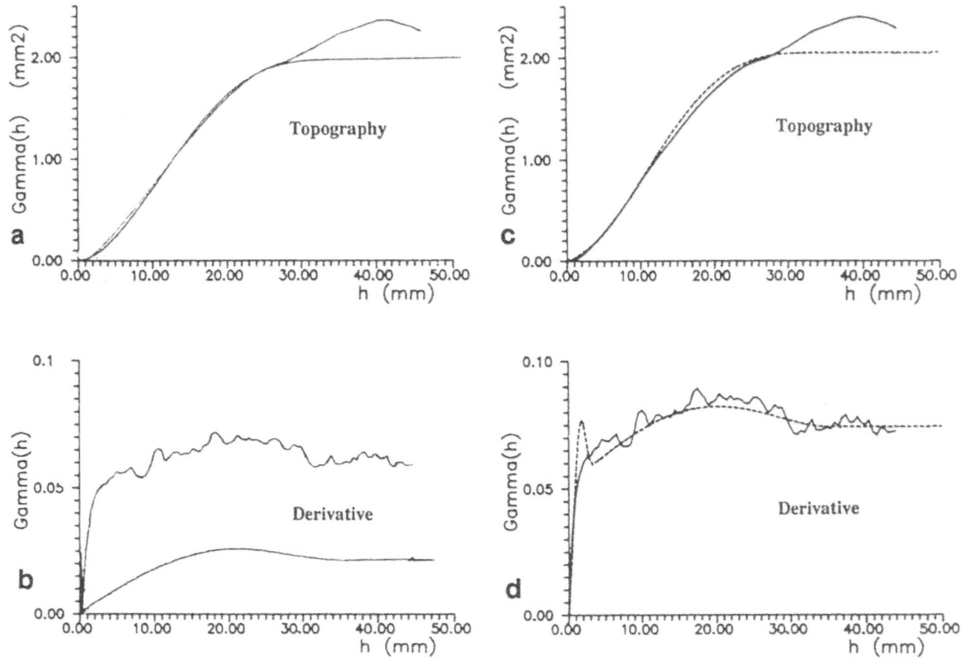

Fig. 2. Variography of a fracture profile and its first derivative : (a,b) with a single cubic model ; (c,d) with two nested cubic models.

MODELLING OF BOTH FRACTURE SURFACES AND APERTURE

Consider a fracture sample that has been taken by means of a core perpendicular to the average fracture plane. Let x be a point in the average fracture plane (two coordinates), and z be the axis normal to this plane. The topography of the lower and upper surfaces of the fracture are described in terms of (realizations of) two random functions $Z_1(x)$ and $Z_2(x)$. The aperture then is the random function $Z_3(x)=Z_2(x)-Z_1(x)$. Because of this relation, Z_1, Z_2 and Z_3 are not independent.

Moreover, the surfaces Z_1 and Z_2 are commonly similar. Before studying a more general case, we should examine a very simple example of the usefulness of a simultaneous study of elevations and aperture.

Case of a constant displacement

Suppose that Z_1 and Z_2 are equal up to a translation :

$$Z_2(x) = Z_1(x+a) + m$$

where a and m are the horizontal and vertical components of the relative displacement of the upper surface compared with the lower one.

Denoting $C(h)=Cov(Z_1(x), Z_1(x+h))=Cov(Z_2(x), Z_2(x+h))$ as the covariance function common to both surfaces and $g(h)=C(0)-C(h)$ as their variogram, such a displacement can be easily detected on the cross-covariance $C_{12}(h)=Cov(Z_1(x), Z_2(x+h))$ of the two surfaces, since $C_{12}(h)$ is equal to $C(h+a)$: it is translated from $C(h)$, with its maximum value for $h=a$ instead of $h=0$. This displacement can also be detected on the aperture variogram $g_{33}(h)$, as it is easy to see that :

$$g_{33}(h) = 2\ g(a) + 2\ g(h) - g(h-a) - g(h+a)$$

When $h>>a$, $g(h)$ can usually be considered as linear between h-a and h+a, so that the last three terms cancel out and $g_{33}(h)$ is equal to its sill value $2\ g(a)$. Figure 3 shows the aperture variogram $g_{33}(h)$ linked with a linear variogram $g(h)$. Along the translation direction, it has a range equal to the translation a. Its behaviour below this range is also linear, but with twice the slope of the variogram of the topography. The behaviour is similar along the other directions, but with a smoother transition to the sill. This type of variogram can be found on actual fractures.

The aperture variogram is thus characteristic of the movement that created the fracture and of the topography of the fracture surfaces. There is usually no sense in studying it independently of the surfaces. Conversely, the joint study of the three variables enables determination of the absolute value of the horizontal offset and its direction. A look at the data is enough to obtain its orientation.

It is obvious that in case of a constant displacement, a simple examination of the data could provide a. But in practice the data are affected by drifts, the displacement is not the same in the centre of the fracture and near its border, and abrasion has occurred.

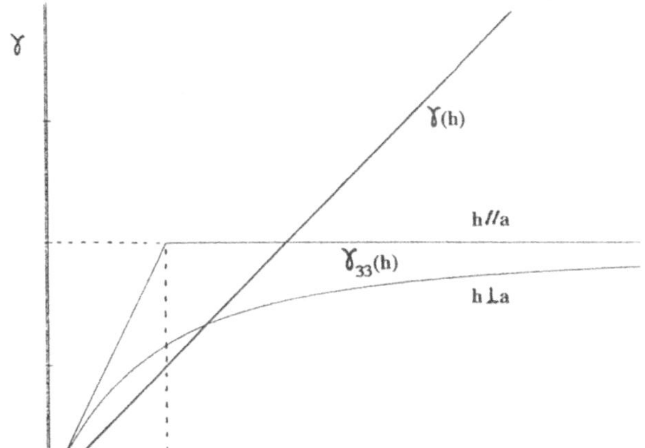

Fig. 3. Variogram g_{33} of the aperture due to a relative translation a of the fracture sides, for a linear variogram $g(h)$.

General case

In the general case, the coregionalizations of several stationary random functions $Z_i(x)$ are usually studied by means of direct and cross-covariances

$$C_{ij}(h) = Cov(Z_i(x), Z_j(x+h))$$

or direct and cross-variograms

$$g_{ij}(h) = \frac{1}{2} E[Z_i(x+h)-Z_i(x)][Z_j(x+h)-Z_j(x)]$$

Variograms can be expressed from covariances, according to the relation :

$$g_{ij}(x,y) = \frac{1}{2} [2C_{ij}(0)-C_{ij}(h)-C_{ij}(-h)]$$

Cross-variograms are doubly symmetrical, for they satisfy :

$$g_{ij}(h) = g_{ij}(-h) = g_{ji}(-h) = g_{ji}(h)$$

whereas cross-covariances usually are not symmetrical, but only satisfy :

$$C_{ij}(h) = C_{ji}(-h)$$

Cross-variograms are therefore less rich than cross-covariances when the latter are not symmetrical, particularly when spatial variations of one parameter are shifted in comparison with those of another one (due, for example, to migration of a chemical element, to delay, or to fault throw), which is precisely the case for the two surfaces of most fractures.

Since the variables here discussed satisfy by definition

$$Z_3(x) = Z_2(x) - Z_1(x)$$

their direct and cross-covariances can be deduced from those of Z_1 and Z_2. Similarly, if we use only variograms (with a possible loss of information, but with fewer bias problems), the direct and cross-variograms of the three variables can be derived from those of Z_1 and Z_2. Because of the symmetry of the cross-variograms, it is also possible, and moreover easier, to take the three direct variograms as basic structure functions, from which the cross-variograms can be deduced by :

$$g_{12}(h) = \frac{1}{2} [\ g_{11}(h) + g_{22}(h) - g_{33}(h)]$$
$$g_{13}(h) = \frac{1}{2} [-g_{11}(h) + g_{22}(h) - g_{33}(h)]$$
$$g_{23}(h) = \frac{1}{2} [-g_{11}(h) + g_{22}(h) + g_{33}(h)]$$

These relationships remain valid if experimental, rather than theoretical, structure functions are considered. It would thus be redundant to consider other theoretical or experimental structure functions than those that were chosen as basic tools.

More precise model for displacement

In practice, because of the measuring method, the vertical displacement is not constant. When the fracture sample is collected from a core normal to the median fracture plane, the two surfaces will be separate. They are also measured separately. The wedging of each half-sample is not perfectly horizontal, so that Z_1 and Z_2 are affected by drifts m_1 and m_2, which are linear functions of x with nearly equal slopes. As a consequence, the vertical component m of the displacement has a linear drift with a small slope :

$$m(x) = \alpha + \beta x$$

Moreover, the horizontal displacement a can also vary because of relaxation, so that it can be considered as a smooth function of x. Finally, each surface may have been affected by dissolution, or the data may be affected by measurement errors. A more precise model is therefore :

$$Z_2(x) = Z_1(x+a(x)) + m(x) + R(x)$$

with

a(x) : relative horizontal displacement, i.e. smooth function of x ;
m(x) : relative vertical displacement, i.e. linear function of x ;
R(x) : residual, considered as a stationary random function.

Z_1 and Z_2 are affected by linear drifts, which may be one order of magnitude higher than m(x). We shall now examine the influence of these various terms.

Displacement with vertical drift : translatogram

In this section, the horizontal displacement a is assumed to be constant, but not the vertical displacement m, and the residual R is assumed to be null. As the experimental covariances and variograms are affected by this slight drift (the drift is small, but so is the aperture), it is useful to reduce its effect. Two approaches can be used :

i) remove a least squares plane from each data set (lower surface and upper surface), in order to get roughly horizontal surfaces ;
ii) define tools that are less sensitive to drift than the usual structure functions.

We suggest a combination of both approaches. The first one poses no problem, so that only the second one will be examined.

Concerning variograms, we have seen above that the generalized variogram enables the filtering of the influence of a polynomial drift. This approach cannot be used for covariances, and covariances of residuals are biased, so that we have to define a specific tool for our problem. The principle for the construction of such a tool is simple ; instead of considering the definition of the cross-variogram

$$g_{ij}(h) = C_{ij}(0) - 1/2\ [C_{ij}(h) - C_{ij}(-h)]$$

which cancels out the asymmetry of the cross-covariance, a pseudo-cross-variogram can be defined by

$$G_{ij}(h) = C_{ij}(0) - C_{ij}(h)$$

which retains this asymmetry. Many functions of this type can be considered. One of them seems to be particularly interesting (see Chilès, 1992, for more details), which is the average of the pseudo-cross-variograms $-G_{31}(h)$ and $G_{32}(-h)$ of the aperture and the surfaces (their theoretical values are equal, since $-C_{31}(h)$ and $C_{32}(-h)$ are equal, as mentioned above) :

$$\tfrac{1}{2}\ [-G_{31}(h)+C_{32}(-h)] = \tfrac{1}{2}\ \mathrm{Cov}(Z_3(x),\ Z_1(x+h)-Z_1(x)+Z_2(x)-Z_2(x-h))$$

Let g_{SA} be this structure function (S for Surface, A for Aperture), which will be called a translatogram (see below) and is defined by :

$$g_{SA}(h) = \tfrac{1}{2}\ \mathrm{Cov}(Z_2(x)-Z_1(x),\ Z_1(x+h)-Z_1(x)+Z_2(x)-Z_2(x-h))$$

When the two surfaces are separated by a horizontal displacement a, one obtains :

$$\begin{aligned} g_{SA}(h) &= C(0) - C(h) - C(a) + C(a-h) \\ &= g(h) + g(a) - g(a-h) \end{aligned}$$

The first expression requires that the covariance C(h) exists, while the second one remains valid if only the variogram g(h) exists.

We call g_{SA} a translatogram in order to differentiate it from the usual cross-variogram, and because it is a powerful tool for the characterization of horizontal translation or displacement. When we consider, as before, the case of a linear variogram (Fig. 4), we see that along the directed direction of translation, the translatogram has a range a and a sill 2 g(a) (like the aperture variogram), but is equal to zero along the opposite direction and displays intermediate sills along the other orientations. g_{SA} keeps all the advantages of g_{33} and moreover makes it possible to define the translation orientation.

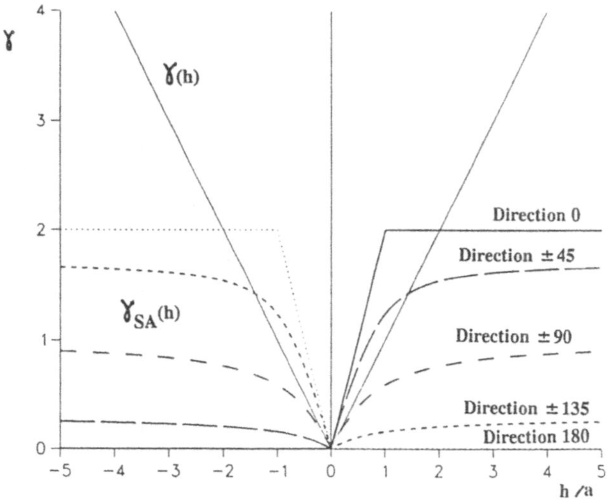

Fig. 4. Translatogram associated with a relative translation a and a linear variogram g(h).

The corresponding experimental translatogram (defined as an average of experimental values rather than a centred mathematical expectation) is weakly sensitive to drift. The second part of the covariance is the difference of two increments concerning Z_1 and Z_2, which filter any constant component ; the difference of these increments filters any linear drift that would affect Z_1 and Z_2 simultaneously, and only the linear part of m(x) remains.

At the level of the mathematical expectation, we have :

$$g_{SA}(h) + g_{SA}(-h) = g_{33}(h)$$

where g_{33} is the symmetrized form of g_{SA}. This explains why the translatogram combines the advantages of a cross-covariance and of a variogram, i.e. oriented directional behaviour and good inference properties. Note also that the above relationship is not numerically valid for the experimental structure functions.

Variable horizontal displacement

If a varies slowly, so that it remains nearly constant over distances shorter than h , the previous expressions of the covariances and variograms simply need to be randomized. If an average tendency a_0 remains, the curves where a intervenes display a smoother behaviour around $h=a_0$. Consider for example the variogram g_{33}. Denoting f(a) the distribution function of a over the studied area, and $g_{33}(a;h)$ the expression of g_{33} as a function of a and h, one has :

$$g_{33}(h) = \int f(a)\ g_{33}(a;h)\ da$$

In particular, its sill is $2 \int f(a)\ g(a)\ da$.

Consider now the particular case where the average displacement has no particular orientation. More precisely, simply suppose that f is symmetrical ($f(a)=f(-a)$). Because of the randomization, the cross-covariances become symmetrical. In this case the cross-covariances give no more information than the cross-variograms, which themselves give no more information than the direct variograms, and the expression of the aperture variogram from the variogram of the topography becomes :

$$g_{33}(h) = 2\ [g(h) + g*f(0) - g*f(h)]$$

where $g*f$ denotes the convolution of g by f. Concerning the translatogram, such a case is defined by the relation :

$$g_{SA}(h) = g_{SA}(-h) = \tfrac{1}{2}\ g_{33}(h)$$

For example, if g(h) has a linear behaviour near the origin, with a slope b, the sill of g_{33} will be :

$$2\ b \int f(a)\ |a|\ da = 2b\ E[|a|]$$

If the distribution of a is normal with mean 0 and variance σ^2, this sill of g_{33} will be equal to $\sqrt{8/\pi}\ \sigma$ in 1-D and to $\sqrt{\pi}\ \sigma$ in 2-D (σ is here the circular standard deviation), while the range will correspond to the maximum value of $|a|$ (around 3σ).

Influence of the residual R(x)

We will now consider the case where the residual R(x) is not negligible. R(x) results from components $R_1(x)$ and $R_2(x)$ included in Z_1 and Z_2. If we assume that R_1 and R_2 are independent, the consequences for the variograms are the following :

- g_{11} and g_{22} include the variograms of R_1 and R_2 respectively ;
- g_{33} includes both variograms of R_1 and R_2.

These variograms are commonly reduced to a nugget effect because of measuring errors, but may also include a structured component that is a feature of the fracture itself.

APPLICATIONS

Thanks to this approach, it is possible to understand better how a fracture was created, to simulate it at a large scale, and then to study its hydrological or mechanical behaviour. For example, when the aperture is known at any point - from a void cast or from a geostatistical simulation - it is theoretically possible to calculate flow by assigning to any point a transmissivity as a function of the aperture, and then using a finite difference model. However, this method would be very awkward, and in fact is not suitable, for the aperture is much smaller under stress than what has been measured on the void cast. Gentier (1986) and other authors have observed that, under stress, the water flow usually is along channels. Hence the idea, from the flow-study viewpoint, to represent each fracture by a network of channels or conducting pipes. Billaux (1990) presents the network obtained on an actual fracture by morphological skeletonization (Serra, 1982) : this operation was performed on the void-height image after cutoff at a minimum height value corresponding to the reduction in aperture width under normal stress.

CONCLUSION

The first conclusion is that a proper modelling of the small-scale variations of a fracture surface requires the simultaneous study of the topography of this surface and of the gradient of this topography.

The second conclusion is that the void-space geometry must be studied in relation with the fracture-surface geometry. Cross-covariances are very efficient tools for the study of the geometry of a single fracture, but require strict stationarity of the data. Cross-variograms are less sensitive to lack of stationarity, but are not as powerful in the study of fractures with a shear offset, because of their symmetrized form. The translatogram proposed here, is a simple tool that combines the advantages of cross-covariances (asymmetry) and cross-variograms (good inference), and that enables efficient study of fractures that are mainly characterized by a relative translation of their two surfaces.

These tools can be used for other fracture types, but cannot be universally applied. For example, in the case of alteration or erosion of fracture surfaces due to a mechanical or chemical origin, the study of covariances is not sufficient, and it is necessary to consider at least the bivariate distributions. Taking explicitely into account that the aperture is non-negative, also requires to consider particular random-function models.

ACKNOWLEDGEMENTS

This is BRGM scientific contribution No. 92023. The work was financially supported by a BRGM research grant. Thanks are due to Jane Long from LBL, who provided the opportunity of initiating this work as part of a joint research project between BRGM and LBL.

REFERENCES

Billaux, D. (1990). Hydrogéologie des milieux fracturés. Géométrie, connectivité, et comportement hydraulique. Doctoral thesis, E.N.S. des Mines de Paris, 277 p.

Billaux, D., and S. Gentier (1990). Numerical and laboratory studies for flow in a fracture. In Rock joints, N. Barton and O. Stephansson (Eds), Balkema, Rotterdam, Netherlands, pp. 369-373.

Brown, S.R., and C.H. Scholz (1985). Broad bandwidth of the topography of natural rock surfaces. Journal of geophysical research, Vol. 90, No. B14, pp. 12575-12582.

Chilès, J.P. (1977). Géostatistique des phénomènes non stationnaires. Thesis, Nancy-I University, France, 152 p.

Chilès, J.P. (1979). Le variogramme généralisé. Internal report, N-612, Centre de Géostatistique, Fontainebleau.

Chilès, J.P. (1989a). Three-dimensional geometric modelling of a fracture network. In Geostatistical, sensitivity, and uncertainty methods for ground-water flow and radionuclide transport modeling, B.E. Buxton, (Ed.), Battelle Press, Columbus, Ohio, pp. 361-385.

Chilès, J.P. (1989b). Modélisation géométrique de réseaux de fractures. In Geostatistics, M. Armstrong (Ed.), Kluwer Academic Publishers, Dordrecht, Netherlands, Vol. 1, pp. 57-76.

Chilès, J.P. (1992). Geostatistical modelling of the geometry of a single fracture. Internal report, BRGM, SGN/ISA, Orléans, France.

Durham, W.B., and B.P. Bonner (1992). Closure and fluid flow in discrete fractures. Proceedings of Fractured and jointed rock masses, Lake Tahoe, California, to be published.

Gentier, S. (1986). Morphologie et comportement hydromécanique d'une fracture naturelle dans un granite sous contrainte normale. Doctoral thesis, Université d'Orléans. Document du BRGM, No. 134, BRGM, Orléans, France, 597 p.

Gentier, S., and J. Riss (1990). Quantitative description and modelling of joints morphology. In Rock joints, N. Barton and O. Stephansson (Eds.), Balkema, Rotterdam, Netherlands, pp. 375-382.

Gentier, S., J.P. Chilès, and J. Riss (1991). Analyse et simulation d'une surface de fracture rocheuse. Cahiers de géostatistique, Fasc. 1, ENS des Mines de Paris, pp. 77-87.

Mandelbrot, B.B., D.E. Passoja, and A.J. Paullay (1984). Fractal character of fracture surfaces of metals. Nature, Vol. 308, pp. 721-722.

Matheron, G. (1965). Les variables régionalisées et leur estimation. Masson, Paris, 306 p.

Matheron, G. (1989). The internal consistency of models in geostatistics. In Geostatistics, M. Armstrong (Ed.), Kluwer Academic Publishers, Dordrecht, Netherlands, Vol. 1, pp. 21-38.

Sayles, R.S., and T.R. Thomas (1977). The spatial representation of surface roughness by means of the structure function : a practical alternative to correlation. Wear, Vol. 42, pp. 263-276.

Serra, J. (1982). Image analysis and mathematical morphology. Academic Press, London, 628 p.

Using the Entropy Statistic to infer Population Parameters from Spatially Clustered Sampling

Neil Schofield
FSSI Consultants (Australia) Pty. Limited
P.O. Box 657, Epping NSW 2121, Australia.

Abstract

This paper proposes an alternative to the Polygonal and Cell Declustering Methods for generating representative estimates of population parameters from spatially clustered sample data sets. The proposed Entropy Method uses the criterion of maximizing the difference between the population entropy and the sample entropy to find the *optimal* set of declustering weights. The method capitalizes on the simplicity of the Cell Declustering Method, is comparatively easy to implement for 2-D and 3-D data sets and provides results very similar to the Polygonal Method.

Introduction

An important practical problem may arise in using the naive or equal weighted statistics of many earth science sample data sets as representative estimates of underlying population parameters.

The sampling schemes used particularly in the natural resource industries usually depart significantly from the ideal schemes of random stratified or systematic sampling schemes as described by Ripley, 1981. While sample measurements are usually intended to inform about some spatial domain of interest, often some areas within the domain have been sampled more extensively than others.

In many mining and mineral exploration projects, there is a tendency to cluster sampling in the high grade areas of mineralized domain because these areas are of most interest to the miner or explorer. Accessability of an area can also lead to sample clustering such as occurs in rugged terrains or from fan drill sampling in underground mines.

In 2-D space, sample clustering is easily observed by simply plotting the sample locations. Combined with a contour map, one can quickly gain an overall impression of the nature of the clustering i.e. clustered in high valued areas or random clustering etc. In 3-D space, it is more difficult to determine the nature of the clustering particularly when drill holes are variously directed within the domain of interest.

In the rare instances where perfectly regular or systematic sampling is available, the equal weighted statistics provide the best estimates of the underlying population parameters (provided there is no periodicity to the phenomenon at the scale of sampling). In the more usual case where sampling is preferentially clustered in high valued areas or, in high and low value areas as often occurs in 3-D sampling, a *optimal* set of declustering weights is desired such that the resultant weighted sample statistics provide the most representative estimates of the population parameters.

Various methods have been proposed to generate the *optimal* set of declustering weights and these are discussed by Isaaks and Srivastava 1989 (Chapter 10) and by Deutsch 1988. These methods include;

- **The Polygonal Method**
  In this method, the weight applied to a sample is directly proportional to the area of the Voronoi polygon (Ripley 1981) surrounding it. In areas of clustered sampling, the polygon areas will tend to be smaller and samples will receive correspondingly smaller weights than samples with larger polygonal areas in more sparse sampling. The polygonal approach requires the definition of a boundary around the domain of sampling to restrict the size of the polygons for samples at the edge of the sampled area. While reliable and fast algorithms exist to implement the Polygonal Method in 2-D (Green and Sibson 1978, Tipper 1991), the author is not aware of any 3-D implementation of the method to generate 3-D polyhedra (as suggested by Isaaks and Srivastava), the volumes of which could be used to generate declustering weights.

- **The Cell Declustering Method**
  The Cell Declustering Method was proposed by Journel 1983 and Deutsch 1988 provides a FORTRAN 77 program for implementing the method in 2-D and 3-D together with an assessment of the performance of the method compared to the Polygonal Method and a Global Kriging Method. The Cell Declustering Method works effectively when sampling is known a priori to be clustered in high or low valued areas. For example, if sampling is clustered in high valued areas, one seeks by iteration over a range of cell sizes, the set of weights which provides the minimum value for the global mean of the samples. If sampling is clustered in low values, the set of weights which maximizes the global mean is sought. Each sample receives a weight equal to $\frac{1}{N_i \times C}$ where N_i is the number of samples in the *ith* cell and C is the total number of cells that contain least one sample.

  In situations where sampling is randomly clustered or clustered in high and low valued areas in the domain of interest, one chooses a cell size such that there is approximately one datum per cell in sparsely sampled areas. This calls for some visual judgement which may be realistic for 2-D data sets but is somewhat unrealistic for 3-D situations.

  Important advantages of the Cell Declustering Method are that it is easy to implement in both 2-D and 3-D and does not necessarily require the prior definition of a boundary to the domain as does the Polygonal Method.

- **Other Methods**
  Deutsch 1988 and Isaaks and Srivastava 1989 discuss other alternatives to obtaining declustered statistics of the sample data including discarding the clustering samples and using the global kriging weights. Discarding clustering data can work well if the clustering data can be properly identified but it can often have the disadvantage of ignoring a large volume of important information.

Using the global kriging weights can also work well and has the important advantage of accounting for spatial continuity if a reliable measure of the spatial continuity can be found from the clustered data.

This paper presents an alternative approach to spatial declustering which uses the measure of entropy (Jones 1979) of a sample set or population as a basis for finding the *optimal* set of declustering weights. A 2-D study of a number of clustered sample data sets from the Walker Lake Data Set (Isaaks and Srivastava 1989) shows that the method provides results very similar to the Polygonal Method.

The Entropy Measure H

The entropy of a discrete random variable X is defined as;

$$H(X) = -\sum_{i=1}^{N} p_i \cdot ln\ p_i$$

where N is the total number of possible outcomes of X and p_i is the probability to the *ith* outcome.

$H(X)$ can be interpreted as the amount of uncertainty that exists as to the value of X and/or the average amount of information received when a value of X is observed. Both interpretations are useful in the following discussion.

The Entropy Method

Consider a discrete random variable V defined on a spatial domain D such that all N possible outcomes of V, $v_i, i = 1, \cdots, N$, are associated with the set of equal sized samples making up the domain D. The entropy measure $H(V)$ is defined as

$$\begin{aligned} H(V) &= -\sum_{i=1}^{N} \frac{1}{N} \cdot ln(\frac{1}{N}) \\ &= -\, ln(\frac{1}{N}) \end{aligned}$$

Each outcome has an equal probability of occurrence with all other outcomes and for this set of uniform probabilities, $H(V)$ is a maximum (Jones 1979). For any other set of unequal weights, $H(V) < -ln(\frac{1}{N})$.

Now consider a limited sampling U of V comprising L samples $u_l, l = 1, \cdots, L$ where $L << N$. The entropy of the sample set U may be measured as

$$H(U) = -\sum_{l=1}^{L} w_l \cdot ln\ w_l$$

where $w_l \epsilon [0,1]$ is the weight associated to the sample u_l and $\sum_{l=1}^{L} w_l = 1$. If the U are chosen systematically (at regular spacing) such that the w_l are all equal to $\frac{1}{L}$, the entropy of the regular sampling U_r is

$$\begin{aligned} H(U_r) &= -\sum_{l=1}^{L} \frac{1}{L} \cdot ln(\frac{1}{L}) \\ &= - ln(\frac{1}{L}) \\ &< - ln(\frac{1}{N}) \end{aligned}$$

The choice of $w_l = \frac{1}{L}$ maximizes the value of $H(U)$ and therefore minimizes the value of $H(V) - H(U)$. This is interpreted to mean that a regular sampling U_r of V provides maximum information about V for that number L of samples. For any other non-regular or clustered sample set U_c comprising L samples such that the w_l are not equal to $\frac{1}{L}$, $H(U_c) < H(U_r)$.

The hypothesis proposed here is that the *optimal* set of sample weights w_l for clustered data should be chosen such that $H(V) - H(U_c)$ is a maximum. This approach may be considered as one of maximizing the unknown information about the underlying population V given the sample U_c. In practice, $H(V) - H(U_c)$ can be maximized by minimizing $H(U_c)$ under the same conditions used in the Cell Declustering Method i.e. that the w_l are directly related to the area of influence of the sample value u_l.

The iterative mechanism of the Cell Declustering Method may be used to find the *optimal* set of weights $w_l, l = 1, \cdots, L$ that minimizes $H(U_c)$ and this set of weights can then be used to derive the representative declustered statistics of the sample data set.

Case Studies

The U variable of the Walker Lake data set (Isaaks and Srivastava 1989) has been used as the basis for testing the performance of the Entropy Method to provide declustering weights and representative statistics of the underlying population. The performance of the Entropy Method is compared to that of the Polygonal Method for the same sample data sets.

The Walker Lake data set is a public domain data set derived from digital elevation data from the Walker Lake area near the California - Nevada border. The data set comprises 78000 data on a grid of 260 by 300. The contour map of Figure 1 shows the dominant structural features of the Walker Lake area. In both case studies, a rectangular boundary to the domain was defined at 0 and 260 and 0 and 300.

Two studies were performed to check the performance of the Entropy Method;

1. In the first study, ten sets of samples comprising between about 350 and 500 samples were collected from the Walker Lake data set in such a way that the samples were preferentially clustered in the high valued areas. This was achieved by first selecting a set of 195 samples on a random stratified grid (13 by 15) and

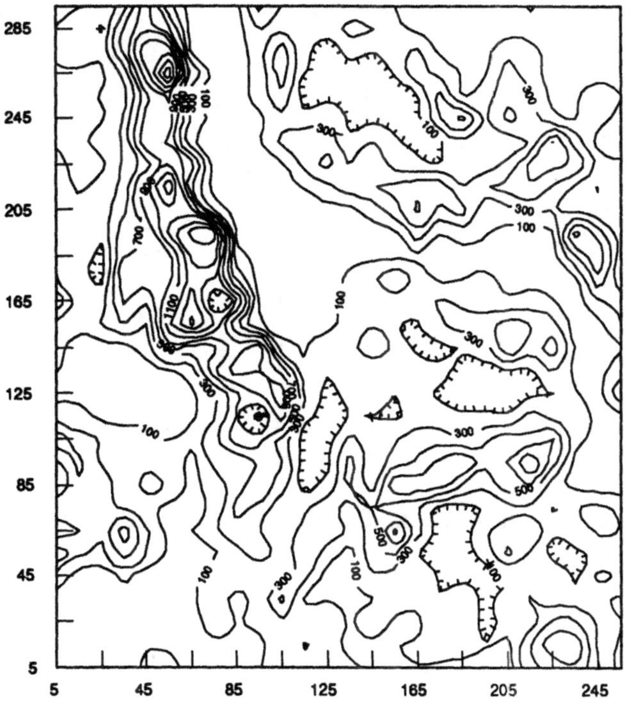

Figure 1: Contour map of the exhaustive average of the Walker Lake variable U within $10 \times 10\ m^2$ areas.

then clustered sampling around the high values found in the initial sample campaign. Figure 2a shows a typical pattern of preferentially clustered sampling.

These ten sample data sets were declustered using both the Polygonal Method and the Entropy Method and the sample statistics compared to the exhaustive statistics (population parameters).

2. The second study is similar to the first except that in each of the 10 sample campaigns, exactly 400 samples were collected using random sampling and producing a pattern of random clustering in the sample data. Figure 2b shows a typical pattern of randomly clustered sampling.

Study 1 - Preferential Clustering

Figure 3 presents q-q plots of the histograms of the sample data sets and clearly demonstrates the adverse effects of clustering in high valued areas on the equal weighted statistics.

Figure 4 presents plots of the Polygonal and Entropy declustered sample means versus the naive sample means. The population mean is shown by the dashed line.

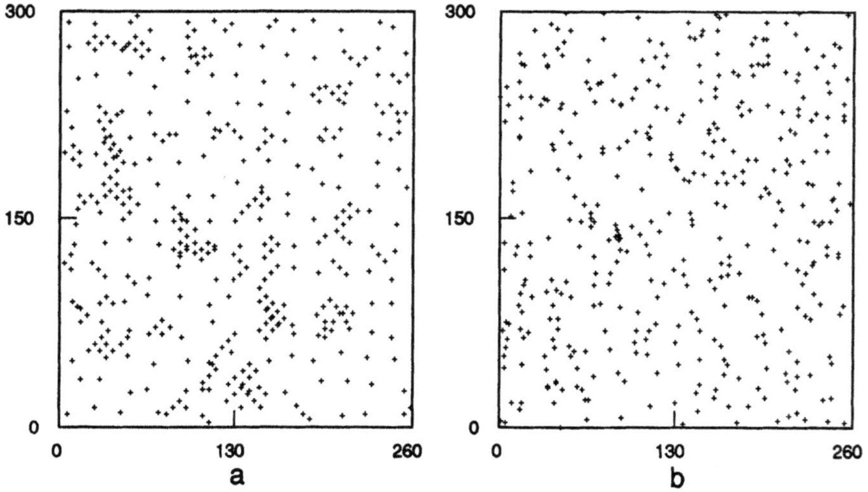

Figure 2: Typical patterns of (a) preferential and (b) randomly clustered sampling.

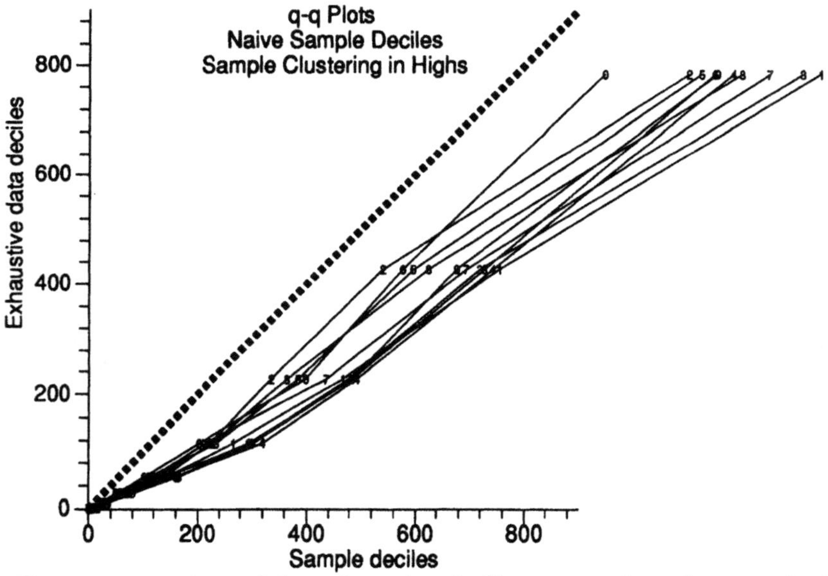

Figure 3: q-q plots of the exhaustive deciles versus the naive sample deciles.

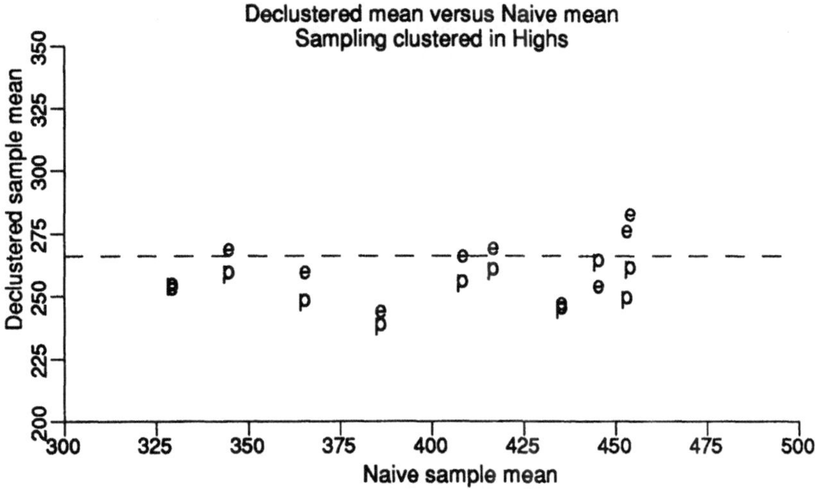

Figure 4: Entropy (e) and Polygonal (p) declustered sample means versus the naive sample means for the 10 sample data sets. The population mean is shown by the dashed line.

Figure 5 presents a similar plot for the declustered and naive sample variance. Both these figures demonstrate the similarity of the result shown by the Polygonal and Entropy Methods and their success in providing a representative mean and variance from the sample data clustered in the high values. In addition, to providing more accurate estimates of the population mean, the spread of the declustered estimates compared to the naive estimates is reduced by a factor of around 3 for the means and 2.6 for the variances.

Figure 6 presents q-q plots of the Polygonal and Entropy declustered sample deciles versus the exhaustive deciles of the U variable. These plots demonstrate the similarity between the declustered histograms generated by the Entropy Method and the Polygonal Method and the improvement brought by declustering over the naive histograms of Figure 3.

Study 2 - Random Clustering

Figure 7 shows q-q plots which compare the naive deciles of the ten randomly clustered data sets to the population deciles. The sampling does not show the extreme bias of the preferentially clustered data sets.

Figure 8 presents plots of the Polygonal and Entropy declustered means and variances versus the corresponding naive statistics of the sample data sets. Again the Entropy Method provides declustered statistics very simular to those of the Polygonal Method and the declustered statistics from both methods show a marginally smaller spread than the naive statistics of the sample data.

Figure 9 shows the q-q plots for Polygonal and Entropy declustered histograms of the 10 sample data sets. The Polygonal and Entropy methods again show similar

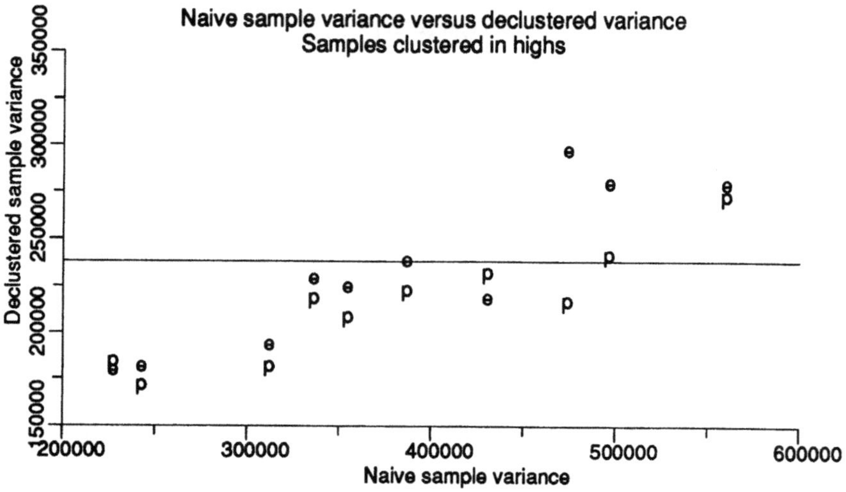

Figure 5: Entropy (e) and Polygonal (p) declustered sample variances versus the naive variances for the 10 sample data sets.

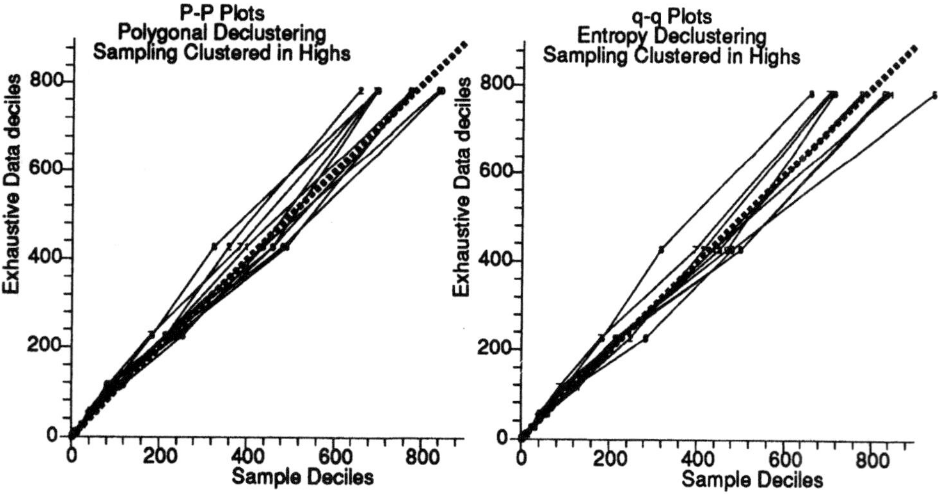

Figure 6: Comparison of the declustered sample histograms generated by the Polygonal and Entropy Methods for sampling clustered in high values.

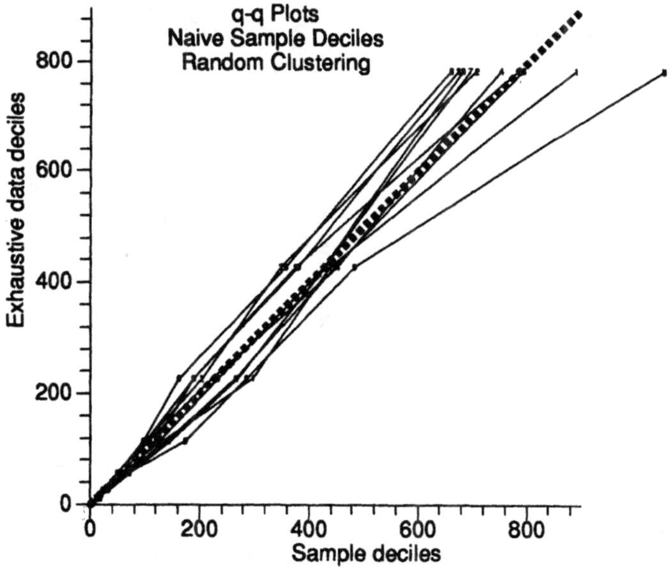

Figure 7: q-q plots of the exhaustive deciles versus the naive sample deciles.

results with a marginal improvement over the naive histograms particularly for the entropy declustering.

Conclusions

The Entropy Method of declustering spatially clustered sampling provides a useful alternative to both the Cell Declustering Method and the Polygonal Method for several reasons;

1. The maximization of the difference between the population entropy and the sample entropy $H(V) - H(U_c)$ provides a consistent basis for finding the *optimal* set of declustering weights independent of the nature of the clustering, i.e. preferential or random clustering.

2. The iterative mechanism of cell declustering using the entropy criterion can be easily implemented for 2-D and 3-D problems.

3. The method produces results very similar to the Polygonal Method.

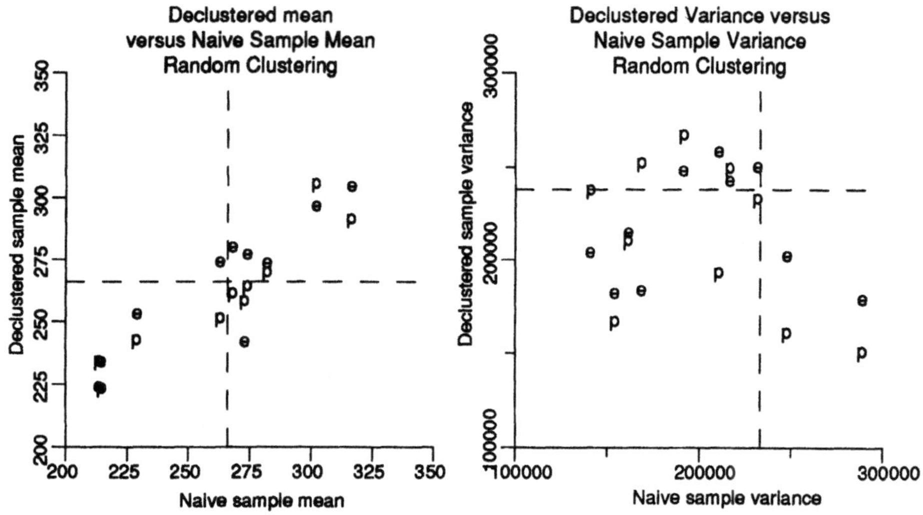

Figure 8: Entropy (e) and Polygonal (p) declustered sample means and variances versus the naive means and variances for the 10 sample data sets. The population parameters are shown by the dashed lines.

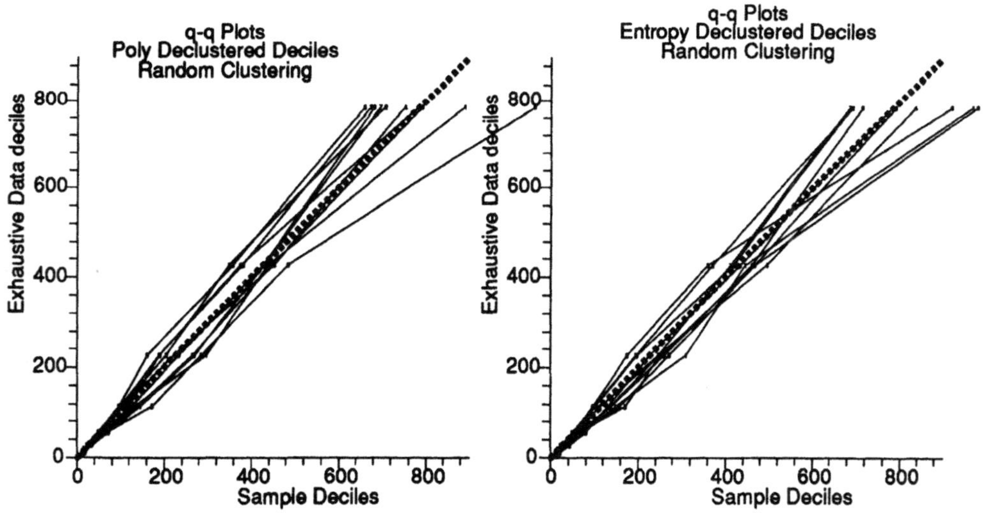

Figure 9: Comparison of the declustered sample histograms generated by the Polygonal and Entropy Methods for randomly clustered sampling.

References

Deutsch, C. (1988) "DECLUS: a FORTRAN 77 program for determining optimum spatial declustering weights", Computer and Geoscience, v.15, no.3, 325-332.

Green, P.J. and Sibson, R. (1978) "Computing Dirichlet tessellations in the plane.", Computer Journal, v.21, 168-173.

Isaaks, E.H. and Srivastava, R.M. (1989) An Introduction to Applied Geostatistics. Oxford University Press, New York.

Jones, D. (1979) Elementary Information Theory. Clarendon Press, Oxford.

Journel, A.G. (1983) "Nonparametric estimation of spatial distributions.", Jour. Math. Geology, v.14, no.3, 445-468.

Ripley, B.D. (1981) Spatial Statistics. John Wiley & Sons, New York.

Tipper, J.C. (1991) "FORTRAN programs to construct the planar Voronoi diagram.", Computers & Geoscience, v.17, no.5, 597-632.

BIASED KRIGING ON THE SPHERE?

BURKHARD SCHAFFRIN
Dept. of Geodetic Science and Surveying
The Ohio State University
1958 Neil Ave.
Columbus, Ohio 43210, U.S.A.

ABSTRACT

It is well known among geodesists that homogeneous-isotropic processes on the sphere which are both, Gaussian and ergodic, do not exist. Thus we have to give up one of the two properties as soon as we consider global phenomena such as the (residual) topography, the disturbing gravity field, etc. In case of a *non-Gaussian* process we are facing problems in deriving the proper distribution functions for our test statistics, beside the fact that such a process is not completely defined by its first two moments. On the other hand, a *non-ergodic* process does not allow us to equivalently express "expectation" and "covariance" as spatial integral over any of its realizations. Hence our usual methods to derive the covariance function (or the semi-variogram) fail to produce consistent - or even unbiased - estimates.

This may turn out to be not as critical as it seems if we only could control the *mean square error* of the predicted phenomenon. For this purpose we introduce, in a second step, *homBLIP (Best homogeneously LInear Prediction)* as a biased alternative to Ordinary Kriging with a slightly reduced mean square error, thereby using new types of *homeograms*. By a similar approach of searching for *biased predictors with smaller mean square error risk*, we may also attempt to improve Universal Kriging in a later stage.

1. INTRODUCTION AND SOME BASIC RESULTS

In geodetic science typical processes such as the (residual) topography - including the Sea Surface Topography - or the disturbing gravity field are global phenomena which are to be modelled on a closed surface, for instance the sphere. If we parametrize this sphere of mean radius R by longitude $\lambda \varepsilon [0,2\pi]$ and colatitude $\vartheta \varepsilon [0,\pi]$, we immediately realize its limited extent, thus rendering the customary geostatistical methods - developed for the Euclidean space $\mathbb{R}^d$ - almost obsolete. As just one example we mention that *homogeneity* (i.e., "translation invariance") and *isotropy* (i.e., "rotation invariance") are no longer distinguishable since, on the sphere, both properties refer to the *invariance with respect to rotations*, either around the axis through the particular point or around an axis perpendicular to it.

Let us therefore assume an incremental *scalar-valued* process in the following, defined by

$$x(s): S \to \mathbb{R}; \quad s:=(\vartheta,\lambda), \tag{1.1}$$

where, for every sεS, the value x(s) represents a *random variable* with

$$E\{x(s)\} = \beta(s) \quad \text{as } \textit{expectation}, \tag{1.2}$$

and

$$D\{x(s)\} = E\{[x(s) - \beta(s)]^2\} = \sigma_x^2(s) \tag{1.3}$$

as *variance function.* In addition, we consider the *auto-covariance function*

$$C\{x(s), x(s')\} = E\{[x(s) - \beta(s)]\,[x(s') - \beta(s')]\} = C_x(s,s') \tag{1.4}$$

and introduce the new notion of a *homeogram* for the non-centered covariance function

$$\eta_x(s,s'): = E\{x(s) \cdot x(s')\} = C_x(s,s') + \beta(s) \cdot \beta(s'), \tag{1.5}$$

both being positive-definite functions; cf. *A.G. Journel/Ch.J. Huijbregts (1978, p. 35).*

The *homeogram* (if it exists) represents a measure for the similarity of the process at points s and s'. This is in contrast to the dissimilarity measure as contained in the *variogram*

$$\begin{aligned} 2\gamma_x(s,s'): &= D\{x(s) - x(s')\} = \sigma_x^2(s) + \sigma_x^2(s') - 2C_x(s,s') = \\ &= \eta_x(s,s) + \eta_x(s',s') - 2\eta_x(s,s') \end{aligned} \tag{1.6}$$

which typically forms the basis of the prediction procedure in geostatistics.

We further notice that for the definition of the *homeogram* $\eta_x(s,s')$ - very much alike the variogram $\gamma_x(s,s')$ - only the values of the incremental process x(s) are needed, not their expectations $\beta(s)$. In our examples above, the function $\beta(s)$ represents *biases* of certain reference surfaces such as the "sea level" (i.e., NN) for the topography, or the "geoid" for the gravity field.

In the case of a *homogeneous (-isotropic)* process, of course, simplifications are possible and we obtain the straight-forward relations

$$E\{x(s)\} = \beta \quad \text{independent of } s\varepsilon S, \tag{1.7a}$$

$$C_x(s,s') = C_x(\psi) = h_x(\psi) - \beta^2 \tag{1.7b}$$

$$\text{with } \cos\psi = \cos\vartheta \cos\vartheta' + \sin\vartheta \sin\vartheta' \cos(l'-l) \tag{1.7c}$$

$$D\{x(s)\} = \sigma_x^2 = C_x(0) \quad \text{independent of } s\varepsilon S, \tag{1.7d}$$

$$\gamma_x(s,s') = \gamma_x(\psi) = \sigma_x^2 - C_x(\psi) = \eta_x(0) - \eta_x(\psi), \tag{1.7e}$$

which allow *unbiased estimation* via

$$\hat{\beta} = \frac{1}{4\pi R^2} \int_0^{\pi}\int_0^{2\pi} x(\vartheta,\lambda) \sin\vartheta \, d\lambda d\vartheta, \tag{1.8a}$$

$$\hat{\eta}_x(\psi) = \frac{1}{4\pi R^2} \int_0^{\pi}\int_0^{2\pi} x(\vartheta,\lambda)\, x(\vartheta',\lambda') \sin\vartheta \, d\lambda d\vartheta \tag{1.8b}$$

$$\hat{\gamma}_x(\psi) = \frac{1}{4\pi R^2} \int_0^{\pi}\int_0^{2\pi} [x(\vartheta,\lambda) - x(\vartheta',\lambda')]^2 \sin\vartheta \, d\lambda d\vartheta \tag{1.8c}$$

with (ϑ',λ') fulfilling condition (1.7c) for any given ψ and arbitrary (ϑ,λ). In contrast, the *estimated auto-covariance function*

$$\hat{C}_x(\psi) = \frac{1}{4\pi R^2} \int_0^{\pi}\int_0^{2\pi} \left[x(\vartheta,\lambda) - \hat{\beta}\right]\left[x(\vartheta',\lambda') - \hat{\beta}\right] \sin\vartheta \, d\lambda d\vartheta \tag{1.9}$$

turns out to even be *not asymptatically unbiased*, in general, since unfortunately $\hat{\beta}$ according to (1.8a) will *not be a consistent estimate*, due to the fact that the surface of the sphere is bounded and thus

$$D\{\hat{\beta}\} = \frac{1}{16\pi^2 R^2} \int_0^{\pi}\int_0^{2\pi}\int_0^{\pi}\int_0^{2\pi} C_x(\psi) \sin\vartheta \sin\vartheta' \, d\lambda d\vartheta d\lambda' d\vartheta' \tag{1.10}$$

would not tend to zero (unless the integral itself vanishes). For the same reason, we cannot expect $\hat{\eta}_x(\psi)$ and $\hat{\gamma}_x(\psi)$ to become consistent estimates, even under the assumption of Gaussianity for the transformation of 4th order moments; see also the discussion in *S. Meier/W. Keller (1990, chapter 6.1)*.

The necessary remedy may come from repeated observations of the process, leading to n equations of type

$$y_i(s) = x(s) + e_i(s), \qquad i\varepsilon\{1,\ldots,n\}, \tag{1.11}$$

with the *observational error function* $e_i(s)$ for the respective epoch $t=t_i$. Since this is basically an *error component model*, proper techniques for both prediction and estimation are readily available and could be applied in the same way as for a Random Effects Model; see *B. Schaffrin (1983)*, for instance.

However, in most cases we are confined to *not having more than one realization* of the process x(s) in hand - or sometimes only one observation of it thus making the information even weaker. In order to arrive at consistent estimates, nevertheless, the stronger assumption of *ergodicity* has to be assumed which states that the entire characterizing information about the spatial process should be completely contained in any of its realizations; for more details, we refer to *J.A.R. Blais (1988, p. 66)* and *N. Cressie (1988)*. As a consequence we only mention that the expectation operator can then be replaced throughout by a *surface integral*, thus yielding the *exact integral relations*

$$E\{x(s)\} = \frac{1}{4\pi R^2}\int_0^{\pi}\int_0^{2\pi} x(\vartheta,\lambda)\,\sin\vartheta\; d\lambda d\vartheta = \beta \quad \text{independent of } s\varepsilon S, \tag{1.12a}$$

$$\eta_x(s,s') = E\{x(s)x(s')\} = \frac{1}{4\pi R^2}\int_0^{\pi}\int_0^{2\pi} x(\vartheta,\lambda)\; x(\vartheta',\lambda')\,\sin\vartheta\; d\lambda d\vartheta = \eta_x(\psi) \quad \text{for } \psi=\psi(s,s'), \tag{1.12b}$$

$$C_x(s,s') = \eta_x(\psi) - \beta^2 = C_x(\psi), \quad C_x(0) = \sigma_x^2, \tag{1.12c}$$

$$\gamma_x(s,s') = \eta_x(0) - \eta_x(\psi) = \sigma_x^2 - C_x(\psi), \tag{1.12d}$$

which may be *approximated* by summations over discrete points. *Apparently, every ergodic process on the sphere turns out to be homogeneous-isotropic*; unfortunately, the reverse is not true. Moreover, while for homogeneous Gaussian processes in $\mathbb{R}^d$ ergodicity is implied by the condition

$$\lim_{\|s'-s\|\to\infty} C_x(s'-s) = 0 \tag{1.13}$$

following *R. Christensen (1991, p. 278)*, for instance, *a homogeneous-isotropic process on the sphere which is Gaussian, can never be ergodic.* This is the famous negative result of *S.L. Lauritzen (1973)* which marks the most important difference between processes on unbounded and bounded domains.

Thus, on a sphere, we are facing the following *major difficulties*:

(i) Consistent estimates of the covariance function - and hence of the "plug-in" predictor - exist only under the *assumption of ergodicity* for the process x(s), in which case this process turns out to be *homogeneous-isotropic* and *non-Gaussian.*

(ii) For a *Gaussian process* x(s), even if it is homogeneous-isotropic, the surface average does not represent a consistent estimate of β(s) which causes the corresponding estimate of the covariance function even to be *not asymptotically unbiased.*

Since we cannot hope to generate unbiased predictors at all using biased covariance estimates *in case (ii)*, and since the search for the unbiased predictor with minimum variance leads to optimal results only under normal conditions which are not fulfilled *in case (i)* - as well as in many other cases of practical relevance, cf. *C.G. Constable (1988)*, - we may therefore try to find *biased predictors* that

(i) minimize the *mean square prediction error* instead, and

(ii) use the *homeogram* $\eta_x(s,s')$ exclusively of which an unbiased estimate exists *even for non-ergodic processes.*

Correspondences to the predictors on the basis of (semi-) variograms for which also unbiased estimates exist in this case, are pointed out.

2. A NEW LOOK AT SOME UNBIASED PREDICTORS

Let us first define the model with respect to which we try to *infer a spatial process at a certain location*, namely x(s), from given values of the same (or a related) process at a couple of *sample points*, for instance $x(s_i)$, $i\varepsilon\{1,...,p\}$; these are taken from the original process or - what is more likely to be available - from the *observed process*

$$y(s_i) = x(s_i) + e(s_i), \quad i\varepsilon\{1,...,p\}, \tag{2.1}$$

where $e(s_i)$ denotes the *observational noise* (in discrete points usually).

Assuming *no correlation* between the processes x(s) and e(s), *nor* between the observational errors at different locations e(s) and e(s'), s≠s', we obtain in the case of a *homogeneous-isotropic* process x(s):

$$\text{(i) } s\neq s'\text{: } C_e(s,s') = 0,\ C_{xe}(s,s') = 0 = C_{ex}(s,s') \Rightarrow$$
$$\Rightarrow C_y(s,s') = C_x(\psi) + 0 = C_y(\psi) \tag{2.2}$$

$$\text{(ii) } s=s'\text{: } C_{xe}(s,s) = 0 = C_{ex}(s,s),\ C_e(s,s) = \sigma_e^2(\vartheta,\lambda) \Rightarrow$$
$$\Rightarrow C_y(s,s) = C_x(0) + C_e(s,s) = \sigma_x^2 + \sigma_e^2(\vartheta,\lambda) \tag{2.3a}$$

$$\text{(ii)' } s=s' \text{ and } e(s) \text{ also homogeneous-isotropic} \Rightarrow$$
$$\Rightarrow C_y(s,s) = \sigma_x^2 + \sigma_e^2 = \sigma_y^2 := C_y(0)$$
(compare with the so-called "nugget effect") (2.3b)

$$\text{(iii) For } \textit{all} \text{ s and s': } C_{xy}(s,s') = C_x(s,s') + C_{xe}(s,s') = C_x(\psi) \tag{2.4}$$

where again

$$\cos\psi = \cos\vartheta\cos\vartheta' + \sin\vartheta\sin\vartheta'\cos(\lambda'-\lambda) \tag{2.5}$$

holds true. Since our goal is to *spatially predict* x(s) at certain locations s, using the sample values $y(s_i)$ or $x(s_i)$ for $i\varepsilon\{1,...,p\}$, respectively, our *general model for homogeneous-isotropic processes* on a sphere reads:

$$\left.\begin{array}{ll} y(s_i) = x(s_i) + e(s_i) & \text{for all } i\varepsilon\{1,...,p\} \\ E\{e(s_i)\} = 0,\ D\{e(s_i)\} = \sigma_e^2 & \text{for all } i\varepsilon\{1,...,p\} \\ E\{x(s)\} = \beta = E\{y(s)\} & \text{independent of } s\varepsilon S \\ D\{x(s)\} = \sigma_x^2 = D\{y(s)\} & \text{independent of } s\varepsilon S \\ C\{x(s), e(s_i)\} = 0 & \text{for all } s\varepsilon S \text{ and } i\varepsilon\{1,...,p\} \end{array}\right] \tag{2.6}$$

In the rare case of a *known value* for β, we may apply the principle of *Best inhomogeneously LInear Prediction (inhomBLIP)* in order to minimize the mean square prediction error in the *entire* class of inhomogeneously linear predictors:

$$MSPE\{\tilde{x}(s)\} := E\{[\tilde{x}(s) - x(s)]^2\} = \min_{\alpha,\delta} E\{[\alpha^T y + \delta - x(s)]^2\} \tag{2.7}$$

where the p x 1 vector y is defined by

$$y: = [y(s_1),...,y(s_p)]^T \tag{2.8}$$

and a is an unknown p x 1 vector, δ an unknown scalar, both to be determined as functions of the arbitrary location s. This approach, in fact, represents "*least-squares collocation*" in its simplest form as developed by *H. Moritz (1970)*, and eventually leads to the well known result

$$\widetilde{x}(s) = \beta + \kappa(s)^T \cdot K^{-1}(y - \tau \cdot \beta) \tag{2.9}$$

where we have introduced the p x 1 vector

$$\begin{aligned} \kappa(s): &= [C_x(\psi_1),\ldots,C_x(\psi_p)]^T && \text{for all } s \varepsilon S, \\ &= [C_y(\psi_1),\ldots,C_y(\psi_p)]^T && \text{for all } s \not\in \{s_i \mid i=1,\ldots,p\}, \end{aligned} \tag{2.10a}$$

the p x p matrix

$$K: = [C_y(\psi_{ij})] = \sigma_e^2 \cdot I_p + [C_x(\psi_{ij})], \tag{2.10b}$$

and the p x 1 "summation vector"

$$\tau: = [1,...,1]^T; \tag{2.11}$$

ψ_i and ψ_{ij} define the angles between s_i and s, or between s_i and s_j, respectively.

Although it was originally not required here, *inhomBLIP* turns out to be *weakly unbiased* in the sense:

$$E\{\widetilde{x}(s)\} = \beta + \kappa(s)^T \cdot K^{-1} \cdot [E\{y(s_i)\} - \beta] = \beta \quad \text{for the } \textit{given} \text{ value } \beta. \tag{2.12}$$

Hence, the *mean square prediction error* results as

$$MSPE\{\widetilde{x}(s)\} = E\{[\widetilde{x}(s) - x(s)]^2\} = D\{\widetilde{x}(s) - x(s)\} = \sigma_x^2 - \kappa(s)^T \cdot K^{-1} \cdot \kappa(s), \tag{2.13}$$

and thus smaller than the original variance σ_x^2. Of course, this procedure is nothing else but "*Simple Kriging*".

We also mention that the use of the *semi-variogram* $\gamma_x(\psi)$ instead of the covariance function $C_x(\psi)$ leads to the representations

$$\gamma(s): = [\gamma_x(\psi_1),\ldots,\gamma_x(\psi_p)]^T = \tau \cdot \sigma_x^2 - \kappa(s), \tag{2.14a}$$

$$\Gamma: = [\gamma_x(\psi_{ij})] = \sigma_e^2 I_p + \sigma_x^2 \cdot \tau\tau^T - K, \tag{2.14b}$$

for the *new* p x 1 vector $\gamma(s)$ and the p x p matrix Γ after applying the relation (1.7e) to (2.10a,b). This does *not* mean that we are entitled to simply replace the covariance function everywhere by the (negative) semi-variogram, due to the *inequality*

$$(\Gamma - \sigma_e^2 \cdot I_p) \cdot K^{-1}\kappa(s) = (\sigma_x^2 \cdot \tau\tau^T - K) \cdot K^{-1}\kappa(s) = \\ = \tau \cdot \sigma_x^2 \cdot [\tau^T K^{-1}\kappa(s)] - \tau \cdot \sigma_x^2 + \gamma(s) \neq \gamma(s) \tag{2.15}$$

with the only *exception* when

$$\tau^T K^{-1}\kappa(s) = 1 \tag{2.16}$$

holds true for the p x 1 "summation vector" τ which is unlikely to be fulfilled.

On the other hand, the value for β will be *unknown* in most cases (though a good guess might be available). Then we better apply the principle of *Best homogeneously Linear weakly Unbiased Prediction (homBLUP)* to minimize the mean square prediction error in the *reduced* class of homogeneously linear predictors which are simultaneously "weakly unbiased":

$$MSPE\ \{\tilde{\tilde{x}}(s)\} = E\{[\tilde{\tilde{x}}(s) - x(s)]^2\} = D\{\tilde{x}(s) - x(s)\} = \\ = \min_{\alpha} (E\{[\alpha^T y - x(s)]^2\} \mid E\{\alpha^T y\} = \beta) = \\ = \underset{\alpha,\nu}{\text{stationary}}\ E\{[\alpha^T y - x(s)]^2\} + 2\nu(\tau^T\alpha - 1) \tag{2.17}$$

where n denotes a "Lagrange multiplier" and a is another unknown p x 1 vector, both to be determined as functions of the arbitrary location s. This approach has been proved to be *scale-error resistant* and was therefore named "*robust collocation*" by *B. Schaffrin (1986; 1989)*. It eventually leads to the result

$$\tilde{\tilde{x}}(s) = b + \kappa(s)^T \cdot K^{-1} \cdot (y - \tau \cdot b) \tag{2.18}$$

with the same p x 1 vector $\kappa(s)$ and the same p x p matrix K as in (2.10a,b), but with an additional *random scale factor* b, defined through the ratio

$$b = (\tau^T K^{-1}\tau)-1 \,.\, (\tau^T K^{-1} y) \quad \text{independent of } s\varepsilon S \tag{2.19a}$$

with

$$E\{b\} = \beta \quad \text{and} \quad D\{b\} = (\tau^T K^{-1}\tau)^{-1}. \tag{2.19b}$$

Thus we immediately confirm *weak unbiasedness* because of the implication

$$E\{y\} = \tau \cdot \beta = \tau \cdot E\{b\} \Rightarrow E\{\tilde{\tilde{x}}(s)\} = E\{b\} + 0 = \beta = E\{x(s)\}. \tag{2.20}$$

The corresponding *mean square prediction error* is then readily obtained by

$$MSPE\ \{\tilde{\tilde{x}}(s)\} = D\{\tilde{x}(s) - x(s)\} = MSPE\ \{\tilde{x}(s)\} + (\tau^T K^{-1}\tau)^{-1}[\tau^T K^{-1}\kappa(s) - 1]^2. \tag{2.21}$$

which is slightly larger than (2.13), but still below σ_x^2.

It is easily seen that this procedure generates "*Ordinary Kriging*" since (2.18) can be transformed into

$$\tilde{\tilde{x}}(s) = \left[\kappa(s)^T - \nu(s) \cdot \tau^T\right] K^{-1} y = \alpha(s)^T \cdot y \tag{2.22a}$$

with

$$\nu(s) := \left(\tau^T K^{-1} \tau\right)^{-1} \cdot \left[\tau^T K^{-1} \kappa(s) - 1\right] \tag{2.22b}$$

such that the *combined system* eventually reads

$$\begin{bmatrix} K & \tau \\ \tau^T & 0 \end{bmatrix} \begin{bmatrix} \alpha(s) \\ \nu(s) \end{bmatrix} = \begin{bmatrix} \kappa(s) \\ 1 \end{bmatrix} \tag{2.23}$$

which nicely corresponds to the systems given by *A.G. Journel/Ch.J. Huijbregts (1978, p. 306)* or *R. Christensen (1991, p. 268-271)*, respectively.

Again we may introduce $\gamma(s)$ and Γ as in (2.14a,b) using the *semi-variogram* $\gamma_x(\psi)$; and now a simple replacement of the covariance function $C_x(\psi)$ by $-\gamma_x(\psi)$ is allowed indeed, due to the fact that the relation

$$\begin{aligned}(\Gamma - \sigma_e^2 I_p) \cdot \alpha(s) - \gamma(s) &= (\sigma_x^2 \cdot \tau\tau^T - K) \cdot \alpha(s) - \tau \cdot \sigma_x^2 + \kappa(s) = \\ &= \tau \cdot \sigma_x^2 \left[\tau^T \alpha(s) - 1\right] + \left[\kappa(s) - K\alpha(s)\right] = 0 + \tau \cdot \nu(s)\end{aligned} \tag{2.24}$$

does necessarily hold, hence leading to the *alternative system*

$$\begin{bmatrix} \sigma_e^2 I_p - \Gamma & \tau \\ \tau^T & 0 \end{bmatrix} \begin{bmatrix} \alpha(s) \\ \nu(s) \end{bmatrix} = \begin{bmatrix} -\gamma(s) \\ 1 \end{bmatrix} \tag{2.25}$$

which has been used in geostatistics extensively.

Let us close this chapter with the remark that most of the above derivations are still valid for processes which are *not homogeneous-isotropic* on the sphere, though the expressions become more complicated. However, the real issue is what will happen to the (spatially) predicted process if we use *estimates* for the covariance function $C_x(s,s')$ or the semi-variogram $\gamma_x(s,s')$ instead of their "true" representations. This has been discussed in chapter 1 with the result that *biased predictors* of x(s) which make use of the *homeogram* $\eta_x(s,s')$ only, might turn out to be more suitable, particularly in view of a smaller mean square prediction error.

3. BIASED PREDICTORS WHICH ARE SUPERIOR TO ORDINARY KRIGING

For the derivation of a *biased predictor with reduced mean square prediction error* - as compared with "Ordinary Kriging" alias homBLUP -, we again assume the general model (2.6) for the stochastic process x(s) on the sphere, (which we suppose to be *homogeneous-*

isotropic for the sake of simplicity only). The value of β might be *unknown* though a guess β_0 should be available, for instance taken from (2.19a).

In complete analogy to chapter 2, we now apply the principle of *Best homogeneously LInear Prediction (homBLIP)* in order to minimize the mean square prediction error in the *enlarged* class of all homogeneously linear predictors, let them be weakly unbiased or not. Unlike the situation within the class of inhomogeneously linear predictors where we found the *identity between inhomBLIP and inhomBLUP*, here it turns out that the minimum is reached indeed for a homogeneously linear predictor that is *biased*, thus we expect homBLIP $\neq$ homBLUP. Our target function is then given through

$$MSPE\{\tilde{\tilde{x}}(s)\} = E\{[\tilde{\tilde{x}}(s) - x(s)]^2\} = \min_{\alpha} E\{[\alpha^T y - x(s)]^2\} \tag{3.1}$$

where the p x 1 vector y is defined as in (2.8), and α is an unknown p x 1 vector to be determined as a function of the arbitrary location s. After short derivations, the necessary *Euler-Lagrange conditions*

$$H \cdot \alpha(s) = \eta(s) \tag{3.2}$$

with the new p x 1 vector

$$\eta(s) := [\eta_x(\psi_1), \ldots, \eta_x(\psi_p)]^T \quad \text{for all } s \varepsilon S, \tag{3.3a}$$

and the p x p matrix

$$H := [\eta_x(\psi_{ij})] + \sigma_e^2 I_p. \tag{3.3b}$$

Since H is a *positive-definite* matrix - considering that the homeogram $\eta_x(\psi)$ is a positive-definite function -, the conditions (3.2) are also sufficient so that the *predictor of type homBLIP* is eventually obtained as

$$\tilde{\tilde{x}}(s) = \eta(s)^T \cdot H^{-1} y \tag{3.4}$$

with the *mean square prediction error*

$$MSPE\{\tilde{\tilde{x}}(s)\} = \eta_x(0) - \eta(s)^T \cdot H^{-1} \cdot \eta(s). \tag{3.5}$$

To compare the results with inhomBLIP (2.9-13) and with homBLUP (2.18-21), we return to the definition (1.5) of the *homeogram* $\eta_x(\psi)$ which formally allows the decomposition of $\eta(s)$ and H according to

$$\eta(s) = \kappa(s) + \tau \cdot \beta^2, \tag{3.6a}$$

$$H = K + \tau\tau^T \cdot \beta^2, \tag{3.6b}$$

and hence the further transformation of homBLIP into

$$\begin{aligned}\tilde{\tilde{x}}(s) &= \left[\kappa(s) + \tau\cdot\beta^2\right]^T\left(K + \tau\tau^T\cdot\beta^2\right)^{-1}y = \\ &= \left[\kappa(s) + \tau\cdot\beta^2\right]^T\left[K^{-1} - K^{-1}\tau(\beta^{-2} + \tau^TK^{-1}\tau)^{-1}\tau^TK^{-1}\right]y = \\ &= \kappa(s)^TK^{-1}\left[y - \tau(\beta^{-2} + \tau^TK^{-1}\tau)^{-1}(\tau^TK^{-1}y)\right] + \\ &\quad + \beta^2\left[1 - \tau^TK^{-1}\tau(\beta^{-2} + \tau^TK^{-1}\tau)^{-1}\right](\tau^TK^{-1}y) = \\ &= \bar{b} + \kappa(s)^T\cdot K^{-1}\cdot(y - \tau\cdot\bar{b}).\end{aligned} \tag{3.7}$$

Here $\bar{b}$ denotes the new (biased) *random scale factor*, defined through the ratio

$$\bar{b} = (\beta^{-2} + \tau^TK^{-1}\tau)^{-1}\cdot(\tau^TK^{-1}y) \quad \text{independent of } s\varepsilon S \tag{3.8a}$$

with

$$E\{\bar{b}\} = \beta\cdot\left[1 - (1 + \beta^2\cdot\tau^TK^{-1}\tau)^{-1}\right] \neq \beta, \tag{3.8b}$$

$$\begin{aligned}D\{\bar{b}\} &= (\beta^{-2} + \tau^TK^{-1}\tau)^{-2}(\tau^TK^{-1}\tau) = \\ &= \beta^2(1 + \beta^2\cdot\tau^TK^{-1}\tau)^{-1}\left[1 - (1 + \beta^2\cdot\tau^TK^{-1}\tau)^{-1}\right],\end{aligned} \tag{3.8c}$$

and the mean square error

$$MSE\{\bar{b}\} = D\{\bar{b}\} + (E\{\bar{b}\} - \beta)^2 = \beta^2\cdot\left[1 - (1 + \beta^2\cdot\tau^TK^{-1}\tau)^{-1}\right]. \tag{3.8d}$$

The corresponding transformation of the *mean square prediction error* (3.5) leads to the expression

$$\begin{aligned}MSPE\{\tilde{\tilde{x}}(s)\} &= \eta_x(0) - \left[\kappa(s) + \tau\beta^2\right]^T(K + \tau\tau^T\beta^2)^{-1}\left[\kappa(s) + \tau\beta^2\right] = \\ &= \sigma_x^2 + \beta^2 - \left[\kappa(s) + \tau\beta^2\right]^T\left[K^{-1} - K^{-1}\tau(\beta^{-2} + \tau^TK^{-1}\tau)^{-1}\tau^TK^{-1}\right]\left[\kappa(s) + \tau\beta^2\right] = \\ &= \left[\sigma_x^2 - \kappa(s)^TK^{-1}\kappa(s)\right] + \beta^2 + \beta^2(1 + \beta^2\cdot\tau^TK^{-1}\tau)^{-1}\left[\tau^TK^{-1}\kappa(s)\right]^2 - \\ &\quad - 2\beta^2\left[\tau^TK^{-1}\kappa(s)\right]\cdot\left[1 - (\tau^TK^{-1}\tau)\cdot\beta^2(1 + \beta^2\cdot\tau^TK^{-1}\tau)^{-1}\right] - \\ &\quad - \beta^4(\tau^TK^{-1}\tau)\left[1 - (\tau^TK^{-1}\tau)\cdot\beta^2(1 + \beta^2\cdot\tau^TK^{-1}\tau)^{-1}\right] = \\ &= MSPE\{\tilde{x}(s)\} + \beta^2(1 + \beta^2\cdot\tau^TK^{-1}\tau)^{-1}\left[\tau^TK^{-1}\kappa(s) - 1\right]^2,\end{aligned} \tag{3.9}$$

and after comparison with (2.13) and (2.21) to the important *inequality*

$$MSPE\{\tilde{x}(s)\} \leq MSPE\{\tilde{\tilde{x}}(s)\} \leq MSPE\{\tilde{\tilde{x}}(s)\} \tag{3.10}$$

which shows the improvement over "*Ordinary Kriging*" while still retaining the value of b as an unknown. Note that the appearance of β in the formulas (3.7-9) is caused by the use of the covariance function only and thus somewhat artificial. When making use of (3.4) in connection with (3.5) *solely the homeogram* $\eta_x(\psi)$ enters the calculations. It can be estimated *unbiasedly* without the knowledge of β under far milder conditions than the covariance function.

Because of space limitations we do not further investigate the *equivalent use of the semi-variogram* $\gamma_x(\psi)$ here to generate homBLIP $\tilde{\bar{x}}(s)$, as suggested by *N. Cressie (1991, p. 178)*. Also the performance of homBLIP $\tilde{\bar{x}}(s)$ in numerical examples remains to be studied; however, this is beyond the scope of the present paper.

4. CONCLUSIONS AND OUTLOOK

For stochastic processes on the sphere which are typical for global applications such as in geodetic science, we face the major problem of *non-ergodicity* or *non-Gaussianity* even under the assumption of homogeneity and isotropy. This behavior is caused by the fact that the surface of the sphere is *bounded.* To overcome difficulties in the *unbiased* (or only asymptatically unbiased) estimation of the *covariance function*, we introduced the non-centralized version of it as the *homeogram* which is certainly unbiasedly estimable under much milder conditions. In a way, it parallels the semi-variogram, but simultaneously allows the *improvement of "Ordinary Kriging"* (in terms of the mean square prediction error) at the expenses of a *small bias*, - which might be entering through the traditional procedure anyway!

These results have been derived so far for the "true" *homeogram*; an analysis of the "plug-in" predictors when estimated homeograms are used, should be done in the future and might soon lead to quite similar results as in *R. Christensen (1991, chapter 6.5)*. Another critical point requiring further studies is undoubtedly the *robustness* of the predictor against modelling errors. First results for *dynamic* - not spatial - processes can be found in *M.H.J. Gruber (1990, chapter IX)* and should be extended to the case considered in this paper.

5. REFERENCES

Blais, J.A.R. (1988): Estimation and Spectral Analysis, The University of Calgary Press, Calgary, Canada.

Christensen, Ronald (1991): Linear Models for Multivariate, Time Series, and Spatial Data, Springer-Verlag, New York, Berlin etc.

Constable, C.G. (1988): "Parameter estimation in non-Gaussian noise", Geophys. J. *94*, 131-142.

Cressie, Noel (1988): "Spatial prediction and ordinary Kriging", Math. Geology *20*, 405-421.

Cressie, Noel (1991): "Statistics for Spatial Data", Wiley, New York etc.

Gruber, Marvin H.J. (1990): Regression Estimators - A Comparative Study, Academic Press, Boston etc.

Journel, A.G. and Huijbregts, Ch. J. (1978): Mining Geostatistics, Academic Press, London etc.

Lauritzen, S. (1973): The probabilistic background of some statistical methods in physical geodesy, Publication of the Danish Geodetic Institute, No. 48, Kopenhagen.

Meier, S. and Keller, W. (1990): Geostatistik (in german), Akademie-Verlag, Berlin.

Moritz, Helmut (1970): "A generalized least-squares model", Studia Geophys. et Geodaet. *14*, 353-362.

Schaffrin, Burkhard (1983): "Variance covariance component estimation and the adjustment of heterogeneous repeated measurements" (in german), Publication of the German Geodetic Commission, Series C, No. 282, Munich.

Schaffrin, Burkhard (1986): "New estimation/prediction techniques for the determination of crustal deformations in the presence of geophysical prior information", Tectonophysics *130*, 361-367.

Schaffrin, Burkhard (1989): "An alternative approach to robust collocation", Bulletin Géodésique *63*, 395-404.

MULTIVARIATE GEOSTATISTICS: BEYOND BIVARIATE MOMENTS

FELIPE B. GUARDIANO and R. MOHAN SRIVASTAVA[1]

Applied Earth Sciences Department, Stanford, CA, USA, 94305

Traditionally, geostatistical models are conditioned only on univariate and bivariate statistics such as the sample histogram and covariance or indicator covariances. Higher order sample statistics such as three, four, multi-point covariances, as obtained, for example, from a training image, would improve considerably stochastic images if they could be reproduced.

An extended concept of indicator kriging allows the production of images that honor any number of multiple-point indicator covariances representing multiple-event experimental frequencies. The technique is applied to the stochastic imaging of phenomena sparsely sampled with complex features that cannot be captured by two-point statistics.

INTRODUCTION

In geostatistics, characterization of uncertainty about a spatially distributed phenomenon is done through conditional simulation. Conditional simulation consists of generating realizations of a random function (RF) conditioned by certain structural information, usually a univariate cumulative distribution function (cdf) and moments of the bivariate distribution such as the attribute covariance or some indicator covariances. Higher order statistics, such as the trivariate, quadrivariate, multiple-point distributions, are imposed by the implicit or explicit RF model, and by conditioning the realizations to local data. There are situations where available data allows the inference of these higher order statistics and one would wish to impart those statistics to the stochastic images. The algorithm proposed here allows for the explicit reproduction of such multiple-point statistics, generating realizations that reflect more accurately the spatial features of the underlying phenomenon.

Consider as an introductory example, the top image of Figure 1. It shows a section of a cross-bedded aeolian sandstone imaged with black and white pixels. The black pixels correspond to very fine sediments. Their tight horizontal connectivity would create low vertical permeability flow barriers in a reservoir. Note the typical dune-like curvilinear shape of the black pixel patterns. The white pixels correspond to

[1]also with FSS International, Vancouver, BC

coarser higher permeability sediments. An accurate appraisal of the flow behaviour of such sedimentary sequences would require a good characterization of the spatial arrangement of the fines.

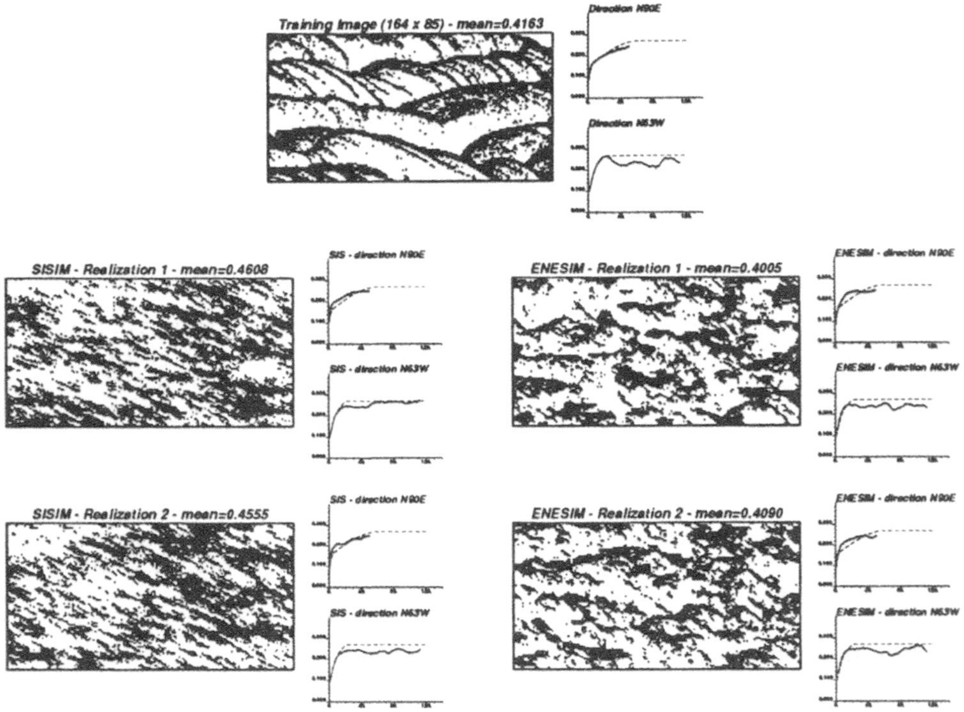

Figure 1: *Stochastic imaging of cross-bedded sand. The reference image (*165 × 84 *pixels) is shown on the top. The two images on the left are realizations produced by sequential indicator simulation (SISIM); the two images on the right are produced by the new proposed extended normal equations simulation (ENESIM) algorithm. All four images are shown to reproduce correctly the univariate and bivariate indicator statistics of the training image.*

The training image at the top of Figure 1 comprises 164 x 85 = 13,940 pixels, 42% of them being black. 73 samples, i.e. about one half of a percent of the training image, were collected at random to be used as conditioning data (sample proportion of black pixels is 47%). In addition, the exhaustive indicator semivariograms in the directions EW (*horizontal*) and N63W were retrieved from the training image. The latter direction corresponds to that of the longer right limb of the dune-shaped black pixels.

Although simplistically binary, the essential features of the training image defy reproduction using the traditional two-point statistics (here an indicator variogram model). The two lower left images of Figure 1 are reproductions honoring the previous univariate and bivariate indicator statistics; yet they fail to *look like* the training image, in particular the characteristic curvilinear dune shape has not been reproduced. These two lower left images are two independent realizations of an indicator RF model $I(\mathbf{u})$, defined as:

$$I(\mathbf{u}) = \begin{cases} 1 & \text{if } \mathbf{u} \text{ is a black pixel} \\ 0 & \text{if not} \end{cases} \tag{1}$$

and conditioned by the 73 sample data in addition to a variogram modeled from the two previous exhaustive indicator variograms. These realizations were obtained from sequential indicator simulation (SISIM) using routine `sisim` of the public domain software GSLIB (Deutsch and Journel, 1992).

Clearly, more than two-point statistics are necessary to account for the curvilinear dune-shaped pattern of black pixels seen on the training image. The same 73 sample data were used to produce the two independent realizations shown at the lower right of Figure 1. Here, the proposed extended normal equations (ENESIM) algorithm was used. Reproduction of the univariate (proportion of black pixels) and the indicator variograms is similar to that achieved by SISIM images. However, the ENESIM images start capturing the dune shape. The additional information provided to the ENESIM simulation algorithm is precisely multiple-point indicator covariances directly read from the training image, as described hereafter.

Note that the goal of any simulation is not to reproduce the training image. The objective is to generate realizations that account for (are constrained by) information deemed relevant to the underlying phenomenon. The training image itself is not the underlying phenomenon. It is assumed, however, that both share some structural properties such as specific multiple-point covariances.

BAYES' RELATION

The indicator formalism for evaluation of conditional probabilities, Journel and Alabert (1989), is actually a particular case and an approximation of a much more general and powerful concept of extended normal equation including Bayes' relation itself, Solow (1986). Consider any particular yet unobserved event A_0 whose conditional probability is to be assessed. The conditioning information is constituted by n data events A_α with $\alpha = 1, \ldots, n$. The event A_0 could be as simple as $Z(\mathbf{u}) \leq z$, indicating whether a particular unsampled continuous attribute value z($\mathbf{u}$) is valued below threshold z or not, or it can involve multiple events at different locations such as a string of high values $A_0 : \{Z(\mathbf{u}) > z; \mathbf{u} \in \mathit{string}\ L\}$. Similarly each of the data events A_α may be as simple as a single-location datum value $A_\alpha : \{Z(\mathbf{u}_\alpha) = z_\alpha\}$, or it can be as complex as the set of all data values $S(n) : \{Z(\mathbf{u}_\alpha) = z_\alpha, \alpha = 1, \ldots, n\}$.

More precisely, let the random variables A_0, A_α, and $S(n)$ be binary random variables, set to 1 if the corresponding event occurs, to 0 otherwise. The conditional probability of the event A_0 given the n data events $A_\alpha = 1$, $\alpha = 1, \ldots, n$, is equal to the conditional expectation of the indicator random variable A_0:

$$\begin{aligned} Prob\{A_0 = 1 \mid A_\alpha, \alpha = 1, \ldots, n\} &= E\{A_0 \mid S(n) = 1\} \\ &\equiv [A_0]^*_{SIK} = E\{A_0\} + \lambda[1 - E\{S(n)\}] \end{aligned} \tag{2}$$

with $[A_0]^*_{SIK}$ being the simple indicator kriging (SIK) estimate of A_0; $S(n) = \prod_{\alpha=1}^{n} A_\alpha$ is the single indicator of the global data event with $S(n) = 1$ if and only if all elementary data events occur simultaneously, $A_\alpha = 1, \forall \alpha = 1, \ldots, n$. The system of normal equations is reduced to a single equation:

$$\lambda.Var\{S(n)\} = Cov\{A_0, S(n)\} \tag{3}$$

with:

$$\begin{aligned} Var\{S(n)\} &= E\{S(n)\}.[1 - E\{S(n)\}] \\ E\{S(n)\} &= Prob\{S(n) = 1\} = Prob\{A_\alpha = 1, \alpha = 1, \ldots, n\} \\ Cov\{A_0, S(n)\} &= E\{A_0.S(n)\} - E\{A_0\}.E\{S(n)\} \\ E\{A_0.S(n)\} &= Prob\{A_0 = 1, S(n) = 1\} \end{aligned}$$

Hence:

$$\begin{aligned} E\{A_0 \mid S(n) = 1\} &= E\{A_0\} + \frac{E\{A_0.S(n)\} - E\{A_0\}.E\{S(n)\}}{E\{S(n)\}} \\ &= \frac{E\{A_0.S(n)\}}{E\{S(n)\}} = \frac{Prob\{A_0 = 1, S(n) = 1\}}{Prob\{S(n) = 1\}} \end{aligned} \tag{4}$$

as given by Bayes' relation.

Indicator kriging with a single global data event is exact inasmuch as Bayes' postulate, itself at the basis of the notion of conditional probability. All that is required is inference of the probabilities of the ratio (4) or, equivalently, of the two indicator covariances $Var\{S(n)\}$ and $Cov\{A_0, S(n)\}$ required by the normal equation (3).

In the case of the exercise depicted in Figure 1, $A_0 = I(\mathbf{u})$ is the indicator of a black pixel at any unsampled location $\mathbf{u}_0$, and $S(n) = \prod_{\alpha=1}^{n} A_\alpha = 1$ represents the set of all 73 indicator data A_α all set to 1, i.e. such that:

$$A_\alpha = A(\mathbf{u}_\alpha) = \begin{cases} I(\mathbf{u}_\alpha) = 1 & \text{if } I(\mathbf{u}_\alpha) = 1 \\ 1 - I(\mathbf{u}_\alpha) = 1 & \text{if } I(\mathbf{u}_\alpha) = 0, \forall\, \alpha = 1, \ldots, n \end{cases} \tag{5}$$

Inference of the probability $E\{A_0.S(n)\} = E\{I(\mathbf{u}_0).\prod_{\alpha=1}^{n} A_\alpha\}$ calls for scanning the training image to evaluate the frequency of occurrence of the joint event $\{I(\mathbf{u}) = 1, A(\mathbf{u}+\mathbf{h}_\alpha) = 1, \alpha = 1, \ldots, n\}$ with $\mathbf{h}_\alpha = \mathbf{u}_\alpha - \mathbf{u}_0$, for the given geometric configuration of the $(n+1)$ locations $\mathbf{u}$, $\mathbf{u} + \mathbf{h}_\alpha$, $\alpha = 1, \ldots, n = 73$. Similarly, the probability $E\{S(n)\}$ would be inferred by the corresponding[2] frequency of occurrence of the global data event $\{\prod_{\alpha=1}^{n} A(\mathbf{u} + \mathbf{h}_\alpha) = 1\}$.

[2]The reader may have associated the choice of a particular training image to the decision of stationarity. For consistency, it is essential that the two experimental frequencies $E\{A_0.S(n)\}$ and $E\{S(n)\}$ be inferred from exactly the same set of $(n+1)$-tuples locations $\{\mathbf{u}, \mathbf{u}+\mathbf{h}_\alpha, \alpha = 1, \ldots, n\} \subset$ training image. This common set is called the *eroded* area of the training image by the specific configuration of the $(n+1)$-tuples considered. The larger n, the larger the distances $\mathbf{h}_\alpha = \mathbf{u}_\alpha - \mathbf{u}_0$, the smaller the eroded training image left for the inference process. The previous problem of consistency is at the root of questions of positive definiteness when modelling indicator covariances as required by normal equations of type (3).

SIMULATION

Once the relative frequency (4), i.e. $E\{A_0 \mid S(n) = 1\}$, is inferred, one can draw by Monte Carlo a realization for A_0. For example, if $E\{A_0 \mid S(n) = 1\} = 0.6$, the simulated value for A_0 is 1 whenever the uniform $[0, 1]$ random number drawn is lesser than or equal to 0.6. Whatever the value simulated (0 or 1) for A_0 at location $\mathbf{u}_0$, it becomes a datum event for the next node $\mathbf{u}_0'$ to be simulated according to the sequential simulation principle, Deutsch and Journel (1992). Therefore, the global data event $S(n+1)$ for the next node $\mathbf{u}_0'$ is of dimension $n+1$. If N nodes are to be simulated with n original elementary data events, the global data event dimensions grows gradually from n for the first node being simulated to $(n + N - 1)$ for the last node.

REDUCING THE INFERENCE PROBLEM

Although the general indicator kriging expression (2) or (4) is exact, it is not practical as soon as the number n of elementary data events becomes large, unless the training image is correspondingly very large. Indeed, as n grows, or if the global data event $S(n)$ is complex, it is unlikely to find on the training image enough $(n+1)$-tuple events such as $S(n) = 1$ to allow inference of the frequencies $E\{A_0.S(n)\}$ and $E\{S(n)\}$. The solution consists of:

- either reducing the number n and complexity (specificity) of the data events, which amounts to reduce the amount of information conditioning the probability $E\{A_0 \mid S(n) = 1\}$
- or model both frequencies $E\{A_0.S(n)\}$ and $E\{S(n)\}$ as functions of, say, the n parameters $\mathbf{h}_\alpha$ of the data configuration. This amounts to interpolate or extrapolate all needed frequencies from a few actually calculated frequencies.

The second alternative is that traditionally used when a two-point Z-covariance $Cov\{Z(\mathbf{u}), Z(\mathbf{u}+\mathbf{h})\}$ or an indicator covariance $Cov\{I(\mathbf{u}), I(\mathbf{u}+\mathbf{h})\}$ are modeled by fitting a curve through actually calculated experimental covariance values. Unfortunately, the modelling option is limited to low order covariances (univariate and bivariate), thus does not allow capturing high order statistics from the data. The first solution is that considered to generate the ENESIM images at the right of Figure 1.

THE ENESIM ALGORITHM

The ENESIM simulation algorithm proceeds as follows:

1 - Define a random path visiting all unsampled nodes of the simulation grid, in the case of Figure 1 all (13,940-73)=13,867 nodes discretizing the training image excluding the 73 conditioning data.

2 - At each node $\mathbf{u}_i$, retain the nearest $\frac{n}{4}$ conditioning indicator data (black or white) in each quadrant, for a total of n set to 16 in this exercise. Scan the

training image for all 17-tuples reproducing the node $\mathbf{u}_i$-to-16 data configuration, and evaluate the conditional probability (4). That conditional probability is retained only if there are at least a minimum number min (fixed at 20 for this exercise) of identical 17-tuples configurations. If not, the furthest away datum is dropped, reducing the data size to 15, and the training image is scanned again[3], etc ..., until the minimum required number of identical data configurations is found.

3 - From the conditional probability (4) draw a simulated value (black or white) for node $\mathbf{u}_i$. That simulated value is then coded as a datum value to be used for the conditional simulation of all subsequent nodes.

4 - Move to the next node $\mathbf{u}_{i+1}$ along the random path and repeat steps 2 and 3.

Loop until all the nodes are informed with a simulated value. One stochastic image has been generated. If another image is needed, reiterate the entire process from step 1.

The algorithm can be easily modified to retain preferentially specific data configurations (e.g. those revealing the characteristic dune shape) rather than the closest $\frac{n}{4}$ data in each quadrant. It can also be adapted to categorical indicator variables that can take more than 2 outcomes; beware though that if there are K categories and n data locations retained there may be up to K^n joint realizations of the n data values, e.g. $K^n = 1.5 \times 10^{11}$ for $K = 5$ and $n = 16$. Fortunately, if the original conditioning data are consistent with the training image statistics the actual data configuration and values retained at any node will be found often enough in the scanning process.

SIMULATION OF PORE GEOMETRY

The training image for this second case study, see Figure 2, is a digitized thin section of an oil bearing sand with shale clasts of different sizes and orientation (Kramers et al, 1989). Three families of shale sizes can be distinguished in this image. The shale are represented as black pixels and the white pixels represent the porous sand matrix through which fluids can flow.

The top image of Figure 2 serves as the phenomenon to be simulated. It comprises 116 x 59 = 6,844 pixels averaging 43% black. 47 pixels (black or white) are sampled from this top image for conditioning the stochastic imaging process, averaging 54% black.

Unlike the example of Figure 1, and as would be the case in most practical applications, the training image (second from top in Figure 2) is different from that to be simulated. However, both images share common structural traits if only because they were taken from the same sand facies. Indicator variograms for the two E-W and N-S directions were extracted from the training image and modeled.

The two lower left images of Figure 2 are two independent SIS realizations honoring the training indicator variogram model and the 47 conditioning data. They fail to

[3]Actually, all the frequencies corresponding to any subset of the initial 16 data configuration are embedded in the scanning made for that initial configuration.

reproduce the clast aspect of either the training or unknown image and would yield a much too low effective permeability: the black pixels are not disconnected into separate clasts.

The two lower right images of Figure 2 are two independent realizations of the ENESIM algorithm conditioned to the same 47 data but building from multiple-point statistics taken from the training image. The ENESIM implementation is identical to that described for Figure 1. The clast aspect of the training and unknown images is better reproduced, hence the flow characteristics (e.g. permeability) of those images would also be better reproduced. Note that the two-point indicator variogram model, although not explicitly considered by the ENESIM algorithm, are well reproduced.

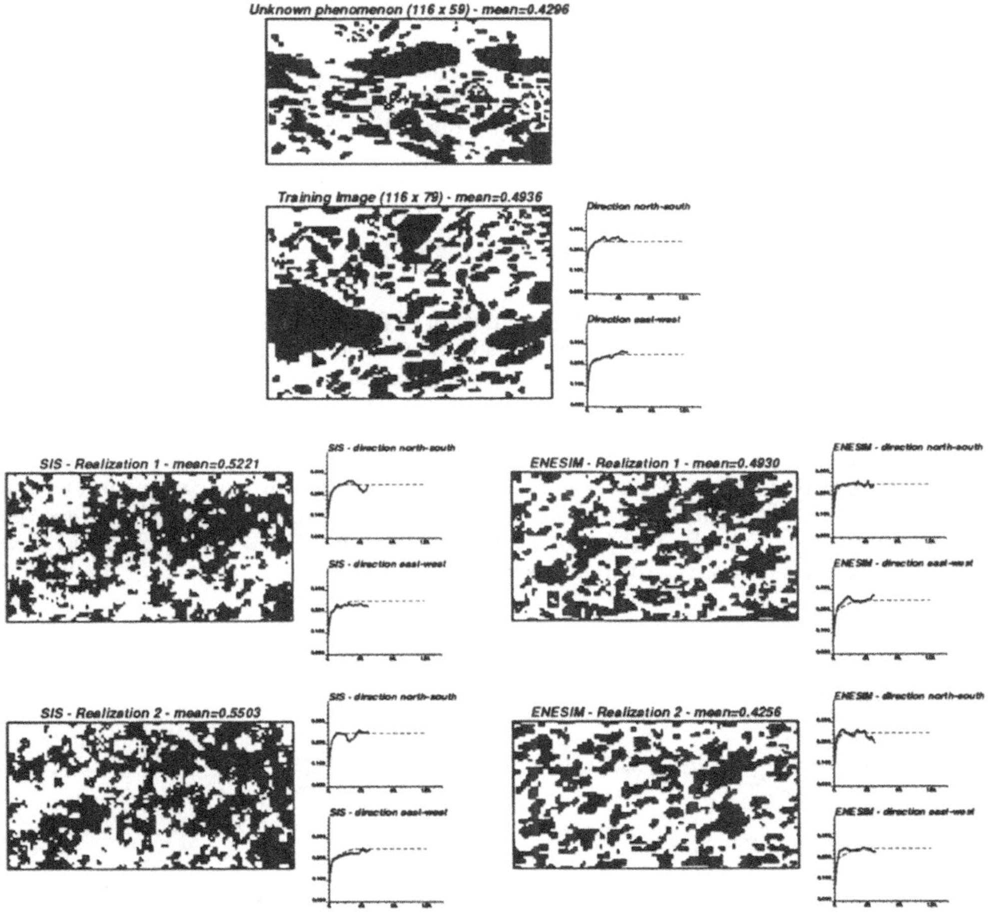

Figure 2: *The reference image to be reproduced is shown on the top. Below it is a training image taken from the same lithofacies. The two images at the lower left are realizations produced by SIS; the images at the lower right are produced by ENESIM. The experimental variograms shown to the right of each image correspond to directions north-south and east-west.*

Figure 3: *Both the conditioning data and the multiple-point statistics are taken from the same training image (second from the top on Figure 2) resulting in a much better reproduction of that training image.*

If the 47 conditioning data had been taken from the training image from which the multiple-point statistics are inferred, then a much better reproduction is achieved as shown by the two independent realizations of Figure 3. Once again, the reproduction of the training image indicator variograms (not shown) is excellent.

SIMULATION OF FRACTURE GEOMETRY

This third case study relates to the simulation of fracture distribution in carbonate rock. The image at the left of Figure 4 has been generated by an algorithm for physical crack propagation from initial (Poisson) seeds, Olson and Pollard (1989). That image has been digitized by 200 x 200 binary pixels (black or white) of which 22% were black. 200 conditioning data were taken at random averaging 23% black. These 200 data were used to condition the ENESIM realization at the right of Figure 4. The ENESIM realization averages 21% black and its indicator variograms in the two major directions EW and NS are shown to reproduce correctly the model fit from the exhaustive variograms derived from the training image. Once again, the ENESIM implementation is identical to that described for Figure 1.

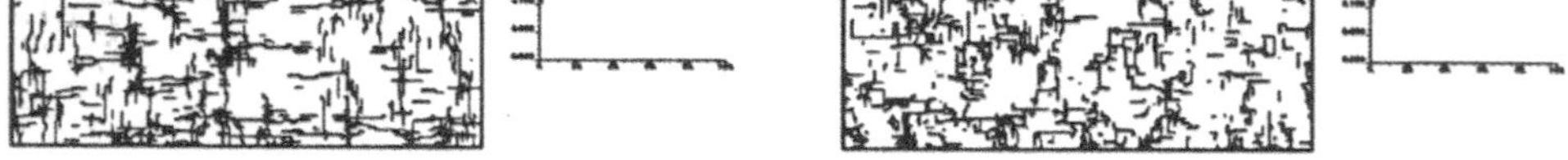

Figure 4: *The training image was generated by physical crack propagation. 200 data were extracted from it to condition the ENESIM realization shown at the right.*

One reason for the poor reproduction of the longer fracture lengths is the limitation to the nearest data found in each neighborhood: quickly in the sequential simulation algorithm these conditional data are very close to the node being simulated and fail to reflect the longer fracture lengths. A smarter search strategy could have been considered such as a neighborhood elongated along the north-south and/or east-west directions.

It is worth noting that the generation of the ENESIM realization using a prototype non-optimized code was about 10 times faster, cpu-wise, than the generation of the training image using the physical growth equations. If that training image is deemed representative of any other section of a 3D volume (isotropy), the ENESIM algorithm can easily expand that 2D training information into multiple 3D realizations, a task now beyond the capability of physical growth simulators.

Recall that the ENESIM simulation algorithm is exact, in that it honors the data values at their exact locations, no matter the density or configuration of these data. This exactitude condition is sometimes difficult to fulfill with Boolean algorithms which, in the case of Figure 4, would amount to drop at random fracture segments according to some prior length distribution. Such Boolean approach would have no control on the resulting indicator variogram.

EXTENDED NORMAL EQUATIONS

The single normal equation or Bayes' relation approach of equations (3) and (4) requires scanning a training image for frequencies related to a single global data event denoted $S(n)$ in these equations. Often such a training image is not available or is not extensive enough to allow reliable inference. One may then decompose that global data event $S(n)$ into simpler (less specific) information components whose related frequencies are easier to assess.

The global data event $S(n) = 1$ can be seen as anyone of 2^n joint realizations of the n elementary binary data events $A_\alpha, \alpha = 1, \ldots, n$, where A_α can take either values 0 or 1. The conditional probability (2) of event $A_0 = 1$ takes the more general expression (Cox, 1972; Journel and Alabert, 1989):

$$\begin{aligned} Prob\{A_0 = 1 \mid A_\alpha = a_\alpha, \alpha = 1, \ldots, n\} \;&\equiv\; [A_0]^*_{SIK} \\ = \lambda_0 + \sum_{\alpha_1=1}^{n} \lambda^{(1)}_{\alpha_1} a_{\alpha_1} + \sum_{\alpha_1=1}^{n} \sum_{\alpha_2>\alpha_1}^{n} \lambda^{(2)}_{\alpha_1\alpha_2} a_{\alpha_1} a_{\alpha_2} & \\ + \sum_{\alpha_1=1}^{n} \sum_{\alpha_2>\alpha_1}^{n} \sum_{\alpha_3>\alpha_2}^{n} \lambda^{(3)}_{\alpha_1\alpha_2\alpha_3} a_{\alpha_1} a_{\alpha_2} a_{\alpha_3} + \cdots + \lambda^{(n)} \prod_{\alpha=1}^{n} a_\alpha & \qquad (6) \end{aligned}$$

The 2^n coefficients (IK weights) λ_0, $\lambda^{(1)}_{\alpha_1}$, $\lambda^{(2)}_{\alpha_1\alpha_2}$, $\lambda^{(3)}_{\alpha_1\alpha_2\alpha_3}$, …, $\lambda^{(n)}$ are determined by an extended system of 2^n normal equations which establishes the orthogonality of the error vector $A_0 - [A_0]^*_{SIK}$ to the 2^n combination of data events $\underline{1}$, A_{α_1}, $A_{\alpha_1}\, A_{\alpha_2}$, …, $\prod_{\alpha=1}^{n} A_\alpha$, see Projection Theorem (Luenberger, 1969). For example, consider the first normal equation expressing orthogonality of the error vector to the constant unit vector $\underline{1}$, otherwise known as unbiasedness condition:

$$< A_0 - [A_0]^*_{SIK}, \underline{1} >= E\{A_0\} - \lambda_0 - \sum_{\alpha_1=1}^{n} \lambda^{(1)}_{\alpha_1} E\{A_{\alpha_1}\} - \ldots - \lambda^{(n)} E\{\prod_{\alpha=1}^{n} A_\alpha\} = 0 \quad (7)$$

This equation allows rewriting the ccdf (6) into the more familiar IK format:

$$\begin{aligned} Prob\{A_0 = 1 \mid A_\alpha = a_\alpha, \alpha = 1, \ldots, n\} \quad - \quad E\{A_0\} \\ = \sum_{\alpha_1=1}^{n} \lambda^{(1)}_{\alpha_1} [A_{\alpha_1} - E\{A_{\alpha_1}\}] + \sum_{\alpha_1=1}^{n} \sum_{\alpha_2>\alpha_1}^{n} \lambda^{(2)}_{\alpha_1\alpha_2} [A_{\alpha_1}.A_{\alpha_2} - E\{A_{\alpha_1}.A_{\alpha_2}\}] + \cdots \\ + \lambda^{(n)} [\prod_{\alpha=1}^{n} A_\alpha - E\{\prod_{\alpha=1}^{n} A_\alpha\}] \end{aligned} \quad (8)$$

Expression (8) identifies expression (4), i.e. Bayes' relation, when the global data realization $S(n)$ corresponds to all switches a_α set to 1.

The traditional IK approximation amounts to retaining only the $(n+1)$ first terms of the expansion (6) or the n first terms of expansion (8). What is lost is the information carried by multiple joint data events such as $A_{\alpha_1} A_{\alpha_2}$ or $A_{\alpha_1} A_{\alpha_2} A_{\alpha_3}$. What is gained is easier inference: all that is needed are (traditional) two-event indicator covariances such as $E\{A_0.A_\alpha\}$ or $E\{A_\alpha.A_\beta\}$, $\forall\, \alpha, \beta = 1, \ldots, n$.

Availability of the full (and exact) expansion (8) allows retaining any number of combinations (products) of data events up to all $(2^n - 1)$ of them, depending on their information content and on the feasibility of the inference of the corresponding indicator covariances. For example, one may retain the partial IK estimate:

$$\begin{aligned} E^*\{A_0 \mid S(n) = 1\} \quad &= \quad E\{A_0\} + \sum_{\alpha=1}^{n} \nu^{(1)}_\alpha [a_\alpha - E\{A_\alpha\}] \\ &+ \quad \nu^{(2)}_{12} [a_1 a_2 - E\{A_1.A_2\}] \end{aligned} \quad (9)$$

if the joint occurrence $A_1 A_2$ (equal to 1 or 0) is deemed an information critical to the event A_0 to assess. The estimator (9) requires inference of the three-event indicator covariances $E\{A_0.A_1.A_2\}$ and $E\{A_\alpha.A_1.A_2\}$, $\alpha = 3, \ldots, n$, in addition to the traditional two-event indicator covariance $E\{A_\alpha.A_\beta\}$, $\alpha, \beta = 1, \ldots, n$.

The main contribution of the general indicator formalism is the possibility of considering any combination of data events of any type. For example, the elementary data events A_α can be:

- hard measurements made on the primary variable, $A_\alpha : \{Z(\mathbf{u}_\alpha) = z_\alpha\}$
- constraint intervals, $A_\alpha : \{Z(\mathbf{u}_\alpha) \in (a_\alpha, b_\alpha]\}$
- categorical data, $A_\alpha : \{\mathbf{u}_\alpha \in \textit{facies } k\}$

- prior (pre-posterior) probability distribution values such as $A_\alpha : \{Prob\{Z(\mathbf{u}_\alpha) \leq z \mid Y(\mathbf{u}_\alpha) = y_\alpha\}$ where the distribution of the primary value $Z(\mathbf{u}_\alpha)$ is constrained by the information brought by the collocated secondary datum $Y(\mathbf{u}_\alpha) = y_\alpha$. That prior local cdf could be obtained by calibration from a scattergram of $z - y$ values, Zhu and Journel (1992).

- any multiple-event (point) information such as $A_\alpha : \{Z(\mathbf{u}_\alpha + \mathbf{h}_\beta) \leq z_\beta, \beta = 1, \ldots, n_\alpha\}$

The limitation is not theoretical (the corresponding extended normal equations are easily written), it lies in the inference of the corresponding indicator covariances. The larger the number (n) of elementary data events and the more complex these data events (including soft information), the larger the training image or data set needed for inference. Problems of inconsistency arise if that inference is done from possibly different and partially inconsistent training data sets. These problems can be fixed in various *ad hoc* ways; this is the price to pay for using more and more relevant information.

CONCLUSIONS

Reconstruction of sparsely sampled phenomena must capitalize on ancillary structural (pattern) information that can be collected from similar fields better sampled (a training image). There is much more valuable information to collect from such a training image than two-point statistics most often limited to a mere covariance/variogram function (linear correlation). Important characteristic features of training images are often curvilinear and involve multiple points; the modeling and exportation of such features are well beyond the flexibility of traditional random function models.

The extended normal equations (indicator kriging) formalism allows capturing such multi-point statistics and using them for deriving more accurate probability distributions for the unknown events to be assessed. Conditioning can be done on very complex data events involving several locations jointly, and hard as well as soft measurements. The only requirement is that the relation (covariance) between these data events and the event to be assessed can be inferred from the training image. A more accurate conditional probability distribution reflects immediately into more accurate stochastic images or conditional simlations.

The algorithm calls for a brute-force, computer-intensive approach to inference and image reconstruction (stochastic imaging) characteristic of many modern engineering and mathematical fields.

REFERENCES

1 - Cox, D. R. (1972) *The analysis of multivariate binary data*, in Applied Statistics, **21** (2): 113-120.

2 - Deutsch, C. V. and Journel, A. G. (1992) *GSLIB: Geostatistical software library and User's Guide*, to be published by Oxford University Press.

3 - Journel, A. G. and Alabert, F. A. (1989) *Non-Gaussian data expansion in the earth sciences*, Terra Nova, **1** (2): 123-134.

4 - Kramers, J. W., Bachu, S., Cuthiell, D. L., Prentice, M. E. and Yuan, L. P. (1989) *A multidisciplinary approach to reservoir characterization: the Provost Upper Manville B Pool*, in Journal of Can. Pet. Tech., **28** (3): 1-11.

5 - Luenberger, D. (1969) *Optimization by vector space methods*, Wiley & Sons, 326 pages.

6 - Olson, J. and Pollard, D. (1989) *Inferring paleostresses from natural fracture patterns: A new method*, in Geology, **17**: 345-348.

7 - Solow, A. (1986) *Mapping by simple indicator kriging*, Math Geology, **18** (3): 335-352.

8 - Zhu, H. and Journel, A. G. (1992) *Formatting and integrating soft data: Stochastic imaging via the Markov-Bayes algorithm*, in Proc. of 4th International Geostatistical Congress (this volume).

PARAMETRIC SURFACE ESTIMATION

E. A. Yfantis
Computer Science Department
University of Nevada, Las Vegas, 89154

G. T. Flatman
Environmental Protection Agency, EAD Division
P.O. Box 93478
Las Vegas, Nevada, 89193-34783-34783-3478

F. Miller
Desert Research Institute
P.O. Box 19040
Las Vegas, Nevada, 89132-0040

ABSTRACT

A parametric surface estimation algorithm is examined. The algorithm is a perfect interpolator. The points surrounding the point to be estimated are weighted according to the length of their paths from the point to be estimated, and not their Euclidean distance from that point. The algorithm is capable of estimating surfaces that are not functions and twist, turn, and fold, into the three dimensional space in any direction. The big advantage of this family of algorithms is that they do not require the process, the data came from, to satisfy the intrinsic hypothesis, or be second order stationary. Furthermore, they do not require equal distance between sampling points or continuity of the first or second derivatives. From the computational point of view they do not require matrix inversion. This family of methods is therefore robust. Given any set of points in the three dimensional space we show that this family of interpolators converges and always produces a surface. The disadvantage of this method is that, due to the lack of strict assumptions, it is difficult to calculate the error of estimation. Under the assumption of stationarity we calculate the error of the estimate, produced by interpolating with our method. Thus under the assumption of stationarity our method can be compared with kriging.

KEY WORDS: Deterministic Interpolators, Parametric Splines

INTRODUCTION

The problem of surface interpolation is a very important one, with applications in Computer Aided Design (CAD), Geosciences, Environmental Sciences, Geography, all areas of Engineering, Health Sciences, etc. Since the algorithm we introduce here is deterministic the adjective parametric does not refer to probability distributions, but parameters in the mathematical sense, or computer graphics sense. Over the past twentyfive years many perfect interpolation algorithms (algorithms producing curves or surfaces passing through the data), for curve and surface estimation have been developed. These algorithms can be classified into three categories: deterministic, stochastic, and chaotic. The deterministic algorithms can be found in the literature under the name splines (Barnhill and Riesenfeld '74), (Barnhill '85), (Barnhill and Boehm '83), (Barsky '88), (Akima '70), (Bartels et al '87), (Beck et al '86), (Coons '74), (de Boor '87), (de Casteljau '86), (Farin '83 '88), (Foley '86 '87 '87), (Lane '88), (Nielson '86 '87), (Salkauskas '84), (Sapidis '87), (Schumaker '81), (Shirman and Sequin '90), (Gregory '74), and others. The stochastic algorithms can be found in the literature usually under the name kriging (Journel '89), (David '77), (Olea '77), (Matheron '76), (Borgman and Frahme '76), (Yfantis et al. '87), (Journel and Huijbregts '78), (Armstrong '84), (Myers '82), (Davis '88), (Delfiner '76), (Englund '90), and others. The third class of interpolators, namely the ones based on the theory of chaos and fractals was developed recently (Carpenter '86), (Yfantis et al '88) and others.

The deterministic interpolation algorithms for surface estimation require that the domain of the input data points constitutes a rectangular grid, and all the grids have to be equal. Also biquadratic, surface estimation algorithms require continuity of the first derivative, and bicubic splines require continuity of the second derivative. Kriging requires that the data is sampled from a random process which is wide sense stationary or satisfies the intrinsic hypothesis, or if there is a trend, the residuals after removing the trend are second order stationary. Interpolation algorithms based on the theory of chaos give unpredictable answers, the generated curve or surface is continuous but has derivative nowhere, and due to their random nature every time the method is applied to the same set of data it gives different results. The method presented here leads to a deterministic perfect interpolation algorithm for curve and surface estimation. This algorithm does not require continuity of the first or second derivative, and does not require the data to satisfy the intrinsic hypothesis, a hypothesis that is impossible to prove for a given set of data. The aim of a surface estimation algorithm can be the interpolation of the surface itself, or of an attribute defined on this surface. The proposed method performs both interpolations.

THE PARAMETRIC INTERPOLATOR

Let $(x_{ij}, y_{ij}, z_{ij}, w_{i,j}), i, j \in N$ be data points in the four dimensional space, where x,y,z, denote the space coordinates of a point and w the attribute of interest measured at the point (x,y,z). Now we are looking at the x-coordinates of the patch only (the equations of the y, and z coordinates can be obtained similarly). The equation expressing the x-coordinate of the (i,j) patch will be denoted by $X_{ij}(t,u), i, j \in N, 0 \leq t \leq 1, 0 \leq u \leq 1$ and is defined by $x_{ij}, x_{i+1,j}, x_{i,j+1}, x_{i+1,j+1}$. The equation of the patch $X_{ij}(t,u)$ can be found if we find the equation of the curve $X_{ij}(t,0), 0 \leq t \leq 1$, defined by $x_{ij}, x_{i+1,j}$, and the curve $x_{ij}(0,u), 0 \leq u \leq 1$, defined by $x_{ij}, x_{i,j+1}$, and then we take their tensor product. The equation of the curve, introduced here, passing through $x_{ij}, x_{i+1,j}$, is:

$$X_{ij}(t,0) = 0.25(t^2 - t)x_{i-1,j} + 0.25(4 - 3t - t^2)x_{i,j} + \quad (1)$$
$$0.25(5t - t^2)x_{i+1,j} + 0.25(t^2 - t)x_{i+2,j}, 0 \leq t \leq 1, 2 \leq i \leq n - 2$$

For i = 1, and i = n-1, namely the first and the last patches are:

$$X_{1j}(t,0) = 0.5(2-3t+t^2)x_{1j} + (2t-t^2)x_{2j} + 0.5(t^2-t)x_{3j} \quad 0 \le t \le 1 \tag{2}$$

and

$$X_{n-1j}(t,0) = 0.5(t^2-t)x_{n-2j} + (1-t^2)x_{n-1j} + 0.5(t^2+t)x_{nj}, \tag{3}$$
$$0 \le t \le 1 \tag{4}$$

The curve $X_{ij}(0,u)$ passing through x_{ij}, x_{ij+1} is defined in a similar way. Thus for an interior surface patch (a patch that has neighboring patches on all sides), the equation is:

$$X_{ij}(t,u) = (0.25)^2 T_1 B_1 U_1,\ 0 \le t,u \le 1 \tag{5}$$

where
$T_1 = [t^2-t, 4-3t-t^2, 5t-t^2, t^2-t]$

$$B_1 = \begin{bmatrix} x_{i-1,j-1} & x_{i-1,j} & x_{i-1,j+1} & x_{i-1,j+2} \\ x_{i,j-1} & x_{i,j} & x_{i,j+1} & x_{i,j+2} \\ x_{i+1,j-1} & x_{i+1,j} & x_{i+1,j+1} & x_{i+1,j+2} \\ x_{i+2,j-1} & x_{i+2,j} & x_{i+2,j+1} & x_{i+2,j+2} \end{bmatrix} \quad U_1 = \begin{bmatrix} u^2-u \\ 4-3u-u^2 \\ 5u-u^2 \\ u^2-u \end{bmatrix}$$

$X_{ij}(0,0) = x_{ij}, X_{ij}(1,0) = x_{i+1,j}, X_{ij}(0,1) = x_{i,j+1}, X_{ij}(1,1) = x_{i+1,j+1}.$

Non interior patches have one or more sides with no neighbors. Non interior patches with one side having no neighbors are for example: the $X_{1j}(t,u), X_{n-1,j}(t,u)$, where j is an index corresponding to an interior edge and 1, n are indices corresponding to edges with no neighbors, also $X_{i1}(t,u)$ and $X_{i,(k-1)}(t,u)$, where k corresponds to an edge with no neighbors. The equations of these patches are :

$$X_{1j}(t,u) = (0.25) T_2 B_2 U_1,\ 0 \le t,u \le 1 \tag{6}$$

where
$T_2 = [0.5(2-3t+t^2), 2t-t^2, 0.5(t^2-t)]$

$$B_2 = \begin{bmatrix} x_{1,j-1} & x_{1,j} & x_{1,j+1} & x_{1,j+2} \\ x_{2,j-1} & x_{2,j} & x_{2,j+1} & x_{2,j+2} \\ x_{3,j-1} & x_{3,j} & x_{3,j+1} & x_{3,j+2} \end{bmatrix}$$

$$X_{n-1,j}(t,u) = (0.25) T_3 B_3 U_1,\ 0 \le t,u \le 1 \tag{7}$$

where
$T_3 = [0.5(t^2-t), (1-t^2), 0.5(t^2+t)]$

$$B_3 = \begin{bmatrix} x_{n-2,j-1} & x_{n-2,j} & x_{n-2,j+1} & x_{n-2,j+2} \\ x_{n-1,j-1} & x_{n-1,j} & x_{n-1,j+1} & x_{n-1,j+2} \\ x_{n,j-1} & x_{n,j} & x_{n,j+1} & x_{n,j+2} \end{bmatrix}$$

$$X_{i,1}(t,u) = (0.25) T_1 B_4 U_2,\ 0 \le t,u \le 1 \tag{8}$$

where

$$B_4 = \begin{bmatrix} x_{i-1,1} & x_{i-1,2} & x_{i-1,3} \\ x_{i,1} & x_{i,2} & x_{i,3} \\ x_{i+1,1} & x_{i+1,2} & x_{i+1,3} \\ x_{i+2,1} & x_{i+2,2} & x_{i+2,3} \end{bmatrix} \qquad U_2 = \begin{bmatrix} 0.5(2-3u+u^2) \\ 2u-u^2 \\ 0.5(u^2-u) \end{bmatrix}$$

$$X_{i,k-1}(t,u) = (0.25)T_1B_5U_3, \; 0 \le t,u \le 1 \tag{9}$$

where

$$B_5 = \begin{bmatrix} x_{i-1,k-2} & x_{i-1,k-1} & x_{i-1,k} \\ x_{i,k-2} & x_{i,k-1} & x_{i,k} \\ x_{i+1,k-2} & x_{i+1,k-1} & x_{i+1,k} \\ x_{i+2,k-2} & x_{i+2,k-1} & x_{i+2,k} \end{bmatrix} \qquad U_3 = \begin{bmatrix} 0.5(u^2-u) \\ 1-u^2 \\ 0.5(u^2+u) \end{bmatrix}$$

Finally there are patches with more than one edge having no neighbors. Such are the $X_{11}(t,u)$ and $X_{n-1,k-1}(t,u)$. The equations of these patches are:

$$X_{11}(t,u) = T_2B_6U_2, \; 0 \le t,u \le 1 \tag{10}$$

where

$$B_6 = \begin{bmatrix} x_{11} & x_{12} & x_{13} \\ x_{21} & x_{22} & x_{23} \\ x_{31} & x_{32} & x_{33} \end{bmatrix}$$

$$X_{n-1,k-1}(t,u) = T_3B_7U_3, \; 0 \le t,u \le 1 \tag{11}$$

where

$$B_7 = \begin{bmatrix} x_{n-2,k-2} & x_{n-2,k-1} & x_{n-2,k} \\ x_{n-1,k-2} & x_{n-1,k-1} & x_{n-1,k} \\ x_{n,k-2} & x_{n,k-1} & x_{n,k} \end{bmatrix}$$

The parametric equations for y and z, and w are similar to those for x. The only difference is that we replace the x's with y's or the x's with z's or w's. Note that our estimation method does not require systematic sampling (equal intervals between neighboring sampling points), continuity of the first derivative, or the assumption that the underlying process satisfies the intrinsic hypothesis, or is wide sense stationary. Furthermore, it does not require finding the inverse of a matrix as kriging and splines require. The calculations therefore are fast, and a surface is always generated. Finally, the generated surface does not have to be a function as in the case of kriging.

THE MEAN SQUARE ERROR OF THE PARAMETRIC INTERPOLATOR UNDER THE ASSUMPTION OF STATIONARITY

Now we are looking at the attribute only. From equation (5) we obtain:

$$W^*_{ij}(t,u) = \frac{1}{16}[w_{i-1j-1}(t^2-t) + w_{ij-1}(4-3t-t^2) + w_{i+1j-1}(5t-t^2) +$$

$$\begin{aligned} &w_{i+2j-1}(t^2-t)](u^2-u)+[w_{i-1j}(t^2-t)+w_{ij}(4-3t-t^2)+w_{i+1j}(5t-t^2)+ \\ &w_{i+2j}(t^2-t)](4-3u-u^2)+w_{i-1j+1}(t^2-t)+w_{ij+1}(4-3t-t^2)+ \\ &w_{i+1j+1}(5t-t^2)+w_{i+2j+1}(t^2-t)](5u-u^2)+w_{i-1j+2}(t^2-t)+ \\ &w_{ij+2}(4-3t-t^2)+w_{i+1j+2}(5t-t^2)+w_{i+2j+2}(t^2-t)](u^2-u) \end{aligned} \tag{12}$$

If we assume that the mean of the stationary process is μ and the variance is σ^2 then:

$$E(W^*_{ij}(t,u))=\mu \tag{13}$$

So our estimator has the same mean as the process it estimates. If $w_0 = W_{ij}(t,u)$ is the true value of the process at the point $(X_{ij}(t,u), Y_{ij}(t,u), Z_{ij}(t,u))$ (notice that the (t,u) could all be considered as a way of reparametrization, or as a way of defining a new coordinate system), then if a square sampling design is assumed in order to make the calculation of the mean square error both manageable and also comparable with kriging with square sampling design, the mean square error at that point is:

$$E(W^*_{ij}(t,u)-w_0)^2=F+G-H-K \tag{14}$$

Where:

$$F=\sigma^2+\frac{1}{16}\sigma^2(u^4-2u^3+7u^2-6u+4)(t^4-2t^3+7t^2-6t+4) \tag{15}$$

$$\begin{aligned} G=\frac{1}{128}\{&[-t^4-u^4+2(t^3+u^3)-17(t^2+u^2)+16(t+u)]R(h) \\ &+2[t(t^2-1)(2-t)+u(u^2-1)(2-u)]R(2h)+[(t^2-t)^2+(u^2-u)^2]R(3h)+ \\ &+[32tu(4-3t-t^2)(4-3u-u^2)+8u(t^2-t)(5t-t^2)(4-3u-u^2)+ \\ &8t(4-3t-t^2)(5u-u^2)(u^2-u)+2(t^2-t)(5t-t^2)(u^2-u)(5u-u^2)]R(\sqrt{2}h) \\ &+[16(t^2-t)(u^2-u)(2+u-u^2)(2+t-t^2)+4(t^2-t)(2+t-t^2)* \\ &(5u-u^2)(4-3u-u^2)]R(\sqrt{5}h) \\ &+8(t^2-t)(2+t-t^2)(u^2-u)(2+u-u^2)R(2\sqrt{2}h) \\ &+[8t(4-3t-t^2)(u^2-u)^2+8u(t^2-t)^2(4-3u-u^2)+ \\ &2(t^2-t)^2(u^2-u)(5u-u^2)+2(t^2-t)(5t-t^2)(u^2-u)^2]R(\sqrt{10}h) \\ &+4[(t^2-t)(2+t-t^2)(u^2-u)^2+(t^2-t)^2(u^2-u)(2+u-u^2)]R(\sqrt{13}h) \\ &+2(t^2-t)^2(u^2-u)^2R(3\sqrt{2}h)\} \end{aligned} \tag{16}$$

$$\begin{aligned} H=\frac{1}{8}\{&[R(w_{i-1j-1}-w_0)(t^2-t)+R(w_{ij-1}-w_0)(4-3t-t^2)+ \\ &R(w_{i+1j-1}-w_0)(5t-t^2)+R(w_{i+2j-1}-w_0)(t^2-t)](u^2-u)\} \end{aligned} \tag{17}$$

$$\begin{aligned}K = \frac{1}{8}\{[R(w_{i-1j} - w_0)(t^2 - t) + R(w_{ij} - w_0)(4 - 3t - t^2)\\ +R(w_{i+1j} - w_0)(5t - t^2) + R(w_{i+2j} - w_0)(t^2 - t)](4 - 3u - u^2)\\ +[R(w_{i-1j+1} - w_0)(t^2 - t) + R(w_{ij+1} - w_0)(4 - 3t - t^2) +\\ R(w_{i+1j+1} - w_0)(5t - t^2) + R(w_{i+2j+1} - w_0)(t^2 - t)](5u - u^2)\\ +[R(w_{i-1j+2} - w_0)(t^2 - t) + R(w_{ij+2} - w_0)(4 - 3t - t^2) +\\ R(w_{i+1j+2} - w_0)(5t - t^2) + R(w_{i+2j+2} - w_0)(t^2 - t)](u^2 - u)\}\end{aligned} \tag{18}$$

The R used in the above equations denotes the autocovariance or autocorrelation function of the random process W, which under the assumption of stationarity depends on the length of the path joining the corresponding points (instead of the Euclidean distance). If the sampling design is square then the mean square error attains its maximum value when u = t = 0.5, or when the point to be estimated is in the center of the square. By substituting in the above equation, the maximum mean square error is:

$$\begin{aligned}E(W_{ij}^*(0.5, 0.5) - w_0)^2 = \frac{1937}{256}\sigma^2 + \frac{1}{128}\{\frac{63}{8}R(h) - \frac{9}{4}R(2h)\\ +\frac{1}{8}R(3h) + \frac{7614}{256}R(\sqrt{2}h) - \frac{405}{64}R(\sqrt{5}h)\\ +\frac{81}{32}R(2\sqrt{2}h) + \frac{63}{64}R(\sqrt{10}h) - \frac{9}{32}R(\sqrt{13}h) + \frac{1}{128}R(3\sqrt{2}h)\}\\ -\frac{1}{32}R(h\sqrt{4.5}) + \frac{9}{16}R(h\sqrt{2.5}) - \frac{81}{8}R(h\sqrt{0.5})\end{aligned} \tag{19}$$

The mean square error of the estimator derived by using kriging and the same neighborhood as in our parametric estimator, is:

$$\begin{aligned}E(W_{ij}^*(u, v) - w_0)^2 = \sigma^2 + \sum_{m=1}^{4}\sum_{l=1}^{4}\sum_{j=1}^{4}\sum_{i=1}^{4} a_{ij}a_{ml}R(w_{ij} - w_{lm})\\ -2\sum_{i=1}^{4}\sum_{j=1}^{4} a_{ij}R(w_{ij} - w_0)\end{aligned} \tag{20}$$

R is the autocorrelation function of W, and the argument is the length along the path. The degree of closeness of the two mean square errors depends on the autocorrelation function or semivariogram of the process, and the zone of influence.

CONCLUSIONS

A robust interpolation method, given a set of data in the three dimensional space, is presented. The estimation algorithm is deterministic, and does not require: equal sampling intervals, continuity of the derivatives, inverting large matrices, or inverting any matrices at all, the assumption of the intrinsic hypothesis or stationarity (which are difficult to verify) or the restriction that the data points are part of a function (not to have more than one z for some x,y). The method introduced here is relatively easy to implement on the computer, and to apply to a given set of data. The mean square error of our method was calculated based on the assumption that the underlying process is wide sense stationary.

Acknowledgment The authors would like to thank Dr. J. P. Chiles for his fine review and his valuable suggestions. They would also like to thank Dr. A. Soares for his encouragement and enormous energy that has exerted so that this conference will pass from the conceptual stage to reality.

NOTICE

Although the research described in this article has been supported by the United States Environmental Protection Agency, it has not been subjected to Agency review and therefore does not necessarily reflect views of the Agency and no official endorsement should be inferred.

REFERENCES

Akima, H. (1970), A new method of interpolation and smooth curve fitting based on local procedures, journal ACM 17 (4), 589-602.

Armstrong, M. (1984), Problems with universal kriging, Mathematical Geology 16, 101-108.

Barnhill, R. (1985), Surfaces in Computer Aided Geometric Design, A Survey with New Results, Computer Aided Geometric Design 2, 1-17.

Barnhill, R. and Riesenfield, W., eds., (1974), Computer Aided Geometric Design, Academic Press, New York.

Barnhill, R and Boehm, W., eds. (1983), Surfaces in Computer Aided Geometric Design, North-Holland, Amsterdam.

Barsky, B. (1988), Computer Graphics and Geometric Modeling using Beta-Splines, Springer Verlag, New York.

Bartels, R., Beatty, J. and Barsky, B. (1987), An Introduction to Splines for use in Computer Graphics and Geometric Modeling, Morgan Kaufmann, Los Altos, Ca.

Beck, J., Farouki, R. and Hinds, J. (1986), Surface Analysis methods, IEEE Computer Graphics and Applications, 6 (12), 18-36.

Borgman, L. E. and Frahme, R. B. (1976) Multivariate properties of bentonite in Northwestern Wyoming, in Advanced Geostatistics in the Mining industry, D. Reidel publ., Dordretch, 381-390.

Carpenter, L. C. (1986), Computer Rendering of Fractal Curves and Surfaces, ACM Siggraph, Dallas, August 18-22, 9-15.

Coons, S. (1974), Surface Patches and B-Spline Curves, In R. Barnhill and R. Riesenfeld, editors, Computer Aided Geometric Design, Academic Press.

David, M. (1977), Geostatistical Ore Reserve Estimations, Elsevier, Amsterdam.

Davis, B. (1988), Uses and Abuses of Cross Validation in Geostatistics, Mathematical Geology, 19 (3) 241-249.

de Boor, C. and Hollig, K. (1987), B-Splines without divided differences, In G. Farin, editor, Geometric Modeling-Algorithms and New Trends, 21-27, SIAM.

de Casteljau, P. (1986), Shape Mathematics and CAD, Kogan Page, London.

Delfiner, P. (1976), Linear Estimation of Nonstationary Spatial Phenomena, In Guarascio, M., Advanced Geostatistics in the Mining Industry, 49-68, D. Reidell, Boston.

Englund, E. J. (1990), A Variance of Geostatisticians, Mathematical Geology, 22 (4), 417-456.

Farin, G. (1988), Curves and Surfaces for Computer Aided Geometric Design, Academic Press, San Diego.

Farin, G. (1983), Smooth interpolation to scattered 3D data, in: [Barnhill and Boehm '83].

Foley, T. (1986), Local Control of Interval Tension Using Weighted Splines, Computer Aided Geometric Design, 3 (4) 281-294.

Foley, T. (1987), Interpolation with Interval and Point Tension controls Using Cubic Weighted v-Splines. ACM Transactions on Mathematical Software, 13 (1) 68-96.

Foley, T. (1987), Weighted Bicubic Spline Interpolation to Rapidly Varying Data, ACM Transactions on Graphics, 6 (1) 1-18.

Gregory, J. (1974), Smooth interpolation without twist constraints, in: [Barnhill and Riesenfeld '74].

Journel, A. G. (1989), Fundamentals of Geostatistics in Five Lessons, Short Course in Geology, American Geoph. Union Publ., Washington D. C.

Journel, A. G. and Huijbregts, G. T. (1978), Mining Geostatistics, Academic Press, London.

Lane, D. (1988), Interpolating Curves and Surfaces Using Spline Functions, Master's Thesis, Computer Science Dept. University of Nevada, Las Vegas.

Matheron, G. (1976), A Simple Substitute to Conditional Expectations, The disjunctive kriging, In Guarasciom, Advanced Geostatistics in the Mining Industry, NATO ASI, D. Reidel Publ., Dordrecht, 221-236.

Myers, D. (1982), Matrix Formulation of Co-kriging, Mathematical Geology, 17 (6), 625-644.

Nielson, G. (1986), Rectangular nu-Splines, IEEE Computer Graphics and Applications, 6 (2), 35-40.

Nielson, G. (1987), Coordinate free Scattered Data Interpolation, In L. Shumaker, editor, Topics in Multivariate Approximation, Academic Press,San Diego.

Olea, R. A. (1984), Systematic Sampling of Spatial Functions, Kansas Geological Survey, University of Kansas, Lawrence, Kansas, Series on Spatial Analysis, 7, 1-57.

Salkauskas, K. (1984), C^1 Splines for Interpolation of Rapidly Varying Data, Rocky Mountain journal, 14 (1), 239-250.

Sapidis, N. (1987), Algorithms for Locally Fairing B-Spline Curves, Master's Thesis, University of Utah.

Schumaker, L. (1981), Spline Functions, Basic Theory, J. Wiley and sons, New York.

Shirman, L. A. and Sequin, C. H. (1990), Local surface interpolation with shape parameters between adjoining Gregory Patches, Computer Aided Geometric Design 7 (5), 375-388.

Yfantis, E. A., Flatman, G. T. and Behar, J. V. (1987), Efficiency of Kriging Estimation for Square, Triangular, and Hexagonal Grids, Mathematical Geology, 19 (3), 183-207.

Yfantis, E. A., Flatman, G. T. and Englund, E. J. (1988), Simulation of Geological Surfaces Using Fractals, 20 (6), 667-673.

How large a sample is needed to estimate the regional variogram adequately?

R. WEBSTER[1] & M. A. OLIVER[2]

[1] *ETH Zürich, Grabenstrasse 3, 8952 Schlieren, Switzerland.*
[2] *School of Geography, The University, Birmingham B15 2TT, England*

Summary

There is little to guide the geostatistical practitioner on the size of sample needed to estimate the variogram adequately. It is often estimated from few data, but because confidence limits cannot be determined analytically from a single set of data its precision is unknown. Approximate confidence intervals can be found numerically by Monte Carlo simulation. A large field of values is created using a plausible model of the variogram. It is then sampled many times, and the observed variogram of each sample is computed. An experimental sampling distribution of the variogram is constructed from which percentiles and confidence limits can be obtained. Our experiments suggest that variograms computed on fewer than 50 data are of little worth and that at least 100 data are needed. For a normally distributed isotropic variable 150 data should suffice, while one derived from 225 data will usually be reliable.

Introduction

The variogram is crucial for geostatistical estimation and for defining the spatial variation of a property in a region, yet it is often determined from few data. There are many examples in the literature of variograms computed from samples of less than 50 data, and some from fewer than 30. Such variograms are unreliable because the estimated semivariances embody large errors, and the apparent confidence with which they are used is not justified.

In kriging, for instance, even though estimates are fairly stable, their estimation variances, and hence confidence limits, are sensitive to the precision of the variogram. As it happens, we rarely know how reliably the regional variogram of a property is estimated in any particular instance. Furthermore, there is no easy way of incorporating the error in the variogram into the final kriging variance. Clearly it is important to estimate the variogram well at the outset.

We already know something of the robustness of determining the variogram with different estimators and how it is affected by the distribution of the population (Cressie & Hawkins, 1980). Although we also know intuitively that generally the larger the sample from which the variogram is computed the more precise it will be, provided that the sampling interval is appropriate, we do not know how many measurements are needed to achieve any desired precision. Studies in time-series analysis (Jenkins & Watts, 1968) show that different realizations of the same random process can give very different correlograms and that large samples (300 or more) are needed to identify spectral peaks with reasonable confidence.

There are situations, especially in mining and petroleum exploration, where only few data are available, and to obtain more may be impossible or too expensive. The computed variograms often appear erratic, and in many instances the principal cause is that the data are too few. Some investigators in other fields, where the constraints on sampling are much less severe, have attempted to justify taking such small samples on the grounds that each semivariance is calculated from many comparisons. They may be misled into computing confidence limits from the classical formula, which is based on independent samples, when neither the sample values nor the estimated semivariances are independent of one another .

Figure 1 shows what can happen when the sample is small. The six variograms in it were obtained by taking six samples of 49 points from a single field of values using a 7×7 square grid with an interval of 15 units and random origins. The variation is very large. The original exponential variogram, which had been used to create the field, and which could be reproduced faithfully in the exhaustive experimental variogram, had no nugget variance. However, anyone fitting models to three of the variograms in Fig. 1 would almost certainly extrapolate to a positive intercept on the ordinate. By increasing the sample to 225 points on a 15×15 grid at an interval of 7 units and again selecting the origin at random we obtained the six variograms in Fig. 2. These are much more similar to one another, especially at the shorter lags, and they match fairly closely the original variogram in the generator. Results such as those in Fig. 1 give little confidence and show clearly the dangers of undersampling when estimating variograms.

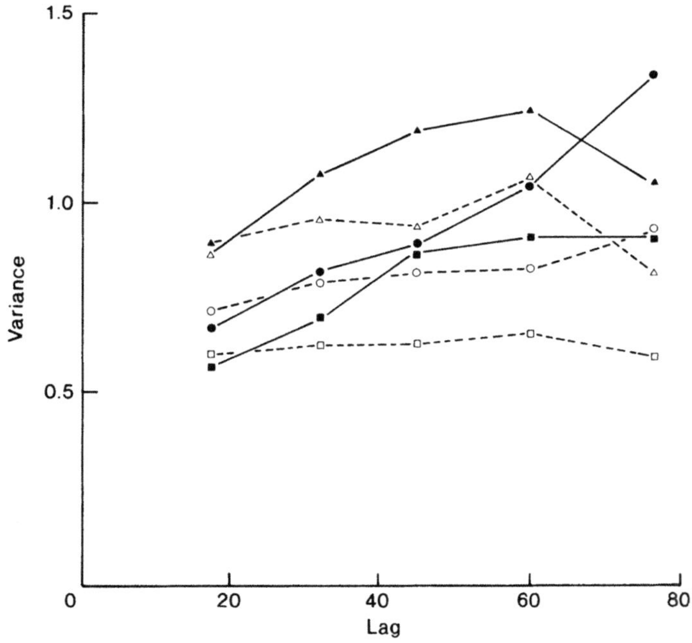

Figure 1. Variograms obtained by sampling a field generated from an exponential variogram, $\gamma(h) = \{1 - \exp(-h/16)\}$, with 49 points on a 7×7 grid at 15-unit intervals.

Geostatisticians often ask themselves, as well as being asked by others: how large a sample is needed to obtain a satifactory estimate of the variogram? The question has no easy answer unfortunately. There is no simple formula for calculating confidence limits on variograms, and it has seemed unprofitable to seek an algebraic solution (Matheron, 1989, p. 107). Some form of Monte Carlo technique is needed to determine them.

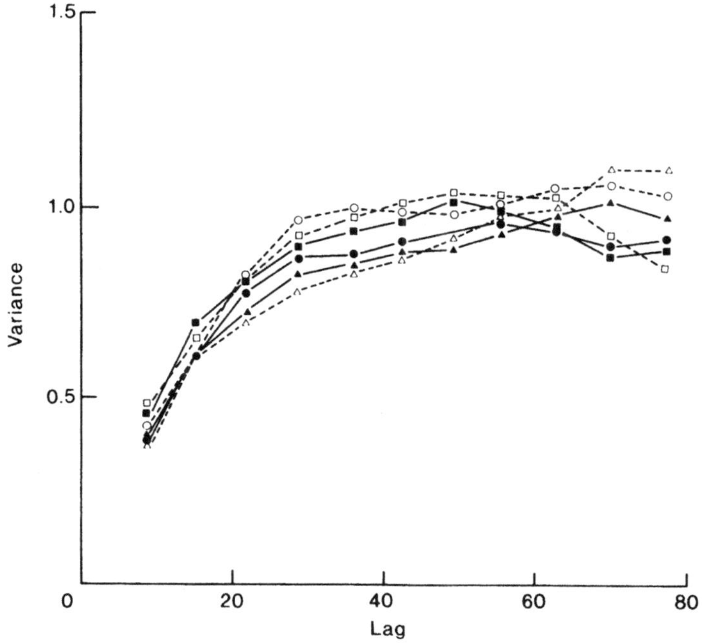

Figure 2. Variograms obtained by sampling the same exponential field as in Fig. 1 with 225 points on a 15 × 15 grid at intervals of 7 units.

Which variogram?

Matheron (1965) and Muñoz-Pardo (1987) have studied fluctuations in the variogram, and they distinguish fluctuations *sensu stricto* arising from different realizations of a generating process and those arising from repeated sampling of a single realization of the process, i.e. the sampling error. This raises the question of precisely which variogram we wish to estimate, and we discuss this next.

To describe a region geostatistically we sample it on a particular support and compute a variogram. This is the *observed variogram* to which we usually fit a smoothly varying mathematical function; the *fitted model*. Our sample values and the observed variogram that we derive from them represent an underlying process whose variogram is unknown: they represent just one outcome from it. We might then determine the fluctuation to expect from the process by simulating numerous sets of values from the particular model and computing the observed variogram from each set. The quantiles of the observed values for each lag would be reasonable estimates of the confidence intervals. Essentially this was the approach taken by McBratney &

spherical variogram in which the nugget variance was 1/3 of the total variance. Their formulae are respectively:

$$\gamma(h) = c\{1 - \exp(-h/r)\} , \tag{1}$$

and

$$\begin{aligned} \gamma(h) &= c_o + c\left\{\frac{3h}{2a} - \frac{1}{2}\left(\frac{h}{a}\right)^3\right\} \quad \text{for } 0 < h \le a, \\ \gamma(h) &= c_o + c \quad \text{for } h > a, \\ \gamma(0) &= 0, \end{aligned} \tag{2}$$

where, in the usual notation, $\gamma(h)$ is the semivariance at lag distance h for the isotropic case, c_o is the nugget variance, c is the sill of the spatially structured variance, and r and a are the distance parameters of the models.
Their coefficients were

model	nugget c_o	sill c	distance r	a
exponential		1	16	
spherical	0.3333	0.6667		50

The distance parameter, a, in the spherical model is the range of spatial dependence. The exponential variogram has an effective range of $3r$, and so the two models that we used are approximately equivalent in this respect.

Their integral ranges are different, however. The integral range, A, can be expressed in the form

$$A = \int_0^\infty \rho(h)\mathrm{d}h = \int_0^\infty \left\{\frac{c_0 + c - \gamma(h)}{c_0 + c}\right\} \mathrm{d}h, \tag{3}$$

where $\rho(h)$ is the correlogram (Lantuéjoul, 1991). The integral range of the exponential model with no nugget is 16, the same as its distance parameter, but that of the spherical variogram we used is only 12.5 because of its nugget component. The larger nugget tends to increase the errors of estimating the semivariances for any given sampling design and lag.

We checked that the outcomes of the simulations had the properties expected. Both had variances close to 1: they were 0.9808 for the exponential field and 1.1110 for the spherical, and slight skew: 0.17 and 0.22 respectively. Their exhaustive observed variograms were close to the underlying ones from which they were generated.

Sampling and computing. Having created the fields we then distinguished two situations in which we might sample. In the first we consider the case where the

Webster (1986) and Taylor & Burrough (1986), though in different contexts. Their results showed very wide intervals except at the shortest lags. Note, however, that the observed variation was in the generating process, and was not a sampling effect. Shafer and Varljen (1990) also attempted to compute confidence intervals for a single realization using the jackknife method.

In geostatistical practice, as distinct from theory, or indeed time-series analysis, we rarely have more than one realization. In any particular case we have a single limited region or volume of soil, rock, water or air, and within it a single realization. Our task is to describe what is there. Provided the region is homogeneous it will have a variogram, the *regional variogram*, for a particular support, and this is the variogram that we need to know or to estimate in order to proceed further. Following Matheron (1965) and others since (eg, Journel & Huijbregts,1978; Davis & Borgman, 1982) we shall call this the *local variogram*. We should also like to know how big a sample to take to estimate it adequately. Matheron (1965) did give a formula, Equation (5) below, to provide what he called a first approximation to an estimation variance of the local variogram. Muñoz-Pardo (1987) pursued Matheron's approach for bounded variograms. He showed by numerical integration that the estimation variance depended on the effective range of the variogram in relation to the size of the region as well as on the size of sample, i.e. the number of observations.

Determining the confidence limits

One way of determining confidence limits on the variogram for a particular region obtained by sampling it on a given support would be to take many more samples of the same size on that support, compute the variogram of each sample, and record the spread of the observed semivariances. Except perhaps in remote sensing where there are abundant data from scanners this is hardly practicable. It is not necessarily sensible either, since once data have been obtained they can be used to compute a single observed variogram, which will be the most precise. The alternative is to simulate a dense field of values from a plausible variogram and to sample it. This has the advantage, in principle, that we know the variogram to start with, though the particular realization will depart from it more or less. This is what we did.

Our study is based on our experience as soil scientists and on feasible practice in soil survey. The problem is almost always two-dimensional. We have a region of interest, and we want to know how some property of the soil varies in it. The soil is accessible almost everywhere, and sampling is neither so expensive that we cannot afford to sample adequately nor so cheap that we can oversample lightly. We can sample the whole of a region soundly to represent it.

The simulated fields. We created large autocorrelated isotropic random fields, each consisting of 65 536 $(= 256 \times 256)$ values on a square grid. We used the method of turning bands (45 of them) with a normal distribution of errors (Journel & Huijbregts, 1978). In the examples that we report here there were two such fields, one generated by an exponential variogram with no nugget variance, and one by a range of the variogram is a substantial proportion of the distance across the region of interest. To simulate this we took a subset of the whole field covering approximately 120 × 120 units, i.e. a square with side somewhat more than twice the range of the variogram. We then sampled the subset using square grids with following sizes and intervals.

Size	25	49	100	144	225	400
Interval	20	15	10	8	7	5

For the last four we sampled exhaustively with no value in the field being used twice. This gave 100, 64, 49 and 25 samplings respectively. For the first two we repeated the sampling 100 times, again without using any value more than once.

The second situation that we envisaged is where the range of the variogram is only a small proportion of the distance across the region. Sampling the whole field on a regular grid would give little information on the structured part of the variogram in these circumstances unless the grid were very dense (and therefore very expensive). A sensible solution here is to sample along transects spread over the region. We did this and selected starting points and directions at random, once more ensuring that no point was sampled twice. We chose the same combinations of sample size and sampling interval as before, so that, for example, with a sample size of 25 there were five transects each of length 5.

Observed variograms. We computed the variogram for each sample using the usual computing formula:

$$\hat{\gamma}(\mathbf{h}) = \frac{1}{2m(\mathbf{h})} \sum_{i=1}^{m(\mathbf{h})} \{z(\mathbf{x}_i) - z(\mathbf{x}_i + \mathbf{h})\}^2, \tag{4}$$

where $z(\mathbf{x}_i)$ are the simulated values at $\mathbf{x}_i$, $\mathbf{h}$ is the lag, $= h$ for isotropic variation, and $\hat{\gamma}(\mathbf{h})$ is the observed semivariance at that lag. Except along the rows and columns of the grid the true lag is rounded to the nearest whole interval of the sampling grid, and this has a smoothing effect on the variogram. Nevertheless, this is what happens in practice and which we wish to illustrate.

Figure 3a shows the observed semi-variances and the 90% confidence limits for each sample size on square grids for the exponential variogram. The graphs show clearly how the confidence intervals narrow as the size of the sample increases. A sample size of 144 would seem to give acceptable confidence, that for 225 shows some small improvement, while for 400 points the variogram is estimated with great precision. The six separate graphs can be summarized in one if we plot the standard deviation of the observed semivariances against size of sample, Fig. 4. Clearly, the standard deviation decreases and the 90 percentiles narrow as the size of sample increases. We can also judge from these graphs approximately what size of sample to use to achieve some particular standard deviation or confidence interval. We then performed the sampling along transects for the same set of sample sizes. Figure 3b shows the results. There is somewhat more sampling fluctuation than for the grid

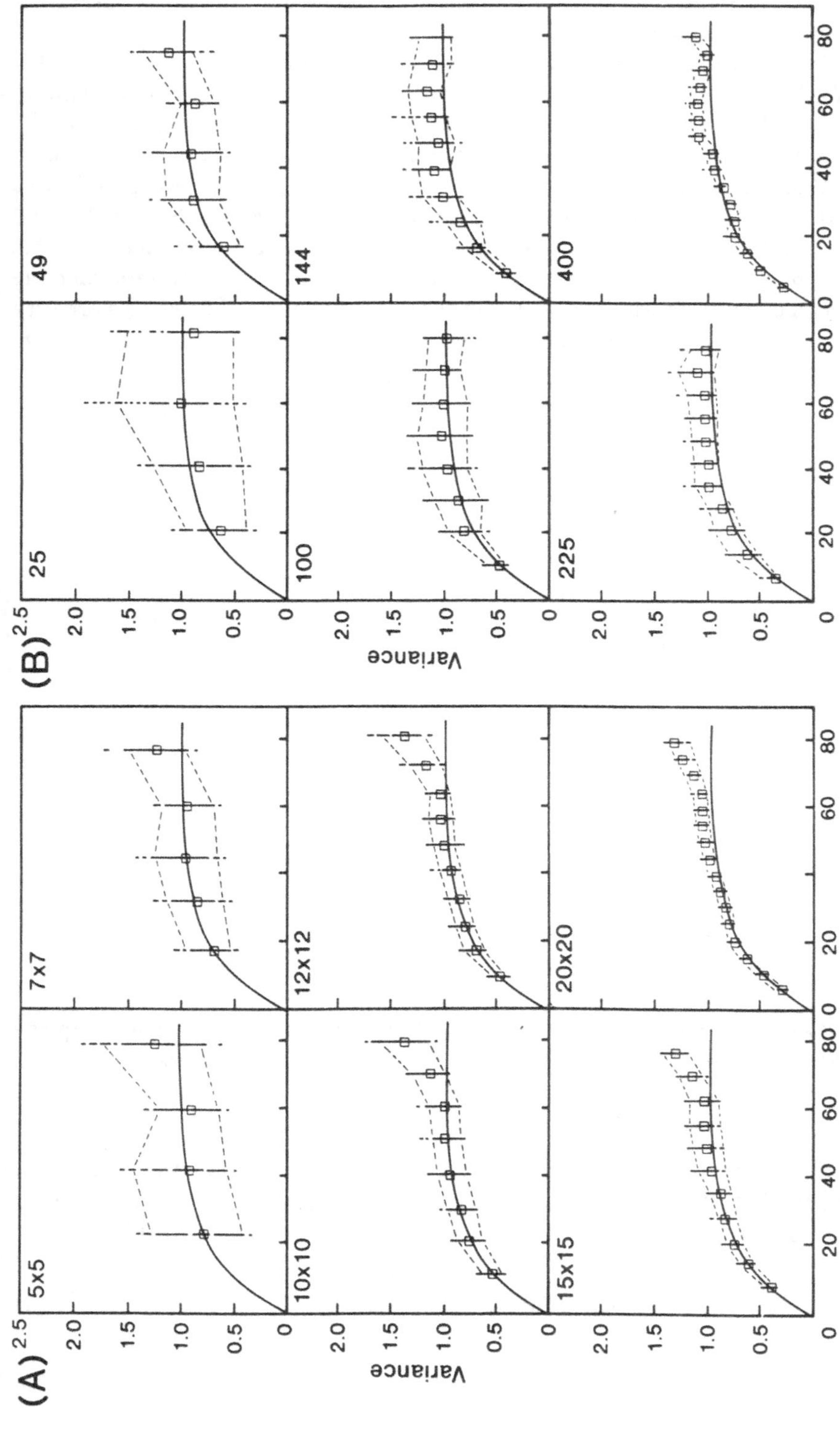

Fig. 3. Observed semivariances and the 90% confidence intervals for samples (a) on a square grid, and (b) along transects computed from the exponential field generated from the variogram $\gamma(h) = \{1 - \exp(-h/16)\}$.

sampling, but otherwise the sequence of narrowing confidence limits with increasing size of sample is the same.

Figure 5a shows the observed values and 90 percentiles for the variogram, $\gamma(h) = 0.33 + 0.67sph(50)$, with the same range of sample sizes on square grids. A variogram, such as this, with an appreciable nugget variance is fairly typical, hence these results are perhaps more relevant. In general the confidence intervals are wider, and in particular those for a sample of 100. To attain a reasonably precise estimate in this situation at least 144 measurements would be needed, while a sample of 400 would give a very precise estimate of the variogram. Figure 6 again summarizes these graphs. Figure 5b shows the observed variograms and their 90 percentiles by sampling the same field along transects with the same range of sizes as before. The graphs are very similar to those in Fig. 5a except that the confidence intervals are somewhat wider.

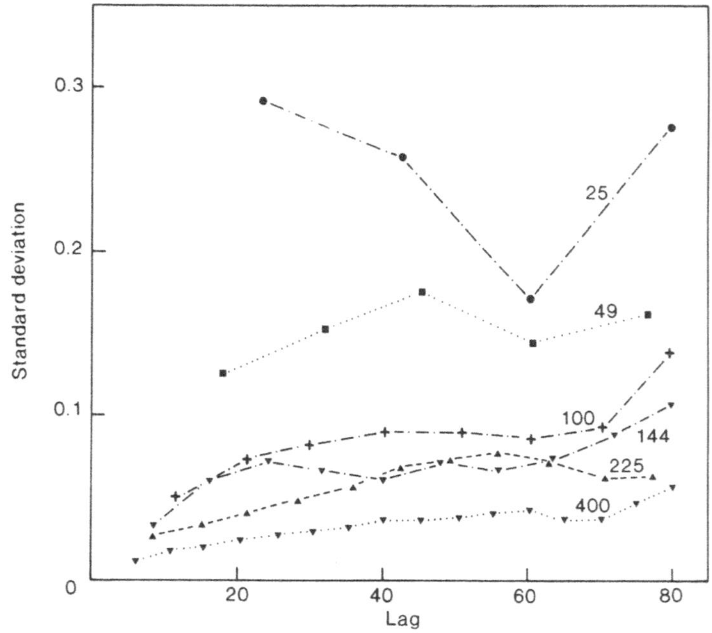

Figure 4. Standard deviations of the observed semivariances against size of samples on square grids from fields generated with the exponential variogram.

Discussion and Conclusion

Matheron (1965) devoted a chapter of his book to fluctuation in the variogram. For a normally distributed isotropic variable with a variogram he gave the following approximation for the estimation variance of the local semivariance, $\gamma_R(\mathbf{h})$:

$$\mathrm{var}[\hat{\gamma}_R(\mathbf{h})] \approx 4\gamma(\mathbf{h})\frac{N}{N'}\sigma_N^2. \tag{5}$$

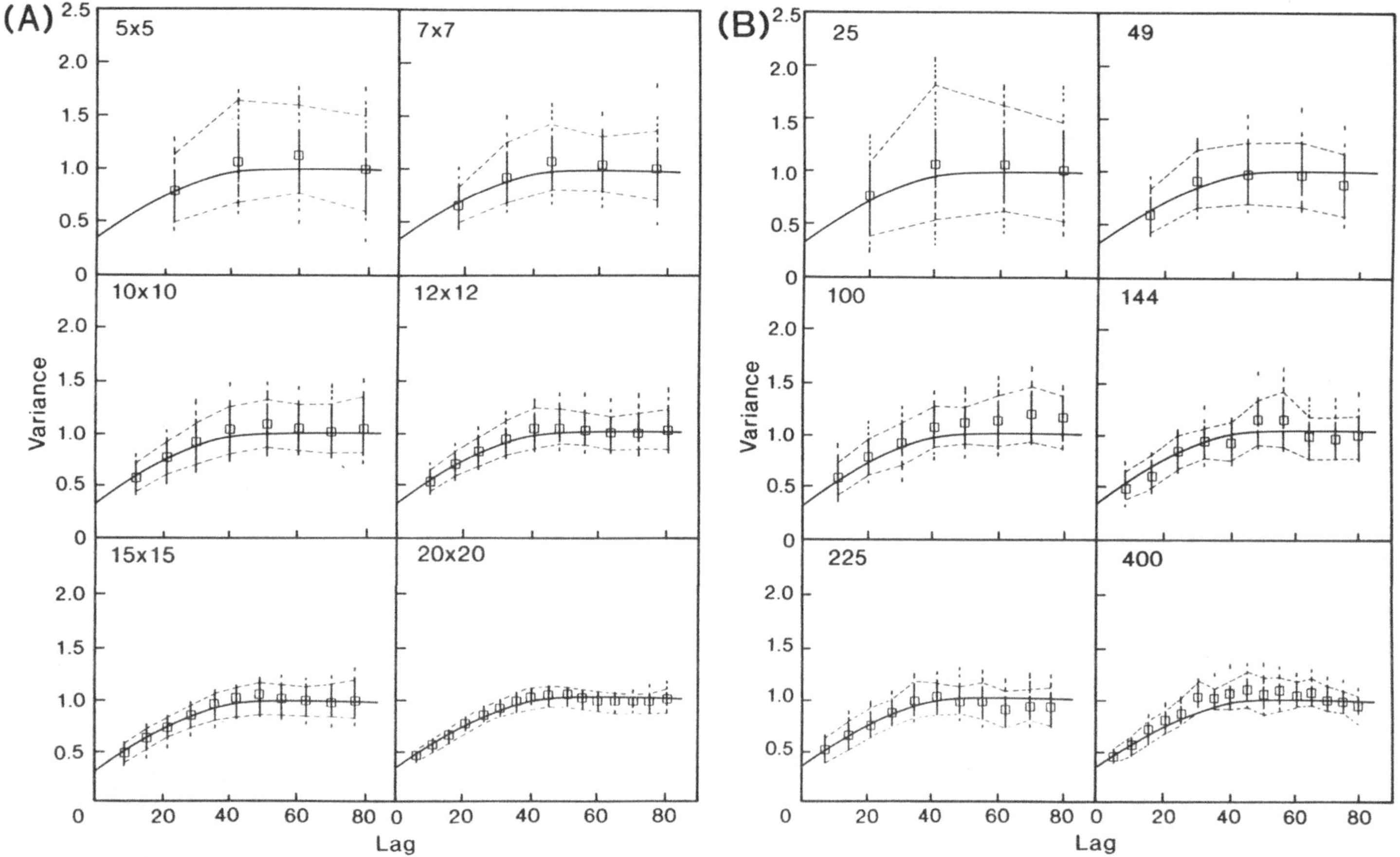

Fig. 5. Observed semivariances and the 90% confidence limits for samples (a) on square grids, and (b) along transects from the field generated with the spherical variogram $\gamma(h) = 0.33 + 0.67sph(50)$.

Here $\gamma(\mathbf{h})$ is the value of the underlying variogram at lag $\mathbf{h}$, $\hat{\gamma}_R(\mathbf{h})$ is the estimate of the local semivariance at that lag, and σ_N^2 is the variance of the mean in the region estimated from N observations. Matheron describes N' as the number of points effectively utilized in estimating $\gamma_R(\mathbf{h})$. It is the number of paired comparisons in the intersection of the region with itself when translated by the vector $\mathbf{h}$. Journel & Huijbregts (1978) summarize this result, and they advise practitioners that a minimum of 30 to 50 paired comparisons should suffice. We have for long judged this to be too few. Following practice in time-series analysis we have recommended a minimum of 100 observations for regular sampling in one dimension and computing the variogram to no more than one fifth of the run, giving a minimum of 80 comparisons for an estimate. As above, this is a matter of judgement, but no more. In two dimensions, however, there is confusion because many practitioners forget or fail to realise that $\mathbf{h}$ is a vector with direction as well as distance. They average their estimates of $\gamma_R(\mathbf{h})$ over all directions, and they readily obtain the minimum of 50 comparisons with few data, and this gives them a false sense of security.

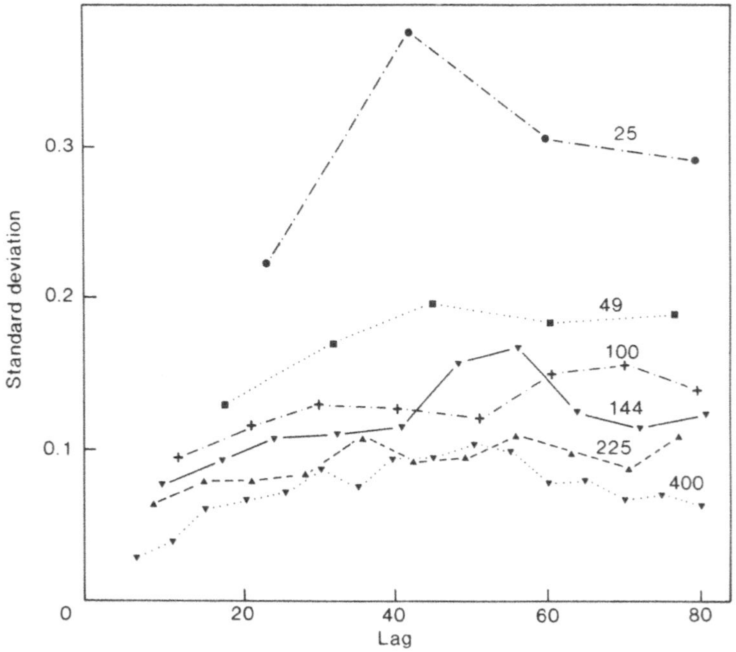

Figure 6. Standard deviations of observed semivariances against sample size for transects for across the field with the spherical variogram.

Muñoz-Pardo (1987) derived the following expression for the estimation variance of a semivariance:

$$\mathrm{var}[\hat{\gamma}_R(\mathbf{h})] = \frac{1}{2S'^2}\int_{S'}\int_{S'}\{f(\mathbf{x},\mathbf{y},\mathbf{h})\}\mathrm{d}\mathbf{x}\mathrm{d}\mathbf{y} + \frac{1}{2N'^2(\mathbf{h})}\sum_{i=1}^{N'(\mathbf{h})}\sum_{j=1}^{N'(\mathbf{h})}\{f(\mathbf{x}_i,\mathbf{x}_j,\mathbf{h})\}$$

$$-\frac{1}{N'(\mathbf{h})S'}\sum_{i=1}^{N'(\mathbf{h})}\int_{S'}\{f(\mathbf{x}_i,\mathbf{x},\mathbf{h})\}d\mathbf{x}, \quad (6)$$

where

$$f(\mathbf{x},\mathbf{y},\mathbf{h}) = \{\gamma(\mathbf{x}-\mathbf{y}+\mathbf{h})+\gamma(\mathbf{x}-\mathbf{y}-\mathbf{h})-2\gamma(\mathbf{x}-\mathbf{y})\}^2$$

for any values of i and j. In Equation (6) $\hat{\gamma}_R(\mathbf{h})$ denotes the estimated value of the regional variogram, i.e. the local variogram, at lag $\mathbf{h}$, S' is the area of intersection when the region is translated by the vector $\mathbf{h}$ and N' is the number of sampling points in the intersection, and $\mathbf{x}$ and $\mathbf{y}$ are two points that describe the region independently.

Muñoz-Pardo went on to solve the equation by numerical integration. He illustrated his approach with two examples that can reasonably be compared with ours. He assumed a square region in which some property had either a spherical or exponential variogram without nugget and with an effective range 1/2.215 times the side of the region. He sampled it with a 7×7 square grid. For the exponential case this matched almost exactly the situation we describe in our first example, and the results were very similar. For his spherical variogram, which had no nugget and approximately the same effective range (50.6), Muñoz-Pardo obtained somewhat smaller estimation variances, and this accords with our experience with that model (Webster & Oliver, 1992). The reason is that the integral range is larger—18.75 for a range of 50, and 18.97 for a range of 50.6, which is what Muñoz-Pardo used—than the 16 of the exponential model. He then derived confidence limits from the variances assuming a normal distribution of errors. His confidence intervals were a lot less than ours, however, and seem too small. His assumption of normal errors for such a small sample could be partly the cause.

Our results show that with samples of only 25 and 49 the confidence intervals are almost certainly too wide for the observed variogram to represent the realization reliably. The precision obtained with a sample of 100 might be acceptable in some circumstances, and 150 might suffice in many. Samples of 225 gave good precision, and in most applications this number would almost certainly be adequate where variation is isotropic. We believe that practitioners should aim to sample at some 150 to 200 points in a region to estimate the variogram of that region. Where the variation is anisotropic, however, larger samples are likely to be needed.

The results also show how important it is to interpret N' correctly. The unsuspecting practitioner, taking N' to be the number of paired comparisons, would have felt secure with 49 points on a square grid (84 to 240 comparisons with our program), and 100 sampling points would have seemed extravagant (180 to 774 comparisons).

As expected from Equation (5), our results show that where there is spatial dependence the shorter is the lag the narrower are the confidence intervals. This is desirable where the variogram is to be used for kriging, and the estimates at longer lags and in the sill values in particular are less important. It suggests that there is little practical value in trying to obtain equal numbers of comparisons at all lags, despite the appealing ingenuity of schemes that do that. However, if a user needs even greater precision near the origin then so be it: he may need more comparisons at lags shorter than our shortest and a more elaborate scheme than the ones we used here.

Acknowledgements

We thank Mr D. Renard for his help with the simulation while R. Webster was at the Ecole des Mines de Paris, and Mrs J. Dowling for preparing the figures.

References

CRESSIE, N. & HAWKINS, D. M. 1980. Fitting variogram models by weighted least squares. *Mathematical Geology* **17**, 563-586.

JENKINS, G. M. & WATTS, D. G. 1968. *Spectral Analysis and its Applications.* Holden-Day, Oakland, California.

JOURNEL, A. G. & HUIJBREGTS, C. J. 1978. *Mining Geostatistics.* Academic Press, London.

LANTUÉJOUL, C. 1991. Ergodicity and integral range. *Journal of Microscopy* **161**, 387-403.

MATHERON, G. 1965. *Les variables régionalisées et leur estimation.* Masson, Paris.

MATHERON, G. 1989. *Estimating and Choosing.* Springer Verlag, Berlin.

MCBRATNEY, A. B. & WEBSTER, R. 1986. Choosing functions for semivariograms of soil properties and fitting them to sampling estimates. *Journal of Soil Science* **37**, 617-639.

MUÑOZ-PARDO, J. F. 1987. *Approche géostatistique de la variablité spatiale des milieux géophysiques.* Thèse Docteur-Ingénieur, Université de Grenoble et l'Institut National Polytechnique de Grenoble.

SHAFER, J. M. & VARLJEN, M. D. 1990. Approximation of confidence limits on sample semivariograms from single realizations of spatially correlated random fields. *Water Resources Research* **26**, 1787-1802.

TAYLOR, C. C. & BURROUGH, P. A. 1986. Multiscale sources of spatial variation in soil. III Improved methods for fitting the nested model to one-dimensional semi-variograms. *Mathematical Geology* **18**, 811-821.

WEBSTER, R. & OLIVER, M. A. 1992. Confidence intervals on variograms for samples of different sizes. In *Proceedings of the 2nd Codata Conference on Geomathematics and Geostatistics,* eds P. A. Dowd and J. J. Royer. *Sciences de la Terre, Série Informatique Géologique* **31**, 11-23.

DEPTH, DIP AND GRADIENT

D. Renard (*) and P. Ruffo (**)

(*) Centre de Géostatistique
Ecole des Mines
35 rue Saint–Honoré
77305 Fontainebleau – France

(**) AGIP S.p.A.
P.O. Box 12069
20120 Milano – Italy

ABSTRACT

This article deals with the problem of depth mapping from depth and dip data. The standard geostatistical procedure consists in cokriging Z_x^* at each point Z_x using the depths Z_α and the gradients $(\nabla Z)_\beta$ known at the experimental points. This procedure is rather demanding insofar as it involves a gradient which implies that the depth formation is twice differentiable. In addition it may not be the most realistic one. The gradient (i.e. a point measure) is never physically accessible; the quantity which is measured is an average dip over a given support. This average dip can be approximated using the formula $\nabla(Z * p)$ with the result that it is possible to cokrige the depths using these average dip data.
This article sets out to compare these local and regional procedures. The influence of the support size on the mapping results will be addressed.

INTRODUCTION

When producing the (two–dimensional) map of the top of a reservoir ("surface"), we may use several types of data:

- depth values measured at the wells
- dip and azimuth values derived from Dipmeter measurements
- time surface relative to the seismic horizon closest to the geologic marker, converted into depth through a time–to–depth conversion procedure.

An affordable reconstruction of a geologic structure should use all these information together.

At present only well depths and seismic horizons are used in a routine way, because of the lack of an automated procedure allowing the use of the dipmeter results. In this paper, we will concentrate on the two first types of data. Note that, to maintain generality, we assume that these data are not necessarily defined at the same point.

Dipmeter data are measured in a well using a specific tool (figure 1) that measures the conductivity of geologic formations crossed by the well.

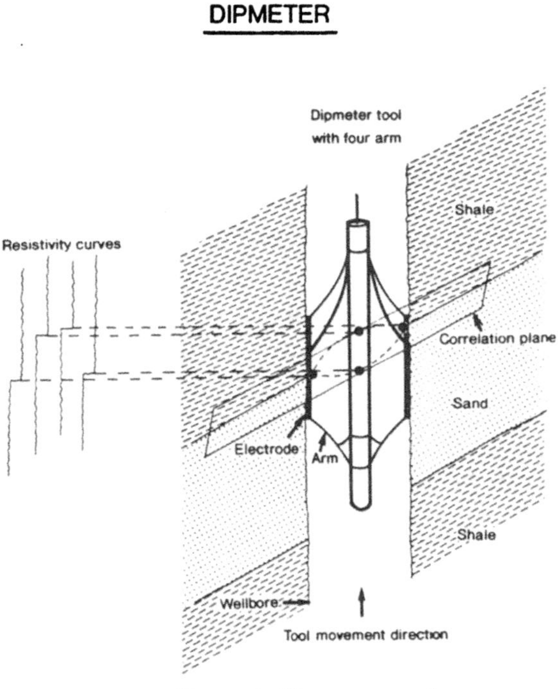

Figure 1: Dip meter tool

Through a specific processing, the resistivity curves (usually 4) are cross–correlated and values of dip and azimuth are estimated at regular depth steps (figure 2).

The precision of the Dipmeter tool is given on average as ± 0,2 degrees for the Dip and ±4 degrees for the Azimuth, while the quality of the measurement is a function of several parameters. Among these parameters, the most important are the type of mud present in the wellbore and the wellbore itself (regular or not, deviated or not). Generally the quality increases with the dip of the surface, therefore gentle dips (and their azimuths) are more uncertain.

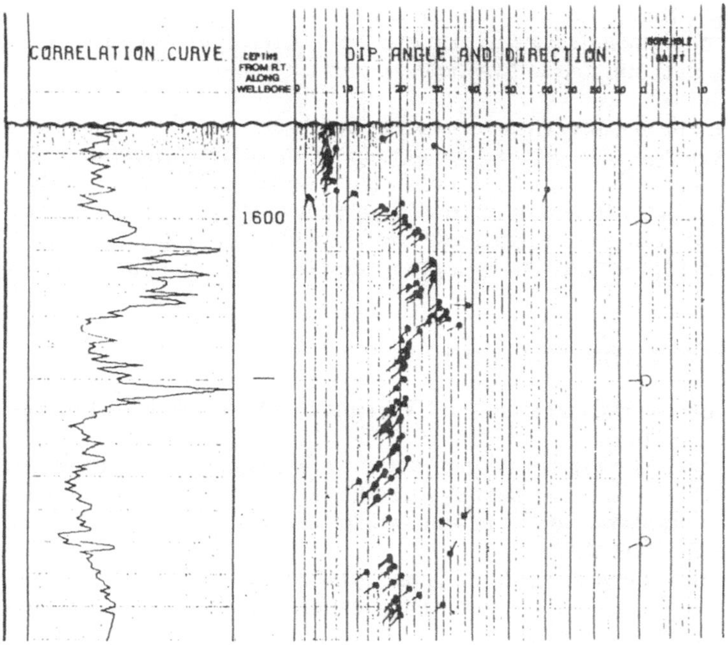

Figure 2: Dip and Azimuth measurements

The lateral extent of the support on which the surface gradient is measured is the same as that of the well depth measurement, while, strictly speaking, the *vertical* extent of this support (along the wellbore) is a function of the correlation interval.
Consequently they can be considered as punctual, compared to the corresponding quantities coming from the seismic data.
However, if the gradient measured is almost constant for several formations in a certain strip of the wellbore, this may give the gradient the meaning of *average slope* of the surface over an area centered at the well location. But this interpretation may vary from well to well.
These two hypotheses will be described together with their limitations in the structural model.

EXISTING TECHNIQUES

The problem of mixing gradients and depth data is a general concern that has been dealt with many times in literature. As the gradient and depth data are usually defined at the same location, the first intuitive solution is to replace the gradient by a set of points (duplicates: usually two points in the dip direction, one upstream, one downstream) and use all these points (true and duplicates) in a kriging procedure [1]. But when no depth data is available together with the gradients, the method fails: although this hypothesis may not be realistic in well logging, it can make sense when gradients denote structural information assigned by geologists.
Instead one could think of adding a constraint in the kriging system so that the weights attached to the duplicates sum up to zero in order to filter out the unknown depth data.

In both cases, the number of duplicates as well as their distance to the depth data are parameters that must be carefully determined. In particular one should worry when the duplicates corresponding to two close gradients overlap.

A second method consists in processing depth and gradient information in a single cokriging system which requires the simultaneous identification of the simple and cross variograms. Note that, this time, depth and gradient may not be defined at the same point. In [2], Haas suggested to use the gaussian variogram for testing the methodology and the linear variogram regularized over a segment for the cartography of a dolomite top. While in [3] the gaussian variogram was chosen for the joint study of geopotential and winds.

DEPTH AND GRADIENT COKRIGING

Let us set up the cokriging system. This part will be developed in 1−D in order to simplify the notation.
We denote by $Z$the depth variable that we will assume to be a differentiable random variable (IRF−k) characterized by its generalized covariance $K(h)$. Its gradient $G$is a IRF−(k−1) associated with the generalized covariance $-K''(h)$. This requires that $K(h)$ is twice differentiable.
The information consists of depth Z_a and gradients G_i measured at wells. If we consider the set of weights $\{\lambda^a, \nu^i\}_{Z}$, the cokriging estimate of the depth data using depth and gradient data will be written:

$$Z^* = \lambda^a Z_a + \nu^i G_i \tag{1}$$

We must first ensure that $(Z^* - Z_0)$ is an authorized linear combination of order k; that is any monomial $f^\ell(x)$of degree $\ell \leq k$is filtered out by $\{\lambda^a, \nu^i\}_Z$:

$$\lambda^a f_a^\ell + \nu^i \frac{df_i^\ell}{dx} - f_0^\ell = 0 \tag{2}$$

Note that, as $f^\ell(x) = x^\ell$, then $\frac{df^\ell(x)}{dx} = \ell x^{\ell-1} = \ell f^{\ell-1}(x)$ for $\ell > 0$, and $\frac{\partial f^0(x)}{\partial x} = 0$
Then the conditions (2)become:

$$\begin{cases} \lambda^a f_a^\ell + \nu^i \ell f_i^{\ell-1} = f_0^\ell & \forall \ell,\ 0 < \ell \leq k \\ \sum_a \lambda^a = 1 \end{cases} \tag{2'}$$

Expanding $Var(Z^* - Z_0)$and minimizing it against each $\lambda^a$and each $\nu_i$under the constraints of (2'), and introducing the Lagrange multipliersμ_ℓ, we derive the cokriging system:

$$\begin{bmatrix} K_{a\beta} & K'_{i\beta} & f_\beta^\ell \\ K'_{aj} & -K''_{ij} & \ell f_j^{\ell-1} \\ f_a^\ell & \ell f_i^{\ell-1} & 0 \end{bmatrix} \begin{bmatrix} \lambda^a \\ \nu_i \\ \mu_\ell \end{bmatrix} = \begin{pmatrix} K_{0\beta} \\ K_{0j} \\ f_0^\ell \end{pmatrix}, \begin{pmatrix} K'_{0\beta} \\ -K''_{0j} \\ \ell f_0^{\ell-1} \end{pmatrix} \tag{3}$$

The first right−hand side is used for depth estimation while the second is for gradient estimation.
$K(h)$and $K''(h)$are symmetrical, whereas $K'(h)$is anti−symmetrical; thus the global matrix is symmetrical. It is regular assuming that the $f^\ell(x)$functions are algebraically

independent. In addition, concerning the first basic drift monomial function, if no depth data are available, then the corresponding column of the cokriging system is uniformly zero and the matrix is singular. This is common sense as it is impossible to determine fully a function, knowing only its derivative, due to the indeterminacy of the integration constants.
The first right−hand side gives the weights $\{\lambda^a, \nu^i\}_Z$ for the depth and the corresponding variance:

$$\sigma_Z^2 = K(0) - \lambda^\beta K_{0\beta} - \nu^j K'_{0j} - \mu_\ell f_0^\ell$$

Similarly, replacing the first right−hand side by the second one, gives the weights $\{\lambda^a, \nu^i\}_G$ for the gradient estimation and the corresponding variance:

$$\sigma_G^2 = -K''(0) - \lambda^\beta K'_{0\beta} + \nu^j K''_{0j} - \mu_\ell \ell f_0^{\ell-1}$$

The limitation of the depth and gradient cokriging comes from the choice of the structural model which requires a structure at least twice differentiable, such as the gaussian or the cubic variograms, the h^3 or the h^5generalized covariances.

DIP INFORMATION

The previous constraints may be very restrictive. Moreover, it may be interesting to consider the dip meter information as the average of the surface measured on a given support. The dip is then the derivative of the depth variable convoluted by a weighting function. To ensure derivability, it seems advisable to use the gaussian weighting function:

$$G^\xi = \frac{d}{d\xi}(Z * p) \text{ with } p(u,v) = \frac{1}{2\pi ab}\exp\left(-\frac{u^2}{2a^2} - \frac{v^2}{2b^2}\right)$$

where a and b characterize the dimensions of the support along its main axes. These two parameters add flexibility to the model.
The cokriging system (3)requires the specification of quantities such as:

$$Cov\{Z_a, G_i^\xi\} = \iint Cov\{Z_a, Z_{i-(u,v)}\} \frac{\partial}{\partial\xi} p(u,v)dudv$$

$$Cov\{G_i^\xi, G_j^\eta\} = \iint Cov\{Z_i, Z_{j-(u,v)}\} \frac{\partial^2}{\partial\xi\,\partial\eta} P(u,v)dudv$$

where ξ and η stand for x or y, and where $P = p * p$

STRUCTURAL IDENTIFICATION

Let us first recall that if the depth variable is a IRF−k characterized by its generalized covariance $K(h)$, the gradient components are IRF−(k−1) with the generalized covariances $\partial^2K/\partial x^2$ (respectively $\partial^2K/\partial y^2$). For the sake of simplicity, we will use the generic term *variogram* to designate covariances, variograms or generalized covariances. We will designate by γ_Z the variogram for the depth and by γ_{G_X} (resp. γ_{G_Y}) the variogram of the gradient component along X (resp. Y).
The identification of the variograms must be performed on a multivariate basis, considering simultaneously the simple and cross variograms of the depth and the gradient components.
Let us assume that the depth variable is isotropic, then the variogram of the gradient components are necessarily anisotropic: in the direction β, the shape of the variogram

component calculated in the direction α depends on the angle $\beta - \alpha$. We must therefore compare the variogram of the X gradient component calculated along X with the variogram of the Y gradient component calculated along Y. Moreover if the depth variogram is bounded, the gradient variograms should reach the same sill in all directions.

As an example, we will assume that γ_Z is a gaussian isotropic variogram (range $d=33$; sill $C=1$). The variogram γ_{G_X} shows a hole effect when calculated along the X direction (0 degree) and presents a gaussian shape when calculated along the Y direction (90 degrees).

Figure 3a shows the shape of γ_{G_X} calculated for angles varying regularly from 0 to 90 degrees, by step of 10 degrees. Figure 3b shows γ_{G_X} for the same angles when γ_Z is a cubic variogram ($d=100$).

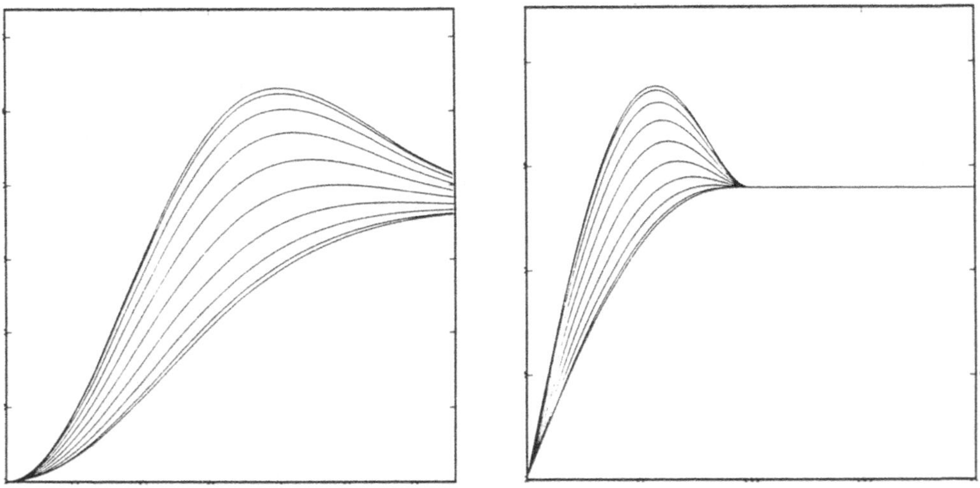

Figure 3: (a) Gaussian variogram – (b) Cubic variogram

The next result concerns the gradient considered as the derivative of the regularized variable. On each figure, we represent γ_{G_X} calculated for different values of the regularization extension ($a=b=1,5,10,20,30,40$) and the punctual gradient (as in figure 3) for two angles: 0 degree in figure 4 and 90 degrees in figure 5.

We can check that the punctual gradient corresponds to the limit of the derivative of the regularized variable when the support size tends towards zero.

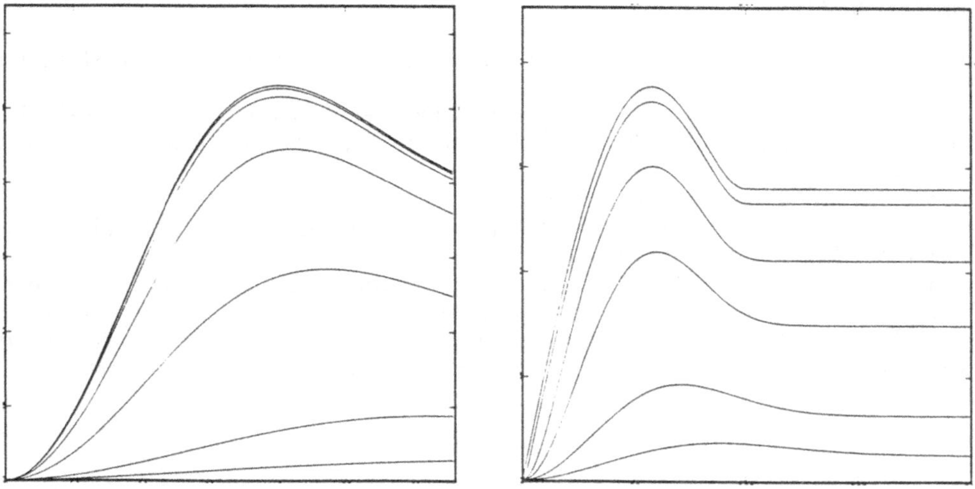

Figure 4: (a) Gaussian variogram – (b) Cubic variogram

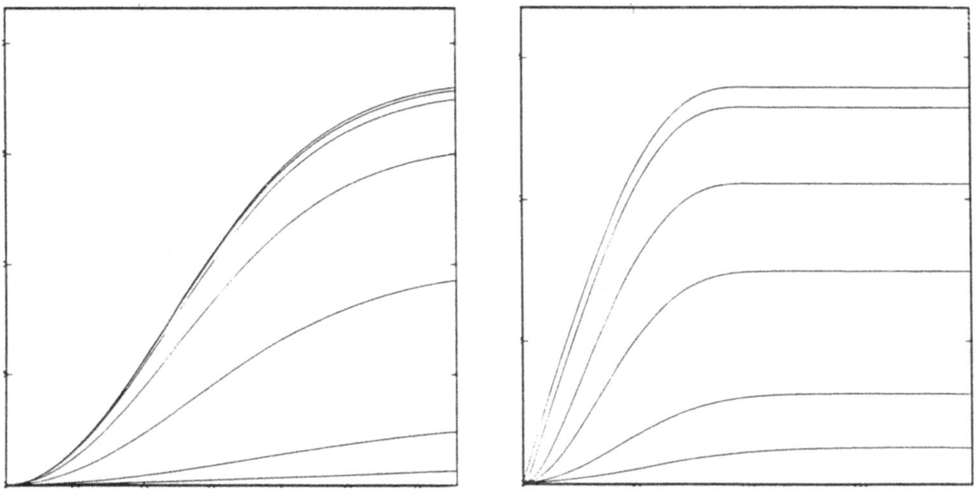

Figure 5: (a) Gaussian variogram – (b) Cubic variogram

PRACTICAL CASE

The data set corresponds to a reservoir constituted by a sequence of finely stratified sand and clay layers.
These sediments reveal a turbiditic origin and have been deposited in a foredeep basin. Due to compression a well defined anticline structure was developed with orientation NW–SE.
The zone of interest covers a 5km by 6km rectangle.
Several levels were interpreted from seismic data. Among these the one that has been chosen for this example represents the top of a gas producing sand layer.
Figure 6 shows the corresponding depth map, obtained from the seismic time map through time–to–depth conversion, and subsequently tied to the 19 wells that have reached this top. Thirteen of them have dipmeter information that has been converted into slope components along X and Y. These are posted with arrows proportional to the gradient modulus (figure 7)

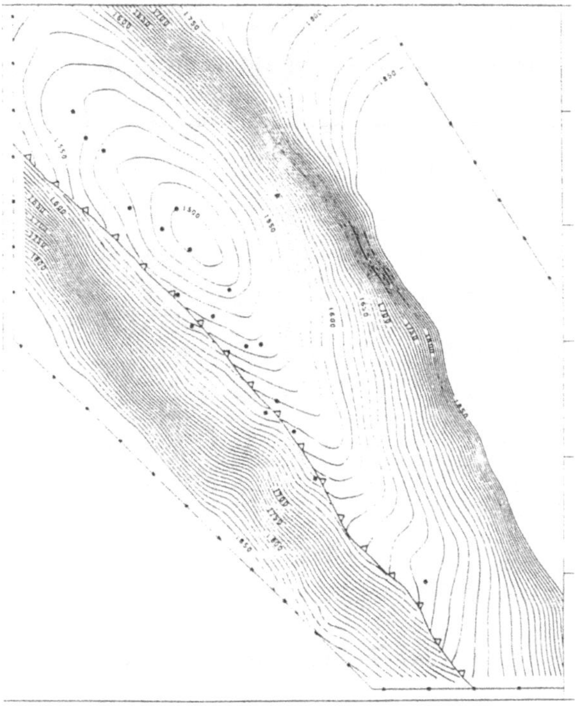

Figure 6: Seismic depth map

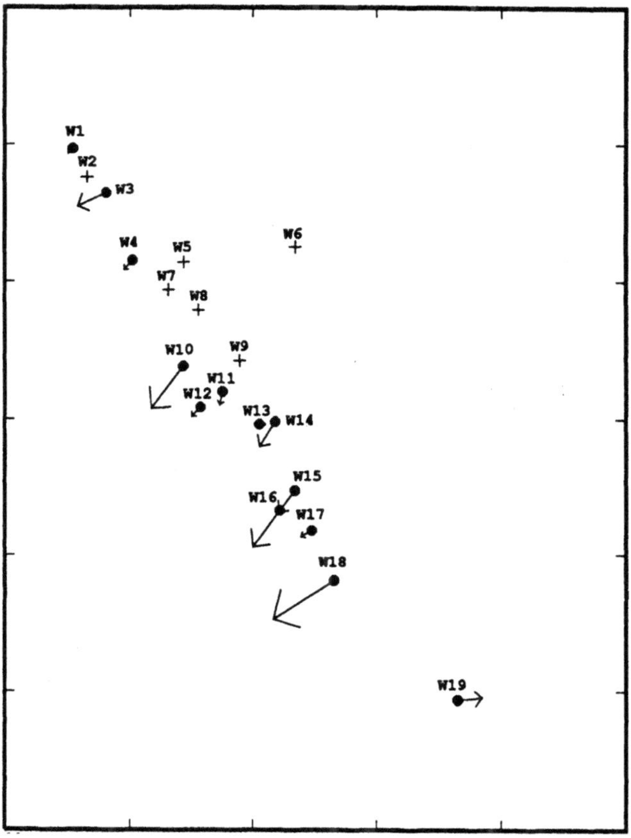

Figure 7: Well and gradient location
● for gradient and depth information
+ for depth information only

Structural Identification

As most of the wells mainly belong to a narrow band lying NW–SE, the structure inference is not easy. The stationarity along the transverse direction must be assumed. The variograms can be calculated only in the longitudinal direction.
Moreover the points located on the top of the structure do not provide gradient information as the measure was not performed.

Kriging without gradients

The first map is calculated using only the depth information. The top of the structure stretches from NW to SE, following the preferential sampling pattern, with two marked flanks in the northern part. The slope then vanishes in the southern part. Finally the well W19 is the center of a local hollow.
This map (figure 8) does not reflect the elongated structure of the seismic depth map that lies in the NW–SE orientation, gently dipping towards the south. In particular the abrupt western edge does not show up. We also see that the map does not honor the largest gradient values around wells W15, W16 and W18.

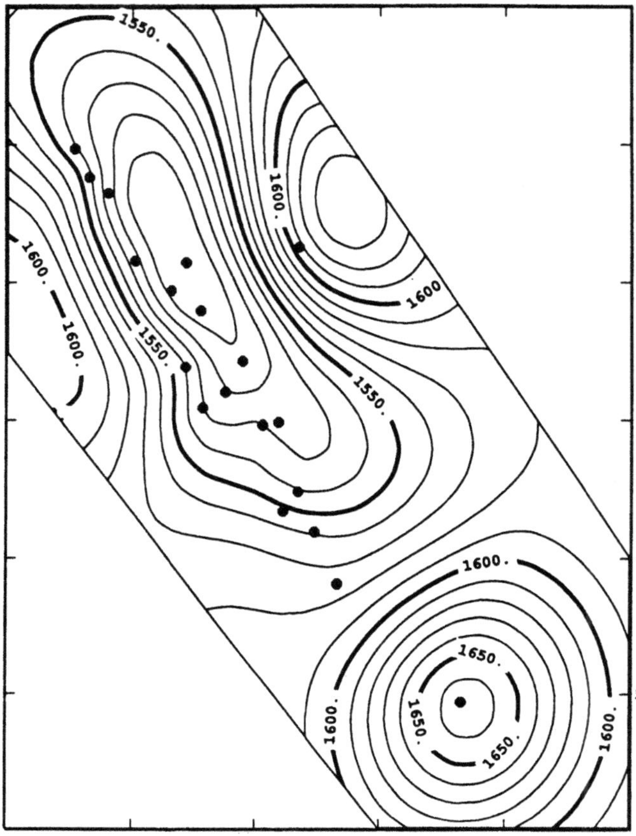

Figure 8: Kriging depth without gradient

Kriging with gradients

Due to the cubic (twice differentiable) variogram, we can use both depth and gradient information through a cokriging procedure. Here the gradient is considered as the derivative of the surface.

Figure 9 presents an elongated structure with two tops around 1500m, separated by a narrow saddle. The main difference with the previous map concerns the evidence of the presence of a fault (the one shown in the seismic map) corresponding to the slope in the south–west part of the structure.

The same slope in the north–east part cannot be interpreted in the same way due to the lack of data in that area.

Besides the kriging with gradient shows that the structure is more elongated towards south–east, as in the seismic, and that therefore a strong slope is created between wells W18 and W19.

Finally we note that the well W19 is still located on a local hollow, which does not correspond to the depth seismic map. But we can argue that W19 is about 90m deeper than W18, that it is isolated and that it presents a gradient oriented in the inverse direction. As a matter of fact, from the seismic map, we understand that W19 is separated from W18 by the fault. Moreover noticing that this fault is a reverse one, we understand that, using only the well data available, the result of the figure 9 is plausible.

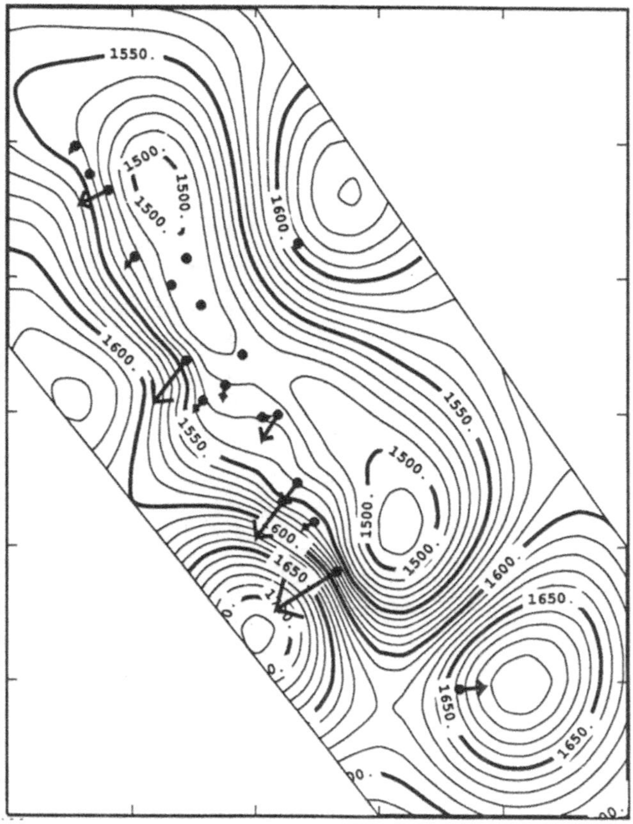

Figure 9: Kriging depth with gradient

Kriging with dip

The map presented in figure 10 results from considering the gradient information as regional dip values. As mentioned before, there is no way to determine the extension parameters for the regularization function with the current data set. We have arbitrarily chosen these: $a=b=50m$ (to be compared with the range of the depth variogram $d=1400$).

The structure still presents its two tops. The northern one remains almost unchanged. On the other hand, the southern top is strongly modified. Its highest point has moved northwards, and is now located on the continuation of the gradient of wells W15 and W16. In the opposite direction, the hole produced is 50m deeper than before, reinforcing the slope of the western edge of the structure. We should also notice the slight modification around wells W11, W12, W13 and W14 where the gradient values are constantly small but they vary in direction.

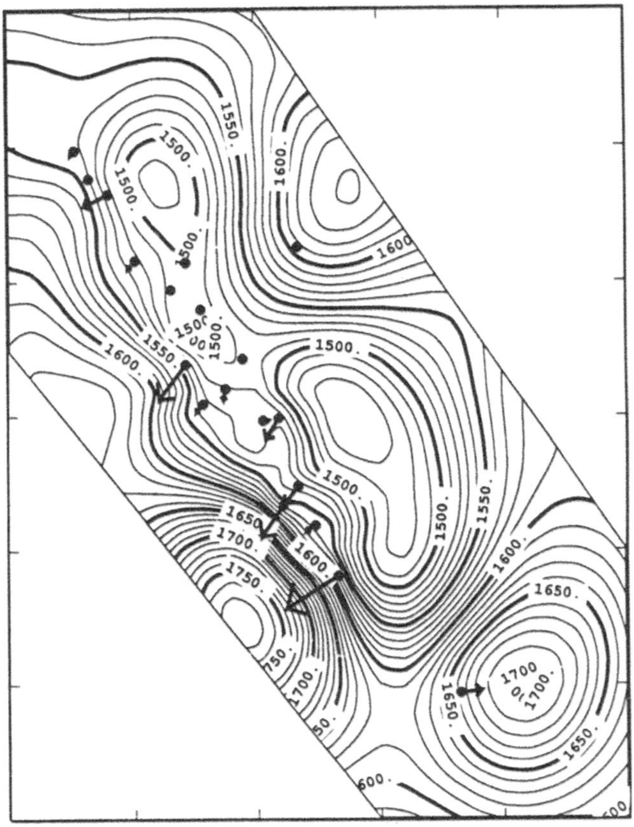

Figure 10: Kriging depth with dip

CONCLUSION

The drawback of any cokriging method is the need to derive a multivariate structure. In this particular case, this is made easier as depth and gradients are linked through formal equations, but it requires the depth variogram to be twice differentiable. Instead, considering the gradient values as a measurement of regional dip allows us to use any type of variogram and offers an additional flexibility through the choice of the regularization extension.

The use of gradient measurements, when available and valuable, together with the depth data, is certainly a step towards the use and integration of all available information in the reconstruction of a depth structure. The results are encouraging and at the same time shows clearly that we will have to include the seismic information.

ACKNOWLEDGEMENTS

The authors would like to thank AGIP S.p.A. for permission to present these data.

REFERENCES

[1] Y. SHIYI. Essai d'utilisation du gradient. Centre de Géostatistique. June 1983
[2] A. HAAS, Ch. LANTUEJOUL M. de VATHAIRE Applications du Cokrigeage. Mathématiques Appliquées. May 1972
[3] J.P. CHILES, P. CHAUVET, P. DELFINER. Analyse des champs météorologiques par Krigeage. Centre de Géostatistique. July 1975.

SIMULATION OF RANDOM FUNCTIONS ON LARGE LATTICES

Henning Omre, Knut Sølna and Håkon Tjelmeland

Norwegian Computing Center
P.O.Box 114, Blindern
N-0314 Oslo
Norway

Abstract. Random functions are in frequent use in applications of spatial statistics. Gaussian and Gaussian intrinsic random functions are differentiated, and the screening sequential algorithm for generating realizations from them are defined. The algorithm is based on the general sequential algorithm and Markov properties for random functions. For exponential and linear variogram functions the algorithm is shown to be exact for the one-dimensional case, while empirical evaluations show that it is highly reliable also in two-dimensional cases. For fractal random functions, the screening sequential algorithm is significally more reliable than the frequently used random midpoint displacement and successive random addition algorithms. The processing requirements for the algorithm is independent of the actual variogram function and linear in number of lattice nodes — both favorable characteristics.

1 Introduction

Stochastic models are in frequent use for describing physical phenomena. In applications like geology, meteorology and pollution monitoring, spatial or spatial/temporal models must be used. Gaussian and fractal random functions have proven to be useful models. The questions being posed concerning these phenomena are usually so complex that one has to rely on stochastic simulation techniques for addressing them. The representation of these phenomena are normally on lattices in higher dimensions, lattices with up to 10^8 nodes. Hence one has to be able to generate reliable realizations of Gaussian and fractal random functions on extremely large lattices. This article addresses this problem.

The size of the lattice, and hence the dimension of the multivariate equivalent, excludes the straight forward approach of simulation from the corresponding multi-Gaussian distribution. Several other procedures for simulation have been suggested. Important references are: Matern (1963), Journel (1974), Borgman et al. (1984), Barnsley at al. (1988), Journel and Alabert (1988).

The current article concerns simulation from a stochastic model without conditioning on observations in particular locations, since the objective is to evaluate the reliability of the algorithm for reproducing a given spatial dependence structure. Moreover, the comparison with the familiar simulation algorithms for fractal random functions should be done unconditionally since the latter is defined for this case only. Extensions to the conditional case can be done by sacrificing some processing efficiency. In applications with few observations, like petroleum reservoir description for example, a two stage approach to conditioning as defined in Journel and Huijbregts (1978) can be very efficient also.

In this article the screening sequential algorithm for simulating random functions is defined and evaluated. It is based on the general sequential algorithm and Markov properties of random functions, see Ripley (1987) and Adler (1981). In spatial dimension two and higher, one has to rely on approximative exact algorithms to obtain sufficient processing efficiency. Accepting an approximation makes it even more complicated and important to: provide an intuitive justification; evaluate the properties empirically; and to constrain the class of models for which the approximation is valid. This seems often to be forgotten.

The authors do not claim that the spatially defined screening sequential algorithm is completely new, although no published work defining exactly the same algorithm has been found. Algorithms of the same flavor are numerous, however, and it is not wise to claim novelty in an area where so much trial and error has been made. The objective of the article is, however, to present a thorough justification for an algorithm we have found very reliable and processing efficient, for a relatively large class of spatial dependence structures. The class includes fractal random functions and a comparison with familiar algorithms for simulating them are done.

2 Model definition and properties

Consider a random function,

$$\{\mathcal{Z}(x)\;; x \in \mathcal{D}\}$$

with $\mathcal{Z}(\cdot)$ being a random variable with realization $z(\cdot) \in \mathcal{R}^1$; and x being a reference location in the bounded domain $D \subset \mathcal{R}^m$, with $m = 1, 2, 3$ in this study.

2.1 Model classes

Two classes of random variables will be considered:

- Gaussian random functions - defined by $\sum_{i=1}^{n} \eta_i \mathcal{Z}(x_i)$ being Gaussian for all $\{\eta_1, \eta_2, \ldots, \eta_n\}$; all location configurations $(x_1, x_2, \ldots x_n) \in \mathcal{D}$; and all integer values of n. The special case $\sum_{i=1}^{n} \eta_i = 0$ is included since constants are considered as degenerate cases of Gaussian distributions, see Mardia et al.(1979). The

Gaussian random function being considered here are stationary, homogeneous and isotropic, with,

$$\begin{aligned} E\{\mathcal{Z}(x)\} &= 0.0 \\ \mathrm{Var}\{\mathcal{Z}(x)\} &= \sigma^2 \\ \mathrm{Cov}\{\mathcal{Z}(x'), \mathcal{Z}(x'')\} &= \sigma^2 \rho(|x' - x''|) \\ \mathrm{Var}\{\mathcal{Z}(x') - \mathcal{Z}(x'')\} &= 2\sigma^2 \gamma_0(|x' - x''|) \\ &= 2\sigma^2[1 - \rho(|x' - x''|)] \end{aligned}$$

and $\rho(\cdot)$ being the correlation function; and $\gamma(\cdot)$ being the variogram function. A valid model require the correlation function, $\rho(\cdot)$, to be non-negative definite ensuring non-negative variances for all linear combinations of random variables. Consequently, $\rho(\Delta)$ has to be bounded and symmetrical around $\Delta = 0$.

- Gaussian intrinsic random functions - defined by $\sum_{i=1}^{n} \eta_i \mathcal{Z}(x_i)$ being Gaussian for $\sum_{i=1}^{n} \eta_i = 0$; for all location configurations $(x_1, x_2, \ldots, x_n) \in \mathcal{D}$; and all integer values of n. This corresponds to the IRF-0 definition in Delfiner (1979) extended by assuming Gaussianity. The Gaussian intrinsic class is larger than, and includes, the Gaussian class, since the former requires Gaussianity for a smaller set of weights $\{\eta_1, \eta_2, \ldots, \eta_n\}$ which is included in the requirement for the latter. In particular, the univariate characteristics need not be Gaussian. In this study, the term Gaussian intrinsic random function will be used for functions in the intrinsic class not being Gaussian random functions.

  The Gaussian intrinsic random functions being considered here are stationary, homogeneous and isotropic, with:

$$\begin{aligned} E\{\mathcal{Z}(x)\} &= 0.0 \\ \mathrm{Var}\{\mathcal{Z}(x') - \mathcal{Z}(x'')\} &= 2 \cdot \gamma(|x' - x''|) \end{aligned}$$

  A valid model require the negative variogram function, $-\gamma(\cdot)$, to be conditional non-negative definite, ensuring non-negative variances for all linear combinations of random variables with weights adding to zero. A consequence of this is that $\gamma(\cdot)$ need not be bounded and hence the univariate variance need not be finite.

2.2 Model characteristics

The following characteristics are relevant for the study:

- Conditional random variables - defined for an unobserved arbitrary location $x_0 \in \mathcal{D}$ given observations in n arbitrary locations $(x_1, x_2, \ldots x_n) \in \mathcal{D}$. For both the Gaussian and Gaussian intrinsic random functions, the conditional distributions will be Gaussian,

$$\mathcal{Z}(x_0)|\mathcal{Z}(x_1) = z_1, \ldots, \mathcal{Z}(x_n) = z_n \rightsquigarrow \mathcal{N}(\mu_{0|1,\ldots,n}, \sigma^2_{0|1,\ldots,n})$$

with $\mathcal{N}(\mu, \sigma^2)$ indicating a Gaussian distribution with:

$$\begin{aligned}\mu_{0|1,\ldots,n} &= E\{\mathcal{Z}(x_0)|\mathcal{Z}(x_1) = z_1, \ldots, \mathcal{Z}(x_n) = z_n\} \\ \sigma^2_{0|1,\ldots,n} &= \text{Var}\{\mathcal{Z}(x_0)|\mathcal{Z}(x_1) = z_1, \ldots, \mathcal{Z}(x_n) = z_n\}\end{aligned}$$

The parameters, $(\mu_., \sigma_.)$, are determined differently under the two models, however. For Gaussian random functions the parameters are:

$$\begin{aligned}\mu_{0|1,\ldots,n} &= \underline{\psi}_{0n}^T \Psi^{-1} \underline{z} \\ \sigma^2_{0|1,\ldots,n} &= \sigma^2 - \underline{\psi}_{0n}^T \Psi^{-1} \underline{\psi}_{0n}\end{aligned}$$

with $\underline{\psi}_{0n}$ being $1 \times n$ -dimensional with elements $\sigma^2\rho(|x_0 - x_i|); i = 1, \ldots, n$; Ψ being $n \times n$-dimensional with elements $\sigma^2\rho(|x_i - x_j|); i, j = 1, \ldots, n$; and $\underline{z}$ being $1 \times n$-dimensional with elements $z_i; i = 1, \ldots, n$. This corresponds to the simple kriging predictor and prediction variance, see Journel and Huijbregts (1978). For Gaussian intrinsic random functions the parameters are:

$$\begin{aligned}\mu_{0|1,\ldots,n} &= \underline{\gamma}_{0n}^T \Gamma^{-1} \underline{z}^+ \\ \sigma^2_{0|1,\ldots,n} &= \underline{\gamma}_{0n}^T \Gamma^{-1} \underline{\gamma}_{0n}\end{aligned}$$

with $\underline{\gamma}_{0n}$ being $1 \times (n+1)$ -dimensional with elements $-\gamma(|x_0 - x_i|); i = 1, \ldots, n$ plus 1 in last row; Γ^{-1} being $(n+1) \times (n+1)$-dimensional with elements $-\gamma(|x_i - x_j|); i, j = 1, \ldots, n$ plus 1's in last column and row except for element $((n+1), (n+1))$ being equal to 0 ; and $\underline{z}^+$ being $1 \times (n+1)$ - dimensional with elements $z_i; i = 1, \ldots, n$ plus 0 in last row. This corresponds to the ordinary kriging predictor and prediction variance, see Journel and Huijbregts (1978).

- General simulation algorithm - defined on a lattice of nodes covering $\mathcal{D}$. Denote the set of lattice node locations $\mathcal{L}_D$, the number of nodes $\mathcal{N}_D = \#\mathcal{L}_D$; and hence the countable set of random variables representing the random function:

$$\{\mathcal{Z}(x);\ x \in \mathcal{L}_D\}$$

This can be seen as a multivariate random variable. Consider an arbitrary ordering of the random variables in the lattice, $\{\mathcal{Z}_j; j = 1, \ldots, \mathcal{N}_D\}$. The following relation follows from the definition of conditional probability:

$$\begin{aligned}&\text{Prob}\{\mathcal{Z}_1 = z_1, \mathcal{Z}_2 = z_2, \ldots, \mathcal{Z}_{\mathcal{N}_\mathcal{D}} = z_{\mathcal{N}_\mathcal{D}}\} = \\ &\quad \text{Prob}\{\mathcal{Z}_{\mathcal{N}_\mathcal{D}} = z_{\mathcal{N}_\mathcal{D}} | \mathcal{Z}_1 = z_1, \mathcal{Z}_2 = z_2, \ldots, \mathcal{Z}_{\mathcal{N}_\mathcal{D}-1} = z_{\mathcal{N}_\mathcal{D}-1}\} \\ &\quad \times \text{Prob}\{\mathcal{Z}_{\mathcal{N}_\mathcal{D}-1} = z_{\mathcal{N}_\mathcal{D}-1} | \mathcal{Z}_1 = z_1, \mathcal{Z}_2 = z_2, \ldots, \mathcal{Z}_{\mathcal{N}_\mathcal{D}-2} = z_{\mathcal{N}_\mathcal{D}-2}\} \\ &\quad \times \ldots \times \text{Prob}\{\mathcal{Z}_2 = z_2 | \mathcal{Z}_1 = z_1\} \times \text{Prob}\{\mathcal{Z}_1 = z_1\}\end{aligned}$$

The sequential simulation algorithm under a multi-Gaussian model,

$$\begin{aligned} \mathcal{Z}_1 &\rightsquigarrow \mathcal{N}(0.0, \sigma^2) \\ \mathcal{Z}_2 | \mathcal{Z}_1 = z_1 &\rightsquigarrow \mathcal{N}(\mu_{2|1}, \sigma^2_{2|1}) \\ &\vdots \\ \mathcal{Z}_{\mathcal{N}_D} | \mathcal{Z}_{\mathcal{N}_D - 1} = z_{\mathcal{N}_D-1}, \ldots, \mathcal{Z}_1 = z_1 &\rightsquigarrow \mathcal{N}(\mu_{\mathcal{N}_D|\mathcal{N}_D-1,\ldots,1}, \sigma^2_{\mathcal{N}_D|\mathcal{N}_D-1,\ldots,1}); \end{aligned}$$

is an immediately consequence of the decomposition above, and it is frequently used for stochastic simulation, see Ripley (1987). This entails generating z_1 from the marginal distribution; z_2 from the conditional distribution given the value obtained for z_1; etc. For generating z_i, a matrix of dimension $(i-1)\times(i-1)$must be decomposed, and the algorithm may be extremely resource requiring if $\mathcal{N}_D$ is large.

For Gaussian random functions the sequential simulation algorithm, as described above, can be used directly, since the marginal distribution is Gaussian, and so are all conditional distributions. The parameter values to be used are defined in the previous sub-section.

For Gaussian intrinsic random functions the picture is slightly more complicated, since the marginal distribution is not necessarily Gaussian. Assume that the value z_1 is generated by a suitable probability distribution centered at 0.0. Then all conditional distributions will be Gaussian and the sequential algorithm, as described above, can be used. The parameter values to be used are defined in the previous sub-section. The fact that realizations of Gaussian intrinsic random functions based on the sequential, or any other, algorithm can be said to be conditional to the value of z_1 obtained, will be ignored in the following discussion. Note also that whenever one or more observations are available, the problem will disappear. The objective of this sub-section is to define a completely general simulation algorithm, and the sequential algorithm, with the reservations made for the Gaussian intrinsic random functions, provide this.

2.3 Special cases

The two classes of random functions, Gaussian and Gaussian intrinsic, contain several sub-classes with particular characteristics. A presentation and discussion of some of the sub-classes relevant for this study follows:

- Markov property - appears as very different for the one-dimensional case, $x \in \mathcal{D} \subset \mathcal{R}^1$, and higher dimensional cases, $x \in \mathcal{D} \subset \mathcal{R}^m$ with $m = 2,3$. This is caused by the lack of ordering in the reference space in higher dimensions.

  In one dimension with $\mathcal{D} \subset \mathcal{R}^1$, define an unobserved arbitrary location $x_0 \in \mathcal{D}$ and observations in n arbitrary locations $(x_1, x_2, \ldots, x_n) \in \mathcal{D}$. Assume, without loss of generality, that the set of locations is ordered increasingly and that $x_i < x_o < x_{i+1}$. The one-dimensional Markov property is defined by:

$$\begin{aligned}&\text{Prob}\{\mathcal{Z}(x_0) = z|\mathcal{Z}(x_1) = z_1, \mathcal{Z}(x_2) = z_2, \ldots, \mathcal{Z}(x_n) = z_n\}\\&\quad = \text{Prob}\{\mathcal{Z}(x_0) = z|\mathcal{Z}(x_i) = z_i, \mathcal{Z}(x_{i+1}) = z_{i+1}\}\end{aligned}$$

Hence the conditioning is only dependent on the closest observations on each side of location x_0, and they are screening the influence of the other observations. For Gaussian and Gaussian intrinsic random functions, it is sufficient that;

$$\begin{aligned}&E\{\mathcal{Z}(x_0)|\mathcal{Z}(x_1) = z_1, \mathcal{Z}(x_2) = z_2, \ldots, \mathcal{Z}(x_n) = z_n\}\\&\quad = E\{\mathcal{Z}(x_0)|\mathcal{Z}(x_i) = z_i, \mathcal{Z}(x_{i+1}) = z_{i+1}\}\\&\text{Var}\{\mathcal{Z}(x_0)|\mathcal{Z}(x_1) = z_1, \mathcal{Z}(x_2) = z_2, \ldots, \mathcal{Z}(x_n) = z_n\}\\&\quad = \text{Var}\{\mathcal{Z}(x_0)|\mathcal{Z}(x_i) = z_i, \mathcal{Z}(x_{i+1}) = z_{i+1}\}\end{aligned}$$

since the conditional distributions are dependent on the two first moments only. For Gaussian random functions defined on $D \subset \mathcal{R}^1$ of arbitrary size, the Markov property is present for the exponential correlation function only, see Adler (1981). The exponential correlation function is:

$$\rho(|x' - x''|) = \exp\{-\alpha|x' - x''|\} \ ; \ \alpha \geq 0$$

with α being a parameter related to extent of spatial correlation. In Adler (1981) a couple of other correlation functions with Markov property are also listed, but for them the size of $\mathcal{D}$ is linked to the extent of spatial correlation. This is considered to be too restrictive for our study. For Gaussian intrinsic random functions defined on $D \subset \mathcal{R}^1$ of arbitrary size, the Markov property is present for the linear variogram function, see Matheron (1971). The linear variogram function is;

$$\gamma(|x' - x''|) = \tau|x' - x''| \ ; \ \tau \geq 0$$

with τ being a scale parameter. Note that this variogram function is unbounded and that it is closely related to Brownian motion in one dimension.

In higher dimensions with $\mathcal{D} \subset \mathcal{R}^m$ with $m = 2, 3$, the Markov property is not uniquely defined due to the lack of ordering in the reference space $\mathcal{D}$. A natural extension of the one-dimensional definition would, however, be: The conditional probability for $\mathcal{Z}(x_0)$ given the observations on a closed surface surrounding x_0 and several other observations outside this surface, is identical to the conditional probability given the observations on the surface only. A definition along these lines will be used here, although it is recognized that the conditioning on observations on a closed surface will not be possible whenever the representation is on a lattice $\mathcal{L}_D$. In higher dimensions approximations must be used anyhow. For Gaussian random functions defined on $D \subset \mathcal{R}^m$ with $m = 2, 3$, the Markov definition above yields only trivial results entailing no stochasticity, see Adler (1981). If, however, conditioning is done on both the value and on the derivative perpendicular to the closed surface, a meaningful Markov definition, named pseudo-Markov, can be obtained, see Adler (1981). Correlation functions having Markov properties for $D \subset \mathcal{R}^1$ tends to have pseudo-Markovian properties in higher dimensions also. For Gaussian intrinsic random functions on $\mathcal{D} \subset \mathcal{R}^2$ it can be shown that the de Wijsian variogram function, $\gamma(\Delta) = \log(|\Delta|)$, would yield Markov properties as defined here, see Matheron (1971). The fact that this variogram function is not a valid variogram function for point support causes problems, and is the reason for not evaluating it further here. For $\mathcal{D} \subset \mathcal{R}^3$ non-valid variogram functions having Markov properties may also be identified, but these are ignored here for the same reason.

- Self affine, or fractal, random functions - defined through the scaling characteristics of the random function. According to Feder (1988) the definition of self affine processes is,

$$\mathcal{Z}(cx) = c^\beta \mathcal{Z}(x) \ ; \ 0 < \beta < 1$$

with c being the enforced scaling factor; and β being a constant for all c. Contrary to the definition of self similar processes, the variability of self affine processes will change with scale. The change should be according to the model described above. For Gaussian random functions the self affine property cannot exist, since the correlation function is bounded, and hence the scaling cannot be arbitrary large. For Gaussian intrinsic random functions it is obvious that the variogram function,

$$\gamma(|x' - x''|) = \tau |x' - x''|^{2\beta} \ ; \ 0 < \beta < 1$$

defines a self affine process. If the parameter β is defined to be the Hurst exponent, H, the corresponding random function is often termed fractal random function, see Feder (1988). The self affine property holds for $D \subset \mathcal{R}^m$ with $m = 1, 2, 3$. Note that for $D \subset \mathcal{R}^1$ with $\beta = 1/2$ the special case of Brownian motion having Markov properties appear.

3 Screening sequential simulation algorithm

The random function $\{\mathcal{Z}(x); x \in \mathcal{D}\}$ will be represented by random variables on a lattice covering $\mathcal{D}, \{\mathcal{Z}(x); x \in \mathcal{L}_D\}$, with $\mathcal{L}_D$ being the set of locations of lattice nodes; and $\mathcal{N}_D = \#\mathcal{L}_D$ being the number of nodes. The algorithm will be defined in an unconditional setting given the spatial correlation or variogram function, hence without available observations.

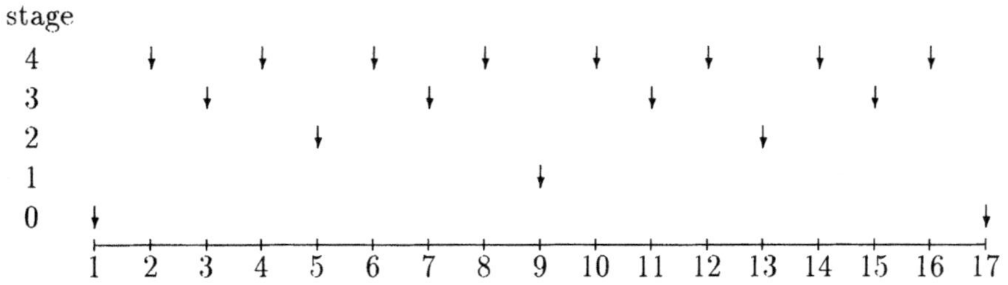

Figure 1: Sketch of halfing procedure used in the simulation algorithm

The justification of the algorithm appears from the sequential simulation algorithm and the Markov properties for $D \subset \mathcal{R}^1$, both defined in the previous section. Consider a Markovian Gaussian or Gaussian intrinsic random function for $D \subset \mathcal{R}^1$, and denote the set of random variables on the lattice $\{\mathcal{Z}_i; i = 1, \ldots, \mathcal{N}_D\}$. Without loss of generality, assume that the set is ordered increasingly in location on $D \subset \mathcal{R}^1$ and that $\mathcal{N}_D = 2^n + 1$ with n being an arbitrary integer. It follows immediately from the assumptions above and the discussion in the previous section that the following holds and fills the lattice;

$$\begin{aligned} \mathcal{Z}_1 &\rightsquigarrow \mathcal{N}(0, \sigma^2) \\ \mathcal{Z}_{\mathcal{N}_D} &\rightsquigarrow \mathcal{N}(\mu_{\mathcal{N}_D|1}, \sigma^2_{\mathcal{N}_D|1}) \end{aligned}$$

$$\mathcal{Z}_{(2k+1)2^{n-i}+1} | \mathcal{Z}_{k2^{n-i+1}+1} = z_{\cdot}, \mathcal{Z}_{(k+1)2^{n-i+1}+1} = z_{\cdot} \rightsquigarrow \mathcal{N}(\mu_{\cdot|\cdot,\cdot}, \sigma^2_{\cdot|\cdot,\cdot})$$
$$\text{for } k = 0, \ldots, 2^{i-1} - 1 \text{ and } i = 1, \ldots, r$$

with the dot sub-scripts being obvious from the context. The recursive expression above defines a halfing procedure as sketched in Figure 1. Firstly, at the initiation stage, the two end values are generated by the sequential algorithm. At stage one, the value half way between is generated by conditioning on the end values. At stage two, values half way between the assigned values are generated conditioned on the two closest values. At stage three, another halfing takes place and so on. The Markov property entails that conditioning on the closest two values is sufficient and this screening effect gives name to the algorithm. The algorithm can be considered as a zooming procedure with the conditioning values on each side getting increasingly closer with increasing stages.

From the assumptions made, it is clear that this is an exact simulation algorithm for a Gaussian random function on $D \subset \mathcal{R}^1$ with exponential covariance function, $\rho(\Delta) = \sigma^2 \exp\{-\alpha|\Delta|\}$. For Gaussian intrinsic random functions on $D \subset \mathcal{R}^1$ with linear variogram function, $\gamma(\Delta) = \tau|\Delta|$, it is also exact, except for the initiation problem discussed in relation to the sequential algorithm in the previous section. The screening sequential algorithm in one dimension, as described above, is of course extremely efficient since conditioning on only the two closest values is required. Moreover, the identical configuration appearing on each stage of the generation procedure makes it necessary to determine $\underline{\psi}_{0n}\Psi^{-1}$ and $\underline{\gamma}_{0n}\Gamma^{-1}$ only once on each stage. Finally, for the linear variogram function it is known that the weights are scale invariant hence their values are known a priori to be equal to 0.5 on each conditioning value. This is recognized as the familiar simulation procedure for Brownian motions and for fractal random functions with $H = 0.5$, corresponding to the linear variogram function, see Feder (1988).

Note that exactness is obtained only for the one-dimensional case, $D \subset \mathcal{R}^1$, and Gaussian and Gaussian intrinsic random functions with exponential covariance functions and linear variogram functions respectively. In particular, the fractal random functions, being self affine, with Hurst exponent different from 0.5 will not be exactly simulated not even in one dimension. This is of course a far too restrictive class of random functions to be useful in practice. The sequential simulation algorithm presented in the previous section is exact also for higher dimensions for all random functions considered here, but it is far too resource consuming for practical applications. Hence one has to rely on approximatively correct simulation algorithms. The remainder of this article is devoted to definition and evaluation of the screening sequential algorithm in higher dimensions and for models for which it is not exact. The objective is to identify an algorithm providing a reasonable trade off between conceptual and implementation simplicity, processing efficiency, and precision.

The screening sequential simulation algorithm for higher dimensions, $D \subset \mathcal{R}^m; m = 2, 3$, and for an arbitrary covariance or variogram function is defined as follows:

- Define a set of directions $\{\phi_i; i = 1, \ldots, \mathcal{N}_\phi\}$ in the lattice $\mathcal{L}_D$. These directions should span the space reasonably uniformly, and will intersect any surrounding surface.

- Initiate the values $\mathcal{Z}(\cdot)$ in the border area of $\mathcal{L}_D$, defined as the set of lattice locations on the border and its neighbors, by the sequential simulation algorithm. Recall that this algorithm is exact, but for higher dimensional cases not the entire border area need be initiated.

- Step for $t = 1, 2, \ldots, n$:
  Use a halfing procedure for densifying the lattice for each dimension of $D \subset \mathcal{R}^m$ in each step. Determine the pdf $\mathcal{N}(\mu_{\cdot|\cdot}, \sigma^2_{\cdot|\cdot})$ for the actual lattice node by conditioning on the two closest available values in each of the directions $\{\phi_i; i = 1, \ldots, \mathcal{N}_\phi\}$. Assign the actual lattice node a value by simulation from $\mathcal{N}(\mu_{\cdot|\cdot}, \sigma^2{}_{\cdot|\cdot})$. Recall that pseudo-Markov properties in higher dimensions require conditioning both on the value and the derivative on the surrounding surface, hence the two closest values are used.

- The values in the $(2^n + 1)_1 \times \ldots \times (2^n + 1)_m$ lattice is generated.

The screening sequential algorithm draws on the properties of the sequential simulation algorithm and the insight in pseudo-Markov properties in higher dimensions. The screening sequential algorithm is of course only approximative in higher dimensions since the surrounding surface is represented by the $\mathcal{N}_\phi$ intersections with the directions only. A clever implementation, using the high symmetry of the algorithm, requires a few matrix decompositions on each step only. Hence the algorithm will appear as extremely processing efficient. Note in particular that the processing requirements will be independent of the correlation structure since the screening effect is assumed.

The screening sequential algorithm is based on the sequential Gaussian algorithm and the Markov properties. The algorithms used in Alabert (1989) and the SIS-algorithm defined in Journel and Alabert (1988) are developed for the conditional case, and are also based on the sequential algorithm. These algorithms run through the lattice in a random path however, and the dimensionality of the conditioning is reduced by using only the closest conditioning values. This is done in line with the 'local' kriging tradition justified through robustness with regard to local departure from stationarity. The screening sequential algorithm differs from the ones above by efficiently using the screening effect provided by the Markov properties. The values are generated in a sequence which reproduces the same configurations of conditioning values and zoom into smaller scales. The justification of the algorithm from approximate Markov properties simplifies the identification of models for which the algorithm is acceptably reliable. The screening sequential algorithm should rather be seen as an extension of the random midpoint displacement algorithm used for generating self affine, or fractal, random functions, see Barnsley et al. (1988). This algorithm is also based on the sequential algorithm and it draws on the Markov properties for justifying conditioning on only the closest values. Moreover, it is defined such that the configuration of conditioning values remain unchanged during the procedure. The random midpoint displacement algorithm appears as somewhat ad hoc, however. This will be more thoroughly discussed in the next section. It is worth noting that the screening sequential algorithm is extended to cover not only fractal random functions, but also classes of Gaussian random functions.

4 Empirical evaluations and discussion

The empirical evaluation is performed along two directions:

- generalization to higher dimensions. For Gaussian and Gaussian intrinsic random functions with exponential and linear variogram functions respectively, the presence of approximate Markov properties for $\mathcal{R}^2$ are evaluated.

- extension of the class of correlation and variogram functions. The generalization will be to other commonly used correlation functions and to other fractal dimensions for the Gaussian intrinsic model. The study is performed for $\mathcal{D} \subset \mathcal{R}^2$. For self affine, or fractal, random functions, a comparison with the commonly used random midpoint displacement algorithm is also included.

Two different simulation algorithms are involved in the study:

- screening sequential algorithms is used in most of the cases. It is defined in the previous section. In the study the directions denoted ϕ are $\{i\pi/4; i = 1, \ldots, 8\}$. Hence 16 observations are used in the conditioning in the two dimensional case.

- random midpoint displacement algorithm as described in Barnsley et al. (1988), frequently used in practice, is used as a comparison for fractal cases. This algorithm can be considered as a simplified version of the screening sequential algorithm in the sense that only the four or six main directions for ϕ are used in two and three dimensions respectively. Moreover, only the closest values in each direction is used and the same weights are assigned to each of the values to obtain the conditional distribution for all stages in the procedure. To avoid severe anisotropy effects, the directions used are rotated $\pi/4$ in two dimensions, and correspondingly in three dimensions, for certain configurations. The initiation described in Barnsley et al. (1988) appears as arbitrary. The initiation step is crucial when simulating fractal random functions since correlation appear for all distances. The initial values have considerable impact through the conditioning on values later to be generated. To provide anything like a fair comparison, the initiation procedure presented in Barnsley et al. (1988) is replaced by the sequential algorithm.

The study in two dimensions are done on a regular lattice of $\mathcal{L}_D \in \{x_{ij} | i,j = 1, \ldots, 129\}$. The shortest distance between two lattice nodes is used as the unit. For each case 100 realizations are generated. The traditional variogram estimator is used:

$$\hat{\gamma}(h) = \frac{1}{2\mathcal{N}_h} \sum_{(i,j)(k,l) \in E_h} (\mathcal{Z}(x_{ij}) - \mathcal{Z}(x_{kl}))^2$$

with $E_h : \{(i,j),(k,l) | |x_{ij} - x_{kl}| = h\}$ and $\mathcal{N}_h = \#E_h$ Recall that the correlation function, $\rho(h)$, can be expressed as a variogram function $\gamma_0(h) = (1 - \rho(h))$.

The empirical results are presented in quantile-plots, for example Figure 3. The horizontal axis is the lag in the variogram function and the vertical one is the variogram value. For each lag the variogram value along one of the main directions is estimated for each of the realizations. The hatched line indicates the average of the estimates of the variogram for each lag. The dotted lines indicates the 0.2, 0.4, 0.6 and 0.8 quantiles for the estimates for each lag. At lag 20, 60 and 100, the histograms for the estimates of the variogram values at the respective lags are displayed. The target variogram function is presented by a solid line.

In Figure 3 and 4, the results from the empirical study of the screening sequential algorithm on Gaussian random functions for $\mathcal{D} \subset \mathcal{R}^2$ are displayed. The exponential variogram function with $\alpha = 0.2$ and $\alpha = 0.02$ are used in the two figures. The former represents a random function with correlation structure having much shorter extent than $\mathcal{D}$, while the latter has long correlation structure. There appears to be no significant bias for any of the cases. The estimation variance seems to increase considerably with increasing correlation structure. This is not surprising since then higher correlation between values in the lattice nodes appear and this redundancy causes the total information content to be less. From plots not reported here, the estimates appear as almost perfectly isotropic. Recall that for exponential variogram functions, the screening sequential algorithm is exact for $D \subset \mathcal{R}^1$, although sampling uncertainty will occur of course. The results in Figure 3 and 4 suggest that the screening sequential algorithm is, practically spoken, exact for $\mathcal{D} \subset \mathcal{R}^2$ also, at least for relatively short correlation structures.

In Figure 5, the results from the empirical study of the screening sequential algorithm on Gaussian intrinsic random functions for $\mathcal{D} \subset \mathcal{R}^2$ are displayed. The linear variogram function is used in the figure. There appears to be no significant bias, but the estimation variance is relatively large. This is, as previously discussed, caused by the long, or in this case infinite, correlation structure. The slight anisotropy in the estimates is considered to be sampling uncertainty only. Recall that for linear variogram functions the screening sequential algorithm is exact for $D \subset \mathcal{R}^1$. The algorithm seems highly reliable for $\mathcal{D} \subset \mathcal{R}^2$ also.

The screening sequential algorithm is considered to be reliable for Gaussian and Gaussian intrinsic random functions with exponential and linear variogram functions for $\mathcal{D} \subset \mathcal{R}^m$ with $m = 1, 2$. This is interpreted to be caused by the Markov property being present for $\mathcal{D} \subset \mathcal{R}^1$, playing approximately over to the two-dimensional case also. Preliminary empirical evaluations for $\mathcal{D} \subset \mathcal{R}^3$ leaves the same impression for that case as well.

In Figure 6 and 7, the results for Gaussian random functions with correlation functions being non-Markovian for even $D \subset \mathcal{R}^1$ are displayed. Two variogram functions are evaluated: the spherical model,

$$\gamma_o(\Delta) = \begin{cases} \frac{3}{2}(\frac{\Delta}{r}) - \frac{1}{2}(\frac{\Delta}{r})^3 & ; \; 0 < \Delta < r \\ 1.0 & ; \; r < \Delta \end{cases} \quad ; r = 20$$

and the second-order exponential model

$$\gamma_o(\Delta) = 1 - \exp\{-\alpha|\Delta|^2\} \; ; \alpha = 0.005$$

The former is frequently used and appears as exact for a three-dimensional spherical moving average operator on a white noise lattice. The latter constitutes an extreme case for Gaussian random functions, resulting in an infinitely times derivable realization. The results for $\mathcal{D} \subset \mathcal{R}^2$ for the spherical model in Figure 6, suggests a fairly distinct downward bias and relatively large estimation variance in the estimates. The results display no anisotropy. It appears as the screening sequential algorithm creates realizations according to the exponential variogram model closest to the spherical one specified. The spherical model can of course be reasonably well represented by linearly combining realizations of exponential models. Direct simulation is not possible since linear combination of exponential models do not exhibit Markov properties, even for $\mathcal{D} \subset \mathcal{R}^1$. The results for the second-order exponential model for $\mathcal{D} \subset \mathcal{R}^2$ in Figure 7, seem promising without noticeable bias although with relatively large estimation variance. No anisotropy can be identified. The experience from other tests is, however, that whenever the correlation structure is extended, the correlation between neighboring values in the lattice increases, and this will cause numerical instability. All algorithms based on the sequential algorithm will face these problems.

In Figure 8 and 9, the results for $\mathcal{D} \subset \mathcal{R}^2$ for Gaussian intrinsic random functions with variogram functions being non-Markovian for even $D \subset \mathcal{R}^1$ are displayed. The following class of self affine, or fractal, random functions are evaluated:

$$\gamma(\Delta) = \tau|\Delta|^{2H} \ ; \ H = 0.2 \ ; \ H = 0.8$$

The former represents a fairly erratic random function, while the latter represents a fairly smooth one. The results in Figure 8 and 9, suggest that no bias is present, although the estimation variance is fairly large. This is caused by the long correlation structure as previously discussed. No anisotropy can be identified. These results are considered to be encouraging.

In Figure 10 through 12, the results from applying the random midpoint displacement algorithm for $\mathcal{D} \subset \mathcal{R}^2$ on the fractal model for $H = 0.5, H = 0.2$ and H=0.8 are displayed. The results in Figure 10 through 12 should be compared to the results in Figure 5, 8 and 9, displaying the corresponding results for the screening sequential algorithm. The random midpoint displacement algorithm seems to provide severely biased results which are significantly less reliable than the ones from the screening sequential algorithm. Moreover, the initiation step in the random midpoint displacement algorithm was improved from what is usually done, not doing so would have provided even more striking differences. Other reasons for the difference are: the random midpoint displacement algorithm only relying on the four major directions in the lattice makes it vulnerable; and the fact that it assigns the same weights to all values makes it vulnerable for non-symmetric patterns in border areas.

In practical applications, the processing requirements are of major concern when representations on large lattices are needed. The screening sequential algorithm has processing requirements being independent of the actual variogram function and linear in the number of lattice nodes. These are favorable characteristics. In Figure 2 the processing requirement of the screening sequential and the random midpoint displacement algorithms, are displayed in a log-log plot. The latter algorithm is considered to be immensely fast. Note that the screening sequential algorithm has similar properties as the random midpoint displacement algorithm, although multiplied by a factor of approximately 1.75. The fact that the screening sequential algorithm is

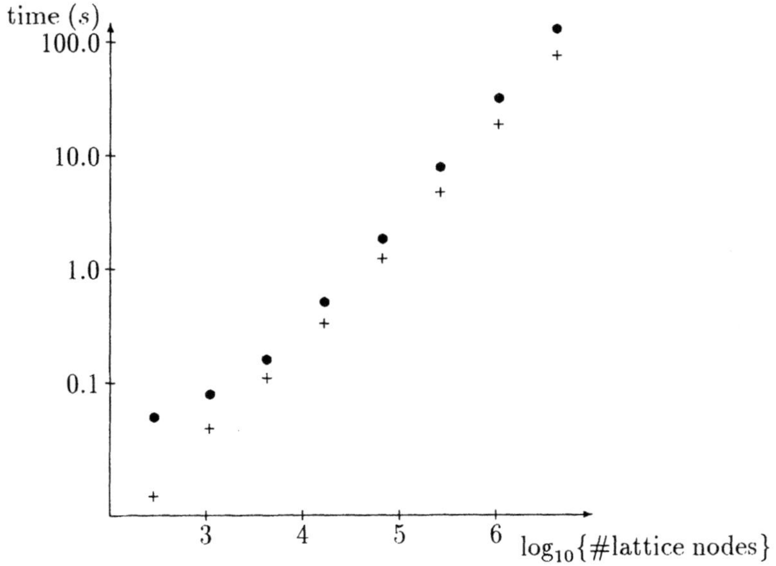

Figure 2: Processing requirements for the screening sequential algorithm (•) and for the random midpoint displacemant algorithm (+) for different grid sizes

both more general and considerably more reliable should justify the additional processing time in practical applications. Note that a two dimensional lattice of size $2049 \times 2049 = 4198401$ can be generated in 2 min. 8 sec. on a IBM Risc 6000/320. A lattice of size approximately 10^9 would require 8 h. 30 min., and considerable memory, of course, on the same computer.

5 Conclusions

The screening sequential algorithm for simulation of random functions is defined and evaluated. It is based on the general sequential simulation algorithm and approximate Markov properties in random functions. For Gaussian and Gaussian intrinsic random functions with exponential and linear variogram functions respectively, i.e. $\gamma_o(\Delta) = 1 - \exp\{-\alpha|\Delta|\}$ and $\gamma(\Delta) = \tau|\Delta|$, the screening sequential algorithm is exact in one dimension. This is due to the Markov properties for these models in one dimension. In higher dimensions, the Markov property is not present exactly, and the reliability of the algorithm must be evaluated empirically. The empirical study shows that the screening sequential algorithm is reliable also in the two-dimensional case for the models mentioned above. Moreover, for self affine, or fractal, models, $\gamma(\Delta) = \tau|\Delta|^{2H}$ with $0.2 \leq H \leq 0.8$ the screening sequential algorithm does also appear as reliable. Consequently, the simulation algorithm is reliable for models with variogram functions being both bounded and unbounded - or "with and without sill". Empirical evaluations of the same variogram functions in three dimensions seem to provide similar conclusions. This provides good opportunities for modeling many real phenomena.

For Gaussian random functions with spherical variogram functions, the algorithm provides relatively strongly downward biased results, and the general shape is not reproduced. It appears like the closest exponential variogram function is reproduced. With second-order exponential variogram functions, the algorithm appears as reasonably reliable, unless the correlation structure is long and numerical instability occurs. The screening sequential algorithm should for both these cases, be used with care.

The random midpoint displacement algorithm, frequently used for generation of fractal random functions, is also empirically evaluated. The results are significantly less reliable than for the screening sequential algorithm. In applications concerning scientific computing, the random midpoint displacement algorithm is not considered to be sufficiently reliable. The authors' experience is that the successive random addition procedure, frequently used in the fractal community, is even less reliable than the random midpoint displacement algorithm, and hence should be used with care. Procedures in the frequency domain tend to be very processing requiring in higher dimensions.

The processing requirements for the screening sequential algorithm are independent of the correlation structure and linear in the number of lattice nodes. The algorithm has processing characteristics similar to the random midpoint displacement algorithm, although with a constant 1.75 times higher. The latter is considered to be an immensely processing efficient simulation algorithm. The screening sequential algorithm required 2 min. 8 sec. to generate a lattice of size $2049 \times 2049 = 4198401$ on an IBM Risc 6000/320.

6 Acknowledgment

The study is financed by Norsk Hydro a.s., Statoil a.s. and Norwegian Computing Center.

References

Adler, R.J. (1981): *The geometry of random fields.* John Wiley & Sons, New York, 280p.

Alabert, F.G. (1989): "Constraining Description of Randomly Heterogeneous Reservoirs to Pressure Test data: a Monte-Carlo Study," paper *SPE* 19600

Barnsley, M.F., Devaney, R.L., Mandelbrot, B.B., Peitgen, H.O., Saupe, D. and Voss, R.F. (1988): *The Science of Fractal Images.* Springer–Verlag, New York, pp. 47-57

Borgman, L.E., Taheri, S.M. and Hagan, R.L. (1984): "Three-Dimensional, Frequency-Domain Simulations of Geological variables." In Verly et. al.: *Geostatistics for Natural Resources Characterization,* Part 1, pp. 517-541.

Delfiner, P. (1979): *The Intrinsic Model of Order k.* Note for Ecole d'Eté 1979. Ecole des Mines, Fontainebleau, France.

Journel, A.G. (1974): "Geostatistics for conditional simulations of ore bodies." Econ. Geol., 69(5), 673-687.

Journel, A.G. and Alabert, F.G. (1988): "Focusing on Spatial Connectivity of Extreme-Valued Attributes: Stochastic Indicator Models of Reservoir Heterogeneities," paper *SPE* 18326, presented at 1988 Annual Technical Conference and Exhibition, Houston, Tx, Oct. 2-5.

Journel, A.G. and Huijbregts, C.J. (1978): *Mining Geostatistics.* Academic Press, 600p.

Feder, J. (1988): *Fractals.* Plenum Press, New York

Mardia, K.V., Kent, J.T. and Bibby, J.M. (1979): *Multivariate Analysis.* Academic Press.

Matern, B. (1963): *Spatial Variations.* 2nd Edition 1986, Lecture notes in Statistics # 36, Springer Verlag.

Matheron, G. (1971): *The Theory of Regionalized Variables and Its Applications.* Ecole des Mines, Fontainebleau, France.

Ripley, B.D.(1987): *Stochastic Simulation,* Wiley and Sons.

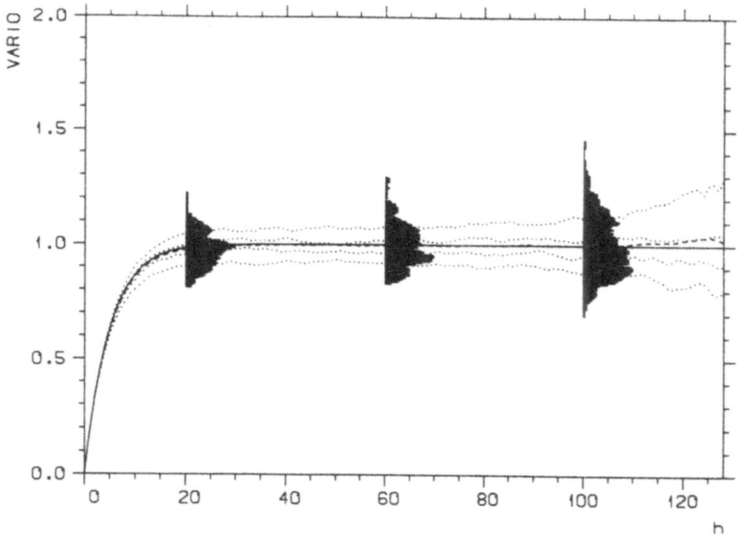

Figure 3: Screening sequential algorithm in two dimensions; exponential variogram function with $\alpha = 0.2$

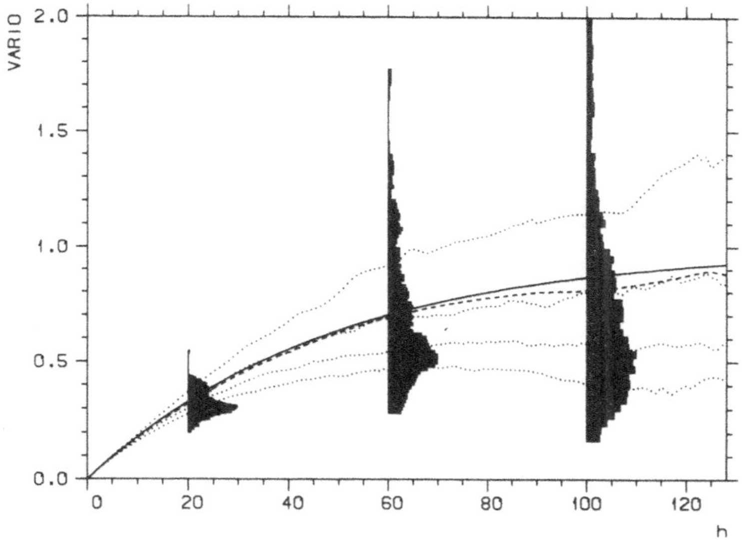

Figure 4: Screening sequential algorithm in two dimensions; exponential variogram function with $\alpha = 0.02$

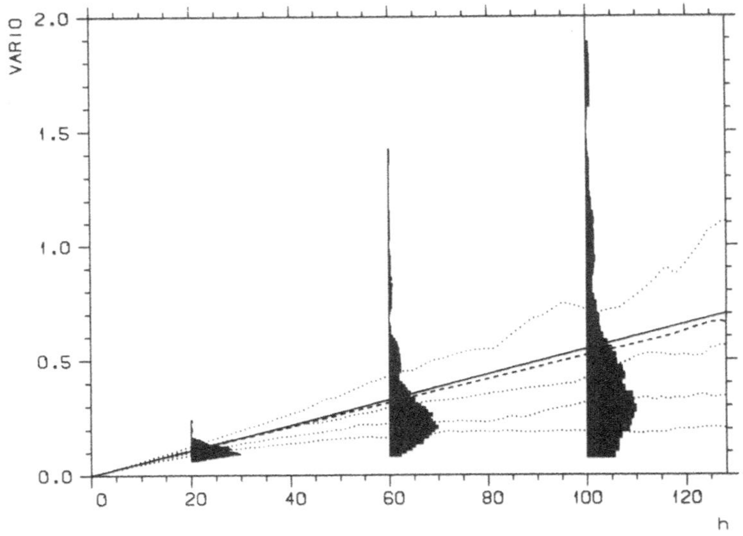

Figure 5: Screening sequential algorithm in two dimensions; fractal variogram function with $H = 0.5$; linear variogram function

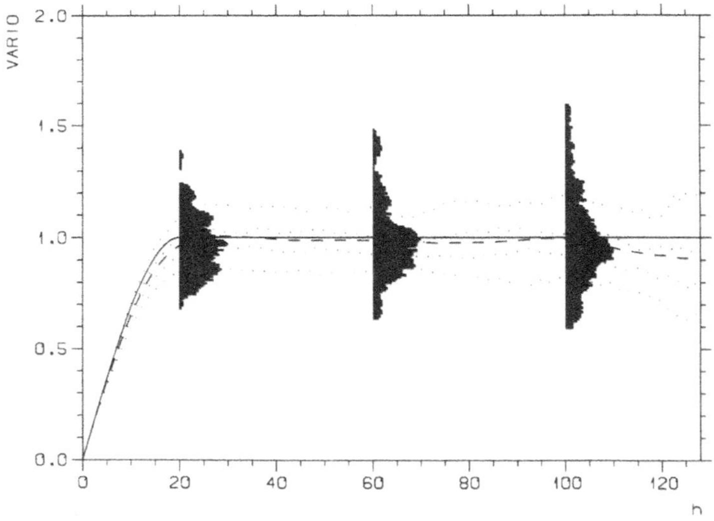

Figure 6: Screening sequential algorithm in two dimensions; spherical variogram function with $r = 20$

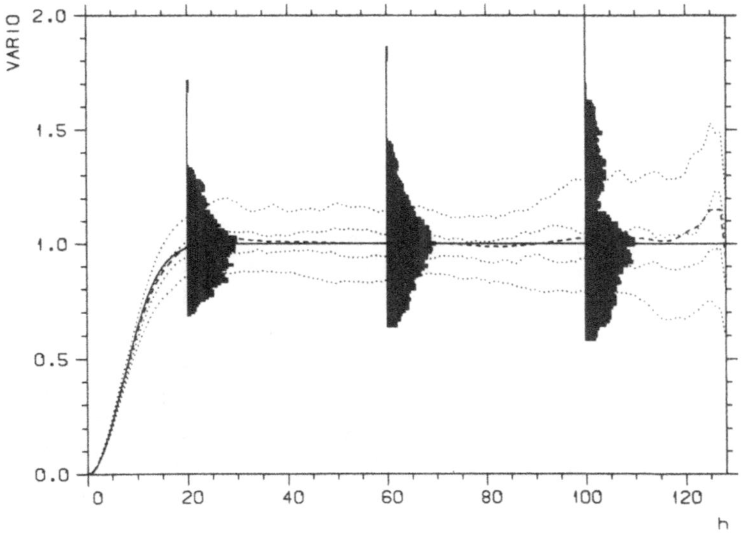

Figure 7: Screening sequential algorithm in two dimensions; second-order exponential variogram function with $\alpha = 0.005$

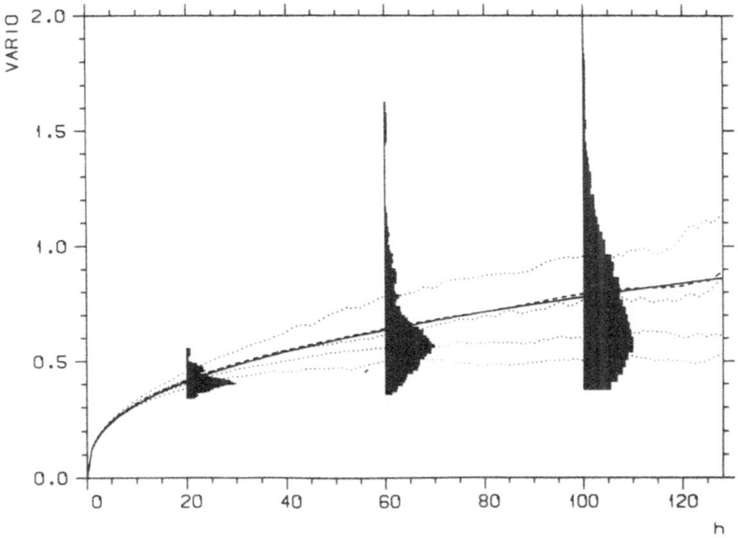

Figure 8: Screening sequential algorithm in two dimensions; fractal variogram function with $H = 0.2$

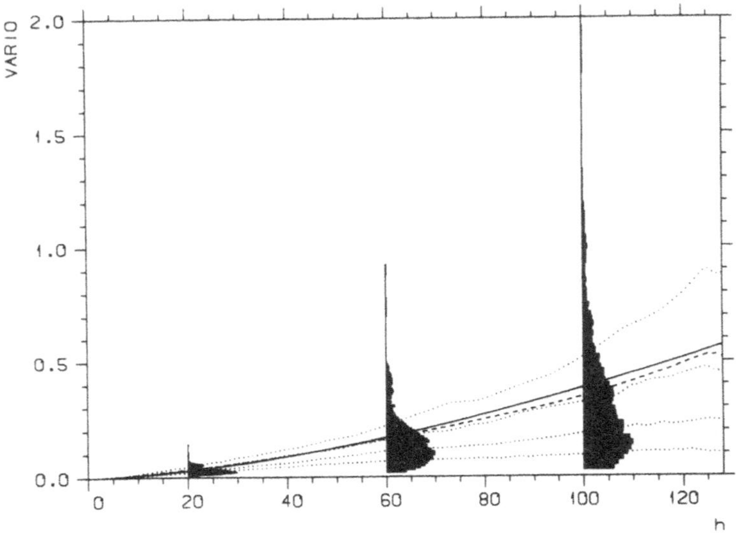

Figure 9: Screening sequential algorithm in two dimensions; fractal variogram function with $H = 0.8$

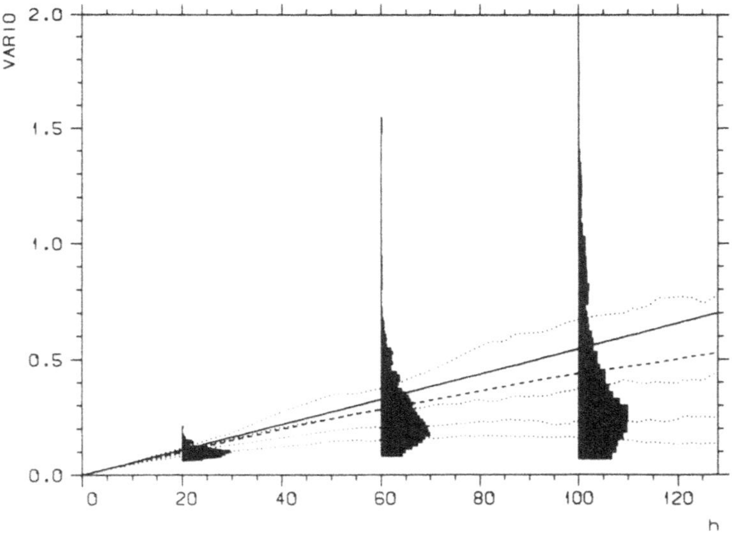

Figure 10: Random midpoint displacement algorithm in two dimensions; fractal variogram function with $H = 0.5$; linear variogram function

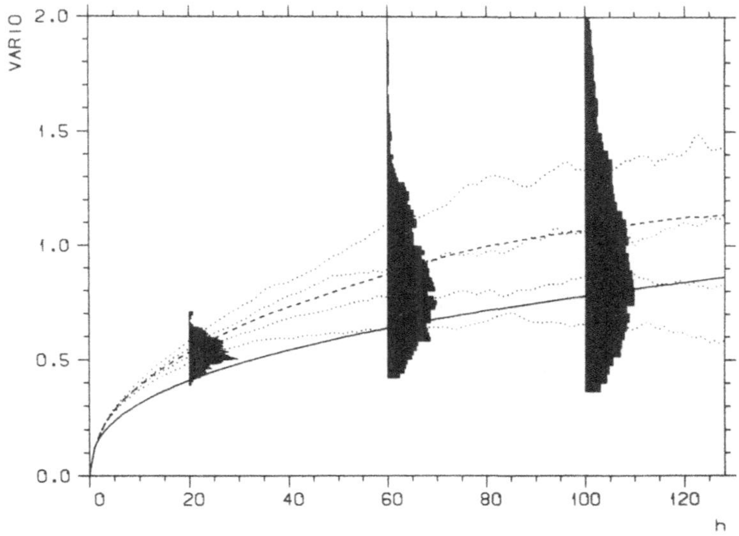

Figure 11: Random midpoint displacement algorithm in two dimensions; fractal variogram function with $H = 0.2$

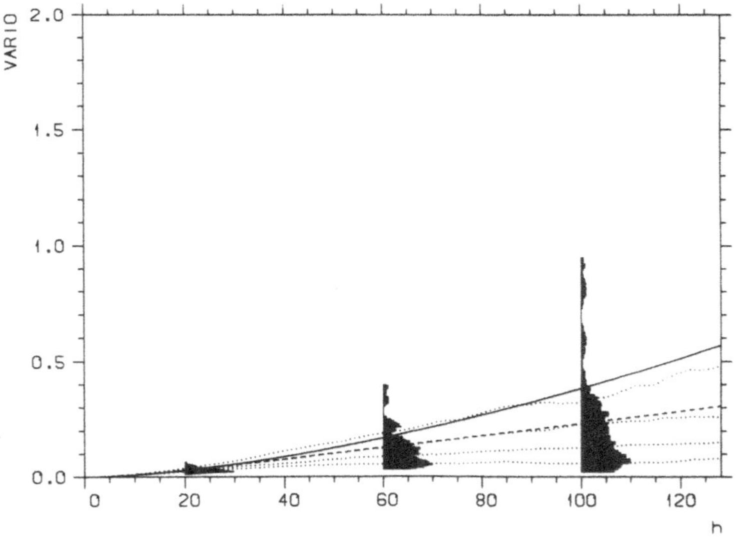

Figure 12: Random midpoint displacement algorithm in two dimensions; fractal variogram function with $H = 0.8$

Conditioning a Gaussian model with inequalities

Xavier Freulon and Chantal de Fouquet
Centre de Géostatistique
Ecole des Mines
35 rue Saint-Honoré
77305 Fontainebleau
France

ABSTRACT : Functions with mixed discrete/continuous distribution, random sets or coregionalizations between a function and a random set can be modelled using gaussian random functions. In these cases, the conditions at experimental points are expressed in terms of inequalities: $1 \leq i \leq n$, $a_i \leq X_i \leq b_i$. The simulation with composite constraints (equalities and inequalities) is performed in two steps: firstly we ensure that the inequality constraints are validated, secondly we perform a classic conditional simulation. To validate the inequalities, we propose two methods: the first one is based on the acceptance-rejection technique but can be used only to validate very few inequalities; in the second, we sample the stationary distribution of a Markov chain. This latter case is illustrated with a few examples.

1 Gaussian model and conditional simulations

Let $X = (X_i)_{1 \leq i \leq p}$ be a standard multigaussian random vector. We study the case where the n ($n \leq p$) first components of the vector to be simulated belong to fixed sets:

$$1 \leq i \leq n \ , \quad X_i \in B_i \ .$$

A conditional simulation of X is therefore a realization of the conditional probability distribution that respects these constraints. In practice, the two following cases are of common use: B_i is reduced to a single point (equality constraint) or B_i is a finite reunion of intervals (inequality constraint).

Equality constraint: $(X_1, \ldots, X_n) = (x_1, \ldots, x_n)$

If $i > n$, X_i can be written as the sum of its simple kriging by the coordinates $(X_j)_{1 \leq j \leq n}$ and an independent residual. The simulation of these residuals can be done using the *classic* conditioning kriging[5].

Inequality constraint: $(X_1, \ldots, X_n) \in B_1 \times \cdots \times B_n$

The conditional density function of the vector X to be simulated is the product of two terms,

$$\frac{g(x_1, \ldots, x_n) 1\!\!1_{B_1 \times \cdots \times B_n}(x_1, \ldots, x_n)}{\int_{B_1 \times \cdots \times B_n} g(u_1, \ldots, u_n) du_1 \ldots du_n} g(x_{n+1}, \ldots, x_p | x_1, \ldots, x_n)$$

where g denotes the n-dimensional Gauss density function of the vector $(X_1, \ldots, X_n)$ and $g(\cdot|x_1, \ldots, x_n)$ the conditional density function of the vector $(X_{n+1}, \ldots, X_p)$ when $(X_1, \ldots, X_n) = (x_1, \ldots, x_n)$. Hence the conditional simulation is performed in two steps: the n first components are simulated with the inequality constraint, then the other components are simulated with the value $(x_1, \ldots, x_n)$ as a condition.

Composite constraint:
$(X_1, \ldots, X_{n'}) = (x_1, \ldots, x_{n'})$ and $(X_{n'+1}, \ldots, X_n) \in B_{n'+1} \times \cdots \times B_n$

If $i > n'$, X_i can be written as the sum of its simple kriging by the coordinates $(X_j)_{1 \leq j \leq n'}$ and an independent residual. Therefore, the simulation of X under the composite constraint is achieved with the simulation of the residuals with only inequality constraints.

In the last two examples, we have to simulate a standard Gaussian vector under inequality constraints. Later on, a truncated Gaussian vector will denote a real random vector whose joint density function is

$$x \in I\!\!R^n \quad f(x) = \frac{g(x) 1\!\!1_B(x)}{\int_B g(u) du}$$

where $B = B_1 \times \cdots \times B_n$ — from now on this set is referred to as the constraint B — and g is the joint probability function of a n-dimensional standard Gaussian vector. This function is defined by the symmetric positive definite $n \times n$ matrix Σ:

$$g(x) = (2\pi)^{-\frac{n}{2}} |\Sigma|^{-\frac{1}{2}} \exp\{-\frac{1}{2} x' \Sigma^{-1} x\} .$$

In this paper, we shall present two methods to simulate such a density function: the first one is based on the acceptance-rejection technique and the second is an iterative method using a Markov chain.

2 Simulation of truncated Gaussian variables

2.1 Simulation by acceptance-rejection technique

The acceptance-rejection technique, due to von Neumann[10], can be used to generate a real random variable — a scalar or a vector — when its density function,

f, is majorized by another density function, h, for which samples are available: $x \in I\!R^n$, $f(x) \leq C\,h(x)$. This technique is well suited to the two following cases: the simulation of a scalar truncated Gaussian variable and the simulation of a truncated Gaussian vector.

2.1.1 A scalar truncated Gaussian variable

The constraint which the Gaussian variable has to verify is a finite union of disjoint intervals $B = \bigcup_i I_i$. As the density function g can be majorized on every interval by a positive and summable function f_i — $x \in I_i$, $g(x) \leq f_i(x)$ — the density function of the truncated Gaussian variable is majorized on whole B:

$$\frac{g(x)\mathbb{1}_B(x)}{\int_B g(u)du} \leq C \sum_i \varpi_i \frac{f_i(x)\mathbb{1}_{I_i}(x)}{\int_{I_i} f_i(u)du} \quad \text{with} \quad \left\{ \begin{array}{lcl} C & = & \frac{\sum_j \int_{I_j} f_j(u)du}{\int_B g(v)dv} \\ \varpi_i & = & \frac{\int_{I_i} f_i(u)du}{\sum_j \int_{I_j} f_j(u)du} \end{array} \right.$$

We choose the functions $(f_i)_i$ so that explicit and easy computations are available for the discrete distribution $(\varpi_i)_i$ and for the inverse of the cumulative function of the density $f_i(x)\mathbb{1}_{I_i}(x)/\int_{I_i} f_i(u)du$. Thus, the majorizing density function can be simulated by the inverse technique. Two kinds of inequalities have been used to majorize the Gaussian density function (figure 1): $x \in I\!R$, $\exp(-\frac{x^2}{2}) \leq 1$ and $0 < a \leq |x|$, $\exp(-\frac{x^2}{2}) \leq |x|/a \, \exp(-\frac{x^2}{2})$.

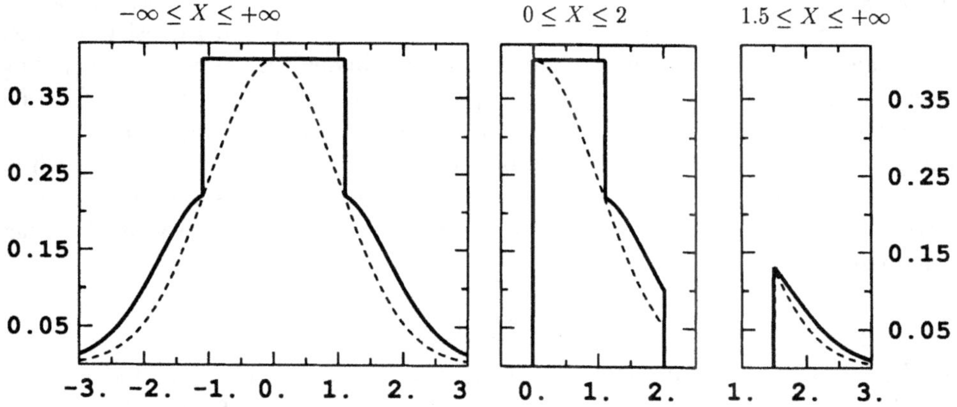

Figure 1: Some majorations of the Gauss density function
---- Gauss density function
—— majorizing function

Therefore, the algorithm to simulate a scalar truncated Gaussian variable using the acceptance-rejection technique is:

(i) *Sample a random interval $\underline{i}$ from the discrete distribution $(\varpi_i)_i$.*

(ii) *Sample the variables X and U from distributions whose densities are respectively,*

$$\frac{f_{\underline{i}} 1\!\!1_{I_{\underline{i}}}(x)}{\int_{I_{\underline{i}}} f_{\underline{i}}(u)du} \quad and \quad 1\!\!1_{]0,1[}(u) .$$

(iii) *If $U f_{\underline{i}}(X) \leq g(X)$, the value of X is kept, else return to (i).*

2.1.2 A truncated Gaussian vector

In order to simulate a truncated Gaussian vector using the acceptance-rejection technique, we choose as a majorizing function the restriction of the Gaussian density function to a set A including the constraint B. If $B \subset A$, we have:

$$\frac{g(x) 1\!\!1_B(x)}{\int_B g(u)du} \leq \frac{\int_A g(v)dv}{\int_B g(u)du} \frac{g(x) 1\!\!1_A(x)}{\int_A g(u)du} .$$

To choose a convenient subset A, we represent the coordinates of the Gaussian vector as linear combinations of independent Gaussian variables: $X = LT$ where L is a $n \times n$ matrix with $\Sigma = LL'$ and T is a vector of independent Gaussian variables. To simulate X under the constraint $B = \{x \in I\!R^n \ , \ a_i \leq x_i \leq b_i\}$ is equivalent to simulating T under the constraint $B^* = \{t \in I\!R^n \ , \ a_i \leq \sum_j l_i^j t_j \leq b_i\}$. Now, we simulate T with the acceptance-rejection technique; the majorizing density function used is the Gaussian truncated on the smallest n-dimensional hypercube including B^*: $A^* = [\alpha_1, \beta_1] \times [\alpha_2, \beta_2] \times \cdots \times [\alpha_n, \beta_n]$. The bounds of A^* are determined by $\alpha_i = \min_{(t_1,\ldots,t_n) \in B^*} t_i$ and $\beta_i = \max_{(t_1,\ldots,t_n) \in B^*} t_i$. When the covariance matrix is decomposed by the Cholesky factorization, the peculiar geometry of the polyhedron A^* can be used to compute its bounds easily, since the matrix L is a lower triangular matrix. The coordinates of T are independent, the majorizing density function can be factorized:

$$\frac{g(t) 1\!\!1_{A^*}(t)}{\int_{A^*} g(u)du} = \prod_{i=1}^{n} \frac{g(t_i) 1\!\!1_{[\alpha_i, \beta_i]}(t_i)}{\int_{\alpha_i}^{\beta_i} g(u_i)du_i} .$$

Therefore, in order to simulate this density, all the coordinates are simulated independently from each other using the acceptance-rejection technique for a scalar variable presented earlier.

The algorithm to simulate a truncated Gaussian vector using the acceptance-rejection technique is:

(i) *Determine* L *and* A^*.

(ii) *Simulate* T, *truncated on* A^*.

(iii) *If* $X = LT \in B$, *the value is kept, else return to (ii).*

In fact, the computation of A^* is not compulsory since the *a priori* Gaussian distribution ($A^* = I\!R^n$) can be used. But using a truncated distribution as a majorizing function reduces the average number of rejections and therefore leads to an important reduction in the computer time needed, as can be seen on figure 2. However, as time increases steadily as the dimension of the simulated vector increases, this algorithm tends to be extravagant and should be limited in practice to small vectors where $n \leq 10$.

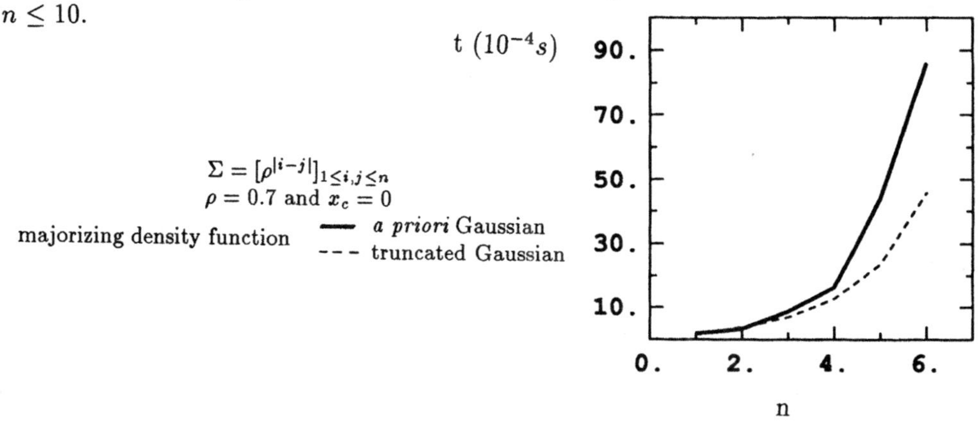

Figure 2: Average computer time to simulate a truncated Gaussian vector by the acceptance-rejection technique versus the dimension of the vector

2.2 The iterative method

In the iterative method, the distribution of the model is not simulated at once. Starting with an initial simulation sampled from an arbitrarily chosen distribution, we modify it at random until we get a suitable vector. More formally, we construct series of random vectors $(X^{(\nu)})_{\nu \geq 0}$ whose marginal distributions converge toward the joint distribution of a Gaussian vector truncated on B. These series constitute a stochastic process whose transitions occur at discrete times; the set of possible states is the constraint B. The random process commonly used is a Markov chain: it is defined by an initial distribution p_0 and stationary transition probabilities P; the computation of the series only requires that the current iteration need be stored in computer memory, as the distribution of $X^{(\nu)}$ is independent of all previous states except its immediate predecessor $X^{(\nu-1)}$. These techniques first used by Metropolis and his collaborators[9] in statistical mechanics are now widely used in image analysis to produce simulations of Markov random fields[4].

2.2.1 Transition probabilities

A simple way to modify a vector is to change only one of its coordinates. Let P_i denote the transition probability which modifies the i-th coordinate; P_i depends only on the current value of the vector. We define an iteration as a sequential change of every coordinate of the vector. Thus its transition probabilities are: $\forall(x, \mathcal{A}) \in B \times \mathcal{B}_B$ ($\mathcal{B}_B$ denotes the σ-algebra of the Borel set of B),

$$\begin{aligned} P(x, \mathcal{A}) &= P\{X^{(\nu+1)} \in \mathcal{A} | X^{(\nu)} = x\} \\ &= \int dP_1(x, u^{(1)})\, dP_2(u^{(1)}, u^{(2)}) \cdots dP_n(u^{(n-1)}, u^{(n)}) \mathbb{1}_{\mathcal{A}}(u^{(n)}) \,. \end{aligned}$$

In the case of a truncated Gaussian vector, a convenient system of transition probabilities has been propounded by Geman S. and Geman D. [4]: we substitute for the value of the i-th coordinate a value sampled from its conditional distribution when the value of the other coordinates is fixed, or, more formally,

$$X_i^{(\nu+1)} = X_i^{SK}{}_{(X_1^{(\nu+1)},\ldots,X_{i-1}^{(\nu+1)},X_{i+1}^{(\nu)},\ldots,X_n^{(\nu)})} + \sigma_{SK}(i)\, U_i$$

where X_i^{SK} is the simple kriging of X_i by the other coordinates of the vector, $\sigma^2_{SK}(i)$ is the kriging variance, and U_i is a truncated Gaussian variable such that $X_i^{SK} + \sigma_{SK}(i)\, U_i \in B_i$.

Under mild assumptions, B is a compact set, the transition probability is ergodic ant its unique stationary distribution is the truncated Gaussian distribution, i.e. for every initial distribution p_0, the distributions of $X^{(\nu)}$ converge toward the truncated Gaussian distribution (proof in appendix). If B is not a compact set, we have not proved the uniqueness of the stationary distribution but in practice non uniqueness should correspond to pathological cases. Besides, as no theoretical criterion is available to forecast the speed of the convergence, this technique will be studied on a few examples.

2.2.2 The initial distribution

An easy way to sample an initial vector which verifies the constraint is to ignore the correlations between the coordinates. They can then be simulated independently from each other. A second method using the correlations is the incremental method[6]: let the i first coordinates of the vector be simulated, the $(i+1)$-th coordinate can be split into two terms : $X_{i+1}^{(0)} = X_{i+1}^{SK}{}_{(X_1^{(0)},\ldots,X_i^{(0)})} + R_{i+1}$; the residual R_{i+1} is a scalar Gaussian variable simulated so that the inequality constraint on the $(i+1)$-th coordinate is verified. But this device does not take into account the inequalities which the coordinates to be simulated have to verify, thus the vector actually simulated is not a truncated Gaussian vector.

3 Some examples of truncated Gaussian variables

To test a sampling technique, we compute some statistics of the simulation, the first two moments of which are available. As the variance of these statistics is generally small, the model is accepted if the value of the statistics does not significantly differ from their mean. As we cannot compute the explicit value of the first two moments of the basic statistics calculated on truncated Gaussian variables, we randomize the constraint and the truncated Gaussian vector becomes an ordinary multigaussian vector. We specify first how we randomize the constraint: the vector to be simulated, $X = (X_i)_{1 \leq i \leq n}$, is a Gaussian vector. For any i, we choose an interval C_i. The randomized truncated Gaussian vector is simulated in two steps: a) we simulate a random vector $\tilde{X}$ following the same distribution as X. The randomized constraint is

$$B = (B_i)_{1 \leq i \leq n} \text{ with } B_i = \begin{cases} C_i & \text{if } \tilde{X}_i \in C_i \\ C_i^c & \text{else} \end{cases}$$

b) then we simulate a Gaussian vector X truncated on B. As $P\{\tilde{X} \in B\} = \int_B g(u)\, du$ and the density function of the truncated Gaussian vector on B is $\mathbb{1}_B\, g(x) / \int_B g(u)\, du$, the density function of the randomized truncated Gaussian vector is

$$\sum_B P\{\tilde{X} \in B\} \frac{\mathbb{1}_B(x)\, g(x)}{\int_B g(u)\, du} = g(x)\ .$$

If we suppose the simulation of X to be correct, the test of the Gaussian model on X allows the evaluation of the algorithm which simulates a truncated Gaussian vector. But, as we randomize the constraint, we just evaluate an average speed of convergence.

To use the iterative method to simulate a truncated Gaussian vector, the kriging of the coordinate i has to be performed with the neighbourhood $\{1,\ldots,i-1,i+1,\ldots,n\}$. A simple way to get the kriging weights, $(\lambda_i^j)_{j \neq i}$, and the variance, σ_i^2, is to invert the covariance matrix, $\Sigma^{-1} = S = [s^{ij}]_{1 \leq i,j \leq n}$:

$$1 \leq i \leq n \quad \begin{cases} & \sigma_i^2 = 1/s^{ii} \\ j \neq i & \lambda_i^j = -s^{ij}/s^{ii} \end{cases}$$

This method is applicable when the dimension of the simulated vector is not bigger than a few hundreds. In some cases, properties of the covariance matrix can be used to reduce the kriging neighbourhood. On the real line, if the process is markovian, i.e. its stationary covariance is exponential, two points on both sides of the point to be simulated constitute the kriging neighbourhood. In a two or three-dimensional space, similar Markovian properties are induced by a factorizable exponential covariance[1];

[1] $\rho(h) = \exp\{-(\frac{|h_x|}{a_x} + \frac{|h_y|}{a_y} + \frac{|h_z|}{a_z})\}$ where h_x, h_y et h_z are the components of h and a_x, a_y, a_z three scaling factors.

the only non null kriging weights are the weights of the nearest neighbours[8]. In more general cases, if the stationary covariance of the Gaussian model to be simulated tends toward zero, the kriging weights also converge toward zero as the distance to the point to be kriged increases. So we may use a moving neighbourhood, $V_i \subset \{1,\dots,i-1,i+1,\dots,n\}$ (if $j \notin V_i$, $\lambda'^j_i = 0$), but the actual simulated vector is the *auto-normal* model[1], which is defined by its conditional density function:

$$1 \le i \le n \quad g(x_i|(x_j)_{j\neq i}) = \frac{1}{\sqrt{2\pi}}\frac{1}{\sigma_i}\exp\{-\frac{1}{2}(\frac{x_i - \sum_j \lambda'^j_i x_j}{\sigma_i})^2\}$$

If $\lambda'^j_i/\sigma_i^2 = \lambda'^i_j/\sigma_j^2$, the joint density function is $g(x) = (2\pi)^{-\frac{n}{2}}|S_*|^{\frac{1}{2}}\exp\{-(x'S_*x)/2\}$ where the $n \times n$ matrix, S_*, is defined by

$$s_*^{ij} = \begin{cases} 1/\sigma_i^2 & \text{if } i = j \\ \lambda'^j_i/\sigma_i^2 & \text{else} \end{cases}$$

When for any i, the kriging neighbourhood is $\{1,\dots,i-1,i+1,\dots,n\}$, we verify that $S_*^{-1} = \Sigma$. But a smaller neighbourhood does not ensure either the verification of the symmetry $\lambda'^j_i/\sigma_i^2 = \lambda'^i_j/\sigma_j^2$ or the positive definiteness of the matrix S_*. Nevertheless, on various examples, the iterative technique using a moving neighbourhood reproduces efficiently the modelled covariance. Besides, this technique can be used whether the points are located on a regular grid or not.

3.1 1-dimensional simulations

A stationary Markov process is sampled along a regular grid. The covariance matrix of the simulated vector is $\Sigma = [\rho^{|i-j|}]_{1\le i,j\le n}$. Given a threshold, x_c, we define the conditioning indicator to be $(\mathbb{1}_{X_i \ge x_c})_{1\le i\le n}$. Statistics — mean and variance — and the experimental variograms are calculated on the first iterations of the Markov chain. Figure 3 presents the experimental variance as a function of the rank of the iteration: the number of iterations needed for the experimental variance to reach the theoretical value does not depend on how the initial vector is simulated, but does depend heavily on the correlation between the coordinates of the vector. The first points of the experimental variograms (figure 4) are correct after a very small number of iterations, but the experimental sill of the variogram, following the variations of the experimental variance, slowly reaches the theoretical value. In this case, the value of the experimental variance seems to be a good criterion for evaluating the time when the chain reaches its stationary state.

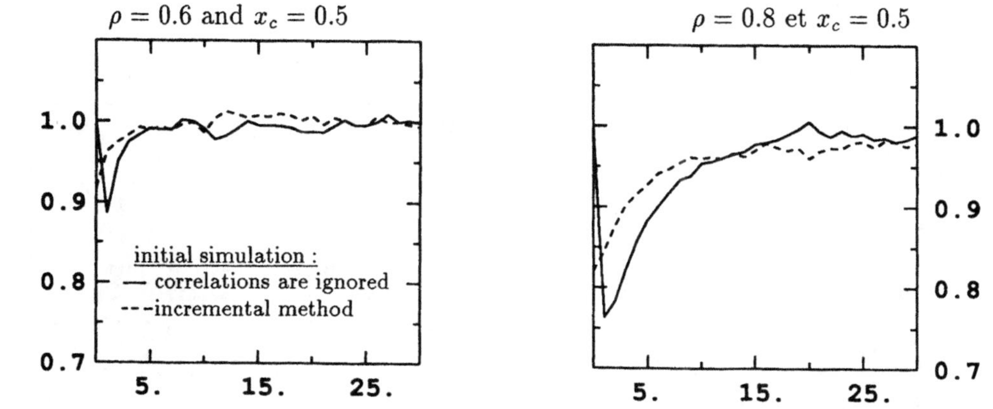

Figure 3: Experimental variance versus the rank of the iteration — Gemans' method

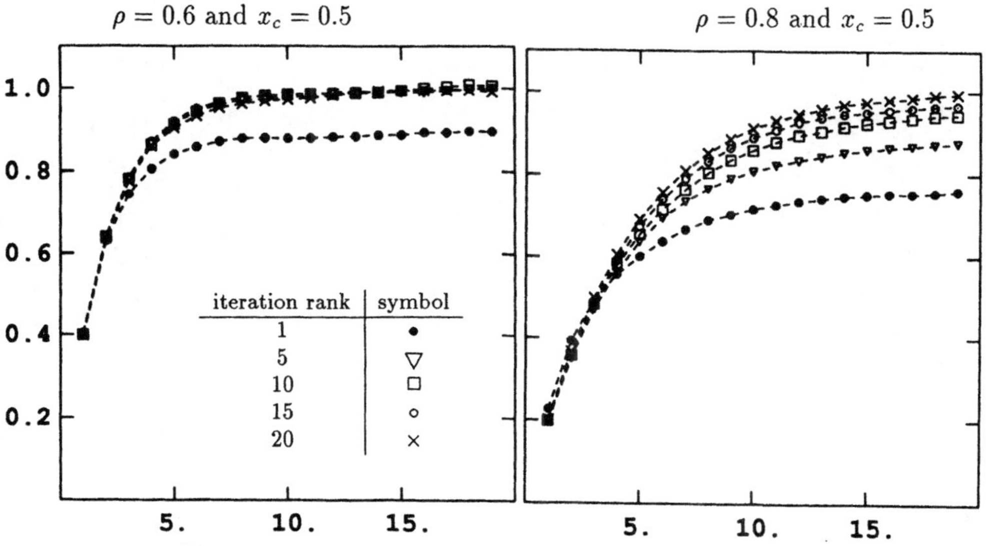

Figure 4: Experimental variograms - Gemans' method

3.2 2-dimensional simulations

Three 200 × 200 Gaussian images are simulated under the constraint defined by the indicator $\mathbb{1}_{X_i \geq 0.67}$ — the proportion of the black pixels is therefore 0.25. They are presented on figure 5: the first one has a factorizable exponential covariance, the second an isotropic exponential covariance — these two exponential covariances coincide on the axes but differ elsewhere — and the third an isotropic spherical covariance. The iterative device is stopped when the experimental variance stabilizes to the theoretical value: 200 iterations are needed in the first case, 100 in the second case and 300 in the third case. For the last two images, a 9×9 moving neighbourhood is used. In all cases, the experimental variograms comply with the models (figure 5).

Figure 5: Conditional simulations by the indicator $\mathbb{1}_{X \geq 0.67}$

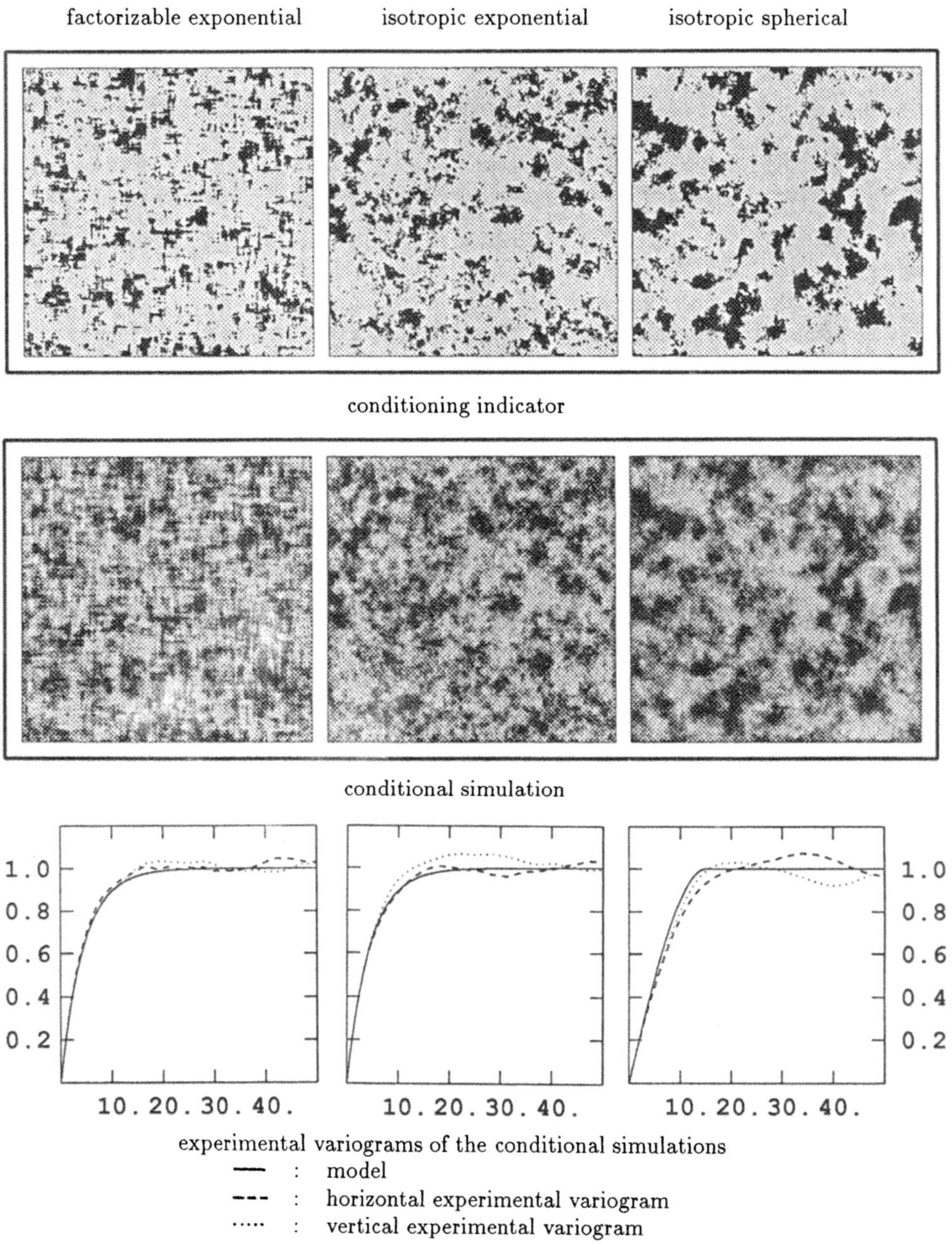

4 Conclusion

The simulation of a Gaussian model subject to inequality constraints is performed in two steps: firstly we validate the inequalities using the iterative method; the experimental variance seems a convenient criterion for convergence. Secondly we perform a *classic* conditional simulation. As the inequality constraint is induced by the observations, the dimension of the truncated Gaussian vector is generally far smaller than the number of the nods of the grid to be simulated. The second step of the simulation is done by a less time consuming device, such as the Turning Band method and conditioning kriging[5].

There is no reason why the studied variables should be Gaussian; they may be modelled as the transformed of a Gaussian variable. For example, discrete geological variables have been modelled[8, 3] by a step function: the value of the discrete variable defines an inequality constraint on the associated Gaussian variable. Inequalities also arise when the measures are blurred by a uniform additive noise. Finally, the simulation of a Gaussian vector under inequality constraints appears in the conditional simulation of substitution random functions[7]. These examples show that conditioning a Gaussian model with inequalities widens the set of natural phenomena to which such a model can be applied. So far, this model was used only when the variable, modelled using a one-to-one univariate transform, was exactly known at every experimental point.

Appendix

If B is a compact set, the transition probability defined in 2.2.1 is ergodic and its unique stationary distribution is the Gaussian distribution truncated on B.

Proof: We prove first that the truncated Gaussian distribution is a stationary distribution of the transition distribution P, i.e. its density function verifies the equation:

$$\forall \mathcal{A} \in \mathcal{B}_B \ , \quad \int f(x)\, dx\, P(x, \mathcal{A}) = \int_{\mathcal{A}} f(x) dx \ .$$

Let $\overline{x_i} = (x_j)_{j \neq i}$ and $f(x_i|\overline{x_i})$ be the conditional density function of X_i when $\overline{X_i} = \overline{x_i}$. The coordinate transition probability can then be written as

$$P_i(x, \mathcal{A}) = \int_{\mathcal{A}} \delta_{(\overline{x_i} - \overline{y_i})}\, f(y_i|\overline{x_i})\, dy$$

where $\delta(\cdot)$ is the Dirac measure on $I\!R^{n-1}$. By the Fubini theorem, we get

$$\int f(x)\, dx\, P_i(x, \mathcal{A}) = \int f(x)\, dx\, \delta_{(\overline{x_i} - \overline{y_i})}\, dy_i\, f(y_i|\overline{x_i}) = \int_{\mathcal{A}} f(y)\, dy \ .$$

This equation implies that for every measurable function h,

$$\int f(x) dx dP_i(x, y) h(y) = \int f(x) dx h(x) \ .$$

$$\text{Thus, } \int dx f(x) P(x, \mathcal{A}) = \int f(x) dx dP_1(x, u^{(1)}) \int \ldots \int dP_n(u^{(n-1)}, u^{(n)}) \mathbb{1}_{\mathcal{A}}(u^{(n)})$$
$$= \int f(x) dx dP_2(x, u^{(2)}) \int \ldots \int dP_n(u^{(n-1)}, u^{(n)}) \mathbb{1}_{\mathcal{A}}(u^{(n)})$$
$$= \int_{\mathcal{A}} f(x) dx \; .$$

So the truncated Gaussian distribution is a stationary distribution of P.

To prove the ergodicity of the transition probability, we note that the transition probability distribution is absolutely continuous with respect to the Lebesgue measure on B and that its density function is continuous and strictly positive. So if B is a compact set, then the greatest lower bound is reached on B and is strictly positive. In this case, the transition probability is ergodic[2] and then its unique stationary distribution equals the truncated Gaussian distribution. ∎

References

[1] Besag, J. (1974) "Spatial interaction and statistical analysis of lattice systems". *J. R. Statist. Soc. B*, 36:192–236.

[2] Doob, J. L. (1953) *Stochastic processes.* John Wiley & Sons, Inc., New York.

[3] Freulon, X. (1992) *Conditionnement du modèle gaussien par des inégalités ou des randomisées.* thèse de docteur en géostatistique, ENSMP, Fontainebleau, France.

[4] Geman, S. and Geman, D (1984) "Stochastic relaxation, gibbs distribution and the bayesian restoration of images". *I.E.E.E. transactions : Pattern Analysis and Machine Intelligence*, 6:721–741.

[5] Journel, A. G. (1977) *Simulations conditionnelles de gisements miniers.* thèse de docteur-ingénieur, Université de Nancy-I, Nancy, France.

[6] Langlais, V. (1990) *Estimation sous contraintes d'inégalités.* thèse de docteur en géostatistique, ENSMP, Fontainebleau, France.

[7] Lantuéjoul, Ch. (1992) "Substitution random functions". In A. Soares, editor, *4th International Geostatistics Conference*, Dordrecht Holland, Kluwer Academic Press.

[8] Matheron, G. et al. (1988) "Simulation conditionnelle à trois faciès dans une falaise de la formation du brent". Etudes Géostatistiques V - Séminaire C.F.S.G. sur la géostatistique 15-16 juin 1987, Fontainebleau. *Sciences de la Terre*, 28:213–249.

[9] Metropolis, N. et al. (1953) "Equations of state calculations by fast computing machines". *J. Chem. Phys.*, 21:1087–1091.

[10] von Neumann, J. (1951) Various techniques used in connection with random digits. *National Bureau of Standards Applied Mathematics Series*, 12:36–38.

GEOSTATISTICS: ROADBLOCKS AND CHALLENGES

ANDRÉ G. JOURNEL
Stanford Center for Reservoir Forecasting, Stanford, CA, USA, 94305

ABSTRACT

Misconceptions about the role of random function models may have distracted geostatistical research into unrewarding debates about model properties and led some practitioners into using more complex algorithms when simpler ones would have sufficed.

Typical of such roadblocks are concerns about stationarity, ergodicity, model consistency when random function models have no intrinsic siginificance. They are but data analysis tools and, as such, can be bent, retooled or simply discarded.

The challenge proposed to geostatistical theory is to data-charge its algorithms to account for an ever increasing variety of information, of both structural nature (multiple-event dependence beyond traditional two-point covariances) and local nature (hard and soft data). Last, like recent developments in simulated annealing, these algorithms need not be probabilistic.

I - INTRODUCTION

Geostatistical research and development is no longer concentrated in a few academic centers, it has spread into the hands of numerous consultants and research and development industrial groups who are accountable for results and time spent. Consultants and engineers are well known for their ingenuity and boldness when it comes to finding a solution in a limited time with limited resources. They may have no hesitation in welding together machines (models), fixing the leaks, trying it out under multiple stresses, then getting from it an approximate solution to their problem. Geostatisticians would learn by understanding why such combinations work, grasping the essence of their success then enlarging the scope of their own models. This is how applied mechanics works, blending intuition, ingenuity, experimentation and rigorous analysis. Experience is proving that geostatistics is progressing the same way with considerable input from the day-to-day practitioners.

In this review paper, some old problems are revisited in the light of the present practice of very fast conditional simulation. Concerns about stationarity and ergodicity, about whether to use a non-stationary model, are mostly due to improper terminology and can easily be put to rest. The long standing roadblock of block distributions has been sidetracked by fine scale simulations: it suffices to observe these distributions.

Then, the paper focuses on the essence of any geostatistical investigation: the determination of a model of spatial uncertainty. The recent surge of new algorithms for stochastic imaging indicates an evolution from theoretically-derived to algorithmically-derived random function models. The former are usually more internally consistent, while the latter are more externally consistent in the sense that they account for more of the available data.

The last section of the paper introduces some of the most promising new conditional simulation algorithms which account for multiple-point statistics beyond the traditional two-point covariance/variogram. The non-probabilistic algorithm of simulated annealing stands out by capitalizing on the revolution in computing brought about by modern graphic workstations.

II - STATIONARITY AND ERGODICITY

Statements such as "the data are non-stationary" reflect a misunderstanding of the concept which may lead to time wasted in "testing for stationarity".

Stationarity and ergodicity are properties of the probabilistic model not attributes of either the data or phenomenon under study[1]. Rather than hypotheses - a term which implies the possibility of testing and invalidation - stationarity and ergodicity are model decisions necessary to perform inference from data. However, the resulting random function (RF) model can be proven *a posteriori* inappropriate if it fails to deliver the application intended.

A full definition of ergodicity[2] is not necessary. It suffices to mention the characteristic Ergodic Theorem which states that moments and distributions of a stationary and ergodic RF can be inferred from single realization statistics. In most applications, only a single realization is available, thus a RF model *must* be stationary and ergodic - at least with regard to the moments being inferred - to allow any data-based inference.

It is the decision of stationarity which allows pooling together observations taken at different locations in space and/or time, and it is the decision of ergodicity that allows inference of the model moments and distributions from the single realization statistics. Often measurements of a spatial phenomenon taken at different times are considered as different realizations of the RF model. In fact, such a phenomenon is distributed in space *and* time, and implicit stationarity in time of the model is required to interpret a single realization in space and time as multiple realizations in space. As more information becomes available or if the study goal changes, one may change model or decide on different pooling alternatives.

In an application mode, rapping about stationarity and ergodicity is most often futile. So are concerns about "existence of moments". If a particular moment, say, the covariance, is needed for the algorithm intended, then a RF model with such moment is adopted and there is no need for further a priori justification. The only important question is whether this moment can be reasonably inferred from available data.

III - NON-STATIONARY MODELS

Although there cannot be inference unless the RF model possesses some form of stationarity, practitioners have always been uneasy about pooling together data taken at different locations, particularly when the pool definition is large due to data sparsity. Therefore, very early in the development of geostatistics, every effort was made to render the RF model robust with regard to inhomogeneity in the data. All these implementation steps amount to inconsistencies with regard to the prior decision of stationarity.

The most extensive and successful violation of the stationary model concept is the practice of ordinary kriging within moving data neighborhoods. This practice amounts to keeping the stationary covariance inferred over a large pool of data, yet considering a locally variable (thus non-stationary) model mean.

The ordinary kriging unbiasedness condition (sum of weights set to one) amounts to re-estimating at each location $\mathbf{u}$ the mean value $m(\mathbf{u})$ from the neighborhood data. Note that because the search neighborhood varies with $\mathbf{u}$, the notation $m(\mathbf{u})$ is more accurate than the more traditional "stationary" notation m. It also indicates that ordinary kriging is already a "non-stationary" algorithm that can handle inhomogeneity in the data, and indeed this is the very source of the success and durability of ordinary kriging when applied within *moving* data neighborhoods.

Later non-stationary models, such as underlying the twin algorithms of universal kriging and intrinsic random function (IRF) kriging, amounted to expanding the functional form of the trend model:

$$m(\mathbf{u}) = a_0(\mathbf{u}) + \sum_{l=1}^{L} a_l(\mathbf{u})\, f_l(\mathbf{u}) \;, \tag{1}$$

where the $f_l(\mathbf{u})$ are known functions of $\mathbf{u}$ and the $a_0(\mathbf{u})$, $a_l(\mathbf{u})$'s are unknown coefficients to be re-estimated locally. Unless the component functions $f_l(\mathbf{u})$ carry a physical significance and estimates of the specific coefficients a_l are needed (e.g. for a filtering of some trend components), there is no practical advantage nor theoretical ground for the decomposition (1).

In interpolation conditions where the location $\mathbf{u}$ being estimated is within the correlation range of the neighborhood data, the ordinary kriging estimate does not differ significantly from any universal kriging estimate. In other words, in such conditions the decomposition (1) is inconsequential[3]. One is then better off using the simpler and more robust ordinary kriging algorithm.

In extrapolation conditions, universal kriging or IRF-kriging amounts to fitting by generalized least squares the specific decomposition (1) from within the data, then extrapolating the result beyond the data correlation range. Then, clearly, it does matter which functions $f_l(\mathbf{u})$ are being extrapolated. The critical problem is not which algorithm has been used to fit the model (1) but which functions $f_l(\mathbf{u})$ are used for extrapolation.

IV - NON-LINEAR KRIGING AND BLOCK DISTRIBUTIONS

Non-linear kriging algorithms are actually traditional linear krigings applied to some non-linear transform of the data, be it a logarithmic transform, a normal score transform or an indicator transform.

Two problems triggered the development of non-linear kriging in the mid-1970's:

(1) - the extreme variability of some attributes such as gold or uranium grades which made variogram inference difficult

(2) - the problem of evaluation of block grade distributions.

In the presence of extreme variability it is tempting to work on a smooth transform of the data, such as the logarithm. Experimental variograms of log-data are better behaved, easier to interpret and model. Unfortunately, it is not the log-attribute which is sought after but the attribute itself. A naive inverse transform (e.g. antilog of the log-attribute estimate) in addition to yielding a biased estimator is extremely sensitive to local inhomogeneity in the data (errors are exponentiated!). What is gained in structural analysis ease and robustness by working on a smooth transform of the data is lost at the backtransform stage. This explains why the early 1980's surge in lognormal kriging applications has essentially faded away.

What has remained from this line of research are non-linear kriging applications that do not require any back transform. Essentially, non-linear kriging is now used *not* to estimate local attribute values but to evaluate conditional distribution functions for those unsampled attribute values. These conditional distributions are then used not so much to provide "better" estimates for the unsampled values but to generate conditional simulations (stochastic imaging), see hereafter sections V and VI.

With regard to the problem of evaluation of block grade distributions, there are two schools of thought:

(i) - one that strives to define analytically the block probability distribution conditional to information on smaller data support

(ii) - one that is content with defining that same conditional block distribution through a series of simulated block values.

In the second case, generation of block conditional distributions is seen as the generation of a series of very fine scale (point) conditional simulations. The simulated values are then averaged within block volumes to generate simulated block values. The problem with that approach is one of CPU time and computer storage, a problem increasingly vanishing with modern computers. Note that this second approach does not require that the block value be a linear average of the point values, any non-linear averaging algorithm could be considered, e.g. a geometric or power average for effective block permeability values.

V - CONDITIONAL DISTRIBUTIONS

What differentiates probabilistic estimation from more traditional interpolation/extrapolation algorithms is the modeling of uncertainty. The kriging variance was a selling point for the kriging algorithm until its shortcoming was better understood:

~ the kriging variance being data values - independent is but a ranking index of data configurations.

~ the kriging variance is specific to a single point or block being estimated. A series (or a map) of kriging variances does not inform about joint spatial uncertainty, say, the uncertainty about an estimated feature involving several points.

The kriging variance has been often misused for locating additional samples. It would lead to adding samples where there are few and independently of prior data values, not where the additional data would reduce most the uncertainty that counts[4]. The solution is to determine a full, data values-dependent, measure of local uncertainty such as the conditional distribution defined below or, better, consider the measure of joint spatial uncertainty provided by conditional simulations, see next section VI.

The conditional cumulative distribution function (*ccdf*) is the probability distribution of the unsampled value updated by the information carried by the neighboring data values:

$$F(\mathbf{u}; z|(n)) = \text{Prob}\{Z(\mathbf{u}) \leq z|n \text{ data values }\} \tag{2}$$

where $Z(\mathbf{u})$ is the random variable modeling uncertainty about the unsampled value $z(\mathbf{u})$, the notation $|(n)$ being read as "conditional" to the n data values $Z(\mathbf{u}_\alpha) = z_\alpha$ retained.

The *ccdf* measure of uncertainty is intrinsic to the data available (n) and to the prior model of spatial dependence adopted; it is not dependent on the particular estimator $Z^*(\mathbf{u})$, i.e. on the particular massaging of the data, retained.

Parametric approach

In a parametric approach a particular distribution model is assumed for the *ccdf* and its parameters, usually its mean and variance, are determined from the data. The most convenient and therefore most used *ccdf* model is the Gaussian model with mean and variance equal to the simple kriging (SK) mean and variance of either $Z(\mathbf{u})$ or its normal score transform. The gaussianity of the *ccdf* is a property of the RF model chosen, it does not generally arise from either the data or the physics of the phenomenon under study.

Of all bivariate distribution models sharing the same covariance the bivariate Gaussian model is that which maximizes bivariate entropy[5]. Some call on that characteristic and on Shannon's principle of maximum entropy to justify their choice of the Gaussian model. The flaw in such an argument is that the principle of maximum entropy should be applied to the output distribution, that of $\varphi(Z(\mathbf{u}))$, not that of $Z(\mathbf{u})$ itself. Indeed, a *ccdf* model such as (2) or a series of stochastic images, are rarely an end goal; instead these images are input to a process or transfer function φ which yields an output $Y = \varphi(Z(\mathbf{u}))$. The entropy to be maximized is not that of the input RF $Z(\mathbf{u})$ but that of the output process $Y = \varphi(Z(\mathbf{u}))$.

The Gaussian RF model shares with non-mixture isofactorial models a very specific property which disqualifies them from being "non-committal" a priori models: extreme-valued realizations of such models are increasingly less spatially connected[6], there cannot be clusters or bands of extreme values. Moreover this "destructuration" property is symmetric around the median cutoff, i.e. extreme highs are equally disconnected as are extreme lows. Realizations of such models would be non-conservative input distributions to transfer functions φ sensitive to spatial connectivity of highs and lows, such as flow simulators in reservoir engineering, remediation processes in pollution control, drilling targets in mining exploration.

Yet, there is no shame in adopting a congenial model as long as the consequences of such a model choice are understood.

Indicator (Bayes') approach

The conditional probability (2) can be written according to Bayes' relation, and using indicator RF notations:

$$F(\mathbf{u}; z|(n)) = \frac{E\{I(\mathbf{u}; z) \cdot \Pi_{\alpha=1}^{n} I_\alpha\}}{E\{\Pi_{\alpha=1}^{n} I_\alpha\}} \tag{3}$$

with: $I(\mathbf{u}; z) = 1$ if $Z(\mathbf{u}) \leq z$, $= 0$ if not, being the indicator of the event to be assessed, the I_α's are the indicator of the n data events, e.g. $I_\alpha = 1$ if $Z(\mathbf{u}_\alpha)$ is in a class containing the datum value z_α, = 0 if not.

Then: $E\{I(\mathbf{u};z) \cdot \Pi_{\alpha=1}^{n} I_\alpha\} = \text{Prob}\{Z(\mathbf{u}) \leq z,\ Z(\mathbf{u}_\alpha) = z_\alpha, \alpha = 1,\ldots,n\}$

$$E\{\Pi_{\alpha=1}^{n} I_\alpha\} = \text{Prob}\{Z(\mathbf{u}_\alpha) = z_\alpha, \alpha = 1,\ldots,n\}$$

Note that simple indicator kriging (IK) considering the single indicator datum $J = \Pi_{\alpha=1}^{n} I_\alpha$ identifies Bayes relation (3). Indeed[6, 7]:

$$I^*(\mathbf{u};z) - F(z) = \lambda[J - E\{J\}] \tag{4}$$

with: $F(z) = E\{I(\mathbf{u};z)\} = \text{Prob}\{Z(\mathbf{u}) \leq z\}$
and the kriging weight λ given by the single normal equation:

$$\lambda \text{ Var } \{J\} = \text{Cov } \{I(\mathbf{u};z),\ J\}, \text{ i.e. } \lambda = \frac{E\{I(\mathbf{u};z) \cdot J\} - F(z) \cdot E\{J\}}{E\{J\} \cdot [1 - E\{J\}]}$$

Finally, for $J = 1$:

$$F(\mathbf{u};z|(n)) = I^*(\mathbf{u};z) = \frac{E\{I(\mathbf{u};z) \cdot J\}}{E\{J\}}, \text{ as in relation (3).}$$

Similarly, it can be shown that the IK estimator (4) provides the *exact* (Bayes) result for the case $J = 0$.

Either IK estimator (4) or Bayes relation (3) calls for inference of the multiple events probabilities (actually they are indicator covariances) $E\{I(\mathbf{u};z) \cdot J\}$ and $E\{J\}$. Such probabilities could be inferred from corresponding experimental frequencies obtained by scanning a training image[8], which amounts to trading the formal *stationary RF* model for a discrete and extensive set of indicator realizations. The only conditions are:

(i) - the training image is representative of the actual dependence between data and the event to be assessed, i.e. the implicit decision of stationarity is appropriate

(ii) - the simultaneous $(n+1)$-event $\{Z(\mathbf{u}) \leq z,\ Z(\mathbf{u}_\alpha) = z_\alpha,\ \alpha = 1,\ldots,n\}$ for the specific geometry $\{\mathbf{u}, \mathbf{u}_\alpha, \alpha = 1,\ldots,n\}$ is observed often enough to infer its probability.

If that last condition is not met, then the training image can be enlarged which amounts to relaxing the decision of stationarity, or the number n of data retained is made smaller, or the data events are made less specific, e.g., by considering the information $Z(\mathbf{u}_\alpha) \leq z$ instead of $Z(\mathbf{u}_\alpha) = z_\alpha$. The last two options amount to reducing the information retained to assess (update) the probability of the event $Z(\mathbf{u}) \leq z$. The larger the data set considered and the more specific the information retained, the larger must be the training image if assessment of the *ccdf* is to be data-based as in (3) or (4). One alternative is to rely on non-data-based high order properties of a RF model. In the case of a Gaussian RF model, the multiple point probabilities $E\{I(\mathbf{u};z) \cdot J\}$ and $E\{J\}$ are derived from the model *not* from actual data.

Another alternative, which stands between the purely data-based Bayes approach (3) and the almost purely model-based Gaussian approach, is offered by the concept of extended IK or extended normal equations developed in[7, 8], and summarized hereafter.

Extended normal equations

To alleviate the inference load, one could replace the global information $J = \Pi_{\alpha=1}^{n} I_\alpha$ defined in relation (4) by any number of lesser products of its elementary components I_α. The cost is information loss and the possibility of inconsistencies.

For example, if the specific joint data event $I_{\alpha_1} I_{\alpha_2} I_{\alpha_3}$ is deemed critical to assess the occurrence of the event $I(\mathbf{u}; z)$, one might consider the *ccdf* model provided by the extended IK estimator:

$$F^*(\mathbf{u}; z|(n)) = F(z) + \sum_{\alpha=1}^{n} \lambda_\alpha \left[I_\alpha - E\left\{I_\alpha\right\}\right] + \nu_{123}\left[I_{\alpha_1} I_{\alpha_2} I_{\alpha_3} - E\left\{I_{\alpha_1} I_{\alpha_2} I_{\alpha_3}\right\}\right] \quad (5)$$

where the $(n+1)$ data events retained are the n elementary data events I_α and the single product $I_{\alpha_1} I_{\alpha_2} I_{\alpha_3}$.

The $(n+1)$ weights $\lambda_\alpha, \alpha = 1, \ldots, n$, and ν_{123} are determined by an extended set of $(n+1)$ normal equations, see[7, 9] and Appendix:

$$\left\{ \begin{array}{lll} \sum_{\beta=1}^{n} \lambda_\beta \operatorname{Cov}\left\{I_\beta, I_\alpha\right\} + \nu_{123} \operatorname{Cov}\left\{I_{\alpha_1} I_{\alpha_2} I_{\alpha_3}, I_\alpha\right\} & = & \operatorname{Cov}\left\{I(\mathbf{u}; z), I_\alpha\right\}, \alpha = 1, \ldots, n \\ \sum_{\beta=1}^{n} \operatorname{Cov}\left\{I_\beta, I_{\alpha_1} I_{\alpha_2} I_{\alpha_3}\right\} + \nu_{123} \operatorname{Var}\left\{I_{\alpha_1} I_{\alpha_2} I_{\alpha_3}\right\} & = & \operatorname{Cov}\left\{I(\mathbf{u}; z), I_{\alpha_1} I_{\alpha_2} I_{\alpha_3}\right\} \end{array} \right. \quad (6)$$

As opposed to the *ccdf* model (4), the determination of model (6) requires only the inference of covariances involving 2, 3, and 4 points, instead of $(n+1)$.

Consistency problems

Because the model (5) does not use the single complete global information $J = \Pi_{\alpha=1}^{n} I_\alpha$, it may not share the strict consistency of Bayes' relation (3) or (4). The *ccdf* model (5) may not be valued in $[0, 1]$, and it may not be a non-decreasing function of the threshold value z. Such order relation deviations can be corrected a posteriori. These corrections can be seen as ad hoc by some; however, if one accepts that there is no "true" *ccdf*, that all *ccdf*'s are but models including the exact Bayes' expression (3), order relation corrections can be seen as an integral part of the *ccdf* modeling process.

Order relation problems are the necessary price to pay for any approach that does not have access to either a full internally consistent RF model (e.g. Gaussian) or an extensive training image allowing consistent inference of all required indicator covariances (experimental frequencies of occurrence).

One can adopt an iron-clad internally consistent RF model, but such models are typically inflexible allowing very little structural information to be accounted for, a single covariance function in the case of a Gaussian model. Such *internally consistent* models are thus, most often, *maximally externally inconsistent* in the sense that they are inconsistent with information they cannot account for. If that information is deemed critical for the problem at hand, one may opt for a richer model that can be more "data-charged", such as the models provided by extended normal equations, even if such models may not be fully internally consistent. Engineers would accept continuously tuning a model towards a solution rather than adopting a shining model that is likely to fail the job.

VI - STOCHASTIC IMAGING

The most vigorous development in geostatistical theory and practice of the late 1980's has been in the field of conditional simulations also referred as stochastic imaging. In little less than five years, the vintage algorithm of turning band simulations has been outclassed by many other Gaussian and non-Gaussian algorithms including *LU* decomposition, random fractals, sequential Gaussian and sequential indicator simulation, truncated Gaussian fields, Markov point processes, Boolean algorithms, simulated annealing, a not yet exhaustive list.

Sequential simulation

For simulations of both categorical and continuous variables, the sequential simulation principle[10] is particularly flexible and straightforward of application. The sequential simulation principle consists of drawing simulated nodal values from successive *ccdf*'s conditioned to the original data *and* to all previously simulated values, thus guaranteeing that the simulated values reproduce the input covariance model(s). The simulation progresses along a random path from one node to another until the stochastic image is completed at the resolution required. The entire process with another random path is repeated if another stochastic image is needed.

The *ccdf*'s can be generated by any algorithm, either using a prior parametric multivariate model (e.g. Gaussian), or building from a more data-rich multiple indicators model. The latter model allows hard and soft data of different types to be used for conditioning and accounts for their respective reliability.

Probability field simulations

Rather than conditioning the *ccdf*'s by all prior simulated values, which requires establishing these *ccdf*'s anew from one realization (stochastic image) to another, Srivastava[11] has suggested that reproduction of the spatial continuity be imposed through spatial correlation of the random numbers drawn to generate the simulated attribute value. More precisely, the probability (p-) field simulations algorithm consists of[12]:

(1) - Generation of the *ccdf*'s at all nodes of the simulation grid, considering only the original data (both hard and soft). This is done only once(!) no matter the number of realizations required. This explains the remarkable speed of the procedure.

(2) - Generation of as many *unconditional* realizations as necessary of the p-field or *RF* $T(Z(\mathbf{u})) = F^{-1}(Z(\mathbf{u}))$, where $F(z)$ is the stationary marginal *cdf* of the *RF* $Z(\mathbf{u})$ modeling the variable of interest. Because these realizations are unconditional, they can be generated by any very fast algorithm such as fractal Brownian motion, or more general spectral techniques using fast Fourier transform. Let $t_l(\mathbf{u}) \in [0,1], \mathbf{u} \in A$, be the $l^{\underline{th}}$ realization of that p-field over the simulation grid A.

(3) - Let $F(\mathbf{u};z|(n)),\ \mathbf{u} \in A$, be the *ccdf*'s generated in step (1) at each node $\mathbf{u}$ of the grid A, the $l^{\underline{th}}$ realization (stochastic image) of the Z-field is simply $z_l(\mathbf{u}) = F^{-1}\{\mathbf{u};t_l(\mathbf{u})|(n)\},\ \mathbf{u} \in A$.

The originality of Srivastava's proposal lies in the dissociation of the process of conditioning performed in step (1) and that of reproduction of spatial correlation obtained by considering correlated random number $t_l(\mathbf{u})$ to draw from the *ccdf*'s.

Simulated annealing

Possibly, the most significant development in stochastic imaging of the last few years has been the introduction of stochastic relaxation techniques[13, 14] in the field of conditional simulations. Except for simulations with extended normal equations[8] which require extensive training images for inference, there is presently no geostatistical algorithm that allows capturing more than two-point statistics. Traditional covariances or variograms, including indicator covariances, are two-point statistics in the sense that they involve the same variable at two different locations, as in $\text{Cov}\{Z(\mathbf{u}), Z(\mathbf{u}+\mathbf{h})\}$, or two different variables at the same or at two different locations as in the cross-covariance $\text{Cov}\{Z(\mathbf{u}), Y(\mathbf{u}+\mathbf{h})\}$.

There may be significant and reliable information involving more than two points or two variables at a time such as

- a particular *curvilinear* feature of extreme-valued attribute values characterized by the K-point indicator covariance (non-centered).

$$E\{I(\mathbf{u}_1)\cdot I(\mathbf{u}_2),\ldots,I(\mathbf{u}_K)\} = \text{Prob}\{Z(\mathbf{u}_1) > z, Z(\mathbf{u}_2) > z,\ldots,Z(\mathbf{u}_K) > z\}, \quad (7)$$

  where $I(\mathbf{u})$ is the indicator of $Z(\mathbf{u})$ exceeding threshold z. Note that the K locations $\mathbf{u}_1,\ldots,\mathbf{u}_K$ need not be aligned, their geometric configuration may correspond to, say, a dune-shaped configuration of fine grain deposition in an eolian sand[8].

- a particular *non*-linear average of attribute values

$$Y(\mathbf{u}_\alpha) = \phi(Z(\mathbf{u}')) \ , \ \mathbf{u}' \in \text{neighorhood of } \mathbf{u}_\alpha, \quad (8)$$

such as provided by, e.g., a well test at location $\mathbf{u}_\alpha$[14].

When building stochastic images of the *RF* $Z(\mathbf{u})$, one may want to impose structural information of type (7) in addition to the traditional two-point covariance model(s), and local information of type (8) in addition to hard or soft data associated with a single location $\mathbf{u}_\alpha$.

The algorithm starts from either a pure random noise image $\{z_l(\mathbf{u}),\ \mathbf{u} \in A\}$ with the correct marginal *cdf* $F(z)$ or, better, from a prior stochastic image reproducing both $F(z)$ and some measure of two-point dependence, then the values $Z_l(\mathbf{u})$ are shuffled or perturbed in such a way that

(1) the prior statistics $F(z)$ and $C(\mathbf{h})$ are modified the least possible

(2) the additional information of type (7) and/or (8) is accounted for approximately.

By repeating the process starting from different prior images $\{z_l(\mathbf{u}), \mathbf{u} \in A\}, l = 1,\ldots,L$, one generates a distribution of data-rich stochastic images. The fact that one does not know or control fully the implicit *RF* $Z(\mathbf{u})$ underlying such a set of "annealed" images is of no great concern: it bears repeating that *RF* models have no intrinsic value, they are usually not generators of the physical phenomenon under study, they are but expendable models of uncertainty (our own uncertainty *not* that of the phenomenon).

VII - CONCLUSIONS

In the 1960's, the development of geostatistics was made possible with the advent of digital computers. The recent surge in new algorithms for stochastic imaging (hundreds of realizations over millions of nodes) was triggered by the workstation revolution. Strict prior considerations about the *RF* model were useful in a time of computer shortage when one had to rely almost entirely on model properties. Such considerations may now be counterproductive. The computing revolution allows brute-force trials and errors (e.g. the simulated annealing concept), progressive fine-tuning of multiple models and algorithms until an "acceptable" result is obtained.

Exactly how such an "acceptable" result has been reached need not be immediately understood, the important point is that it can be repeated. The underlying generating *RF* model (an elusive and often misleading concept) can be analysed a posteriori from its realizations. In other words, the algorithm defines the *RF*, not the reverse; or, even better, the concept of an underlying *RF* model is dropped altogether.

References

[1] **Myers, D. E.** *(1989)* - To be or not to be ... stationary? That is the question, Math. Geol. **21**(3): 347-362.

[2] **Blum, J. R.** *(1982)* - Ergodic theorems. Entry in Encyclopedia of Statistical Sciences, ed. Kotz and Johnson, Wiley & Sons, **2**: 541-545.

[3] **Journel, A. G.** and **Rossi, M.** *(1989)* - When do we need a trend model in kriging?, Math. Geol. **21** (7): 715-739.

[4] **Srivastava, M.** *(1987)* - Minimum variance or maximum profitability?, in CIM **80** (901): 63-68.

[5] **Christakos, G.** *(1990)* - A Bayesian/maximum entropy view of the spatial estimation problem, in Math. Geol. **22**(7): 763-777.

[6] **Journel, A. G.** and **Alabert, F** *(1989)* - Non-Gaussian data expansion in the earth sciences, in Terra Nova **1** (2): 123-134.

[7] **Solow, A.** *(1986)* - Mapping by simple indicator kriging, in Math. Geol. **18** (3): 335-352.

[8] **Guardiano, F.** and **Srivastava, M.** *(1992)* - Multivariate geostatistics: beyond two-point moments, in Proc of 4th Int. Geostat. Congress, in this volume.

[9] **Luenberger, D.** *(1969)* - **Optimization by vector space methods**, Wiley & Sons, 326p.

[10] **Deutsch, C.** and **Journel, A. G.** *(1992)* - GSLIB: **Geostatistical software library and Users' guide**, Oxford University Press.

[11] **Srivastava, R. M.** *(1991)* - P-fields conditional simulations, FSS International internal report.

[12] **Froidevaux, R.** *(1992)* - Probability field simulation, in Proc. of the 4th Geostat. Congress, in this volume.

[13] **Geman, S.** and **Geman, D.** *(1984)* - Stochastic relaxation, Gibbs distributions and the Bayesian restoration of images, in **IEEE Transactions on Pattern Analysis and Machine Intelligence**, PAMI-**6** (6): 721-741.

[14] **Deutsch, C. V.** *(1992)* - Conditioning stochastic reservoir models to well test information, in Proc. of 4th Int. Geostat Congress (this volume).

APPENDIX

Extended normal equations

For reasons of simplicity, the indicator formalism and equations (3) to (6) were presented for the case of a continuous variable $Z(\mathbf{u})$ with data events $I_\alpha : \{Z(u_\alpha)$ in class containing value $z_\alpha\}$ related to the same attribute.

However, the event (A_0) to assess can be anything, e.g., A_0 could be related to the occurrence of a particular lithofacies k at the single location $\mathbf{u}$ or over a set of locations $\mathbf{u} \in S$, i.e. $A_0 = 1$ if and only if S is entirely within facies k. Similarly, the elementary data events A_α could be anything, related to a single location as hard datum $A_\alpha : \{Z(\mathbf{u}_\alpha) = z_\alpha\}$ or a constraint interval $A_\alpha : \{Z(\mathbf{u}_\alpha) \in (a_\alpha, b_\alpha]\}$, or related to several locations $A_\alpha : \{Z(\mathbf{u}') > z, \ \forall \, \mathbf{u}' \in S(\mathbf{u}_\alpha)\}$.

A *ccdf* model of type (5), not retaining the global data event $J = \Pi_{\alpha=1}^{n} A_\alpha$ for reasons of inference difficulty, would be written using the more general set of notations:

$$\text{Prob}^* \{A_0 = 1 | A_\alpha = a_\alpha, \alpha = 1, \ldots, n\} = [A_0]^*_{IK} = E\{A_0\} + \sum_{\alpha=1}^{n+1} \lambda_\alpha [a_\alpha - E\{A_\alpha\}] \quad (9)$$

with a_α being the binary realization (0 or 1) of the random event A_α. The notation $A_{n+1} = A_{\alpha_1} A_{\alpha_2} A_{\alpha_3}$ is used for the specific product of 3 elementary data events A_α deemed critical to the assessment of the (unknown) event A.

The normal equations determining the $(n+1)$ weights λ_α can be interpreted as the condition for the estimator $[A_0]^*_{IK}$ to reproduce the input probabilities $E\{A_0 | A_\alpha = 1\}$ which have been inferred either from a training image[8] or from a prior multivariate model, say Gaussian if all events A_0, A_α relate to a continuous variable. Indeed, the orthogonality of the random error vector $A_0 - [A_0]^*_{IK}$ to each of the random data event A_α is written[9]:

$$< A_0 - [A_0]^*_{IK}, A_\alpha >= 0 \quad , \ \forall \, \alpha = 1, \ldots, n+1 \quad (10)$$

Since: $< A_0, A_\alpha >= E\{A_0 A_\alpha\} = \text{Prob}\{A_0 = 1, A_\alpha = 1\}$,

and: $< [A_0]^*_{IK}, A_\alpha >= E\{[A_0]^*_{IK} A_\alpha\} = E\{[A_0]^*_{IK} | A_\alpha = 1\} \cdot \text{Prob}\{A_\alpha = 1\}$,

the orthogonality conditions (10) are rewritten:

$$E\{[A_0]^*_{IK} | A_\alpha = 1\} = \text{Prob}\{A_0 = 1 | A_\alpha = 1\} \quad , \ \forall \alpha = 1, \ldots, n+1 \quad (11)$$

In words, the normal equations ensure that in expected value the model probabilities $[A_0]^*_{IK}$ are consistent with each of the $(n+1)$ input probabilities $E\{A_0|A_\alpha = 1\} = \text{Prob}\{A_0 = 1|A_\alpha = 1\}$. However, that consistency does not entail necessarily consistency with the global data event $J = \Pi^n_{\alpha=1} A_\alpha$, i.e. $E\{[A_0]^*_{IK}|J = 1\}$ is usually different from $E\{A_0|J = 1\}$, unless the global data event has been considered explicitly as in the IK expression (4).

Once again, which data event to retain in the expression of the conditional probability model (4) or (9) is strictly a question of inference ability.

RANDOM FUNCTIONS AND FRACTURE STATISTICS MODELS

Dominique JEULIN
Centre de Géostatistique, Ecole des Mines de Paris
35 rue St–Honoré, 77305 FONTAINEBLEAU, France

Abstract

For reliability analysis of materials, it is necessary to be able to predict the probability of fracture of a specimen, as a function of :

- a fracture mechanism assumption, relying on a local critical state,
- the loading conditions and the geometry of the specimen,
- a probabilistic model for the spatial distribution of the critical criterion,
- a model for the change of scale.

Two different fracture criteria are introduced : the weakest link assumption and a critical damage model. They involve two different kinds of change of support, namely by a convolution and by an inf. operation.

Examples of random function models are presented, for which a theoretical calculation of coherent fracture statistics at different scales is possible, and examples of scaling laws are given.

INTRODUCTION

In various domains, there is a need for probabilistic models to predict the changes of scale effects in fracture statistics of heterogeneous media.The potential field of applications of these models is wide : in addition to the fracture of materials (for the nuclear industry, aeronautics, etc.), where some of them are already implemented, they can be used to solve various problems, such as soils and rock mechanics, fracture processes in earthquakes and in dam stability, etc.

The purpose of this paper is to introduce some types of models which enable us to calculate the probability of fracture of a given specimen, according to its loading conditions, and to its local load bearing capacity. A more complete presentation is provided in [1–3].

A fracture can be initiated in a material when the applied loading conditions are such that a fracture criterion (considered as an intrinsic property of the material) is reached. Various fracture criteria may be used. The most common is the critical stress σ_c. It may be a critical deformation ε_c (which is equivalent to the previous one for a linear elastic medium). For media undergoing a partial degradation before the fracture propagation, the criterion can be a critical damage D_c (which is a nonlocal property).

Whatever the fracture criterion, its critical value never remains constant in heterogeneous media : at a small (usually microscopic) scale, spatial fluctuations of the critical value are observed, so that local drops in the strength may be encountered ; moreover, stress concentration may appear, as a result of the spatial variations of the physical properties of the components of materials. In any case, the microstructural heterogeneities are enhanced for fracture properties, resulting into a high dispersion of the strength of many materials. To account for this experimental evidence, probabilistic models are required. To study the fracture of the same material under various sollicitations (size and shape of a part, various boundary conditions, etc.), a complete random function model must be introduced, as will be illustrated later with some specific models.

We will first introduce the principles of fracture statistics models, and then review the main results obtained from two kinds of fracture criteria : the weakest link approach, and a critical damage assumption.

Principle of fracture statistics models

A first step in the model is to select a fracture criterion, which reflects the physical aspects of the fracture process. Most often, it is based on a stress criterion, like the derivations presented below.

For a given specimen, its fracture will occur as long as the fracture criterion is satisfied. This event involves :

- the knowledge of the microscopic critical stress field $\sigma_c(x)$. The experimental determination of σ_c at every point of a material is out of reach of the present experiments, so that the introduction of models is necessary. We will use random function models with a point support.
- the determination of the applied stress field $\sigma(x)$, which depends on the local behaviour and on physical properties. When using a general probabilistic approach, $\sigma(x)$ is a random function (that should be correlated to the critical field $\sigma_c(x)$), which depends on the boundary conditions and on the microstructure. For instance, in the case of an elastic medium, $\sigma(x)$ is the solution of the Navier partial derivative equations with a random elasticity tensor. In the framework of the so-called "local approach" [4–5], which we will follow here, an important simplification is introduced to make the problem tractable : $\sigma(x)$ is computed for an equivalent homogeneous medium, and there is no stress redistribution during the fracture process. With this assumption, $\sigma(x)$ is a deterministic field (depending on the geometry of the problem), which is compared in each point x of the structure to a random threshold $\sigma_c(x)$.

Two families of fracture criteria deriving from the local comparison of the fields $\sigma(x)$ and $\sigma_c(x)$ were introduced [1–3]. For the case of random fields, they enable us to calculate the probabilities of fracture of a given specimen.

• The weakest link assumption is common for brittle materials (like ceramics) and for cleavage fracture in steels [4–6] : the material breaks, as soon as the critical value $\sigma_c(x)$ is reached by $\sigma(x)$ at a single point x. Fracture will occur when :

$$\exists x, \sigma(x) \geq \sigma_C(x) \tag{1}$$

• One can relax the previous assumption, to account for the fracture to initiate at several potential locations. The "critical damage threshold" assumption may be well suited for progressive fractures or for ductile fractures, despite the fact that, in the present study, we will disregard the redistribution of a stress during the progression of the fracture process.

As will be seen later, from a formal point of view, the two families of criteria, together with the simplifications introduced, imply a change of support of the field $\sigma_c(x)$ by the infimum operator $\wedge$ (weakest link assumption), or by a convolution operation (critical damage criterion). Random function models for which these change of support effects can be predicted, are appropriate candidates for use in fracture statistics.

WEAKEST LINK MODELS

Formulation of the weakest link assumption

It is quite common in materials sciences to use the weakest link assumption : the fracture of a specimen B occurs when a fracture can be initiated at any point x in B, as expressed in relation (1). This situation corresponds to the initiation of an unstable crack, that will propagate immediately, as can be seen in brittle materials. In that case, no competition between several crack fronts, and so subcritical crack growth is accounted for.

The weakest link assumption can be reformulated as follows : given a finite specimen B_y (a compact set), (implanted at point y in an infinite medium, there is no fracture of B, as long as we have :

$$\{\sigma_C(x) - \sigma(x-y) \geq 0, \ \forall x \in B_y\} \tag{2}$$

For a random medium, the probability of no fracture is given by :

$$P\{\text{no failure of } B_y\} = P\left\{ \bigwedge_{x \in B_y} (\sigma_C(x) - \sigma(x-y)) \geq 0 \right\} \tag{3}$$

In relation (2), the applied stress field σ does not depend on the local variations of σ_c if we make the homogeneous equivalent medium assumption. In addition, we will consider stationary random functions for σ_c, so that the probability P in (2) is invariant by translation of B.

If $H_{\sigma_C}(\sigma)$is the set of points y for which (2) is satisfied, (3) can be rewritten as :

$$P(\sigma) = P\{y \in H_{\sigma_c}(\sigma)\} \tag{4}$$

Given a stationary random function for $\sigma_C(x)$, it is generally not possible to know the probability law (3) or (4) in a closed form. However specific models were proposed.

The Boolean random Varieties in the weakest link model

For particular classes of models, the Boolean random Varieties, but also the Dead Leaves Varieties [1–3], theoretical calculations of the non–fracture probability can be made in a closed form. For these models, it is possible to derive theoretical models of change of scale including size and shape effects (concerning the specimen B), and also microstructure effects. Analytical results are available for the following sress fields :

- a uniform stress field $\sigma(x) = \sigma$ for a cubic specimen with volume $V = L^3$
- the stress field developed at the tip of a macroscopic crack in elastic–plastic materials.

We will recall here the general results and illustrate them for the uniform stress field.

The Boolean random Varieties can be constructed as follows :

- in $\mathbb{R}^n$, n Poisson linear Varieties, with dimension k ($k = 0,\ldots, n-1$) can be defined [7]. A geometrical interpretation of the varieties is the following : consider a Poisson point process $\mathcal{P}_k(\omega)$ with intensity $\theta_k(d\omega)\mu_{n-k}(dx)$ on the linear varieties of dimension $(n-k)$, with the orientation ω, and containing the origin (μ_{n-k} is the Lebesgue measure in $\mathbb{R}^{n-k}$). In each point $x_i(\omega)$ of the process, a variety of dimension k, $V_k(\omega)_{x_i}$, orthogonal to direction ω, is located. A realization of the variety of dimension k is obtained as

$$V_k = \bigcup_{x_i(\omega)} V_k(\omega)_{x_i}$$

For fracture statistics, we operate in $\mathbb{R}^3$, so that we consider the following Poisson varieties : points ($k = 0$), lines ($k = 1$), and flats ($k = 2$). In addition, we will limit ourselves to isotropic models, but anisotropic structures (with appropriate $\theta_k(d\omega)$) can be used when necessary.

- in a second step, we consider independent primary random functions $Z'(x)$, lower semi–continuous, with a subgraph admitting compact sections almost surely and with $Z'(x) \leq \sigma_m$. Starting with a homogeneous medium with $Z(x) = \sigma_m$, we build the Boolean random Variety of dimension k as follows :

$$Z(x) = \wedge\{Z'(x - y_i)\ ; y_i \in V_k x_i\ ; x_i \in \mathcal{P}_k\} \tag{5}$$

In $\mathbb{R}^3$, we obtain random grain defects (for $k = 0$), random fiber defects (for $k = 1$) and random strata defects (for $k = 2$). It can be shown that, for the Boolean Varieties, the probability of fracture, (3) can be expressed as :

$$P\{\text{no failure of B}\} = \exp - \theta\ \overline{\mu}\ (H^c_{Z'}(\sigma - \sigma_m)) \tag{6}$$

In (6), μ is equal to volume V (for grains), $(\Pi/4)S$ (S being the surface of the area) for fibers, or to the integral of mean curvature M for strata (with

$M = \frac{1}{2}\int_S \left(\frac{1}{R_1} + \frac{1}{R_2}\right) dS$, R_1 and R_2 being the main radii of curvature over the surface area S of the set $H^c_{Z'}(\sigma - \sigma_m)$. It is assumed in (6) that $H^c_{Z'}(\sigma - \sigma_m)$ is a convex set for the case of fiber and strata defects.

A slightly more general formulation of (6) can be obtained, if we start from a family of primary random functions Z'_t, implanted on random Poisson varieties $V_k(t)$ with intensity $\theta(t)$ (t being a time parameter):

$$P\{\text{no failure of B}\} = \exp - \int_{\mathbb{R}} \bar{\mu}(H^c_{Z'_t}(\sigma - \sigma_m))\theta(dt) \tag{7}$$

The general properties of these models for fracture statistics applications are the following :

- By construction, the fracture stress allows correlations at a microscale (depending on the size of the support of the defects $Z'_t(x)$).
- When considering the fracture strength at a macroscale (namely the scale of the specimen B), the following effects are expected to be observed : size effect (through the set of points $H^c_{Z'_t}(\sigma - \sigma_m)$, depending both on B and on Z'_t ; shape effect for the specimen B ; microstructure effect (grains, fibres or strata give different weights to the probability of failure, through V, S or M).

Some examples of Boolean Varieties

We develop now some specific examples of models, useful for applications, based on two separate constructions, and with $\sigma_m = +\infty$.

- First we consider defects Z'_t with a random closed set support X'_0. The critical stress of the defect σ_c is a random variable Z', independent of X'_0, implanted in space with the intensity $\theta(\sigma)$. Since we are interested in brittle fracture in traction, we restrict ourselves to the scalar case with $\sigma > 0$ (σ being here the local maximal principal stress). In the case of a uniform stress field σ over the specimen B, the distribution of the fracture strength σ_R of B can be deduced from equation (7) as :

$$P\{\sigma_R \geq \sigma\} = P\{\text{no failure of B}\} = \exp\left(-\bar{\mu}\,(X'_0 \oplus \check{B})\phi(\sigma)\right) \tag{8}$$

In (8), $\bar{\mu}$ is the average over the realizations of the random closed set X'_0; $X'_0 \oplus \check{B} = \{x; B_x \cap X'_0 \neq \emptyset\}$, and

$$\phi(\sigma) = \int_0^{\sigma} \theta(t)\,dt \tag{9}$$

In practice, any increasing non−negative function $\phi(\sigma)$ can be used in (8). We illustrate this with well−known examples.

- For defects with intensity $\theta(\sigma) = m\theta(\sigma - \sigma_0)^{m-1}$ with $m > 1$ and $\sigma > \sigma_0$, we get $\phi(\sigma) = \theta(\sigma - \sigma_0)^m$, and σ_R obeys a Weibull distribution, with $\sigma_u^m = 1/\theta$:

$$P\{\sigma_R \geq \sigma\} = \exp - \bar{\mu}(X'_0 \oplus \check{B})\left(\frac{\sigma - \sigma_0}{\sigma_u}\right)^m \qquad (10)$$

This distribution is very popular amongst praticioners of fracture statistics, as can be seen from the numerous references mentioned in [8–9]. The classical version of the distribution is limited to point defects X'_0, where the more severe defects (with a low σ_c) are much sparser than the less critical defects. This model is expected to be suited to high grade materials with a thorough control of defects.

From equation (10), it is easy to calculate the expectation $E[\sigma_R]$ and the variance of σ_R, as a function of $V = \bar{\mu}(X'_0 \oplus \check{B})$:

$$E[\sigma_R] = \sigma_0 + V^{-1/m}\sigma_u\Gamma(1 + 1/m) \qquad (11)$$

$$D^2[\sigma_R] = \sigma_u^2(V)^{-2/m}(\Gamma(1 + 2/m) - \Gamma^2(1 + 1/m)) \qquad (12)$$

where $\Gamma(x)$ is the Eulerian function $\Gamma(x) = \int_0^\infty t^{x-1} \exp - t \; dt$.

When $\sigma_0 = \sigma$, the coefficient of variation of σ_R, $D[\sigma_R]/E[\sigma_R]$ does not depend on V, which can be easily checked from experimental data :

$$D[\sigma_R]/E[\sigma_R] = \frac{(\Gamma(1 + 2/m) - \Gamma^2(1 + 1/m))^{1/2}}{\Gamma(1 + 1/m)} \qquad (13)$$

Equations (11) and (12) express a scale effect for the expectation and the variance of the strength, which always decrease with the size of the specimen, but differently according to the kind of Boolean Variety. This scale effect is also observed on the median σ_M of the fracture strength :

$$\sigma_M = \sigma_0 + K\, V^{-1/m} \qquad (14)$$

Similarly, we can use a bimodal Weibull distribution, where there is a superposition of defects from two Weibull populations (with parameters (θ_1, σ_{01}, m_1), and (θ_2, σ_{02}, m_2)) ; in that case,

$$\phi(\sigma) = \theta_1(\sigma - \sigma_{01})^{m1} + \theta_2(\sigma - \sigma_{02})^{m_2} \qquad (15)$$

- For defects with intensity $\theta(\sigma) = \frac{\theta}{\sigma_u}$, the distribution of strength is exponential, and can be considered as a particular Weibull distribution, with $m = 1$.

$$P\{\sigma_R \geq \sigma\} = \exp - (\theta V \sigma/\sigma_u) \qquad (16)$$

with expectation $\frac{\sigma_u}{\theta V}$, variance $\left(\frac{\sigma_u}{\theta V}\right)^2$ and coefficient of variation 1.

- For defects with intensity $\theta(\sigma) = \frac{\theta}{(m - 1)\sigma^m}$ for $\sigma \geq \sigma_0 > 0$, we have $\phi(\sigma) = \frac{1}{\sigma_0^{m-1}} - \frac{1}{\sigma^{m-1}}$ when $m > 1$.

• For defects with intensity $\theta(\sigma) = \theta/\sigma$ for $\sigma \geq \sigma_0 > 0$, we get

$$P\{\sigma_R \geq \sigma\} = (\sigma_0/\sigma)^{\theta V} \tag{17}$$

for $\sigma \geq \sigma_0$ and $V = \overline{\mu}(X'_0 \oplus \check{B})$, as previously.

Contrary to the Weibull model, the last two populations of defects are dominated by defects with very low critical stress σ_c.

The distribution (17) is well known as the distribution of Pareto [10]. Its expectation and variance are infinite for $\theta V \leq 1$ and $\theta V \leq 2$ respectively, namely for specimens B with a low size.

When $\theta V > 1$, we have

$$E[\sigma_R] = \frac{\sigma_0 \theta V}{\theta V - 1} \tag{18}$$

For $\theta V > 2$,

$$D^2[\sigma_R] = \frac{\sigma_0{}^2 \theta V}{(\theta V - 2)(\theta V - 1)} \tag{19}$$

and

$$D(\sigma_R)/E(\sigma_R) = \sqrt{\frac{\theta V - 1}{\theta V(\theta V - 2)}} \tag{20}$$

The last properties, given by equations (18–20), decrease with the size of the specimen B. The size effect is higher for Pareto's model than for Weibull's model.

• When $\phi(\sigma)$ is a sigmoidal function, a saturation effect is observed on the cumulative distribution of defects :

$$\phi(\sigma) = \theta \left[1 - \exp\left(-\left(\frac{\sigma - \sigma_0}{\sigma_u}\right)^m\right)\right] \qquad \text{for } \sigma \geq \sigma_0. \tag{21}$$

Such a distribution was observed for the fracture of carbon fibers [11].

▪ A second family of Boolean Varieties models uses defects made of random closed sets with a constant critical stress σ_c as previously, but the value of the σ_c depends on the realization of the primary grain X'_0, as follows. We consider random grains X'(d) (homothetics with ratio d of random grains X' = X'(1)). For each grain, $\sigma_c = \psi(d)$. In fracture mechanics, it is common to assume that $\psi(d) = K/\sqrt{d}$, each grain X'(d) behaving like a crack with length d. For the model, with X'(u) = uX', we have, in the case of a homogeneous stress field σ :

$$P\{\sigma_R \geq \sigma\} = \exp - \int_{K^2/\sigma^2}^{\infty} \theta(u)\mu(X'(u) \oplus \check{B})du \tag{22}$$

In equation (22), we can use Steiner's formula [7, 12] when the sets B and X'(u) are convex :

for Boolean grains

$$\overline{V}(X'(u) \oplus B) = V(B) + \frac{S(B)}{4\pi}uM + \frac{M(B)}{4\pi}u^2S + Vu^3 \tag{23}$$

for Boolean fibers

$$\overline{S}(X'(u) \oplus B) = S(B) + \frac{M(B)}{4\pi}uM + u^2S \tag{24}$$

for Boolean strata

$$\overline{M}(X'(u) \oplus B) = M(B) + uM \tag{25}$$

In equations (23–25), V, S, and M are the average volume, surface area, and integral of mean curvature, over the realizations of the random set X'.

Depending on the choice of $\theta(u)$ (i.e. of the grain size distribution of the model), various distributions can be obtained. When $\mu(B) \to \infty$, we recover asymptotically Weibull's distribution for $\theta(u) = \theta u^{-m}$ (with $m > 4$ for grains, $m > 3$ for fibers, and $m > 2$ for strata), and Pareto's distribution for $\theta(u) = \theta/u$.

As is illustrated by the examples, many fracture strength distributions can be derived from the Boolean Varieties models. They induce various scaling laws, concerning the decrease with the size of the specimen B of the average strength, and of the dispersion of the macroscopic strength. The decrease of the strength towards the minimal strength σ_0 is expected since the larger B, the higher the probability to find the most critical defect, in the context of the weakest link model. Such scaling laws should be useful for selecting appropriate models from fracture data obtained at different scales.

It is also clear from the examples that various models can lead to the same fracture strength distributions, when restricting ourselves to a given sollicitation (and mainly to asymptotic results). Different conditions of sollicitations should be used to select a proper model.

CRITICAL DAMAGE MODELS

When relaxing the weakest link assumption, it is possible to propose critical damage models, based on a critical volume fraction, or on a critical intensity, depending on the nature of the defects [3].

Fracture statistics models with a critical volume fraction

We consider in a bounded specimen B a stress field $\sigma(x)$ applied to a material with the critical stress field $\sigma_c(x)$. We define the indicator function $1_{\sigma_c}(x)$ of the domain D as :

$$1_{\sigma_c}(x) = 1 \quad \text{if} \quad \sigma_c(x) < \sigma(x), \quad \text{else } 0.$$

We have $\mu_n(D) = \int_B 1_{\sigma_c}(x)dx$. We assume that B will not fail, as long as $\frac{\mu_n(D)}{\mu_n(B)} < p_c$, where p_c is a parameter of the model. This parameter indirectly accounts for percolation effects, since when p_c increases, the domain D (where the breakage is expected to accur) progressively invades the specimen B.

In our probabilistic approach, $\sigma_c(x)$ is assumed to be a realization of a random function, and therefore $1_{\sigma_c}(x)$ is the indicator function of the random set D. We have

$$P\{\text{no failure of B}\} = P\left\{\frac{\mu_n(D)}{\mu_n(B)} < p_c\right\} \tag{26}$$

Solving (26) as a function of the probabilistic properties of $\sigma_c(x)$ and of the geometry of B (and of $\sigma(x)$) requires finding the solution of a change of support problem. Since no general solution of (26) is known, appropriate models, or appropriate simplifications, must be used, as is proposed below.

It is easy to calculate the average and the variance of the random variable $\mu_n(D)$, depending on $\sigma_c(x)$. We have

$$E[\mu_n(D)] = \int_B F(\sigma(x), x)dx \tag{27}$$

where $F(\sigma(x), x) = P\{\sigma_c(x) < \sigma(x)\}$. Even when σ is constant, the distribution may depend on x in B for a nonstationary random function $\sigma_c(x)$). The geometrical covariogram of D, $K_\sigma(h)$, is given by :

$$K_\sigma(h) = \int_{B \cap Bh} E[1_{\sigma_c}(x)1_{\sigma_c}(x+h)]dx = \int_{B \cap Bh} F_2(\sigma(x), \sigma(x+h))dx \tag{28}$$

where $F_2(\sigma(x), \sigma(x+h) = P\{\sigma_c(x) < \sigma(x), \sigma_c(x+h) < \sigma(x+h)\}$ is the bivariate distribution of σ_c. The variance $D^2[\mu_n(B)]$ is obtained as :

$$D^2[\mu_n(B)] = \int_{R^n} K_\sigma(h)dh - \left[\int_B F(\sigma(x))dx\right]^2 \tag{29}$$

From now on, we consider large specimens B and random functions $\sigma_c(x)$ for which the Central Limit Theorem can be applied.

Asymptotically, $\mu_n(D)$ is a Gaussian variable with average and expectation given by relations (27) and (29). Therefore, we get :

$$P\{\text{no failure of B}\} = \frac{1}{\sqrt{\pi}} \int_{-\infty}^{\frac{p_c - m}{\sqrt{2}D}} \exp(-y^2)dy \tag{30}$$

where $m = E[\mu_n(D)]/\mu_n(B)$ and $D^2 = D^2[\mu_n(D)]/\mu_n(B)^2$.

In the case of a stationary random function $\sigma_c(x)$, and a uniform stress field σ over B, we get $m = F(\sigma)$ and

$$D^2 = m(1 - m)\ A(\sigma)/\mu_n(B) \tag{31}$$

Relation (31) can be applied when the integral range $A(\sigma)$ of the random set, deduced from $\sigma_c(x) < \sigma$ is different from 0 and remains finite. We have :

$$A(\sigma) = \int_{R^n} \frac{F_2(\sigma, \sigma, h) - F^2(\sigma)}{F(\sigma)(1 - F(\sigma))} dh \tag{32}$$

with , $F_2(\sigma, \sigma, h) = P\{\sigma_c(x) < \sigma,\ \sigma_c(x + h) < \sigma\}$.

We can also use the following empirical expressions for the variance D^2 [13] :

$$D^2 = F(\sigma)(1 - F(\sigma))B(\sigma)/|\mu_n(B)|^{\alpha} \tag{33}$$

with $\alpha < 1$ for $A(\sigma) = 0$ and $\alpha > 1$ for $A(\sigma) = \infty$.

From law (30), general properties of the fracture probabilities can be easily deduced :

• For a given stress field $\sigma(x)$, P{fracture} decreases with the threshold p_c ; we have P{fracture} = 1/2 for $m = p_c$.

• For a uniform stress field σ, the median macroscopic strength σ_M is obtained from $F(\sigma_M) = p_c$. There is therefore no size effect for σ_M in this model, which can be checked experimentally. The median σ_M increases with p_c, while the variance $D^2(\sigma_R)$ decreases with V(B). When $V(B) \rightarrow \infty$, the fracture strength σ_R converges to the constant $\sigma_R = F^{-1}(p_c)$: a deterministic fracture behaviour results on a large scale, towards a fracture strength depending both on $F(\sigma)$ and on p_c.

Fracture statistics models with a critical density of defects

The previous models cannot be applied to a material containing defects with a zero measure (points, thin microcracks, etc.). An alternative criterion is a critical number density, θ_c. For a given applied stress field $\sigma(x)$, $N(\sigma)$ is the random number of defects where $\sigma_c(x) < \sigma(x)$. We assume that the fracture will occur when $N(\sigma) > n$, with $n = \theta_c\ \mu_n(B)$ (n increases with the critical density θ_c, and with the volume of B).

For a Boolean Random Function (Boolean Variety with k = 0), we allow the superposition of defects and $N(\sigma)$ follows a Poisson distribution with parameter $\theta' = \theta\ \overline{V}(H^c_{Z'}(\sigma - \sigma_m)) = n\ \theta_1$.

• When $n < 1$ (for specimens B with a small size), we recover the weakest link criterion, and P{no failure of B} $= \exp(-\theta') = p_0$.

• When $n \geq 1$, we get :

$$P\{\text{no failure of B}\} = p_0 + p_1 ... + p_m + ... + p_n = 1 - F_n(\theta_1) \tag{34}$$

In the case of point defects and a uniform stress field over B, we have $\theta' = \theta V(B)\ \phi(\sigma)$ (where $\phi(\sigma)$ is an increasing positive function of σ, as earlier) and

$\theta_1 = \frac{\theta}{\theta_c}\phi(\sigma)$. In that case, (and approximately for large B), θ_1 does not depend on the geometry of the specimen B.

In relation (34), $p_m = P\{N(\sigma) = m\} = \frac{(\theta_1 n)^m}{m!}\exp - n\theta_1$. When $\phi(\sigma)$ is derivable, the distribution $F_n(\theta_1)$ admits a density $f_n(\theta_1)$ and a mode $\left(\frac{n}{e}\right)^n \frac{1}{(n-1)!}$ for σ such that $\frac{\theta}{\theta_c}\ \phi(\sigma) = 1$ (where e = exp 1). This mode is unique when $\phi(\sigma)$is strictly monotonous. When increasing the size of B such that $n \rightarrow \infty$, the pdf $f_n(\sigma)$ converges toward a Dirac distribution : the fracture strength of the material becomes a constant value σ_R, such that $\frac{\theta}{\theta_c}\ \phi(\sigma_R) = 1$. This behaviour is very similar to the previous model, based on a critical volume function. It can also be shown [3] that, unlike the weakest link model, the large specimens are less sensitive to the most severe defects (with low σ_c).

CONCLUSION

The approach proposed in this paper and in [3] enabled us to introduce a wide range of models useful for Fracture Statistics : using various fracture criteria, brittle behaviour or progressive fracture initiation can be described. For each fracture criterion, various random function models can be proposed and can be tested, either on a macroscopic scale (from experimental data of mechanical testing result obtained in various loading conditions, it is possible to identify the parameters of a given model and to check its validity), or on a microscopic scale (after the defects have been located, a statistical description of their spatial distributions can be obtained by image analysis ; this is useful for testing the geometrical models).

From a practical point of view, models with few parameters (at least 2) can be built. Empirical scaling laws can be useful for selecting proper models, but are not sufficient in general and must be completed by fracture strength distributions for various geometries of loading. After a given model has been identified, it can be used to predict fracture statistics in complex loading situations, using for instance Finite Element Calculations to determine the applied stress field $\sigma(x)$ [14] : from the simplification given by decoupling the two stress fields $\sigma(x)$ and $\sigma_c(x)$, closed form results can be obtained without simulating the random field $\sigma_c(x)$. Some of these models were applied to various materials : ceramics [14], carbon fibers [11], steels [6]. Recent theoretical developments include a competition between several fracture criteria [3, 15].

REFERENCES

[1] Jeulin, D. (1989), "Morphological models of change of scale in brittle fracture statistics", CG Report N−18/89/G, Ecole des Mines de Paris, Fontainebleau. Accepted for publication in *Strength and Fracture.*

[2] Jeulin, D. (1989), "Random fields models for fracture statistics", Actes du 32ème Colloque de Métallurgie, INSTN, *Ed. de la Revue de Métallurgie*, n° 4, 3ème trimestre 1990, pp. 99−113.

[3] Jeulin, D. (1991), *Modèles morphologiques de structures aléatoires de changement d'échelle*, Thèse de Doctorat d'Etat ès Sciences Physiques, Université de Caen.

[4] Pineau, A. (1981), *Proc. of the 5th Int. Conf. on Fracture*, Cannes, D. François ed., Vol. 2, p. 533.

[5] Beremin, F.M. (1983), *Metall. Trans. A.*, 14A, p. 2277.

[6] Piques, R., Borit, F., Jeulin, D., Pineau, A. (1990), *Rupture fragile et fragilisation intergranulaire de l'acier des cuves de réacteurs à eau sous pression (acier MND5)*, Centre des Matériaux Report, Ecole des Mines de Paris, Evry.

[7] Matheron, G. (1975), *Random sets and integral geometry*, J. Wiley & Sons, N.Y.

[8] Harter, H.L. (1978), "A bibilography of extreme−value theory", *International Statistical Review*, 46, 279−306.

[9] Kittl, P., Diaz, G. (1988), *Res. Mechanica*, 25, pp. 99−207.

[10] Johnson, N.L., Kotz, S.(1970), in : *Continuous univariate distributions*, Vol. 1, J. Wiley, N.Y., Chap. 20−21.

[11] Baxevanakis, C., Jeulin, D., Valentin, D. (1992), "Fracture statistics of single fiber composite specimens". To appear in *Composites Sciences and Technology.*

[12] Serra, J. (1982), *Image analysis and mathematical morphology*, Academic Press, London.

[13] Lantuéjoul, C. (1991), "Ergodicity and integral range", *Journal of Microscopy*, Vol. 161, Pt 3, March 1991, pp. 387−403.

[14] Berdin, C., Baptiste, D., Jeulin, D., Cailletaud, G. (1991), "Failure models for brittle materials", *Fracture Process in Concrete, Rock and Ceramics*, ed. by G.J.M. van Mier, J.G. Rots and A. Bakker, E. & F.N. Spon, London, pp. 83−92.

[15] Jeulin, D. (1992), "Morphological models for fracture statistics". To appear in *Proc. CMDS7*, N.H. Anthony Ed., Special Issue of Materials Science Forum, Trans. Tech. publications.

SPATIAL FILTERING UNDER THE LINEAR COREGIONALIZATION MODEL

[1]BOURGAULT GILLES and [1]MARCOTTE DENIS

1 École Polytechnique de Montréal, Dépt. Génie Minéral, C.P. 6079, Succ. A, Montréal (Quebec) CANADA, H3C 3A2

ABSTRACT. Under the linear coregionalization model (LCM), p-variable observation vectors are interpreted as resulting from linear combinations of underlying orthogonal elementary variables at k different scales of variation. An approach is presented where the LCM parameters are estimated directly by an iterative procedure that seeks to simultaneously maximise the fit of elementary variograms to all the experimental variograms and cross-variograms. The multivariate variogram calculated with the Mahalanobis metric is shown to be helpful in determining the number and type of elementary variograms. Once the LCM coefficients are estimated, underlying elementary variables are estimated by ordinary factorial cokriging.

Two examples are presented and results are compared with those of a principal components analysis (PCA). The first example concerns a simulated data set with noise added; the second, a case study of exploration for nickel-copper deposits in the Labrador trough. It is shown that spatial factorial analysis provides results more easily interpretable in geological terms than PCA.

1. INTRODUCTION

It is essential to take the spatial information into account in order to explain the multivariate structure of coregionalized data. This is achieved, here, with the application of the multivariate variogram to the linear coregionalization model. The results of our approach are compared with those of a Principal Components Analysis (PCA), which does not take into account spatial information. A simulated data set and a case study are used to compare both analyses.

2. THEORETICAL REVIEW

We are interested in simplifying the problem of studying p variables, by replacing them with k ($k \leq p$) underlying elementary variables. We propose that the k underlying variables have unit variance and null cross-variograms. This is written as a linear model (Journel and Huijbrets, 1978; Matheron, 1982; Myers, 1983, 1988; Wackernagel, 1985, 1988; Bourgault and Marcotte, 1991):

$$\mathbf{Z(x)}_{1\times p} = \mathbf{X(x)}_{1\times k}\ \mathbf{A}_{k\times p}\ ,\ k \leq p. \tag{1}$$

where

Z(x): vector of p measured variables at the sample point located in **x**.
X(x): vector of k elementary variables (not directly measurable).
A : matrix (kxp) which contains the coregionalization's parameters associating the measured to the elementary variables.

As a consequence of the elementary variables'properties we can write all variograms and cross-variograms associated with Z(**x**) variables as linear combinations of the variograms associated with X(**x**) variables (Borgman and Frahme, 1976; Davis and Greenes, 1983; Myers, 1983, 1988):

$$\overline{\gamma}_Z(\boldsymbol{h}) = \boldsymbol{A}^t\ \overline{\gamma}_X(\boldsymbol{h})\ \boldsymbol{A} \tag{2}$$

where

$\overline{\gamma}_Z(\boldsymbol{h})$: matrix (pxp) of the variograms and cross-variograms, at lag **h**, associated with the measured variables.
$\overline{\gamma}_X(\boldsymbol{h})$: diagonal matrix (kxk) of the variograms, at lag **h**, associated with the elementary variables.

Under second-order stationarity, an exactly similar formula can be established for the spatial covariance matrices. Bourgault and Marcotte (1991) have demonstrated that the multivariate variogram, calculated using the pseudo-inverse of covariance matrix as a metric, is equal to the sum of the elementary variograms. In practice this metric can be found by a singular-value decomposition of the covariance matrix. The metric used is then:

$$\mathbf{C}^- = \mathbf{V}\ \mathbf{D}^{-1}\ \mathbf{V}^t \tag{3}$$

where

C : covariances matrix (pxp) with rank=k.
D : diagonal matrix (kxk) of non-null eigenvalues.
V : matrix (pxk) of eigenvectors associated with non-null eigenvalues.

The multivariate variogram is written as:

$$2G(\boldsymbol{h}) = E[(\boldsymbol{Z(x)} - \boldsymbol{Z(x+h)})\ \boldsymbol{C}^-\ (\boldsymbol{Z(x)} - \boldsymbol{Z(x+h)})^t] = 2\sum_{q=1}^{k}\gamma_{X_q}(\boldsymbol{h}) \tag{4}$$

where

$\gamma_{X_q}(\boldsymbol{h})$: elementary variogram associated with X_q(**x**).

Therefore, it is easy to recognize the k elementary variograms by breaking the multivariate variogram down into elementary structures associated with different spatial scales. Once the number and type of elementary variograms is identified, an estimate of the matrix **A** can be obtained directly by modelling the p(p+1)/2 variograms and cross-variograms of Z(**x**) variables with linear combination of the elementary variograms (eq. 2). This ensures semi-positive definiteness (or positive definiteness if k=p) (eq. 2) (Journel and Huijbrets, 1978; François-Bongarçon, 1981; Wackernagel, 1985; Galli and Wackernagel, 1987; Myers, 1988). Further details on this algorithm are given in Bourgault and Marcotte (1991).

In the presence of white noise the linear model of coregionalization is modified to:

$$\mathbf{Z(x)} = \mathbf{X(x)}\ \mathbf{A} + \mathbf{E(x)} \tag{5}$$

where

$\mathbf{E(x)}$: vector of p independent white noise with null mean.

The covariance matrix **C** is now of full rank, and the metric to calculate the multivariate variogram is the Mahalanobis metric $\mathbf{C}^{-1}$ (Harff and Davis, 1990; Bourgault and Marcotte, 1991). Thus, equation (2) can be rewritten as:

$$\overline{\gamma}_Z(\boldsymbol{h}) = \boldsymbol{A}^t\, \overline{\gamma}_X(\boldsymbol{h})\ \boldsymbol{A} + \boldsymbol{D}_E(1-\delta(\boldsymbol{h})) \tag{6}$$

where

$\mathbf{D}_E$: diagonal matrix (pxp) of variances associated with white noise.

$\delta(\mathbf{h})$: Kronecker delta equal to 1 for $\mathbf{h}=\mathbf{0}$, otherwise equal to 0.

The automatic adjustment of variograms and cross-variograms by linear combinations of elementary variograms also allows the estimation of $\mathbf{D}_E$ by adding p nugget effects to the k elementary variograms (Bourgault and Marcotte, 1991). These p nugget effects appear only on the $Z(\mathbf{x})$ variograms and not on their cross-variograms. A nugget effect can also be considered as an elementary structure of the multivariate variogram. This means that a cross-correlated micro-structure is present in the data. This cross-correlated micro-structure could either be interpreted as a significant signal or simply as a correlated noise component. For example, with sparse geochemical data, one would expect that mineralization should appear almost as an impulse simultaneously affecting many variables. On the other hand, in remote sensing, a very short scale structure could identify a correlated noise stemming from atmospheric variations or from variations in the position and/or orientation of the multi-spectral sensor.

Under the second-order stationarity hypothesis, the vector $\mathbf{X(x)}$ can be estimated by ordinary factorial cokriging, as we can write:

$$E[\mathbf{Z(x)}] = E[\mathbf{X(x)}]\ \mathbf{A} + E[\mathbf{E(x)}]$$
$$\mu_Z = \mu_X\ \mathbf{A} \tag{7}$$

Note that here it is not necessary that $\mu_Z = \mathbf{0}$ and $\mu_X = \mathbf{0}$ as is usually assumed in factorial kriging. The estimator is written as:

$$\boldsymbol{X}^*(\boldsymbol{x}) = \sum_{\alpha=1}^{n} \boldsymbol{Z}(\boldsymbol{x}_\alpha)\ \boldsymbol{\Gamma}_\alpha \tag{8}$$

The non-bias equation, $E[\boldsymbol{X}(\boldsymbol{x})^*] = \boldsymbol{\mu}_X \boldsymbol{A} \sum_{\alpha=1}^{n} \boldsymbol{\Gamma}_\alpha = \boldsymbol{\mu}_X$, is respected if:

$$\sum_{\alpha=1}^{n} \boldsymbol{\Gamma}_\alpha = \boldsymbol{A}^t(\boldsymbol{A}\ \boldsymbol{A}^t)^{-1} \tag{9}$$

The ordinary factorial cokriging system is written in matrix form, according to Myers (1982), as:

$$\left.\begin{aligned} \sum_{\beta=1}^{n} \overline{\gamma}_Z(\alpha,\beta)\ \boldsymbol{\Gamma}_\beta + \overline{\mathbf{v}} &= \boldsymbol{A}^t\, \overline{\gamma}_X(x,\alpha) \\ \sum_{\alpha=1}^{n} \boldsymbol{\Gamma}_\alpha &= \boldsymbol{A}^t(\boldsymbol{A}\boldsymbol{A}^t)^{-1} \end{aligned}\right\} \tag{10}$$

where
Γ_α : matrix (pxk) of cokriging weights.
$\bar{\mathbf{v}}$: matrix (pxk) of Lagrange multipliers.

The squared norm of $(\mathbf{X(x)}^* - \mathbf{X(x)})$ is given by:

$$\sigma^2(\boldsymbol{x}) = Tr\{\sum_{\alpha=1}^{n} \bar{\gamma}_X(x,\alpha)\,\mathbf{A}\,\boldsymbol{\Gamma}_\alpha\} + Tr\{\,\bar{\mathbf{v}}^t\mathbf{A}^t(\mathbf{A}\mathbf{A}^t)^{-1}\} \quad \mathbf{(11)}$$

It is minimal when the cokriging weights are obtained by equation (10).

3. SIMULATED DATA

A 1-D linear coregionalization is simulated (see details in Table 1) for three Z(**x**) variables using two X(**x**) variables; the variograms and cross-variogram of X(**x**) are shown in Figure 1. Piecewise linear variograms with ranges 9 (X_1) and 40 (X_2) were simulated by a dilution process (Serra, 1968). The pseudo-periodic behaviour is probably an artefact of the uniform random-numbers generator used (Sharp, 1982). A gaussian white noise with variance equivalent to 50% of the original Z(**x**) variables'variances has been added to the 200 simulated data points (equally spaced on a line). Figure 2 shows the variograms and cross-variogram for the X^*(**x**) estimated by ordinary factorial cokriging (Marcotte, 1991) at the sampling points (note that the estimated values were rescaled to unit variance). Figure 3 shows the variograms and cross-variograms of the principal components. The first two component variograms present similar structures with a high nugget effect, whereas the third component shows a pure nugget effect. This indicates that PCA failed to discriminate between the noise and the X(**x**)'s signal (despite their small spatial cross-correlations). Contrary to this, the X^*(**x**) variables show variograms and cross-variogram almost identical to the original simulated X(**x**).

4. CASE STUDY

The data come from the Ministère de l'Energie et des Ressources du Québec (Beaumier, 1987) and were earlier studied by Marcotte and Fox (1990). A total of 1314 lake sediment samples from the Schefferville area in northern Quebec, were analysed for Cr, Cu and Ni. Data was standardized before processing. Figure 4 shows the simplified geology of the study area. Figure 5 shows the multivariate variogram, calculated using the Mahalanobis metric. A geometric anisotropy is observed that follows the known stratigraphy (Dimroth, 1978). The experimental multivariate variogram can be broken down into two elementary variograms: a nugget effect (punctual scale) and an anisotropic spherical variogram (regional scale, with minor range a_{N58E}=7500 m and major range a_{N32W}=13500 m). Figure 6 shows the automatic variogram and cross-variogram adjustments for Cr, Cu and Ni in the N58E direction by linear combinations of the two elementary variograms. Table 2 shows the $\mathbf{A}^*$ matrix and the PCA results obtained with these data.

The elementary variables were estimated by ordinary factorial cokriging at the sampling points. Figure 7, for the X^*(**x**), and Figure 8, for the

PCA, show the variograms along and across the stratigraphy. The $X^*(\mathbf{x})$'s variograms reproduce well the structures recognized on the multivariate variogram (Fig. 5). On the other hand, PCA variograms are all spherical with a nugget effect. Even the last component (PCA3) which explains only 6% of the total variance shows a structure with a range. It would thus be hazardous to reject this last component solely because of its low variance contribution. The cross-variograms for the $X^*(\mathbf{x})$ variables and for the PCA components (not shown) indicate weak spatial correlations.

The map associated with the anisotropic spherical structure ($X_2^*(\mathbf{x})$) is shown in Figure 9. We recognize a regional pattern for this structure with negative values belonging to the sedimentary group (see Fig. 4) and high positive values associated with the ultramafic rocks within an igneous group. This map discriminates very well between poor and rich backgrounds in Cr, Cu and Ni, and furnishes a clear geological picture. The map associated with the micro-structure is shown in Figure 10. This elementary variable is interpreted as a common punctual variation (with this sampling pattern) that may be associated with mineralization. The 12 higher values of this elementary variable correspond well with the known showings (Dugas, 1970) and are mainly associated with the ultramafic rocks.

The map of the first principal component is not presented, it is very similar to our regional map (Figure 9). The variables'loading on the first eigenvector are all positive (Table 2) as was the case for our regional $X_2^*(\mathbf{x})$. The loadings on the second eigenvector are positive for Cu and Ni and negative for Cr. The 12 higher positive values of this component (Figure 10) mainly belong to the sedimentary group and show no association with known mineralization. This component is possibly a mathematical artefact of PCA related to the orthogonality constraint on the eigenvectors.

DISCUSSION AND CONCLUSION

For a simulated data set, with a significant added noise, PCA failed to correctly identify the underlying model, whereas our approach succeeded. In effect, the variograms of the first two principal components appear as a mixture of the two simulated structures. Also, they both present a strong nugget effect due to the added noise. On the other hand, factorial cokriging estimates reproduce almost exactly the simulated variograms. With real data, our approach led to the estimation of underlying variables which were more easily interpretable and more closely related to the known geology and mineralization. Our estimates of underlying variables show the same scale of variation that was modelled on the multivariate variogram. Although the PCA components are weakly spatially cross-correlated, they cannot be associated to a unique scale of variation as are our $X^*(\mathbf{x})$. For this reason, it will be more difficult to explain the PCA components in geological terms than the $X^*(\mathbf{x})$ variables. Frequently, the first principal component will capture the regional scale variations because of its importance compare to local scale variations. The picture is not so clear however for the other

components. The regional scale of variation could be present for most components, even the component with the least variance (Figure 8; see also Marcotte (1990) and Marcotte and Fox (1990)). The orthogonality between PCA vectors dictates that factors with similar coefficients will be merged on the same component even when acting at different scales. On the other hand, our approach enables the estimation of factors with similar coefficients but acting at different scales (Table 2; Figure 7). It succeeds in discriminating single scale sources of variation (Figures 2 and 7) and determining their importance for each measured variable. Contrary to PCA, our approach is sensitive to permutations of data locations. Finally, the multivariate Mahalanobis variogram helps identify the number and scale of significant elementary variables.
Applied to the Schefferville data, our approach and PCA led to the identification of similar regional patterns of variation. However, the anomalies defined by our approach followed more closely the known mineralized indices than did those obtained by PCA (Figure 10).

TABLE 1. Simulated linear coregionalization with noise added (50%) to the $Z_i(x)$. Results of coregionalization and PCA analysis.

$$Z_1(x) = 1.0\,X_1(x) + 4.0\,X_2(x) + 2.92\,E_1(x)$$
$$Z_2(x) = -3.0\,X_1(x) + 2.0\,X_2(x) + 2.55\,E_2(x)$$
$$Z_3(x) = 2.0\,X_1(x) - 1.0\,X_2(x) + 1.58\,E_3(x)$$

$X_1(x)$: 200 samples on a line, local structure, see variogram in Figure 1.
$X_2(x)$: 200 samples on a line, regional structure, see variogram in Figure 1.
$E_i(x)$: independent noises, N(0,1).
$X_i(x)$ was simulated by a dilution process as proposed by Serra (1968)

linear coregionalization results

$$A' = \begin{bmatrix} 2.23 & -2.53 & 1.53 \\ 3.74 & 2.83 & -1.76 \end{bmatrix} \begin{matrix} X_1(x) \\ X_2(x) \end{matrix}$$
$$\qquad Z_1(x) \quad Z_2(x) \quad Z_3(x)$$

eigenvectors

	PCA1	PCA2	PCA3
$Z_1(x)$	.796	-.6049	.0225
$Z_2(x)$	.5385	.7245	.4303
$Z_3(x)$	-.2766	-.3304	.9024

eigenvalues

$$\begin{bmatrix} 31.2451 & 0 & 0 \\ 0 & 16.7406 & 0 \\ 0 & 0 & 3.3835 \end{bmatrix}$$

TABLE 2. Linear coregionalization and PCA results on Schefferville data.

STARTING SOLUTIONS FOR COREGIONALIZATION ANALYSIS

$A_0 = [0]_{2x3}$ $E_0 = [0]_{1x3}$

ESTIMATES

$$A^* = \begin{bmatrix} 0.27 & 0.34 & 0.41 \\ 0.32 & 0.48 & 0.62 \end{bmatrix} \begin{matrix} X_1(x) \\ X_2(x) \end{matrix}$$

(columns: *Cr*, *Cu*, *Ni*)

individual nugget effect: $E^* = [\ 0.21\ 0.22\ 0.08]$

$X_1(x)$: elementary variable, common nugget effect (micro-structure).
$X_2(x)$: elementary variable, anisotropic spherical variogram (regional structure).

EIGENVECTORS

	PCA1	*PCA2*	*PCA3*
Cr	0.57	-0.76	0.33
Cu	0.57	0.65	0.51
Ni	0.59	0.10	-0.80

EIGENVALUES

$\lambda_1 = 2.5173$ (84%) $\lambda_2 = 0.3065$ (10%) $\lambda_3 = 0.1762$ (6%)

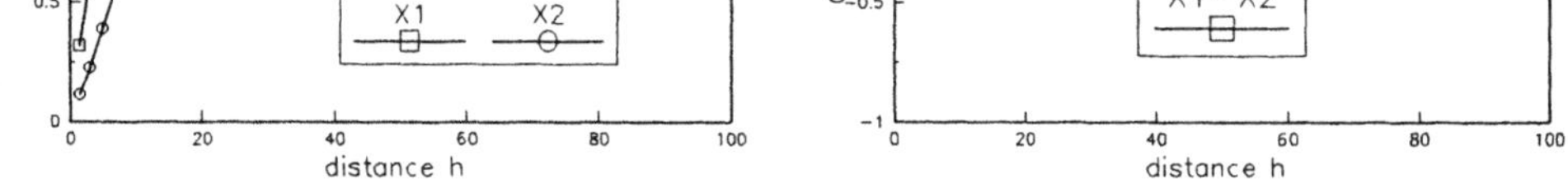

Figure 1. (A) Variograms of underlying variables for simulated linear coregionalization. (B) Cross-variogram for the two underlying variables.

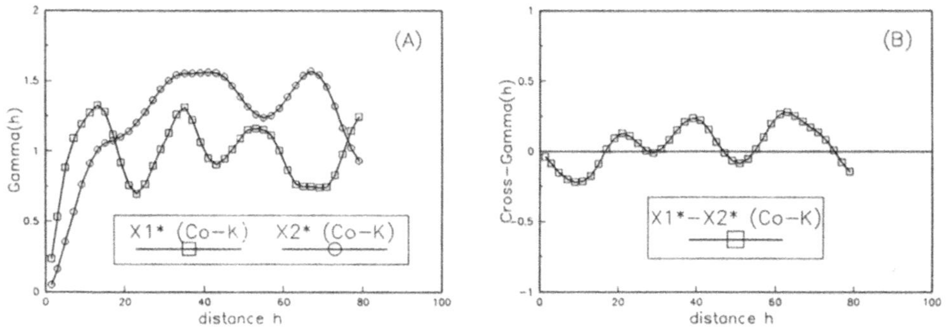

Figure 2. Variograms (A) and cross-variogram (B) for elementary variables estimated by ordinary factorial cokriging on simulated data.

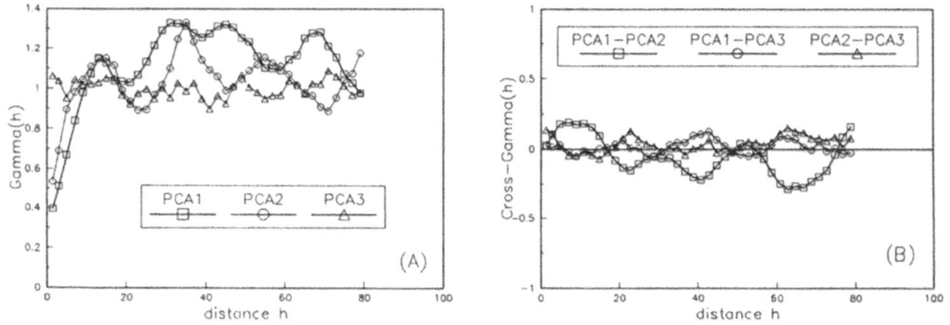

Figure 3. Variograms (A) and cross-variograms (B) for principal components on simulated data.

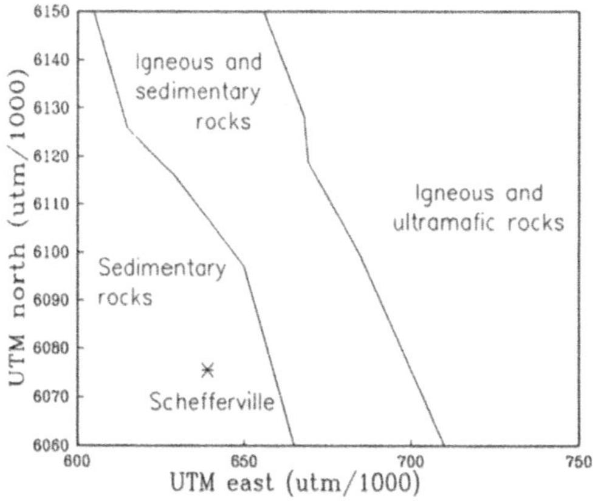

Figure 4. Simplified geology of Schefferville area.

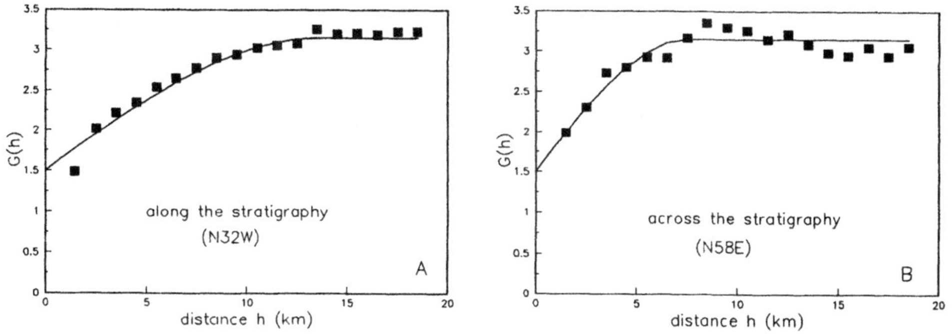

Figure 5. Geometric anisotropic multivariate variogram, calculated with Mahalanobis metric, on Schefferville data.

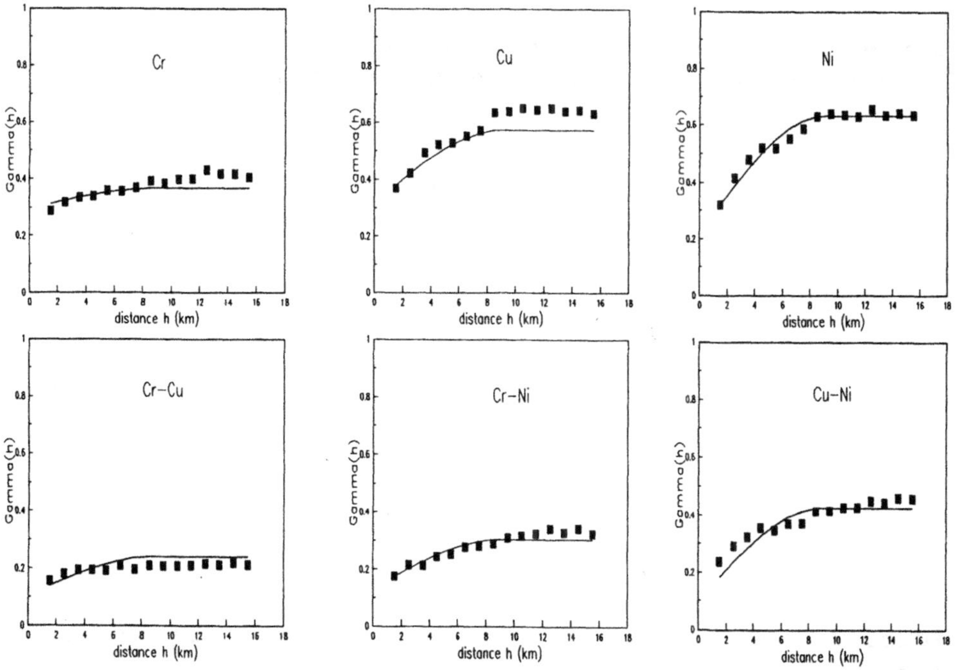

Figure 6. Automatic adjustments of variograms and cross-variograms (in minor range direction) by linear combinations of two elementary variograms (Schefferville data): a common nugget effect + a geometric anisotropic spherical scheme.

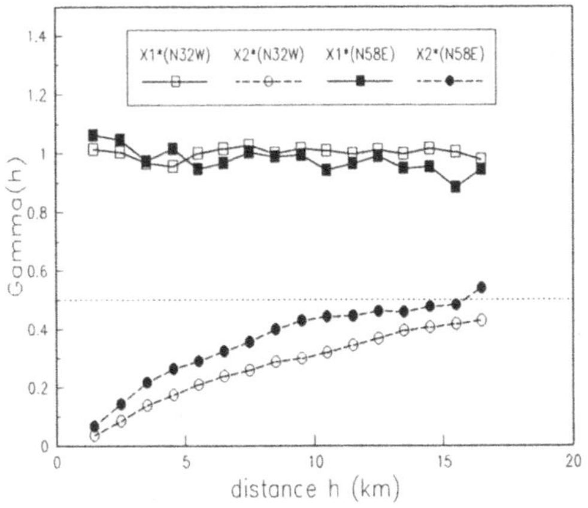

Figure 7. Variograms for $X^*(x)$ variables estimated by ordinary factorial cokriging of Schefferville data.

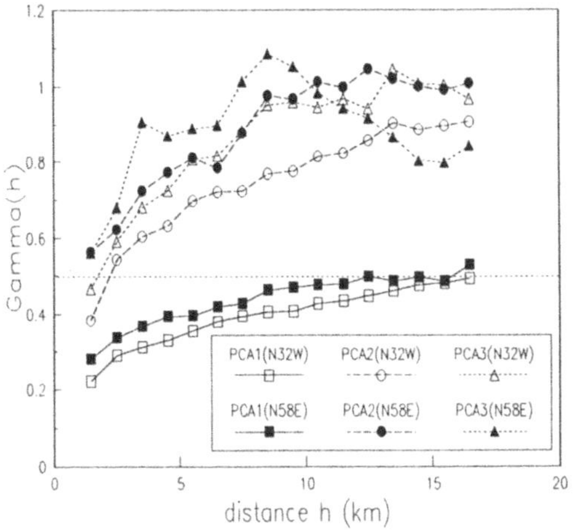

Figure 8. Variograms for PCA components of Schefferville data.

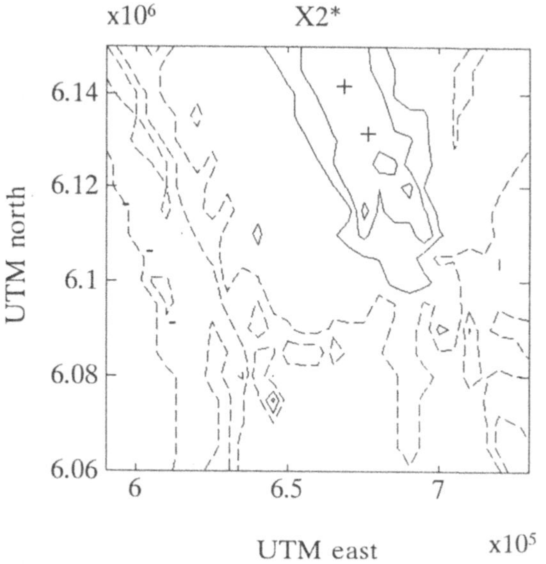

Figure 9. Isocontour map for regional elementary variable $X_2^*(x)$ of Schefferville data (positive values are outlined by continuous lines and negative by dashed ones).

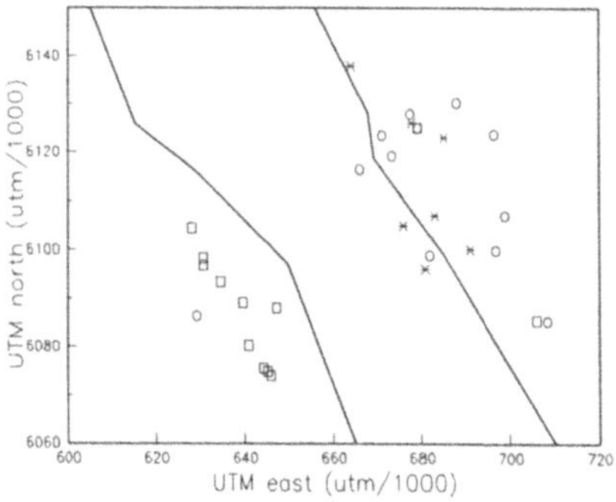

Figure 10. Thelve higher values for elementary variable $X_1^*(x)$ (circles) and for second principal component PCA2 (squares) of Schefferville data. Asterix locate the known Cu-Ni mineralization.

REFERENCES

Beaumier, M. (1987) "Géochimie des sédiments de lac de la région de Schefferville", MB 87-32, M.E.R.Q.

Borgman, L.E. and Frahme, R.B. (1976) "A case study: multivariate properties of bentonite in northeastern Wyoming", in: M. Guarascio et al. (eds.), Advanced Geostatistics in the Mining Industry, Reidel-Dordrecht, pp.381-390.

Bourgault, G. and Marcotte, D. (1991) "Multivariable variogram and its application to the linear coregionalization model", Math. Geol. 23, 7, 898-927.

Davis, B.M. and Greenes, K.A. (1983) "Estimation using spatially distributed multivariate data: an exemple with coal quality", Math. Geol. 15, 2, 287-300.

Dimroth, E. (1978) "Labrador trough area", Geol. Rep. 193, Min. of Energy and Ressources, Quebec.

Dugas, J. (1970) "Metallic mineralization in part of the Labrador trough", Spec. Pap. 5, Min. of Energy and Resources, Quebec.

François-Bongarçon, D. (1981) "Les corégionalisations le cokrigeage", Cahiers du Centre de Géostatistique et de Morphologie Mathématique de Fontainebleau, École Nationale Supérieure des Mines de Paris, C-86.

Galli, A. and Wackernagel, H. (1987) "Multivariate geostatistical methods for spatial data analysis", in E. Diday et al. (eds.), Data Analysis and Information, V.5, North Holland,.

Harff, J. and Davis, J.C. (1990) "Regionalization in geology by multivariate classification", Math. Geol. 22, 573-588.

Journel, A.G. and Huijbregts, Ch.J. (1978) "Mining Geostatistics", Academic Press.

Marcotte, D. (1990) "Lake Sediments in the Manicouagan Area: Multivariate Analysis and Variography Used to Enhance Anomalous Response", In Statistical Applications in the Earth Sciences, F.P. Agterberg and G.F. Bonham-Carter (eds), GSC paper 89-9, 349-355.

Marcotte, D. (1991) "Cokriging with Matlab", Computer and Geosciences 17, 9, 1265-1280.

Marcotte, D. and Fox, J.S. (1990) "The Schefferville area: multivariate analysis and variography used to enhance interpretation of lake sediment geochemical data", Journ. of Geoch. Explor. 38, 247-263.

Matheron, G. (1982) "Pour une analyse krigeante des données régionalisées", N-732, Centre de Géostatistique et de Morphologie Mathématique, Fontainebleau.

Myers, D. (1982) "Matrix formulation of co-kriging", Math. Geol. 14, 3, 249-257.

Myers, D. (1983) "Estimation of linear combination and co-kriging", Math. Geol. 15, 5, 633-637.

Myers, D. (1988) "Multivariate geostatistics for environmental monitoring", Sciences de la Terre 27, 411-427.

Serra, J. (1968) "Les Structures Gigognes: Morphologie Mathématique et Interprétation Métallogénique", Mineralium Deposita 3, 135-154.

Sharp, W.E. (1982) "Stochastic Simulation of Semivariograms", Math. Geol. 14, 5, 445-456.

Wackernagel, H. (1985) "L'inférence d'un modèle linéaire en géostatistique multivariable", Thèse, École Nationale Supérieure des Mines de Paris.

Wackernagel, H. (1988) "Geostatistical techniques for interpreting multivariate spatial information", in C.F. Chung et al. (eds.), Quantitative Analysis of Mineral and Energy Ressources, Reidel-Dordrecht, pp.393-409.

GRAVITY DATA TRANSFORMATION BY KRIGING

DENIS MARCOTTE MICHEL CHOUTEAU
Département de génie minéral, École Polytechnique
C.P. 6079, Succ. A, Montréal, Québec, Canada, H3C-3A7

ABSTRACT

The gravimetric model of covariance is used to derive formulas for gravity data transformation by kriging. The transformed kriged surface obeys the potential field equation automatically. A deterministic model is sampled and estimated by spectral methods and by ordinary and simple kriging with different number of neighbours. Results are compared and advantages of kriging are outlined.

1. Introduction

Geophysical data transformation has many objectives such as: standardization of data presentation (e.g. pole reduction with magnetic data), noise filtering, smoothing, enhancement of anomalies, correction for different elevation of data, and determination of depth of sources.

These objectives are attained by transformations of the original data in either the frequency or the spatial domain. Routinely used in the fifties and sixties, spatial domain transformations fell progressively into disuse with the development of the very efficient FFT algorithm. The complexity of the FFT algorithm of only NlogN, where N is the number of grid points, explains this fact. Also, the convolution theorem, that is, a convolution in the spatial domain corresponds to a product in the frequency domain, also favours the frequency approach.

The spatial transformations, especially when done by kriging, may still be very useful for the following reasons:

i. Geophysical data, especially gravimetric data, are rarely sampled on a regular grid. An interpolation phase is thus necessary which attenuates the computational advantage of the frequency approach. Moreover, when the interpolation is performed by kriging, the desired transformations can be estimated simultaneously in the spatial domain at a very low marginal cost.

ii. Whatever the interpolation method used, the frequency content of the signal will be more or less affected because of the spectral characteristic of the interpolator. The problem is that the frequency modifications will vary from place to place in the field depending on the density and distribution of data. To take this effect into account, frequency analysis should be made on a local basis instead of globally; however, this is not possible. On the other hand, kriging is usually performed locally although the covariance function is estimated globally.

iii. In the spectral approach, a white noise uniformly affects all the frequencies, and is difficult to identify and to separate from interesting high frequency variations. In the spatial domain, however, the noise appears as a nugget effect which is filtered out automatically by kriging.

iv. Data collected at different altitudes will show different frequency signatures. The frequency domain approach being global, it can not take this problem fully into account. In this case, the problem with kriging is to estimate the covariance function. This can be done if sufficient data are available near an arbitrary reference altitude. Once the covariance function is known, it is very easy to account for data at different altitudes in the kriging system.

v. Frequency method operates under a second-order stationarity hypothesis over the entire field. Kriging necessitates only local stationarity hypothesis. This hypothesis can even be relaxed to allow for drift of the mean (FAI-k) or to allow for variability proportional to the local mean (relative or logarithmic variograms).

vi. With frequency domain method, there is no measure of precision of the estimate. With kriging, the kriging variance provides an average measure of precision for interpolation performed with the same data configuration. A kriging variance is computable for any desired transformation and could help determine data transformations that may be estimated precisely.

vii. Frequency domain transformation are usually univariate. It is however simple to generalize kriging interpolators to take into account multivariate information. This could be used, for example, with magnetic data and/or topographic data to improve estimates performed on gravity data. Moreover, it is possible to take into account functional relationship that may exist between variables in the (cross)variograms modelling. An example of such a functional link exists between newtonian (or gravitational) and magnetic potential field.

This paper presents the kriging systems and equations required to perform common gravity data transformation. A simulated example is used to evaluate the performance of kriging compared to frequency domain transformation with data on a regular grid. When data is irregularly spaced, the influence of the necessary interpolation phase on the frequency domain transformation is evaluated.

2. Methods

Kriging equations and notation

Let

$$Z(\mathbf{x}) = P(\mathbf{x}) + Y(\mathbf{x}) \quad (1)$$

where

$\mathbf{x}$	=	a vector of coordinates; in 2-D, $\mathbf{x}$: (x,y)
$Z(\mathbf{x})$	=	the function evaluated at location $\mathbf{x}$.
$P(\mathbf{x})$	=	a polynomial term expressing the local mean.
$Y(\mathbf{x})$	=	a zero mean stochastic component.

Let us consider the general situation where a given linear operator $L(Z(\mathbf{x}))$ on $Z(\mathbf{x})$ is to be estimated from the available data $Z(\mathbf{x}_i)$, i=1...n. We assume here that $Z(\mathbf{x})$ is continuous enough to allow $L(Z(\mathbf{x}))$ to be defined everywhere. An estimate of the desired transformation $L(Z(\mathbf{x}_0))^*$ is obtained by:

$$L(z_0)^* = \boldsymbol{\lambda}^T \mathbf{z} \quad (2)$$

with:

$$\mathbf{z} = \begin{pmatrix} \mathbf{z}_i \\ 0 \end{pmatrix} \quad (3)$$

$\mathbf{z}_i$ is the (nx1) data vector, i.e.$[Z(\mathbf{x}_1), Z(\mathbf{x}_2)...Z(\mathbf{x}_n)]'$.
$\mathbf{0}$ is a vector of zeros. Its length equals the number of monomials in P(x).

Allowing for a possible drift of the local mean, the kriging system can be written:

$$\mathbf{K}\,\boldsymbol{\lambda} = \mathbf{k}_0^L \quad (4)$$

with:

$$\mathbf{K} = \begin{pmatrix} \mathbf{K}_{ij} & \mathbf{X} \\ \mathbf{X}^T & 0 \end{pmatrix} \quad (5)$$

$$\mathbf{k}_0^L = \begin{pmatrix} \mathbf{k}_{0i}^L \\ \mathbf{x}_0^L \end{pmatrix} \quad (6)$$

where $\mathbf{X}$ is the matrix of monomials in x_i and y_i, i=1...n. Under stationarity assumption, X reduces to a vector of ones (ordinary kriging), or does not appear (simple kriging).

$\mathbf{x}_0^L$ is the vector of monomials in x_0 and y_0, linearly transformed by L.

$\mathbf{K}_{ij}$ is the covariance matrix (or the generalized covariance when a drift is present) between observations, i,j=1...n.

$\mathbf{k}_{0i}^{L}$ is the vector of covariances between observations and the linear operator at the point to be estimated.

The kriging estimator is unbiased and minimizes the estimation variance $Var(Z_0-Z_0^*)$. The estimation variance is then written

$$\sigma_k^2 = \sigma_L^2 - \boldsymbol{\lambda}^T \boldsymbol{k}_0^L \tag{7}$$

where:

σ_k^2 is the kriging variance.
σ_L^2 is the a priori variance of the linear operator on Z.

More detailed presentation of kriging can be found in standard textbooks such as Matheron (1971) and Journel and Huijbregts (1978). All the preceding equations can easily be generalized to include the multivariate case, that is, cokriging (Matheron, 1971; FrançoisBongarçon, 1981; Myers, 1982).

It is interesting to formulate the kriging interpolator and the kriging system in a dual form (Davis and Culhane, 1984; Royer and Vieira, 1984; Galli et al., 1984).

Rearranging equations 2 and 4 gives:

$$L(z_0)^* = \boldsymbol{z}^T \boldsymbol{K}^{-1} \boldsymbol{k}_0^L = \boldsymbol{b}^T \boldsymbol{k}_0^L \tag{8}$$

with:

$$\boldsymbol{K}\boldsymbol{b} = \boldsymbol{z} \tag{9}$$

also, because of the linearity of the operator:

$$L(Z_0)^* = L(Z_0^*) \tag{10}$$

Equation 8 shows that the interpolated surface possesses an analytic expression:

$$L[Z(\boldsymbol{x})]^* = L[P(\boldsymbol{x})] + \sum_{i=1}^{n} b_i \, L[k(|\boldsymbol{x}-\boldsymbol{x}_i|)] \tag{11}$$

The kriging interpolator is thus the sum of a slowly varying, usually low order (0 or 1) polynomial plus a linear combination of n basis functions each associated to an observation. These basis functions depend on both the covariance and the desired linear operator. The weights, b_i, are invariant under the different transformations; they are completely determined by the observed values and the covariance function.

They thus need be estimated only once. Also, from equation 10, it is seen that the best estimator (in the sense of minimum estimation variance) of any (defined) linear operation on Z(x) is obtained by applying this linear transformation to the kriged surface $Z(x)^*$. This is a remarkable coherence property of kriging which justifies applying frequency domain transformations to the kriged surface.

Adapted model of covariance

Chilès and Guillen (1984), demonstrated that under the hypothesis particular to Spector's model for the distribution of gravity sources (Spector and Bhattacharyya, 1966; Spector and Grant, 1970), the resulting covariance model would be:

$$K(h) = \frac{C\,2t}{[h^2 + 4t^2]^{1/2}} \qquad (12)$$

where the range parameter "t" corresponds to the average depth of the sources, and C is the variance or, equivalently, the asymptotic sill of the corresponding variogram. This covariance model possesses some remarkable properties. The upward continuation operator which relates the observed field at height "t" to the field at height "t+s", is:

$$w_s(h) = \frac{s/2\pi}{[h^2+s^2]^{3/2}} \qquad (13)$$

The field measured at Z_{t+s} is obtained by a convolution product:

$$Z_{t+s}(x,y) = w_s(h) * Z_t(x,y) \qquad (14)$$

The covariance between two points at the height (t+s) can thus be written:

$$K_{t+s}(h) = w_s(h) * w_s(h) * K_t(h) \qquad (15)$$

Applying the convolution theorem to solve this equation gives:

$$K_{t+s}(h) = \frac{C_1\;2(t+s)}{[h^2+4(t+s)^2]^{1/2}} \qquad (16)$$

This is similar to (12), with $C_1 = C\ t/(t+s)$ and range t+s. Thus, the covariance function that would be obtained at height (t+s) can be deduced entirely from the covariance obtained at t. Similarly, the covariance between a point at height t and a point at height (t+s) is given by:

$$K_{t,t+s}(h) = \frac{C_2 \quad (2t+s)}{[h^2+(2t+s)^2]^{1/2}} \tag{17}$$

The sill of this cross-covariance is simply $C_2 = C\ 2t/(2t+s)$, and the range is t+s/2. Knowing the covariance function at height t, the calculation of (17) is straightforward. Also, we have, as expected, the obvious ordering between sills and ranges:

$$C_t \geq C_{t,t+s} \geq C_{t+s} \tag{18}$$

$$a_t \leq a_{t,t+s} \leq a_{t+s} \tag{19}$$

Finally, the covariance between a point at height t+s and another point (horizontally h apart) at height t+s+q is:

$$\frac{2\ C\ t}{[h^2+(2t+2s+q)]^{1/2}} \tag{20}$$

This latter formula enables the different altitudes of the points to be taken into account in the construction of the kriging system. All that is required is the knowledge of the covariance gravimetric model at a reference height t. Note that this "correction" is required even if the gravimetric data are Bouguer anomalies (i.e. corrected for free air, plateau and topographic correction) as noted by Naudy and Neumann (1965).

In all the preceding equations, we assumed s,q>0. In fact, these relations hold true also for negative s and q, as long as $|s|,|s+q|<t$. Thus, downward continuation is also easily achieved by kriging with the gravimetric model of Chilès and Guillen (1984).

3. Results and discussion

Gravity data transformations

In this section, we shall develop, in a 2-D context, the above general results for some widely used gravity data transformations. We shall assume, for simplicity of notation, that P(x)=m although results obtained could be generalized to higher order drift polynomials.

Table 1 summarizes our results. The first column gives the desired transformation, the second column gives the a priori variance of this transformed variable, and the third column gives the resulting expression for the right member of the kriging system (4) when the "gravimetric" model of covariance of Chilès and Guillen (1984) is used. It is interesting to note from these results that the estimates of second horizontal derivatives and second vertical derivatives obey the Laplacian equation of a potential field ($\nabla^2 U = 0$).

Table 1. Formulas to solve kriging systems (equations 4 and 7 in the text) for the various transformations. The available data is at a reference height t, the transformation is estimated at level t+s.

Transformation	A priori variances	k_{0i}^L with "gravimetric" covariance
Interpolation	C	$\frac{C\,2\,t}{\left[h_{0i}^2+4\,t^2\right]^{1/2}}$
Upward (s>0) or downward (s<0,-s<t) continuation	$\frac{C\,t}{(t+s)}$	$\frac{C\,2\,t}{\left[h_{0i}^2+(2t+s)^2\right]^{1/2}}$
First derivative (along x), at height t+s	$\frac{Ct}{4\,(t+s)^3}$	$\frac{-2\,C\,t\,(x_0-x_i)}{\left[h_{0i}^2+(2t+s)^2\right]^{3/2}}$
First vertical (upward) derivative at height t+s	$\frac{Ct}{2\,(t+s)^3}$	$\frac{-2\,C\,t\,(2t+s)}{\left[h_{0i}^2+(2t+s)^2\right]^{3/2}}$
Second derivative (along x), at height t+s	$\frac{3\,Ct}{16\,(t+s)^5}$	$\frac{6\,Ct\,(x_0-x_i)^2-2Ct\,[h_{0i}^2+(2t+s)^2]}{[h_{0i}^2+(2t+s)^2]^{5/2}}$
Second vertical derivative at height t+s	$\frac{3\,Ct}{2\,(t+s)^5}$	$\frac{4\,C\,t(2t+s)^2-2\,C\,t\,h_{0i}^2}{\left[h_{0i}^2+(2t+s)^2\right]^{5/2}}$
Spector decomposition		Factorial kriging; see Chilès and Guillen (1984).

One way to evaluate the performance of a spatial estimator for the different transformation is to compare its amplitude response surface to the theoretical amplitude response corresponding to the desired transformation. Figure 1 illustrates that the circular nature of the theoretical response surface associated with the vertical gradient transformation is well reproduced by kriging.

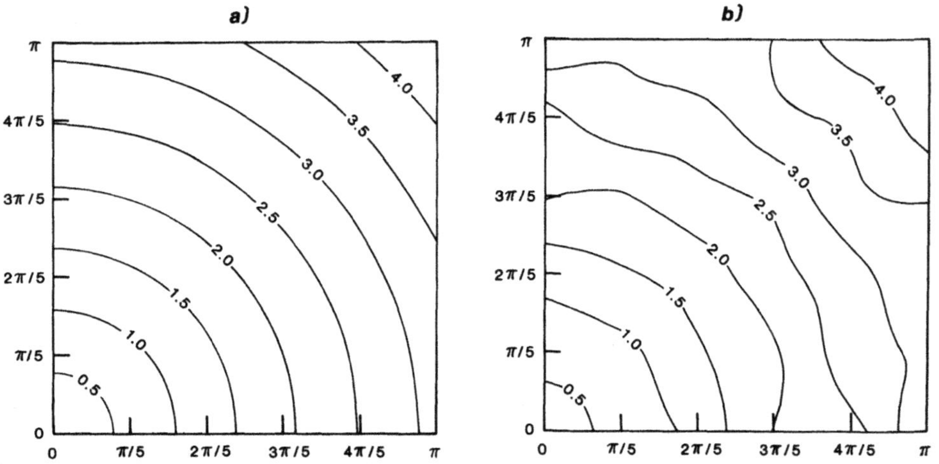

Figure 1. Amplitude response of ideal filter (left) and simple kriging filter (right). Sources at one unit depth. Regular 11 x 11 grid with one unit spacing between grid points.

A simulated example

A 2-D deterministic gravity field is modelled from four prisms located at shallow and intermediate depths using the algorithm of Nagy (1966). The emplacement of prisms and the corresponding gravity field are shown in Figure 2. A grid of 129 x 129 is formed. The spectral method is applied to this grid to obtain first and second vertical derivatives that will be considered as the "real surfaces" for these transformations. Two different "sample" data are extracted; one on a regular (32 x 32) grid, and the other on an irregular grid obtained by random selection of 300 nodes from the regular grid. Kriging and frequency domain transformation for first and second vertical derivatives are performed with both sample data and are compared to the actual surface. In the irregular grid case, an estimation to a regular (32 x 32) grid is obtained by an "inverse of the distance" weighting (with the four nearest neighbours) prior to frequency domain transformation.

The experimental and modelled variograms for both data are shown in Figure 3. It should be noted that the "sill" of the variogram is asymptotically reached at a "practical range" of 40*t. Ordinary and simple kriging were performed locally with 9, 25 and 49 neighbours.

second derivative, particularly in the viscinity of the two very thin prisms. For the vertical gradient, with the regular grid, simple kriging and the spectral method gave better results than ordinary kriging. With the irregular grid, simple kriging gave the best results, followed by ordinary kriging and the spectral method. This illustrates how the interpolation phase could seriously hampered the quality of results obtained by the spectral method.

An increase in neighbourhood improves ordinary kriging but does not significantly affect simple kriging. Results for ordinary kriging were better with the irregular grid (300 sample points) than with the regular grid (1024 sample points). This rather surprising result that may be due to a particular positioning of the regular grid relatively to the modelled prisms. In effect, the modelled range of the covariance function (4.3) with the regular grid seems to reflect mainly the intermediate depth prism (5.0) and very little the shallow ones (0.1). On the other hand, the covariance obtained with the irregular grid presents a range (2.45) that is a compromise between shallow and intermediate depths. The better performance of simple kriging compared to ordinary kriging is probably related to the inverse screen effect (Rivoirard, 1996) of ordinary kriging: the weights of external neighbours are increased due to the implicit estimation of the global mean with the neighbourhood data.

Table 2. Correlations between estimates and real surface for the vertical gradient (dz) and the second vertical derivative (dz^2). OK (ordinary kriging), SK (simple kriging). All: internal estimated grid (900 points); part: subset of estimated grid points at 1.5 or more units from the two very thin prisms (676 grid points).

Estimator	Regular grid data				Irregular grid data			
	dz		dz^2		dz		dz^2	
	all	part	all	part	all	part	all	part
Spectral method	.87	.86	.18	.15	.65	.68	.06	.14
OK (9 points)	.67	.67	--	--	.69	.74	--	--
OK (25 points)	.72	.73	.02	.60	.81	.88	.03	.58
OK (49 points)	.75	.83	.02	.53	.85	.92	.03	.57
SK (9 points)	.96	.96	--	--	.88	.95	--	--
SK (25 points)	.85	.90	.02	.60	.88	.96	.03	.58
SK (49 points)	.84	.92	.02	.52	.89	.96	.03	.57

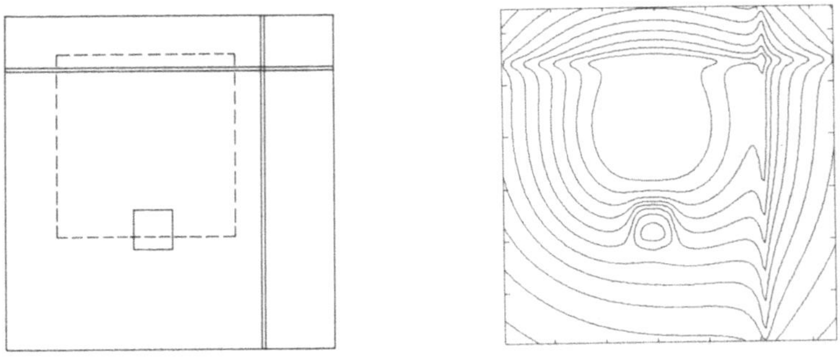

Figure 2. Left: prisms modelled, Right: corresponding gravity field. Shallow elements (depth of 0,1 unit) in solid lines. Intermediate depth prisms (5 units) in hatched lines. The two very thin prisms and the big square have positive density contrast. The small square has negative density contrast. The field is 26 units x 26 units large.

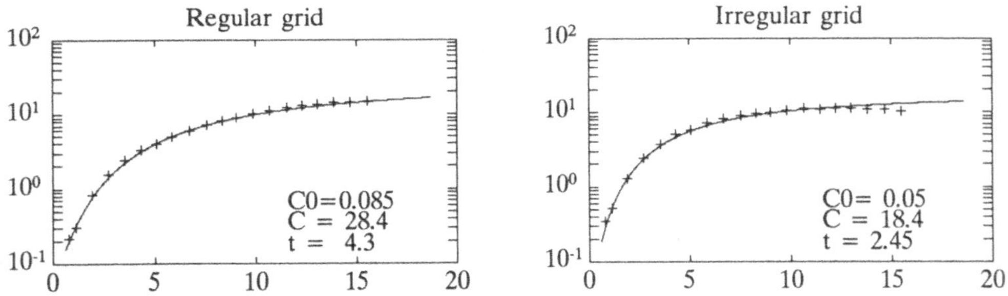

Figure 3. Experimental and modelled variograms. Left, with the regular sample grid (1024 points); right, with the irregular sample grid (300 points). Note that vertical scale is logarithmic to facilitate visualization at short distances.

Table 2 presents the correlations between the actual surface and the estimates, for vertical gradient and second vertical derivative obtained by the two methods with the two different sampling schemes. The comparison is made on the 30 x 30 internal estimated grid points in order to avoid edge effects. Correlations are also calculated for the subset of grid points located at 1.5 units or more from the two very thin prisms. In all cases, the second vertical derivative was poorly estimated. This is because the sampling grid is not fine enough to provide precise estimation of the

This simulated example is not deemed particularly favourable to kriging because of its deterministic nature opposed to the stochastic characterization of the geostatistical approach. We recall that the gravimetric model of covariance stems from the hypotheses particular to the Spector's model. These hypotheses are not met in our simulation. Also, the "real surfaces" were obtained by spectral methods on a finer grid. Albeit, results obtained by kriging are comparable to those obtained by frequency analysis, especially when simple kriging is performed. The interpolation phase required to obtain a regular grid before performing spectral analysis seems to be an important element for the quality of the results. If the interpolation is to be done by kriging, because of its good properties, then the estimation of the transformed values could also be obtained simultaneously by kriging at a very low marginal cost and with a comparable precision. From the computational point of view, this would greatly simplify the conception and execution of software, as a single technique (kriging) instead of two (interpolation followed by frequency analysis) can be used for both interpolation and transformation. As a bonus, the kriging variance may help in judging the appropriateness of a given transformation.

4. Summary and conclusion

The equations to estimate various common data transformations have been derived. From these, it was noted that the kriged estimates for second derivatives obey the Laplacian equation of a potential field.

The model of covariance for gravity developed by Chilès and Guillen (1984) according to Spector's model for the distribution of the sources seems to be applicable under a wider range of situations than first expected by those authors. On a deterministic modelled field, this covariance provided a very good fit to the experimental variogram. Simple kriging gave better estimations of vertical gradients than its spectral counterpart or ordinary kriging. Moreover, estimates obtained by simple kriging were relatively insensitive to the neighbourhood size. Results for the second derivative were unsatisfactory in all cases. This emphasizes the difficulty in estimating the second derivative and throws doubt on its usefulness compared to the vertical gradient.

In real world applications, the data is seldom, if ever, sampled on a regular grid. Accordingly, a preliminary interpolation phase is required before proceeding to data transformation by spectral methods. In this context, kriging appears as a strong contender to the present two-phase approach in terms of the quality of results, the simplification of the code and the versatility of the method.

Acknowledgments

This research was financed by Department of Energy, Mines and Resources (research grant 224 4 92) and NSERC (CRD 133234 and OGP0105603).

References

Chilès, J.P. and Guillen, A. (1984) "Variogrammes et krigeages pour la gravimétrie et le magnétisme", Sciences de la Terre, 20, 455-468.

Davis, M.W. and Culhane, P.G. (1984) "Contouring very large datasets using kriging", in G. Verly et al. (eds.), Geostatistics for Natural Resources Characterization, NATO-ASI, 122, Series C, Reidel Publishing Co., Dordrecht, pp. 599-620.

François-Bongarçon, D. (1981) "Les corégionalisations, le cokrigeage", Centre de Géostatistique et de Morphologie Mathématique, Fontainebleau, C-86, 26 p.

Galli, A., Murillo, E. and Thomann, J. (1984) "Dual kriging - its properties and its uses in direct contouring", in G. Verly et al. (eds.), Geostatistics for Natural Resources Characterization, NATO-ASI, 122, Series C, Reidel Publishing Co., Dordrecht, pp. 621-634.

Matheron, G. (1970) "La théorie des variables régionalisées et ses applications", Cahiers du Centre de Morphologie Mathématique, fascicule 5, 212 p.

Myers, D.E. (1982) "Matrix formulation of co-kriging", Math. Geology, 14, 3, 249-257.

Nagy, D. (1966) "The gravitational attraction of a right rectangular prism", Geophysics, 21, 2, 362-371.

Naudy, H. and Neumann, R. "Sur la définition de l'anomalie de Bouguer et ses conséquences pratiques", Geoph. Prosp., 13, 1-11.

Rivoirard, J. (1984) "Looking for a kriging plan in a stockwerk deposit", in G. Verly et al. (eds.), Geostatistics for Natural Resources Characterization, NATO-ASI, 122, Series C, Reidel Publishing Co., Dordrecht, pp. 621-634.

Royer, J.J. and Vieira, P.C. (1984) "Dual formalism of kriging", in G. Verly et al. (eds.), Geostatistics for Natural Resources Characterization, NATO-ASI, 122, Series C, Reidel Publishing Co., Dordrecht, pp. 691-702.

Spector, A. and Bhattacharyya, B.K. (1966) "Energy density spectrum and autocorrelation function of anomalies dues to simple magnetic models", Geophys. Prosp., 14, 3, 242-272.

Spector, A. and Grant, F.S. (1970) "Statistical models for interpreting aeromagnetic data", Geophysics, 35, 293-302.

THE SPATIAL VARIABILITY OF PREDICTION ERRORS

Paul Switzer
Stanford University
Statistics Department
Stanford, CA 94305 USA

INTRODUCTION

This paper is about the estimation of error in spatial problems, how we do it and how we might do it. It is also about our methods of interpreting error estimates to our scientific audience. We have two specific objectives in mind. The first objective is to elucidate the role of averaging or pooling, i.e., the role of statistics, in the reporting of error estimates. Our discussion of pooling is largely philosophical and is focused on the proper interpretations that should be given to error estimates. The second objective is to emphasize the importance of viewing errors as a spatially varying field. Here we will have some specific suggestions which affect how we usually estimate errors in geostatistical problems.

Errors in geostatistical problems arise from incomplete observation of a function on a geographic domain. Geostatisticians have developed methods for efficiently using the observed part to make predictions regarding the unobserved part. More importantly, geostatisticians have provided a framework for assessing and reporting the errors associated with predictions, arguably the major achievement of the field.

The points raised in this paper have been discussed and debated on other occasions both within the geostatistics community and among statisticians in general. In terms of geostatistical practice, the main point will be that recognizable regions of high and low variability are typical of spatial data and should be routinely recognized when reporting estimates of prediction errors. Error estimates are more useful and easier to interpret if they are strongly tied to the local variability of the data rather than the global variability which depends on sometimes arbitrary definitions of the spatial domain.

The purpose of this paper is not to propose specific algorithms for regionalization of error estimates. Rather, the purpose is to draw attention to the utility of routine regionalization of error estimates. The examples were chosen to illustrate, as simply as possible, the main points of the paper.

ERROR AVERAGING

Let Z(x) and Z(x) denote the true and predicted values respectively at location x. Z(x) may be a point value at x or another local value such as a block average centered at x. We take the prediction error at location x to be the difference D(x) = |Z(x)-Z(x)|. Since the true Z(x) is not known at the time of the prediction, the error D(x) also is not known.

At the heart of the statistical method is the replacement of the unknowable and particular error D(x) by a some sort of distribution of errors. Conceptually, this distribution is created by grouping or pooling errors associated with similar prediction problems. The pooled error distribution is summarized using one or more statistics such as the mean or the mean square. We would then report the same mean value repeatedly as a statistical substitute for the particular errors attached to particular predictions in the same group. The heterogeneity of the pooled error distribution might also be reported in terms of an appropriate statistic such as its standard deviation.

The terminology "averaging set" will be used here to refer to the chosen grouping of similar prediction problems. If the averaging set is chosen appropriately then information about the pooled error distribution may be available even though information about particular errors is not. In the most common form of geostatistical practice the averaging sets correspond to fixed spatial configurations of observations relative to the prediction point.

The critical step is the specification of the averaging set, i.e., what are the comparable prediction problems which should be grouped together and be treated alike for purposes of error estimation? It is important to note that there are many possible "correct" estimates of a prediction error corresponding to different choices for the averaging set attached to the given prediction problem. Unless it is clearly understood from the context, any error report should contain an explicit statement regarding the statistical averaging which underlies the report.

The distinctions among alternative statistical error estimates are precisely the distinctions in the choice of averaging sets. What criteria should govern the choice of averaging sets? Arguably, a better averaging set is one for which the dispersion is smaller. The quest is then for averaging sets which are believed to be relatively homogeneous for values of the prediction error D. We want the differences between the particular errors and the reported statistical errors to be small, i.e., we should do our best to distinguish prediction problems from one another. One important distinction is related to regionalization of prediction errors.

LOCAL MODULATION OF PREDICTION ERRORS

Prediction errors show spatial structure on two different scales. The first, which we call local modulation, relates to the spatial configuration of the data in the prediction neighborhood; we are likely to have smaller prediction errors closer to data points. The larger scale spatial structure of prediction errors derives from the changing variability statistics of the underlying field rather than from the spatial configuration of the observations. We are likely to have smaller interpolation errors where there is less variability in the underlying field. This we call the regional modulation of prediction errors.

Describing the local modulation of the spatial variability of prediction errors has been one of the principal contributions of classical geostatistics. Specifically, squared differences in the expression for a particular prediction error, D(x), are replaced by global averages of "similar" squared differences. Here, "similar" usually means similar spatial separation and this provides the rationale for variography. The restriction to neighborhood prediction means that only relatively short distances are involved. Furthermore, when predicted values are linear combinations of observed values in a neighborhood, with coefficients summing to one, then the prediction error D(x) for either point prediction or block prediction can always be expressed in terms of squared differences between point values. [It is not necessary to invoke more abstract concepts of stationarity and unbiasedness.]

An advantage of the variogram approach to spatial structural modeling for short distances is that it is insensitive to larger scale drift and the effects of low frequency modulations which may become apparent as the domain get larger. This robustness property is not shared, for example, by autocorrelations and the global variance. Typically, both the autocorrelations and the global variance will increase as the domain for global statistics gets larger.

However, since the variogram is a global statistic, the same relative spatial configuration of neighborhood data produces the same error estimate regardless of the position of the neighborhood within the domain. By itself, the global variogram for the Z(x) field cannot provide information regarding the regional modulation of prediction errors.

In the usual geostatistical approach the statistical averaging set consists of point pairs with similar spatial separation. But there other possibilities for forming statistical averages. For example, we may wish to distinguish different orientations of point pairs, as in non-isotropic variography. We may also wish to have different variograms for zones of high and low values of the field. If we have auxiliary covariate information such as rock type or grain size we may want to use similarity of the covariate as an averaging set either instead of or in addition to spatial separation. Finally, if we have monitoring data with repeated observations of the field at fixed monitoring locations, then we can use time averages of observed differences for monitor site pairs, together with a model for extending these averages to unmonitored locations.

Clearly, different choices of the statistical averaging set will lead to different estimates of a particular prediction error. While the estimates are different, they are all nevertheless correct in the context of their respective averaging. For example, the choice between isotropic or non-isotropic variography is simply a choice of specificity. This underlines the importance of reporting estimation errors together with the averaging context from which they are derived. Misunderstanding regarding the averaging context sometimes leads to confusion in the interpretation of the reported prediction error.

While there is not a unique approach to error estimation because of the arbitrary choice of averaging sets, the richness of the available information should guide the specification of averaging sets. If a cruder form of averaging is used where a more refined specification can be supported by the available data, then there may be a lost opportunity to provide more informative error estimates which are more descriptive of the particular prediction problem. Still, a lost opportunity is not really the same as a mistake if care is taken always to specify the averaging set.

REGIONAL MODULATION OF PREDICTION ERRORS

A global variogram is useful for declustering of clustered data but it lacks the specificity needed for estimating prediction errors. Of course, global variograms are a useful data-analytic tool for exhibiting the scales of spatial variability. On the other hand it is local variability based on data neighborhoods which provides the means to extract the regional pattern of prediction errors. Estimates of prediction errors should be capable of responding to and exhibiting regional spatial pattern.

There are several ways in which prediction error estimates can be made regionally specific. The most straightforward way is to allow for a regionalized proportional scaling factor, say C(x), for the variogram, in the spirit of Haas (1990) or Host, Omre, and Switzer (1991). It is then natural to use neighborhoods to estimate the scaling factor locally. A suitable quadratic form such as the sample variance, using only observations in the neighborhood of x and the globally fitted variogram, can be used to provide an estimate of C(x).

The virtues of neighborhoods have been routinely recognized for construction of the predictor itself as a means of adapting the prediction to local changes in mean level. But the importance of neighborhoods to the estimation of the prediction error have not been systematically exploited in the applications of geostatistics. Thus the reported prediction error estimates have a global quality which does not reflect conditions of local variability.

In some cases it may happen that the introduction of a covariate serves as a surrogate for regional changes in variability patterns. For example, stratification of variography according to rock type is sometimes an obvious choice. Then the reported estimates of prediction errors will be specific to rock type. This stratification assumes that the covariate has a spatially coherent pattern and thereby the error estimation process is enriched.

A third method for introducing regional modulation in error estimates is to scale the variogram by a factor which depends on the mean level of the field as described by Journel and Huijbregts (1978, p.187) and Cressie (1991, p.64), for example. Sometimes even a simple proportional variogram scaling will capture the essential heterogeneity of the estimation error. But the magnitude of estimation errors is not always so simply tied to mean levels. For example, density or concentration fields which span a broad dynamic range will have the smallest prediction errors in neighborhoods where the concentrations are either very low or very high. Similar remarks apply to the spatial heterogeneity of the variability of precipitation fields.

A fourth method would be to do a structural analysis on squared differences of the data. This method is particularly appropriate to regularly spaced data, but could be adapted to irregular data. The idea is to define a new field

$$S(x) = \text{Average}[Z(x\text{-}h/2)\text{-}Z(x+h/2)]^2 ,$$

where the average is over all vectors h of unit length, say. A fitted S variogram could then be used to smoothly interpolate the S field by Kriging, from which one gets the regional modulation of the scaling constant of the Z variogram.

Even when we allow for regional modulation of prediction errors, there is still a large but spatially unstructured component of the prediction error. Since, in principle, the unstructured component of the prediction error cannot be estimated, our estimates of the prediction error can recover only the spatially structured part of the error field.

SOME DETAILS FOR GRIDDED DATA

In order to focus on the regional modulation of prediction errors we consider the following two prototypical situations where a fixed configuration of neighborhood data removes the local modulation component of the prediction error. The first situation is the problem of predicting the center point of a grid cell where the data are on a regular grid and the predicted value is taken to be the simple average of the data at the corners of the grid cell. The second situation is the problem of predicting the block average of the 2x2 block centered at the central point of a grid cell; once again, the predicted value is taken to be the simple average of the data at the corners of the grid cell. The spatial variability of the prediction error derives only from the spatial modulation of variability of the underlying field. These simplified "regular" problems should be kept in mind for the discussion which follows.

Turning first to the problem of interpolating the central point of the neighborhood at location x, the squared prediction error may be written as

$$D2(x) = 2 * T2(x) - S2(x) * (1\text{-}1/N) \text{ , where}$$

$$2 * T2(x) = \text{Average}\{Z(xi)\text{-}Z(x)\}^2 \text{ over xi ,}$$

$$2 * S2(x) = \text{Average}\{Z(xi)\text{-}Z(xj)\}^2 \text{ over xi and xj .}$$

Here N is number of observations in the neighborhood, and xi,xj are distinct observation locations within the neighborhood, and Z(x) is the spatial field we are trying to predict. This is an exact expression for the prediction error at location x and does not involve any modeling or hypotheses.

The value of S2 is computable from the data and is, in fact, just the sample variance of the observations in the neighborhood of x. On the other hand, the particular value of T2 must be replaced by a statistical estimate. The following argument suggests that S2 may be a reasonable, but somewhat conservative, estimate of T2.

Global variography may indicate significant spatial autocorrelation at distances on the scale of the prediction neighborhood together with a large global variance for the Z(x) field. However, within a neighborhood, both the apparent spatial autocorrelation and the apparent variation of the Z(x) field are typically much smaller. To a rough approximation, Z(x) may be viewed as a locally stationary random function with unimportant correlation except at the shortest distances.

The preceding argument may be used to justify a conservative approximation of T2 by S2. This leads to an estimate of the prediction error magnitude at location x given by

$$\text{estimate } D2(x) = S2(x) * (1+1/N) .$$

This naive estimate of the prediction error is proportional to the local standard deviation of the data. The proportionality factor is exactly appropriate to the prediction problem with iid sampling from a fixed distribution. Not only is this naive estimate likely to be conservative, i.e. too large, but it is indifferent to the location of the interpolation point x within the neighborhood and would certainly be inappropriate to prediction error estimation close to an observed datum.

One alternative to the naive estimate for the unobserved component T2 of the true prediction error derives from an approach taken by Loader and Switzer (1992). Let W(x) denote the vector W1(x),..,WN(x) of ordinary kriging weights for neighborhood prediction at location x, corresponding to some variogram, and let D2i be the vector $\{Z(xi)\text{-}z(xj)\}^2$, j=1..N. Then the Loader-Switzer approach gives

$$\text{estimate } \{Z(xi)\text{-}Z(x)\}^2 = W(x)' * D2i .$$

For prediction at the center of the square grid, the kriging weights are all 1/N. The implied estimate of T2(x) is S2(x)*(1-1/N), and the corresponding prediction error estimate is

$$\text{estimate } D2(x) = S2(x) * (1\text{-}1/N) .$$

This approach can be used also for prediction error estimates at arbitrary locations with arbitrary neighborhood configurations because of its use of a kriging weight vector. Since the kriging weights are invariant to scale changes in the variogram, we do not explicitly estimate a local scale factor for the variogram.

Of course, we can explicitly estimate a local scale factor for a variogram. For example, with a linear variogram model. Then an estimate for T2(x) is obtained by a linear combination of the observed values of $\{Z(xi)\text{-}Z(xj)\}^2$ where the weights are determined by the geometry of the neighborhood relative to the prediction location x. When x is the midpoint of a data square, then the four squared differences at unit distance would each have coefficient 0.707/6 while the two diagonal squared differences would each have coefficient 0.500/6.

True prediction errors exhibit substantial short scale variability on top of the regional modulation. Therefore, it can be argued that estimated prediction errors should be spatially smoothed. This point is taken up again briefly in the next section.

For the second problem of predicting a block average from neighborhood data which are not highly clustered, the squared prediction error can be expressed as

$$D2(x) = 2 * TB2(x) - S2(x) * (1-1/N) - B2(x), \text{ where}$$

$$2 * TB2(x) = \text{average}\{Z(xi)-Z(xu)\}^2 \text{ over xi and xu,}$$

$$2 * S2(x) = \text{average}\{Z(xi)-Z(xj)\}^2 \text{ over xi and xj,}$$

$$2 * B2(x) = \text{average}\{Z(xu)-Z(xv)\}^2 \text{ over xu and xv.}$$

Here, xu and xv are locations in the block being predicted and xi and xj are distinct data locations.

Once again, S2 is computable from the data. An argument similar to that above suggests that TB2 might be conservatively estimated by S2. Finally, B2 is the "exhaustive" variance for the block at location x and it, too, might be estimated by the variance, S2, of the data within the block. Putting these naive estimates together produces an estimate of the magnitude of the block prediction error:

$$\text{estimate } D2(x) = S2(x)/N .$$

This naive estimate is the same as the one appropriate to the estimation of the population mean from N iid sample values. In general, prediction errors for block averages or a global average should not be too different from those obtained under classical iid assumptions provided the data are not highly clustered. Refinements similar to those discussed for the point prediction problem are possible here as well.

THE UNCERTAINTY OF PREDICTION ERRORS

The true prediction errors D(x) exhibit variability unrelated to coherent spatial structure. We may consider that the actual error at location x is drawn from a distribution centered at M(x), say. It is M(x), rather than D(x), that we can hope to recover.

For discussion purposes imagine that the difference between the unobserved Z(x) and its estimated value is approximately a Gaussian random variable with standard deviation M(x). Then the error measure D(x) is a random variable equivalent to $M(x)*|Q(x)|$ where Q(x) is a zero-mean Gaussian variable. This implies that the coefficient of variation for the prediction error, D(x), is approximately 0.75, large enough to make the regional structure of D(x) not always apparent. For example, it will be difficult to distingush a doubling of M(x) based on the value of D(x).

On the other hand, if we want to estimate only the regional pattern of M(x), i.e., regioanlly averaged values of M(x), then the corresponding regioanlly smoothed values of D(x) will have enhanced precision. This argues for regional smoothing of the estimated pointwise prediction errors. However, regional smoothing of estimaed prediction errors is not quite the same as estimating pointwise prediction errors using larger neighborhoods. In particular, the variance of the data in a larger neighborhood will involve data-pairs with larger spatial separation, whereas smoothing of variances derived from the small nieighborhoods does not introduce inter-point differences with larger spatial separation. Since the the true prediction error involves only short inter-point distances, the use of data variances from larger neighborhoods seems less appropriate than the smoothing of dta variances from small neighborhoods.

The severity of regional smoothing of estimated prediction errors depends on the desired spatial resolution for prediction errors weighed against the reduction in the irreducible error uncertainty that comes with more smoothing. Some guidance might be obtained through a structural analysis of the pointwise error estimates.

ILLUSTRATIONS

The first illustration is the simple case of a one-dimensional and unilateral equally spaced time series of length 100. It is chosen to exhibit aspects of regional modulation of prediction errors in an uncomplicated way and is not meant to be typical of geostatistical problems. The data at time t are the observed values up to time t-1. For the sake of discussion suppose that the predicted value at time, t, is the average of the observed values of the time series between t-1 and t-5. Figure 1 shows the underlying time series, the predicted time series, and the time series of the true magnitudes of the prediction errors. The underlying time series has fairly rich spatial structure with autocorrelations of 0.79 at lag 1, decreasing slowly to 0.30 at lag 13.

The goal is to estimate the prediction error magnitude at time t, using the historical information. A special feature of this illustration is that we have available the true prediction errors up to time t-1.

The classical geostatistical approach to this problem would be to do a structural analysis of the available time series data and fit a variogram as a function of time lag. From the variogram we would calculate the estimation variance, i.e., the estimated squared prediction error, which would correspond to a data configuration of five consecutive observed points followed by the prediction point. Since this configuration is same for all t, the calculated prediction error would be the same for all t.

In this example the true prediction errors are progressively revealed. In principle, we can use classical geostatistical modeling of the observed time series of prediction errors up to time t-1 and use Kriging methods to estimate the prediction error at time t. However, as a simplification we use a simple unweighted average of the recently observed prediction errors between t-1 and t-10. The important distinction, vis-a-vis the classical geostatistical approach of the preceding paragraph, is that we would now have a time-varying estimate of the prediction error.

Another approach to time-varying error estimation would be to simply calculate the variance of the of the original data in the time window between t-1 and t-10, say. The window variance is then multiplied by a fixed factor to obtain a time-varying error estimate, along the lines described in the earlier section on gridded data. This differs from the approach of the last paragraph in that we are basing the time-varying error estimate on the local variance of the underlying observed time series rather than on the local mean of the observed error series. The importance of this

FIGURE 1a. UNDERLYING TIME SERIES

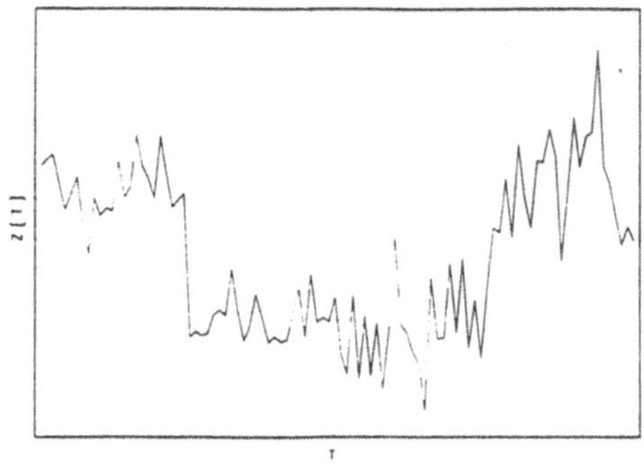

FIGURE 1b. PREDICTED TIME SERIES

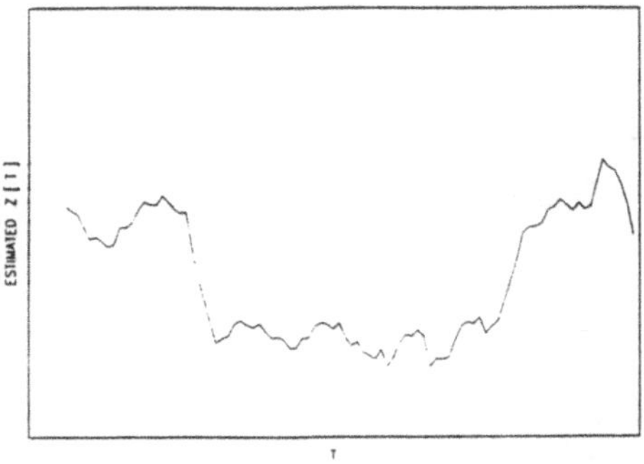

FIGURE 1c. PREDICTION ERROR TIME SERIES

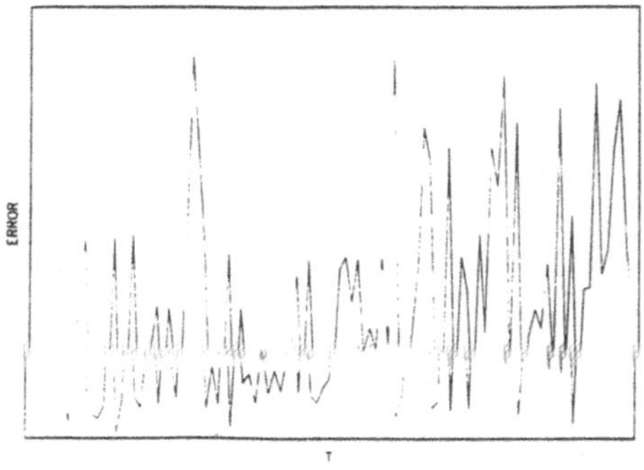

FIGURE 2a. SMOOTHED PREDICTION ERRORS

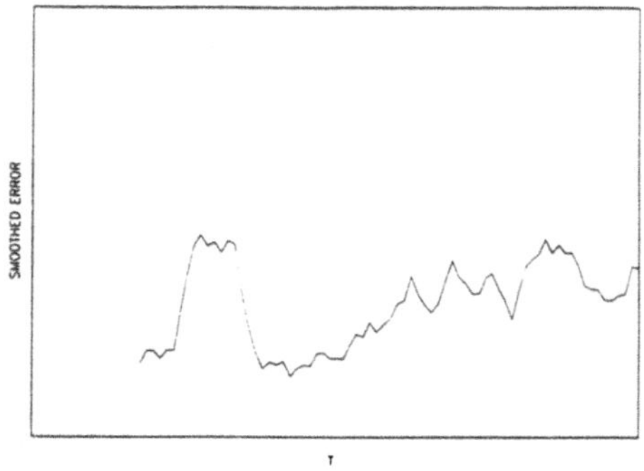

FIGURE 2b. ESTIMATED ERRORS – I

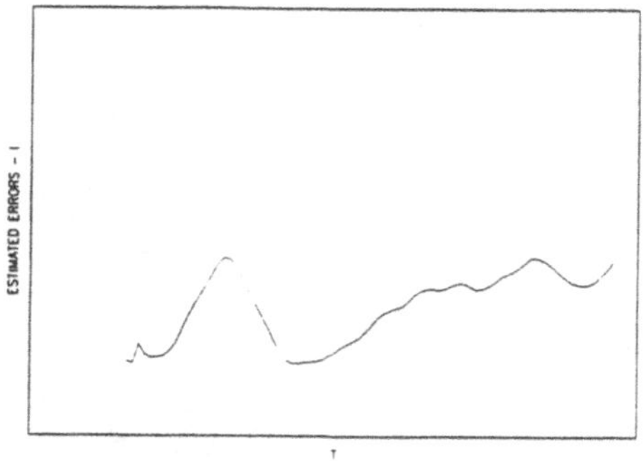

FIGURE 2c. ESTIMATED ERRORS – II

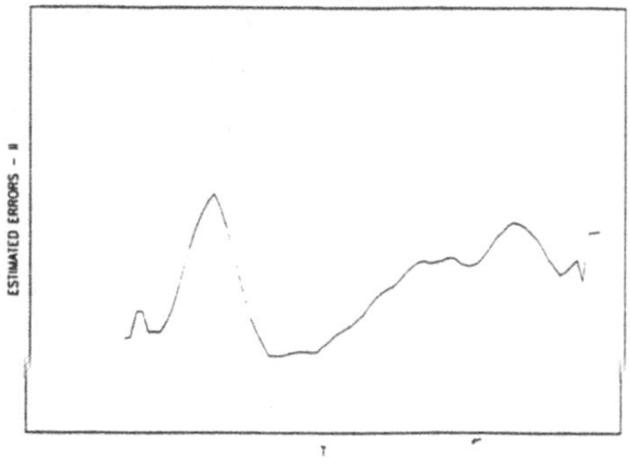

FIGURE 3a. DATA VALUES

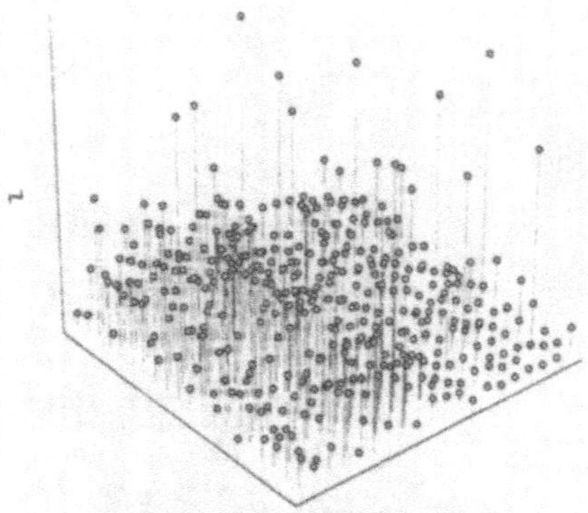

FIGURE 3b. SMOOTHED PREDICTION ERRORS

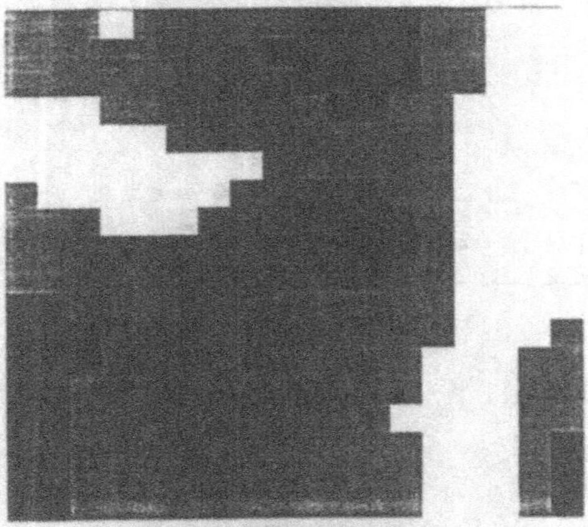

FIGURE 3c. ESTIMATED PREDICTION ERRORS

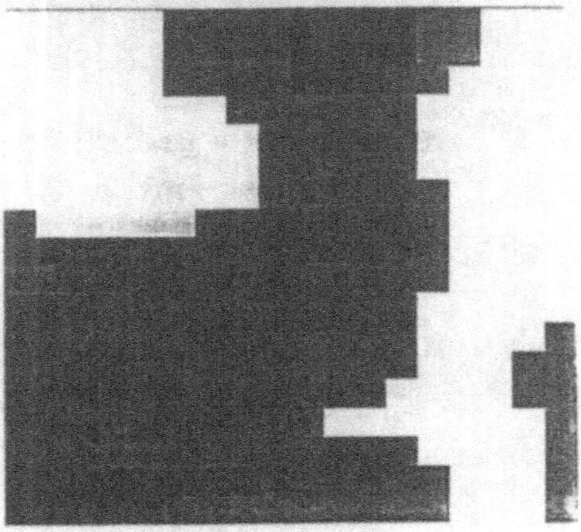

last approach to time-varying error estimation is that the historically observed errors are not used and, therefore, this approach is more easily generalized to spatial problems where observed errors at other locations are typically not available for prediction at location x.

Figure 2 shows the time series of the actual prediction error magnitudes together with the two time series of the estimated error magnitudes according to whether the prediction errors up to time t-1 are available or not. A nine-point smoother has been used to help suppress the randomness of the prediction error and to bring out the regional pattern of prediction errors, as described in the last section. The pattern of the estimated prediction errors follows the pattern of the true prediction errors. By comparison, the time series of estimated prediction errors using a global variogram is necessarily a global constant.

However, a time-varying prediction error estimate in place of the constant prediction error estimate used in practical geostatistics will not necessarily provide a large improvement when viewed pointwise because of the large spatially unstructured component of the error field D. The large improvement is apparent only when viewed on a regional scale, that is when the goal of error estimation is the "mean" error field M as described in the last section. Thus the contemporaneous cross-correlations between the unsmoothed series of true prediction errors and estimated prediction errors are 0.08 and 0.11, respectively, for the two local estimation procedures described above. However, these contemporaneous cross-correlations rise to 0.51 and 0.57 for the smoothed versions of these time series. [The unilateral one-sided nature of the estimation creates a delay effect which tends to reduce the contemporaneous cross-correlation.]

The first example of one-step ahead prediction in an equally-spaced time series was chosen because we have several clear alternatives regarding the estimation of prediction errors and the localization of error estimates seems quite natural.

The second example is based on a two-dimensional 20x20 grid. The data are represented by the graph in Figure 3a. As an exercise, the value at each grid location, not on the boundary, is predicted by the average of the four nearest data. The magnitudes of the true prediction errors at each grid location were smoothed using a two-dimensional spline smoother and then discretized to three levels. The resulting simplified pattern of true prediction errors is shown in Figure 3b.

The estimate of the prediction error magnitude at each grid location was formed by applying the same two-dimensional spline smoother to the four-point standard deviations for each of the prediction neighborhoods. The calculated standard deviations were scaled by the constant factor 0.7, reflecting the discussion in Section 4. Once again, the pattern of true prediction errors is reflected in the smoothed estimates.

CONCLUSIONS

Estimates of prediction errors attached to individual locations are actually estimates of spatially averaged prediction errors. The averaging context should always be made explicit. Common geostatistical practice treats error estimation in global averaging framework. However, we can often distinguish regions of high and low variability which should be used to modulate prediction error estimates. Local estimates of variability can be computed from the same neighborhood data used to derive the predicted value. But regional error patterns are easily obscured by large random fluctuations in the true prediction errors. Therefore, neighborhood-based estimates of variability should be smoothed and reported error estimates should refer to a regional averaging scale.

REFERENCES

Cressie, N. (1991) *Statistics for Spatial Data*, Wiley, New York.

Haas, T.C. (1990) "Lognormal and moving window methods of estimating acid deposition", *J. American Statist. Assn.*, vol. 85, pp.950-963.

Host, G., Omre, H. & Switzer, P. (1991) "Spatial prediction of air pollution from space/time observations", Technical Report, Norwegian Computing Center, Oslo.

Journel, A. & Huijbregts, C. (1978) *Mining Geostatistics*. Academic Press, New York.

Loader, C. and Switzer, P. (1992) "Spatial covariance estimation for monitoring data. In *Statistics in Environmental and Earth Sciences* (A. Walden & P. Guttorp, eds), pp. 72-90. Edward Arnold, London.

RELATIONS BETWEEN THE INDICATORS RELATED TO A REGIONALIZED VARIABLE

Jacques RIVOIRARD
Centre de Géostatistique
Ecole des Mines de Paris
35 rue Saint–Honoré
77300 Fontainebleau - France

0. INTRODUCTION

In the seventies, in order to make predictions for future selective mining, geostatistics had to deal with the problem of estimating reserves above cut–off grades. Since then practitioners have used techniques based on, for example, multigaussian distributions, disjunctive kriging (cokriging of indicators), or more simply kriging of indicators. The indicators above a threshold define the ore at this cut–off or, in a non mining context, define the geometric set of values above the cut–off.

During the past decade the interest in geometric problems has increased, in particular for simulations. So we have seen the development of techniques for simulating random functions $Z(x)$ from indicators $I[Z(x) < z]$ at different thresholds z, or conversely, techniques for simulating sequential geological facies obtained by thresholding a multigaussian random function.

In many such cases several thresholds and indicators corresponding to different random sets related to the variable under study have to be handled simultaneously. These sets depend on each other and their mutual arrangement is an important structural characteristic of the variable. In this paper we propose simple tools to describe this arrangement.

1. SOME THEORY

Thresholding a Random Function (RF) gives Random Sets (RS), the structure of which is related to the structure of the RF. Here we will only characterize structures from pairs of points (x, x+h). The RF Z(x) is supposed to have stationary bivariate distributions (Z(x), Z(x+h)) with covariance $\sigma(h)$. The cut–off z divides the space into the RS of points with value $\geq z$ and its complementary set. The indicator $I[Z(x) \geq z]$ (equal to 1 on the first set, 0 on the other) has the mean $E(I[Z(x) \geq z]) = P[Z(x) \geq z] = T(z)$.

Its covariance

$$\sigma_z(h) = \mathrm{Cov}\{I[Z(x) \geq z], I[Z(x+h) \geq z]\}$$

$$= P[Z(x) \geq z, Z(x+h) \geq z] - T(z)^2$$

and its variogram $\gamma_z(h) = 0.5\,\mathrm{E}\{\,I[Z(x) \geq z] - I[Z(x+h) \geq z]\,\}^2$

$$= 0.5\,(P\{\,I[Z(x) \geq z] \neq I[Z(x+h) \geq z]\,\})$$

i.e. $0.5\,(\,P[Z(x) < z, Z(x+h) \geq z] + P[Z(x) \geq z, Z(x+h) < z]\,)$

are functions of the bivariate distribution (Z(x), Z(x+h)). Thus, it is possible to explicitly compute the structure of this indicator under the hypothesis of a given bivariate distribution (such as bigaussian). Note that the indicator $I[Z(x) \geq z]$ and its complement $I[Z(x) < z] = 1 - I[Z(x) \geq z]$ have the same variogram and covariance.

Generally the structure of the indicator $I[Z(x) \geq z]$ changes when the cut–off z varies: apart from the variance which acts as a multiplicative factor, the shape changes as we will see in examples. Recall (e.g. Matheron 1982) that summing the indicator variograms for all cut–offs gives the order 1 variogram of Z(x)

$$\int \gamma_z(h)dz = \frac{1}{2}\,\mathrm{E}\,\,|Z(x+h) - Z(x)|$$

The cross covariance between indicators at cut–offs z and z' is

$$\sigma_{z\,z'}\,(h) = \mathrm{Cov}\,(\,I[Z(x) \geq z]\,,\, I[Z(x+h) \geq z']\,)$$

$$= P[Z(x) \geq z, Z(x+h) \geq z'] - T(z)\,T(z')$$

It is not necessarily symmetric in h. Knowing these covariances for all cut–offs is equivalent to knowing the bivariate distributions (Z(x), Z(x+h)) for all distances h. Summing these covariances gives the covariance of Z(x)

$$\int\!\!\int \sigma_{z\,z'}(h)\;dz\;dz' = \sigma\,(h)$$

Unlike the covariance, the cross variogram mixes the cases +h and –h

$$\gamma_{z\,z'}(h) = 0.5\;\; E[I(Z(x+h) \geq z) - I(Z(x) \geq z)][I(Z(x+h) \geq z') - I(Z(x) \geq z')]$$

or, if $z \leq z'$: $0.5\,(\,P[Z(x) < z, Z(x+h) \geq z'] + P[Z(x) \geq z', Z(x+h) < z]\,)$

We will see (next section) that the way the RS are arranged for two cut–offs $z \leq z'$ can be conveniently described using the conditional probabilities

$$P\;[Z(x+h) \geq z' \mid Z(x+h) \geq z,\; Z(x) < z] \quad \text{denoted} \quad P_h\;[\rightarrow\geq z' \mid \rightarrow\geq z]$$

$$P\;[Z(x+h) < z \mid Z(x+h) < z',\; Z(x) \geq z'] \quad \text{denoted} \quad P_h\;[\rightarrow< z \mid \rightarrow< z']$$

The first one is the probability that, going from a point x with value $<z$ to a point x+h with value $\geq z$, the second value is $\geq z'$. In other words, it is the probability with which, getting into the domain of values $\geq z$, one meets a value $\geq z'$. The second one is the probability with which, getting into the values $<z'$, one meets a value $<z$. The first probability describes the "border effects" within the domain $\geq z$, the second one the border effects within the domain $< z$.

In practice, such probabilities can be computed from pairs (Z(x), Z(x+h)) of points separated by h. They are expected to be equal for +h and –h only if the distributions of pairs are symmetric in h. This will be supposed here from now on. Then the cross variogram can be written $(z \leq z')$

$$\gamma_{zz'}(h) = P[Z(x) \geq z', Z(x+h) < z]$$

whereas the simple one becomes

$$\gamma_z(h) = P[Z(x) \geq z, Z(x+h) < z]$$

We then get $(z \leq z')$ $$P_h\,[\rightarrow \geq z' \mid \rightarrow \geq z] = \gamma_{zz'}(h) / \gamma_z(h)$$

$$P_h\,[\rightarrow < z \mid \rightarrow < z'] = \gamma_{zz'}(h) / \gamma_{z'}(h)$$

So comparing the cross variogram between two indicators to their simple variograms gives our conditional probabilities directly. All these results are not original, but seem to have been insufficiently exploited for structural analysis as well as for the choice of estimation or simulation techniques. We will now see what these probabilities physically mean, then how they can be used.

2. SOME TYPES OF ARRANGEMENTS

Some schematic figures will help in understanding. In each case the RF Z(x) takes values 1, 2, 3, 4 with given probabilities (the names : mosaic, residual, and diffusion will be justified in section 4.).

In the mosaic model (Figure 1), the space is divided into compartments. Each compartment is given one of the four values, according to their probabilities, and independently from the other compartments. So, when going from a point x with value 1 to a point x+h with value $\neq 1$, this point takes any of the other values 2, 3, 4 without preference. Going into the domain C2 of points ≥ 2, the probability to reach a value ≥ 3 does not depend on h either. In the same way, going into the domain <3, the probability to reach a value <2 does not depend on h. To sum up, probabilities such as $P_h\,[\rightarrow \geq 3 \mid \rightarrow \geq 2]$ and $P_h\,[\rightarrow < 2 \mid \rightarrow < 3]$ are constant in h.

In a diffusion–type model (Figure 2), when leaving the domain of points 1, we meet first values 2. When leaving the 2's, there is a transition to 1 or 3, depending on whether we are going upwards or downwards. Probabilities like $P_h\,[\rightarrow \geq 3 \mid \rightarrow \geq 2]$ or

P_h [$\rightarrow$< 2 | $\rightarrow$< 3] are not constant. They start at 0 (no contact between 1 and 3) and then increase with h.

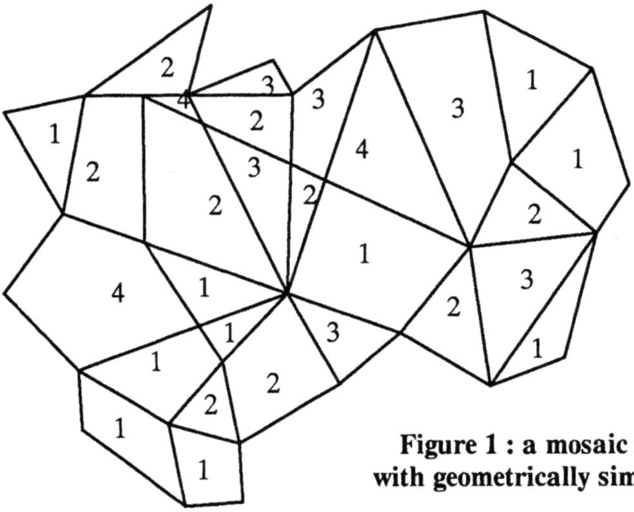

Figure 1 : a mosaic model with geometrically simple tiles

It is possible to imagine less strict transitions, Figure 3. Imagine that the values ≥ 3 are anywhere within the domain C2 of points ≥ 2, but are rarer within the borders of this domain. Leaving the 1's, we will meet, not always but still preferentially, values 2. The probability P_h [$\rightarrow$≥ 3 | $\rightarrow$≥ 2] increases with the first distances h, but starts from a positive value, expressing a possible contact between values 1 and values ≥ 3.

Imagine now – **model with indicator residuals**, Figure 4 – that within the domain C2 of values ≥ 2, the values ≥ 3 are randomly distributed no matter the proximity of the frontier of C2. There are no border effects within C2: leaving the 1's, we meet values 2 or ≥ 3 without any preference: P_h [$\rightarrow$≥ 3 | $\rightarrow$≥ 2] is independent of h. There is no longer any tendency to transit at value 2 when going from the 1's domain to the ≥ 2 values. Similarly, the values 4 being distributed randomly and without border effects within the domain ≥ 3, there is no tendency to transit through 3 when entering the ≥ 2. However there is a hierarchy in this model, and what happens when entering the domain of values < a cut-off is not the same as when entering the domain ≥ a cut-off. Here, leaving C3, the probability P_h [$\rightarrow$< 2 | $\rightarrow$< 3] to reach a value 1 depends on h : it is small for small h (because the frontiers of C3 are more in contact with values 2), and increases afterwards.

Many other arrangements may be thought of, e.g. that the values 4 are distributed randomly and without any border effects within the domain C3, but that these values ≥ 3 themselves avoid the borders of C2. Or that C3 is preferentially located on the right parts of C2, etc. Here we will only consider the simple types.

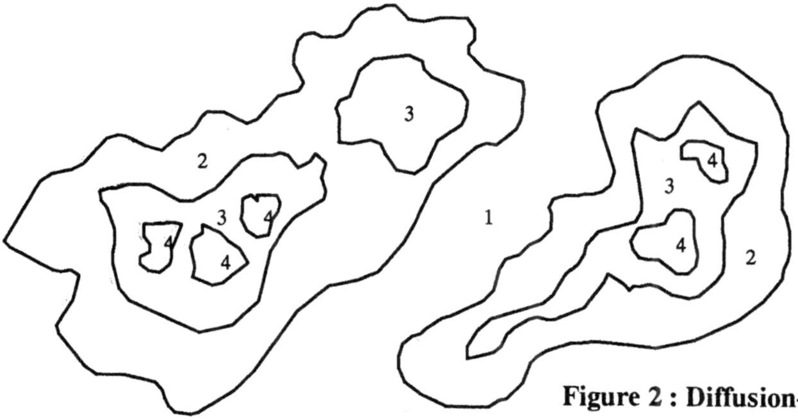

Figure 2 : Diffusion–type model

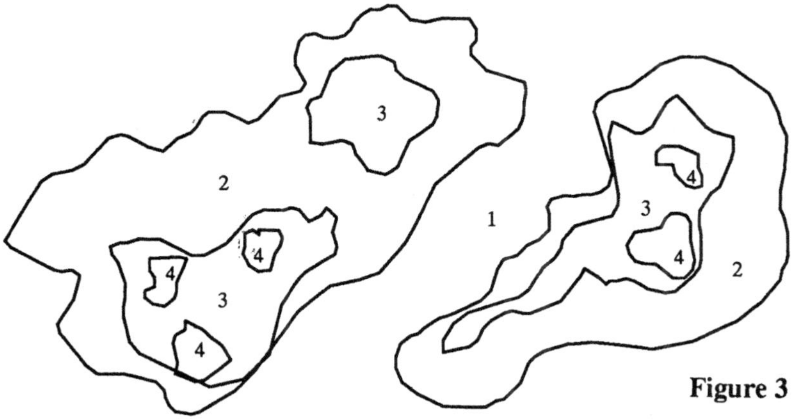

Figure 3

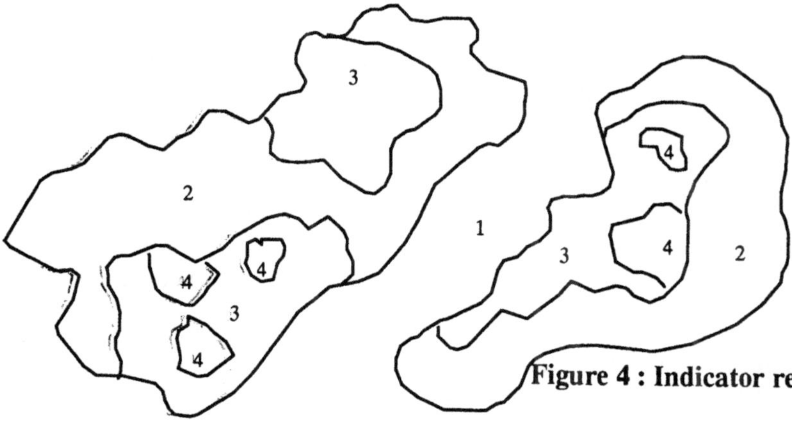

Figure 4 : Indicator residual model

In short, and in term of indicator variograms :

If the cross variograms of indicators $\gamma_{z\,z'}(h)$ are proportional to the variograms of the lower threshold indicator $\gamma_z(h)$, it is the indicator residual model. The sets of points $\geq$ the different thresholds are nesting without border effects one within the other.

If the cross variograms are proportional to the variograms of the larger cut–off, it is also an indicator residual model, but in the other direction ("decreasing" instead of "increasing"). Now it is the sets with values lower than the different cut–offs which are nesting one within the other without border effects.

If the ratios $\gamma_{z\,z'}(h)/\gamma_z(h)$ are constant in h whatever the thresholds z and z', we have the mosaic model. There is no border effect whether increasing or decreasing.

Lastly, if the ratios $\gamma_{z\,z'}(h)/\gamma_z(h)$ increase with h, there are border effects, increasing and decreasing, like in the diffusion–type models.

However, when the cut–offs z and z' correspond to close probabilities, their cross and simple variograms are of course close to each other, even in a diffusion–type model. Moreover for quite distinct cut–offs, there are in practice always border effects, more or less well marked. Taking them into account is a matter of appreciation and the above indications have to be moderated. The best is to look at real examples.

3. REAL EXAMPLES

Example 1

These data are herring acoustic densities, which were provided by K. Foote and I. Rottingen (Laboratory of Marine Research, Bergen, Norway). Petitgas (1991) has made a geostatistical study of them. Each value is in fact the regularized density over a 1 nautical mile segment along parallel transects. In the units that were used, 37% of the 986 values exceed $10^2=100$, 12% $10^3=1000$, and 3% $10^4=10000$.

The fluctuations that can be seen on the variograms of the corresponding indicators I2, I3, I4 show that the structures are not well known: Figure 5, distances in nautical miles. However there is a clear destructuration of the simple variograms when the threshold increases. The cross variogram between two indicators, e.g. I2 and I3, looks like I2, although it is not similar. The ratio between these two (the probability to exceed 1000 when entering the >100 domain) shows a rapid increase, which indicates small border effects within this domain. Petitgas chose to neglect these effects and used an indicator residual model to obtain a disjunctive kriging map (see next section).

In comparison, the ratio of the cross variogram between I2 and I3 on the variogram of I3 (the probability to be lower than 100 when entering the <1000 domain) shows much more important border effects, which conforms to the increasing model with indicator residuals.

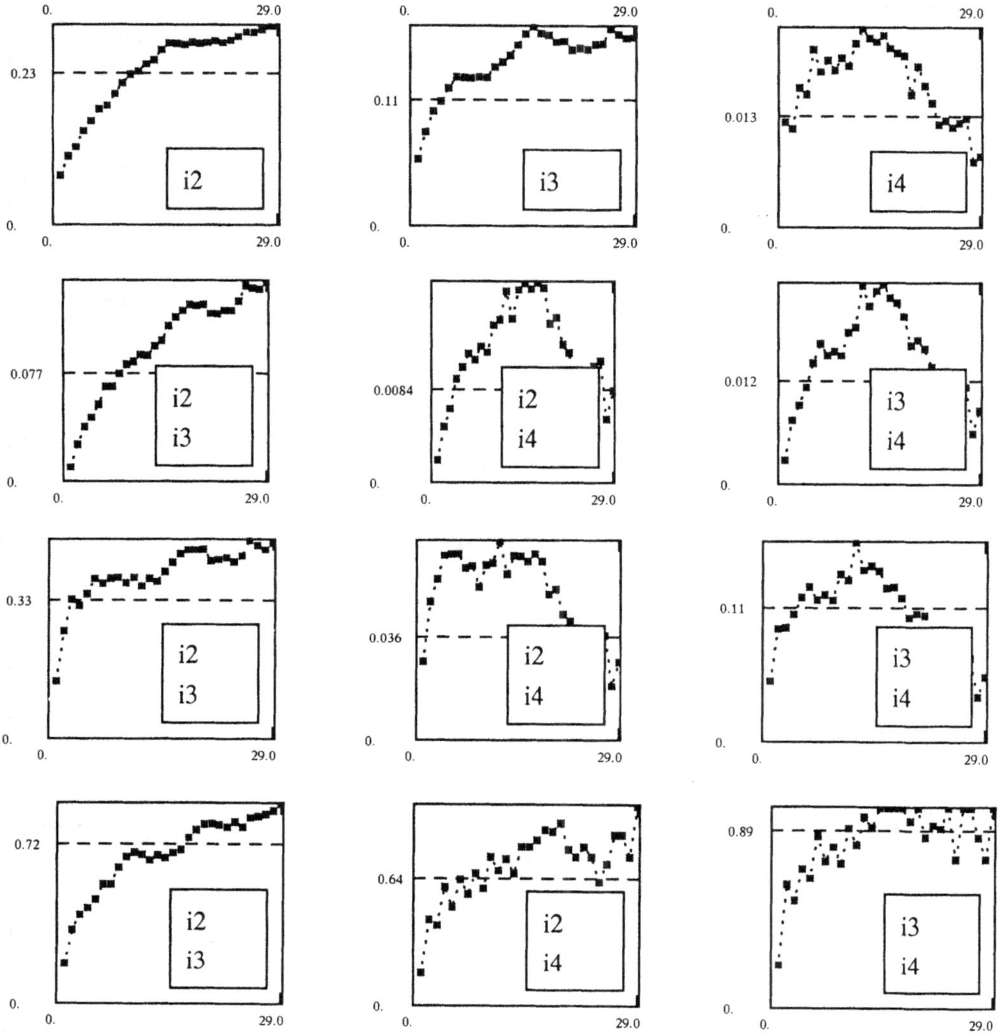

Figure 5 : thresholds 10^2, 10^3, 10^4 on fish densities.
From top to bottom:
- variograms of indicators;
- cross variograms of indicators;
- conditional probabilities upwards: $P_h\ [\rightarrow \geq 10^3 \mid \rightarrow \geq 10^2]$, ...;
- conditional probabilities downwards: $P_h\ [\rightarrow < 10^2 \mid \rightarrow < 10^3]$, ...

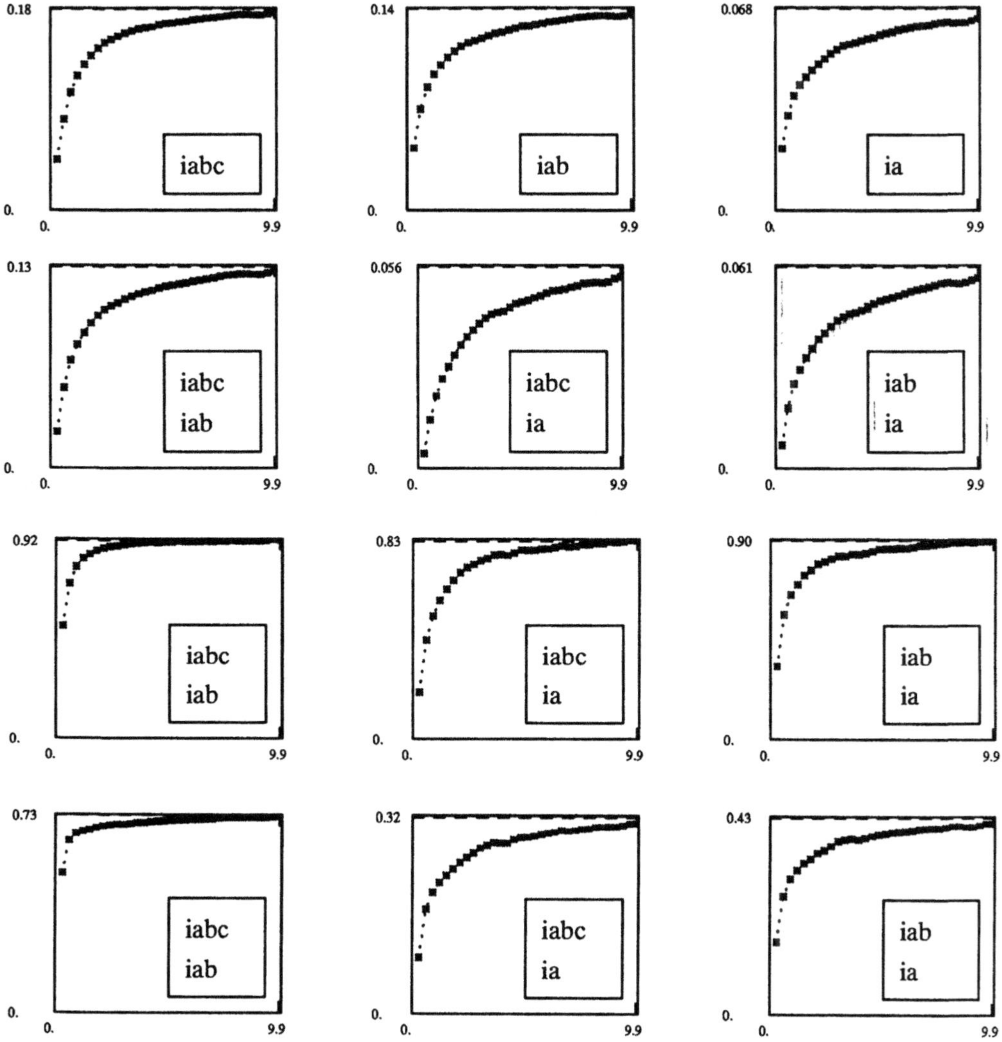

Figure 6 : cumulated rock facies: ABC, AB, A.

From top to bottom:

– variograms of indicators;

– cross variograms of indicators;

– conditional probabilities upwards: $P_h\ [\rightarrow AB \mid \rightarrow ABC]$, ...;

– conditional probabilities downwards: $P_h\ [ABC \rightarrow \mid AB \rightarrow]$, ...

Example 2

These rock facies data were provided by the Direction Production & Developpement Total. They correspond to increasing granulometries: shales (denoted D), shaly sandstones C, sandstones B, sandstones A, and appear in a fluvio-deltaic sedimentation with probabilities 77%, 6%, 10%, 7% (cumulated: 100%, 23%, 17%, 7%). The data are known along wells, along which the variograms of the cumulated indicators A, AB, ABC are calculated. They show beautiful structures (Figure 6).

The cross variogram AB–ABC is very close to the simple variograms of AB and ABC. This is because AB and ABC corresponds to close probabilities (17 and 23%). The facies C being little represented, the conditional probability to reach AB when entering ABC, or to leave ABC when leaving AB, increases rapidly with the distance. The cross variograms A–AB and A–ABC are more regular at the origin than the corresponding simple ones. The conditional probabilities increase regularly with the distance, showing well marked border effects within the domains AB and BCD. The facies B is a transition between A and CD. These data have been modelled by a diffusion-type model (thresholded gaussian model, see further).

4. USING THE RESULTS : THE CHOICE OF TECHNIQUES

We have seen the descriptive tools that are the variograms of indicators, simple and cross, as well as the conditional probabilities. Let us now look at the estimation or simulation, according the different types encountered in 2. (see Rivoirard 1990 for more details on estimation). Of course the indicators are not independent: if $Z(x) \geq 3$, then $Z(x) \geq 2$. The dependence between $I[Z(x) = 2]$ and $I[Z(x) = 3]$ also holds between neighboring points x and x+ h, as generally $Z(x+h)$ has less chance to be ≥ 3 if $Z(x)$ is itself < 2.

The mosaic model with independent values

We have already seen how to build this model from a mosaic (random partition of the space). Let $\varrho(h)$ be the probability for two points h apart, x and x+ h, to belong to the same compartment. If x and x+h belong to the same compartment, then $Z(x)=Z(x+h)$. If not, $Z(x)$ and $Z(x+h)$ are independent. So, whatever the functions f and g, we have

$$E[\ f(Z(x))\ g(Z(x+h))] = E[f(Z(x))\ g(Z(x))]\ \varrho(h) + E[f(Z)]\ E[g(Z)]\ (1-\varrho(h))$$

hence a covariance

$$Cov\ [f\ (Y(x)),\ g\ (Y(x+h))] = Cov\ [f\ (Y(x)),\ g\ (Y(x))]\ \varrho\ (h)$$

which is proportional to $\varrho(h)$. Thus the simple (take $f=g$) or cross covariances between any two functions of $Z(x)$ are proportional to $\varrho(h)$ (to $1-\varrho(h)$ for variograms). In particular this is so for the indicators, which are intrinsically correlated (Matheron 1965).

The consequence for the estimation is the following. Cokriging the indicators is reduced to their separate kriging, with weights that are the same for all indicators. Moreover, as any function f of $Z(x)$ (taking the values $f_1, f_2, \ldots$ when $Z(x) = 1, 2, \ldots$) can be written as a linear combination of the indicators

$$f[Z(x)] = f_1 \; I(Z(x) = 1) + f_2 \; I(Z(x) = 2) + \ldots$$

the resulting estimation of $f[Z(x)]$, i.e. its disjunctive kriging, is nothing but its kriging.

To sum up this is the model in which the separate kriging of indicators, as recommended by Journel (1982), finds its theoretical justification. Basically the reason for any indicator to be estimated separately comes from the elementary property: knowing $I(Z(x) = 2)$, no matter the exact value of $Z(x)$ when one wants to estimate $I(Z(x+h) = 2)$.

The model with orthogonal indicator residuals

Its construction comes directly from the properties seen in section 2. Starting from independent RS A_1, A_2, A_3, A_4, we put $Z(x) = 1$ if $x \in A_1$; $Z(x) = 2$ if $x \notin A_1$ and $x \in A_2$; $Z(x) = 3$ if $x \notin A_1$, $x \notin A_2$ and $x \in A_3$; $Z(x) = 4$ if $x \notin A_1$, $x \notin A_2$ and $x \notin A_3$. The indicator

$$I(Z(x) \geq j+1) = \prod_1^j I(x \notin A_i) = I(Z(x) \geq j) \quad I(x \notin A_j)$$

depends only on the A_i up to j. Geologically this process makes one think of successive and partial erosions. After each of them the eroded parts are filled in with a material, the value of which is smaller and smaller (other hierarchical values can of course be imagined).

For the estimation, the indicators can be factorized with the residuals of the regressions between successive indicators (hence the name): Rivoirard 1989. These residuals need only to be kriged to obtain the cokriging of the indicators and then the disjunctive kriging of any function of $Z(x)$. In this hierarchical model, the estimation of $I(Z(x) \geq j)$ depends only on the indicators for cut–offs $\leq j$ (the cut–offs higher in the hierarchy).

The way to build the model is similar to the technique proposed by Alabert (1987) to sequentially simulate the indicators $I[Z(x) < z_k]$ corresponding to a thresholded RF $Z(x)$. Alabert simulates each indicator using the values of this indicator, either known at the data points, or deduced at points from previously simulated indicators (if $Z(x) < 1$, then $Z(x) < 2$ for this point x). Although the order of the sequence is important and may generate bias, Alabert has no direct criteria for choosing this order. As a matter of fact, a sequential simulation of indicators $I[Z(x) < z_k]$ in the order z_{i_1}, z_{i_2}, ..., z_{i_n} corresponds to a building from independent RS A_1, A_2, ... with

$$I(Z(x) \geq z_{i_p}) = I(Z(x) \geq z_{i_{p-1}}) \; I(x \notin A_p)$$

or

$$I(Z(x) < z_{i_p}) = I(Z(x) < z_{i_{p-1}}) \; I(x \notin A_p)$$

according to whether z_{i_p} is larger or smaller than $z_{i_{p-1}}$. It is an indicator residual model with a particular hierarchy. We expect, for $p<q$, to have a cross variogram between $I(Z(x) < z_{i_p})$ and $I(Z(x) < z_{i_q})$ proportional to the variogram of $I(Z(x) < z_{i_p})$. This gives a criterion to choose such a sequential processing of indicators.

Remark : Even with a continuous distribution, the indicator residual model is theoretically made of compartments with constant values. Then it is also a mosaic model (in the distinction made by Matheron 1989, page 309), but where the value of a compartment is not independent of its size, nor of the neighboring compartments.

The diffusion–type models

In probability, physics, or earth science, diffusion processes correspond to phenomena with gradual variations. There is a transition through neighboring values (but this may be rapid, in particular for some ranges of values). In these models the RS obtained by thresholding are closely linked. Whereas estimating an indicator uses only the values of this indicator at the data points in the mosaic model, and in the residual model it uses also the values of the hierarchically higher indicators, here it depends on the values for all indicators.

In the multigaussian model for instance, the most common diffusion model, the estimation of $I(Z(x) < z)$ depends on the kriging of the variable Z(x) itself, then on all the indicators at the data points. The fact that we know the multivariate and then the conditional distributions, and that the residuals of the regressions are independent from the conditioning values, makes this model suited for estimation, simulation, and above all for conditional simulation. As the variable under study is rarely normal, this is usually considered, for this model, as being transformed from a normal variable. Thresholding a gaussian RF is a particular transformation.

Besides the gaussian model, there are other diffusion models, based on a given statistical distribution (e.g. gamma, Hu 1988), or built on empirical distributions (Matheron 1984, Lajaunie et Lantuéjoul 1989). These models are suited, with the flexibility of an additional transformation, to various distributions (discrete, or with large peaks). As multivariate and conditional distributions are not workable, the estimation is performed through the disjunctive kriging technique (which needs only bivariate hypotheses and is also available in the gaussian case).

Remark: It is possible to imagine models with border effects which are not strictly diffusive (transition by neighboring values). Example: a mosaic where the value is the average, within each compartment, of a given RF.

5. CONCLUSION

Looking at the cross as well as simple variograms of indicators is a convenient way to study the arrangement between the RS obtained by thresholding a RF with stationary and symmetric bivariate distributions. The conditional probabilities that can be deduced can help choosing a mosaic model without border effects, an indicator residual model (no border effects upwards or downwards), or a diffusion-type model (border effects upwards and downwards).

These are simple models, and more sophisticated ones may be needed sometimes. For instance a geological environment resulting from combined diffusion and erosion (mixed diffusion-residual models). Or a sedimentary process where the granulometry gradually decreases, but sometimes increases by jumps (there is no symmetry in h vertically).

References

Alabert, F. (1987) "Stochastic imaging of spatial distributions using hard and soft information", M. Sc. Thesis, Stanford University.

Hu, Lin-Ying (1988) "Mise en oeuvre du modèle gamma pour l'estimation de distributions spatiales", Thèse de Doctorat de l'Ecole des Mines de Paris.

Journel, A. (1982) "The indicator approach to estimation of spatial distributions", Proceedings of the 17th APCOM, T. B. Johnson and R. J. Barnes Editors, New York, pp. 793-806.

Lajaunie, Ch. et Lantuéjoul, Ch. (1989) "Setting up the general methodology for discrete isofactorial models", Proceedings, 3rd International Geostatistics Congress, Avignon, 5-9 Sept. 1988 : "Geostatistics", M. Armstrong Editor, Kluwer Academic Publ., Dordrecht, Holland.

Matheron, G. (1965) Les variables régionalisées et leur estimation, Thèse de Doctorat d'Etat, Ed. Masson, Paris.

Matheron, G. (1976) "A simple substitute for conditional expectation: the disjunctive kriging", Proceedings, NATO ASI : "Advanced Geostatistics in the Mining Industry", Rome, Oct. 1975, pp. 221-236.

Matheron, G. (1982) "La destructuration des hautes teneurs et le krigeage des indicatrices", Centre de Géostatistique, Ecole des Mines de Paris.

Matheron, G. (1984) "Une méthodologie générale pour les modèles isofactoriels discrets", Sciences de la Terre n°21, pp. 1-78.

Matheron, G. (1989) "Two classes of isofactorial models", Proceedings, 3rd International Geostatistics Congress, Avignon, 5-9 Sept. 1988 : "Geostatistics", M. Armstrong Editor, Kluwer Academic Publ., Dordrecht, Holland.

Petitgas, P. (1991) "Contributions géostatistiques à la biologie des pêches maritimes", Thèse de Géostatistique, Ecole des Mines de Paris - 211p.

Rivoirard, J. (1989) "Models with orthogonal indicator residuals", Proceedings, 3rd International Geostatistics Congress, Avignon, 5-9 Sept. 1988 : "Geostatistics", M. Armstrong Editor, Kluwer Academic Publ., Dordrecht, Holland.

Rivoirard, J. (1990) "Introduction to disjunctive kriging and nonlinear geostatistics", Centre de Géostatistique, Ecole des Mines de Paris, 89 p. : bibliogr.

GETTING BETTER QUALITY SEISMIC SECTIONS

Z. ZHANG and A. GALLI

Centre de Géostatistique
Ecole des Mines de Paris
35 rue Saint–Honoré
77305 Fontainebleau – France

ABSTRACT

With the development of reservoir geophysics, one of the major problems is no longer just to determine the external geometry of the reservoir, but also to assess its internal variability. The final aim is to extract lithological and petrophysical information from seismic data. This raises a lot of problems. Two of them are the scale of seismic measurement and the fact that generally seismic amplitudes are noisy.

One of the models currently used by geophysicists is the following one :

The amplitudes are the vertical convolution of the reflexion coefficients with the signal. The noise is not correlated with amplitudes nor is it correlated horizontally.

In this paper we will show how geostatistics can work under these hypotheses, mainly to filter out noise from seismic information. After presenting the method, we will discuss how to test the model and to infer the parameters. Studies have been carried out on one seismic section. The model appears to be a good approximation, and we show different ways to infer the parameters so as to have a good confidence in the result.

INTRODUCTION

Seismic measurements are generally associated with a noise whose origin is very diverse. This noise makes both the interpretation of the section and the precise target of the horizons difficult. Moreover for subsequent processing, this element of interference must be eliminated before or during the processing.

Numerous techniques are available for filtering noise, most of them related to Fourier analysis. In this paper we propose using a purely spatial technique based on variogram decomposition and spatial estimation of "factors". This technique called factorial kriging analysis was developed by MATHERON G. (1982). Since this period factorial kriging analysis has been used mainly in geochemistry [SANDJIVY L. 1984 ; WACKERNAGEL H., 1986], and gravimetry [GALLI A., 1984, CHILES J.P. and GUILLEN A., 1984]. It has also been applied to image analysis by MA Y.Z. and ROYER J.J. (1988) and DALY C. et al (1989, 1991) and in relation with Fourier analysis it can be found in SANDJIVY L. and GALLI A. (1985).

The first step with factorial kriging analysis is to decompose the experimental variogram in a sum, each term being allotted to a factor. This is the crucial step, because most of the further results depend on this identification. In the 70's a method was proposed by DASH B.P. and OBAIDULLAH K.A. to identify these parameters by using a very simple model where amplitudes are supposed to be dependent only on time, and so do the noise with the assumption of non correlation of noise from one trace to the other. This technique has been used with success by the authors and others [RICHARD V., 1988]. However this method assuming that the only difference from one trace to another even if used for adjacent traces can over−estimate the variance of noise. Moreover it is difficult to verify the hypothesis needed, and this method does not work in the case of non stationarity. Here we propose a specific way to make this decomposition in the scope of a specific 2D model of random function. The model used here is a factorized model of covariance, well adapted to data presenting an overall stratification, thus allowing a more precise determination of the vertical variogram of noise which is a difficult point because, as we will see, the experimental vertical variogram is strongly oscillating, not showing clearly the sum of two structures. This model and the additional usual hypothesis can be tested. After presenting the properties of this model we apply it to a real case study, taking care to check its validity and its stability.

THE MODEL

The following model is usually examined for post stack seismic trace : the measured amplitude (AM) is the result of the convolution of the reflexion coefficients (CR) in time, by the signal (S) plus a noise (N) :

(1) $$AM(x,t) = (CR * S)(x,t) + N(x,t)$$

where x is the horizontal distance and t the time.

In what follows, we shall call A (x, t) the real amplitude:

$$A(x,t) = (CR * S)(x,t)$$

So the model is :

$$AM(x,t) = A(x,t) + N(x,t) \tag{2}$$

If we have only one seismic section at our disposal (which is the case here), or few wells, this model is too general for filtering the noise. Therefore we shall begin by making two additional hypotheses which are standard in seismic :

1. H1) the noise has a zero mean and is not correlated to the real amplitude.
2. H2) the noise is independent from one trace to another. It is only correlated vertically (that is in the same trace).

FILTERING METHOD

Under the hypothesis H1 filtering the noise can be carried out using the method known as factorial kriging analysis (MATHERON G., 1982).

Let us recall this method in the monovariable case which interests us, assuming stationarity with a zero mean.

In kriging analysis we construct an estimator N* of the noise by a linear combination of the amplitudes measured at the neighboring points of the point to be estimated:

$$N^*(x,t) = \sum_{\alpha=1}^{\alpha=n} \lambda^\alpha(x,t)\, AM(x_\alpha, t_\alpha) \tag{3}$$

The weights λ^a ($a = 1$ *to* n)are obtained by minimizing the variance of estimation:

$$VAR(N(x,t) - N^*(x,t))$$

which leads to the usual system of factorial kriging analysis:

$$\sum_{\alpha=1}^{\alpha=n} \lambda^\alpha \, Cov\left(AM(x_\alpha, t_\alpha), AM(x_\beta, t_\beta)\right) = Cov\left(N(x,t), N(x_\beta, t_\beta)\right) \tag{4}$$

$$\forall \quad \beta = 1 \text{ to } n$$

In this way we see that as soon as the covariances of the noise and the measured amplitudes are known, the estimation is made by solving a linear system, where the size is the number of points used for the estimation.

DETERMINING THE COVARIANCES

Let us recall that in the model (2)

$$AM(x,t) = A(x,t) + N(x,t)$$

as A and N are independent, the covariance of the measured amplitudes C_{AM} is :

(5) $$C_{AM}(h_x, h_t) = C_A(h_x, h_t) + C_N(h_x, h_t)$$

where $C_A(h_x, h_t)$ = the covariance of the real amplitudes, $C_N(h_x, h_t)$= the covariance of the noise.

As a variogram :

(6) $$\gamma_{AM}(h_x, h_t) = \gamma_A(h_x, h_t) + \gamma_N(h_x, h_t)$$

Consequently, in practice, it suffices to be able to fit the experimental variogram of the measured amplitudes by a sum of variograms, some for the real amplitudes and the others for the noise.

This direct method seems possible in certain cases where the variogram is plainly fitted by a sum of several quite different variograms that, for physical reasons, correspond to each of the quantities we want to estimate (SANDJIVY L., 1984), (GALLI A. et al, 1984) etc.. This only seems possible here in non vertical directions, as under the hypothesis H2 we have:

(7)

$$\gamma_{AM}(h_x, h_t) = \gamma_A(h_x, h_t) + \delta_{h_{x=0}}\gamma_N(0, h_t) + (1 - \delta_{h_{x=0}}) C_N(0, 0)$$

where $\delta_{h_{x=0}} = 1$ if $h_x = 0$

$\delta_{h_{x=0}} = 0$ otherwise

If $h_t = 0$ (horizontal direction) in (7) :

(8) $$\gamma_{AM}(h_x, 0) = \gamma_A(h_x, 0) + (1 - \delta_{h_{x=0}}) C_N(0, 0)$$

On the horizontal variogram of the measured amplitudes we could therefore assign the nugget effect to the noise. Besides, the horizontal and diagonal variograms of the noise are none other than this pure nugget effect.

However, along the vertical direction, this direct method seems very problematic because the vertical variogram of the measured amplitudes oscillates (figure 1).

We know that this behaviour is partly due to the shape of the seismic signal, but it can also be partly allotted to the reflexion coefficients or even to the noise.

This method can only be considered if the signal and the vertical variogram of the reflexion coefficients are known. Or if we accept the hypothesis, common in seismic, that this variogram is a pure nugget effect vertically, it is then possible to obtain the vertical variogram of the real amplitudes. We can then fit the experimental vertical variogram of the real amplitudes and then after substraction fit the variogram of noise. Here, we shall use a special type of model : factorized models. These make it possible to treat the noise filtering problem without having data on the signal and on the reflexion coefficients if we only have seismic sections at our disposal.

FACTORIZED MODEL

We assume here that the covariance of the reflexion coefficients C is of the type

(9) $$C(h_x, h_t) = C^H(h_x)\, C^V(h_t)$$

It's easy to show that in this case the covariance of the real amplitudes is also factorized in

(10) $$C_A(h_x, h_t) = C^H(h_x)\, C_A{}^V(h_t)$$

with $$C_A{}^V(h_t) = C^V * \overline{S}(h_t)$$

where $$\overline{S}(t) = \int S(t+\tau)\, S(\tau)\, d\tau$$

If we use now the hypotheses H1 and H2, we obtain the following model

$$C_{AM}(h_x, h_t) = C^H(h_x)\, C_A^V(h_t) + \delta_{h_{x=0}}\, C_N(0, h_t)$$

from where the property of this model.

For $h_x \neq 0$

(11) $$\frac{C_{AM}(h_x, h_t)}{C_{AM}(h_x, 0)} = \frac{C_A{}^V(h_t)}{C_A{}^V(0)} = \frac{C_A(0, h_t)}{C_A(0, 0)}$$

Formula 11 shows us that the relation between the diagonal and horizontal covariances of the measured amplitudes does not depend on the horizontal distance and is equal to the vertical covariance (reduced) of the real amplitudes. This formula gives not only a way to identify the variogram of noise but is also a powerful test of the hypotheses we use.

From formula 11 we have

$$\frac{C_{AM}(h_x, h_t)}{C_{AM}(h_x, 0)} = \frac{C_{AM}(0, h_t) - C_N(0, h_t)}{C_{AM}(0, 0) - C_N(0, 0)}$$

We deduct from it

$$C_N(0, h_t) = C_{AM}(0, h_t) - \frac{C_{AM}(h_x, h_t)}{C_{AM}(h_x, 0)}\left(C_{AM}(0, 0) - C_N(0, 0)\right)$$

(12)

Using this formula we shall calculate the vertical covariance (or variogram) of the noise, from the covariances (or variograms) of the measured amplitudes.

We have just presented the kriging analysis system and the formulae for obtaining the covariances. We shall now apply them to a real case.

APPLICATION EXAMPLE

The data

The section we will use comes from a carbonate reservoir. As we said previously, usually seismic sections show a well developed overall stratification. However this stratification is far from being along the horizontal direction, so it is very important before starting work to use a reference marker for flattening in order to be able to make the good correlation between traces. The section after flattening is shown in figure 13.

The first geostatistical analyses of the section have led to a division of this section into several zones in terms of the time variograms. For the present study, we are going to use a part of the section taken between zone 3 and zone 7. This zone includes 400 traces with 86 samples per trace. The distance between each trace is 20 m, the vertical sampling is 2ms.

The means of the measured amplitudes along the vertical and horizontal directions have been calculated and presented in figures 2 and 3.

It is clear that in terms of mean the variation of the amplitudes is much stronger from one time to another (figure 3) than from one trace to another (figure 2). Notice also that the mean variation of the amplitudes along the trace show erratic and approximately zero mean fluctuations on top of an overall variation. In terms of variogram, this is seen in the difference of the sill levels, notably that the horizontal variogram of the measured amplitudes (figure 4) has a much lower sill than the vertical variogram (figure 1). In order to get rid of this difference of variation, or at least reduce it, we have transformed the data in the following way:

$$\text{AMT}(x, t) = \text{AM}(x, t) - \text{MXAM}(t) \quad (13)$$

with $\text{MXAM}(t) = \frac{1}{N} \sum_x \text{AM}(x, t)$

where $AMT(x, t)$ is the transformed measured amplitude, $MXAM(t)$ is the mean of the $AM(x, t)$ in terms of the time shown in figure 3. It is easy to see that this mean value is the mean value of the real amplitudes, and will not bear any influence on the noise.

Structural analysis

The variograms (horizontal, vertical and diagonal with fixed h_x) of the measured amplitudes have been calculated and presented respectively on figures 4,1, 5 and 6.

In these figures, the following points should be pointed out :

1) the horizontal variogram (figure 4) has a linear shape at the origin with a nugget effect.

2) the vertical variogram (figure 1) has a cubic behaviour at the origin then a hole effect where the first maximum is reached towards 10 ms.

3) The diagonal variograms with fixed hx (figure 5 and 6) have an oscillating shape. As the hx increase, they approach the data variance. In order to understand this phenomenon, let us look at formula 11 which is :

$$\gamma_{AM}(h_x, h_t) = \gamma_{AM}(h_x, 0) + \frac{C_{AM}(0, 0) - \gamma_{AM}(h_x, 0)}{C_A(0, 0)} \gamma_A(0, h_t)$$

So, if h_x fixed, $\gamma_{AM}(h_x, h_t)$ is equal to $\gamma_A(0, h_t)$ to the nearest constant, which depends on the distance h_x. As $\gamma_A(0, h_t)$ oscillates, so does $\gamma_{AM}(h_x, h_t)$ with fixed hx.

Calculation of the vertical variogram of the noise

After transposition in terms of variogram : formula 12 reads

$$\gamma_N(0, h_t) = \gamma_{AM}(0, h_t) - \frac{C_A(0,0)}{C_{AM}(0,0) - \gamma_{AM}(h_x, 0)} [\gamma_{AM}(h_x, 0) - \gamma_{AM}(h_x, h_t)]$$

(14)

The data variance is $C_{AM}(0, 0) = 0{,}00182$.

The variograms $\gamma_N(0, h_t)$ were calculated according to the above formula (formula 14) with the variograms $\gamma_{AM}(h_x, 0)$, $\gamma_{AM}(0, h_t)$ and $\gamma_{AM}(h_x, h_t)$ shown in figures 4,1,5 and 6. Theoretically, they should be identical.

Figure 7 represents the $\gamma_N(0, h_t)$ calculated for h_x= 1dx, 2dx, 3dx, 4dx, and 5dx with dx being the distance between two traces (20 meters). Figure 8 shows these variograms for hx at respectively 10,30, 40 and 50 times the distance between two traces.

In figures 6 and 7 we note that when h_x is small, the $\gamma_N(0, h_t)$ are practically identical especially when ht is small. For the large distances of h_x and h_t the experimental variograms are less reliable so the variograms $\gamma_N(0, h_t)$ are less significant. We also see the effect of a non stationarity on the horizontal variogram for large distances. Consequently it is advisable to take the variogram $\gamma_N(0, h_t)$ calculated at $h_x = d_x$ as the vertical variogram of the noise.

We note that this vertical variogram of the noise (figure 7) presents a hole effect with a cubic behaviour at the origin at the scale of the vertical variogram of the measured amplitudes, its first maximum being reached towards 8 ms.

Fitting the variograms

$$\gamma_{AM} = \gamma_A + \gamma_N$$

The vertical variogram of the noise was fitted as the sum of a cubic model and a cosine exponential model. (figure 9).

The horizontal and diagonal variograms of the noise are fitted with a pure nugget effect with 0.0004 as variance.

The vertical variogram of the real amplitudes :

was also fitted by the sum of a cubic model and of a cosine exponential model (figure 10).

The sum of $\gamma_N(0, h_t)$ (figure 9) and $\gamma_A(0, h_t)$ (figure 10) gives the fit of the vertical variogram of the measured amplitudes (figure 11).

We see that the vertical variogram (figure 11) of the measured amplitudes $\gamma_{AM}(0, h_t)$is made up of three structures, a cubic model and two cosine exponentials. 33.3% of this cubic model is due to the noise and 66.6% to the real amplitudes. The cosine exponential with a sill of 28ms and a period of 16ms is allotted to the noise, whereas that with a sill of 32ms and a period of 22ms to the real amplitudes.

The horizontal variogram of the real amplitudes is fitted by three spherical models having different sills. (Figure 12).

Noise estimation

– *Selecting the neighborhood*

The estimation will be carried out in moving neighborhood of nh traces and nv times around the point to be estimated. These numbers nh and nv are taken as small as possible.

It can be noted that at the estimation point, the weight is very strong and positive, and that on the same line the other points also have a strong weight but are negative.

The results

We present the results for the two neighborhoods.

Figures 14 and 15 show the noise and resulting amplitudes estimated with a 5x5 neighborhood, while figures 16 and 17 show the noise and resulting amplitudes estimated with a 7x7 neighborhood.

CONCLUSION

The method used here seems to be promising for filtering noise on seismic sections. It allows a check of hypotheses, a good control of the variogram of noise, and so a reasonable confidence on the results. The importance of choosing a horizon for flattening has to be stressed. This gives the possibility of good correlations parallel to the stratification, and of using factorized variograms. This choice has to be precise because it is one of the key points in order to attribute the nugget effect to noise. It is not difficult to extend this methodology in order to take into account well information, and more than one seismic section. Another natural extension that could be considered is the deconvolution with filtering by kriging or cokriging which is the next step envisaged.

ACKNOWLEDGMENTS

We wish to thank H. BEUCHER, J. RIVOIRARD (Centre de Géostatistique) and V. RICHARD (Institut Français du Pétrole) for participating in the many fruitful discussions.

REFERENCES

CHILES, J.P. and GUILLEN, A. (1984) "Variogrammes et krigeage pour la gravimétrie et le magnétisme", Intern. colloquim on Comput. in Earth Sci. édité par Royer JJ in Sciences de la Terre, Série Inf. Géol. n°20, 445–468.

DALY, C., LAJAUNIE, Ch., JEULIN, D. (1989), "Application of multivariate kriging to the processing of noisy images", in Proc. 3rd Intern. Geostatistics Congress, Sept. 5–9, 1988, Avignon, *Geostatistics*, ed. by. M. ARMSTRONG, Vol. 1, Kluwer Academic Publ., 749–760.

DALY, C., "Applications de la Géostatistique à quelques problèmes de filtrage" Thèse de Docteur en Géostatistique, Paris School of Mines, 235 p. (1991).

DASH, B.P. and OBAIDULLAH, K.A. (1970)," Determination of signal and noise statistics using correlation theory", Geophysics, 35, 24–32.

GALLI, A. et al. (1984), "Factorial kriging analysis : a substitute to spectral analysis of magnetic data". G. Verly et al., Geostatistics for Natural Resources Characterization, Reidel Publishing Company, 543–557.

MA, Y.Z. (1987), "Filtrage géostatistique des images numériques", Thèse Doctorat Institut National Polytechnique de Lorraine, 143 p.

MA, Y.Z., ROYER, J.J. (1988), "Local Geostatistical filtering . Application to remote sensing", in Geomathematics and Geostatistics analysis applied to space and time dependent data. Edited by B. Namyslowska–Wilcztnska–J.J. Royer. (Mars 1988).

MATHERON, G. (1982), "Pour une analyse krigeante des données régionalisées", Rapport interne, CGMM, N–732, Fontainebleau, 22 p.

RICHARD,, V. and BRAC J. (1988), "Wavelet Analysis using Well Log Information", Annual Meeting of the Society of Exploration Geophysicists, Anaheim, California (October 1988).

SANDJIVY, L. (1984), "The factorial kriging analysis of regionalized data", in Verly G. et al. (eds) Geostatistics for Natural Resources Characterization, 559–571, Reidel, Dordrecht.

SANDJIVY, L. and GALLI, A. (1985), "Analyse krigeante et analyse spectrale", Sciences de la Terre, Série Inf. Géol., n°21, 115–124.

WACKERNAGEL, H. (1986), "Geostatistical Techniques for Interpreting Multivariate Spatial Informations", NATO, A.S.I. on Statistical Treatment for Earth Resources, Il Cioco, Italy.

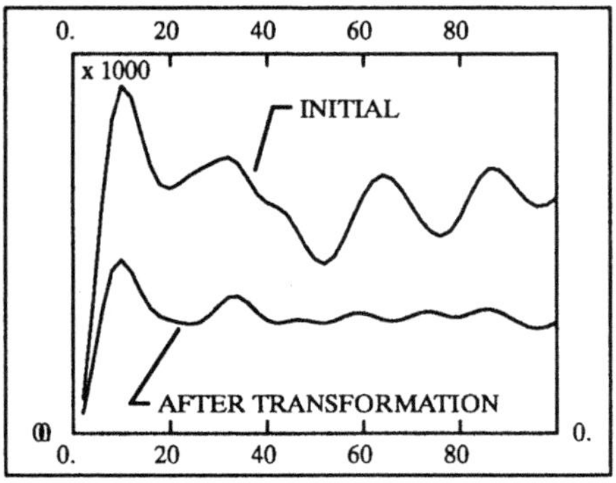

Fig 1 : Vertical variograms of the measured amplitudes

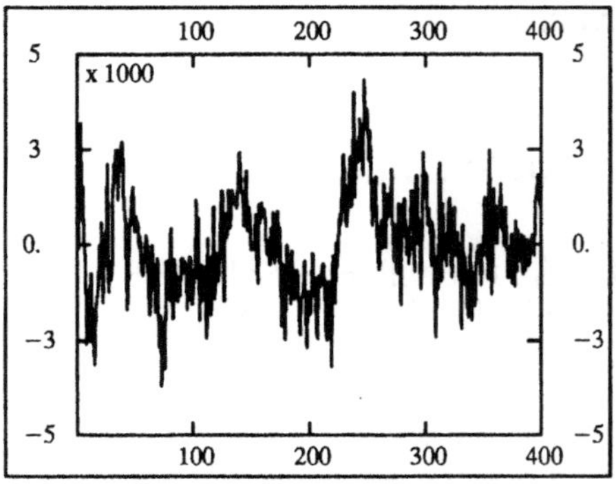

Fig 2 : Mean of the measured amplitudes versus trace

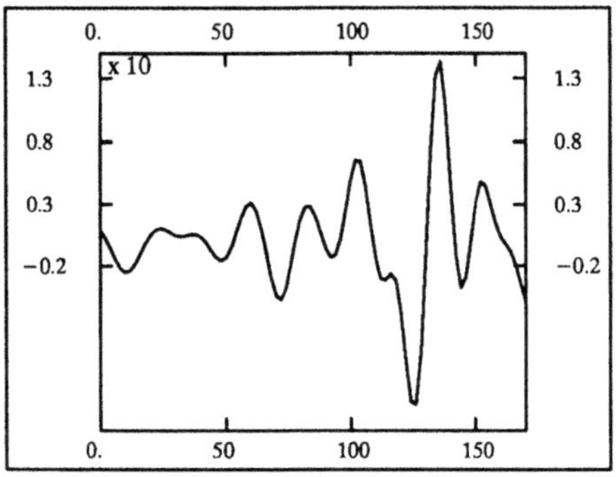

Fig 3 : Mean of the measured amplitudes versus time

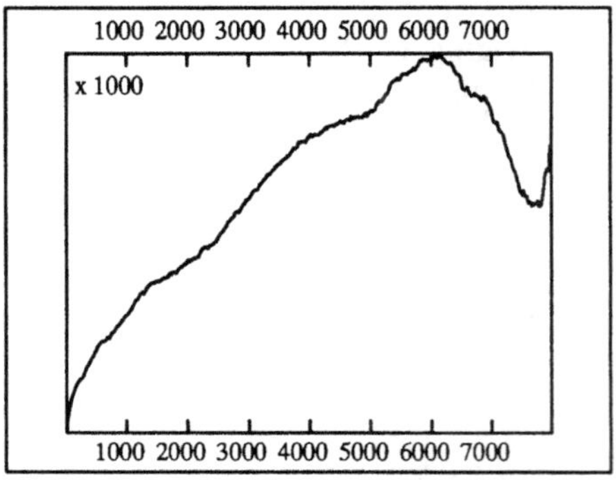

Fig 4 : Horizontal variogram of the measured amplitudes

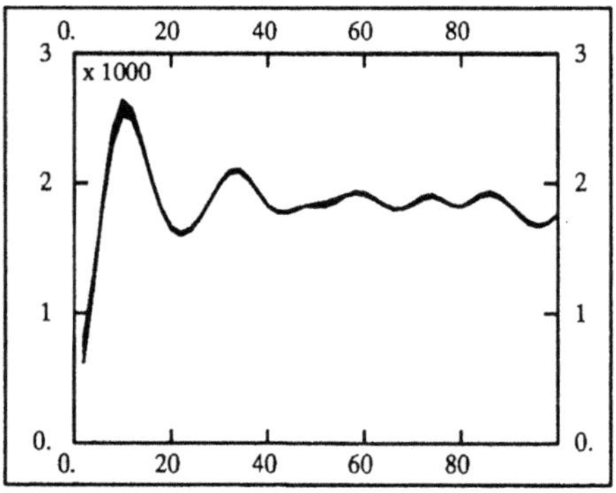

Fig 5 : Diagonal variograms of the measured amplitudes with hx=1,2,3,4,5dx

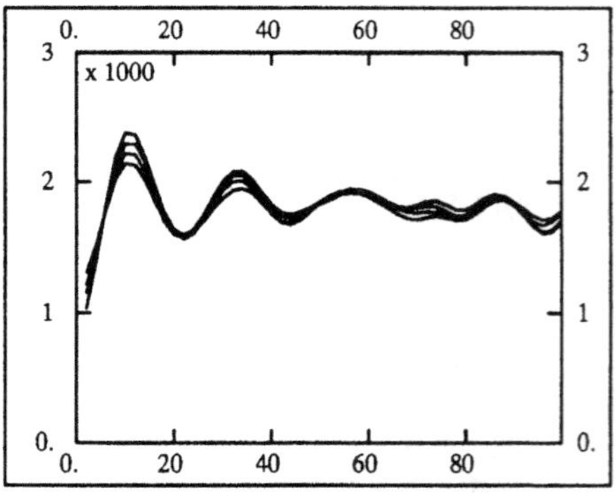

Fig 6 : Diagonal variograms of the measured amplitudes with hx=10,30,40,50dx

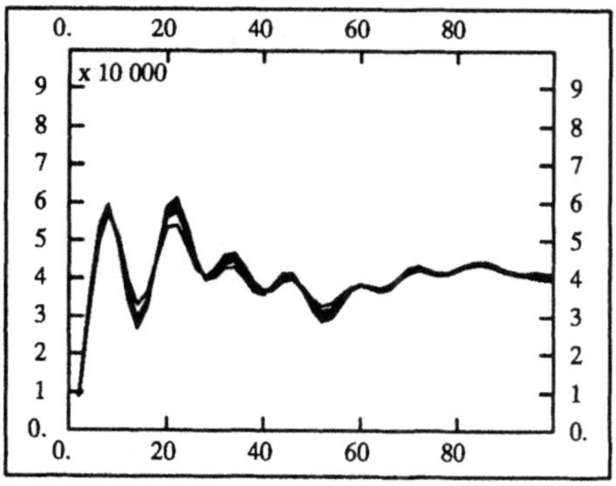

Fig 7 : Vertical variograms of the noise at hx=1,2,3,4,5dx

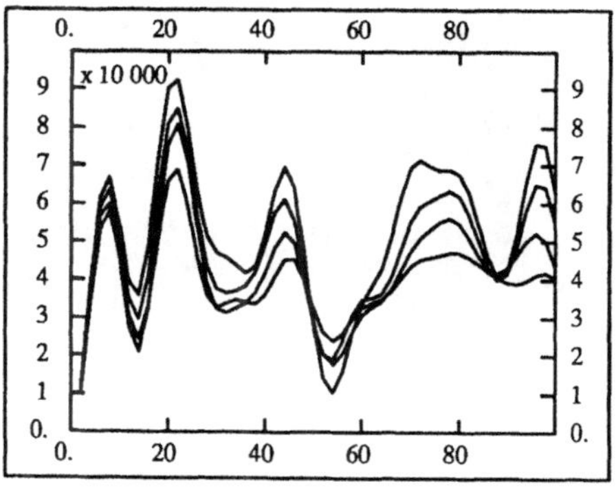

Fig 8 : Vertical variograms of the noise at hx=10,30,40,50dx

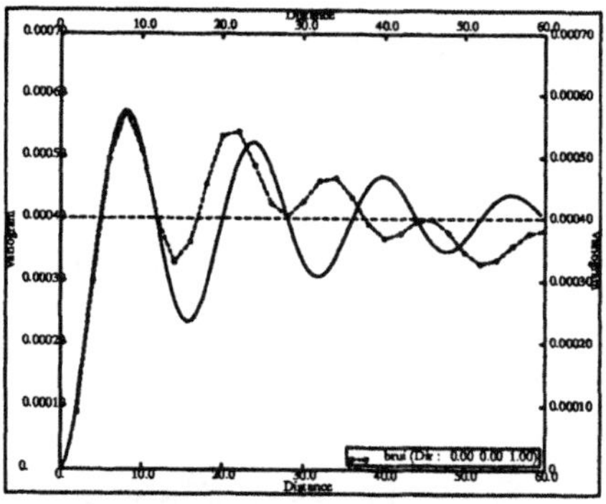

Fig 9 : Fitting of experimental vertical variogram of the noise

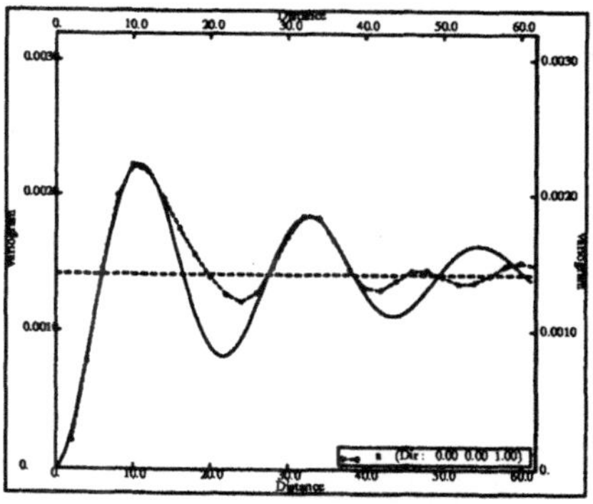

Fig 10 : Fitting of experimental vertical variogram of amplitudes

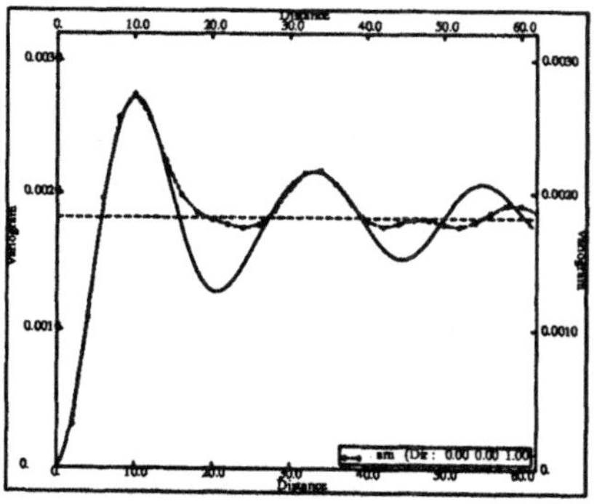

Fig 11 : Fitting of experimental vertical variogram of the measured amplitudes

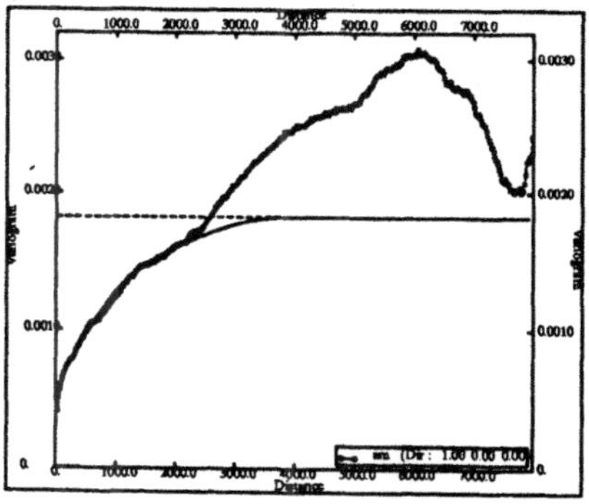

Fig 12 : Fitting of experimental horizontal variogram of the measured amplitudes

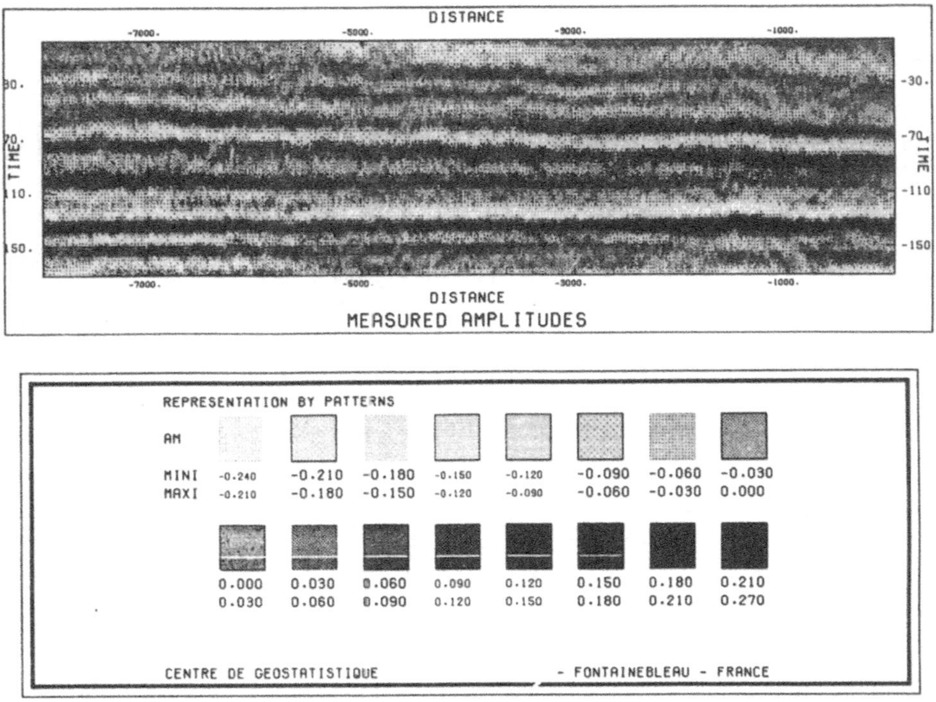

Fig 13 : Section of the measured amplitudes

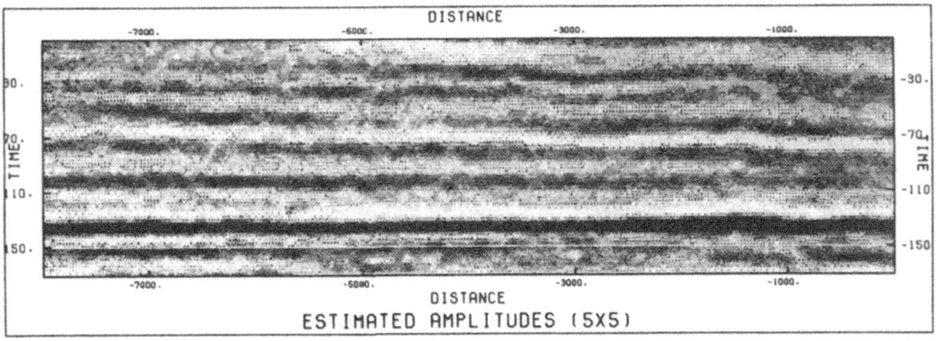

Fig 14 : Section of the estimated amplitudes (5x5)

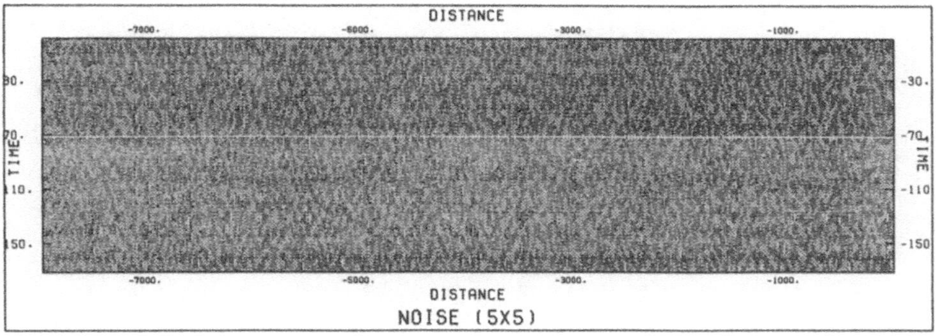

Fig 15 : Section of the noise (5x5)

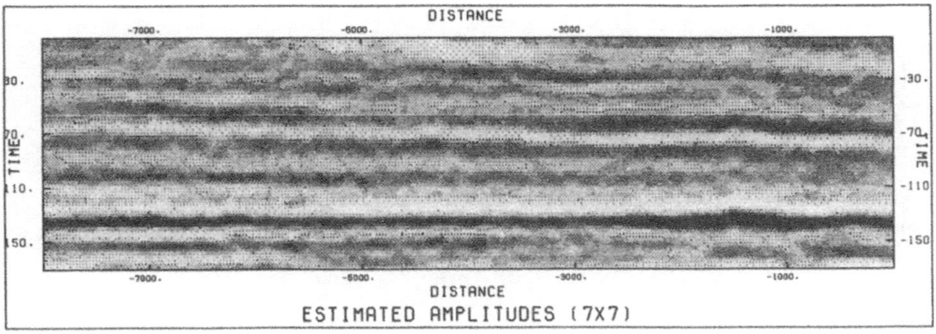

Fig 16 : Section of the estimated amplitudes (7x7)

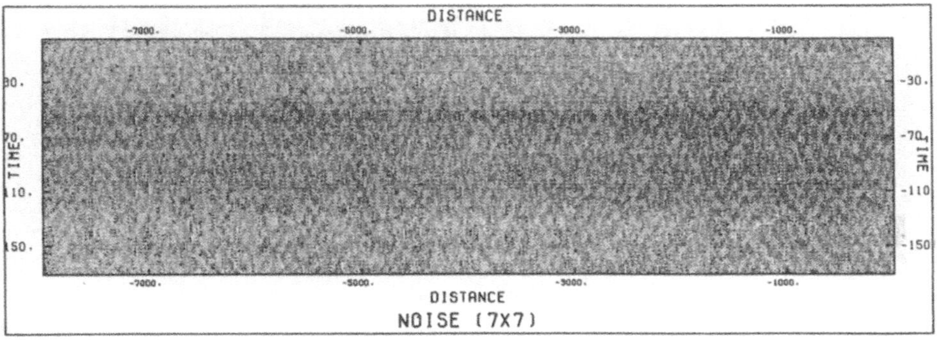

Fig 17 : Section of the noise (7x7)

COMPARISON OF SEVERAL METHODS OF LITHOFACIES SIMULATION ON THE FLUVIAL GYPSY SANDSTONE OF OKLAHOMA

Valérie Langlais, James Doyle
BP exploration
5151 San Felipe
Houston, Texas, 77210, U.S.A.

Introduction

This paper compares simulations of lithofacies in a fluvial environment generated using three different simulation techniques. The methods used are: Simulated Annealing using a training image, Sequential Indicator Simulation, and the Threshold Gaussian Functions. They have in common the use of two-point statistics. The goal of such simulation techniques is to be able to provide equiprobable images of the sedimentary heterogeneity of a subsurface reservoir of this type. It is desirable that the images retain the main geological features of this environment while representing the local variability that may be smoothed in reservoir models developed deterministically from conceptual geological model. The geological heterogeneities can then be assessed for their impact on reservoir performance. After a description of the geological environment, and a brief presentation of the three techniques, an example of a 2D simulation of the outcrop is described, followed by a 3D example.

Geological information

The Gypsy sandstone is an informal name for the lower-most sandstone of the upper Carboniferous Vamoosa Formation of the northern Oklahoma. (Greig 1959; Ford 1978). Regionally, the Gypsy was deposited on a stable cratonic platform which received coarse clastics shed from the south and southeast by the Ouachita and Arbuckle uplifts. Ford (1978) characterized the Gypsy on outcrop as an alluvial system that varies down depositional dip from conglomeratic alluvial-piedmont to fine-grained alluvial-plain deposits. At our principal study site, located in the fine-grained alluvial plain of Ford's study, the Gypsy comprises a multilateral and stacked sequence of meanderbelt channel sandbodies. These sandbodies incise and are surrounded by floodplain deposits of mudstone, siltstone, and fine-grained sandstone.

At the main study site, six channel sandbodies are exposed on parallel walls of a 300m roadcut where the fluvial architecture of the Gypsy was mapped on a photomosaic base. A grid of twenty-two shallow borings was cored adjacent to the surface exposure to extend the fluvial architecture to three dimensions. Channel sandbodies vary in size from 200m wide and 10m thick at the base to 60m wide and 3m thick at the top of the Gypsy. Erosional surfaces defining the bases of the channel sandbodies could be traced on the borehole grid, but lower order features could not be correlated.

For Gypsy channel-fill sequences, five sandstone lithofacies are defined in addition to mudstone and siltstone. Channel sandbodies are variable in vertical succession of lithofacies, but the most common sequence upward from the channel base consists of a lag of granule-to-cobble-size mudclast conglomerate (MC) to trough cross-bedded sandstone (XB) to planar-laminated sandstone (PB) to ripple-laminated fine sandstone (RP) and siltstone (SH) at the top. These lithofacies were mapped on the outcrop photomosaic and described from the shallow cores. Both the map and core descriptions were digitized for use in the stochastic modelling that is the subject of this paper. Geologic models based on the outcrop use deterministic mapping of channel sandbodies together with stochastic representation of facies. The initial focus is on channel sandbody 2 selected because of its continuity at the study site (Fig. 1). All the simulations are performed after eliminating structural dip.

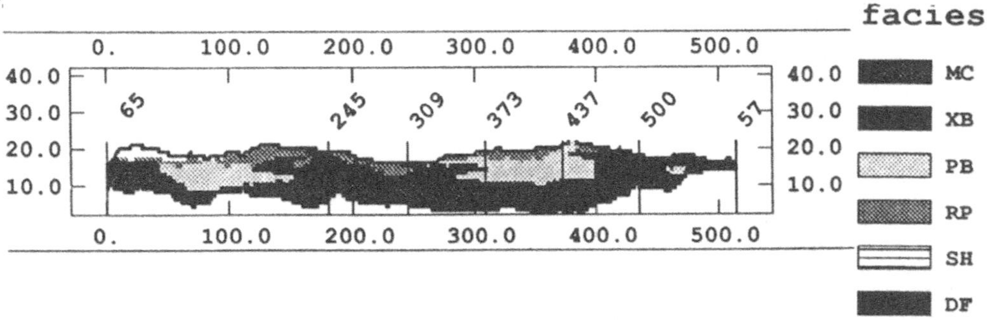

Figure 1: Channel sandbody 2 on the outcrop. (Units are in feet)

1 Methods and parameters

1.a Simulated annealing using a training image. (Farmer,1992, Doyen,1989)

Rather than using a framework of probabilities as do the other methods, simulated annealing reproduces the statistics of two-point histograms. The histograms are computed from a training image assumed to be of the same geological environment. From a random image, each point is visited, following a random path. An objective function is calculated for each possible state of this point. The state which minimizes the objective function is kept, this could lead to a local minimum. An iteration is defined when all points have been visited. The statistics used here are:

$N_{ij}(\vec{u})$ *number of transitions from lithofacies i to lithofacies j along* $\vec{u}$ ($N_{ij}^{tim}(\vec{u})$ for *the training image*)

$N_i(z_k)$ *number of lithofacies i at elevation* z_k ($N_i^{tim}(z_k)$ for *the training image*)

σ_1^2 and σ_2^2 *are the different weights given the different statistics*

The function used to define the notion of closeness to the input statistics is:

$$E = \sum_{\vec{u}} \sum_{i} \sum_{j} \frac{(N_{ij}(\vec{u}) - N_{ij}^{tim}(\vec{u}))^2}{2\sigma_1^2} + \sum_{k} \sum_{i} \frac{(N_i(z_k) - N_i^{tim}(z_k))^2}{2\sigma_2^2} \quad minimum$$

Actually these transition statistics reflect the same information as the experimental variograms and cross variograms of the training image.

By honoring the data, this simple conditioning will lead to discontinuities at wells. Such discontinuities do not have a large importance in the global statistics, but need to be eliminated. By splitting the function into two terms, one involving pairs with well data, the other without, with different weights, these discontinuities can be decreased.

If other statistics should be honored, they can be added in the objective function , e.g. proportions along any directions, information derived from seismic. Each term of the objective function can have its own weight if for any reasons some emphasis should be put on some statistics (more confidence, more importance for the ultimate goal).

1.b Sequential Indicator Simulation (S.I.S.) (Journel,1983, Alabert,1987)

With simple kriging (SK) of indicator functions of each lithofacies $I_i(x)$, a probability distribution is estimated at point x and used to assign a lithofacies at this point. This point will act as a data point for other points. Points are visited following a random path. Simple kriging does not guarantee that these estimators $I_i^*(x)$ belong to [0,1] and that their sum equals 1. Some corrections need to be done to satisfy these requirements. The first correction puts all probability values between 0 and 1: The second and last correction forces their sum

$$I_i^{**}(x) = I_i^*(x) \text{ if } I_i^*(x) \in [0,1], \quad I_i^{**}(x) = 0 \text{ if } I_i^* < 0, \quad I_i^{**}(x) = 1 \text{ if } I_i^*(x) > 1$$

to equal1.. This last estimator is used to assign lithofacies. These modifications of the initial

$$I_i^{***}(x) = I_i^{**}(x) / \sum_{i=1}^{n} I_i^{**}(x)$$

estimator $I_i^*(x)$ can introduce bias to some extent.

In the Gypsy sandstone, lithofacies show in the vertical direction, in that they have a preferred sequence within each channel sandbody. As the ultimate goal is a reservoir simulation, it could be important to keep this feature, because the lithofacies differ in reservoir properties. Therefore the following assumption has been made: a given lithofacies has the same spatial correlation, only the proportion (or mean) differs with the vertical location z (Langlais,1992).Then the estimation of $I_i^0(x,y,z_0)$ is a cokriging (with a particular coregionalisation model) of $I_i^k(x,y,z_k)$ which is the sum of $I_i(x,y,z_k)$ and p_i^k, where $I_i(x,y,z_k)$ is a stationary random function of expectation zero and p_i^k the proportion of lithofacies i at elevation z_k. This option is particularly useful when the conditioning data are not able to force this vertical variation. Note that by doing this, the possibility of inferring the variogram model from the experimental variogram is lost along the vertical direction. Also note that, if for a given elevation the proportion equals 0 or 1, this estimator does not necessarily reproduce these values, but these values can be forced by the program. The input parameters of this method are : the variogram models of each of the indicator functions and the vertical proportions (if used).

This method can be improved by using a simple cokriging of the indicator functions instead of a kriging. This will take into account the spatial correlation between the different lithofacies but will increase the difficulty of having a consistent model of coregionalisation. It offers the possibility of taking into account soft information.

1.c The threshold gaussian functions. (Matheron, 1987,Galli and Ravenne,1990)

This method uses a transformation of the variable lithofacies into a gaussian function. A simulation of this function is performed followed by a back transformation. The transformation is done according to vertical proportions and spatial correlation of lithofacies. It implies an orderly relationship between lithofacies. As the gaussian function is continuous, for a lithofacies sequence L_1, L_2, L_3, going from L_1 to L_3 will implied a transition through L_2. Of course a short spatial correlation implies abrupt transitions. The use of a single variogram model for the gaussian function could lead to some difficulties when experimental variograms of lithofacies are very different (e.g. different directions of anisotropy).

In the software (HERESIM) used here, some choices have been made to reduce computing time.The covariance of the gaussian variable is a factorized exponential model. This implies that the data set should reflect a strong vertical anisotropy. The exponential choice reduces the neighbourhood (conditioning data) by taking advantage of the screen effect of this model (Journel,1978).

As the proportions of lithofacies vary in vertical direction, thresholds used for the transformation are not constant vertically. It is also possible to take into account lithofacies proportions varying in 3D.

2 Comparison in 2D on a small subset of the outcrop

2.a Data set

On the outcrop, the central part of the exposure of channel sanbody 2 (Fig.1) was first studied in an area 515ft (157m) wide by 24ft (6.7m) high. For this 2D section an extensive floodplain mudstone was selected based on examination of lithofacies along the vertical direction and used to eliminate the structural dip.

2.b Criteria of comparison.

The comparison criterion used is limited by the scope of the project which does not include yet flow simulation. Correct ergodicity has not been checked and only global statistics have been considered.The first way to compare different simulations is to make for each a visual assessment of the similarity to the outcrop geology. Second the match between the input parameters (variograms, transition statistics, proportions) and the simulated parameters can be checked. Error terms are computed by comparing the observed and simulated lithofacies proportions. For each simulation, the following expressions have been computed (E_g and E_v are the errors for global and vertical proportions):where n_i is the input occurence of lithofacies i (ns_i for simulation), $n_i(z_j)$ input occurence of lithofacies at elevation z_j ($ns_i(z_j)$ for simulation), l number of lithofacies, $nvert$ is the number of nodes along the vertical. Proportions smaller than 3 grid blocks have been regrouped together.

$$E_g = \sum_{i=1}^{l} \frac{(n_i - ns_i)^2}{n_i} \qquad E_v = \sum_{j=1}^{nvert} \sum_{i=1}^{l} \frac{(n_i(z_j) - ns_i(z_j))^2}{n_i(z_j)}$$

2.c Simulated annealing using a training image.

For the 2D case, the method is used in a particular and favorable context, the training image is the expected result. The neighbourhood, which is a list of vectors $\vec{u}$ describing the statistics used in the simulation, has been chosen in order to cover different distances along X and Z axis and reflects the main feature of these statistics.

In order to avoid discontinuity at well locations, the major parameter to adjust is the weighted ratio between pairs with and without well data . Using a ratio of 20 leads to a reasonable image (continuity at wells). The result (Fig.2) reproduces almost exactly the expected image (the outcrop). The variability between simulations is extremely small; the specification of the model is so strong as to preclude variation. As the criteria for the simulation is to match the input statistics, the two errors E_g and E_v are very small. For E_g, the minimum is 0.03, the maximum 0.19, the mean 0.10, the variance 0.005, the coefficient of variation 0.65. For E_v, the minimum is 3.6, the maximum 22, the mean 9.3, the variance 36, the coefficient of variation 0.64.

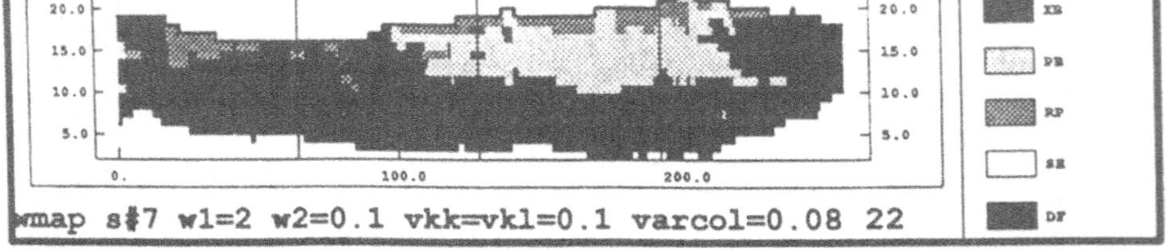

Figure 2: A simulation of channel 2 with simulated annealing.

2.d Sequential Indicator Simulation.

The non stationarity of the outcrop lithofacies distribution is quite strong (Fig.1). Experimental variograms were mainly used to fit a range of correlation on each direction for each lithofacies. The models used are spherical with a range between 140. and 90. in the East-West direction and betwenn 5. and 6. in the vertical direction. Figure 3 shows that the scattered conditioning data do not enable this method to reproduce the vertical organization of the outcrop.

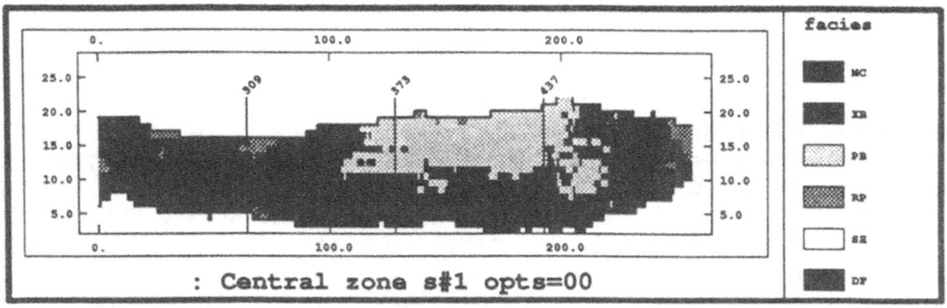

Figure 3: simulation with no vertical lithofacies proportions.

This result clearly reflects the need to take into account the vertical lithofacies proportions of the data set (Fig.4).

The variability in the reproduction of the proportions is quite high. For E_g , the minimum is 29, the maximum 1562, the mean 432, the variance 182414, the coefficient of variation 1.For E_v , the minimum is 321, the maximum 2392, the mean 1025, the variance 330555, the coefficient of variation 0.56.There is no guarantee of having a "good" reproduction of global and vertical proportions; these statistics can vary significantly between simulations. The first correction made on the initial estimator introduced a bias, but there is no way to avoid this step.

The variograms of the simulations (Fig.5) also reflect some dispersion. By modeling the variograms for each lithofacies, there is no guarantee of having a consistent model which could lead in difficulty in reproducing the input variograms. The introduction of the vertical proportions does not allow proper inference from vertical experimental variograms. In addition within this data set, one can question the assumption that the vertical lithofacies proportion curve is stationary.

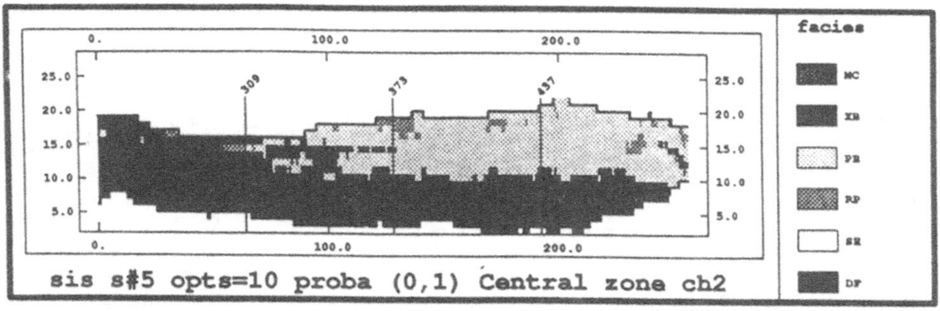

Figure 4 : a simulation of the outcrop with S.I.S. with vertical proportions

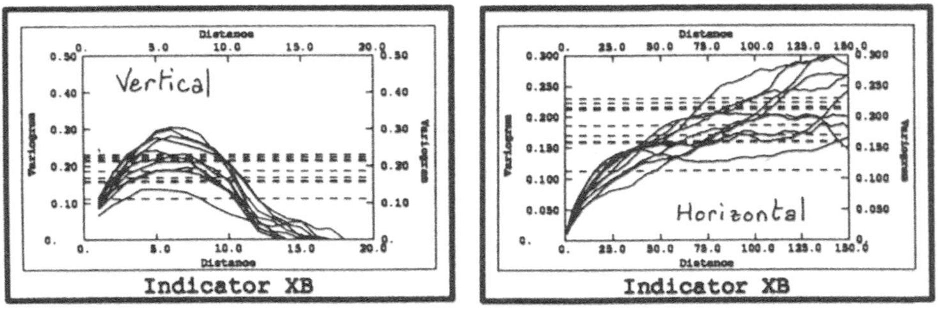

Figure 5: Variograms of 10 simulations for XB.

2.e The threshold gaussian functions.

This method first requires specification of the vertical sequence. In the Gypsy channel sandbodies the most common upward lithofacies sequence is: MC, XB, PB, RP and SH which corresponds to a change in bedforms from channel lag to top of the point bar.

Simulations using this method produce the following error terms: for E_g , the minimum is 21, the maximum 338, the mean 145, the variance 11709, the coefficient of variation 0.74.For E_v , the minimum is 312, the maximum 1063, the mean 664, the variance 11709, the coefficient of variation 0.34.

The global proportions dispersion is quite high, whereas there is a good reproduction of the vertical proportions. The fact that PB has an observed preferential location (Fig.1) is not completely compatible with the same vertical proportions everywhere (Fig 6).

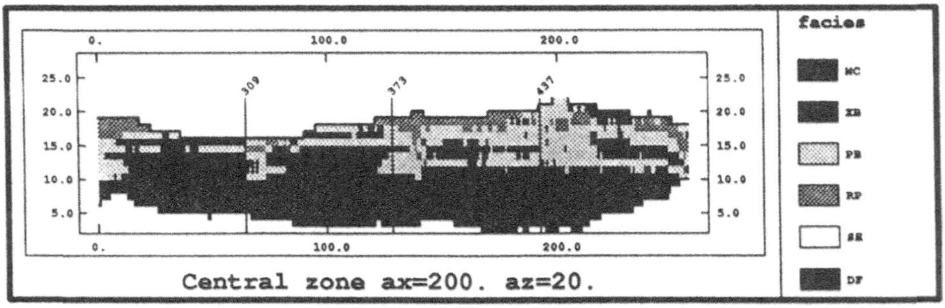

Figure 6 : a simulation of the outcrop with the threshold gaussian functions

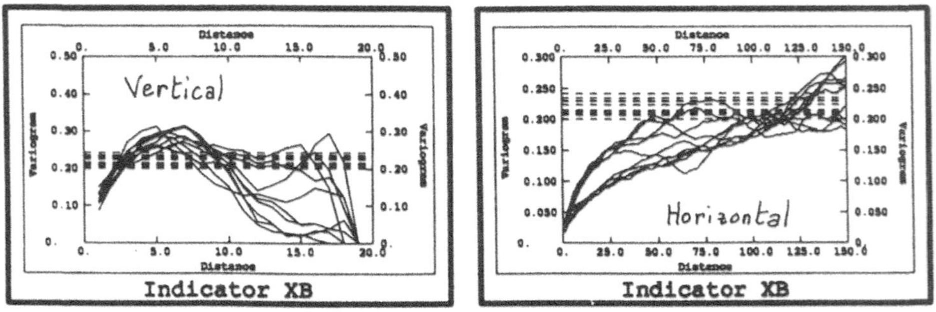

Figure7: Variograms of simulations for XB.

3 3D example.

In the 3D case study, the dimensions of the channel sandbodies are assumed to be known. They have been modeled by the geologist using the outcrop and 22 bore holes behind the outcrop. Only simulations of small scale heterogeneities within channel sandbody 2 have been considered here.

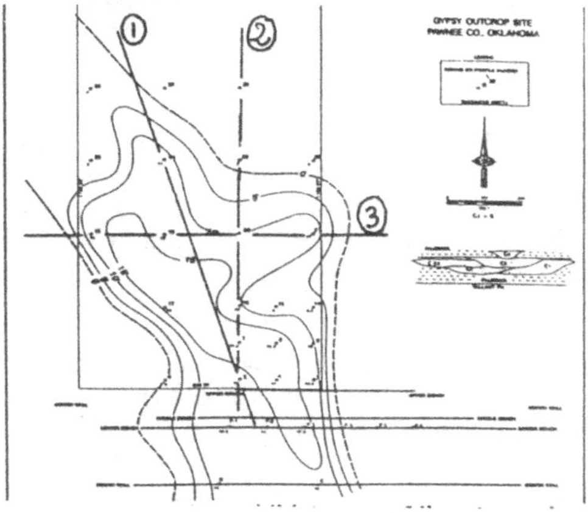

Figure 8 : location of the data and thickness of the channel sandbody 2

On the entire area covered by channel sandbody 2, no one suitable reference datum is present everywhere. The reference level mapped by the geologist has been used.

To specify the input parameters, the main source of quantitative information is the outcrop; for the other directions some choices have to be made. For the axis of anisotropy, the main axis of the channel has been chosen. The horizontal anisotropy ratio chosen is 4. For the method of simulated annealing, the input statistics are the data (outcrop and bore holes) on which the parameters are computed (global proportions, vertical proportions, transition statistics). It is still possible to check the match between the input parameters and the parameters of the simulations. However now the visual comparison is slightly different. The questions are: "do the simulations look geological reasonable and are they consistent with the geological interpretation of the environment ? ".

3.a. Simulated annealing using a training image. (Figs. 9,12)

Using the outcrop and the bore holes as a training image is not enough to obtain a geological image. The sections parallel to the outcrop are satisfactory but perpendicular sections display undesirable randomness due to the lack of statistics along this direction.

To overcome this problem, one could introduce additional statistics in another direction. This can be done by using the outcrop with a distortion to introduce anisotropy. A complete training image may also be generated by sampling a variogram model and using it as statistics. This will raise the problem of choosing a consistent model of variograms.

The results from several simulations with training image in two perpendicular directions are extremely close to each other as was the case in the 2D example. The lack of fluctuations in the generated statistics could be a mojor drawback if it percludes the generation of a wide spectrum of flow simulations.

Due to the trial and error process, this is by far the slowest of the three methods in computation. To simulate 16608 grid nodes took 4hr. 34min. of cpu time on a workstation SUN sparc2. To simulate 1000000 points which might be desirable in characterization of a large reservoir one will use 20 days of cpu time. Although it is a research code which has not been designed for good time performance, it is likely to be time consuming by nature of the algorithm.

3.b. Sequential Indicator Simulation. (Figs. 10,12)

At a first glance, the perpendicular sections to the outcrop obtained from simulations with S.I.S. look better than the ones obtained with simulated annealing, mainly because statistics are available in any directions.Global and vertical proportions are not very well reproduced as seen on figure 12. For example XB are found twice as often in the upper layers than expected. This may result from the use of a biased estimator for the conditional probabilities.To simulate 16608 grid nodes took 24mn of cpu time. For a simulation of 1000000 points one will need 25 hours of cpu time.

3.c. The threshold gaussian functions. (Figs. 11,12)

The reproduction of the input parameters (global and vertical proportions) reflects more stability than S.I.S. In all simulations these parameters are consistently reproduced. As expected the sequence of lithofacies is reproduced, perhaps more often than desirable. The simulation took 40 seconds of cpu time, 1000000 points will only require 10mn, this is mainly due to the choice of the covariance model (section 1.c).

Conclusions

It seems possible to reproduce the geology with all methods except with simulated annealing when taking only the outcrop and the bore holes in the training image. The variability of the simulations are good with S.I.S. and HERESIM with a better respect for proportions in the latter. Computational efficiency are extremely different, to say the least. The next step is assessing the effect of representing the geological detail in flow. It implies simulation of petrophysical variables then a scaling up to provide an input to flow simulator. History matching, when production data is available, may be a way to select "good" simulations among the different ones obtained.

Bibliography

Alabert, F.G., (1987). Stochastic Imaging of Spatial Distribution Using Hard and Soft Information. M.Sc. Thesis, Stanford U., 197p.

Doyen P.M. and al, (1989). Seismic Discrimination of Lithology in Sand/Shale Reservoirs, a Bayesian Approach, Expanded abstract., Proc. of 59th annual SEG meeting, Dallas.

Farmer C., (1992). Numerical Rocks, "The mathematics of oil recovery", Clarendon Press, Oxford.

Ford, G.W., (1978). Genesis and trend of the lowermost unit of the Vamoosa formation in parts of the northeastern and central Oklahoma: Stillwater, Oklahoma, Oklahoma State U.

Galli A. and al., (1990). "Outcrop studies and geostatistical modeling of a Middle Jurassic Brent Analogue". Proc. of the European Oil and Gas Conference, Palermo, Sicilia,.

Greig, P.B., (1959). geology of Pawnee County Oklahoma: Oklahoma Geological survey Bulletin 83, 188p

Journel, A.G. (1983). "Nonparametric Estimation of Spatial Distribution", Journal of Mathematical Geology, Vol.15, No. 3, pp. 445-468.

Journel, A.G., and Huijbregts, Ch.J. (1978). Mining Geostatistics, 2nd printing (revised), Academic Press, London, 600pp.

Langlais, V., (1992). (to be published) Small scale heterogeneities simulations on the Gypsy sand stone of Oklaoma. IFP conference, Scarborough, Yorkshire, U.K.

Matheron, G. and al. (1987). "Conditional Simulation of the geometry of fluvio-deltaic reservoirs", SPE 167533.

Acknowledgements

We would like to thank Philippe Doyen, Georges Verly, Alain Galli and Christian Ravenne for sharing with us their experiences of the different techniques of simulations.

Finally, we would like to thank the I.F.P. and the Centre de Géostatistique which gave us the opportunity to use their software HERESIM on the Gypsy data set.

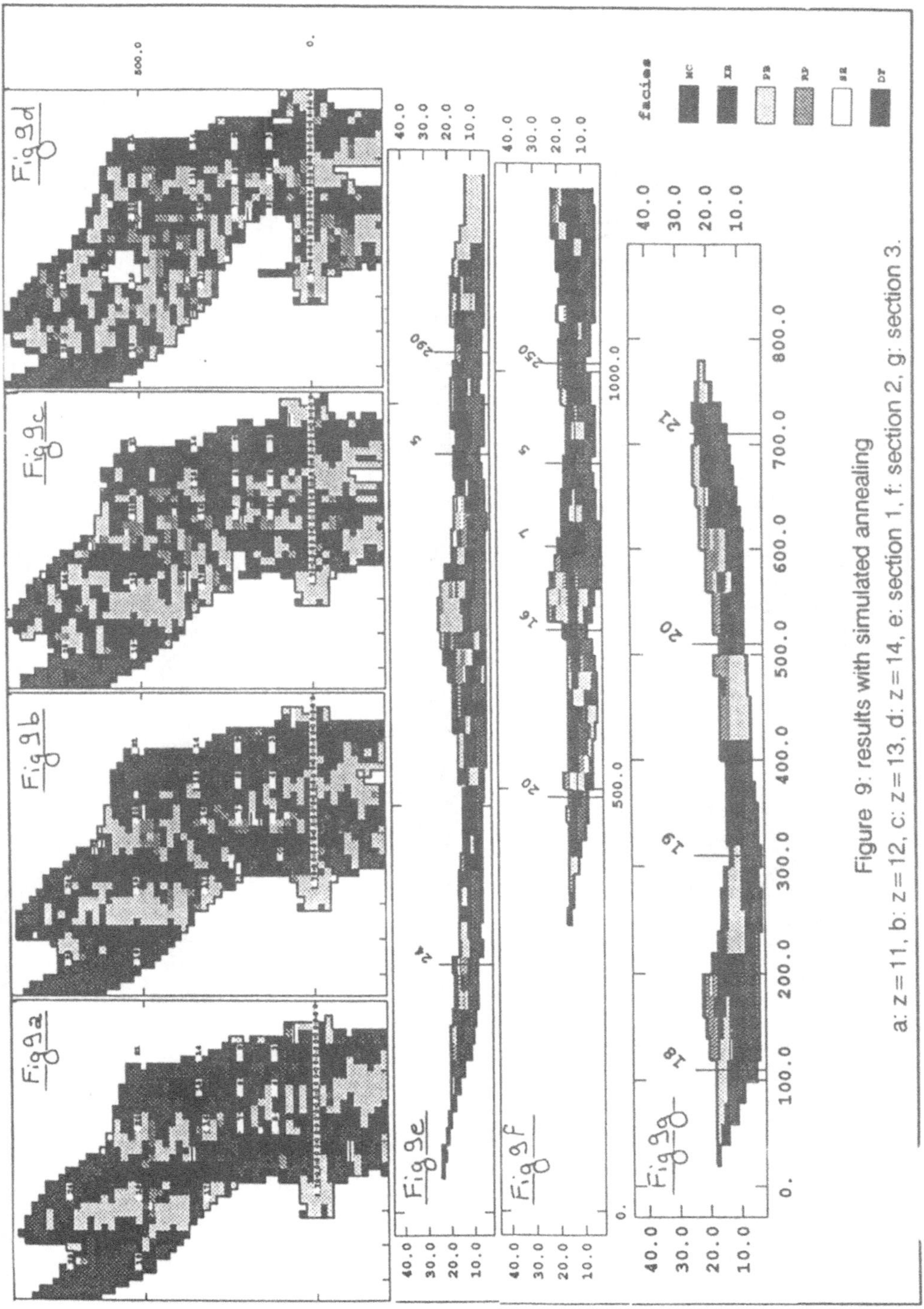

Figure 9: results with simulated annealing

a: z=11, b: z=12, c: z=13, d: z=14, e: section 1, f: section 2, g: section 3.

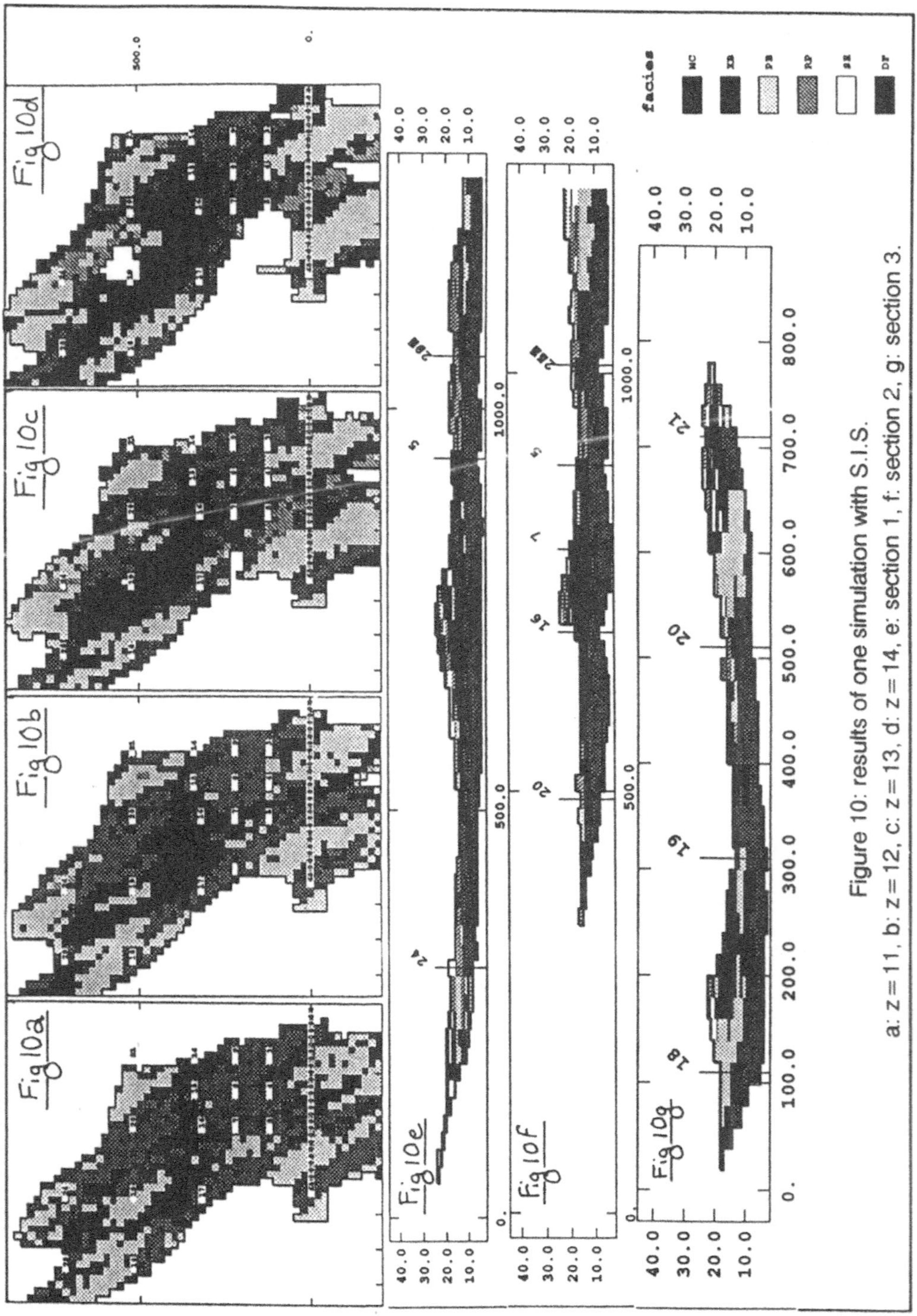

Figure 10: results of one simulation with S.I.S.

a: $z = 11$, b: $z = 12$, c: $z = 13$, d: $z = 14$, e: section 1, f: section 2, g: section 3.

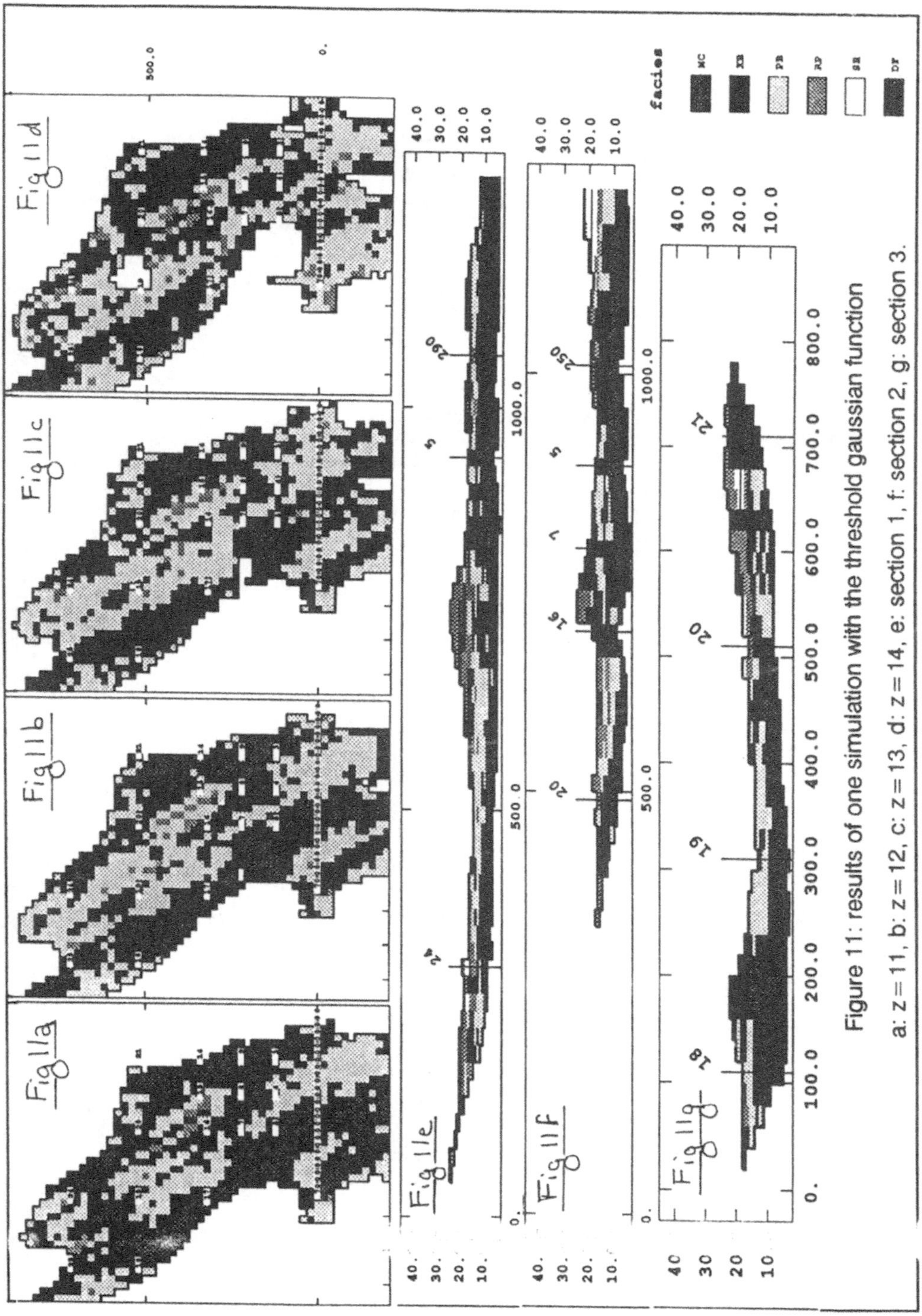

Figure 11: results of one simulation with the threshold gaussian function

a: z = 11, b: z = 12, c: z = 13, d: z = 14, e: section 1, f: section 2, g: section 3.

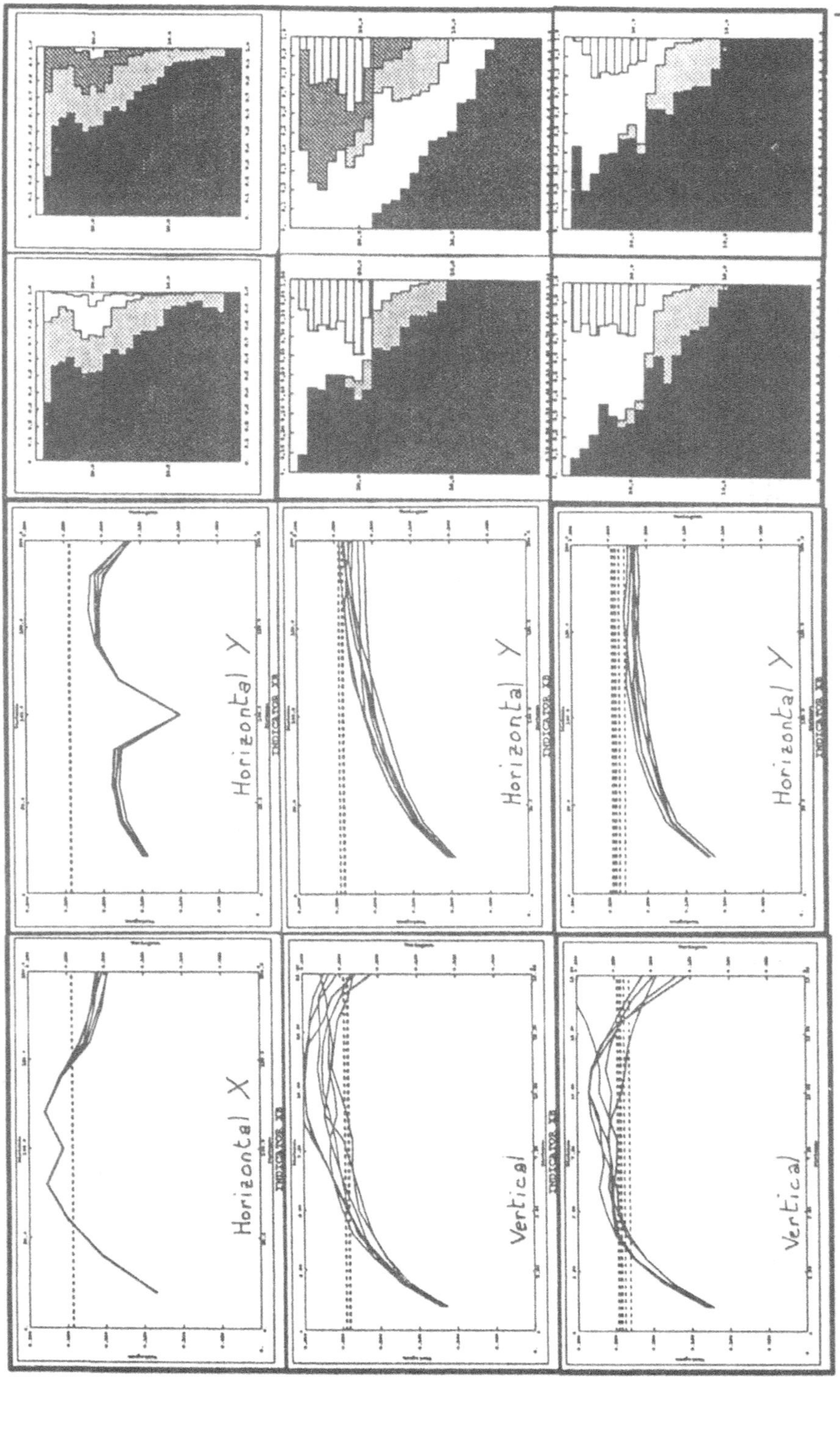

Figure 12: variograms for different simulations. vertical proportion curves for 2 simulations
From top to bottom: simulating annealing, S.I.S., threshold gaussian functions.

GEOMATHEMATICAL MODELLING OF A TURBIDITIC RESERVOIR IN S.E. HUNGARY

GEIGER, J.[1], WEROWSKY, V.P. [2]
[1] MOL RT. Hungarian Oil Institute and Laboratory, PO Box 43 1311 Budapest
[2] MOL RT. Nagyalfold Oil and Gas Production Co., PO Box 86, 5000 Szolnok

1. Introduction

The Algyo Field is the largest hydrocarbon accumulation in Hungary and consists of several oil and gas bearing reservoirs. The upper members of these reservoirs developed in a Pannonian delta system due to the complex lateral shifting and prograding phases (Geiger and Revesz 1987, Geiger 1988). The lower reservoirs are regarded as turbidity deposits (Fig.1). Earlier studies suggested two types of turbidity deposits: deep basin turbidities and prodelta turbidity fans.
The AP-13 reservoir is a complex turbiditic rock body penetrated by about 150 wells (Fig.2). This reservoir with a large gas cap is the biggest in the sequence of the Lower Pannonian sediments of Algyo. After 15 years of exploitation (at 10% recovery which is only 30% of the expected production), the reservoir studies show serious problems: (1) Because of the high pressure decline which had occurred, probably through the cement failures of some wells, productivity decreased. (2) The productivity rates varied extremely in the reservoir and the water injection method was not be able to increase the production in some part of the rock body. (3) The lateral distributions of the better and the poorer developed reservoir units were not be able to understand on the basis of the previous sedimentological model.
This situation urged to work out a new sedimentological model for this reservoir, which was expected to show the most probable facies framework, to outline the possible lateral and vertical distributions of the members of this facies framework, to clarify the reservoir characters of these members by using the porosity, permeability and carbonate content data measured on core samples, and to reveal the best continuity directions for the log porosity, log-permeability and gross reservoir thickness data.

2. The Analysis Framework

The first step was the development of a macro-sedimentological stochastic model for the sedimentation history (Fig.3). In the second step the lateral variation in the depositional system can be taken into account by building the sedimentological interpretations of the sporadic core samples and grain size analyses into this conceptual model (Fig.3). In this way the vertical appearance of the sedimentary facies contained by the model may be extended to the third dimension according to the Walter facies rule. This synthesis was the third step (Fig.3). The fourth step was the geostatistical analysis of the porosity, permeability and the gross thickness data derived from the geophysical log interpretations (Fig.3). Finally, as the fifth step, the sedimentological model was compared with the reservoir characters and tendencies (Fig.3).

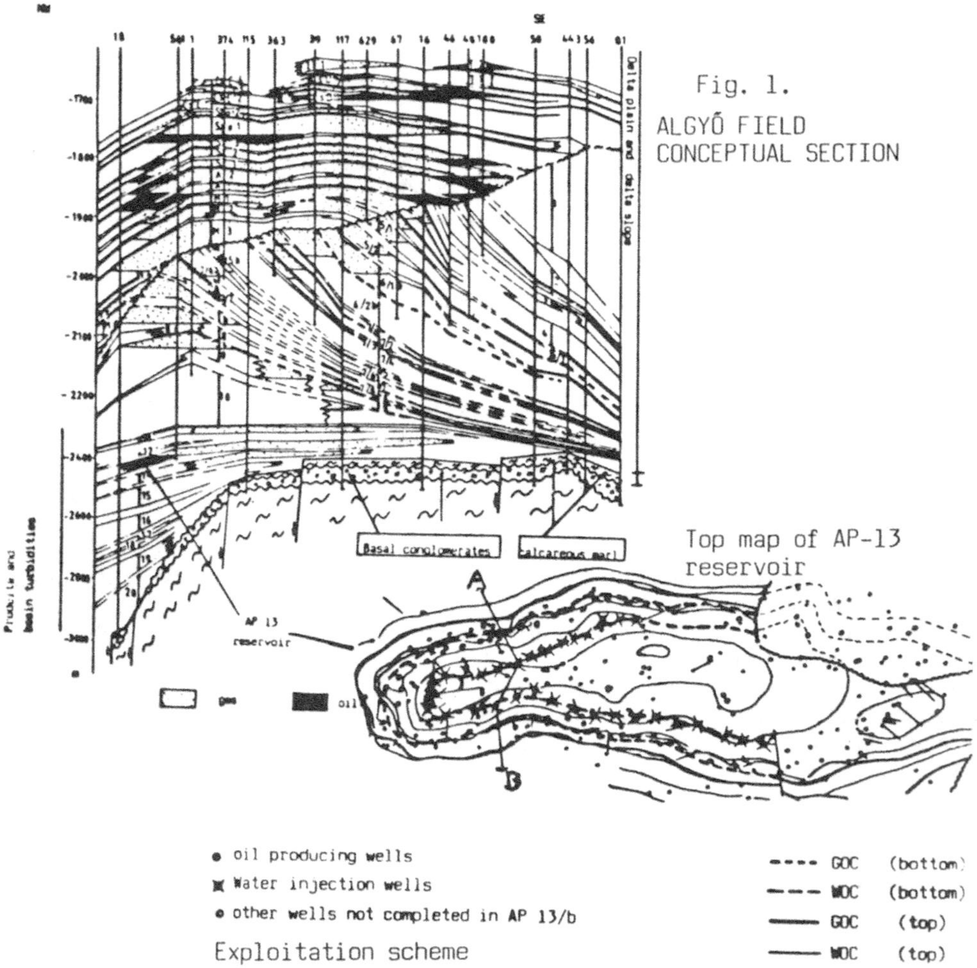

Fig. 1.
ALGYŐ FIELD
CONCEPTUAL SECTION

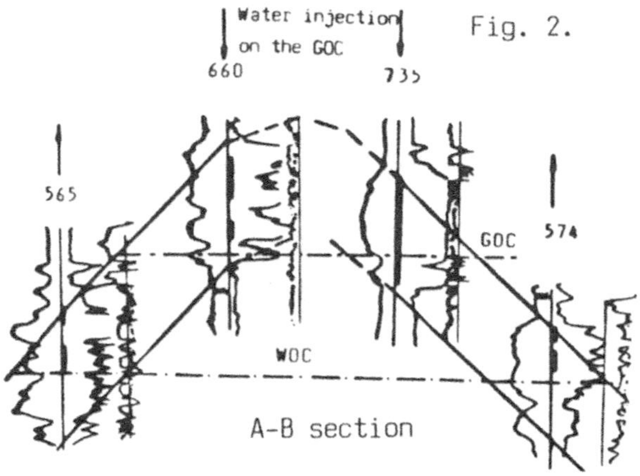

Fig. 2.

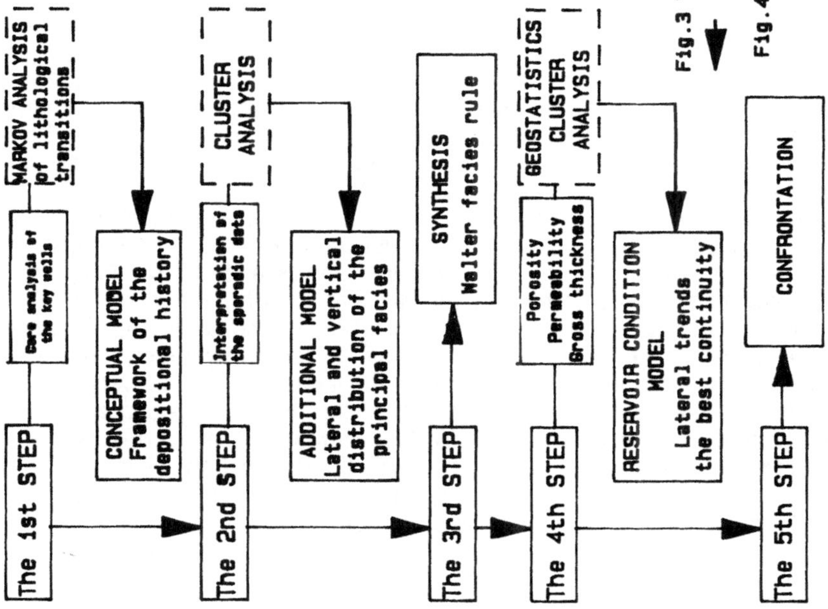

Fig.3 The modeling strategy

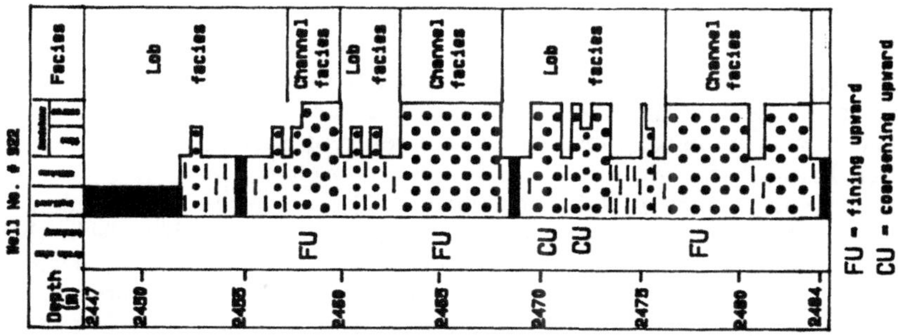

Fig.4 A key-well section

2.1. The macro-sedimentological model

The key wells are defined as wells penetrating the whole reservoir thickness with 100% core recoveries mostly in the NW part of the reservoir (Fig.2, Fig.4). This part shows the most heterogeneous development within the reservoir rock body, so more information was expected from here.
In order to built a meaningful conceptual model sequence, the random and the deterministic phenomena of the sequences must be differentiated from each other (Potter and Blakely 1968). The lithological units appearing above each other because of the deterministic processes will reflect the general depositional history through time. This model sequence was determined by using the embedded Markov chain method of the vertical lithological transitions suggested by Krumbein and Dacey (1969). This method subdivides the section into thickness intervals which have a uniform lithological composition. Each lithological units, regardless of its thickness, forms a step in the Markov chain. So, the diagonal of the transition matrix have only zero-elements. In this work we made several attempts for using the alternative method: coding of lithological states at equal thickness interval (Schwarzacher 1975). However it was found that the sedimentary structures with several scales did not make meaningful any of these efforts. In this way almost all information that is of stratigraphic value has lost for us. But the embedded method can be accepted from the sedimentological point of view because it reflects the most probable facies transition series.
In the structuring phase the number of states were stepwise reduced to obtain the most significant transition matrix in the sense of sedimentology and simultaneously it was expected to satisfy a chi-square significance test for the first order Markov process (Billingsley, 1961) Eight sedimentation units were used as states: homogeneous argillaceous marl; sandy coarse siltstone with parting lamination; coarse siltstone with sandy lenses; very fine sandstone and cross-laminated very fine and fine sandstone; homogeneous fine sandstone and cross-laminated fine sandstone (Fig.5). The difference matrix of the Markov analysis indicates those facies transitions which can be regarded more or less "deterministic" events, i.e. the transition probabilities are not smaller than the corresponding observed probabilities (Potter and Blakely, 1967; Gingerich, 1969) This difference matrix is represented by the facies relationship diagram (Fig.5). It shows a main depositional frame cycle, which starts at the horizontal laminated silty argillaceous marl (this is the starting facies member of the AP-13 reservoir) above which appears the homogeneous fine sandstone facies, then comes the laminated fine and very fine sandstone member, the unbedded fine sandstone and the chain terminates again at the horizontal laminated argillaceous marl facies (Fig.5). The symmetric nature of this modal frame cycle is suggested by the bi-directional arrows. So the flow pattern shows an initially high then gradually decreasing transport energy, which is followed by an initially low then gradually high energy period. This general process is occasionally interrupted by three other more or less quiet water sedimentation types characterized by the three siltstone member appearing in the inside subcycles (Fig.5).
The modal frame cycle can be drawn as a vertical rock column starting at the laminated argillaceous marl (Fig.6). So, it can be stated that the initial fining upward lithofacieses of the frame cycle may represent the domination of the lower channel facieses in the key well sequence (Fig.4, Fig.6). In the modal cycle the upper coarsening upward series correspond to the lobe facies domination in the upper parts of the key well sequence (Fig.4, Fig.6). Because of these correspondences, the idealized lithofacies succession obtained by the Markov chain analyses can be laterally extended in a hypothetical transport direction on the basis of the Walter facies rule (Fig.6). Moreover, in this way the theoretical lateral equivalent of the vertical repetition can also be sketched (Fig.6). So, an initial channelized period (i.e. decreasing energy) and a lobe prograding period (i.e. increasing energy) have been outlined by this model. Consequently the geological

subdivision of this reservoir can be done by identifying and correlating the channelized and lobe sequences from well to well. By restriction this result to the key well sequence in Fig.4, at least two main prograding cycles can be identified during the accumulation of the total reservoir thickness.

2.2. Multivariate analysis of the core data

The next question is: are there any laterally significant different ways of deposition of these main sedimentary facieses within the reservoir. The recognition of depositional processes causing the accumulation of core samples can be done by the interpretation of the clusters containing the sets of core samples deposited under similar circumstances (Park, 1974). Because of the present lithification stage of sandstone diagenezis, the available 200 grain size distributions were supposed to characterize the original depositional conditions. Therefore the weight percents of the grain size fractions as a multidimensional vector characterized the corresponding core sample. For transforming these sample vectors into the multidimensional sphere with unique radius a regular vector normalizing method was used. These normalized multidimensional vectors were classified by using an r-hierarchal Q-mode cluster analysis. The similarity coefficient applied was the cosine of the angles between any two multidimensional vectors. For the reduction the centroid method was used (Gower, 1967).

Fig.7 shows the result of this classification method. The determinations of the groups and subgroups were derived partly from the interpretation of the R-mode factor analysis performed for the grain size classes of the samples belonging to every particular clusters (Joreskog et al. 1976), and partly from the corresponding sedimentary structures of the available hand specimens. The maximum-likelihood method was used to compute the factor loadings matrix and for assessing the goodness of fit of the factor model a chi-square statistical measure was applied after Joreskog et al (1976). In the interpretation of any R-mode unrotated factor loadings matrix the factors were traced back to such independent depositional processes which can be suggested on the basis of the logical associations showed by the variables appearing with "considerable" factor loadings. It has been proved that these sedimentological interpretations are close to those which can be obtained from the sedimentary structures (Geiger, 1984).

The reservoir characterizations were carried out only for the end-members of the genetic hierarchy , i.e. for the B, A(1/1), A(1/2) and A2 groups (Fig.7). The main reservoir rocks belong to the sandy turbidite facies (A group on Fig.7) within which the A(1/1) class has the best porosity and the smallest clay and carbonate content values (Fig.7). The other types of turbidities can be characterized by smaller effective porosity and higher carbonate content (Fig.7). The sediment of the traction current origin have the smallest effective porosity and the highest carbonate contents (Fig.7).

The significance of the above classification was tested and accepted by performing a discriminant analysis between the clusters defined. So as to avoid redundancies, the discriminant analyses were counted such grain size class variables which appeared in different factors of the rotated factor matrix (varimax-method was used).

After putting back the samples to their real stratigraphic positions, it has been shown that the lower third part of the AP-13 reservoir is characterized by the lateral coexistence of undisturbed argillaceous marl, siltstone with "traction carpet" origin and sand laden turbidity currents with increasing transport energy (Fig.8). The middle third part comprises fine and coarser grained sandstones accumulated by turbidity currents and a few quite water argillaceous marls (Fig.8). The uppermost third consists of deposits similar to the lower third but the sand content is much more on the top (Fig.8). The reservoir characters fitted to the end-clusters have shown that these conditions laterally can be expected to become worse in the theoretical transport direction (Fig.8). However in a particular location they may improve from the bottom to the top of the reservoir (Fig.8).

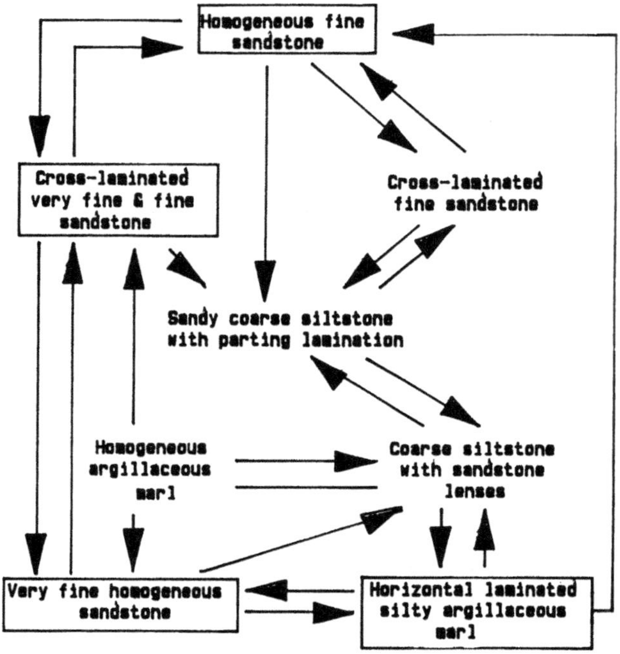

Fig.5 Facies relationship diagram

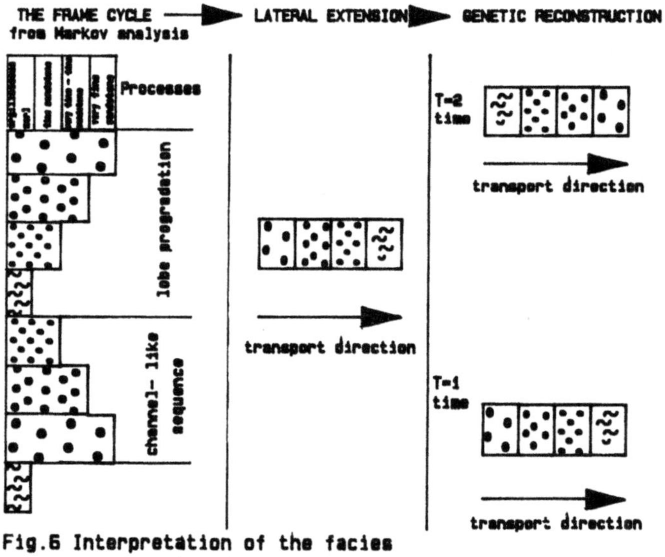

Fig.6 Interpretation of the facies relationship diagram

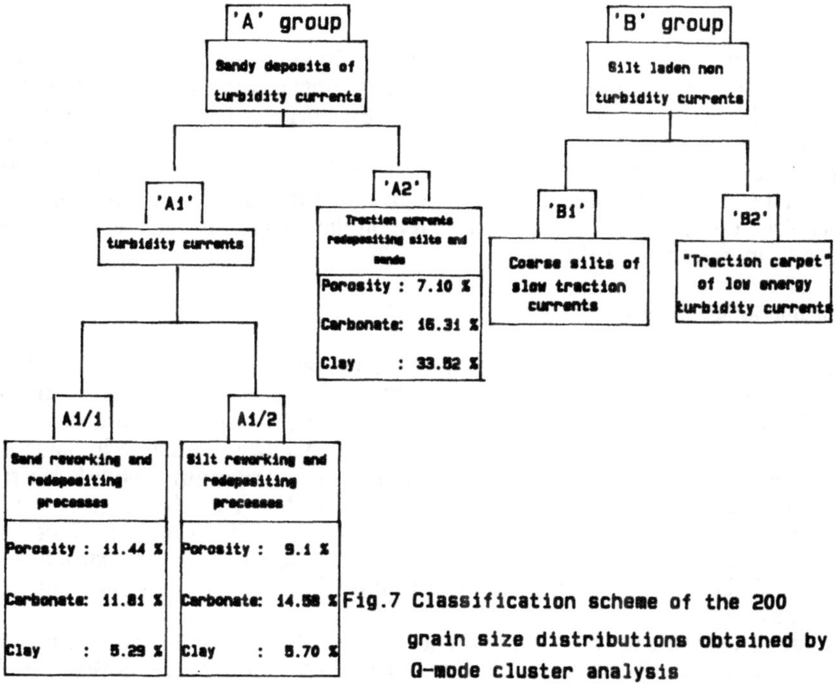

Fig.7 Classification scheme of the 200 grain size distributions obtained by Q-mode cluster analysis

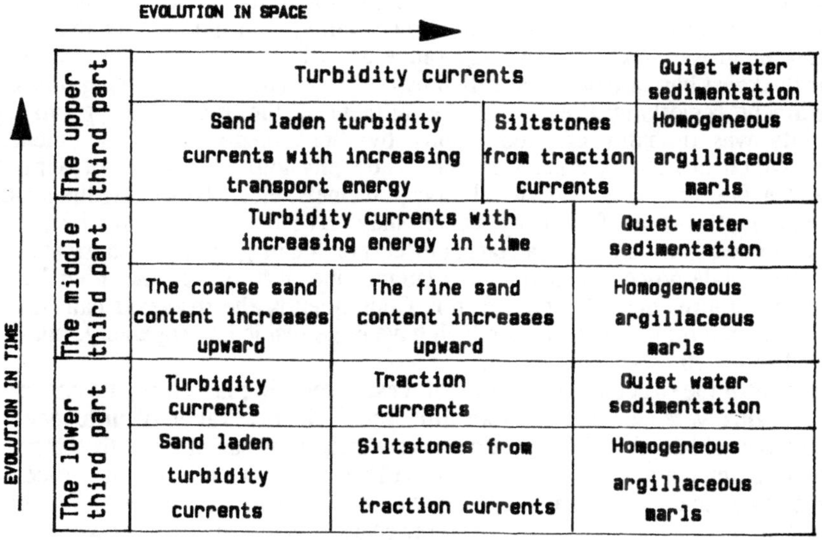

Fig.8 Synthesis of the sedimentological evolution

2.3. Synthesis

The third step of the modelling strategy was to combine the results of the first two steps. Thus, it is clear that the sedimentary facies belonging to the main modal cycle have a wide lateral distribution. However, the subcycles merge partly into the laminated and partly into the homogeneous argillaceous marl facieses. Moreover this phenomena becomes more common from the NW to the SE. It is also evident that the recurrence of an individual sedimentary facies occurs under a little bit coarser textural conditions. The transport direction from NW to SE is probable on the basis of the tendency of the inside subcycles to merge from NW to SE.

2.4. Geostatistical analyses

The fourth step was the geostatistical analysis of geophysical porosity, permeability and gross-thickness data averaged for each well. The semi-variograms calculated for the N-S, NE-SW, E-W, and SE-NW direction made possible to understand the anisotropic nature of the reservoir (Journel and Huijbregts 1978). This anisotropy must be in harmony with the main transport direction. The omi-derectional semi-variograms show large range of influence for these parameters (Fig.9, Fig.10): 4900 m for gross thickness (Fig.9), 1000 m for permeability (Fig.10), and only 380 m for porosity (The average real well spacing is 150 m). In these cases the starting distance was 300 m, i.e. the structure of the variability of the tested parameters within this distance is constant, consequently homogeneous.

The numerical experiments have shown that a smaller homogeneity radius (in the above sense) gives a worse modelling possibilities. This experience suggests that (1) the real reservoir heterogeneity is fairly large on a small horizontal scale; (2) the sizes of the individual morphogenetic units within this lower-fan system not smaller than 300 m.

In the case of gross thickness, the large range of influence obtained expresses that the reservoir thickness basically was controlled by the large scale morphologic processes. The local sediment reworking events did not influenced the rock body geometry. The permeability was probably effected mainly by those medium-scale sedimentological processes which are characteristics in the inside physiographic units of a prograding middle-lower fan series (e.g. individual lobes, channels, etc.). However, the small range of influence in the case of porosity values may have been caused by local sediment reworking processes which have had larger effects on the original petrophysical characters than the large scale processes such as low lateral lobe shifting. On the other hand, local interpretation failures may also be a factor. Consequently the reservoir can be expected to built up by such rock genetic units which have large inside heterogeneity, but they may not differ from each other as a whole.

In the analysis of the structural anisotropy of these reservoir characters, the corresponding semi-variograms were calculated in four directions. The best range of influence for each parameters was obtained in NW-SE direction which is in harmony with the previously established sedimentary transport direction (Fig.11). After the kriging processes, the resulting contour maps as well as those showing the estimated averages above the blocks with 1,000 m of size suggest hat the reservoir characters tested becomes worse from NW to SE (Fig.12).

2.5. The lateral distribution of reservoir conditions

By averaging the log-porosity, permeability, water saturation and net sand thickness data for each well four dimensional vectors can be obtained which characterize the reservoir conditions for the whole reservoir thickness within a circle around the well.

In the interpretation of the above geostatistical analyses the following consequency was

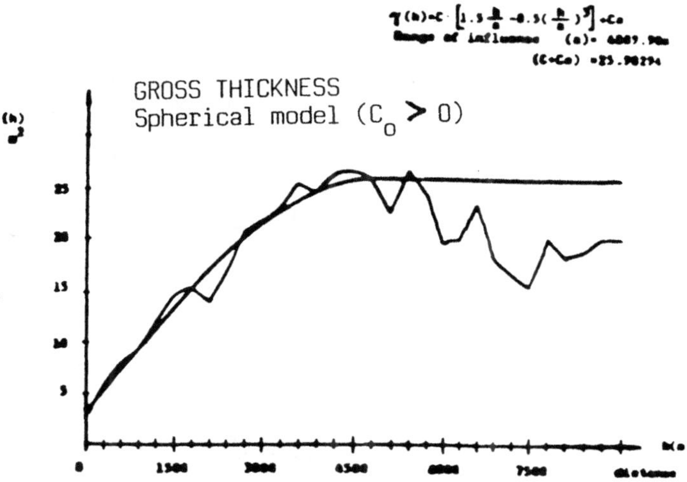

Fig. 9.

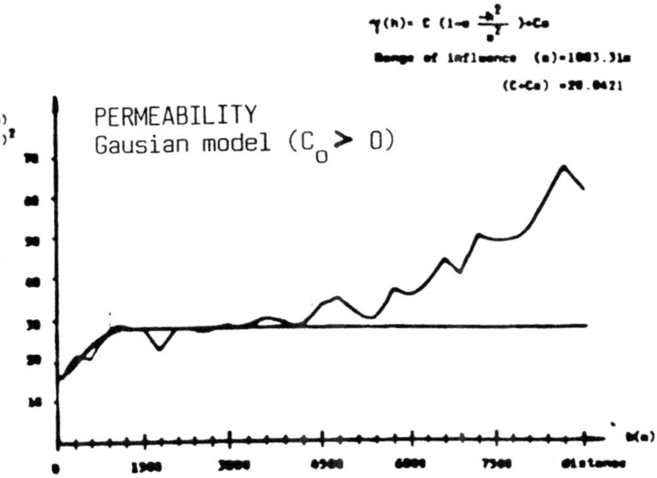

Fig. 10/a.

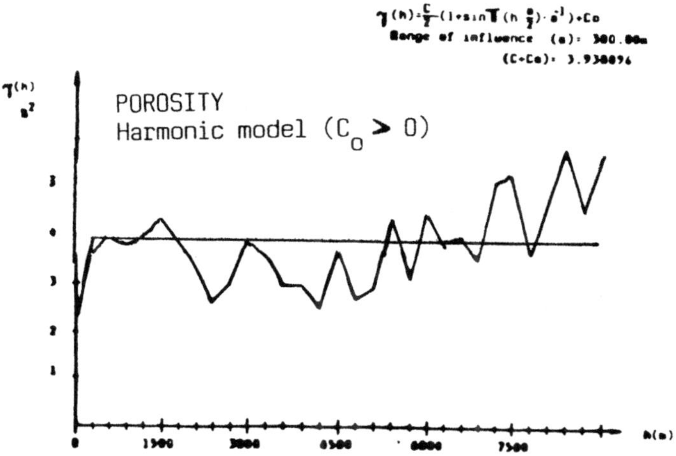

Fig. 10/b.

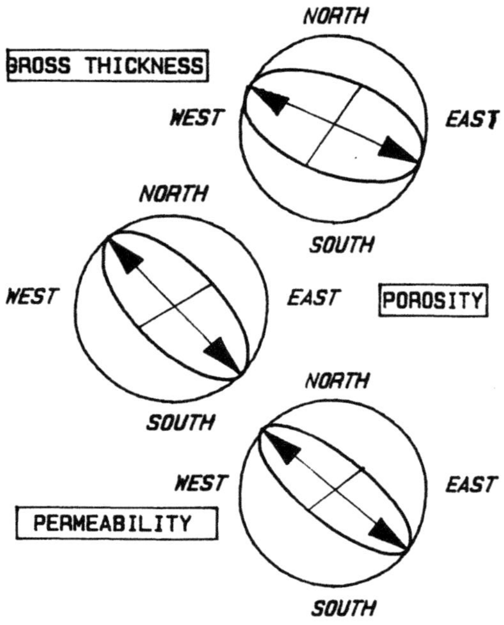

Fig.11 Anisotropy ellipses

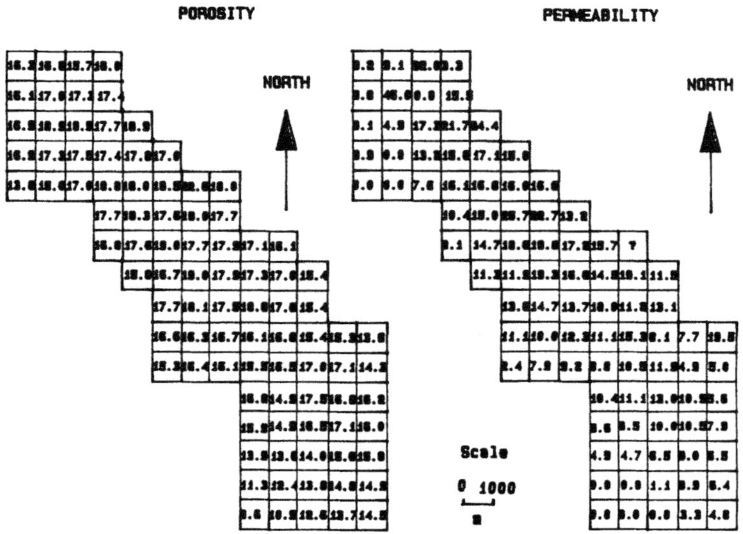

Fig.12 The results of the block-kriging

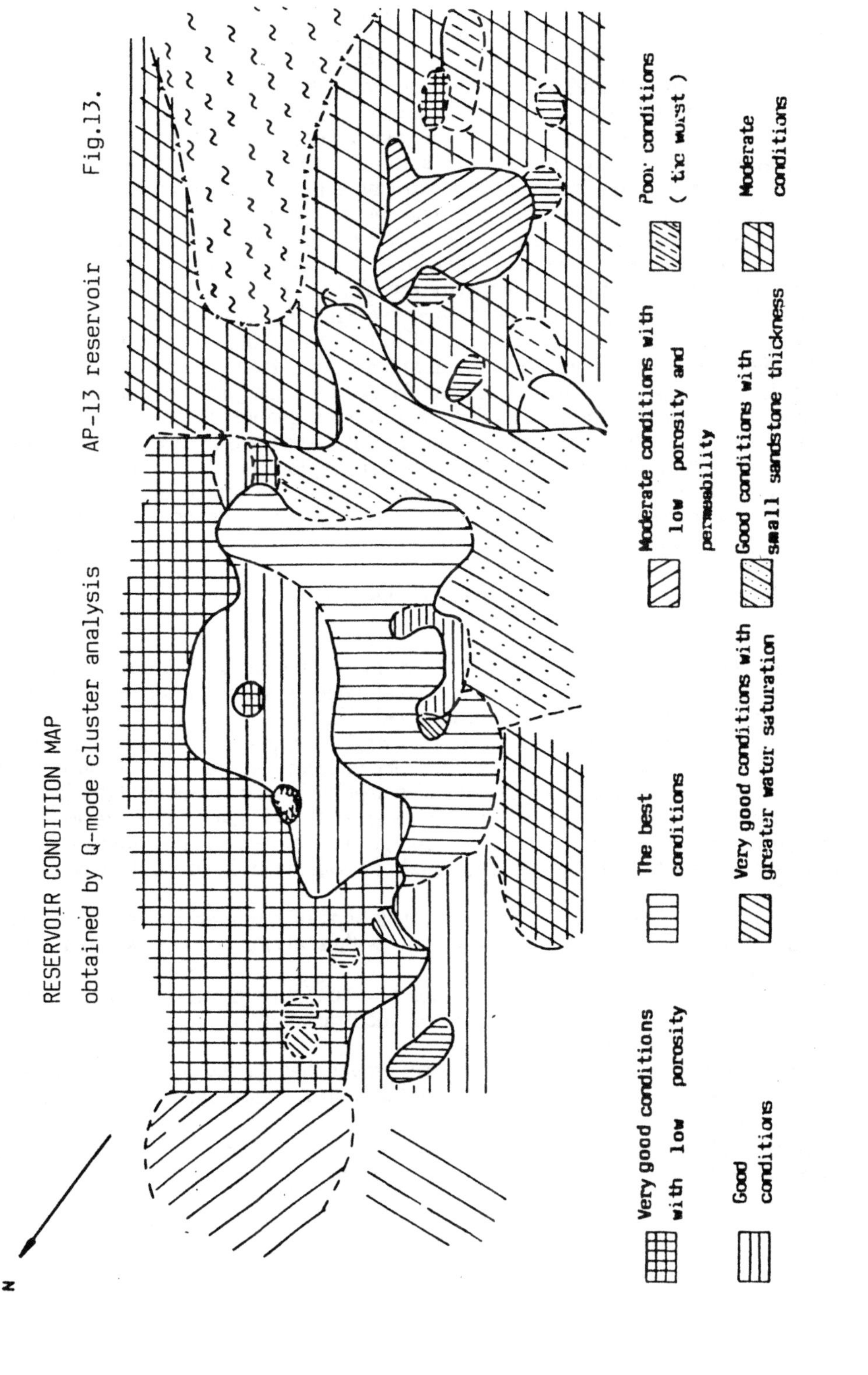
RESERVOIR CONDITION MAP
obtained by Q-mode cluster analysis
AP-13 reservoir
Fig.13.
N
Very good conditions with low porosity
The best conditions
Moderate conditions with low porosity and permeability
Poor conditions (the worst)
Good conditions
Very good conditions with greater water saturation
Good conditions with small sandstone thickness
Moderate conditions

derived:"...the reservoir can be expected to built up such rock genetic unots which have large inside heterogeneity, but they may not differ from each other as a whole". In this hypothesis two important questions should be answered: Which are the most similar parts of the reservoir rock body? What is the content of this similarity?
The first problem can be solved by using a Q-mode cluster analyses of the above mentioned four dimensional vectors in the same manner as it is done in the classification of the grain size data. In this case the definition of the end-clusters on the dendogram carried out when the classification was significant in the sense of discriminant analysis and in the sense of production experience as well. In this way eight clusters have been mapped (Fig.13). Each can be regarded as more or less homogeneous inside, but they show significant differences among each other in the four dimensional vector space.
For the characterization of the similarity revealed the following method was used: For each parameter the cluster showing the best average value is given by index 1, and those having the worst average value an index of 8. In the final qualification these indexes are added. Between the maximum and the minimum indexes the following four categories are defined: very good, good, moderate, poor. Within each category two other subclasses are formed. The areal distributions of these reservoir quality classes clearly show that reservoir conditions become worse from NW to SE , and the best reservoir conditions can be outlined within the central part of the western region (Fig.13). Moreover it can also be seen clearly, that the identified reservoir quality units do not for any continuous zones, but they appear as smaller or larger areas. This fact can be understood on the basis of the lobe shifting and prograding processes.

3. Conclusions

(1.) The sedimentation occurred in a lower mostly unchannelized turbidity fan which developed on or in front of the prodelta area of a delta systems coming from the NW. At least two fan lobe prograding phases have had effects.
(2) By using Markov model and multivariate analyses a genetic sedimentological system was build up in which the defined sedimentary facieses determine significantly different reservoir characters. Because of the coincidence of the Markov model and the depositional genetic units obtained by the multivariate statistical techniques, these facieses can be recognized on the basis of macroscopic geological characters and discriminant functions. These discriminant functions were generated for both the grain size and the petrophysical variables.
(3) The numerical experiences, in the computations of semi-variograms, suggested that the size of the individual morphogenetic units within this fan geomorphology not smaller that 300 m.
(4) The anisotropy ellipses showed that the best continuity direction is the NW-SE axis which is in harmony with the axis of the transport direction suggested by the sedimentological interpretations.
(5) The results of the block-kriging processes proved that the reservoir conditions become better form NW to SE, so the sedimentary transport direction must be from NW to SE.
(6) The areal distributions of the different reservoir condition units obtained by a cluster analysis can be understand by taking into account the lobe prograding processes. These units can be easily identified by using the corresponding discriminant function system.

REFERENCES

Billingsley, P. (1961) " Statistical methods in Markov chains", Annals Math.Statistics 32, 12-40.

Geiger, J.(1984) "Geomathematical analyses and interpretation of grain size distributions in the lithification stage of sandstone diagenezis", PhD Thesis, Attila Jozsef University, Szeged.

Geiger, J.(1988) "Geomathematical interpretation system and its entropy set for the reconstruction of sedimentary basin accumulation on the Great Hungarian Plain", COGEODATA-IAMG Symposium on Computer Applications in Resource Exploration 21-23 July 1988 Espoo (Helsinki), Abstracts, 13-15.

Geiger, J., Revesz, I.(1987) "Genetic model of Post-Sarmatian sedimentation in the Great Hungarian Plain", VIIIth Congress of the RCMNS, Budapest 12-22 September 1985, Proceedings, 245-251.

Gingerich, P.D. (1969) "Markov analysis of cyclic alluvial sediments", Jour. Sed.Petr. 39, 330-332.

Gower, J.C. (1967) " A comparison of some methods of cluster analysis", Biometrics 23, 623-637.

Journel, A.G. and Huijbregts, Ch.J. (1978) "Mining geostatistics", Academic Press, London

Joreskog, K.G. and Klovan, J.E. and Reyment, R.A. (1976) "Geological factor analysis", Elsevier Scientific Publisher Co., Amsterdam

Krumbein, W.C. and Dacey, M.F. (1969) "Markov chains and embedded Markov chains in geology", Mathematical Geology 58, 1246-1266.

Park, R.A. (1974) "A multivariate analytical strategy for classifying paleoenvironments", Mathematical Geology 6, 333-351.

Potter, P.E. and Blakely, R.F. (1968) "Random process and lithological transitions", Journal of Geology 76, 154-170.

Revesz, I. (1984) "Post-Miocene delta formations in the Southeastern part of the Great Hungarian Plain", 27th International Congress, Moscow, Abstracts VII, 223.

Schwarzacher, W (1975) "Sedimentation models and quantitative stratigraphy", Elsevier Publ. Co, Amsterdam

CONSTRAINING STOCHASTIC IMAGES TO SEISMIC DATA
Stochastic Simulation of Synthetic Seismograms

Louis-Jérôme Bortoli[†]
Francois Alabert[‡], André Haas[‡], André Journel[†]

[†]Stanford Center for Reservoir Forecasting, Stanford, CA 94305, USA
[‡]Elf Aquitaine, Avenue Larribau, 64018 Pau Cedex, France

Abstract

Selection of conditional simulations of acoustic variables is done by matching the simulated values with a series of actual seismic vertical sections. This is done by forward convolution of the simulated acoustic models. The proposed algorithm allows an iterative adjustment of the convolved simulated vertical section to the original seismic data: first impedances along a vertical trace are simulated, then this impedance trace is convolved into a synthetic trace at the seismic scale. An acceptance/rejection criterion based on a correlation function is applied to compare the synthetic with the actual trace. If accepted, the algorithm proceeds and builds the synthetic image by simulating an impedance trace at the next node, otherwise another simulation of the same trace is drawn. Particular attention is given to calibration of the synthetic seismograms at well locations and calibration of acoustic variables to porosity at the log scale. Once the adjustment of the vertical section is obtained, the stochastic impedance image -i.e. the adjusted section before convolution- is easily converted into a porosity image. This image can directly be used for interpretation or flow simulator processing. The method is developped and tested on a real field, a shaly/sand formation covered by an 3D seismic survey.

1 Introduction

Imaging the subsurface, and the modelling of reservoir properties such as lithology, porosity, permeability or fluid saturations, represents a major challenge for petroleum geologists and geophysists. As a result of very high costs of drilling, only a minimum number of exploration and appraisal wells can be justified. Therefore lateral variations of heterogeneous reservoir properties cannot be accurately predicted from only well data. The integration of 3D seismic data, beyond its traditional use in mapping large scale subsurface structures, may significantly improve the modelling of petrophysical properties.

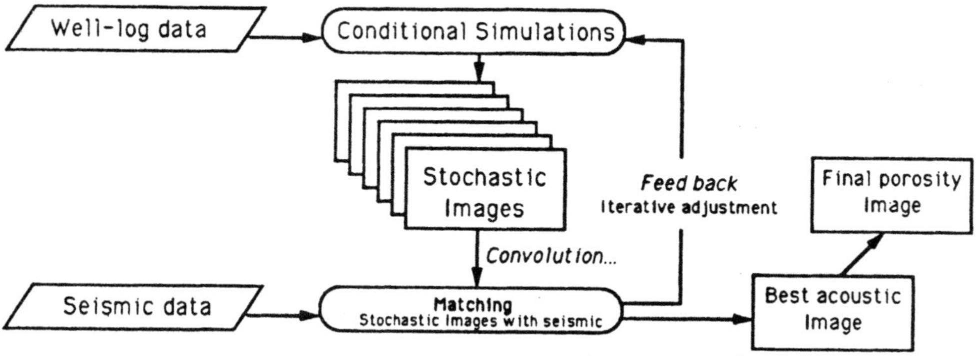

Figure 1: The posterior matching approach.

Seismic data integration has been approached in different ways. One way is to combine large scale "soft" seismic data and small scale "hard" well data in one set of prior conditioning data through algorithms such as straight cokriging[4], the external drift concept[9] or indicator coding of seismic data[1, 8, 12]. Another way is to derive an acoustic model from seismogram inversion through deconvolution of seismic amplitudes in reflectivity sequences. However, acoustic parameters, such as velocity, density, impedance or reflectivity, derived from band-limited and noise-contaminated seismic data are non unique, they have only large scale resolution (decametric versus decimetric for well data) and are difficult to relate to the petrophysical parameters relevant to reservoir production.
This study investigates a third way: to perform a stochastic imaging of an acoustic-related variable based only on well data. From each equiprobable realization, a synthetic seismogram is derived, and the synthetic seismogram that best matches the actual seismic data indicates the most realistic image. This method can be seen as a "posterior matching": the actual 3D seismic data are used as an exhaustive reference data set to select from various prior stochastic images (see Figure 1).

The approach proposed presents an important advantage. Seismic and well data are considered at different steps of the processing: hence, the approach does not call for combining data taken at different scales. As opposed to inversion techniques, the final best acoustic model is derived only from well data at a scale allowing a good calibration between acoustic and petrophysical properties.
As compared to methods integrating seismic data as prior conditioning data, the approach presented does not require a prior probabilistic calibration step, either Markov-Bayes calibration[8, 12] or affine mapping[1]. Such calibration is critical, and often very difficult to establish[1]. The calibration required by the method proposed here is deterministic: it consists of determining a wavelet such that synthetic and actual seismograms match at well locations.

The parameters of the calibration are then used to convolve the stochastic acoustic images, as obtained by the sequential Gaussian simulation (SGS) algorithm, into synthetic seismic images of the reservoir. Among the equiprobable realizations, one

[1]Such methods are usually not able to deal with 3D seismic amplitudes since there is often a weak correlation between amplitudes and any acoustic or petrophysical parameter.

can select the one that best matches the actual seismic data; but, such a selection may be tedious. Therefore an algorithm to feed back the simulation process with the results of the posterior matching has been developed.

This paper mainly focuses on the elaboration of such an optimization algorithm which allows a dynamic integration of 3D seismic data. The principle of the algorithm is to apply the posterior matching approach (Figure 1), trace by trace instead of simulating at once the whole stochastic image. It allows an iterative adjustment of the convolved simulated vertical section to the original seismic data: first impedances along a vertical trace are simulated, then this impedance log is convolved into a synthetic trace at the seismic scale. An acceptance/rejection criterion based on a correlation coefficient is applied to compare the synthetic with the actual trace. If accepted, the algorithm proceeds and builts the synthetic image by simulating an impedance trace at the next node; otherwise another simulation of the same trace is drawn.

Last, through a calibration between acoustic and petrophysical properties (i.e. relation between impedance and porosity at the log scale), the single optimized simulated acoustic image is transformed into the petrophysical image which can be interpreted and used for further processing.

2 Geostatistical Modelling and Simulation

Exploratory data analysis:
The methodology has been developed using data from a shaly sandstone reservoir. The study area, named *Eldorado*, is about 2 km EW x 7 km NS and informed by 7 wells, see Figure 2. The reservoir interval is characterized by a low sand/shale ratio with unknown extension and direction of sand deposition.

The whole *Eldorado* zone has been surveyed by a 3D seismic of 100 *Lines* $\times$ 30 CDP^2. The seismic horizon at the base reservoir has been picked on a 3D workstation and transformed into a flat event so that the base of the reservoir occurs at apparent constant seismic time of 100 ms. The window retained for the study is about 100 ms thick, see Figure 2.

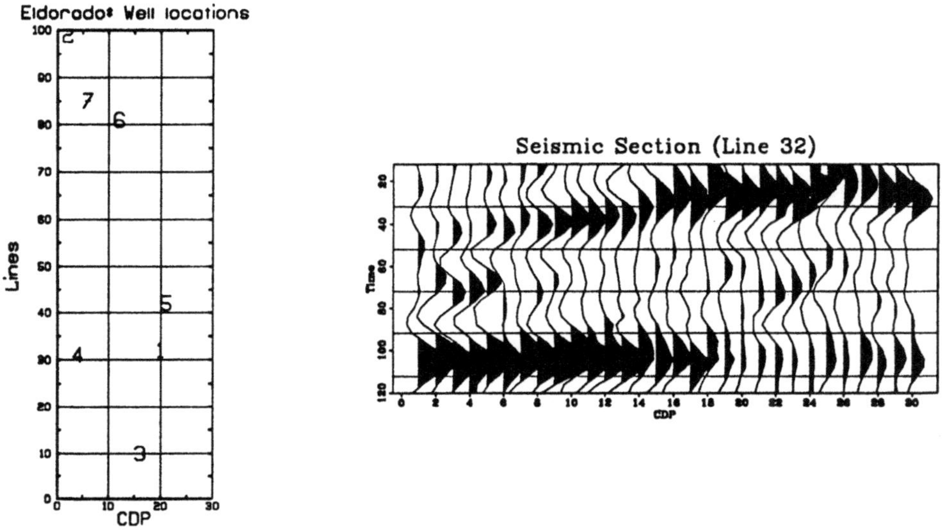

Figure 2: Seismic grid and well locations - Typical seismic line.

The following logs were retained to characterize the reservoir acoustic and petrophysical properties:

- Sonic transit time, expressed in ms/ft and equal to the inverse of the p-wave interval velocity V_p;
- Bulk density ρ, expressed in ton/m^3;
- Acoustic impedance $\rho\ V_p$, calculated from sonic and density log data;
- Porosity ϕ, expressed as a fraction: this is the petrophysical variable to model.

The acoustic impedance is retained as the variable to simulate because the derivative of its logarithm is used directly in the definition of the synthetic amplitude:

$$Amplitude = wavelet \star d\ln(\rho V_p)$$

[2]Common Depth Point.

To build a reservoir model consistent with the 3D seismic format, the following transformations are applied to the coordinates:

- A depth-to-time transformation based on a rough calibration (Z_{TVD}[3] v. Z_{TWT}[4]) is applied to the vertical coordinate[5].
- Wells are relocated to the nearest nodes on the seismic grid[6].
- Acoustic well-logs (sonic and density logs) are resampled from the original 15 cm sampling to the simulation interval of discretization chosen[7].
- The stratigraphic transformation applied to the well-log data,

$$z' = Z_{TWT} - Z_{TWT\ at\ base\ reservoir} + 100\ ms,$$

amounts to take the base of the reservoir for reference level[8], and to keep the true thickness of the formation for each well[9].

Variogram analysis:

The purpose of this analysis is to model the prevailing spatial correlation (ranges, anisotropies...) through a variogram model to be used as an input for geostatistical simulation. Since the simulation random function model used is Gaussian, the model of spatial correlation needed is that of the normal score transforms of the impedances. The vertical density of well-log data after resampling is still high. So the 7 wells over the *Eldorado* zone allow building a reliable vertical variogram model.

Horizontal information density is typically sparse at early stages of production. The average distance between two wells is one kilometer; hence, characterization of horizontal correlations using only well-log data is very hazardous. Horizontal ranges of correlation and anisotropy, must be determined from information provided by seismic data and geological interpretation.

The main direction of continuity (NNW-SSE) shows a range of 30 units (i.e. $30 \times 75m = 2.25km$), while in the orthogonal direction the horizontal range is only 15 units[10].

The analytical expression of the 3D variogram model retained is given below; the system of coordinates is based on the seismic grid: the unit in x-East is a CDP interval (75m), the unit in y-North is a line interval (75m) and z' is the vertical coordinate.

$$\gamma(h'_x, h'_y, h'_z) = Spherical(\sqrt{(\frac{h'_x}{30})^2 + (\frac{h'_y}{15})^2 + (\frac{h'_z}{25})^2}$$

[3]True vertical depth.

[4]Two-way travel time.

[5]Working directly in time avoids the need for a posterior depth-to-time transformation after the simulation.

[6]The system of coordinates is: x-CDP, y-Line, z'.

[7]An initial resampling allows to derive the spatial model of correlation at the simulation scale which is not the initial scale of the data.

[8]To be consistent with the seismic 3D grid of data, the reference level is put at the same arbitrary level: 100 ms.

[9]This transformation accounts for the main structural features of the field (dipping, erosion...).

[10]Comparing seismic horizontal variograms for different planes (isochrones) does not reveal any significant variability of the anisotropy directions and ratio.

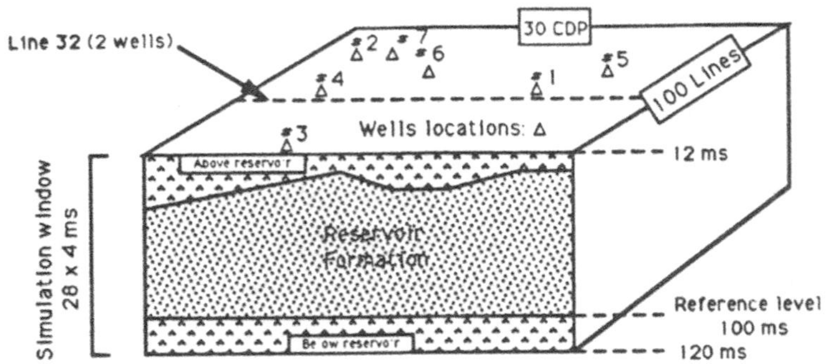

Figure 3: Definition of the simulation grid: 30 CDP × 100 Lines × 108 ms.

Simulation grid:

To allow comparison with the seismic data set, the horizontal simulation grid is identified to the regular grid of the 3D seismic survey; it includes 100 Lines × 30 CDP, see Figure 3. Choosing a vertical grid amounts to decide the simulation vertical resolution. The sampling rate of the seismic data is 4 ms which gives, for a window of 108 ms, 28 samples. The original sampling rate of the well-data is about 15 cm. For simplicity, the vertical simulation lag retained is the seismic lag, i.e. an interval of discretization of 4 ms[11].

3 Convolution of the Stochastic Images

Calibration phase:

The calibration consists of comparing the synthetic trace along the conditioning wells (convolution of the reflectivity sequences from well data) with the actual collocated trace from the 3D survey.
This is a critical phase, a fail/pass step, for the feasibility of the study. In the eventuality that the calibration at well locations is poor, comparison of actual versus simulated seismic traces cannot be expected to be good between the wells. The calibration is expected to satisfy the following equation for each well:

$$S_{eismic} = w \star R_{eflectivity} + N_{oise},$$

where S_{eismic} is the actual seismic trace and $R_{eflectivity}$ is the reflectivity sequence from well-log data (derivative of the log of acoustic impedance).

At every well location, a wavelet w is estimated, then an "average" of these w is used to convolve the whole reservoir. There are several ways to evaluate the wavelet that provides the best fit between synthetic and actual seismograms. A traditional approach consists of building the synthetic seismograms at well locations with different *a priori* wavelets until a maximum correlation is obtained at each location. This method was used for the 7 wells of the *Eldorado* zone. Different correlation measures were tested to calibrate synthetic and actual seismograms. Other factors such as well location or the existence of static shifts can significantly affect the calibration. The

[11]The choice of a vertical simulation lag must be related to the particular use of the final stochastic image, a flow simulator for instance.

best relocation and the best shift for each well were automatically determined. The average wavelet retained is a ricker2, zero phase, 30Hz, normal polarity plus an average static shift of the simulated traces of +4 ms over the entire zone, see Figure 4.

The following table provides the Pearson's linear correlation coefficients attained for the seven wells using the above wavelet with or without static shift and relocation, see Figure 4 for the synthetic and actual seismograms for well # 1 and # 2. The calibration is considered as satisfactory whenever all coefficients of correlation are higher than 0.65.

Calibration at well locations: correlation coefficients between synthetic and actual seismograms (7 wells):

WELL	w/ shift	w/ shift	w/o shift	w/o shift
	w/ reloc.	w/o reloc.	w/o reloc.	w/ reloc.
#1	**0.70**	0.63	0.76	0.79
#2	**0.73**	0.47	0.43	0.66
#3	**0.80**	0.63	0.50	0.81
#4	**0.90**	0.78	0.39	0.57
#5	**0.68**	0.55	0.55	0.70
#6	**0.74**	0.74	0.62	0.64
#7	**0.67**	0.35	0.39	0.57

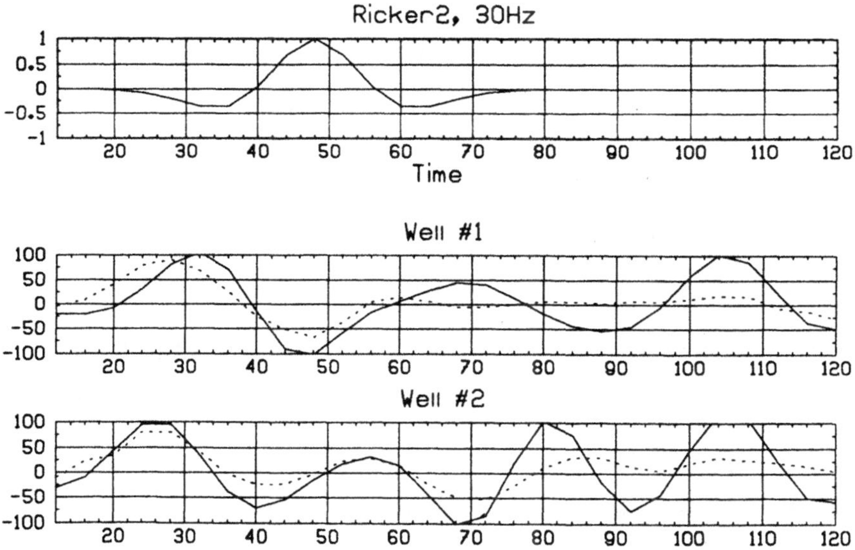

Figure 4: Wavelet retained for convolution,
Well # 1 and # 2: actual (dotted line) and synthetic seismograms.

4 Optimization

The proposed algorithm performs an iterative adjustment of the convolved simulated vertical sequences of impedances to the actual 3D seismic traces. Instead of simulating the whole 3D grid of acoustic impedance at once, the impedances are simulated trace by trace. Each simulated trace is convolved into a synthetic seismogram and compared to the collocated actual trace. In case of "acceptable matching", the simulation proceeds to another node on the horizontal seismic grid. Otherwise another impedance trace is drawn at that same location. The final result is one single 3D "optimal realization" of impedance.

Optimization algorithm:
A vertical sequence of nz simulated impedances is defined as a random vector, Z constituted of nz random variables. Z is simulated at every node (x, y) of the horizontal seismic grid (CDP, Lines) according to the following algorithm:

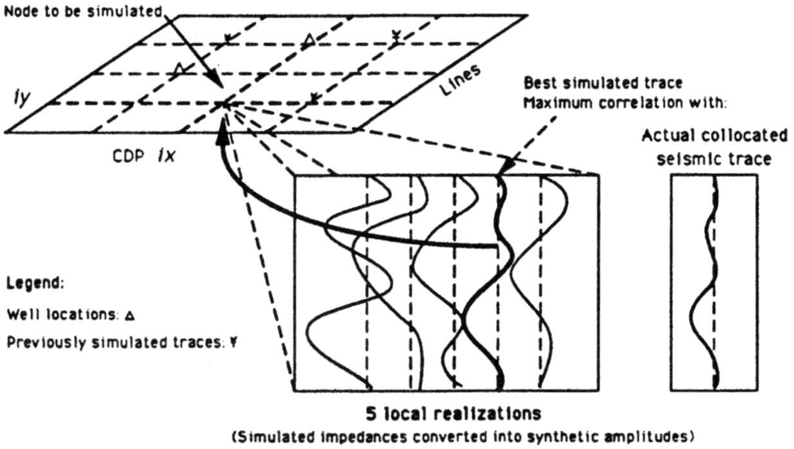

Figure 5: Local optimization process.

1. Define a random path through all nodes (x, y) to be simulated.
2. For each node (x, y) perform a local optimization (see Figure 5):
   - Determine the conditional cumulative distribution function (ccdf) by solving a simple kriging system.
   - Draw $lsim$ realizations (Z_i^{isim}) from that ccdf.
   - Compare them to the actual collocated seismic trace (matching function).
   - Retain the best trace that then becomes a conditioning data.
3. Go to the next node.

To obtain another "global optimization", a new random path through the grid is defined; new simulated traces are retained only if the fit is better than that achieved during the previous run.

The inputs of the optimization program are:

- the conditioning wells relocated on the simulation grid (i.e. seismic grid).
- the parameters of the convolution (wavelet and static shift) to build the synthetic simulated seismic traces;
- the 3D actual seismic amplitudes.

The main output is an optimal impedance image (3D grid of simulated values) to be transformed into an image of the petrophysical variable of interest. Other outputs allow assessing the quality of the optimization achieved. Section by section, the simulated synthetic images can be visually compared with the reference seismic data. In addition, a 2D correlation map provides a quantitative measure of the correlation attained at every location between the best synthetic and the actual seismic trace, see Results: Figure 7 and Figure 8.

Aside from the parameters of the sequential Gaussian simulation algorithm (essentially a covariance model), the optimization program calls for specific parameters such as:

- *lsim*: number of traces drawn at any location to select the best one (local optimization);
- *nsim*: number of times all the nodes should be revisited along a new 2D random path (global optimization).

The global optimization process protects the optimal image from biases that may be associated to particular simulation paths. Note that if *lsim* and *nsim* are set to 1, the optimization is equivalent to a traditional Gaussian simulation with every trace being simulated at once.

Validation:
The results shown hereafter indicate that the optimization algorithm succesfully integrate the seismic data. However, to be valid it should also insures reproduction of the input spatial correlation model of impedance normal scores, and consistency of the optimal acoustic image with the original impedance well-data. Since deconvolution does not give a unique solution, a matching synthetic trace could come from many different impedance sequences. Therefore a selection based on similarity between synthetic and actual seismic data does not necessarily ensure a realistic impedance image. However, the sequential approach proposed does ensure that every realization of impedance trace is consistent with the calibrated well-data.

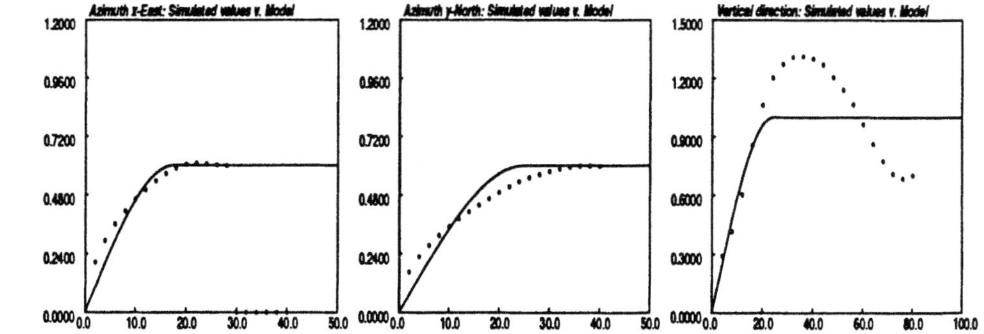

Figure 6: Impedance variogram reproduction after final optimization ($nsim$=10, $lsim$=50).

The global reproduction of the covariance model (see Figures 6) is expected to be good since:

- vertically: each trace is simulated at once, therefore the optimization (selection) process does not perturb the vertical variogram reproduction;
- horizontally: because the 2D simulation path is random and because the impedance horizontal variogram was modelled from the seismic data, the reproduction should be reasonable.

With regard to program efficiency, the algorithm speed is of the same order as that of the sequential Gaussian simulation algorithm (SGS). For each trace, the vertical sequence of simple kriging systems is solved only once, and then any number *lsim* of local realizations can be drawn from the same set of ccdfs. As for the operations involved to build the synthetic amplitudes and to compare them to the actual seismic data, their CPU requirement is relatively unimportant.

Results:

To analyse the performances of the optimization algorithm, a section (Line 32) was extracted from the simulated 3D grid. With *nsim* and *lsim* set to 1, the impedance simulated image is a straight realization from SGS. As *lsim* increases, the image significantly converges towards the actual seismic reference image. The following table provides a measure of the performance of the optimization algorithm for that section: the average of the Pearson's linear correlation coefficients between synthetic and actual seismic traces of Line 32 increases from *0.28* with no optimization to *0.83* with *lsim=100* local realizations.

Means and variances of correlation coefficients(Line 32, 30CDP):

nsim	lsim	Mean	Variance 10^{-2}	nsim	lsim	Mean	Variance 10^{-2}
1	1	0.28	13.5				
1	10	0.68	1.50	2	5	0.70	1.86
1	50	0.78	0.86	5	20	0.86	0.67
1	100	0.83	0.70	2	100	0.86	0.46
1	150	0.84	0.50				

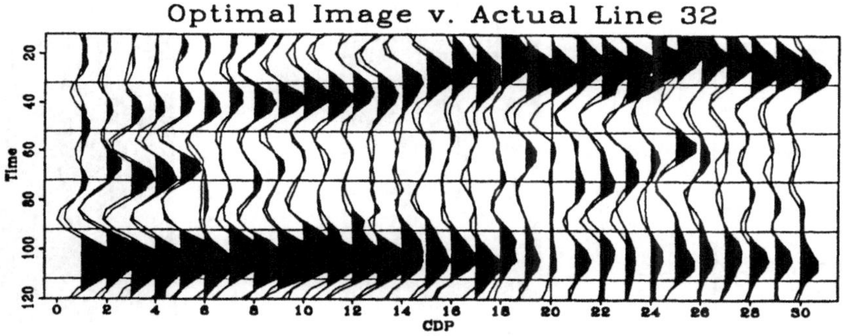

Figure 7: Superposition of the synthetic and actual Line 32 (note the excellent match).

The level of similarity, achieved through the optimization process, is demonstrated by superposing the synthetic realization of Line 32 with *nsim*=10 and *lsim*=50 on top of the actual seismic line, see Figure 7. To assess the quality of the optimization for a whole 3D realization, a 2D correlation map is provided, see Figure 8.

5 Conclusion

A single 3D acoustic image of the reservoir, constrained *a posteriori* by 3D seismic data with a satisfactory correlation level, has been obtained. This optimal image is easily transformed into a porosity image through the acoustic-petrophysical calibration performed on well-data[12]. Such 3D porosity-image, produced at a scale defined by the user, can directly be used for reservoir modelling and flow simulation. The final porosity section corresponding to the best synthetic image on the Line 32 example appears very realistic, see Figure 9. A scatterplot of these simulated porosity values versus the original porosity well-data is given in the same figure.

This paper has presented a feasible approach to integrate seismic data into porosity mapping: the acoustic images generated honors both well and seismic data. The method comprises two components: simulation of reservoir properties conditional to hard well data using the sequential Gaussian algorithm and integration of 3D seismic amplitudes through an iterative selection of simulated impedance traces.

In practice, the results are found to be sensitive to the quality of the calibration at well locations as well as to the acoustic-petrophysical relation used to derive the final porosity image. The results could be improved by making the convolution and the acceptance/rejection criterion of a simulated trace dependent on the location (by using the specific results of the calibration at the closest well).

The method does not aim at providing any measure of uncertainty, instead it yields a unique optimal image. However, the correlation map of synthetic and actual

[12]The relation $\phi = f(\rho V_p)$ proposed in this paper was empirically derived from simple multilinear regression from well data at the log scale.

$$\phi = 0.6546 - 0.0636\rho V_p\ , \quad with\ R^2 = 60\%.$$

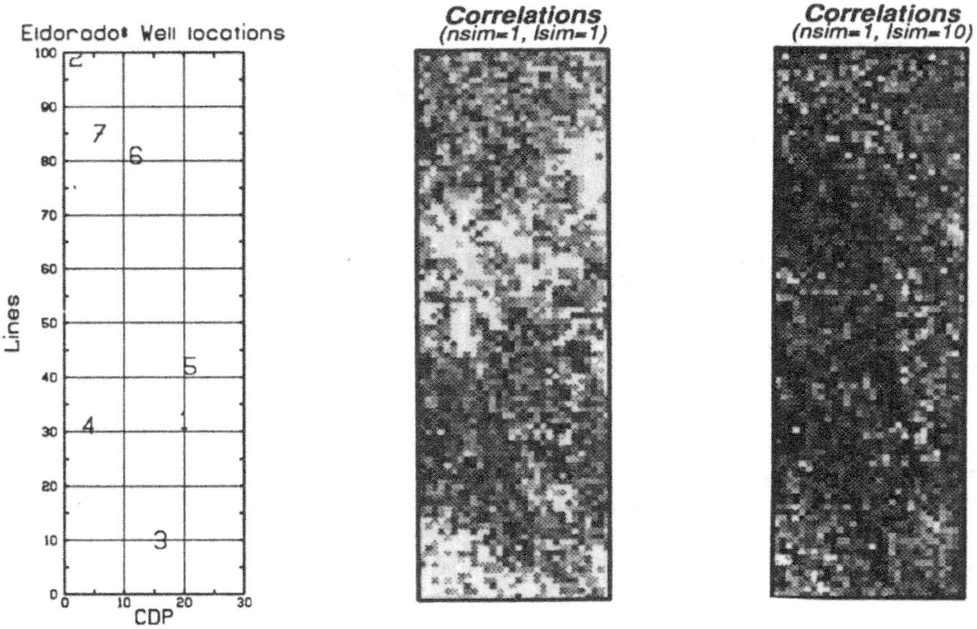

Figure 8: 2D correlation maps: *nsim* and *lsim* set to 1 (no optimization) result in a poor level of similarity between synthetic and actual traces. The darker zones indicate better correlations around the wells. With $lsim = 10$ and $nsim = 1$, the correlation map becomes darker indicating a better match.
Wells are left as white points. Scale: White= no correlation - Black= maximum correlation

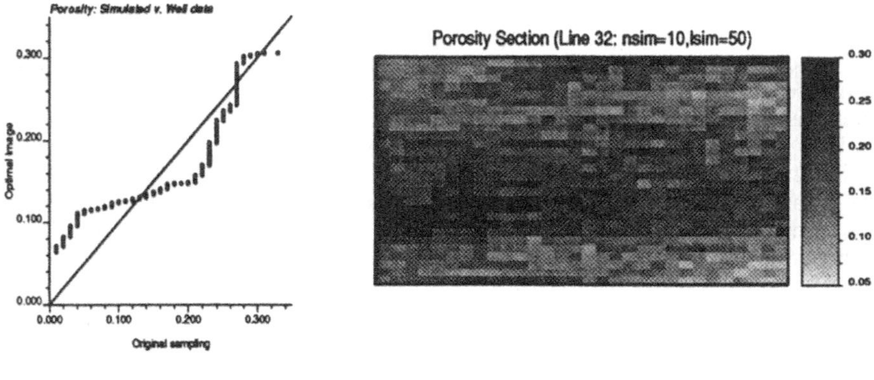

Figure 9: Final porosity image (Line 32 extracted from the 3D optimal realization): the boundaries of the reservoir are indicated by low-porosity zones. The lower west part of the reservoir is featuring very high porosities.

seismic traces provides an indication of uncertainty: areas where high correlations between simulated values and corresponding seismic data are obtained after few iterations give confidence in the optimal image and in the prior calibration; areas where correlations remain poor indicate seismic characters not fully accounted for, or locally inconsistent with the well data.

References

[1] F. Alabert, S. Thadani and A.G. Journel. *An integrated geostatistical/pattern recognition technique for the characterization of reservoir variability.* SEG Annual Meeting, New Orleans, 1987.

[2] F. Alabert. *Stochastic Imaging of Spatial Distributions Using Hard and Soft Information.* Master's thesis, Stanford University, Stanford, CA, 1987.

[3] C. Deutsch. and A.G. Journel. *GSLIB: Geostatistical Software Library and User's guide,* to be published by Oxford University Press, New York, 1992.

[4] P. Doyen. *Porosity from seismic data: A geostatistical approach.* Geophysics, 53 (10), pp 1263-1275, October 1988.

[5] J. Gomez-Hernandez. *A stochastic approach to the simulation of block conductivity fields conditioned upon data measured at the short scale.* PhD thesis, Stanford University, Stanford, CA, 1991.

[6] E. Isaaks and R. Srivastava. *An introduction to applied geostatistics.* Oxford University Press, New York, 1989.

[7] A.G. Journel. *Fundamentals of Geostatistics in Five Lessons.* Short course in Geology, vol.8, AGU publ.,Washington, D.C., 40p., 1989.

[8] A.G. Journel and H. Zhu. *Integrating soft seismic data: Markov-Bayes updating, an alternative to cokriging and traditional regression.* In Report 3, Stanford Center for Reservoir Forecasting, Stanford, CA, 1990.

[9] A. Marechal. *Kriging seismic data in presence of faults.* In G. Verly et al., editors, Geostatistics for natural resources characterization, pages 271-294, Reidel, Dordrecht, Holland, 1984.

[10] W.J. Ostrander, *Plane-wave reflection coefficients for gas sands at non-normal incidence.* Geophysics, 49 (10), pp 1637-1648, October 1984.

[11] S.C. Key and S.C. Smithson. *New approach to seismic-reflection event detection and velocity determination.* Geophysics, 55 (8), pp 1057-1069, August 1990.

[12] H. Zhu. *Modeling Mixture of Spatial Distributions with Integration of Soft Data.* PhD thesis, Stanford University, Stanford, CA, 1991.

A NEW APPROACH FOR PERMEABILITY CLASSES SIMULATION IN UNDERSAMPLED OIL RESERVOIRS

A. Costa e Silva(*), H. Garcia Pereira(**), A. Soares(**), J. Almeida(**), A. Guterres(**)

(*) PARTEX - Companhia Portuguesa de Serviços
(**) CVRM - IST, Av. Rovisco Pais 1096 Lisboa Codex

ABSTRACT

In the characterization of permeabilities in an oil reservoir, two main problems are identified: i) due to shortage of "hard" permeability data, "soft" information provided by other sources should be taken into account. ii) The non-additive nature of the variable does not allow for direct spatial inference.

To overcome jointly these usual difficulties, a new methodology, combining Correspondence Analysis and Multi-phase Simulation, is proposed. With the Correspondence Analysis technique, used as a qualitative regression tool, a grade of membership of each sample to a permeability class is assigned, constituting a soft data set. Based on the categorized hard and soft information, the all set of permeability classes is obtained, via a multi-phase simulation technique, reproducing, in the final maps, the volume proportions of each class and the contiguity architecture of the all set, as revealed by the multi-phase covariances.

The methodology was tested in a case study regarding an extensively recognized oil reservoir.

INTRODUCTION

The classical procedures used in oil industry for permeability estimation in undersampled wells rely usually on empirical relationships between permeability and one or two "explicative" variables, namely porosity and irreducible water saturation (Timur, 1968). In fact, these procedures can not satisfactorily account for more than a restricted portion of the useful range of permeabilities, being useless for predicting the occurrence of extreme values, which are of great importance for understanding the fluids flow in the reservoir, since low permeability zones define flow barriers and high values define preferential paths.

The research work reported in this paper was supported by a grant of JNICT (Junta Nacional de Investigação Científica e Tecnológica), in the framework of the project GEOMETRIC AND STOCHASTIC SIMULATION OF ENERGETIC FLUIDS, developed by CVRM.

On the other hand, classical geostatistical techniques are not straightforward to apply on raw permeability data, given the non-additive nature of the variable. An approach, relying on geological controls for kriging conditioning, was recently developed by Rumhy et al., 1991, to cope with some of the above mentioned problems.

To overcome jointly the usual difficulties stemming both from the shortage of hard permeability data and from non-additivity problems, a new methodology, combining Correspondence Analysis (Benzécri, 1980) and Multi-phase Simulation (Soares,1992a), was developed in this paper.

Under the assumption that expert reservoir engineering knowledge allows the splitting of the experimental permeability range into useful classes (being the extreme classes significant for production planning purposes), Correspondence Analysis, used as a "qualitative regression" tool on the available quantitative and qualitative "explicative" variables, provides soft information on the grade of membership of each formation interval to the above defined permeability classes, for the undersampled wells. This soft information is then merged with hard data, categorized by an indicator into the same classes, to be used as an input for the multi-phase simulation technique, which can cope with non-additivity, at this stage, since it deals only with frequency data in ordered categories.

METHODOLOGY

Given a set of wells, some of which are undersampled for permeability but contain other explicative variables, the objective of the proposed methodology is to produce a simulated image of the reservoir architecture, reflecting the spatial distribution of permeability classes and taking into account, jointly, soft data in the undersampled wells and hard permeability values.

This approach includes two main steps - Qualitative Regression and Multi-phase Simulation -, which are described in the sequel.

. Qualitative Regression

Even though permeability is usually linked to other available explicative variables (porosity, facies, lithologies, fluids saturation), it is not straightforward to derive some form of multiple regression permitting to calculate the missing permeability values. In fact, apart from the non-additivity problem, also the set of explicative variables is usually non-homogeneous, containing numerical and categorical data. Assuming that the non-additivity issue is approached by encoding permeability into significant classes, also the non-homogeneity problem can be treated by the same procedure, categorizing all explicative variables and performing a qualitative regression of permeability classes on the explicative variables categories.

The qualitative regression technique relies on the supplementary projection properties of Correspondence Analysis (Cazes, 1982) and includes the following steps :

i) Select an uncored well where the permeability is to be estimated and categorize their q explicative variables into p classes. A Boolean matrix S (n1 x p) is obtained by taking each formation interval (denoted in the sequel by the sample x=1,....,n1) and assigning it to a certain class of each explicative variable.

ii) From the set of estimator wells (those that contain hard permeability values), choose a sub-set of estimators, according to the criterion of minimizing the χ^2 distance between the explicative variables' categories, calculated for the contingency table crossing the uncored well with the estimator wells. Once selected the estimator wells, with a total of n cores, two Boolean matrices are constructed : the matrix L (n x p) contains the indicator values referring to the categorized explicative variables and the matrix P (n x N) contains indicator values referring to the N permeability classes.

iii) Once defined the matrices L, P and S, project as supplementary variables the columns of P and, as supplementary individuals, the rows of S, onto the p-q engeinvectors given by the Correspondence Analysis of matrix L. Those projections are arranged under the form of two new matrices PSUP (N x p-q) and SSUP (n1 x p-q), which give the coordinates of permeability classes and samples, respectively, in the same space of the factorial axes.

iv) Calculate de distance d(x,i) from each sample to each permeability class in the space of the factorial axes using matrices SSUP and PSUP and derive the grade of membership gm(x,i) of sample x to the permeability class i :

$$gm(x,i) = [1/(N-1)]\,[1 - d(x,i)/d(x)] \qquad (1)$$

where :

N is the number of permeability classes

d(x,i) is the distance from sample x to permeability class i

$$d(x) = \sum_i d(x,i)$$

.<u>Multi-phase</u> <u>Simulation</u>

To simulate multi-phase structures derived from a categorical variable - such as a set of lithofacies, a set of classes of non-numerical variables, etc.- Suro-Perez (1991) proposed the Sequential Indicator Simulation, using the cokriging of indicators corresponding to the set of phases. This method requires the knowledge of the covariance models of all individual phases as well as the cross-covariances between the phases. This is the main limitation of the method, in particular in petroleum applications where the most common situations, characterized by the scarcity of data, do not allow inferring even the covariances of all individual phases.

In this paper it is proposed a methodology of simulation of indicator vectors and direct conditioning to hard and soft data by a multi-phase kriging technique (Soares, 1992,b) which requires only the inference of one global model - the multi-phase covariance.

The following steps resume the method:

i) - The available information is composed, in this particular case, by hard indicator vectors $I_i(x)$, associated to each phase -class of permeability- X_i :

$$I_i(x) \quad \begin{matrix} 1 \text{ if } x \in X_i \\ 0 \text{ if } x \in X_j \end{matrix} \qquad \text{for all phases } i \neq j$$

and by soft grades of membership of x to the class of permeability X_i, given by gm(x,i), calculated by equation (1). In the sequel, both hard indicator and soft grade of membership are denoted by $I_i(x)$.

Multi- phase covariances - probability of two points, separated by a distance vector h, to belong to the same phase X_i , i=1,N- are calculated:

$$C(h) = E\{ \sum_i [I_i(x) \cdot I_i (x+h)] \} \qquad \text{or the corresponding variogram}$$

$$\gamma (h) = 1/2 \, E\{ \sum_i [I_i(x) - I_i (x+h)]^2 \} \tag{2}$$

ii) - After the simulation of a Gaussian variable y(x) - by using the turning bands method (Journel and Huigbregths, 1978)- this is truncated by the N Gaussian cutoffs equivalent to the cumulative experimental indicator mean of the phases (Journel and Isaaks,1984, Matheron et al., 1987):

$$Is_i(x) = \begin{matrix} 1 \text{ if } yc_{i-1} < y(x) <= yc_i \\ 0 \text{ otherwise} \end{matrix} \tag{3}$$

The corresponding Gaussian cutoff are :

$$yc_i = G^{-1} (1-m_i)$$

with the cumulative experimental mean $m_i = \sum_{j=1}^{i} m_j$, being m_j the experimental mean of phase xj and G the standard Gaussian probability distribution function.

It is worth noting that, in the simulation of the Gaussian variable y(x) , it is required a variogram model, correspondent to the multi-phase variogram, which can be obtained by the following methods : an hermitian expansion of the Gaussian cumulative probability function (Matheron et all,1987) or by an heuristic approximation proposed by Journel et all, 1990, that can be easily generalized for the multi-phase covariances, by interpreting it as a sum of individual indicator covariances. In both approaches the relation between the multi-phase variogram and the Gaussian variogram can take into account the several thresholds yc_i, but is still a one to one relation, which means that to one Gaussian model corresponds only one multi-phase model.

iii) - In the conditioning step, the multi-phase kriging is applied to hard indicator data and soft data. The conditioned map $Ip_i(x)$ is calculated using the classical relation:

$$Ip_i(x)=I^*_i(x) + \epsilon\ (x) \tag{4}$$

The conditional simulated map Ip_i is the estimated map I^*_i plus a deviation $\epsilon(x)$:

$$\epsilon(x)=[\ Is_i(x)\ -Is_i^*(x)]$$

with the two multi-phase kriging estimates :

$$I^*_i(x) = \sum_\alpha \lambda_\alpha\ I_i\ (x_\alpha)$$

$$Is_i^*(x) = \sum_\alpha \lambda_\alpha\ Is_i\ (x_\alpha) \qquad i=1 \text{ to N phases}$$

iv) The final step consists on the calculation of the morphology of the simulated multi-phase set. The resulting conditioned values $Ip_i(x)$, have the meaning of the probability of x to belong to each phase X_i. The objective of this step is to allocate the points x to each phase Xi, based on the previous probability vectors, such that the volume proportions of each phase, as inferred from the experimental samples, are honored in the final multi-phase map. Soares (1992,b) proposed the following algorithm consisting on two steps: i) the vectors $Ip_i(x)$ are sorted into descending order for each individual phase, ii) each point x is allocated to the phase i corresponding to the highest value $Ip_i(x)$ and such that the proportion of these points identifies the experimental volume proportion of X_i. Note that, following the decreasing order, one point x that is allocated to the phase i cannot be part of phase j if $Ip_i(x)>Ip_j(x)$ even if the volume proportion of phase j is not reached yet.

CASE STUDY

. Statement of the problem

Given the oil field depicted in Fig. 1 (Pereira et al, 1990), it is required to simulate the spatial behavior of permeability in a selected zone A, based on core data available in a set of wells, in which there is a group where no permeability values were measured (denoted soft in Fig. 1). This set of wells - 1, 8, 19, 25, 37, 52, - contains information on other variables, linked, to a certain extent, with permeability.

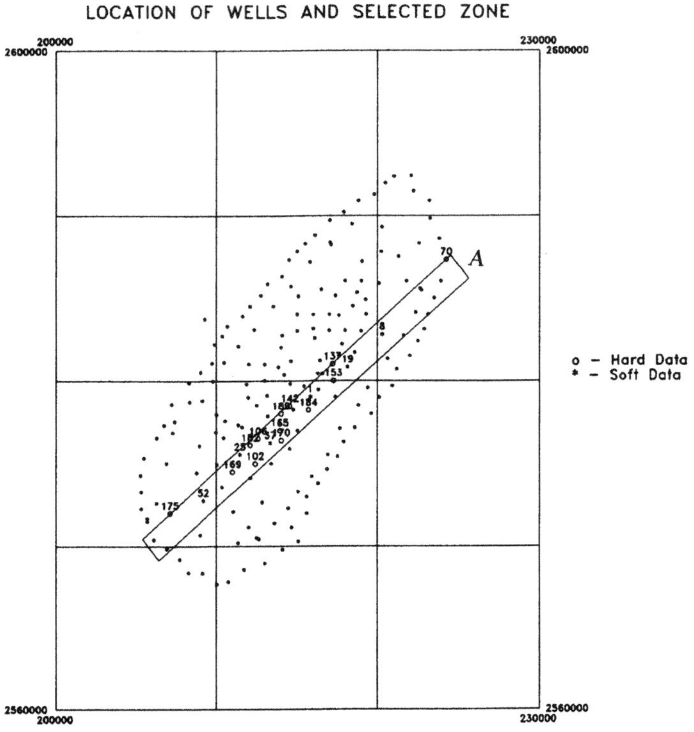

Fig. 1 - Location of wells in the oil field and selected zone A

The soft data on permeability can be derived from the available explicative variables, which are non-homogeneous : some are expressed by numerical values (Depth, porosity, water and oil saturations) and other are nominal (lithologies - limestone, clay, dolomite). Provided that all auxiliary variables are categorized, by splitting numerical data into classes, it is possible to apply Correspondence Analysis, producing a file containing the grade of membership of each formation interval (sample) to a certain phase (permeability class), for the undersampled wells. Merging this file with hard permeability indicators for the remaining wells, the required input for multi-phase simulation is obtained.

. Estimation of soft permeability values

According to the methodology previously outlined, the procedure for the estimation of soft permeability values in 5 classes (0/10, 10/30, 30/60, 60/150, >150 md) was applied to each one of the undersampled wells, as detailed in the sequel.

For a given well 1, the set of wells containing permeability values was sorted by increasing order of the χ^2 distance measured in the contingency table relating the categories of explicative variables in well 1 and in each one of the possible estimators (cf. Fig.2). Hence, a sub-set of 3 wells was selected as exhibiting the maximum similarity with the well to be estimated (cf Fig. 3 - wells 10, 193, 137).

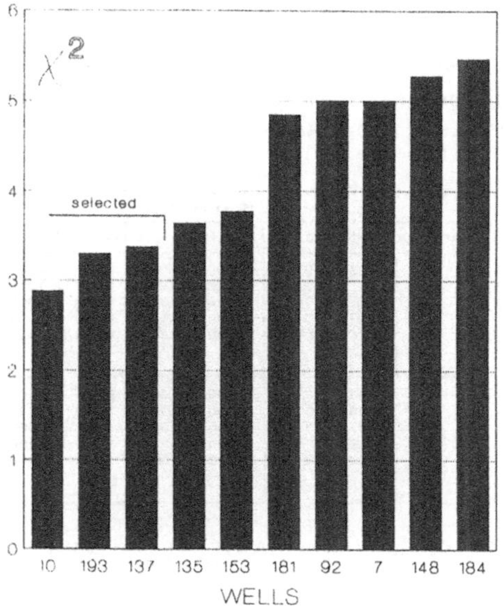

Fig. 2 - Sorting by similarity with well 1

In Fig. 3 it is shown a comparison of the explicative variables histogram for the well 1 and for the set of estimator wells (10+137+193), providing an average χ^2 distance of 3.19.

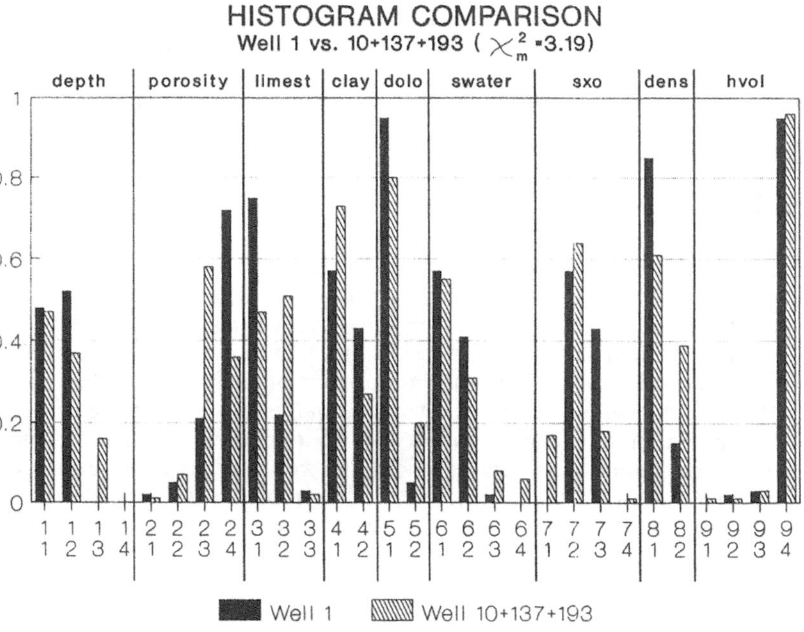

Fig. 3 - Histogram comparison for the explicative variables (well 1 vs. 10+137+139)

The same methodology described as qualitative regression was used for wells 8, 19, 25, 37, 52, producing a matrix containing the grade of membership to the permeability classes, by applying equation (1). This matrix, coupled with the sharp indicator data for the wells where permeability was measured, constitutes the input, sketched in Fig. 4, for the multi-phase simulation technique.

		PERMEABILITY CLASSES (md)				
DATA TYPE	DEPTH	0 - 10	10 - 30	30 - 60	60 - 150	› 150
SOFT	7486	0.000	0.000	0.000	0.500	0.500
	7487	0.000	0.000	0.290	0.710	0.000
	7488	0.000	0.000	0.320	0.680	0.000
	7489	0.000	0.000	0.590	0.410	0.000
	7490	0.000	0.000	0.590	0.410	0.000
	⋮					
HARD	7445	0.000	0.000	1.000	0.000	0.000
	7446	0.000	0.000	0.000	1.000	0.000
	7447	0.000	0.000	0.000	1.000	0.000
	7448	0.000	0.000	0.000	1.000	0.000
	7449	0.000	0.000	0.000	1.000	0.000
	⋮					

Fig. 4 - Format of the input matrix for the multi-phase simulation

.Structural Analysis: multi-phase variograms

Since the morphology of the dome representing the oil reservoir is extremely regular, the original Cartesian coordinates of the samples were previously transformed into stratigraphic coordinates according to the geological stratification (Journel,1989), before computing the variograms.

The multi-phase variograms were calculated using (2). Fig. 5 shows the horizontal omni-directional multi-phase variogram, fitted by two exponential structures with ranges: a1=600m. and a2=4000 m.. The corresponding two structures in vertical direction are a1=6 vertical units and a2=45 vertical units, shown in Fig. 6.

It is worth noting that only for two classes, the calculation of individual phase variograms was succeeded, which means that any method that calls for the covariance models of each individual phase (or cross-covariance) is not applicable.

.Simulation of Classes of Permeability

To calculate the Gaussian variogram models, the approximation proposed by Journel and Posa, 1990, was used: Each structure of the multi-phase variogram corresponds to a Gaussian spherical variogram with a range *a* and is approximated by a sum of two exponential structures with practical ranges *a* and *3a* and the sill-parameters as a function of the volume proportions of each phase. After simulating the standard Gaussian values , using the turning bands method , in selected area A, they were truncated in five classes, according to equations (3), reproducing the indicator simulated maps. The simulated maps were conditioned directly to the experimental coded vectors using (4), and the resulting probability maps were transformed in the multi-phase images as described in point (iv) of multi-phase simulation methodology.

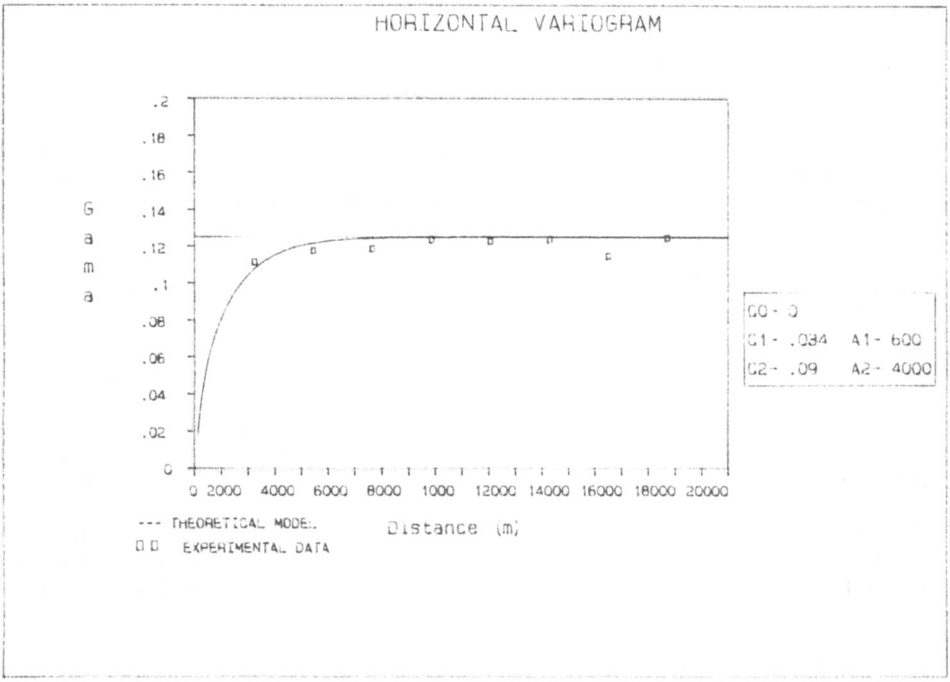

Fig. 5 Horizontal omni-direccional experimental variogram and theoretical model

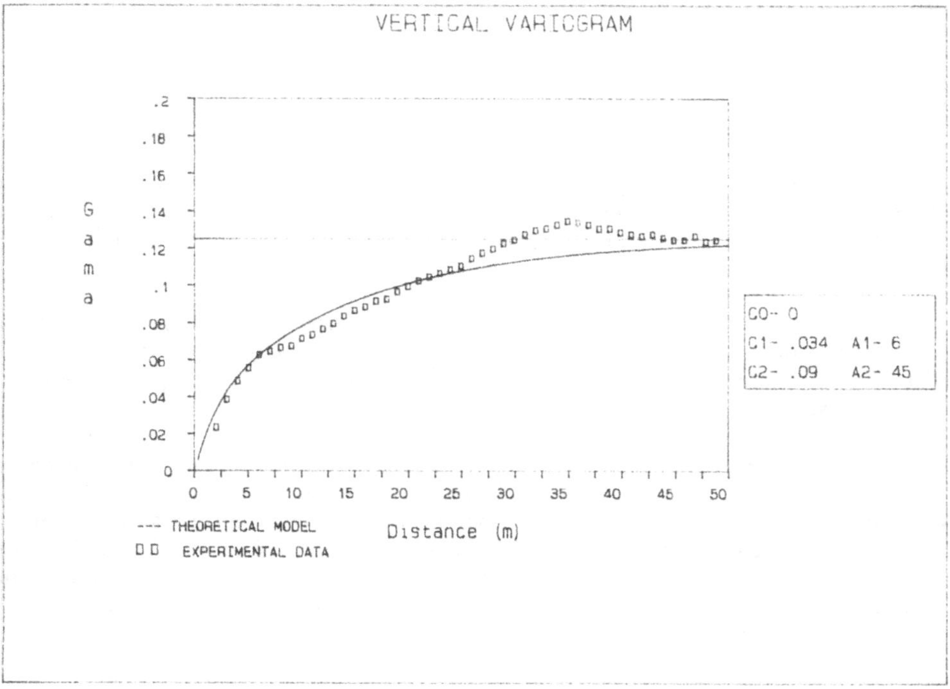

Fig. 6 Vertical experimental variogram and theoretical model

.Conditional Simulated Images

The spatial dispersion of the classes of permeability can be visualized in the conditional simulated images of Fig. 7 showing a sequence of 4 cross sections. The results allow the visualization of the internal architecture of the reservoir, specially the distribution of the extreme K classes that condition the fluids flow. Two high permeability zones -preferential paths for the the fluids movement, -are visible in the central part and in one flank. A potential fluid barrier-a low permeability class- can also be visualized in the other flank.

Fig 8 shows in detail the first cross-section with the coordinates transformed to the original dome geometry.

.Some remarks :

- In the area A the multi-phase variograms are satisfactorily reproduced in the final simulated images (Fig 9) . To transform the indicator variograms into Gaussian variograms some practical approximations were used (Journel and Posa, 1990).

- The final simulated multi-phase image fits the architecture of classes of permeability: due to the ordered truncation after the Gaussian simulation, they generally follow the ordered sequence from the low to the highest class, which is a very convenient effect for this application.

- The conditioning effect of the soft information depends on the grade of membership of each class: the soft information is more conditioning as it is closed to hard indicator data and, vice-versa, the conditioning effect is less evident as the degree of membership to the classes does not differentiate them clearly.

CONCLUSIONS

. The application of the proposed methodology to a case study provided simulated images of the internal architecture of the reservoir, specially for the extreme permeability classes which can be visualized as potential paths or barriers for the fluids flow.

. For udersampled wells in permeability, it is possible to derive a grade of membership to permeability classes by using correspondence analysis as a qualitative regression technique, providing soft information to be inputted into the multi-phase simulation technique.

. The Multi-phase simulation is a method that allows to simulate categorical variables such as lithological phases, classes of permeability (or any other non-addictive variable), reproducing the global multi-phase variogram and individual volume proportions of each phase. It is an adequate method for petroleum applications where, normally, the scarcity of data does not allow the individual covariances and cross covariances inference, but a global multi-phase covariance.

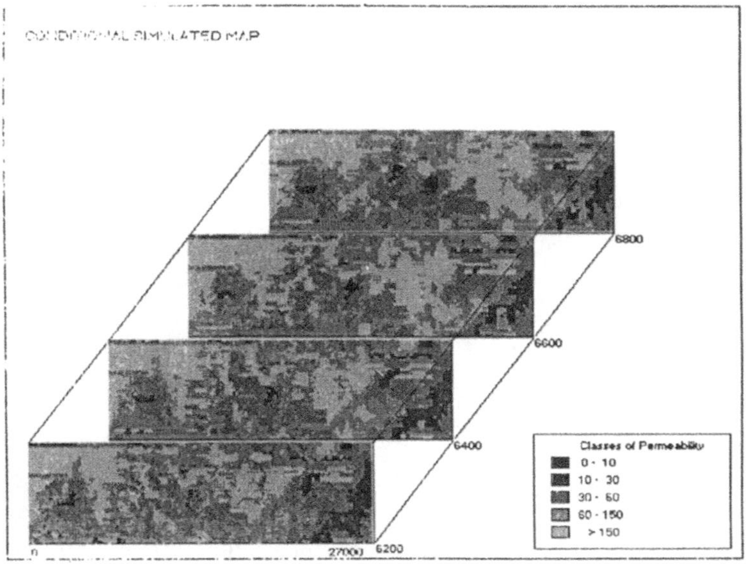

Fig. 7 Four sequenced cross sessions in the area A with simulated classes of permeability

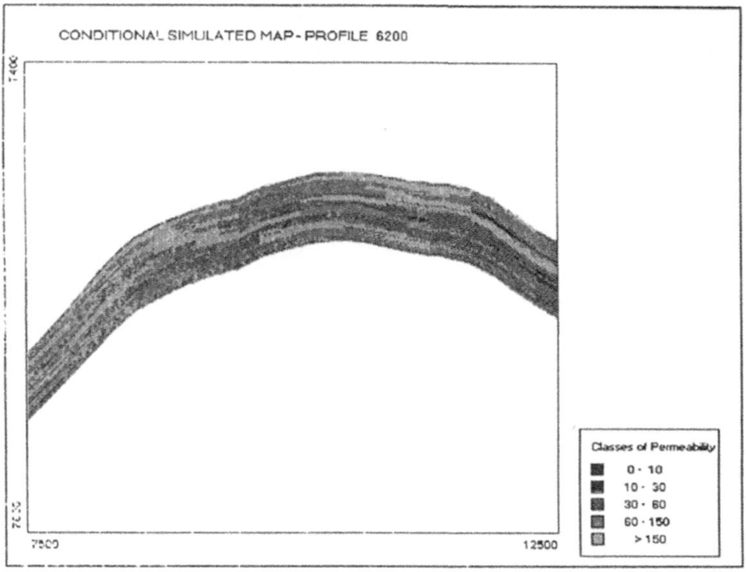

Fig. 8 Detail of the first cross- session of Area A where the coordinates were transformed back to the dome geometry

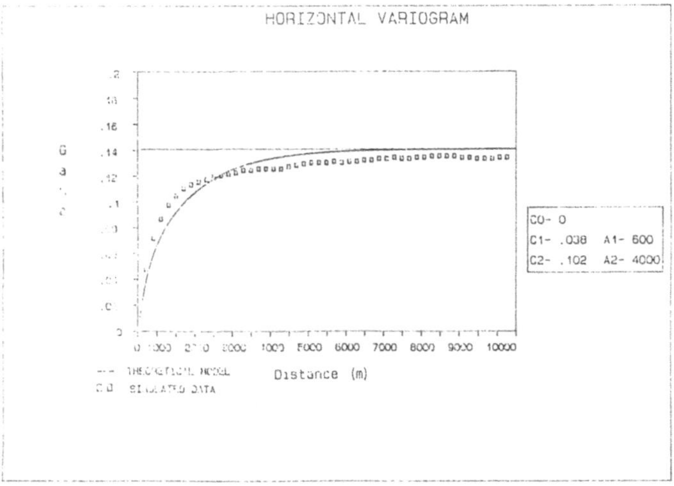

Fig 9 . Horizontal multi-phase variogram of simulated permeability classes and the theoretical model.

REFERENCES

Benzécri, 1980 : Pratique de l'analyse des données, Dunod, 424 p.

Cases, P., 1982 : Note sur les éléments supplémentaires en analyse des correspondences, *Les cahiers de l'analyse des données,* vol VI, n. 1, p. 9-23

Journel, A., Huijbregts, C., 1978, Mining Geostatistics, Academic Press, 600 p.

Journel ,A., Isaaks, E., 1984, " Conditional Indicator Simulation : Application to a Saskatchewam uranium deposit", *Mathematical Geology,* Vol. 16, No. 7, p. 658- 718.

Journel, A.G., Gomez-Hernandez J.J., 1989, Stochastic Imaging of the Wilmington Clastic Sequence. SPE paper # 19857.

Journel, A.G., Donato Posa, 1990, Characteristic Behavior and Order Relations for Indicator Variograms, *Mathematical Geology,* Vol. 22, No. 8, p. 1011- 1025.

Matheron,G., Beucher H., Fouquet C., Galli A., Guerrilot D., Ravenne C., 1987, Conditional Simulation of the Geometry of Fluvio-Deltaic Reservoirs. SPE paper # 16753.

Pereira, H., Silva, A., Soares, A., Ribeiro, L., Carvalho, J., 1990 : Improving reservoir description by using geostatistical and multivatiate data analysis techniques, *Mathematical Geology,* Vol. 22, No. 8, p. 879- 913.

Rumhi, M. , Archer, J., Daltaban, S., 1991 : A synergistic approach to characterization of reservoir permeability : a conditional kriging method, SPE paper 21446, p. 915-924

Soares A., 1992 a, : Multi-Phase Simulation of a Set of Lithoclasses. Internal report of CVRM.

Soares 1992,b, : Geostatistical Estimation of Multi-Phase Structures. *Mathematical Geology,* Vol. 24, No. 2 .

Suro-Perez, ., Journel A., 1990: Simulation of Lithofacies. Proceedings of 2nd European Conference on the Mathematics of Oil Recovery. pp- 3-10, Techn. Pub., Paris.

Timur, A., 1968 : An investigation of permeability, porosity and residual water saturation relashionships, Proc. SPWLA Ninth Annual Logging Symposium, New Orleans, Paper K

INTRODUCING MORE GEOLOGY IN STOCHASTIC RESERVOIR MODELLING

OLIVIER DUBRULE, ELF AQUITAINE PRODUCTION,
Avenue Larribau, 64018 Pau Cedex, FRANCE

ABSTRACT

More and more often, detailed probabilistic models of petrophysical parameters are used as input to dynamic simulation of flow in petroleum reservoirs. For example, variogram-based geostatistical techniques honour well data, together with elementary statistical parameters, such as proportions of various facies or spatial correlations. Other techniques have also been developed, which take into account other kinds of geological constraints, such as the geometry and relative position of geological units. New quantitative reservoir characterisation approaches now compromise between the advantages of various stochastic techniques, by applying the most appropriate method at each modelling stage. Multi-step approaches, where the stochastic modelling of genetic units is followed by the representation of internal lithological variations, and finally, of permeability distributions, are gaining popularity. These approaches and their geological assumptions are reviewed, together with the current industry efforts towards gathering more quantitative geological data to use as input to these stochastic models.

INTRODUCTION

A realistic modelling of heterogeneities is prerequisite to a reliable forecast of flow behaviour during production.
Stochastic reservoir modelling techniques present two characteristics (Omre et al, 1988):
1. they make the generation of realistic three-dimensional heterogeneity models possible.
2. not one single model, but a family of "possible" models can be generated, which all honour the available data. By running flow simulation on a number of these models, the impact of geological uncertainty on flow prediction can be evaluated.

A number of reviews of probabilistic approaches for heterogeneity modelling have been published (Farmer, 1988, Dubrule, 1989, Haldorsen and Damsleth, 1990, Ripley, 1992). The goal of the present review is to emphasize the geological assumptions used by various approaches.

The current industry interest in stochastic reservoir modelling has resulted in a cross-fertilisation between techniques of various origin. Variogram-based methods were originally used for mining applications, but many interesting applications have now resulted from their use in reservoir modelling. Another group of techniques, called "object-based" in a broad sense (Dubrule, 1989) originate from the developments of stochastic geometry (Stoyan et al, 1987). The third group of approaches, very promising but still presenting the smallest number of applications, are based on multi-point statistics or other generalisations of the variogram function. Among this last group, Markovian methods are adapted from statistical physics or image analysis.

After a few years of development and testing, the complementarity between these various approaches is becoming clearer. Each technique has its use, depending on the phase of development of the oil field, on the geological depositional environment, and on the parameters that are modelled.

THE MULTI-STEP MODELLING APPROACH

With a multi-step approach, modelling of large-scale heterogeneities is followed by the representation of smaller scale variations (Fig. 1).

Step 1: Definition of the layering

The layering often reflects significant changes in depositional environment. Such changes will affect the distribution pattern of geological heterogeneities in space, the statistical parameters used to characterise heterogeneities, or the stochastic approach used. For instance, the modelling approach will not be the same in a "layer-cake" reservoir formation where sand continuity is good, or in a "labyrinth" reservoir where sand arrangement is complex and well-to well correlation is difficult (Weber and Van Geuns, 1990).

Step 2: Modelling genetic units

Identification of genetic units, and correlation of these units between wells, can significantly contribute to the reduction of uncertainties at unsampled locations. Genetic units are often associated with permeability heterogeneities, and their delineation can be a major step towards a reliable flow prediction. Fluvial channels, mouth bars deposited at a delta front, or turbidite lobes and channels are examples of genetic units. Delineation of these units appears to be most important in labyrinth-type reservoirs (Weber and Van Geuns, 1990). In other types of reservoirs presenting a layer-cake arrangement (i.e. braided channels), it may be more relevant to model the distribution of units associated with permeability barriers (shales, cemented intervals).

Step 3: Modelling lithological facies variations

Lithology variations can be significant within genetic units. See for instance the work of the HERESIM group (Rudkiewicz et al, 1990), who has modelled the distribution of 15 different lithologies (such as sandstone, argillaceous sandstone, clay...) within fluvial units of the Scalby Formation (Ravenscar Group) in England. Lithology variations can have a significant impact on flow, especially if they occur at large scale.

Step 4: Distributing petrophysical properties in space

To the reservoir engineer, the final objective of stochastic modelling is to reliably represent the spatial distribution of petrophysical properties, for use in flow simulation: modelling of genetic units and lithological variations are only steps towards this goal. Once the depositional/lithological framework has been defined, permeability variations can be associated with each lithology class.

Examples of applications of each of the four above steps are presented below. In some of these applications, all four steps have been combined in sequence, whereas other approaches have skipped one or two steps (see Table 1).

GENETIC UNIT MODELLING

In the following, it will be assumed that the reservoir has been subdivided in layers (step 1). For probabilistic modelling, a proper layering scheme is of crucial importance, and should identify intervals that are "statistically homogeneous": the term "homogeneous heterogeneity" is another expression of statistical homogeneity: it simply means that a few statistical parameters (size of sand bodies, variogram, fractal dimension, Markov transition probabilities...) are representative of the heterogeneity pattern and can be considered as the statistical "signature" of each layer. Boundaries between layers are often associated with changes in depositional environment.

Object-based techniques

Object-based methods, which represent genetic units as objects distributed in space, are the most widely used at this stage of modelling. For each unit, a variety of input parameters, such as shape, size (through width/thickness ratios) or orientation can be taken into account. Constraints on the relative position of units in space can also be honoured.

Examples of applications exist for the modelling of sand and shales in the Hassi-Messaoud field, Algeria (Delhomme and Giannesini, 1979), the distribution of channels within fluvial reservoirs (Clemetsen et al, 1990, Henriquez et al, 1990, Hirst et al, 1992), and the distribution of crevasses and channels within the fluvio-deltaic Ness formation (Gundeso and Egeland, 1990). A similar approach has also been applied to different types of units in a North Sea reservoir of coastal marine origin (L.M. Fält et al, 1991), in a North Sea fan delta complex (Damsleth et al, 1990) and in a coastal plain sequence of the Cano Limon field, Colombia (Dubrule et al, 1990). A large number of papers have also documented the use of object-based methods for representing the spatial distribution of small ("stochastic") shales (Haldorsen et al, 1987, Hoiberg et al, 1990).

Indicator Simulation

This approach quantitatively characterises the size of various units by means of the indicator variogram. Typically, the indicator variogram is not the best tool for characterising the geometry of genetic units: in a recent paper, Journel and Deutsch (1992) have shown that very different heterogeneity "images" could have the same indicator variogram. However, the flexibility of the simulation algorithms used with indicator simulation, and their easy conditioning by various types of data, has justified the use of this approach for genetic units modelling.
Examples exist for turbidite settings (Alabert and Massonnat, 1990) or for modelling of shales (Journel and Gomez-Hernandez, 1989). MacDonald et al, 1991, have also used it in conjunction with object-based techniques for representing various facies in shallow marine and coastal plain environments in the Gullfaks field.

Proportion curves

The HERESIM group (Rudkiewicz et al, 1990) has proposed an approach, illustrated on cases of reconstructing lithological variations along outcropping cliffs from a limited number of data, which is based on proportion curves. A vertical proportion curve gives, as a function of depth, the proportion of various lithologies. Strong variations along the proportion curves can help characterize boundaries between genetic units.

Two-Dimensional Markov Fields

This approach, based on 2D Markov fields, has been proposed to describe thickness variations of large-scale permeability barriers. Statistical relationships between thicknesses measured at various group of points are characterised by a Markov formalism, representing probabilities of transition from one thickness value to another. Another original feature of this approach is that it can be used to take into account probabilities of correlation of genetic bodies between wells. Hoiberg et al, 1990 have applied this approach to shales, whereas Omre et al, 1990, have applied it to calcite cemented layers.

LITHOFACIES MODELLING

Lithofacies modelling may consist for instance of representing the spatial distribution of shale, sand, and shaly sandstone. Most published applications have used indicator simulation for this purpose. In some examples, lithofacies variations are modelled within genetic units, in others, genetic unit modelling is skipped.

Indicator Simulation

Indicator simulation was used by Alabert and Massonnat, 1990, for modelling reservoir quality variations within turbiditic lobes and channels. The HERESIM Group also used it in conjunction with proportion curves for generating lithological variations within channels in a fluvio-deltaic environment (Rudkiewicz et al, 1990). The Stanford School has proposed a wide range of applications of indicator simulation for lithofacies modelling. One example is the simulation of the sand/shale distribution in the Wilmington clastic sequence (Journel and Gomez-Hernandez, 1989).

Markov Random Fields and Multiple-Point Statistics

Markov Random Fields have experienced successful applications in such fields as physics or image analysis. A number of authors have proposed to use this approach for petroleum applications (see for instance Farmer, 1988, or Ripley, 1992). Markov models use multi-point statistics, instead of the two-point statistics used by indicator variogram simulations. The hope is that, using Markov random fields, more constraints on the relative position of various lithofacies will be satisfied than with indicator simulation. The formalism is well developed, but problems of statistical inference of model parameters have reduced the number of applications so far.

Other interesting attempts have been made by Deutsch, 1992 and Guardiano and Srivastava, 1992, to constrain geostatistical simulations by multiple-point statistics rather than by indicator variograms. It is still too early to evaluate the relations between these various techniques, and their potential for petroleum applications.

Two-Point Histogram

This approach, proposed by Farmer, 1992, is another attempt to characterise facies transitions by means of a better tool than the indicator variogram: instead of differences between values of various lithofacies, geological architecture is characterised by a function $N_{ij}(h)$ which measures, for each vector h, the probability of transition from facies i to facies j. Doyen, 1990, has applied this approach to the modelling of a sand-shale distribution using high-resolution seismic data as a constraint.

PETROPHYSICAL VARIATIONS MODELLING

For these applications, the variogram-based geostatistical approach is by far the most popular. Damsleth et al, 1990, Alabert and Massonnat, 1990 have used it for modelling internal permeability variations within genetic units. Ringrose et al, 1992, have applied it to the modelling of small-scale heterogeneities, with the objective of testing the influence of these heterogeneities on flow.
Fractal models can be regarded as a special case (they correspond to power-law variograms) of geostatistical conditional simulation. A wide number of applications of fractal models exist. Hewett, 1986, or Perez and Chopra, 1991, have applied these models to the direct modelling of permeability or porosity variations, without the intermediate step of genetic units or lithofacies modelling. Lucia and Fogg, 1990, have also used a direct approach, modelling permeability variations in a carbonate reservoir using a standard variogram-based conditional simulation.

SIMULATION ALGORITHMS

The most widely used simulation algorithms can be subdivided into two main classes: those, such as the variogram-based ones, which directly construct a realisation satisfying the input statistical parameters and those which, starting from an initial distribution, modify it sequentially until it satisfies these input parameters.

Direct Approaches

One of the reasons for the success of indicator simulation techniques is the speed of the simulation algorithms: the approach developed by the HERESIM Group (Rudkiewicz et al, 1990) or the Sequential Indicator Simulation (SIS) method proposed by the Stanford School (Alabert, 1987) are both very fast.
The HERESIM technique consists of generating a gaussian random function, which is truncated into a number of intervals using calculated treshold values: each indicator value is then associated with one of these intervals. The simulation of the gaussian random function is obtained by adding a random error to a kriged estimate. A number of limiting assumptions on the variogram model simplify the kriging system.
With the SIS algorithm, the probability of occurence of each indicator value at any given location is first estimated, using Indicator Kriging as an approximation.

Then, using this probability as input, the indicator value is randomly drawn by Monte-Carlo simulation. This approach is used sequentially for each location.

The method of successive random additions (Hewett, 1986) is also a direct method commonly used with fractal simulations.

Iterative Approaches

Iterative approaches have been more recently introduced than direct approaches, but have proved very powerful, because of the experience already acquired with these algorithms in fields such as physics or image analysis. Most iterative approaches are inspired from Metropolis algorithm (Ripley, 1987).

For instance, simulations of marked-point processes based on death/birth algorithms (Ripley, 1987, Stoyan et al, 1987) have brought a lot of flexibility to the simulation of object-based models. Hoiberg et al, 1990 give a detailed explanation of the method used to model the spatial distribution of small shales using death/birth algorithms. Note that Clemetsen et al, 1990, use death/birth algorithms for representing channel-belt distributions, but propose a direct sequential approach for generating channel distribution within a channel-belt.

However, in spite of the new modelling algorithms which are proposed, conditioning by well data still appears to pose problems with object-based techniques when the size of genetic units becomes as large as the well spacing.

The use of iterative algorithms is not limited to object-based methods. Applications which are very similar in philosophy to birth/death processes, exist for models based on two-point histograms or Markov random fields: starting from an initial realisation, locations are selected at random from this realisation, and their value is changed. If this change improves the optimality criterion, the change is accepted, otherwise it is rejected. This approach has been used by Doyen, 1990. Farmer, 1992 also proposes a technique based on iterative modification of an initial model. Instead of changing pixels one at a time, this approach sequentially interchanges values at randomly selected pairs of pixels. A simulated annealing approach is used to determine an appropriate sequence of transpositions. Other iterative algorithms, similar in philosophy to the ones above, have been proposed. The Gibbs sampler (Ripley, 1987) has been applied by Fält et al, 1991, whereas Hoiberg et al, 1990 use Flinn's spin exchange approach for generating large shales. The latter approach works for genetic units that correlate between wells, and probabilities of correlations can be taken into account.

DERIVING LATERAL CONTINUITY INFORMATION FROM VARIOUS SOURCES

The above shows that a variety of techniques exist for stochastic modelling. The choice of which approach to use, and of the input statistical parameters (variograms, size of genetic units, Markov transition probabilities..), should be such that a geologically realistic distribution of heterogeneities is obtained.
In the case of mature reservoirs, the number of wells may be sufficient to infer lateral continuity information from the wells. For instance, Journel and Gomez-Hernandez, 1989 or Alabert and Massonnat, 1990 have been able to calculate horizontal indicator variograms on mature fields. However, in most cases, the well spacing is too large to provide any useful statistical information about lateral continuity. Therefore, other means of gathering quantitative information must be used.
The most common approach is to derive quantitative lateral information from analog outcrops (a large number of publications is available. See Bryant and Flint, 1992, for a review). Such outcrop studies should clearly identify, from the beginning, the scale of variation that is quantified: quantitative information about the lateral extent of genetic units will usually require larger outcrops than quantification of lithology variation within units (Fig. 2). Outcrop studies of the HERESIM group (Rudkiewicz et al, 1990) have focussed on this latter objective. At a smaller scale, petrophysical plug measurements and minipermeameater data will often be used to associate permeability data with various lithologies (Step 4, see for instance Tyler et al, 1991).
In the absence of any geological information about lateral continuity, results from pressure tests have proved to be useful, although great care should be taken in analysing them. Alabert and Massonnat, 1989 have shown how well tests could provide information about the width of turbiditic sand bodies.
Horizontal wells, the number of which is rapidly increasing, are a very promising source of information, since they provide information on lateral continuity at the metric scale within the reservoir itself (no need to use quantitative information from geological "analogues"). Perez and Chopra, 1991 have used horizontal wells to determine horizontal fractal dimension in a braided streams complex.
Another promising avenue is the use of process (or genetic) simulations, where the depositional process is modelled in the computer using a deterministic approach, and where the result of this simulation is used to infer statistical patterns of variability, to be provided as input to stochastic simulations. Teztlaff, 1991 runs deterministic simulations to infer two-point histograms, Clemetsen et al,

1990 use outputs from deterministic models (Bridge and Leeder, 1979) to constraint the relative position of sand bodies in their simulation. A new trend of stochastic models also incorporates constraints mimicking the process that generated the sand bodies (Gundeso and Egeland, 1990, Henriquez et al, 1990, Hirst et al, 1992, Hu et al, 1992).
Also worth-mentioning is the direct use by Falt et al, 1991 in the simulation process of vertical cross-sections drawn by the geologist. These cross-sections are both used as data in the modelling and as a basis for calculating statistical parameters characterising the geology.

USING INDIRECT MEASUREMENTS TO CONSTRAIN THE MODEL

Lateral continuity information is quantified through input parameters of the stochastic model (variograms, width/thickness ratios..). However, together with well data, seismic and dynamic data are most of the time available. In favourable cases, these data can provide information about the spatial distribution of heterogeneities. One of the difficulties in using such data is that numerical models used for translating an heterogeneity model into a seismic response or a pressure drawdown test are not straightforward. For seismic data, the simplest form of modelling is the convolution of acoustic impedance contrasts, whereas for well tests, the transformation is more complicated, ranging from numerical approximations, often regarded as too simplistic, to computer-intensive flow simulations. In a number of applications, seismic data have been succesfully incorporated into the stochastic modelling process (see for instance Doyen, 1990, Bortoli et al, 1992), whereas Deutsch has recently proposed a new approach, based on simulated annealing, to constrain geostatistical realisations by well-test results.

CONCLUSION: AVENUES FOR FUTURE RESEARCH

The above developments show that stochastic reservoir modelling is currently a "hot" research topic. In spite of this, a few areas can be mentioned where a substantial research effort is still required.
Simulation algorithms: In spite of the great improvements brought by techniques such as birth/death processes in the simulation of object-based models, conditioning by wells in cases where objects are the same size or greater than the

well spacing remains a problem. It can also be argued that promising techniques, such as Markov random fields or two-point histograms deserve more practical tests

Better use of indirect information A few applications report the use of seismic data for constraining stochastic models. The use of seismic and dynamic data (pressure tests, production data) deserves to be better investigated.

Better geological constraints: A great variety of stochastic approaches is now available for distributing heterogeneities in space. These approaches will generate realistic models only if the reservoir geologist is involved early in the choice of the approach and of the statistical parameters to use. It is also important to have the objective and scale of interest of quantitative outcrop studies clearly identified from the start. If geologists make the effort of understanding the philosophy of stochastic models, and if geostatisticians try and make their techniques easier to understand to non-specialists, more of the geologists' experience will be accounted for and quantified, leading to more realistic stochastic models.

ACKNOWLEDGEMENTS

The authorisation of Elf Aquitaine Production to present this paper is gratefully acknowledged, together with the precious help from Elf Aquitaine Production's "Service Information et Documentation" and Drafting Department.

REFERENCES

Alabert, F.G., 1987, Stochastic Imaging of Spatial Distributions using Hard and Soft Information, M.Sc. Thesis, Stanford University, 197 p.

Alabert, F.G. and G.J. Massonnat, 1990, Heterogeneity in a Complex Turbiditic Reservoir: Stochastic Modelling of Facies and Petrophysical Variability, SPE Paper 20604, p. 775-790.

Bortoli, L.J., F. Alabert and A. Haas, 1992, Constraining Stochastic Images to Seismic Data, this volume.

Bridge, J.S., and Leeder, M.R., 1979, A Simulation Model of Alluvial Stratigraphy, Sedimentology 26, p. 617-644.

Bryant, I.D., and S.S. Flint, 1992, Quantitative Clastic Reservoir Geological Modelling: Problems and Perspectives, to be published in IAS Special publication 15, Quantitative Description and Modelling of Clastic Hydrocarbon Reservoirs and Outcrop Analogues, Flint and Bryant eds.

Clemetsen, R., A.R. Hurst, R. Knarud and H. Omre, 1990, A Computer Program for Evaluation of Fluvial Reservoirs, North Sea Oil and Gas Reservoirs II, Graham and Trotman, Buller et al eds, p. 373-385.

Damsleth, E., C.B. Tjolsen, H. Omre, and H.H. Haldorsen, 1990, A two-stage Stochastic Model Applied to a North Sea Reservoir, SPE Paper 20605, p. 791-802.

Delhomme, A.E.K. and J.F. Giannesini, 1979, New Reservoir Description Techniques improve Simulation Results in Hassi-Messaoud Field, Algeria, SPE Paper 8435.

Deutsch, C.V., 1992, Annealing Techniques Applied to Reservoir Modeling and the Integration of Geological and Engineering (Well Test) Data, PhD Thesis, Stanford University, 304 p .

Doyen, P.M., 1990, Statistical Method for Estimating Lithology from Seismic Data, US Patent Number 4926394, 15 May 1990.

Dubrule, O., 1989, A Review of Stochastic Models for Petroleum Reservoirs, in Geostatistics, Kluwer publ., Armstrong ed, p. 493-506.

Dubrule, O., S. Flint, A. Paardekam and M.C. Budding, 1990, Three-Dimensional Reservoir Geological Modelling, Abstracts Volume, 13th International Sedimentological Congress, 26-31 August 1990, Nottingham, England.

Fält, L.M., A. Henriquez, L. Holden and H. Tjelmeland, 1991, MOHERES, a program system for simulation of Reservoir Architecture and Properties, Proceedings of the 6th European EOR Symposium, Stavanger, Norway, May 21-23, 1991.

Farmer, C.L. 1988, The Generation of Stochastic Fields of Reservoir Parameters with Specified Geostatistical Distributions, Mathematics of Oil Production, Oxford Science Publications, Clarendon Press, Oxford, p. 235-252.

Farmer, C.L, 1992, Numerical Rocks, in Mathematics of Oil Recovery, King ed, Oxford University Press.

Fielding, C.R. and R.C. Crane, 1987, An Application of Statistical Modellling to the Prediction of Hydrocarbon Recovery Factors in Fluvial Reservoir Sequences, Recent Developments in Fluvial Sedimentology, Soc. Econ. Pal. Miner. Spec. Pub., 39, p. 321-327.

Geehan, G.W., T.F. Lawton, S. Sakurai, H. Klob, T.R. Clifton, K.F. Inman, K.E. Nitzberg, 1986, Geologic Prediction of Shale Continuity, Prudhoe Bay Field, Reservoir Characterisation, Academic Press Inc, Orlando, Fl., Lake and Carroll eds, p. 63-82.

Guardiano, F.B. and M.R. Srivastava, 1992, Multivariate Geostatistics: Beyond Bivariate Moments, this volume.

Gundeso, R. O. and O. Egeland, 1990, SESIMIRA - A New Geological Tool for 3-D Modelling of Heterogeneous Reservoirs, in North Sea Oil and Gas Reservoirs, II, Graham and Trotman, London, Buller et al eds, p. 363-371.

Haldorsen, H.H., D.M. Chang and S.H. Begg, 1987, Discontinuous Permeability Barriers: A Challenge to Engineers and Geologists, in North Sea Oil and Gas Reservoirs, II , Graham and Trotman, London, Buller et al eds, p. 127-151.

Haldorsen, H.H. and E. Damsleth, 1990, Stochastic Modelling: Journal of Petroleum Technology, April 1990, p. 404-412.

Henriquez, A., Tyler K.J., and Hurst A., 1990, Characterization of Fluvial Sedimentology for Reservoir Simulation Modeling, SPEFE Journal, September 1990, p. 211-216.

Hewett, T.A., 1986, Fractal Distributions of Reservoir Heterogeneity and Their Influence on Fluid Transport, SPE Paper 15386

Hirst J.P.P., C.R. Blackstock and S. Tyson, 1992, Stochastic Modelling of Fluvial Sandstone Bodies, to be published in IAS Special publication 15, Quantitative Description and Modelling of Clastic Hydrocarbon Reservoirs and Outcrop Analogues, Flint and Bryant eds.

Hoiberg J., H. Omre and H. Tjelmeland, 1990, Large-Scale Barriers in Extensively Drilled Reservoirs, 2nd European Conference on the Mathematics of Oil Recovery, p.31-41, Technip, Paris.

Hu, L.Y., Ph. Joseph and O. Dubrule, 1992, Random Genetic Simulation of the Internal Geometry of Deltaic Sandstone Bodies, accepted for presentation at the SPE Fall Meeting, October 1992.

Journel, A.G. and J.J. Gomez Hernandez, 1989, Stochastic Imaging of the Wilmington Clastic Sequence, SPE Paper 19857.

Journel, A.G. and C.V. Deutsch, 1992, Entropy and Spatial Disorder, Stanford Center for Reservoir Forecasting, Report 5, May 1992.

Lucia, F.J. and G.E. Fogg, 1990, Geologic/Stochastic Mapping of Heterogeneity in a Carbonate Reservoir, Journal of Petroleum Technology, October 1990, p. 1298-1303.

MacDonald A.C., T.H. Hoye, P. Lowry, T. Jacobsen, J.O. Aasen, A.O. Grindheim, 1991, Stochastic Flow Unit Modelling of a North Sea Coastal-Deltaic Reservoir, Proceedings of the 6th European EOR Symposium, Stavanger, Norway, May 21-23 1991, p.41-52.

Omre, H., K.B. Halvorsen, L. Holden, and J. Hoiberg, 1988, Reservoir Heterogeneity, Geological Description and Effects on Fluid Flow - Models for Heterogeneity: Proceedings from the SPOR Seminar, Stavanger, Norway, Sept 20-21, p. 296-308.

Omre, H., K. Solna and H. Tjelmeland, 1990, Calcite Cementation: Description and Production Consequences, SPE Paper 20607, p. 811-823.

Perez, G. and A.K. Chopra, 1991, Evaluation of Fractal Models to Describe Reservoir Heterogeneity and Performance, SPE Paper 22694, p. 387-398.

Ravenne, C. and H. Beucher, 1988, Recent Developments in Description of Sedimentary Bodies in a Fluvio-Deltaic Reservoir and their 3D Conditional Simulations, SPE Paper 18310.

P.S. Ringrose, G.E. Pickup, J.L. Jensen and K. Sorbie, 1992, The Use of Correlation Statistics for Modelling Immiscible Displacements in Petroleum Reservoirs, Paper to be presented at the 3rd European Conference on the Mathematics of Oil Recovery, 17-19 June 1992, Delft, Netherlands .

Ripley, B.D., 1992, Stochastic Models for the Distribution of Rock Types in petroleum Reservoirs, to be published in Statistics in the Environmental and Earth Sciences, Griffin, Walden and Guttorp eds.

Ripley, B.D, 1987, Stochastic Simulation, Wiley & Sons Inc.

Rudkiewicz, D. Guérillot and A. Galli, 1990, An Integrated Software for Stochastic Modelling of Reservoir Lithology and Propery with an Example from the Yorkshire Middle Jurassic, in North Sea Oil and Gas Reservoirs, II , Graham and Trotman, London, Buller et al eds, p. 399-406.

Stoyan, D., W.S. Kendall and J. Mecke, 1987, Stochastic Geometry and its Applications, Wiley & Sons, Inc.

Tetzlaff, D.M., 1991, The Combined Use of Sedimentary Process Modelling and Statistical Simulation in Reservoir Characterization, SPE Paper 22759.

Tyler, N., M.D. Barton and R.J. Finley, 1991, Outcrop Characterization of Flow Unit and Seal Properties and Geometries, Ferron Sandstone, Utah, SPE Paper 22670.

Weber, K.J., 1986, How Heterogeneity affects Oil Recovery, Reservoir Characterisation, Academic Press Inc, Orlando, Fl., Lake and Carroll eds, p. 487-544.

Weber, K.J. and L.C. van Geuns, 1990, Framework for Constructing Clastic Reservoir Simulation Models, Journal of Petroleum Technology, October 1990, p. 1248-1297.

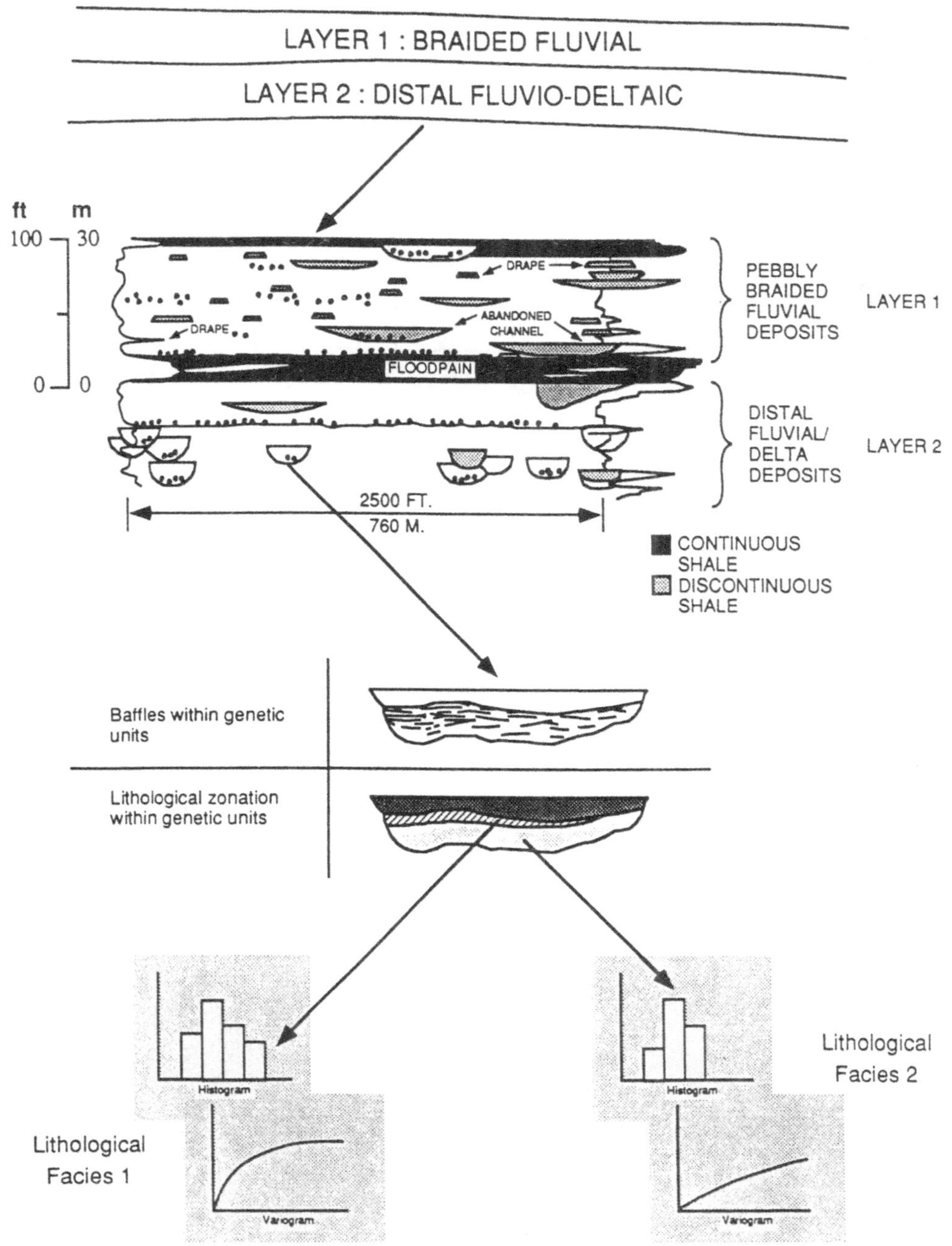

Figure 1 : Illustration of a step-by-step quantitative modelling approach
(Inspired from Geehan et al 1986, Weber 1986, Damsleth et al 1990)

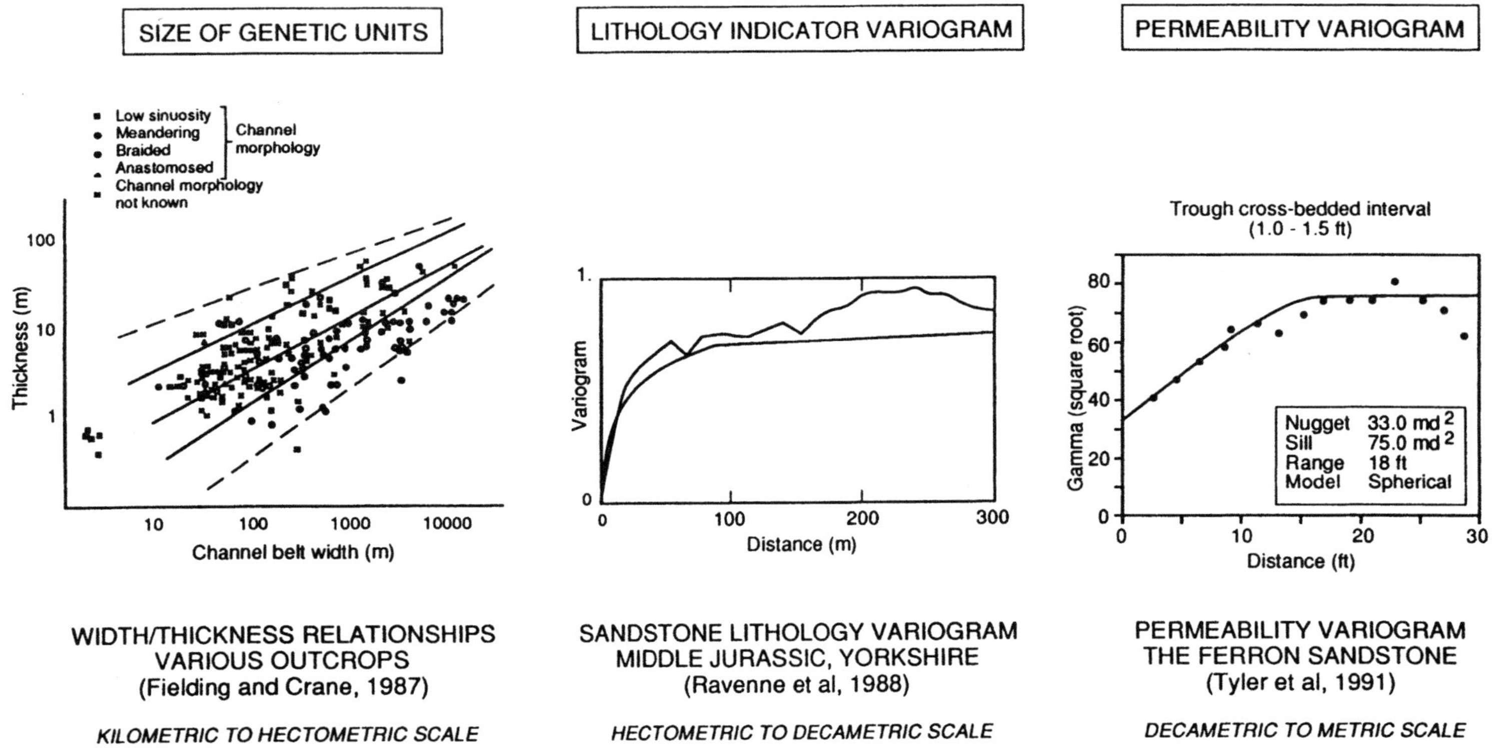

Figure 2 : Outcrop data can be used to derive quantitative information at different scales

Modelling Step / Depositional Environment	Genetic Unit Modelling	Internal Lithology Variations	Permeability Variations
Fluvio-Deltaic Outcrop (Rudkiewicz et al, 1990)	Vertical proportion curves delineate the main genetic units (calculated on the outcrop).	Indicator Simulation. Large number of lithofacies. Variogram calculated on outcrop.	Modelling of permeability variations within each reservoir lithofacies. Same variogram range as for lithofacies.
Fan-Delta Reservoir (Damsleth et al, 1990)	Object-Based approach. Mouth-bars and channels. Size derived from analog outcrops.		Direct modelling of permeability variations within sand bodies. Variogram based on educated guess.
Coastal-Marine Reservoir (Fait et al, 1991)	Object-Based approach. Tidal inlets, wash-over bodies. Size derived from hand-drawn fence diagrams.		Direct modelling of permeability variations within sand bodies. Sub-range fractal variogram calculated along-hole.
Turbiditic Reservoir (Alabert & Massonnat, 1990)	Indicator Simulation. Turbiditic lobes and channels. Variogram calculated from well data and well tests.	Indicator Simulation. Reservoir/non-reservoir facies. Variogram based on educated guess.	Permeability variations within reservoir facies. Vertical variogram from wells. Educated guess for horizontal variogram.

Table 1: Examples of published multi-step stochastic modelling approaches

AN INTEGRATED APPROACH TO RESERVOIR STUDIES USING STOCHASTIC SIMULATION TECHNIQUES

Alberto S. Almeida*°, Thomas Tran*, Paulo R. Ballin*°

*Stanford Center for Reservoir Forecasting, Stanford, CA, USA, 94305
° Petrobras, Petroleo Brasileiro S.A.

Abstract - In this methodological paper, geostatistical techniques are applied to synthetic reservoir modeling in three different steps: (1) high-resolution numerical modeling of a depositional environment; (2) integration of seismic data to improve the reservoir petrophysical characterization; and (3) a probabilistic approach for assessing uncertainty in reservoir performance prediction. This study shows that, seismic data does improve the petrophysical description of the reservoir, that stochastic simulation allows evaluating the impact of the uncertainties inherent in reservoir modeling.

1 Introduction

Accurate reservoir performance prediction is key to proper reservoir management. In recent years, much effort has been dedicated to improving reservoir management through better reservoir modeling (Haldorsen and Damsleth, 1990), more flexible flow simulators (Mattax and Delton, 1990), the development of enhanced recovery schemes (Lake, 1989). All these improvements require a better characterization of the porous media heterogeneities which explains the growing interest in geostatistical techniques (Journel 1990).

The survey of oil and gas fields share common characteristics: 1) the high drilling costs result in a small number of wells, providing few hard data; 2) the data are a by-product of production drilling; they do not come from planned sampling campaigns; 3) wells are preferentially located in high-pay zones, thus providing data biased towards those zones; 4) seismic data, although more abundant, have poor spatial resolution and limited information content; 5) geological data, although non-quantitative, are of paramount importance in sketching the geometric architecture of the reservoir; and 6) production and well test data must be matched.

Geostatistics has brought forward some exciting approaches to these challenges, most importantly, algorithms for data integration and for appreciating the impact of uncertainty in the reservoir description. Traditional low-pass interpolation algorithms such as kriging and cokriging are now limited to volumetric calculations where the drawback of spatial variability smoothing is less serious. The major contribution of geostatistics has been the introduction and development of the concept of stochastic imaging of reservoir heterogeneities. Conditional simulations have demonstrated the ability to honor various data, of different types and accuracies, including qualitative geological information, yet avoiding the smoothing effect of traditional interpolators (Journel and Alabert, 1990; Doyen, 1988). Stochastic reservoir modeling provides multiple equiprobable reservoir models all data-intensive, rather than a single, smooth, usually data-poor, deterministic model. Experience has shown that these data-intense stochastic reservoir models yield a better history match of production data, yet providing a measure of uncertainty in prediction of future performance (Alabert and Massonat, 1990).

This paper, methodological in nature, presents a typical suite of geostatistical interventions in reservoir modeling and performance forecast. More precisely, three different areas of applications are adressed:

- **Geological modeling** - or the building of a high-resolution numerical model of a given depositional environment.
- **Reservoir data integration** - A methodology for integrating seismic data to improve stochastic images of the spatial distribution of petrophysical properties in the reservoir.
- **Reservoir study** - A probabilistic approach for performing reservoir performance forecasts.

2 Numerical reservoir modeling

The creation of the *Stanford-1* data base provides a pedagogical example of how a numerical model of reservoir geometry and petrophysical properties can be built (Almeida et al. 1992). Although this example does not relate to any actual reservoir, it does borrow many features and statistics typical of sandstone turbiditic reservoirs as found in the North Sea (Heriter et. al 1979).

The *Stanford-1* data base consists of a 3-dimensional grid of geological and petrophysical information. The grid comprises 150x150x86= $1,935,000$ nodes informed with facies type, thickness, depth, porosity, permeabilities, and saturations. This grid may represent the quasi- exhaustive information that would be provided by 150x150= $22,500$ cored wells on a regular 100 ft. x 100 ft. grid, and sampled every 2 ft. along the vertical. Thus, the 3D volume informed by this data base is approximately 15,000 ft. x 15,000 ft. x 172 ft.

The generation of *Stanford-1* proceeded in seven steps: 1)Simulation of the reservoir geometry; 2)Simulation of porosity; 3) Simulation of horizontal permeability ; 4) Simulation of vertical permeability; 5) Final structural setting; 6)Simulation of water saturation; and 7) simulation of seismic velocity.

The geological model was idealized as a deep-sea turbiditic fan with three depositional and one erosional events (see Figure 1):

- the first event corresponds to a deep sea deposition of fine sediments generating the (bottom) first layer, a continuous mudstone layer.

- the second event corresponds to a turbiditic sequence of six successive layers. These layers are not as continuous as the mudstone layer and, in many places, they may be completely absent.
- the third event is an erosional surface caused, for example, by an up-lifting of the area or a marine regression, which eroded the top of the turbiditic sequence.
- the fourth event corresponds to the last (upper) layer. Over the previous erosional surface coarse sediments due to marine transgression were deposited. This layer could be interpreted as a beach-type deposit producing a sandstone with varying proportions of shale increasing upward.

The generation of the geometries related to these four events was performed through a series of two-dimensional unconditional simulations of layer thicknesses, using the sequential Gaussian algorithm (Deutsch and Journel, 1992). The reason for using this technique is that a sedimentary layer boundary prior to any faulting is expected to have the extreme continuity characteristic of Gaussian models.

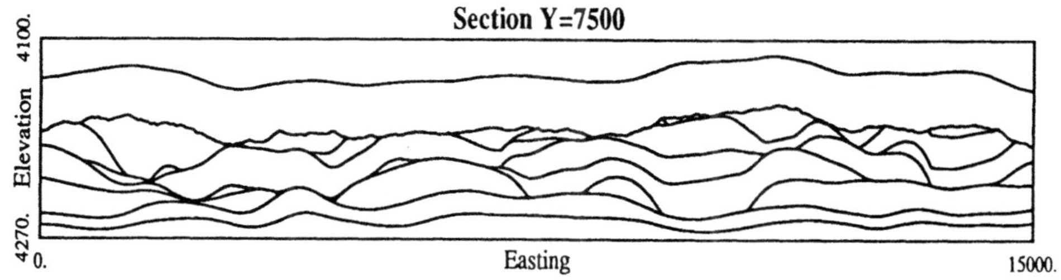

Figure 1: Cross-section showing layers geometry

The following procedure was used to simulate the reservoir geometry: 1) First, the normal score transform of each layer thickness is simulated over the 150 x 150 horizontal grid, assuming a specific covariance model consistent with the the corresponding conceptual geological model. For instance, layer 1 thickness is very continuous with a N60°E direction of maximum continuity, while the thicknesses of layer 2 to 6 within the turbiditic sequence were less continuous yet sharing the same anisotropy direction. All layer thicknesses were simulated independently of each other. A different model could account for cross correlation between layers. Also, in real practice, such simulations are made conditional to well lithofacies data.

2)Next, the simulated normal score values are back-transformed to some predefined thickness histogram. For instance, for layer 1 (mudstone) the thickness histogram is lognormal with mean 28 ft. and coefficient of variation 0.5.

3) Starting from a base surface, each layer thickness was stacked upon the previous one, determining the elevation of the eight top surfaces.

4) Some surfaces were smoothed using an moving window averaging technique.

5) To set the reservoir at a realistic depth, an arbitrary value of 4100.00 ft was added to the elevation of each node.

Figure 1 shows a typical section of the geometry of the resulting depositional model.

Within the previously generated geometric architecture, porosity was simulated in 3-D again using an unconditional sequential Gaussian algorithm. Porosity was simulated independently from one layer to another. Histograms of actual core porosity were used for the back transforms. Simulations were performed using stratigraphic coordinates relative to the bottom surface of each layer (Journel and Gomez-Hernandez 1989), to ensure stratigraphic continuity rather than strict horizontal or vertical continuity. Once again, in practice, it would be easy to make such a simulation conditioned to porosity data available along wells.

Following the simulation of porosity, horizontal and vertical absolute permeability values were simulated and assigned to all nodes of the petrophysical grid (150x150x86). The approach used consists of drawing each horizontal permeability nodal value from a specific probability distribution conditional to the porosity value simulated at this node. These conditional distributions were taken from an actual experimental scattergram of core horizontal permeability vs. core porosity available from an actual turbiditic reservoir. A better, but much more CPU intensive, approach would have been to consider the probability distribution of permeability given all surrounding porosity and permeability data rather than only the collocated porosity value. Such approach could be implemented using either a Gaussian model for the joint distribution of permeability/porosity, or an indicator model (Suro-Perez 1991). This approach was not implemented for *Stanford-1* because no local permeability data were available.

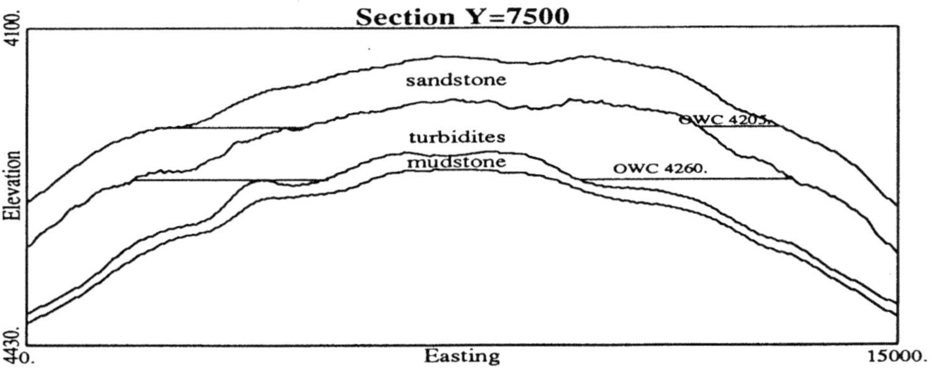

Figure 2: Cross-section showing the main sequences and corresponding OWC

Last, the vertical permeability at each node was drawn from the probability distribution conditional to the collocated simulated value for horizontal permeability. Again, an actual experimental scattergram of horizontal vs. vertical permeability was used for this purpose.

Up to this point the *Stanford-1* geometry consists of a vertical stack of 8 layers, defined over a 3-dimensional grid. In order to simulate water and hydrocarbon saturations, it is necessary to define the setting of the reservoir structure. The reservoir structure was obtained by bending the original rectangular geometry with a regular paraboloid surface. Other more complex structural settings could have been considered by adding a higher order trend surface to the original rectangular structure.

The next step was the simulation of water saturation. The turbiditic and the sandstone sequences were considered as two isolated oil reservoirs, while a constant value of $S_w = 100\%$ was given to all nodes in the mudstone layer. The separation of the two oil reservoirs is marked by an impervious layer corresponding to a paleosoil generated during the period of sub-areal conditions. As a consequence, two different hydrocarbon water contacts (HWC) were defined for the two reservoirs. In both reservoirs, a value of $S_w = 100\%$ was attributed to all points below the HWC , and for the points above the HWC a value S_w decreasing towards the top of the sequence was attributed, utilizing experimental capillary data. Figure 2 shows a cross section of the entire depositional environment where the three main sequences as well as the two water-oil contacts can be seen.

The last step in the generation of *Stanford-1* reservoir was the simulation of synthetic seismic velocity mimicking a dense 3-D seismic survey. Seismic data such as compressional velocity is a multivariate function of reservoir properties such as thickness, density, porosity, mineral content, etc. However, the purpose of this exercise is not generation of seismic data but its usage to improve reservoir description. An experimental scattergram of porosity vs. compressional velocity measured in laboratory samples at ultrasonic frequencies (0.5-1.5 MHz) provided the following linear regression (Klimentos, 1991): $V_p(km/s) = 5.87 - 6.99 * \phi - 3.33 * C$, where the clay content (C) was supposed here to be constant and equal to 15.0%.

This regression was perturbed with random coefficients allowing the generation of a velocity $V_p(\mathbf{x})$ at any node $\mathbf{x}$ where the simulated porosity value is $\phi(\mathbf{x})$: $V_p(\mathbf{x}) = \beta(\mathbf{x}) - \xi(\mathbf{x}) * \phi(\mathbf{x})$, where $\beta(\mathbf{x})$ and $\xi(\mathbf{x})$ are independent normal random deviates with respectives means, 5.37 and 6.99 and respective standard deviations, 0.30 and 0.20. This results in a scattergram and a cofficient of correlation -0.57 consistent with the experimental porosity -velocity scattergram. Note that in actual practice, seismic data are given.

Fault simulation

Seismic data are traditionally used to map reflective horizons such as layer transition or major discontinuities such as faults or fractures. Two major faults striking NS were introduced in the data base. These faults were simulated as two pseudo-planes crossing the entire turbiditic sequence, and generating two sealing horizons with low porosity values varying from 1.50 to 3.50% and correspondingly low permeability values (horizontal permeability varies from 1.0 to 3.0 md in the vicinity of the fault planes) and higher seismic velocity. Figures 3a and 3b show grayscale representations of typical porosity and seismic velocity over a horizontal section. Dense seismic data would allow the detection of such faults while sparse well data may miss them.

The *Stanford-1* data base is now considered complete. It comprises a 150 x 150 x 86 grid with each node informed with lithofacies type, depth, porosity, horizontal and vertical permeabilities, saturation and seismic velocity.

3 Modeling with sparse sampling

Availability of an "exhaustive" reference data base allows checking any model based on a sparse sampling extracted from it. It allows the comparison various sampling strategies and evaluating their impact on reservoir performance prediction.

For the present exercise, seven wells were assumed available (see dots on Figure 3a). Along those wells, the porosity, permeability and saturations data were read directly from the data base. In addition, as may be the case in an actual reservoir, the dense 3-D seismic data were assumed available, i.e. all 1,935,000 gridded seismic velocity data. The porosity spatial distribution is then mapped using first only the seven wells data, then in addition the seismic velocity data. The intention is to evaluate the information brought by the seismic and how it improves reservoir performance prediction. All predictions are checked against the reference results built from the exhaustive data base.

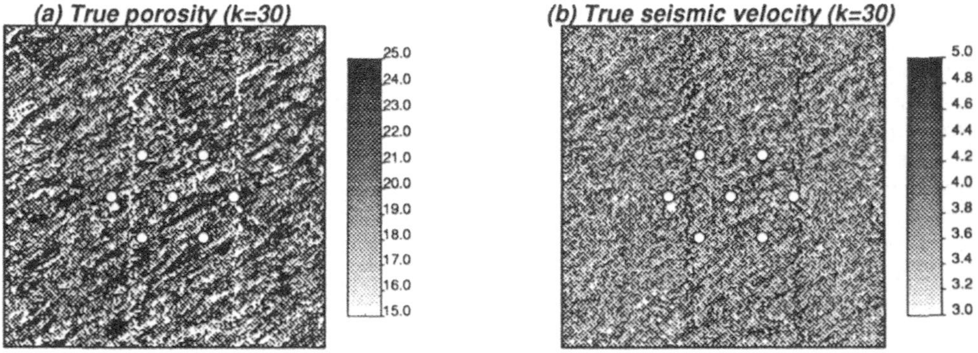

Figure 3: Grayscale of a) porosity and b) seismic velocity

Using only well data

Stochastic imaging of the porosity spatial distribution using only the seven well data was performed using the sequential Gaussian algorithm. Since the seven wells were not enough to provide a reasonable variogram modeling, the variogram model used was assumed to be known from others similar reservoirs. In this study, the model was taken from the exhaustive data set. The model used was spherical with anisotropy in the direction N60E, range of 1100 ft., horizontal anisotropy ratio of 1 to 10 and vertical to horizontal anisotropy ratio 1 to 100. Simulation was performed over the sole turbiditic sequence in 3D and made conditional to the seven wells porosity data. One realization (stochastic image) is shown on Figure 4a. The various realizations were checked and reproduce reasonably well the global spatial statistics (histogram and variogram) of the reference porosity distribution, except for the discontinuities generated by the fault.

Using seismic data

The exhaustive seismic data was used in addition to the seven wells data to improve stochastic imaging of the porosity distribution. This merger of data of different types calls for the concept of cokriging.

A straight cokriging approach (Isaaks and Srivastava, 1989), or a kriging with a trend model made proportional to the seismic data (Marechal, 1984), were options. They were not retained because of the smoothing effect associated with any kriging as opposed to stochastic simulation,and because kriging does not deliver any useful measure of uncertainty. Instead the algorithm used for stochastic imaging of porosity given both porosity and seismic data is sequential Gaussian simulation. More precisely:

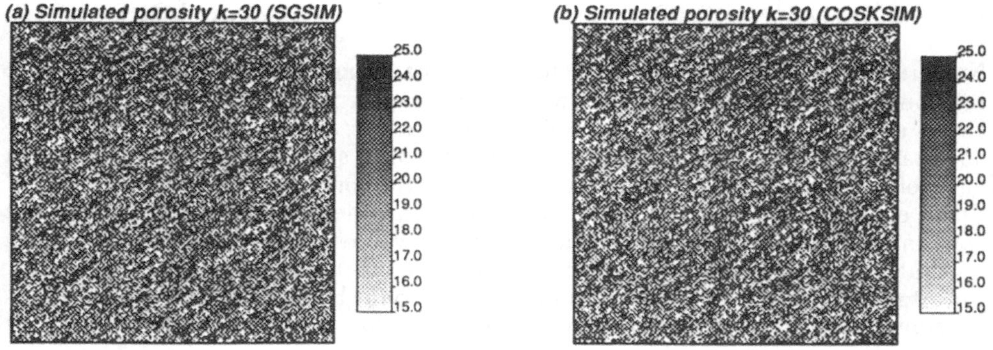

Figure 4: Grayscale of simulated porosity values: a)considering just well data and b) integrating seismic velocity

1. In the space of normal score transforms of both porosity and seismic data, the Gaussian conditional distribution of any porosity value is obtained by a reduced simple cokriging technique, called collocated cokriging (Xu et al.,1992). This technique amounts to retaining only the secondary datum, in this case seismic velocity, closest to the location where the primary variable (porosity) is to be simulated, e.g., the collocated seismic velocity datum. This approximation besides reducing the calculation effort involved in the cokriging procedure, makes the cokriging matrix system more stable when the secondary data is very dense with regard to the primary data grid.

2. The burden of cross-covariance inference is considerably reduced by adopting for the cross-covariance of the normal score seismic vs. porosity data a Markov type model:

$$\rho_{s\phi}(h) = \rho_{s\phi}(0).\rho_{\phi}(h)$$

   where the subscripts s and ϕ, represent respectively seismic velocity and porosity. This model, is a consequence of the statement: "The closest datum, e.g. the collocated datum, screens the influence of further away data", (Zhu, and Journel,1992).

3. Drawing from successive conditional distributions for normal scores of porosity, a stochastic image of the porosity spatial distribution is built.

Figure 4a gives a grayscale representation of one realization (stochastic image) of the porosity spatial distribution over one horizontal section. For Figure 4a, the only conditioning data originated from the seven wells. Figure 4b gives a realization over the same horizontal section but now conditioned to both the seven wells data and the seismic. Besides an improvement in local accuracy, the major difference between the two images is the reproduction of the two faults in Figure 4b, obtained through using seismic data.

4 Probabilistic reservoir performance forecasts

Numerical reservoir models and/or stochastic images are not end goals; they are to be input to specific flow simulators for reservoir performance prediction. Traditionally, a single numerical model is used which does not allow assessing the response uncertainty due to spatial uncertainty. The concept of stochastic imaging provides a mean of varying the input numerical model under the constraint of the available data. These equiprobable stochastic reservoir models can then be processed through the flow simulator, yielding a distribution of response values. That distribution provides a quantitative measure of output uncertainty due to the input spatial uncertainty. Uncertain parameters of the flow simulator itself could be made variable, thus allowing simultaneous assessment of output uncertainty due to imperfect modeling of the flow process.

In this section, the uncertainty of a reservoir response related to the turbiditic sequence of *Stanford-1* is assessed. This exercise analyses the effect of spatial heterogeneities on the performance of a small reservoir under natural depletion subject to a strong water influx. Seven wells were drilled in a seven-spot pattern with an average spacing of 3000 ft, see Figure 3.

The probabilistic approach used to perform this study consisted of the following steps:

- Generate the stochastic numerical model

  The first step is the generation of several sets of equiprobable images of the petrophysical characteristics of the reservoir, more precisely, porosity, horizontal permeability and vertical permeability. These images can not be generated independently since the petrophysical variables are intercorrelated. The joint simulation algorithm used in this exercise was the same used for the generation of the reference data base. These images were generated over a dense grid comprising 150 x 150 x 12= 270,000 nodes at a horizontal spacing of 100 ft x 100 ft. and average vertical spacing of 10 ft. (see Figure 4b for an example).

  With present flow simulators and computing power it would not be practical to run a complex multiphase 3D flow simulation on, say, 100 such high resolution numerical models. Coarsening of the high resolution grid provided by geostatistical stochastic imaging is needed.

- Upscale the images

  Coarsening calls for averaging the simulated petrophysical properties over grid blocks. No matter the size and geometry of the blocks, averaging of volumetric properties such as porosity and saturations poses no difficulty: the averaging process is straightforward linear. The difficulty resides in averaging dynamic properties such as permeabilities. On this subject, a large literature is available (Kasap and Lake, 1989; White and Horne, 1987; Deutsch, 1989; Gomez-Hernandez and Journel, 1990; King, 1990; Hewett and Behrens, 1990).

The approach adopted here is the power average proposed by Deutsch (1989). A coarse grid of 10 x 10 x 12 with average block size of 1500 ft. x 1500 ft. x 10 ft. was selected. The simulated scalar permeabilities values K_i falling into a given block are averaged according to a prior calibrated power average ω ; the resulting effective block permeability is K_{eff} such as :

$$K_{eff}^{\omega} = \frac{1}{N} \sum_{i=1}^{N} K_i^{\omega}$$

For this study, values of $\omega = 0.8$ and $\omega = 0.2$ are retained for upscaling horizontal and vertical permeability, respectively. Recall that $\omega = -1.0$ corresponds to an harmonic averaging, $\omega = 0$ to a geometric averaging and $\omega = +1.0$ to an arithmetic averaging.

Because the upscaling in this exercise was limited to a rather homogeneous single-facies turbidite, there was no need to consider multiple sets of averaging powers or permeability tensors. In practice, when multiple lateral facies are present, the averaging power ω is calibrated from small scale flow simulations using high resolution stochastic images (Deutsch, 1989).

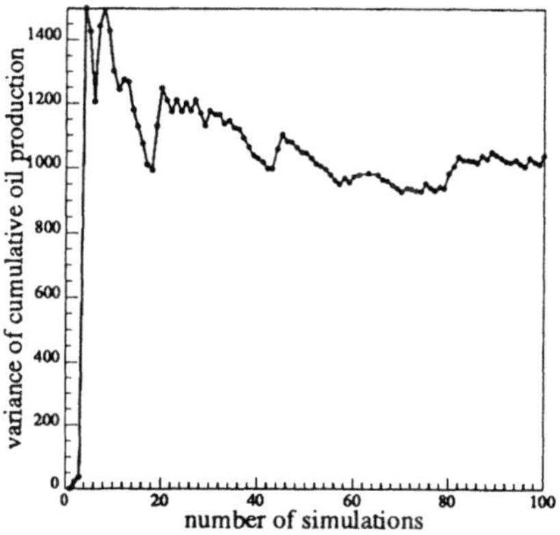

Figure 5: Variance of cum. oil production versus the number of images

- Rank the images

At this point it is necessary to define how many stochastic images should be considered for a correct assessment of uncertainty after processing through the flow simulator. Ballin (1992) suggests an approach that consists of analysing the fluctuations of statistics of a given reservoir response as the number of upscaled stochastic images processed through the flow simulator increases. In this present exercise the variance was used. The number of stochastic images to retain is that for which the variance stabilizes.

The calibration reservoir response should be that most appropriate for the case under study. For instance, if the limitation in the field production is the amount of gas to be processed, the reservoir response should be the gas-oil ratio. One can also weight several reservoir responses according to their importance for the case under study.

Figure 5 shows the profile of the variance of the field cumulative oil production after 7 years as a function of the number of coarse grid images. As can be seen the oscillations stabilize after 100 runs. Therefore, only 100 high resolution numerical models were generated by stochastic imaging. These images were generated conditional to the seven wells data and to the seismic data; then they were upscaled into 100 coarse grid block models.

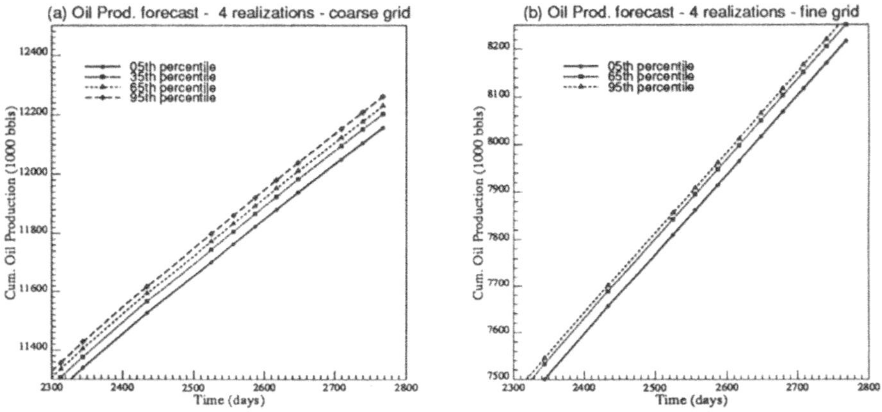

Figure 6: Cumulative oil production forecast: a) coarse grid runs, b) fine grid runs

Repetitive flow simulation was performed considering these 100 coarse grid numerical models. The field cumulative oil production is the response variable used for ranking the 100 numerical models. Figure 6a shows the four profiles of this response variable for the numerical models ranked 5th, 35th, 65th, and 95th.

- Obtain the distribution of the reservoir response

  Comparing the set of data in this exercise with previous calibration done to check the rank preserving assumption (Ballin, 1992), it was considered that the ranking of images done from coarse grid models is preserved for flow simulation using finer grid models. This limits the number of runs using a high resolution grid (50 x 50 x 12) considered reasonable for describing the flow dynamics. The three models chosen are the previously ranked 5th, 65th and 95th percentiles. Then, the reservoir responses obtained from these fine grid models are classified as the 5th, 65th, and 95th percentiles.

Figure 6b shows the cumulative oil production profiles for these three fine grid models. This figure shows that the uncertainty in predicting the field cumulative oil production is small. This small uncertainty reflects the extreme homogeneity of the

Stanford-1 turbiditic sequence. Short scale variability at the scale of the variogram ranges (1100 ft.) does not greatly affect the flow response values. The single major heterogeneities in *Stanford-1* are the two faults (see Figure 4b), which were correctly mapped from the seismic velocity data. In practice, the flow simulation may involve several heterogeneous layers (mudstone, turbidites, clean and shaly sandstone) with complex fracturations that could be both open and sealed. The resulting uncertainty in reservoir performance prediction would then be much higher. Based on the rank preserving assumption, these three response values are considered to be the 5th, 65th, and 95th percentiles of the distribution of responses using fine grid models. Interpolation between these three points provides the cumulative probability distribution shown in Figure 7a. For example the median estimate for cumulative production after 7 years is read as 8.2 x 10^6 bbls.

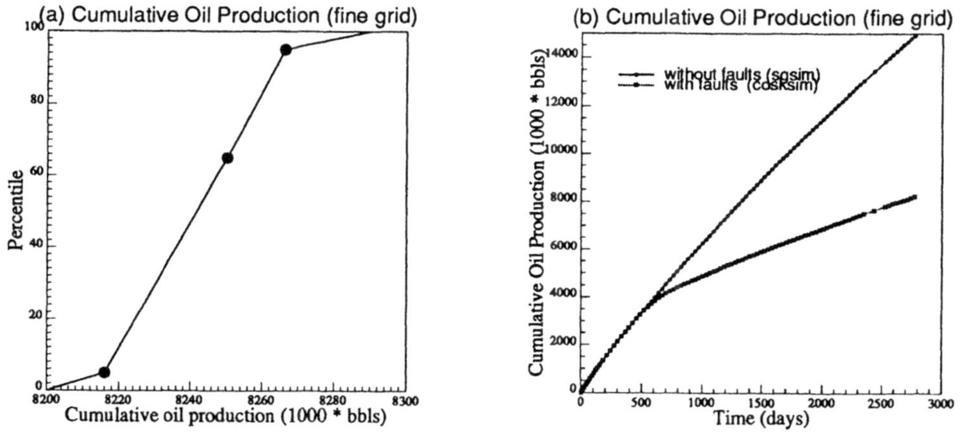

Figure 7: a) cumulative distribution of the cumulative oil production; and b) comparison between two cumulative oil production forecasts: considering just the well data and considering well data and seismic information

Figure 7b compares the two cumulative oil production forecasts obtained from: 1) stochastic models using only the well data and 2) stochastic models using both well and seismic data (the faults are identified). The two forecasts are clearly differents showing the importance of using all information available.

5 Conclusions

This study demonstrates the potential of geostatistical techniques in:

- modeling a complex depositional environment and geological architecture
- integrating (soft) seismic with (hard) well data
- assessing the impact of petrophysical modeling uncertainty on reservoir performance forecast.

More specifically, this synthetic case study shows that :

1. once the geometric architecture (flow piping) of the reservoir is determined, intra-facies variability of petrophysical variables (porosity, permeability) has little impact on the precision (estimation variance) of the forecasts

2. seismic data are most useful in determining large scale heterogeneities such as faults and abrupt changes in facies. They do contribute to more accurate determination of petrophysical properties, but as stated above, within an homogeneous lithofacies these local details have little impact on the forecasts. What really counts is the lithofacies distribution.

3. Gaussian-related stochastic models with their maximum entropy property tend to generate very homogeneous and look-alike intra-facies distributions for the petrophysical parameters. This may lead to an over-optimistic assessement of actual uncertainty.

6 References

1. Alabert, F., and Massonat G. J., (1990) "Heterogeneity in a Complex Turbiditic Reservoir: Stochastic Modelling of Facies and Petrophysical Variability", SPE paper #20604, 65th Ann. Tech. Conf. of SPE, New Orleans.

2. Almeida, A. S., Guardiano F., and Cosentino L. (1992) "Generation of The Stanford-1 Turbiditic Reservoir", *Stanford Center for Reservoir Forecasting - Report 5*, Stanford.

3. Ballin, P. R. (1992) *Approximation of Flow Simulation for Uncertainty Assessement*, PhD thesis, Stanford University, Stanford.

4. Deutsch, C. V. (1989) "Calculating Effective Absolute Permeability in Sandstone/Shale Sequences", *SPE Formation Evaluation*, September 1989.

5. Deutsch, C. V., and Journel, A. (1992) *GSLIB: Geostatistical Software Library User's Guide*, Stanford Center for Reservoir Forecasting, Stanford.

6. Doyen, P. M. (1988) "Porosity from Seismic Data: A Geostatistical Approach", *Geophysics*, October 1988.

7. Haldorsen, H. and Damsleth, E. (1990) "Stochastic Modelling", *JPT*, April 1990.

8. Heriter, F., Lossel, and P., Wathne, E. (1979) "Frigg Field - Large Submarine -Fan Trap in Lower Eocene Rocks of the Viking Graben, North Sea, Giant Oil and Gas Fields of the Decade 1968 - 1978", AAPG.

9. Hewett, T. A., and Behrens R. A. (1990) "Considerations Affecting the Scaling of Displacements in Heterogeneous Permeability Distributions", SPE paper #20739, 65th Ann. Tech. Conf. of SPE, New Orleans.

10. Isaaks, E., and Srivastava, M. (1989) *An Introduction to Applied Geostatistics*, Oxford University Press, New York.

11. Journel, A., and Gomez-Hernandez, J. (1989) "Stochastic Imaging of the Wilmington Clastic Sequence", SPE paper #19857, 64th Ann. Tech. Conf. of SPE.

12. Journel, A. (1990) "Geostatistics for Reservoir Characterization", SPE paper #20750, 65th Ann. Tech. Conf. of SPE, New Orleans.

13. Journel, A., and Alabert, F. (1990) "New Method for Reservoir Mapping", *JPT*, February 1990.

14. King, P. R. (1989) "The Use of Renormalization for Calculating Effective Permeability", *Transport in Porous Media 4*, pages 37-58, Kluwer Academic Publishers.

15. Klimentos, T. (1991) "The effects of porosity-permeability-clay content on the velocity of compressional waves", *Geophysics*, December 1991.

16. Lake, L. W. (1989) *Enhanced Oil Recovery*, Prentice Hall, Englewood Cliffs.

17. Marechal, A. (1984) "Kriging Seismic Data in Presence of Faults", *Geostatistics for Natural Resources Characterization*, G. Verly et al., editors, pages 271-294, Reidel, Dordrecht, Holland.

18. Mattax, C. C., and Dalton, R. L. (1990) "Reservoir Simulation", *JPT*, June 1990.

19. Suro-Perez, V. (1991) *Indicator Kriging and Simulation Based on Principal Component Analysis*, PhD thesis, Stanford University, Stanford.

20. Xu, W., Tran T., Srivastava, R. M., and Journel, A. G. (1992) "Integrating Seismic Data in Reservoir Modeling: The Collocated Cokriging Alternative", SPE paper #24742, 67th Annual Tech. Conf. of SPE, Washington D.C. (in press).

21. Zhu, H. and Journel, A. G. (1992) "Formatting and integrating soft data: Stochastic imaging via the Markov-Bayes algorithm", in Proc. of 4th. Int. Geostat. Congress (this volume).

BAYESIAN KRIGING FOR SEISMIC DEPTH CONVERSION OF A MULTI-LAYER RESERVOIR

Petter Abrahamsen
Norwegian Computing Center
P.O.Box 114, Blindern
N-0314 Oslo
Norway

Abstract. A stochastic model for a petroleum reservoir with L seismic subsurfaces is presented. The seismic velocities within each layer is described by linear regression models and Gaussian random fields. Seismic interpretation errors are modeled as Gaussian random fields. Intercorrelations between all subsurfaces and all velocity fields are taken into consideration. This simplifies the handling of deviating wells and ensures consistent prediction and prediction variances for all L subsurfaces and L velocity fields. Bayesian kriging is used for prediction of subsurfaces and velocity fields.

In the limit corresponding to exact prior knowledge, the Bayesian method is equivalent to cokriging with $2L$ covariables. In the limit corresponding to no prior knowledge, the method is equivalent to a combination of universal kriging and cokriging with $2L$ dependent regression models and $2L$ covariables.

1 Introduction

The large scale geometry of a petroleum reservoir is usually described by a set of geological subsurfaces separating almost homogeneous layers. Available information for the depth to the subsurfaces are precise depth observations in wells, and less precise information from seismic travel times. The travel times are recordings of the time used by a sonic pulse from the surface to a reflecting subsurface. The importance of the seismic travel times are their lateral coverage. The travel times are usually available on a fine-meshed grid which allows an almost continuous — but inexact — description of the lateral depth trends. On the other hand, the almost exact well measurements are available in just a few locations possibly several kilometers apart. The challenge is therefore to combine the exact well measurements with the general trends from the travel times. This requires the simple kinematic relation between depth, the velocity of sound, and travel time: $z = v \cdot t$. In the next section, spatial stochastic models for v and t are established, and the resulting model for z is considered. Using kriging techniques, predictions for the velocity field and the depth are available. The process of calculating the depth to a seismic reflector from the measured travel times is known as depth conversion.

In most applications several seismic reflectors are observed above the reservoir layers under study. The velocity of sound changes significantly at seismic reflectors. Thus, a separate model for the velocity field in every layer should be used to obtain the best results. The traditional approach to depth conversion is to predict the thickness of each layer independently. This paper proposes a method which considers all subsurfaces and velocity fields in one consistent model.

A general problem — at least in off-shore applications — is the very limited number of wells. The lack of data must be compensated by prior information, so Bayesian statistics is necessary. This ensures stable and reasonable results for any number of well observations — including zero.

The paper is organized as follows: the next section describes the stochastic models, Section 3 reviews Bayesian kriging and outlines how the current problem is organized to fit into the linear kriging machinery. Section 4 gives an example which illustrates the properties of the proposed method. Section 5 close the discussion by some final remarks.

2 Stochastic model for depth conversion

The fine-meshed grids containing the travel times are regarded as continuous 2-dimensional fields. The recorded travel times to subsurface l is denoted $t_l(\mathbf{x})$, where $\mathbf{x} \in \mathbf{R}^2$ is a lateral reference. The average vertical velocity of sound within a layer is called the interval velocity and is denoted $v_l(\mathbf{x})$. If both the travel times and the interval velocities are exactly known, the depth to subsurface L, $z_L(\mathbf{x})$, is found as the sum of the interval thicknesses, $\Delta z_l(\mathbf{x})$, above:

$$z_L(\mathbf{x}) = \sum_{l=1}^{L} \Delta z_l(\mathbf{x}), \tag{1}$$

where $\Delta z_l(\mathbf{x}) = v_l(\mathbf{x})\Delta t_l(\mathbf{x}) = v_l(\mathbf{x})\,(t_l(\mathbf{x}) - t_{l-1}(\mathbf{x}))$ The crucial notational convention is that properties associated with intervals are given the name of the subsurface below. The Δ-operator is used for the difference between a property at subsurface $l-1$ and subsurface l, i.e. associated with layer l. Note that by definition $z_0(\mathbf{x}) = t_0(\mathbf{x}) = 0$ for any $\mathbf{x}$.

2.1 Seismic travel times

The seismic travel times, $t_l(\mathbf{x})$ are usually collected on a regular grid sometimes as fine-meshed as 12.5m×25m, so the lateral resolution appears very fine. However, the measured travel times are an averaged sum of reflections from a considerable area depending on the depth to the reflector. Thus the travel time map, $t_l(\mathbf{x})$, should be regarded as a smoothed indirect measurement of the depth to the reflector. There are also physical limits to the vertical resolution of the seismic sound signals due to strong attenuation of short wavelengths. These two measuring errors are modeled as a space dependent random field, $R_l^t(\mathbf{x})$, The 'true' unsmoothed travel times to reflector l is therefore:

$$T_l(\mathbf{x}) = t_l(\mathbf{x}) + R_l^t(\mathbf{x}); \qquad \mathbf{x} \in \mathbf{R}^2. \tag{2}$$

2.2 Interval velocities

Within a homogeneous layer l the interval velocity usually varies laterally due to different average rock density and mineralogy. The lateral trend is a linear model:

$$V_l(\mathbf{x}) = \sum_{p=1}^{P_l} A_l^p g_l^p(\mathbf{x}) + R_l^v(\mathbf{x}); \qquad \mathbf{x} \in \mathbf{R}^2, \tag{3}$$

where A_l^p are partly known coefficient parameters, $g_l^p(\mathbf{x})$ are known space dependent regression functions, and $R_l^v(\mathbf{x})$ is modeled as a Gaussian random field with zero expectation. The regression functions are typically interval velocities from stacking velocities or functions of interpreted travel times. The part of Equation (3) excluding the residual is called the interval velocity trend, and is denoted by the symbol $\tilde{V}_l(\mathbf{x})$.

In the Bayesian approach the coefficient parameters are regarded as multi-Gaussian distributed random variables. The trained geophysicist should assign prior probability distributions to these variables, that is essentially, expectations and variances.

2.3 Depth to subsurfaces

By definition of interval velocity, the thickness of a layer l outlined by two seismic reflectors is:

$$\Delta Z_l(\mathbf{x}) = V_l(\mathbf{x})\Delta T_l(\mathbf{x}) = \left(\tilde{V}_l(\mathbf{x}) + R_l^v(\mathbf{x})\right)\left(\Delta t_l(\mathbf{x}) + \Delta R_l^t(\mathbf{x})\right). \tag{4}$$

The residuals are generally small compared to the trends so a reasonable simplification is to ignore products of residuals:

$$\Delta Z_l(\mathbf{x}) \approx \left(\tilde{V}_l(\mathbf{x}) + R_l^v(\mathbf{x})\right)\Delta t_l(\mathbf{x}) + \tilde{V}_l(\mathbf{x})\Delta R_l^t(\mathbf{x}).$$

Reorganizing by using $\Delta R_l^t(\mathbf{x}) = R_l^t(\mathbf{x}) - R_{l-1}^t(\mathbf{x})$, $R_0^t(\mathbf{x}) = 0$, and $\Delta\tilde{V}_{l+1}(\mathbf{x}) = (\tilde{V}_l(\mathbf{x}) - \tilde{V}_{l+1}(\mathbf{x}))$ gives the depth to subsurface L:

$$\begin{aligned} Z_L(\mathbf{x}) &= \sum_{l=1}^{L} \Delta Z_l(\mathbf{x}) \\ &= \sum_{l=1}^{L} \left(\tilde{V}_l(\mathbf{x}) + R_l^v(\mathbf{x})\right)\Delta t_l(\mathbf{x}) + \sum_{l=1}^{L-1} \Delta\tilde{V}_{l+1}(\mathbf{x})R_l^t(\mathbf{x}) + \tilde{V}_L(\mathbf{x})R_L^t(\mathbf{x}). \end{aligned}$$

The velocity changes, $\Delta\tilde{V}_{l+1}(\mathbf{x})$, are usually small compared to $\tilde{V}_L(\mathbf{x})$. Noting that these velocity changes are multiplied by residuals and using the fact that the sum of two residuals are dominated by the largest, suggests that simplifying by ignoring the 'velocity-change' terms has minor implications. This simplification reduce the stochastic model for depth to subsurface L to:

$$Z_L(\mathbf{x}) = \sum_{l=1}^{L} \left(\tilde{V}_l(\mathbf{x}) + R_l^v(\mathbf{x})\right) \Delta t_l(\mathbf{x}) + R_L^z(\mathbf{x}). \tag{5}$$

The time residual is replaced by the depth residual according to $R_L^z(\mathbf{x}) = \tilde{V}_L(\mathbf{x})R_L^t(\mathbf{x})$. The depth residual is modeled as a Gaussian random field with expectation zero.

Two important aspects should be recognized from Equation (5): 1) The uncertainty in all interval velocities above subsurface L contributes to the uncertainty in the depth to subsurface L. 2) The uncertainty in the travel times to a subsurface above subsurface L does not contribute to the uncertainty in the depth to subsurface L. This is a consequence of removing the 'velocity-change' terms.

A full specification of the stochastic model for the depth to the subsurfaces requires a full specification of the residual Gaussian random fields and the prior multi-Gaussian distribution for the coefficient parameters. Independence among different residuals and the coefficient parameters is assumed:

$$\mathrm{Cov}\{R_l^z(\mathbf{x}), R_{l'}^z(\mathbf{x}')\} = 0 \qquad \text{for } l \neq l', \tag{6a}$$

$$\mathrm{Cov}\{R_l^v(\mathbf{x}), R_{l'}^v(\mathbf{x}')\} = 0 \qquad \text{for } l \neq l', \tag{6b}$$

$$\mathrm{Cov}\{R_l^z(\mathbf{x}), R_{l'}^v(\mathbf{x}')\} = 0 \qquad \text{for any } l \text{ and } l', \tag{6c}$$

$$\mathrm{Cov}\{A_l^p, R(\mathbf{x})\} = 0 \qquad \text{for any residual.} \tag{6d}$$

The expectations of the residuals are assumed zero everywhere. The variance and correlation function must be specified for every residual:

$$\mathrm{Var}\{R_l^{z/v}(\mathbf{x})\} = \left[\sigma_l^{z/v}(\mathbf{x})\right]^2 \qquad \text{for } l = 1, \ldots, L \tag{7a}$$

$$\mathrm{Corr}\{R_l^{z/v}(\mathbf{x}), R_l^{z/v}(\mathbf{x}')\} = \rho_l^{z/v}(\mathbf{x}, \mathbf{x}') \qquad \text{for } l = 1, \ldots, L. \tag{7b}$$

Finally consider a vector $\mathbf{A}$ containing all $P = \sum_{l=1}^{L} P_l$ coefficient parameters, A_l^p. The prior distribution for this multi-Gaussian vector must also be specified in terms of P expectations, $E\{\mathbf{A}\} = \boldsymbol{\mu}_0$, and a $P \times P$-dimensional covariance matrix, $\mathrm{Var}\{\mathbf{A}\} = \Sigma_0$.

2.4 Unification of depth and velocity models

Standard linear methods are used for predictions. The multi-layer model must therefore be formulated as a linear regression model with a spatial dependent Gaussian residual:

$$Z(\mathbf{x}) = \sum_{p=1}^{P} A_p f_p(\mathbf{x}) + R(\mathbf{x}) = \mathbf{f}(\mathbf{x}) \cdot \mathbf{A} + R(\mathbf{x}). \tag{8}$$

All P coefficient parameters are organized in the column vector $\mathbf{A}$ and the corresponding regression functions are organized in the P-dimensional row vector $\mathbf{f}(\mathbf{x})$. First observe that Equation (5) — the model for depth to subsurfaces — has this form: the residual is $R(\mathbf{x}) = \sum_{l=1}^{L} R_l^v(\mathbf{x})\Delta t_l(\mathbf{x}) + R_L^z(\mathbf{x})$, and the sum of regression functions is: $\sum_{p=1}^{P} A_p(\mathbf{x}) f_p(\mathbf{x}) = \sum_{l=1}^{L} \sum_{p=1}^{P_l} A_l^p g_l^p(\mathbf{x})\Delta t_l(\mathbf{x})$, where $P = \sum_{l=1}^{L} P_l$ is the total number of coefficient parameters. The interval velocity model, Equation (3), has exactly the form of Equation (8). The objective however, is to consider depth and velocity simultaneously so a common regression model must be used. The idea is to multiply Equation (3) by $\Delta t_l(\mathbf{x})$ to obtain a thickness having length units. Note that this thickness is not the true layer thickness since the depth residuals are ignored. Once the velocities are 'transformed' into thicknesses, a common regression model for both depths and velocities is possible. Although this is the key to simultaneous prediction of all subsurfaces and interval velocities, it is mainly a matter of notation. Details are therefore given in Appendix A.

3 Bayesian kriging

Two different kinds of data are considered: depth observations from every subsurface and velocity observations from every interval. The number of observations from subsurface and interval l is denoted N_l^z and N_l^v respectively. The lateral locations of the observations are in principal arbitrary but the position to observations from vertical wells obviously coincide. The observations are considered exact.

Travel times, $t_l(\mathbf{x})$, and regression functions, $g_l^p(\mathbf{x})$, are assumed known for every $\mathbf{x} \in \mathbf{R}^2$.

3.1 Posterior distribution for coefficient parameters

A posterior multi-Gaussian distribution for the coefficient parameters in the velocity models is assessed from a prior distribution and the available depth and velocity observations.

Consider a column vector, $\mathbf{Z}$, of all the $N = \sum_{l=1}^{L} (N_l^z + N_l^v)$ depth and velocity observations (rescaled by Δt), and a vector $\mathbf{R}$ of the corresponding N unobserved residuals: $\mathbf{Z}^T = [Z_1(\mathbf{x}_1), \ldots, Z_N(\mathbf{x}_N)]$, and $\mathbf{R}^T = [R_1(\mathbf{x}_1), \ldots, R_N(\mathbf{x}_N)]$, where '$T$' is used for transposed. In this notation, Equation (8) becomes:

$$\mathbf{Z} = F\mathbf{A} + \mathbf{R}. \tag{9}$$

The $N \times P$-dimensional 'design-matrix' F is constructed from the known space-dependent regression functions. Every row corresponds to an observation and every column corresponds to a particular regression function $f_p(\mathbf{x})$.

The prior $N \times N$-dimensional covariance matrix for all the observations are: $K_z = \text{Var}\{\mathbf{Z}\} = F\Sigma_0 F^T + K$, where $F\Sigma_0 F^T$ is the prior contribution from the trend, and K is the kriging matrix: $K = \text{Var}\{\mathbf{R}\}$. The Bayesian prediction and prediction variance for the coefficient parameters are the expectations and the covariances of the posterior multi-Gaussian distribution given by:

$$\hat{\boldsymbol{\mu}}_b = E\{\mathbf{A}|\mathbf{Z}\} = \boldsymbol{\mu}_0 + \Sigma_0 F^T K_z^{-1}(\mathbf{Z} - F\boldsymbol{\mu}_0) \tag{10}$$
$$\hat{\Sigma}_b = \text{Var}\{\mathbf{A}|\mathbf{Z}\} = \Sigma_0 - \Sigma_0 F^T K_z^{-1} F \Sigma_0. \tag{11}$$

This is a standard result from linear regression analysis found in most textbooks on Bayesian statistics such as Berger [2].

3.2 Kriging

Define the prior variance of $Z(\mathbf{x})$ at an arbitrary location, say $\mathbf{x}$, and the prior covariances between $Z(\mathbf{x})$ and the observation vector, $\mathbf{Z}$:

$$k_z(\mathbf{x}) = \text{Var}\{Z(\mathbf{x})\} = \mathbf{f}(\mathbf{x})\Sigma_0 \mathbf{f}^T(\mathbf{x}) + k(\mathbf{x}) \tag{12}$$
$$\mathbf{k}_z(\mathbf{x}) = \text{Cov}\{Z(\mathbf{x}), \mathbf{Z}\} = F\Sigma_0 \mathbf{f}^T(\mathbf{x}) + \mathbf{k}(\mathbf{x}), \tag{13}$$

where $k(\mathbf{x}) = \text{Var}\{R(\mathbf{x})\}$ and $\mathbf{k}(\mathbf{x}) = \text{Cov}\{R(\mathbf{x}), \mathbf{R}\}$. Using these definitions the Bayesian kriging predictor and the corresponding prediction variance are:

$$Z_b^*(\mathbf{x}) = E\{Z(\mathbf{x})|\mathbf{Z}\} = \mathbf{f}(\mathbf{x}) \cdot \boldsymbol{\mu}_0 + \mathbf{k}_z(\mathbf{x}) K_z^{-1}(\mathbf{Z} - F\boldsymbol{\mu}_0) \tag{14}$$
$$\sigma_b^2(\mathbf{x}) = \text{Var}\{Z(\mathbf{x})|\mathbf{Z}\} = k_z(\mathbf{x}) - \mathbf{k}_z(\mathbf{x}) K_z^{-1} \mathbf{k}_z^T(\mathbf{x}). \tag{15}$$

This result is found in Omre and Halvorsen [3] and in Omre, Halvorsen, and Berteig [4].

4 Example and discussion

Sections 2.2 and 2.3 describes stochastic models for interval velocities and depth to subsurfaces. Section 2.4 shows how these models can be regarded as a single linear regression model with a correlated residual which is handled by standard estimation and prediction techniques as well as Bayesian prediction techniques. This 'unification' of the models determines correlations between separate subsurfaces and between subsurfaces and interval velocities. Consequences of these dependencies on predictions obtained by kriging will be illustrated.

4.1 Data

Data is taken from a Norwegian off-shore petroleum reservoir. The data are slightly manipulated to maintain confidentiality. Two subsurfaces — Top and Base — are considered.

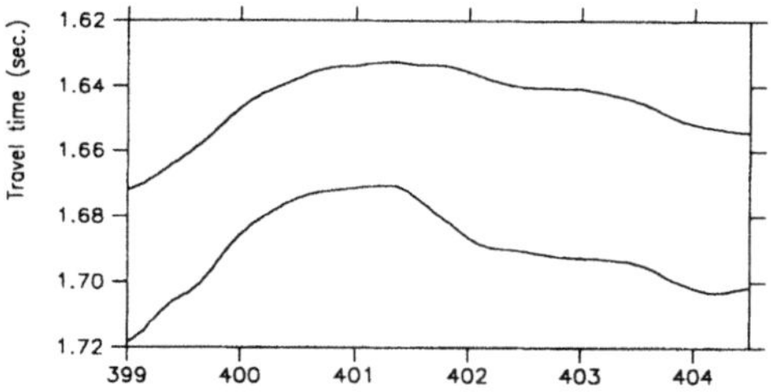

Figure 1: Recorded travel times, $t_l(\mathbf{x})$, to Top and Base.

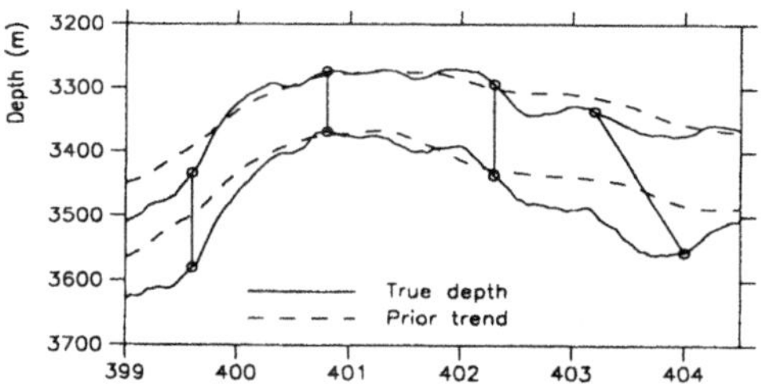

Figure 2: Possible true depth to Top and Base. Dashed lines are prior guesses. Depth observations are shown as small circles.

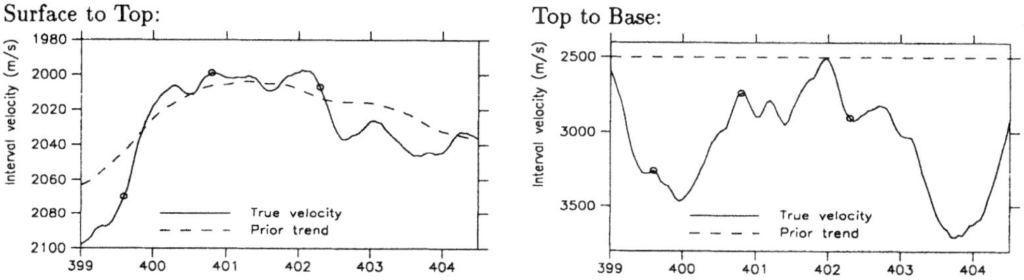

Figure 3: Possible true interval velocities. Dashed lines are prior guesses. Velocity observations are shown as small circles.

Figure 1 shows a 5.5 kilometer long cross-section of the travel time maps in the west-east direction. Figure 2 shows a cross section of the two subsurfaces. Four wells — three vertical and one deviating — are indicated by small circles connected by solid lines. To obtain depths from travel times, interval velocities must be specified. Figure 3 shows cross-sections of the possible true velocity fields and the six velocity observations from the vertical wells. Available observations are listed in Table 1.

4.2 Stochastic models

The stochastic model for Top is:

$$Z_{\text{Top}}(\mathbf{x}) = V_{\text{Top}}(\mathbf{x}) t_{\text{Top}}(\mathbf{x}) + R^z_{\text{Top}}(\mathbf{x}) \tag{16}$$

$$V_{\text{Top}}(\mathbf{x}) = A^1_{\text{Top}} + A^2_{\text{Top}} \left(t_{\text{Top}}(\mathbf{x}) - 1.63\text{s}\right) + R^v_{\text{Top}}(\mathbf{x}). \tag{17}$$

Note that $\min_{\mathbf{x}}\{t_{\text{Top}}(\mathbf{x})\} \approx 1.63\text{s}$. The two coefficient parameters have simple interpretations: A^1_{Top} is the average interval velocity to the crest of the geological structure and A^2_{Top} determines the increase (or decrease) in interval velocity at the flanks. So A^2_{Top} controls the curvature of the geological structure.

The stochastic model for Base is:

$$Z_{\text{Base}}(\mathbf{x}) = V_{\text{Top}}(\mathbf{x}) t_{\text{Top}}(\mathbf{x}) + V_{\text{Base}}(\mathbf{x}) \left(t_{\text{Base}}(\mathbf{x}) - t_{\text{Top}}(\mathbf{x})\right) + R^z_{\text{Base}}(\mathbf{x}) \tag{18}$$

$$V_{\text{Base}}(\mathbf{x}) = A^2_{\text{Base}} + R^v_{\text{Base}}(\mathbf{x}). \tag{19}$$

Well	Subsurface/ interval	X-coor. (km)	Depth (m)	Velocity (m/s)
1	Top	400.8	3273	1999
	Base		3368	2739
2	Top	402.3	3293	2007
	Base		3436	2903
3	Top	399.6	3433	2070
	Base		3580	3254
4	Top	403.2	3335	
	Base	404.0	3556	

Table 1: Available observations. Note that interval velocity observations are only available in vertical wells.

Residual	σ	a (m)	ρ
$R^z_{\text{Top}}(\mathbf{x})$	4m	1800	Spherical
$R^v_{\text{Top}}(\mathbf{x})$	12m/s	600	Gaussian
$R^z_{\text{Base}}(\mathbf{x})$	8m	2000	Spherical
$R^v_{\text{Base}}(\mathbf{x})$	300m/s	600	Gaussian

Table 2: Specifications for the Gaussian random fields. The covariance functions are: $\text{Cov}\{R(\mathbf{x}), R(\mathbf{x}')\} = \sigma^2\rho(|\mathbf{x}-\mathbf{x}'|; a)$.

The stochastic models for Top and Base include four residuals. They are modeled as Gaussian random fields with zero expectation. Table 2 gives the specifications used in the computations. The spherical correlation function is defined by $\rho(r;a) = 1 - \frac{3}{2}\frac{r}{a} - \frac{1}{2}\left(\frac{r}{a}\right)^3$ for $r < a$ and 0 else. The Gaussian (2. order exponential) correlation function is defined by $\rho(r;a) = \exp(-3(r/a)^2)$.

A full specification of the stochastic model requires prior guesses on the coefficient parameters. The expectations and standard deviations are found in Table 3.

General figure caption. Cross-sections of predicted depths and velocities are shown in Figures 4 to 8. Predictions, $Z^*(\mathbf{x})$, are shown as solid lines and prediction variances, $\sigma^2(\mathbf{x})$, are shown as dotted lines of $Z^*(\mathbf{x}) \pm \sigma(\mathbf{x})$. Well observations are shown as small circles. Observations from the same well is connected by a solid line. The horizontal axis are in the west-east direction and spans approximately 5.5 kilometers. The vertical scale are meters and meters per second for depth and velocity predictions respectively.

4.3 "Traditional" versus proposed method

The traditional approach to predicting the depth to subsurfaces in a layered media is to start by predicting the uppermost subsurface. Prediction of the second subsurface is done by adding the predicted thickness of the intermediate layer to the uppermost subsurface. Prediction variances are obtained for the first subsurface and the intermediate layer. The problem however, is to evaluate the prediction variance for the second subsurface. Simple addition of the two prediction variances would give the correct answer if the first subsurface and the thickness of the intermediate layer where uncorrelated. This is not so: the thickness of the intermediate layer between

Traditional method: Proposed method:

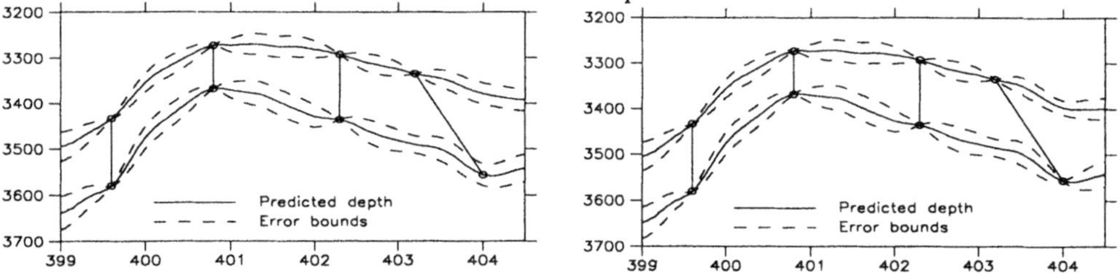

Figure 4: Predicted depth to Top and Base. Notice the failure of the traditional approach at the deviating well.

Table 3: Expectation μ, standard deviation σ, and correlations between the coefficient parameters for 0, 1, 2, 3, and 4 wells.

	Bayes estimates										GLS estimates (universal kriging)					
# of wells	0 (prior)		1		2		3		4		2		3		4	
# of obs.	0		2+2		4+4		6+4		8+6		4+4		6+4		8+6	
Parameter	μ	σ	μ	σ	μ	σ	μ	σ	μ	σ	μ	σ	μ	σ	μ	σ
A^1_{Top} m/s	2000	50	1993	12	1993	10	1997	9	1996	9	1991	25	1978	14	1987	11
A^2_{Top} m/s^2	1500	750	1493	749	1499	731	2269	625	2402	469	1701	3609	4050	1169	2992	610
A^1_{Base} m/s	2500	500	2675	257	2772	195	2864	186	2963	157	2821	212	2826	208	3000	166
Corr$\{A^1_T, A^2_T\}$	0		−0.25		−0.49		−0.70		−0.77		−0.94		−0.88		−0.84	
Corr$\{A^1_T, A^1_B\}$	0		0		0		0.01		−0.02		0		0.20		0.01	
Corr$\{A^2_T, A^1_B\}$	0		0		0		−0.16		−0.06		0		−0.30		−0.09	

Top and Base is:

$$
\begin{aligned}
\Delta Z_{\text{Base}}(\mathbf{x}) &= Z_{\text{Base}}(\mathbf{x}) - Z_{\text{Top}}(\mathbf{x}) \\
&= V_{\text{Base}}(\mathbf{x})\,(t_{\text{Base}}(\mathbf{x}) - t_{\text{Top}}(\mathbf{x})) + R^z_{\text{Base}}(\mathbf{x}) - R^z_{\text{Top}}(\mathbf{x}),
\end{aligned}
$$

which is (anti-)correlated to $Z_{\text{Top}}(\mathbf{x})$ due to the common residual $R^z_{\text{Top}}(\mathbf{x})$. Ignoring negative correlations causes to large prediction variances. Figure 4 compares the traditional approach to an approach where all correlations are considered. Both are produced by Bayesian kriging. The results are almost identical except from at the right flank. The prediction variance for Base fails to be zero at the well observation. The very strong intercorrelations between $\Delta Z_{\text{Base}}(\mathbf{x})$ and $Z_{\text{Top}}(\mathbf{x})$ in the vicinity of the well is ignored in the traditional approach leading to wrong results.

A different, but less dramatic effect, is that the rightmost observation of Base should have an impact on the prediction of Top. This is obviously not true for the traditional approach since Top is predicted independently of Base. The proposed method however, gives a deformation of the prediction for Top directly above the observation. Notice also a small local reduction in the prediction variance.

4.4 Bayesian kriging

Bayesian kriging is a necessity to obtain reasonable results in some applications; the lack of well observations in off-shore applications make universal kriging unreliable.

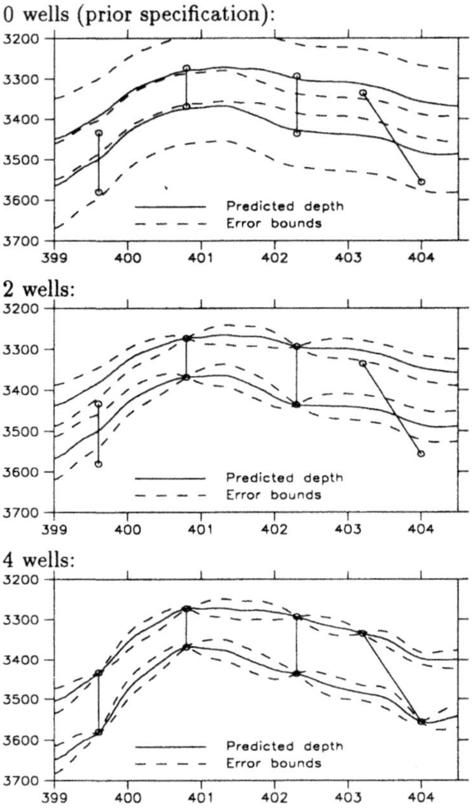

Figure 5: Predicted depth to Top and Base by Bayesian kriging.

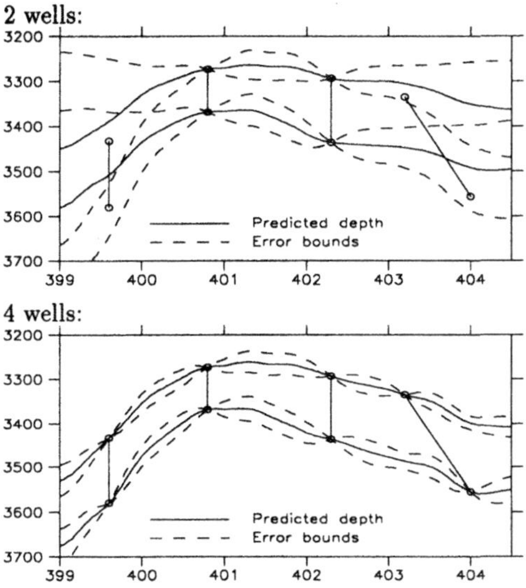

Figure 6: Predicted depth to Top and Base by universal kriging.

Figures 5 and 6 show predictions by Bayesian and universal kriging respectively. The predictions are conditioned on different numbers of wells.

The difference between Bayesian and universal kriging is the method for obtaining estimates for the coefficient parameters. Both methods use simple kriging for local adaptions in the vicinity of observations. Table 3 contains Bayesian and GLS estimates for the coefficient parameters. The estimation of A^1_{Top} and A^1_{Base} are very successful even when a single well is used. The reason is that the specified variance for the residuals are quite small compared to the prior variances for the coefficient parameters. The acceleration parameter, A^2_{Top}, controlling the curvature, is not successfully estimated before any of the deep observations in well 3 or 4 are included.

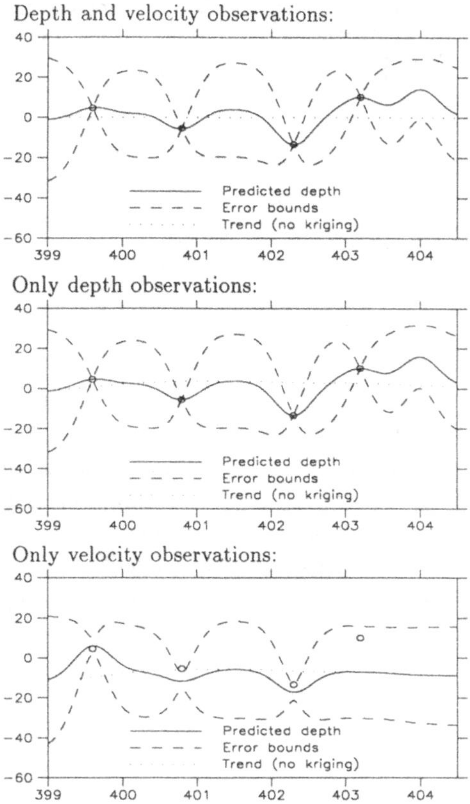

Figure 7: Bayesian depth predictions for Top. The trend for predictions with both depth and velocity data has been subtracted to exaggerate the local influence of the data.

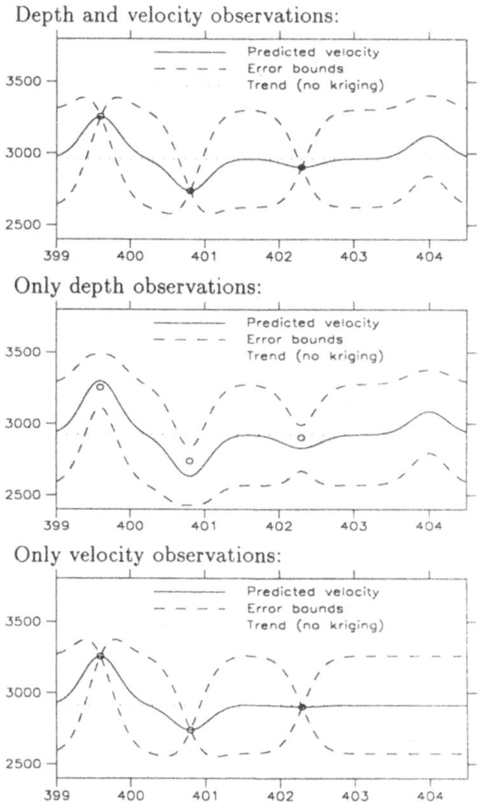

Figure 8: Bayesian velocity predictions for interval between Top and Base.

4.5 Use of velocity information

The impact of velocity observations on depth predictions are quite small due to strong intercorrelations. To visualize this Figure 7 shows depth predictions for Top conditioned on 1) all observations 2) depth observations alone 3) velocity observations alone. To exaggerate the variations the predicted trend obtained by using all data are subtracted from all the predictions. The two upper figures are almost identical indicating minor influence from velocity observations. Notice especially the 'hump' on the right side due to the depth observation of Base below. The reduction in uncertainty at this location is seen quite clearly. The prediction in the lower figure do not interpolate the depth observation since it is conditioned on velocity observations alone.

Figure 8 shows the predicted interval velocity in the layer between Top and Base.

Table 4: Generalized least squares estimates for expectation μ, standard deviation σ, and correlations of the coefficient parameters for all observations, depth observations, and velocity observations.

GLS estimates (universal kriging)						
Observations:	Depth+velocity		Depth		Velocity	
# of observations	8+6		8		6	
Parameter	μ	(σ)	μ	(σ)	μ	(σ)
A^1_{Top} (m/s)	1987	(11)	1990	(11)	1984	(11)
A^2_{Top} (m/s^2)	2992	(610)	2880	(621)	2915	(638)
A^1_{Base} (m/s)	3000	(166)	2975	(204)	2965	(173)
$\text{Corr}\{A^1_{\text{Top}}, A^2_{\text{Top}}\}$	-0.84		-0.84		-0.80	
$\text{Corr}\{A^1_{\text{Top}}, A^1_{\text{Base}}\}$	0.01		-0.01		0	
$\text{Corr}\{A^2_{\text{Top}}, A^1_{\text{Base}}\}$	-0.09		-0.10		0	

The influence from the depth observation of Base in the deviating well is seen as a hump and a small reduction in prediction variance.

Table 4 shows the corresponding estimated coefficient parameters. The more sensitive GLS estimates are shown. The general observation is that adding velocity data has minor implications since depth and velocity data are highly correlated.

Note however that A^1_{Base} is determined more precisely using the 3 relevant velocity observation than the 4 relevant depth observations. The reason is that the 6 velocity observations splits into two sets of uncorrelated observations while all the depth observations are strongly correlated.

5 Final remarks

A new method for depth conversion of seismic travel times to L reflecting subsurfaces has been presented. The method has the following characteristics:

- it is a combination of cokriging with $2L$ covariables and L dependent linear regression models for velocity and depth trends,
- predictions are possible by Bayesian or universal kriging,
- the predicted depths and velocities are conditioned on all observations from the N subsurfaces and N velocity fields,
- depth predictions interpolate the relevant depth observations, and velocity predictions interpolate the relevant velocity observations,
- prediction variances for all predictions are available,
- using deviating wells are trivial,
- estimates of coefficient parameters in the velocity models are based on all observations.

Conditional simulation based on the same model is described in Abrahamsen and Omre [1]. Simulation is necessary for the assessment of uncertainty in volumetric predictions.

References

[1] P. Abrahamsen and H. Omre. Horizon 4.0 — geometrical module, user guide and theory. NR-note SAND/01/1992, Norwegian Computing Center, P.O.Box 114 Blindern, N-0314 Oslo, Norway, 1992.

[2] J. O. Berger. *Statistical Decision Theory and Bayesian Analysis.* Springer series in statistics. Springer-Verlag Inc., New York, 2. edition, 1985.

[3] H. Omre and K. B. Halvorsen. The bayesian bridge between simple and universal kriging. *Math. Geol.*, 21(7):767–785, 1989.

[4] H. Omre, K. B. Halvorsen, and V. Berteig. A bayesian approach to kriging. In M. Armstrong, editor, *Geostatistics*, volume 1, pages 109–126. Kluwer Academic Publishers, 1989.

A Unified notation for depth to subsurfaces and interval velocities

We intend to show that the regression models:

$$(20) \qquad V_l(\mathbf{x}) = \sum_{p=1}^{P_l} A_l^p g_l^p(\mathbf{x}) + R_l^v(\mathbf{x})$$

$$(21) \qquad Z_L(\mathbf{x}) = \sum_{l=1}^{L} \left(\tilde{V}_l(\mathbf{x}) + R_l^v(\mathbf{x})\right) \Delta t_l(\mathbf{x}) + R_L^z(\mathbf{x})$$

can be written as a single regression model:

$$Z(\mathbf{x}) = \mathbf{f}(\mathbf{x}) \cdot \mathbf{A} + R(\mathbf{x}).$$

The exercise is mainly a change in notation.

It will prove convenient to use a new equivalent set of regression functions:

$$f_l^p(\mathbf{x}) = \Delta t_l(\mathbf{x}) g_l^p(\mathbf{x}).$$

Multiplying Equation (20) by $\Delta t_l(\mathbf{x})$ and introducing new regression functions gives:

$$(22) \qquad V_l(\mathbf{x})\Delta t_l(\mathbf{x}) = \sum_{p=1}^{P_l} A_l^p f_l^p(\mathbf{x}) + R_l^v(\mathbf{x})\Delta t_l(\mathbf{x})$$

$$(23) \qquad Z_L(\mathbf{x}) = \sum_{l=1}^{L} V_l(\mathbf{x})\Delta t_l(\mathbf{x}) + R_L^z(\mathbf{x}).$$

Note that Equation (22) is equivalent to Equation (20) since $\Delta t_l(\mathbf{x})$ is considered a known function.

All coefficient parameters will be organized in a $P = \sum_{l=1}^{L} P_l$-dimensional column vector starting with the P_1 parameters for the first interval and so on:

$$\begin{aligned} \mathbf{A}^T &= (A_1, \dots, A_P) \\ &= \left([A_1^1, \dots, A_1^{P_1}], \dots, [A_L^1, \dots, A_L^{P_L}]\right). \end{aligned}$$

The square brackets simply emphasize the initial grouping of parameters. The space dependent regression functions $f_l^p(\mathbf{x})$ are organized accordingly in a P-dimensional row vector such that the trends can be written

$$\begin{aligned} \tilde{V}_l(\mathbf{x})\Delta t_l(\mathbf{x}) &= \mathbf{f}_l^{[v]}(\mathbf{x}) \cdot \mathbf{A} \\ \tilde{Z}_l(\mathbf{x}) &= \mathbf{f}_l^{[z]}(\mathbf{x}) \cdot \mathbf{A}. \end{aligned}$$

The crucial point is to replace all $f_l^p(\mathbf{x})$'s irrelevant for a certain surface by zeroes:

$$\begin{aligned} \mathbf{f}_l^{[v]}(\mathbf{x}) &= \left(0, \dots, 0, [f_l^1(\mathbf{x}), \dots, f_l^{P_l}(\mathbf{x})], 0, \dots, 0\right) \\ \mathbf{f}_l^{[z]}(\mathbf{x}) &= \left([f_1^1(\mathbf{x}), \dots, f_1^{P_1}(\mathbf{x})], \dots, [f_l^1(\mathbf{x}), \dots, f_l^{P_l}(\mathbf{x})], 0, \dots, 0\right). \end{aligned}$$

Introducing generic forms:

$$(24) \qquad Z(\mathbf{x}) = \begin{cases} Z_L(\mathbf{x}); & \text{for depth to subsurface } L \\ V_L(\mathbf{x})\Delta t_L(\mathbf{x}); & \text{for velocities in } L-1 \text{ to } L \end{cases}$$

$$(25) \qquad \mathbf{f}(\mathbf{x}) = \begin{cases} \mathbf{f}_L^{[v]}(\mathbf{x}); & \text{for } Z(\mathbf{x}) = Z_L(\mathbf{x}) \\ \mathbf{f}_L^{[z]}(\mathbf{x}); & \text{for } Z(\mathbf{x}) = V_L(\mathbf{x})\Delta t_l(\mathbf{x}) \end{cases}$$

$$(26) \qquad R(\mathbf{x}) = \begin{cases} R_L^z(\mathbf{x}) + \sum_{l=1}^{L} R_L^v(\mathbf{x})\Delta t_L(\mathbf{x}); & \text{for } Z(\mathbf{x}) = Z_L(\mathbf{x}) \\ R_L^v(\mathbf{x})\Delta t_L(\mathbf{x}); & \text{for } Z(\mathbf{x}) = V_l(\mathbf{x})\Delta t_l(\mathbf{x}) \end{cases}$$

Using these gives the single regression model

$$(27) \qquad Z(\mathbf{x}) = \mathbf{f}(\mathbf{x}) \cdot \mathbf{A} + R(\mathbf{x}),$$

for both depths to subsurfaces and velocities. The parameter vector, $\mathbf{A}$, is independent of the interpretation of Z, $\mathbf{f}$, and R.

Equation (27) describes $V_l(\mathbf{x})\Delta t_l(\mathbf{x})$ rather than $V_l(\mathbf{x})$ and the same applies to the Bayesian kriging equation. This implies that the velocity observations must be multiplied by $\Delta t_l(\mathbf{x})$ when used in the kriging equation. The result — the predicted surface — must therefore be divided by $\Delta t_l(\mathbf{x})$ to obtain the interval velocity field.

For the N observations of the surfaces Equation (27) can be written

$$(28) \qquad \mathbf{Z} = F\mathbf{A} + \mathbf{R}.$$

where $\mathbf{Z}$ is the N observations organized as a column vector and $\mathbf{R}$ is the corresponding unknown residuals. The $N \times P$-dimensional matrix F contains the corresponding regression functions. Every row of F is the P-dimensional $\mathbf{f}(\mathbf{x})$ vector for the corresponding observation.

INDICATOR SIMULATION OF PETROPHYSICAL ROCK TYPES

CHRISTOPHER J. MURRAY
Stanford Center for Reservoir Forecasting
Department of Applied Earth Sciences
Stanford, CA 94305-2225 USA

ABSTRACT. Because the reservoir properties of various rock types differ considerably, it is better to simulate the distribution of reservoir properties in two steps: simulate the rock types, then simulate the reservoir properties, conditional on the rock types. However, identification of the rock types themselves can be difficult, due to the interaction of depositional processes and diagenesis, and the variable amount and quality of data available from well to well. This paper presents a methodology using cluster analysis for identification of petrophysical rock types in a training data set of cored wells. Analysis of the cores is then used for geologic verification of the rock types. Discriminant function analysis is used to extend the rock type classification to wells without core data. The spatial distribution of the rock types is then modeled and simulated in three dimensions using indicator principal components.

1. INTRODUCTION

Petroleum reservoirs result from the complex interaction of several geologic processes. The interaction of erosion, deposition, and diagenesis controls the distribution of reservoir properties. Although those properties exist in a deterministic fashion beneath the ground, only a limited sample of well log and core data are available from the reservoir. Reservoir flow simulations, which can be used to examine the uncertainty attached to the future performance of the reservoir, require a complete grid of the important reservoir properties, and stochastic simulation provides one way to generate such a grid (Haldorsen and Damsleth, 1990). Because the reservoir properties of various rock types can be quite different, some researchers have proposed that the geologic architecture of the reservoir be simulated first, then infilled with reservoir properties (Suro-Perez and Journel, 1990; Journel and Gomez-Hernandez, 1989).

Petrophysical rock types can be defined as classes of rock characterized by important differences in reservoir properties, especially porosity and permeability. Petrophysical rock types cannot be defined solely on the basis of the original depositional environments, because diagenesis often alters the original reservoir properties. Also, in most cases, core data is only available for a limited number of wells, so the petrophysical rock types must be identifiable using well log data. This paper presents a methodology for defining the rock types using multivariate statistical analysis of well log and core data. The spatial distribution of those rock types is then simulated in three dimensions using indicator principal component simulation (Suro-Perez and Journel, 1990).

2. GEOLOGIC SETTING

The Lower Cretaceous Muddy Formation is the major producing formation in the study area, which consists of a portion of Amos Draw Field, in the northern Powder River Basin of Wyoming. The study area is a square measuring 30,000 feet (9,144 m) on a side (Fig. 1). The Rozet sandstone member of the Muddy Formation, which is the reservoir at Amos Draw, is a progradational sandstone deposited during a regression of the Cretaceous Sea (Odland et al., 1988). After deposition of the Rozet, a further drop in sea level led to the development of an unconformity that truncates the Rozet member. Relief on the post-Rozet unconformity, and sub-aerial weathering that developed below the unconformity, directly control the permeability distribution and production at Amos Draw (Odland et al., 1988; Chambers, 1986). The Rozet at Amos Draw has an average of 12-14 feet of net pay, average porosity of 12-18 percent, and low permeability, averaging 1-2 millidarcies (VonDrehle, 1985).

3. IDENTIFICATION OF PETROPHYSICAL ROCK TYPES

One difficulty that is common to many reservoir studies is the scarcity of core data. Core data (porosity and permeability) were available for 7 of 22 sample wells in the study area (Fig. 1). Well log data (gamma ray, density, and resistivity) were available for the rest of the wells. Only three of the cores were available for description of the depositional environments and thin-section analysis. Therefore, it was desirable to develop a two-step method for identifying the rock types. In the first step, the important rock types were defined on the cored wells, so that the geologic framework of the rock types could be examined in detail, and hopefully understood in terms of the depositional and diagenetic settings in which the rock types developed. Once the rock types were defined in the cored wells, the classification was extended to wells for which well log data alone were available. The multivariate techniques of cluster analysis and discriminant function analysis provided tools for performing that two-step analysis.

3.1 Rozet Depositional Environments

Three Rozet cores in the study area were measured and described (Fig. 2). The interpretations of the depositional environments within the Rozet were based on the comparison of sedimentary structures, grain size, and stratigraphic relationships with published depositional models (Berg, 1986). Five lithofacies were identified in the described cores, including four depositional lithofacies and an altered zone (Fig. 2). From bottom to top the depositional lithofacies were: lower shoreface, middle shoreface, upper shoreface/foreshore, and backshore. Berg (1986) includes a detailed description of those depositional environments and their properties. The depositional lithofacies were overlain by an altered zone that appears to be a paleosol developed beneath the Rozet unconformity (Odland et al., 1988). Examination of the petrophysical data indicated substantial overlap between the reservoir properties of the five lithofacies, apparently caused by diagenetic alteration of the original depositional lithofacies.

3.2 Identification of Petrophysical Rock Types by Multivariate Analysis

Although it might have been possible to identify the depositional lithofacies in the non-cored wells using detailed well log correlations, that approach was not pursued because of the variation in their reservoir properties. As an alternative, cluster analysis and discriminant function analysis (Davis, 1986; Dillon and Goldstein, 1984) were used to identify petrophysical rock types having different reservoir properties.

Agglomerative hierarchical cluster analysis was used for identification of "natural clusters" of samples with similar properties (Dillon and Goldstein, 1986). The cluster analysis was performed on cored wells. The Euclidean distance provided the similarity measure for the cluster analysis, which was performed using the average linkage algorithm in the CLUSTER program (SAS Institute Inc., 1988). The variables used in the analysis were the gamma ray, bulk density, resistivity, grain density, core porosity, and the logarithm of the core permeability. All variables had been standardized to have a mean of zero and a standard deviation of one.

Over 180 samples were used in the analysis, so display of the dendrograms from the cluster analysis is impractical here. Instead, the results of the cluster analysis are displayed on a canonical discriminant plot (Fig. 3). Canonical discriminant analysis summarizes the between-class variation for grouped observations in a manner similar to the way in which principal component analysis summarizes total variance (SAS Institute Inc., 1988). Three petrophysical rock types were identified by the cluster analysis, and the canonical coefficients indicate good separation between the rock types (Fig. 3). Because cluster analysis is sensitive to the choice of clustering algorithm, the analysis was repeated using the complete linkage algorithm (farthest neighbor) and Ward's algorithm (Dillon and Goldstein, 1986). Over 95 percent of the samples were placed in the same clusters by all three clustering algorithms.

After completion of a cluster analysis, the user must determine if the classification of samples makes sense geologically. In this study, the stratigraphic, petrographic, and petrophysical properties of the clusters were examined. The three clusters correspond to reservoir rock, rock type 3; non-reservoir rock, rock type 1; and an intermediate rock type, rock type 2 (see Fig. 4). The reservoir rock occurs in clean, well-sorted sandstones of the middle shoreface, and has excellent porosity and permeability. The intermediate rock type has low gamma ray values and low densities indicating clean sandstones with good porosity. However, the permeability and resistivity of the intermediate rock type are low. Most of the intermediate rock type occurs in the middle and upper shoreface depositional environments. Thin-section analysis indicates most of the intermediate sandstones were originally porous and permeable, but have been damaged by weathering and burial diagenesis.

After identifying the three petrophysical rock types in the cored wells using cluster analysis, and determining the geologic character of the rock types, discriminant function analysis extended the classification of the rock types to the wells which did not have core data. Discriminant function analysis (DFA), unlike cluster analysis, requires a training data set in which the classes of objects are known (Davis, 1986; Dillon and Goldstein, 1986). DFA identifies a linear combination of the variables that maximizes the distance between the class vector means, while minimizing the within-class variance. The DFA was performed using the SAS procedure DISCRIM (SAS Institute Inc., 1988), with the cored wells as a training set, and the logged wells as the target set to be classified. Cross-validation of the DFA on the cored wells indicated an eight percent misclassification rate when the rock types were classified using only the well log data.

The final output from the rock type classification process was a foot by foot determination of the rock type in each of the 22 wells, a total of 645 samples. The non-reservoir rock, rock type 1, constituted 71 percent of the observations. Rock type 2, the intermediate rock type, comprised 17 percent, and the remaining 12 percent of the samples were rock type 3, the reservoir rock.

4. SIMULATION OF THE ROCK TYPES

The method chosen for simulation of the rock types was Indicator Principal Component Simulation (Suro-Perez and Journel, 1990). This technique relies on the orthogonal nature of principal components in an attempt to account for the spatial cross-correlation between rock types. The first step in an indicator principal component (IPC) study is an analysis of the correlation matrix of the rock type indicators for a given lag, often chosen as a lag of zero. After the principal components are determined, the variogram of each principal component must be calculated and modeled.

The indicator principal component simulation program used for this study was IPCSIM (Deutsch and Journel, 1992). This program uses the sequential simulation algorithm, conditioning each simulated node by the nearby data, as well as all previously simulated nodes in the area (Deutsch and Journel, 1992). Each principal component is kriged independently during the simulation of a grid node, based on the assumption of orthogonality of the principal components. Back-transformation of the kriged principal components provides the conditional distribution of the rock types for that node, from which the simulated rock type is drawn at random (Suro-Perez and Journel, 1990). This process is repeated until all nodes are simulated.

Since the lithofacies indicators constitute a closed system, only two principal components are needed to characterize the three petrophysical rock types identified in this study. The scores for the two principal components (see below) indicate that the non-reservoir rock (type 1) is accounted for by the first indicator principal component (IPC1), while the reservoir rock (type 3) is mostly represented by the second principal component (IPC2). The scores for rock type 2, the intermediate rock type, are split between the first and second indicator principal components.

Principal Component Scores

	IPC 1	IPC2
Rock type 1	-0.8036	-0.1448
Rock type 2	0.5072	-0.6235
Rock type 3	0.2764	0.7683

Because the principal component transformation was performed at lag zero, a check was performed to ensure that the orthogonalization remains valid at other lags. The low cross-correlation between the indicator principal components indicates the assumption is reasonable in this case (Fig. 5).

4.1 Variogram Modeling

For the variogram analysis and the simulations, the data location coordinates were transformed to stratigraphic coordinates (Journel and Gomez-Hernandez, 1989), based on an average Rozet thickness of 30 feet (9.14 meters). Fig. 6 presents the experimental and model variograms for the two indicator principal components in the vertical direction. These vertical variograms are well-behaved and easily modeled, with ranges of 12 and 6 feet, respectively. The horizontal variograms could not be modeled because there were only 22 wells in the study area, with an average well spacing of 2640 feet (805 meters). The horizontal variogram model was derived from geologic studies of the distribution of productivity at Amos Draw (Murray, 1992), and included a horizontal to vertical anisotropy ratio of 2500:1, a direction of maximum continuity of N60E, and a 3:1 anisotropy ratio in the horizontal direction.

4.2 Conditional Simulations

Several simulations of the 3 petrophysical rock types were generated using the Indicator Principal Component algorithm. Each three-dimensional simulation contained 108,000 grid cells, with a cell size of 500 feet by 500 feet by 1 foot (152 by 152 by 0.3 m). The simulations capture several important geologic features present at Amos Draw (Figs. 7 and 8). The marginal and reservoir rock types (2 and 3, respectively) are concentrated in the middle of the Rozet, with most of the reservoir rock encased in the intermediate rock type. Also, the intermediate rock type is thicker above the main reservoir than it is below. The simulations honored the well data and the proportions of the three rock types. In the simulation shown in Figs. 7 and 8, the proportions of the rock types were: 74 percent of rock type 1, 16 percent of rock type 2, and 10 percent of rock type 3 (the input proportions were 71, 17, and 12 percent, respectively).

The horizontal continuity of the rock types appears reasonable and reproduces the range and anisotropy ratio of the input variogram model. The vertical variograms calculated from the simulation (Fig. 9) show the presence of high relative nuggets compared to the input models, for which the nuggets were zero. Examination of the simulations also indicate that the vertical transition frequencies between the rock types do not match those seen in the well bores. This may indicate that IPCSIM did not fully capture the cross-correlation between the rock types. A post-processing step using simulated annealing (Deutsch and Journel, 1992) was applied in subsequent work, resulting in a better reproduction of the facies transition frequencies and the vertical variogram model (Murray, 1992).

5. CONCLUSIONS

Diagenesis severely altered the reservoir properties of sandstones within the study area, so simulation of the depositional lithofacies would not be sufficient to characterize the reservoir. As an alternative, cluster analysis provided a method of identifying petrophysical rock types in cored wells. Discriminant function analysis was then used to extend the classification of the rock types to wells that did not have core data. This two-step process identified three rock types with different reservoir properties.

Three-dimensional simulations of the petrophysical rock types were generated using the Indicator Principal Components algorithm. The simulations could be used directly for analysis of the effect of reservoir heterogeneity on recovery. By infilling the simulated rock types with reservoir properties (porosity, permeability, and saturation) the simulations could be used in reservoir fluid flow simulations. This infilling could be accomplished by generating conditional Gaussian simulations (Deutsch and Journel, 1992) of the reservoir properties for each rock type.

One major uncertainty in this study concerns the horizontal continuity of the rock types. In industrial studies performed for decision making, sensitivity analyses should be conducted using simulations generated with several different horizontal to vertical anisotropy ratios. In addition, analogous surface outcrops and seismic data might be used to better constrain the horizontal continuity of the petrophysical rock types between the wells.

ACKNOWLEDGEMENT

The author thanks Amoco Production Co. for providing the data used in this study.

REFERENCES

Berg, R. R., 1986, Reservoir sandstones: Prentice Hall, Engelwood Cliffs, NJ, 481 p.

Chambers, L. D., 1986, Hydrocarbon trapping mechanism for the Muddy Sandstone reservoir, Amos Draw Field, Powder River Basin, Wyoming: M.Sc. thesis, Univ. of Colorado, Boulder, CO, 223 p.

Davis, J. C., 1986, Statistics and data analysis in geology, 2nd ed.: John Wiley and Sons, New York, 646 p.

Deutsch, C. V., and A. G. Journel, 1992, GSLIB: Geostatistical software library and user's guide: Oxford University Press, New York, in press.

Dillon, W. R., and M. Goldstein, 1984, Multivariate analysis: John Wiley and Sons, New York, 587 p.

Haldorsen, H. and E. Damsleth, 1990, Stochastic modeling: Journal of Petroleum Technology, p. 404-412.

Journel, A. G., and J. Gomez-Hernandez, 1989, Stochastic imaging of the Wilmington clastic sequence, *in* Omega, Formation evaluation and reservoir geology, SPE paper #19857, p. 591-606.

Murray, C. J., 1992, Geostatistical applications in petroleum geology and sedimentary geology: Ph.D. dissertation, Stanford University, in press.

Odland, S. K., Patterson, P. E., and Gustason, E. R., 1988, Amos Draw Field: A diagenetic trap related to an intraformational unconformity in the Muddy Sandstone, Powder River Basin, Wyoming: Wyoming Geol. Assoc. Guidebook, 39th Ann. Field Conf., p. 147-160.

SAS Institute Inc., 1988, SAS/STAT user's guide, Release 6.03: SAS Institute Inc., Cary, NC, 1028 p.

Suro-Perez, V. and A. G. Journel, 1990, Simulation of lithofacies, *in* Guerillot and Guillon, eds. Proceedings of 2nd European Conference on the Mathematics of Oil Recovery: Technip Publ., Paris, p. 3-10.

VonDrehle, W. F., 1985, Amos Draw Field, Campbell County, Wyo.: Wyoming Geol. Assn., 36th Field Conf. Guidebook, pp. 11-31.

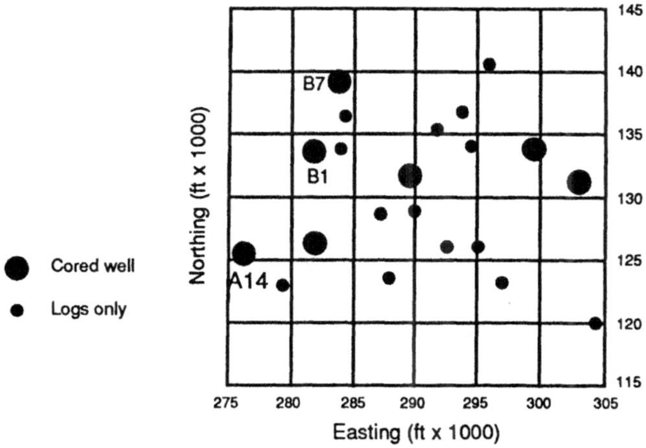

Fig. 1 Sample Location Map, Amos Draw 3D Study Area
B1 = Amoco Bertolet #1, B7 = Amoco Bertolet #7,
A14 = Davis Andy #14

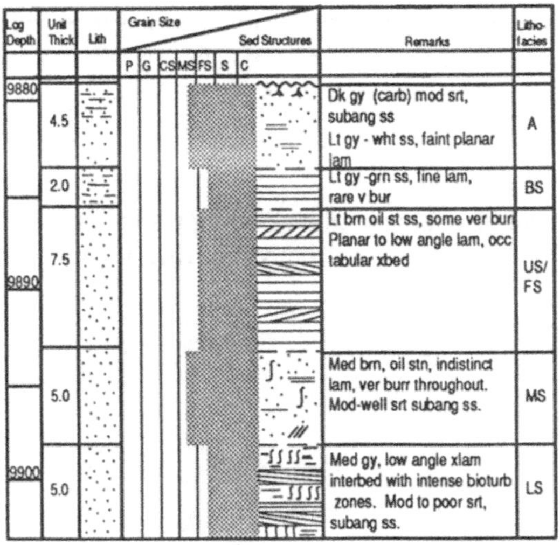

Fig. 2 Geologic Core Description, Rozet Member of Muddy Formation, Amoco Bertolet #1, Amos Draw Field

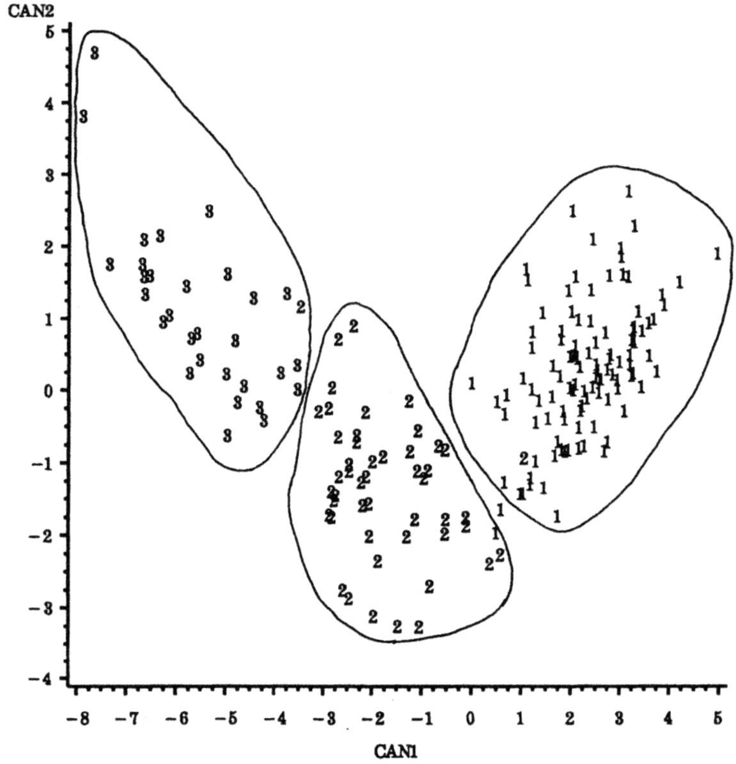

Fig. 3 Canonical discriminant plot showing results of cluster analysis of well log and core data. Three petrophysical rock types were identified.

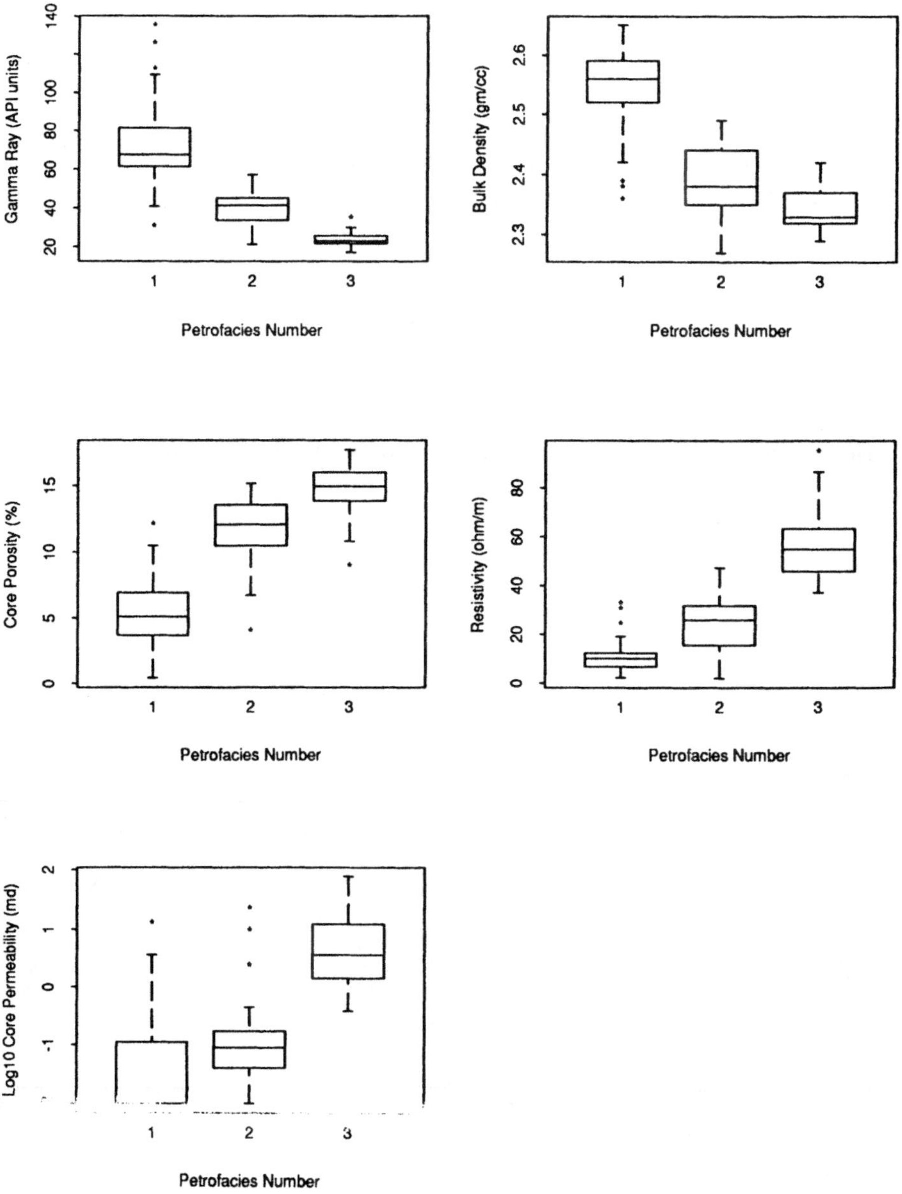

Fig. 4 Box plots of 5 well log and core properties for the 3 petrophysical rock types

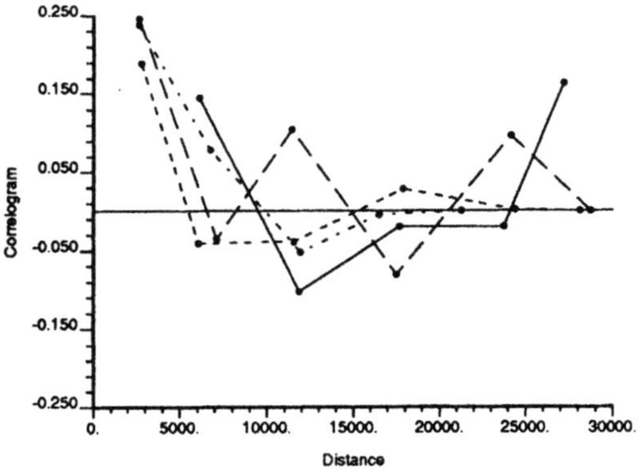

Fig. 5 Horizontal cross-correlogram between the two principal components

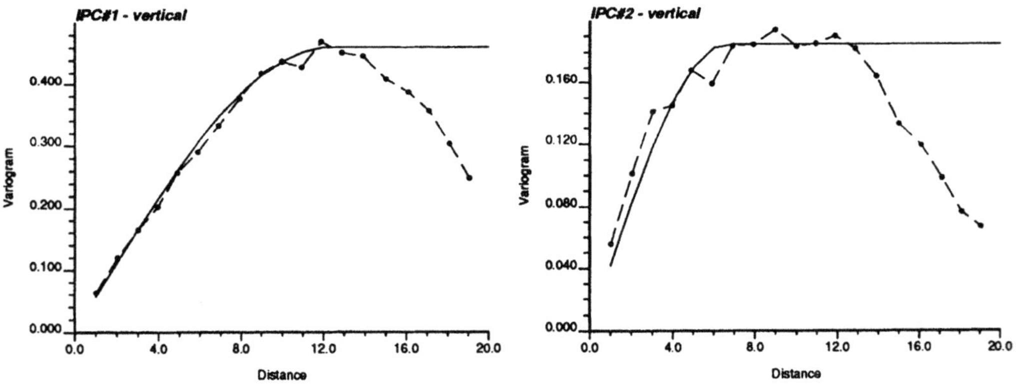

Fig. 6 Vertical variograms for the two principal components. Dashed lines show the experimental variograms, solid lines are the models fit to the variograms.

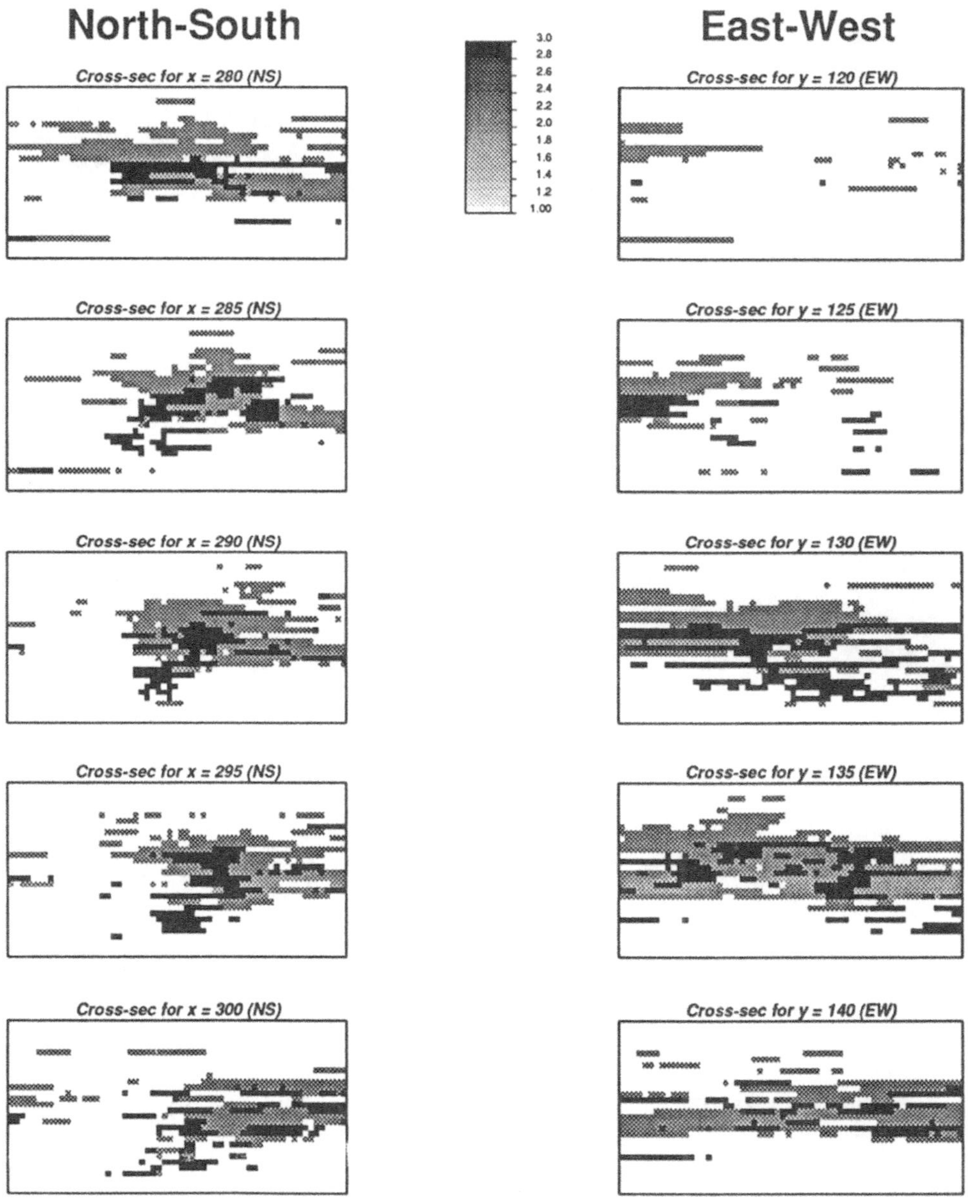

Fig. 7 Vertical cross-sections from a 3D simulation of the 3 petrophysical rock types. Cross-sections are spaced 5,000 ft (1,524 m) apart. Each cross-section is 30,000 ft (9,144 m) across and 30 ft (9.14 m) high. Rock type 3 is the best reservoir rock. The x and y coordinates are linked to Fig. 1.

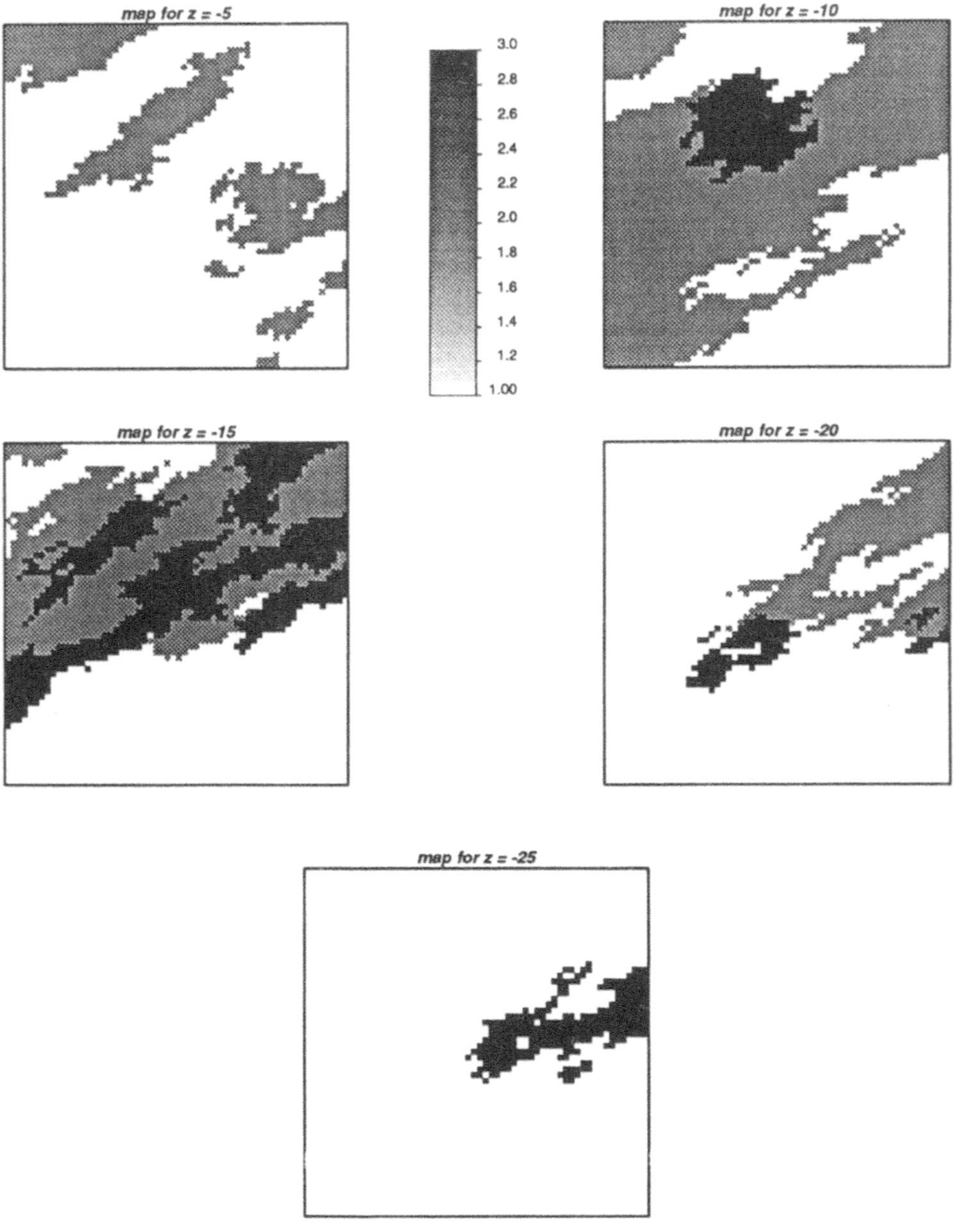

Fig. 8 Horizontal maps showing 5 levels of the 3D simulation of the three rock types. Maps are spaced 5 ft (1.5 m) apart. Each map is 30,000 ft (9,144 m) on a side. The z coordinate is measured down from the top of the Rozet Member.

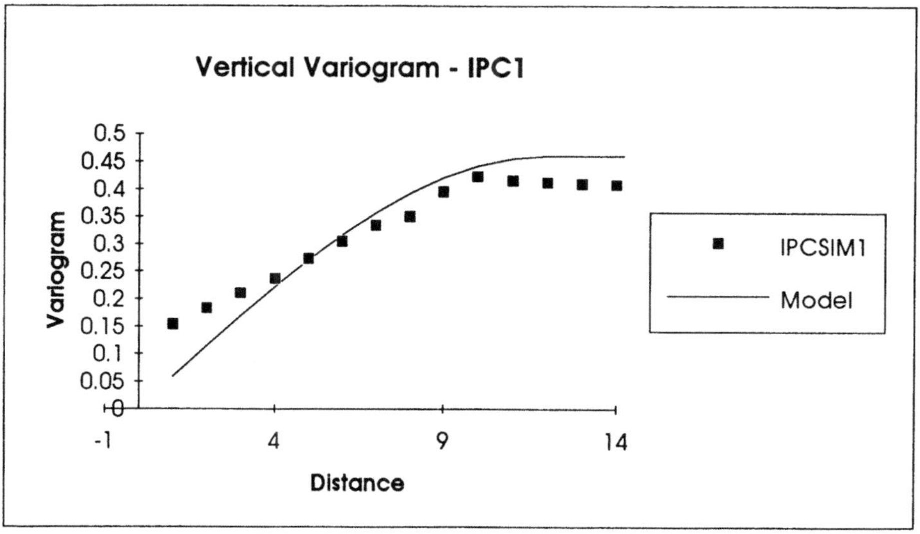

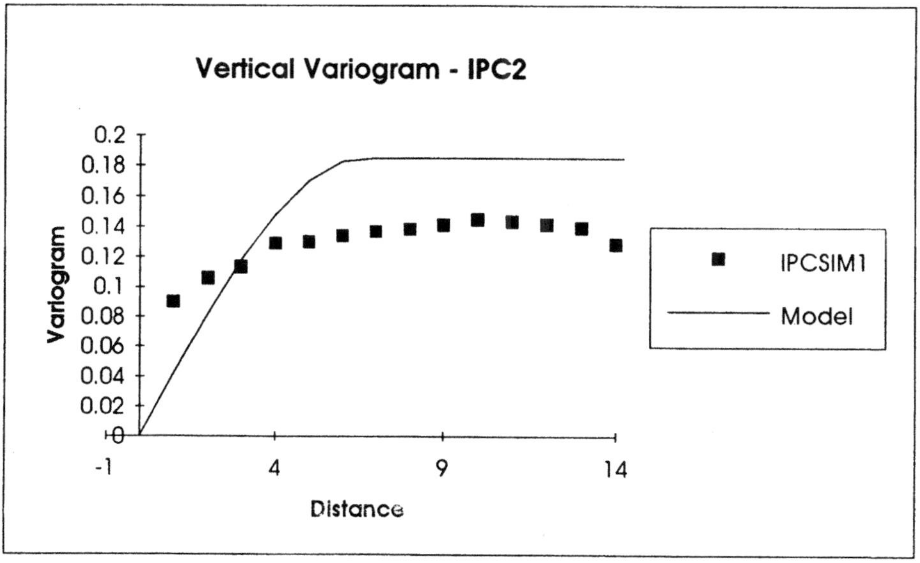

Fig. 9 Reproduction of the variogram model by IPCSIM

SEQUENTIAL MODELLING OF RELATIVE INDICATOR VARIABLES: DEALING WITH MULTIPLE LITHOLOGY TYPES

ROUSSOS DIMITRAKOPOULOS[1] and MICHEL DAGBERT[2]

[1] McGill University, Dept. of Mining and Metallurgical Eng., 3480 University St., Montreal, Qc, CANADA H3A 2A7
[2] GEOSTAT Systems International Inc., 4385 St. Hubert Str., #1, Montreal, Qc, CANADA H2J 2X1

ABSTRACT
This paper deals with the use of relative indicators as the means to model qualitative variables such as multiple lithology types. Relative indicator variables are defined sequentially, each within the space of the preceding. As a result, indicator type modelling approaches can be implemented to sequentially generate models of qualitative variables, using both indicator kriging and indicator conditional simulation. Two examples are given, one in estimating the lithologies of a mineral deposit and one in simulating the lithofacies of a reservoir.

INTRODUCTION

Qualitative characteristics such as lithologic types, play an important role in understanding geological phenomena of interest. For instance, ore mineralization and grade distributions are most often controlled by multiple rock types. In a similar fashion, petroleum reservoirs and corresponding rock properties are controlled by sedimentary lithofacies variations and architecture. It is therefore necessary to consider alternative techniques that can be used for the modelling of qualitative (categorical) variables, both in the sense of smooth estimated models as well as simulated images.

In estimating qualitative variables, most of the existing attempts are based on indicator kriging (Journel, 1983). Along this path, more recent developments include morphological kriging (Soares, 1990), which combines indicator kriging and spatial orientation data in dealing with two qualitative variables. The method can also be generalized to accommodate a series of variables (Soares, 1992). Other approaches include the combination of geostatistics with geometric modelling techniques (Mallet, 1989) and the use of boolean type algorithms (Haldorsen et al., 1988). Developments in the context of conditional simulation (Journel and Huijbregts, 1978; Christakos, 1992) include a variety of indicator type simulations (Journel and Isaaks, 1984; Journel and Alabert, 1990; Soares and Costa e Silva, 1991). The latter have been developed to handle two qualitative characteristics, or deal with additional variables based on principlal component indicator kriging (Suro-Perez and Journel, 1990). A method based on the conditional simulation of random sets has also been developed (Matheron et al., 1988).

Although the above developments provide some alternatives, they may be limited by the number of variables that can be modelled, require data that is not always available, may not provide satisfactory solutions, or may be based on distributional assumptions that are difficult to verify. Nevertheless, it seems imperative to search for alternative frameworks, allowing the modelling of sequences of qualitative spatial variates. In this context, the concept of relative (nested) indicator kriging (Dagbert, 1990) is explored and adapted to the problem in hand. In modelling continuous variables, relative indicator kriging is an indicator type technique which has been used to estimate conditional probability distributions, based on Bayes relation. An important aspect of the technique is the introduction of relative indicator variables and the corresponding relative indicator variograms. Using both these concepts, any number of qualitative variables of the same physical type may be numerically coded, analyzed and modelled sequentially, by considering the one as a proportion of the remaining.

In the present paper, the basic ideas in relative indicator kriging are used for the analysis and modelling of spatial qualitative physical varieties and, specifically, lithology types. Basic relations are outlined first for estimation and practical aspects in estimating lithologies are shown in the modelling of a hard-rock mineral deposit. Subsequently, the conditional simulation is examined in the context of relative indicators and its application illustrated in generating images of sedimentary lithofacies in a petroleum reservoir. Finally, conclusions and suggestions for further work are given.

KRIGING RELATIVE INDICATOR VARIABLES

Relative indicator kriging in modelling conditional distributions

Given a stationary spatial random function $\{Z(x),\ x \in D\}$ with univariate and bivariate probability distribution functions $F(z)=P[Z(x)\leq z]$ and $F(z,z')=P[Z(x)\leq z,\ Z(x+h\leq z']$, respectively, a stationary indicator random function is defined, such that

$$I(x;z) = \begin{cases} 1 & \text{if } Z(x)\geq z \\ 0 & \text{otherwise} \end{cases} \tag{1}$$

and $D=\{x: I(x;z)=1\}$. Then, the cumulative conditional distribution function (ccdf) $F(x;z|(N))$ can be estimated from

$$[1\text{-}F(x;z|(N))]^* = P^*[Z(x)\geq z\,|\,(N)]$$

$$= [I(x;z)|Z(x_a)=z_a,\ a=1,\ldots,N]^* = \textstyle\sum_{a=1}^{N} \lambda_a(x_a;z)\ I(x_a;z) \tag{2}$$

where the weights $\lambda_a\ (x_a;z)$ are derived using the ordinary kriging equations.

Given a series of k threshold values z_i with $z_1<\ldots<z_k$, field D may be partitioned into relative sub-fields, such that $D_k\subset\ldots\subset D_1$ and $D_i = \{x:\ x \in D_{i-1},\ I_i(x;z_i)=1\}$ with $i=2,\ldots,k$. In this case and for $i>1$, the ith indicator RF is defined relative to the previous one and is, apparently, different from the indicator RF corresponding to the same threshold value in the conventional indicator kriging (Journel, 1983). As a result, eq. 2 can not be used as in indicator kriging.

However, if one recalls Bayes relation, then eq. 2 may be rewritten to express relative indicator kriging, such that

$$[1-F(x;z_i|(N_i))]^* = \prod_{j=1}^{i} [I_j(x;z_j)|(N_j)]^* \tag{3}$$

with

$$[I_i(x;z_i)|(N_i)]^* = P^*[Z(x)\geq z_i|x \in D_{i-1}, (N_i)] = \sum_{a=1}^{N_i} \lambda_a^i(x_a;z_i)\, I_i(x_a;z_i) \tag{4}$$

where the weights $\lambda_a^i(x_a;z)$ are derived using the ordinary kriging equations, using the corresponding set of N_i data, subset of the full data set N.

The apparent advantage of relative indicator kriging is the reduction of order relation problems that may arise in the estimation of a ccdf, when the latter is estimated using conventional indicator kriging.

Estimating qualitative spatial variables

Consider field D, within which a sequence of related qualitative variables $\{q_i, i=1,...k\}$ are recognized. Similarly to the approach in the previous section, D may be partitioned based on q_i, such that $D_k \subset ... \subset D_1$. Then, a sequence of k-1 indicator variables may be defined, the one relative to the preceding, such that

$$I_i(x) = \begin{cases} 1 & \text{if } x \in D_i \\ 0 & \text{if } x \in \text{remaining} \end{cases} \tag{5}$$

The above binary coding may then be interpreted as the probability $P[x \in D_1|(N_1)]$ for q_1 and $P[x \in D_i | x \in D_{i-1}, (N_i)]$, i=2,...k-1 for the remaining qualitative variables. Note that this is similar to the related interpretation of indicator variables (e.g. Solow, 1986), except that it is applied within each of the sub-fields D_i.

Generally, each relative indicator in eq. 5 may be estimated from

$$[I_i(x)|(N_i)]^* = \sum_{a=1}^{N_i} \lambda_a^i(x_a)\, I_i(x_a) \tag{6}$$

where the weights $\lambda_a^i(x_a)$ may be derived using ordinary kriging and, each time, the corresponding set of N_i data.

In practice, the estimation of a sequence of qualitative variables proceeds in a sequential fashion. First, the estimation of the first variable and the transformation of probabilities to binary indicators is performed. Then, the second variable is modelled within the domain outlined from the previous step and, again the estimated probabilities are transformed to binary indicators. The process is repeated until all the qualitative variables appear within the domain being modelled.

Remarks

A general analysis and discussion on the estimation of ccdf's using the concept of relative indicators is beyond the scope of the present study. However, several aspects relevant to the estimation of sequences of qualitative variables need to be addressed.

The ordering of the qualitative variables within a sequence is of critical importance to the resulting model. Evidently, different sequencing of the same variables will result in different models. The proper sequencing can only be decided upon the physical characteristics of the phenomenon under study. For example, the sequencing of rock types in a mineral deposit can be decided on the basis of a geological interpretation. Considering that different interpretations may exist, it seems reasonable that a modelling technique should produce different results for different interpretations.

There is no unique way to transform estimated probabilities to binary values. Common sense, however, suggests that the relative proportions of qualitative variables should be respected. In this context, a solution previously suggested (Soares, 1990) may be followed. Accordingly, the l estimated probabilities within a domain are ordered in descending order and the first p values, p = l x m, are considered to belong to one quality while the rest belong to the remaining. m is relative indicator mean derived from either a global estimation or inferred from the data.

To perform ordinary kriging as required in eqs. 4 and 6, the corresponding variogram is required. Generally, the variogram of the ith relative indicator is

$$\gamma_I^i(h) = 1/2 \ \mathrm{Var}\{[I_i(x+h)-I_i(x)] \mid x, x+h \in D_i\} \tag{7}$$

and it is estimated using the standard experimental variogram calculation formula and the number of pairs within the corresponding sub-field. Clearly, caution should be taken in inferring variogram models when only a few data of one qualitative variable are available relative to the remaining.

It can be shown that the approach described above is identical to estimating the probabilities of a point in the domain of consideration to belong to each of the qualitative variables in the sequence, using a straight forward adaption of eq. 3. It can also be shown that these probabilities will sum up to 1.

Additional relevant comments can be found in the following section.

SEQUENTIAL ESTIMATION OF ROCK TYPES IN A MINERAL DEPOSIT

The sequential estimation of qualitative variables may be demonstrated in an application in a gold deposit occurring in a greenstone belt. Data from 19 drill holes in a small part of a mine, 20m by 75m by 20m, were available. Given a geological interpretation and rock type descriptions in each of the drill holes, five main lithologies were identified on 1m length composits and properly coded in a sequence as A, B, C, D and E. Based on this sequence, four relative indicator variables were defined. I_1 for rock type A and the remaining, I_2 for rock type B and the remaining excluding A, I_3 for C and the remaining excluding A and B, and I_4 for D and the remaining i.e. E.

In the first step, experimental variograms were calculated along strike, downdip, and perpendicular to the bedding. The experimental variograms and spherical models fitted are shown in Figure 1. The characteristics of the variogram models are given in Table I. Note that all variograms exhibit long ranges along

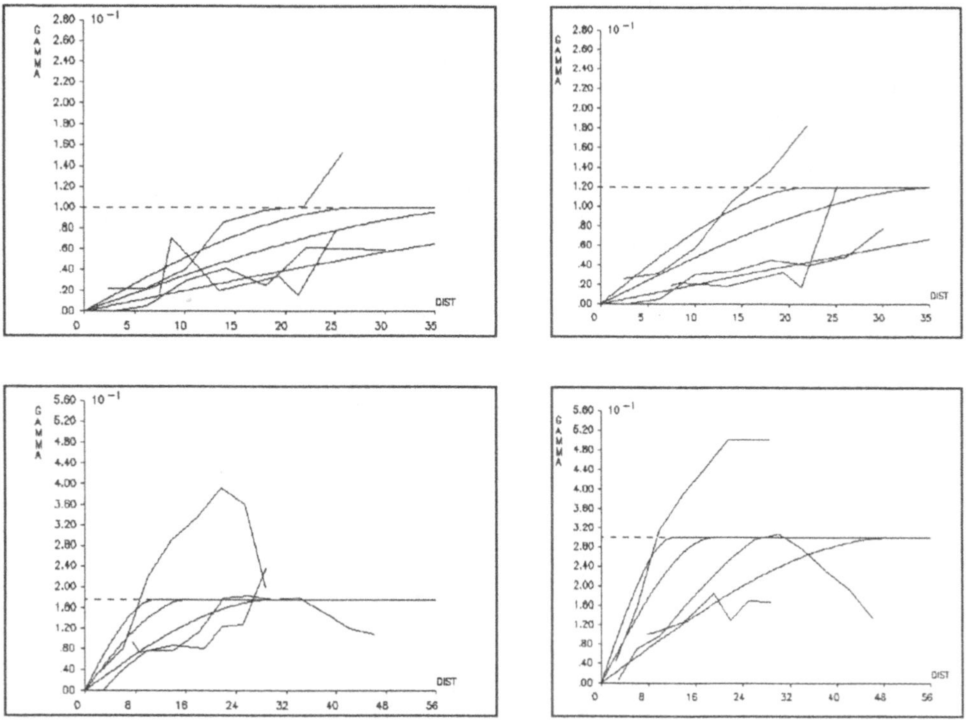

Figure 1. Relative indicator variograms of lithology types in a gold deposit: (a) rock type A; (b) rock type B and the remaining; (c) rock type C and the remaining; (d) rock type D and E.

TABLE I. Variogram models of lithology types (all models are spherical).

	Along strike	Downdip	Perpendicular to bedding
I_1			
Rng.	90m	80m	15m
Sill	.11	.11	.11
I_2			
Rng.	75m	70m	25m
Sill	.10	.10	.10
I_3			
Rng.	90m	85m	20m
Sill	.12	.12	.12
I_4			
Rng.	30m	25m	10m
Sill	.18	.18	.18

strike and intermediate ranges downdip. This was found to be consistent with the geology of the deposit.

Using the relative indicator variograms, the lithology indicators were estimated. First I_1 was kriged on a grid of size 1m by 10m by 5m and the estimated probabilities were transformed as described previously and using the indicator means calculated from the data. Having defined the boundary between rock type A and the remaining, I_2 was kriged and transformed to define the boundary between rock type B and the remaining. The same step was repeated for I_3 and I_4 to produce the final model of the lithologies of the deposit. A section of this model is shown in Figure 2.

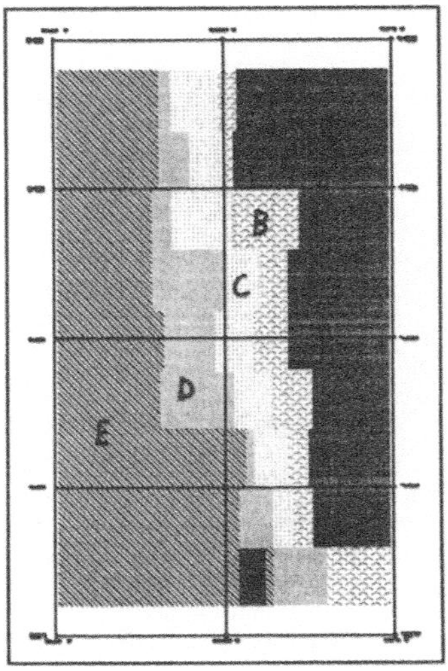

Figure 2. Section from a mineral deposit showing estimated lithology types.

SEQUENTIAL RELATIVE INDICATOR CONDITIONAL SIMULATION

Implementation based on Gaussian simulations

The modelling of series of qualitative variables in a sequential fashion, the one relative to the remaining, may be examined in a similar fashion as in the previous section and in the context of conditional simulation. The basic idea for simulating two qualitative variables described as a single indicator function, has been previously suggested (Soares and Costa e Silva, 1991). The same approach may now be restated, as appropriate, for a sequence of relative indicator variables.

Consider the conditionally simulated realization of a variable $\{j_i(x), x \in D_i\}$ in its classical form (Journel and Huijbregts, 1978), i.e.

$$j_{ics}(x) = j_{iK}^{*}(x) + [j_{iS}(x) - j_{iSK}^{*}(x)] \tag{8}$$

where $j_{iS}(x)$ is a simulated realization of values 0 and 1, generated as described in Journel and Isaaks, 1984); $j_{iK}^{*}(x)$ and $j_{iSK}^{*}(x)$ are kriged estimates using actual and simulated data, respectively, with values 0 and 1. If the conditionally simulated values $j_{ics}(x)$ are transformed to 0's and 1's using the transformation applied previously when estimated probabilities from eq. 6 were transformed, then an indicator realization $i_{ics}(x)$ is generated. The later realization will reproduce the available indicator data and their mean and variance, while it is expected to closely approximate their indicator variogram.

If a sequence of qualitative variables is defined as in the previous sections, then this approach can be sequentially applied to the corresponding sequence of relative indicator variables. The final image of the sequence will then reproduce the available data, relative proportions and approximate relative indicator variograms.

Implementation based on non-Gaussian simulations

As an alternative to Gaussian conditional simulation, the sequential indicator simulation (SIS) algorithm (Alabert, 1987; Journel and Alabert, 1990) may be considered. Although it remains to examine a general adaption of SIS to the idea of relative indicators, it appears that a simple implementation can proceed similarly as before. First, SIS can be used to conditionally simulate the first qualitative variable and the remaining. Then, within the space of the remaining the second variable and the remaining can be simulated, and so on, until all variables are present. Again and following the properties of SIS, the available data, relative proportions and relative indicator variograms will be reproduced.

Remarks

From a practical point of view, it is always appropriate to assess the quality and acceptability of a conditionally simulated image in terms of relative indicator variograms. This is particularly important if a Gaussian based approach has been chosen, given the related approximations.

An attractive feature of the present approach, similar to conventional indicator ones, is that relative indicator variograms preserve their interpretive character. This is particularly important in cases of widely spaced data.

It remains to be explored how different types of soft information can be used in a generalized framework. In a first approach, however, prior information may be coded each time a qualitative variable vs the remaining is simulated, as probabilities in]0,1[.

GENERATING IMAGES OF LITHOFACIES IN A PETROLEUM RESERVOIR

Conditional simulation as previously described, is applied in this section to generate an image of the sedimentary lithofacies in part of a petroleum reservoir. The

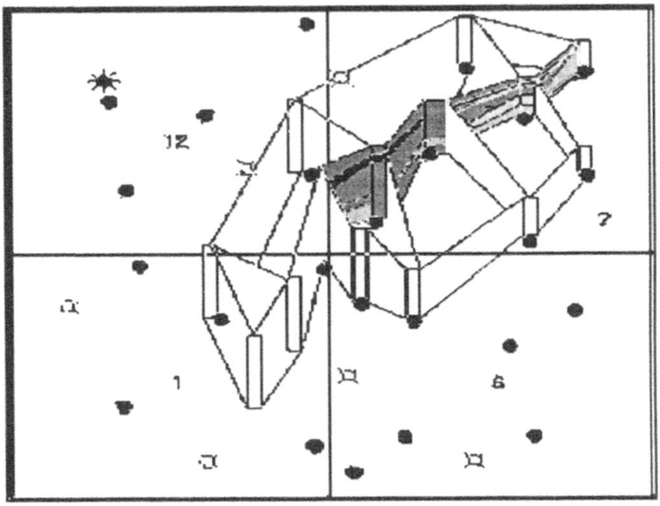

Figure 3. Wells and geological interpretation of the petroleum resevoir in this study.

available data come from 14 wells where sedimentary facies have been identified and correlated from well to well, given the geological interpretation of the deposit as shown in Figure 3. Six facies have been identified in a sequence according to the geology of the deposit and coded as A, B, C, D, E and F. Note that facies A are non-reservoir facies. Five relative indicator variables were consequently defined, with the first, I_1, involving facies A and the remaining, the second I_2 including facies B and the remaining (excluding A), and so on.

Experimental average horizontal and vertical variograms for each of the relative indicators were calculated and models fitted. Experimental variograms and fitted models are shown in Figure 4, while the parameters of the models are given in Table II.

TABLE II. Variogram models of lithofacies in a petroleum reservoir (all models are spherical).

	I_1	I_2	I_3	I_4	I_5
Rng. Hor.	1000m	1200m	750m	900m	800m
Rng. Vert.	30m	6m	10m	18m	15m
Sill	.245	.035	.095	.230	.150
Nugget	.050	.001	.050	.060	.030

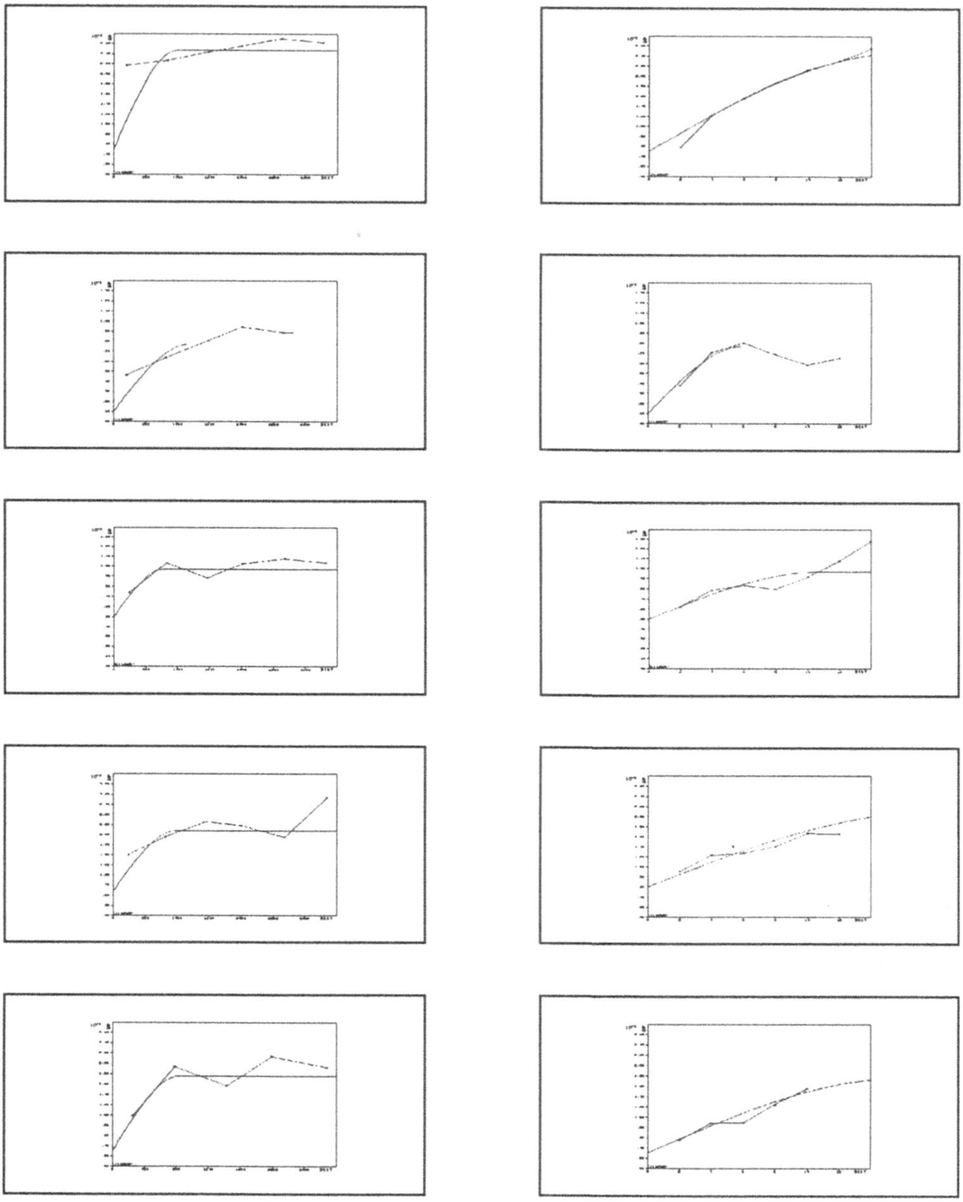

Figure 4. Variograms of the five reservoir lithofacies relative indicators, horizontal and vertical are left and right, respectively.

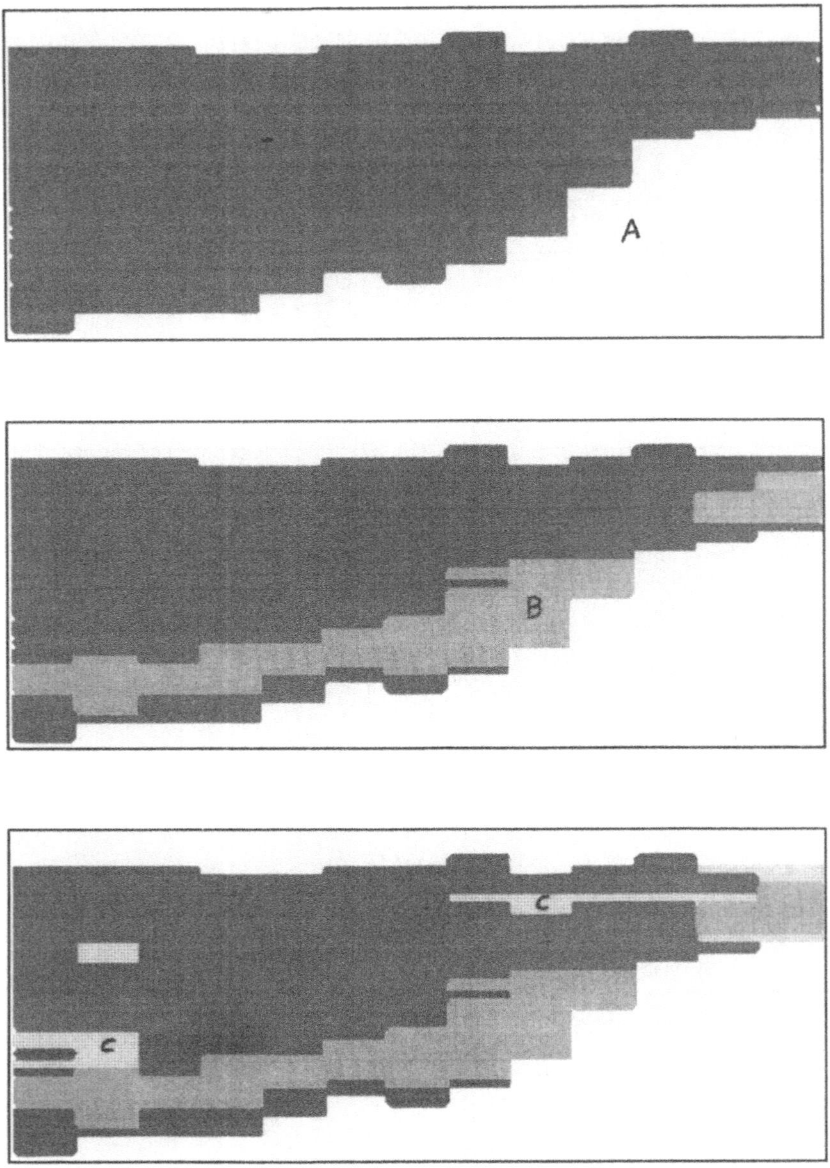

Figure 5. Images of the sequentially generated three first lithofacies in the reservoir in this study.

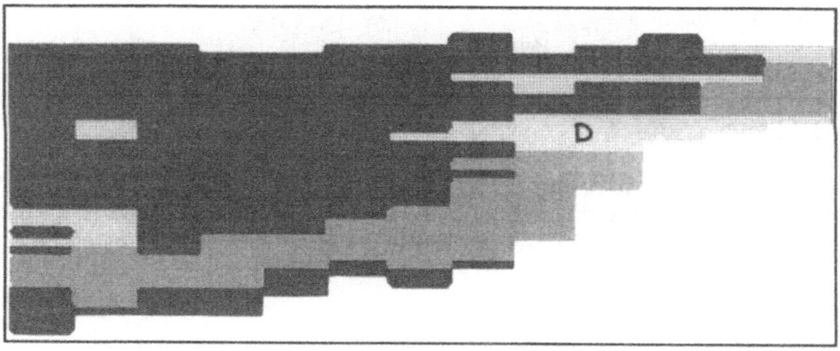

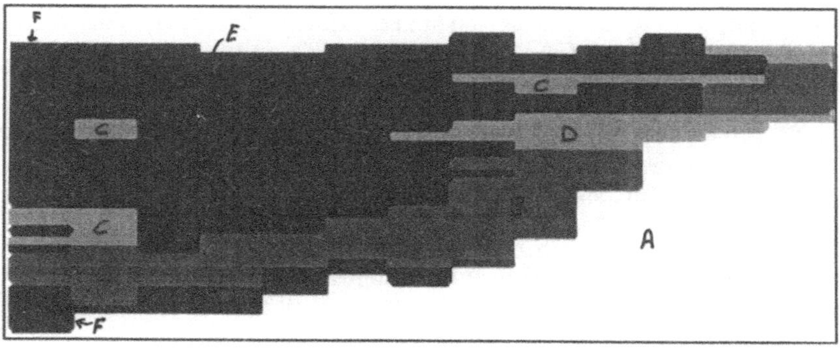

Figure 6. Images of the sequentially generated fourth, fifth and last lithofacies in the reservoir in this study.

The conditional simulation of the five relative indicators was performed as described and using blocks of 200m by 200m by 2m. The mean of each indicator was used in the transformation step. To provide an insight to the sequential process of generating the final image of the reservoir lithofacies, Figures 5 and 6 show the "sequential filling" of the reservoir grid blocks with corresponding lithofacies.

Note that the reproduction of relative indicator variograms was verified.

CONCLUSIONS

The concept of modelling using relative indicator variables has been shown both in the context of estimation and conditional simulation. The idea has been used to estimate rock types in a mineral deposit as well as conditionally simulate reservoir lithofacies.

The sequential modelling of qualitative variables using relative indicators and the characteristics of sequences of relative indicator variables need further exploration and research.

ACKNOWLEDGMENTS

Acknowledgments are in order to B. Robbins, J. Clark and J. Boulanger for coopeation and data; and R.Olea and T.Hewett for their reviews. Funding was provided from NSERC of Canada operating grant OPG 0105803 to RD.

REFERENCES

Alabert, F. (1987) Stochastic Imaging of Spatial Distributions using Hard and Soft Information, M.Sc. thesis, Applied Earth Sciences, Stanford University, Stanford, CA.

Christakos, G. (1992) Random Field Models in Earth Sciences, Academic Press, New York, NY.

Dagbert, M. (1990) " Nested indicator approach for ore reserve estimation in highly variable mineralizations", 92nd General Annual Meeting of the Canadian Institute of Mining, Ottawa.

Haldorsen, H.H., Brand, P.J. and Macdonald, C.J. (1988) "Review of the stochastic nature of reservoirs", in S.Edwards and P.R.King (eds.), Mathematics of Oil Production, Oxford Science Publications, Oxford.

Journel, A. G. (1983) "Non-parametric estimation of spatial distributions", Math. Geology, 15, 445-468.

Journel, A.G. and Alabert, F. (1990) "New method for reservoir mapping", JPT, February 1990, 212-218.

Journel, A.G. and Isaaks, E. (1984) "Conditional Indicator simulation: Application to a Saskatchewan uranium deposit", Math. Geology, 16, 685-718.

Journel, A.G. and Huijbregts, Ch. (1978) Mining Geostatistics, Academic Press, New York, NY.

Mallet, J.L., 1989, "Geometric modelling and geostatistics", in M. Amstrong (ed.), Geostatistics, 2, 737-747.

Matheron, G., Beucher, H., de Fouquet Ch., Galli A. and Ravenne Ch. (1988) "Simulation conditionelle a trois facies dans une falaise de la formation du Brent", Sci. de la Terre, Ser. Inf., 28, 213-249.

Soares, A. (1990) "Geostatistical estimation of orebody geometry: Morphological kriging", Math. Geology, 22, 787-802.

Soares, A. (1992) "Geostatistical estimation of multi-phase structures", Math. Geology, 24, 149-161.

Soares, A. and Costa e Silva, A. (1991) "Geostatistical simulation of the geometry of a shale/limestone sequence in an oil reservoir", SPE paper # 21633.

Solow, A. (1986) "Mapping by simple indicator kriging", Math. Geology, 18, 335-352.

Suro-Perez, V. and Journel, A.G. (1990) "Stochastic simulation of lithofacies for reservoir characterization", in Mathematics in Oil Production, O.Guillon and D. Guerillot (eds.), Technip Editions.

COMBINING FIBRE PROCESSES AND GAUSSIAN RANDOM FUNCTIONS FOR MODELLING FLUVIAL RESERVOIRS

Frode Georgsen and Henning Omre
Norwegian Computer Center
P.O.Box 114, Blindern
N-0314 OSLO
Norway

Abstract. *A combination of fibre processes, a special version of marked point processes, and Gaussian random functions is applied to build a 3D model for fluvial reservoir. A fluvial reservoir consists of a set of channel-belts, each containing several river-beds.*

The directions of the channel-belts are represented by fibres. Geological studies provide prior knowledge about the directions. Combining this information, according to Bayesian rules, with the well-observations, gives the posterior probability distributions of the directions. The joint distribution of the fibres includes spatial interaction.

The individual river-beds inside a channel-belt are modelled by correlated 1D Gaussian random functions. These Gaussian functions define the location, thickness and width of the river-beds. A fairly complex conditioning scheme is used for reproducing the conditioning pattern usually defined by geologists.

The simulation technique is based on the Ripley-Kelly algorithm for birth/death processes, and a standard filter procedure for the 1D Gaussian random functions. Examples are presented.

1 Introduction

Heterogeneities in the characteristics of petroleum reservoirs have considerable impact on fluid flow, and hence on the recovery potential. Reservoir units with low permeability may create barriers for fluid flow, and units with high permeability may create transition zones for injected water. Consequently, reliable models for heterogeneities in reservoirs are prerequisits for comprehensive reservoir evaluation. Stochastic models for reservoir characteristics with associated stochastic simulation algorithms have proven to be well suited for this purpose, see Dubrule [5], Haldorsen and Damsleth [7] and Omre [10].

In this paper, a stochastic model and an associated simulation algorithm for the facies distribution in reservoirs of fluvial origin are presented. This type of reservoirs will frequently contain a background of non-permeable sediments superimposed by sediments from river-beds of good reservoir quality. Hence modelling the geometry of the river-beds will be the key to a reliable reservoir model. One river-bed will

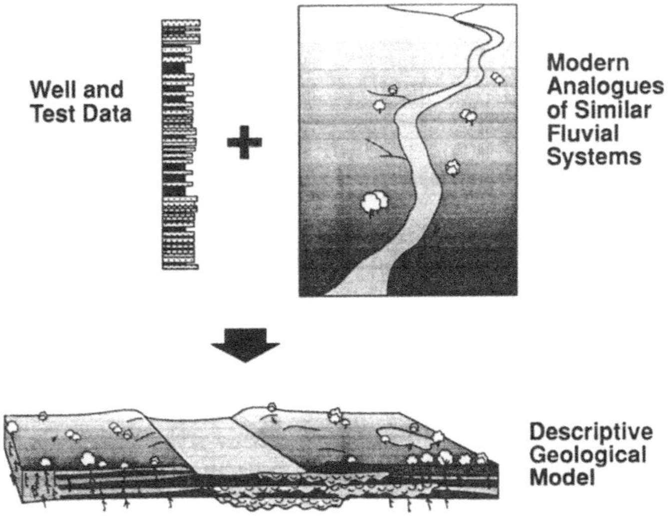

Figure 1: Schematic illustration of the geology in a fluvial reservoir as given in Stanley et al.[13]

normally show a sinous behaviour in a main direction through the reservoir and its change in location will usually be small in subsequent years. This constitutes a channel-belt. In some years major changes in location will take place due to avulsion caused by for example severe flooding. This will be the origin of a new channel-belt and some spatial repulsion from previously created belts will occur. In Figure 1, a schematic illustration of the main characteristics of fluvial reservoirs is presented. Many attempts have been made to model this behaviour stochastically. Among 3-dimensional models it is worth mentioning: Haldorsen et al. [6], Augedal et al. [2], Gundesø and Egeland [1], Meling et al. [9] and Clemetsen et al. [4].

All these models are based on the theory of marked point processes, but on very different levels of sophistication. Common to all of them, however, is the lack of ability to reproduce the river-beds observed in a large number of wells. Hence conditioning on observations creates problems. In real cases in the North Sea, the number of wells can be in the range of 20-50 at the production stage. Moreover, production tests may identify that the same river-bed is observed in more than one well. This complicated conditioning pattern must be handled by a reliable stochastic model for reservoir characterization. The model defined in this article is created for solving this problem.

2 Stochastic model and parameter estimation.

The geological characteristics of the reservoir are represented by a stochastic model. General knowledge about fluvial reservoir provides the basis for the prior model, while reservoir specific observations are used to update this prior guess to obtain a

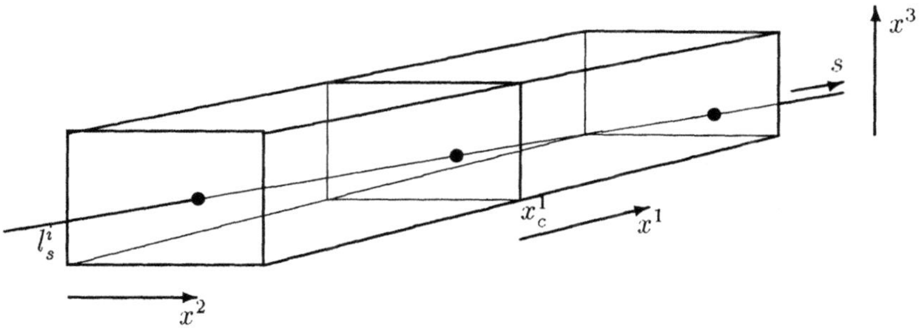

Figure 2: Principal direction line for one channel-belt, intersecting the reservoir box.

posterior model for the characteristics. This posterior model is used for simulating realizations of the reservoir.

2.1 Prior model.

The simulation volume is assumed to be a rectangular reservoir-box in 3D, defined in a (x^1, x^2, x^3)-system. The river-beds are grouped together and organized in different channel-belts. Each of these channel-belts has a principal direction given by a line. For channel-belt number i, this is given by l_s^i, see Figure 2. This line is parametrized by $l_s^i : \{(x^1, x^2, x^3) : (x_c^1, \theta_1^i, \theta_3^i) + s \cdot (1, \theta_2^i, \theta_4^i)\}$, with $s \in (-\infty, \infty)$ being the argument along the line; and x_c^1 being the midpoint of the reservoir-box in the x^1-direction. The x^1-direction is identified as the most probable direction for a channel-belt. l_s^i is fully specified by the parameter-vector $\vec{\theta^i} = (\theta_1^i, \theta_2^i, \theta_3^i, \theta_4^i)$, with θ_1^i and θ_3^i being the x^2- and x^3-coordinates respectivally, of the intersection point between l_s^i and the plane $x^1 = x_c^i$; and θ_2^i and θ_4^i being the slopes of l_s^i projected into the (x^1, x^2)-plane and the (x^1, x^3)-plane respectivally.

Consider an arbitrary channel-belt i, and let everything related to it be referenced by s along the principal direction l_s^i. Then the position of the channel-belt, at reference s, is defined by a multi-dimensional vector:

$$\vec{U^i}(s) = \{H_j^i(s), V_j^i(s), W_j^i(s), T_j^i(s); j = 1, \ldots, M^i\}, \tag{1}$$

with M^i being the number of river-beds in the channel-belt; $H_j^i(s)$ being the horizontal sinuousity of the center line of river-bed j; $V_j^i(s)$ being the vertical sinuousity of the center line of river-bed j; $W_j^i(s)$ being the width of river-bed j; and $T_j^i(s)$ being the thickness of river-bed j, see Figure 3.

The vector $\vec{U^i}(\cdot)$ is defined as a $4M^i$-dimensional correlated Gaussian function, with a one-dimensional reference along l_s^i. The univariable properties of this Gaussian function, is given by:

$$\begin{array}{llll} E\{H(s)\} & = 0 & ; \; Cov\{H(s), H(s')\} & = \sigma_H^2 \rho_H(|s - s'|) \\ E\{V(s)\} & = 0 & ; \; Cov\{V(s), V(s')\} & = \sigma_V^2 \rho_V(|s - s'|) \\ E\{T(s)\} & = \mu_T & ; \; Cov\{T(s), T(s')\} & = \sigma_T^2 \rho_T(|s - s'|) \\ E\{W(s)\} & = \mu_W & ; \; Cov\{W(s), W(s')\} & = \sigma_W^2 \rho_W(|s - s'|), \end{array} \tag{2}$$

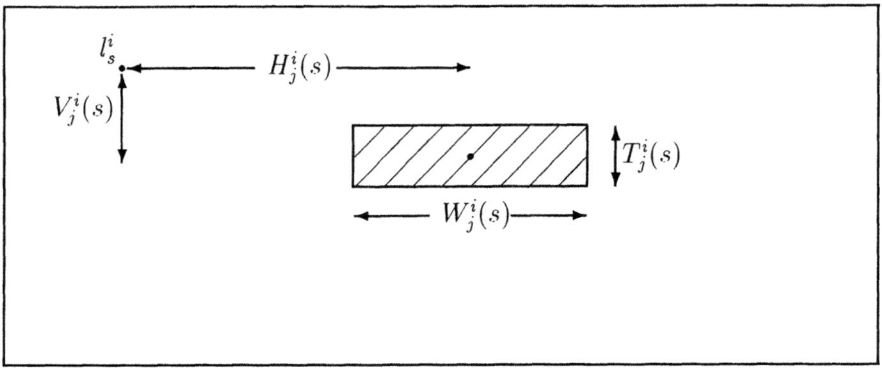

Figure 3: Cross-section of the reservoir perpendicular to the principal direction of the channel-belt, containing one river-bed (hatched).

where $\sigma_\cdot^2$ is the variance of the variable, while $\rho_\cdot(\cdot)$ is the spatial correlation-function of the variable. The parameters μ_T and μ_W are the user-specified expected thickness and width of the river-beds respectively.

The multivariable properties of the Gaussian function is given by the $(4M^i) \times (4M^i)$ covariance matrix Σ_U, which entails the so-called intrinsic coregionalization. Here both the covariance between identical variables in different river-beds, different variables in the same river-bed and different variables in different river-beds can be specified. For this complex multi-dimensional correlation-structure to be consistent, a positive semi-definite matrix is required.

So far the parameters $\vec{\theta}$ have been looked upon as being constant and user specified. However there are reasons to believe that the geologists only have some general, prior understanding of the most probable principal directions of the channel-belts from regional geological studies. This justifies a stochastic interpretation of the parameters, $\vec{\Theta}$, according to the Bayesian paradigm, see Berger [3]. By assigning prior distributions to $\vec{\Theta}$, one will obtain a fully specified model for channel-belts not observed in wells. This means that expression (1) can be rewritten in a Bayesian setting, as:

$$\{\vec{U}^i(s)|\vec{\Theta}^i = \vec{\theta}^i\} = \{H_j^i(s), V_j^i(s), W_j^i(s), T_j^i(s); j = 1, \ldots, M^i|\vec{\Theta}^i = \vec{\theta}^i\}.$$

To specify the model further the probability distribution of $\vec{\Theta}$ is needed. This is defined to be a multi-Gaussian distribution:

$$\vec{\Theta}^i \sim P_{\vec{\Theta}}(\vec{\theta}^i) \sim N(\vec{\mu}_\Theta, \Sigma_\Theta),$$

with $\vec{\mu}_\Theta$ being the expectation; and Σ_Θ being the covariance matrix. It is reasonable to consider this prior distribution to be equal for all channel-belts.

Summarizing the definitions above, gives the conditional probability distribution for the channel-belt given the principal direction:

$$\{\vec{U}^i(s)|\vec{\Theta}^i = \vec{\theta}^i\} \sim P_{\vec{U}^i|\vec{\Theta}^i}(\vec{u}^i(s)|\vec{\theta}^i),$$

which is a $4M^i$ - dimensional Gaussian probability distribution with parameters given by expression (2). Combining this with the Gaussian prior distribution for $\vec{\Theta}^i$, gives the probability distribution for the channel-belt, as,

$$\vec{U}^i(x) \sim P_{\vec{U}^i}(\vec{u}^i(x)) = \int_\Omega P_{\vec{U}^i|\vec{\Theta}^i}(\vec{u}^i(s)|\vec{\theta}^i)P_{\vec{\Theta}^i}(\vec{\theta}^i)d\vec{\theta}^i, \tag{3}$$

with Ω being the domain in which $\vec{\theta}^i$ can vary. Note that the location-reference for the channel-belt, $\vec{U}^i(x)$, must be in the reservoir x-system of axis, while its conditional counterpart is defined in the channel-belt s-system of axis. This is caused by the fact that the prior $P_{\vec{\Theta}}(\cdot)$ is defined in the x-system. The change of reference system also causes $\vec{U}^i(x)$ not to be exactly Gaussian in spite the fact that Gaussian distributions are conjugate priors in linear settings like here. The reasons for defining $\vec{U}^i(s)|\Theta^i$ in the s-system is to make the realizations geologically trustworthy. Note that this change of axis does not cause problems for stochastic simulation, but it will make the parameter estimates slightly non-optimal however. This will be discussed later.

So far only one channel-belt has been considered, but the reservoir generally consists of several channel-belts. Assume there are B channel-belts in the reservoir; then the interesting distribution is of course the joint probability distribution for the whole ensemble of channel-belts ($\vec{U}(x) = (\vec{U}^i(x); i = 1, \ldots, B)$), defined by $P_{\vec{U}}(\vec{u}(x))$.

It is reasonable to assume that the channel-belts interact, since they are deposited sequentially according to certain geological "rules". In this model the interaction is only through the principal directions of the belts, i.e. $\{l_s^i; i = 1, \ldots, B\}$, or equivalently; the parameters $\{\vec{\Theta}^i; i = 1, \ldots, B\}$. These principal directions are looked upon as fibres in 3D, and the ensemble of such fibres constitutes a fibre process according to Stoyan et al.[14] and Clemetsen et al. [4]. In this context the following prior joint probability distribution for $\vec{\Theta} = (\vec{\Theta}^i; i = 1, \ldots, B)$ may be defined;

$$P_{\vec{\Theta}}(\vec{\theta}) = C_0 \cdot \exp\{-[\sum_{i=1}^{B} b^\Theta(\vec{\theta}^i) + \sum_{i=1}^{B}\sum_{j=1}^{B} c^\Theta(\vec{\theta}^i, \vec{\theta}^j)]\}, \tag{4}$$

with C_0 being a normalizing constant; $b^\Theta(\vec{\theta}) = (\vec{\theta} - \vec{\mu}_\theta)^T \Sigma_\theta^{-1}(\vec{\theta} - \vec{\mu}_\theta)$ being the equivalent of the prior Gaussian marginal distribution $P_{\vec{\Theta}}(\vec{\theta})$; and $c^\Theta(\cdot, \cdot)$ being the pairwise interaction between the principal directions of the channel-belts. The form of the interaction function $c^\Theta(\cdot, \cdot)$ can be chosen according to the spatial pattern the fibres are supposed to show. When the fibres represent channel-belts in a fluvial reservoir, geological knowledge suggests to use a function where the fibres repulse each other, see Clemetsen et al. [4].

The collection of all channel-belts is denoted $\vec{U}(x) = (\vec{U}^i(x); i = 1, \ldots, B)$ and the following probability distribution is obtained:

$$\vec{U}(x) \sim P_{\vec{U}}\{\vec{u}(x)\} = \int_\Omega P_{\vec{U}|\vec{\Theta}}(\vec{u}(x)|\vec{\theta})P_{\vec{\Theta}}(\vec{\theta})d\vec{\theta} = \int_\Omega P_{\vec{\Theta}}(\vec{\theta}) \prod_{i=1}^{B} P_{\vec{U}^i|\vec{\Theta}^i}(\vec{u}^i(s^i)|\vec{\theta}^i)d\vec{\theta}. \quad (5)$$

The last equality comes from the assumption that the channel-belts only influence each other through their principal directions. Expression (5) cannot easily be analytically evaluated due to the interaction in the fibre process and the changes in reference system. The probability distribution can however be explored through stochastic simulation based on the Ripley -Kelly algorithm, see Ripley [12], and standard simulation algorithms for Gaussian random functions.

At this stage a stochastic model for the fluvial reservoir is defined and prior assignments of probabilities to all parameters are made based on general characteristics of fluvial reservoir. In the following section the prior model will be updated by reservoir specific observations obtained in the wells in order to define a posterior model.

2.2 Posterior model.

Observations from wells give additional information about the parameters in the model and about the exact location of river-beds. Denote the number of wells in the reservoir by N. The observations appear in vertical wells at $\{x_k = (x_k^1, x_k^2); k = 1, \ldots, N\}$. In each well the intersections with river-beds are identified, i.e. center-location $x_k^{3,i,j}$ and thickness $t_k^{i,j}$ for river-bed j, in channel-belt i, in well k.

Consider one channel-belt with principal direction l_s^i. For notational simplicity, assume that l_s^i is horizontal, i.e. $\theta_4^i = 0.0$. Then the well-locations $\{x_k; k = 1, \ldots, N\}$ can be transformed to the system defined by l_s^i to obtain $\{s_k^i = (s_k^{1,i}, s_k^{2,i}); k = 1, \ldots, N\}$, with $s_k^{1,i}$ being the reference along l_s^i, and $s_k^{2,i}$ being the reference along the horizontal normal to l_s^i. For each well k and river-bed j, identify,

$$\tau_j^i(s_k^i) = \begin{cases} 0 & \text{if well } k \text{ does not penetrate river-bed } j. \\ t > 0 & \text{if well } k \text{ penetrates river-bed } j, \text{ with thickness } t = t_k^{i,j}. \end{cases}$$

$$\gamma_j^i(s_k^i) = \begin{cases} \infty & \text{if well } k \text{ does not penetrate river-bed } j. \\ c & \text{if well } k \text{ penetrates river-bed } j, \text{ with } c = x_k^{3,i,j} - \theta_3^i, \text{ being the} \\ & \text{depth of the river-bed observation relative to the principal} \\ & \text{direction of the channel-belt.} \end{cases}$$

This defines the set of available observations, which has to be pre-identified for each i, j at all wells:

$$\vec{o}^i : \{\tau_j^i(s_k^i), \gamma_j^i(s_k^i); k = 1, \ldots, N; j = 1, \ldots, M^i\}.$$

Now the posterior version of expression (3) can be identified as:

$$\{\vec{U}^i(x)|\vec{O}^i = \vec{o}^i\} \sim P_{\vec{U}^i|\vec{O}^i}(\vec{u}^i(x)|\vec{o}^i) = \int_\Omega P_{\vec{U}^i|\vec{\Theta}^i,\vec{O}^i}(\vec{u}^i(s)|\vec{\theta}^i, \vec{o}^i)P_{\vec{\Theta}^i|\vec{O}^i}(\vec{\theta}^i|\vec{o}^i)d\vec{\theta}, \quad (6)$$

where:

- $P_{\vec{U}^i|\vec{\Theta}^i,\vec{O}^i}(\vec{U}^i(s)|\vec{\theta}^i,\vec{o}^i)$ is the probability distribution of a $4M^i$-dimensional Gaussian random function. For each element in $\vec{o}^i$ one has, either; (i) the well penetrates the channel-belt: $\tau_j^i(s_k^i) > 0$, which gives an inequality constraint, $s_k^{2,i} \in \{H_j^i(s_k^{1,i}) \pm \frac{W_j^i(s_k^{1,i})}{2}\}$, and the equality constraints $\tau_j^i(s_k^i) = T_j^i(s_k^{1,i})$ and $\gamma_j^i(s_k^i) = V_j^i(s_k^{1,i})$; or (ii) the well does not penetrate the channel-belt: $\tau_j^i(s_k^i) = 0$, which gives the inequality, $s_k^{2,i} \notin \{H_j^i(s_k^{1,i}) \pm \frac{W_j^i(s_k^{1,i})}{2}\}$. The inequality conditionings are hard to obtain analytically, but simulation procedures based on rejection sampling are efficient. The equality conditioning can be honored analytically along the lines of conditional simulation.

- $P_{\vec{\Theta}^i|\vec{O}^i}(\vec{\theta}^i|\vec{o}^i)$ is the posterior distribution of the parameters of the principal direction of the channel-belt. This distribution is defined by $b^{\Theta|O}(\vec{\theta}^i|\vec{o}^i)$, which is the posterior analog to $b^{\Theta}(\vec{\theta}^i)$ from expression (4). This term can be found analytically by making three assumptions; (i) the conditioning is only performed on the wells penetrating the river-beds, i.e. on the set $\{s_k^i : \tau_j^i(s_k^i) > 0; k = 1,\ldots,N; j = 1,\ldots,M^i\}$; (ii) the wells hit the river-beds at their center-line; and (iii) the vector $\vec{U}^i(x)$ is a Gaussian random function in the reservoir x-system of axis, which from the discussion above is shown not to be exactly true. None of the assumptions above are considered to be crucially violated in the modelling of fluvial reservoirs. These assumptions justify the application of the methodology of Bayesian kriging, see Omre and Halvorsen [11], for defining the posterior distribution $b^{\Theta|O}(\cdot|\cdot)$.

To obtain the posterior version of the probability distribution in expression (5), the joint posterior distribution of the model parameters is needed, $P_{\vec{\Theta}|\vec{O}}(\vec{\theta}|\vec{o})$, with $\vec{O} = (\vec{O}^i; i = 1,\ldots,B)$. This is defined according to expression (4) by using $b^{\Theta|O}(\vec{\theta}^i|\vec{o}^i)$ as defined above and $c^{\Theta|O}(\cdot,\cdot|\cdot) = c^{\Theta}(\cdot,\cdot)$. The latter entails that the observations do not influence the spatial interaction relation for principal directions of channel-belts.

This defines the posterior probability distribution of all river-beds in the reservoir:

$$\{\vec{U}(x)|\vec{O} = \vec{o}\} \sim P_{\vec{U}|\vec{O}}(\vec{u}(x)|\vec{o}) = \int_{\Omega} P_{\vec{\Theta}|\vec{O}}(\vec{\theta}|\vec{o}) \prod_{i=1}^{B} P_{\vec{U}^i|\vec{\Theta}^i,\vec{O}^i}(\vec{u}^i(s^i)|\vec{\theta}^i,\vec{o}^i)d\vec{\theta}, \qquad (7)$$

As already mentioned, the analytical evaluation of the probability density function above is not possible but fast simulation techniques can be used to sample from this distribution, as described in the next section.

3 Simulation algorithm

In order to generate realizations of $\{\vec{U}(x)|\vec{O} = \vec{o}\}$ from expression (7), it is necessary to find a reliable stochastic simulation algorithm. The algorithm should ensure that; (i) every realization reflects the stochastic properties of the model; (ii) every realization reproduces all observations in the reservoir; and (iii) all realizations are stochastic independent of each other. In addition, performance time is an important subject, due to the size of the reservoir.

The simulation procedure defined for this model consists of two parts:

A. Simulate the fibre process defining the principal directions of all channel-belts given the observations. This is done according to $P_{\vec{\Theta}|\vec{O}}(\cdot|\cdot)$.

B. For each channel-belt simulate the individual river-beds, given the principal directions and the observations. This is done according to $P_{\vec{U}^i|\vec{\Theta}^i,\vec{O}^i}(\cdot|\cdot,\cdot)$, $i = 1,\ldots,B$.

The realization obtained in part A will be used as conditioning values in part B, hence generating realizations according to expression (7).

Simulation in part A is based on the concept of spatial birth-and-death processes, see Ripley [12], and the algorithm is due to Ripley and Kelly:

- Compute the number of channel-belts, B, from the specified sand/gross ratio for the reservoir or by an iterative procedure.
- Generate B principal directions, parameterized by $\vec{\theta} = (\vec{\theta}^1,\ldots,\vec{\theta}^B)$, randomly, such that $P_{\vec{\Theta}|\vec{O}}(\vec{\theta}|\vec{o}) > 0$.
- Perform the following iteration a specified number of times, for example $4 \times B$ times:

  Draw a number k uniformly from $\{1,\ldots,B\}$.

  Remove $\vec{\theta}^k$ from the ensemble.

  $*$ Generate a new family $\vec{\theta}^k_*$ randomly.

  Accept $\vec{\theta}^k_*$ with probability $p_{\vec{\Theta}^k|\vec{\Theta}_{-k},\vec{O}}(\vec{\theta}^k_*|\vec{\theta}_{-k},\vec{o})$ with $\vec{\Theta}_{-k}$ being $\vec{\Theta}$ excluding $\vec{\Theta}^k$.

  If not accept go to $*$.

  End of iteration.

In practical situations one may have to perform this procedure several times, using simulated annealing techniques, to optimize the number of channel-belts. Note that the probability involved in this algorithm, can be written as:

$$p_{\vec{\Theta}^k|\vec{\Theta}_{-k},\vec{O}}(\vec{\theta}^k_*|\vec{\theta}_{-k},\vec{o}) = \exp\{-[b^{\Theta|O}(\vec{\theta}^k_*|\vec{o}) + \sum_{i\neq k} c^{\Theta|O}(\vec{\theta}^k_*,\vec{\theta}^i|\vec{o})]\},$$

which is relatively easily obtainable since it does not involve a normalizing constant.

Part B is based on an algorithm for simulation of Gaussian random functions, see Journel and Huijbregts [8], and it is conditioned on the realization of the parameters $\vec{\Theta} = (\vec{\Theta}^1, \ldots, \vec{\Theta}^B)$ obtained above:

- For each channel-belt, generate the values of $\{\vec{u}^i(s_k^{1,i}); k = 1, \ldots, N | \vec{\theta}^i, \vec{o}^i\}$, i.e. the values of the variables for all river-beds at locations corresponding to the wells. This is done by first identifying a conditional multi-Gaussian distribution, where the conditioning is on the equality constraints stated in section 2.2. Then realizations are generated from this conditional distribution until the inequalities stated in section 2.2 are satisfied. The values obtained are used for conditioning in the next step.

- For each channel-belt generate a realization of the multi-Gaussian random function $\{\vec{u}^i(\cdot)|\vec{\theta}^i, \vec{o}^i\}$ on a discrete grid along l_s^i. This is done by using the filter method for unconditional simulation and conditioning by linear combinations as described in Journel and Huijbregts [8].

The resulting realizations of river-beds $\{\vec{u}(x)|\vec{o}\} = \{\vec{u}^i(x); i = 1, \ldots, B|\vec{o}\}$ generated from the probability distribution in expression (7) constitute one possible outcome of the reservoir.

4 Examples

In this section the method presented in the previous sections is illustrated by examples. The presentation emphasizes the following features; the location and behaviour of the river-beds; the conditioning on well-observations; and the interaction between channel-belts. The first examples are performed in 2D to simplify the visualization, although the method described generally applies in 3D. These schematic examples are all performed on synthetic data. At the end of the section, a 3D example from a reservoir with real data is presented.

Figures 4 and 5 shows the results from simulating one channel-belt, with three river-beds, and with four different levels of conditioning. The four simulations have been performed with identical prior parameters listed in Table 1(a). In the simulations the practical range of correlation for both $H(s)$ and $W(s)$ has been set to 50. In Figure 6, the posterior marginal probability density function for the slope of the principal direction of the channel-belt, Θ_2^i, is shown for each of the four examples, to illustrate the changes as the number of wells increases.

Figure 4(a) shows the unconditional case, a realization $\vec{u}_a^i(x)$ from the distribution $P_{\vec{U}^i}(\cdot)$ given in expression (3). In this case, as Figure 6(a) shows, the posterior distribution for Θ_2^i is identical to the prior. In Figure 4(b), two wells have been drilled in the reservoir, both penetrating all river-beds. This plot shows a realization $\vec{u}_b^i(x)$ from the distribution $P_{\vec{U}^i|\vec{O}^i}(\cdot|\vec{o}_b^i)$ given in expression (6), with $\vec{o}_b^i$ being the set

All units are in meters			
(a)River-beds in Figures 4 and 5			
Res. length	200	σ_H	40
Res. width	100	$\rho_H(r)$	$1 - e^{\{-3(\frac{r}{50})^2\}}$
$E(\Theta_1)$	0.0	μ_W	10
$sd(\Theta_1)$	40.0	σ_W	2
$E(\Theta_2)$	0.0	$\rho_W(r)$	$1 - e^{\{-3(\frac{r}{50})^2\}}$
$sd(\Theta_2)$	$\frac{1}{2\sqrt{3}}$	$Cov(\Theta_1, \Theta_2)$	0.0
(b)Principal directions in Figure 6			
Res. length	200	$E(\Theta_2)$	0.0
Res. width	100	$sd(\Theta_2)$	0.18
$E(\Theta_1)$	0.0	$Cov(\Theta_1, \Theta_2)$	0.0
$sd(\Theta_1)$	20.0		

Table 1: Prior parameters for the simulations shown in Figure 4, 5 and 6

of observations from the two wells. Figure 6(b) shows the decreased variation in the posterior distribution for Θ_2^i and also the posterior expectation. In Figure 4(c) two more wells have been included, also penetrating all river-beds. The realization $\vec{u}_c^i(x)$ from the distribution $P_{\vec{U}^i|\vec{O}^i}(\cdot|\vec{o}_c^{\,i})$, with $\vec{o}_c^{\,i}$ being the set of observations from the four observations, is shown. By forcing all river-beds to reproduce the well observations the variability of $H^i(s)$ in the model is reduced. The density function in Figure 6(c) also becomes slightly narrower. The last plot, Figure 4(d), shows the inclusion of two more wells not penetrating any of the river-beds. The realization $\vec{u}_d^i(x)$ from the distribution $P_{\vec{U}^i|\vec{O}^i}(\cdot|\vec{o}_d^{\,i})$, with $\vec{o}_d^{\,i}$ being the set of observations from all six wells, is shown. Note from Figure 6(d), that Θ_2^i has the same posterior distribution as in case (c), since non-penetrating wells are not accounted for. The conditioning on two non-penetrating wells, both lying in the normal path of the river-beds, enforce larger dispersion to avoid intersection with the wells.

In Figure 5 just another set of realizations based on the same model is displayed, and the same features can be observed.

Figure 7 shows the principal directions of eight channel-belts; two of them being conditioned on six wells each; the others being unconditioned. The prior parameters used are presented in Table 1(b). Figure 7(a) shows the principal directions without interaction. It is a realization $\vec{\theta}|\vec{o}$ from the distribution $P_{\vec{\Theta}|\vec{O}}(\cdot|\cdot)$, being the conditional version of expression (4) as discussed in section 2.2, with the interaction function $c^{\Theta|O}(\cdot,\cdot|\cdot) \equiv 0.0$. Figure 7(b) shows a realization of the principal directions based on $\vec{\theta}|\vec{o}$ from $P_{\vec{\Theta}|\vec{O}}(\cdot|\cdot)$ with considerable spatial repulsion, i.e. $c^{\Theta|O}(\cdot,\cdot|\cdot) \gg 0.0$ for channel-beltss being close. The interaction function used, $c^{\Theta|O}(\cdot,\cdot|\cdot)$, is easily implemented although too complicated to present here. It includes two parameters; δ being the minimum allowed spatial distance between two principal directions, in this case $\delta = 0.3$; and ρ being the range of interaction between two principal directions,

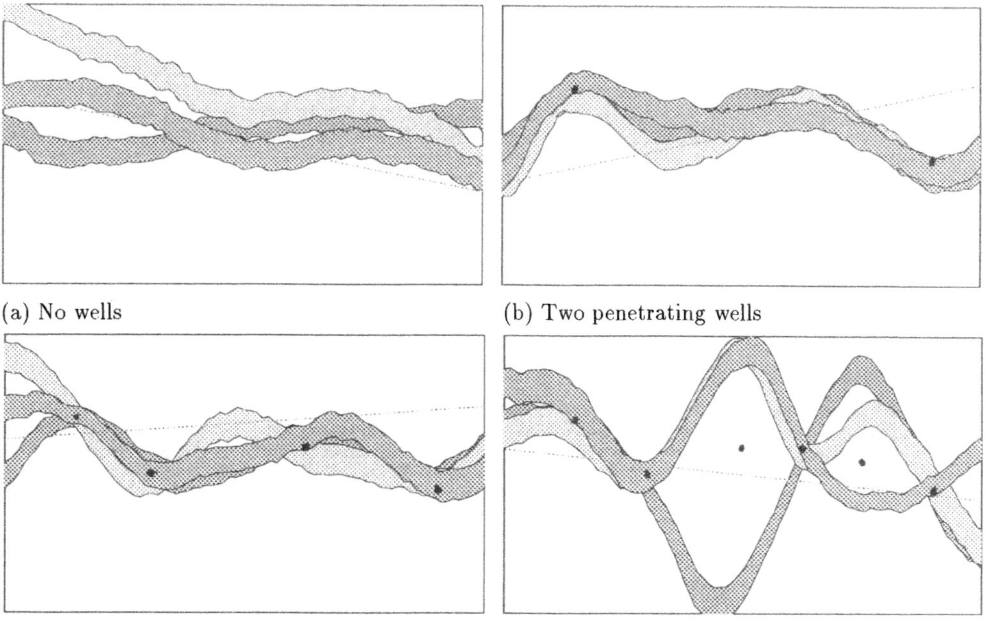

(a) No wells (b) Two penetrating wells

(c)Four penetrating wells (d)Four penetrating, two non-penetrating wells

Figure 4: Four examples of one channel-belt with three river-beds simulated in 2D.

in this case $\rho = 3.0$. In 2D the principal directions tend to end up relatively parallel. In 3D, however, one tends to have relatively parallel directions for equal depths, but considerable dispersion between depths.

Figure 8 shows parts of two vertical cross-sections of one simulated realization from an off-shore reservoir. Figure 8(a) shows a cross-section parallel to the main direction of river-beds, while Figure 8(b) shows a cross-section perpendicular to the main direction. On both plots note the clustering in channel-belts due to the interaction between the principal directions.

An important aspect of the model presented, is the processing time of the simulations. A simulation of a full off-shore reservoir conditioned on 10 wells, consisting of 30 channel-belts and 206 river-beds, with 100 iterations in the simulation algorithm took about 30 minutes on a DEC-station 5000/25.

5 Closing Remarks

A stochastic model for the facies distribution in a fluvial reservoir has been presented. The combination of fibre processes and Gaussian random functions has made it possible to meet the challenges introduced by conditioning on a large number of wells, especially the need for allowing the same river-bed to appear in more than one well. Reliable and fast simulation algorithms makes the model useful for evaluation of real

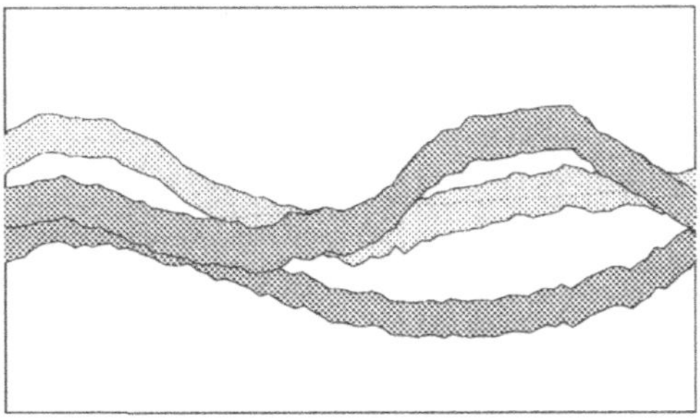

(a) No wells

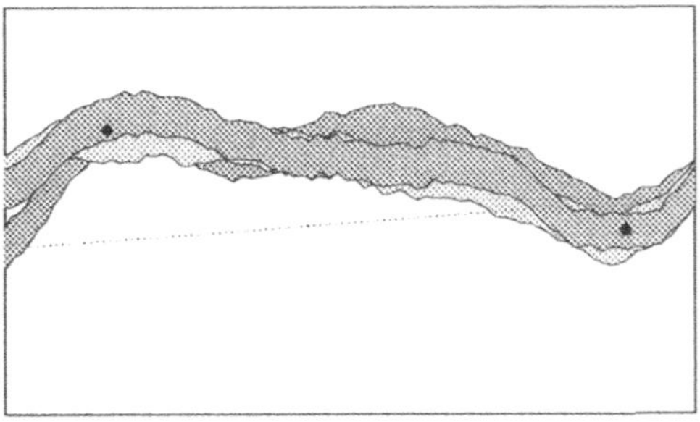

(b) Two penetrating wells

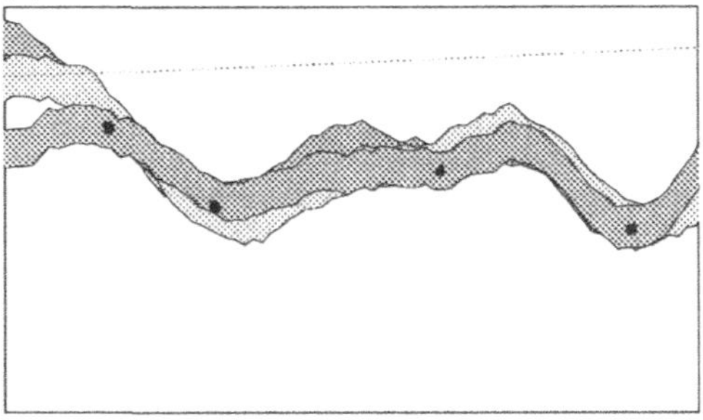

(c)Four penetrating wells

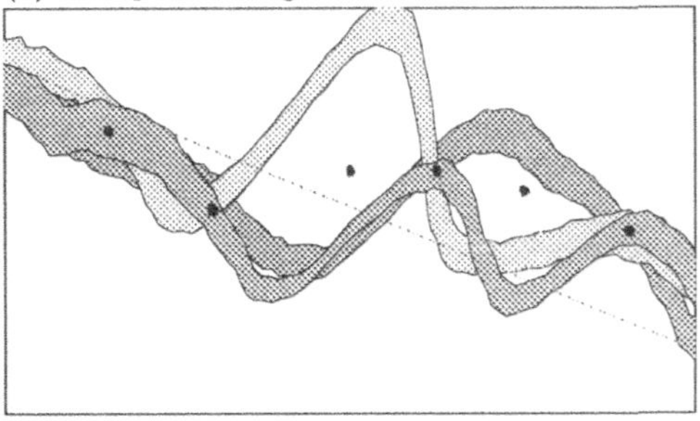

(d)Four penetrating, two non-penetrating wells

Figure 5: Four examples of one channel-belt with three river-beds simulated in 2D.

reservoirs.

The model presented will typically constitute one part of a larger system for reservoir evaluation. Other parts being for instance: Structural models, models for petrophysical variables in the facies, and techniques for change of scale to prepare input for the reservoir production simulator.

Further development will be done in the direction of several facies, not only sand and non-permeable background. These facies should be closely related to the river-beds, for example crevasses. Again, living up to complex conditioning patterns constitutes the main challenge.

Some draw-backs of the model should be mentioned; wide river-beds can give inconsistency between observations and model; difficult conditioning schemes may cause the simulation algorithm to be very time-consuming; and the model is not easily generalized to several facies.

6 Acknowledgments

The research has benefited from financial support from Saga Petroleum a.s. and Statoil a.s. [13]

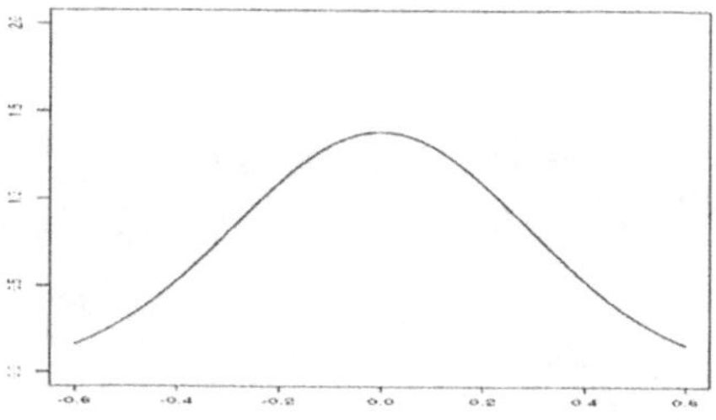

(a)No wells, $m = 0.0, sd = 0.289$

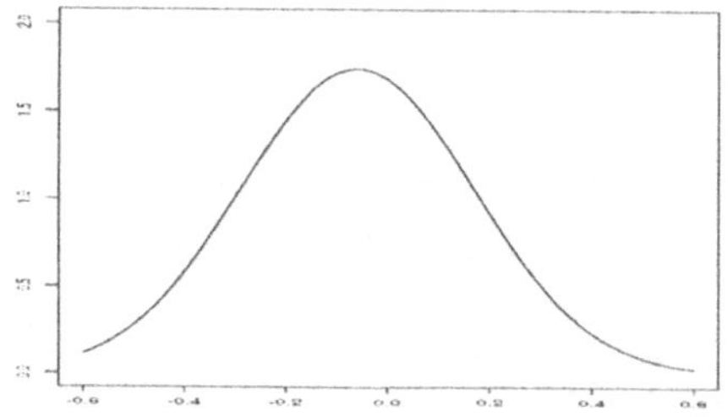

(b)Two penetrating wells, $m = -0.06, sd = 0.229$

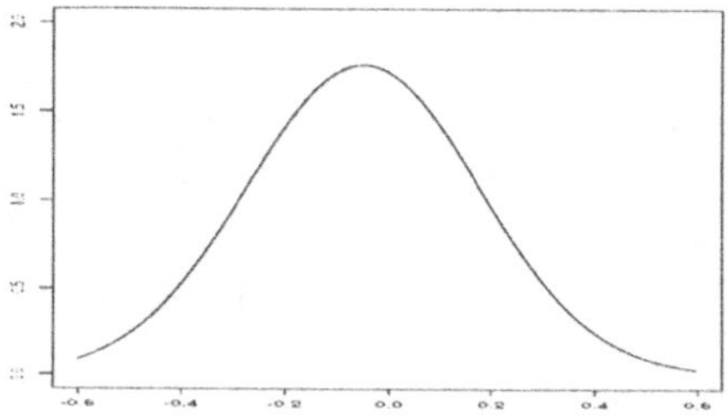

(c)Four penetrating wells, $m = -0.05, sd = 0.226$

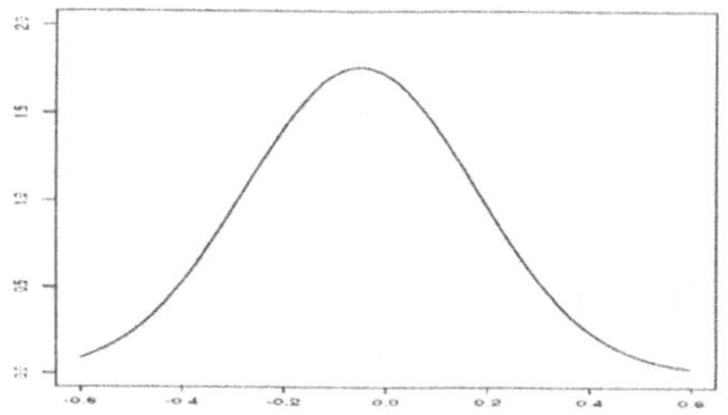

(d)Four penetrating, two non-penetrating wells, $m = -0.05, sd = 0.226$

Figure 6: Posterior probability distribution for the slope of the principal direction of the channel-belt, $P_{\Theta_2^i|\vec{O}^i}(\cdot|\cdot)$. m is the mean, sd the standard deviation.

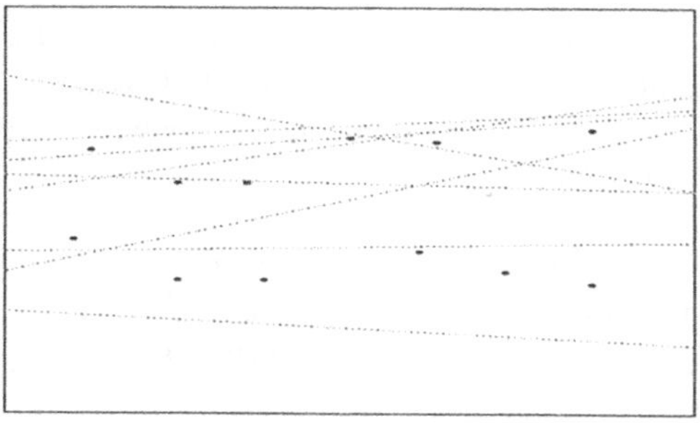

(a)No interaction

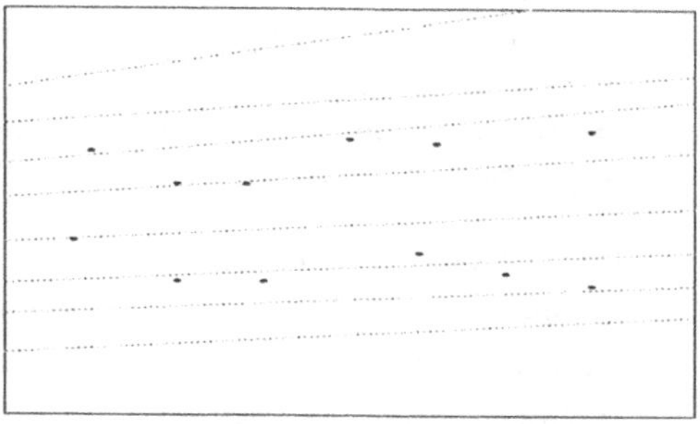

(b)Repulsion

Figure 7: The principal directions of eight channel-belts in 2-D, with and without interaction

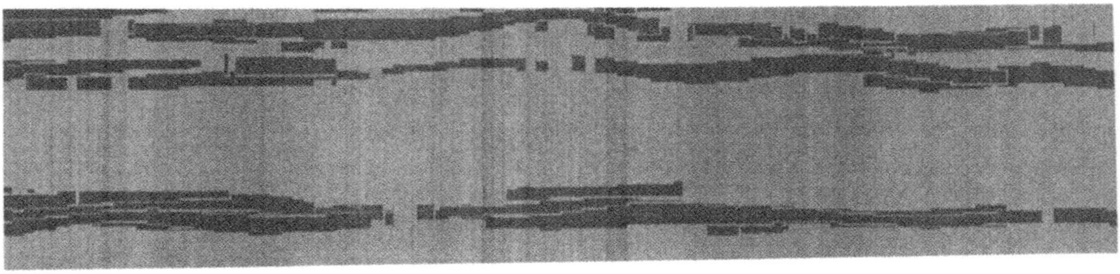

(a)Cross-section parallel to main-direction of river-beds.

(b)Cross-section perpendicular to main-direction of river-beds.

Figure 8: Cross-sections from a simulated realization with real data.

References

[1] R. Gundesø and O. Egeland. Sesimira – a new geological tool for 3-d modeling of heterogeneous reservoirs. In A. T. Buller, E. Berg, O. Hjelmeland, J. Kleppe, O. Torsæter, and J. O. Aasen, editors, *North Sea Oil and Gas Reservoir – II*, pages 363–371. Graham & Trotman, May 1989.

[2] H. O. Augedal, K. O. Stanley, and H. Omre. Sisabosa, a program for stochastic modelling and evaluation of reservoir geology. In *Proceedings from Conference on Reservoir Description and Simulation with Emphasis on EOR, Oslo*, Oslo, Norway, September 1986. IFE.

[3] J. O. Berger. *Statistical Decision Theory and Bayesian Analysis.* Springer series in statistics. Springer-Verlag, New York, 2. edition, 1985.

[4] R. Clemetsen, A. R. Hurst, R. Knarud, and H. Omre. A computer program for evaluation of fluvial reservoirs. In A. T. Buller, E. Berg, O. Hjelmeland, J. K-leppe, O. Torsæter, and J. O. Aasen, editors, *North Sea Oil and Gas Reservoir – II*, pages 373–385. Graham & Trotman, May 1989.

[5] O. Dubrule. A review of stochastic models for petroleum reservoirs. In *Meeting on Quantification of Sediment Body Geometries and Their Internal Heterogeneities, London*. British Society of Reservoir Geologists, March 1988.

[6] H. H. Haldorsen, P. J. Brand, and C. J. MacDonald. Review of the stochastic nature of reservoirs. In S. Edwards and P. King, editors, *Mathematics in Oil Production*, pages 109–209. Oxford Science Publications, Oxford, 1988.

[7] H. H. Haldorsen and E. Damsleth. Stochastic modeling. *Journal of Petroleum Technology*, pages 404–412, April 1990.

[8] A. G. Journel and Ch. J. Huijbregts. *Mining Geostatistics*. Academic Press, London, 1978.

[9] L. M. Meling, P. O. Mørkeseth, and T. Langeland. Production forecasting for gas fields with multiple reservoirs of limited extent. In *1988 SPE Annual Technical Conference and Exhibition*, Houston, Texas, October, 1988. SPE 18287.

[10] H. Omre. Stochastic models for reservoir characterization. In J. Kleppe and S. M. Skjæveland, editors, *Recent Advances in Improved Oil Recovery Methods for North Sea Sandstone Reservoirs*. Norwegian Petroleum Directorate, Stavanger, Norway, 1991.

[11] H. Omre and K. B. Halvorsen. The Bayesian bridge between simple and universal kriging. *Mathematical Geology*, 21(7):767—786, 1989.

[12] B. D. Ripley. *Stochastic Simulation*. John Wiley & Sons, New York, 1987.

[13] K. O. Stanley, K. Jorde, N. Ræstad, and C. P. Stockbridge. Stochastic modelling of reservoir sand bodies for input to reservoir simulation, snorre field, northern north sea, norway. In A. T. Buller, E. Berg, O. Hjelmeland, J. Kleppe, O. Torsæter, and J. O. Aasen, editors, *North Sea Oil and Gas Reservoir – II*, pages 91–101. Graham & Trotman, May 1989.

[14] D. Stoyan, W. S. Kendall, and J. Mecke. *Stochastic Geometry and its Applications*. John Wiley & Sons, New York, 1987.

Indicator Principal Component Kriging: The Multivariate Case

Vinicio Suro-Pérez
Dirección de Exploración y Producción, PEMEX
Av. Marina Nacional 329, Torre Ejecutiva Piso 8, 11311 México, D.F.

Abstract

A multivariate non-parametric technique is presented to model conditional distributions. This algorithm is an extension of Indicator Principal Component Kriging and allows considering simultaneously several variables or attributes as conditioning data. Specific linear combinations of binary random variables, called indicator principal components, are determined such that their spatial crosscorrelation is negligible. It is shown that this linear transformation ranks the indicator principal components according to their spatial autocorrelation. The approach proposed is evaluated for the specific case of two biGaussian variables jointly biGaussian. The exact conditional distributions derived from this biGaussian model are compared to the corresponding models derived from the approach proposed. Next, a case study dealing with a 2-variable field illustrates the performance of this approach under different sampling strategies of each variable.

I. Introduction

This paper deals with the generalization of Indicator Principal Component Kriging (IPCK) (Suro-Pérez and Journel, 1991) to the multivariate case. The goal is the modeling of conditional distributions under conditioning data belonging to different variables. There is a practical interest on that topic. For example, the ore reserve evaluation of a polymetalic mine demands consideration of the different metals of economical interest, taking into account their possible spatial dependence. The modeling of an oil reservoir exhibits similar behavior. There are a number of variables such as porosity and permeability, spatially related, which need to be inferred in the reservoir for forecasting the oil production. Both examples, show the need of modeling such spatial dependence between variables, if any, and use it to estimate or simulate the behavior of one or both variables at the unsampled locations. One traditional solution consists of using cokriging (see Journel, 1989), a multivariate generalized linear regression, to obtain an estimate of any single variable as a linear combination of sample values on any number of other variables. However, just as any regression or kriging, cokriging tends to smooth out the spatial variability of the variable being estimated, which could be detrimental for applications which require retention of the observed spatial variability. Also cokriging, in general, does not provide any measure of uncertainty of the estimate produced, less a measure of joint spatial uncertainty.

A solution around some of the limitations of cokriging is provided by the concept of conditional probability distributions. Rather than providing a single estimate for the unsampled location, one builds a model for the probability distribution of its possible outcomes. This probability distribution model provides an uncertainty model about the unsampled value; one can derive from it estimated values, e.g., the expected value akin to the (least square error) cokriging estimate, or draw from it realizations of the unsampled value (conditional simulation).

II. Statement of the Problem

Consider several locations in an oil reservoir or ore deposit. Each one of these locations has been sampled considering K different properties or attributes. A probabilistic model for the joint variability of these K attributes consists of K stationary and interdependent random functions $Z_k(\mathbf{x})$, $k = 1, \cdots, K$, one for each sampled property. This formulation allows uncertainty modeling through the conditional probability distribution:

$$Prob\{Z_{k_0}(\mathbf{x}) \leq z | Z_k(\mathbf{x}_{\alpha_k}) = z_k(\mathbf{x}_{\alpha_k}) \ \forall \alpha_k \in \{n_k\},\ k = 1, \cdots, K\} \tag{1}$$

with $Z_{k_0}(\mathbf{x})$ being the k_0^{th} variable of the the K variables sampled at $\{n_k\}$ locations, $\mathbf{x}_{\alpha_k}$ is a sample location for the kth variable and z is a threshold value.

Various assumptions and approximations are necessary to determine the probability (1). One possible assumption is to consider that the dependence between the K random functions can be modeled by a multiGaussian distribution (Anderson, 1984), with parameters (esentially the covariance matrix) inferred from the data. Other approach consists of a non-parametric approach which captures more of the original data statistics when building a model for the probability distribution (1). Rather than the mere covariance between variables, the non-parametric approach has the potential to consider the full bivariate distribution of any two of the K variables considered. However, its drawback is the burden of inferring these additional statistics calling, usually, for more and better data.

The non-parametric approach considered here is based on a prior binary transform of any variable $Z_k(\mathbf{x})$ into a series of indicator variables, one per threshold value, defined as:

$$I_k(\mathbf{x}; z_l) = \begin{cases} 1 & \text{if } Z_k(\mathbf{x}) \leq z_l \\ 0 & \text{otherwise} \end{cases} \tag{2}$$

with z_l being a threshold value applied to the random function $Z_k(\mathbf{x})$. The reason for such binary transform is that the probability distributions of $Z_k(\mathbf{x})$ are directly related to expected values of the binary variables $I_k(\mathbf{x}; z_l)$. Indeed:

$$E[I_k(\mathbf{x}; z_l)] = Prob\{Z_k(\mathbf{x}) \leq z_l\} = p_{kl} \tag{3}$$

with p_{kl} being the proportion of $Z_k(\cdot)$ values below the threshold z_l. Also the non-centered covariance of the indicator transform is exactly the bivariate cdf of the corresponding two original variables:

$$E\{I_k(\mathbf{x};z_l)I_{k'}(\mathbf{x}+\mathbf{h};z_{l'})\} = Prob\{Z_k(\mathbf{x}) \leq z_l, Z_{k'}(\mathbf{x}+\mathbf{h}) \leq z_m\} \tag{4}$$

Therefore, inference of the various indicator auto(cross)covariances amounts to infer directly from data the full bivariate cdf's between any pair of variables $Z_k(\mathbf{x})$ and $Z_{k'}(\mathbf{x})$. This is as opposed to the multiGaussian model where these bivariate cdf's are assumed a priori Gaussian (after normal score-transform). In this regard, the indicator approach makes more use of the data statistics provided that they are enough and reliable. Once again, one advantage is the source of a disadvantage, since the indicator approach calls for a more complete thus more difficult inference from the data available.

The conditional probability (1) being in fact an expected value, it should be estimated by least squares regression techniques, i.e. by kriging using indicator covariances of the type (4). An indicator cokriging estimate using all information available on the K variables would be written:

$$I_m^*(\mathbf{x};z_{k'}) = p_{mk'} + \sum_{k=1}^{K}\sum_{l=1}^{L_k}\sum_{\alpha=1}^{n_k} \lambda_{k\alpha}^l [I_k(\mathbf{x}_\alpha;z_l) - p_{kl}] \tag{5}$$

with L_k being the number of threshold values for the variable $Z_k(\mathbf{x})$ and the weights $\lambda_{k\alpha}^l$ obtained by solution of a system of linear equations (cokriging system). Unfortunately, this conditional distribution model (5) requires inference of $(\sum_{k=1}^{K} L_k)^2$ indicator auto(cross)covariances, invalidating (5) as a practical alternative to model conditional distributions.

III. Multivariate Indicator Principal Component Kriging

Consider that the number of threshold values L_k for each variable $Z_k(\mathbf{x})$ is the same, say L, and that these values correspond to the same quantiles, i.e.,

$$E[I_k(\mathbf{x};z_l)] = E[I_{k'}(\mathbf{x};z_l)] \ \forall\ k,k'$$

The notation z_l associated to the indicator $I_k(\cdot;\cdot)$ corresponds to the lth quantile of the random variable $Z_k(\cdot)$. For the sake of simplicity, the notation z_l does not include any reference to the corresponding variable, which is given by the subindex of the indicator variable. Further, consider that at each location $\mathbf{x}$, the entire set of variables $Z_k(\mathbf{x})$ is present, hence the notation n_k is reduced to the same number of samples n. With these considerations, an indicator vector considering the L thresholds is:

$$\mathbf{I}_k(\mathbf{x}) = [I_k(\mathbf{x};z_1)\cdots I_k(\mathbf{x};z_L)]^T$$

and an indicator vector considering the K variables is:

$$\mathbf{I}(\mathbf{x}) = [\mathbf{I}_1^T(\mathbf{x})\cdots \mathbf{I}_K^T(\mathbf{x})]^T \tag{6}$$

The covariance matrix corresponding to the indicator vector (6) is written, for lag $\mathbf{h}'$,

$$\mathbf{\Sigma}_I^K(\mathbf{h}') = \begin{bmatrix} \mathbf{C}_I^{11}(\mathbf{h}') & \cdots & \mathbf{C}_I^{1K}(\mathbf{h}') \\ \mathbf{C}_I^{21}(\mathbf{h}') & \cdots & \mathbf{C}_I^{2K}(\mathbf{h}') \\ \vdots & \ddots & \vdots \\ \mathbf{C}_I^{K1}(\mathbf{h}') & \cdots & \mathbf{C}_I^{KK}(\mathbf{h}') \end{bmatrix} \tag{7}$$

with $\mathbf{C}_I^{kk'}(\mathbf{h}')$ being the covariance matrix between the indicator vectors $I_k(\mathbf{x})$ and $I_{k'}(\mathbf{x}+\mathbf{h})$. The dimension of (7) is $(KL \times KL)$.

III.1 Indicator Principal Components

The indicator covariance matrix $\mathbf{\Sigma}_I^K(\mathbf{h})$ can be expressed as a matrix multiplication of an orthogonal matrix $\mathbf{A}$ and a diagonal matrix $\mathbf{D}$ (Golub and Van Loan, 1983), i.e.,

$$\mathbf{\Sigma}_I^K(\mathbf{h}) = \mathbf{A}\mathbf{D}\mathbf{A}^T \tag{8}$$

This decomposition exists if and only if the indicator covariance matrix is symmetric. For example, the choice $\mathbf{h} = 0$ yields a symmetric matrix $\mathbf{\Sigma}_I^K(0)$ which can be determined fully from elementary statistics of the K original variables $Z_k(\mathbf{x})$. This simple choice is considered hereafter.

The orthogonal matrix $\mathbf{A}$ defines the indicator principal components (ipc), in number $(KL \times KL)$ equal to the dimension of the indicator vector $\mathbf{I}(\mathbf{x})$ as defined in (6):

$$\mathbf{Y}(\mathbf{x}) = \mathbf{A}^T\mathbf{I}(\mathbf{x}) \tag{9}$$

As consequence of transformation (9), the ipc covariance matrix $(\mathbf{\Sigma}_Y^K(\mathbf{h}))$ is diagonal for $\mathbf{h} = 0$ since decomposition (8) was done at such lag. For lags different from zero, the key hypothesis underlying the indicator principal component approach is that the elements of the ipc vector (9) are uncorrelated for all lags $\mathbf{h}$, i.e., a diagonal ipc covariance matrix for all $\mathbf{h}$.

III.2 Conditional Probability Distribution Models

The lack of crosscorrelation between ipc's $Y_k(\mathbf{x})$ $(k = 1, \cdots, KL)$ allows estimation of each ipc without need of considering cokriging. Indeed, $Y_k(\mathbf{x})$ is estimated by kriging, i.e.,

$$Y_k^*(\mathbf{x}) = \sum_{\alpha=1}^{n} \lambda_{k\alpha} Y_k(\mathbf{x}_\alpha) \quad k = 1, \cdots, KL$$

and conditional distribution models for each variable and for each cutoff are obtained by an inverse transformation applied on the estimated ipc vector, i.e.,

$$\mathbf{I}^*(\mathbf{x}) = \mathbf{A}\mathbf{Y}^*(\mathbf{x}) \tag{10}$$

with,

$$\mathbf{Y}^*(\mathbf{x}) = [Y_1^*(\mathbf{x}) \cdots Y_{KL}^*(\mathbf{x})]^T$$

Each element of the indicator vector $\mathbf{I}^*(\mathbf{x})$ is an estimate of the conditional probability (1), for example,

$$Prob_{IPCK}(Z_k(\mathbf{x}) \le z_l |\ Z_k(\mathbf{x}_\alpha)\ \ \alpha = 1, \cdots, n\ ;\ \ k = 1, \cdots, K) = I_k^*(\mathbf{x}; z_l)$$

The notation $Prob_{IPCK}(\cdot)$ refers to a conditional distribution model obtained by using IPCK.

Remarks:

- Similar to the univariate case (Suro-Pérez and Journel, 1991), the first ipc autocovariances show a strong spatial correlation. The autocovariances for the last indicator principal components show lack of spatial correlation.
- For ipc's with spatial correlation $Y_k(\mathbf{x})$, the corresponding estimate $Y_k^*(\mathbf{x})$ requires solution of a system of linear equations with dimension equal to $n+1$.
- One limitation of this approach is that all K variables must be sampled equally at all n data locations, as was made explicit in expression (6). The ipc's (9) can only be defined considering the existence of the full indicator vector.
- Just like indicator kriging and univariate IPCK, multivariate IPCK does not ensure that the corresponding conditional distribution models satisfy the following order relations,

$$I_k^*(\mathbf{x}; z_l) \ge I_k^*(\mathbf{x}; z_{l'})\quad l > l'\ \ \forall\, k \tag{11}$$

$$I_k^*(\mathbf{x}; z_l) \in [0, 1] \tag{12}$$

Therefore, IPCK-derived conditional probability models must be corrected to obtain consistent conditional distributions.

III.3 The Case of Undersampling

This is a common case observed in practice. Unfortunately, there is no easy answer to that problem in the framework of IPCK. Consider for example a reduced number of variables, say $K = 2$, and suppose that there are n_1 locations where the variable $Z_1(\mathbf{x})$ has been sampled. Similarly, there are n_2 sampled locations for the variable $Z_2(\mathbf{x})$. As before, the threshold values z_l will correspond to the same l-quantile for both variables.

Indicator covariance matrices for both variables can be defined, i.e., $\mathbf{C}_I^{kk}(\mathbf{h})\ \ k = 1, \cdots, 2$, and also indicator crosscovariance matrices, $\mathbf{C}_I^{12}(\mathbf{h})$, $\mathbf{C}_I^{21}(\mathbf{h})$. Notice that if $\mathbf{h} = 0$ the indicator covariance matrices are symmetric and equal. The indicator crosscovariance matrix is not symmetric in general but the transpose of $\mathbf{C}_I^{12}(0)$ is equal to $\mathbf{C}_I^{21}(0)$. There are special cases, isofactorial distributions (Beckmann, 1973) for example, which show that the indicator crosscovariance matrix is symmetric.

Once again, the final goal is to obtain a conditional distribution model by using a simpler procedure than indicator cokriging. To achieve that, consider the problem of finding an orthogonal matrix $\mathbf{A}$ such that:

$$\mathbf{C}_I^{11}(0) = \mathbf{A}\mathbf{D}^1\mathbf{A}^T \tag{13}$$

with $\mathbf{D}^1$ being a diagonal matrix. For this problem, the solution provided by the spectral decomposition suffices. However, if the problem is modified to find $\mathbf{A}$ such that expression (13) is satisfied and also that indicator crosscovariance matrix can be factorized as:

$$\mathbf{C}_I^{12}(0) = \mathbf{A}\mathbf{D}^{12}\mathbf{A}^T \tag{14}$$

with $\mathbf{D}^{12}$ being also a diagonal matrix. There is no reason to expect joint satisfaction of (13, 14) with the same orthogonal matrix $\mathbf{A}$. Therefore, it is required an orthogonal matrix $\mathbf{A}$ such that:

$$\begin{aligned} \mathbf{A}\mathbf{C}_I^{11}(0)\mathbf{A}^T &\approx \mathbf{D}^1 \\ \mathbf{A}\mathbf{C}_I^{12}(0)\mathbf{A}^T &\approx \mathbf{D}^{12} \end{aligned} \tag{15}$$

A method to achieve such diagonalization is the FG-algorithm (Flury and Gautschi, 1986). This matrix $\mathbf{A}$ can be obtained by this procedure and the approximation level of the diagonalization (15) can be checked.

III.3.1 Conditional Distribution Models in the Case of Undersampling

Knowledge of matrix $\mathbf{A}$ honoring (15) allows construction of the following two ipc vectors:

$$\begin{aligned} \mathbf{Y}^1(\mathbf{x}) &= \mathbf{A}^T\mathbf{I}_1(\mathbf{x}) \\ \mathbf{Y}^2(\mathbf{x}) &= \mathbf{A}^T\mathbf{I}_2(\mathbf{x}) \end{aligned} \tag{16}$$

with

$$\mathbf{Y}^k(\mathbf{x}) = [Y_1^k(\mathbf{x}) \cdots Y_L^k(\mathbf{x})]^T \quad k = 1, 2$$

Ipc covariance and crosscovariance matrices are assumed to be approximately diagonal for all $\mathbf{h}$, i.e.

$$\mathbf{C}_Y^{kk}(\mathbf{h}) = \mathbf{A}^T\mathbf{C}_I^{kk}(\mathbf{h})\mathbf{A} \rightsquigarrow \textit{diagonal} \quad k = 1, 2$$

$$\mathbf{C}_Y^{kk'}(\mathbf{h}) = \mathbf{A}^T\mathbf{C}_I^{kk'}(\mathbf{h})\mathbf{A} \rightsquigarrow \textit{diagonal} \quad k \neq k',\ k = 1, 2\ , k' = 1, 2$$

Runing a few experimental crosscovariances allows checking the previous approximations. If the assumptions are well justified, each ipc is estimated by a cokriging estimator, i.e.,

$$Y_l^{k'*}(\mathbf{x}) = \sum_{k=1}^{2}\sum_{\alpha=1}^{n_k} \lambda_{k\alpha} Y_l^k(\mathbf{x}_\alpha)$$

To ensure unbiasedness, the weights $\lambda_{k\alpha}$ must satisfy,

$$\sum_{k=1}^{2}\sum_{\alpha=1}^{n_k}\lambda_{k\alpha}=1 \tag{17}$$

This expression (17), together with the system of linear equations below provides the weights,

$$\sum_{k''=1}^{2}\sum_{\alpha=1}^{n_{k''}}\lambda_{k''\alpha}C_{Y_l}^{k''k}(\mathbf{x}_\alpha-\mathbf{x}_\beta)-\mu=C_{Y_l}^{k'k}(\mathbf{x}-\mathbf{x}_\beta)\ \ \beta=1,\cdots,n_k\ \ k=1,2 \tag{18}$$

with

$$C_{Y_l}^{k''k}(\mathbf{x}_\alpha-\mathbf{x}_\beta)=Cov(Y_l^{k''}(\mathbf{x}_\alpha),Y_l^{k}(\mathbf{x}_\beta))$$

The conditional distribution models are obtained by a back-transformation:

$$\begin{aligned}\mathbf{I}_1^*(\mathbf{x}) &= \mathbf{A}\mathbf{Y}^{*1}(\mathbf{x})\\ \mathbf{I}_2^*(\mathbf{x}) &= \mathbf{A}\mathbf{Y}^{*2}(\mathbf{x})\end{aligned}$$

with $\mathbf{Y}^{*k}(\mathbf{x})$ being the estimated indicator principal component vector whose elements have been estimated by using expressions (17, 18).

Remarks:

- This multivariate version of IPCK does not require modeling the entire set of ipc autocovariances and crosscovariances. Indeed, only the first autocovariances and crosscovariances are correlated. Therefore, the last ipc's are estimated by an arithmetic average.
- Practical implementation of this method is limited to approximations (15) and to the number of variables K. A large K implies larger systems of equations and a much more constrained modeling of autocovariances and crosscovariances through the linear model of coregionalization (Journel and Huijbregts, 1978, p. 171).

IV. A Gaussian Model

To study the approximation level of the ipc covariance matrix under isofactorial bivariate distributions, consider two random variables $Z_1(\mathbf{x})$ and $Z_2(\mathbf{x})$ spatially correlated and crosscorrelated between them. Assume that the bivariate distribution of any pair $Z_1(\mathbf{x}),Z_1(\mathbf{x}+\mathbf{h})$, $Z_2(\mathbf{x}),Z_2(\mathbf{x}+\mathbf{h})$ and $Z_1(\mathbf{x}),Z_2(\mathbf{x}+\mathbf{h})$ is biGaussian with autocorrelograms $\rho_i(h), i=1,2$, i referring to the variable $Z_i(\cdot)$, and crosscorrelogram $\rho_{12}(h)$. The correlograms are assumed identical for either $Z_1(\cdot)$ or $Z_2(\cdot)$ and defined by:

$$\rho_i(h_x,h_y)=\begin{cases}1-Sph_a(\sqrt{h_x^2+4\cdot h_y^2}) & \text{if } \sqrt{h_x^2+4\cdot h_y^2}\le 20\\ 0 & \text{if } \sqrt{h_x^2+4\cdot h_y^2}>20\end{cases}\qquad i=1,2 \tag{19}$$

The crosscorrelogram is,

$$\rho_{12}(h_x, h_y) = \begin{cases} 0.8 - 0.8Sph_a(\sqrt{h_x^2 + 4 \cdot h_y^2}) & \text{if } \sqrt{h_x^2 + 4 \cdot h_y^2} \leq 20 \\ 0 & \text{if } \sqrt{h_x^2 + 4 \cdot h_y^2} > 20 \end{cases} \tag{20}$$

with the notation $Sph_a(h)$ referring to an spherical variogram with range $a = 20$.

The assumption of biGaussianity together with the spatial variability of $Z_1(\cdot)$, $Z_2(\cdot)$ and $Z_1(\cdot)Z_2(\cdot)$, as given by correlograms and crosscorrelograms, allows obtention of the indicator covariance matrix (7). The retained number of threshold values K is three and they correspond to the three quartiles. Therefore, the dimension of matrix (7) is six and it will be used to obtain the orthonormal matrix $\mathbf{A}(0)$ and investigate the level of approximation of the ipc covariance matrix under the biGaussian model. Other isofactorial distributions could be used, however results from Suro-Pérez (1992) show that the cumulative bivariate distribution does not present significant variations considering different isofactorial models.

At the level of crosscovariances, the ipc covariance matrix at lag $|\mathbf{h}| = 5$ along the $x-$direction is:

$$\Sigma_Y^2(5) = \begin{bmatrix} 0.4243 & 0.0000 & 0.0000 & 0.01365 & 0.0000 & 0.0000 \\ 0.0000 & 0.0490 & 0.0000 & 0.0000 & 0.0000 & 0.0000 \\ 0.0000 & 0.0000 & 0.0484 & 0.0000 & 0.0000 & -0.0098 \\ 0.0137 & 0.0000 & 0.0000 & 0.00264 & 0.0000 & 0.0000 \\ 0.0000 & 0.0000 & 0.0000 & 0.0000 & 0.0099 & 0.0000 \\ 0.0000 & 0.0000 & -0.0098 & 0.0000 & 0.0000 & 0.0030 \end{bmatrix}$$

and for $|\mathbf{h}| = 10$ along the same direction is:

$$\Sigma_Y^2(10) = \begin{bmatrix} 0.2041 & 0.0000 & 0.0000 & 0.00848 & 0.0000 & 0.0000 \\ 0.0000 & 0.0077 & 0.0000 & 0.0000 & 0.0000 & 0.0000 \\ 0.0000 & 0.0000 & 0.0218 & 0.0000 & 0.0000 & -0.0052 \\ 0.0084 & 0.0000 & 0.0000 & 0.0005 & 0.0000 & 0.0000 \\ 0.0000 & 0.0000 & 0.0000 & 0.0000 & 0.00177 & 0.0000 \\ 0.0000 & 0.0000 & -0.0052 & 0.0000 & 0.0000 & 0.0013 \end{bmatrix}$$

Both matrices are well approximated as diagonal matrices. The crosscovariances between first and fourth, and first and sixth ipc's appear significant as compared to the autocovariance level of the fourth and sixth ipc, however the actual crosscorrelation is negligible and therefore, cokriging of those components will be well approximated by kriging. Also, observe that some ipc crosscovariances are equal to zero, similar to the univariate case (Suro-Pérez and Journel, 1991). Once again, symmetry and persymmetry of the indicator covariance matrix yields such behavior.

IV.1 Simulation of Gaussian Fields

Two Gaussian fields have been generated using a procedure based on the Cholesky decomposition of the covariance matrix (Kennedy and Gentle, 1980, p. 294). The autocorrelogram for both random fields corresponds to (19). The crosscorrelogram is

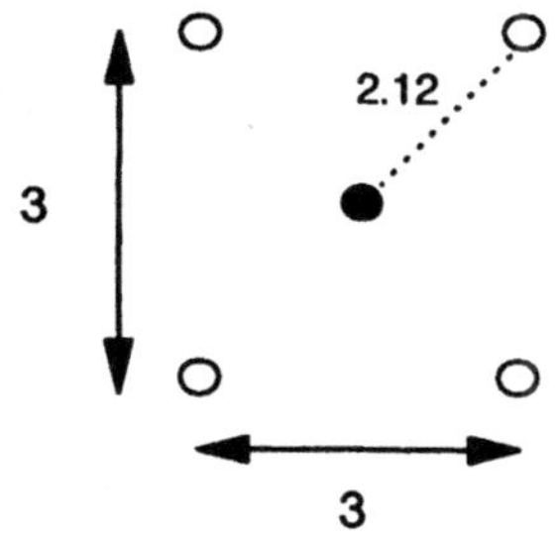

Figure 1: Conditioning data configuration. Given four conditioning locations fully sampled, the conditional distribution is obtained at the center of the conditioning data.

Table 1: Modeling of conditional distributions. $\bar{F}_{q_p}$ is the arithmetic average of the difference between the Gaussian conditional distribution and the IPCK conditional distribution. q_p refers to the pth quantile of each variable.

	$F_{q_{0.1}}$	$F_{q_{0.3}}$	$F_{q_{0.5}}$	$F_{q_{0.7}}$	$F_{q_{0.9}}$
$Z_1(\mathbf{x})$	0.070	0.0052	0.0050	0.0092	-0.0346
$Z_2(\mathbf{x})$	0.0244	0.0588	0.0208	-0.0815	-0.0139

defined by (20). In both cases (19, 20), the correlation range is $a = 10$. Since both fields are Gaussian, the conditional distribution given the configuration shown in Figure 1, is Gaussian and obtained by simple cokriging (Anderson, 1984). This conditional distribution then is compared with the conditional distribution obtained by the multivariate version of IPCK. In this exercise, four locations fully sampled are considered, although the IPCK undersampling implementation (16) has been used. The considered cutoffs correspond to the deciles of both variables. Table 1 shows the average of 361 differences between the Gaussian conditional distribution and the conditional distribution obtained by IPCK, at the nine cutoffs. Figure 2 shows scattergrams for $q_{0.2}$, median and $q_{0.8}$ cutoffs. It is observed a conditional bias for the extreme cutoffs ($q_{0.2}, q_{0.8}$), though all cutoffs are globally unbiased. This behavior is typical of indicator methods, IPCK included, since the outliers influence is reduced at the very moment of the indicator definition: extreme values are coded as 1's or 0's for all cutoffs, with no distinction of how high are the highest values or how low are the lowest values. Gaussian methods, as compared to indicator methods, are sensitive to outliers, since the conditional mean is a linear combination of sample values, outliers or not.

Figure 3 shows how the conditional bias is corrected by removing those distributions where outliers ($|Z_i(\mathbf{x})| \leq 2.0$) are one or more of the four conditioning data. These scattergrams (Figures 2 and 3) show that for the first cutoffs, the IPCK probabilities are lower than the Gaussian probabilities if there are negative outliers. If there are positive outliers, the last cutoffs show IPCK probabilities higher than the Gaussian probabilities.

V. Case Study: Walker Lake Data Set

The Walker Lake Data Set (Isaaks and Srivastava, 1989) is used here to test

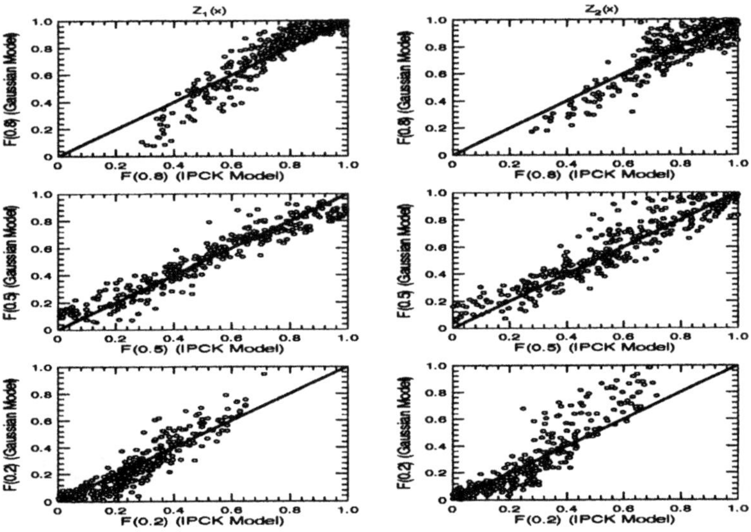

Figure 2: Scattergrams between Gaussian conditional distributions and IPCK models. Observe that the IPCK estimates are globally unbiased, with conditional bias towards the extreme cutoffs.

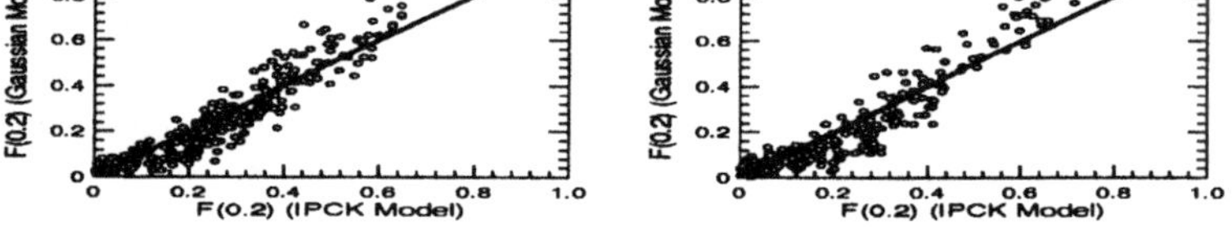

Figure 3: Scattergrams between Gaussian conditional distributions and IPCK models. No outliers are present in the conditioning data.

Table 2: Basic Statistics Walker Lake Data Set: $Z_1(\cdot)$, $Z_2(\cdot)$ variables $\bar{x}$ is the arithmetic average; s^2 is the variance; $q_{0.25}$, $q_{0.5}$ and $q_{0.75}$ are the first quartile, median and third quartile respectively; *min* and *max* are the minimum and maximum values. These values have been obtained from the exhaustive 19800 available data.

	$\bar{x}$	s^2	$q_{0.25}$	$q_{0.50}$	$q_{0.75}$	*min*	*max*
$Z_1(\mathbf{x})$	194.13	127570.2	4.324	35.22	229.66	0.0	5505.92
$Z_2(\mathbf{x})$	181.76	32161.5	23.97	129.37	295.80	0.0	985.53

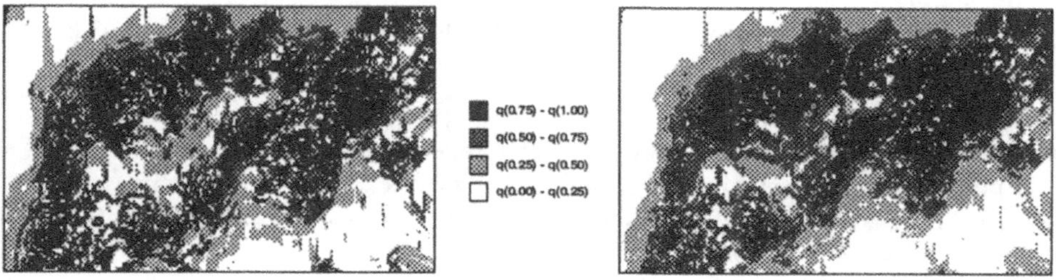

Figure 4: Walker Lake Data Set. There are 19800 data, 110 rows and 180 columns. (left) $Z_1(\mathbf{x})$. (right) $Z_2(\mathbf{x})$.

the applicability of the multivariate version of IPCK. This data set consists of two elevation related variables, say $Z_1(\cdot)$ and $Z_2(\cdot)$. The variable $Z_1(\cdot)$ corresponds to the variance of 25 data, equally spaced, in a block of 5×5. $Z_2(\cdot)$ is a function of the local mean and variance of each block. Thus, $Z_2(\cdot)$ is related to $Z_1(\cdot)$. Table 2 shows basic statistics of these variables and Figure 4 shows maps of both variables. The gray scaling has been selected considering the quartiles as threshold values.

V.1 Undersampling Case

The following case to consider is when one of the variables has not been sampled at the same locations than the other. Suppose that one of the variables has been densely sampled and the other only at fewer locations. Figure 5 shows an example of such a situation where sampling locations for both variables are located differently. There are 62 samples for the $Z_1(\cdot)$ variable and 260 locations, regularly spaced, for $Z_2(\cdot)$. The variable $Z_2(\cdot)$ in this context is used as ancillary data that helps to build conditional distribution models for $Z_1(\cdot)$. For example, $Z_2(\cdot)$ could be seismic information and $Z_2(\cdot)$ information derived from well logs.

Two ipc vectors are defined (16) considering as threshold values the deciles of each variable. The orthonormal matrix $\mathbf{A}$ used to obtain both ipc vectors has been computed from the exhaustive data sets. This matrix $\mathbf{A}$ yields approximately a diagonal ipc covariance matrix. Figure 6 shows some ipc auto(cross)correlograms defined for $Z_1(\cdot)$ and $Z_2(\cdot)$. Note that there is a ranking of the spatial correlation defined by the ordering of the ipc's. The crosscorrelations for different ipc show a negligible correlation level as

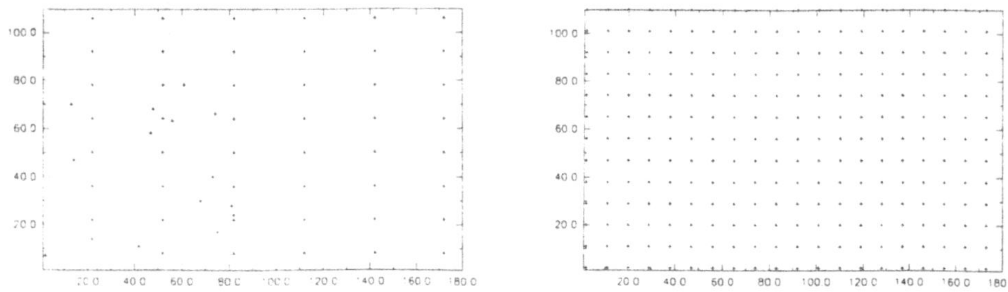

Figure 5: Conditioning data to be used for the case of unequal sampling. (left) 62 locations where variable $Z_1(\cdot)$ has been sampled. (right) 260 locations where variable $Z_2(\cdot)$ has been sampled.

compared to the autocorrelation levels. Therefore, estimation of the first $Y^1(\cdot)$ ipc's, say $Y_l^1(\mathbf{x})$ for $l = 1, \cdots, 2$, demands cokriging involving $Y_l^2(\mathbf{x}), l = 1, \cdots, 2$. Estimates of other ipc's, say $Y_l^1(\mathbf{x})$, $l = 3, \cdots, 9$, can be obtained by an arithmetic average since the spatial correlation appears negligible.

Models of conditional distribution for $Z_1(\cdot)$ are obtained at 660 locations using as conditioning data the available information of both variables. Figure 7 shows a contour map of the probability of exceeding the ninth decile. Note the vales above 0.9 in the right side. Nine conditional probability models indicate as extremely likely to find outcomes above the ninth decile. The true values extracted from the exhaustive data set show that indeed eight out of nine values are above the ninth decile.

VI. Conclusions

Use of the multivariate version of IPCK on the Walker Lake Data Set has shown its applicability for modeling conditional distribution when the conditioning data pertain to different attributes. It has been shown that multivariate IPCK requires reasonable amounts of computer time and that the modeling effort is limited to few ipc autocovariances and crosscovariances. This is an important feature which is not shared for other non-parametric multivariate techniques, if any, for modeling conditional distributions.

Two implementations of multivariate IPCK have been presented. The first case considers that all locations $\mathbf{x}$ have been sampled fully with all attributes. A conditional distribution model, under this context, requires only kriging for a limited number of ipc's. The second case considers that attributes have been sampled at different locations. This features implies that a conditional distributional model should be obtained by doing cokriging in a few ipc's. However, even when this case is more complex that the full sampling case, its practical implementation only requires modeling of a limited number of ipc autocovariances and crosscovariances.

A comparison on a simulated Gaussian field, between IPCK conditional distributions and the corresponding, theoretically exact, Gaussian distribution, has shown that IPCK estimates are conditionally biased around the extreme cutoffs. This behavior is yield by outliers presence in the conditioning data. Gaussian methods are sensible to their

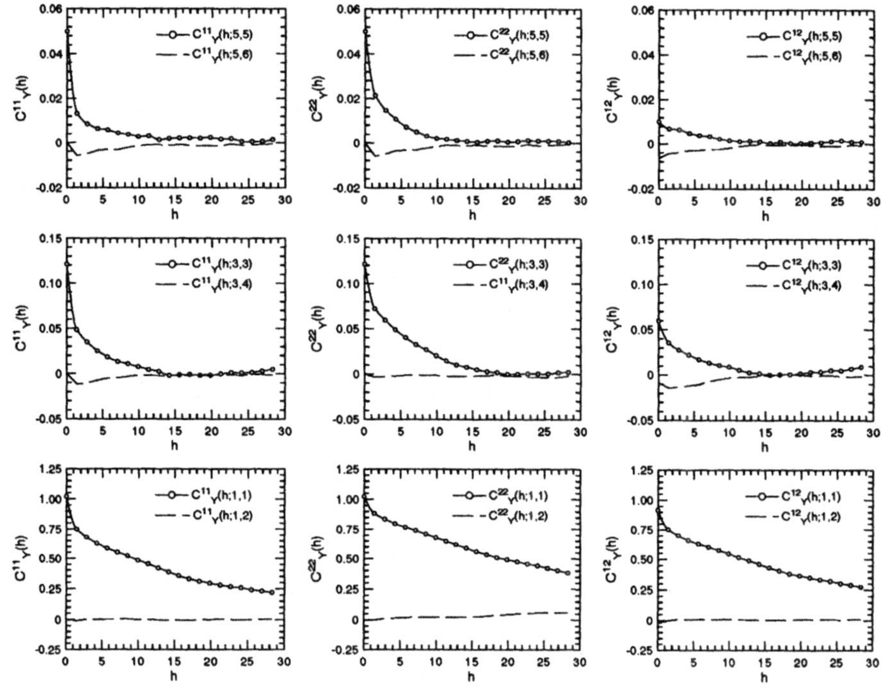

Figure 6: Ipc auto(cross)covariances for the Walker Lake Data Set. Observe that all ipc crosscovariances are negligible as compared to the ipc autocovariances. Also, observe that most of the spatial correlation is captured in the first ipc's.

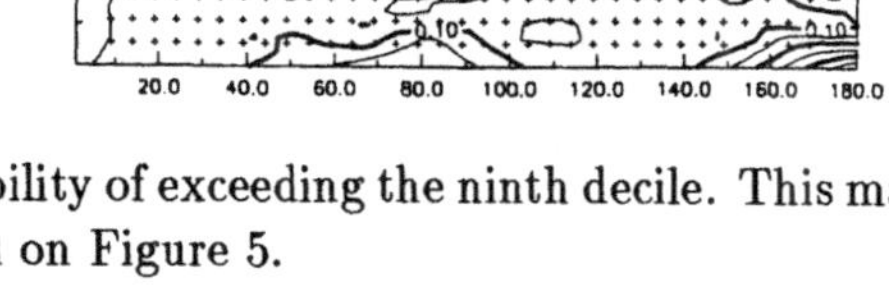

Figure 7: Probability of exceeding the ninth decile. This map was built considering unequal sampling showed on Figure 5.

presence. Multivariate IPCK, as any method based on indicators, reduces their influence.

VII. References

- Anderson, T. W., 1984, **Multivariate Statistical Analysis**, Wiley and Sons, New York, 675 pp.
- Beckmann, P., 1973, **Orthogonal Polynomials for Engineers and Physicists**, The Golem Press, 280 pp.
- Flury, B. N. and W. Gautschi, 1986, An algorithm for simultaneous orthogonal transformation of several positive definite symmetric matrices to nearly diagonal form, SIAM J. Sci. Stat. Comput., V. 7, p. 169-184.
- Golub, G. and Van Loan Ch., 1983, **Matrix Computations**, The John Hopkins University Press, 476 pp.
- Isaaks, E. and M. R. Srivastava, 1989, **An Introduction to Applied Geostatistics**, Oxford University Press, New York, 561 p.
- Journel, A. G., 1989, **Fundamentals of Geostatistics in Five Lessons**, Short Course in Geology, V. 8, American Geophysical Union publ., Washington, D. C., 40 pp.
- Kennedy, W. J. and Gentle, J. E., 1980, **Statistical Computing**, Marcel Dekker, New York, 591 pp.
- Suro-Pérez, V. and A. G. Journel, 1991, Indicator principal component kriging, Math. Geol., V. 25, No. 5, p. 759-788.
- Suro-Pérez, V., 1992, Estimation and Stochastic Simulation using Indicator Principal Component Kriging, Ph. D. Thesis, Stanford University.

MODELLING RESERVOIR HETEROGENEITY WITH FRACTALS

THOMAS A. HEWETT
Petroleum Engineering Department, Stanford, CA, USA 94305-2220

The properties of random fractals and fractional noises are reviewed. The intermittency of fractional noises is shown to provide a good model for the spatial intermittency of property variations seen in well logs and is consistent with the temporal intermittency of hydrologic records. An anisotropic fractal model is proposed for constructing conditional simulations of geologic property distributions which match measured data where available, mimic the character of variations in the interwell region, and reproduce the observed bedding characteristics and univariate statistics of available data. These simulations can be constructed with arbitrary resolution and can be used to evaluate the effects of reservoir heterogeneity on oil recovery processes.

INTRODUCTION

In recent years there has grown an increasing appreciation of the need to understand the influence of the non-uniform distributions of flow properties characteristic of natural porous media on the dynamics of oil recovery processes. In particular, it is recognized that the heterogeneous distributions of permeability typically found in oil reservoirs exert a controlling influence on the channeling and early breakthrough of injected fluids, the sweep efficiency of those fluids, the distribution of bypassed oil which is available as a target for infill drilling or enhanced oil recovery processes, and the evolution of phase behavior in multi-component miscible and semi-miscible enhanced oil recovery processes.

Numerical methods for simulating the performance of oil displacement processes have reached a high level of sophistication. The equations which are solved in these simulators are general enough to account for a wide variety of physical mechanisms which can affect the dynamics of oil displacement. The validity of these equations for describing displacement processes in porous media has been verified by comparison with laboratory experiments.

The primary application of these simulators is for the prediction of reservoir performance at scales much larger than the laboratory scale. Owing to the limitations of current computer resources, the geometric distribution of reservoir flow properties is typically discretized into computational grid blocks which are large compared to the scale at which the flow properties are defined and measured in the laboratory. This means that each block must be assigned an effective average value for each of the relevant flow properties, and these averages must include the effects of property variations at scales finer than the scale of the grid block. In addition, the use of coarse grid blocks introduces a "numerical dispersion," due to discretization errors in the finite difference approximations to the equations, which is generally much larger than properly scaled physical dispersion.

In order to properly account for the effects of heterogeneous distributions of flow properties on reservoir processes, it is necessary to model the variation of properties at the scale at which they are defined and measured in the laboratory. When data specifying variations at this scale are unavailable, synthetic property distributions can be used to evaluate the influence of heterogeneity on flow processes (Hewett, 1986; Hewett and Behrens, 1990). These synthetic fields should match the data available at wells and should mimic the character of property variations in the interwell region which can influence the predicted performance of flow processes.

A variety of approaches is available for producing synthetic heterogeneous fields. Most are based on reproducing statistical measures of correlation and variation observed in the available data and propagating variations with similar statistical characteristics to the interwell region. The aim in constructing these simulation models is to produce a realistic representation of the reservoir property distributions which affect fluid flow processes at an arbitrary scale of resolution. The properties required for fluid flow modeling are the permeabilities, relative permeabilities, porosities, fluid saturations, and capillary pressures. In practice, the controlling influences of variable permeability and porosity on fluid displacements are usually all that are modeled.

Although field scale models at this scale of resolution cannot be accommodated in current reservoir simulators, simulations on local portions of a detailed model can be used to evaluate the impact of heterogeneity on flow processes and the effect of averaging flow properties in coarser gridded models. These models also can be used to provide the input to streamtube methods for scaling up the effects of heterogeneity, or methods for deriving effective flow properties for coarse grid block models (Hewett and Behrens, 1991).

This paper will focus on an approach based on a fractal model of interwell property variations for producing simulation models. This approach uses techniques drawn from the field of geostatistics to reproduce the marginal distributions of property values and to make the resulting simulations conditional to measured data. The synthetic variations of properties are generated using the Weierstrass-Mandelbrot representation of random fractals which is based on the power law scaling of variations in fractal distributions.

RANDOM FRACTALS AND FRACTIONAL NOISES

Mandelbrot's fractal geometry provides a mathematical framework for describing the irregular and complex shapes found in nature. Fractals are characterized by variations at all scales (and are, therefore, undifferentiable) and have partial correlations over all scales. Every attempt to divide such a geometry into smaller, more uniform, regions results in the resolution of even more structure.

The geometries of fractal distributions are characterized by their intermittent or "spotty" nature; they do not fill space. This characteristic is quantified by a parameter, D, called the fractal dimension of the distribution. The statistics of fractal distributions are strongly affected by the mismatch between the fractal dimension of the distribution and the Euclidean dimension of the space available for it to fill. This is quantified by the fractal co-dimension, $H = d - D$, where d is the Euclidean dimension of the embedding space (Mandelbrot, 1983).

The properties of correlated random fields are the subject of the field of geostatistics. The primary tool used in geostatistics is the variogram, or variance of increments. The variogram, $2\gamma(h)$, measures the mean square variability of a random function $Z(x)$ as a function of the spatial separation, or lag, between two points within it. The variogram for random fractals is given as (Mandelbrot, 1983)

$$2\gamma(h) = V_H h^{2H} \tag{1}$$

where H is the co-dimension defined above, and is restricted to the range $0 < H < 1$. This form for the variogram has no range, reflecting the presence of variations at all scales in a true fractal. As a practical matter, fractal distributions will approximate geologic property distributions only over a finite range of scales between some effective outer scale and some effective inner scale. Several examples of a random fractal with zero mean and unit variance are shown in Figure 1. These were generated using the same sequence of random numbers, but different values of H, in the Weierstrass-Mandelbrot function, to be described later. As H increases from zero toward unity, the persistence of correlations increases and the contributions of the smallest scales of variation (highest frequencies) decrease. Increasing values of H result in fewer crossings of the axis and the function becomes more regular. Random fractals with $H > 0.7$ bear a strong resemblance to the topography of the Earth and have been used to simulate mountainous terrains (Voss, 1985).

Although random fractals described by Equation (1) have received the most attention in the literature, distributions described by a related function may actually have more relevance to the characterization of geologic property distributions in sedimentary systems. This type of random function is referred to as "fractional noise," and is just the derivative of functions described by Equation (1), after smoothing with a moving window filter to remove the highest frequencies which otherwise make fractals undifferentiable (Mandelbrot and Van Ness, 1968). This function is stationary and has a semi-variogram given as

$$\gamma(h,\delta) = V_H \delta^{2H-2}\left\{1-\frac{1}{2}\left[\left|\frac{h}{\delta}+1\right|^{2H} - 2\left|\frac{h}{\delta}\right|^{2H} + \left|\frac{h}{\delta}-1\right|^{2H}\right]\right\} \tag{2}$$

where δ is the length of the moving window filter. This variogram model asymptotes to a sill value of $V_H\delta^{2H}$ for large h. Samples of fractional noises are more erratic than samples of random fractals. The behavior of fractional noises for $H \to 1$ is the same as the behavior of random fractals as $H \to 0$. Near the origin, the behavior of the variogram for fractional noises is identical to that for random fractals. When $h/\delta < 1$, Equation (2) is approximately equal to $V_H\delta^{-2}\, h^{2H}$. Thus, the behavior of random fractals can be modeled as fractional noise with a smoothing interval much larger than the domain of observation.

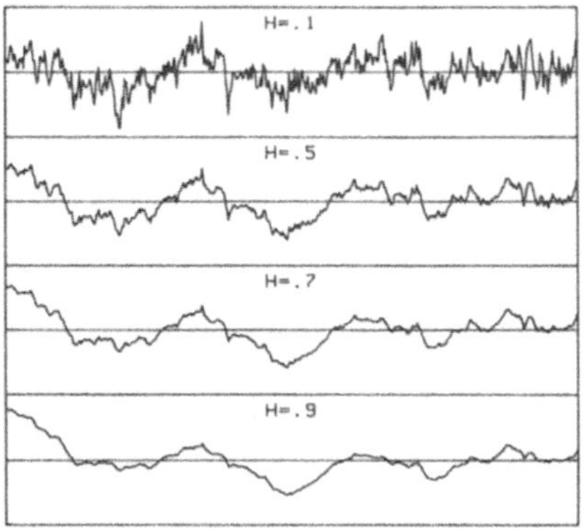

Figure 1: Traces of normalized random fractals with different values of the co-dimension, H.

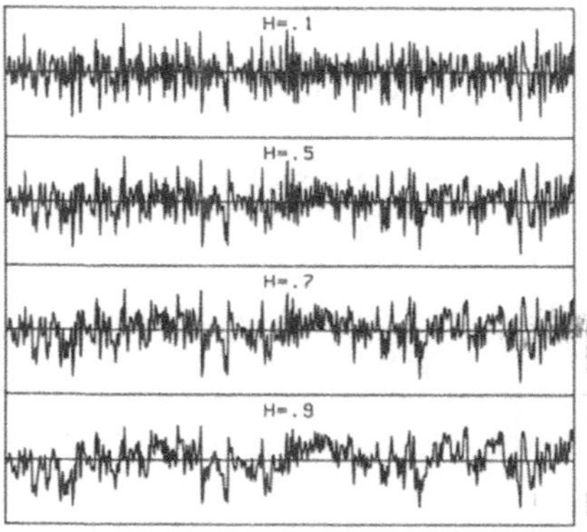

Figure 2: Traces of normalized fractional noises calculated from the increments of the traces in Figure 1.

The normalized fractional noises calculated from the increments of the fractals shown in Figure 1 are shown in Figure 2. The characteristics of fractional noises depend strongly on the value of H, with the ranges $0 < H < 1/2$, $H = 1/2$, and $1/2 < H < 1$ having distinctly different properties. This can be seen by considering the correlation of successive samples, given as $2^{2H-1} - 1$ (Hewett, 1986; Mandelbrot, 1983). When $H = 1/2$, the correlation is zero and the sequence corresponds to a completely uncorrelated "white noise." When $H < 1/2$ the correlation of successive values is negative. When $H > 1/2$ the correlation between successive samples is positive. This leads to a persistence of variations in local regions of all scales and a less dense set of zero crossings. The strength of this persistence, or the tendency of like values to cluster together, is measured by H. Samples of geologic property variations observed in well logs typically show the persistence of fractional noises with $H > 0.7$ (Hewett, 1986).

The persistence of variations in fractional noises with $H > 1/2$ means that long samples are required to draw reliable conclusions about the statistics of the population from which they are drawn. As an example, the standard deviation of the mean of a sample of fractional noise of length s, $\sigma_{m,s}$, is given by (Mandelbrot and Wallis, 1969a)

$$\sigma_{m,s} = s^{H-1}\sigma \tag{3}$$

where σ is the standard deviation of an infinite sample. When $H = 0.9$, the standard deviation of the mean of a sample of 1000 values is still half as big as the standard deviation of individual samples from the infinite population.

One of the first geophysical records to be described by the correlation structure of fractional noise is the hydrologic data of the Nile River analyzed by Edwin Hurst in his planning for the Aswan High Dam (Hurst, 1951). In examining the sequence of annual river flows over a period of several thousand years, he concluded that the span of interdependence of annual flows exceeded the length of the historical record and could for all practical purposes be regarded as infinite. Standard statistical tests are not well suited to detecting this type of long run interdependence between samples. An alternative introduced by Hurst (Hurst, 1951) and refined by Mandelbrot and Wallis (Mandelbrot and Wallis, 1969b) looks at the behavior of the accumulated variations from the mean. If $Z(t)$ is the record of a sample after removing the mean, then the accumulation of variations is given as

$$Z^*(t) = \sum_{u=1}^{t} Z(u) \tag{4}$$

If sequential values in the sample are completely uncorrelated, the range of the accumulated variations will grow as the square root of the sample length (Feller, 1951). By the range, we mean the maximum range of excursions of the accumulated variations above and below their mean value of zero. In samples of fractional noise drawn from hydrologic data, the range of accumulated variations was observed to increase in proportion to the H power of the sample length, where $0.5 < H < 1$, indicating a partial clustering of like values at all scales.

Since we are interested in detecting the clustering tendency of like variations in different length subsamples, and different subsamples of the total record can have different means and variances, the appropriate statistic for identifying the clustering tendency in fractional noises is the rescaled range, where each subsample is normalized to have zero mean and unit variance when the range of accumulated variations is calculated. With $Z^*(t)$ defined above, the range of a subsample extending from a position t to $t + s$, after removal of the mean, is given by (Mandelbrot and Wallis, 1969b)

$$\begin{aligned} R(t,s) = & \max_{0<u<s} \left\{ Z^*(t+u) - Z^*(t) - (u/s)\left[Z^*(t+s) - Z^*(t)\right]\right\} \\ & - \min_{0<u<s} \left\{ Z^*(t+u) - Z^*(t) - (u/s)\left[Z^*(t+s) - Z^*(t)\right]\right\} \end{aligned} \tag{5}$$

The subtraction of the trend in $Z^*(t)$ in each of the terms above accounts for the removal of the subsample mean from the calculation of the range. To account for the different variances of subsamples, this sequential range is divided by the standard deviation of the sample, $S(t,s)$. The asymptotic behavior of the rescaled range, $R(t,s)/S(t,s)$, is indicative of the long run correlations in the sample. For uncorrelated sequences, or sequences with a finite range of correlations

$$\lim_{s \to \infty} s^{-\frac{1}{2}} R(t,s) \,/\, S(t,s) = \text{constant}$$

For fractional noises,

$$\lim_{s \to \infty} s^{-H} R(t,s) \,/\, S(t,s) = \text{constant}$$

and the rescaled range asymptotically obeys an s^H scaling law. The value of H for a fractional noise can be determined by calculating the value of the rescaled range for a large number of subsamples with different lengths and starting points within the record and plotting $R(t,s)/S(t,s)$ vs. s on logarithmic coordinates. The slope of the line formed at large values of s gives the appropriate value of H.

When this procedure was applied to the Nile River discharge data, a value of $H = 0.91$ was obtained. Similar analyses of hydrologic data from other rivers showed that most have H in the range $0.7 < H < 0.9$ (Mandelbrot and Wallis, 1969b). The same type of analysis has been applied to precipitation data directly with similar results, but the short records available (< 200 years) don't permit a confident conclusion about the presence of long run interdependence. In order to better test the idea of long run correlations in weather variations, it has been necessary to analyze long fossil geophysical records which are related to annual precipitation. These include measurements of tree ring thicknesses and lake varve thicknesses. In each case, values of $H > 5$ have been found (Mandelbrot and Wallis, 1969b).

Considering the intermittent character of hydrologic data over long time periods, it would be surprising not to find similar variations in spatial records of sedimentation, given that the driving forces for many sedimentation processes are related to the magnitudes of hydrologic flows. As an example, consider the sequence of 1732 values, normalized to zero mean and unit variance and shown in Figure 3, taken from a spherically focused log (SFL) run in a well from a braided fluvial environment. This log is a high resolution resistivity log which shows a good correlation with permeability in this formation.

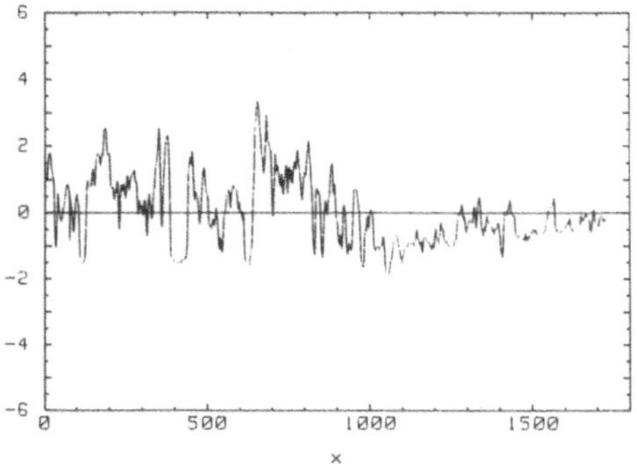

Figure 3: Normalized sequence of values from a spherically focused log run in a well from a braided fluvial environment.

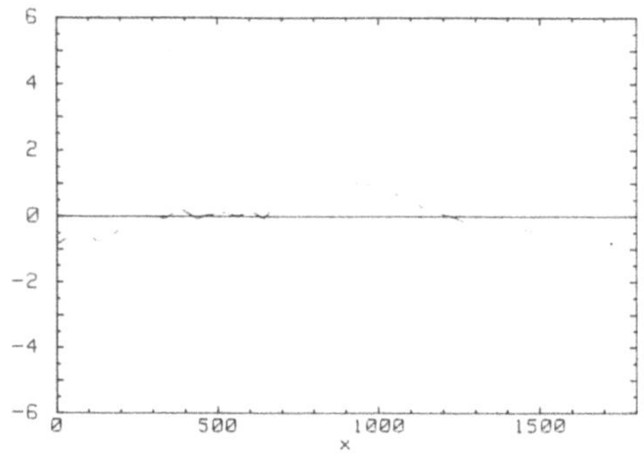

Figure 4: Normalized sequence of the accumulated variations from the mean for the trace in Figure 3.

The normalized accumulated variations required for the calculation of the rescaled range are shown in Figure 4. The behavior of the rescaled range is shown in Figure 5. The symbols represent the average of the rescaled range for a large number of subsamples for each lag. The slope for lags greater than 20 was determined by linear regression to be $H = 0.90$. A line with slope $H = 0.5$ is also shown for reference. Clearly, this trace shows long run correlations with no signs of breaking to a square root behavior for large lags.

The variogram for this same trace is shown on logarithmic coordinates in Figure 6 along with the variogram model for fractional noise given by Equation (2). The value of using logarithmic coordinates is that the fractional errors in fitting the model near the origin and for large lags can both be seen in the same plot. The value of $H = 0.90$ determined from the rescaled range analysis was used in plotting the model. The values of the smoothing interval, δ, and the sill value, $V_H\delta^{2H-2}$, were adjusted to provide the best "eyeball" fit to the experimentally determined points. The value of the sill required to fit the measured variogram was 1.65, indicating that the sample variance underestimates the sill by 40 percent.

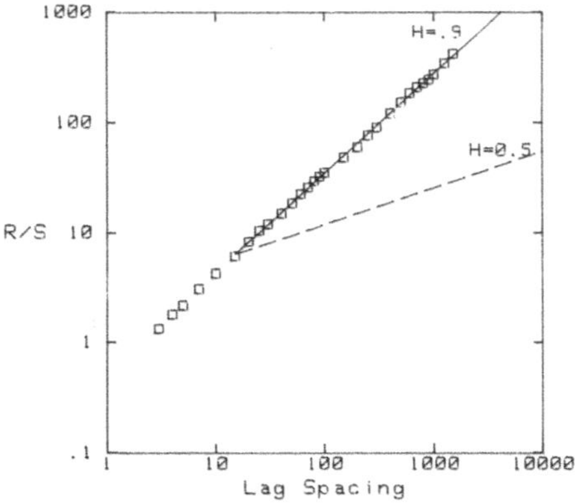

Figure 5: The rescaled range of the accumulated variations in Figure 4 as a function of the lag spacing.

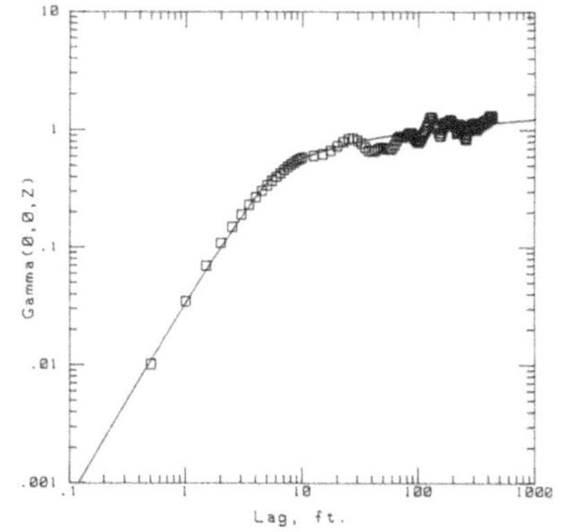

Figure 6: The variogram for the trace in Figure 3 on logarithmic coordinates.

The fit of the model is very good for small lags and shows oscillations around the model for large lags. These oscillations are a consequence of the "fluctuation variance" which results from calculating a variogram with data from a finite domain which is not small compared to the total range of correlations (Journel and Huijbregts, 1978). The quality of the fit on a linear plot is shown in Figure 7. The behavior of the variogram near the origin is shown in Figure 8. The quality of the fit near the origin where the fluctuation variance is smallest is excellent. Attempts to fit this data with other variogram models proved fruitless since their basic shapes could not accommodate the nature of variations for both large and small lags.

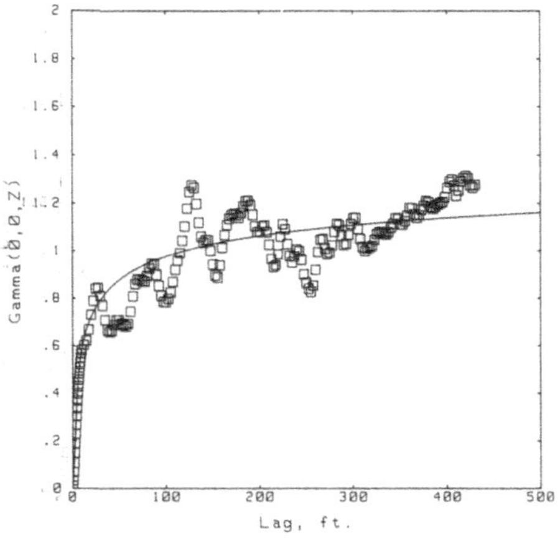

Figure 7: The variogram for the trace in Figure 3 on linear coordinates.

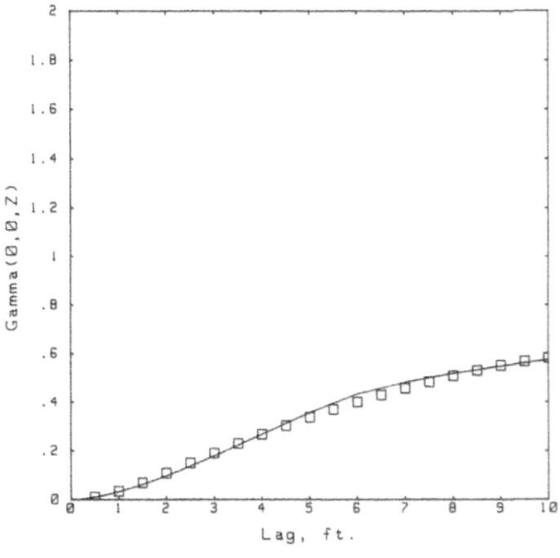

Figure 8: The behavior near the origin of the variogram in Figure 7.

Measurements of the correlation structure of lateral variations of geologic properties are more difficult to evaluate. This is a consequence of the sparser sampling of data due to the wide spacing of wells and the irregular spacing of the sample locations. In the absence of reliable estimates of the lateral correlation structure, one can appeal to the relationship observed between hydrologic data and the observed correlations in well log data. If δ is the length of the record of an annual cycle of storms in the geologic column, then the corresponding length in lateral directions would be expected to be much larger, given the stratified nature of sedimentary deposits. The behavior of fractional noises with large lateral mixing lengths is the same as that of random fractals with the same value of H. Thus, a reasonable model for both the lateral and vertical variations which is consistent with the observed behavior of hydrologic records is fractional noise with a large horizontal to vertical geometric anisotropy, or equivalently, a zonal anisotropy (Journel and Huijbregts, 1978) with the behavior of fractional noise in the vertical and random fractals in the horizontal. This will be referred to as the anisotropic fractal model.

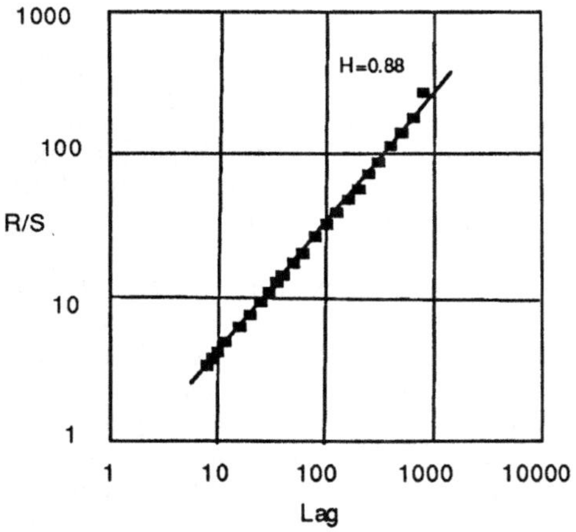

Figure 9: R/S analysis of density log from carbonate vertical Well 1.

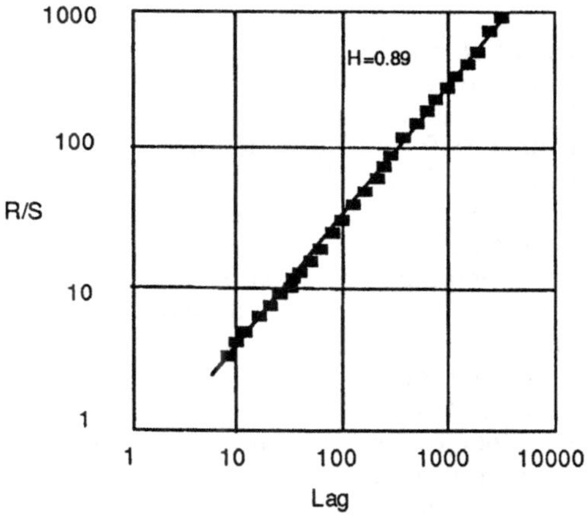

Figure 10: R/S analysis of density log from carbonate horizontal Well 2.

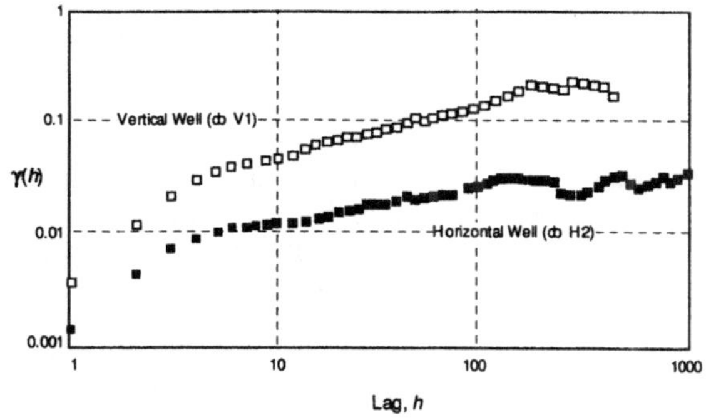

Figure 11: Variograms of vertical Well 1 and horizontal Well 2 plotted on logarithmic coordinates.

Experimental validation of this model has been provided recently by measurements of horizontal and vertical well log data from wells in the same formation (Crane and Tubman, 1990). Both showed evidence of the long run correlations of fGn when analyzed with the rescaled range. The vertical data had an intermittency coefficient of $H = 0.88$. The horizontal data had a similar value of $H = 0.89$, indicating that the same model form is valid in both directions. This is shown in Figures 9 and 10. Variograms calculated on the two well logs showed a similar form, but different variance levels at large lags. Neither experimental variogram showed a clear sill for the largest lags measured. These are shown plotted on logarithmic coordinates in Figure 11. The characteristics of fGn with similar values of H have also been observed in other reported analyses of horizontal well log data (Hardy, 1991; Perez and Kelkar, 1991).

CONDITIONAL SIMULATION OF FRACTAL DISTRIBUTIONS

With a model for the variogram, there are a variety of techniques for constructing realizations of random fields which have spatial correlations matching those given by the variogram. Simulations of fractal distributions are particularly well suited to spectral methods. This results from the power law form of the variogram which, according to the generalized Wiener-Khintchine theory (Mandelbrot, 1967), corresponds to a power law spectral density. The form of the spectral density corresponding to Equation (1) is

$$S_Z(f) = C_H 1 / f^{\beta} \tag{6}$$

where $\beta = 2H + 1$. The construction of a random fractal can be accomplished in a straightforward fashion by constructing a Fourier series with random amplitudes drawn from a Gaussian distribution with zero mean and unit variance and random phases drawn from a uniform distribution in $[0,2\pi]$. By filtering the amplitudes to obey Equation (6), a fractal with any desired value of H can be constructed (Voss, 1988). The resulting simulation can then be scaled to have the mean and variance of the distribution being simulated.

When the simulation is only required on a regular grid, fast Fourier transform (FFT) methods can be used to calculate the distribution. An alternative approach to constructing random fractals based on a generalization of the Weierstrass-Mandelbrot random function can be calculated for an arbitrary configuration. This approach, described in detail in the Appendix, is based on a Fourier series with a geometric sequence of spatial frequencies obeying Equation (6). It has the advantage of providing an analytical representation of the simulation which can be used to calculate values at arbitrary locations. The process of conditional simulation takes one of the random fields produced by any of the above methods and alters it to pass through the data we have measured, while at the same time preserving the overall correlation structure of the field (Journel and Huigbrets, 1978). These conditional simulations are not unique because they contain random components which represent the variability of the function at scales smaller than those we have actually been able to sample. However, because we constrain them to honor the same variogram and pass through the same data values, they all have a similar appearance.

Conditional simulations generated using the Weierstrass-Mandelbrot function will be Gaussian in character, because the variability is produced by adding a number of random components together. When we are interested in simulating fields with marginal distributions which are not Gaussian, such as distributions of permeability, efforts should be made to preserve the marginal distribution as well as the character of spatial correlations. This can be accomplished by working with normal score transforms of the data (Journel and Huijbregts, 1978).

The anisotropic fractal model of property distributions proposed here is based on the idea that the same sequence of driving forces for sedimentation processes acted over large areas which were contemporaneously exposed to these processes. Deformation of strata subsequent to burial must be accounted for separately. This can be accomplished by constructing the simulations in a stratigraphic coordinate system which follows the bedding planes of the sediments (McArthur, 1988).

As an example of this procedure, four SFL logs along a 113 ft. thick by 186 ft. long cross-section are shown in Figure 12. The center log is a subsample of the log previously analyzed. Eight markers are defined in each well and have been kriged to the interwell region to define the stratigraphic horizons between wells. These are used to define the stratigraphic grid shown, where the thickness of the grid blocks is proportional to the local thickness between adjacent markers, and the layers of grid blocks follow horizons which maintain a proportional spacing between the markers. The well log values were then kriged onto the grid nodes using a fractal variogram with the value of $H = 0.9$ derived from the previous analysis. The resulting distribution of values is shown in Figure 13, where the kriged data is shown in the form of pseudo-well logs. The slight mismatch at the well locations is due to the fact that the grid locations have a coarser vertical resolution and do not coincide with the well log sample points but have been connected up as a continuous trace. A contour plot of the same distribution, in Figure 14, shows that kriging produces a smoother distribution of horizontal variations in the interwell region than is typically observed in outcrops.

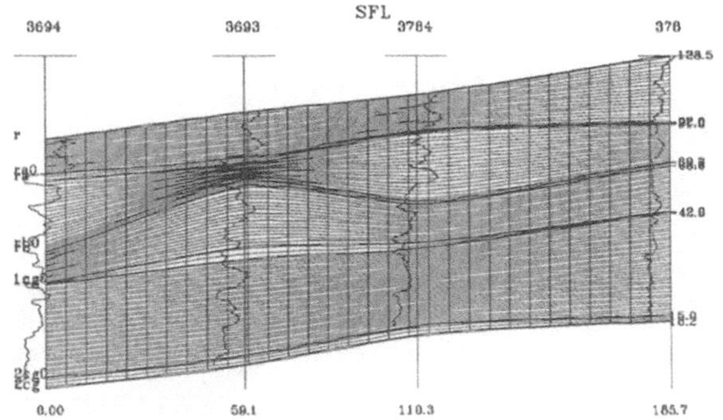

Figure 12: Well log traces, correlated markers, and stratigraphic grid for a 113 ft. thick cross-section to be simulated.

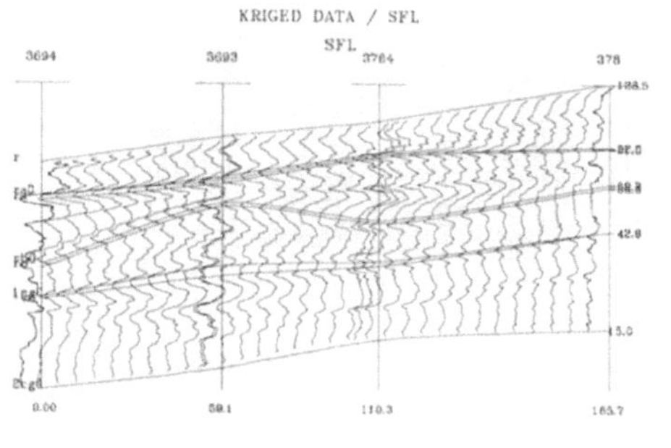

Figure 13: Kriged well log values shown as pseudo-well logs.

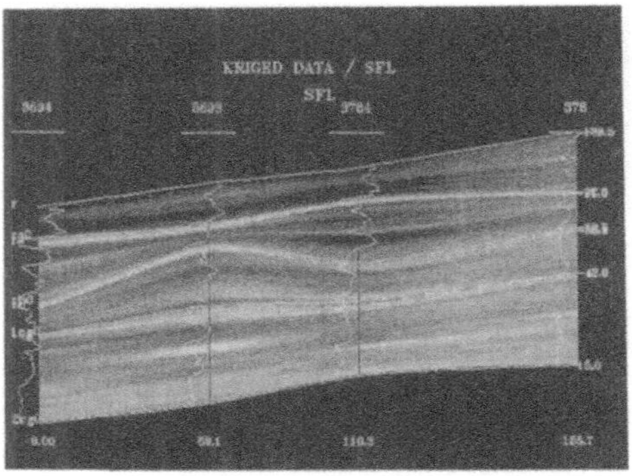

Figure 14: Contour plot of kriged well log values.

A conditional anisotropic fractal simulation was then constructed using the procedure described in the Appendix on the same grid and well log locations. The resulting pseudo-well logs after inversion of the transforms are shown in Figure 15. A contour plot of this distribution is shown in Figure 16. Here, the correlated small scale variations produce a cross-section with a variable character more typical of outcrops.

Conditional simulations of areal distributions of properties can also be constructed. As an example, consider the kriged surface through the distribution of transmissibilities (permeability times thickness) sampled in 29 wells shown in Figure 17. It can be seen that kriging produces smooth variations of properties between sample points. A conditional simulation through these same data is shown in Figure 18.

These simulations can be used as input to reservoir simulation models to evaluate the impact of unsampled heterogeneity in the interwell region on oil recovery processes. Simulations of flow through multiple realizations of these fields can provide an assessment of the uncertainty in process performance predictions which remain due to uncertainty in the distribution of properties due to sparse sampling.

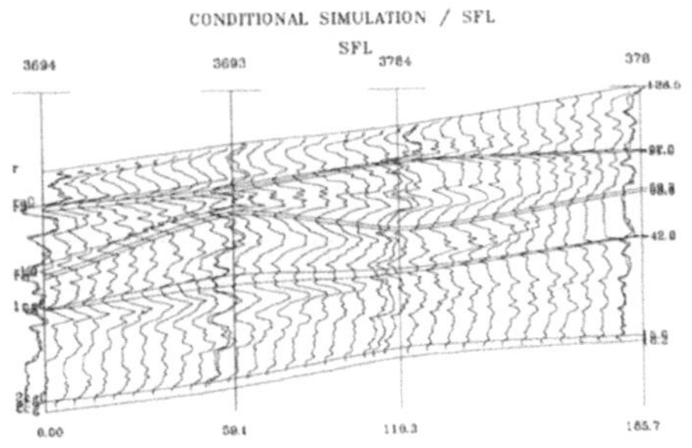

Figure 15: Conditional simulation of the well logs in Figure 12 shown as pseudo-well logs.

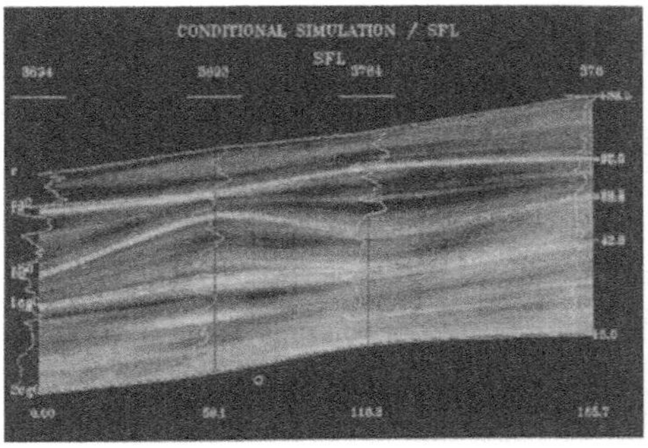

Figure 16: Contour plot of the conditional simulation.

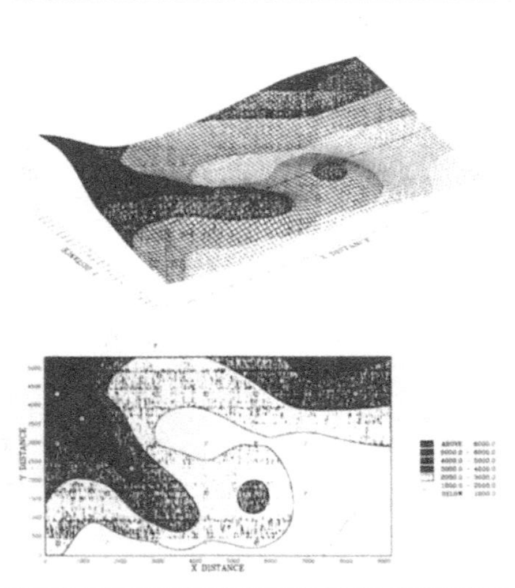

Figure 17: Distribution of transmissibilities obtained by kriging the 29 transmissibility data. © 1988 Society of Petroleum Engineers.

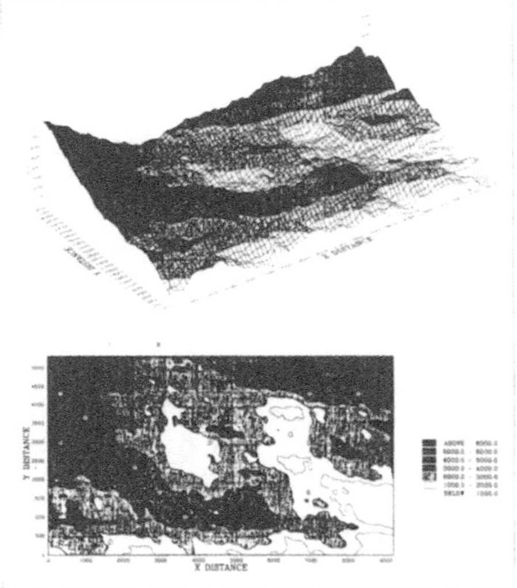

Figure 18: Conditional simulation of the transmissibility distribution. © 1988 Society of Petroleum Engineers.

CONCLUSIONS

1. An anisotropic fractal model of geologic property distributions in sedimentary environments matches the spatial intermittency of variations observed in well log data and is consistent with the temporal intermittency of hydrologic records and analyses of horizontal well log data.

2. Conditional simulations of geologic property distributions can be constructed which honor measured data and simulate unsampled heterogeneity with an anisotropic fractal correlation structure. Construction of the simulation on a stratigraphic grid accounts for the deformation of strata subsequent to burial and the use of Gaussian transforms preserves the marginal distribution of the data in the simulated values.

3. The construction of spatial distributions of reservoir flow properties which preserve the spatial and univariate statistics of the data at arbitrary scales of resolution can be used to evaluate the impact of different scales of heterogeneity on flow processes.

APPENDIX

The Weierstrass-Mandelbrot Random Fractal Function

A variety of techniques are available in the literature for constructing random fractal distributions (Voss, 1988). The Weierstrass-Mandelbrot fractal function is based on a Fourier representation with a spectral density satisfying Equation (8) with an infinite range of frequencies (Berry and Lewis, 1980). The algorithm presented here limits the range of frequencies to those corresponding to an outer scale of variation, L_{max}, and an inner scale of variation, L_{min}.

An isotropic three-dimensional surface depending on two space coordinates and having a fractal dimension $D = 3 - H$ can be calculated as

$$Z_{wm}(X,Y) = \sum_{n=o}^{N} \sum_{m=o}^{M} A(m,n)\left(\frac{K_{oo}}{K_{mn}}\right)^{H} \sin\left[K_{x,m} X + K_{y,n} Y + \phi(m,n)\right] \qquad \text{(A-1)}$$

where the wave numbers are given by

$$K_{mn} = \left(K_{x,m}^2 + K_{y,n}^2\right)^{1/2}$$

$$K_{x,m} = K_{x,o}/r^m$$

$$K_{y,n} = K_{y,o}/r^n$$

$$K_{x,o} = 2\pi/L_{x,\max}$$

$$K_{y,o} = 2\pi/L_{y,\max}$$

The $A(m,n)$ are Gaussian random variables with zero mean and the same variance for all m and n, and the $\phi(m,n)$ are random phases uniformly distributed in $[0,2\pi]$. The amplitude of the variations can be scaled to any desired value by suitable choice of the variance of the $A(m,n)$ distribution. Here, r ($r < 1$) is the ratio of successive spatial frequencies, or the ratio of successive nested scales of variation. Changes in r affect the "lacunarity" of the distribution, without affecting the fractal dimension (Mandelbrot, 1983; Voss, 1988). Values of $r < 1/2$ have most of the variability concentrated in a few clumps. The simulations in this study used $r = .707$, so that every other term in the series halves the scale of variations. H is the fractal co-dimension ($H = 3 - D$) and controls the amplitude of variations at every scale. The number of terms which must be calculated depends on the range of scales and the scale ratio; i.e., $M = \log(L_{x,\min}/L_{x,\max})/\log(r)$ and $N = \log(L_{y,\min}/L_{y,\max})/\log(r)$. For the simulations used in this study, $L_{x,\max} = L_{y,\max}$ was set to 4 times the maximum dimension of the simulation grid, and $L_{x,\min} = L_{y,\min}$ was set to half the grid spacing.

In terms of the spectral density, $Z_{wm}(x,y)$ contains only discrete frequencies $f_n = C/r^n$, each component having a mean square amplitude proportional to $1/f^{2H}$ in a bandwidth proportional to f (equal spacing on a logarithmic scale) (Voss, 1988). Thus, the spectral density (mean square amplitude per unit bandwidth) is

$$S_Z(f) = C\,1/f^{2H+1}$$

This is the power law behavior required by Equation (8).

Equation (A-1) can be modified to produce an anisotropic distribution with the characteristics of random fractals in X and the characteristics of fractional noise in Y by simply differentiating the expression above with respect to Y. The resulting variations with Y will tend to appear more irregular than data observed in well logs because of the finite volume of investigation of logging tools.

REFERENCES

Berry, M.V., and Lewis, Z.V. (1980) "On the Weierstrass-Mandelbrot Fractal Function," Proc. Roy. Soc., London, A, 370, 459–484.

Crane, S.D., and Tubman, K.M. (1990) "Reservoir Variability and Modeling with Fractals," SPE 20602, presented at the 65th Ann. Tech. Conf. of SPE, New Orleans.

Feller, W. (1951) "The Asymptotic Distribution of the Range of Sums of Independent Random Variables," Ann. Math. Stat., 22, 427–432.

Hardy, H.H. (1991) "Fractal Analysis of Core Photographs with Application to Reservoir Characterization," NIPER/DOE 3rd International Reservoir Characterization Technical Conference, DOE CONF 911125, NIPER-558/2, Nov. 3–5.

Hewett, T.A. (1986) "Fractal Distributions of Reservoir Heterogeneity and Their Influence on Fluid Transport," SPE 15386, 61st Ann. Tech. Conf. of SPE, New Orleans.

Hewett, T.A., and Behrens, R.A. (1990) "Conditional Simulation of Reservoir Heterogeneity with Fractals," SPE Formation Evaluation, Sept., 217–225 (originally SPE 18326, 63rd Ann. Tech. Conf. of SPE, Houston, 1988).

Hewett, T.A., and Behrens, R.A. (1991) "Scaling Laws in Reservoir Simulation and Their Use in a Hybrid Finite Difference/Streamtube Approach to Simulating the Effects of Permeability Heterogeneity," in L.W. Lake, H.B. Carroll, Jr. and T.C. Wesson (eds.) *Reservoir Characterization II*, pp. 402–441, Academic Press, San Diego.

Hurst, H.E. (1951) "Long Term Storage Capacity of Reservoirs," Trans. ASCE, 116, 770–808.

Journel, A.G., and Huijbregts, Ch.J. (1978) *Mining Geostatistics*, Academic Press, New York.

Mandelbrot, B.B. (1967) "Sporadic Random Functions and Conditional Spectral Analysis: Self-Similar Examples and Limits," Proc. Fifth Berkeley Symp. Math. Stat. and Prob., 3, (eds. LeCam and Neyman), Univ. of Cal. Press, 155–179.

Mandelbrot, B.B. (1983) *The Fractal Geometry of Nature*, Freeman, New York.

Mandelbrot, B.B., and Van Ness, J.W. (1968) "Fractional Brownian Motions, Fractional Noises, and Applications," SIAM Review, 10(4) 422–437.

Mandelbrot, B.B., and Wallis, J.R. (1969a) "Computer Experiments with Fractional Gaussian Noises, Part I, Averages and Variances," Water Resources Research, 5(1).

Mandelbrot, B. B., and Wallis, J.R. (1969b) "Some Long-Run Properties of Geophysical Records," Water Resources Research, 5(2).

McArthur, G.J. (1988) "Using Geology to Control Geostatistics in the Hellyer Deposit," Math. Geo., 20(4), 343–366.

Perez, G. and Kelkar, M. (1991) "Assessing Distributions of Reservoir Properties Using Horizontal Well Data," NIPER/DOE 3rd International Reservoir Characterization Technical Conference, Nov. 3–5.

Voss, R.F. (1985) "Random Fractal Forgeries," Proc. of the NATO A.S.I. Fund. Algorithms in Comp. Graphics, Ilkley, U.K.

Voss, R.F. (1988) "Fractals in Nature: From Characterization to Simulation," in *The Science of Fractal Images*, (eds. Peitgen and Saupe), Springer-Verlag, New York.

On the Connectivity of two Random Set Models: The Truncated Gaussian and the Boolean.

Denis ALLARD & Heresim Group
Centre de Géostatistique
Ecole des Mines
35 rue Saint-Honoré
77305 Fontainebleau
France

ABSTRACT. The connectivity of two 2-D random sets, namely the truncated gaussian model and the boolean model, is evaluated by simulation. It appears that several parameters of the random sets (porosity, integral range) affect connectivity. The critical values of some of these parameters are brought out.

1 Introduction

To give numerical models of the geology of reservoirs which include the maximum of the available information is one of the major aims of today's petroleum geostatistics. These models are used as input to numerical flow simulations, and allow the computation of histograms of the characteristics studied, such as recoverable reserves, flow, pressure etc., and finally help the engineers in their decision making. In geostatistics, a geological facies in a reservoir is considered as the realization of a random set, and stochastic simulations can provide multiple realizations of this random set. The realizations thus obtained respect the experimental proportion and covariance, but little is known about their connectivity.

The probability that two points are connected and the set of points connected to a fixed point are characteristics which cannot be derived from the proportion and the covariance. Actually, these characteristics involve the entire realization, and must depend on the entire multivariate law. In particular, they cannot depend only on the covariance —a bivariate characteristic—, but also on the model of random function.

The aim of this article is to evaluate by simulation some connectivity characteristics of two models commonly used in the petroleum world: the truncated gaussian model and the boolean model. After having specified the mathematical framework and briefly recalled the two random sets studied here, we shall try to bring out what is common and what distinguishes them from the connectivity point of view.

2 Mathematical framework and definitions

The 4-connectivity on the square lattice

Most of the time, simulations are made on a regular grid defined by a lattice on which neighbourhood relations must be defined. Let us take an example: on the square lattice we can either consider that each point has 4 neighbours —this is the 4-connectivity— or add the 4 points in the corners as neighbours —that is the 8-connectivity. Other lattices with other connectivities are possible in 2-D like the triangular lattice —6 neighbours—, the honeycomb lattice —3 neighbours— etc.. All the characteristics studied will depend on the kind of lattice and the kind of connectivity chosen. In the following, only the square lattice with the 4-connectivity (simple square lattice) will be considered as it is the most often used in petroleum applications.

On the square grid $\mathcal{S}$ of size $L \times L$, let S be a random subset of $\mathcal{S}$. S can be considered as the pores or the permeable phase, and $S^c = \mathcal{S} \backslash S$ the grains or the impermeable phase. Two points x and x' of S are connected (notation $x \leftrightarrow x'$) if there exists a sequence in S of neighbouring points between x and x'. A subset $C \subset S$ is connected if $x \leftrightarrow x'$ for any x and x' of C, and the connected components are the largest connected subset of S. So, far from being a bivariate characteristic, the existence of a connection between two points involves not only the entire simulated field S, but also the relations of neighbourhood. If any connected component appears to intersect two opposite boundaries we say that the corresponding connected component is a percolating one, and that the percolation of the permeable phase holds. Last, we can define a connectivity function, τ, as the probability that two points are connected (i.e. they belong to the same connected component), knowing that they both belong to the permeable phase S.

$$\forall x, y \in \mathcal{S}, \ \tau(x, y) = P(x \leftrightarrow y \mid x, y \in S)$$

Given a lattice, this function, and the distribution of the connected components depend on the probability P on $(\mathcal{S}, \mathcal{P}(\mathcal{S}))$. Thus we have to define our models of stochastic reservoirs.

The truncated gaussian model (TGM)

This model, today commonly used for the modeling of reservoirs, allows geostatistical studies and conditional simulations of lithologic sequences (MATHERON et al., 1988 [5]) and of random sets obtained by truncation of regionalized variables such as a grade, a permeability etc. (FREULON, 1992 [1])

Assume Y to be a stationary multigaussian random function with covariance ρ. On $\mathcal{S}$, the threshold y_c defines the random set S represented by its indicator function Z:

$$S = \{x \in \mathcal{S} : Y(x) \geq y_c\} \quad Z(x) = \begin{cases} 1 \text{ if } Y(x) \geq y_c \\ 0 \text{ if } Y(x) < y_c \end{cases}$$

$Z(x)$ is a stationary random function :

$$\begin{aligned} E[Z(x)] &= p &&= 1 - G(y_c) \\ Cov(Z(x), Z(x+h)) &= C(h) &&= g(y_c)^2 \sum_{n \geq 1} \frac{H_{n-1}^2(y_c)\rho^n(h)}{n} \end{aligned}$$

where g and G are the gaussian density and distribution function, and where H_n is the normalized Hermite polynomial of degree n. To simplify computations we have chosen the factorized covariance model:

$$Cov(Y(x), Y(x+h)) = \rho(h) = \rho(h_1, h_2) = e^{-\frac{|h_1| + |h_2|}{a}}$$

where a is the range of the covariance of Y in the two main directions.

As we want to compare two models when their covariances vary, we must choose a criterion to compare them. A natural characteristic is the integral range A, defined as the integral of $C(h)$ over $\mathbb{R}^2$: $A = \frac{1}{C(0)} \int_{\mathbb{R}^2} C(h)\, dh$ (MATHERON, 1989 [6]; YAGLOM, 1986 [8]; LANTUEJOUL, 1990 [3]). In the case of a stationary random function, the integral range is asymptotically proportional to the estimation variance of the mean value over a domain V, provided V large enough and $0 < A < +\infty$. Physicists call it the correlation area, and it can be seen as the scale of the phenomenon. The integral range of a truncated gaussian random set,

$$A_t = 4a^2 \frac{g(y_c)^2}{G(y_c)(1 - G(y_c))} \sum_{n \geq 1} \frac{1}{n} \left(\frac{H_{n-1}(y_c)}{n} \right)^2$$

is equal to the product of the square of the gaussian covariance range, multiplied by a function of the threshold. It is symmetric around $p = 1/2$ where it is maximal with $A_t \simeq 0.7 \times 4a^2$; it equals 0 for $p = 0$ and $p = 1$ (fig. 1).

The discrete boolean model with fixed primary grain (BM)

The boolean model, well adapted to the idea of randomly arranged objects and of honouring a geometry, has for this reason a great ability to simulate the geometry of various phenomena such as petroleum reservoirs, mineralized lens or microscopic porous media. An application to the stochastic modeling of underground reservoirs facies can be found in HALDORSEN, 1987 [2], whereas there is a theoretical presentation in MATHERON, 1967 [4]. In order to compare it to the truncated gaussian model, which can only be simulated on a grid, we shall use the discrete version of this model.

On $\mathcal{S}$, let W be a $l \times l$ square primary grain, located at the origin, and W_x be its translation at $x \in \mathcal{S}$. At each point x of $\mathcal{S}$, W_x is located with the probability $p_0 \in [0,1]$ independently of all other points. Let S_0 be the set of all points on which a grain has been set up. The random set S is then defined as the union of all primary grains, and is represented by its indicator function Z:

$$S = \bigcup_{x \in S_0} W_x \qquad Z(x) = \begin{cases} 1 \text{ if } x \in S \\ 0 \text{ if } x \notin S \end{cases}$$

$Z(x)$ is a stationary random function :

$$\begin{aligned} p &= 1 - q_o^{l^2} \\ C(h) &= \left(q_0^{l^2 + (|h_1| + |h_2|)l - |h_1||h_2|} - q_0^{2l^2} \right) \mathbb{1}_D(h) \end{aligned}$$

where $q_0 = 1 - p_0$ and $D = \{-l, \cdots, l\}^2$ is the domain where C is not equal to 0. The integral range.

$$A_b = 4l^2 \, \frac{1-p}{p} \sum_{i \geq 2} \frac{(-\ln(1-p))^{i-1}}{i.i!}$$

is proportional to the area of the primary grain for a fixed proportion p. It is no longer symmetric, but monotonically decreasing from $p = 0$ where $A_b = l^2$ to $p = 1$ where $A_b = 0$ (fig. 1). For the choice of $l = 5$ and $a = 2.7$, we can see on the same figure that the integral ranges of both models are almost equal around $p = 0.5$.

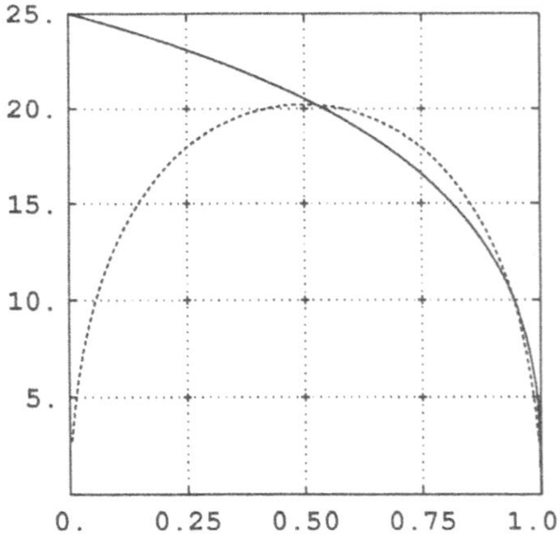

Figure 1: Integral range versus proportion for TGM with $a = 2.7$ (dashed line) and BM with $l = 5$ (solid line).

3 Results of simulations

Introduction

The principle of the computation will be as follows. On a square grid of size $L \times L$, 250 independent simulations are carried out with a covariance range $a \in \{0, 1, 2, 2.7, 3, 3.5\}$ for the gaussian random function, and a grain size $l \in \{1, 2, 3, 5, 7\}$ for the boolean model. Hence, the simulations depend on three parameters: the proportion p, the integral range A and the grid size L.

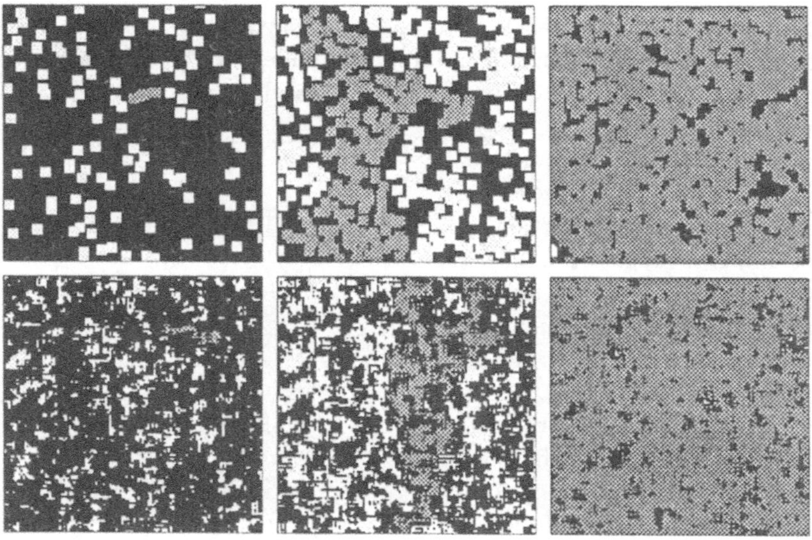

Figure 2: Random sets on a 128×128 grid. The biggest connected component is in grey. Top : boolean model with $l = 5$; the proportion p is from left to right 0.15, 0.5525, 0.85. Bottom : truncated gaussian model with $a = 2.7$; p is from left to right 0.15, 0.3975, 0.85

Both models of random sets for three different proportions are depicted on fig. 2. One can observe that for a small proportion the random set is constituted of many, small, widely scattered connected components (left), whereas for a large proportion almost all of the connected components have merged into a huge cluster covering the entire field, except some small holes (right). Between these two extremes a connected component of much bigger extension than the others joins two opposite boundaries for the first time at a particular value of the proportion called the percolation threshold P_c (center). To describe this general evolution, some characteristics of the simulations have been chosen: the percolation threshold P_c, the number of connected components

$N_{cc}(p)$, the relative area occupied by the biggest connected component in the permeable phase $V_{max}(p)$ and the connectivity function $\tau_p(h)$. All these characteristics are random variables. They all depend not only on the parameters of the random set but also on the dimension of the grid.

The number of connected components

A first description can be made using the number of connected components. From $p = 0$ where $N_{cc} = 0$, let the proportion increase. At first time, new connected components are created in the many empty spaces, and N_{cc} increases. Little by little, different connected components come in contact, tending to decrease N_{cc}. These two antagonistic effects find a balance for a particular proportion where N_{cc} reaches its maximum. Beyond this value, N_{cc} decreases and is equal to 1 for $p = 1$. On fig. 3, $\overline{N}_{cc}$, the mean value of N_{cc} on 250 simulations divided by the area of the simulation field, is represented for three different models of random sets.

In correlated random sets ($\triangle$ and $\circ$), many very small connected components are eliminated and $\overline{N}_{cc}$ decreases significantly. For a given proportion ($p = 0.5$) and integral range ($A \simeq 20$), the gaussian model ($\triangle$) generates three times more connected components than the boolean one ($\circ$). However, for very large proportions ($p \geq 0.75$), $\overline{N}_{cc}$ tends to 1 identically for all models. This fact corresponds exactly to the general invasion of the percolating cluster in the permeable phase. As the number of connected components is not suitable to illustrate this invasion, we have introduced the relative area occupied by the biggest connected component, $V_{max}(p)$.

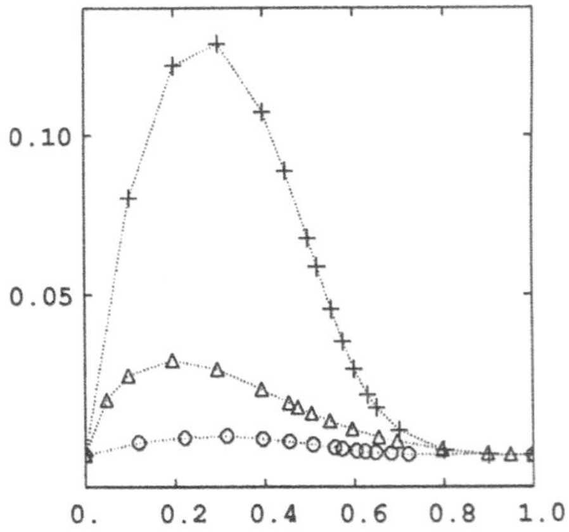

Figure 3: $\overline{N}_{cc}$ versus the proportion on 700 x 700 grid, for three models: without correlation (+), truncated gaussian model with $a = 2.7$ ($\triangle$) and boolean model with $l = 5$ ($\circ$).

The biggest connected component

As seen above, starting at a given proportion value, one of the components individualizes, grows quicker than any other and is responsible for the percolation of the permeable phase. Fig. 4 represents the mean value $\overline{V}_{max}$ of V_{max} on the same set of simulations, versus the proportion. $\overline{V}_{max}$ increases suddenly around a critical proportion (generally between $p = 0.5$ and $p = 0.6$) in a second order phase transition. This transition can be characterized by two values: the inflexion point or the abscissa $\hat{p}_c$ for which $\overline{V}_{max} = 0.5$ on one hand, and the "width" Δp of the transition defined by a percentile deviation on the other hand. Besides the proportion which varies from $p = 0.4$ to $p = 0.7$, there are three different ways to make A and L vary: L remains constant while A is varying (fig. 4a), A remains constant and L varies (fig. 4b), and L^2/A remains constant while L and A are varying together (fig. 4c).

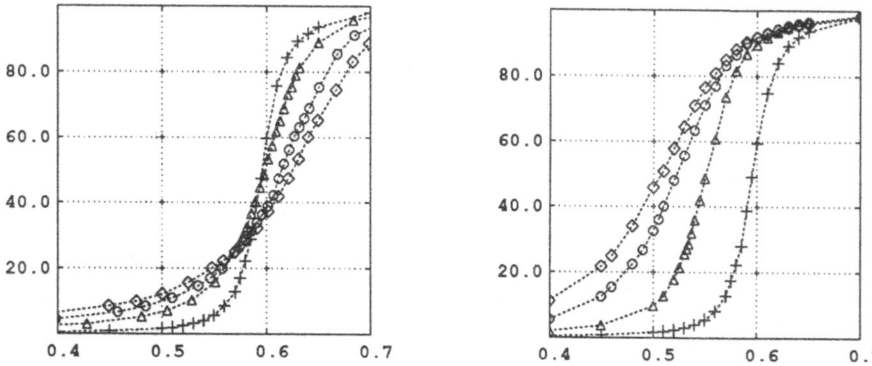

Figure 4a: $\overline{V}_{max}$ versus the proportion: $L = 200$. *Left:* BM with l successively equal to 1(+), 3($\triangle$), 5($\circ$), 7($\diamond$). *Right:* TGM with a successively equal to 0(+), 1($\triangle$), 2($\circ$), 3($\diamond$).

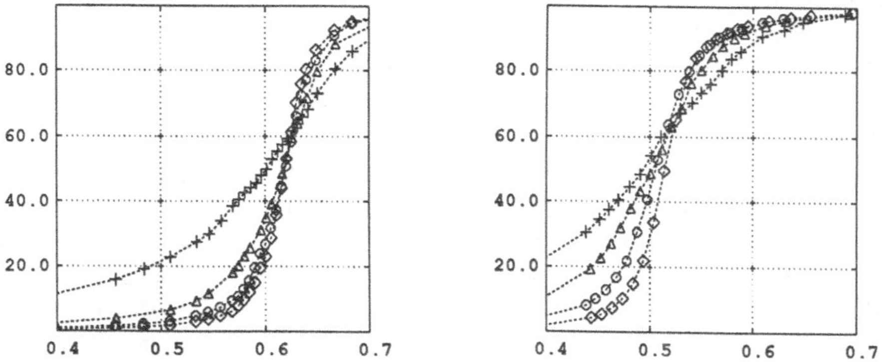

Figure 4b: $\overline{V}_{max}$ versus the proportion: A=20. *Left:* BM with $l = 5$; L is successively equal to 100(+), 300($\triangle$), 500($\circ$), 700($\diamond$). *Right:* TGM with $a = 2.7$; L is successively equal to 100(+), 200($\triangle$), 400($\circ$), 700($\diamond$).

For a constant grid size (fig.4a), $\hat{p}_c$ increases with l in the boolean model, whereas it decreases with a in the truncated gaussian model; Δp increases with both a and l. For a constant integral range ($A = 20$) $\hat{p}_c$ increases and Δp decreases in both models with L. For a constant ratio $L^2/A = 12\,500$ (fig.4c), $\hat{p}_c$ decreases in the TGM and increases in the BM with L, whereas Δp remains constant for those variations.

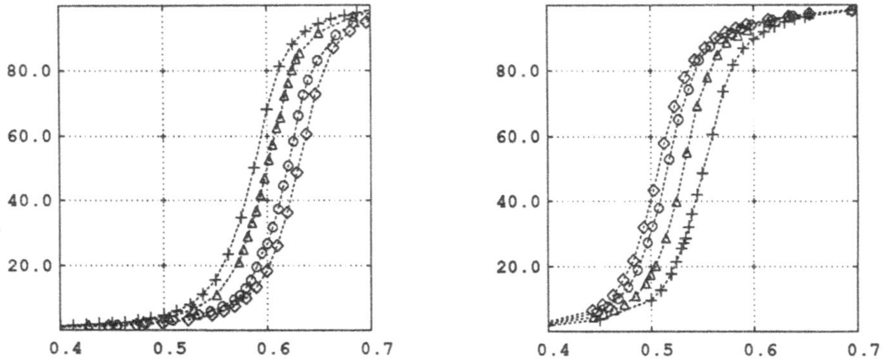

Figure 4c: $\overline{V}_{max}$ versus the proportion: $L^2/A = 12\,500$ *Left:* BM, the ratio $L/l = 100$ being kept constant; l is successively equal to 2(+), 3($\triangle$), 5($\circ$), 7($\diamond$). *Right:* TGM the ratio $L/a = 200$ being kept constant; a is successively equal to 1(+) 1.5($\triangle$), 2.7($\circ$), 3.5($\diamond$).

In other words, Δp is a decreasing function of the scale factor L^2/A, which controls the transition. $\hat{p}_c$ increases always with the grid size; it increases also with l in the boolean model, whereas it decreases with a in the truncated gaussian model. This two variables characterize the invasion of the simulation field by a huge connected component assigned to become the percolating one. Thus, $\hat{p}_c$ should be related to the percolation threshold, and Δp must find its equivalent in the percolation phenomenon.

The percolation threshold

For a realization of the gaussian random function Y, the lowest value of the proportion for which two opposite boundaries of the grid $\mathcal{S}$ are put in contact by a connected component belonging to S is called the percolation threshold, P_c. With a similar construction, we define the percolation threshold for the boolean model. 250 simulations give 250 samples of the random variables P_c, on which the distribution function $F(p) = P(P_c < p)$ can be computed. Hence, $F(p)$ is the probability that the simulation percolates at proportion p. $F(p)$ behaves exactly like $\overline{V}_{max}(p)$ when a, l and L are varying, as it can be seen on fig. 5. On this figure, $F(p)$ is represented with the same choice of parameters as those on fig. 4a, with which it has to be compared.

The mean value $\overline{p}_c = \mathbf{E}\{Pc\}$ (fig. 6) is always an increasing function of the grid size L, but the main point is that $\overline{p}_c$ increases with l in the boolean model, whereas it decreases with a in the truncated gaussian model —as $\hat{p}_c$ does.

A log-log plot of the variance $\sigma^2(P_c)$ versus L^2/A (fig. 7) suggests a power-law relation

$$\sigma^2(P_c) = \frac{K}{(L^2/A)^\alpha} \tag{1}$$

with $\alpha = 3/4$. This result needs some comments:

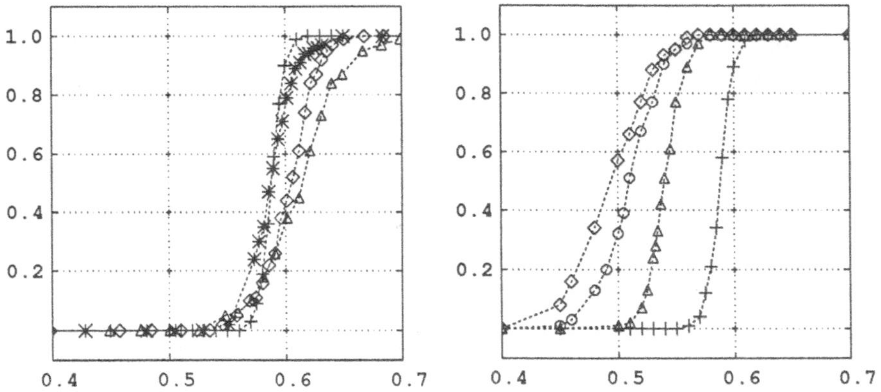

Figure 5: Distribution function $F(p) = P(P_c < p)$: $L = 200$. Left: BM with l successively equal to $1(+)$, $3(\triangle)$, $5(\circ)$, $7(\diamond)$. Right: TGM with a successively equal to $0(+)$, $1(\triangle)$, $2(\circ)$, $3(\diamond)$.

- The value $\alpha = 3/4$ of the exponent is known from the percolation theory (STAUFFER, 1987[7]) which deals with uncorrelated random sets [1]. According to the simulations, it is also valid for both models and for any choice of their parameters.

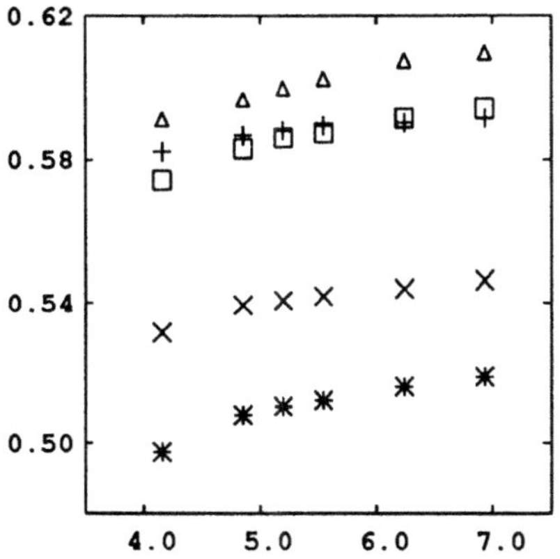

Figure 6: $\overline{p}_c$ versus $\ln L$ for different models of random sets, TGM with $a = 0(+)$, $a = 1(\times)$, $a = 2(*)$; BM with $l = 3(\square)$, $l = 5(\triangle)$.

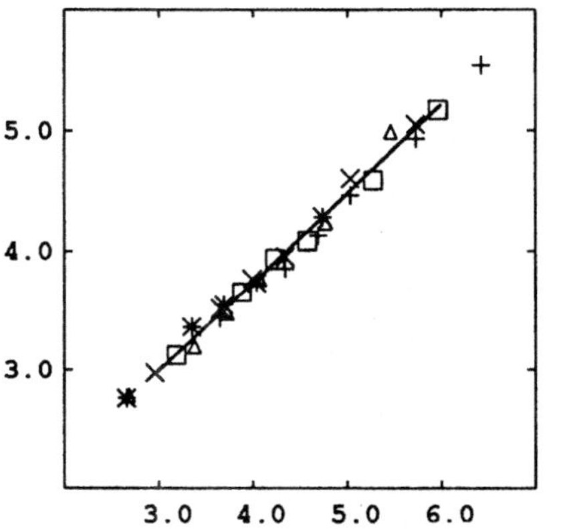

Figure 7: $-\ln \sigma^2(P_c)$ versus $\ln \frac{L^2}{A}$ for different model of random sets. TGM with $a = 0(+)$, $a = 1(\times)$, $a = 2(*)$; BM with $l = 3(\square)$, $l = 5(\triangle)$.

[1]On uncorrelated random sets, it is believed by the percolation theory that this exponent is dependent on the dimension of the space: $\alpha(d = 3) = 1.136$, but is independent on the type of lattice.

- Assuming equation (1) to be valid when the grid size becomes very large, we have $\sigma^2(P_c) \to 0$ and thus $P_c = p_\infty$ a.s. for $L \to \infty$: on an infinite square grid, the percolation threshold is almost surely (i.e. with probability 1) equal to a particular value p_∞.

In this section, it has been shown that if the variance of the percolation threshold is completely controlled by the scale factor L^2/A, the percolation threshold cannot be easily predicted. Its relation with L and A is much more complicated than a power law like (1).

The connectivity function

The connectivity function $\tau(h) = P(x \leftrightarrow x+h \mid x, x+h \in S)$ has been calculated on 50 simulations of a 512×512 grid, with an important restriction: in order to avoid edges effect, we have only considered couples of points both belonging to the 200×200 central square.

It has been observed that this function is isotropic for both models and whatever the parameters, in spite of the covariance anisotropy of the models and the anisotropy of the lattice (cf fig. 8). Hence, $\tau_p(h)$ can be written as a function of the distance $|h| = \sqrt{h_1^2 + h_2^2}$.

$\tau_p(|h|)$ is a decreasing function, with $\tau_p(1) = 1$. Indeed, if we have $x \leftrightarrow x + h'$, there necessarily exists at least one h such that $|h| \leq |h'|$ with $x \leftrightarrow x + h$ and $x + h \leftrightarrow x + h'$. Thus $P(x \leftrightarrow x + h) \geq P(x \leftrightarrow x + h')$ if $|h| \leq |h'|$. If $|h| = 1$, x and $x + h$ are neighbours, and if they both belong to S they are obviously connected. The way $\tau_p(|h|)$ decreases with $|h|$ depends on the proportion p and is illustrated on fig. 9.

- If $p \ll \overline{p}_c$, $\tau_p(|h|)$ quickly decreases and converges towards 0 for large $|h|$, since all connected components are of small size.
- If $p \gg \overline{p}_c$, $\tau_p(|h|)$ decreases till it reaches a sill $\geq \overline{V}^2_{max}$. This can be explained as follows: for short distances two connected points of S can either belong to a small connected component, or belong the the percolating one. But for long distances, two points are connected only if they both belong to the biggest connected component : for large $|h|$, $P(x \leftrightarrow x + h \mid x, x + h \in S) \simeq P(x \in C_{max}$ and $x + h \in C_{max}) \geq P(x \in C_{max}) \times P(x + h \in C_{max}) \simeq \overline{V}^2_{max}$, according to a standard inequality on increasing events.
- around $\overline{p}_c$, there is a transition domain, where $\tau_p(|h|)$ tends neither to 0 nor to $\overline{V}^2_{max}$ as $|h|$ becomes very large.

To illustrate this, let us define a connectivity range $D = \int \tau_p(h)\, dh$. D is an area, which can be seen as the scale of the connectivity phenomenon. $\tau_p(h)$ being an isotropic function, D can be assumed to be the area of a disc with diameter $\xi = \sqrt{\frac{4D}{\pi}}$. ξ is a characteristic length of the connected component, and is called the connectivity length.

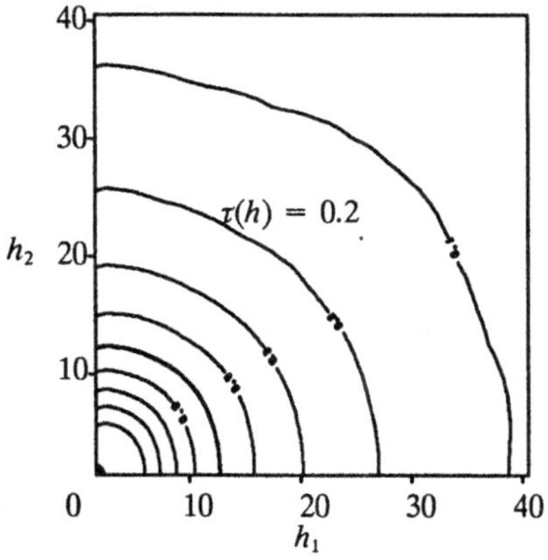

Figure 8: Iso-probability line for $\tau_p(h_1, h_2)$; BM with $l = 5$. $p = 0.5$.

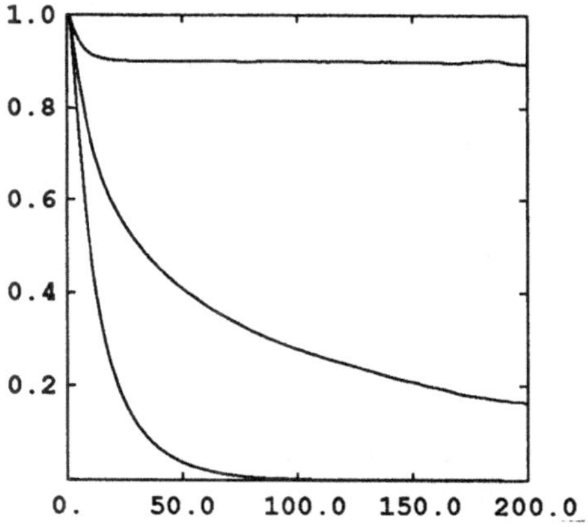

Figure 9: $\tau_p(|h|)$ for truncated gaussian model, $a = 2.7$ for $p = 0.4, 0.5, 0.6.$: $\overline{p}_c = 0.507$

Fig. 10 shows ξ versus $p < p_c$ for three different models. One can see that ξ diverges when $p \to p_c$ because of the invasion of the biggest connected component. On an infinite grid, we would have $\xi \to \infty$ for $p \to p_c$, with the existence of an infinite cluster at the percolation threshold. For uncorrelated random sets, ξ diverges sharply around the percolation threshold. Increasing the correlations leads to a smoothing of this divergence. Boolean model and truncated gaussian model have approximatively the same connectivity length when the difference $\overline{p}_c - p$ between the percolation threshold and the proportion is the same.

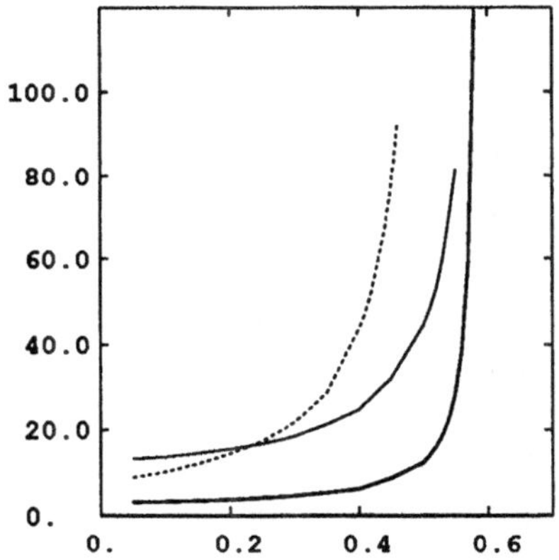

Figure 10: ξ versus p for three different models: uncorrelated random set (double line), TGM ($a = 2.7$-dashed line) and BM ($l = 5$-solid line).

4 Conclusion

The main facts reported in this paper can be summarized as:

(1) The number of connected components N_{cc} depends strongly on the model: for a same proportion and integral range, N_{cc} is three times larger for the truncated gaussian model than for the boolean model.

(2) The percolation threshold varies with the model: it decreases with the covariance range a in the truncated gaussian model, whereas it increases with the square size l in the boolean model.
(3) The variance of the percolation threshold follows a power relation independent of the model; the scale factor L^2/A controls this variance.
(4) The connectivity function is isotropic for both models, in spite of the anisotropy of the covariances and of the lattice.
(5) The divergence of the connectivity function is similar for both models when the difference $(\overline{p}_c - p)$ is considered.

In conclusion, in 2-D simulations, only few practical consequences follow the choice of the model. An extension to rectangular grid, rectangular primary grains and covariance with different ranges in the main directions should be undertaken to generalize this results. 3D simulations must also be studied, in spite of the computer-time cost. Last but not least, how connectivity information could be taken into account in conditional simulations is to be found.

Acknowledgements I gratefully thank Ch. Lantuéjoul for his helpful comments and suggestions. Research support came from the HERESIM GROUP and the IFP.

References

[1] Freulon, X. and de Fouquet, Ch. (1992) *"Conditionning Gaussian Model with Inequalities. Application to Random Functions or Random Sets."*, Poceedings of the 4th International Geostatistics Conference, Troia, sept. 1992, Ed. Soares A.

[2] Haldorsen, H. H. and MacDonald, C. J. (1987) *"Stochastic Modeling of Underground Reservoirs Facies (SMURF)"*, 62nd Technical Conference and Exhibition of the Society of Petroleum Engineers, Dallas 1987, 575-589.

[3] Lantuéjoul, Ch. (1991) *"Ergodicity and Integral Range"*, Journal of Microscopy, vol, 161-3, 387-404.

[4] Matheron, G. (1967) *Eléments pour une théorie des milieux poreux* , Masson & Cie Editeurs, Paris.

[5] Matheron, G., Beucher, H., de Fouquet, Ch., Galli, A., Guerillot, D. and Ravenne, C. (1987) *"Conditional Simulations of the Geometry of Fluvio-Deltaic Reservoirs"* 62nd Technical Conference and Exhibition of the Society of Petroleum Engineers, Dallas 1987. 591-600

[6] Matheron, G. (1989) *Estimating and Choosing*, Springer-Verlag, Berlin.

[7] Stauffer, D. (1985) *Introduction to the Percolation Theory*, Taylor & Francis, London.

[8] Yaglom, A. M. (1987) *Correlation Theory of Stationary and Related Random Functions*, Springer Series in Statistics, New-York.

SEMI-MARKOV RANDOM FIELDS

Håkon Tjelmeland and Lars Holden
Norwegian Computer Center
P.O.Box 114, Blindern
N-0314 OSLO
Norway

Abstract. *The spatial distribution of sedimentary facies in petroleum reservoirs is of major concern due to its influence on the flow regime in the reservoir, and thereby the production of oil and gas. A class of stochastic models, called semi-Markov Random Fields, for the facies distribution in petroleum reservoirs is presented. Semi-Markov Random Fields are a sub-class of Markov Random Field models. The model parameters in semi-Markov Random Fields have, in contrast to the usual Markov Random Fields with local neighborhood, a clear physical interpretation as size distributions of homogeneous volumes and border relations. Semi-Markov Random Fields can be regarded as a three dimensional generalization of the well known semi-Markov chain in one dimension. Simulation results both from semi-Markov Random Fields and Markov Random Fields with local neighborhood are presented.*

1 Introduction

The spatial distribution of sedimentary facies in petroleum reservoirs is of major concern due to its influence on fluid flow, and thereby the production of oil and gas. Two classes of phenomena can be defined, see Omre (1991). The first class is event phenomena representing occurrence of smaller facies units in a background facies. Examples are small shale units in a background matrix of sand and river beds in a non-permeable background matrix. The second class is mosaic phenomena representing a packing of different facies with no facies constituting a background. This class can be divided into two, dependent on if there are a natural ordering in the facies or not. If there is a natural ordering in the facies, truncated continuous random functions can be used, see Rudkiewicz et al. (1990). If there are no natural ordering in the facies Markov Random Fields (MRF) is a natural choice for modeling of the facies architecture.

2 Markov Random Fields

In this section some notation will be introduced and the definition of MRFs will be given. The notation and the definition of MRF follows the one used in Geman and Geman (1984). We will restrict ourselves to consider MRFs on regular grids in two or three dimensions. Let N be the number of cells or pixels in the grid, and assume that they are labeled in some manner by integers $s = 1, 2, \ldots, N$, and let $\mathcal{S}$ be the set of all pixels:

$$\mathcal{S} = \{1, 2, \ldots, N\}. \tag{1}$$

Each pixel shall be assigned one out of K facies. We will further assume that all combinations of facies assignments have a positive probability. This means that the minimal sample space is:

$$\Omega_o = \Lambda^N. \tag{2}$$

with $\Lambda = \{1, 2, \ldots, K\}$. An arbitrary facies assignment of $\mathcal{S}$ will be denoted $\boldsymbol{X} = (X_1, X_2, \ldots, X_N)$ where X_s is facies in pixel number s. MRFs are defined with respect to a *neighborhood*. For each pixel $s \in \mathcal{S}$ a neighborhood $\mathcal{G}_s$ is defined, which defines $\mathcal{G} = (\mathcal{G}_1, \mathcal{G}_2, \ldots, \mathcal{G}_N\}$, with properties:

1. $\mathcal{G}_s \subseteq \mathcal{S}$ for all $s \in \mathcal{S}$
2. $s \notin \mathcal{G}_s$ for all $s \in \mathcal{S}$
3. $s \in \mathcal{G}_r \Leftrightarrow r \in \mathcal{G}_s$ for all $r, s \in \mathcal{S}$

If $r \in \mathcal{G}_s$ then pixel r is said to be a neighbor of pixel s. Then, the first condition simply says the all neighbors of s are in $\mathcal{S}$. The second condition says that a pixel cannot be a neighbor of itself, while the last condition says that neighborhood is symmetric; if pixel r is neighbor of pixel s, then pixel s also has to be a neighbor of pixel r. Given a neighborhood $\boldsymbol{\mathcal{G}} = (\mathcal{G}_1, \mathcal{G}_2, \ldots, \mathcal{G}_N)$ a subset $C \subseteq \mathcal{S}$ is said to be a clique if all distinct pairs of pixels in C are neighbors:

$$s \in \mathcal{G}_r \text{ for all } s, r \in C,\ s \neq r. \tag{3}$$

With the notation introduced above, $\boldsymbol{X}$ is defined to be a MRF with respect to a neighborhood $\boldsymbol{\mathcal{G}}$ if:

1. $P\{\boldsymbol{X} = \omega\} > 0$ for all $\omega \in \Omega_o$
2. $P\{X_s = \omega_s | X_{\mathcal{S}\setminus\{s\}} = \omega_{\mathcal{S}\setminus\{s\}}\} = P\{X_s = \omega_s | X_{\mathcal{G}_s} = \omega_{\mathcal{G}_s}\}$
   for all $s \in \mathcal{S}$ and for all $\omega = (\omega_1, \omega_2, \ldots, \omega_N) \in \Omega_o$.

The first condition simply says that all facies combinations must have a non-zero probability, while the second condition is the Markov assumption; The conditional probability for the facies in pixel s given the facies in all other pixels is actually only dependent on the facies assigned to pixels that are neighbors of pixel s. One should observe that if $\mathcal{G}_s = \mathcal{S} \setminus \{s\}$ for all $s \in \mathcal{S}$, then condition 2 always will be fulfilled.

The Hammersley–Clifford theorem (see Besag, 1974) states that if $\boldsymbol{X}$ is a MRF with respect to a neighborhood $\boldsymbol{\mathcal{G}}$ then the most general form of the point probability function is:

$$\pi_o(\omega) = P\{\boldsymbol{X} = \omega\} = \frac{1}{Z} \cdot \exp\left\{-\sum_{C \in \mathcal{C}} V_C(\omega)\right\}, \omega \in \Omega_o \tag{4}$$

with Z being a normalizing constant; $\mathcal{C}$ being the set of all cliques C under the neighborhood $\mathcal{G}$; and $V_C(\cdot)$ being arbitrary functions on Ω_o with the property that $V_C(\omega)$ only depends on those coordinates ω_s of ω for which $s \in C$. The V_C-functions are often called potential functions. The Hammersley–Clifford theorem further states that the local characteristics, $P\{X_s = \omega_s | X_r = \omega_r, r \in \mathcal{G}_s\}$, corresponding to (4) are:

$$P\{X_s = \omega_s | X_{\mathcal{G}_s} = \omega_{\mathcal{G}_s}\} = \frac{1}{Z_s} \exp\left\{- \sum_{C:s \in C} V_C(\omega)\right\} \tag{5}$$

with Z_s being a normalizing constant; $V_C(\cdot)$ being potential functions and where the sum is over all cliques C which contain the pixel s.

When using MRFs in practice it is common to use a local neighborhood; each pixel s has only a few neighbors which are located spatially close to the center pixel s. Frequently used local neighborhoods are first and second order neighborhoods. In a 2D grid first order neighborhood, $\mathcal{G}^1 = (\mathcal{G}_1^1, \mathcal{G}_2^1, \ldots, \mathcal{G}_N^1)$, means that each pixel has the four nearest pixels as neighbors, i.e.:

$$\mathcal{G}^1_{(i,j)} = \{(i-1,j), (i+1,j), (i,j-1), (i,j+1)\} \tag{6}$$

with a pragmatic approximation at the border, while in a second order neighborhood in a 3D grid each pixel will have 26 neighbors. In a similar manner one can introduce neighborhoods of higher orders.

The major justification for using a local neighborhood comes from the computing resources required to simulate or estimate from the model. With a large neighborhood simulation and estimation will be far too computer intensive to be usable with the computers that exist today, unless the potential functions are carefully chosen.

2.1 Simulation algorithms

Several simulation methods for MRF models are described in the literature. The most commonly used is probably the Gibbs sampler algorithm (see Geman and Geman, 1984), but also Metropolis algorithms (see Ripley, 1987) are in frequent use. Both of these are special cases of a more general class of simulation algorithms described in Hastings (1970). The idea behind the algorithms is to construct a Markov chain with Ω_o as sample space and with $\pi_o(\omega)$ as limiting distribution.

Hastings described the following procedure for construction of such a Markov chain:

- Let $Q = (q_{\omega\eta})$ be a transition matrix which defines an irreducible Markov chain on Ω_o. The Q matrix should have a kind of symmetry in that $q_{\omega\eta} > 0 \Leftrightarrow q_{\eta\omega} > 0$ for all $\omega, \eta \in \Omega_o$.
- Based on this Q matrix, define a new Markov chain on Ω_o with transition matrix $P = (p_{\omega\eta})$ given by:

- $p_{\omega\eta} = q_{\omega\eta} \cdot \rho_{\omega\eta}$ with $\rho_{\omega\eta} = \frac{\alpha_{\omega\eta}}{1+\frac{q_{\omega\eta}}{q_{\eta\omega}} \cdot \frac{\pi_o(\omega)}{\pi_o(\eta)}}$ if $q_{\omega\eta} \neq 0$ and $\omega \neq \eta$

  where $\alpha_{\omega\eta}$ may be chosen freely under the two constraints:
  - $\alpha_{\omega\eta} = \alpha_{\eta\omega}$ for all $\omega, \eta \in \Omega_o$
  - $0 < \alpha_{\omega\eta} \leq 1 + \frac{q_{\omega\eta}}{q_{\eta\omega}} \cdot \frac{\pi_o(\omega)}{\pi_o(\eta)}$ for all $\omega \neq \eta$; $\omega, \eta \in \Omega_o$
- $p_{\omega\omega} = 1 - \sum_{\eta \neq \omega} p_{\omega\eta}$

It is then straight forward to show that the Markov chain with transition matrix P will have $\pi_o(\cdot)$ as limiting distribution, see for example Hastings (1970). This gives an iterative simulation algorithm where each iteration consists of first to draw one potentially new state with probabilities as specified by the Q-matrix and then accept this potentially new state with probability given by $\rho_{\omega\eta}$.

For both the Gibbs sampler and the Metropolis algorithm one chose the $\alpha_{\omega\eta}$'s as large as possible:

$$\alpha_{\omega\eta} = \min\left\{1 + \frac{q_{\omega\eta}}{q_{\eta\omega}} \cdot \frac{\pi_o(\omega)}{\pi_o(\eta)}, 1 + \frac{q_{\eta\omega}}{q_{\omega\eta}} \cdot \frac{\pi_o(\eta)}{\pi_o(\omega)}\right\} \tag{7}$$

such that

$$\rho_{\omega\eta} = \min\left\{1, \frac{q_{\eta\omega}}{q_{\omega\eta}} \cdot \frac{\pi_o(\eta)}{\pi_o(\omega)}\right\} \tag{8}$$

If one restricts the Q-matrix to be symmetric, $q_{\omega\eta} = q_{\eta\omega}$, for all $\omega, \eta \in \Omega_o$, one gets:

$$\rho_{\omega\eta} = \min\left\{1, \frac{\pi_o(\eta)}{\pi_o(\omega)}\right\} \tag{9}$$

which is the transition probabilities used in the Metropolis algorithm. If one instead choose the following Q-matrix:

$$q_{\omega\eta} = \begin{cases} \frac{1}{N} \cdot P(X_s = \eta_s | \boldsymbol{X}_{\mathcal{G}_s} = \omega_{\mathcal{G}_s}) & \text{if } \omega \text{ and } \eta \text{ differ mostly in one pixel position} \\ 0 & \text{else} \end{cases} \tag{10}$$

one gets $\rho_{\omega\eta} = 1$ for all $\omega, \eta \in \Omega_o$ and the potential new state will always be accepted. This corresponds to the Gibbs sampler algorithm.

2.2 Simulation examples

In this section simulation examples from MRFs with second order neighborhood will be presented. The simulations have been done with a Metropolis algorithm with transition matrix Q given by:

$$q_{\omega\eta} = \begin{cases} \frac{1}{N \cdot K} & \text{if } \omega \text{ and } \eta \text{ differ mostly in one pixel position} \\ 0 & \text{else} \end{cases} \tag{11}$$

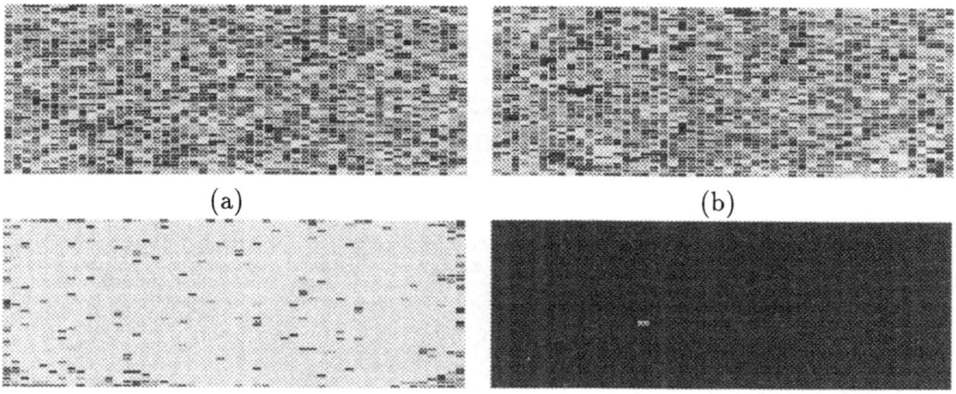

Figure 1: Cross sections (50 × 50 pixels) through four simulation result when using a Metropolis algorithm to simulate from a MRF with first order neighborhood. The β-parameter were: (a) $\beta = 0.10$, (b) $\beta = 0.22$, (c) $\beta = 0.23$ and (d) $\beta = 0.3$.

with $K = 10$ facies, on a grid with $N = 50 \times 5 \times 50$ pixels and $10.000 \times N$ iterations, i.e. for each pixel potentially new facies will be drawn 10.000 times. In the model there is only pairwise interaction, i.e. all potential functions not corresponding to cliques with two pixels are identical to zero. The potential functions for cliques with two pixels are:

$$V_{\{r,s\}}(\omega) = \begin{cases} \beta & \text{if } \omega_r = \omega_s \\ 0 & \text{else} \end{cases} \tag{12}$$

with $\beta > 0.0$ being a parameter specifying the degree of homogeneity. In figure 1 simulation results for four different values of β are plotted. There seems to be a threshold value for β somewhere in the interval $[0.22, 0.23]$. For β-values lower than this threshold the result appear as close to independence, while for β-values higher than the threshold, even for β-values only slightly higher, one facies dominate totally. It should be no surprise that one facies dominate totally for β-values larger than some threshold. For a first order neighborhood, with $N = 2$ facies and an infinitely large 2D-grid it can be shown theoretically that one facies always will dominate for $\beta > 1.76$, see Kindermann and Snell (1980). We have simulated realizations with several different local neighborhoods and with different potential functions and our experience is that the general picture is the same if one iterate until convergence; the realizations appear either close to independent noise or one facies dominate totally. This means that MRFs with local neighborhood are not suitable to model facies distributions in petroleum reservoirs. Note that our simulations are unconditional simulations or conditioned on a small number of observations only. In other

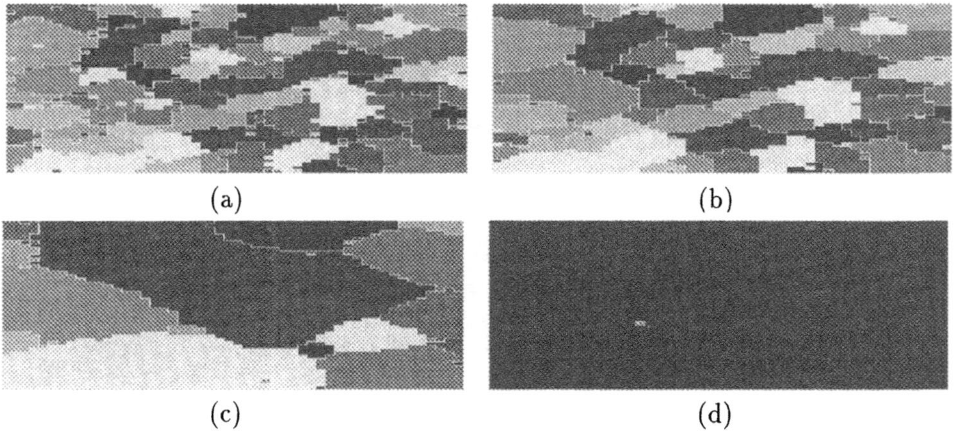

Figure 2: Cross sections (50×50 pixels) through four simulation result when using a Metropolis algorithm to simulate from a MRF with first order neighborhood and with $\beta = 0.40$ after (a) $50 \times N$ iterations, (b) $100 \times N$ iterations, (c) $1000 \times N$ iterations and (d) $10000 \times N$ iterations.

applications such as image analysis with a lot of data present MRFs with local neighborhood have been used with success, see for example Geman and Geman (1984). Note also that for β-values well above the threshold value realistic facies distributions can be obtained if the iteration is stopped before convergence is reached. This can be seen in figure 2 which shows simulation results with the same model as described above and with $\beta = 0.4$ after $500 \times N$, $1000 \times N$, $5000 \times N$ and $10000 \times N$ iterations. A disadvantage by using realizations which have not converged is that it is difficult to interpret the model parameters and thereby to control the size of homogeneous volumes or bodies. Parameter estimation will also be difficult.

3 Semi-Markov Random Fields

As shown in the simulation examples in the last section it seems difficult to control large scale characteristics in a MRF by using a local neighborhood only. In this section we will describe a sub-class of MRFs called semi-Markov random fields (semi-MRF). This name has been chosen because the model class can be seen as a 2 and 3D generalization of the well known semi-Markov process in 1D. Semi-MRFs have large neighborhoods, but the potential functions are chosen such that required computer resources for simulation from the model should not be much larger than for a MRF with local neighborhood. Furthermore, at least for modeling of facies distribution in a petroleum reservoir, the parameterization in semi-MRFs seems more natural than using a local neighborhood only. Typical parameters in a semi-MRF will be size distributions of homogeneous areas or volumes and border relations.

The most natural way to define the point probability distribution for a semi-MRF is the following: Given a realization ω split the pixels into the minimal number of homogeneous connected bodies, and denote these bodies $B_1, B_2, \ldots, B_{n_B}$ with n_B number of bodies. Also let c_i be the facies in body B_i for $i = 1, 2, \ldots, n_B$. Note that then both the bodies $B_i; i = 1, 2, \ldots, n_B$, the corresponding facies $c_1, c_2, \ldots, c_{n_B}$ and number of bodies n_B will be functions of the realization $\boldsymbol{X}$. The point probability for ω is then:

$$\pi_o(\omega) = P\{\boldsymbol{X} = \omega\} = \frac{1}{Z} \cdot \exp\left\{\sum_{i=1}^{n_B} U_1(B_i, c_i) + \sum_{i=1}^{n_B} \sum_{j=i+1}^{n_B} U_2(B_i, c_i, B_j, c_j)\right\} \tag{13}$$

with Z being a normalizing constant; $U_1(\cdot, \cdot)$ being a function describing the form and size of a connected homogeneous body and $U_2(\cdot, \cdot, \cdot, \cdot)$ being a function describing interaction between pairs of connected homogeneous bodies. Note that $U_1(\cdot, \cdot)$ and $U_2(\cdot, \cdot, \cdot, \cdot)$ are not potential functions as defined for MRFs since they are not related to a specific clique. Note also the equivalence between equation (13) and the point probability distribution of marked point processes with pairwise interaction, see for example Stoyan et al. (1987). An important difference is that in equation (13) the number of bodies n_B will be stochastic which is not true for marked point processes. Another difference is that the bodies B_i in equation (13) can not overlap and must fill the space, and this will give a kind of implicit interaction between bodies which one does not have for marked point processes.

Since the class of semi-MRFs is a subset of the class of MRFs, an alternative way to specify the model class is to specify the neighborhood $\mathcal{G}$ and the corresponding potential functions. A semi-MRF has a global neighborhood which means that all pixels are neighbors to each other:

$$\mathcal{G}_s = \mathcal{S} \setminus \{s\} \text{ for all } s \in \mathcal{S}. \tag{14}$$

This means of course that there will be an enormous number of cliques and a corresponding number of potential functions to be specified. The relation between the potential functions and the $U_1(\cdot, \cdot)$ and $U_2(\cdot, \cdot, \cdot, \cdot)$ functions can be expressed by:

$$V_C(\omega) = \begin{cases} U_1(B_i, c_i) & \text{if } C = B_i \cup R(B_i) \\ U_2(B_i, c_i, B_j, c_j) & \text{if } C = B_i \cup R(B_i) \cup B_j \cup R(B_j) \\ 0 & \text{otherwise} \end{cases} \tag{15}$$

with B_i, B_j, c_i and c_j being defined relative to the given realization ω as in equation (13); and $R(B)$ being the outer rim of body B defined by:

$$R(B) = \left(\bigcup_{r \in B} \mathcal{G}_r^1\right) \setminus B. \tag{16}$$

Note that since semi-MRFs are a subset of MRFs, the simulation algorithms described in section 2.1 can be used to simulate from semi-MRFs. Furthermore, with a clever implementation, the computer resources should not be much larger than for MRFs with local neighborhood. The latter should at least be true per iteration, the convergence rate is still an open question.

3.1 Simulation examples

In this section two examples of simulation from semi-MRFs will be presented. In these simulation a $U_2(\cdot,\cdot,\cdot,\cdot)$ function equal to zero has been used, but still there will be an implicit interaction between the bodies since they are not allowed to overlap and must fill the space. The $U_1(\cdot,\cdot)$ function used in the simulations is given by:

$$U_1(B,c) = U_x(l_x(B),c) + U_y(l_y(B),c) + U_z(l_z(B),c) + U_v(B,c) \qquad (17)$$

with $l_x(B)$, $l_y(B)$ and $l_z(B)$ being definitions of the lengths of body B in x-, y- and z-directions respectively; $U_x(l_x,c)$, $U_y(l_y,c)$ and $U_z(l_z,c)$ being functions related to the length distribution of a body B of facies c in x-, y- and z-direction respectively; and $U_v(B,c)$ being a function related to the volume of B and will be discussed later. The parametric form for $l_p(B)$ and $U_p(l_p,c)$ with p being x,y or z are:

$$l_p(B) = \sqrt{\frac{\sqrt{20}}{V_B} \cdot \int_B (p-\overline{p})^2 \mathrm{d}x\mathrm{d}y\mathrm{d}z} \qquad (18)$$

$$U_p(l_p,c) = \frac{(l_p - \mu_{p,c})^2}{\sigma_{p,c}^2} \qquad (19)$$

with V_B being the volume of B. Note that:

- with $U_v(\cdot,\cdot) = 0$ the corresponding 1D model will be a semi-Markov process with normal distributed waiting times with mean $\mu_{p,c}$ and standard deviation $\sigma_{p,c}$ inside each facies and arbitrary transition between facies.
- for bodies being ellipsoids with axis parallel to coordinate axis, the lengths defined by (18) will equal the true lengths of the ellipsoid.

In figure 3 plots from two different realizations are given. For all of them the model was as described above and with $\mu_{p,c} = 10$ and $\sigma_{p,c} = 1$ for all the $K = 10$ facies used. Note that then the most probable length and the most probable height of the bodies are 10 pixels which corresponds to 1/5 of the length and height respectively of the plots in figure 3. The simulation was done on a $50 \times 10 \times 50$ grid and the simulation algorithm used was the one described in section 2.2. In the model for figure (a) $U_v(B,c) = 0$ were used and the realization is after $1000 \times N$ iterations. This model corresponds, as already mentioned, to a direct 3D-generalization of a semi-Markov process with Gaussian distributed waiting times and arbitrary transition between

(a)

(b)

Figure 3: Cross sections (50 × 50 pixels) through two simulation results when using a Metropolis algorithm to simulate from semi-MRF with (a) $U_v(B,c) = 0$ (b) $U_v(B,c) \neq 0$

facies. The lengths of the bodies in the realization, measured as given by equation (18), correspond satisfactory to what one should expect from the model. In the figure one can observe that the bodies are very little compact and is obviously not suitable for modeling of the facies distribution. A term which gives higher probability to more compact bodies is necessary. In the model for figure (b) the following $U_v(\cdot,\cdot)$ function were used:

$$U_v(B,c) = \frac{(V_B - \frac{\pi}{6} l_x(B) l_y(B) l_z(B))^2}{\sigma_v^2} \tag{20}$$

with $\sigma_v = 6$ and the realization is after $1000 \times N$ iterations Note that this $U_v(\cdot,\cdot)$ function gives highest values to bodies were the volume of the body is close to the volume of an ellipsoid with corresponding lengths and should give more compact bodies. As can be seen in the figure, the bodies are more compact than in (a), but the lengths of the bodies, measured as given by equation (18), are much smaller than one should expect from the model parameters. We believe this is due to very slow convergence of the simulation algorithm used. Under the iterations bodies can grow larger in two ways; bodies can grow larger pixel by pixel, but then other bodies must become smaller and vanish, or several bodies can merge together to one. In the algorithm both these events have very low probabilities for occurring; bodies have very small probability to become smaller and vanish since the model specifies very low probability for small bodies; and bodies will have have very small probability for merging because when merging they will have a form which is very little compact and the model specifies very low probabilities for bodies not being compact. The fundamental problem with the algorithm is that only one pixel can be changed in each iteration. All algorithms used in practice so far changes only one or two pixels in each interaction even though the general algorithm described in section 2.1 may change arbitrary many pixels in each iteration.

3.2 Modified simulation algorithm

A simulation algorithm which change more than one pixel in each iteration is required and a preferable simulation algorithm would accept hole bodies to be put in or taken out. To specify a transition matrix, Q, which allow this is difficult if bodies are not allowed to overlap and if the bodies have to fill the space for each iteration. One could imagine a simulation algorithm operating on an extended model which allow overlap between bodies and pixels without any facies assigned. As a generalization of the simulated annealing algorithm (see Geman and Geman, 1984 or Winkler, 1990), the probability for realizations with overlap between bodies and pixels without any facies assigned could gradually be reduced toward zero.

The extended model on which the model should operate could be defined by:

- Each pixel can be assigned non, one or several facies, i.e.

$$\omega_s \subseteq \Lambda \tag{21}$$

- All combinations of assignments have a positive probability. Then the minimal sample space for the extended model is:

$$\Omega = \{\omega = (\omega_1, \omega_2, \ldots, \omega_N) | \omega_s \subseteq \Lambda, s = 1, 2, \ldots, N\} \tag{22}$$

- The point probability function is given by:

$$\pi_T(\omega) = \frac{1}{Z_T} \cdot \exp\left\{ - \sum_{C \in \mathcal{C}} V_C'(\omega) - \frac{\sigma(\omega)}{T} \right\}, \omega \in \Omega \tag{23}$$

  with Z_T being normalizing constant; $T > 0$ being a parameter which in connection with simulated annealing often is called the temperature; $V_C'(\omega) = V_C(\omega)$ for all $\omega \in \Omega_o$; $\sigma(\omega) = 0$ for all $\omega \in \Omega_o$; and $\sigma(\omega) > 0$ for all $\omega \in \Omega \setminus \Omega_o$. Note that this implies that the point probability function for the extended model will converge toward the point probability function for the original model when the temperature, T, goes to zero:

$$\lim_{T \to 0} \pi_T(\omega) = \begin{cases} \pi_o(\omega) & \text{for } \omega \in \Omega_o \\ 0 & \text{for } \omega \in \Omega \setminus \Omega_o \end{cases} \tag{24}$$

Note also that this extended model will be a MRF with respect to a global neighborhood, where $\frac{\sigma(\omega)}{T}$ corresponds to a potential function for the clique which consists of all pixels. This implies that simulation algorithms within the class described in section 2.1 can be used to simulate from the model. The modification in the simulation algorithm is that the temperature T is replaced by a temperature function $T(t)$ which decrease toward zero as the iteration number t increase.

Simulated annealing will converge if the temperature function goes 'slow enough' toward zero, see for example Geman and Geman (1984). Our assumption is that this will be true also when the potential functions are not zero, at least as a good approximation.

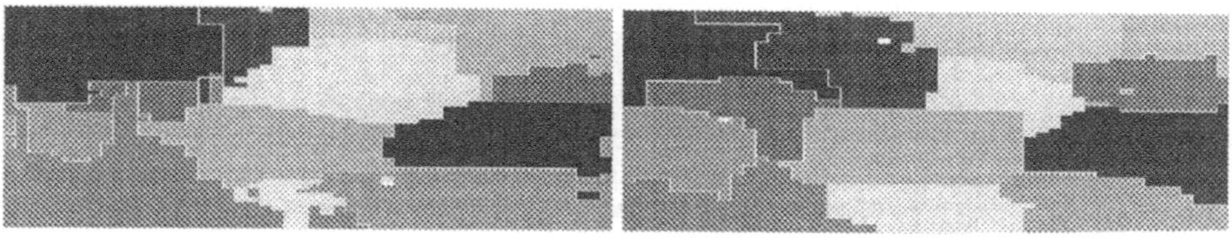

Figure 4: Two parallel cross sections (50 × 50 pixels) through one simulation result when using a modified simulation algorithm to simulate a from semi-MRF

3.3 Simulation examples

In this section one simulation example from a semi-MRF with the type of simulation algorithm described in the previous section will be presented. The model is exactly the same as presented in section 3.1. The $V_C'(\cdot)$ functions used in the extended model is the one corresponding to (13), but where the bodies B_i now may overlap and do not need to fill the space. The $\sigma(\cdot)$ function used gave smaller probability to realizations with overlap and/or space:

$$\sigma(B) = \sum_{s=1}^{N} |\#\omega_s - 1| \tag{25}$$

with $\#\omega_s$ being number of facies assigned to pixel s. An exponential decrease in the temperature, $T(t) = T_0 \cdot a^{\frac{t}{N}}$ with $T_0 = 10.0$ and $a = 0.995$, and iteration continued until $T(t) = 0.0001$. Note that this annealing schedule is too fast to guarantee convergence, see Geman and Geman (1984), but using an annealing schedule which guaranteed convergence would have required too much computer resources. In the Q-matrix used four types of transitions had positive probabilities: adding one pixel to a body, removing one pixel from a body, putting in a new body and removing a body. Transitions where several bodies were merged together to one or where one body were split into several bodies had zero probabilities.

In figure 4 two parallel cross sections through one realizations is given. The realization was simulated on a 50 × 20 × 50 grid. The model and the simulation algorithm used is as described above with $\mu_{p,c} = 10$ and $\sigma_{p,c} = 1$ for $p = x, y$ and z for all the $K = 10$ facies and with $\sigma_v = 6$. Remember that with this $\mu_{p,c}$, the most probable lengths and heights are 1/5 of the length and height of the plots in figure 4. When comp aring figure 4 with figure 3 one can observe that the modified simulation algorithm gives realizations which are more consistent with what one should expect from the model parameters. Note that the model and the model parameters used to produce the realizations shown in figure 3(b) and in figure 4 in the same, only the simulation algorithms differ. If the simulation algorithm used to produce the realization shown in figure 3(b) had been run for enough iterations, the result would have appeared as in figure 4, but the number of iterations necessary is too large to be performed in practice.

4 Closing Remarks

In this paper a description of Markov Random Fields is given together with some simulation examples. In particular two sub-classes of Markov Random Fields have been evaluated; MRFs with local neighborhood and Semi-Markov Random Fields. We have not been able to find any MRF with local neighborhood which gave realizations with realistic sizes of homogeneous volumes. It should be noted that all our simulations were unconditioned or conditioned on a limited number of observations only, which will be the typical situation when modeling facies distribution in petroleum reservoirs. In applications with a lot of data, for example image analysis, Markov Random Fields with local neighborhood have been used with success, see for example Geman and Geman (1984).

Simulation algorithms where only one pixel can change value in each iteration seems to converge too slow for semi-MRFs. A simulation algorithm which is a combination of the simulation algorithm given in Hastings (1970) and simulated annealing is proposed. This algorithm gives positive probabilities for taking in or removing hole bodies in addition to changing the value of one pixel. It remains to prove that this algorithm really converge toward to specified distribution, but the simulations done are encouraging. More work is necessary both in better simulation algorithms for semi-MRFs and for a better understanding of the model.

All the simulation examples presented in this paper are without conditioning data. All the simulation algorithms used can easily be extended to reproduce observed data; for simulation algorithms changing only one pixel in each iteration the pixels corresponding to a observation can just be fixed to the facies observed, for the modified simulation algorithm presented in section 3.2 it can be done by putting a penalty for inconsistency between observations and realization into the $\sigma(\cdot)$ function. But our experience is that conditioning, with respect to point observations or with respect to global properties such as the percentage of each facies, often result in a much slower convergence of the algorithms.

References

Besag, J.: "Spatial interaction and the statistical analysis of lattice systems [with discussion]," *J. Royal Statist. Soc B, (1974), Vol. 36*,192-236.

Geman, S. and Geman, D.: "Stochastic relaxation, Gibbs distributions, and the Bayesian restoration of images," *IEEE Trans. (1984) PAMI 6*, 721-741.

Hastings, W.K.: "Monte Carlo sampling methods using Markov chains and their applications," *Biometrika (1970) 57*, 97-109.

Kindermann, R. and Snell, J.S.: *Markov Random Fields and their Applications*. American Mathematical Society, 1980.

Omre, H.: "Stochastic Models for Reservoir Characterization", *Recent Advances in Improved Oil Recovery Methods for North Sea Sandstone Reservoirs*, Kleppe, J. and Skjæveland, S.M. (eds.); To appear.

Ripley, B.D.: *Stochastic Simulation.* Wiley, New York (1987).

Rudkiewicz, J.L., D. Guerillot, D., Galli, A. and Group Heresim: "An Integrated Software for Stochastic Modelling of Reservoir Lithology and Property with an Example from the Yorkshire Middle Jurassic," *North Sea Oil and Gas Reservoirs - II*, A.T. Buller et al (ed); NIT, Graham & Trotman; (1990) 399-406.

Stoyan, D., Kendall, W.S. and Mecke, J.: Stochastic Geometry and its Applications, John Wiley & Sons Inc., 1987.

Winkler, G.: "An Ergodic L^2-theorem for Simulated Annealing in Bayesian Image Reconstruction," *J. Appl. Prob. 28*, 779-791.

A Sensitivity Analysis of Geostatistical Parameters on Oil Recovery Predictions

Udo G. Araktingi
Chevron Oil Field Research Company
1300 Beach Boulevard
La Habra, California 90631

ABSTRACT

The effect of different spatial covariance parameters; range, covariance model, anisotropy direction and ratio on important reservoir engineering parameters such as breakthrough time, fractional flow of oil and oil recovery as a function of pore volumes injected is examined. The study is conducted for situations where the range of spatial continuity of permeability is smaller than the well spacing, meaning it cannot be detected through the calculation of an experimental variogram. This is a relatively common occurrence for even well developed oil reservoirs. The impact of the different parameters on the flow behavior is analyzed using a hybrid finite-difference/streamtube approach for the fluid displacement simulations. The permeability and porosity grids are generated using both Gaussian based conditional simulation techniques and non-parametric methods such as indicators. Some of the results indicate that Gaussian simulation techniques are not suitable for fluid flow predictions while the choice of the variogram type (exponential, Gaussian or spherical) is unimportant.

INTRODUCTION

Increasingly severe economic constraints and the realization of the importance of reservoir heterogeneity on fluid displacements processes have highlighted the need for more accurate reservoir description to improve the design and planning of successful recovery projects. The use of spatial statistics to incorporate different data types (well logs, seismic, expert knowledge) in an unbiased and consistent framework has allowed the generation of quantitative reservoir models which can be input directly into flow simulators. Reservoir properties, such as permeability and porosity, are spatially correlated in some fashion. Geostatistics provides a way to measure that spatial continuity and use it to build a reservoir model. The spatial correlation of each property is measured by calculating the experimental variogram or covariance of the available data. Once the experimental variogram is modeled using an analytical function, numerous conditional simulation techniques exist that transform that measure of spatial correlation into a large number of equiprobable reservoir descriptions. Such realizations can be sampled and tested for their fluid flow response to gain a better understanding of the remaining level of uncertainty due to the insufficient amount of conditioning well data.

Thus, the determination of the correct variogram model appears crucial. All variogram models are defined by an analytical type, a range, a sill, an anisotropy ratio and direction and a distinctive behavior at distances shorter than the range. It is this section of the variogram that needs to be modeled most carefully since it will have the most impact, through the kriging weights, on the resulting permeability fields. This study will concern itself with the effect of the parameters mentioned above on fluid flow predictions.

RESERVOIR SPATIAL CONTINUITY

A typical variogram for reservoir permeability is shown in Fig. 1.

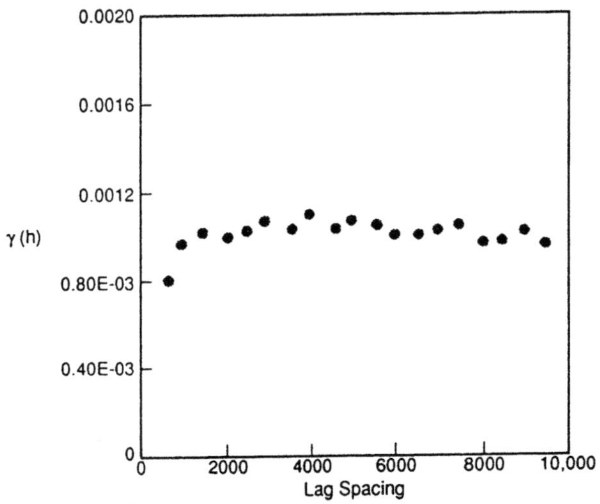

Figure 1
Typical shape of an oil reservoir permeability variogram

It indicates that for separation distances greater than approximately 1000 ft., permeability becomes uncorrelated since the variogram fully reaches the data variance (sill value). A variogram value exists for a 500 ft. lag, however it has often little statistical significance since it is usually based on very few data pairs. This situation, where very little well data is available over distances shorter than the range occurs more often than not. Even in developed oil fields, well spacing is usually on the order of 1000 feet or more except for heavy oil reservoirs where the nature of the recovery process necessitates closer spacing. Thus, many alternate variogram forms exist that will fit the experimental variogram shown in Fig. 1. These fitted variograms will vary by their type, range, anisotropy ratio and direction. The usual solution to this problem is to generate reservoir descriptions using different variograms until the geologist approves of the "look" of the simulation. The variogram for that particular simulation is deemed the "correct" one. Unfortunately, in most cases, reasonably different variograms still produce reservoir models that are acceptable to the geologist. The other alternative is to use information derived from a pertinent outcrop study. In recent years, several outcrop studies[1,2] using mini-permeameters to measure permeability directly, have been conducted to collect such data over very small intervals (≤1 ft.). The resulting experimental variograms typically show extremely short ranges (≤50 ft.) which means that fluid displacements in reservoirs with interwell distances on the order of 1000 feet should be extremely efficient[3]. However, such ideal displacements happen infrequently meaning that these observed ranges in the outcrop data reflect the data being collected over a relatively short scale as compared to the flow scale. Thus, these measurements are only detecting the very short scale structure in a series of nested structures. These studies, although cheaper than gathering and analyzing well core data on an information/cost basis, are labor intensive and extending the scale over which they are currently being conducted will be even more demanding. Eventually, studies of outcrop exposures of most basic depositional features will be available to provide the appropriate patterns of spatial variation. In the meantime, it appears that outcrop data cannot be generally used to generate reservoir descriptions for flow simulation studies. Thus, a clear need exists to understand the effect of the variogram parameters on the resulting flow displacements and the parameters (breakthrough time, fractional flow and oil recovery) that describe them.

METHODOLOGY

An area, consisting of two water injectors and two oil producers, from the reservoir for which the variogram in Fig. 1 was derived, was chosen to simulate fluid displacements to investigate the effect of the variogram parameters on flow behavior. In addition, the 25 neighboring wells, as shown in Fig. 2, were used to condition the different realizations generated for the study.

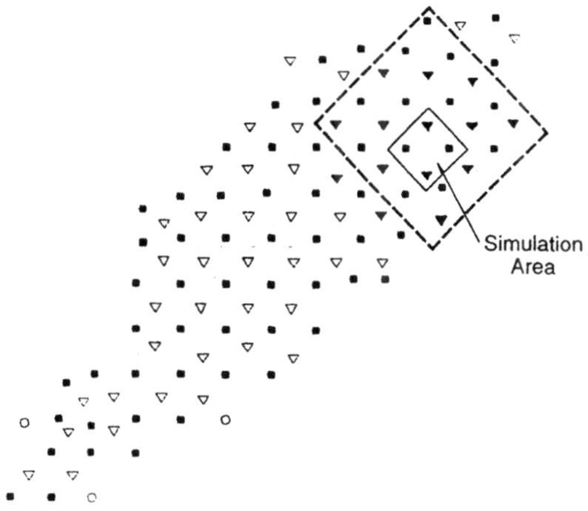

Figure 2

Study area located within square. Conditioning wells are located within – – – area.

These wells are located in a West Texas carbonate reservoir and the zone under study is a series of dolomitized grainstones.

The variogram for permeability, obtained from a functional transform of porosity logs, is shown in Fig. 1 and clearly indicates that the range is less than 1000 ft. Three types of variograms (exponential, Gaussian and spherical) with different ranges that fit that particular variogram of permeability were used and are identified in Table 1. Given the asymptotic character of the exponential and Gaussian models, the ranges listed in Table 1 all correspond to the same practical range of correlation.

Table 1
Parameters of Variograms Used in Figures 4-12

Name	Type	Range (ft)	Anisotropy Ratio	Direction	Figures Used In
S1	Spherical	900	2	65°	5, 7, 8
S2	Spherical	450	1	NA	5, 7, 8, 11, 12
S3	Spherical	225	1	NA	5, 7
S4	Spherical	600	1	NA	11
S5	Spherical	900	2	0°	12
E1	Exponential	300	2	65°	4, 9
E2	Exponential	150	1	NA	4, 9
E3	Exponential	75	1	NA	4
G1	Gaussian	520	2	65°	6, 10
G2	Gaussian	260	1	NA	6, 10
G3	Gaussian	130	1	NA	6

Three ranges and two levels of anisotropy were used for each variogram type. To investigate the effects of variogram type, flow behavior through conditional simulations obtained with equivalent variograms, meaning different variogram types but equal practical ranges, are compared. The type of displacement considered is a waterflood with a mobility ratio of 2.

Flow Simulations

The waterflood simulation were modeled using a streamtube simulator using the hybrid approach[4,5]. The following gives a brief description of this approach:

1. Once the area under investigation has been determined, realizations of two-dimensional grids of *kh* (cumulative permeability thickness) and ϕh (cumulative porosity thickness) values are generated.

2. Using these grids and the appropriate fluid properties, injection and production rate data, the solution of the convection equation determines the location of the flow streamlines and associated streamtubes.

3. Very detailed cross-sections of ϕ and k values between wells are also generated. A finite-difference waterflood simulation model is run to derive the characteristics solutions relating fractional flows of water and oil to the pore volumes injected.

4. Finally, these fractional flow curves are mapped onto the streamtubes to estimate the 3D performance of the area under study.

Streamtube calculations assume single phase, incompressible flow in the 2D areal model with the streamtube geometry remaining essentially fixed in time and no crossflow occurring between streamtubes. As a result, this approach is not suited for addressing energy related issues such as pressure maintenance, depletion or buildup in a reservoir and the modeling of significant near wellbore effects such as coning and cusping.

However, this method has been applied very successfully for the estimation of recoveries from many different types of enhanced recovery processes ranging from simple waterfloods to miscible displacements[5]. It also provides a reservoir management screening tool for project design and process. The overall success of this approach stems from the use of very detailed geologic representations in the areal and vertical sense which are not currently possible with conventional 3D coarse-grid models. A 50 x 95 grid was used for the cross-sections while the areal grid was made up of 2500 cells. The average number of log values per well for this zone was 95, meaning that the vertical grid had the same resolution as the available data. Concurrently, the breaking up of the 3D problem into two more tractable 2D flow problems makes this method extremely fast with today's cheap computer power (a true 3-D model with the equivalent resolution would require more than 11 x 10^6 cells). Thus, the computational efficiency makes it feasible to process a great number of conditional simulations in a reasonable amount of time.

RESULTS

Both Gaussian techniques, Gaussian sequential simulation (GSS) and the LU[6] method, and a non-parametric method, sequential indicator simulation[7] (SIS), were used to generate the property grids needed for the flow simulations. The results were broken up based on the conditional simulation technique used for the reservoir modeling. First, the results obtained with the Gaussian approach are described.

Three sets of runs were made each using a different variogram type (exponential, spherical and Gaussian) and the GSS method. The final 3D fractional flow of oil and the cumulative recovery as a function of pore volumes injected (PVI) are plotted in Figs. 3, 4 and 5. A total of fifteen realizations were run for each set. Within each set, the range of the variogram was varied (Table 1). Surprisingly, the flow behavior in all three sets appears for all practical purposes, independent of all the fitting parameters considered. Variogram range, type and anisotropy direction do not have a significant impact on the effluent and recovery curves. Breakthrough time is similarly unaffected indicating the absence of any preferential flow paths or flow barriers, even though some of the realizations were obtained using large variogram ranges. In addition, very little spread can be seen between realizations. For example, breakthrough times obtained with the exponential model range from 0.27 to 0.29 PVI. Also, at 1.0 PVI the difference in recoveries range from 64% to 66% of oil in place. This spread is very small considering that no conditioning data was available at distances shorter than the variogram ranges used. It should be noted that the differences in the wiggles in the fractional flow curves between 0.8 and 1.6 PVI in Figs. 4 and 5 are due to the wells inadvertently not being completed in the top layer for the latter set of realizations. Very similar flow behavior in all respects is observed when using the LU approach to generate the different realizations. Again, the various variogram parameters do not have an effect on flow behavior with the spread between realizations being very small, as shown in Fig. 6. An additional ten realizations were made using a Gaussian variogram (G1 in Table 1) with a 20% nugget and a 50% nugget, and showed the same results as in Fig. 5.

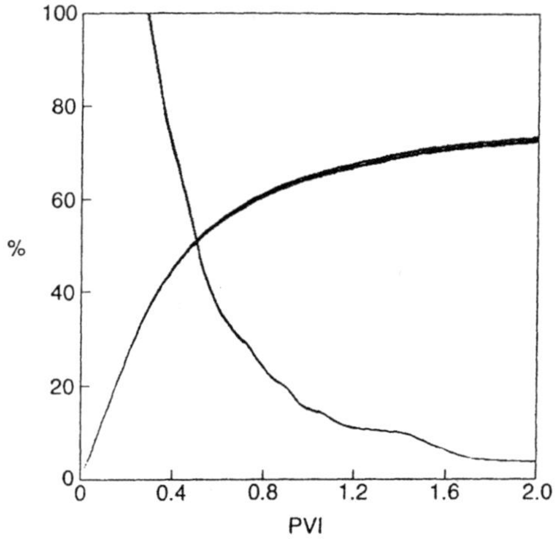

Figure 3

Flow results for GSS realizations using variograms E1, E2 and E3 (total of 15 realizations). BT = 0.29 PVI ±0.01, Recovery (at 1.0 PVI) = 65% ±1%

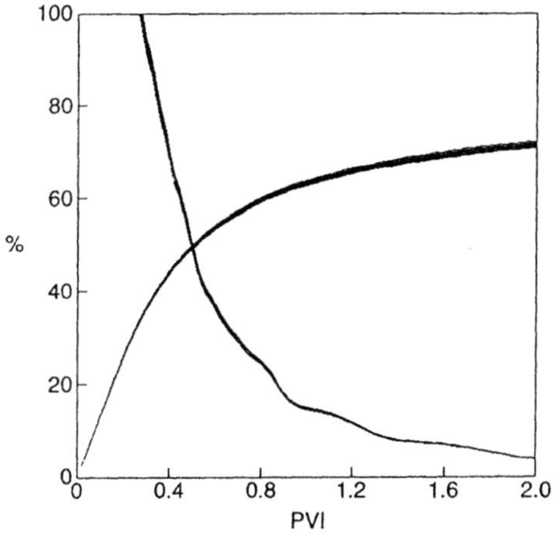

Figure 4

Flow results for GSS realizations using variograms S1, S2 and S3 (total of 15 realizations). BT = 0.28 PVI ±0.01, Recovery (at 1.0 PVI) = 65% ±1%

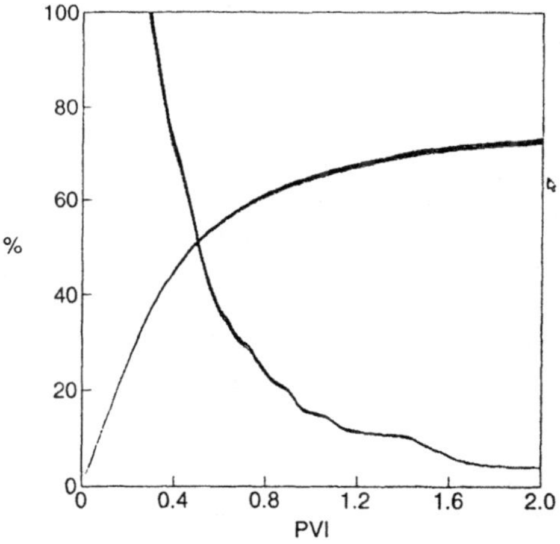

Figure 5

Flow results for GSS realizations using variograms G1, G2 and G3 (total of 15 realizations). BT = 0.28 PVI ±0.01, Recovery (at 1.0 PVI) = 65% ±1.0%

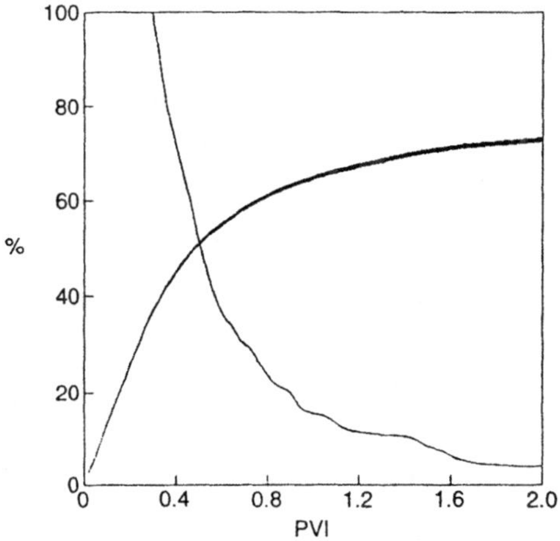

Figure 6

Flow results for LU realizations using variograms S1, S2 and S3 (total of 15 realizations). BT = 0.28 PVI ±0.01, Recovery (at 1.0 PVI) = 65% ±1.0%

For the SIS realizations, five thresholds were used to break up the permeability distribution which for this specific case covers five orders of magnitude (0.01-110.0 md with thresholds chosen at 0.05, 0.51, 2.15, 3.79, and 10.77 md). Using the identical variogram for all cutoffs, which corresponds to median indicator kriging,[9] results in flow behavior similar to the Gaussian results except that the spread in recovery at 1.0 PVI is 64.5% ± 2.0% and breakthrough occurs at 0.29 PVI ± 0.02 PVI. As before, the variogram parameters do not affect recovery or effluent behavior significantly. Three additional sets of runs were conducted in which the range of the highest threshold was significantly larger (900 ft.) than the ranges of the other cutoffs (all 450 ft.) In addition, the anisotropy direction for the highest cutoff was also significantly different than for the other cutoffs where isotropic variograms were used. The flow results obtained are shown in Figs. 7, 8 and 9 with each figure corresponding to one of the different variogram types (exponential, spherical and gaussian).

The magnitude of the range of breakthrough times increases significantly with values ranging from 0.07 to 0.22 PVI. Also, unlike the Gaussian simulations, earlier breakthrough times confirm that the high permeability streaks input through the indicator variograms are indeed being honored in the realizations. A persistent result is that the choice of the variogram type does not alter significantly the recovery or effluent behavior. The differences in flow behavior between the Gaussian and the non-parametric model based realizations will now be discussed.

DISCUSSION

The use of either the GSS or the LU method to generate quantitative reservoir descriptions assumes that the random function describing the multivariate distribution of permeability is Gaussian. Such Gaussian models impart to their realizations a high level of spatial disorder (entropy) or noise. Consequently, a fluid flow process through such a medium will result in a very efficient, piston-like displacement independently of the variance of the permeability distribution. This was observed in the flow simulations. The efficiency of the displacement process will improve with increasing spatial disorder resulting in erroneously high oil recoveries. Of all the random functions available, the Gaussian model maximizes such spatial disorder[8]. A high entropy model is contrary to the notion of an oil reservoir which could simply be stated as a porous medium possessing an overall permeability and porosity structure capable of commercially producing oil. Thus, even if the range indicated by the experimental variogram is less than the well spacing, this does not mean that the permeability is randomly distributed in space which is how the Gaussian generated realizations used in this study behave in terms of fluid flow. Checking the bivariate normality[9] of oil fields where the well spacing is small enough to provide a well defined variogram at small distances shows that the assumption of an underlying Gaussian random function does not hold.

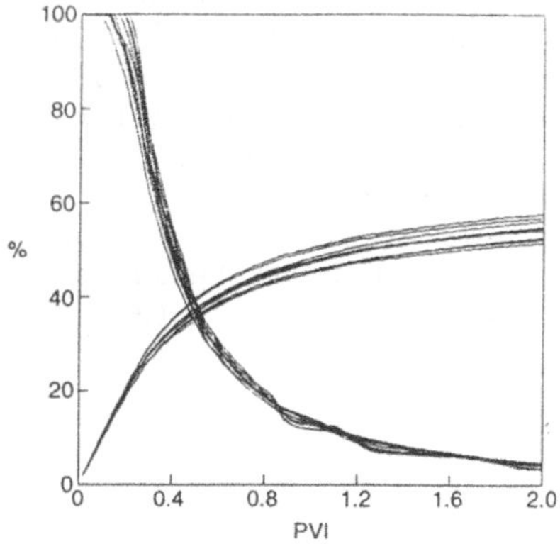

Figure 7

Flow results for SIS realizations using variogram S1 for highest cutoff and S2 for all others (total of 10 realizations). BT = 0.145 PVI ±0.075, Recovery (at 1.0 PVI) = 47.5% ±4%

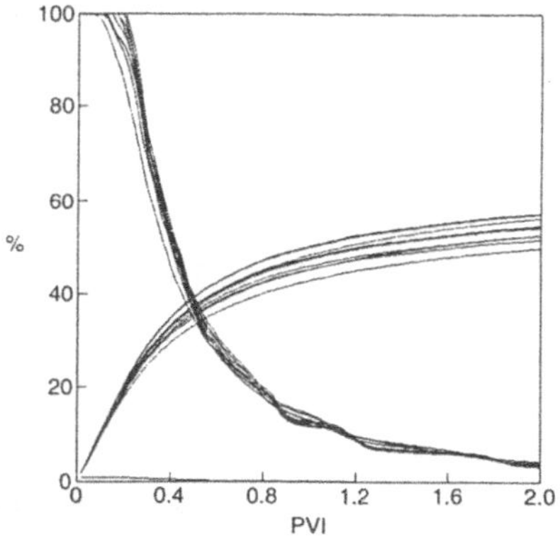

Figure 8

Flow results for SIS realizations using variogram E1 for highest cutoff and E2 for all others (total of 10 realizations). BT = 0.14 PVI ±0.07, Recovery (at 1.0 PVI) = 48.5% ±5.5%

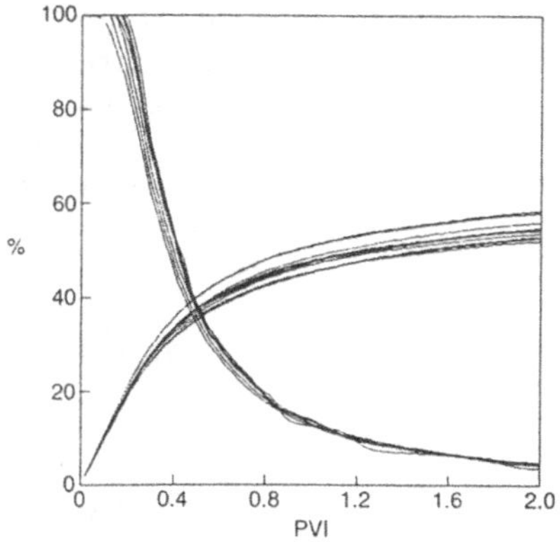

Figure 9

Flow results for SIS realizations using variogram G1 for highest cutoff and G2 for all others (total of 10 realizations). BT = 0.135 PVI ±0.065, Recovery (at 1.0 PVI) = 48% ±4%

On the other hand, non-parametric realizations provide the flexibility to distinguish between the spatial behavior of low permeability values and high permeability values. Such extreme values are the most crucial since they usually have the highest impact on process performance and fluid distribution. In the cases where the range of the high values is considerably larger than the ranges of the remaining cutoffs, a more reasonable spread of recoveries and breakthrough times more in expectation with the small amount of conditioning data is observed. Reducing the high cutoff range from 900 feet to 600 feet, as shown in Fig. 10, produces later breakthrough times and improved recoveries while also slightly decreasing the level of uncertainty, indicating a trend towards the Gaussian flow results.

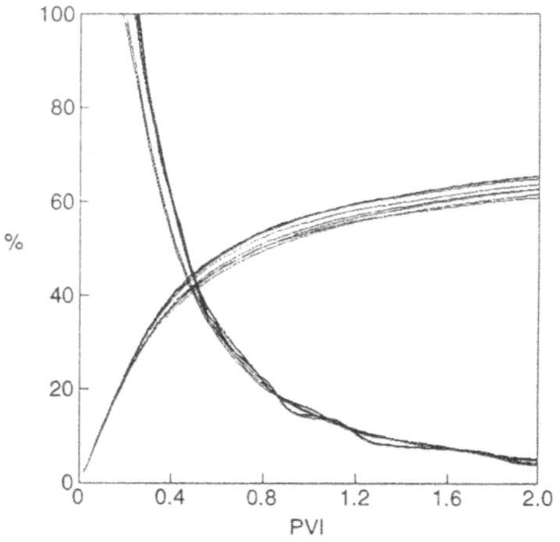

Figure 10

Flow results for SIS realizations using variogram S4 for highest cutoff and S2 for all others (total of 10 realizations). BT = 0.22 PVI ±0.04, Recovery (at 1.0 PVI) = 55% ±2.5%

As noted in the previous section, dropping that range to the other cutoffs' range indeed produces flow performance similar to the results obtained with the Gaussian realizations. Expectedly, a Gaussian model implies that the ranges specific to high and low values become increasingly shorter[9], meaning that a persistent high permeability streak or a prolonged low permeability zone cannot exist. This explains the extreme similarity of the flow behavior observed in Figs. 3 through 6. The Gaussian simulation only considers an average $C(k)$ of all indicator covariances and cross-covariances $C_I(h;k,k')$[7].

$$C(h) = \int_{-\infty}^{+\infty} \int_{-\infty}^{+\infty} C_I(h;k,k')\, dkk'$$

This will bias the permeability field toward the median cut-off for which $C_I(0;k,k')=0.25$ is maximum. The effect of range and anisotropy of the low permeability values was also investigated by running a set of simulations where the range of those values was set to

900 feet. The anisotropy was defined so that these low values should impede the flow between injectors and producers. The results, shown in Fig. 11, indicate again that the range plays a crucial role. There is less crossflow occurring between layers because the more continuous low permeability values will tend to compartmentalize the fluid flow resulting in a more efficient displacement.

The impact of the choice of anisotropy direction and ratio is not as clear because it depends on the injector/producer well layout with respect to the anisotropy orientation. For the simple well arrangement studied here, orienting the flow barriers in the direction of the streamlines gives average recovery levels close to the Gaussian results but with a much larger spread in recovery behavior.

The choice of variogram model used had a small effect on the visual appearance of the conditional simulations but did not contribute to any significant differences in flow

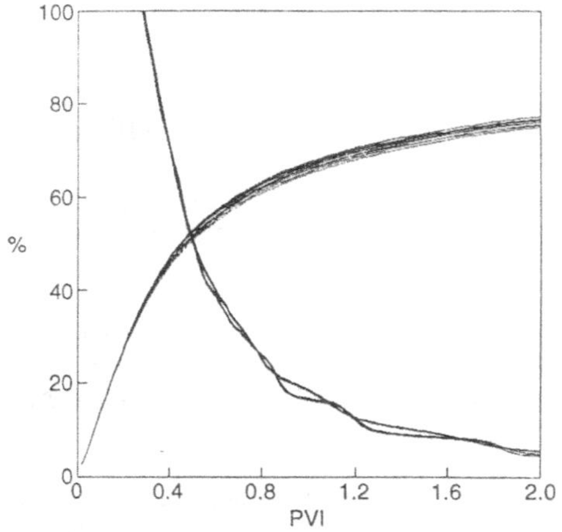

Figure 11

Flow results for SIS realizations using variogram S5 for lowest cutoff and S2 for all others (total of 10 realizations). BT = 0.30 PVI ±0.03, Recovery (at 1.0 PVI) = 67% ±2%

behavior. The flow simulator, in this case the hybrid approach, is acting as a smoothing function that, for the fluid properties used in this study, ignores such subtle differences. It should also be remembered that this level of smoothing is process dependent, meaning that while waterfloods are not sensitive to the choice of variogram function, this may not be true for other recovery processes such as miscible displacements or steamfloods.

CONCLUSIONS

1. The most important variogram parameter affecting flow behavior is the range. This clearly underscores the need to derive a good estimation of the range whether it be from a related property measurement such as seismic impedance derived porosity, or from direct measurements off an outcrop face over an appropriate distance (i.e., a typical interwell spacing).

2. The impact of the anisotropy ratio from the simulations performed in this study was not clear and will need further research.

3. The choice of the variogram type (exponential, spherical and Gaussian) is not important in terms of fluid flow predictions.

4. Indicator simulation techniques appear better suited than Gaussian methods to generate the reservoir descriptions needed for flow simulation. The indicator approach allows the level of uncertainty resulting from the lack of information about the interwell reservoir description to be adequately quantified.

REFERENCES

1. Grant, C. W., Goggin, D. J., and Harris, P. M., "Permeability Variation in Carbonate Parasequences and its Effect on Viscous Dominated Flow Behavior", GSA abstract, San Diego, CA, Oct. 21-24, 1991.

2. Tyler, N., Barton, M. D. and Finley, R. J., "Outcrop Quantification of Flow-Unit Sand Properties and Geometries, Ferron Sandstone, Utah", presented at 75th AAPG annual convention, Dallas, TX, April 7-10, 1991.

3. Araktingi, U. G. and Orr, F. M., "Viscous Fingering in Porous Media", SPE 18095 presented at the 63rd Annual Technical Conference of SPE, Houston, TX, Oct. 2-5, 1988.

4. Hewett, T. A. and Behrens, R. E., "Sealing Laws in Reservoir Simulation and Their Use in a Hybrid Finite Difference/Streamtube Approach to Simulating the Effects of Permeability Heterogeneity", Reservoir Characterization II, Academic Press, 1991, pp.402-941.

5. Tang, R. W., Behrens, R. E. and Emanuel, A. S., "Reservoir Studies with Geostatistics to Forecast Performance", SPERE (May 1991) 253-258.

6. Alabert, F. G., "The Practice of Fast Conditional Simulations through the LU Decomposition of the Covariance Matrix", Mathematical Geology, Vol. 19, No. 5, 1987.

7. Alabert, F. G., "Stochastic Imaging of Spatial Distribution using Hard and Soft Information", M. Sc. Thesis, Stanford University, Stanford, CA, 197 pp. (1987).

8. Jones, D. S., Elementary Information Theory, Clarendon Press, Oxford p.137 (1979).

9. Deutsch, C. W. and Journel, A. G., "GSLIB: Geostatistical Software Library, User's Guide", Oxford University Press, New York, 1992.

CONDITIONING RESERVOIR MODELS TO WELL TEST INFORMATION

CLAYTON V. DEUTSCH
Stanford Center for Reservoir Forecasting, Stanford, CA, USA, 94305

Pressure transient well test information has largely been ignored in stochastic reservoir modeling. Well testing informs the effective absolute permeability of some region around the well. This information does not resolve local details of the spatial distribution of permeability; however, it does constrain the average permeability. This paper presents an approach, based on simulated annealing, that will condition stochastic reservoir models to such well test-derived effective properties.

The volume and type of averaging informed by the well test must first be calibrated by forward simulating the well test on stochastic reservoir models that are consistent with the geological interpretation, core data, well log data, and seismic data. Stochastic reservoir models are then constructed with simulated annealing where one component of the global objective function relates to honoring the well test-derived average permeability.

INTRODUCTION

The concept underlying stochastic reservoir modeling is to construct numerical models of the reservoir properties that are consistent with all relevant data. Applying a flow simulator to multiple numerical models allows an appreciation for the uncertainty in the reservoir response. The accuracy and precision of the output response distribution requires integrating a maximum amount of relevant prior information, e.g., core plug measurements, well log data, surface and borehole geophysical interpretations, geological interpretations, spatial information gathered from similar reservoirs, properties inferred from well tests, ... This paper is concerned with integrating this last source of data, i.e., effective permeabilities inferred from pressure transient well tests, into stochastic models of absolute permeability.

Conventional conditional simulation techniques such as Gaussian [6,18], fractal [10], or indicator [13] models have the ability to account for local conditioning data (core and well log data), a global histogram, and varying amounts of spatial information in the form of two-point variogram/covariance models. Variants of these techniques based on cokriging [7] or some type of trend model [17] allow geophysical information to be integrated into the resulting realizations. However, none of these techniques allow the integration of well test-derived effective permeabilities.

Simulation techniques based on marked point processes [9] are well-suited to simulating spatial phenomenon that are characterized by a repetition of easily characterized shapes. The resulting spatial structure is implicitly controlled by the placement of the digitized or analytical shapes. Conditioning to local data and a global histogram is achieved by disallowing inconsistencies at data locations and controlling the number of objects placed in the realization. The integration of geophysical interpretations and well test-derived effective permeabilities is not possible with these latter techniques.

A well test-derived effective permeability does not directly inform the smaller scale permeability values near the well bore. However, it does constrain a complex non-linear average of the small scale values. Many flow processes are directly influenced by the average flow characteristics as informed by well test-derived effective permeabilities. Historically, well test-derived permeabilities were used in homogeneous reservoir models for reservoir performance forecasting [20]. Two inadequacies of such homogeneous models are that they do not allow an assessment of uncertainty and they do not allow for the important influence of small scale permeability heterogeneities. Stochastic reservoir models allow the small scale permeability heterogeneities to be accounted for. However, as mentioned above, current stochastic modeling techniques do not account for the information carried by average properties measured by well tests.

The problem of conditioning stochastic models to well test-derived effective permeability is difficult because the average is non-linear. Conventional simulation techniques allow data of different volumetric supports only when the averaging is linear, see page 513 [14].

One way to achieve this conditioning would be to discard all models that do not yield a forward simulated well test response close enough to the actual measured pressure response. Such selection procedures may be practical when building models with a single well test; however, it is not practical in presence of multiple well test interpretations, some with more advanced multirate tests which inform a number of different permeability averages near the well. In general, a prohibitively large number of realizations would be required to find a few that simultaneously match all well test data.

The algorithm proposed in this paper is to generate realizations with *simulated annealing* [15] that honor the measured well test effective permeabilities in addition to conventional data such as the local conditioning data, a histogram, and a variogram/covariance. Application of simulated annealing requires that the generation of a stochastic realization be posed as an optimization problem. The two-part objective function in this optimization problem consists of the deviation from the desired statistical/geological description plus the deviation from the well test effective permeabilities. The deviation from the measured well test results could be known through forward simulation of the well test on all candidate realizations. Again, it would be impractical to forward simulate the well test after each (set of) perturbation(s) as demanded by simulated annealing. Therefore, the forward simulation must be replaced by a more easily calculated numerical approximation. A non-linear power average [3,5,16] of the small scale permeability values has been adopted for this purpose.

More precisely, the well test is first interpreted to provide the effective permeability near the well bore. Second, the type of averaging, as quantified by an averaging power, and the volume of averaging must be calibrated. The averaging power depends essentially on the connectivity of the extreme permeability values. The volume of averaging depends on the duration of the well test. Third, the initial stochastic realizations are altered by systematically changing the elementary grid block permeability values so that the previously calculated effective permeability and the variogram model are honored. Calculation of that effective permeability can be done very fast after each (set of) perturbation(s) using the previously calibrated power average.

The following presentation of the algorithm will clarify the implementation details and acknowledge a number of limitations. One limitation is that the method imposes the well test results without accounting for uncertainty in the underlying well test interpretation. A second limitation is that the method requires that the full well test response be summarized by a single non-linear weighted power average. The severity of these limitations may be judged through numerical experimentation.

QUANTIFYING WELL TEST-DERIVFED EFFECTIVE PERMEABILITY

Pressure transient well tests are performed by generating some flow rate impulse in the reservoir (e.g., start production, change the flow rate, stop production, ...) and measuring the pressure response. Well test interpretation consists of interpreting the pressure response by using some appropriate mathematical model to relate the pressure response (output) to flow rate history (input) [11]. Provided that the mathematical model is appropriate the model parameters can be associated to certain reservoir parameters. Of particular interest is the effective absolute permeability k_e.

To apply the envisaged optimization technique it is necessary to translate this well test effective k_e into a more easily calculated property while retaining the flexibility to differentiate a wide variety of heterogeneous systems encountered in practice. The power averaging formalism is used to model the non-linear averaging of absolute permeabilities [5]. The assumption is that the elementary permeability values average linearly after a non-linear power transformation, i.e.,

$$\overline{k}(\omega) = \left[\frac{1}{N}\sum_{\mathbf{u}_i \in V} k(\mathbf{u}_i)^{\omega}\right]^{\frac{1}{\omega}} \tag{1}$$

Where $\overline{k}(\omega)$ is the ω-power average permeability of the N permeability values $k(\mathbf{u}_i)$, $i = 1, \ldots, N$, at locations $\mathbf{u}_i$ within the volume of interest V. The power ω varies between the bounding values of -1 and 1 corresponding to the harmonic and arithmetic averages respectively (the geometric average is obtained for ω=0.0). The idea is to use relation (1) and to calibrate the averaging volume V and averaging power ω for each particular reservoir.

To define the appropriate averaging volume it is necessary to consider the portion of the pressure response used to derive the well test effective permeability k_e. In practice, k_e is obtained by interpreting the pressure response during the time at which the response resembles infinite-acting radial flow. Early-time effects such as wellbore storage and late-time boundary effects are not considered in the interpretation. It is possible to define an inner radius r_{min} and an outer radius r_{max} that correspond to the limits of infinite acting radial flow since the pressure response, at any time t, may be related to block permeabilities within a time-dependent radius-of-drainage $r(t)$. Consider a typical pressure response shown on the Miller-Dyes-Hutchinson (MDH) plot at the bottom of Figure 1. The pressure response between 1 hour and 10 hours is used to derive an estimate of the effective permeability $k_{welltest}$. The inner and outer limits of the shaded region (on the schematic illustration of the reservoir) correspond to the radius-of-drainage at 1 hour and 10 hours respectively.

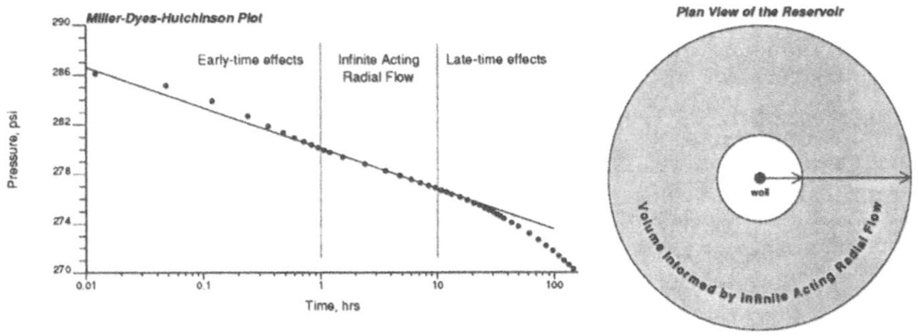

Figure 1: A schematic illustration of the volume measured by a given well test interpretation. The inner and outer limits of the shaded region, on the schematic illustration of the reservoir, correspond to the radius-of-drainage at 1 hour and 10 hours (the limits of infinite acting radial flow).

The time interval during which the pressure response resembles infinite-acting radial flow is easily determined by standard interpretation techniques. Evaluating the radius-of-drainage $r(t)$ at the time limits is not as straightforward; depending on the arbitrary definition chosen for $r(t)$, the radius can change by as much as a factor of 4. It will be necessary to calibrate the radius-of-drainage $r(t)$ by repeated flow simulations. The block permeabilities contributing to the pressure response measured up to time t are approximately enclosed by a circular volume centered at the well defined by a time-dependent radius $r(t)$ written as [3,12,21]:

$$r(t) = A\sqrt{\frac{k_e t}{\phi \mu c_t}} \tag{2}$$

where A is a constant, k_e is the reservoir permeability around the well, ϕ is the porosity, μ is the fluid viscosity, and c_t is the total compressibility. Depending on the definition chosen for the radius-of-drainage the value of A varies from 0.023 to 0.07 (for oil field units). Alabert [3], in the context of evaluating the averaging volume of a well test, and for specified levels of discretization and test durations, found an optimal A_{opt} value of 0.010 in oil field units.

The averaging power ω describes the type of averaging within the volume $V(A)$. In many cases, this averaging power is close to the geometric average ($\omega = 0$). For practical test durations and for complex heterogeneous permeability distributions, the type of averaging can differ significantly from the geometric average.

The constant A and the averaging power ω are calibrated with the following procedure (see also [3,4]):

1. Generate n_s (20-100) multiple realizations of the permeability field with relevant statistical properties.

2. Forward simulate a well test, with conditions as close as possible to those used in the field to arrive at k_e, on each realization to obtain n_s pressure response curves.

3. Deduce an effective permeability $\overline{k}_i, i = 1, \ldots, n_s$ from each pressure curve using established well test interpretation techniques [11].

4. Compute average permeabilities $\overline{k}(A,\omega)_i, i = 1,\ldots,n_s$ for A values between the practical bounding limits of 0.001 and 0.020 and for ω values between practical bounding limits of -0.5 and 0.5.

5. Choose the pair (A_{opt}, ω_{opt}) that yields the closest agreement between the reference $\overline{k}_i, i = 1,\ldots,n_s$ values and the approximate $\overline{k}(A_{opt}, \omega_{opt})_i, i = 1,\ldots,n_s$ values.

After establishing appropriate A_{opt} and ω_{opt} values the *goodness* of the power average approximation can be checked by plotting a scatterplot of the $\overline{k}(A_{opt}, \omega_{opt})_i$ values versus the well test-derived $\overline{k}_i$ values.

The weighted non-linear power average (1) is proposed as a computationally simple replacement for the full 3-D well-test response. In practice, a well test response is interpreted to yield an estimate of the effective permeability k_e and the averaging volume parameter A_{opt} and power ω_{opt} are calibrated for the particular geological setting. The next step is to impose the well test-derived effective permeability k_e, i.e., the appropriate ω-average $\overline{k}(A_{opt}, \omega_{opt})$, on stochastic models. The technique of simulated annealing is considered for this simulation.

SIMULATED ANNEALING

In the "annealing" approach to stochastic simulation there is no explicit random function model, rather, the creation of a simulated realization is formulated as an optimization problem to be solved with a stochastic relaxation or "annealing" technique. The first requirement of this class of methods is an objective (or energy) function which is some measure of difference between the desired spatial characteristics and those of a candidate realization. The essential feature of annealing methods is to iteratively perturb (relax) the candidate realization and then accept or reject the perturbation with some decision rule. The decision rule is based on how much the perturbation has brought the candidate image closer to having the desired properties. One possible decision rule is based on an analogy with the metallurgical process of annealing, hence the name *simulated annealing*. Technically the name "simulated annealing" only applies to those stochastic relaxation methods based strictly on simulated annealing [2,15]; however, through common usage the name "annealing" is used to describe the entire family of methods that are based on the principle of stochastic relaxation.

Annealing is the process where a metallic alloy is heated, without leaving the solid phase, so that molecules may move positions relative to one another and reorder themselves into a low energy crystal (or grain) structure. The probability that any two molecules will move relative to one another is known to follow the Boltzmann distribution. Simulated annealing is the application of the annealing mechanism of perturbation (swap the attribute values assigned to two different grid node locations) with the Boltzmann probability distribution for accepting perturbations.

At first glance this approach appears terribly inefficient. For example, millions of perturbations may be required to arrive at an image that has the desired spatial structure. However, these methods are more efficient than they might seem as long as few arithmetic operations are required to update the objective function after a perturbation; virtually all conventional *global* spatial statistics (e.g., a covariance) may be updated locally rather than globally recalculated after a local perturbation. Also, the power average representation of the well test k_e developed earlier is easily updatable.

The objective function is defined as some measure of difference between a set of reference properties and the corresponding properties of a candidate realization. The reference properties could consist of any *quantified* geological, statistical, or engineering property. In the context of this paper the reference properties consist of traditional two-point variogram/covariance functions and the well test derived effective permeability. Thus the objective function could be written as:

$$O = \lambda_1 \cdot \sum_{i=1}^{i=n_{\mathbf{h}}} \left[C^{reference}(\mathbf{h}_i) - C^{realization}(\mathbf{h}_i)\right]^2 + \lambda_2 \cdot \sum_{j=1}^{j=n_K} \left[k_{e_j}^{reference} - k_{e_j}^{realization}\right]^2 \tag{3}$$

where O is the objective function, $n_{\mathbf{h}}$ is the number of covariance lags $\mathbf{h}_i$ considered important, $C^{reference}(\mathbf{h}_i)$ is the ith reference covariance value, $C^{realization}(\mathbf{h}_i)$ is the ith covariance value taken from the candidate realization, n_K is the number of well test-derived effective permeabilities to be reproduced, $k_{e_j}^{reference}$ is the jth well test-derived effective permeability, $k_{e_j}^{realization}$ is the jth effective permeability calculated from the candidate realization using power averages, and λ_1 and λ_2 are relative weights to ensure that the covariance contribution has the same importance as the well test contribution.

The starting image is a 3-D array, $z(\mathbf{u}_i), i = 1, \ldots, N$, or permeability values. The annealing methodology to achieve a realization l, $z^{(l)}(\mathbf{u}_i), i = 1, \ldots, N$, with a low objective function (3) is as follows:

1. Establish the reference components in the objective function: the reference covariance values $C^{reference}(\mathbf{h}_i), i = 1, \ldots, n_{\mathbf{h}}$ are based on experimental data or on an analytical model. The reference effective permeabilities $k_{e_j}^{reference}, j = 1, \ldots, n_K$, are the direct result of applying standard interpretation techniques to the pressure transient response measured in the field.

2. Generate an easily constructed initial realization: the initial realization is either the output of a more conventional stochastic simulation algorithm or the initial realization could be generated by assigning each nodal value at random from the stationary univariate distribution $F(z)$.

3. Compute the realization components in the objective function: the covariance values $C^{realization}(\mathbf{h}_i), i = 1, \ldots, n_{\mathbf{h}}$, and the effective permeabilities $k_{e_j}^{realization}$, $j = 1, \ldots, n_K$, are calculated from the candidate realization.

4. Compute the objective function O based on the reference and the realization statistics, see (3).

5. Perturb the realization $z(\mathbf{u}_i), i = 1, \ldots, N$, to generate a new realization $z_{new}(\mathbf{u}_i)$, $i = 1, \ldots, N$ by swapping the permeability value at any two locations $\mathbf{u}_i$, $\mathbf{u}_j$, $i \neq j$, $0 < i,j \leq N$.

6. Update each component in the objective function and recompute the objective function O_{new} with the perturbation. The change to the objective function is $\Delta O = O_{new} - O_{old}$.

7. The perturbation is accepted or rejected on the basis of a specified decision rule. One approach would be to accept all helpful perturbations $\Delta O \leq 0$ and to reject all disruptive perturbations $\Delta O > 0$. This choice, which corresponds to a steepest descent approach, can lead to a local minima. The essential contribution of simulated annealing is a prescription for when to accept or reject a given perturbation. The acceptance probability distribution is given by:

$$P\{accept\} = \begin{cases} 1, & \text{if } \Delta O \leq 0 \\ e^{\frac{-\Delta O}{t}}, & \text{otherwise} \end{cases} \tag{4}$$

   All favorable perturbations ($\Delta O \leq 0$) are accepted and some unfavorable perturbations are accepted with an exponential probability distribution. The parameter t of the exponential distribution is analogous to the "temperature" in annealing. The higher the temperature the more likely an unfavorable perturbation will be accepted.

   Accepting the perturbation causes the image, $z(\mathbf{u}_i), i = 1, \ldots, N$, and the objective function O to be updated.

8. When the objective function O gets close to zero then the realization is considered finished since it now honors both the reference covariance and the well test data; otherwise, return to step 5 and continue the perturbation process.

The idea is to start with an initially high temperature parameter t (say the initial objective function $t = O$) and lower it by some multiplicative factor λ (say 0.1) when enough perturbations have been accepted (K_{accept}=10 times the number of grid nodes in the system N) or too many have been tried ($K_{max} = 100 \cdot N$). The algorithm is stopped when efforts to lower the objective function become sufficiently discouraging [19].

One remaining issue is to establish the weights λ_1 and λ_2 applied to each component in the objective function. The purpose behind these weights is to have each component play an equally important role in the global objective function. Without any weighting the component with the largest units would dominate the objective function. The weights λ_1 and λ_2 are established so that, in average, each component contributes equally to a change in the objective function ΔO. That is, each weight λ_c is inversely proportional to the average change of that component objective function:

$$\lambda_c = \frac{1}{\overline{|\Delta O_c|}}, \quad c = 1, \ldots, 2 \tag{5}$$

In practice, the average change of each component $\overline{|\Delta O_c|}$ can not be computed analytically; however, it can be numerically approximated by evaluating the average change of M (say 1000) independent perturbations:

$$\overline{|\Delta O_c|} = \frac{1}{M} \sum_{m=1}^{M} |O_c^{(m)} - O_c|, \quad c = 1, \ldots, 2 \tag{6}$$

where $\overline{|\Delta O_c|}$ is the average change for component c, $O_c^{(m)}$ is the perturbed objective value, and O_c is the initial objective value. Each of the M perturbations $m = 1, \ldots, M$ arises from the swapping mechanism employed for the annealing simulation.

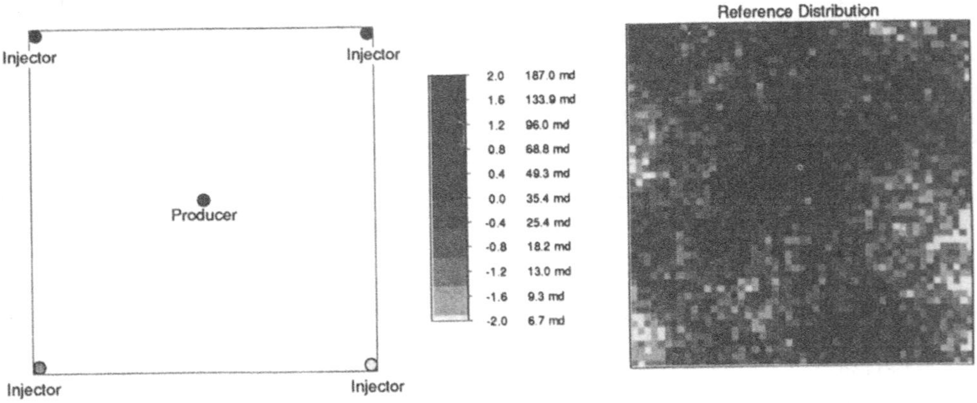

Figure 2: The reference distribution of permeability considered as the true reservoir for the purposes of obtaining the well test response and interpreted effective permeability.

All of the elements needed for integrating well test data are now in place. The resulting realizations obtained after going through the simulated annealing procedure with the objective function (3) are conditional to both the initial geological/statistical description and the well test-derived permeabilities. The following example illustrates how the methodology is implemented in practice.

AN EXAMPLE APPLICATION

Consider a 2-D example where the the block horizontal absolute permeabilities are known to follow a lognormal distribution $F(z)$ with a mean of 50.0 md and a variance of 2500 md^2. The distribution of log permeability is characterized by an exponential normal scores covariance model $C_Y(\mathbf{h})$ with an effective range of 25 grid block units. A 51 unit by 51 unit realization of this lognormal permeability field was generated to serve as a reference distribution, see Figure 2. The permeability at the central producer and the four injectors are known. Interpretation of a simulated drawdown response yields an estimated effective permeability of 102.1 md.

Simulated annealing is capable of generating realizations conditional to the distribution model $F(z)$, the covariance model $C_Y(\mathbf{h})$, and the five conditioning data. One hundred such realizations were generated by annealing with the objective function:

$$O = \sum_{i=1}^{i=n_{\mathbf{h}}} \left[C^{reference}(\mathbf{h}_i) - C^{realization}(\mathbf{h}_i) \right]^2$$

where $n_{\mathbf{h}}$ corresponds to the most compact arrangement of 18 lags, see top of Figure 3. Annealing allows this objective function to go to zero, i.e., the resulting realizations provide an excellent reproduction of the covariance. The first two realizations are shown on Figure 3.

These 100 initial realizations reproduce the measured permeability at the five well locations. However, a forward simulation of a well test at the producing well location does not yield the 102.1 md value obtained from the reference image. Figure 4 shows a histogram of the 100 permeability values. These 100 effective permeability values were used to calibrate the parameters A_{opt} and ω_{opt} that provide the best numerical approximation to the well test result. The average permeabil-

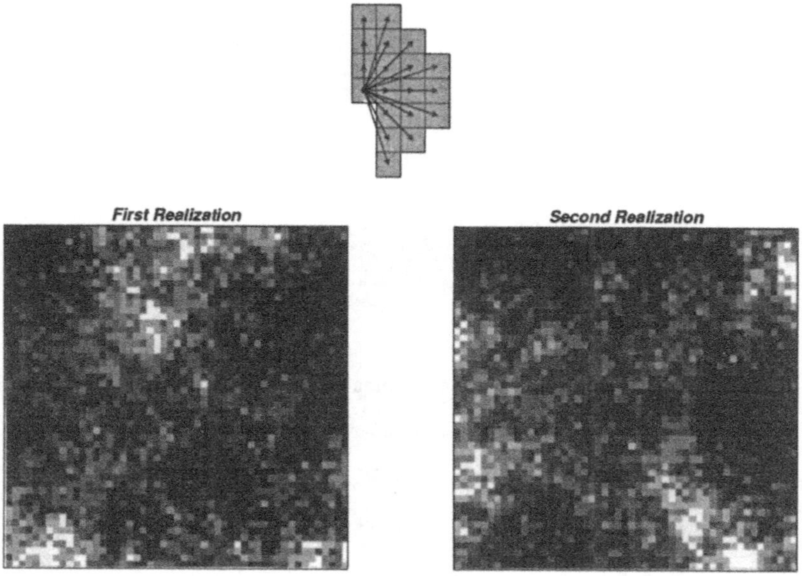

Figure 3: The lag vectors used to generate realizations of permeability and the first two realizations.

ities $\overline{k}(A,\omega)_i, i = 1,\ldots,n_s$ for A values between the practical bounding limits of 0.001 and 0.020 and for ω values between practical bounding limits of -0.5 and 0.5 were computed using the 100 initial realizations. The criteria for an optimal pair (A_{opt}, ω_{opt}) was to minimize the mean normalized absolute deviation and the mean normalized error [3] defined as:

$$mNAD(A,\omega) = \sum_{i=1}^{i=n_s} \frac{|\overline{k}(A,\omega)_i - \overline{k}_i|}{\overline{k}_i} \tag{7}$$

$$mNE(A,\omega) = \left|\frac{\sum_{i=1}^{i=n_s} \overline{k}(A,\omega)_i - \sum_{i=1}^{i=n_s} \overline{k}_i}{\sum_{i=1}^{i=n_s} \overline{k}_i}\right| \tag{8}$$

$mNAD$ measures correlation and mNE measures the bias between the power average approximation and the true well test values. The optimal pair $A_{opt} = 0.003$ and $\omega_{opt} = 0.10$ was found to minimize both error terms.

Figure 5 shows a scatterplot of the power average numerical approximation and the true well test-derived effective permeability. Note the lack of any bias (the average effective permeabilities are 58.3 md in both cases) and the excellent correlation of 0.98. The physical volume informed by the well test can be determined by the constant A_{opt} and knowledge of the time limits of the infinite acting portion of the pressure response: $r_{min} = 1.0$ grid units (70.1 feet) and $r_{max} = 4.6$ grid units (320.0 feet).

The annealing simulation now uses a two part objective function (3). Annealing is able to generate realizations that lower this objective function to zero. The first two realizations are shown on Figure 6.

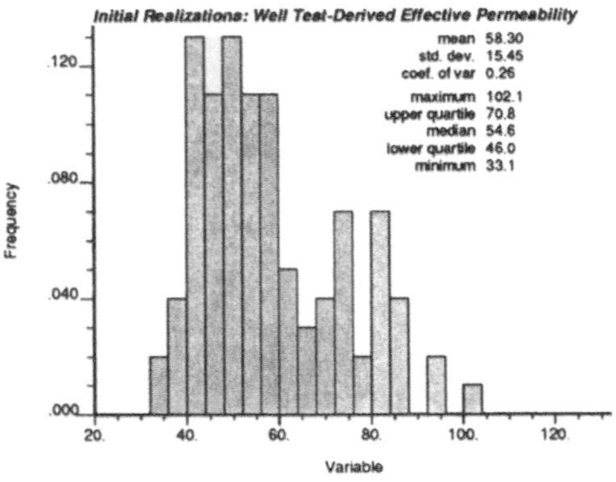

Figure 4: The histogram of 100 effective permeability values from the initial realizations.

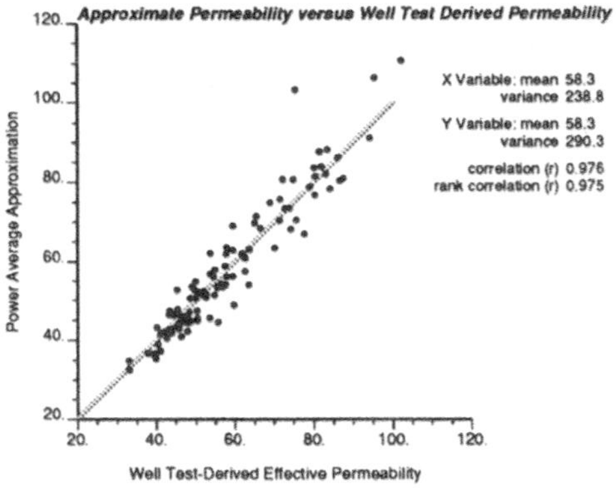

Figure 5: A scatterplot of the power average approximation and the true well test-derived effective permeabilities for all 100 initial realizations.

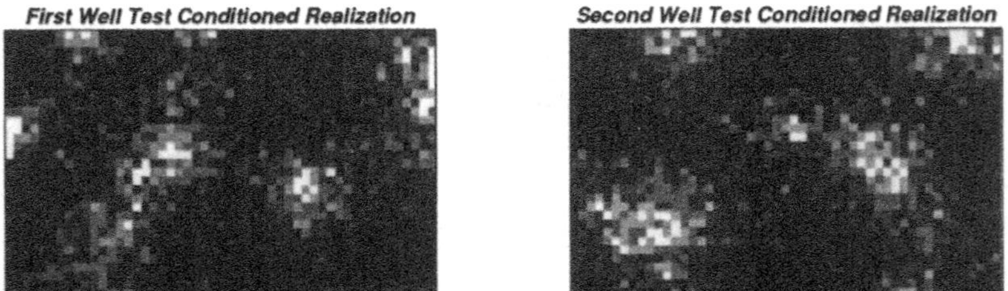

Figure 6: The first two realizations conditional to the five well data, the histogram, the covariance, and the well test result.

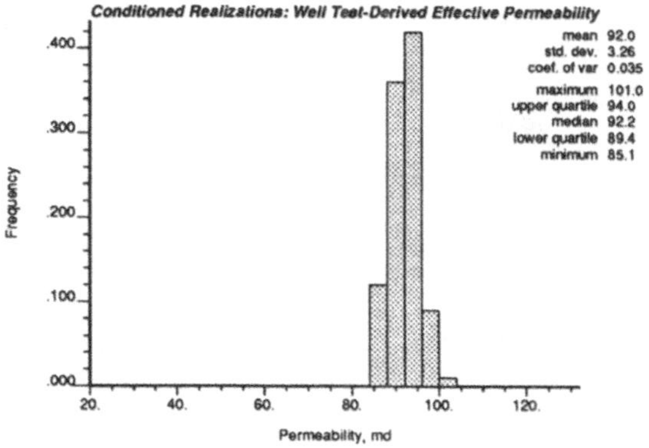

Figure 7: The histogram of 100 effective permeability values conditional to the well test result.

To verify that the calibrated power average is a fair approximation to the actual well test result a full well test was forward simulated on the final 100 realizations. Figure 7 shows the distribution of the 100 well test derived effective permeability values before and after post processing. Note that the distribution of effective permeabilities, after post processing, is much closer to the reference value of 102.2 md.

This illustrates that well test effective permeabilities can be imposed on stochastic realizations to a fair degree of approximation. What has not been shown is that accounting for well test-derived effective permeability actually helps predicting future reservoir performance. To illustrate the improvement in future prediction all 201 realizations (the reference, the 100 initial, and the 100 well test conditioned realizations) have been associated to the five spot injection/production pattern shown on the left of Figure 2 with an injector at all corner locations and a central producer. All variables except the block absolute permeabilities have been held constant and the performance of each realization has been simulated with Eclipse [1].

Two response variables are presented here: 1) the breakthrough time, i.e., the time of first water arrival at the producer, and 2) the final oil in place after 50 *years* of simulated production (note that the units are important only in a relative sense). The reference image yielded a breakthrough time of 0.81 time units and a fractional final oil in place 0.38. The histograms of values obtained before and after conditioning to the well test data on Figure 8. The results are summarized on Table 1. Note that the well test conditioned distributions are better centered around the reference value and the uncertainty as measured by the 95% probability interval ($q_{0.975} - q_{0.025}$) has been reduced.

REMARKS AND CONCLUSIONS

A methodology has been presented to integrate well test-derived properties into stochastic reservoir models. The methodology consists of generating realizations with simulated annealing. The objective is to minimize deviations from the initial variogram/covariance model and yet honor a numerical approximation of the well test-derived effective permeability. The power average numerical approximation is useful because it would not be feasible to re-run a flow simulation after each perturbation called for by the annealing technique.

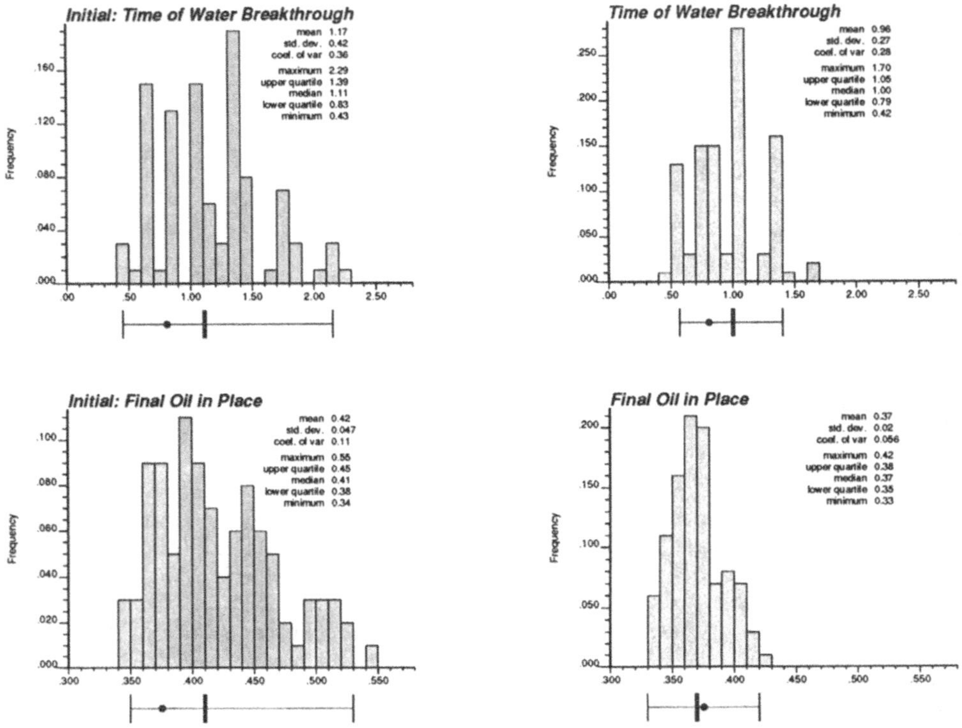

Figure 8: The histograms of the breakthrough time and the final oil in place obtained by running flow simulations on the realizations before (to the left) and after (to the right) integrating the well test data. The black dot in the lower box plot is the reference value, the thicker vertical line is the median, and the outermost vertical lines represent the 95% probability interval.

Response Variable		95% Probability Interval $q_{0.025}$	$q_{0.5}$	$q_{0.975}$	percent reduction
Breakthrough time	reference		0.81		
	before	0.45	1.11	2.15	
	after	0.57	1.00	1.40	-51%
Final oil recovery	reference		0.38		
	before	0.35	0.41	0.53	
	after	0.33	0.37	0.42	-50%

Table 1: Summary of the breakthrough time and final water flow rate before and after conditioning to include the well test result. The reference values of 0.81 and 0.38 are shown under the median ($q_{0.5}$) column, the 95% probability intervals as established from the 100 flow simulations are shown by the 0.025 and 0.975 quantiles ($q_{0.025}$ and $q_{0.975}$), and the percent reduction refers to the width of the 95% probability interval.

An example shows how the methodology could be implemented in practice. The significant improvement brought by accounting for the well test data was demonstrated in terms of distributions of output uncertainty (after performing a waterflooding flow simulation).

The annealing methodology presented to integrate well test data is quite general. The method has the potential to integrate many disparate data as long as these data can be quantified to enter a global objective function. For example, multiple-point statistics could be used to input curvilinear geological structures and seismic data could be incorporated by adding another component to the global objective function [4,8]. Other sources of data that could be incorporated are the results of multiple rate and tracer tests.

References

[1] *ECLIPSE Reference Manual.* Exploration Consultants Limited, Highlands Farm, Greys Road, Henley-on-Thames, Oxon, England, November 1984.

[2] E. Aarts and J. Korst. *Simulated Annealing and Boltzmann Machines.* John Wiley & Sons, New York, NY, 1989.

[3] F. Alabert. Constraining description of randomly heterogeneous reservoirs to pressure test data: a Monte Carlo study. In *SPE Annual Conference and Exhibition, San Antonio*, Society of Petroleum Engineers, October 1989.

[4] C. Deutsch. *Annealing Techniques Applied to Reservoir Modeling and the Integration of Geological and Engineering (Well Test) Data.* PhD thesis, Stanford University, Stanford, CA, 1992.

[5] C. Deutsch. Calculating effective absolute permeability in sandstone/shale sequences. *SPE Formation Evaluation*, 343–348, September 1989.

[6] C. Deutsch and A. Journel. *GSLIB: Geostatistical Software Library.* to be published by Oxford University Press, New York, NY, 1992.

[7] P. Doyen. Porosity from seismic data: a geostatistical approach. *Geophysics*, 53(10):1263–1275, 1988.

[8] P. Doyen and T. Guidish. Seismic discrimination of lithology: a Bayesian approach. In *Geostatistics Symposium*, Calgary, AB, May 1990.

[9] H. Haldorsen and E. Damsleth. Stochastic modeling. *J. of Pet. Technology*, 404–412, April 1990.

[10] T. Hewett. Fractal distributions of reservoir heterogeneity and their influence on fluid transport. 1986. SPE paper # 15386.

[11] R. Horne. *Modern Well Test Analysis.* Petroway Inc, 926 Bautista Court, Palo Alto, CA, 94303, 1990.

[12] P. Johnson. The relationship between radius of drainage and cumulative production. *SPE Formation Evaluation*, 267–270, March 1988.

[13] A. Journel and F. Alabert. New method for reservoir mapping. *J. of Pet. Technology*, 212–218, February 1990.

[14] A. Journel and C. J. Huijbregts. *Mining Geostatistics.* Academic Press, New York, NY, 1978.

[15] S. Kirkpatrick, C. Gelatt Jr., and M. Vecchi. Optimization by simulated annealing. *Science*, 220(4598):671–680, May 1983.

[16] G. Korvin. Axiomatic characterization of the general mixture rule. *Geoexploration*, 19:267–276, 1981.

[17] A. Marechal. Kriging seismic data in presence of faults. In G. Verly et al., editors, *Geostatistics for natural resources characterization*, pages 271–294, Reidel, Dordrecht, Holland, 1984.

[18] G. Matheron, H. Beucher, H. de Fouquet, A. Galli, D. Guerillot, and C. Ravenne. Conditional simulation of the geometry of fluvio-deltaic reservoirs. 1987. SPE paper # 16753.

[19] W. Press, B. Flannery, S. Teukolsky, and W. Vetterling. *Numerical Recipes.* Cambridge University Press, New York, NY, 1986.

[20] H. Ramey Jr. Advances in practical well test analysis. In *65th Annual Technical Conference and Exhibition*, pages 665–676, Society of Petroleum Engineers, September 1990.

[21] H. van Poollen. A hard look at radius of drainage and stabilization-time equations. *Oil and Gas Journal*, 139–147, September 1964.

RESERVOIR CHARACTERIZATION WITH SEISMIC DATA USING PATTERN RECOGNITION AND SPATIAL STATISTICS

SURESH G. THADANI
Stochastic Systems International
1524 Mockingbird,
Plano, Texas 75093, USA

ABSTRACT. This paper addresses the problem of utilizing seismic data to augment (sparse) well control information in constructing reservoir descriptions, that are accompanied by robust measures of uncertainty. A hybrid approach is presented that combines statistical pattern recognition techniques, with spatial statistics techniques for the stochastic imaging of reservoir properties. The pattern recognition techniques are used to calibrate the seismic data to well-log (geological) data, so as to generate *soft* control information (such as inequality constraints), on these properties in the neighborhoods of the wells. Stochastic Imaging techniques are then used to image Indicator representations of these soft estimates, and thus quantify the uncertainty associated with the spatial variability of the reservoir properties. The soft estimates serve to constrain the stochastic images in between the wells, while facilitating the inference of the underlying spatial covariance models. The paper presents test results of an application of the hybrid methodology to a dolomitic reservoir.

1. INTRODUCTION

In recent years there has been considerable interest in the possibility of utilizing seismic data to augment sparse well control information in reservoir characterization problems, while also providing for the robust quantification of uncertainty. This problem poses significant challenges: seismic data has limited bandwidth and resolution [1,21,22], and contains only indirect information about reservoir properties (such as porosity, thickness, ...). The limited "information content" implies that seismic data can realistically be expected to yield only "soft" information on reservoir properties. The indirect nature of the data makes it necessary to establish and exploit the crucial relationships between the seismic and reservoir property (geological) "spaces". These relationships could in principle be established directly by utilizing techniques such as Cokriging [2], and indeed, CoSimulation [20]. However, the inference of the underlying multivariate spatial auto/cross correlation models requires significant control information and modeling effort - such approaches become problematic when well control is sparse.

This paper presents a hybrid methodology that was developed to address the above concerns, particularly in sparse well settings. This methodology uses (multivariate) statistical pattern recognition techniques to calibrate the seismic to the geological space. This "Soft Inversion" process *generates* local soft control information away from the wells, in the form of inequality constraints (or more generally, distributions) on *average* reservoir properties. Stochastic Imaging techniques are then used to image Indicator representations of these soft estimates, and thus quantify the uncertainty associated with the spatial variability of the reservoir properties. The soft estimates are generated at *most* seismic sample locations neighboring the wells - they help constrain the stochastic images, while facilitating the inference of the spatial correlation models directly in the geological space.

The Soft Inversion methodology is described in Section 2 of this paper. In section 3, the Soft Inversion process is interpreted from the (random function) viewpoint of Spatial Statistics. Also, this section provides a brief overview of Stochastic Imaging techniques suited to imaging the soft estimates. Section 4 describes results of an application of the methodology to a dolomitic reservoir sampled by 3D seismic and well-log data. Finally, Section 5 summarizes the major conclusions of this research and outlines directions for future work.

2. SOFT INVERSION

As noted above, the object of Soft Inversion is to infer local soft estimates (eg. inequality constraints, distributions) on reservoir properties by calibrating seismic to geological data. This calibration is based on assumed linear (more precisely, affine) functional relationships between the seismic data and *average* reservoir properties, that are consistent with available linearized "physics based" models [1,18,21]. In particular, these affine relationships are utilized in only an *implicit* sense (ie. "least committal" in the jargon of maximum entropy methods). The multivariate Pattern Recognition techniques used for Soft Inversion, do not *directly* exploit spatial correlation information. Rather, the goal is to *generate* approximate control information away from the wells, whose *implicit* spatial correlation characteristics may be utilized by Stochastic Imaging techniques such as Indicator simulation. As will be shown in section 3.1, the transition from the random vector Pattern Recognition models to the random function Stochastic Imaging models is possible under certain stationarity assumptions.

Soft Inversion consists of three main steps: representation, mode estimation, and statistical calibration. In the following sections, these steps will be described with reference to the example reservoir whose spatial survey map is illustrated in figure 1. As indicated in this figure, the reservoir interval is sampled by logs from 12 wells, and also by a 3D seismic survey within the enclosed polygonal area. Typically each well will have a suite of well-logs sampled at 1/2 foot vertical intervals (and less frequently, core measurements), while the seismic survey may have several (hundred) thousand *stacked* [21,22] seismic traces (each "trace" consisting of seismic amplitudes as a function of seismic travel time) at a spatial sampling of roughly 100' square. A typical seismic line from this example is illustrated in Figure 2, with the 200' thick reservoir interval located (in the 60 mSecs time interval) between the upper and lower seismic horizons.

2.1 REPRESENTATION

In the representation step, the seismic traces and the wells are each described by a set of attributes or "features". The feature vector 's', representing each seismic trace, may include the sequence of trace amplitudes within the reservoir interval, the instantaneous attributes of the trace [4], the trace power spectra [5], trace locations, etc. Similarly, each well may be represented by the *multidimensional distribution* ("scatter plot") of its well-logs (and when available, core measurements), within the reservoir interval. The well feature vector 'w' can be derived from a discrete sampling of this distribution, or from its selected moments such as the mean, covariance, skewness, kurtosis, etc. From the viewpoint of Pattern Recognition, the seismic and well features are each modeled as random vectors with associated probability distributions F(s) and F(w) respectively - the corresponding data sets are considered the result of *independent* sampling from these distributions.

The important point regarding the selection of these two sets of vector attributes is to ensure that they are indeed *informative* about the reservoir properties of interest and mutually consistent. In addition to statistical tests such as canonical correlation / spatial factor analysis [6,7], seismic modeling techniques may be used to confirm the utility of these multivariate representations [1]. Also, in practice, principal components analysis may be employed to identify feature redundancies, and thus reduce the intrinsic dimensionality of these representations.

2.2 MODE ESTIMATION

The result of the representation step is the set of seismic trace feature vectors S, and well feature vectors W - these two sets define the seismic and geological feature vector spaces. In general, the joint seismic-geological {S-W} space distribution will be *multimodal* - the underlying hypothesis being that these modes represent groups of similar seismic traces ("seismic facies"), and corresponding homogeneous geological facies. Soft Inversion is designed (as far as possible) to perform the seismic-geological calibration *separately* over each mode (facies) of this distribution. This mode by mode calibration facilitates the application of *linear* (or more precisely, *affine*) model based estimation techniques. Figure 5 illustrates this piecewise calibration process.

The modes of the joint S-W distribution may, in principle, be estimated directly in the joint S-W space by using multivariate cluster analysis algorithms [8]. In practice, since the sample size in the seismic space far exceeds that in the geological space, the cluster analysis is applied *separately* to the two spaces - as noted earlier, the seismic modes indicate traces of "similar character", while the geologic modes represent wells with "similar" logs. After cluster analysis, the seismic/geological clusters are evaluated for spatial and feature correspondence. Cluster pairs that "correspond" are assumed to be indicative of modes of the joint S-W distribution. In the case of sparse well coverage, geological cluster analysis may not be possible - the seismic clusters provide the main evidence of the S-W distribution modes.

It should be noted in passing that the cluster analysis algorithms do not directly exploit spatial correlation information - such information does however, exist *implicitly* in the feature representations. As an example, figures 3 and 4 illustrate clustering results for the seismic traces from the reservoir of figure 1, in the spatial and feature domains respectively. All four clusters exhibit spatial continuity - more precisely, seismic traces that are "close" (ie. "similar") in the seismic feature space are also likely to be "close" in the physical coordinate space. Indeed, the clusters represent a spatial partition of the data set into zones with *distinct* feature statistics such as the mean, covariance, etc.

As noted earlier, after cluster analysis, the seismic and geological clusters are *tested* for both spatial and feature correspondence. The spatial correspondence is evaluated by examining seismic/well cluster pairs for spatial overlap ie. whether the spatial region ("extent") containing all of the seismic cluster traces, overlaps significantly with the spatial region containing all of the wells in the well cluster. The feature correspondence between each seismic - well cluster pair is evaluated by means of either classical canonical correlation analysis [6], or spatial factor analysis [7,9,18]. Such techniques evaluate the multivariate correlation between spatially collocated seismic and geological feature vectors within each cluster pair, and also identify linear combinations of features ("canonical factors") that maximally correlate. Canonical correlation analysis assumes spatial independence, while spatial factor analysis incorporates spatial correlation information. A useful fallout of such analyses is the potential to flag clusters (geological facies) where seismic inversion is *not justified* by the available evidence. In the case of sparse well coverage, each seismic cluster is associated only with those well(s) that fall within its spatial extent. Geophysical modeling may be used when possible to further validate these cluster-cluster or cluster-well associations.

The final result of the mode estimation step is a set of seismic-geological cluster (well) pairs that exhibit significant spatial and feature correspondence. These pairs are hypothesized to represent modes of the joint S-W distribution, and are calibrated using the linearized statistical calibration techniques to be described below.

2.3 STATISTICAL CALIBRATION

Several strategies for Statistical Calibration are possible, depending on the extent of well control that is available. We consider first the case when well coverage is sparse - that is, well clusters containing significant numbers of wells are not available.

2.3.1 SPARSE WELL COVERAGE

In this case it is possible to calibrate only those seismic clusters that have wells *within* their spatial extent. Figure 6 illustrates a seismic cluster in the seismic feature space, and the *vertical* sample distribution (in the geological space) of a single calibrating well that falls within its spatial extent. For the purpose of explanation, we assume that each well is represented by the mean vector (vertical sample *averages*) of its vertical sample (marginal) distributions and that the seismic and geological spaces have dimensions N and M respectively. The extension to statistics other than sample averages is discussed in reference [18].

In Soft Inversion, the primary assumption is that the seismic attributes s, and the well attributes w (ie. the mean vector) are *approximately* related by an Affine transformation:

$$w \sim A\,s + b \tag{1}$$

where

A is a M x N matrix,
b is a N x 1 vector.

This Affine transformation may be viewed as the first order term of the Taylor expansion of w in terms of s. When considering parameters such as acoustic impedance, it can readily be shown that this approximation is consistent (see Appendix I) with linearized wave propagation models, based for instance on the Born approximation [1,18,21,22,23], with the matrix A representing (primarily) an averaged version of the seismic wavelet. In contrast to traditional seismic inversion however, the method developed here does not require the rather delicate estimation of this wavelet or indeed of the affine transform {A,b}. In a more general sense, the *validity* of approximation (1) is enhanced if we constrain its application to *unimodal* ("Gaussian like") subsets of the joint S-W parameter space. As noted earlier, this is in fact the goal of the mode estimation step.

Assuming sample independence, equation (1) implies that

$$E\{w\} = A\,E\{s\} + b \tag{2}$$
$$C(w) = A\,C(s)\,A' \tag{3}$$

where

E{x} = m(x) represents the expected value of the vector x
C(x) = the covariance matrix of x given by E {(x-m) (x-m)'}
A' = the matrix transpose of A.

Since the matrices C(w) and C(s) are symmetric, while A is in general nonsquare and nonsymmetric, equation (3) is under-determined for the matrix A, and thus admits an arbitrarily large set of solutions for A.

Let the generalized inverse [24] of A be denoted by the (N x M) matrix 'a'. Then the following properties hold:

$$A\,.\,a\,.\,A = A \qquad a\,.\,A\,.\,a = a \tag{4}$$

Equations (3) and (4) imply that

$$A\,.\,a\,.C(w) = A\,.\,a\,.\,A\,C(s)\,A' = A\,C(s)\,A'$$

$$A\,.\,a = A\,C(s)\,A'\,.\,c(w) = I \tag{5}$$

where c(.) represents the inverse of C(.), and I the identity matrix.

Also, the general solution to equation (1) is given by [24]:

$$s = a \cdot w - a \cdot b + [I - aA] \cdot z \quad (6)$$

$$m(s) = a \cdot m(w) - a \cdot b + [I - aA] \cdot z$$

where

z is an arbitrary (N x 1) vector.

Equation (6) implies that

$$C(s) = a \cdot C(w) \cdot a' \quad (7)$$

Therefore

$$A \cdot C(s) = C(w) \cdot a'$$

$$c(w) \cdot A = a' \cdot c(s)$$

$$c(w) = a' \cdot c(s) \cdot a \quad (8)$$

Now any seismic feature vector s lies on the surface of an ellipsoid in the seismic feature space:

$$(s - m(s))' \; [c(s)] \; (s - m(s)) = k(s) \quad (9)$$

where k(s) is a constant.

Substituting equation (6) in equation (9) we obtain

$$(w - m(w))' \; [a' \; c(s) \; a] \; (w - m(w)) = k(s) \quad (10)$$

or equivalently, using equation(8):

$$(w - m(w))' \; [c(w)] \; (w - m(w)) = k(s) \quad (11)$$

Equation (11) states that the corresponding well feature vector w (formed by applying the affine transform to s) also lies on the surface of an ellipsoid in the geological space (Fig 6). This "target" ellipsoid is *independent* of the specific affine transformation used. Its size increases as 's' moves away from the mean of the seismic space distribution.

The ellipsoid surface represented by equation (6) represents the locus of all possible geological feature values corresponding to the seismic feature vector s. In particular, the *distribution* of possible feature values is uniform and takes non-zero values only on the surface of this ellipsoid. This (symmetric) distribution represents the *soft estimate* of the geological parameters corresponding to the seismic trace s. A more conservative soft estimate in the form of inequality constraints, can be constructed simply by picking the *extremities* of the ellipsoid in the geological space coordinates. However, in this case, we must be careful to note that the inequality constraints for each geological parameter are not independent, these parameters being constrained to lie on the surface of equation (6) above. The *width* of these inequality constraints is zero when the vector s coincides with the mean of the seismic cluster, and increases as s moves away from the mean (see equation (2)). These inequality constraints *implicitly* contain spatial correlation information, based as they are on seismic (and when available, well) clusters that contain such information (see remarks in section 2.2, mode estimation).

REMARKS:

We conclude this section with three remarks:

1. The soft estimates can strictly be derived only if the mean and covariance of the vector w are available. However, in the case of sparse well coverage, we have access only to the vertical sample distribution of well-log sample values, not the distribution of *sample averages* w. Nevertheless, the statistical calibration procedure outlined above is applied directly to this vertical sample distribution. This approach is based on the *covering hypothesis*: i.e. the distribution of the vector w of vertical sample averages is "enclosed" by the *vertical sample distribution*. The covering assumption is illustrated in figure 5. Extensive testing with real data suggests that this assumption is reasonable, provided that the reservoir zone sampled by the calibrating well is *representative* of the geological facies identified by the seismic cluster (seismic facies). Specifically, experience indicates that the closer the "well trace" (ie. the average trace in the neighborhood of the well) is to the mean of the seismic cluster, the greater the confidence that may be placed on the covering hypothesis. The price paid for the covering assumption is that the "soft" estimates that we obtain will be conservative - however, preliminary tests with real data indicate that, if the w distribution is primarily related to the vertical sample distribution by a scale factor, then spatial correlation models inferred from these estimates will be "correct" to within a constant (possibly estimable} scale factor [18].

2. Each soft estimate may be *weighted* by a measure of the degree to which the calibrating well *represents* (in the sense of remark 1 above) the seismic facies. As noted above, one such measure is the distance of the "well-trace" from the seismic cluster mean. When only a single calibrating well is available, *consistency* of the soft estimates requires that the well-trace be close to the seismic cluster mean, since then the inequality constraints have ~zero width and are centered at the mean property values at the well (see equation (2) above). When the well trace is not close to the seismic cluster mean, the mean of the soft estimates is likely to be biassed. In this case, in practice, it is desirable to further split the seismic cluster into *subclusters* (either by reclustering or by using pattern classification techniques [18]), and calibrate only the subcluster whose mean is indeed close to the well-trace. When several such representative calibrating wells are available for a seismic (sub)cluster, each (sub)cluster sample may be calibrated *separately* to each of these wells - the representation weights may be used to construct a weighted average (using representation measures normalized to [0,1] over the set of wells) of the soft estimates derived from each well. An ancillary result of such a weighting

process is that the seismic (sub)cluster mean now maps to a *weighted mean* over the calibrating wells.

3. The *reliability* of the soft estimates is of course governed by the validity of the affine approximation, and of the covering hypothesis mentioned above. Both the reliability and consistency of the soft estimates depend on the degree to which each calibrating well is *representative* of the seismic facies. It is appropriate to reemphasize that the validity of each of these assumptions is enhanced by the mode estimation step, since it limits their applicability to homogeneous subsets of the joint parameter space. It should be also be pointed out that the soft estimates do not *explicitly* use the affine approximation - indeed, the technique has a "maximum entropy" flavor in that the estimates are least committal with respect to the affine mapping. Finally, it may be noted that it is in fact the above assumptions that "substitute" for the lack of spatial cross-correlation information, particularly with sparse well coverage.

2.3.2 EXTENSIVE WELL COVERAGE

We now consider the case when we have a cluster of several wells that has significant "correspondence" (both spatially, and in the feature space) with the seismic cluster. The representation measure weighted approach outlined in section 2.4.1 is still applicable. However, if the number of wells in this cluster is deemed adequate to estimate the mean and covariance of the sample averages w, the affine calibration process can be applied directly to these averages [16,17]. Since the w distribution is used directly, the covering hypothesis is not required, and this approach will produce tighter (albeit less conservative) soft estimates as compared to those obtained from single well calibrations.

It may be noted in passing that with extensive well coverage, it becomes possible to apply Cokriging / CoSimulation directly to each seismic/geological cluster pair in the joint S-W space, albeit with significant structural analysis and computational effort [12,18]. Alternately, classical multivariate regression may be employed to produce both single valued estimates and parametric confidence intervals - at the expense of ignoring spatial correlation information [19].

3. STOCHASTIC IMAGING

Soft Inversion allows the inference of distributions (and/or inequality constraints) on reservoir properties "away from the wells", even when well coverage is sparse. As noted earlier, this inference process for the most part, does not directly exploit spatial correlation information. In the hybrid methodology presented here, Stochastic Imaging techniques are employed for precisely this purpose, while allowing for the quantification of uncertainty. The soft estimates have two potential uses for Stochastic Imaging: first, they offer the possibility of constraining the stochastic images between the wells; second, they provide crucial "samples" for structural analysis, particularly when well coverage is sparse. Also, since these estimates are directly in terms of the reservoir parameters, the problem of modeling cross-correlations *between* the seismic-geological spaces is effectively sidestepped.

In this section we first review the conditions under which results based on the random vector models underlying Soft Inversion are *compatible* with the random function models underlying Stochastic Imaging. This is followed by a review of two Stochastic Imaging algorithms that are suited to imaging the soft estimates.

3.1 COMPATIBILITY OF SOFT INVERSION WITH STOCHASTIC IMAGING

In Soft Inversion, the underlying statistical model is based on the distributions F(s) and F(w), of the seismic and geological space feature vectors. The data samples in these two spaces are assumed to be drawn *independently* from these two distributions. The basic strategy is to partition the joint S-W space into homogeneous modes, and then perform the statistical calibration separately over each mode, to produce the soft estimates on reservoir properties.

In order to use these soft estimates in Stochastic Imaging, we need to make the (conceptual) transition from a single random vector model, to a random vector function (ensemble of random vectors) model. Such a transition may be effected, by making the following two hypotheses:

(1) The spatial extent of each seismic cluster CL(j) defines a separate region of (2nd order) stationarity. Each sample in the seismic cluster is now viewed as a *realization* of a separate random vector.

(2) Within this region, each seismic random vector has the (stationary) first order distribution F{s / CL(j)} as determined from the seismic cluster analysis, with corresponding (stationary) first order distribution F{w / CL(j)} for the collocated geological random vector.

Under these stationarity hypotheses, each random vector in the seismic cluster may still be hypothesized to be related to the geological vector by an affine transformation and the soft estimates be derived exactly as before. Note that although each seismic feature vector may be related to its collocated geological vector by *distinct* affine mappings, this will not affect the soft estimates as they are transparent to specific mappings, and depend only on the mean and covariance of the two conditional distributions referred to above.

To summarize, *compatibility* of the Soft Inversion results with the Stochastic Imaging algorithms can be ensured by requiring that each seismic cluster be imaged separately under its own stationarity hypothesis.

3.2 STOCHASTIC IMAGING ALGORITHMS

There are at least two stochastic imaging algorithms that could utilize the soft estimates: Indicator Simulation [10], and the Markov_Bayes algorithm [3]. In the case of Indicator simulation, the inequality constraints are coded as vectors of "hard" (ie. 0/1/undefined) Indicators. In the case of the Markov-Bayes algorithm, the (soft estimate) distributions on reservoir parameters and associated representation weights are used to construct approximations [18] to the prior (or "pre-posterior") cdfs, which are coded as "soft" indicators (ie. in the closed interval [0,1]).

3.2.1 INDICATOR SIMULATION

The technique of Indicator simulation [10], generates approximations to the local conditional expectations, (and therefore to the local Bayesian posterior distribution), by simple kriging of the "hard" indicator data. The image samples are drawn sequentially across the simulation grid from these local distribution approximations. The technique is quite flexible, but does require considerable effort for structural analysis, particularly with several cutoffs; also the computational burden is not negligible. These problems may be alleviated somewhat by the use of principal component methods such as those proposed by V. Suro-Perez [11], and by the use of direct Bayesian imaging algorithms [12,18]. As with most stochastic imaging algorithms, Indicator simulation does assume weak stationarity of the underlying random process; also, as has recently pointed out by Matheron [13], the internal consistency of indicator covariance models is not guaranteed. With regard to stationarity, it may be pointed out that the process of mode estimation in Soft Inversion may serve to (implicitly) partition the data set into zones that might better honor this assumption. In practice, the problem of internal consistency can be kept manageable by using heuristic "order relations" corrections as proposed by Journel [14].

3.2.2 THE MARKOV-BAYES ALGORITHM [3]

As noted above, it is possible to construct approximations [18] to the prior cdf at each seismic trace location using the results of Soft Inversion. In general, when well coverage is extensive, such prior cdfs might be also inferred directly from well-logs and/or local qualitative/expert information. These cdf's may be coded as "soft" indicators in the interval [0.1]. The recently developed Markov-Bayes algorithm allows for the stochastic imaging of both hard indicators and prior cdfs coded as "soft" indicators. This is accomplished by treating the hard and soft indicators as separate RF's that are jointly simulated by using CoKriging to estimate the local conditional distribution approximations. By using the so-called "Markov hypothesis" that "hard" indicators prevail over collocated "soft" indicators, the effort required for inference of the hard-soft spatial cross covariances is significantly reduced, although this process does still require significant amounts of collocated hard and soft indicator data. The formulation of separate RF's for hard and soft indicators, also makes it possible to update the prior cdf's based on neighboring hard and soft indicator information. The algorithm as presently formulated is univariate - generalizations to vector seismic and geological parameters appear to be possible.

Both approaches, Indicator Simulation and the Markov-Bayes algorithm, should preferably be applied (depending on the sample size, and on stationarity criteria) to each seismic-geological cluster (well) pair in sequence, for the reasons summarized in section 3.1 above. However, for the results reported in this paper, only the Indicator Simulation approach has been evaluated on the example reservoir application, and has produced satisfactory results even though it has been applied to the entire seismic data set, rather than to each cluster pair. This evaluation is described in the next section.

4. APPLICATION OF METHODOLOGY

The Soft Inversion / Stochastic imaging methodology has been evaluated on at least three different reservoir description problems to date. In this section we describe the results of an application of the methodology to mapping porosity in a dolomitic reservoir [15,16], that was sampled by a dozen wells and by a 3D seismic survey (see Figure 1). The reservoir interval in this example was ~ 200' thick, while exhibiting significant lateral variability in porosity. The dozen wells available each had density, P-wave velocity, and interpreted porosity logs available over the reservoir interval. The seismic data was available in the form of ~ 100,000 stacked traces, and a tracked horizon dilineating the top of the reservoir was available. The reservoir interval spanned 60 mSecs on the seismic section (see Figure 2).

Each seismic trace was *represented* by 15 amplitude samples over the 60 mSec interval. A principal components analysis on the seismic trace data revealed 6 significant principal components with the first three components capturing more than 85% of the data variability. A cluster analysis using density estimation techniques, detected 4 major clusters. These clusters are illustrated in the spatial domain in figure 6, and in the principal component domain in Figure 7. As noted earlier, these four clusters exhibit significant lateral continuity. The green cluster primarily represents the dolomitic facies, while the black cluster identifies known faults in this reservoir zone. Each of the red, blue, and green clusters contains at least one well in its spatial extent. In particular, the average seismic traces in the neighborhood of these calibrating wells were "close" to the red and green cluster means. The affine calibration technique was used to construct *soft* estimates on average porosity (and velocity , density, acoustic impedance) at each of the seismic trace locations in these clusters. Figures 8 and 9 illustrate spatial maps of the porosity soft estimates coded by width of the inequality constraints for the red and green clusters. In particular, there are a significant number (~1000) relatively narrow (< 5% porosity) bounds as coded in black in these figures. The accuracy of these bounds (ie. whether the "true" value falls within the constraint interval), has been verified at a limited number of "blind" test wells.

The small number of wells precluded the inference of reliable "hard" data spatial covariance models. In contrast, the large number of soft estimates does indeed provide crucial control information in between the wells. The soft porosity estimates were coded into 5 Indicator levels or cutoffs. Figure 10 illustrates spatial correlograms for the cutoffs at 9%, 11%, 13%, and 15% porosity. The indicator correlograms exhibit significant correlation lengths of ~10000' for the first three cutoffs - the 15% cutoff indicates far less spatial correlation than the first three cutoffs.

Figures 11 illustrates two realizations (using the SIS algorithm, [10]) of average porosity using highly approximate indicator covariances based on just the "hard" data from 12 wells. The variability in the two realizations is significant, indicating lack of sufficient constraining information. Figures 12 illustrates the same two realizations using the soft data - the variability of the images is significantly reduced particularly within the polygonal area dilineating the seismic survey and associated soft estimates. The average over 100 realizations for both hard and soft data runs are illustrated in Figures 13-

once again, the soft data realizations have clearly identified the known high porosity zones in the middle of the seismic survey area. Figure 14 illustrates a maximum probability map - ie. at each trace location, the porosity value corresponding to the maximum of the emperical local histogram (derived from 100 porosity realizations) is displayed. Figure 15 illustrates maps of the probability that porosity at a given trace location falls within a specified interval. (It should be noted in passing that both figures 14 and 15 illustrate characteristics of the emperical *local* porosity histograms - this is to be contrasted with characteristics of the *joint* distribution of any spatial subset of porosity sample values). The subset of these figures corresponding to the use of soft data, confirms the presence of the high porosity zones , while showing significant detail particularly within the polygonal seismic survey area.

5. CONCLUDING REMARKS

The test results reported here and known to have been conducted in other studies elsewhere, confirm the utility of the Soft Inversion / Stochastic Imaging methodology. Based on these tests, it is reasonable to conclude that these methods are both computationally feasible and of significant practical value, provided that the seismic/petrophysical data are indeed representative of the reservoir properties of interest. Specifically, the methodology does indeed (a) generate *useful* local soft control information on average reservoir properties in between the wells, while (b) providing for the assessment of uncertainty via stochastic imaging; in particular, the task of structural analysis is reduced to that of modeling spatial correlations *only* in the geological space. In addition to its *multivariate* capability, the methodology has the critical advantage of flagging seismic clusters that fail the correspondence tests, and thus identify subsets of the seismic data where *inversion is not justified* by the available evidence. On the debit side, one of the primary current limitations of the methodology is the lack of adequate performance measures on the soft estimates, particularly in sparse well settings - such performance measures might be derived from additional a priori information based on the underlying physics / geology.

Future research directions include extensions of Soft Inversion to non-linear mappings and to pre-stack seismic data [18,21], the development of robust performance measures in sparse well settings, further evaluation of the feasibility of using soft indicators generated by Soft Inversion in stochastic imaging, multivariate generalizations of the Markov-Bayes algorithm using Pattern Recognition concepts, and further incorporation of the physics into the estimation process.

6. ACKNOWLEDGEMENTS

Early development of these techniques occured during the author's tenure at British Petroleum (see [15]) in collaboration with Francois Alabert (ELF) and Prof. Journel (Stanford University). The author wishes to express his gratitude to these individuals for several illuminating and enjoyable discussions, to Francois for his work in generating some of the early test simulation results, and to BP for providing the positive environment that facilitated the development of these techniques. Some of the test results in this paper were first presented at the Stanford SCRF meeting in June 1989, and at the SEG annual meeting in November 1987.

REFERENCES:

1. Oldenburg, D. W., Levy, S., (1983), Recovery of the Acoustic Impedance from reflection seismograms, Geophysics, vol 48, pp 1318-1337.

2. Doyen, P.M., Porosity from Seismic Data, (1988), A geostatistical approach, Geophysics, vol 53, pp 1263-1275.

3. Journel, A.G., and Zhu, H., May 1990, Markov-Bayes updating, an alternative to cokriging and traditional regression, Stanford Center for Reservoir Forecasting.

4 Tanner, M.T., and Sheriff, R.E., Application of amplitude, frequency, and other attributes to stratigraphic and hydrocarbon detection, AAPG Memoir No. 26, C.E.Payton, Ed.

5. Khattri, K.N., Sinhval, A., Sinhval, H., and Awasthi, A.K., Seismic Indicators of Stratigraphy, Handbook of Geophysical Exploration, vol 20, Klaus Helbig and Sven Treitel, Editors, pp225-26.

6. Kshirsagar, A.M., Multivariate Analysis, (1972) M. Dekker, New York.

7. Grunsky, E.C., Spatial Factor Analysis: A technique to assess the spatial relationships of multivariate data, Statistical Applications in the Earth Sciences, Geological Survey of Canada, p 329-347.

8. Hartigan, J., Clustering Algorithms,(1975), Wiley, New York.

9. Royer, J.J., Proximity Analysis, A method for multivariate geodata processing: Sciences de la Terre, Informatique Geologique, no. 20, pt. 1: Universite de Nancy, France, pp 223-243.

10. Journel, A.G., and Alabert, F., (1988) Focussing on spatial connectivity of extreme valued attributes: stochastic indicator models of reservoir heterogeneities, S.P.E. paper no. 18324.

11. Suro-Perez, V., Joint Characterization and simulation of multiple reservoir properties: the IPCK Approach, (May 1991) Stanford Center for Reservoir Forecasting.

12. Thadani, S.G., A direct Bayesian algorithm for the Soft Inversion of Seismic Data, (1992) EAEG , Paris, France.

13. Matheron, G., The Internal Consistency of Models in Geostatistics Geostatistics, (1989) Kluwer Academic Publishers, vol 1, pp 21-38.

14. Journel, A.G., Comment on "Positive Definiteness Is Not Enough" (1992) Math. Geology, vol 24, no. 1.

15. Thadani, S.G., Alabert, F., Journel, A.G., An integrated Geostatistical /Pattern Recognition technique for the characterization of reservoir spatial variability, SEG 1987, New Orleans, La.

16. Alabert, F. A., Thadani, S.G., Reservoir Porosity Mapping using Soft Data inferred from seismic: Two applications of the Indicator Simulation technique, Stanford Center for Reservoir Forecasting, Report no. 2, May 1989.

17. Fournier, F., Extraction of Quantitative Geological Information with Multidimensional Statistical Analyses, SEG 1989, Dallas, Texas.

18. Thadani, S.G., Reservoir Characterization using Pattern Recognition and Spatial Statistics, paper in preparation for publication in Geophysics, 1992.

19. Fournier, F., Derain, J., A Statistical Methodology for Calibrating a 2D Seismic data set, EAEG 1992, Paris, France.

20. Hernandez, J.,G., Joint Sequential Simulation of MultiGaussian Fields, 4'th Internatrional Geostatistics Congress, TROIA, Portugal, 1992.

21. Aki, K and Richards, P.G., Quantitative Seismology, Theory and Methods, (1980) Volume II, W.H.Freeman and Co, San Francisco.

22. Waters, K.H., Reflection Seismology, (1981) Wiley-Interscience, New York.

23. Santosa, F., and Symes, W,W., Linear Inversion of Band-Limited Reflection Seismograms, (1986) SIAM Journal of Scientific and Statistical Computing, Volume 7, No. 4.

24. Golub, G.H., and Van Loan, C.F., Matrix Computations, (1983) John Hopkins University Press, Baltimore, Marland.

APPENDIX I

Relationship of the Affine Mapping to linearized models of seismic data

In linearized 1D models of wave propagation derived, for instance, from the Born approximation, the seismic data (say, in a specific time interval) is expressed by the discretized matrix equation [23]:

$$d \sim G \cdot m + h \qquad (1)$$

where

d = (N x 1) vector of seismic "data" samples in the (time) interval of interest

m = (M x 1) vector of "model" parameter values (say, acoustic impedance)
h = (N x 1) constant vector representing the contribution of subsurface zones outside the interval of interest
G = (N x M) matrix "kernel" (seismic wavelet in the case of the 1D convolution model [21,22]

In general the matrix kernel G is rank deficient - as noted by Backus and Gilbert [21] and others, this implies that with linear estimators, only certain unique "averages" of the model parameters 'm', may reliably be extracted from the data. These forms of these unique averages may be derived if detailed knowledge of the matrix kernel G is available.

However, in practical reservoir characterization applications, G is both inadequately defined and varies spatially, so that the above "averages" are not readily accessible. Following the Backus-Gilbert approach, we would still like to identify "averages" that are estimable over the *class of possible kernels* G. Indeed, the affine mapping model that was presented in section 2 is based on the hypothesis that one such average is the simple arithmetic average of the model parameters in the reservoir interval. In this appendix we show that, provided the linearized model of equation (1) is a valid approximation, the affine mapping model can in principle be expected to identify the arithmetic average reasonably well, without requiring specific knowledge of G, and even when G is rank deficient. In this sense, the affine model is "consistent" with the linearized models.

Consider the affine mapping model as an estimator of the arithmetic average:

$$Avg(m) \sim A' . d + b \tag{2}$$

where

A is a (N x 1) vector and b is a (M x 1) vector

$$Avg(m) = f' . m \tag{3}$$

$$f' = (M \times 1) \text{ row vector } = [1/M \ 1/M \ \ldots \ 1/M]$$

Equations (1), (2), and (3) imply that

$$A' \ [G\, m + H] + b \sim f' . m \tag{4}$$

Equating coefficients of like terms in equation (10) we have:

$$b = - A' H$$

$$G'A = f \tag{5}$$

Since G is in general rank deficient, we consider the generalized inverse estimate for A [21,24]:

$$A \sim g(G') . f$$

where

$$G' = U . Q. V'$$

$g(G')$	=	$U' . q . V$
Q	=	(p x p) diagonal matrix of the 'p' nonzero eigenvalues of G' and 'q' its inverse
U	=	(N x p) orthogonal matrix
V	=	(M x p) orthogonal matrix

Substituting this estimate of A in equation (1) we have:

$$Avg(m) \sim f' . g(G') . G. m$$

$$= f' V V' m \qquad (6)$$

$$R = V . V' \qquad (7)$$

In the parlance of linear inverse theory, the matrix $R = V.V'$ is known as the "resolving kernel" or "resolution matrix [21].

At one extreme, if G is full rank then $V.V' = I$ (the identity matrix), and the affine model can identify the arithmetic average exactly.

At the other extreme, if G is of rank = 1, it can be shown [21] that the resolving kernel degenerates to the ~ constant matrix

$$R = \{rij\}$$

$$rij \sim 1/M \qquad \text{for all } i,j = 1,2, \ldots, M \qquad (8)$$

Substituting this matrix in equation (6) shows that once again the arithmetic average may be recovered with negligible error.

In the intermediate cases as rank(G) increases from 1 to M, the rows of the resolution matrix construct weighted (convex) averages of the model parameters via the product [VV'. m]. Each row of weights may be shown to be a discrete approximation to a delta function [21] - these approximations progressively "narrow" as rank(G) approaches M (ie. when R becomes the identity matrix). The affine model in this case produces *arithmetic averages of these weighted averages* via equation (6). Since the weighted averages are dominated by the value at the middle of the averaging window, these arithmetic averages are expected to be close to the arithmetic averages of the model parameters.

In summary, when linearized wave equation models are a valid approximation, the affine model can be expected to provide robust (if not exact) estimates of the *arithmetic averages* of the model parameters, independent of the specific underlying kernel G and even when G is rank deficient. Indeed, these remarks are valid for any linearized model (1D/3D) relating reservoir properties to seismic response (for instance, those relating porosity to seismic response).

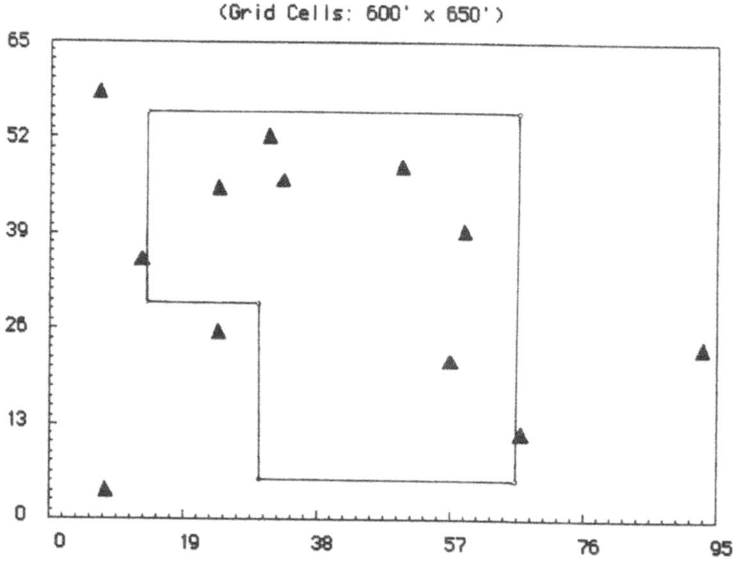

Fig. 1: Reservoir survey area

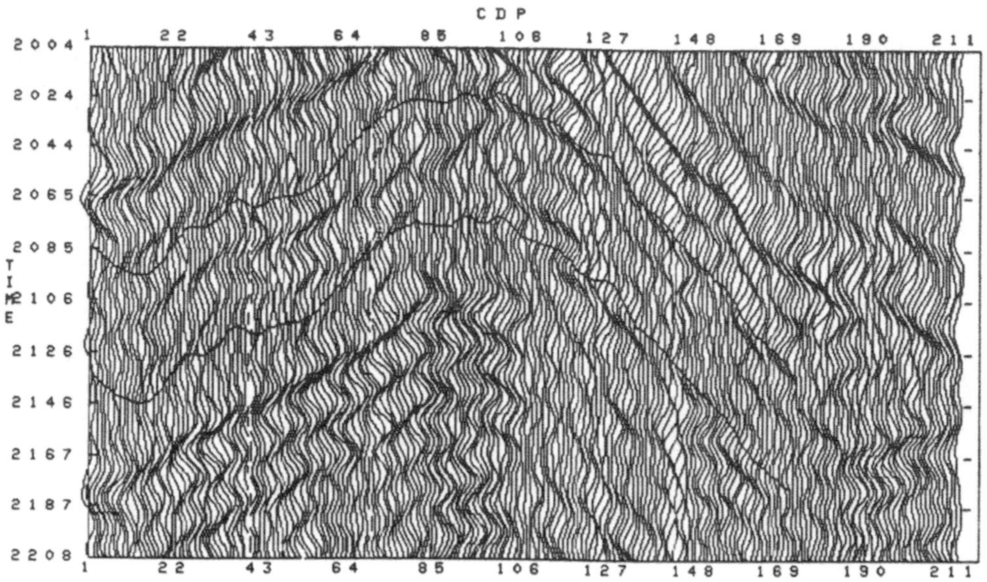

Fig. 2: A seismic line with tracked reservoir horizons

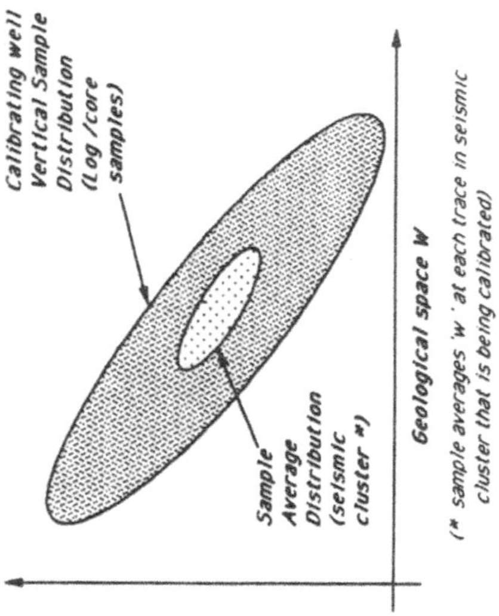

Fig. 7 Soft Inversion: The Covering Hypothesis

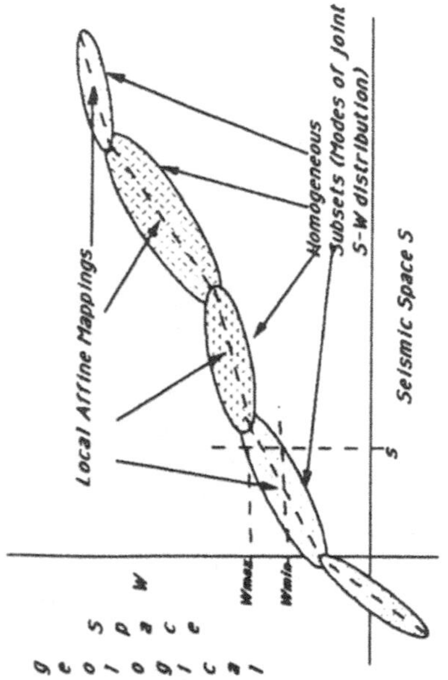

Fig. 5. Soft Inversion:
A piecewise calibration approach

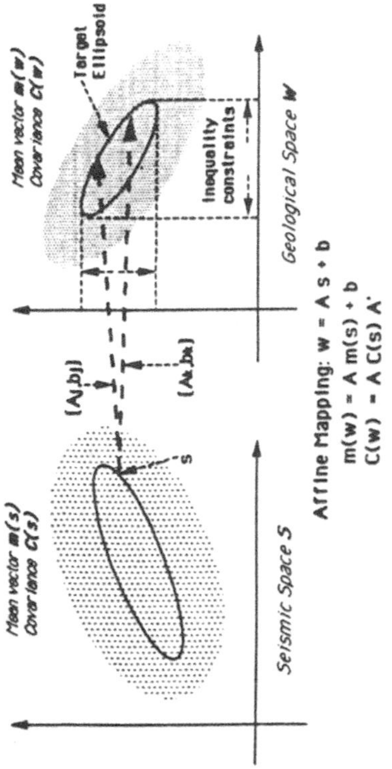

Figure 6: Statistical Calibration with Affine Mappings

[(Aj,bj), and (Ak,bk) are two possible solutions of the mean and covariance equations]

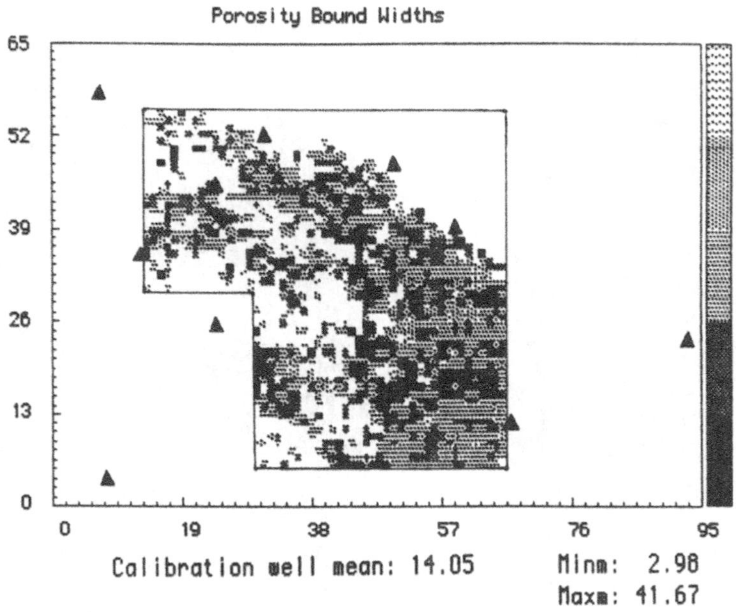

Fig. 8: Green (dolomitic) cluster calibration

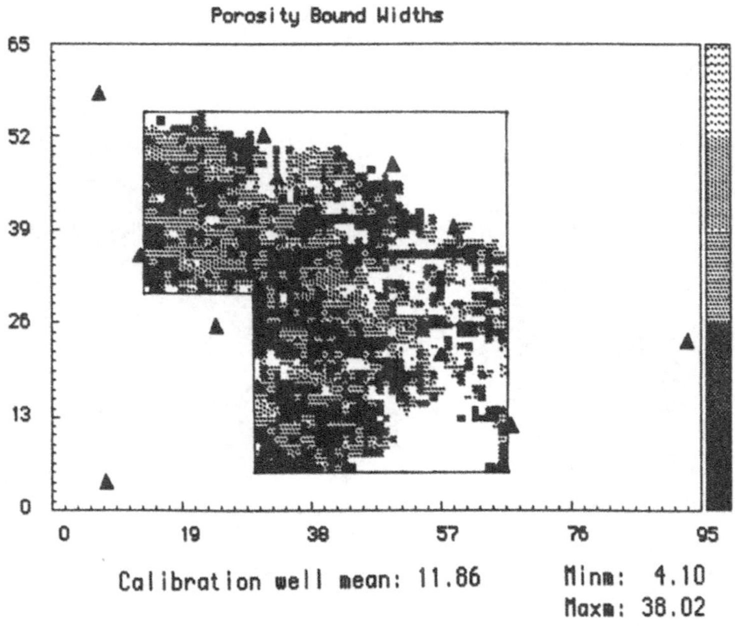

Fig. 9: Red Cluster Calibration

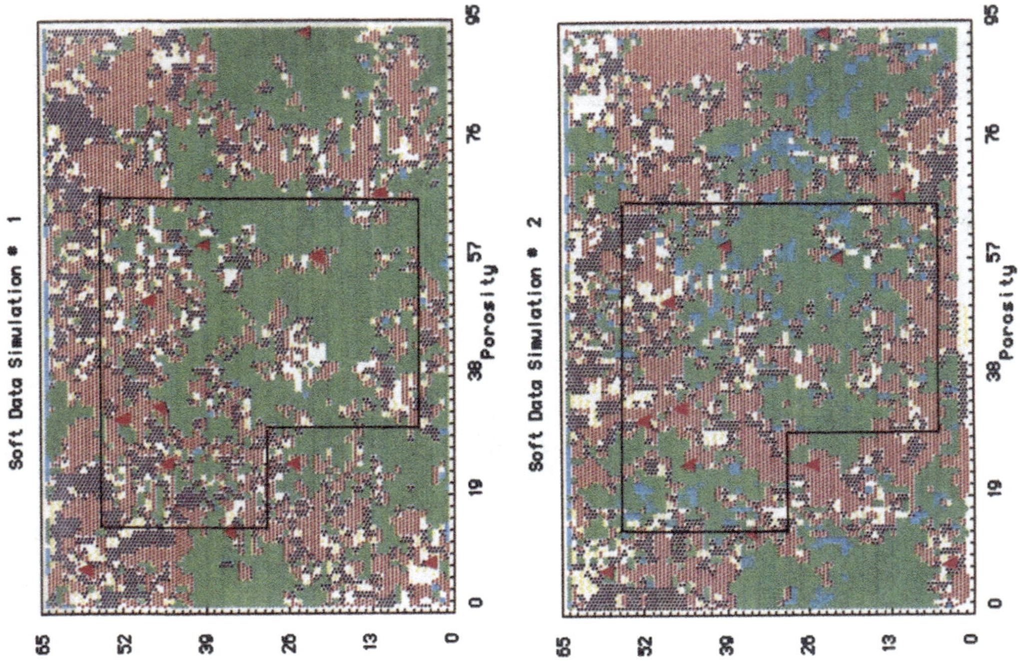

Fig. 12: Indicator simulations: hard and soft data

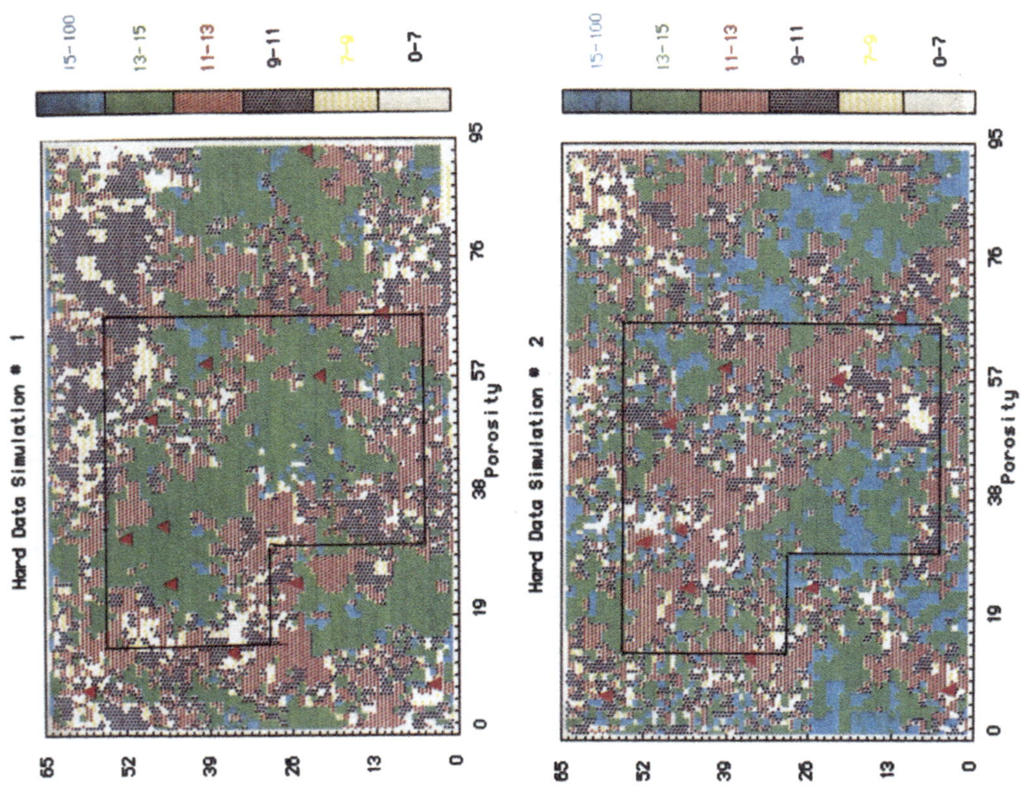

Fig. 11: Indicator simulations: hard data only

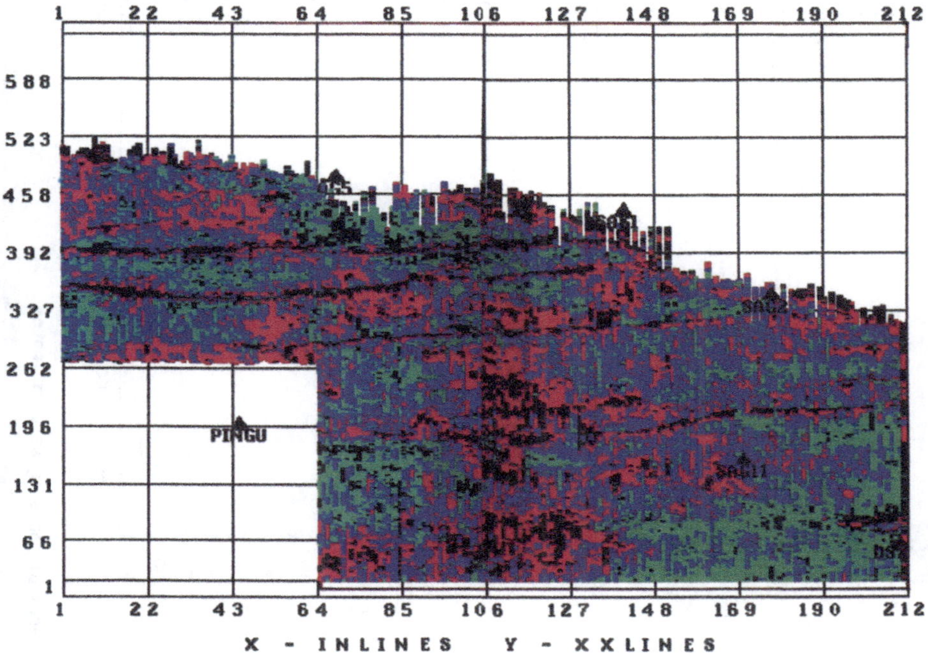

Fig. 3: Spatial cluster map of seismic data

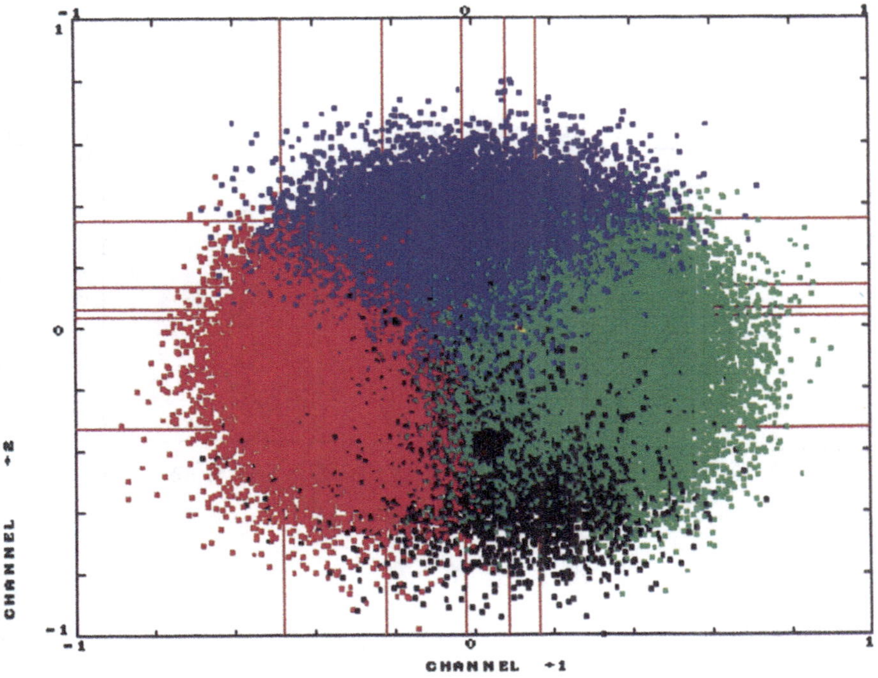

Fig. 4: Feature space cluster map of seismic data
(first two principal components)

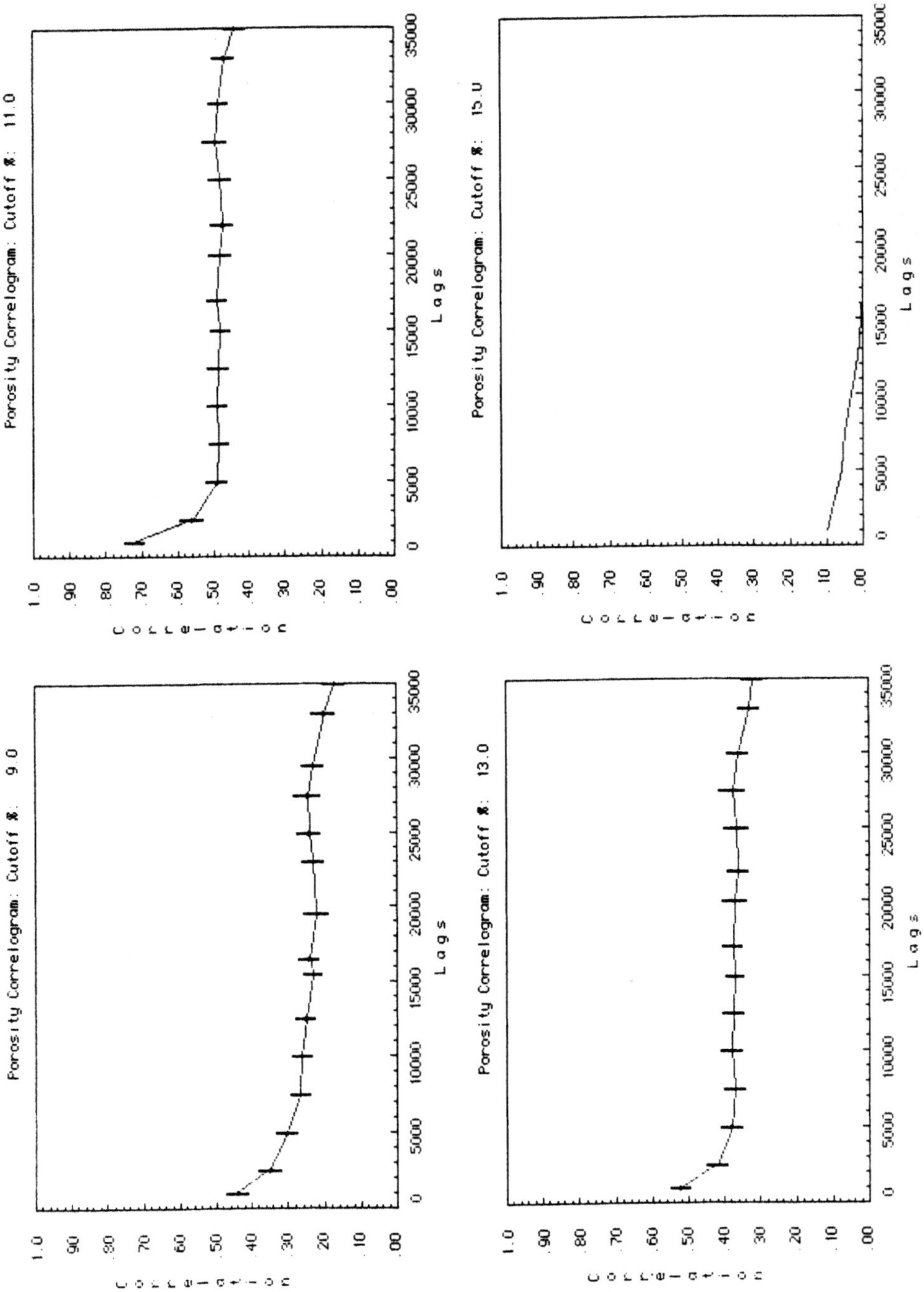

Fig. 10: Indicator correlograms of porosity soft estimates

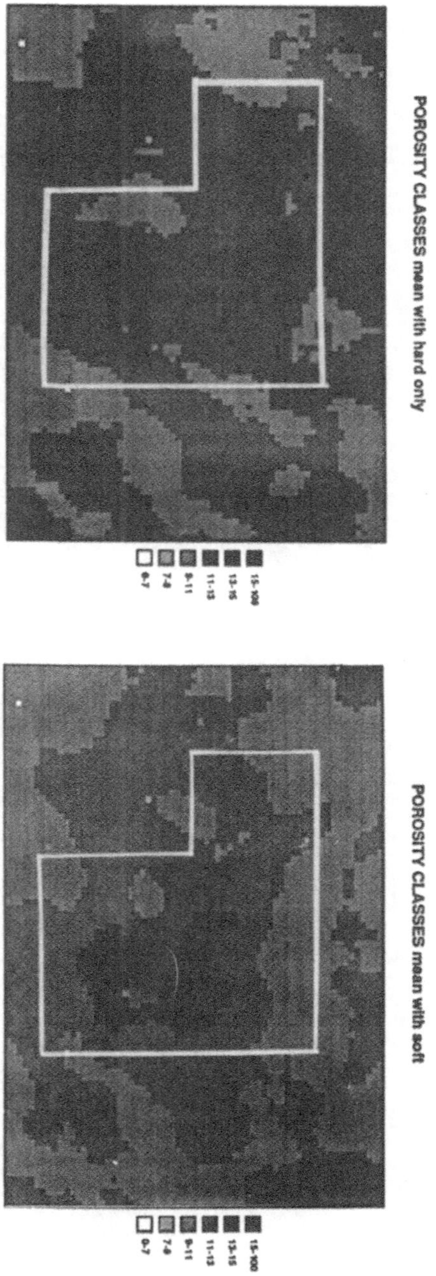

Fig. 13: Indicator simulations: Average (kriged) map

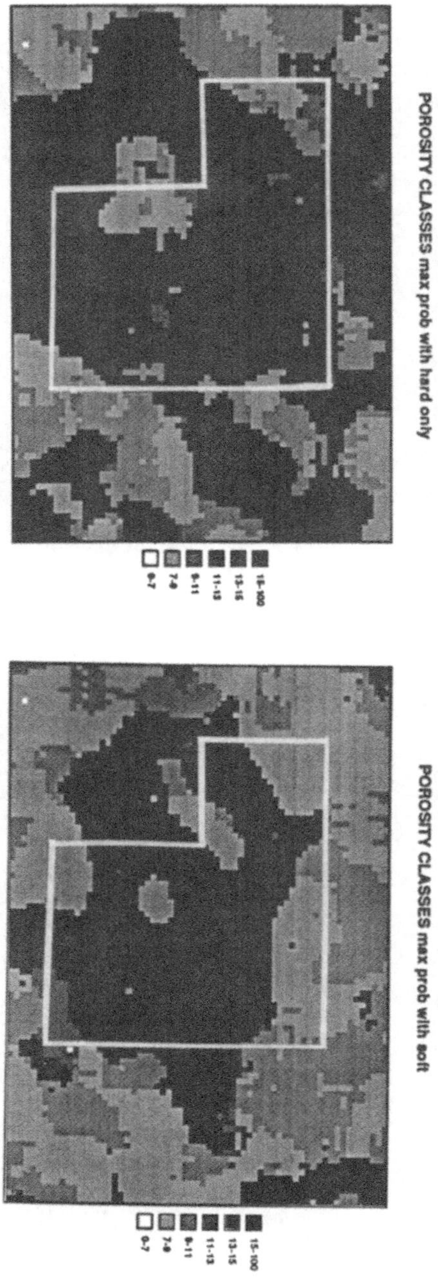

Fig. 14: Indicator simulations: Maxm probability map

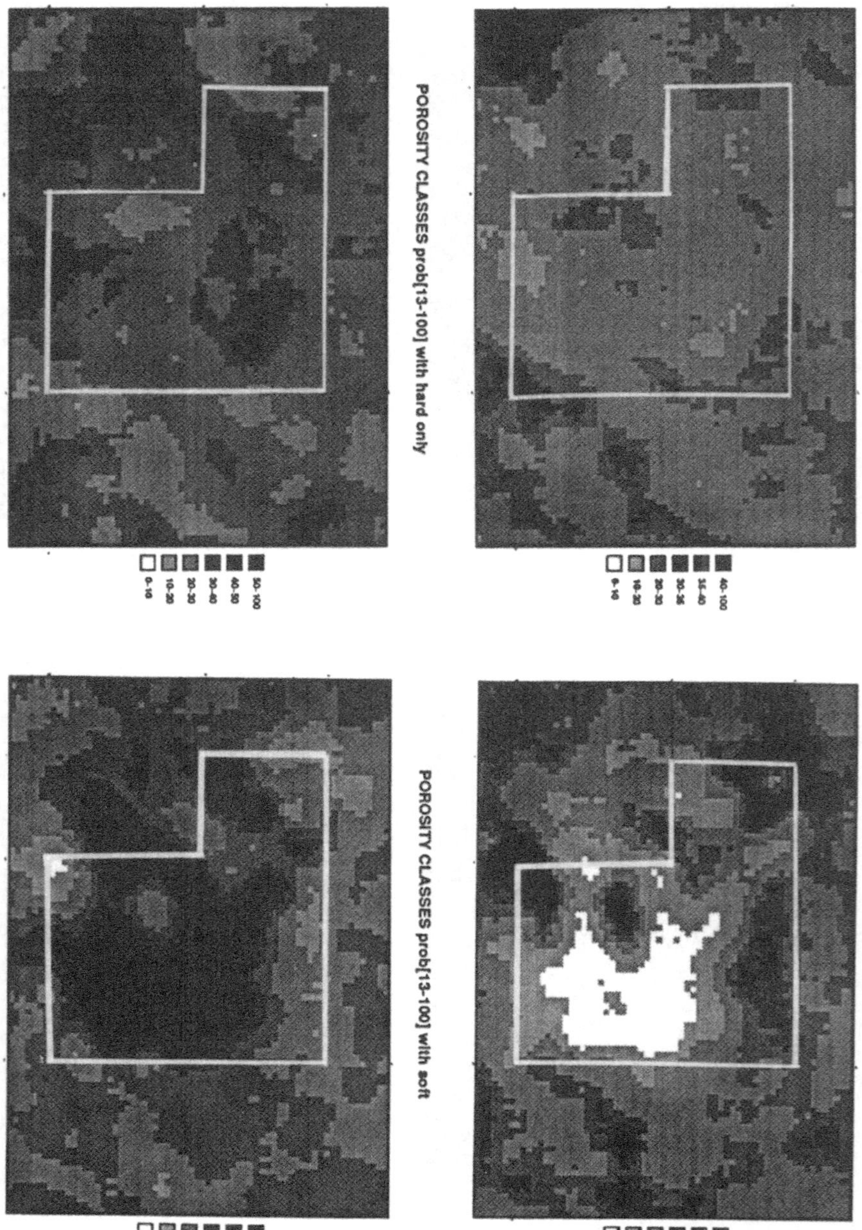

Fig. 15: Indicator simulations: Probability maps

Sequential Gaussian Cosimulation: A Simulation Method Integrating Several Types of Information

G.W. Verly

BP Research Centre, Chertsey Road, Sunbury-on-Thames
Middlesex, TW16 7LN, UK.

Summary

An important task in reservoir description is the integration different types of information in order to improve predictions of recoverable reserves. Sequential gaussian cosimulation is a method which produces a series of equiprobable models of one or more continuous properties by incorporating different types of information. Two applications of the method are: 1) the joint simulation of several properties such as horizontal and vertical permeabilities; and 2) the simulation of one property, e.g. porosity, in the light of two types of information, e.g. core measurements and seismic data. The method is simple and flexible. It reproduces the frequency distributions, and the auto- and cross-correlation functions of the properties involved. It also honours the property values where they are known. In this paper, sequential gaussian cosimulation is presented and illustrated by means of a case study. Some practical implementation aspects are also discussed.

1. Introduction

An important problem in the petroleum industry is the integration of several types of data when modelling reservoir heterogeneities. There are two aspects to the problem: firstly, heterogeneities must be reproduced, and secondly, information from different sources must be incorporated.

The first aspect, reproducing heterogeneities, has been the subject of research for several years. The most popular solution is stochastic simulation because it reproduces spatial variability. Several methods have been suggested to generate discrete or continuous variables, such as lithotypes or porosities. These methods include: sequential indicator simulation (Alabert, 1987; Journel and Alabert, 1988); truncated random functions (Isaaks, 1984; Matheron et al., 1987), sequential gaussian simulation (Isaaks, 1990; Verly, 1991); and boolean models (Haldorsen, 1988).

The second aspect, integrating data, is a more recent research topic. Some of the suggested solutions include: external drift (Marechal, 1984); bayesian kriging (Omre, 1987); maximum a posteriori estimation (Doyen et al., 1989); pattern recognition (Thadani et al., 1987); and Markov Bayes updating(Journel and Zhu, 1990).

Sequential gaussian cosimulation (SGCS), an extension of sequential gaussian simulation, is proposed as a mean for simulating models of several continuous variables. It reproduces the distributions and the auto- and cross-variograms of the variables together with their values where they are known. Two important applications are: 1) using seismic absolute acoustic impedance to simulate porosities; and 2) simulating jointly vertical and horizontal permeabilities. The method is simple.

It is also integrated in the sense that important aspects of the simulation – multiple correlation structures, anisotropies, conditioning, regular or irregular grids – are all combined in a single process. Finally, it is general in the sense that it can be applied to different types of problems. All it needs in practice are correlated variables.

The concept of cosimulation is not new in geostatistics. The usual approach is ,however, arduous. Firstly, independent isotropic fields are generated. Anisotropy is then built in by compressing or stretching the simulated grids. Multiple correlation structures are generated by adding grids. Cross-correlation results from multiplication. Finally conditioning is imposed by kriging. SGCS, because it is simple and integrates all these steps in one process, has a better chance of being applied routinely.

In this paper, the theory and practice of SGCS are described. The advantages and disadvantages are also discussed. The method is then used to simulate porosities using absolute acoustic impedance in a real North Sea reservoir.

2. Theory

Consider a set a K variables, $[z_k(x), k = 1 \text{ to } K]$ where x denotes a location in space. If $K = 2$, $z_1(x_0)$ and $z_2(x_0)$ represent for example the acoustic impedance and the porosity at location x_0. The problem amounts to generating spatial models of these variables such that: 1) some statistical properties of the variables taken individually or jointly are reproduced; and 2) the values of the variables are reproduced where they are known.

Geostatistics interprets the K variables $[z_k(x), k = 1 \text{ to } K]$ as realizations of K random variables, denoted by $[Z_k(x), k = 1 \text{ to } K]$. This set of random variables in space constitutes K random functions, also denoted by $[Z_k(x), k = 1 \text{ to } K]$. Within this framework, the problem amounts to generating joint realizations of the K random functions such that: 1) the distribution and variogram of each random function are reproduced; 2) the cross-variograms between the random functions are reproduced; and 3) the sample values are honoured.

The first step consists in assuming that each random function $Z_k(x)$ is strictly stationary and that there exists a random function $Y_k(x)$ such that (Journel and Huijbregts, 1978):

$$Y_k(x) = \phi_k[Z_k(x)], \quad \text{and} \quad Y_k(x) \sim N(0,1),$$

where the gaussian transform ϕ_k is a one-to-one function. The transformed variables are referred to as normal scores. By construction, the random functions $[Y_k(x), k = 1 \text{ to } K]$ are strictly stationary and standard normally-distributed. It is also assumed that they are joint multivariate normally-distributed. As a consequence, the full joint probability law of the set $[Y_k(x), k = 1 \text{ to } K]$ is known if the mean, auto- and cross-covariance functions are known. By construction, the means are zero. The auto- and cross-covariance functions are linked to the auto- and cross-variograms via the relationship:

$$C_{kk'}(h) = C_{kk'}(0) - \gamma_{kk'}(h), \quad k = 1 \text{ to } K, \quad k' = 1 \text{ to } K,$$

where $C_{kk'}(h)$ and $\gamma_{kk'}(h)$ respectively are the covariance and cross-variogram between the two variables z_k and $z_{k'}$ separated by h.

The second step is sequential simulation. It is performed as follows:

- Select at random a location, say x_0, which has not been simulated yet.
- Using the surrounding normal scores, both measured *and* already simulated, compute the conditional distribution of the random vector at location x_0. This random vector consists of the random variables $[Y_k(x_0), k = 1 \text{ to } K]$ excluding those which have been sampled.
- Draw a realization from the conditional distribution. This realization constitutes a vector of simulated normal scores.
- Continue until there are no more locations to simulate.

Once all of the locations have been processed, the simulated normal scores are back-transformed using the inverse of the gaussian transforms.

The remainder of this section describes how the conditional distribution of the random vector is computed, and how a realization is obtained from it.

The K random functions $[Y_k(x), k = 1 \text{ to } K]$ are joint multinormally distributed. This implies that the random vector consisting of 1) the random variables surrounding location x_0; and 2) the random variables to be simulated at x_0, is multinormally-distributed. For simplicity, consider the following notation:

- X_1 is the vector whose entries are the random variables surrounding location x_0. This vector contains at most $N \times K$ entries, where N is the number of "samples" surrounding x_0, and K is the number of variables. In most cases, the number of entries will be less than NK because not all of the K variables are sampled at the N locations. By "samples", we mean both measured and already simulated variables.
- X_2 is the random vector whose entries are the random variables to be simulated at x_0. This random vector contains at most K entries because some of the variables might have been measured.
- X is the random vector such that: $X^T = (X_1^T, X_2^T)$.

The multivariate distribution of the random vector X is:

$$X \sim N(\mu, C),$$

where the mean vector μ is zero by construction, and the entries of the covariance matrix C are defined using the auto- and cross-covariance functions $[C_{kk'}(h); k, k' = 1 \text{ to } K]$. Partitioning X into X_1 and X_2, and the mean vector and covariance matrix accordingly leads to:

$$X = \begin{pmatrix} X_1 \\ X_2 \end{pmatrix} \sim N\left(\mu = \begin{pmatrix} 0 \\ 0 \end{pmatrix}, C = \begin{pmatrix} C_{11} & C_{12} \\ C_{21} & C_{22} \end{pmatrix}\right).$$

The conditional distribution of X_2, given the known values of X_1 (x_1), is multivariate normal such that (Anderson, 1958):

$$(X_2 | X_1 = x_1) = X_{2|1} \sim N\left(\mu_{2|1} = C_{21}C_{11}^{-1}x_1, \quad C_{22|1} = C_{22} - C_{21}C_{11}^{-1}C_{12}\right), \quad (1)$$

where $\mu_{2|1}$ and $C_{22|1}$ are the conditional mean vector and covariance matrix respectively. A realization of the conditional random vector $X_{2|1}$ is (Rubinstein, 1981):

$$x_{2|1} = L_{22|1}r_2 + C_{21}C_{11}^{-1}x_1, \quad (2)$$

where r_2 is a vector of independent standard gaussian values, and $L_{22|1}$ is the lower triangular matrix of the LU decomposition of $C_{22|1}$:

$$C_{22|1} = L_{22|1} U_{22|1}.$$

Davis (1987) has shown that Equation (2)can be simplified to:

$$x_{2|1} = L_{22}^{-1} r_2 + A_{21} L_{11}^{-1} x_1, \tag{3}$$

where L_{22}^{-1}, A_{21}, and L_{11} are obtained from the LU decomposition of the covariance matrix C:

$$C = \begin{pmatrix} C_{11} & C_{12} \\ C_{21} & C_{22} \end{pmatrix} = LU = \begin{pmatrix} L_{11} & 0 \\ A_{21} & L_{22} \end{pmatrix} \begin{pmatrix} U_{11} & B_{12} \\ 0 & U_{22} \end{pmatrix}.$$

3. Practice

In practice, SGCS is not much different from sequential gaussian simulation, or multigaussian kriging (Isaaks, 1990; Verly, 1986). A brief description is given in this section. We will see that the method is based on a number of necessary but justifiable approximations.

The first step is exploratory data analysis. This step is necessary to detect any potential problems, such as errors in the dataset and lack of stationarity. Severe lack of stationarity is worthy of special attention. For example, zones with different statistics are processed separately, and trends are subtracted prior to the study and added afterwards. Note that if there are enough conditioning data, it might be possible to keep the trend as it is. Exploratory data analysis also provides the variable distributions, with declustering if necessary (Journel, 1982).

K gaussian transforms $[\phi_k, k = 1 \text{ to } K]$ are computed by matching the variable cumulative distributions to a cumulative standard normal distribution (Journel and Huijbregts, 1978):

$$G(y) = F_k(z) \Leftrightarrow y = \phi_k(z) = G^{-1}\left[F_k(z)\right] \Leftrightarrow z = \phi_k^{-1}(y) = F^{-1}\left[G(y)\right],$$

where $F_k(\cdot)$ is the cumulative frequency distribution of the kth variable, and $G(\cdot)$ is the cumulative standard normal distribution. Each gaussian transform is stored as a series of bounds (z_i, y_i). Special treatment is given to the first and last few bounds in order to impose minimum and maximum z-values, and to avoid the problem of infinite y-values (Verly, 1986). Note that if there are no 'spikes' of identical z-values, the gaussian transforms are one-to-one functions. If there are such 'spikes', they must be eliminated using a 'de-spiking' procedure (Verly, 1986).

The next step is to assume that the K normal score random functions $[Y_k(x), k = 1 \text{ to } K]$ are joint multinormally-distributed. There is no test of multinormality in the case of a unique, irregular and incomplete sampling, which is the usual situation. However, several checks exist for binormality, which is a necessary but not sufficient condition for multinormality (Verly, 1986).

The next step is to compute the auto- and cross-variograms of the normal score variables, and to fit a model of linear coregionalization (Journel and Huijbregts, 1978; Isaaks and Srivastava, 1989). If a good variogram-fitting program is available, and if all of the variables have been sampled at a reasonable number of locations, this task is relatively easy. In some cases however, at least one of the variable may be significantly undersampled with respect to the other variables and some assumptions have to be made for the corresponding auto- and cross-variograms.

The final step, simulation, is automatic. A few practical points concerning the program are worth mentioning. The sequential algorithm relies on the concept of screen effect (Isaaks and Srivastava, 1989). Indeed, it quickly becomes impractical to use all of the existing values to compute the conditional distribution at a new location. It is then necessary to restrict the conditioning to the values in the immediate neighbourhood of the location. Although no systematic tests have been done, it seems that retaining the 16 closest values per variable, i.e. 2 per variables per octant if an octant search (Isaaks and Srivastava, 1989) is used, gives good results.

In theory, any path through the locations to be simulated can be used. In practice, a random rather than regular path is chosen in order to avoid possible spreading of artifacts in the results. As a consequence, it is not possible to speed up the simulation because there is no regularity in the configuration of the conditioning values. The computation of the conditional vectors and matrices of Equations (1), (2), and (3) have to be repeated for each new location. However, it is possible to obtain several realizations "for the price of one" if the same random path is chosen, and if different sets of random numbers are used in Equations (2) or (3).

If the simulation is performed on a regular grid, the concept of "intermediate" grids has several advantages; in particular when very long variogram ranges have to be reproduced. In this approach, each grid is nested within the next one, with the last one being the grid to be simulated. The simulation starts on the coarsest grid. When this has been completed, the simulation continues on the next coarsest grid, and so on, until the finest grid is simulated. A random path within each grid is followed. The number of intermediate grids depends on the variogram ranges, and on the final grid spacing. An advantage of this scheme is greater speed of execution and less memory requirement. Indeed, the program can be optimized in terms of sample search, covariance tables, etc. Another advantage is more stability in the simulation in the sense that, as long as one grid has been completed, the locations of the conditioning values (at least the simulated ones) is regular, thus preventing the possibility of strong clustering.

4. Discussion

If the model is consistent with the data, SGCS reproduces for a given set of variables:

- a standard normal distribution for each transformed variable;
- the variograms of the transformed variables;
- the cross-variogram between the transformed variables;
- the values of the transformed variables where they are known;

By back-transforming the simulated values, the distributions and known values of the original variables are reproduced. It is assumed that the auto- and cross-variograms of the original variables (which have not been computed) are also reproduced.

SGCS has several advantages, most notably simplicity and generality. Simplicity is most important when it comes to transferring the technique to operational centres. SGCS is simple because it can be explained in a short space of time to somebody who has some knowledge of the variogram and of simulation. Generality is important because different types of problem can be solved. This generality stems from the fact that as long as there is some correlation between the variables, cosimulation can be used with some success. A note of caution, however, is necessary. The solution reproduces the spatial correlation between the variables. This does not necessarily imply consistency with the physics underlying the correlation.

The method is also integrated. No special manipulation is necessary to create anisotropies, enforce conditioning, or simulate irregular grids. The many computational advantages in simulating regular grids makes irregular grids not very attractive, especially since the support of the simulated values stays the same. Only the spacing between the simulated values is irregular.

The method is not limited to two variables: three or more variables can be considered. The price is a more complex coregionalization model, and increased computation time. Whatever the number of variables of interest, two different approaches are possible. The first improves the simulation of one (or several) variable(s) using the information from the other variables. An example is simulating porosities using absolute acoustic impedance from seismic. The second simulates directly all of the variables, e.g. *jointly* simulating vertical and horizontal permeabilities.

When incorporating seismic information, SGCS does not need a full grid of seismic values. It can, for example, simulate porosities where there is no seismic information, or simply fill the gaps by simulating the missing seismic information along with the porosities. This is an advantage over methods which generate values only where seismic information is located.

SGCS reproduces the degree of correlation between the variables. Consider, for example, the case of two variables. If they are independent, their realizations are independent, as if generated by two separate sequential gaussian simulations. If their dependence is perfect, their two realizations are identical. Any intermediate case is also possible. This is an advantage over the external drift method which does not incorporate the degree of correlation of the variables.

It is also possible to create realizations of different variables which are correlated and show different scales of variability. For example, it is possible to simulate porosities correlated with, but not as smoothed as, impedance values.

There is no significant difference in applying the method in 2D or 3D, in the sense that modelling the auto- and cross-variograms is fairly easy in both cases. This remains true if information is missing for one variable or along one direction. Indeed, the variogram is an intuitive tool and lack of information can often be compensated by personal expertise.

The method also has some limitations. The first is the assumption of stationarity. This is not a problem if the domain can be divided into different stationary zones. As mentioned earlier, a trend can be subtracted prior to the study and added afterwards. Although this is an acceptable approximation, it is not obvious to do this for correlated variables with dependent trends. In some cases, however, it seems that an obvious trend can be kept as long as there are enough conditioning data.

A second limitation is the hypothesis of multinormality. In the case the multinormal hypothesis appears unreasonable, other non parametric techniques, such as indicator cosimulation or Markov Bayes updating migh be preferable.

A second limitation is the screen effect. What happens if the screen effect is not good enough? It probably does not happen very often, but it might happen. More work is needed here.

Another limitation is in the model of coregionalization for the auto- and cross-variograms. Any positive-definite model can be used. In practice, a model of linear coregionalization is used (Journel and Huijbregts, 1978; Isaaks and Srivastava, 1989). Such models are not always easy to fit and approximations are often needed. A good variogram modelling program is therefore useful.

Finally, the algorithm is computer intensive, especially if more than two variables are considered. For example, SGCS is much slower than the Turning Band method if there are no conditioning data. The Turning Band method however loses its computational advantage when conditioning is needed. It is also less practical when anisotropies have to be reproduced. There exist several ways of speeding up the sequential process. One is to use a special succession of grids resulting in regular configurations of conditioning values, whislt maintaining the advantage of a random path. Another is to simulate, at each location, a first variable; update the solution ; simulate the second variable; and so on, until all the variables are simulated.

5. Case Study

SGCS has been applied on a real North Sea reservoir. For reasons of confidentiality, the name and location of the reservoir can not be disclosed. Distances and data values have been modified. The problem is as follows. Measurements representing absolute acoustic impedance (AAI) of a layer are available at more than 30,000 locations (Figure 1a). These measurements are called seismic AAI. Average porosity and AAI measurements of the layer are also available at 20 well locations. These measurements are called well porosities and AAI. Note that well AAI values are not obtained from well logs; they are seismic AAI data which happen to be located at well locations. AAI and porosity are negatively correlated. The objective is to simulate a map of porosities using all of the available AAI and porosity values.

Figure 1a shows a definite trend at the scale of the reservoir. A similar trend can be seen on a map showing the porosity values. However, because there are many conditioning AAI values, and because there is good correlation between AAI and porosity, the trend need not be subtracted prior to the study.

Figures 1b and 1d show the histograms of the AAI and porosity values, reespectively. The AAI histogram is very well defined because AAI is known almost everywhere. By contrast, the porosity histogram is poorly defined.

Two gaussian transforms have been computed: one using the seismic AAI, and the other using the well porosities. The porosity histogram has been smoothed prior to being transformed in order not to reproduce the jagged shape of the experimental curve. Note that the need for a gaussian transform is questionable. It is clear from Figure 1a that at the scale of the study, the phenomenon does not appear stationary. This means that the non-normal shape of the AAI histogram can be interpreted as a local fluctuation of a phenomenon within a much larger area. The joint distribution of the transformed AAI and porosity values has been assumed multinormal without further checks.

Experimental variograms using AAI normal scores and an anisotropic model have been fitted. Figure 2a shows the experimental variograms along 4 directions together with the model. The 4 directions include those of best and worst continuity, and two intermediate cases. Because the anisotropy main directions do not coincide with the seismic grid, it was not possible to compute variograms along directions corresponding to the same model. This is why the two models representing the intermediate directions are not superimposed. Figure 2a shows that the experimental variograms and the model are in good agreement up to a distance of 150 to 200 units. After that, the agreement deteriorates because the model does not reproduce the hole effect (Journel and Huijbregts, 1978; Isaaks and Srivastava, 1989). This will have no significant bearing on the result since there are many conditioning data. Note that the total sill of the model is larger than 1, the variance of the transformed AAI. This is acceptable since the variance of the transformed AAI is local and does not represent the variance of the model.

The auto- and cross-variograms of the well AAI and porosity normal scores have been computed. As expected, the results (Figures 2b, 2d, and 2f) are less reliable. In fact, no experimental value for these Figures could be computed with more than 7 pairs. However, the anisotropies observed in the seismic AAI variogram (Figure 2a) are reproduced. Rescaling the seismic AAI variogram model by a factor of 1.5 results in an "acceptable" fit for the well AAI variogram. This well AAI model in turn served as a basis for fitting the well porosity auto- and cross-variograms. It can be seen from Figures 2b, 2d, and 2f that the three models are of poor but equal quality. Better fits are not possible because there is not enough data. The fitted well auto- and cross-variograms are representative of the statistics of 20 well samples. It is known that the "global" AAI variogram, i.e. the seismic AAI model, is 1.5 smaller than the well AAI variogram. It is assumed that the other global auto- and cross-variogram are also 1.5 smaller than their well counterparts. The model of coregionalization is then defined.

One realization of the AAI/porosity model has been obtained using SGCS. The auto- and cross-variogram of the simulated normal scores are shown in Figures 2c, 2e, and 2g. The histograms of the simulated AAI and porosities (back-transformed normal scores) are shown in Figures 1c and 1e. The simulated AAI and porosity maps are shown in Figure 1f and 1g.

Figures 2c, 2e, and 2g show that the model of coregionalization is very well reproduced. The histogram of the simulated AAI (Figure 1c) is almost identical to the histogram of the original AAI measurements. This is due to the large amount of conditioning values. It is difficult to compare the histogram of the simulated porosities (Figure 1e) with the original but unreliable distribution (Figure 1d). However, the fact that the simulated histogram is positively-skewed makes sense since the AAI histogram is negatively-skewed and their correlation is negative. Note that, because the simulated field is small with respect to the observed structures, and because there are few porosity measurements, the simulated porosity distribution probably fluctuates a lot from one realization to another. This aspect of the problem has not been covered since only one realization has been simulated.

Figures 1f and 1g show the maps of simulated AAI and porosities. As expected, the simulated AAI map is almost identical to the original map, except that the gaps have been filled with simulated values. The porosity map is negatively correlated with the AAI map. The porosity map does not show any short scale variability because this was not modelled in the coregionalization.

SGCS has provided a very reasonable map of porosities using actual data. Additional maps can be obtained with different random numbers. More importantly, additional maps can be obtained by changing some of the model parameter values (coregionalization, gaussian transform, etc.), which are known to be subjective.

6. Conclusions

In this paper, sequential gaussian cosimulation (SGCS), a stochastic simulation method for integrating several variables, has been described. The method reproduces the distributions, and the auto- and cross-variograms of the variables, together with the values of the variables where there are known. The method assumes that the random functions of the variables are strictly stationary and joint multinormally distributed after transformation.

SGCS is "general" in the sense that all it needs in practice are correlated variables. It can be used in two important types of application. The first one is to improve the simulation of one variable using information about another variable. The second one is to simulate several variables at the same time.

The method has been applied on a real reservoir. A 2D porosity map has been simulated using 20 porosity samples and more than 30,000 absolute acoustic impedance values. The results obtained are very good.

The concept of cosimulation within a multinormal framework is not new in geostatistics. The advantage of SGCS over existing methodology is its simplicity. Not only is the theory simple, but also it integrates all aspects of simulation — multiple structures, anisotropies, regular or irregular grids, conditioning — in a single process. SGCS has other advantages. It is relatively easy to replace missing information, for example a variogram or a cross-variogram, using expert knowledge. Multiple realizations are obtained at no extra cost if the same random path is used.

The method has some limitations. The most important one is the hypothesis of stationarity which, in fact, is inherent to most other methods. Practical solutions amount to dividing the area into stationary zones or subtracting the trend. Note that a trend does not seem to be a problem when there are many conditioning values for one of the variables. Another limitation is that there is no guarantee that the physics underlying the correlation between the variables is honoured, even if the spatial auto- and cross-correlation of the variables are reproduced. This might be a problem especially where seismic information has to be incorporated.

In conclusion, SGCS is a simple and attractive method for incorporating different types of data. The method has several advantages and some limitations. The results obtained so far on actual data are satisfactory. It appears therefore that SGCS has good potential for future use.

Acknowledgements

The author would like to thank the British Petroleum Co. plc for permission to publish this paper. He is also grateful to Colin Daly and Giles Watts for useful discussions.

References

Alabert, F.G. (1987). *Stochastic Imaging of Spatial Distributions Using Hard and Soft Information*, M.Sc. Thesis, Stanford U., 197 pp.

Anderson, T.W. (1958). *An Introduction to Multivariate Statistical Analysis.* Wiley, New York. 374 pp.

Davis, M.W. (1987). "Production of Conditional Simulations via the LU Triangular Decomposition of the Covariance Matrix", *Journal of Mathematical Geology*, Vol. 19, No. 2, pp. 91–98.

Doyen, P.M. , Guidish, T.M., and de Buyl, M.H. (1989). "Seismic Discrimination of Lithology in Sand/Shale Reservoirs: A Bayesian Approach". Exp. Abs., Proc. Soc. Expl. Geoph. 59th An. Int. Meeting, Dallas, Texas.

Haldorsen, H. H., Brand, P. G., and Macdonald, C.J. (1988). "Review of the Stochastic Nature of Reservoir", *Mathematics in Oil Production*, Ed. Edwards S., King P.R., Clarendon Press, Oxford, pp. 109-209.

Isaaks, E.H. (1984). "Indicator Simulation: Application to the Simulation of a High Grade Uranium Mineralization", *Geostatistics for Natural Resource Characterization*, Part 2, Eds. Verly et al., D. Reidel Publ. Co., pp. 1057-1069.

Isaaks, E.H., Srivastava, R.M. (1989). *An Introduction to Applied Geostatistics*, Oxford University Press, New York, 561 pp.

Isaaks, E.H. (1990). *The Application of Monte Carlo Methods to the Analysis of Spatially Correlated Data*, Ph.D. Thesis, Stanford U., 213 pp.

Journel, A.G., and Huijbregts, Ch.J. (1978). *Mining Geostatistics*, 2nd printing (revised), Academic Press, London, 600pp.

Journel, A.G. (1982). "Indicator Approach to Spatial Distribution.", *Proc. of the 17th APCOM Int. Symp.*, Eds. Johnson, T.B., and Barnes, R.J, Soc. of Min. Eng. of AIME, Littleton, CO, pp. 793-806.

Journel, A.G., Alabert, F.G. (1988). "Focusing on spatial connectivity of extreme-valued atrributes: stochastic indicator models of reservoir heterogeneities", SPE 18324, *Ann. Tech. Conf. and Exib. of SPE, Houston, TX*. pp. 621–632.

Journel, A.G., and Zhu, H. (1990). "Integrating Soft Seismic Data: Markov-Bayes Updating, an Alternative to Cokriging and Traditional Regression", Stanford Center for Reservoir Forecasting, Report No. 3.

Marechal, A. (1984). "Kriging Seismic Data in Presence of Faults", *Geostatistics for Natural Resource Characterization*, Part 1, Eds. Verly et al., D. Reidel Publ. Co., pp. 271-294.

Matheron, G., Beucher, H., de Fouquet, C., Galli, A., Guerillot, D., and and Ravenne, C. (1987). "Conditional Simulation of the Geometry of Fluvio-Deltaic Reservoirs", SPE 16753.

Omre, H. (1987). "Bayesian Kriging - Merging Observations and Qualified Guesses in Kriging", *Journal of Mathematical Geology*, Vol. 19, No. 1, pp. 25–39.

Rubinstein, R.Y. (1981). *Simulation and the Monte Carlo Method*. Wiley, New York. 272 pp.

Thadani, S.G., Alabert, F.G., and Journel, A.G. (1987). "An Integrated Geostatistical/Pattern Recognition Technique for the Characterization of Reservoir Spatial Variability", SEG expanded abstracts.

Verly, G. (1986). "Multigaussian Kriging - A Complete Case Study" *Proc. of the 19th APCOM Int. Symp.*, Ed. Ramani, R.V., Soc. of Min. Eng. of AIME, Littleton, CO, pp. 283–298.

Verly, G. (1991). "Sequential Gaussian Simulation: A Monte Carlo Method for Generating Models of Porosity and Permeability". *Special Publication No. 3 of EAPG - Florence 1991 Conference*, Ed. Spencer, A.M., To be published.

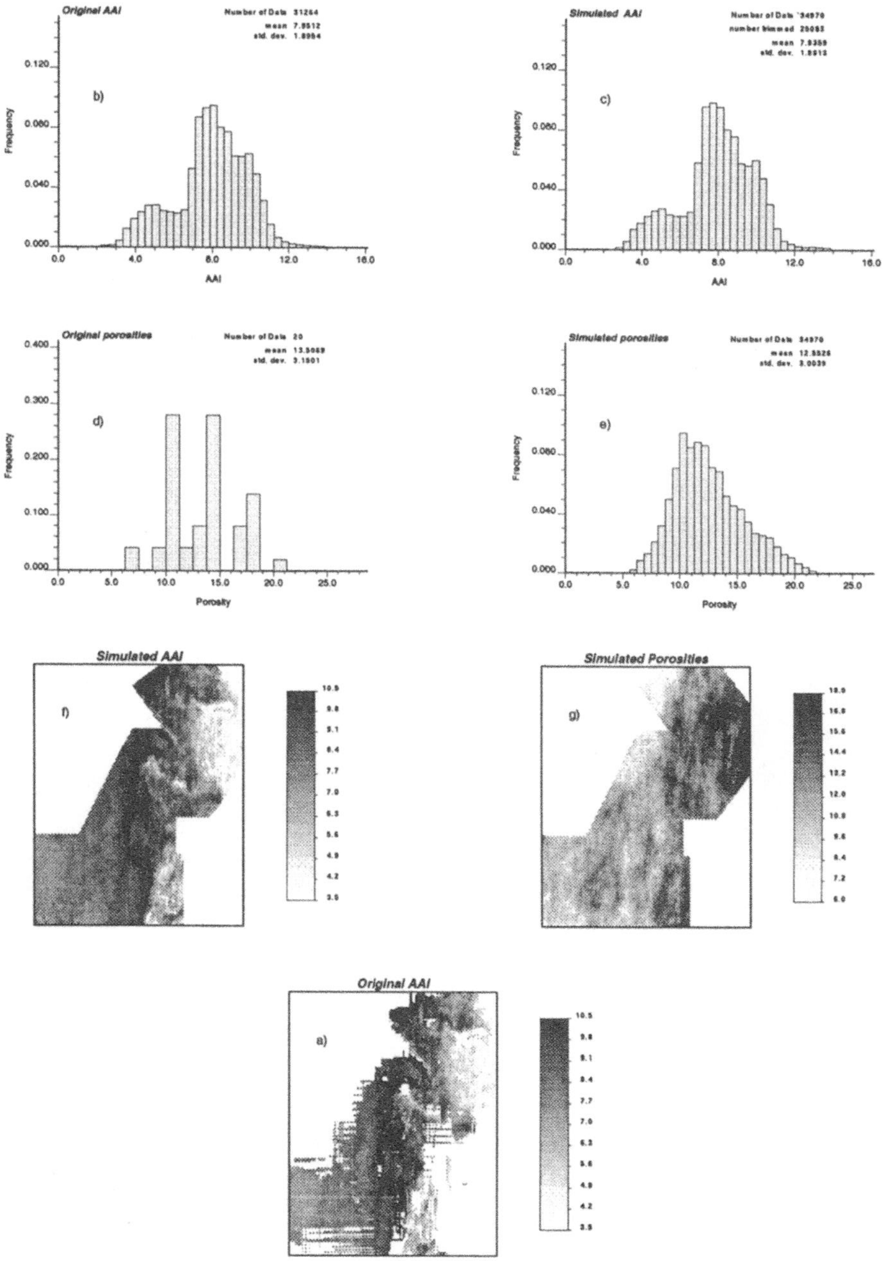

Figure 1 a) Original absolute acoustic impedance (AAI) map. b & d) Original AAI and porosity distributions. c & e) Simulated AAI and porosity distributions. f & g) Simulated AAI and porosity maps. The grey scale minimum and maximum values have been chosen to enhance the contrast of the maps. As a consequence, they do not correspond to the histogram minimum and maximum values.

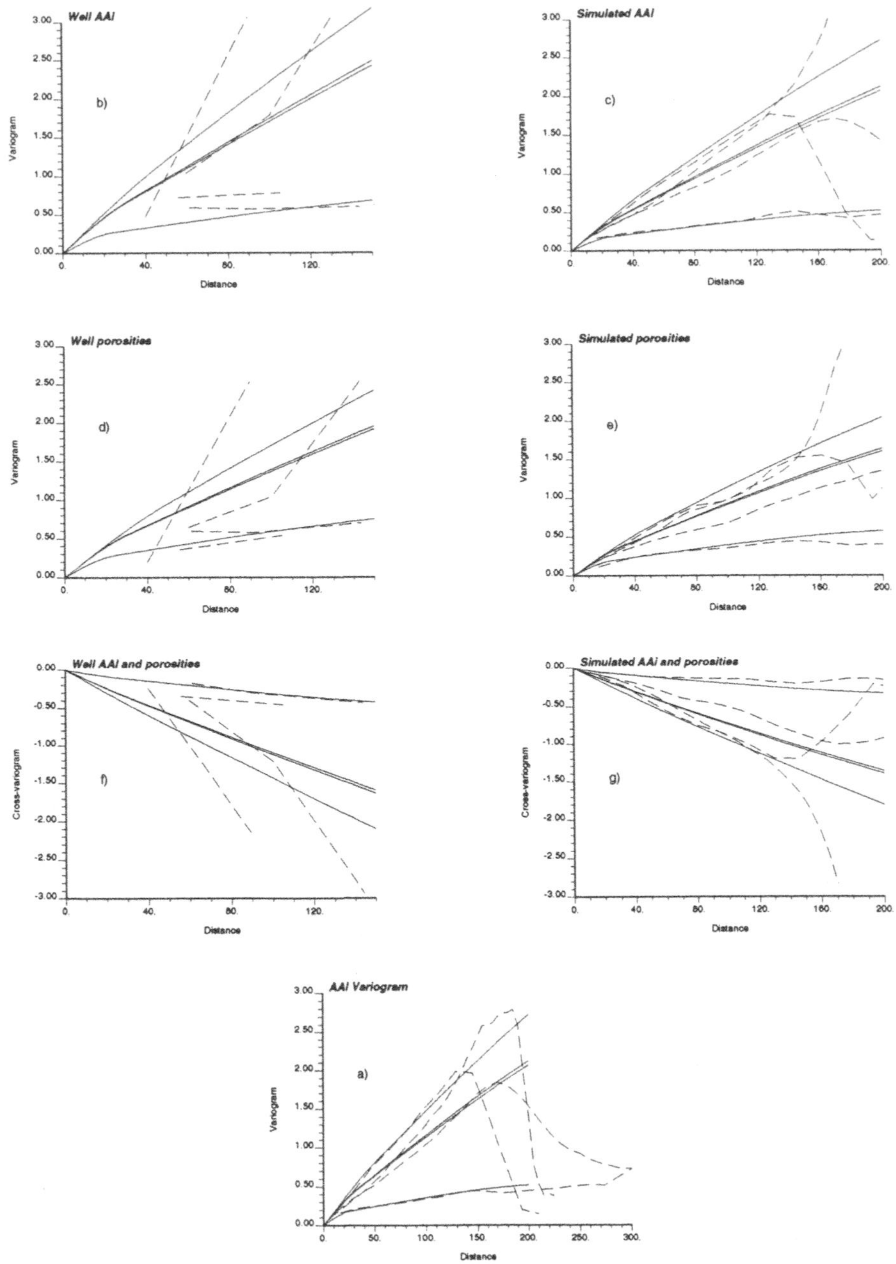

Figure 2 Experimental (dashes) and simulated (dashes)auto- and cross-variograms, together with the coregionalization model (full line). Variograms are shown along the directions of best and worst continuity, and two intermediate cases. See the text for more explanations. a) AAI Variogram from seismic. b & d) AAI and porosity variograms from wells. f) AAI and porosity cross-variogram from wells. c & e) AAI and porosity variograms from simulation. f) AAI and porosity cross-variogram from simulation.

INCLUDING A REGIONAL TREND IN RESERVOIR MODELLING USING THE TRUNCATED GAUSSIAN METHOD

H. Beucher*, A. Galli*, G. Le Loc'h*,
C. Ravenne**
and HERESIM Group

(*) Centre de Géostatistique
Ecole des Mines
35 rue Saint-Honoré
77305 Fontainebleau - France

(**) IFP
1&4 av. bois Préau
92506 Rueil-Malmaison - France

ABSTRACT

When studying the geometry of petroleum reservoirs, we have to integrate different kinds of data (wells and seismic) as well as geological knowledge. Usually, wells show the fine vertical variation of lithofacies, but are less precise about horizontal variations. On the another hand, seismic data give a good idea of the horizontal evolution but we lose information on the vertical variability. A realistic reservoir model has to take this into account.

Until now, we have been able to model vertical variability with HERESIM. Here we show that the truncated gaussian model can be used to model horizontal variability or both horizontal and vertical variability. This can be achieved by using both horizontal and vertical non stationary proportions with a stationary gaussian random function. We present the extension of the method in detail, focusing on how to perform structural analysis, which is possible because the gaussian is still stationary, even if the lithofacies show strong non stationarities. Thus we work with coherent non stationary models for the simple and cross variograms of lithofacies. This methodology is illustrated on examples.

INTRODUCTION

For modelling heterogeneous reservoirs, we have chosen the truncated gaussian method[1] [2]. This method has already given satisfactory images in fluvial deposits and has great flexibility, as we will present in this paper. This method considers the geological image of the reservoir as a realization of a random set. The spatial analysis of the geometry of the lithotypes is based on two tools :variograms and proportion distributions. On these tools, the stationary or non−stationary hypothesis has to be chosen. Then, under the good one previous choice, the lithofacies structural analysis is made. The results of this analysis are the parameters of the underlying gaussian function which is the base of the simulation. In this paper we present the structural analysis −experimental calculations and fits in terms of gaussian functions− when the lithofacies can be considered as a stationary variable or not. We will see the importance of the choice of an appropriate model, evident on the simulations but also on the structural analysis which can give overestimated ranges.

STRUCTURAL ANALYSIS USING THE TRUNCATED GAUSSIAN METHOD

Experimental tools

The geostatistical tools used in this method are:

- First,the proportion curves of each of the various lithotypes along different directions.These curves are computed according to a reference level chosen with geological criteria for each stratigraphic unit.
  Along the vertical axis, these curves summarize the sequence deposits. On the horizontal plane, the distribution of the lithotypes show the global behavior of the deposits. These curves and displays are the first tools that show if we can work in the stationary framework or if we have to consider non stationary models.
- Then the simple and cross−variograms of the indicator functions of the lithotypes. There is, for each studied direction, as much simple variograms as lithotypes. Throughout the presentation, we shall call variogram the following mean, even in the non−stationary case:

$$\gamma\ (h,u)\ =\ 1/2E[A(u)\ -\ A(u\ +\ h)]^2$$

With : $A(u)$ = *indicator function of lithofacies* A *at 3D point* u $\{u\ =\ (x,y,z)\}$

E = *mathematical expectation*

h = *calculation distance along the chosen direction*

In the stationary case this variogram depends only on h, whereas in the non stationarity case, it also depends on the location of the point u. This causes the problem of statistical inference because basically we have one measure per data point. Nevertheless, note that in all the cases, the values of the variogram belong to the interval [0.,.5].

The model

The indicator function of one lithotype is developed with Hermite polynomials[3][4].

$$A(u) = f_0(s(u)) + f_1(s(u))H_1[Y(u)] + f_2(s(u))H_2[Y(u)] + \ldots$$

$$\text{With } Y(u) = \textit{Underlying Gaussian Function}$$

$$A(u) = \textit{indicator of } Y(u) < s(u)$$

$$f_n(s(u)) = \textit{function of the thresholds } s \textit{ at point } u$$

$$f_0(.) = G(.)$$

$$f_n(.) = H_{(n-1)}(.) * g(.)/\sqrt{(n)} \qquad n \geq 1$$

$$H_n[.] = \textit{Hermite polynomials}$$

The calculation of the variogram, in terms of gaussian function gives:

$$(u,h) = 1/2E[f_0(s(u)) + \sum_{n=1}^{nh} f_n(s(u))H_n[Y(u)] - f_0(s(u+h)) - \sum_{n=1}^{nh} f_n(s(u+h))H_n[Y(u+h)]]^2$$

that is:

$$\gamma\ (u,h) = 1/2[P_a(u) + P_{a(u+h)}] - P_a(u)P_a(u+h) - \sum_{n=1}^{nh} f_n(s(u))f_n(s(u+h))C^n(h)$$

$$P_a(u) = \textit{Probability of lithofacies } A \textit{ at point } u$$

$$C(h) = \textit{Stationary covariance of the Gaussian function}$$

Fitting

To perform the statistical inference, we have averaged in the same way experimental and model variogams. For instance horizontal variograms correspond to individual variograms at given z level averaged on all the levels.

The corresponding fits need:

- the variogram of the underlying gaussian function which is always stationary.
- the experimental thresholds for the truncation, given by the proportions of each lithotype depending on the 3D location.

As the whole lithofacies are obtained by truncation of only one gaussian function, they are interdependent. The used parameters (ranges, proportion maps..) have to fit, as best as possible, all the simple and cross–variograms. This consistency do not ensure that this method is available for all data sets. These lithofacies correlations on this model correspond to data constraints in sequential deposit.

EXAMPLES

Horizontal and vertical Stationary case

In this case the expectation of each indicator function is constant. The heterogeneity of the reservoir depends only on the spatial correlation. This means that the proportions of each lithotype vary slowly around a constant value.

This case is illustrated on real data which are a part of Long Nab cliff [2]. The geological environment of this reservoir located in Yorkshire, is fluvio–deltaic. To clarify the presentation, its heterogeneity is represented by only two lithotypes. On the chosen levels the proportion of one lithotype and obviously of the other, which is the complementary of the image, varies weakly around a constant value for the vertical direction (figure 1). On the same levels, the horizontal proportions do not vary around a constant but appear as an almost black and white image (figure2).Although the 3D stationarity hypothesis is valid as it is confirmed by the horizontal and vertical variograms.

Note that when there are only two lithotypes, there is only one variogram for each direction to represent the spatial variability of the reservoir. The fits in this stationary case are satisfactory even if they do not honor all the oscillations.(figures 3,4)

On actual data, a 3D stationarity of the lithotypes is quite improbable. This hypothesis can be used for a part of the reservoir inside a homogeneous unit. But generally it is necessary to take into account the vertical non stationarity as we show it in the next paragraph.

Vertical non stationarity (ordinary case)

In this case, the expectation of each lithotype depends on the level. This corresponds in sedimentary units to the fact that the lithofacies are sequentially distributed along the vertical axes. This sequence is summarized on the vertical proportion curves. On the corresponding example which is another part of Long Nab cliff drawn with 3 lithotypes, it is obvious that the vertical stationarity cannot be assumed (figure 5); however horizontal stationarity hypothesis can be accepted(figure 6). This corresponds to most of the cases we have been working on.

The corresponding vertical variograms show also a non stationarity. The fits performed with the corresponding non stationarity proportion curves are clearly more precise (figure 7a) than with a stationary hypothesis (figure 7b).

Even if this hypothesis is often sufficient, sometimes this approximation does not correspond to the geological environment. Then it is necessary to analyze the variability of the lithotypes under a more general model.

3D non stationarity (general case)

When the lithofacies proportions are clearly non stationary on the whole reservoir, the expectation of each lithotype depends on the 3D location. To obtain a good fit in the chosen model and then a good simulation, it is necessary to take into account these variations. This is the case for instance, when a lithotype is present in a part of the reservoir and disappears in the other part (alluvial fan) or when the observation scale is such that only one meander is present in fluvial deposits.

Before showing results on real data, we will present a study on simulations. This is needed to see clearly what is the feasibility of the method, what we can expect at best to reproduce. This gives also guidelines for real cases: how is the fit sensitive to the gaussian range, to the proportions we use...

Simulated example

For this example, we have simulated a 2D vertical section 40m high, 2000m long (figure 8) with, for the gaussian ranges, 200 meters horizontally and 4 meters vertically. The

3 lithotypes present a 3D non stationarity. The global proportion curves (figures 9,10) summarize these 3D non−stationarity. Assuming the hypothesis of stationarity along the horizontal axis, which corresponds to Pa depending only on the vertical location, the "better" fits are obtained with a gaussian range of 1000 meters which is larger than the range used for the simulation (figure 11). Conversely, if we use the horizontal variations of the proportions, P_a varying along the two directions, the fits are correct with the input range and the behavior is honor up to large distances (figure 12). We also see that a horizontal range of 400meters does not give a good fit at small distances (lithotype 1).

It is obvious that non stationarity fits are better but what happens on the simulations? Using the two extremity "wells" as conditioning data, we have computed a simulation under horizontal stationarity [P_a depends on z] with the fitted gaussian range of 1000 meters (figure 13). When comparing this simulation with the data (2D non stationary simulation [P_a depends on x and z] and gaussian range of 200 meters (figure 14)), a major difference appears in the location of the lithotypes 1 and 3 which are spread on the whole section.The result shows clearly that in that case you need a non stationary model to obtain a realistic simulation.

Real data

When studying a real reservoir, of course you do not have all the previous information. But the proportion curves and the variograms indicate the possible choice of stationarity or not, depending on the direction. The case presented here is part of an alluvial fan deposit . .On the three lithotypes represented in this zone, two are only present in the extremities of the section when the third one appears on the whole section (figures 15 to 16). As in the previous simulation the 3D non−stationarity approach seems more appropriate even if the second lithofacies appears stationary. As in the previous chapter, we compare the horizontal fits performed under the 2 hypothesis:

- Horizontal stationarity: it is impossible to reconstitute the behavior of all the 3 variograms. The "better" gaussian range (1000 m) which can be obtained smooths the behavior (figure 17).
- Horizontal non−stationarity: the shapes of the three simple variograms are better honored (figure 18). We can note also that in this example the value of the gaussian range under stationary hypothesis is greater than under non stationary hypothesis (500 m).

Obviously in this example, to perform the reservoir simulation, we need the 3D distribution of the lithofacies proportions .These maps have to be made using conditioning values, already used for fitting, added to geological hypothesis on the deposit. This can be performed for instance using seismic interpretation.

CONCLUSION

We have shown that non stationarity (the most frequent case in practice) has to be taken into account not only for the final simulation but also for the structural analysis. On the last example, we have seen that the fit with a stationary model of a non stationary experimental variogram leads to an overestimated range for a poor result.

Non stationary structural analysis has clearly to be done. This analysis is more difficult to do because it mixes the influence of the proportions and the variograms. However further experiments seem to show that:

- One can have an idea of the range to use, making a stationary fit which respects the shape of the first points (not the values)
- If we have to estimate proportions using well data and geological information, it is important they keep the meaning of proportions, which is done by kriging. A mean value has to be estimated with a point neighborhood reasonably extended.
- When performing the fits, proportions incompatible with the data show up clearly.

To sum up, the truncated gaussian method, because the model is specified through a stationary gaussian function which also insures consistency, is very flexible and allows us to take into account additional information(coming from seismic or/and from geological knowledge) for the modelling.

REFERENCES

[1] Matheron G., Beucher H., de Fouquet Ch., Galli A., Conditional simulation of the geometry of fluvio–deltaic reservoirs. SPE paper n° 16753, SPE 1987 Annual Technical Conference and Exhibition, Dallas, Texas, 27–30 September 1987. Vol. X : Formation, Evaluation and Reservoir Geology.

[2] de Fouquet Ch., Beucher H., Galli A., Ravenne Ch., Conditional simulation of random sets : application to an argilaceous sandstone reservoir. In : *Geostatistics* ed. by M. Armstrong, Proc. 3rd International Geostatistics Congress, Avignon, 5–9 Sept. 1988, Dordrecht, Holland : Kluwer Academic Publ., 1989, Vol. 2, pp. 517–530.

[3] Matheron G., A simple substitute for conditional expectation: the disjonctive kriging, 1976, Proceedings, NATO ASI: "advanced geostatistics in the mining Industry" Rome.

[4] Rivoirard J., Introduction to disjunctive kriging and nonlinear geostatistics, 1990, Centre de Géostatistique ENSMP Fontainebleau

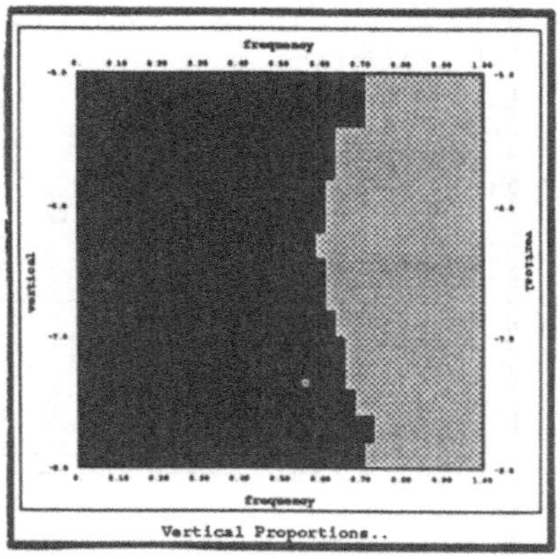

Figure 1: Vertical proportion curves in a simple case

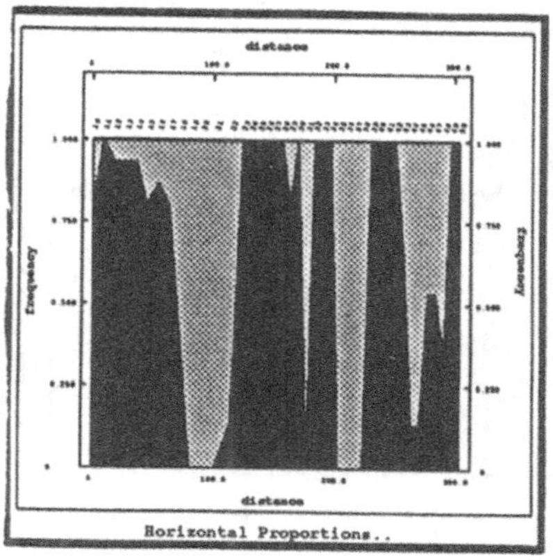

Figure 2: Horizontal proportion curves in a simple case

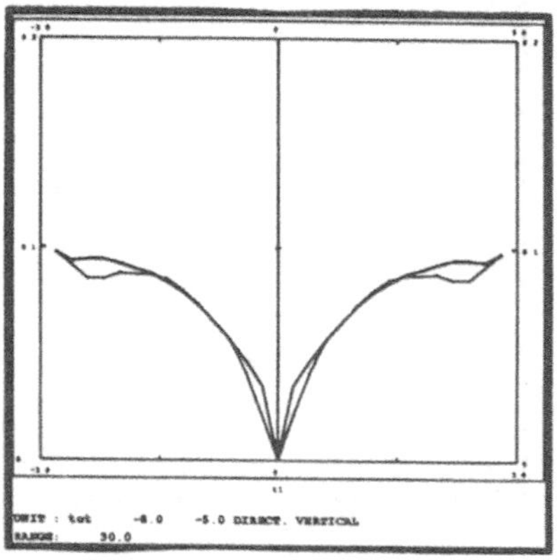

Figure 3: Fitted vertical variogram

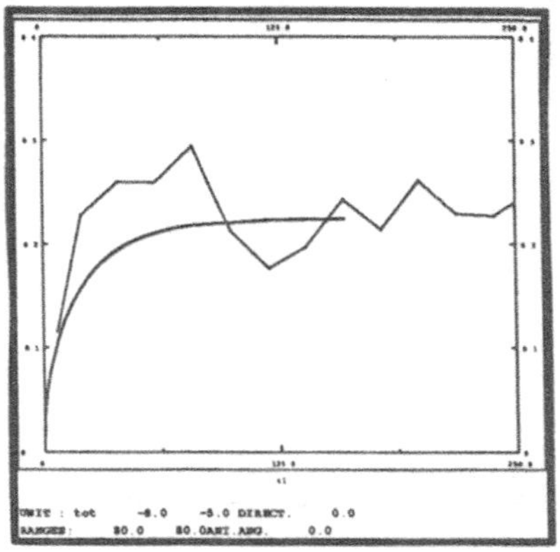

Figure 4: Fitted horizontal variogram

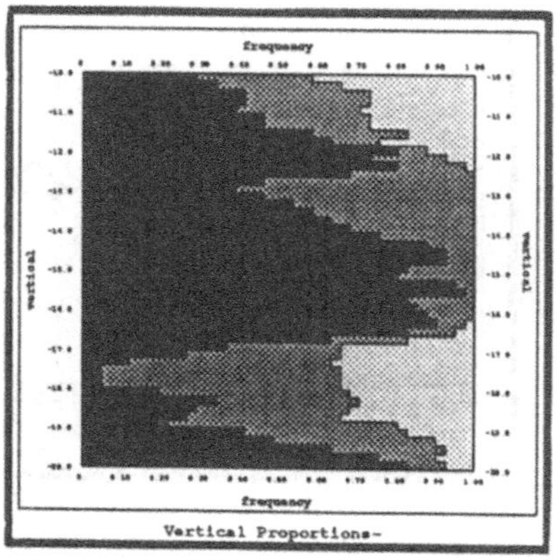

Figure 5: Vertical proportion curves in an ordinary case

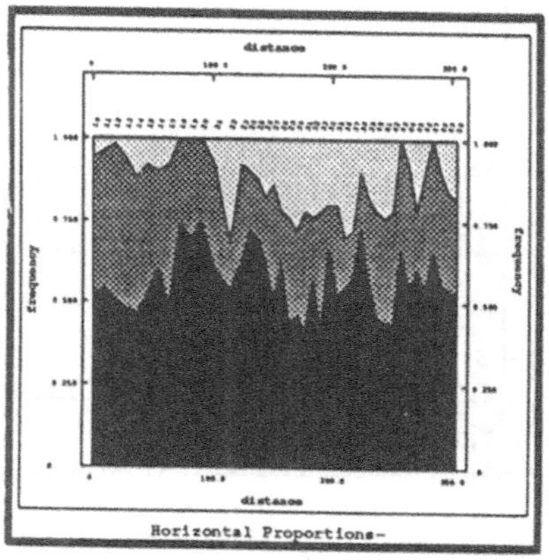

Figure 6: Horizontal propotion curves in an ordinary case

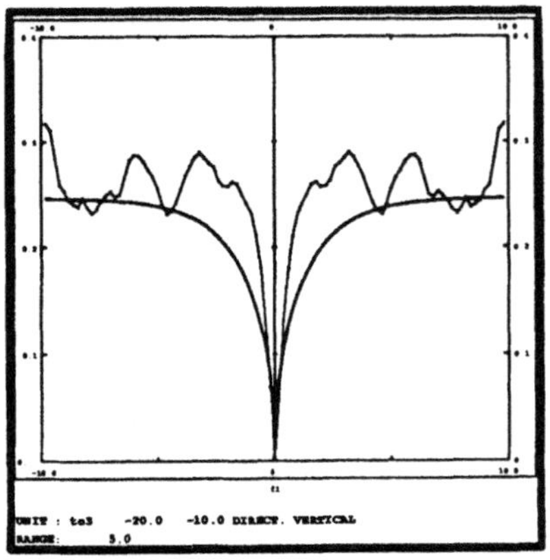

Lithotype 1

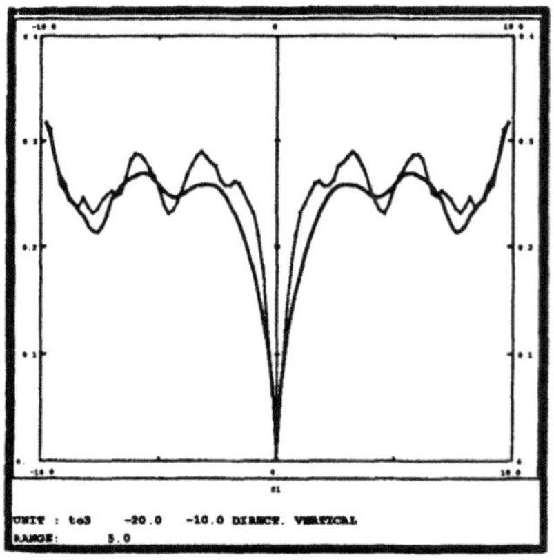

Lithotype 1

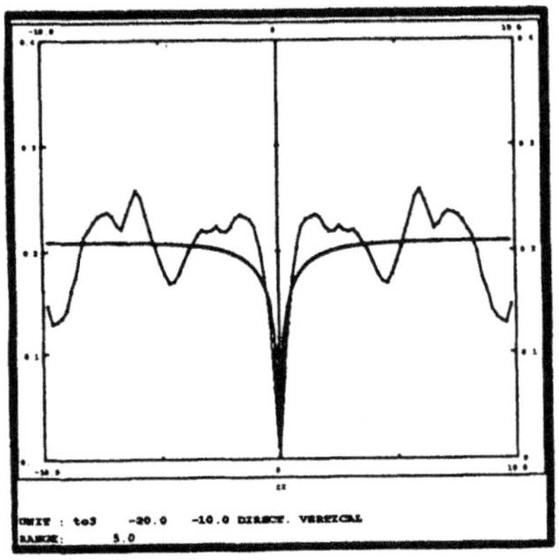

Lithotype 2

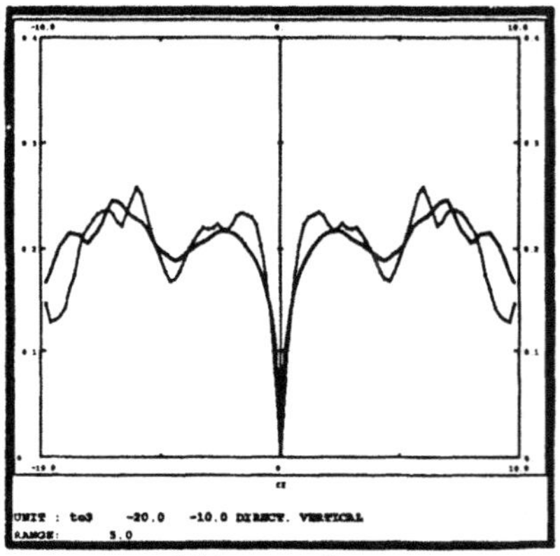

Lithotype 2

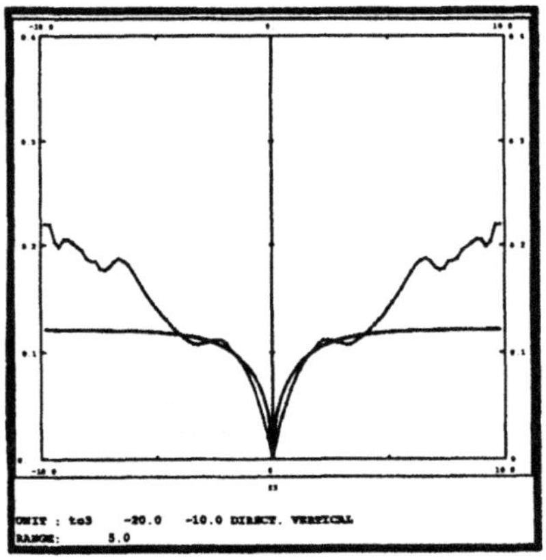

Lithotype 3

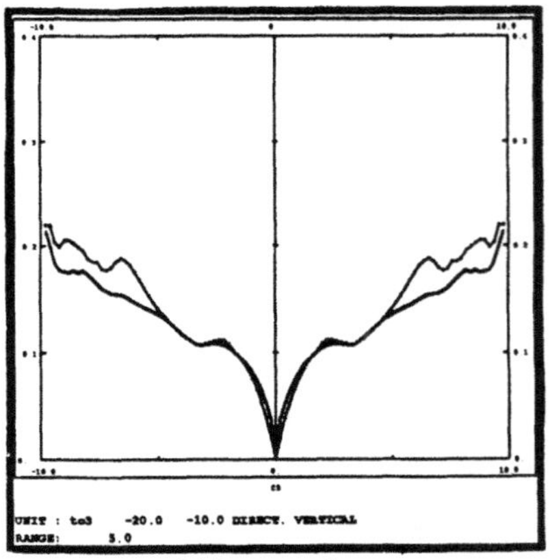

Lithotype 3

Figure 7a: Fitted vertical variograms with stationary hyphothesis

Figure 7b: with non stationary hyphothesis

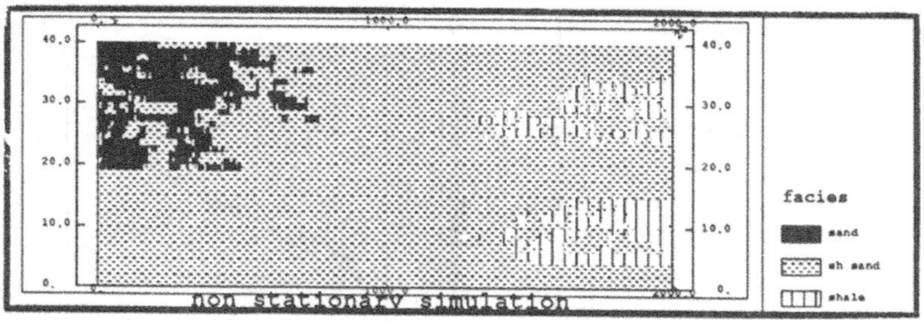

Figure 8: Simulated section

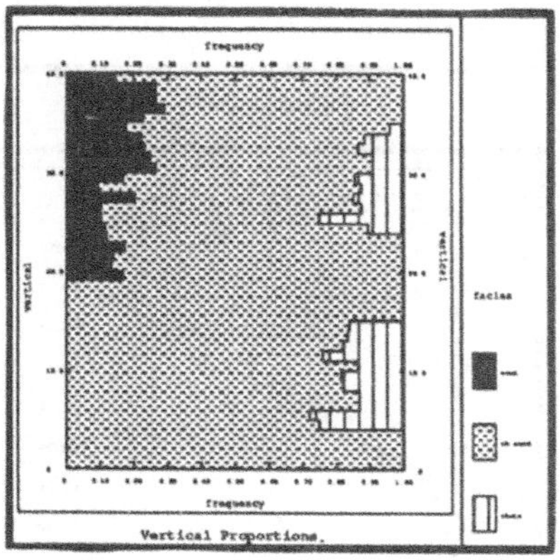

Figure 9: Vertical proportion curves

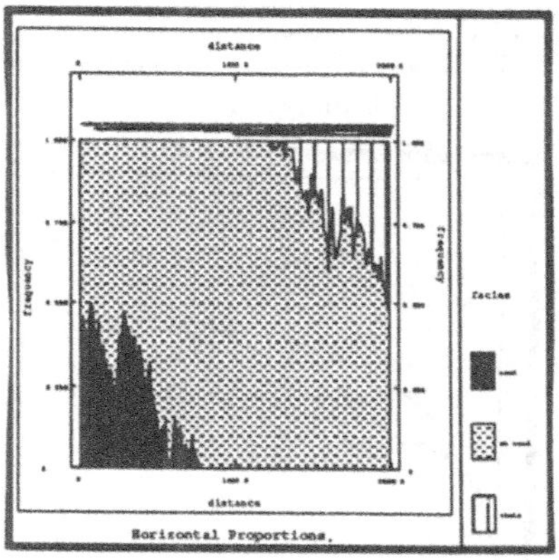

Figure 10: Horizontal proportion curves

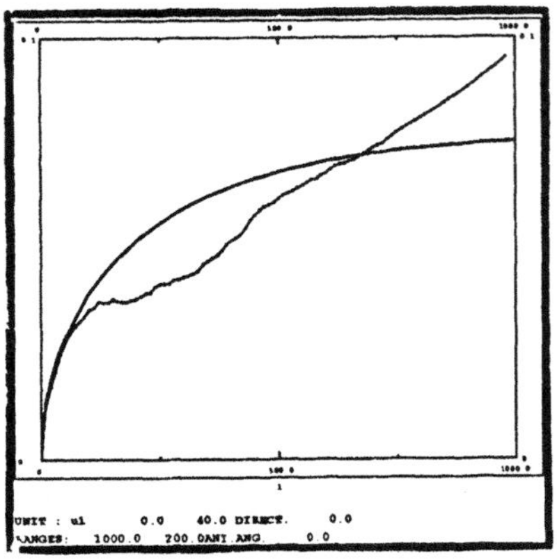

Lithotype 1

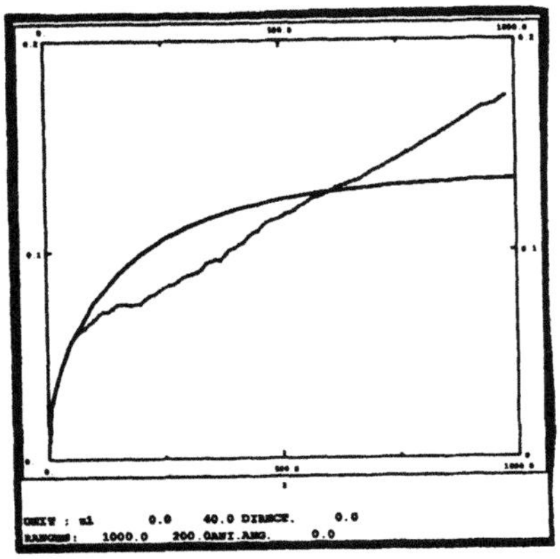

Lithotype 2

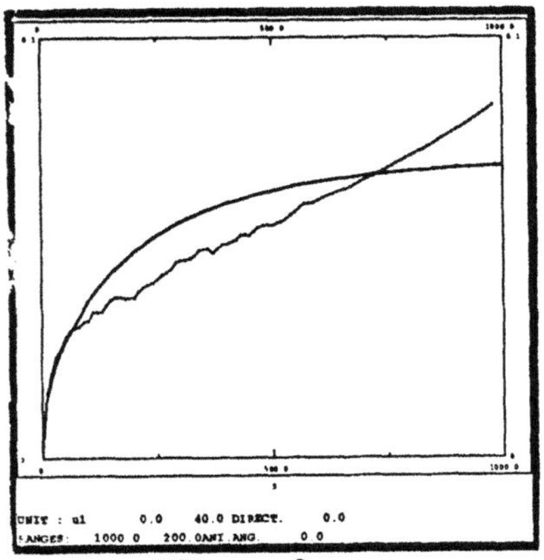

Lithotype 3

Figure 11: Fitted variogams with horizontal stationarity

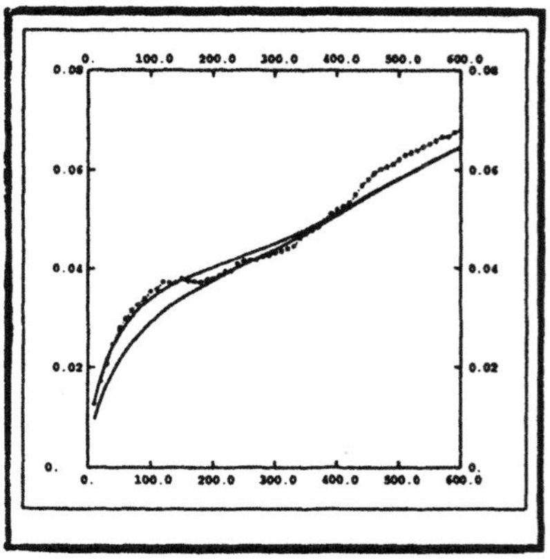

Lithotype 1

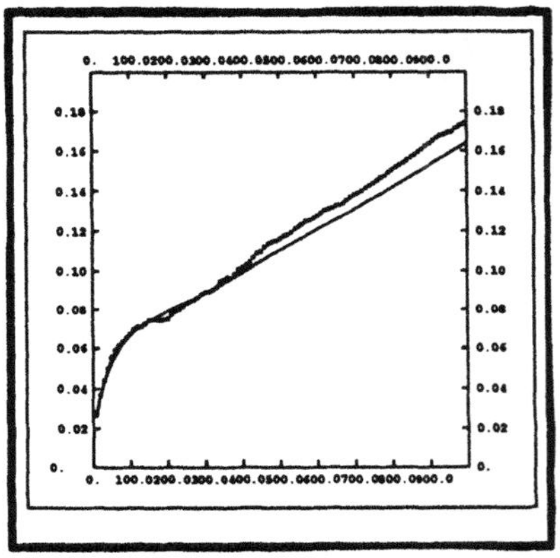

Lithotype 2

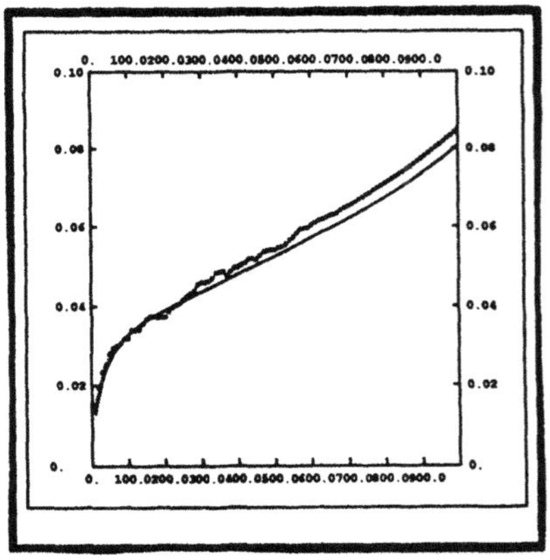

Lithotype 3

Figure 12: Fitted variograms with global non stationarity

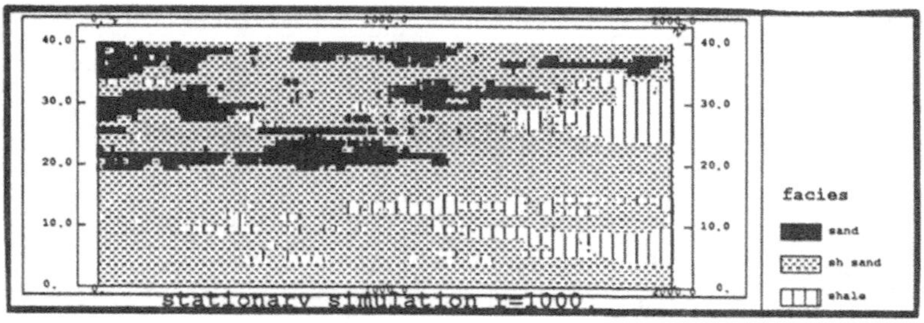

Figure 13:Horizontal stationary simulation - two conditioning wells-1000m range-

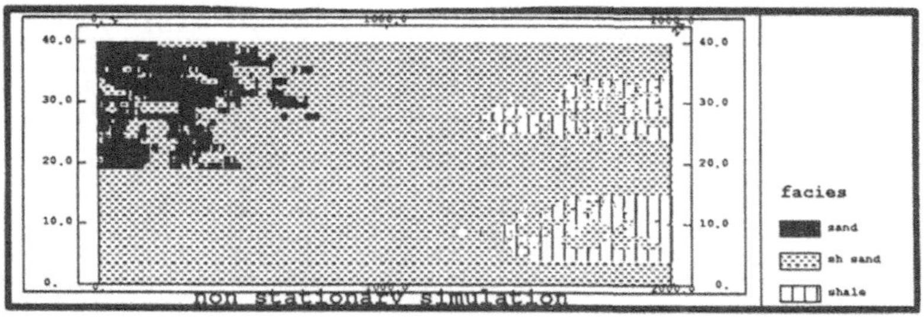

Figure 14: 2D non stationary simulation - 200m horizontal range-

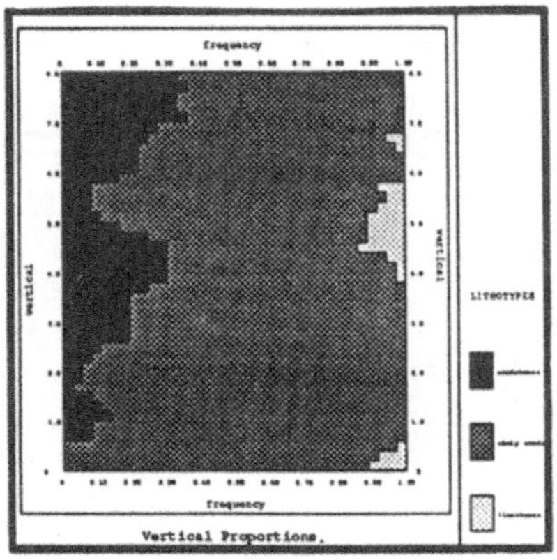

Figure 15: Vertical proportion curves in alluvial fan deposits

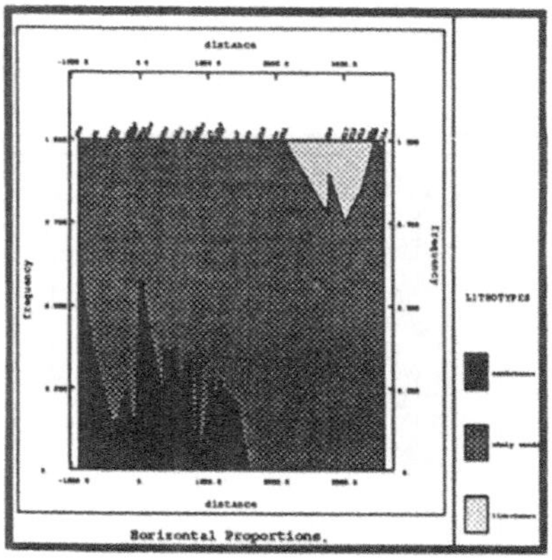

Figure 16: Horizontal proportion curves in alluvial fan deposits

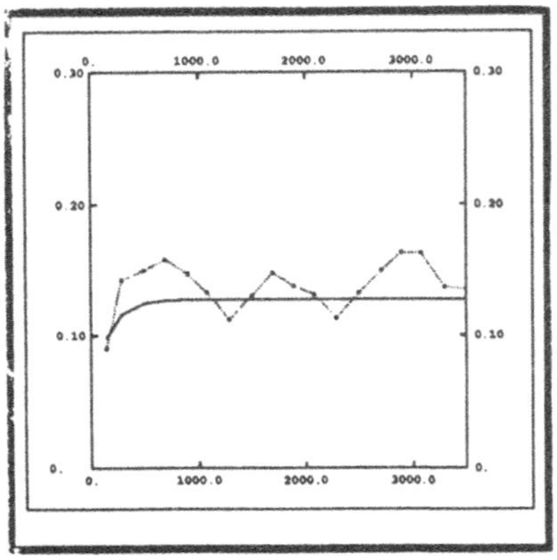

Lithotype 1

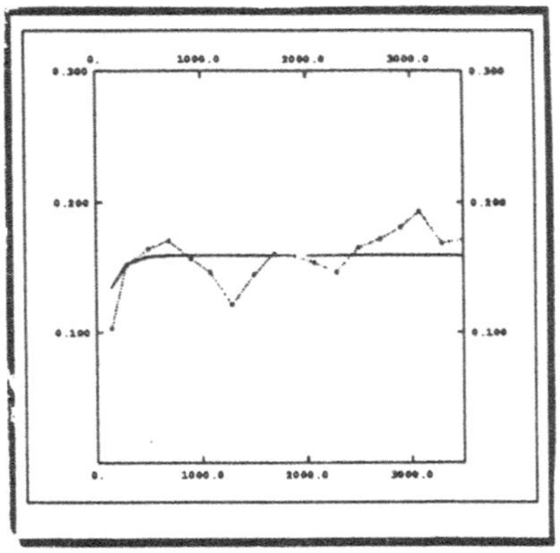

Lithotype 2

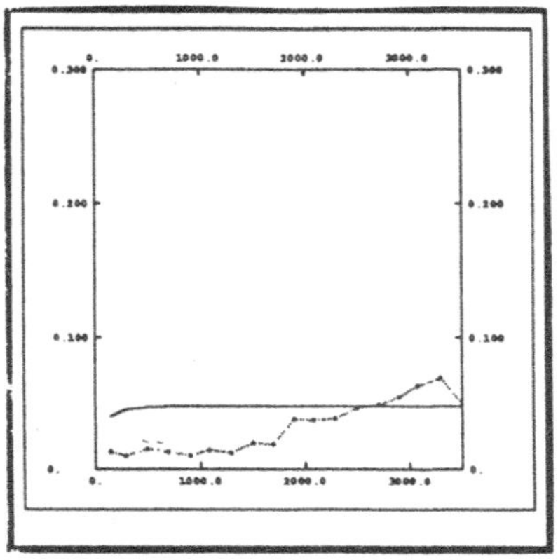

Lithotype 3

Figure 17: Fitted variogams with horizontal stationarity

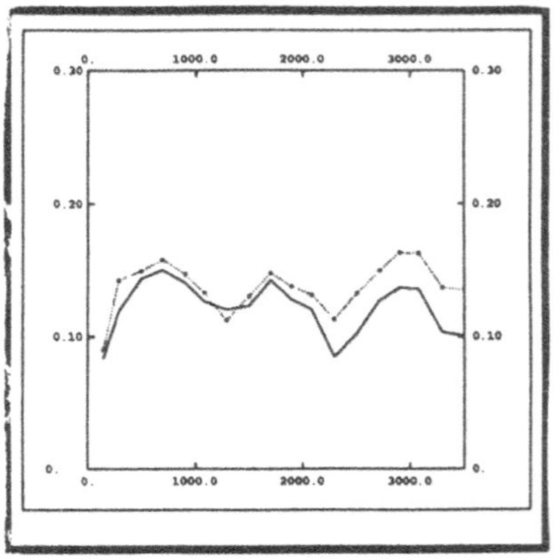

Lithotype 1

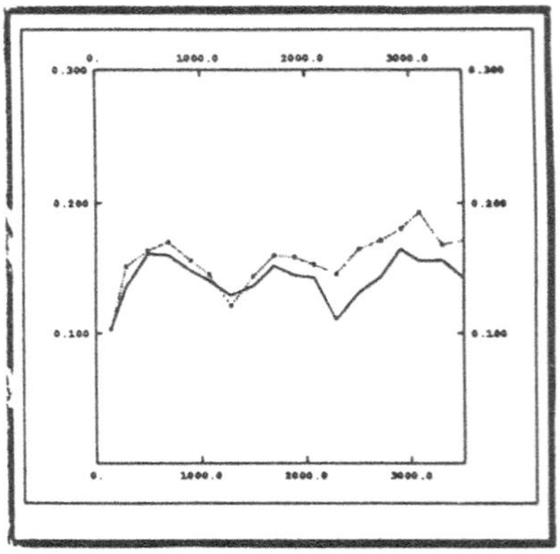

Lithotype 2

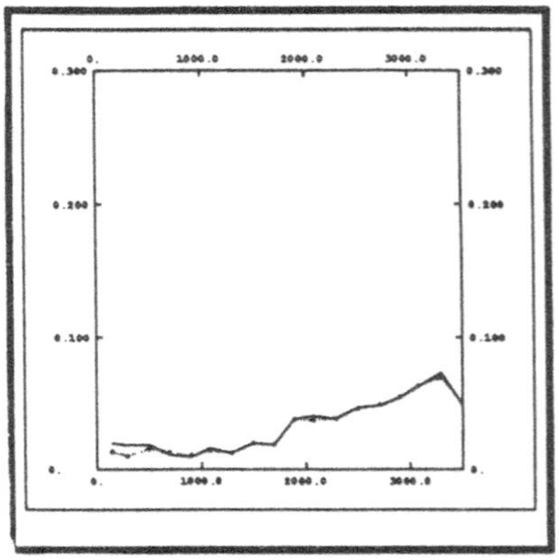

Lithotype 3

Figure 18: Fitted variograms with global non stationarity

Quantitative Geology and Geostatistics

Series Editor: F.M. Gradstein (*Canada*)

1. F. M. Gradstein, F. P. Agterberg, J. C. Brower and W. S. Schwarzacher: *Quantitative Stratigraphy*. 1985 ISBN 90-277-2116-5
2. G. Matheron and M. Armstrong (Eds.): *Geostatistical Case Studies*. 1987 ISBN 1-55608-019-0
3. *Cancelled*
4. M. Armstrong (Ed.): *Geostatistics*. Proceedings of the 3rd International Geostatistics Congress, held in Avignon, France (1988), 2 volumes. 1989 Set ISBN 0-7923-0204-4
5. A. Soares (Ed.): *Geostatistics Tróia '92*, 2 volumes. 1993 Set ISBN 0-7923-2157-X

GPSR Compliance

The European Union's (EU) General Product Safety Regulation (GPSR) is a set of rules that requires consumer products to be safe and our obligations to ensure this.

If you have any concerns about our products, you can contact us on ProductSafety@springernature.com

In case Publisher is established outside the EU, the EU authorized representative is:

Springer Nature Customer Service Center GmbH
Europaplatz 3
69115 Heidelberg, Germany

Batch number: 10823951

Printed by Printforce, the Netherlands

GEOSTATISTICS TRÓIA '92

Volume 2

Edited by

AMILCAR SOARES

Centro de Valarizacao de Recursos Minerais,
Instituto Superior Técnico,
Lisbon, Portugal

Kluwer Academic Publishers

Dordrecht / Boston / London

Library of Congress Cataloging-in-Publication Data

Geostatistics Tróia '92 / edited by Amilcar Soares.
p. cm. -- (Quantitative geology and geostatistics ; v. 5)
"[Papers from] the Fourth Congress Troia '92"--Foreword.
Includes index.
ISBN 0-7923-2157-X (set : acid-free). -- ISBN 0-7923-2155-3 (v. 1 : acid-free). -- ISBN 0-7923-2156-1 (v. 2 : acid-free)
1. Geology--Statistical methods--Congresses. 2. Petroleum--Geology--Statistical methods--Congresses. I. Soares, Amilcar. II. Series.
QE33.2.S82G46 1993
550'.72--dc20 92-46924

ISBN 0-7923-2156-1 (vol. 2)
ISBN 0-7923-2155-3 (vol. 1)
ISBN 0-7923-2157-X (set)

Published by Kluwer Academic Publishers,
P.O. Box 17, 3300 AA Dordrecht, The Netherlands.

Kluwer Academic Publishers incorporates
the publishing programmes of
D. Reidel, Martinus Nijhoff, Dr W. Junk and MTP Press.

Sold and distributed in the U.S.A. and Canada
by Kluwer Academic Publishers,
101 Philip Drive, Norwell, MA 02061, U.S.A.

In all other countries, sold and distributed
by Kluwer Academic Publishers Group,
P.O. Box 322, 3300 AH Dordrecht, The Netherlands.

Printed on acid-free paper

TABLE OF CONTENTS

PETROLEUM

OTHER APPLICATIONS

ENVIRONMENT

HYDROLOGY

SOIL SCIENCES

NATURAL RESOURCES

HUMAN SCIENCES

MINING

FOREWORD

From Frascati to Troia

In the Seventeen years between the first International Geostatistics Institute in Frascati and the Fourth Congress Troia'92, the geostatistical community has grown in number and diversity and has reached recognition yet holding on its trademark. In recent years it has emerged from its original field of application in the mining industry to include applications in petroleum, oceanography, soil, entomological and environmental sciences. The "others" category of past congresses has grown into the most important, and had to be subdivided into many new sub classifications.

Troia'92 has been a most fruitful forum where more than 200 participants from 28 different countries challenged each other's understanding of a still fast moving theory and learnt from unexpected applications from remote fields. Troia'92 will be remembered as the Congress of consolidation when Geostatistics reached its place within the dynamics of the roaring scientific growth of this century end.

The Organizing Committee.

Margaret Armstrong
André Journel
Peter Dowd
Amilcar Soares (General Secretary)

Acknowledgements

TROIA'92 was organized by CVRM - Instituto Superior Técnico (Lisbon).

The Organizing Committee would like to thank the members of the Scientific Committee who had selected the abstracts and reviewed the papers. The quality of the final version of the articles is due to their conscientious work.

The Congress was made possible by the financial aid of the following institutions and companies which have sponsored the technical sessions:

- Commission of the European Communities - DGXII
- SCRF - Stanford Center for Reservoir Forecasting
- SOMINCOR
- SPE - Sociedade Portuguesa de Empreendimentos
- RTZ Corporation PLC
- COGEMA
- EDM - Empresa de Desenvolvimento Mineiro
- AFAIRS - Associação Universidade - Empresa para Formação no Âmbito da Industria e Recursos do Sub-solo.
- INIC - Instituto Nacional de Investigação Científica
- DGGM - Direcção Geral de Geologia e Minas
- FLA - Fundação Luso Americana

The Organizing Committee is deeply obliged to these institutions and companies.

The extraordinary success of the Congress would not have been possible without the dedication of the students of the CVRM - Instituto Superior Técnico.

ANALYSIS OF QUASI-PERIODIC TIME SERIES BY A GEOSTATISTICAL METHOD WITH APPLICATION TO CLIMATOLOGY

H ALLISON and G B DAVIS
CSIRO Division of Water Resources
Private Bag PO, Wembley WA 6014, Australia

ABSTRACT

Confirmation of global climatic change requires analysis of quasi-periodic processes which is usually performed by one-dimensional time-series analysis. In our paper, a two-dimensional spatial analysis in the form of kriging over a moving block of data as described in Journel and Rossi (Math Geology, Vol 21, N.7: 715-39) is applied to time series data. The analysis allows separation of seasonal quasi-periodicity from over-the-years time behaviour and is thus a powerful method for use in climatology or for application to other quasi-periodic time series. This has been achieved by introducing a new morphological transformation "montage", which is an inverse operation to "stripe sampling" by Serra (1982).

The technique has been applied to a 100 year time series of rainfall data for Perth, Western Australia. It was found that the geostatistical method was capable not only of identifying a previously unreported 22 year (a double solar cycle) periodicity in rainfall, but also was able to predict, a year ahead, the heavy rainfall experienced in winter (May-August) 1991 in Perth.

1. INTRODUCTION

Methods of mathematical morphology which have been successfully applied to texture analysis in petrography, wood anatomy, histology and other applications (Serra 1982) has not been applied to climatology. The need to reveal the structure of "objects" in climatology is becoming increasingly important in view of such effects as greenhouse global warming, or, its converse "nuclear winter" (Pittock *et al* 1989).

In this article it is shown that some of the "objects" in climatology permit a new mathematical transformation, which is an inverse transformation to "stripe sampling" by Serra (1982). The "objects" in this study are quasi-periodic time-series, such as solar activity, reflected by sunspot numbers as in Fig 1 or rainfall. Note, however, that it is not sunspot numbers but solar irradiance (Lean 1991) and the solar magnetic cycle (Webb *et al* 1984) with periods of about 11 and 22 years respectively, which may influence climatic processes. There is a considerable controversy on this subject, some claiming a considerable influence of solar cycles on global climate (Newell *et al* 1989), while others are more sceptical (Pittock 1978, 1983). We do not intend to discuss this issue further except to state that the methods of mathematical morphology employed here do reveal a 22 year periodicity in Perth rainfall data.

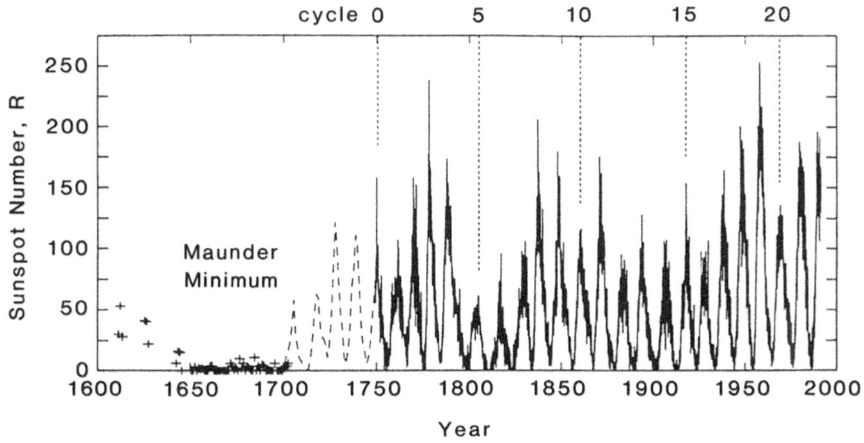

Fig 1 Quasi-periodic solar activity with period of about 11 years (After Lean 1991).

Quasi-periodic time-series are classically analysed by means of spectral representation of stationary processes (Priestley 1981), the power spectrum being capable of detecting hidden periodicies. As an example, a power spectrum of rainfall in Perth is presented on Fig 2, clearly showing, apart from a dominant one year periodicity, some other peaks of about 2.6 years, 8.4 years and 11.2 years.

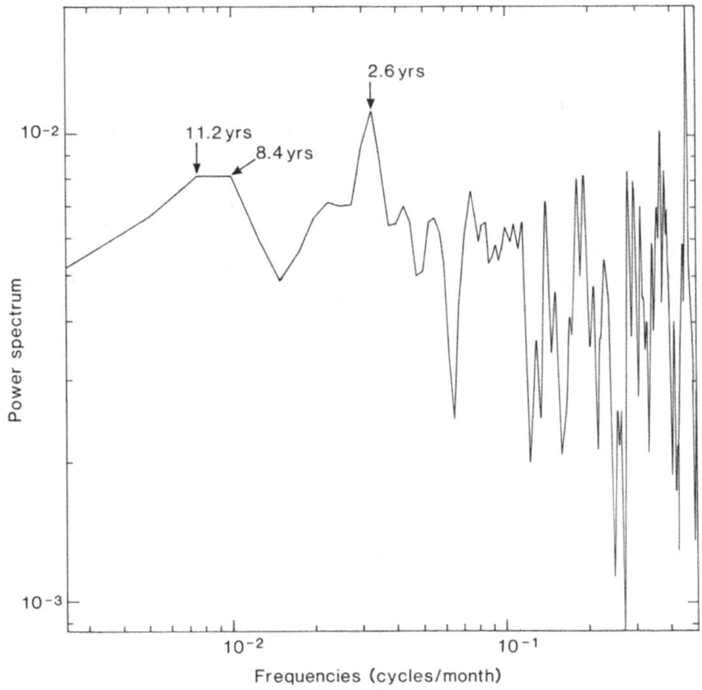

Fig 2 Power spectrum [in mm/month]^2 of Perth rainfall data (1907-1989) (private communication A.B. Pittock and R. Lean, CSIRO Division of Atmospheric Research).

In this article we present an entirely different approach, most closely related to Matheron's (Matheron 1989) preference for analysis of "real" data, rather than Fourier Domain analysis. To operate in the "real" time domain we propose a mathematical construction which converts a quasi-periodic time-series into a "regionalised variable". This is described below as a *montage*.

2. *MONTAGE* OF QUASI-PERIODIC TIME SERIES

A function f(t) is called periodic if there exists a period T such that

$$f(t+T) = f(t) \tag{1}$$

for any time t.

Periodicity implies a complete determinism and predictability of f(t) beyond any interval of length T. A quasi-periodic function however, is assumed random within each periodic interval T. That is, each cycle does not exactly repeat itself at previous and consequent times.

We shall consider such a class of quasi-periodic functions {f(t,T)} which has a periodic support with period T. Subdividing t into a sequence t_1, t_2, $t_3 \ldots t_n$ such that the Euclidean distance d $[t_{i+1} - t_i]$ = T, we project from each point t_i an axis τ_i, orthogonal to t.

The values of the quasi-periodic function at the interval $[t_i, t_{i+1}]$ are assigned to the axis τ_i, forming in this way a two-dimensional *montage* which consists of juxtaposed portions of the original quasi-periodic function. This is illustrated on Fig 3, where a *montage* of monthly rainfall data is shown; the t axis is labelled "years" and the τ_i axis is labelled "months".

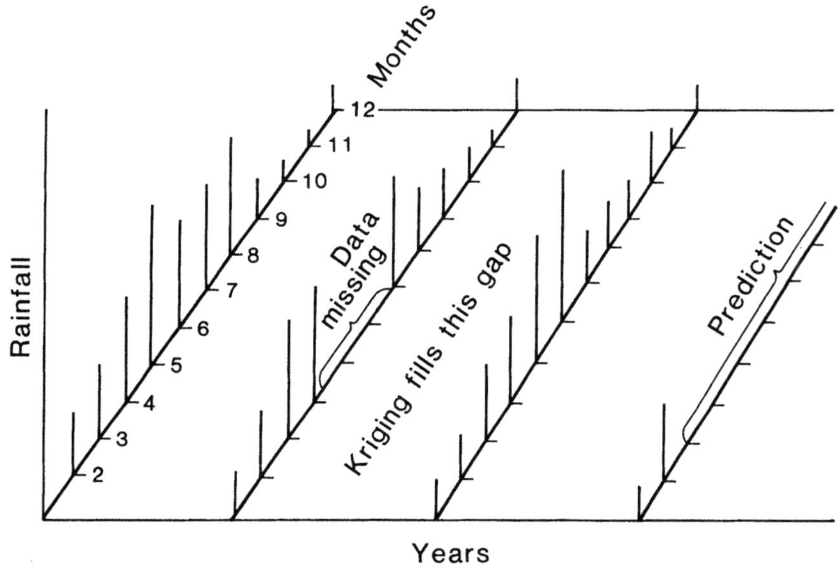

Fig 3 *Montage* of rainfall data into a point-process on the plane.

A *montage* consisting of individual measurements such as shown on Fig 3, represents a two-dimensional point process on the (t,τ) plain (Bartlett 1975). Such a point process possesses a correlation structure in two dimensions, in years and months. The two-dimensional correlation may have a certain structure, specific to regions on the (t,τ) plane, therefore the *montage* which we shall denote now Z (t,τ) is one realisation of a random regionalised variable {Z (t,τ)} to which the geostatistical theory of Matheron (1975) is applicable.

When the original quasi-periodic function is continuous, its *montage* takes the form of stripes, juxtaposed along the t axis periodically with the period T as shown on Fig 4.

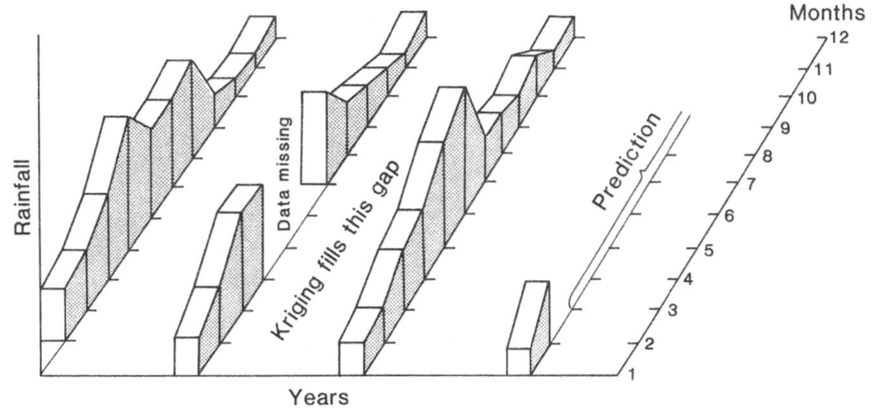

Fig 4 Stripes of the *montage* showing that the *montage* is an inverse transformation of stripe sampling by Serra (1982).

One may interpret the stripes on Fig 4 as akin to the "stripe sampling" by Serra (1982), by which the original quasi-periodic time series can be recovered from its *montage*. It is seen now that there exists a one-to-one correspondence between the original quasi-periodic time series and its *montage*, *ie* montaging and stripe-sampling are inverse mathematical operations.

3. REGIONALISATION OF A *MONTAGE* OF A QUASI-PERIODIC FUNCTION

The data points between the stripes of the *montage* have no physical meaning, however, an artificial surface can be, theoretically stretched over the whole domain (t,τ) by interpolation between the stripes, for example by ordinary kriging. To apply kriging (as in de Marsily 1986), we need firstly to construct a variogram, which in turn, depends on the "distances" (ranges) in the (t,τ) plane. Our experiments with distances on the Perth rainfall data set has shown that variograms obtained display a marked anisotropy. However, by a choice of linear transformation of one coordinate, an isotropic variogram can be obtained.

In particular, by choosing a time interval of one-tenth of a month (called decimonth, abbreviated to dmonth) a near-isotopic variogram was calculated which was approximated, by an exponential model. (Fig. 5). Alternative variogram models could have been fitted to the "experimental" variogram (eg. a parabolic model), however kriging has been shown to be robust even with significant variogram model fitting errors. Yakowitz and Szidarovszky (1985) tested a number of variogram models and found the exponential model performed as well as many others. This is further supported by work of Carr and Myers (1984) and Carr (1990) who established the similarities between kriging and low-pass filtering which leads to smoothed data. The choice of variogram model is therefore not critical since variogram fitting errors will be significantly smoothed by kriging.

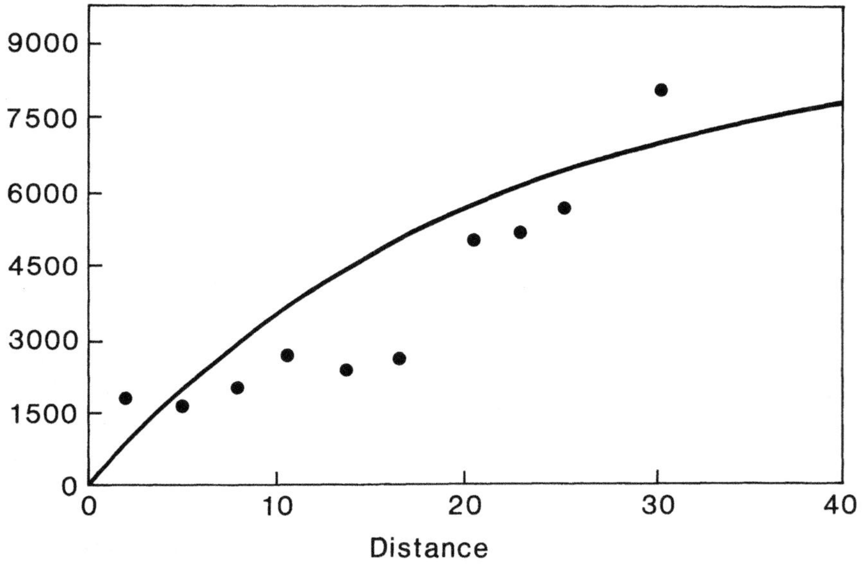

Fig 5 Variogram of the *montage* of Fig 3 for rainfall data.

A regular lattice of data points with axes dmonth and years is shown on Fig 6, where the circle of radius r limits the neighbourhood of a central estimation point and where the inscribed square delineates a moving window used in ordinary kriging.

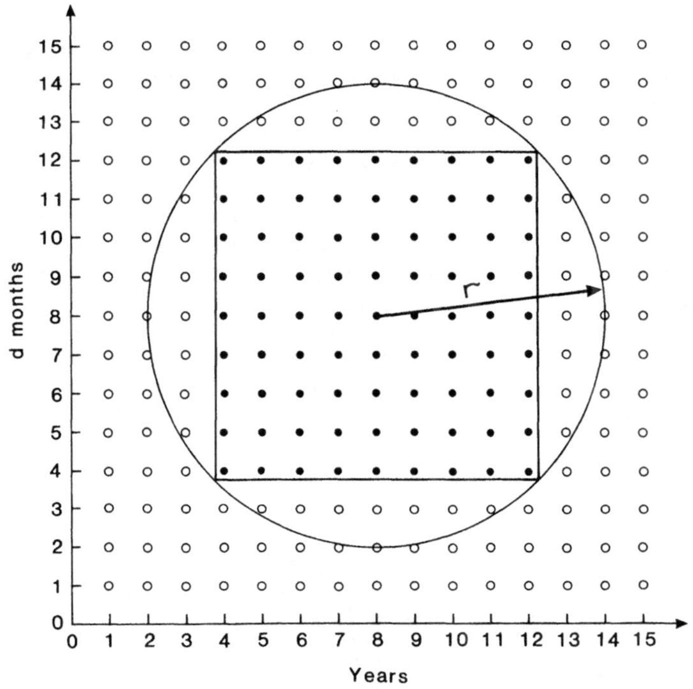

Fig 6 Zone of influence with radius r and the moving window of kriging.

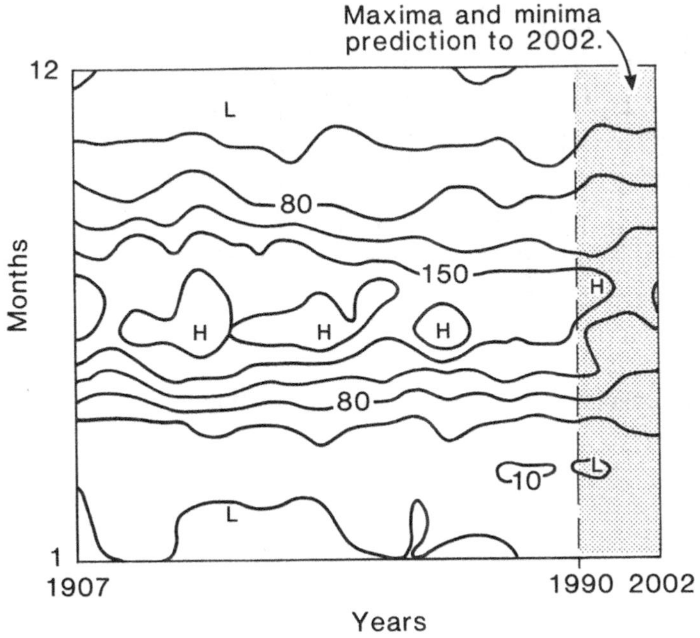

Fig 7 Kriging of the *montage* for rainfall data (1907-1989).

Kriged results are shown in Fig 7. Kriging has not only transformed the point-process *montage* of Fig 3 into a continuous surface, but has also provided an extrapolation beyond 1990. Although there are inaccuracies when extrapolating by any method, winter rainfall in Perth in 1991 was significantly above average, in agreement with the plot on Fig 7.

4. MORPHOLOGICAL CLEANING OF THE *MONTAGE*

The grey-tone picture as on Fig 7 could be subjected to a variety of morphological transformations, such as the "Hit and Miss" transformation (Serra 1982). We have applied median filtering. Our interest was to sharpen the repetitive pattern of "H" on Fig 7 apparently corresponding to a certain periodicity in rainfall. We have used two filtering intervals for median filtering of the original rainfall data: 5 data points and 121 data points; then the *montage* has been constructed, and ordinary kriging applied again. Five data-point median filtering produced results not much different from Fig 7, but 121 data-point median filtering refined the kriged data to that shown in Fig 8.

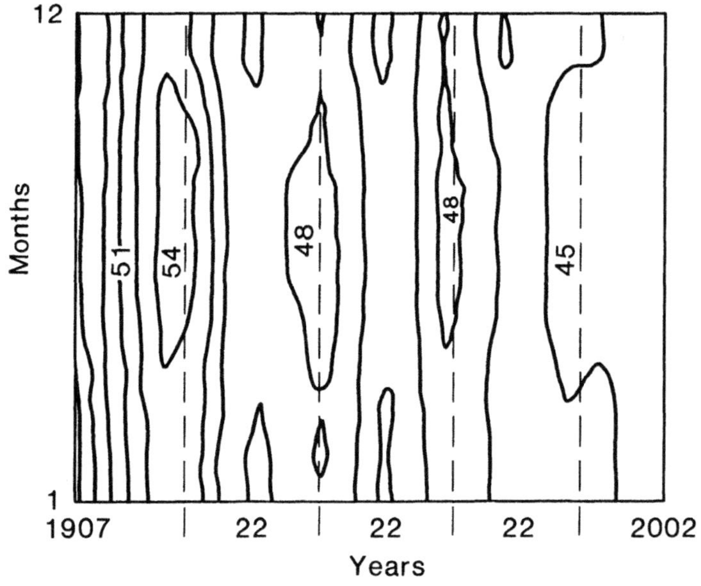

Fig 8 Morphological cleaning by median filtering.

A 22 year periodicity is fairly clearly evident in Fig 8 separated by 22 year intervals. This periodicity was not apparent in the Fourier spectral analysis graph of Fig 2. Newell *et al* 1989 also identified a 22 year solar magnetic cycle.

SUMMARY AND CONCLUSION

1. A *montage* allows transformation of a quasi-periodic time-series to a regionalised variable to which mathematical morphology is applied.

2. Using Perth rainfall as an example of a quasi-periodic process, interpolation, extrapolation and median filtering were performed on the two-dimensional *montage* using years/decimonth as coordinates.

3. Using a two-dimensional variogram in years/dmonth coordinates, the *montage* is kriged to reveal between-years and seasonal correlations, unlike correlations used in classical time-series analysis. This allowed long-term between-year trends to be identified.

4. Although geostatical analysis of the *montage* has revealed periodicities (of 22 years in Perth rainfall) which a traditional spectral Fourier analysis appeared unable to resolve, the methods are considered complementary.

5. The results presented are preliminary, leaving a detailed comparison with other time-series methods to future research.

ACKNOWLEDGMENTS

We are grateful to the Bureau of Meteorology (Perth) for rainfall data. Spectral analysis on Fig 3 is courtesy of Drs B Pittock and R Allan (CSIRO Division of Atmospheric Research, private communication). The computer code for median filtering was written by Mr R Stuart (formerly CSIRO).

REFERENCES

Bartlett, M.S. (1975) The Statistical Analysis of Spatial Pattern. Chapman and Hall, London.

Carr, J.R. (1990) Application of spatial filter theory to kriging: Math Geol. 22, 1063-1079.

Carr, J.R. and Myers, D.E. (1984) Application of the theory of regionalized variables to the spatial analysis of Landsat data: 9th William T. Pecora Memorial Remote Sensing Symposium, Institute of Electrical and Electronic Engineers (IEEE) Computer Society Press, Silver Spring, Maryland, p. 55-61.

Journel, A.G and Rossi, M.E. (1989) When do we need a trend model in kriging? Math. Geol. 21(7) 715-39.

Lean, J. (1991) Variations in the sun's radiative output. Review of Geophysics 29, 4, 505-535.

Marsily, G. de. (1986) Quantitative Hydrogeology. Academic Press, London.

Matheron, G. (1975). The theory of regionalised variables and its applications. Cahiers du Centre de Morphologie Mathematique de Fountainebleau No. 5.

Matheron, G. (1989) Estimating and Choosing. Springer-Verlag, Berlin.

Newell, N.E., Newell, R.E., Hsuing, J. and Zhongziang, W. (1989) Global marine temperature variation and the solar magnetic cycle. Geophys. Res. Let. 16(4) 311-14.

Pittock, A.B. (1978) A critical look at long-term sun-weather relationships. Review of Geophysics and Space Physics Vol. 16 No. 3 pp. 400-420.

Pittock, A.B. (1983) Solar variability, weather and climate: an update. Quart. J.R. Met. Soc. 198, 23-55.

Priestley, M.B. (1981) Spectral Analysis and Time Series, Vol. I and II. Academic Press, London.

Pittock, A.B., Walsh, K. and Frederiksen, J.S. (1989) General circulation model simulation of mild nuclear winter. Climate Dynamics, 3, 191-206.

Serra, J. (1982) Image Analysis and Mathematical Morphology. Vol. 1. Academic Press, London.

Webb, D.F., Davis, J.M. and McIntosh, P.S. (1984) Observations of the re-appearance of polar coronal holes and the reversal of the polar magnetic field. Solar Physics, 92, 109-132.

Yakowitz, S.J. and Szidarovszky F. (1985) A comparison of kriging with non-parametric regression methods. J. Multivariate Anal. 16, 21-53.

Kriging Temperature in Scotland using the External Drift Method

Gordon Hudson
Land Use Division
Macaulay Land Use Research Institute
Craigiebuckler
Aberdeen AB9 2QJ
Scotland

Abstract

Land use research includes the formulation and testing of models for land evaluation, land suitability and crop yield at the regional or district scale. Such models provide the bio-physical inputs necessary to make economic forecasts about the returns from alternative land uses. Modelling the seasonal and spatial variation of climatic variables is necessary for land evaluation and yield prediction. Temperature data are sparse in Scotland, with 149 climate stations covering a domain extending 600 km from north to south and 300 km from west to east. The temperatures are correlated with elevation, which is known everywhere, so it was possible to make use of this relationship to krige temperatures by using elevation as external drift. The underlying hypothesis is that the regression of temperature on elevation is linear. The spatial structures of monthly means of temperature were analysed by computing variograms in several directions. The temperatures are nonstationary in different directions in winter and summer but there appear to be stationary periods in spring and autumn. The underlying variogram of January temperature was used in universal kriging with elevation as external drift. The parameters of the linear regression with elevation may not be constant over the whole domain and using a moving neighbourhood allows for this. Kriging temperatures with elevation as external drift gives estimates of temperature which are closer to reality than those obtained kriging temperatures using only standard drift functions.

Keywords:

january temperatures, underlying variogram, non-stationary geostatistics, external drift, Scotland.

Introduction

Land evaluation for agriculture, forestry or conservation and land suitability mapping for individual crops needs information about the spatial and temporal variation in physical and economic factors. The physical data comes from a variety of sources; chloropleth maps of soil types or bioclimatic regions, isoline maps of topographic contours, and point measurements of soil properties in soil profile pits or of climate variables measured at climate stations. This point data must be interpolated to the nodes of a regular grid to prepare thematic maps for input to a Geographic Information System. The soil point data are usually interpolated in the field by the expert interpretation of a soil surveyor who generates maps of soil types, from which chloropleth maps of associated variables can be derived. Climatic variables can be mapped by expert assessment also (Birse, 1971), but it is becoming more common to use interpolation techniques. The information sources in the form of these thematic maps are compiled by overlay into climate, soil and land inventories, either manually or by using a Geographic Information System, and then matched with crop requirements, management inputs and conservation needs to produce land suitability assessments, following the methods of the F.A.O.(F.A.O., 1984).

Soil and topographic maps provide information about static properties such as soil available water or slope, though spatial interpolation can also give this information. The climate and more specifically soil-climate interactions are dynamic properties which vary throughout the average season and also exhibit differences between seasons referred to as seasonality. The climate variables in Scotland are observed and recorded to facilitate the weather forecasting service provided by the U.K. Meteorological Office and are sited particularly on low ground in agricultural districts or near cities and airports. Rainfall is measured on a denser network than temperature, especially in upland catchments supplying water to hydro-electric power generators and in major river catchments for hydrological studies.

When using climate data for land suitability mapping in temperate, developed countries such as Scotland it is long term averages and seasonal departures from these averages which are of interest, rather than the short term fluctuations referred to as 'weather', although some information on 'uncertainty' can be derived by analysis of year to year fluctuations. There are also requirements for prediction of extreme events, such as strong winds which cause windthrow of timber crops, or intense rainfall events causing soil loss or catastrophic flooding. The primary variables that require to be estimated are temperature, rainfall and wind speed, though solar radiation, humidity, surface roughness and albedo are also needed to model secondary variables such as potential evapotranspiration.

On the global scale, climate is controlled by physical processes which are modelled by general circulation models. At the local scale of a continent or country it may be more appropriate to use a simpler empirical modelling procedure which takes account of local correlations between climate and topography. In this paper the spatial estimation of time-averaged temperatures using kriging with elevation as an external drift function is examined.

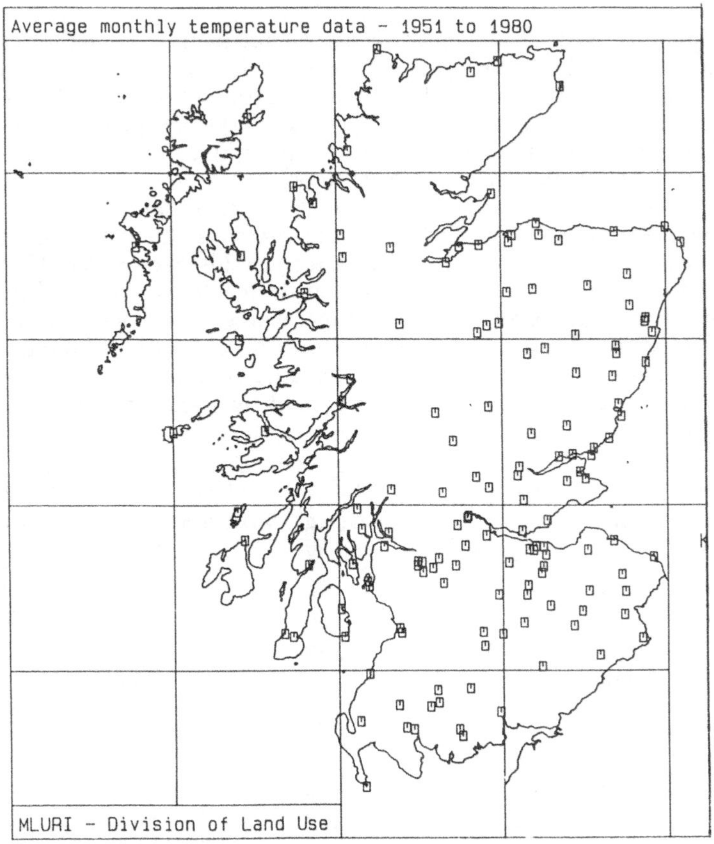

Figure 1: The location of climate stations in Scotland

The study area and the temperature data

The Scottish Office Agriculture and Fisheries Department have commissioned research on land suitability mapping in Scotland, for agricultural and forest crops suited to northern temperate climate zones. This research needs a large number of data sources for overlay within a Geographic Information System. The aim of the work presented here is to investigate a method for the preparation of thematic temperature maps.
Temperatures are recorded several times daily at each climate station and daily means are computed. From these daily readings the arithmetic mean for each month of the year was calculated for the 30-year period, 1951–80. The spatial variability of these 30-year monthly means is examined in this study by using variograms and average temperatures are mapped by kriging. To develop a method for mapping temperatures over the whole country, the computing was simplified by producing point kriged estimates on a regular, square grid with 5 km spacing over the whole of Scotland and aligned east-west and north-south.

Month	Mean temp.	Variance	Correlation with easting	Correlation with northing	Correlation with elev.
January	2.8	1.02	-0.47	0.11	-0.80
February	2.8	0.94	-0.46	0.05	-0.84
March	4.6	0.63	-0.50	-0.02	-0.89
April	6.7	0.55	-0.49	-0.31	-0.78
May	9.5	0.57	-0.54	-0.44	-0.62
June	12.4	0.57	-0.33	-0.54	-0.49
July	13.6	0.58	-0.18	-0.55	-0.49
August	13.6	0.49	-0.28	-0.48	-0.62
September	11.8	0.49	-0.33	-0.34	-0.76
October	9.3	0.58	-0.44	-0.15	-0.82
November	5.4	0.89	-0.46	0.01	-0.80
December	3.9	0.98	-0.52	0.05	-0.77

Table 1: Monthly mean temperatures and correlations with easting, northing and elevation

The 30-year mean temperature data from 145 climate stations (excluding the stations on the Orkney and Shetland Islands) in Scotland (Figure 1) were used in the study. The irregular scatter of the data points is worth noting and the lack of stations in the north west gives small kriging systems with few data points for interpolation. The stations on the Shetland Islands and the Orkney Islands were dropped from the study at this stage since the maritime influence on climate is strong and these stations are too isolated from those of the mainland for the establishment of kriging systems with more than three data points.
The average temperatures for each month are normally distributed when 145 climate stations are considered. The correlation of 30-year mean temperatures with easting (x_1–coordinate), northing (x_2–coordinate) and elevation

is shown in Table 1. There is an annual cycle of temperature ranging from a low of 2.8°C in winter to a high of 13.6°C in summer and a fluctuation in the variance from 1.03°C^2 in winter to 0.59°C^2 in summer. The correlations of temperature with easting, northing and elevation are also seasonal.

Experimental variograms for each month

An initial description of the spatial variation throughout the year was gained by computing omnidirectional variograms for each 30-year monthly mean. The experimental variograms are shown for January, April, July and October in (Figure 2). The variogram sills are periodic though the range remains at a constant distance of about 80 km in April, July and October. In January the variogram continues to rise, behaviour typical of non-stationary data.

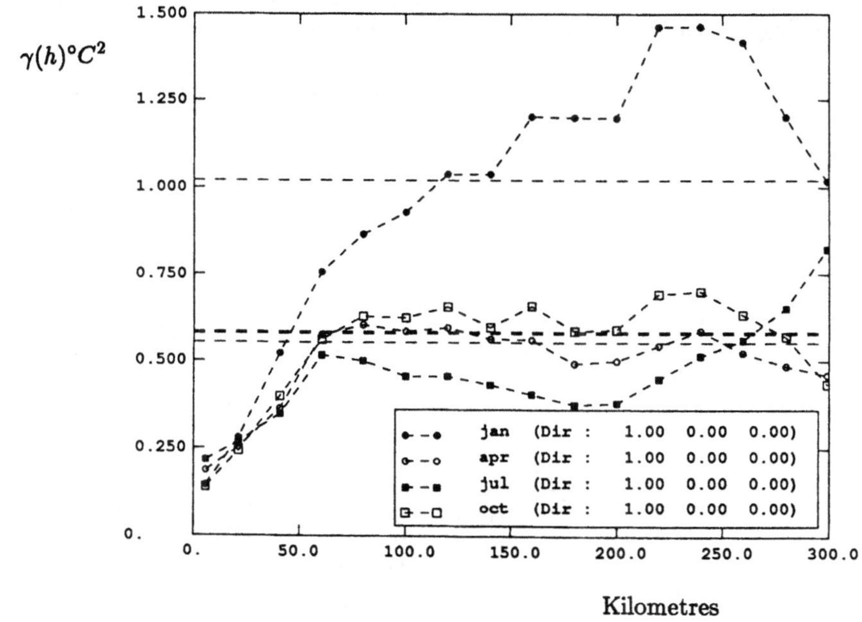

Figure 2: Omnidirectional variograms of 30-year mean temperatures

Considering the annual period to run from January to December, the variogram sill is high in January and February then falls to its lowest in August and September before rising to the high value in November and December. The annual cycle in the variogram sill expresses the average behaviour of the temperature but the weather patterns in individual months may change from year to year (Lamb, 1950). The variograms of this average behaviour in time were
computed in four orientations for each month of the year and revealed dif-

ferent anisotropic structures that are linked to the different weather systems that, on average, prevail over Scotland (Hudson, 1991) in summer and winter. The experimental variograms of January temperatures illustrate the winter spatial structure, and July is typical of the summer spatial structure. The next step was to interpret and model the experimental variograms for January temperatures.

Modelling the variogram of January temperatures

The 30-year mean temperature for January is more strongly correlated with easting (−0.47) than with northing (0.11). The variogram in the east-west direction is strongly influenced by a trend in the data ('drift') and the variance continues increasing to 160 km, approximately half the width of the domain in this orientation. A sill with a value less than the sample variance is reached in the south-north direction at a range of about 80 km (Figure 3). The variograms paired in the intermediate directions are parallel

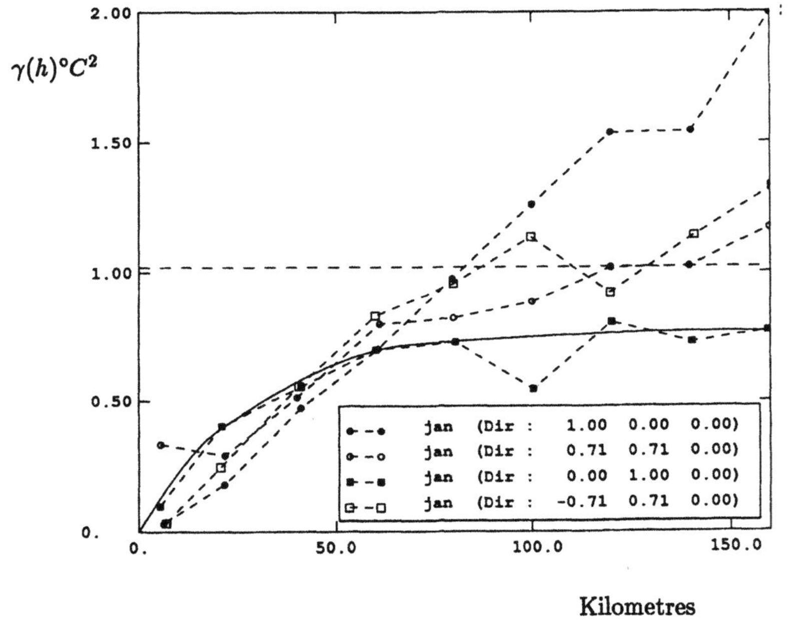

Figure 3: Variograms of 30-year mean January temperatures

which establishes the directions of maximum and minimum trend.
The temperatures in January are regarded as a realization of a random function $T(x_1, x_2)$ composed of a mean function $m(x_1, x_2)$ and a second-order

stationary random function $Y(x_1, x_2)$ with a mean equal to zero, such that

$$T(x_1, x_2) = m(x_1, x_2) + Y(x_1, x_2).$$

The random function $Y(x_1, x_2)$ is called the residual and has a stationary covariance function $C_Y(h)$ which can be used to construct the variogram for $Y(x_1, x_2)$ by the relation

$$\gamma_Y(h) = C_Y(0) - C_Y(h),$$

where $\gamma_Y(h)$ is the variogram of the residual and $C_Y(0)$ and $C_Y(h)$ are the covariances of the residual at lags of zero and h. The mean function $m(x_1, x_2)$ is called the drift and we assume this is a smooth function of the coordinate vector $\mathbf{x}$ and can be represented by a polynomial of order k.
We make the assumption that the theoretical variogram $\gamma_Y(h)$ of the residuals $Y(h)$ is isotropic. Taking the drift as a function of the coordinates:

$$m(x_1, x_2) = a_0 + a_1 x_1 + a_2 x_2 \qquad \text{(linear)}$$

$$m(x_1, x_2) = a_0 + a_1 x_1 + a_2 x_2 + a_3 x_1^2 + a_4 x_2^2 + a_5 x_1 x_2 \qquad \text{(quadratic)}.$$

Assuming the drift is constant in one direction, say x_2, then we have

$$m(x_1, x_2) = a_0 + a_1 x_1 \qquad \text{(linear)}$$

$$m(x_1, x_2) = a_0 + a_1 x_1 + a_2 x_1^2 \qquad \text{(quadratic)}$$

and it follows from these assumptions that the underlying variogram can be identified experimentally in the direction x_2, which corresponds to north to south for January temperatures in Scotland. Generally the drift of a regionalised variable is not constant in a particular direction, and may change within the data domain. Then a more sophisticated approach using generalised covariance functions (Matheron 1973, Chauvet and Galli, 1982) becomes necessary.
The south-north experimental variogram reaches a uniform value (the 'sill') since there is no trend in this orientation in January and is taken as the variogram underlying the trend. Assuming spatial continuity is high in mean monthly temperatures because the large sample support in time (930 days) smoothes the overall variability, the underlying north – south variogram is modelled with a zero nugget effect. A theoretical variogram comprising a linear combination of three spherical models was fitted to the experimental variogram in this direction (Figure 3). The ranges of the nested models are 20, 70, and 150 km with most of the variance attained by about 80 km. This model is used in kriging with and without external drift. Cross-validation, using the climate station data, was used to test the effectiveness of linear and quadratic drift functions and a linear function was found to give the best results.

July mean temperatures

In July there is a south to north trend in temperatures (correlation with northing -0.55), which is much reduced in the east to west direction (correlation with easting -0.18). The variograms in two principal orientations are reproduced (Figure 4) for comparison with those of January. The July variogram does

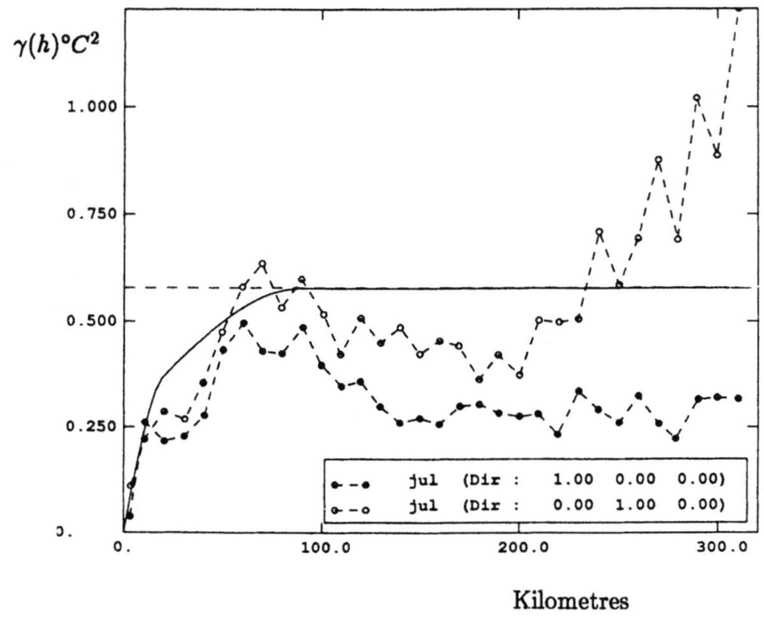

Figure 4: Variograms of 30-year mean July temperatures

not rise steadily in the south to north direction and the effects of the trend are seen mainly in the depression of the sill value to about half the variance in the east-west direction at approximately 100 km lag. The trend in July probably arises from the stable air masses associated with the high pressure systems which prevail over Scotland in summer, together with the long hours of sunshine which particularly affect the south.

Universal kriging of 30-year January mean temperatures

The kriging interpolator is used here to produce an estimate at each node of a regular grid, in this case at points separated by 5 km and where the elevation is known. The nearest climate stations to each grid node, up to a maximum number of 20, and inside a neighbourhood of 80 km radius selected after cross-validation tests are used for kriging. The neighbourhood was selected to have 20 stations in an 80 km radius. The point-kriged map of 30-year mean temperatures, taking no account of the correlation between temperature and elevation, produced a very smooth interpolation. This map, not shown here, was useful in revealing broad regional trends, but masked the local variability caused by the irregular, often mountainous, terrain in Scotland. The knowledge of the correlation of temperature with elevation (Table 1) was used next to improve the estimation of temperature with elevation as external drift.

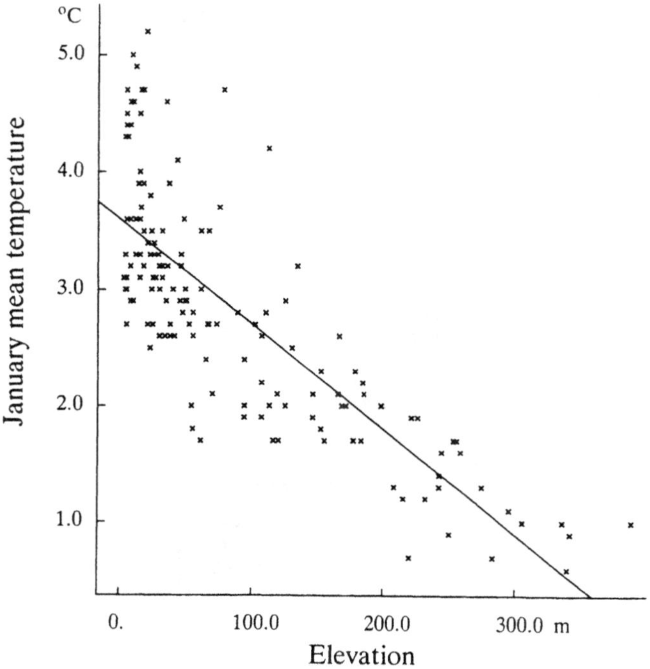

Figure 5: Scatter plot of January temperatures versus elevation at climate stations

Using elevation as external drift to krige January mean temperature

Mean temperatures are controlled by factors which are strongly correlated with spatial location and topography. Incoming solar radiation declines with increasing distance from the equator and the amount received by the land varies with both aspect and slope. Locally, in temperate maritime countries, cool air masses from polar regions exert a variable cooling influence. Thus we have a set of conditions which suggest the use of nonstationary geostatistics to estimate temperatures, taking account of other correlated variables. The correlation between temperature and elevation (Table 1) exists at all seasons though is stronger in winter than summer. The seasonal effects are due to the deterministic periodic variation in solar radiation, overlaid by the effects of various warmer and cooler air mass movements, particularly in winter.
Here we consider January temperatures which have a correlation coefficient of -0.82 with elevation over the whole data field. The correlation with elevation should not be considered as constant over geographic space and this is where

a local interpolator, rather than a global one, offers advantages. The scatter diagram (Figure 5) shows that in January, temperatures are scattered at elevations below about 50 m and approximate a linear function above 100 m. The linear behaviour is assumed to extend to temperatures above 400 m although this hypothesis cannot be confirmed. The scatter diagram of January temperature (Figure 6)kriged with elevation as external drift shows how the correlation with elevation has provided a reasonable map. The kriged raster map of January mean temperature (Figure 7) was produced with a neighbourhood of 80 km and with the nearest 20 data points used in the kriging system. The value of the external drift coefficient varies over the data domain

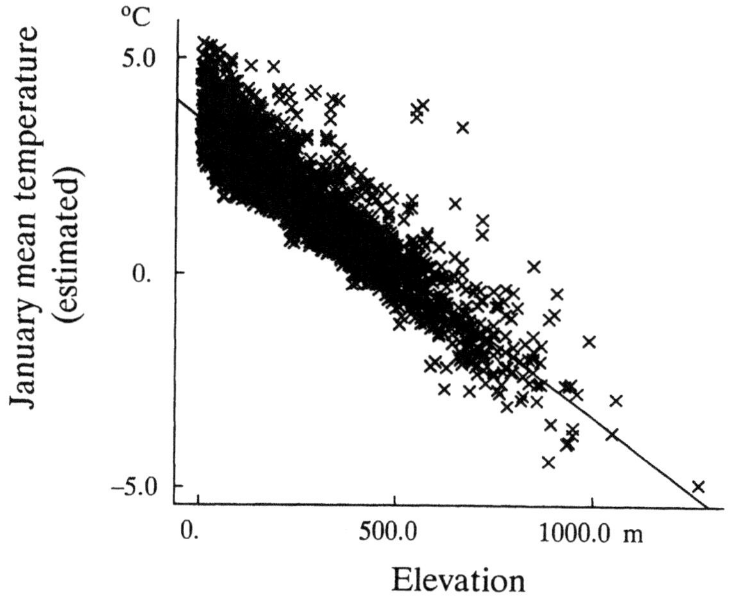

Figure 6: Scatter plot of kriged estimates of January temperatures versus elevation in Scotland

and can be mapped (Hudson and Wackernagel, *in preparation*) to aid interpretation of the kriging. The north-west of Scotland, where there are few climate stations, has a steeper external drift coefficient than the remainder of the country. This is probably due to the coastal site at Benbecula being warmer and the site at Strathconon in a sheltered glen being cooler than expected in an area of sparse data. A numerical classification of site shelter and distance from the sea is potentially useful for estimations by kriging in mountainous terrain. The effect in the north-west could be reduced by enlarging the neighbourhood for kriging, but increasing the number of climate stations in the network would be more useful.

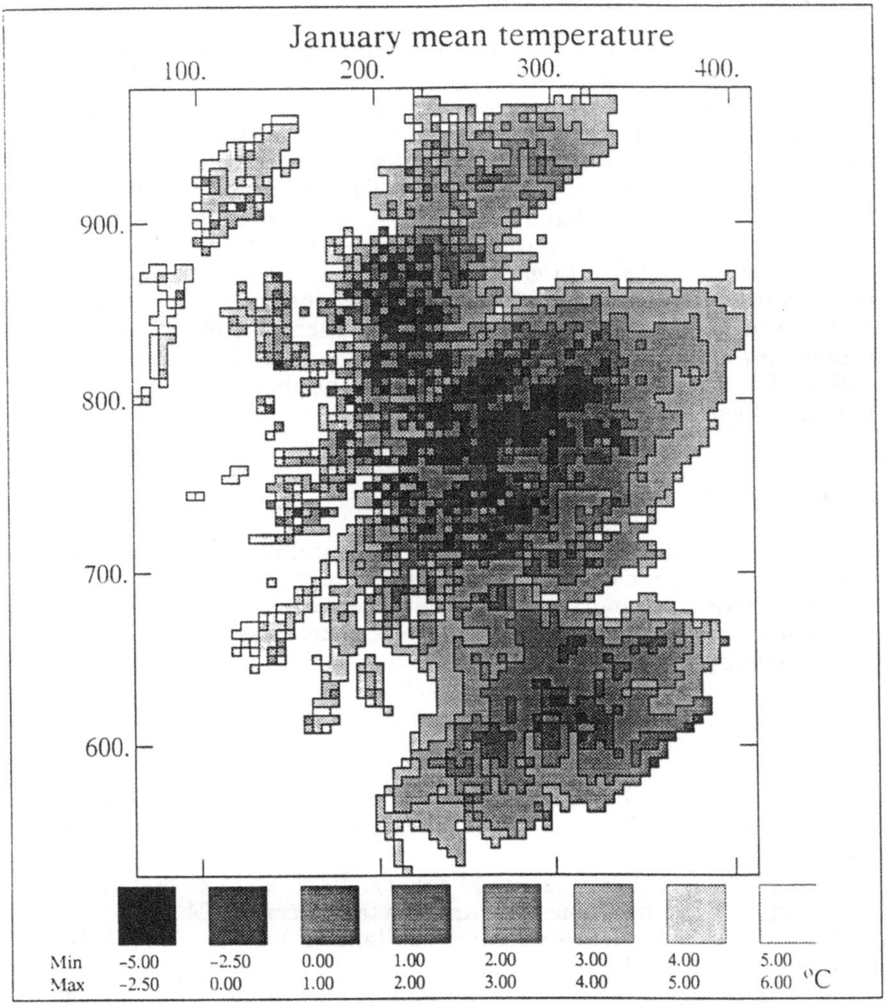

Figure 7: January mean temperatures kriged using elevation as external drift in Scotland

Conclusion

The spatial interpolation of 30-year mean monthly temperatures was explored using nonstationary geostatistics with elevation as external drift. The temperatures were shown to be nonstationary in winter and summer with quasi-stationary periods at the spring and autumn equinoxes and a linear drift function of the coordinates was incorporated in a universal kriging system to estimate January mean temperature. The distribution of climate stations means that there is interpolation in geographic space, but extrapolation in data space because the stations occur only below 400 m.
Temperatures are correlated with elevation which is known with reasonable accuracy everywhere in Scotland. The correlation coefficient differs from place to place in the data field and kriging with external drift, because it uses a local moving neighbourhood, computes an implicit local regression which allows for this variability. The size of the neighbourhood and the number of data points to include are critical.

Acknowledgements

The author wishes to thank the Scottish Office Agricultural and Fisheries Department for support while at the Centre de Géostatistique, Fontainebleau. The work was carried out as part of a course in geostatistics at the Centre. I also thank my supervisor Dr. H. Wackernagel for explaining nonstationary kriging with external drift.

References

Birse, E.L. 1971. The Bioclimatic subregions of Scotland. **3**: Moisture Subdivisions and Thermal Subzones. Macaulay Land Use Research Institute: Aberdeen.
Chauvet, P. and Galli, A. 1982. *Universal Kriging*. Publication C-96, Centre de Géostatistique, Ecole des Mines de Paris, Fontainebleau, p. 94.
F.A.O. 1984. 'Guidelines: Land Evaluation for Rainfed Agriculture', F.A.O. Soils Bulletin **52**. F.A.O., Rome.
Hudson, G. and Wackernagel, H. 1992. 'Mapping temperature using kriging with external drift: theory and an example from Scotland. *In preparation.*
Lamb, H.H. Types and spells of weather around the year in the British Isles: annual trends, seasonal structure of the year, singularities. Q.J.R. Met. Soc., London, **76**, 1950, 393–438.
Matheron, G. 1973. 'The intrinsic random functions and their applications', *Advances in Applied Probability*, **5**, 439–468.

CLIMATOLOGICAL IDENTIFICATION OF THE NON-STATIONARITIES OF INTERMITTENT PRECIPITATION FIELDS.

I. BRAUD, Ph. CROCHET and J.D. CREUTIN
Laboratoire d'étude des Transferts en Hydrologie et Environnement
(LTHE/IMG)
BP 53 X
38041 GRENOBLE Cédex FRANCE

1- Introduction.

In many geophysical sciences, precipitation appears as a key variable. In hydrology, it forms the input of rainfall-runoff models used for flood forecasting or the management of hydroelectric dams. In meteorological or climatological models, its knowledge is necessary on meshes of 30x30 to 100x100 Km^2. Its evolution through the soil and the vegetation determines the mass and heat exchanges between the biosphere and the atmosphere and , consequently the evolution of weather and climates. Therefore, it is necessary to have powerful tools for the interpolation of rainfall fields, or the estimation of mean areal rainfall. However, when the timestep considered decreases (a few hours to a few minutes), the intermittency phenomenon appears and precipitation is thus highly non stationary. Non-stationarities are also present in rainfall cells themselves. Indeed, the physical study of rainfall generation shows that lower intensities than in the center are encountered at the edge of the cells. Prior to any kriging or interpolation, it would be of interest to remove this first order non-stationarity.

When dealing with precipitation fields, several realizations are available, contrarily to mining or hydrogeology. Therefore, the first idea was to estimate climatological means (or ensemble averages) at each gauge and, then to remove them from the rainrate. This approach was fruitful for rainfall accumulations of several hours, when it was raining everywhere (Obled and Creutin, 1986). Unfortunately, with hourly rainfall fields, the obtained drift function was too smooth and was mixing non-zero rainrates and intensities equal to zero (Barancourt, 1990). The use of universal kriging did not prove more useful because the polynomial drift functions did not suit the physical "bell" shape of the rain cells (Chua and Bras, 1982). Furthermore, the available density of the network was not large enough to allow for the robust identification of the degree of the drift (Creutin and Obled, 1982).

In this paper, we propose a method for the estimation of the drift function, more adapted to the structure of rainfall fields and taking into account the intermittency phenomenon. It is based on the definition of a conditional expectation of the rainrates, conditioned by the distance between a point where it rains and the edge of the rain area. This function of the distance exhibits a bell shape which fits the physical observations very well. The definition of the drift function is given in section 2, with the practical method used for its estimation. In section 3, we investigate the non-stationarity of the non-zero rainrate by means of the statistical analysis of the residuals, defined as the differences between the observed rainrates and the value of the drift at the same location, in terms of non-bias, variance and covariance.

All the results discussed here are based on the analysis of a data set of hourly rainfall accumulations collected in 1989 in Niger around Niamey, on an area of about 12000 Km^2 (Roux, 1990; Thauvin and Lebel, 1991). Finally, some perspectives for the practical use of the above defined drift function for point rainfall estimation are discussed in section 4.

2- The drift function and its estimation.

a- *Formulation.*

Let's denote the instantaneous rainrate by $Z(\underline{x}, t)$ where $\underline{x}$ and t refer respectively to the space location and the time. The area under study is A.

For each realization t of the random process $Z(\underline{x}, t)$, the rain area A(t) can be defined using the indicator function of $Z(\underline{x}, t)$:

$$\begin{cases} I(\underline{x}, t) = 0 & \text{if } Z(\underline{x}, t) = 0 \\ I(\underline{x}, t) = 1 & \text{otherwise.} \end{cases}$$

$$A(t) = \{ \underline{x} \in A \,/\, I(\underline{x}, t) = 1 \} \qquad (1)$$

For a given direction u, we introduce the distance $h_u(\underline{x}, t)$:

$$h_u(\underline{x}, t) = \mathrm{Min} \{ \| \underline{x} - \underline{y} \| \,/\, \underline{x} \in A(t), \underline{y} \notin A(t), \underline{x} - \underline{y} = k\,u \} \qquad (2)$$

which is the minimum distance between a point $\underline{x}$ belonging to A(t) and the frontier of A(t), in the u direction.

A conditional expectation of $Z(\underline{x}, t)$ is then calculated as a function of the distance h :

$$D_u(h) = E[Z(\underline{x}, t) \,/\, \underline{x} \in A(t) \text{ and } h = h_u(\underline{x}, t)] \qquad (3)$$

If $D_u(h)$ depends effectively on h, the existence of a systematic drift inside the rain areas is shown.

If $D_u(h)$ is independent of the direction u considered, i.e. isotropic, the u indice can be dropped and an isotropic drift function D(h) is defined :

$$D(h) = E[Z(\underline{x}, t) \,/\, \underline{x} \in A(t), h = h(\underline{x}, t)] \qquad (4)$$

with $$h(\underline{x}, t) = \mathrm{Min} \{ \| \underline{x} - \underline{y} \| \,/\, \underline{x} \in A(t), \underline{y} \notin A(t) \} \qquad (5)$$

$h(\underline{x}, t)$ is thus the distance between $\underline{x}$ and the frontier of A(t).

b- *Identification of the drift function.*

In practice, $Z(\underline{x}, t)$ is only measured at a finite number of locations $\underline{x}_i$, i=1,..,p. The method can be applied either to radar or gauge measurements. Thus $\underline{x}_i$ represents the centre of one pixel or the location of one gauge.

In order to get a robust estimate of $D_u(h)$, we consider classes of distances, centred on h, and of width Δh. If M is the number of realizations available, $D_u(h)$ is estimated by:

$$D^*_u(h) = \frac{1}{N(h)} \sum_{t=1}^{M} \sum_{i=1}^{n(h,t)} Z(\underline{x}_i , t) \tag{6}$$

n(h,t) is the number of locations $\underline{x}_i$ for which $I(\underline{x}_i, t) = 1$ and $h_u(\underline{x}_i, t)$ belongs to $[\ h - \Delta h/2,\ h + \Delta h/2\ ]$ in the realization t and

$$N(h) = \sum_{t=1}^{M} n(h,t) \tag{7}$$

is the total number of values in the class h.

To get an accurate estimate $D^*_u(h)$, the density of the network must be large enough to identify correctly the position of the frontier of A(t). The problem is less crucial when radar data are available. Indeed, we can not exactly use (3) or (6) to compute $h_u(\underline{x}_i, t)$ or $h(\underline{x}_i, t)$ because $I(\underline{x}, t)$ is not known at each location of A. They are estimated with (8) or (9):

$$h^*_u(\underline{x}_i, t) = \text{Min} \{ \, || \, \underline{x}_i - \underline{x}_j \, || \ / \, \underline{x}_i \in A(t), \, \underline{x}_j \notin A(t), \, \underline{x}_i - \underline{x}_j = k\, u, \, j = 1,..p \} \tag{8}$$

$$h^*(\underline{x}_i , t) = \text{Min} \{ \, || \, \underline{x}_i - \underline{x}_j \, || \ / \, \underline{x}_i \in A(t), \, \underline{x}_j \notin A(t), \, j = 1,..p \, \} \tag{9}$$

However, a bias is thus introduced, due to the finite size of the network, as will be seen below.

c- *Estimation of the drift function on the Niger data set.*

The study area is situated in Niger, around Niamey, between longitudes 2°E and 3°E and latitudes 13°N and 14°N on a so-called "square degree" of about 12000 Km^2. This region is characterized by a Sahelian type climate with a rain season lasting from May to October. Rainfall events are mostly associated with squall lines, initiating above Lake Tchad and propagating westwards at wind speeds of about 60 Kmh^{-1}. These squall lines can be schematically described as follows: they are made of a front with convective rainfall associated with heavy rainrates, followed by a stratiform part giving lower intensities.

In the framework of the Epsat-Niger project (Lebel et al., 1992), this site has been instrumented since 1989 with 78 gauges (71 were used in this study, see Fig. 1). This network was designed to cover the area with a regular grid, with a space step of about 12,5 Km. An additional network of 9 gauges was also installed in the center of the domain, with a mean space step of 3 Km, in order to study the structure of rainfall at small time and space steps. Given the fairly flat relief, the centre of the domain can be considered as representative of the whole area.

In this work hourly accumulation rainfall fields are used. 135 hours were recorded in 1989 and are part of 31 events lasting 7 hours in the mean.

The isotropic drift function $D^*(h)$ computed on the data set is presented on Fig. 2. Note that no significant anisotropy was found. This function exhibits a "bell" shape showing clearly a drift of the rainrates which increase from the edge to the centre of the rain areas. The maximum values of $D^*(h)$ is 11 mmh^{-1}, which is much higher than values given by climatological means which are of the order of 3 mmh^{-1} if zero rainrates are taken into account and 6mmh^{-1} if only non-zero values are considered. This maximum value is reached at a distance of about 70 Km. This range can be viewed as half the maximum size of the rain areas.

For large distances, the curve decreases, but this is likely due to the finite size of the study area. Indeed, when the rain area almost completely covers the study area, the frontier is not well described by the network and h can be overestimated (see Fig. 3 for an illustration). Therefore, only the increasing part of the curve is used and modelled using a spherical model given by (10). Furthermore, for large distances, the number of values N(h) in (6) decreases dramatically and the dispersion on the estimate is very large. This provides another argument not to take this part of the curve into account, as in a variogram model.

$$D^m(h) = c + (b - c). \{ 3/2 . (h/a) - 1/2 . (h/a)^3 \} \quad (10)$$

a is the range of the drift for which the maximum value b of $D^*(h)$ is reached. c is the intercept for h=0. The modelled drift function is also given on Fig. 2. The values of the parameters are:

a=72 Km b=11mmh^{-1} c=0.9 mmh^{-1}.

The parameter c is the value of $D^*(h)$ when the distance h moves to zero. In theory, D(h=0) = 0 because, when h moves to zero, Z($\underline{x}$, t) moves to its value at the frontier , i.e. zero. Here c is not strictly equal to zero. This shows the existence of a bias in the drift function estimation, introduced by the finite size of the network. Indeed, (8) or (9) don't provide the exact distance to the frontier of A(t), but the distance to the nearest point outside A(t). h is therefore overestimated. If we denote by h the computed distance, h_1 the true distance to the frontier and dh the bias, we have : $h = h_1 + dh$

When dealing with a regular network, we can approximate dh with:

$$dh = 1/2p . \sum_{i=1}^{p} || \underline{x}_i - Prox(\underline{x}_i) || \quad (11)$$

where $Prox(\underline{x}_i) = \underline{x}_j \quad / \quad ||\underline{x}_i - \underline{x}_j|| = \mathop{Min}_{k=1,k\neq i}^{p} ||\underline{x}_i - \underline{x}_k||$

i.e. half the mean interdistance between two neighbouring gauges of the network.

Thus, the drift function can be corrected by defining a new model:

$$D^{m1}(h) = D^m(h+dh) \quad (12)$$

When dealing with radar or satellite data, this problem is not so significant and dh is probably very small. Note also that (12) is only meaningfull when precise information on the localization of the rain area is available, i.e. when the indicator function I($\underline{x}$, t) is known for each $\underline{x} \in A$. On the other hand, if the only accessible information is the distance to the nearest non-rainy gauge, the initial estimation $D^m(h)$ must be kept in all the calculations , as it is done in section 3.

However, this explanation implies a translation of the curve towards the left, which tends to increase the bias at the origin. Another factor is linked to the finite size of the study area, which tends to underestimate the drift function more and more when the distance considered increases. This implies a lower slope near the origin which can explain the bias (see also Braud et al., 1992).

3- Statistical analysis of the residuals.

In this section, we shall use the above drift function to investigate the non-stationarity of the non-zero rainrates. Thus, we define, at each location where $I(\underline{x}, t) = 1$, the residual $R(\underline{x}, t)$ by:

$$R(\underline{x}, t) = Z(\underline{x}, t) - D(h^*(\underline{x}, t)) \qquad (13)$$

where $h^*(\underline{x}, t)$ is given by (9).

The statistical analysis of $R(\underline{x}, t)$ is then performed in terms of non-bias, variance and covariance.

a- *Non-bias of the residuals.*

In the appendix, it is shown that when the indicator function is known, the definition of the drift function given by (3) or (4) provides an unbiased estimate, i.e. that: $E[\,R(\underline{x}, t)\,] = 0$. However, we saw in section 2 that some errors occur in the estimation of the drift function, and, experimentally, we must check that these errors do not introduce biases.

Globally, they do not because the $R(\underline{x}, t)$ expectation is: $E[\,R(\underline{x}, t)\,] = .025$ mmh^{-1}, i.e. almost zero. However, the global distribution of these residuals is far from being gaussian (Fig. 4). It is highly asymmetric: most of the residuals are negative and correspond to intensities lower than the drift. These negative values are bounded by -b, where b is the maximum value of the drift function. On the other hand, positive residuals are not bounded and can go up to 60 mmh^{-1}.

To go further in the non-bias analysis, the map of the climatological means of the residuals is provided on Fig. 5. It shows that the positive and negative residuals are not randomly spatially distributed. Indeed, the climatological mean decreases from the center to the edges of the domain. Rainrates are thus higher than the drift in the center and lower near the frontiers.This fact can be related to the uncertainties on $D^*(h)$ for large distances described in section 2-c (see also Fig. 3) because the probability of overestimating h^* increases from the center of A to its frontier.

The systematic tendency exhibited here by the residuals shows that the drift function estimation is very sensitive to the density of the network and the size of the study area. The results also suggest that the combination of remote sensing data covering an area larger than the study area (to avoid edge effects) and gauge data would probably give the best accuracy. Indeed, remote sensing provides an accurate localization of the frontier of the rain area, while gauges give a good measurement of the rain intensity.

b- *Variance of the residuals.*

The global standard deviation of the residuals is $\sigma R(\underline{x}, t) = 8.5$ mmh^{-1}. The map of the climatological standard deviation (not shown) does not exhibit any particular tendency, unlike the map of the means.

More interesting is Fig. 6, which provides the square of the residuals as a function of the distance to the nearest point where it does not rain. As $E[\,R(\underline{x})\,]^2 << E[\,R(\underline{x})^2\,]$, the ordinate is almost equal to the variance of the residuals. The curve also exhibits an increasing part for small distances. This proves that the residuals are not stationary at the second order and that the variance of the rainrate (relative to the drift) also increases from the edge to the center of the rain areas. However, the maximum value 132 $(mmh^{-1})^2$ is reached at a=40 Km (instead of 70 Km for the mean) and then, the curve stabilizes for h>a. The same discussion as in section 2-c can be made for the curve at large distances and justifies it not to taken into account.

Let's go further in the analysis and define a standardized residual (st. residual below):

$$RS(\underline{x}, t) = R(\underline{x}, t) / \sigma(h^*(\underline{x}, t)) \qquad (14)$$

where $\sigma(h)$ is the square root of the drift of the square of the residuals :

$$\sigma(h) = \{ E [R^2(\underline{x}, t) \mid \underline{x} \in A(t), h=h(\underline{x}, t)] \}^{1/2} \qquad (15)$$

and $h(\underline{x}, t)$ is given by (5) and practically estimated using (9).

The drift function of $RS(\underline{x}, t)$ is shown on Fig. 7. Contrarily to Fig. 2 and 6, this function can be considered as independent of h. Thus the st. residuals are stationary at the first order. The schematic pattern of the rainfall rate process, deduced from our results is given on Fig. 8. It appears as highly non-stationary.

c- *Structure of the residuals.*

The ensemble variograms of the residuals and of the st. residuals are provided on Fig. 9. Both of them show a well-defined sill at 59 $(mmh^{-1})^2$ for the residuals and 0.76 $(mmh^{-1})^2$ for the st. residuals. The range of the st. residuals is higher than that of the residuals (56 Km against 44 Km). Their nugget effect is also more pronounced. Nevertheless, both residuals show a structure which can be used for their kriging , as will be discussed now.

4- On the use the drift function for point rainfall estimation.

At this stage of the study, two cases must be considered :

i) The indicator function is supposed to be known over the whole domain A. This hypothesis is realistic in the context of future experiments such as the TRMM mission (Simpson et al., 1988). Its major goal is to provide monthly mean areal rainfall on areas of about 10000 Km^2 (the mesh of climatological models). A satellite is to be launched and is expected to give at least qualitative information on rainfall fields, and more specifically on the localization and size of the rain areas. Furthermore, at least on land surfaces, classical meteorological networks will provide values of point rainfall and the two types of information can be combined to interpolate rainfall fields. (On oceans, the problem is more difficult to solve because the indicator function will often be the only information available. Thus, coarser methods must be developed to provide mean areal rainfall (see for example Atlas et al. (1990) or Rosenfeld et al. (1990) for the so-called threshold methods, or Braud et al. (1992) for the use of the drift function discussed here)).

ii) The indicator function and the rainrate are only known at the gauges of a classical network and both the rain area and the non-zero rain intensity have to be estimated.

a- *The indicator function is known over the whole domain A.*

We saw in section 3 that both the residuals and the st. residuals have a well defined structure. However, the st. residuals are more stationary, and their kriging should a priori be prefered. On the other hand, we know that some uncertainties occur in the estimation of the "drift" functions, and to obtain the st. residuals, the residuals have to be divided by $\sigma(h)$. Thus, the performances of the two following possible estimators have to be tested :

$$\begin{cases} Z_1(\underline{x}, t) = [\ D(h(\underline{x}, t)) + R^K(\underline{x}, t)\ ] \cdot I(\underline{x}, t) & (16) \\ Z_2(\underline{x}, t) = [\ D(h(\underline{x}, t)) + \sigma(h(\underline{x}, t)) \cdot RS^K(\underline{x}, t)\ ] \cdot I(\underline{x}, t) & (17) \end{cases}$$

Here the subscript ^{K} denotes a kriged estimate.

As the indicator function is known, no estimation error arises on $h(\underline{x}, t)$. (Here, we don't take into account the errors on the estimation of $D(h)$ or $\sigma(h)$, as it is done in geostatistics with the variogram).

The true rainrate $Z(\underline{x}, t)$ can be decomposed as follows :

$$\begin{cases} Z(\underline{x}, t) = [\ D(h(\underline{x}, t)) + R(\underline{x}, t)\ ] \cdot I(\underline{x}, t) & (18) \\ Z(\underline{x}, t) = [\ D(h(\underline{x}, t)) + \sigma(h(\underline{x}, t)) \cdot RS(\underline{x}, t)\ ] \cdot I(\underline{x}, t) & (19) \end{cases}$$

Thus, the estimation variance is reduced to :

$$\begin{cases} E[\ (Z-Z_1)^2\ ] = E[\ (Z-Z_1)^2\ ] = 0 & \text{if } I(\underline{x}, t) = 0 \\ E[\ (Z-Z_1)^2\ ] = E[\ (R-R^K)^2\ ] & \text{if } I(\underline{x}, t) = 1 \\ E[\ (Z-Z_2)^2\ ] = E[\ (R-RS^K)^2\ ] & \end{cases} \quad (20)$$

In this case, the estimation variance of the residuals or the st. residuals given by kriging is sufficient to derive the estimation variance of $Z(\underline{x}, t)$.

b- *The indicator function is not known over the whole domain A.*

$I(\underline{x}, t)$ must then be estimated where it is not measured. This also implies that $h(\underline{x}, t)$ will also be estimated. As before, two estimates can be proposed:

$$\begin{cases} Z_1(\underline{x}, t) = [\ D(h^*(\underline{x}, t)) + R^K(\underline{x}, t)\ ] \cdot I^*(\underline{x}, t) & (21) \\ Z_2(\underline{x}, t) = [\ D(h^*(\underline{x}, t)) + \sigma(h^*(\underline{x}, t)) \cdot RS^K(\underline{x}, t)\ ] \cdot I^*(\underline{x}, t) & (22) \end{cases}$$

$I^*(\underline{x}, t)$ is defined as in Barancourt et al. (1992) in two steps:

i) Kriging of the indicator. It gives a kriged estimate $I^K(\underline{x}, t)$ which belongs to [0, 1].

ii) A cut-off x is defined to fulfill the non-bias condition, i.e. $E[\ I(\underline{x}, t)\ ] = E[I^*(\underline{x}, t)\ ]$ and $I^*(\underline{x}, t)$ is finally obtained as:

$$\begin{cases} I^*(\underline{x}, t) = 0 & \text{if } I^K(\underline{x}, t) < x \\ I^*(\underline{x}, t) = 1 & \text{if } I^K(\underline{x}, t) \geqslant x \end{cases} \quad (23)$$

Then, $h^*(\underline{x}, t)$ can be computed with (8) or (9), using the estimated indicator function $I^*(\underline{x}, t)$. $D(h^*(\underline{x}, t))$ and $\sigma(h^*(\underline{x}, t))$ are thus obtained and the final stage is the kriging of the residuals (resp. of the st. residuals).

Due to the non-linearity of (23) and the estimation of $h(\underline{x}, t)$, the estimation variance is very hard to derive and compute. Formally, with the Z_1 estimator, it can be decomposed into four terms:

$$\begin{aligned} E[(Z-Z_1)^2] = \; & \text{Prob}[I=1 \mid I^*=0] \,.\, E[(Z>0)^2] + \\ & + \text{Prob}[I=1 \mid I^*=1] \,.\, E[(Z>0 - Z_1>0)^2] + \\ & + \text{Prob}[I=0 \mid I^*=0] \,.\, 0 + \\ & + \text{Prob}[I=0 \mid I^*=1] \,.\, E[(Z_1>0)^2] \end{aligned} \quad (24)$$

Here the notation $Z>0$ (resp. $Z_1>0$) denotes the non-zero rainrate (resp. the non-zero estimated rainrate). The term $E[(Z>0 - Z_1>0)^2]$ combines estimation errors on $h(\underline{x}, t)$ and kriging errors on $R(\underline{x}, t)$ and is thus practically not computable. The problem would be more complicated with the Z_2 estimator because the st. residuals is multiplied by $\sigma(h^*(\underline{x}, t))$.

c- *Discussion.*

Till now, the above techniques have not been tested. The following remarks can however be made. When the indicator function is known, (19) probably provides a lower estimation variance than (18) and the use of Z_2 is preferable. On the other hand, when the indicator function is estimated, the uncertainties introduced on $h^*(\underline{x}, t)$ are likely to lower the advantage of working with the st. residuals, even if their estimation variance is smaller. To get reliable conclusions on the performances of all the proposed estimators, a validation procedure as described in Barancourt et al. (1992) is desirable but has not been undertaken yet, because the emphasis was put on the formulation and analysis of the drift function.

5- Conclusions.

The main goal of this paper was to present the concept of a drift function, based on conditional expectations. The mathematical framework was given with the practical method for the estimation of this function. A case study brought several estimation problems to light when only gauge data are available. It suggests that the combined use of remote sensing data (with a high spatial resolution) for the localization of the rain area and of gauge data for the measurement of the rain intensity would probably improve the accuracy of the estimated drift function.

The proposed drift function was then used to investigate the non-stationarity of rainfall rates. It appeared that, in rain areas, the mean but also the variance were decreasing from the frontier to the center of the cells. Finally, the use of the concept of drift function for point rainfall estimation was presented. It has now to be tested and validated.

Appendix : Non bias of the residuals when the indicator function is known.

Consider the non-zero rainrate $Z(\underline{x}, t)>0$ and define $p(Z>0)$, the probability density function (pdf) of $Z>0$, $p(h)$, the pdf of the minimum distance to the edge h and $p_h(Z>0)$, the conditional pdf of $Z>0$, knowing h. We also set $Z^D(\underline{x}, t) = D(h(\underline{x}, t))$. Thus, we have:

$$p(Z>0) = \int_0^\infty p_h(Z>0) \,.\, p(h) \,.\, dh$$

$$E[Z>0] = \int_0^\infty Z \,.\, p(Z>0) \,.\, dZ = \int_0^\infty Z \,.\, \int_0^\infty p_h(Z>0) \,.\, p(h) \,.\, dh \,.\, dZ$$

$$E[Z>0] = \int_0^\infty p(h) \,.\, \int_0^\infty Z \,.\, p_h(Z>0) \,.\, dZ \,.\, dh$$

$$E[Z>0] = \int_0^\infty p(h) \,.\, E[Z>0 \mid h] \,.\, dh$$

$$E[Z>0] = \int_0^\infty p(h) \,.\, D(h) \,.dh = E[Z^D>0]$$

Furthermore, if we set $q =$ Prob $[I(\underline{x}, t) = 1]$, i.e. the probability that it rains, we have:

$$E[Z] = q.\, E[Z>0] = q \,.\, E[Z^D>0] = E[Z^D]$$

Thus, the residuals defined by (13) are unbiased.

References.

Atlas D., D. Rosenfeld and D.A. Short, 1990 : The estimation of convective rainfall by Area Integrals. 1- The theoretical and empirical basis, *Journal of Geophysical Research*, **95**(D3), 2153-2160.

Barancourt C., 1990 : Etude de l'intermittence et de la variabilité des champs de précipitation par une approche stochastique. Thèse de l'Université Joseph Fourier, Grenoble, France, 271 pp.

Barancourt C., J.D Creutin and J. Rivoirard, 1992 : A method for delineating and estimating rainfall fields, *Water Resources Research*, **28**(4), 1133-1144.

Braud I., Ph. Crochet and J.D. Creutin, 1992: A method for estimating mean areal rainfall using Lagrangian trend functions of the intensities, submitted to *Journal of Applied Meteorology*.

Chua S.M. and R.L. Bras, 1982 : Optimal estimations of mean areal precipitation in regions of orographic influence, *Journal of Hydrology*, **57**, 23-48.

Creutin J.D. and Ch. Obled, 1982 : Objective analysis and mapping techniques for rainfall fields: an objective comparison, *Water Resources Research*, **18**(2), 413-431.

Lebel T., H. Sauvageot, M. Hoepffner, M. Desbois, B. Guillot and P. Hubert, 1992: Rainfall estimation in the Sahel: the EPSAT-NIGER experiment, *Hydrological Sciences Journal* **37**, 1-15.

Obled Ch. and J.D. Creutin, 1986 : On the use of empirical orthogonal functions for mapping meteorological fields. *Journal of Applied Meteorology*, **25**(9), 1189-1204.

Rosenfeld D. , D. Atlas and D.A. Short, 1990 : The estimation of convective rainfall by Area Integrals. 2- The Height-Area Rainfall Threshold (HART) method, *Journal of Geophysical Research*, **95**(D3), 2161-2176.

Roux C., 1990: Rapport de campagne EPSAT-NIGER 1989, ORSTOM, Direction de la Météorologie du Niger, 38 pp.

Simpson J., R.F. Adler and G.R. North, 1988 : A proposed Tropical Rainfall Measurements Mission (TRMM) Satellite, *Bulletin of the American meteorological Society*, **69**(3), 278-295.

Thauvin V. and T. Lebel, 1991: EPSAT-NIGER, Study of rainfall over Sahel at small time steps using a dense network of recording raingauges, *Hydrological Processing*, **5**, 251-260.

Acknowledgements: We thank ORSTOM, which is in charge of the Epsat Niger network, for providing us with the data used in this study.

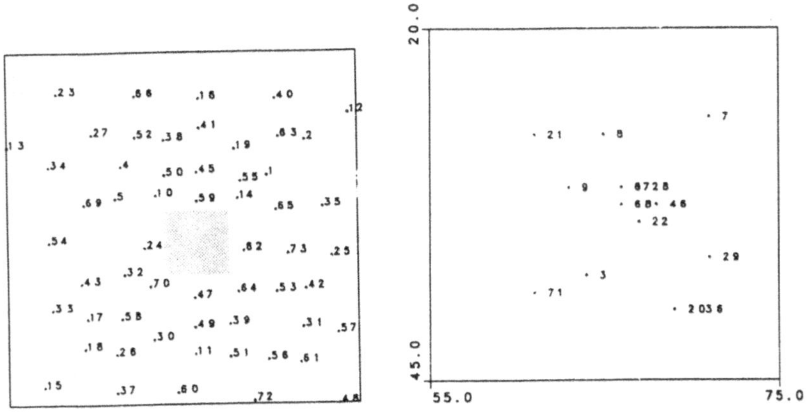

Fig. 1: The Epsat-Niger network in 1989 with the additional network in the centre.

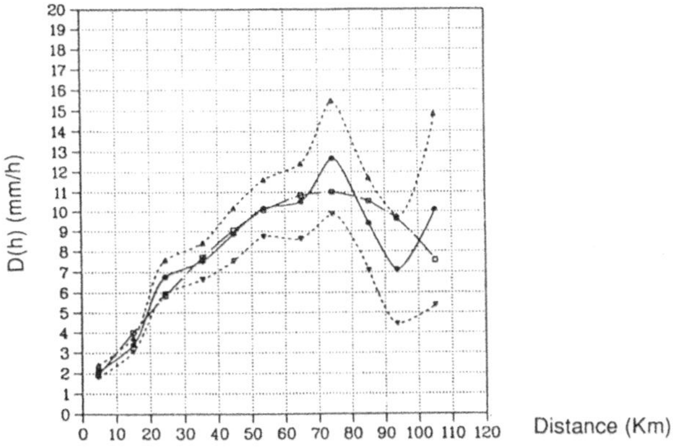

Fig. 2 : Drift fonction of the non-zero rainrate ___ Experimental curve, - - - +/- 2. Standard deviation, __ _ __ The model.

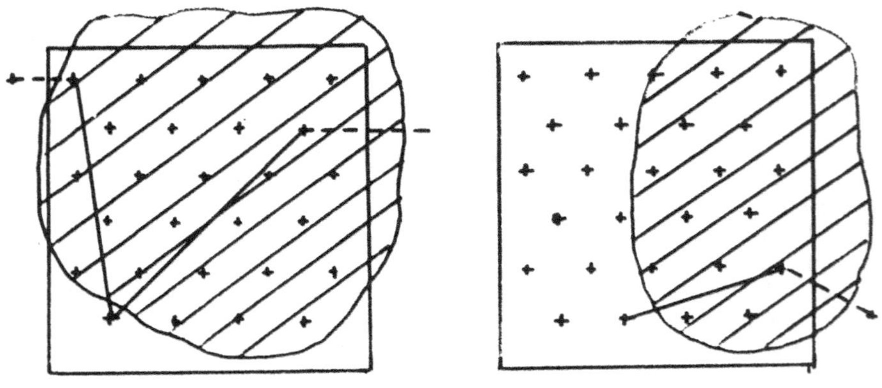

Fig. 3 : Schematical illustration of the bias introduced by the finite size of the study area on the drift function estimation. The rain area is the shaded zone. The full line is the estimated distance h_1 and the dotted line, the value which would have been estimated if the study area was larger.

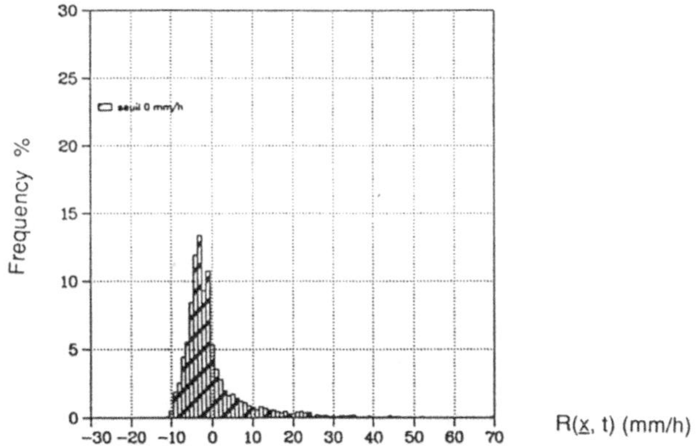

Fig. 4: Global distribution of the residuals.

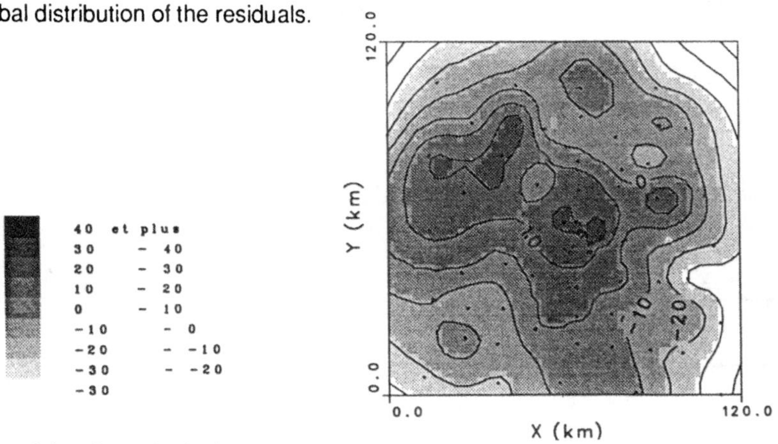

Fig. 5 : Map of the climatological mean of the residuals (values were multiplied by 10 for clarity and are thus in 1/10 mmh^{-1}).

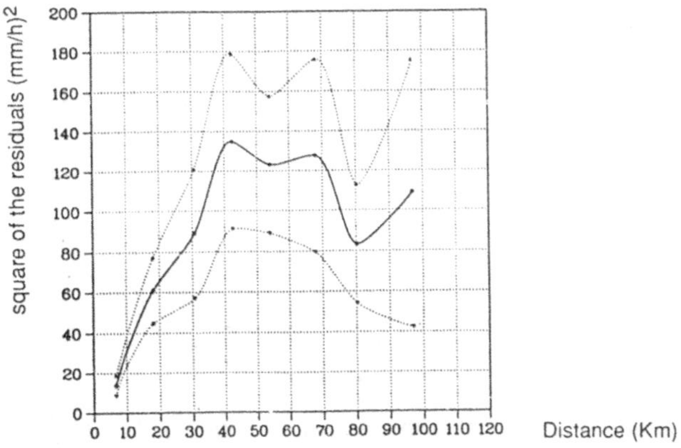

Fig. 6 : ___Square of the residuals as a function of h, - - +/- 2 standard deviation.

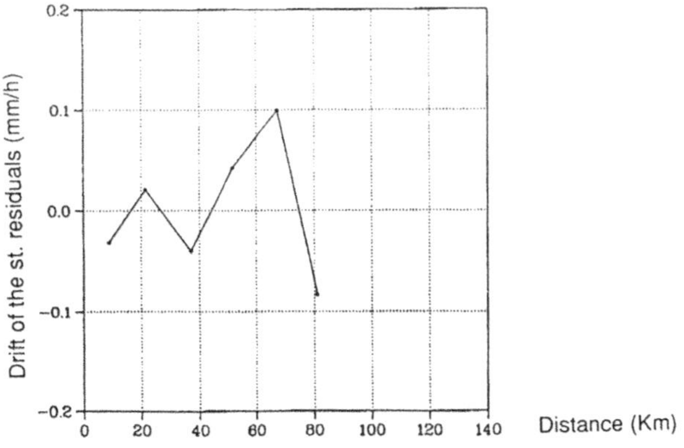

Fig. 7: Drift function of the standardized residuals..

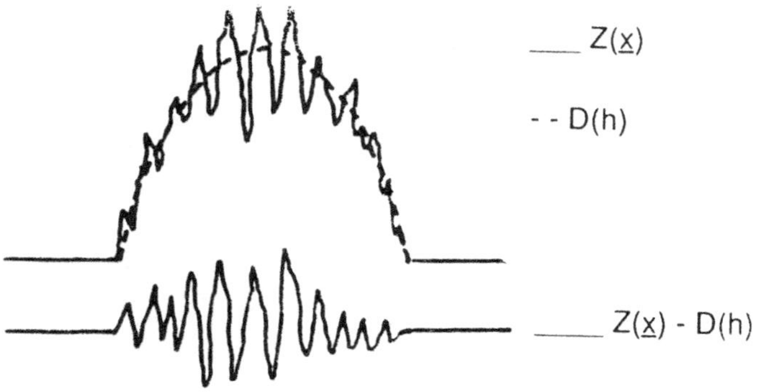

Fig. 8 : Schematical illustration of the pattern of rainfall areas.

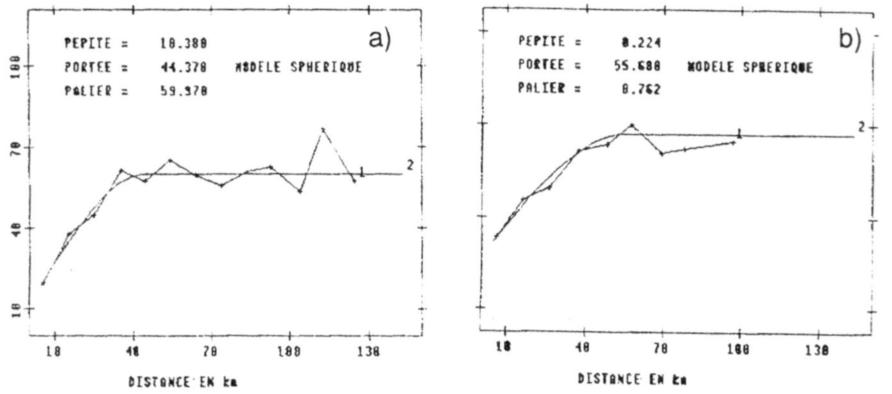

Fig. 9 : Experimental and modelized variograms of a) the residuals : nugget=10. mmh^{-1}, range=44 Km, sill= 60 $(mmh^{-1})^2$; b) the standardized residuals : nugget=.224 mmh^{-1}, range=56 Km, sill= .76 mmh^{-1}.

FORECASTING GYPSY MOTH DEFOLIATION WITH INDICATOR KRIGING

MICHAEL E. HOHN, LINDA S. GRIBKO, West Virginia Geological Survey, P.O. Box 879, Morgantown, WV 26507, USA, and ANDREW M. LIEBHOLD, USDA Forest Service, Northeastern Forest Experiment Station, P.O. Box 4360, Morgantown, WV 26505, USA

ABSTRACT

Outbreaks of the gypsy moth, *Lymantria dispar (L.)*, occur over large areas, but are difficult to predict. Models used previously to forecast defoliation utilized preseason counts of egg masses in a given stand of trees. Three dimensional indicator kriging is applied to the problem of forecasting local probability of outbreak in each of 6,075 2 km by 2 km cells in the state of Massachusetts, USA. From defoliation data representing about thirty years of records, variography showed a temporal persistence of outbreaks to be about 5 years on the average. An omnidirectional, spatial variogram was computed for each of the years since 1984, and found to be sensitive to the number of cells defoliated. At low densities, spatial variograms approached a pure nugget effect, whereas in years of relatively high densities, defoliated cells were clearly clustered. These observations were used to construct and test a three-dimensional simple indicator kriging estimator that uses two years of observed defoliation prior to the year to be forecast, and estimates of global level of defoliation for selecting a variogram and specifying an estimated frequency of defoliation.

Introduction

The gypsy moth, *Lymantria dispar (L.)*, is an example of an eruptive species (Price et al. 1990): population densities vary through several orders of magnitude, often reaching epidemic densities that have spectacular effects on their habitat (i.e., total defoliation of host trees). Gypsy moth populations commonly persist for many years at densities that are so low that it may be difficult to detect any life stages except male moths. Occasionally, for unknown reasons, population densities increase, often to defoliating levels in excess of 6000 egg masses/ha, within only a few generations. These outbreak populations may persist for several years before collapsing.

Outbreaks of the gypsy moth, *Lymantria dispar*, often occur over very large areas but are notoriously difficult to predict. Previous models developed for predicting defoliation have mostly been based upon pre-season counts of egg masses (Gansner and Herrick 1982, Williams et al. 1991, Montgomery 1990. Although there is a significant relationship between egg mass density and subsequent defoliation at the stand level, the variance about this relationship is great at the local level. Lack of more precise methods for prediction of defoliation can have a devastating impact on the efficiency of the gypsy moth management program (Ravlin et al. 1987).

Earlier models developed for predicting outbreaks have relied on data collected at a single point and have largely ignored spatial processes. More than 20 years ago, Campbell (1967) recognized that the dynamics of gypsy moth populations are affected by population conditions in nearby areas; there is often considerable synchronicity in yearly fluctuations in the development of gypsy moth outbreaks (Liebhold and McManus, 1991).

In an attempt to capture these spatial relations in a quantitative way, we have adapted a geostatistical modeling approach to the problem of forecasting gypsy moth defoliation. We have applied Indicator Kriging for modeling the spatial dynamics of gypsy moth outbreaks. Historical defoliation data from Massachusetts were used to parameterize a model that is used to extrapolate defoliation maps into the future.

Data

A geographic information system (GIS), IDRISI, was employed to assemble and collate gypsy moth defoliation data. (Eastman, 1989). IDRISI is a raster-based (grid cell) GIS for capturing, storing, analyzing and displaying geographical data, designed for research applications. A base-map of Massachusetts county boundaries was used to define the study area. This map was generated from latitude and longitude coordinates and these coordinates were projected using the Azimuthal Equal-Distant Projection (Snyder 1987). This projection conserves true distance linearly from a designated point (Boston, Massachusetts). A 2 x 2 km grid cell size was selected as standard for all map layers in the GIS. Each map layer comprised 198 by 93 cells. The grid size was selected because it represented the minimum dependable spatial resolution of the available defoliation.

The Massachusetts Department of Environmental Management monitors gypsy moth defoliation annually in all parts of the state using aerial sketch maps. Maps are sketched during a series of low level reconnaissance flights in late July when defoliation is at its peak. Thirty percent (30%) defoliation is considered the lower threshold for detection from the air. In situations where there is doubt as to the cause of the defoliation, ground checks for the presence of gypsy moth life stages are made. Initially the aerial sketch maps are overlaid on standard U.S. Geological Survey (1:24,000) topographical maps. Subsequently a composite mosaic is generated for the entire state at 1:760,000 scale. Mapping processes vary from region to region and year to year within Pennsylvania, resulting in a strong likelihood of significant data errors from systematic and non- systematic sources.

The likely presence of these errors dictated the course spatial resolution of the data used in this study (2 x 2 km rasters).

To create a uniform set of geographically-referenced defoliation data, the composite maps for the period 1963 to 1989 were first transferred to mylar stable-base sheets. At least four geo- reference points, on clearly recognizable intersections of county boundaries, were accurately located. The prepared maps were then scanned using a digital scanner set at 150 dots per inch resolution. Binary TIFF files from the scanner were converted to ASCII IDRISI 'image' format and saved as IDRISI images or map layers. The transformation of each map layer to a common base map resolution and projection was achieved through a "rubber-sheeting" procedure.

Variogram Analysis

Data for years 1969-1991 were used to calculate a variogram in the time domain to which the following model was fitted by eye (Figure 1):

$$\gamma(h) = C_1\,(1-e^{-h/r_1}) + C_2\,(1-e^{-h/r_2}).$$

where: $C_1 = 0.021$, $r_1 = 0.7$; $C_2 = 0.050$, $r_2 = 1.7$

No attempt was made to model the pseudo-hole effect that results from the historic pattern of gypsy moth outbreak and collapse (Figure 2).

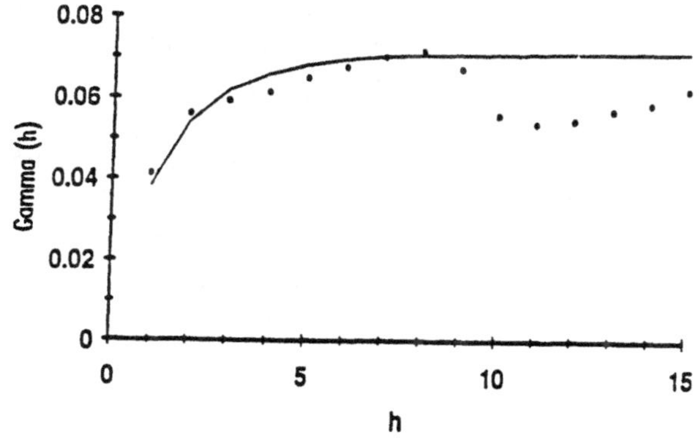

Figure 1. Temporal variogram of gypsy moth defoliation in Massachusetts. Unit of h is year.

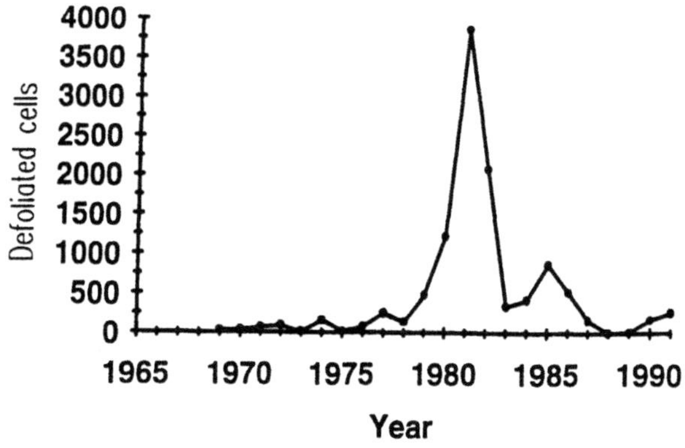

Figure 2. Frequency of gypsy moth defoliation in Massachusetts by year.

From defoliation data for each year, spatial variograms were calculated in four directions. No strong anisotropies were observed in the spatial domain, and therefore, omnidirectional variograms were used throughout the rest of the study. Variogram appearance was found to be dependent in part on the density of defoliation for a given year. Variograms for 1985 and 1986—years with relatively high defoliation rates—variograms reflecting spatial dependence over large distances (Figure 3), whereas the variograms for 1989 and 1990 have very low spatial dependence. The variogram for 1987 is an exception, having a degree of spatial dependence similar to 1985-6, but density of defoliation on the level of 1990.

Each spatial variogram was fitted with the same model used in the time domain; the ratio C_1/C_2 was maintained across all spatial variograms and the single time variogram. During the kriging of a particular year, a geometric model was specified that was anisotropic in the time-space plane (Table 1). In essence, the time variogram was scaled to match the respective spatial variogram. Obviously, a number of options exist for accomplishing this initial scaling of the temporal and spatial variograms. Because estimation variances were not to be computed, only the shape of the variograms needed to be maintained. For two of the years, 1989 and 1990, the first exponential term in the model has the larger range, r_1. The result of these adjustments is a single, spatio-temporal variogram for each year:

$$\gamma(h) = C_1\,(1-e^{-h/R_1}) + C_2\,(1-e^{-h/R_2})$$

$$R_1 = \frac{r_{1s}\; r_{1t}}{\sqrt{r_{1s}^2 \sin^2\phi + r_{1t} \cos^2\phi}}$$

$$R_2 = \frac{r_{2s}\; r_{2t}}{\sqrt{r_{2s}^2 \sin^2\phi + r_{2t} \cos^2\phi}}$$

$r_{1s} = r_1$ *for spatial domain*
$r_{1t} = r_1$ *for time domain*
$r_{2s} = r_2$ *for spatial domain*
$r_{2t} = r_2$ *for time domain*
ϕ = ***azimuth of vector relative to the spatial plane (i.e. $\phi = 90°$ for separation in time, and $\phi = 0°$ for separation in space).***

By constructing a single, anisotropic variogram model, we avoided the possiblity of non-unique solutions of kriging equations which occurs when separate spatial and temporal variograms are used.

Table 1
Variogram constants used in kriging gypsy moth defoliation.

	C_1	r_1	C_2	r_2
Spatial Variograms				
1985	2.10×10^{-2}	1.8	5.00×10^{-2}	20.0
1986	2.10×10^{-2}	1.3	5.00×10^{-2}	40.0
1987	4.14×10^{-3}	1.1	9.86×10^{-3}	10.0
1988	no variogram			
1989	3.55×10^{-4}	3.0	8.45×10^{-4}	0.4
1990	3.55×10^{-3}	9.0	8.45×10^{-4}	0.5
Temporal	2.10×10^{-2}	0.7	5.00×10^{-2}	1.7

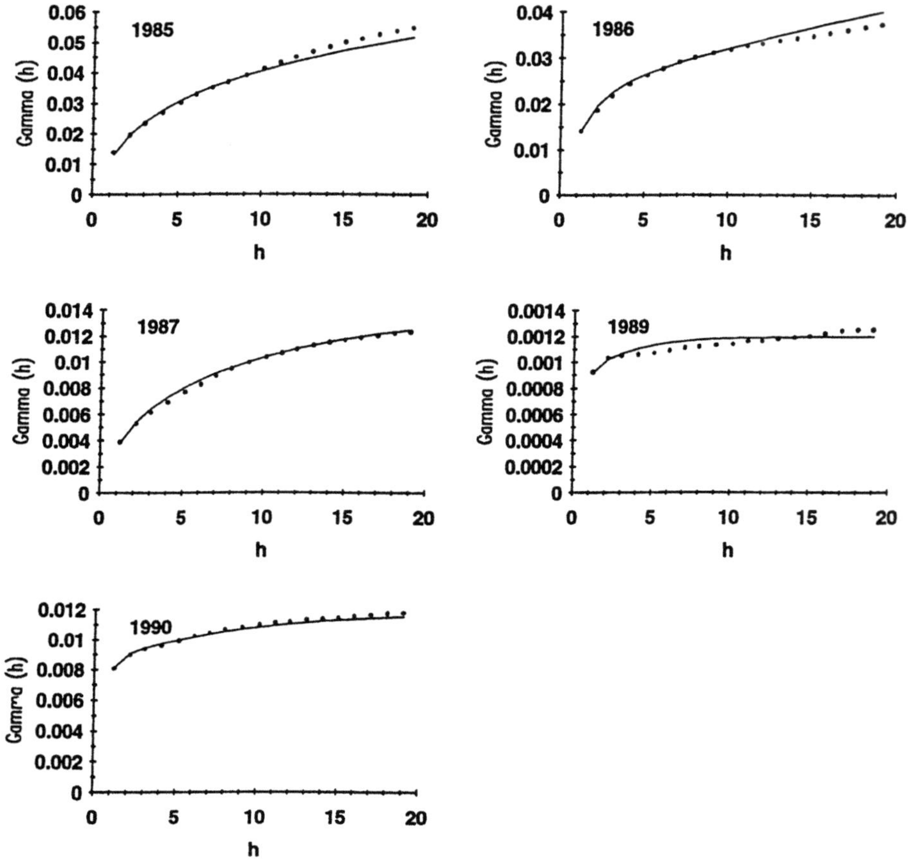

Figure 3. Spatial omnidirectional variograms of gypsy moth defoliation for 1985-1990. Unit of h is 2 km.

Indicator Kriging

The probability of defoliation for each node x_k is $P(x_k)$; note that location x has two spatial components and a time component. A kriged estimate for $P(x_k)$ is a linear combination of observed values of $i(x_k)$ in spatial and temporal proximity to location x_k, i.e.:

$$P^*(x_k) = F^*(x_k)\,[1 - \sum_{j=1}^{n} w_j] + \sum_{j=1}^{n} w_j\, i(x_j),$$

in which $P^*(x_k)$ is an estimate of $P(x_k)$ and $F^*(x_k)$ is an estimate of the expected frequency that $i(x)=1$ (Journel, 1983).

The weights w_j are computed by means of a system of n equations:

$$\sum_{m=1}^{n} w_m\, \rho\,(x_m - x_1) = \rho\,(x_1 - x_k)$$

$$\sum_{m=1}^{n} w_m\, \rho\,(x_m - x_2) = \rho\,(x_2 - x_k)$$

$$\sum_{m=1}^{n} w_m\, \rho\,(x_m - x_n) = \rho\,(x_n - x_k)$$

where:

$$\rho(a-b) = \rho(h) = 1 - \gamma(h)/\sum_{v=1}^{V} (C_v)$$

In all kriging, the search neighborhood was 5 cells wide spatially and 2 cells deep temporally.

Results

To use simple indicator kriging to predict defoliation, one must specify a value of the expected frequency, $F^*(X_k)$ at each location, as well as a variogram model. Statistical or rule-of-thumb methods can be used to predict $F^*(X_k)$, and to suggest which variogram is most appropriate. Prediction of low levels of defoliation would lead one to use a model like that observed in 1989, whereas prediction of widespread defoliation would point to a model like that observed in 1985. An example of a predictor for overall defoliation density in a given year is the number of egg masses observed per unit area of forest (William, et al., 1991; Gansner and Herrick, 1982). Clearly, the effectiveness of this method for estimating local probabilities of defoliation depends upon the ability of the predictor of overall defoliation level.

To test the method, defoliation was estimated for a given year from the two previous years, using a constant value of $F^*(X_k)$ for all locations X_k. $F^*(X_k)$ was set equal to observed frequency of defoliation for all of Massachusetts for the year being predicted. Although this procedure assumes knowledge of $F^*(X_k)$ before the fact, evaluating ways of predicting global defoliation level is beyond the scope of this paper. We are investigating the use of egg mass counts per unit area to predict level of outbreak in the following year.

To save space, only results for selected years are shown. Because 1986 was a year of high defoliation frequency (Table 1) the most appropriate variogram to use is the one for the previous year (Figure 3), resulting in a probability map having compact, well-defined areas of high probability (Figure 8). These areas are centered on cells that were defoliated in 1985. The probability of defoliation was overestimated in an area southeast of the major outbreak in central Massachusetts. However, most high-risk areas correspond with observed defoliation (Figure 9), reflecting the temporal persistence of gypsy moth outbreaks.

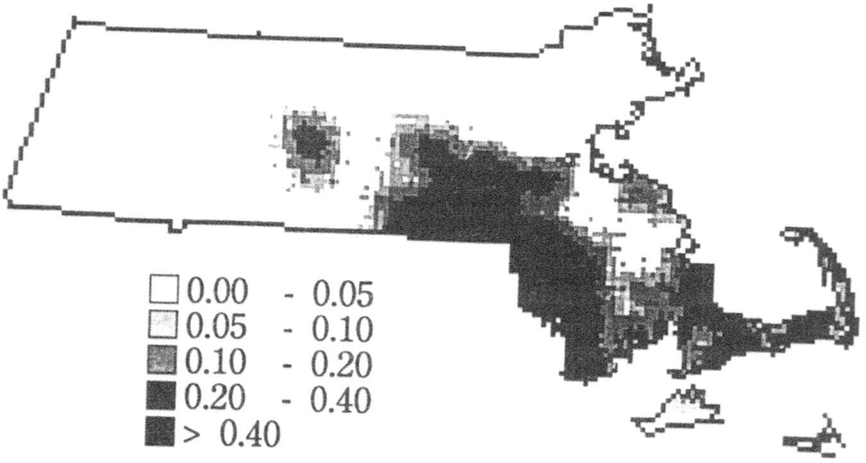

Figure 4. Estimated probability of defoliation in Massachusetts in 1986.

Figure 5. Actual defoliation in Massachusetts in 1986. Black squares indicate presence of defoliation within a 2 km by 2 km area.

In contrast, 1990 displayed low level of defoliation (Figure 7). The combination of a small value for $F^*(X_k)$ and a variogram approaching a pure nugget effect yielded a probability map (Figure 6) with little contrast and small, scattered areas of slightly elevated probabilities that in fact predicted areas of defoliation very poorly.

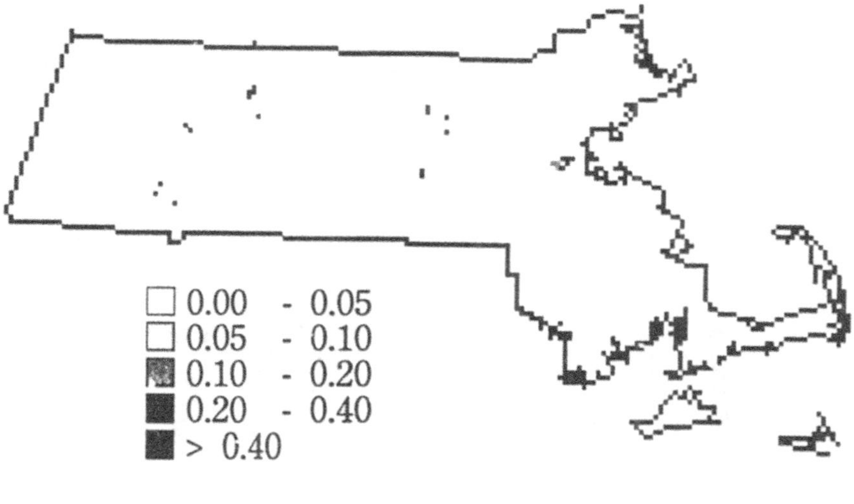

Figure 6. Estimated probability of defoliation in Massachusetts in 1990.

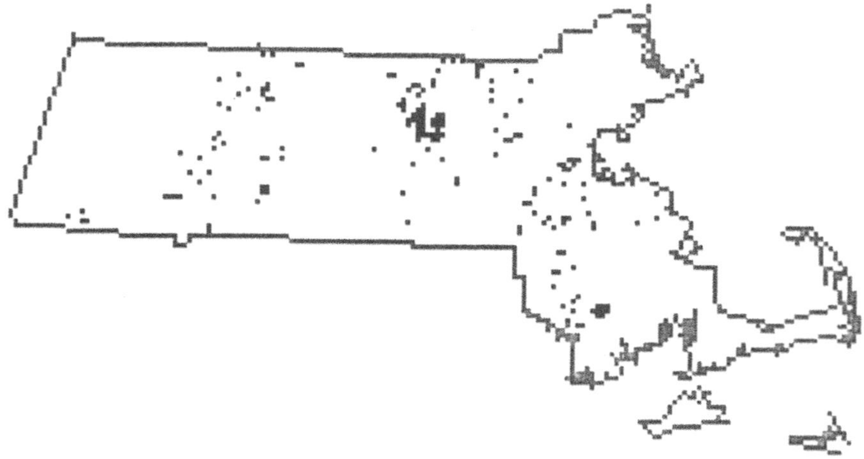

Figure 7. Actual defoliation in Massachusetts in 1990. Black squares indicate presence of defoliation within a 2 km by 2 km area.

1991 exhibited an intermediate level of defoliation (Figure 2). The probability map (Figure 8) calculated with use of the 1985 variogram reflects the clumped nature of actual 1991 defoliation (Figure 9), although some clumps hit and others miss. There exists a strong stochastic component to the spatial/temporal distribution of gypsy moth outbreaks. Nevertheless, most defoliation occurred in areas of higher probabilities.

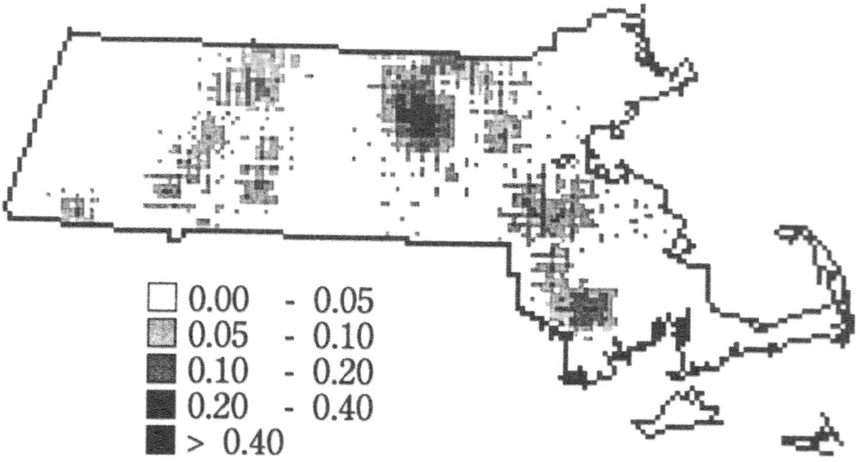

Figure 8. Estimated probability of defoliation in Massachusetts in 1991.

Figure 9. Actual defoliation in Massachusetts in 1991. Black squares indicate presence of defoliation within a 2 km by 2 km area.

In conclusion, the simple indicator estimator does a much better job of predicting areas of defoliation for years in which area-wide levels of defoliation are high. The estimator cannot accurately predict areas of defoliation during lulls and immediately following years of low defoliation. However, the need for accurate predictions is most important during outbreaks.

Obviously any substantial error in prediction of defoliation could have detrimental effects on a gypsy moth management program. If defoliation is erroneously predicted in an area that is not subsequently defoliated, pesticide applications may needlessly be applied, resulting in unnecessary expenditure of funds as well as possible detrimental environmental impacts. Conversely, if predictions erroneously predict no defoliation for an area that subsequently receives substantial defoliation, the management program fails to achieve its goal of minimizing the adverse affects of defoliation. Most current gypsy moth management programs use preseason counts of egg masses as a basis for predicting defoliation but this method is also known to often yield faulty predictions of defoliation. We expect that in the future it may be possible to combine the time-space kriging procedure described here with data from preseason egg mass counts collected on a regional level, to provide more precise predictions of defoliation.

Acknowledgments

We thank the Massachusetts Department of Environmental Management for providing defoliation data, and Eugene Luzader who helped in many aspects of this research. This research was funded in part by funds from the USDA Forest Service Gypsy Moth Research and Development program and by grants 89-37250-4533 and 91-37302-6278 from the USDA CSRS Competitive Grants Program.

References

Campbell, R.W. (1967) "The analysis of numerical change in gypsy moth populations", Forestry Science Management, 15: 32 pp.

Eastman, J.R. (1989) IDRISI: a Grid-Based Geographical Analysis System, Clark University, Worcester.

Gansner, D.A. and Herrick, O.W. (1982) "Predicting the rate of change in timber value for forest stands infested with gypsy moth", U.S. Department of Agriculture Forest Service Research Note NE-311, 3 pp.

Journel, A.G. (1983) "Nonparametric estimation of spatial distributions", Mathematical Geology 15: 445-468.

Liebhold, A.M., and McManus, M.L. (1991) "Does larval dispersal cause the expansion of gypsy moth outbreaks?" Northern Journal of Applied Forestry, 8: 95-98.

Montgomery, M.E. (1990) "Role of site and insect variables in forecasting defoliation by the gypsy moth", in A.D. Watt, S.R. Leather, M.D. Hunter, and N.A.C. Kidd (eds.), Population Dynamics of Forest Insets, Intercept, Andover, pp. 73-84

Price. P.W., Cobb, N., Craig, T.P., Fernandes, G.W., Itami, J.K., Mopper, S., and Preszler, R.W. (1990) "Insect herbivore population dynamics on trees and shrubs: new approaches relevant to latent and eruptive species and life table development", in E.A. Bernays, (ed.), Insect-Plant Interaction, vol. 2 CRC Press, Boca Raton.

Ravlin, F.W., Bellinger, R.G., and Roberts, E.A. (1987) "Gypsy moth management programs in the United States: status, evaluation and recommendations", Bulletin Entomol. Society of America 33, pp. 90-98.

Snyder, J.P. (1987) Map Projections—A Working Manual, United States Geological Survey Professional Paper 1395, 383 pp.

Williams, D.W., Fuester, R.W., Metterhouse, W.W., Balaam, R.J., Bullock R.H., and Chianese, R.J. (1991), "Oak defoliation and population density relationships for the gypsy moth", Journal of Economic Entomology.

CONDITIONAL SIMULATION: PRACTICAL APPLICATION FOR SAMPLING DESIGN OPTIMIZATION

EVAN J. ENGLUND
U. S. Environmental Protection Agency, Environmental Monitoring Systems Laboratory, Las Vegas, Nevada, 89193, U.S.A.

NASER HERAVI
University of Nevada - Las Vegas, Harry Reid Center for Environmental Studies, 4505 S. Maryland Parkway, Las Vegas, Nevada, 89154, U.S.A.

ABSTRACT

Detailed spatial models generated by conditional simulation provide a powerful tool for case-specific optimization of sampling designs. The entire process of sampling, estimation, and decision can be simulated on such a model by a Monte-Carlo approach. Optimization can be based on economic functions or on decision quality constraints rather than simple minimization of estimation variance. Efficient algorithms and 32-bit desktop computers make simulation feasible for routine use. A design solution based on conditional simulation will approach the true optimum only to the degree that the simulations accurately reflect the relevant real-world characteristics. The quality of a simulation depends on a number of factors including the number of conditioning data, the accuracy of the variogram model, and the use of data transformations. The method is illustrated with two examples - one based on the well-known Walker Lake model and the other on an actual case study involving remediation of contaminated soils. For practical purposes, the method appears to be accurate, precise, and robust.

INTRODUCTION

Many authors have applied geostatistics to the problem of spatial sampling design, including Barnes (1989), Burgess et.al. (1981), and Olea (1975). Most such applications involve some variation on minimization of the kriging variance to determine ideal sampling patterns, to identify the best locations for one or more proposed additional samples, or to find the minimum number of samples needed to attain a specified maximum level of error. This approach, however, does not incorporate economics (Srivastava, 1987), nor does it readily permit evaluation of the consequences of decision-making with uncertainty. Rendu (1980) presented a parametric approach to computation of opportunity costs resulting from estimation errors for the normal and log-normal cases.

This paper describes an alternative approach using conditional simulation (Journel, 1974) and presents two examples which illustrate and attempt to validate the method. Although computer-intensive, the conditional simulation approach can deal with true optimization when an economic objective function is available. A wealth of intermediate results permit the investigator to better understand the consequences of the available choices.

THE ENVIRONMENTAL REMEDIATION SCENARIO

Both examples illustrate how conditional simulation might be applied in a remediation situation. The problem involves sampling contaminated soils to select portions of the site with concentrations in excess of a regulatory standard which require remediation. We assume: an initial data set from a representative sub-area of the site; a specified action level applied to a remediation unit (RU) of a specified size appropriate to the remediation method (e.g., removal by front-end loader); and an economic objective function where total cost equals the sum of sampling cost plus remediation costs for all selected RU's, plus the cost of residual contamination in all non-selected RU's.

The decision rule is that if the estimated value of an RU exceeds the action level, it will be remediated. Unit RU remediation cost is constant; unit RU non-remediation cost is proportional to concentration; and the two unit costs are defined to be equal at the action level. The objective is to estimate the number of samples in a single-phase campaign which would result in the lowest total project cost. This defines the optimal sampling density for the remainder of the site. We will use the following notation:

$\mathbf{N_o}$ The optimum number of samples, where the expected value of the total project cost is a minimum.

$\mathbf{N_e}$ An estimate of $\mathbf{N_o}$ from the conditional simulation design procedure.

$\mathbf{n_c}$ The number of initial samples used to condition a simulation (step 2, below).

$\mathbf{n_s}$ The number of simulated samples taken in one iteration of the design procedure (step 4, below).

$\mathbf{n}$ The number of samples to be taken in an actual field sampling campaign.

THE SAMPLING DESIGN PROCEDURE

The procedure is a Monte Carlo resampling scheme which simulates the remediation operation, including data collection, interpolation, and selection.

1) Estimate the variogram model from the initial (conditioning) data set.

2) Generate a detailed simulated site model which is consistent with both the conditioning data and the variogram model. The site model is a dense array of possible sample values.

3) Compute "true" RU values from the site model. Each RU value is the mean of all simulated values within it.

4) Input a value for n_s (selected by trial and error). Generate n_s sample locations by dividing the site model into n_s equal cells, and selecting one random location within each cell. Draw the corresponding simulated sample values from the site model.

5) Estimate mean RU concentrations by block kriging using the variogram model estimated in step 1.

6) Apply the decision rule to the estimated RU values. Evaluate the decisions by comparison to the RU values from step 3. Compute total project cost and other relevant statistics such as numbers of false positives and false negatives, the total quantity of contaminant remediated, etc.

7) Repeat steps 4 - 6, drawing alternate random sets of n_s samples. The number of repetitions necessary to obtain an adequate estimate of the distribution and expected value of total project cost for n_s samples will depend on the variability of the results.

8) Repeat steps 4 - 7 for additional values of n_s, until a cost curve is obtained from which N_e can be determined.

This procedure is not subtle; it relies heavily on brute-force computing. Fortunately, the current generation of desktop computers is capable of running this software. Most of the examples in this paper were run on a 486/50 PC. Typical run times for an entire design sequence, steps 1 - 8 above, were under two hours.

Conditional Simulation

The conditional simulation procedure in step 2 generates a spatial array of simulated values consistent with (conditioned by) known values at measured locations, and with a given variogram model. In this paper, the sequential gaussian simulation (SGSIM) algorithm, (Deutsch and Journel, 1992) is used. This algorithm simulates nodes on a grid in random sequence by first estimating the value at the selected node by kriging with a local neighborhood of conditioning data, and then adding a random component from a normal distribution with zero mean and the kriging standard deviation. Once simulated, values are added to the conditioning set for use in simulating additional nodes.

THE QUALITY OF A SAMPLING DESIGN

How good is the N_e obtained by this procedure? If the procedure is valid and practical, it should be unbiased, precise and robust. That is, the mean N_e should be close to N_o, N_e should converge to N_o as n_c increases, and N_e should be insensitive to small changes in conditioning data, variogram model, or random seed.

CASE STUDY 1: WALKER LAKE

The model

The Walker Lake model used here has been described previously (Englund, 1990, Isaaks and Srivastava, 1989). It contains 19800 values on a 110x180 grid (Figure 1). The data, arbitrarily assigned units of parts per million (ppm), are a reasonable surrogate for contaminated soils. They are highly skewed, discontinuous, and contain a spike of zero values analogous to non-detects. The log histogram is shown in Figure 2 (with zero values set equal to 1/2 the smallest non-zero value) and the exhaustive variogram in Figure 3.

The decision rule and costs

A secondary grid of 198 RU means (Figure 4) is derived from the original model. The decision rule is to remediate an RU when its estimated value exceeds 100 ppm. The cost of remediating any RU is $10,000; the cost of not remediating any RU is proportional to the actual concentration of contaminant in the RU, whether or not it exceeds 100 ppm. The unit sample cost is $500.

The optimum $\mathbf{N_o}$

$\mathbf{N_o}$ was obtained by running the sampling design procedure on the Walker Lake model. Several initial values for $\mathbf{n_s}$ were selected, and the model was resampled 1250 times for each $\mathbf{n_s}$. The process was repeated with additional values of $\mathbf{n_s}$ to "zero in" on the minimum cost. The results are shown in Figure 5; $\mathbf{N_o}$ is approximately 256. This value is itself an estimate subject to some uncertainty. Note that the curve is quite flat in the range 200 - 350. This is helpful in practice, because small errors in estimating $\mathbf{N_o}$ will have only minor impact on total cost. The variability among the 1250 individual cost outcomes for each $\mathbf{n_s}$ emphasizes that a correct estimate of $\mathbf{N_o}$ only guarantees the best *expected* value. The actual outcome from any particular sample of size $\mathbf{n}$ = $\mathbf{N_o}$ might be quite different.

The simulation procedure

The SGSIM algorithm was used with

Figure 1. Shaded map of the Walker Lake model. Darker shades represent higher values.

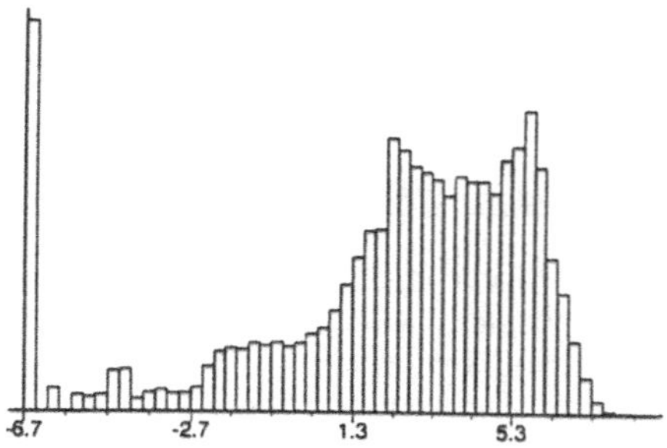

Figure 2. Histogram of log-transformed Walker Lake model.

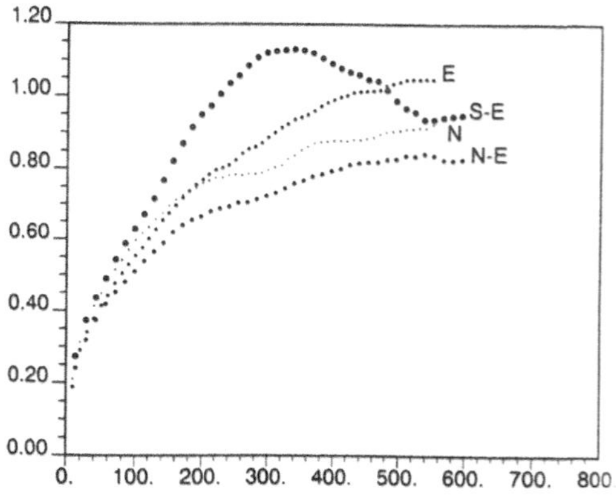

Figure 3. Directional normal-score variograms of the Walker Lake model.

conditioning data sets drawn from the Walker Lake model. Simulations were performed on normal-score transformed data, and back-transformed. The maximum possible simulated value was set at 10,000 (the actual Walker Lake Maximum is 3304). The "linear" option for the upper tail model was used, which results in a uniform distribution between the maximum conditioning value and the maximum simulated value. Variogram models were estimated manually from the transformed conditioning data.

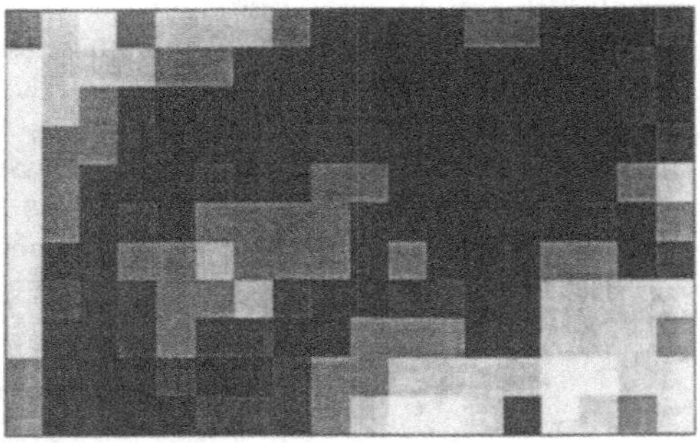

Figure 4. Shaded map of block means from the Walker Lake model.

Accuracy and precision

The accuracy and precision of N_e was evaluated on the Walker Lake model with the SGSIM algorithm by repeating the sampling design process for each of three values of n_c (50, 250, and 1250). Conditioning data sets were drawn independently from the Walker Lake model. Figure 6 shows N_o and the resulting N_e values plotted against n_c. For any specified n_c, there are four sources of variability in this method: A.

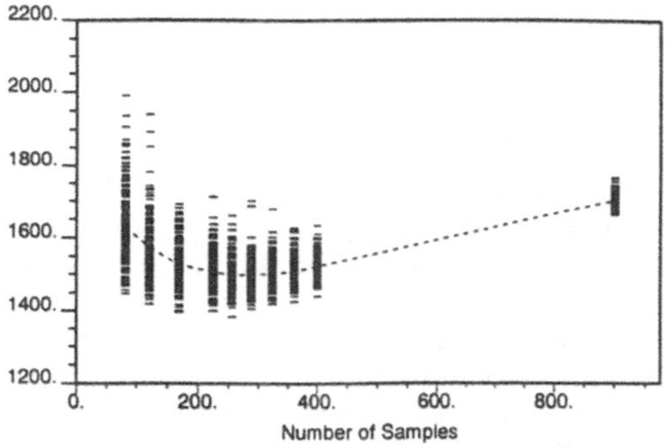

Figure 5. Cost vs. n_s for the Walker Lake model. The dotted line connects the means. N_o is approximately 256 samples.

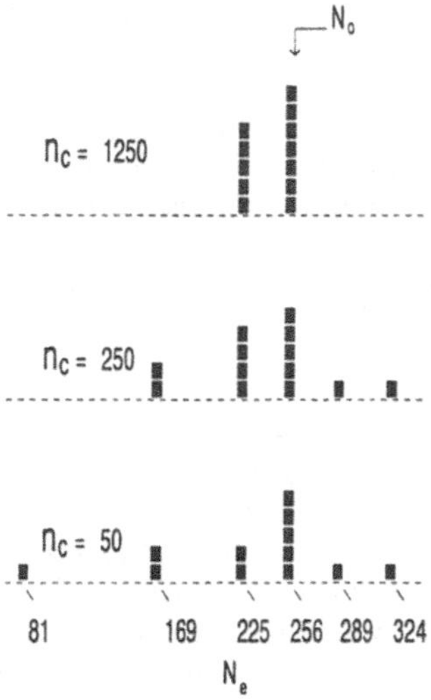

Figure 6. N_e vs. number of conditioning data for several simulations. N_o is approximately 256.

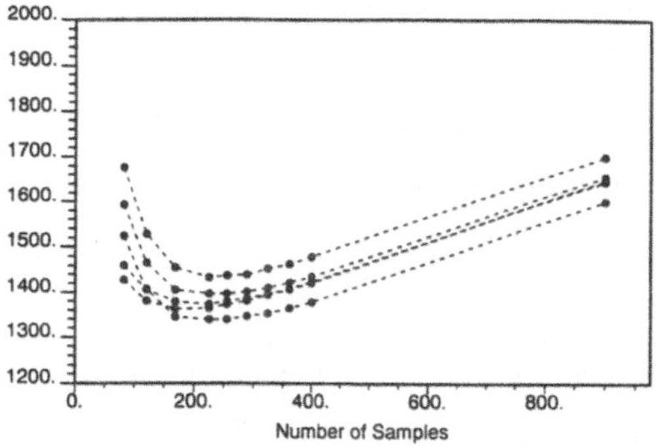

Figure 7. Cost curves from simulations based on a single set of 50 conditioning data, differing only in initial seed value.

variability among conditioning data sets; B. variability among estimated variogram models; C. variability among simulations due to different random seeds; and D. measurement variability in the sampling design process due to different random seeds and the number of iterations. Results for n_c =50 are summarized in Table 1, Case 1,and will be discussed in more detail below. Possible sources of bias in this example are the simulation algorithm, the data transformation, and the method of estimating variogram models. However, the results in Figure 6 do not appear to show any bias of practical significance.

Table 1. Summary sampling design results from simulations with n_c = 50.

Case	Sources of Variance	N_e Mean	N_e Variance
1	A,B,C,D	230	4126
2	C,D	221	668
3	A,C,D	320	1990

Sensitivity to random seeds

Each sampling design run has involved resampling a single site model, i.e., a single realization of the simulation algorithm generated by a single seed value for the random number generator. Would the results be more valid if a separate seed value and corresponding site model were used for each sampling draw? This question was addressed by generating several different site models conditioned by a single set of 50 sample data and the corresponding variogram model. Sampling design results from these site models are shown in Figure 7, and in Table 1, Case 2. Although cost curves for individual site models are higher or lower, reflecting different simulated distributions of contaminant, N_e is relatively constant. Resampling a single site model thus appears to be appropriate for estimating N_o (but note that this would not be true if the objective were to estimate absolute costs associated with N_o). Multiple simulations would add the variability among the curves to that already encountered about a single curve. This would require more iterations to obtain a solution, without necessarily improving accuracy. Nevertheless, any particular seed might lead to an extreme result - repeating the procedure on several different site models would be prudent.

Sensitivity to the conditioning data and the variogram model

The sampling design procedure was rerun for seven of the data sets with n_c = 50, values, using only a single variogram model. The results (Table 1, Case 3) incorporate all sources of variability except the variogram model. By subtraction, the variance components A and B are estimated to be 1322 and 2136, respectively. Although these variance estimates are based on only a few observations, they suggest that the variogram model component, B, is the most significant. Figure 8 illustrates the problem involved in estimating the variogram with sparse data, and reproducing it in conditional simulations. Figure 9 indicates that the method is somewhat more successful at reproducing the frequency distributions.

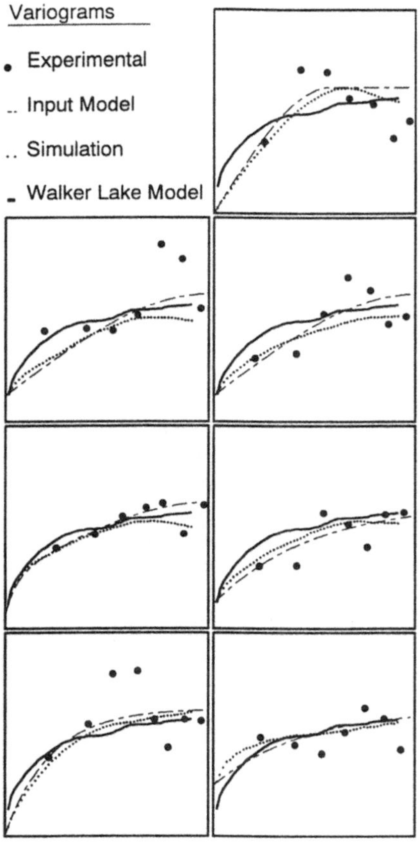

Figure 8. Variograms from seven simulations based on $n_c = 50$, compared with the corresponding experimental variograms, fitted models, and the exhaustive Walker Lake variogram.

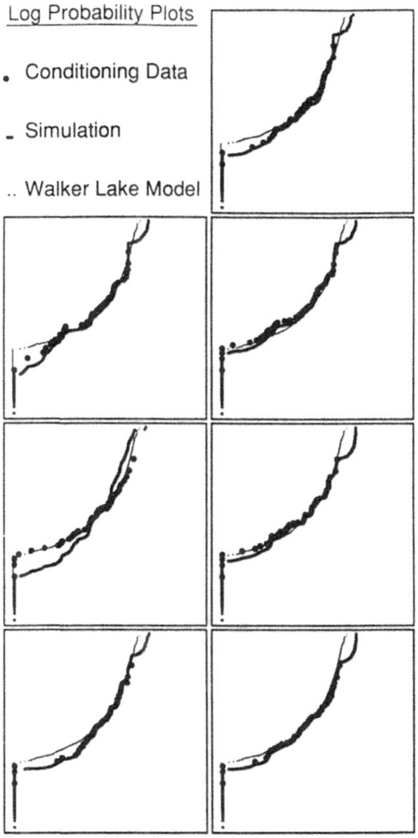

Figure 9. Cumulative probability plots for the seven simulations in Figure 8, the corresponding conditioning data, and the Walker Lake Model. The X-axis is ln(concentration).

CASE STUDY 2: PIAZZA ROAD

At the Piazza Road site in Missouri, dioxin-contaminated oil was sprayed as a dust suppressant on gravel roads. Soil in adjacent pasture land was also contaminated by drainage from the roads. The United States Environmental Protection Agency established a risk-based action level of 1 part per billion (ppb) over an exposure unit (EU) of 5000 sq. ft. (465 m^2) to a depth of 2 in. (5 cm). Soil from EU's estimated to exceed 1 ppb was to be removed and stored for future treatment (RU = EU). High remediation costs ($91,000 per EU) and relatively low sampling costs ($156 per sample) suggested that a more selective removal (RU < EU) might be cost-effective. In an extensive pilot study

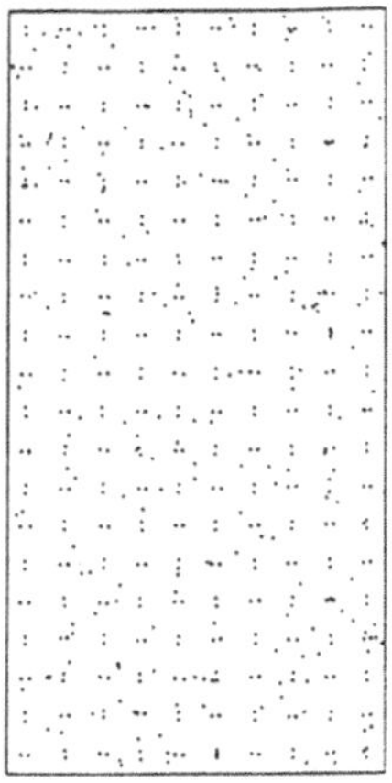

Figure 10. Map of 600 Piazza Road data locations.

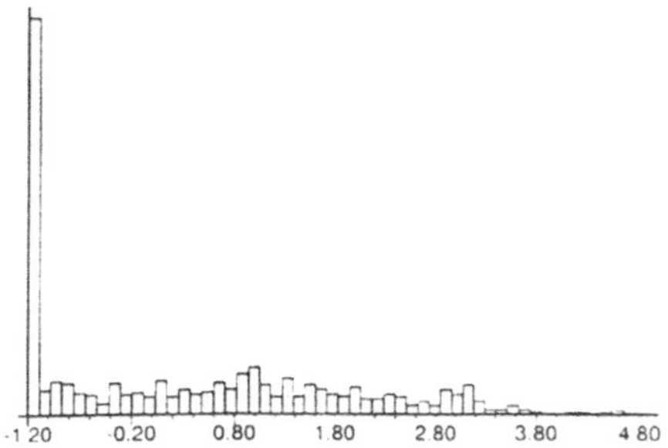

Figure 11. Log histogram of Piazza Road data.

on a representative portion of the site, 600 locations were sampled and analyzed over an area of four EU's to provide information on sampling, analytical, and spatial variability. Ryti, et al (1991) used this information to estimate the RU size and corresponding sampling grid which would minimize total sampling plus remediation costs, subject to pre-defined error limits. The RU was constrained to a minimum practical size of 10 x 10 feet. They coupled the size of the sampling grid to the RU, and recommended a 14 x 14 foot unit size (equivalent to $\mathbf{N_e}$ = 100).

The site remediation has since been completed, and the revised approach reduced total costs more than \$5 million, after allowing for the \$430,000 cost of the pilot study.

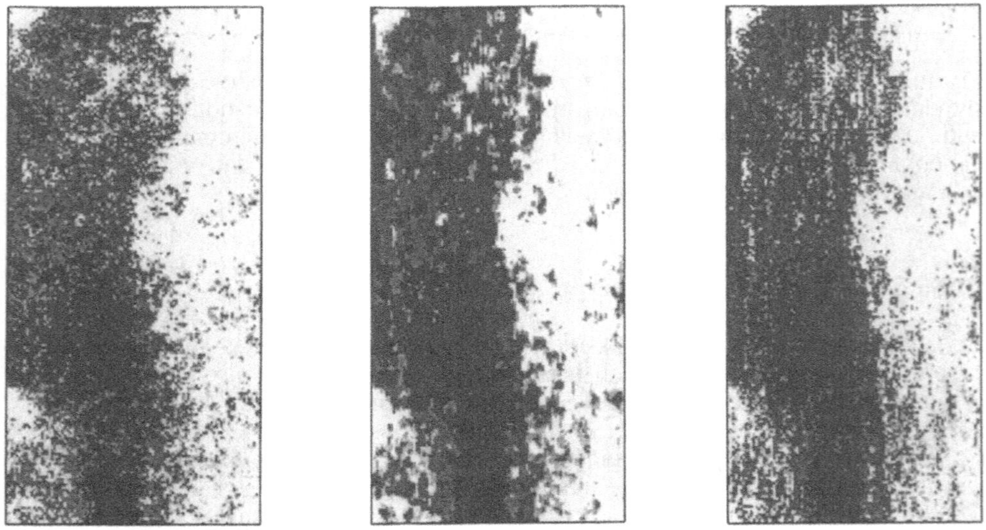

Figure 12. Three Piazza Road simulations based on 600 data.

Although successful in this case, the time and expense involved in such large pilot studies may be prohibitive in many cases. In this example, we will evaluate whether conditional simulations based on a much smaller pilot study could have provided useful results.

Optimization based on the full pilot data set

The 600 data locations are shown in Figure 10, and the histogram in Figure 11. The spike at 0.3 ppb represents non-detects arbitrarily set equal to the detection limit. The earlier results suggest that the main source of variability in the sampling design procedure is the variogram model. To evaluate the level of uncertainty, we simulated the pilot area, conditioned by the 600 data, with each of three alternative variogram models which represented a range of possible interpretations of the experimental variogram. Each of the three simulations (Figure 12) was used in a sampling design run, with the sampling and remediation costs listed above. The three optimization curves (solid lines, Figure 13) have nearly identical $\mathbf{N_e}$ values of 64, 64, and 49 samples.

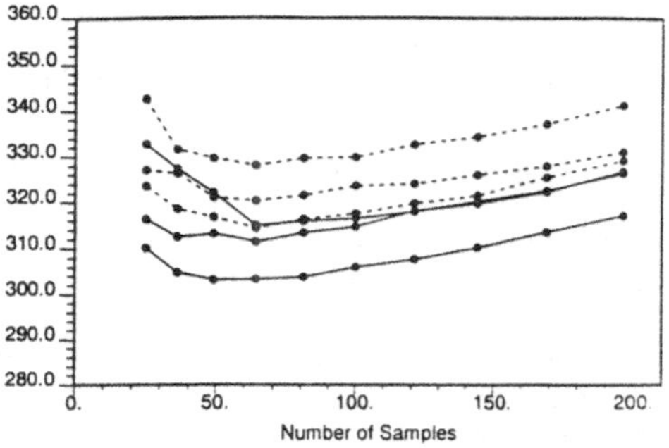

Figure 13. Cost optimization curves from Piazza Road simulations. Solid lines: $\mathbf{n_c}$ = 600; dashed lines: $\mathbf{n_c}$ = 60.

Simulation and optimization with small subsets of the pilot data

The sampling design exercise was repeated with three independent subsets of 60 $\mathbf{n_c}$ data drawn at random from the 600 pilot data. Each variogram was estimated from the subset only. Differences among the three simulations (Figure 14) illustrate the increased

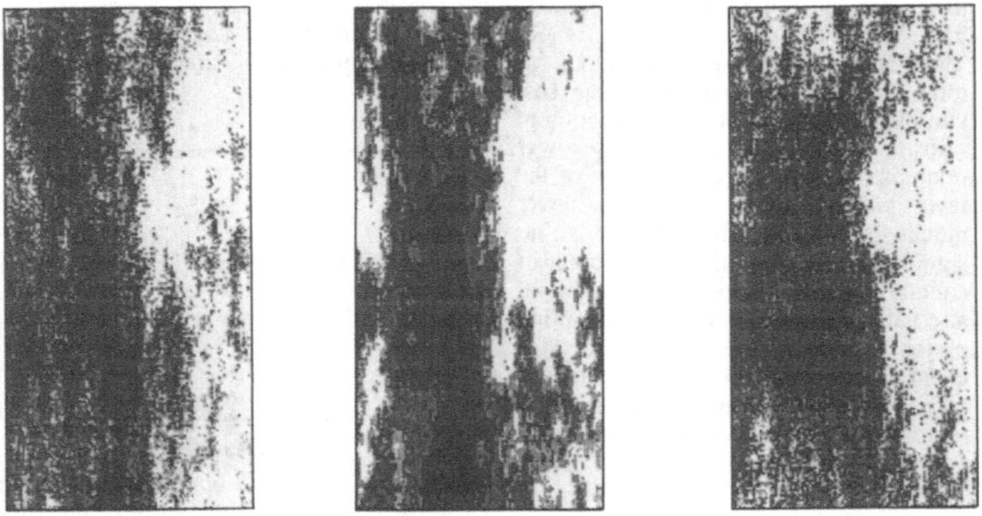

Figure 14. Three Piazza Road simulations based on subsets of 60 data.

variability resulting from fewer data. Nevertheless, the three corresponding cost optimization curves (dashed lines, Figure 13) provide results nearly identical to those obtained with all 600 data, with all three N_e values equal to 64 samples.

Evaluation of sampling alternatives

Suppose an alternative sampling approach such as a portable analyzer could provide data at one-fifth the cost ($31), but with an incremental loss of precision of 25% (relative standard error) compared to the existing method. Would it be advantageous to change methods? If so, how many samples should be taken?. Figure 15 compares sampling designs for the two cases run on the same site model. The upper curve was run as before; the lower curve was run with lower sample cost and with a random error component added to each simulated sample (error from the existing method is already included in the site models via the variogram). Note that more complex, less favorable

Table 2. Comparison of Sampling Alternatives

Point (From Figure 15)	1	2	3
Unit sample cost	$156	$31	$31
Number of samples	64	121	196
Total sampling cost	$9,984	$3,751	$6,076
Remediation cost	$288,789	$284,020	$283,811
Non-remediation cost	$51,268	$51,496	$49,688
Total cost	$350,040	$339,267	$339,555
% False positives	17.1	14.0	13.2
% False negatives	7.4	7.0	6.3
% of total contaminant remediated	97.7	97.6	97.7

error distributions could be used if appropriate. In spite of the added sampling error, the lower sample cost option has lower total costs, with the minimum occurring between 121 and 196 samples. The curve was so flat in this range that 200 iterations were insufficient to resolve the exact minimum. Much of the power of this procedure comes from the large amount of detail that can be obtained in addition to the cost curves. This is illustrated in Table 2 with data about the three points labeled on Figure 15. The reduced total cost from the lower sampling cost alternatives results from a combination of a lower total sampling cost and a lower remediation cost due to reduction of the false positive rate. Although fewer blocks are remediated, there is no significant change in the total amount of contaminant remediated. Evaluation of alternatives is not limited to sampling options. It would be just as easy, for example, to examine the trade-offs associated with changing to a larger remediation unit to obtain economies of scale in remediation.

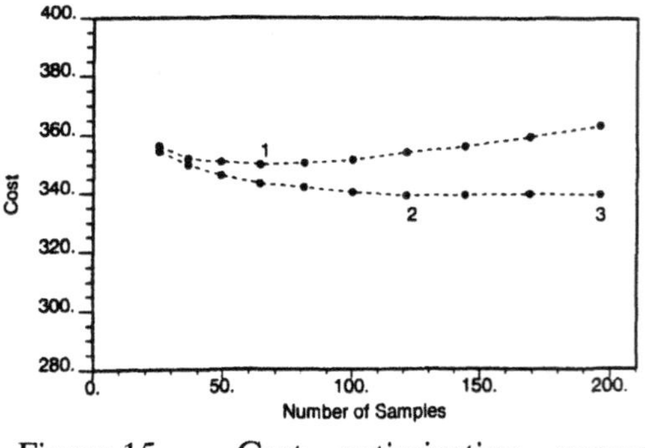

Figure 15. Cost optimization curves comparing two alternate sampling/analytical methods.

These detailed results can help to

evaluate sampling alternatives even when it is not possible to quantify the costs associated with not remediating contaminated material. Examining the relationships among sampling cost, remediation cost, and decision quality from results such as those in Table 2 can effectively quantify the consequences of design choices.

A NOTE ON CONDITIONAL SIMULATION ALGORITHMS AND DATA TRANSFORMATIONS

The objective of this paper was to determine whether resampling a conditional simulation site model has potential as a sampling design tool, particularly in cases where conditioning data are sparse, and the population histogram and variogram are not well known. The SGSIM algorithm used in this study was selected because of its simplicity and efficiency, not because of any perceived theoretical superiority. Other conditional simulation algorithms, such as turning bands (Journel, 1974), LU decomposition (Davis, 1987), and frequency domain (Easley et.al., 1991) could be expected to produce similar results. The indicator simulation algorithm (Alabert, 1987) might deal more easily with truncated data, but it is relatively cumbersome. Evaluation of alternative simulation methods is a topic for further investigation.

The cost model used here is particularly sensitive to the upper tail of the simulated distributions. If the simulated values are much too high, they will tend to exaggerate the cost of false negatives and over-estimate $\mathbf{N_e}$[1]; if the values are too low, they will under-estimate $\mathbf{N_e}$. Because simulated normal scores frequently extend beyond the range of the conditioning data, a model defining the upper tail must be subjectively chosen in order to back-transform the simulation. This choice cannot be avoided - it can be made implicitly as in a log transform, or explicitly, as required by the normal-score transform. Either way, it is a potential source of operator-induced bias in the procedure.

CONCLUSIONS

Monte-Carlo resampling of conditionally simulated site models is a powerful tool for the design of sampling programs. In the cases presented here, which involved less than ideal circumstances, it provided estimates of optimal sampling density that are sufficiently precise, unbiased and robust to be of considerable practical value. The fact that this method can be successfully performed on small computers with relatively small sets of conditioning data makes it potentially routinely applicable in remediation situations, even at small sites.

ACKNOWLEDGEMENT

The authors gratefully acknowledge the work of Allen R. Sparks, who developed and assembled most of the software system used in this investigation.

---

[1] This was the case with the simulations in our original manuscript which were based on log-transformed data. The change to normal-score transforms was made at the suggestion of one of the reviewers.

REFERENCES

Alabert, F.G. (1987), Stochastic imaging of spatial distributions using hard and soft information, MSc thesis, Stanford University.

Barnes, R. (1989) "Sampling design for geologic site characterization", in M. Armstrong (ed.), Geostatistics, Kluwer Academic Publishers, Dordrecht, pp. 809-822.

Burgess, T., Webster, R., and McBratney, A. (1981) "Optimal interpolation and isarithmic mapping of soil properties. IV Sampling strategy", Journal of Soil Science 32, 643-659.

Davis, M. (1987) "Production of conditional simulations via the LU triangular decomposition of the covariance matrix", Mathematical Geology 19, 91-98.

Deutsch, C., and Journel, A. (1992, in press) GSLIB: Geostatistical Software Library, Oxford University Press, New York.

Easley, D., Borgman, L., and Weber, D. (1991) "Monitoring well placement using conditional simulation of hydraulic head", Mathematical Geology 23, 1059-1080.

Englund, E. (1992) "A variance of geostatisticians", Mathematical Geology 22, 417-455.

Isaaks, E. and Srivastava, R. M. (1989) An Introduction to Applied Geostatistics, Oxford University Press, New York.

Journel, A. (1974), "Geostatistics for conditional simulation of ore bodies", Economic Geology 69, 673-687.

Olea, R. (1975) Optimum Mapping Techniques Using Regionalized Variable Theory, Kansas Geological Survey, Lawrence.

Rendu, J.M. (1980) "Optimization of sampling policies: a geostatistical approach", Fourth Joint Meeting MMIJ-AIME 1980, Tokyo.

Ryti, R., Neptune, D. and Groskinsky, B. (1991) "Superfund soil cleanup", Environmental Testing and Analysis, Jan/Feb.

Srivastava, R.M. (1987) "Minimum variance or maximum profitability?", CIM Bulletin, 80:63-68.

NOTICE

Although the research described in this article has been funded in part by the United States Environmental Protection Agency through Cooperative Agreement CR818526 to the Harry Reid Center for Environmental Studies, University of Nevada - Las Vegas, it has not been subjected to Agency review. Therefore it does not necessarily reflect the views of the Agency.

PREDICTING PROBABILITY MAPS OF AIR POLLUTION CONCENTRATION: A CASE STUDY ON BARREIRO/SEIXAL INDUSTRIAL AREA.

Amilcar Soares(*), João Távora(**), Luisa Pinheiro(**), Cristina Freitas(**), José A. Almeida (*)
(*) CVRM- Centro de Valorização de Recursos Minerais
(**) CCRVLT - Comissão de Coordenação da Região de Lisboa e Vale do Tejo

Abstract

Two kinds of problems related to air pollution control in the industrial area of Barreiro/Seixal have been tackled by geostatistical methods: i) One is based on observations made over a large period of time at fixed monitoring stations; two different pollutant structures are identified and split by a threshold value of SO_2. Spatial probability maps for the critical high spikes of SO_2 concentrations are calculated by kriging indicator variables ii) For a given threshold the shape of the plume - high spikes of SO_2 - is determined by transforming the estimated probability maps into morphological maps, fundamental tools for evaluation of the environmental impact on the area.

1 . Introduction

Geostatistics have been used to characterize the spatial dispersion of pollutants, in particular the distribution in time and space of extreme values (see for example Switzer, 1977, Isaaks, 1984). This paper aims at including geostatistics in land use planning by estimating the shape of high spikes of SO_2 and accounting for its intersection with critical areas that should be kept free of pollution, for example schoolyards, hospitals, new urban developments, etc.

In the Barreiro/Seixal industrial area, located in the centre of Portugal, the dispersion of the main airborne pollutant, SO_2 , is characterized by two different concentration structures: one is continuous in time, corresponding to low concentrations and the other has some periodic spikes of high concentration values, resulting from the factories discharges. These two structures are identified and split by calculating the spatial correlation of indicator variables for increasing thresholds: the randomness of the low concentration structure prevails for the low thresholds; whereas above some critical value, the spatial correlation between the monitoring stations emerges, which means that only the high spikes of SO_2 have significant spatial contiguity at the scale of the spacing of the monitors.

The main objective of this study is to preview the impact of the SO_2 pollution in the Barreiro/Seixal area. This implies knowing which important zones within the area are intersected by the pollutant each day.

A factory discharge, in terms of spatial dispersion, is more like a plume - a high SO_2 discontinuity - than a cloud ranging continuously from the high to low values of SO_2. In general, the high spikes of SO_2 must be considered as an independent body from the low values structure and, consequently, to calculate estimated maps of the global SO_2 dispersion, does not make much sense in most air pollution situations.

Thus, in this paper it is proposed to calculate the shape of the plume, i.e. the set of points with SO_2 values above a critical threshold, based on the measurements of each day at a few monitoring stations. Note that the aim is the daily control of the spatial dispersion of extreme values, rather than the knowledge of SO_2 dispersion of the past.

For this purpose daily probability maps are estimated by kriging an indicator variable for one critical threshold. Meteorological conditions, such as wind direction and velocity, are accounted for the structural model: the orientation of the anisotropy ellipse is conditioned to the daily average of the wind direction for the area.

The last step is the transformation of the probability maps into binary maps where the shape of the high concentration of the pollutant is reproduced, as well as the areas with the highest probability of being misclassified.

2. The Data Set

In the Barreiro/Seixal industrial area, the monitoring stations (see fig 1) were installed according to the location of the pollution sources. Hourly values of SO_2 concentration and wind velocity and direction are recorded for all monitors (see CCRL Report, 1989) . For this study the set of daily averages of SO_2 in 1989 for the area of Barreiro was used.

3. Structural Analysis of the Pollutant in Space and Time.

. Indicator Variography of the High Concentrations of SO_2.

In the Barreiro area the spatial dispersion of SO_2 is characterized by a twofold structure: a continuous, in time, low sulphur concentration, combined with occasional high spikes of SO_2 concentration resulting from factories discharges.

As revealed by the variograms, the dispersion of the low SO_2 values does not have any spatial contiguity at the scale of the spacing of the monitoring stations. On the other hand the variograms of transformed indicators, for high cutoff values, show a clear contiguity structure between monitoring stations.

Let us denote by $Z(x,t)$ the average SO_2 concentration of a given day t at the monitoring station located in x, then $I_z(x,t)$ is the corresponding indicator value for the SO_2 threshold z:

$$I_z(x,t)=1 \text{ if } Z(x,t)>=z \quad \text{and} \quad I_z(x,t)=0 \quad \text{if} \quad Z(x,t) < z$$

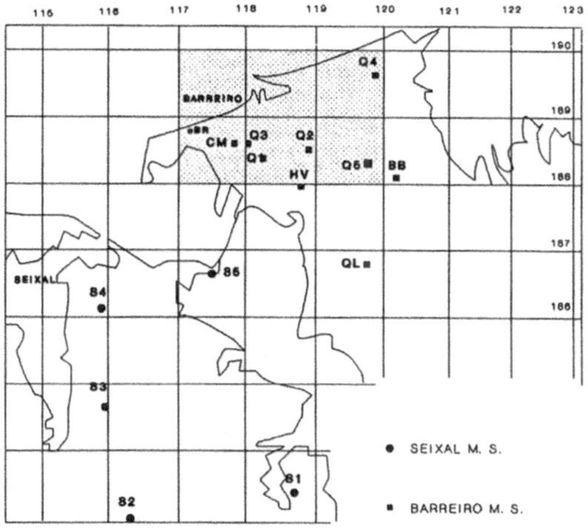

Fig 1. Barreiro/Seixal area showing the monitoring stations.

Averaging the indicator variograms of space-h over N days of a homogeneous period, the variogram which is representative of the area covered by the Ns monitoring stations for that period is obtained:

$$\gamma(h,z)= \frac{1}{2.N.Ns} \sum_{t=1}^{N} \sum_{x=1}^{Ns} [\, I(x,t)-I(x+h,t)]^2$$

Increasing the threshold values the resulting variograms show (see example in fig 2) that the spatial randomness of the low values disappears as the high concentration thresholds increases.

To find the threshold that splits the two populations it is of great importance to define homogeneous periods of time, in which the indicator variograms are calculated.

For the purpose of temporal stationarity some time-series approaches can be used (see for example, Bilonick, 1984). In this study the criteria used to classify homogeneous periods of time was the spatial contiguity and temporal correlation revealed by the variograms of each month: the periods of time are clustered according to the similarity of their variograms in space and time.

Averaging, for all monitoring stations, the indicator variograms corresponding to the pollutant plume, along the time axis, the resulting ranges of the variograms measure the average length of time of the high spikes of SO_2 in that period (see example of fig 3).

$$\gamma(d,z)= \frac{1}{2.Ns.N} \sum_{x=1}^{Ns} \sum_{t=1}^{N} [\, I(x,t)-I(x,t+d)]^2$$

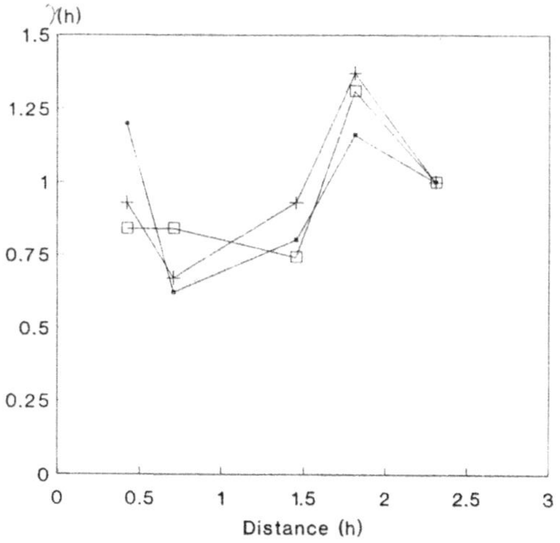

(2a)

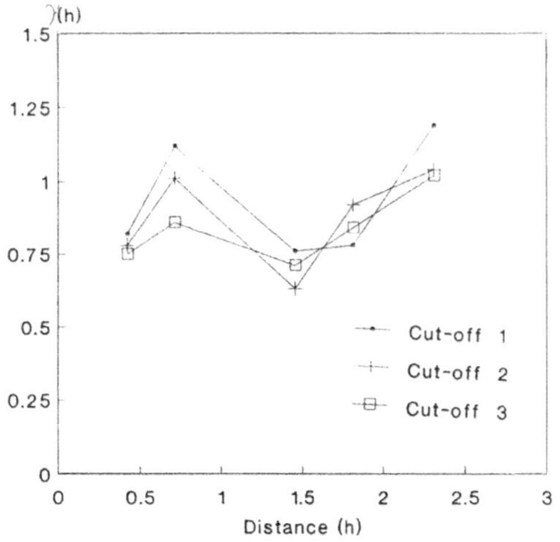

(2b)

Fig 2. - Indicator Variograms for 3 increasing thresholds in August (2a) and October(2b).

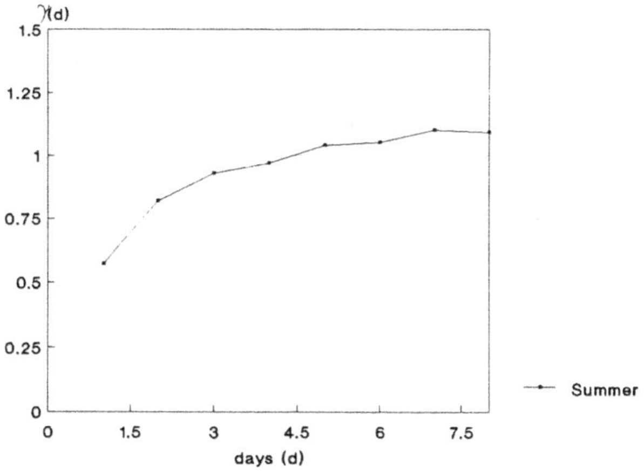

Fig 3 - Indicator Variograms (corresponding to 45 ppm of SO_2) of Time Spacing for a summer period.

In summary, the shape of the high SO_2 dispersion can be estimated after the contiguity measures (variogram or covariances) have been calculated for each homogeneous period of time.

. Meteorological Conditioning

The spatial morphology and the length of time of the pollution plume is strongly influenced by the weather, in particular the direction and velocity of the wind, which are measured with the same period as the SO_2 concentrations.

The wind velocity affects mostly the durability of the plume, which means that, for example, in winter the average length of time of one plume is about two days whereas in summer it is about 4 days.

Calculating the pollution roses - SO_2 concentrations for each wind direction- at the monitoring stations (see fig 4) , we can see the most significant concentrations of SO_2 measured in the classes of wind directions from N/S to NE/SW. The intersection of the main vectors of the different pollution roses of monitoring stations gives us the source point of the pollution plumes, i.e. the spatial location of the factories. Thus the spatial dispersion of the pollution plumes is not isotropic, it depends strongly on the wind direction, which is a parameter that must be accounted in the SO_2 spatial estimation procedure.

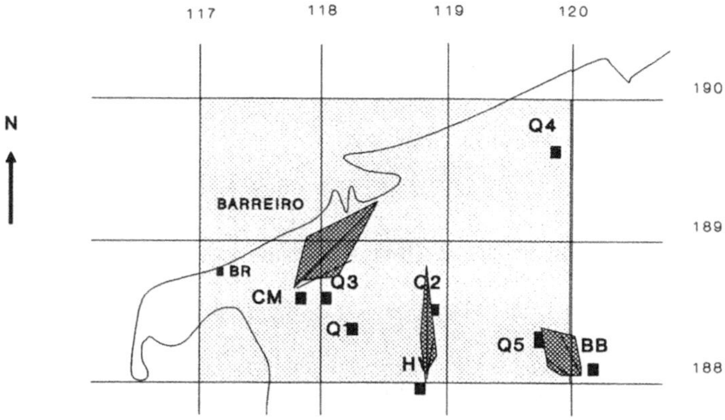

Fig 4 - Pollution roses for 3 monitoring stations during the summer months.

The small number and the spatial location of the monitoring stations does not allow the inference of the possible anisotropic relations for all wind directions. In fact, the variogram anisotropy relation can only be calculated for the NE/SW and NW/SE directions , which correspond to the spatial orientation of the monitoring stations (see example of fig 5).

For other wind directions we assume an anisotropy ellipse with the same axis relation but coincident with the average wind direction of each day. Under this assumption, we have conditioned the stochastic modeling to one important parameter of the physical process of transport of the plume. Nevertheless this is valid for a specific class of wind directions (which is the main class) that carry the plumes from the factories through the area.

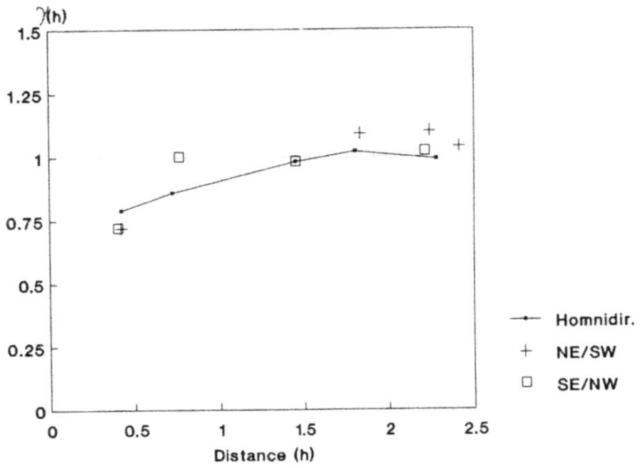

Fig 5 - Indicator Variograms for the 2 main directions and averaged over all directions.

4 . Morphological Estimation of the Pollution Plume

. Estimation of the Probability Maps.

The final objective of the geostatistical modeling is to calculate the shape of the plume, i.e. the set of points with SO_2 values above a critical threshold, based on the measurements of a given day at a few monitoring stations.

To infer, in area A, the characteristics measured at the monitoring stations, we have to assume stationarity of the following parameters for a given homogeneous period of time: the critical threshold value of SO_2 and the spatial contiguity measures - indicator variogram or covariance models- in particular the anisotropy ellipse relations- which implies that evident changes of the covariance parameters in the recent past must be taken into account.

Denote the pollution plume by the two dimensional set $\{X\}$ consisting of all points x in area A with the SO_2 content - $Z(x,t)$ greater than the critical value - the threshold z.

$$\{X\} = \{ x ; I_z(x,t)=1\} \qquad I_z(x,t)=1 \quad \text{if } Z(x,t)>=z$$

The shape of $\{X\}$ is estimated from the experimental values $I(x_\alpha,t)$, known at the monitoring stations at x_α, $\alpha=1,N$ on the day t.

Thus, in a close grid of points, the probability of each point to belonging to the pollution plume - prob $\{x \in X\}$ - was estimated. One linear estimator of that probability is the kriged indicator value at x - $I^*_z(x,t)$, based on the neighborhood values $I_z(x_\alpha,t)$ (Sollow, 1986, Soares,1990) :

$$[\text{ prob}\{x \in X\}]^* = I^*_z(x,t) = \sum_\alpha \lambda_\alpha \, I(x_\alpha,t)$$

To solve the kriging system the anisotropy ellipse of covariances has the main orientation coincident with the average wind direction of that day, thus conditioning the estimator $I^*_Z(x,t)$ on that meteorological parameter.

Fig 6 shows the probability maps in the area A for three successive days in August (6a) and October (6b).

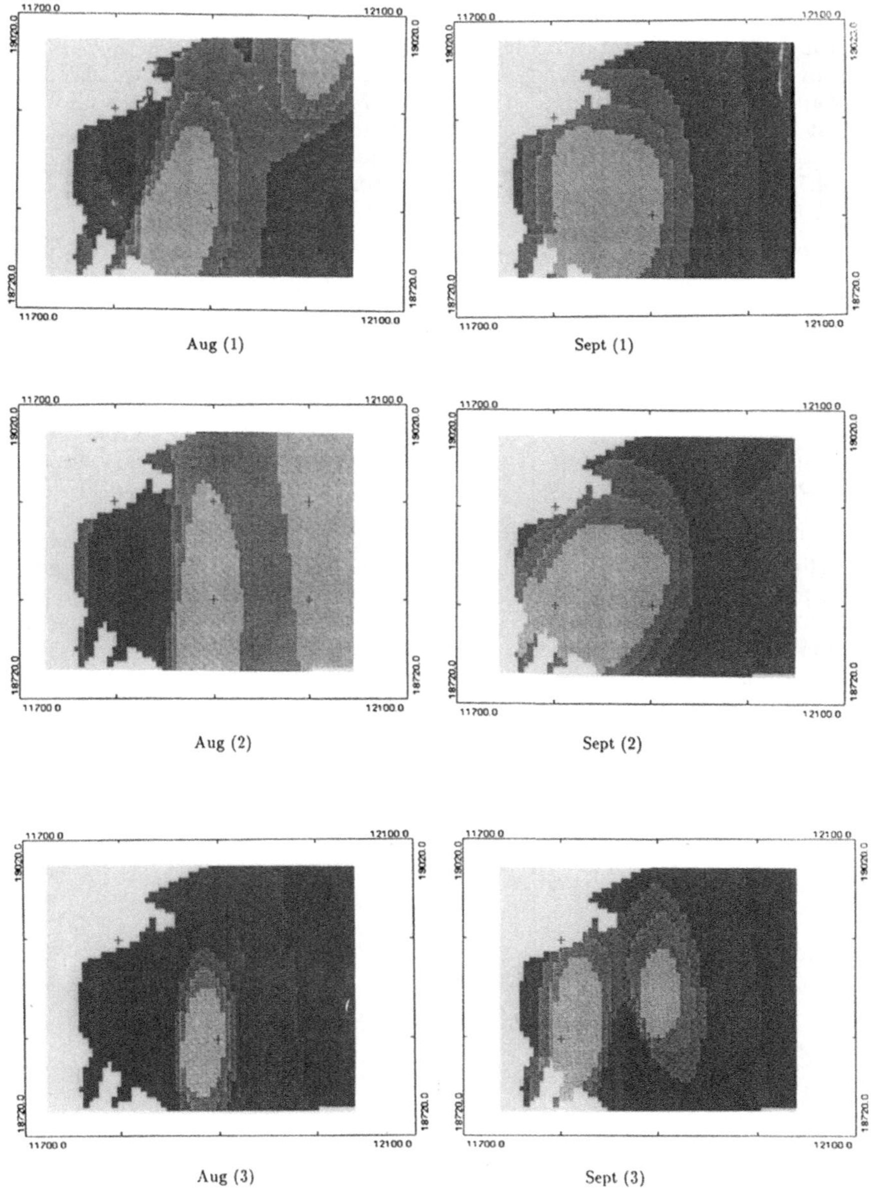

Fig. 6 - Probability maps of 3 successive days of August (6a) and October (6b).

. Calculating the Shape of the Pollution Plume

The last step is the transformation of the probability maps into binary maps where the shape of X is reproduced.

Soares (1990) proposes an algorithm to convert a probability map into a binary map, based on the following simple criterion: denote by m_I the mean of the indicator values of all monitoring stations on a given day, the proportion of unit values in the final binary map must be equal to m_I. Sorting by descending order the total M estimated values $I^*_z(i,t)$ (i=1, M) in area A, the set {X} consists of the first K values - highest probability values of belonging to {X}- such that $K/M=m_I$, the experimental mean. Instead the experimental mean m_I one can use the kriged average $[m_I]^*$, the kriged indicator value for the entire area A.

The set {X} consists of all points x such that:

$$I_z(x,t) = \begin{cases} 1 \text{ if } I^*_z(x,t) >= I^*_z(K,t) \\ 0 \text{ if } I^*_z(x,t) < I^*_z(K,t) \end{cases}$$

where $I^*_z(K,t)$ is the Kth value of the ordered vector $I^*_z(i,t)$

Fig 7 shows the shape of pollution plume in Barreiro area for the previous daily sequences in August and October.

For each day we can see four tones of grey corresponding to the shape of the plume (light grey) , the complementary, i.e. the region definitely outside the plume (black), and two risky areas of misclassification around the threshold:

. the highest probability values (40% of total points) of belonging to the plume - light grey area .

. the next 2 grey tone bands (20% of total points) corresponding to the areas, with the highest probability of being misclassified, around the threshold.

. the lowest probability values (40% of total points) of belonging to the plume - black area.

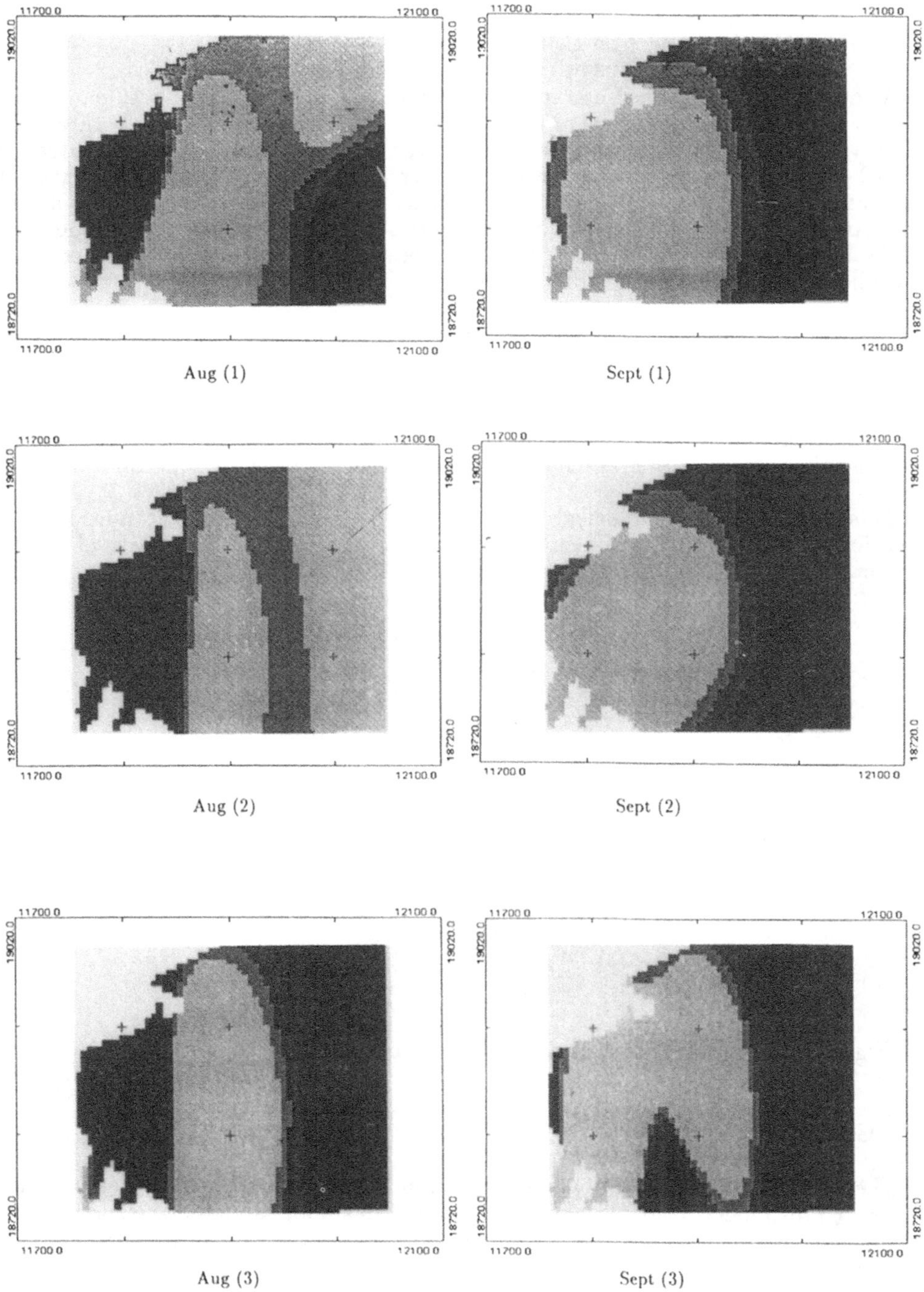

Fig 7 - The shape of plume, the complementary and two risky areas of misclassification around the threshold, for the 3 successive days in August and October.

5. Final Remarks

. Predicting the Impact on Critical Areas.

Although this study is illustrated with shapes of plumes above just one critical threshold, the morphological maps can be calculated for several thresholds. However, in most situations, on a given day, the shape of the high concentrations of SO_2 does not change by increasing or decreasing the thresholds, as long as the indicator values at the monitoring stations remain the same. The pollutant discharges are more like sharp discontinuities than continuous SO_2 clouds, and their spatial characterization can be achieved by the proposed method of morphological estimation.

By overlapping the morphological maps of the plumes with maps of urban, industrial and protected green areas (see example of fig 8)- the impact on critical areas such as hospitals, schoolyards, etc., can be measured and controlled, and risky situations can also be foreseen. For example, assuming that the same values of $I_Z(x_\alpha,t)$ are maintained in the monitoring stations x_α on the next day, we can forecast the impact of the plume shape for eventual changes of wind direction.

This important morphological tool can also be used for land use purposes: the location of new critical areas can be planned according the clean past of a given region.

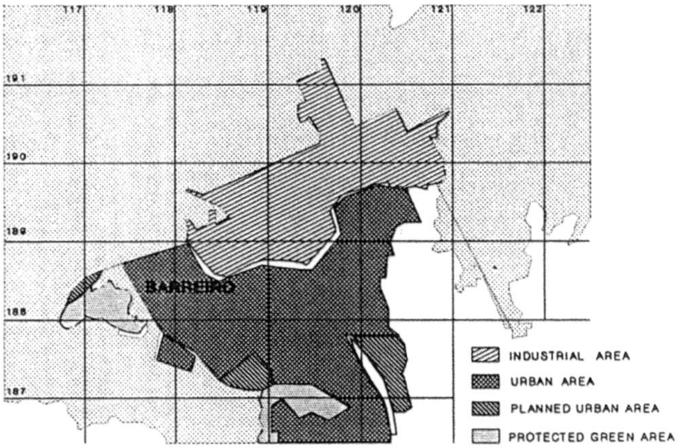

Fig 8 - Geographic information map of Barreiro.

. Geostatistic Models

The spatial estimation of SO_2 concentration is based on stationarity assumptions of variogram models over the entire area and throughout an homogeneous period of time. This means that if one specific day is assumed to belong to the previous homogeneous period of time, it is of great concern to validate the models, used for that period, in the exploratory data analysis , mostly in periods of seasonal transition.

The final, but not less important remark, concerns stationarity in space. One characteristic of the pollution control application fields is the great amount of data in time but measured at only a few monitoring stations. Thus, great care must be taken when analyzing interpolated shapes in areas of scarce data and extrapolated shapes in zones not bounded by the monitoring stations.

References

Bilonick, R., 1984, Risk-Qualified Maps for a Long Term Sparse Network-Mapping the USGS New-York Acid Precipitation Data. G. Verly et al (eds), Geostatistics for Natural Resources Characterization , Part 2, 851-862, pub. Reidel, Dordrecht.

CCRLVT -Internal Report, 1989. Qualidade do ar no Barreiro/seixal.

Isaaks, E., 1984, Risk Qualified Mappings for Hazardous Waste Sites: a Case-Study in Distribution-Free Geostatistics, M.Sc. thesis, Stanford Univ., 85p..

Lajaunie C., 1984, Geostatistical Approach to Air Pollution Modelling.G. Verly et al (eds), Geostatistics for Natural Resources Characterization , Part 2, 851-862, pub. Reidel, Dordrecht.

Soares, A., 1990, Geostatistical Estimation of Orebody Geometry: Morphological Kriging, Math. Geol., Vol.22, pp. 787-802.

Solow, A., 1986, Mapping by Simple Indicator Kriging. Math. Geol., Vol. 18, no 3, pp 335-352.

Switzer, P., 1977, Estimating of Spatial Distributions from Point Sources with Application to Air Pollution Measurement. Bull. Int. Stat. Inst., XLVII (2), pp. 122-137.

MODELLING OF MAP IMPURITIES USING SEQUENTIAL INDICATOR SIMULATION

Marc F. P. Bierkens and Peter. A. Burrough
Department of Physical Geography, University of Utrecht
P.O. box 80.115, 3508 TC, Utrecht, The Netherlands

ABSTRACT

The indicator approach is used to describe the spatial variability of categorical data in soil survey. Indicator kriging is used to estimate the conditional probability on the occurrence of classes at unvisited locations. From the estimated conditional distribution the unknown class at these locations can be predicted. A new definition of map purity is given, which is spatially varying and directly dependent on the classes found in surrounding borings. When maps of categorical data are used for resource analyses, Sequential Indicator Simulation (SIS) can be used to model the propagation of the map impurities. A case study on the mapping of water table classes in the Netherlands is presented to demonstrate the methods.

INTRODUCTION

In soil survey, about 90% of the collected data are of a qualitative nature; they are either ordinal or nominal (Bregt et al., 1991). It is difficult to interpolate or predict these data with regular geostatistical methods because they generally focus on the description and prediction of quantitative variables. Instead, borings of categorical data that are classified equally are grouped together by enveloping polygons. In this way a choropleth map is constructed which serves in a sense as a prediction model for the soil classes at unvisited locations.

The quality of these maps can only be investigated afterwards by taking randomly (in space) sufficient additional borings. The percentage of these borings that is classified to the same class as its enveloping map unit is called the "map purity". There are several disadvantages to this method: First, the practice of manually drawing boundaries around clusters of equally classified borings is, to some extend, subjective. Second, the map purity, a measure of the quality of the map, can be only obtained by additional sampling. Furthermore, the map purity which is thus obtained is spatially invariant (i.e. is a constant) within one map unit and independent of the number of borings taken and the classes found at the boring sites. Obviously, if a class is predicted at an unvisited location, the uncertainty of this prediction should be larger if the surrounding borings all belong to different classes. This uncertainty should decrease when more borings are placed around the location of the prediction. The "classical" method, i.e. predicting categorical data by making choropleth maps and assessing the map purity afterwards, does not account for these effects.

Maps of categorical data are frequently used as input for land resources analyses. For instance, a soil map may be used to assess the suitability of land for a certain type of land use. Usually, its is assumed that the information obtained from these maps is correct. Heuvelink et al. (1989) investigated how errors in quantitative properties, as derived from the soil map, propagate through procedures of resource analysis. However, the way that impurities in input maps of categorical data, i.e. misclassifications, cause errors in the resources analyses is rarely investigated. An important reason for this is the lack of a comprehensive statistical model to predict soil classes and estimate the quality of these predictions.

In this paper indicator kriging (Journel, 1983) is used to estimate the conditional probability on occurrences of classes of categorical data, given the classes found at the original observations. The estimated conditional probabilities are used to predict the unknown classes at unvisited locations. A new definition of map purity is obtained, which is spatially variant and directly dependent on the boring density and the classes found at the observation sites. Sequential indicator simulation (SIS) (Journel and Goméz-Hernández, 1989, Goméz-Hernández and Srivastava, 1990) is used for the generation of alternative equiprobable realisations of classified maps. If some method of resource analysis uses the original map as input, the uncertainties accruing from map impurities can be assessed by performing the analysis on each of the maps generated by SIS.

In the remaining part of this paper, the application of the indicator approach to categorical data is explained. After that, a case study is presented to demonstrate the methods.

THEORETICAL FRAMEWORK

Random fields of categorical variables

Consider a field where m mutually exclusive classes can be found. At each point $\mathbf{x}_0$ a local observation (boring) would be classified to one of the possible classes $\{U_1,\ldots,U_m\}$. Which of the m classes is found at $\mathbf{x}_0$ is not known beforehand. In other words, at a certain location a random variable $N(\mathbf{x}_0)$ is defined which can take any of the values $\{U_1,\ldots,U_m\}$. The probability

$$F_N(U_k;\mathbf{x}_0) = P[N(\mathbf{x}_0) \in \{U_1,\ldots,U_k\}] \qquad (1)$$

is called the (marginal) cumulative probability distribution function (cpdf) of $N(\mathbf{x}_0)$ at location $\mathbf{x}_0$. The function $N(\mathbf{x})$ is then said to be a random field, consisting of categorical variables. If n observations (classified borings) are available the occurrence of a class may be predicted by estimating the conditional probability on each of the m classes, given the n observations. If a classified boring is denoted by n_i, $i\in(n)$, this conditional probability is written as:

$$F_N(U_k;\mathbf{x} \mid n_i;\mathbf{x}_i{:}i \in (n)) = P[N(\mathbf{x}) \in \{U_1,\ldots,U_k\} \mid n_i;\mathbf{x}_i{:}i\in(n)] \qquad (2)$$

The remaining problem is to estimate this conditional probability.

Indicator coding

To estimate the conditional probability (2) first a new random variable is introduced. The indicator random function (indicator RF) or indicator transform $I(U_k;\mathbf{x})$ of $N(\mathbf{x})$ is defined as:

$$I(U_k;x) = \begin{cases} 1 & \text{if } N(x) \in \{U_1,\ldots,U_k\} \\ 0 & \text{if } N(x) \in \{U_{k+1},\ldots,U_m\} \end{cases} \tag{3}$$

The indicator transform has some important properties (Journel, 1983): (1) The expected value of the indicator RF equals the first order cumulative probability distribution function (cpdf) of N(x). (2) The second (non-central) moment of the indicator RF equals the bivariate cumulative distribution function of N(x) and N(x'). It represents the probability that at position x one of the classes $U_1,\ldots,U_k$ is found and at position **x'** one of the classes $U_1,\ldots,U_{k'}$.

Estimation of conditional probabilities
The conditional expected value of $I(U_k;x)$, given a set of n classified borings $\{n_i;x_i : i \in (n)\}$ is given by (Journel, 1983 and Goméz-Hernández, 1989):

$$E[I(U_k;x) \mid n_i;x_i:i \in (n)] = P[N(x) \in \{U_1,\ldots,U_k\} \mid n_i;x_i:i \in (n)] \tag{4}$$

From (4) it can be seen that we can estimate the conditional probability $F_N(U_k;x \mid n_i; x_i, i\in(n))$ by estimating the corresponding indicator conditional expected value $E[I(U_k;x) \mid n_i;x_i : i \in (n)]$ for the same "threshold" U_k ($\forall k= 1,..,m-1$). The exact derivation of this conditional expectation is very cumbersome (Journel & Zhu, 1990). Instead, one may approximate this expected value by computing the expected value of the indicator RF in **x** for threshold U_k, given the indicator transforms $i(U_k;x)$ of nearby data for the <u>same</u> threshold:

$$E[I(U_k;x) \mid i(U_k;x_i) : i \in (n)] \approx E[I(U_k;x) \mid n_i;x_i : i \in (n)]$$

For second order stationary N(x) this conditional expectation is estimated by the following linear combination (Journel, 1983):

$$F_{N\mid(n)}{}^{*}(U_k;x \mid n_i;x_i : i \in (n)) \approx E[I(U_k;x) \mid i(U_k;x_i) : i \in (n)]$$

$$\approx (1 - \sum_{i\in(n)} \lambda_i(U_k;x))\cdot F_N(U_k) + \sum_{i\in(n)} \lambda_i(U_k;x)\cdot i(U_k;x_i)$$

$$k = 1,\ldots, m-1 \tag{5}$$

$F_N(U_k)$ is the marginal distribution of N(x), $i(U_k;x_i)$ is the indicator transform of the observation (classified boring) n_i at location x_i for threshold U_k and $\lambda_i(U_k;x)$ is the corresponding simple kriging weight. These weights are obtained by solving an indicator simple kriging system using the (stationary) indicator covariance function $C_I(U_k;\mathbf{h})$ for the threshold U_k:

$$\sum_{i\in(n)} \lambda_i(U_k;x)\cdot C_I(U_k;x_i - x_j) = C_I(U_k;x - x_j) \quad \forall j \in (n)$$

$$k = 1,\ldots, m-1 \tag{6}$$

Although theoretically simple kriging should be used to estimate the cpfds, equations (5) and (6) may be replaced by an ordinary kriging system. When ordinary kriging together with a search neighbourhood is used the expected value of the indicator RF is re-estimated implicitly at each location. Therefore, it is more robust with regard to the stationarity assumption than the simple kriging estimates.

Inference
From equation (6) it can be seen that in order to estimate the complete conditional cpdf of N(**x**), given the n classified borings, m-1 indicator covariance functions are to be inferred. Usually, one prefers to infer a variogram model instead of a covariance model. For a second order stationary N(**x**) the indicator semi-variogram $\gamma_I(\mathbf{h})$ and the indicator covariance function $C_I(\mathbf{h})$ are related as ($\mathbf{h} = \mathbf{x} - \mathbf{x}'$):

$$C_I(U_k;\mathbf{h}) = C_I(U_k;\mathbf{0}) - \gamma_I(U_k;\mathbf{h}) \quad (7)$$

The indicator variograms can be inferred from the indicator data using the estimator:

$$\gamma_I(U_k;\mathbf{h}) \approx \frac{1}{2n(h)} \sum_{i=1}^{n(h)} [\, i(U_k;\mathbf{x}) - i(U_k;\mathbf{x} + \mathbf{h} \pm \triangle\mathbf{h})]^2 \quad (8)$$

$$k = 1,\ldots,m\text{-}1$$

With n(**h**) the number of pairs of indicator data that are distance $\mathbf{h} \pm \triangle\mathbf{h}$ apart. Indicator covariances and variograms represent bivariate cumulative distribution functions. Therefore, they should obey a number of order relations (inequalities) related to the properties of a bivariate cumulative distribution function. These order relations can be easily checked by using the inequalities presented by Journel and Posa (1990).

Order relation violations
Because conditional cumulative probabilities are estimated the respective estimates $F_{N\,|\,(n)}^{*}(U_k;\mathbf{x})$ should obey the following relations: $F_{N\,|\,(n)}^{*}(U_k;\mathbf{x}) \geq 0$ and $F_{N\,|\,(n)}^{*}(U_k;\mathbf{x}) \leq 1$ and furthermore $F_{N\,|\,(n)}^{*}(U_k;\mathbf{x}) \leq F_{N\,|\,(n)}^{*}(U_l;\mathbf{x})$ if $k < l$. When any of these inequalities are violated the estimates should be corrected. Examples on how to perform these corrections can be found in Journel (1983).

Predicting classes; a new definition of map purity
Once the conditional cpdf at a certain location is known, so is the corresponding probability density function (pdf). From the pdf we can derive any type of estimate to use as a prediction of the unknown class. For real data the conditional expectation (E-type) is often used because it is best in least squares sense. However, when dealing with categorical data such an estimate would not make any sense. Therefore, we propose to use as a best predictor of the unknown class the class whose estimated probability of occurrence is largest (the P-type). This maximum probability, say **P'**, is then defined as the "map purity" and is a measure of the accuracy of the prediction. The resulting probability mass (1-P') is called the "map impurity".
An important property of the conditional cpdf estimated by indicator kriging is its dependence not only on the number of classified borings used, but also on the classes found at these locations. Consequently, this is also true for the map purity as defined above. In conclusion, by adapting the indicator approach we have arrived at a procedure to predict categorical data at unvisited locations, given the classes found in neighbouring borings. Also, the procedure produces an associated map purity which is dependent on the number of surrounding borings and the classes found at these locations.

Simulation
After a map of categorical data has been produced, by means of the above procedure, it may serve as input for land resources assessment. To investigate the uncertainties, due to map impurities, that arise when these maps are used as input for land resources

analyses, Monte Carlo simulation can be applied: Limiting ourselves to conditional simulation, (preferably a large number of) equiprobable realisations of the conditional RF N(**x** | (n)) are generated. These conditional realisations should honour the statistical properties of the RF N(**x**) as well as the classes found at the locations of the borings. Each of these realisations is used as input for the land resources assessment. By statistical evaluation of all of the results we obtain a measure of the uncertainties in the resource analysis that result from the impurities in the input maps of categorical data.

A method that is completely general with regard to the type of variable and its distribution and which can generate conditional realisations directly is called "sequential simulation". Sequential simulation has been applied extensively for the conditional simulation of indicator random functions (Journel and Gómez-Hernández, 1989, Gómez-Hernández and Srivastava, 1990) and also to multigaussian random functions (Gómez-Hernández, 1991). If we define a grid of N points where realisations of the conditional RF N(**x** | (n)) are to be simulated a sequential simulation of the N indicator data proceeds as follows (Journel and Gómez-Hernández, 1989, Gómez-Hernández and Srivastava, 1990):

<u>Sequential indicator simulation algorithm</u>

0) All N points are visiting according to some random path.

1a) Determine with equations (5) and (6) or ordinary kriging the conditional density of $N(\mathbf{x}_1)$, given the (m-1) indicator transforms of the n conditioning data (borings) $\{n_i;\mathbf{x}_i, i \in (n)\}$.

b) Solve any order relation violations that may occur.

c) draw from this conditional density the realisation n_1. This is done by drawing a number between 0 and 1 and determining $n_1 \in \{U_1,\ldots,U_m\}$ from the inverse of $F_{(1 \mid (n))}{}^*(U_k;\mathbf{x}_1)$.

d) Take the (m-1) indicator transforms of n_1 and include them in the conditioning data set.

2a) Determine with equations (5) and (6) or ordinary kriging the conditional density of $N(\mathbf{x}_2)$, given the indicator (m-1) indicator transforms of n_1 and of the n conditioning data (borings) $\{n_i;\mathbf{x}_i, i \in (n)\}$.

b) Solve any order relation violations that may occur.

c) draw from $F_{(2 \mid 1 \cup (n))}{}^*(U_k;\mathbf{x}_2)$ the realisation $n_2 \in \{U_1,\ldots,U_m\}$.

d) Take the (m-1) indicator transforms of n_2 and include them in the conditioning data set.

3) Repeat step 2a) to 2d) until all N values are simulated.

For the generation of a second realisation steps 1 to 3 are repeated using a different random path.

The statistics of the RF N(**x** | (n)) are completely defined by the N-variate (or multivariate) conditional pdf $f_{(N) \mid (n)}(n_i;\mathbf{x}_i, i \in (N) \mid n_j;\mathbf{x}_j, j \in (n))$, n being the number of conditioning data (classified borings). It can be shown that, by construction, the SIS algorithm generates realisations from the multivariate pdf of N(**x** | (n)) (Gómez-Hernández, 1991).

When estimating the conditional cpdfs the (local) first moment and the second moment of the indicator RF are retained, as well as the indicator data at the conditioning points. This means that the maps simulated by SIS reflect the probability of occurrence of the classes, the observed spatial connectivity patterns of the classes and honour the classes found at the observation sites.

A CASE STUDY

Water table classes

The study area, called the "Ooypolder", is located in the east of the Netherlands. For a land sanitation project a very detailed soil survey has been conducted, comprising 2259 borings and a 1:10,000 scale general soil map (Netherlands Soil Survey Institute in 1987).

An important attribute of the soil map is the so called "water table class". Water table classes represent the mean fluctuation dynamics of the water table (Van der Sluijs and de Gruijter, 1985). The water table class is dependent on the mean highest water table (MHW) and the mean lowest water table (MLW), which are measured as a depth in (cm) from the surface. The MHW and the MLW are usually determined from the soil colours, which reflect the degree of oxidation or reduction of the iron in the soil. Table 1 gives the seven main water table classes along with their associated MHW and MLW levels. In this case study the water table class is used to demonstrate the application of the indicator approach to categorical soil attributes.

Experimental set up

The total data set contains 2259 borings at which the water table class was determined. From the 2259 borings 1139 were randomly chosen and kept apart exclusively for validation purposes. The remaining 1120 borings, evenly spread across the area, were used to apply the indicator procedures. To investigate the effect of the number of conditioning data the 1120 data were randomly reduced to 500, 250 and 100 data sets, again evenly spread across the area. The four data sets were each used to: a) infer prior statistics b) predict the water table classes at unvisited locations as well as the associated purities using indicator kriging. c) generate equiprobable maps of water table classes using SIS. d) perform a land use suitability analysis on each of the generated maps.

Table 1: Water table classes in relation to the mean highest (MHW) and mean lowest (MLW) water table (cm below the surface).

	I	II	III	IV	V	VI	VII
MHW	-	-	<40	>40	<40	40-80	>80
MLW	<50	50-80	80-120	80-120	>120	>120	>120

Statistical inference

For each of the four data sets an "a priori" probability mass function or histogram was estimated. Water table class IV was not found and thus not considered. Because 6 different water table classes (WTC) were found, 5 so called "thresholds" could be distinguished. So at each data location a vector of 5 indicator transforms was obtained. For each of the four datasets and five thresholds an (isotropic) experimental indicator variogram was estimated using equation (8). Allowable variogram models were fitted to the experimental variograms. The variogram models were checked on their respective order relations using the inequalities presented by Journel and Posa (1990) and adapted if necessary. Both the a priori probability values and the parameters found for 1120, 500 and 250 data sets of indicator data were very similar. For the 100 data set the probabilities and variogram parameters were somewhat different, which may be caused by the limited amount of data used to estimate them. Due to space limitations only the statistics and variogram parameters of the 500 data set are given here (table 2), with P: probability mass, C_0: the nugget, C: the sill and a: the range.

Table 2: A priori probability mass and Parameters of the indicator variogram models for the set of 500 borings.

WTC	P	type	C_0	C	a (km)
I	0.019	SPH	0.011	0.0178	3.03
II	0.011	SPH	0.018	0.0265	2.19
III	0.097	SPH	0.060	0.119	2.60
V	0.115	SPH	0.110	0.218	2.09
VI	0.484	SPH	0.100	0.144	1.04
VII	0.174	-	-	-	-

SPH = spherical model

Prediction of water table classes

Using ordinary indicator kriging, the variogram models and the four data sets, estimates of the conditional cumulative probabilities were obtained. This was done on a 50*50 meter grid overlying the study area.

From the estimated conditional cpdfs the water table classes were predicted using at each location on the grid the WTC with the highest estimated probability of occurrence (i.e. the P-type estimator). The original choropleth map of water table classes, as obtained in the classical way using all the available borings (Soil Survey Institute, 1987), was rasterized onto the same grid as used in the kriging procedures.

Figure 1 shows this map together with the maximum probability map obtained with the indicator approach using 500 data as well the associated purity map. The maps obtained with the other data sets are not shown, again due to space limitations. Notice the remarkable resemblance between the maps of predicted water table classes (figures 1a and 1b). (For the grid points where the estimated purity is larger than 0.5 about 75% have the same class for both maps.) The differences in the outside river dike lands (the northern edges of the maps) are largest because no conditioning data were available there to predict the water table classes with the indicator approach. Moving from the 1120 data maps to the 100 data map it was found that more and more detail is lost (See Bierkens and Burrough, 1992). Instead, the area is increasingly dominated by WTC VI, the class which has the highest "a priori" probability of occurrence (table 2). This is logical, because if no data are available this is the class most likely to be found.

It can be concluded that, using the indicator approach, maps of water table classes are obtained that are comparable to the traditional choropleth method, but using only 25% of the original amount of data. Also, a direct measure of the resulting map purity is obtained.

The purity maps of the 1120-, 500- and 250 predictions are not much different. Some noticeable reduction in purity was obtained in certain areas when only 100 data were used. Figure 1c shows the purity map obtained with the 500-data set. When this map is compared with the map of predictions (1b) it is clear that the smallest purities occur on the boundaries of more or less homogeneous areas. This can be expected because these are the areas where the unvisited locations are surrounded by borings that are classified differently. It is to these areas that additional sampling efforts should be directed to ensure a certain level of purity.

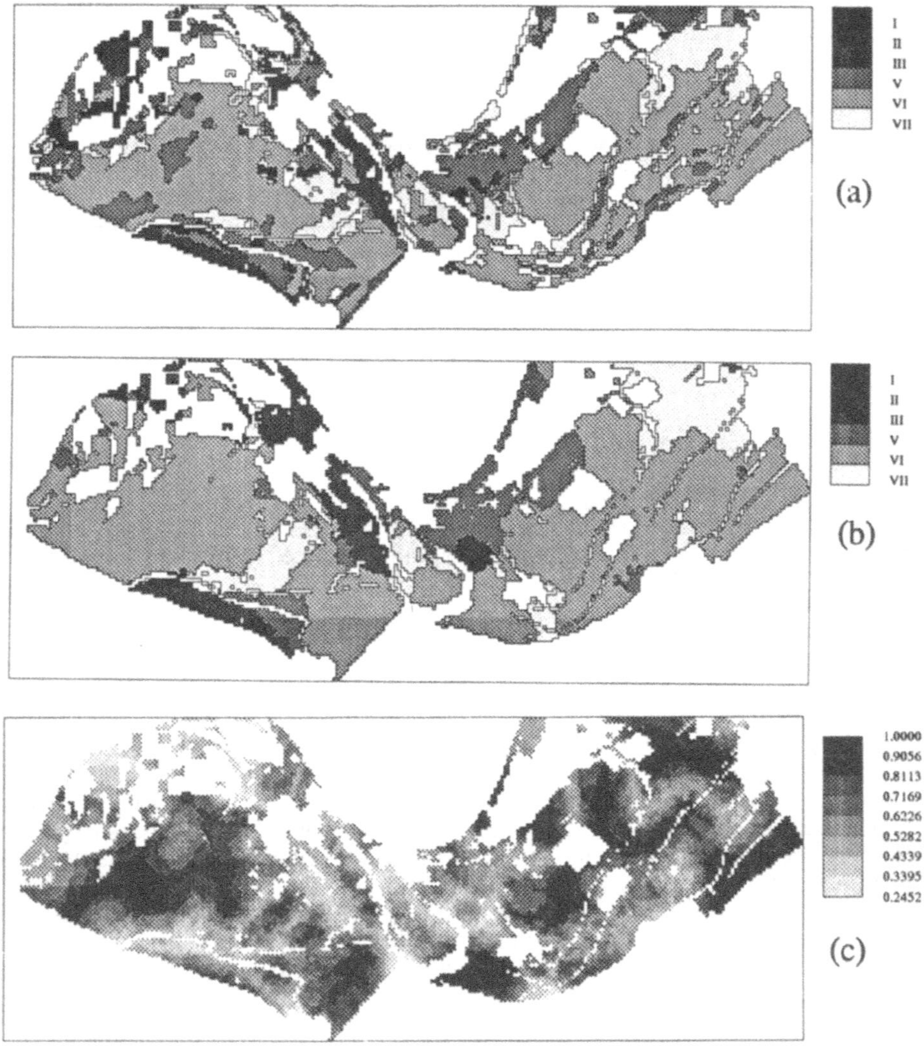

Figure 1: Maps of water table classes: a) rasterized choropleth map of the Soil Survey Institute, b) predicted with the indicator approach from 500 borings, c) map of purities of the predicted water table classes estimated from 500 borings.

Validation

With indicator kriging at each location the conditional cpdf is estimated. However, at the 1139 selected validation points only the realisations (the classes themselves) are known. Consequently, it is impossible to validate at each point individually if the estimation procedure is correct. However, it is possible to obtain a collective measure of correctness for the validation points together: The observed average map purity over all the validation points is given by (N being the number of validation points):

$$AMP_O = \frac{1}{N} \sum_{i=1}^{N} J_i \qquad (9)$$

where

$$J_i = \begin{cases} 1 & \text{if } U_i^* = U_i \\ 0 & \text{if } U_i^* \neq U_i \end{cases}$$

with U_i and U_i^* respectively the observed and predicted class at validation point i. The estimated average map purity over all the validation points is obtained by:

$$AMP_E = \frac{1}{N} \sum_{i=1}^{N} P_i^* \qquad (10)$$

P_i^* being the estimated map purity at validation point i. If the procedure is correct then $AMP_E \approx AMP_O$. In table 3 the estimated and observed average map purities are given for the four different sets of conditioning data. Also given are the relative differences in percent. The results are satisfactory, except for the 100 point data set, which severely overestimates the true purity. A probable cause for this is the inability to estimate proper indicator variograms with only 100 borings.

Table 3: Validation statistics for water table class predictions based on four different sets of borings.

# borings	AMP_E	AMP_O	$(\lvert AMP_E - AMP_O \rvert / AMP_O)*100\%$
1120	0.68	0.66	3.0 %
500	0.67	0.62	8.1 %
250	0.64	0.59	8.5 %
100	0.63	0.51	23.5 %

Simulation and error propagation

Using the ISIM3D program for SIS (Gómez-Hernández & Srivastava, 1990) for the 1120-, the 500-, and 250 dataset, 100 equiprobable realisations of the conditional RF $N(\mathbf{x} \mid (n))$ were generated. The 100 point data set was not considered any longer because of the bad validation results. Figure 2 shows one of the realisations for the 500- data case. Apart from being somewhat noisier, the generated WTC map closely resembles the maximum probability map and the choropleth map of water table classes from the Soil Survey Institute (see figures 1a and 1b).

Each of the realisations was used as input for a land use suitability classification. A suitability classification may consist of deriving suitability ratings for a certain land use from the map of water table classes and the soil map. The classification concerned the suitability for pasture. We considered only the three major suitability classes: class 1: very suitable, no limitations for pasture; class 2: moderately suitable, some limitations for pasture; class 3: poorly suitable, pasture severely limited. The digital map of soil

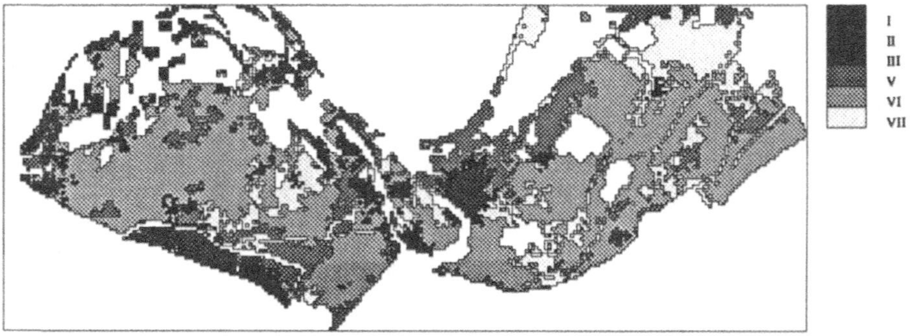

Figure 2: A conditional simulation of water table classes conditioned on 500 borings.

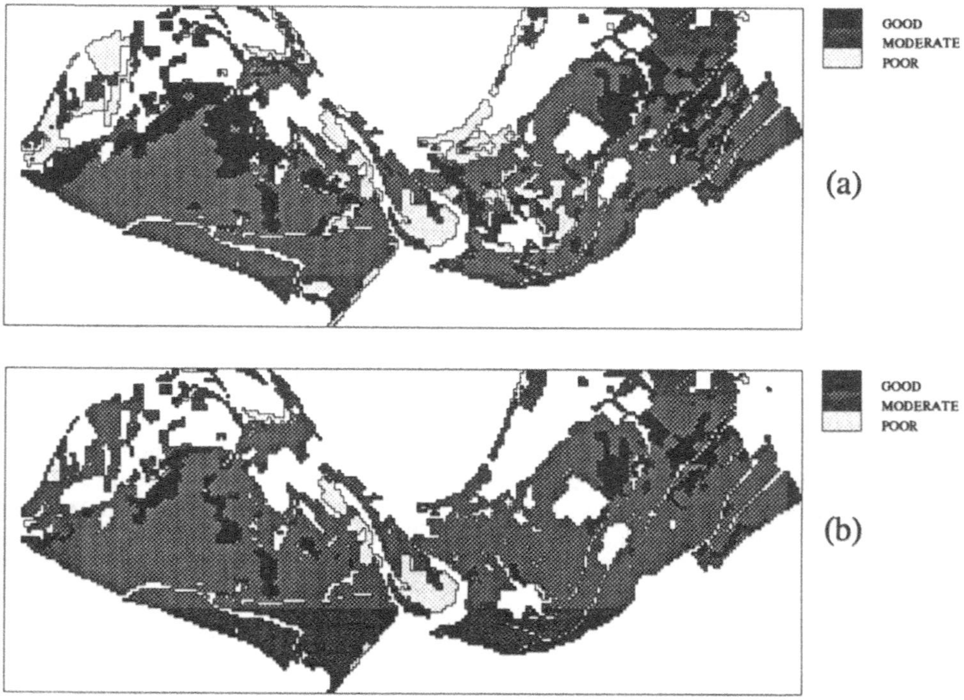

Figure 3: Maps of suitability classes for pasture: a) obtained from the (rasterized) choropleth maps of soil- and water table classes of the Soil Survey Institute, b) predicted with the choropleth soil map and 100 indicator simulations using 500 borings.

classes was combined with the 100 realisations to obtain 100 realisations of suitability maps. From the 100 suitability maps probability distributions were estimated for all the grid points. Proceeding the same way as for the water table classes a map was constructed of the suitability classes with the highest probability of occurrence.
This probability (a suitability map purity) can be seen as a measure for the uncertainty in the suitability analysis caused by the impurities in the map of water table classes. Figure 3 shows the maximum probability map using 500 data (3b) together with the map resulting from combining the choropleth maps of soil- and water table classes (3a). Again, the maps resulting from the indicator simulation resemble the one produced with the choropleth maps. Moreso then for the water table classes the suitability maps produced with the 1120-, 500- and 250 data sets were very much alike (see Bierkens and Burrough, 1992). This is also the case for the purity maps (not shown here). The mean map purity for the three cases is the same and equals 0.95. From these results it can be concluded that for suitability mapping only a small amount of borings is needed to obtain the input map of water table classes. Also, given the large purities of the suitability maps, it can be concluded that the suitability classification as presented here is relatively insensitive to errors made in the input map of water table classes.

CONCLUSIONS AND DISCUSSION

This paper has demonstrated how the indicator approach can be used to predict categorical soil data. Instead of a single class the conditional probability of occurrence was obtained. Using a p-type prediction a new formulation of map purity was obtained which is (a) dependent on the local density and the classes of the surrounding borings and (b) directly available from the prediction procedure. Given this, sequential indicator simulation could be used to generate equiprobable stochastic images of categorical data, conditional to the classified borings. In this way, the errors in a land use suitability analysis that accrue from uncertainties in categorical data could be assessed
The case study demonstrated the method to perform well when validated on an independent set of data. Little difference could be seen between the original map of water table classes that was available from the Soil Survey Institute and the maps produced with indicator kriging and indicator simulation, even when the number of conditioning data was reduced to some 25% of the original data. The estimated map purities were noticeably smaller on the boundaries of more or less homogeneous areas; i.e. areas where adjacent borings are classified differently. It is to these areas that additional sampling should be directed. A suitability analysis for using land as pasture was not very sensitive to impurities in the input map of water table classes.
Although the traditional methods of making choropleth maps are subjective and do not provide a suitable purity measure, they have an important advantage: Because map boundaries are often drawn in the field or from aerial photographs, additional information such as relief and vegetation patterns is included in the mapping procedure. The next task therefore is to find a formal method to include soft information in the indicator approach to handling categorical data. Journel & Zhu (1990) propose a method which is based on converting the soft information to "fuzzy" indicator transforms. By assuming that hard information (borings) always prevails over soft information they arrive at a method that takes the softness of the additional information into account, but does not need the inference of additional (cross)variograms. We propose to further investigate the applicability of this method.

ACKNOWLEDGEMENTS

Dr. J. Jaime Gómez-Hernández is gratefully acknowledged for introducing us to the field of non-parametric geostatistics and for providing the ISIM3D program. A.K. Bregt, The Winand Staring Centre for Integrated Land, Soil and Water Research and "de Landinrichtingsdienst" are kindly acknowledged for providing the data. We also like to thank Victor Jetten for helping with the preparation of the data.

REFERENCES

Bierkens, M.F.P. and Burrough, P.A. (1992) "The indicator approach to categorical soil data; part 2: application to mapping and land use suitability analysis", (submitted).

Bregt A.K., Stoorvogel, J.J., Bouma, J. and Stein, A. (1991) "Ordinal data in soil survey: a Costa Rica Example", Soil Science Society of America Journal (in press).

Gómez-Hernández, J.J. (1991) "A stochastic approach to the simulation of block conductivity fields conditioned upon data measured at a smaller scale", Ph.D. Thesis, Department of Applied Earth Sciences, Stanford University.

Gómez-Hernández J.J. and Srivastava, R.M. (1990) "ISIM3D: An ANSI-C three-dimensional multiple indicator conditional simulation program" Computers and Geosciences 16(4), 395-440.

Heuvelink, G.B.M., Burrough, P.A. and Stein A. (1989) "Propagation of errors in spatial modelling with GIS", Int. J. Geographical Information Systems, 3, 442-452.

Journel, A.G. (1983) "Nonparametric estimation of spatial distributions", Mathematical Geology 15(3), 445-468.

Journel A.G. and Gómez-hernández, J.J. (1989) "Stochastic imaging of the Wilmington clastic sequence", SPE paper 19857, 591-606.

Journel A.G. & Posa, D. (1990) "Characteristic behavior and order relations for indicator variograms", Mathematical Geology 22(8), 1011-1025.

Journel, A.G. & Zhu, H. (1990) "Integrating soft seismic data: Markov-Bayes updating, an alternative to cokriging and traditional regression", Report Stanford Center for Reservoir Forecasting.

Netherlands Soil Survey Institute (1987) "De bodemgesteldheid van het herinrichtingsgebied "Ooypolder"", Report No. 1914 (in Dutch).

Van Der Sluijs, P. and De Gruijter, J.J. (1985) "Water table classes: a method to describe seasonal fluctuation and duration of water tables on Dutch soil maps", Agricultural Water Management 10, 109-125.

Particulate Stack Emissions from Coal Fired Power Stations

I. Clark and M.J. Daly
Geostokos Limited, PO Box 1897, London NW6 1DQ, UK
ESKOM Fuel and Water, PO Box 1091, Johannesburg 2000, RSA

Introduction

Coal fired power stations are generally accounted to be amongst the major potential contributors to atmospheric pollution. Eskom, as one of South Africa's prime producers of power, has been — and remains – committed to maintaining a responsible approach towards all environmental issues, especially air quality. Since the primary generation of power is derived from coal fired Stations serving a heavily populated area, considerable attention has been focussed over the years on the issues of particulate and gaseous stack emission levels.

The levels of particulate and SO_2 emissions are influenced primarily by the characteristics of the coal burnt, but may also be controlled by how (and when) the coal is mined and by the design and operation of the various items of power plant. Many studies have been carried out — particularly in the western world — to investigate the emission levels of plants in operation. That is, monitoring in and around power stations is carried out to quantify the levels of emissions and their impact on the local environment (cf. Cressie et al., 1990, Dixon 1989, Pitelka and Raynal 1989).

The studies described in this paper differ in both nature and environmental impact, in that they are carried out *before* the coal reaches the power station. That is:

- the coal *in situ* is analysed for its potential emission characteristics;
- the mine plan is superimposed on these characteristics to assess the quality of the coal *before burning*;
- potential stack emissions are predicted on a year-to-year basis using different mining scenarios.

This type of study has proved invaluable in the assessment of: the planned mining of coal seams; in the amount of blending and/or quality control sampling which is required during production; in the decision to use or extend the use of washing plants; and — perhaps most important in this context — in the design of the power plant and the control of stack emissions. At the time of writing, three detailed studies have been carried out on particulate stack emissions. In each case, significant changes have been made to either the mine design, the processing of the coal or to the planned gas cleaning equipment.

This paper describes a typical investigation into the potential particulate stack emission characteristics of a coal mine over its planned life and demonstrates how geostatistical methods may be used to control air pollution *before* its inception. Some comments are made on the planned extension of this work to the quantification of potential SO_2 emissions.

Coal Fired Power Stations in the Eastern Transvaal

Eskom is one of South Africa's key producers of power, which it generates primarily from coal fired Power Stations. The majority of these Stations are located in the Eastern Transvaal, relatively close to the heavily populated Pretoria/ Witwatersrand/Vereeniging area (See Figure 1). Because of this juxtaposition, a considerable amount of attention has been directed to the potential problems of air pollution — both particulate and gaseous. The concentration of Stations and their potential impact on air quality is an ongoing subject of intense debate with the licensing authorities, principally the Chief Air Pollution Control Officer (CAPCO).

In the late 1970's, CAPCO designated an area in the Eastern Transvaal — referred to as the air pollution "egg" - within which the level of air pollution could not be increased further. The 1981 limits of this "egg" are shown in Figure 1. The current limits of the controlled area (now referred to as a "blanket") cover the whole of the area in Figure 1. This essentially means that any new Power Station constructed within this area has to adhere to, or improve upon, specific environmental standards for particulate and show predicted gaseous emissions, not yet subject tp legal standards.

The standards or licence limits which are set for particulate emissions have over the years become increasingly strict. Vierfontein Power Station, for example, which was commissioned in the mid-1950's, had a limit of 1020 mg/Sm^3. Highveld Power Station, completed in the early 1960's, had a limit of 396 mg/Sm^3. Matla (1979–1983) has an average limit of 245 mg/Sm^3, whilst the currently developing Power Stations are subjected to limits of 90–100 mg/Sm^3. There has been a consequent need to improve upon plant design and operation to reduce potential levels of particulate emissions. It seems likely that CAPCO will expect improvement of plant performance and installation of control technology to impact positively on the air quality parameters under the "blanket".

Figure 1: RSA Coal Field

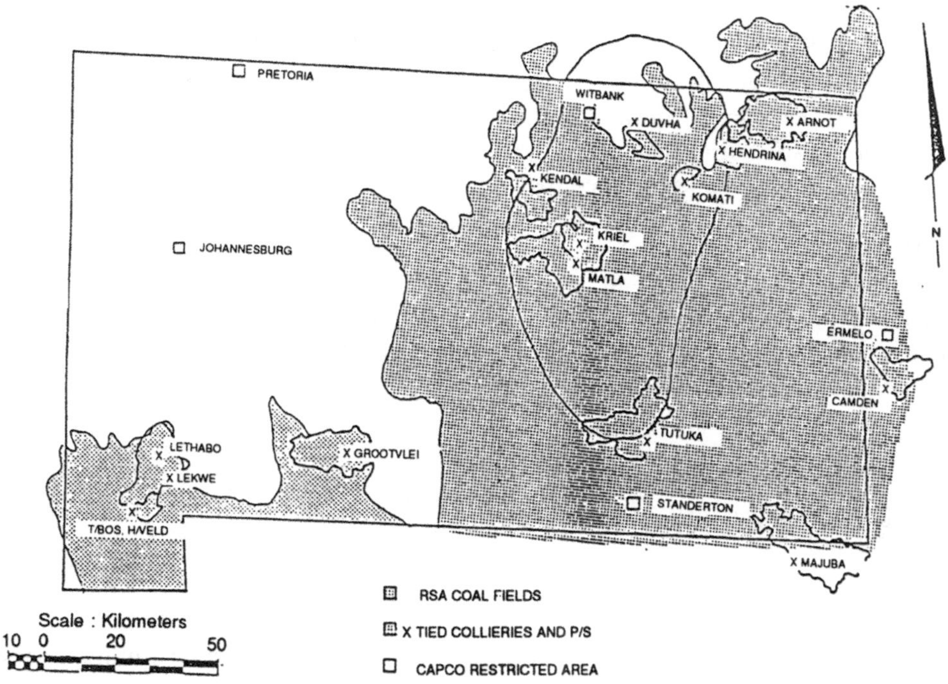

The levels of emission from a power station are influenced principally by the characteristics of the coal burned. However, the method of mining, the mine design and processing (if carried out) affect the quality of the coal delivered to the Station. The design and operation of the power and gas cleaning plant itself has a major impact on the particulates released from the stack. Historically, the power plant design — including precipitators — within Eskom was based upon the descriptive analysis of the geological data as provided in the tender document for that coal field. Often the only data utilised were global averages for coal quality parameters over the field and over the life of the mine. The likely variability of the coal quality over the life of the mine plan was not generally considered in power plant design, until the late 1970's. In addition, geological data and quality predictions supplied by the mining companies often proved unreliable or overly optimistic.

It should, perhaps, be noted that the regulatory levels specified in the CAPCO licences are *continuous* limits. That is, *any* violation of these limits over whatever period is subject to severe penalty. It should also be noted that most mine plans and schedules supplied to Eskom are made on the basis of the mining company's financial year. It is necessary, therefore, to assess whether there is any chance of exceeding the limits during a particular year of operation of the power station — and if so, over what period the limits are likely to be exceeded.

The studies described in this paper were aimed at remedying the short-comings of a 'global' approach to evaluating the potential emission levels, especially as regards the spatial variability of coal quality parameters and the temporal variability introduced by the mining procedures.

Emission Levels

The potential particulate emission level from coal burned depends on various coal quality parameters and on the method of mining and associated factors. These factors can be modified by processing ('washing'). This discussion will be confined to *in situ* characteristics of the coal. The relevant factors are:

- the sulphur content of the coal;
- the ash content of the coal;
- the calorific value of the coal;
- the total moisture in the coal;
- the level of stone contamination introduced by the mining process.

This last factor is generally assumed to be constant for a particular mining method and will have a significant impact on the dust generated when the product is burned. The projected emission level can be calculated from these parameters, using formulae developed by Eskom's Boiler Plant Engineering department for each specific gas cleaning plant. These formulae must contain plant-specific characteristics, such as gas volumes and velocities, dust burdens, input temperatures and precipitator collecting efficiency. The resulting emission level is expressed in mg/Sm^3 — that is, milligrams per standard cubic metre.

In the earliest study (1988) within Eskom, particulate emission levels were calculated for each borehole intersection available in the projected mine area. These values were then used to (arbitrarily) contour the emission levels over the study area, at levels of 20 mg/Sm^3. The mine plan prepared by the colliery was overlaid on this contour plan and typical emission levels read along the centre line of each planned mining panel. It was thus possible to quantify the likely emission level for a specific year of production, or for any specific period after mining had commenced.

Three such studies were undertaken and potential levels of non-compliance were indicated. The results of these (fairly simplistic) studies suggested that a more flexible, computer based approach would be invaluable.

Why Geostatistics?

The relative success of the above simple methodology suggested that spatial and temporal analysis of the mine plan yielded valuable decision making information. Statistical and geostatistical methods were considered for these studies to provide:

- more reliable forecasts of particulate emission levels, in particular for Power plant design purposes;
- results which could be presented and assessed with a greater degree of confidence by the licencing authorities;
- greater flexibility in studying various mining and plant design scenarios;
- information on the interaction between coal quality parameters and trends across the mine area.
- objective confidence levels for the final estimates for each and every scenario.

Geostatistical methods allow us to predict the average values over each mining panel *and* the likely variability of the coal within each panel. Standard errors associated with these estimates may be used to provide confidence levels to assess the reliability of the final projections. Panels can be combined over any planned time period to assess the likely non-compliance with CAPCO limitations.

It is felt that the information which geostatistics can generate will be invaluable in predicting emission levels, identifying time periods of non-compliance and, thus, in modifying production and in negotiating with CAPCO. This information will also aid in planning and in setting of policy.

A Typical Study

The procedure which has been developed over the studies carried out to date will be described *in general* in the following section. Due to confidentiality concerns, it is not possible to publish a detailed case study to illustrate this process. However, the following should give a reasonable idea of the procedure.

The necessary information for such a study consists of:

- Plans and sections of the coalfield, showing borehole collar locations and major geological features;
- Plans showing the location, shape and size of the proposed mining panels and details as to the year in which each is to be mined;
- Data on the elevation and/or dip, width of seam, calorific and ash values, sulphur content, the resulting predicted emission levels and any other relevant variables for the individual boreholes.

Figure 2: Planned mining panels and boreholes

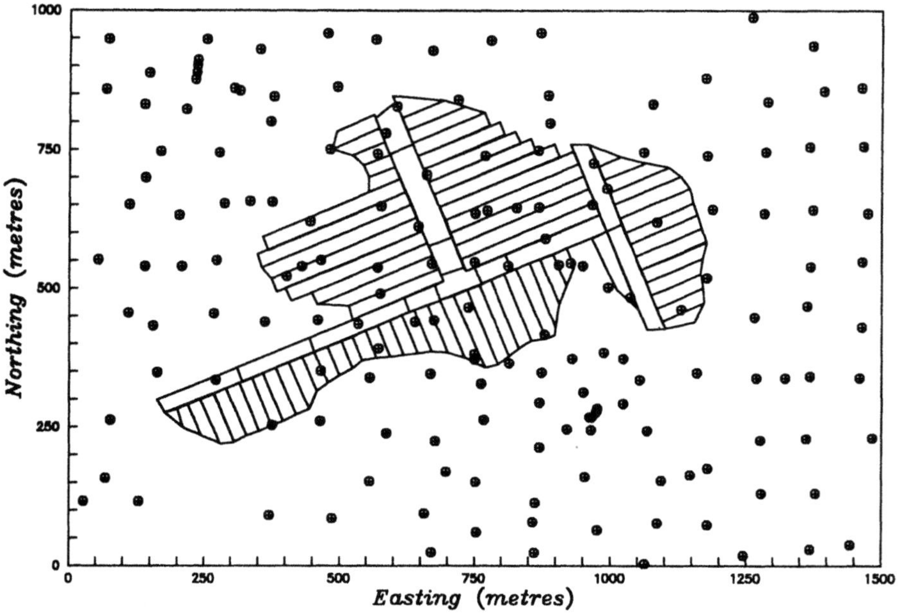

A typical area is shown in Figure 2, with borehole locations and a typical mining panel layout. Panels are generally irregular in shape, conforming to mining considerations and to local geological characteristics. The study is undertaken in various stages:

a. Statistical analysis of the various measurements, with special emphasis on the behaviour of the emission levels. Histograms were constructed for each variable and studied for outliers and inconsistencies.

b. Scatterplots were used to establish relationships amongst the variables and also to identify aberrant behaviour.

c. Trend surface analysis was used to determine large scale 'regional' movements in value for various variables — where appropriate. Residuals from these trend surfaces were inspected for evidence of outliers and/or data inconsistencies.

d. Semi-variogram graphs were constructed to display the spatial continuity or 'predictability' of values from one location to another within the coal seam. The particular variables of interest are the thickness of the seam and the projected emission level. Simple models were fitted to these graphs for estimation purposes.

e. The method of 'cross validation' was used to test the fit of the semi-variogram model and to highlight extreme values or outliers within the sample data set. This analysis also provided an opportunity to study the potential distribution of estimation errors.

f. Data value corrections were made on the basis of these analyses and the whole process repeated once more.

g. Using the distribution and semi-variogram models, values were estimated for each mining panel within the mine plan. These values — weighted by volume of panel — were summarised for each year of mining.

h. Using the panel estimates, two scenarios were constructed:

   a 'best case', where it is assumed that controlled blending will take place at all times — i.e. the yearly average can be maintained throughout the year;

   a 'worst case' scenario, in which it is assumed that no quality control or blending will take place and mining panels will be mined one at a time in sequence.

i. Results from the 'best case' and 'worst case' scenarios are summarised in the form of simple graphs for clarity, showing the likelihood of emission levels exceeding the established permissible CAPCO level.

As a very limited example of the type of intermediate results produced by such a study, Figures 3 and 4 are provided. Figure 3 shows a 'spotted dog' illustrating the particulate emission levels for each borehole. It can be seen that the levels vary considerably over a relatively small area — but still retain a certain spatial continuity, which may be described by a semi-variogram model. Figure 4 shows a typical kriged map of emission levels over the study area. Such a map is produced by Ordinary Kriging of values at points on a grid placed over the study area. This map is, of course, too *smooth* to be used in the 'worst case' scenario described below. However, it gives a good idea of the likely fluctuation in level for (say) mining panels.

Best Case Scenario

Two 'scenarios' are considered in each study. In this first scenario, we assume that the quality of the coal mined in any specific year can be kept constant over that year by carefully controlled blending and, possibly, a quality control sampling campaign carried out during production. Around 40 mining panels are shown in Figure 2. For the 'best' case, we need to estimate the average value of each year's production. Contiguous panels planned to be mined in a specified year are combined into 'polygons' for estimation purposes. For each specified year of production we must estimate the average value over each polygonal area. These individual averages

Figure 3: Emission Level --- mg/Sm³ (OK)

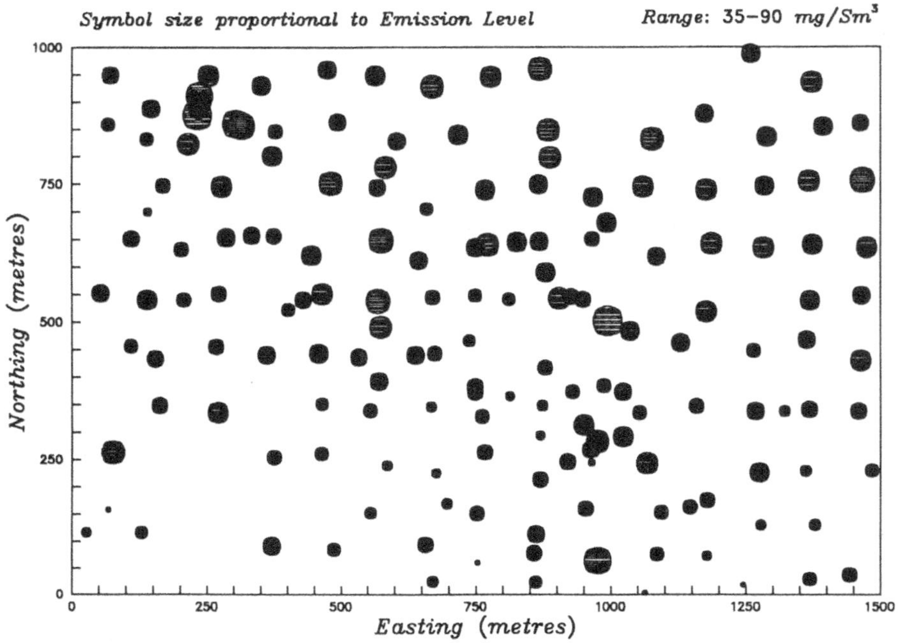

Figure 4: Estimated Emission Level --- mg/Sm³ (OK)

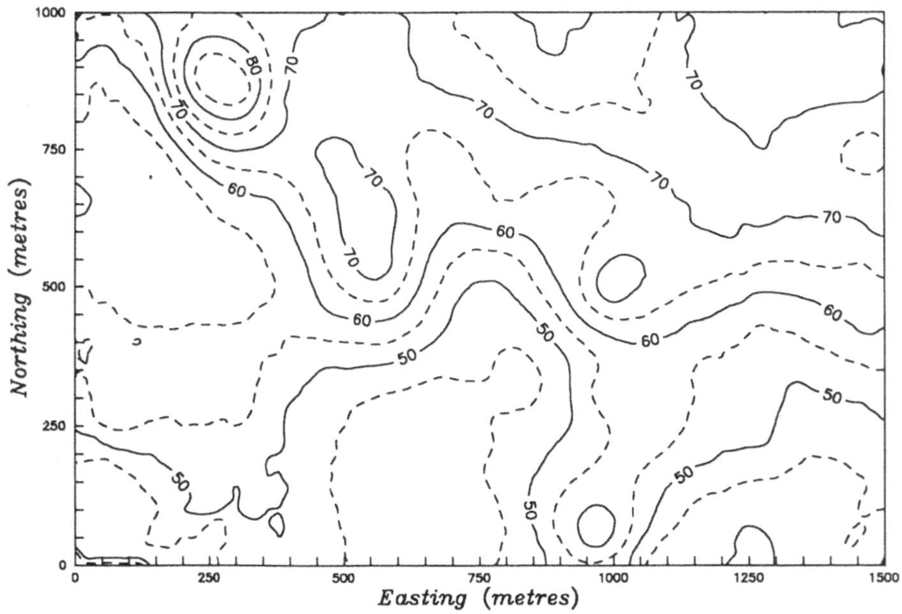

may then be combined to form the yearly average *and* its associated standard error may be calculated.

The average emission level for each year of production, then, is calculated by combining the estimated emission level per panel, the seam thickness per panel and the year of mining as follows:

> The volume mined from each panel is average width × area, corrected for dip if necessary;
>
> The emission level from each panel is weighted by the volume to be mined from that panel;
>
> All panels to be mined in one year are combined by volume and the yearly average calculated;
>
> The standard error of this estimate is found by a similar combination of the individual panel standard errors.

This last stage can only be carried out on the assumption that the areas mined in a particular year are fairly widely scattered over the study area. In most RSA coal mines this assumption seems to be fairly realistic — except, perhaps, in the initial years of mining.

Having produced an average emission level for each year of mining and an associated standard error for this average we may also produce statistics such as *confidence levels* on the results. If we can assume a particular distribution for the estimated values — and for the error of estimation — we can reverse the calculation and find out exactly how probable it is that the yearly mined average will exceed the limit. A cross validation exercise was carried out during each study to verify the likely distribution of estimation errors. In most cases, a Normal distribution could be assumed. In the remainder, lognormal estimation theory was used.

Prior to applying geostatistics, the effective approach was to assume a 'best case' scenario using the overall average for the whole mine. Assuming 'best case' or average coals over the lifetime of a mine — or, at least, a power station — has resulted in non-compliance emissions and, thus, in confrontation with CAPCO. An estimation method which produces better/more data on the likelihood of mining poor quailty coal will allow coal blending or washing to achieve compliance with the CAPCO emission limits.

Worst Case Scenario

In the previous section we assumed that the coal could be mined to the estimated average value — at all times. That is, we assume that the coal can be blended in such a way as to maintain an even feed to the power station over each year of mining. This was viewed as a "best case scenario". Under these assumptions, we might find that the emission levels are unlikely to exceed the CAPCO limits in a particular year.

The 1988 Eskom study considered a "worst case scenario" — that is, supposing that no blending or quality control took place and that the coal was simply mined from one 'end' of the first mining panel to the other 'end' of the last. The variation of values within each mining panel was estimated by taking several points within the panel and estimating the value at each. The proportion of points with values above the CAPCO level was assessed and this figure quoted as the "worst possible case".

In a geostatistical study, we need to estimate (or simulate) the emission value at each individual *point* within each mining panel. In cases where the density of sampling permits estimation without the inherent *smoothing* problems of kriging, a dense grid of points was estimated within each panel. Where drilling is more sparse, simulation methods or 'support' corrections are used to insure the correct variability for the expected emission levels.

We can then assess how many of these *points* are above the acceptable limit and express them as a percentage of the whole mining panel. For each mining panel, then, we will have a 'percentage exceeding CAPCO limit (mg/Sm^3)'. As noted before, each mining panel represents a different mined tonnage which we have approximated by a 'mined volume' — area of mining panel × width. For each year of mining, then, we take all panels to be mined in that year and form a weighted average of the 'percentage' according to the *volume* of coal to be mined.

Figure 5 shows a typical graph illustrating the probability of exceedance for both 'best' and 'worst' cases. We can see the considerable impact that controlled blending would have on the coal produced. Although a high percentage of the coal may exceed the permitted limit, blending with 'better' coal can reduce this well below the required level. This implies that the production quality control sampling should be an important factor in the detailed mine plan.

Using these two 'best' and 'worst' scenarios effectively *brackets* the likely power station performance for coal supplied under a specified mine plan. The difference between these two outcomes in the case illustrated in Figure 5, demonstrates that control of particulate emissions can (in some cases) be controlled by modifying the planning and scheduling of the actual mining — so that better quality coal can be provided for cleaner burning.

Figure 5: Likelihood of exceeding permitted emission level

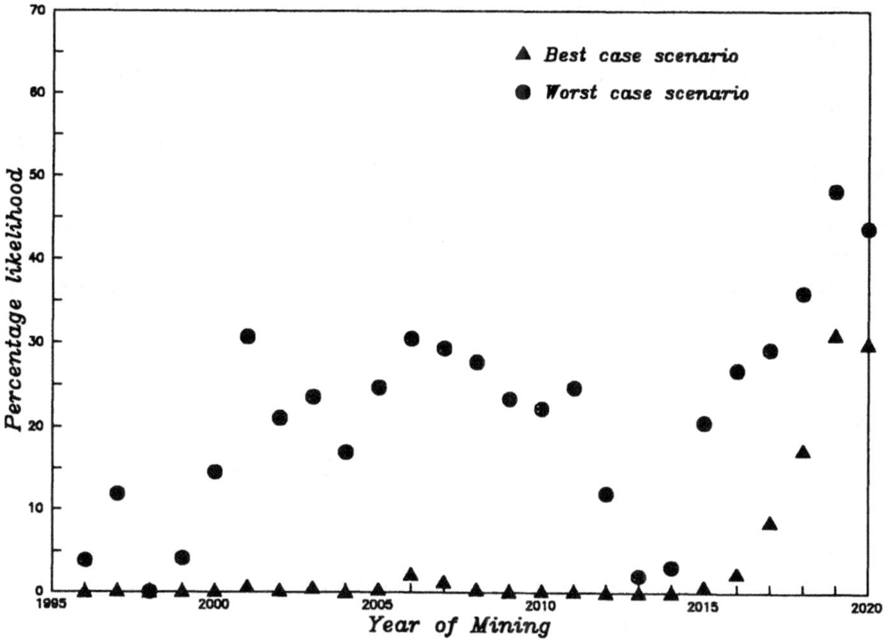

Summary

On the basis of the results of the initial studies it is apparent that being able to predict particulate emission levels from geological data provides an important decision-making tool in terms of power plant design and operation.

This approach enables alternative mine and Power Station development options to be modelled and evaluated from a technical and economic point of view. For example, in one case studied several development options were considered, including:

1. Modification of the mine plan, a reduction in stone contamination and blending between panels which would lead to good quality coal within the CAPCO limits for most of the life of the mine. After this period, SO_3 injection could be retrofitted to improve collecting efficiency so as to maintain acceptable levels of particulate emission.

2. The precipitators could be enlarged or SO_3 injection fitted earlier — putting the onus on the Power Station rather than the mine.

3. A complete change in gas cleaning plant design could be made to reduce stack emission — for example, the fitting of fabric filters.

In a different study, the mine was found to be in danger of non-compliance during almost every year of its life (even in the 'best case'). In this case, the need to fit SO_3 injection is being critically reviewed together with the possible introduction of partial beneficiation of the coal to reduce the dust burden. However, this last option has to be considered very carefully as beneficiation also reduces the inherent sulphur in the coal. Since sulphur acts to reduce the level of particulate emission, a delicate balance must be maintained in any planned beneficiation.

In view of the success achieved in the studies of particulate emission using geostatistics, a similar approach is being adopted for the study of SO_2 emission levels. The study of particulate and SO_2 emission levels in coal fired Power Stations is considered to be of fundamental importance in Eskom's continued efforts to minimise air pollution.

References

Cressie, N., Gotway, C.A. and Grondona, M.O. (1990) "Spatial prediction from networks". *Chemometrics and Intelligent Laboratory Systems*, **7**, 251–271.

Dixon E.C. (1989) "Modeling under uncertainty: Comparing three acid-rain models". *Journal of the Operational Research Society*, **40**, 29–40.

Pitelka, L.F. and Raynal, D.J. (1989) "Forest decline and acidic deposition", *Ecology*, **70**, 2–10.

KRIGING THE RAINFALL IN LESOTHO

M. Armstrong*, G. Chetboun**, P. Hubert***

* Centre de Géostatistique, Ecole des Mines de Paris, Fontainebleau, France
** Lesotho Highland Development Authority (LHDA), Maseru, Lesotho
*** Centre d'Informatique Géologique, ENSMP, Fontainebleau, France.

ABSTRACT

In this paper, three geostatistical kriging techniques are considered for estimating the monthly rainfalls at points or over areas such as catchment basins, in the highland regions of Lesotho. The data available for making these estimates come from the monthly pluviometric recordings made at about 100 raingauges in Lesotho and adjoining parts of South Africa, from 1922 onward. Two problems were encountered: missing monthly values at some raingauges and the sparsity of raingauges in highland areas where estimates are required. Of the three methods tested, one based on using the average summer or winter rainfall as an external drift gave the best results.

INTRODUCTION

The landlocked Kingdom of Lesotho lies about 300km south of Johannesburg in Southern Africa. Unlike the surrounding area, it is endowed with an abundant water supply – particularly in the highlands which receive an annual rainfall of more than 1000mm. Even with full and massive investment in irrigation, hydropower, and other water resources projects, only a fraction of the water could be used. At present the water drains away to the ocean through the Orange–Senqu river system, with no benefit to Lesotho. The challenge of the Lesotho Highlands Water Project is to convert this resource into a source of income for the Kingdom by constructing a series of dams and tunnels to trap the water and divert it to the river system which supplies the densely populated Johannesburg–Pretoria area in neighbouring South Africa. (Fig 1).

In addition to providing hydroelectric power for the kingdom (and thus reducing its need to import energy), the project will provide royalties for Lesotho for all the water diverted. These are expected, in the long term, to exceed any other source of income. As the basic resource for the whole project is the rainfall, accurate estimates of this for highland areas are essential. They have to be provided on a month by month basis for areas such as catchment basins for dams. The data available for making these estimates come from the monthly pluviometric recordings (i.e. monthly accumulations) made at about 100 raingauges in Lesotho and adjoining parts of South Africa, from 1922 onward. Several earlier studies had been carried out using traditional estimation methods. The aim of this study was to see if geostatistics could give better estimates.

CORRELATION BETWEEN RAINFALL AND ELEVATION

The essential difficulty in estimating rainfall comes from the scattered locations of the raingauges. In this case observations have been recorded at some or all of 107 sites since 1922. Seventy–two of these raingauges lie inside Lesotho; the others are nearby in the Republic of South Africa. Most of the gauges in Lesotho lie in the arable low lands to the west around the capital, Maseru. The few that are in highland areas tend to be in the valleys because they can be read regularly. So the underlying problem is to infill the sparsely sampled areas. How can this be be done?

From experience, hydrologists know that rainfall is affected by topography; in general, it is positively correlated with elevation. As maps of Lesotho are readily available, it was hoped that elevation could be used "to interpolate" the spaces between the raingauge sites in the highlands (Fig 2), and to counteract the fact that raingauges tend to be located preferentially at low altitudes. Unfortunately a preliminary study showed virtually no correlation between rainfall and elevation for the data, thereby eliminating the possibility of using elevation as an external drift when kriging the monthly rainfall.

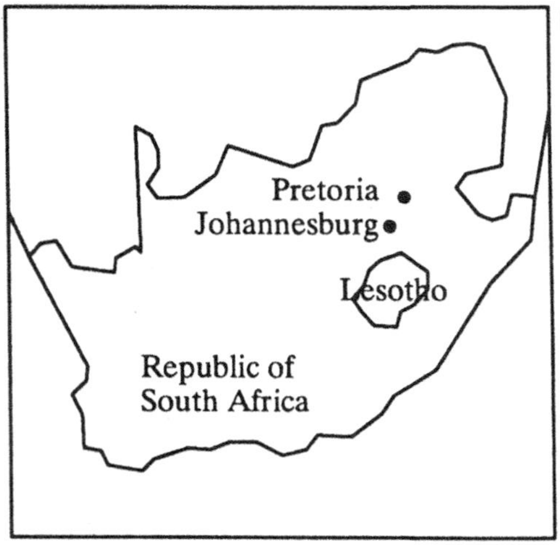

Figure 1: Map of Southern Africa

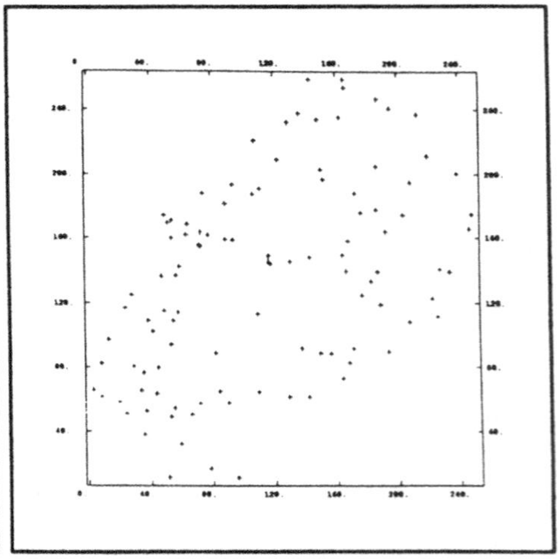

Figure 2: Location of rainguages

BACK TO THE DRAWING BOARD

Having ruled out this tempting possibility, a different approach to the estimation problem was needed. This could only come from a close study of the data. A total of 37056 monthly rainfall readings had been recorded at 107 gauges for periods going back, in some cases, to 1922. The basic statistics on a month by month basis (Table 1) show clear seasonal differences, in particular the means and variances differ from one month to another. Hence the need to standardize the rainfall data to allow comparisons.

	Oct	Nov	Dec	Jan	Feb	Mar	Apr	May	Jun	Jul	Aug	Sep
No	3074	3087	3040	3073	3075	3112	3121	3094	3101	3084	3095	3100
Max	322	618	645	879	849	700	318	628	164	261	26.	352
M	72	93	106	122	111	99	57	29	14	13	22	34
σ	48.6	60.5	64.1	76.4	47.9	66.0	41.3	31.9	18.3	19.5	28.2	39.1

Table 1: Basic statistics (number of readings, maximum value, mean value M, standard deviation, the minimum being zero for all months)

As could be expected for an east coast region in the southern hemisphere where prevailing winds come westward across the ocean, the rainfall is characterized by summer showers from October to April and winter drizzle from May to September. The histograms of the standardized rainfall (i.e. (rainfall – monthly average)/ monthly standard deviation) were of two types: one for summer and another for winter, thus confirming what was known a priori about the climate. Typical examples of a summer month, December, and a winter one, July, are shown in Figure 3.

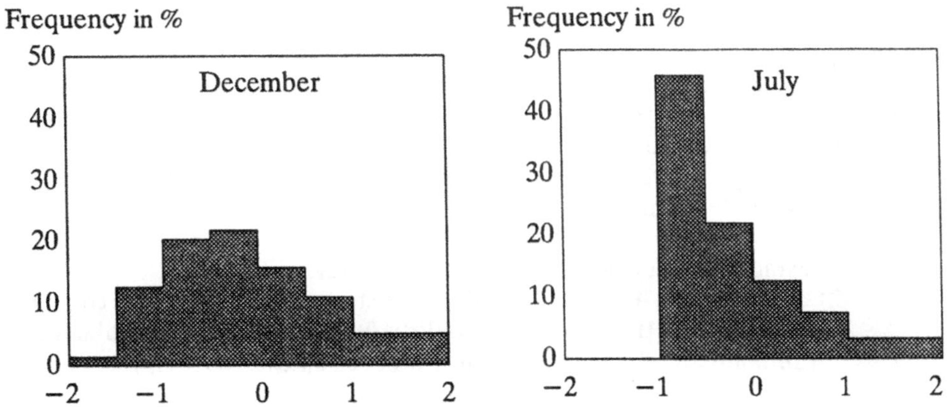

Figure 3: Histograms of the standardised rainfalls for a typical summer month, December, and a typical winter one, July.

OVERVIEW OF THE THREE METHODS TESTED

As was stated earlier, the aim of the project was to use geostatistics to get the best possible estimates of the rainfall over specified areas, based on the available data. As the computing equipment available in Lesotho is limited to PCs, it was out of the question to use all 37056 readings in any one kriging. And in any case, one could question the wisdom of using rainfall measurements from decades earlier to (say) krige the rainfall in the 1980s. Going to the opposite extreme, for any given month, a maximum of 107 observations were available; in general there were far fewer, located mostly around Maseru. Estimates based on so few readings could easily be obtained but would clearly not be the best. However they could serve as a base case against which to test other geostatistical methods. It was felt that better maps and estimates could be obtained by including more information, for example, from the months just before or after the one under study, or from the same month in earlier years.

After some trial and error, in addition to the base case, two geostatistical approaches were developed and tested on the data. One consisted in using the data from the two months just prior to the one in question, in order to give more information on general rainfall, particularly at gauges where the observation for the month under study was missing. In order to ensure that all readings were of the same order of magnitude, the work was carried on the standardised rainfall.

In the other approach we wanted to develop a standard background image of the rainfall at different times of the year. This could be done on a month by month basis, but as it turned out, the pattern of rainfall was very similar for all summer months provided that standardised units are used. The same was true of the winter months. So it was possible to use a single background map for summer and another for winter, which significantly reduces the number of background images from 12 to 2. As with the previous approaches, the calculations were carried out in terms of the standardised rainfall.

In order to be able to compare the three different estimation methods, a test month was chosen – October 1987 because it is a crucial month for design of the dams and is typical or poorly sampled months. Twenty–eight readings were available. As a test, the kriged estimates were calculated for 98 square blocks each 20 x 20 km inside Lesotho using the three methods described above, that is:

1) using the data from the month under study,
2) using the data from the month under study and the previous two months,
3) using the data from the month under study and the appropriate background rainfall map (either summer or winter) as an external drift.

In addition to point and block kriging, a procedure was required for estimating the rainfall in polygonal areas (such as the whole of Lesotho, or the highland area, or individual catchment basins). In all cases the program BLUEPACK 3D was used for kriging.

The first method involved kriging the rainfall for October 87 using only the 28 readings available for that month. Which variogram should be used? One could imagine calculating the experimental variogram for Oct 87, fitting a model and then kriging the data.. Given the limited number of data available, this seems perilous. Secondly, the Lesotho Highland Development Authority (LHDA) needed a reliable and efficient method for kriging any one of the 700 odd months for which data exists, on a PC for input into its hydrological prediction programs. A quick automatic procedure was required. We thereforefore chose to work with standardised data (i.e. mean 0, variance 1) and the corresponding variogram to facilitate comparisons between the data and variograms from different months.

HOW TO STANDARDISE THE DATA ?

The problems of standardising data that vary in both time and space have been faced by many researchers. One has to decide which mean (and standard deviation) to use. For example if we let z_{ijk} denote the rainfall at the ith raingauge during the jth month of the kth year, then we can standardise in any of the following ways:

$$(Z_{ijk} - Z_{\bullet j \bullet})/\sigma_{\bullet j \bullet}$$

$$(Z_{ijk} - Z_{ij \bullet})/\sigma_{ij \bullet}$$

$$(Z_{ijk} - Z_{i \bullet\bullet})/\sigma_{i \bullet\bullet}$$

where * denotes the average over that index.. The first option consists in removing the influence of the average rainfall for that month (for all years and for all raingauges) whereas the second removes the effect of the average rainfall for that month at that raingauge (for all years). So both of these remove the seasonal trend which is our objective, but the second has the added advantage of removing the spatial component of the variability as well. The third option standardises by removing only the spatial component, which is less interesting here. Now if very long time series had been available at all gauges, the second option would be preferable. Unfortunately this is not the case and so the first option was adopted. As the 12 monthly averages were each calculated from over 3000 readings, they are stable over time and need not be updated often.

Switzer (1989) faced a similar dilemma in handling monitoring data on a space–time lattice where the number of monitoring stations was much smaller than the number of observed time points. He developed location–specific covariances. His paper gives an interesting discussion that is of general relevance to space–time estimation.

VARIOGRAMS

Of all the rainfall variograms calculated, three are presented here: the variogram of the raw data as a function of time, and the variograms of the standardised rainfall as a function of time and of space. Figures 4, 5 and 6. The time variogram of the unstandardised data is periodic, as could be expected. A model consisting of a damped cosine was fitted to this. (N.B. This is only authorised for 1–D).

In contrast to this, the time variogram of the standardised data inside Lesotho is not periodic at all. It reaches a sill after a lag of 5 months. This means that the variability left in the standardised data after removing the seasonal trend is uncorrelated for readings 6 months or more apart. One consequence of this is that in the second approach, the data from only the previous one or two months need be used to krige the data from any one month. Provided that standardised data are used, it is not necessary to go back to the same month in earlier years, as one might have imagined. The spatial variograms of the standardised data showed some variability between the four directions, but not enough to be considered anisotropic.

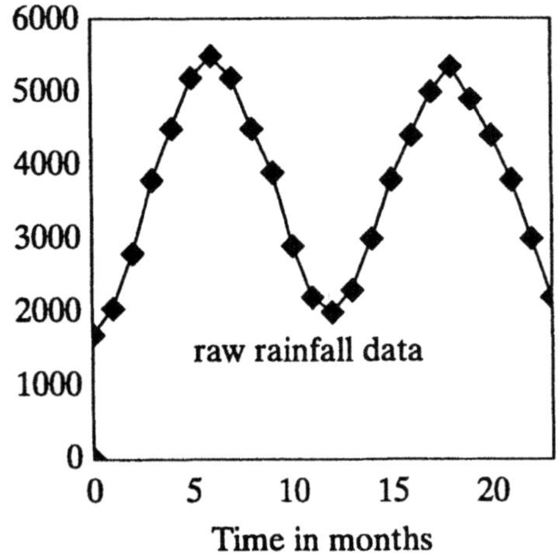

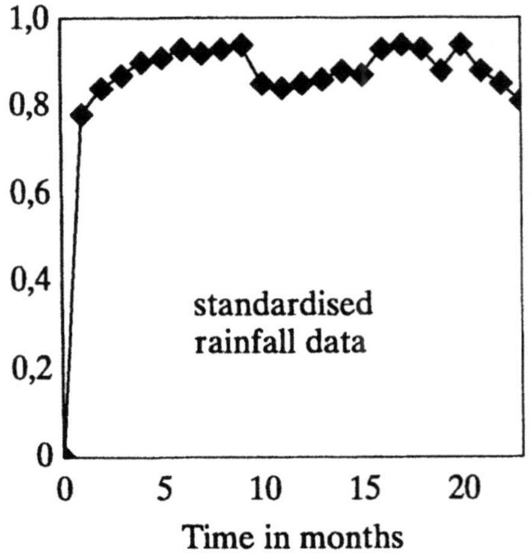

Figures 4 and 5: The rainfall variograms as a function of time; on the left, for the raw data and on the right, for the standardised data.

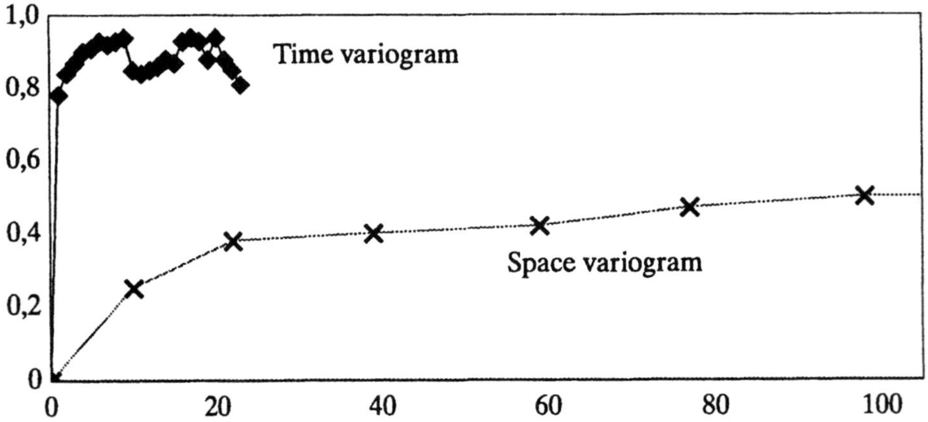

Figure 6: Combined space–time variogram model for standardised data

It is never easy to fit a combined time–space variogram model because there is no natural equivalence between time in months and distance in kilometers. By trial and error, a good fit was obtained with a model consisting of a zero nugget effect, an exponential with a sill of 0.45 and a range of 12.,and a spherical with a sill of 0.47 and a range of 500. The corresponding anisotropy ratios between X, Y and time were 1 : 1 : 11 for the exponential and 0.5 : 0.5 : 500 for the spherical.

For the third approach, maps of the average summer and winter rainfalls were required, and so the corresponding spatial variogram models were needed. No directional differences were found for either data set. A spherical model with a range of 55 km and a sill of 0.16 (and no nugget effect) was fitted to the summertime averages, while a spherical with a range of 25 km and a sill of 0.052 fitted the winter one. The average winter rainfalls were much less variable than those for summer, and were less well structured, as can be seen by comparing the ranges. This lack of structure will be apparent in the contour maps shown later. Having calculated the experimental variograms and fitted models to them, we could proceed to testing the different approaches by kriging (1) the points on a regular 20 x 20 km grid inside, (2) the same block averages on the grid and (3) the average rainfall inside Lesotho.

FIRST METHOD : USING ONLY THE OCTOBER 87 DATA

In this approach we used the standardised rainfall data, but as these are obtained by a linear transformation from the raw data, there is no difference between kriging the raw data using their variogram or kriging the standardised data using the corresponding variogram and then backtransforming. Figure 7 shows the map of the kriged point estimates in standardised units. The striking feature is its lack of structure. The presence of the mountain range in the east of the country has had no apparent effect on the resulting estimates. So the first approach does not seem to infill the sparsely sampled areas in a satisfying way, as one could have guessed.

SECOND METHOD : USING AUGUST AND SEPTEMBER AS WELL AS OCTOBER

When only the October rainfalls were used, kriging the raw data was equivalent to kriging the standardised data. This is no longer true when we want to incorporate information from previous months because their statistical characteristics (mean and variance) are quite different. It is better to use the standardised variable.

Because the time variogram of the standardised data has a range of 5 months (and shows no signs of periodicity), we used only the previous two months data with the data from the month under study. For October 1987, this gave a total of 117 observations compared to 28 for October itself. Clearly this should improve the quality of the estimates as it provides values at sites unavailable in October. It therefore came as no surprise to find that the standard deviations corresponding to the kriging variances were lower for this approach than for the first one. As the same variogram and kriging procedure were used in both cases, it is valid to compare these.

The map of the estimated rainfalls (Figure 8) shows much more structure than did the previous one. As it happened, the month of September 1987 was a very wet one whereas the following month was dry compared to the October average. So when the standardised rainfalls for the three months were used together, the high September values caused some curious high and low values to appear in some areas. Examples include the local high along the SE border with South Africa and the local low on the

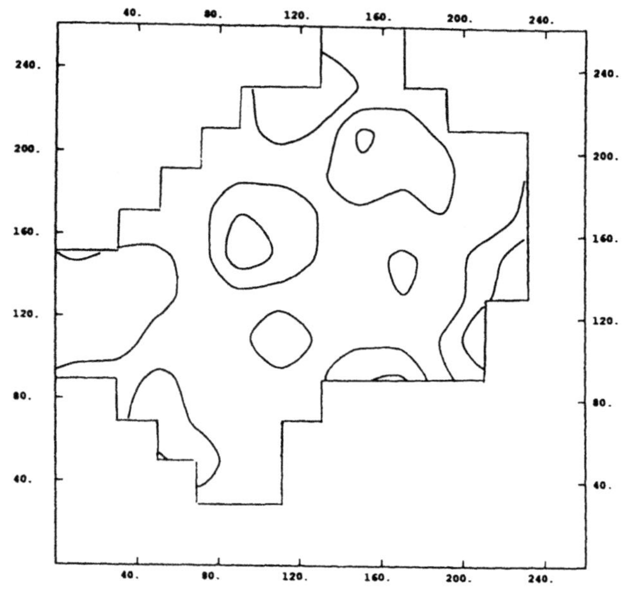

Figure 7: Kriged map of the Oct 87 rainfall obtained using only data for Oct 87

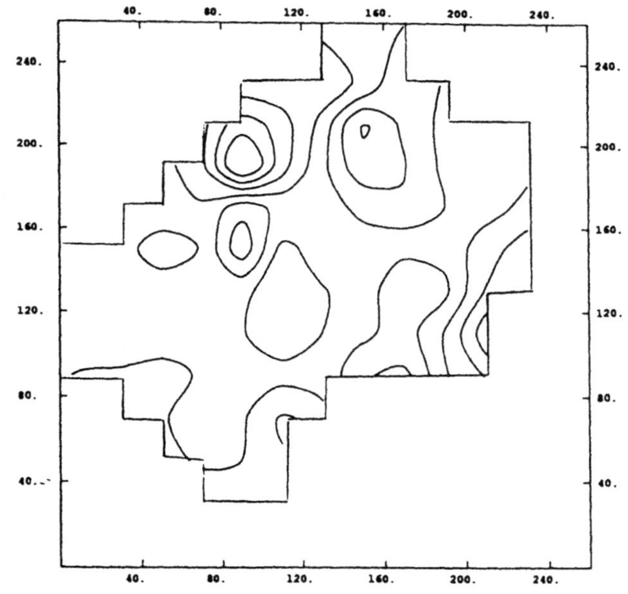

Figure 8: Kriged map of the Oct 87 rainfall obtained using Aug Sep & Oct 87 data

opposite side of the country. So it seems that the data from the previous two months (even after correction for the seasonal trend) are not sufficiently reliable for infilling the unsampled stations. This is why we sought, in the third approach, to find a better "background image" of the summer or winter rainfall.

THIRD METHOD : USING THE SUMMER MAP AS AN EXTERNAL DRIFT

The third approach consists of using the average standardised rainfall for summer as an external drift to describe the overall spatial distribution of the rain in summer. (And similarly for winter.) The theory of the external drift method and its applications are well–known. See, for example, Delfiner et al (1983), Galli and Renard (1986), Galli and Meunier (1987) and Moinard (1987). So the first step is to produce kriged maps of the average summer (and winter) rainfalls. (Figures 9 and 10).

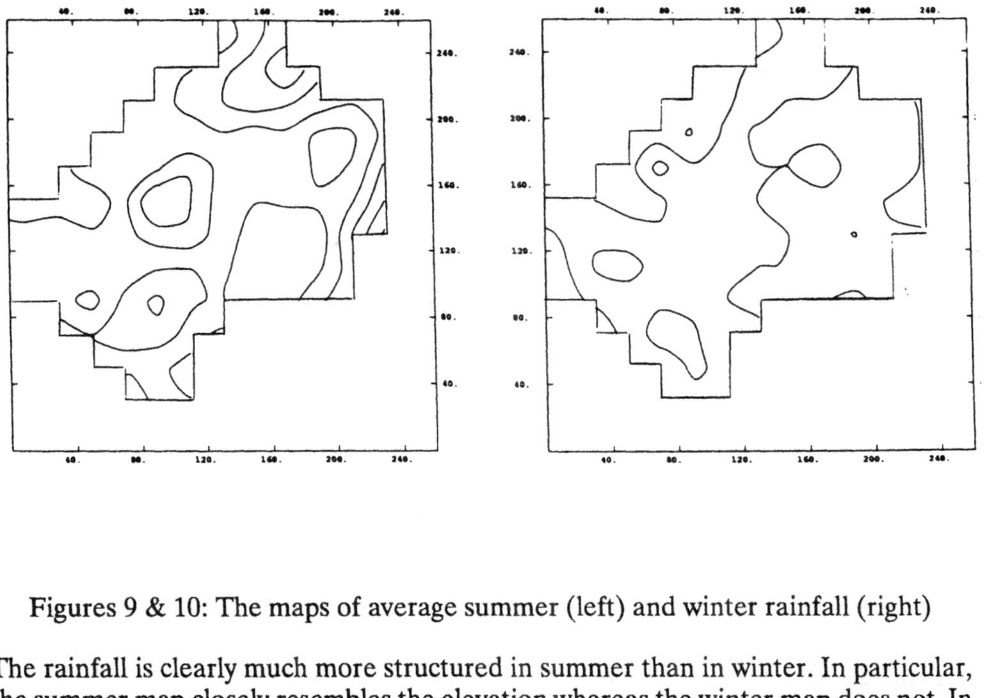

Figures 9 & 10: The maps of average summer (left) and winter rainfall (right)

The rainfall is clearly much more structured in summer than in winter. In particular, the summer map closely resembles the elevation whereas the winter map does not. In fact the correlation coefficients between elevation and rainfall are 0.30 for summer and 0.15 for winter. This dichotomy between summer and winter is one of the reasons why the elevation–rainfall correlation found earlier was so low. Another reason is the change of support. At the beginning of the study individual monthly rainfall readings were considered whereas these maps represent the average rainfall over many years. The effect on the correlation of increasing the support has been described in detail by Chauvet (1987) for a uranium deposit.

In the second step the kriged estimates for October 1987 were obtained using the 28 standardised rainfalls for that month together with the values of the average summer rainfall at all 107 raingauges and at the nodes of the 20 x 20 km kriging grid. Figure 11 shows the map obtained. This is considered as being more satisfactory than the other two because local specialists feel that it gives more plausible estimates in undersampled areas such as the highlands and because it is better structured. This approach is recommended for subsequent work.

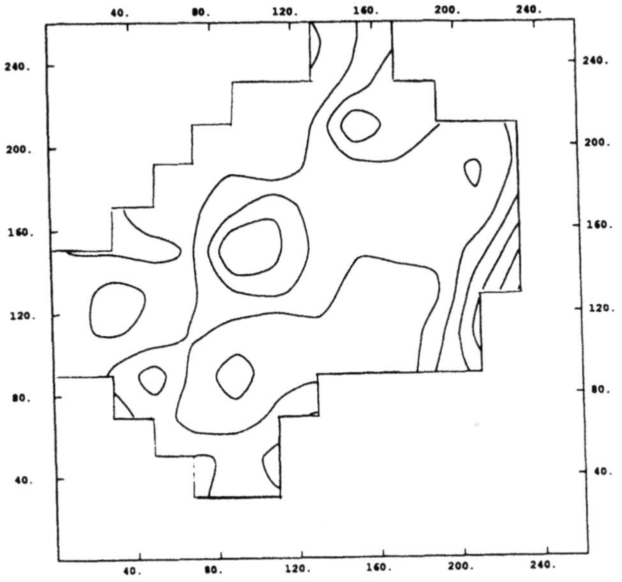

Figure 11: Map of the kriged rainfall for Oct 87 using the average summer rainfall as an external drift

DISCUSSION OF THE METHOD AND THE RESULTS

A 2D or 3D kriging algorithm? The first approach used only the data from the month under study and as such required only a 2D kriging algorithm. In contrast to this, in the second approach, data from three months were used. These could be represented visually as three superimposed planes. So a 3D kriging algorithm is needed. Having said that, although the areas under study are continuous in each plane, the planes are distinct and so one has to be careful in choosing a kriging routine suited to the layered structure. The rainfall data are monthly accumulations The third approach requires an external drift kriging algorithm suited to 2D data.

Statistical comparisons between the results. The simplest way to compare the results obtained using the three methods is by comparing the statistics of the estimates. Firstly, the estimates of the average rainfall over the whole of Lesotho for October 1987 obtained were

(1) 57.5 mm with a standard deviation of 6.2 mm
(2) 54.5 mm with a standard deviation of 6.0 mm
(3) 52.0 mm with a standard deviation of 6.4 mm

These estimates are very similar, given the standard deviation. This suggests that although the three geostatistical methods are based on different assumptions and different data sets, they are mutually compatible. While this is, to a certain extent, reassuring, it does not prove that they are right nor does it allow us to choose from among the three approaches. It would be more satisfying to have external criteria to aid us in making this choice.

External criteria. The ideal way of checking whether an estimation/prediction method is correct is by comparing the estimates with the true values. In mining this can be done, to a limited extent, by mining blocks and comparing the results with the anticipated values. No such test appears to be possible with rainfall. Of course, after the dams have been constructed, far more information will be available as the water starts to bank up. But that will be a little too late for this project, as its aim is to provide input information into the water project.

We therefore sought some other way of either confirming or refuting the geostatistical models being used, and of choosing between them. We wondered if some sort of cross-validation test could be designed, similar to those used for validating variogram models. Late in the study the rainfall readings from three additional sites were received. These had not been used either in establishing the variogram model or in the different kriging procedures, and so could be considered as "external" to the model. The idea was to krige these three points using the initial sets of data and then compare the results to the actual values using the kriging standard deviation as a guide to judge the difference.

But before doing the calculations, we looked in detail at some of the point estimates and their corresponding standard deviations. The most accurately estimated grid node had an estimated rainfall of 38.35 mm with a standard deviation of 15.46. If for argument's sake we consider a 95% confidence interval for the normal distribution, then for this case, it extends from 7.43 mm to 69.2 mm. The adjoining grid node, which is more typical, has an estimate of 55.04 mm with a standard deviation of 33.88, thus giving a confidence interval going from 0 mm to 121.6 mm. As the October 87 rainfalls range from 1.0 mm to 126.0 mm, there is little chance of an experimental value contradicting the prediction. One can argue whether to use common garden normal distribution intervals or something more sophisticated, but the underlying problem persists: the data are sparse and erratic, and so the resulting kriging standard deviations are large.

Initially this may seem disheartening. When the experimental data for close spaced pairs of raingauges were examined more closely, we found that there were often considerable differences between the rainfalls. For example the two gauges at Roma which are a few kilometers apart, gave readings of 69 mm and 126 mm respectively for that month, showing the type of short scale fluctuation possible with this sort of data. (Other tests done by P. Hubert had showed that the rainfall data are stationary in time; in particular, there was no evidence of long term trends or long cyclic variations.)

Final recommendation. In the end we came back to our comments on the form of the kriged maps when choosing the most appropriate estimation method. As the third method produced the map that best corresponded with the impression that water resource experts working in the area have of the local rainfall, it was recommended that it be adopted. The advantage of this procedure is that it gives plausible rainfall maps even when relatively few rainfall observations were available, as was the case in October 87. The kriging procedure is easy to implement routinely on a PC−type computer. As the summer and winter background maps and the variogram were based on several thousand readings spread over more than 60 years, they are considered as being representative of the Lesotho region, and would only have to be update every 5 or 10 years. In this sense, the method proposed seems to be a robust one.

ACKNOWLEDGEMENTS

The authors would like to thank the Lesotho Highland Development Authority for permission to publish the paper, and the French Foreign Affairs Ministry for funding this research project under contract No 41/89.

REFERENCES

Chauvet, P. (1987) "On the estimation of a uranium deposit", In G. Matheron and M. Armstrong (eds) *Geostatistical Case Studies* Reidel and Co Dordrecht pp23−37

Delfiner, P. Delhomme, J.P. and Pelissier Combescure J. (1983) "Application of geostatistical analysis to the evaluation of petroleum reservoirs with well logs." 24th Annual Logging Symposium at Calgary Canada June 27−30.

Galli, A. and Renard D. (1986) BLUEPACK Geostatistical Background, Internal Report Centre de Geostatistique, Ecole des Mines de Paris, Fontainebleau.

Galli, A. and Meunier, G (1987) "Study of a gas reservoir using the external drift method." In G. Matheron and M. Armstrong (eds) *Geostatistical Case Studies* Reidel and Co Dordrecht pp105−120

Moinard, L. (1987) "Application of kriging to the mapping of a reef from wireline logs and seismic data; a case history." In G. Matheron and M. Armstrong (eds) *Geostatistical Case Studies* Reidel and Co Dordrecht pp93−104.

Switzer P. (1989) "Nonstationary spatial covariances estimated from monitoring data", in M. Armstrong (ed), *Geostatistics,* the Proceedings of the 3rd International Geostatistical Congress, Avignon September 5−9, 1988, Kluwer Academic Press Dordrecht Holland pp127−138

PROBABILITY KRIGING AS A DECISION SUPPORT TOOL FOR LOCAL SOIL POLLUTION PROBLEMS

J.P. Okx [1], H. Leenaers [2] and R.M. Krzanowski [3]

[1] TAUW Infra Consult B.V., P.O. Box 479, 7400 AL Deventer, The Netherlands
[2] CSO Consultants for Environmental Management and Survey, P.O. Box 30, 3734 ZG Den Dolder, The Netherhlands
[3] Alberta Research Council, Terrain Sciences Department, P.O. Box 8330, Postal Station F, Edmonton, Alberta, Canada T6H 5X2

ABSTRACT

In a soil remediation project the amount of polluted soil that will be removed seldom equals the total amount of polluted soil. This is because the decision to remove the soil depends not only on the characteristics of the polluted soil body, but also on the current environmental policy and on the technical and economic context of the remediation project. Nevertheless an estimate of the amount of polluted soil is essential.

Probability kriging is a non-parametric geostatistical technique that is suitable for estimating the amount of polluted soil. For volume estimation purposes two pieces of information are essential. Firstly the concentrations of the pollutant with regard to different remedial actions or to different policy-views have to be known. Secondly the level of probability above which some remedial action becomes inevitable must be known. In the described case-study - soil pollution by atmospheric emissions from a pigment factory - probability kriging was used to map the lead contamination in terms of conditional probabilities that given thresholds are exceeded.

It was found that in a large area of about 30 km^2 around the pigment factory the probability of exceeding the 150 mg/kg threshold is over 80 %. The pattern of concentric circles around the factory obtained when estimating the probabilities of exceeding the 300 mg/kg threshold is exactly what is expected in a case of atmospheric deposition.

The potential for working with more than one cut-off level makes the technique an important management tool in decision making.

INTRODUCTION

In a soil remediation project the total volume of polluted soil that has to be removed is the key factor in the cost calculations. These volumes are estimated from the isarithmic maps on which the results of the soil pollution research are presented. However, in practice, the estimated volumes are often exceeded. Normally this is not caused by poor estimation but by not taking into account the errors associated with the estimates. In many cases these errors are not known; simply because the non-statistical interpolation routines used do not provide them. Fortunately there are now

a number of geostatistical techniques available which are not only better estimators than the methods normally used in pollution studies but they also give the errors related to the estimates. The work of Krige (1951) and Matheron (1965) should be considered as the basis for this work.
So the main advantage of geostatistical interpolation techniques, essentially ordinary kriging, is that an estimation variance is attached to each estimate. Unfortunately, unless a Gaussian distribution of spatial errors is called for, an estimation variance falls short of providing confidence intervals and the error propability distribution required for risk assessment. Regarding the characterization of uncertainty, most interpolation algorithms, including kriging, are parametric in the sense that a model for the distribution of errors is assumed. Such models are questionable when used for spatial interpolation errors (Journel, 1987).

Non-parametric geostatistical methods, such as indicator kriging (Journel, 1983) and probability kriging (Sullivan, 1984 and Journel, 1984), which puts the modelling of the uncertainty as priority, do not make use of these assumptions.
Since probability kriging is considered as an improvement of the indicator kriging procedure in the sense that the data is used more completely we decided to use probability kriging.
One of the ways to present the results of the probability kriging procedures is by means of a probability map on which conditional probabilities of exceeding given thresholds are represented. This way of representation makes the technique an important tool in decision making under risk because it communicates the uncertainties to the decision makers (Okx and Kuipers, 1991). After a short description of the method the advantages of the method will be demonstrated in the form of a case study.

PROBABILITY KRIGING

Probability kriging is a non-parametric procedure which does not depend on any particular statistical distribution. It is based on an indicator function $I(x_\alpha;z)$, which is then used for estimating a local distribution. The indicator function is defined in terms of the cut-off value z which is a threshold given by environmental protection regulations. If the values of the property are below the threshold no action is required, but if then certain measures may become necessary.

The indicator function is :

$$I(x_\alpha, z) = \begin{cases} 1 & if \quad z(x_\alpha) > z \\ 0 & else \end{cases}$$

In this article a short description of probability kriging is given and we recommend for further information Sullivan (1984) and Journel (1984). As stated earlier probability kriging is an improvement on indicator kriging. Indicator kriging (see Journel, 1983) does not use all the available information about the variable to be estimated. When the probability of exceeding a certain cut-off value, (P(z)), is estimated only the indicator values of that particular cut-off value (z) are used.
The basic idea is to use the grade information $Z(x_\alpha)$ in addition to the indicator $I(x_\alpha,z)$. The values of $Z(x_\alpha)$ and $I(x_\alpha,z)$ are usually of a different magnitude, which

may give numerical problems. To overcome this problem the grades are replaced by their cumulative distribution function. We will use a variable U(x):

$$U(x_\alpha) = F(Z(x_\alpha))$$

Now the estimated value is:

$$\phi^*(V,X) = \sum_{i=1}^{n} \lambda_\alpha \, I(x_\alpha, z) + \sum_{i=1}^{n} \nu_\alpha \, U(x_\alpha)$$

Weights λ_α and ν_α will be obtained from a standard cokriging system:

$$\sum_{i=1}^{n} \lambda_\alpha \rho_i(x_\alpha - x_\beta, Z) + \sum_{i=1}^{n} \nu_\alpha \rho_{ui}(x_\alpha - x_\beta, Z) + \mu_1 = \rho_i(x_\beta, V, Z)$$

$$\sum_{i=1}^{n} \lambda_\alpha \rho_{ui}(x_\alpha - x_\beta, Z) + \sum_{i=1}^{n} \nu_\alpha \rho_u(x_\alpha - x_\beta, Z) + \mu_2 = \rho_{ui}(x_\beta, V, Z)$$

$$\sum_{i=1}^{n} \lambda_\alpha = 1$$

$$\sum_{i=1}^{n} \nu_\alpha = 0$$

(David, 1988).

CASE STUDY : ATMOSPHERIC DEPOSITION OF LEAD

Source of contamination and sampling scheme

The soil in areas surrounding the town of Eijsden in the Province of Limburg in The Netherlands has been polluted with heavy metals in several ways. Apart from the diffuse inputs through atmospheric deposition and the use of fertilizers containing metals, three site specific sources of heavy metals can be distinguished (Rang et al., 1987).

One source is a pigment producing factory, which was established in Eijsden in 1870. At the beginning of this century the factory emitted large quantities of particles containing barium, zinc and lead into the atmosphere. In 1974 the annual rate of emission was assessed to be 64 tonnes of zinc, 210 tonnes of barium and 94 tonnes of lead. In 1975 the emission stopped almost completely as a result of changes in the production process.

A second site specific source of heavy metal pollution is from roads sealed with zinc oxide cinders, which are a waste product of the pigment industry. Between 1870 and 1975 250,000 tons of cinders were produced. The total length of cinder-sealed roads in the area is about 100 km.

A third source of heavy metals is the river Meuse. Part of the area of investigation is located on the floodplain of the river. The river muds are contaminated with heavy metals as a result of waste disposal by mines and metallurgic industries in the Belgian part of the catchment. Flooding and subsequent deposition of these muds has resulted in a considerable increase in the heavy metal concentrations of the topsoil in the floodplain area.

To estimate the spatial extent of soil pollution by atmospheric emissions from the pigment factory 98 samples were taken from the topsoil in the area. To avoid disturbing the pattern of atmospheric deposition random stratified sampling restricted to areas were the presence of cinders or river-sediments was unlikely was used. The strata were obtained by dividing the area into 1 x 1 km^2 squares. In this case-study we focused on the lead content of the soil because lead turns out to be the most critical element in terms of human-toxicological and ecotoxicological risk. The sampling scheme is given in Figure 1.

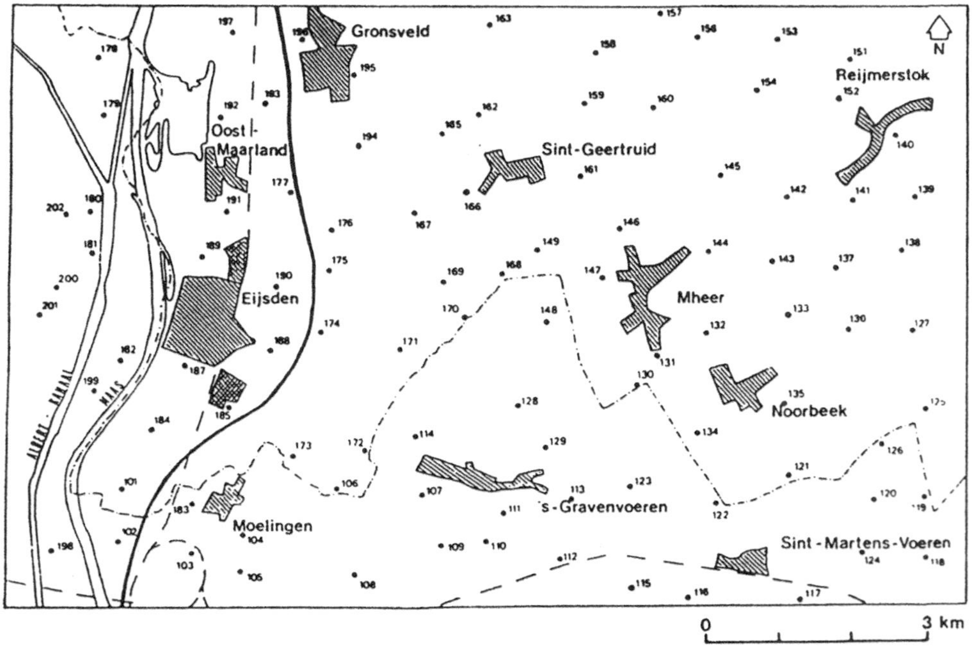

Figure 1. The sampling scheme

Summary statistics

The statistical analysis, which should be the starting point of any geostatistical study, gives us a first impression of the variability of the variable in study. Table 1 shows the most important summary statistics.

Table 1. Summary statistics of the lead content

number of samples	min (mg/kg)	max (mg/kg)	mean (mg/kg)	median (mg/kg)	standard-deviation	coefficient of variation	skew-ness
98	83	5640	248	155	570	2.30	9.0

The large value of 2.30 for the coefficient of variation illustrates the considerable variability in the lead content of the soils in the area. A maximum value of 5640 mg/kg is reached. This means that the soil is extremely polluted locally which calls for remedial action. The minimum value of 83 mg/kg shows that all values are well above the detection limit of 10 mg/kg so there is no zero effect. The frequency histogram (Figure 2) is markedly asymmetric and this is also reflected by the large skewness-value of 9.0. The distribution clearly does not follow any gaussian or lognormal law, a fact which is not unusual for soil pollution data.

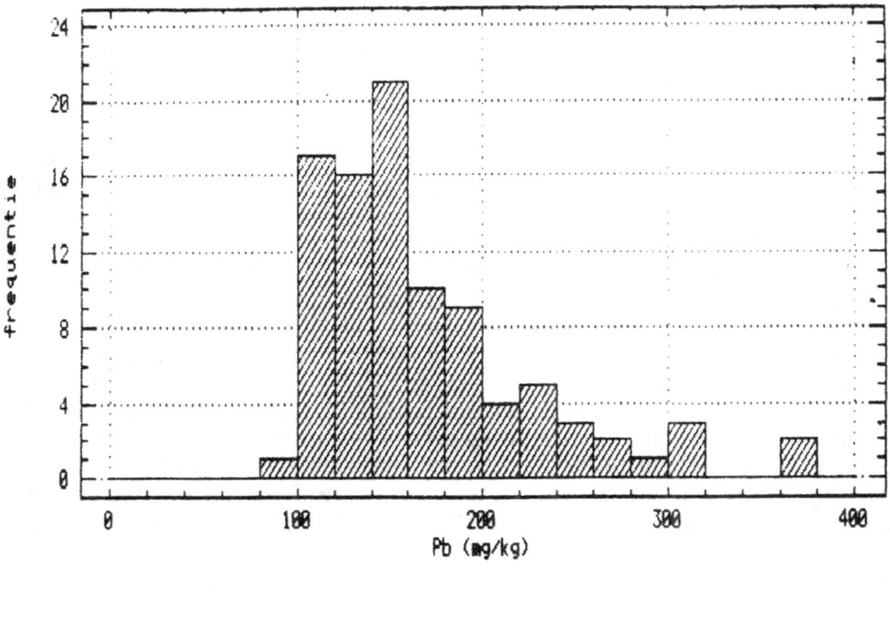

Figure 2. Frequency histogram of lead content

ANALYSIS AND RESULTS

The first step in the probability kriging procedure is to define a number of cut-offs. Different cut-offs could be related to a number of different policy-views. These views can be based on :

- risk assessment
- technical possibilities
- economical context
- political context.

In Table 2 two different cut-offs (z_{ci}) and their corresponding transformations ($U(z_{ci})$) are given. These $U(z_{ci})$ values correspond to the relative cumulative frequency of the distribution function $H(z(x_i))$. The cut-off of 150 mg/kg correspond to the so-called B-value of the Dutch Soil Cleanup Act and the cut-off of 300 mg/kg coincides with the C-value of the same act.

Table 2. Cut-offs z_{ci} and the corresponding transformations ($U(z_{ci})$)

cut off (mg Pb/kg) Z_{ci}	150	300
$U(Z_{ci})$	0.4694	0.9151

Recently, however, the National Institute of Public Health and Environmental Hygiene proposed two new C-values. The first C-value of 162 mg/kg is based on human-toxicology (van den Berg, 1991); it represents the value above which a serious threat to (human) health can be expected. The other C-value is an ecotoxicologically based C-value of 290 mg/kg (Denneman and van Gestel, 1990) which represents the value above which a serious threat to the environment can be expected. Working with these new cut-offs would yield results that are almost identical to those produced with the cut-offs drawn from the Cleanup Act.

Figure 3 shows the experimental values (dots) of the two indicator variograms (I_{150}, I_{300}), the variogram of the transformed sample values, U, and the two cross variograms, UI_{150} and UI_{300}, of the transformed sample values and the indicatorvalues and the solid lines are the fitted spherical models. The experimental variograms I_{150}, U and UI_{150} all start with relatively high values, but they are based on two sample pairs only. Since the parameters of the variogram models are fitted by a simple weighted least squares approximation (Cressie, 1985) these starting points are practically neglected.

In the process of cross variogram fitting the Cauchy-Schwarz inequality :

$$|(\gamma_{ui}(h))| \le \sqrt{(\gamma_u(h) * \gamma_i(h))} \quad for\ all\ h \ge 0$$

was checked to guarantee the positive-definiteness of the kriging matrices (Journel and Huybregts, 1978; Myers, 1982 and 1984; Nienhuis, 1987).

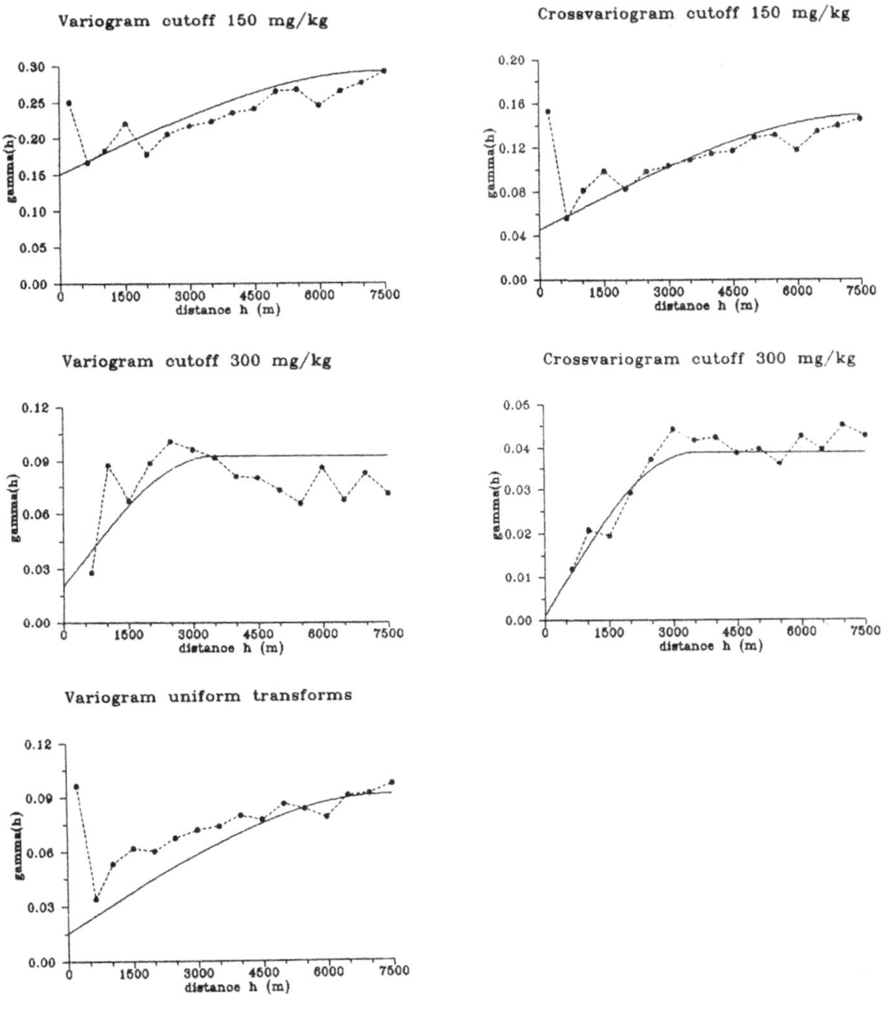

Figure 3. The complete variography

The model parameters (c_0: nugget effect; c_1, a_1: sill and range; c'_0: relative nugget effect) are given in Table 3.

Table 3. Parameters of the spherical models of the different variograms

Variogram	C_0	C_1, a_1	C'_0
I_{150}	0.150	0,293, 7500	0.51
I_{300}	0,020	0.093, 3500	0.21
U	0.015	0.091, 7500	0.16
UI_{150}	0.045	0.150, 7800	0.30
UI_{300}	0.001	0.039, 3500	0.03

The cross variogram models (UI_{150} and UI_{300}) were used for probability kriging which resulted in a number different blocks (125 x 125 km²) filled with the probability that the cut-offs in question (150 mg/kg and 300 mg/kg) is being exceeded. Figures 4 and 5 show the results for the topsoil in the Eijsden-area, the probabilities are represented as percentages.

Figure 4 of the probabilities of exceeding the 150 mg/kg threshold shows that there are two distinct areas. One large area of about 30 km² with probabilities of more than 80 % that the 150 mg/kg threshold is exceeded can be related to the pigment producing factory. This indicates a possible threat to human health. In the rest of the area the probabilities of exceeding the 150 mg/kg threshold are highly variable and range from 10 to 70 %.

In Figure 5 the probabilities of exceeding the 300 mg/kg threshold are mapped and the relation with our factory is even clearer. The pattern of concentric circles around the factory is what you would expect in a case of atmospheric deposition. The area where the probability of exceeding the 300 mg/kg threshold is over 80 % is about 2.4 km².

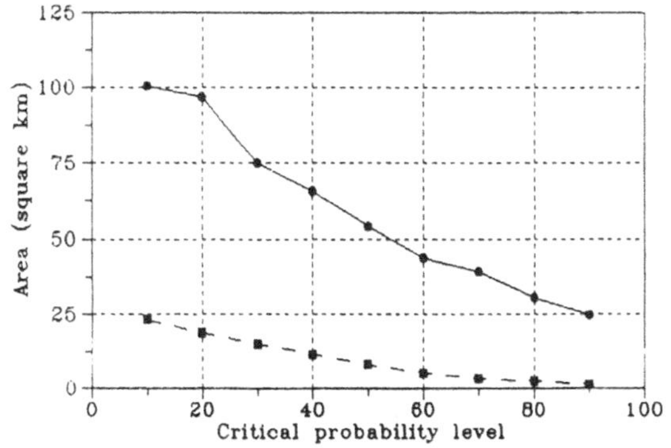

Figure 6. Area in which measures are necessary as a function of cut-off and a critical probability level

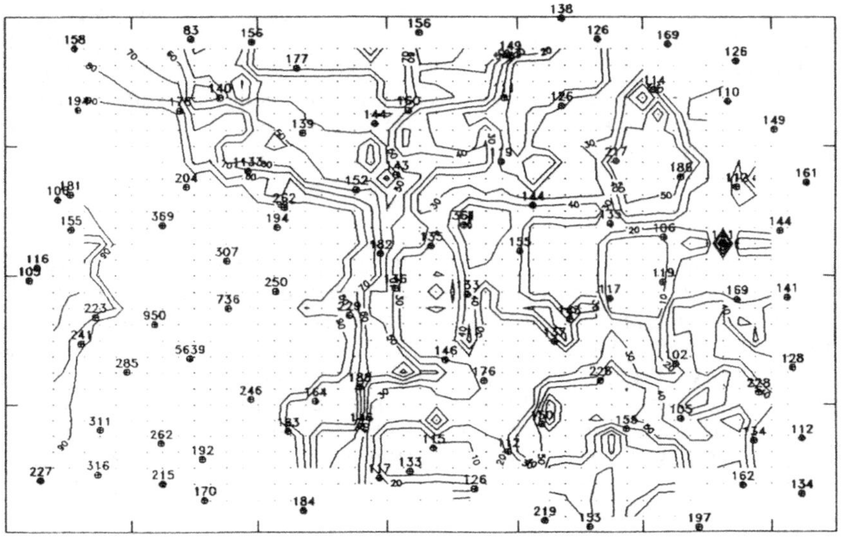

Figure 4. Conditional probabilities of exceedance of the 150 mg Pb/kg threshold

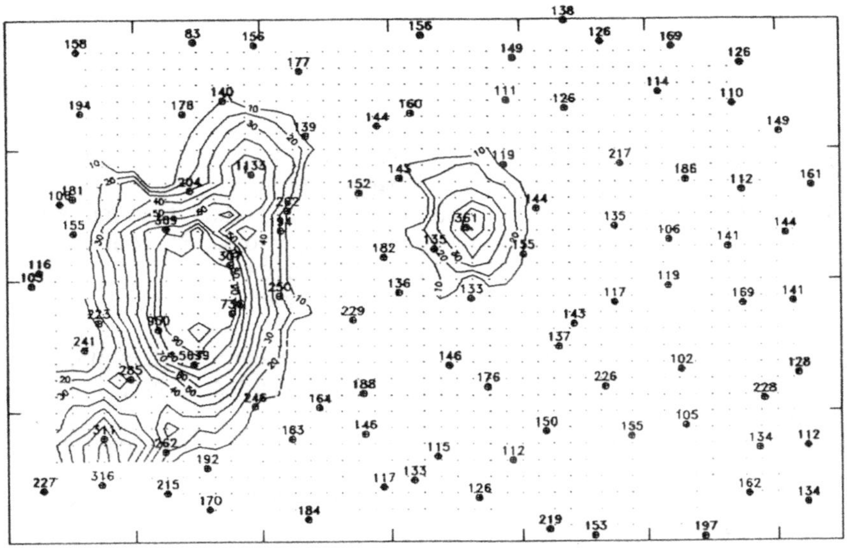

Figure 5. Conditional probabilities of exceedance of the 300 mg Pb/kg threshold

By using a geographical information system we can easily calculate the consequences - in terms of the area for which certain measures becomes necessary - for each possible combination of a cut-off and a critical probability level. This makes the technique an important management tool in the decision making process.
The results of such calculations are given in Figure 6. The solid line shows the relation for the 150 mg/kg threshold and the dotted line for the 300 mg/kg threshold. If we consider the probability levels in Figure 6 as being consumer's risks, then it becomes clear that normally remedial action is a matter of reducing the consumer's risk rather then eliminating it.

DISCUSSION

Probability kriging is, in our opinion, an essential tool in the decision making process. The ability to produce probability maps greatly enhances the possibility of communicating uncertainties to the authorities responsible for soil clean-up operations. At present decision-makers are not used to think in terms of probabilities and risks and it is therefore unlikely that critical probability levels will be specified a priori. The concept of a remedial action that reduces risk instead of eliminating it will force the decision-makers to make formal statements about what should be considered as an acceptable risk. Just as choosing different cut-offs can be based on different aspects such as risk assessment, and economic and political contexts, making statements about acceptable risks is also a matter of taking into account and weighting different aspects.
Information on the local background levels - preferably also in the form of conditional probabilities - is, in our opinion, essential in formulating statements which lead to decisions in which costs and efficiency are well balanced.
The non-parametric nature of this method is yet another advantage when analyzing highly skewed phenomena which are so frequently encountered in soil pollution studies of this kind.

REFERENCES

Akin, H. and Siemes, H. (1988), Praktische Geostatistik. Eine Einführung für den Bergbau und die Geowissenschaften, Springer-Verlag.

Berg, R. van den (1991), Blootstelling van de mens aan bodemverontreiniging. Een kwalitatieve en kwantitatieve analyse, leidend tot voorstellen voor humaan toxicologische C-toetsingswaarden. RIVM, Bilthoven.

Cressie, N. (1985), Fitting variogram models by weighted least squares. Mathematical Geology, Vol.17, No.5, pp.563-586.

Denneman, C.A.J. and Gestel, C.A.M. van (1990), Bodemverontreiniging en bodemecosystemen : voorstel voor C-(toetsings)waarden op basis van ecotoxicologische risico's. RIVM, Bilthoven.

Journel, A.G. and Huybregts, A. (1978), Mining geostatistics. Academic Press, New York.

Journel, A.G. (1983), Nonparametric estimation of spatial distributions, Mathematical Geology, Vol.15, pp.445-468.

Journel, A.G. (1984), The place of non-parametric geostatistics, In :Verly, G. et al. (eds.), Nato Advanced Study Institute Series, pp.307-335, Lake Tahoe 1983, D. Reidel, Dordrecht.

Journel, A.G. (1987), Non-parametric geostatistics for risk and additionalsampling assessment, Stanford.

Krige, D.G. (1951), A statistical approach to some basic mine valuation problems on the Witwatersrand, J. Chem. Metall. Min. Soc. S. Africa, 52 (6), pp.119-139.

Matheron, G. (1965), Les variables regionalisées et leur estimation, Masson, Paris, pp.212.

Myers, D.E. (1982), Matrix formulation of cokriging. Mathematical Geology, Vol.14, No.3, pp.249-257.

Myers, D.E. (1984), Cokriging : new developments. In : Verly, G., David,M., Journel, A.G. and Marechal, A. (eds.), Geostatistics for Natural Resources Characterization (part 1). NATO ASI Series, Series C : Mathematical and Physical Sciences, 122, pp.295-305. D. Reidel Publishing Company, Dordrecht.

Nienhuis, P.R. (1987), CROSSV, a simple FORTRAN 77 program for calculating2-dimensional experimental cross-variograms. Computers & Geosciences, 13, 4, pp. 375-387.

Okx, J.P. and Kuipers, B.R. (1991), 3-Dimensional Probability Mapping forSoil Remediation Purposes. Proceedings of the Second European Conference on GIS, Bruxelles, Belgium.

Rang, M.C., Okx, J.P. and Burrough, P.A. (1987), The use of geostatistical methods within the framework of the dutch soil sanitation operation. In : Duijvenbooden, W. van en Waegeningh, H.G. van (eds.). Proceedings of the Conference of Vulnerability of soil and groundwater to pollutants, TNO Committee on Hydrological Research, The Hague, The Netherlands.

Sullivan, J. (1984), Conditional recovery estimation through probability kriging - Theory and practice, In : Verly, G. et al. (eds.), Nato Advanced Study Institute Series, Lake Tahoe 1983, D. Reidel, Dordrecht.

NON LINEAR APPROACH TO THE MARINE INTRUSION PROCESS MODELLING

DELGADO-GARCIA,J. & CHICA-OLMO,M.
Geostatistics and Remote Sensing Research Group
Dpto. Geodinámica. Univ. de Granada-CSIC
Avda. Fuentenueva s/n
18071-GRANADA (Spain)

Abstract.

In this paper, a practical application of the non linear estimation method of Disjunctive Kriging to the hydrogeochemical characterization of a marine intrusion process in the coastal carbo-detritic aquifer of Aguadulce (Campo de Dalías, Almería, SE Spain) is shown. The marine intrusion has been caused by an intense exploitation of water for the irrigation of the very high perfomance agriculture.

In order to obtain the characterization of the contamination process, an exhaustive campaign of hydrogeochemical analysis with 97 samples has been developped, choosing the chloride content (Cl^-) as an indicator of this intrusion.

The results obtained in ths work demostrate the efficency of this type of estimation method for the probabilistic analysis of geochemical anomalies with many applications in the treatment of hydrogeochemical data and aquifer management.

Keywords. Geostatistics, Hydrogeochemistry, DK, Aquifer Overexploitation.

1. Introduction

Human activities often produce, modification and disturbance in the hydrologic cycle and water circulation. The result is frequently a contamination of the waters, often practically, irreversible and very seldom saveable.

The contamination of the surface water is a very evident process and as such, cause actions and recourses to avoid it or at least to mitigate its effects, can be taken. This does not happen with the subterranean waters, as they are not visible and their exploitation is widely distributed, the actions usually taken arrive frequently late, if they can be produced.

An example of the modification of the natural environmental conditions is the "Campo de Dalias" sector (Almería, SE Spain) (fig.1). This area has suffered important changes related to a strong agricultural "boom" in the past decades. So, this sector has passed from a cereal and leguminous production with poor irrigation (as dry land support) around 3000 m^3/Ha/year in 1961, to a high perfomance intensive production (green houses) with several crops of vegetables per year with elevated consumption (around 13000 m^3/Ha/year), quintupling its production, due to this, it produces a considerable increase in water consumption.

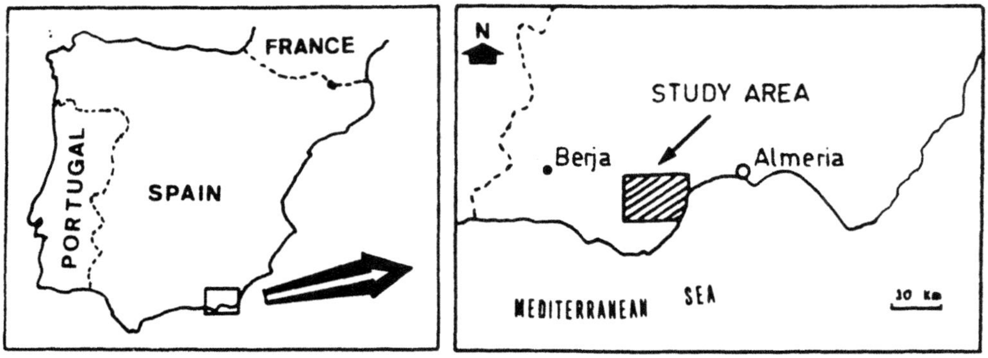

Fig. 1.- Geographical situation of Campo de Dalias area.

The intensive agricultural exploitation, which requires heavy labour, added to a very bad economic situation in this area (Almeria was the lowest *per capita* income province of Spain) has produced a heavy inmigration from the adjacent sectors causing an population increase passing from around 25.000 inhabitants in 1960 to 65.000 in the 80's. To this increase, directly related to the agriculture is necessary to add the summer influx to the several touristic centers of the area which causes an increase of twice or more the regular population.

This population increases originates a logical increase that has produced a deficit in the hydric balance around 50 Hm^3/year (inputs= 50 Hm^3/year; outputs= 100 Hm^3/year), particularly critical in the eastern area (deficit= 30; inputs= 15; outputs= 45). This has produced a very high decrease of piezometric levels.

The decrease of the piezometric levels, and the aquifer's proximity to the sea, have originated the increase of the freshwater-saltwater contact, and thus the deterioritation and aptitude lost of the water for agricultural utilization.

It was necessary to establish the spatial distribution of the contamination caused by marine intrusion, choosing the chloride (Cl^-) contents as an indicator of it. The problem consists in determine the block average value of Cl^-, and the conditional probability of that value being above a determinated limit. The limits are marked by the permissivity of the cultives to the Cl^- contents. The World Healt Organization (WHO) mark 200 and 600 mg/l levels, as proper and admissible values for human use.

The problem is similar to that in mining. As well as in mining too, the problem is compounded by the knowing the conditional probability distributions of the variable that we analysed. The combination of estimation, risk assessment and unknown probability distribution that makes DK specially atractive to the hydrogeochemical data treatment, gives to resources (water) management very important information.

2. Probabilistic analysis of the spatial distribution of contamination

2.1 Local Estimation of Probability Distribution

From a theoretical point of view, the problem to be tackled consists of a local estimation of the distribution probability of the conditional variable $\{Z_v/Z(x_\alpha);\ \alpha= 1 \text{ a } n\}$ where Z_v is the average chlorid contain of the area v and $Z(x_\alpha)$, the experimental data used for the estimate.

It is generally known (Neveu, 1964) that the most absolute means for estimation of the value Z_v, starting with data item Z(x) is the conditional expectation

$$Z^* = E\ [Z_v/Z(x_\alpha);\ \alpha= 1 \ldots n]$$

The statistical inference of Z_v is derived from the prior knowledge of the simultaneous distribution to the n+1 variable $\{Z_v, Z(x_1),\ldots,Z(x_n)\}$ giving rise to important practical problems, that can be resolved by imposing large-scale hypothesis which are difficult to control in numerical terms. This, in turn, undermines the validity of the method used.

2.2 The used method: The Disjunctive Kriging.

Over the last fifteen years various geostatistical methods have been put forward as being capable of estimating the distribution of conditional probability on a local level for a regionalized variable: Multigaussian approach (Verly, 1984), Disjunctive Kriging (Matheron, 1973), Indicator Kriging (Journel, 1982), Uniform Conditioning (Remacre, 1982) and Probabilistic Kriging (Journel, 1984), among others. Each of these involves hypothesis more or less taken for granted in practice. As Alli et al. (1990) have pointed out, these methods can be classified into two categories: parametric and non parametric. The first category can be distinguished by its need for assumptions about the variable distributions.

In this work the Disjunctive Kriging (DK) geostatistical method has been used, classified as part of the group of parametric methods, so that it use the binormality hyphotesis of the variable pairs (Y_0, Y_i) and (Y_i, Y_j).

The method formulated by Matheron (1973,75) as a means of solving mining problems, can be placed, in terms of its actual function between the well known linear kriging and the Conditional Expectation methods. In contrast to the hypothesis involved in the previous method this has the advantage of being practical, it can be used to tackle non linear estimation problems in environmental and hydrogeochemical contexts, as with the case studied here.

The following mathematical equations are based on articles by Matheron (1973, 75) and Journel & Huijbregts (1978). In short, DK is a non linear method to obtain estimates that are applicable to the calculation of functions of a regionalized variable z(x). The estimator has the overall structure:

$$Z_{DK}=\sum f_\alpha [Z(x_i)]$$

This is an estimator made by the combination of functions of each of the $Z(x_i)$. By using the information from the non linear mode it is possible to calculate the probability $Pr[Z_v/z_c]$ or any function of Z_v, $f(Z_v)$.

The most important hypothesis is the binormality of pairs of variables (Y_i,Y_j) (Y_o,Y_i). This hypothesis, which is perhaps one of the weakest points of the method can be avoided in practice, given that normal variables obtained by gaussian anamorphosis: $Y(x) = G^{-1}\ [Z(x)]$ (Marechal, 1980) can be used. Remacre (1984) and Chica-Olmo (1987) have applied different kinds of tests in order to verify by experimental means , the binormality of pairs based on the articles by Matheron (1982).

The estimator for the conditional probability function of the $\{Z_v/Z(x_\alpha);\ \alpha=1,\ldots,n\}$ where Zv is the average chloride content for the area v and $Z(x_\alpha)$ the experimental data used in the estimation is

$$f_{DK}=\sum_{n\geq 0}^{N}\frac{H_n(y)\,H_n^*}{n!}g(y)$$

where the unknowns requiring estimation are the H_n^*, which represent the Hermite coefficients of the function f_{DK}. The other parameters are: g(y), gaussian density functions; $H_n(y)$, the Hermite polynomial of order n for the y value and n! the factorial of n.

The H_n^* coefficients are obtained by kriging from $H_n(y_i)$ calculated in the experimental data x_i:

$$H_n^*=\sum_{i=1}^{N}\lambda_n^i\,H_n(y_i)$$

The λ_n^i are the weights derived of the resolution of the n kriging systems. The systems are formed by the correlations matrix of the estimator ensemble $\{Y_i,Y_j\}$ and between these and the estimated block $\{Y_i,Y_v\}$ with the correlogram, or equivalent variogram, functions: $\rho(h) = E\ \{Y_i,\ Y_{i+h}\}$

$$\forall i \qquad \sum_{i=1}^{N}\lambda_n^i\,\rho_{i,j}^n = \rho_{i,v}^n$$

The expectation of the function f_{DK} is the estimator Z_{DK} of the average value of the block v.

$$Z_{DK}^*=\sum_{n=0}^{N}\frac{\psi_n\,H_n^*}{n!}$$

The kriging error is:

$$\sigma_{DK}^2=\sum_{n=1}^{N}\frac{\psi_n^2}{n!}(1-\lambda_n^\alpha\,\rho_{\alpha,o}^n)$$

It is important to point out that on a strict sense the function f_{DK} is not a true function of density, given that it might not increase with all values of y. Although it is true for $\int f_{DK}(y)\,dy=1$, it cannot always be guaranteed that $\int y^2 f_{DK}(y)\geq 0$ is positive. As such, in extreme situations, negative values for probability can be encountered. This problem does not have important practical importance as it has only occured when y has very large absolute values, having only a slight effects on the tails of distributions.

3. Case Study

The most important objective of the research has been to check the validity of the method used to treat the problem of contamination caused by marine intrusion. The methodology has been applied to the hydrogeochemical data study of the Aguadulce aquifer (Campo de Dalías, SE Spain).

Campo de Dalías is an area situated in the SW sector of the Almeria province (SE Spain), between the slopes bassins of the Adra and Andarax rivers. The sector ocuppies an aproximated surface of 300 Km^2, bordered to the N by the Sierra de Gádor mountains and to the S by the sea. Morphologically, it presents a relatively smooth relief between 300 and 0 meters above sea level with a continuous slope to the sea.

Of the three aquifer systems (Pulido et al., 1987) that provide water to the area (fig.2) we have chosen for the study, the Aguadulce aquifer. The system is situated in the eastern region of the bassin. It is formed by dolomies, calcarenite and miocenic volcanic rocks, and pliocenic sandstone-marl episodies and quaternary materials. This is an unit of great tectonic complexity with a bad known geometry (multilayer formation with lateral facies changes), due to this, it exists a high variability of the hydraulic parameters.

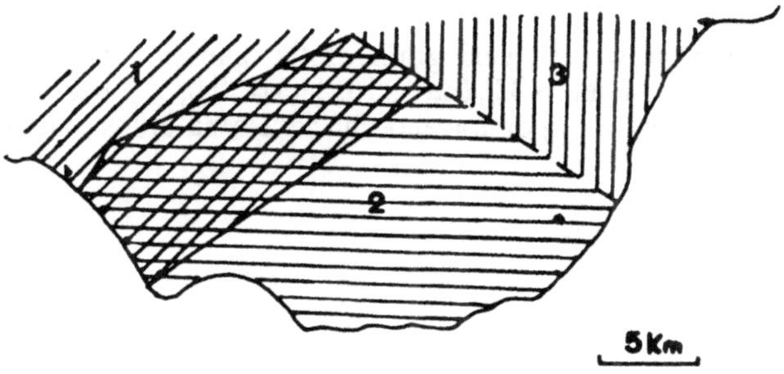

Fig. 2.- Hydrogeology of the Campo de Dalias area (Pulido et al., 1987). Legend 1: Balanegra aquifer, 2: Balerma-Las Marinas aquifer and 3: Aguadulce aquifer.

As we mentioned before, this system has been suffered an intensive overexploitation since the 60's, originating a strong decrease of the piezometric levels, producing marine intrusion processes (in 1980 there was a cone of -20 m depth and 3 Km radius).

For to make the spatial variability clear, several hydrogeochemical control campaigns have been developped, choosing for this example the corresponding with May/June 1986. In this campaign 97 points (fig.3) were sampled inside the system, the results indicate that 75% of the data is above the tolerance for human usage. Among the analysed variables, we chose the chloride content as water quality indicator due to the elevated correlation coeficients (> 0.95) with another parameters: electric conductivity, sodium, magnesium, lithium, among others.

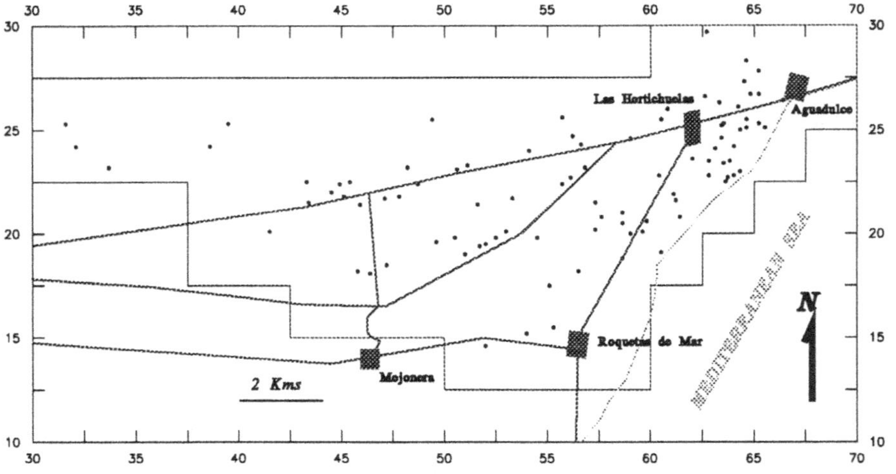

Fig. 3.- Map of localization of case study.

Figure 4 shows the variogram of the gaussian transformed values with a spherical function of parameters C_o (nugget effect)= 0.0, r (range)= 9 and s (sill) = 1.0, the fit has been perfomed by operator and checked by a cross validation process.

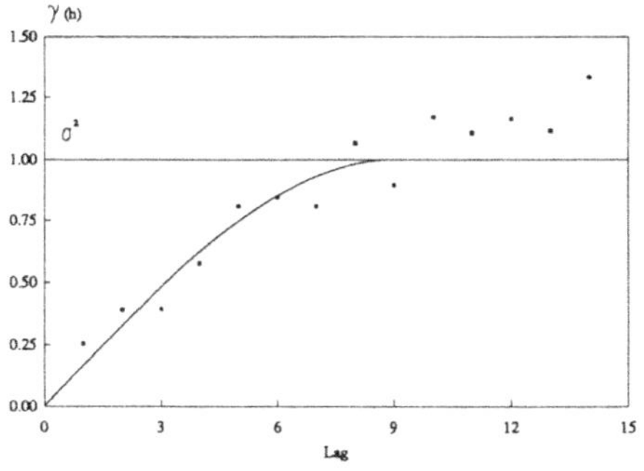

Fig. 4.- Gaussian variogram and fitted model.

The study consists in a DK estimation in blocks of size 1.25x1.25 Km^2, resulting 67 blocks inside the area of the study. The estimated value map is presented in fig. 5 and shows the existence of a high concentration sector where the admissible limits for agricultural use are exceeded in the eastern zone, coinciding with a strong exploitation area.

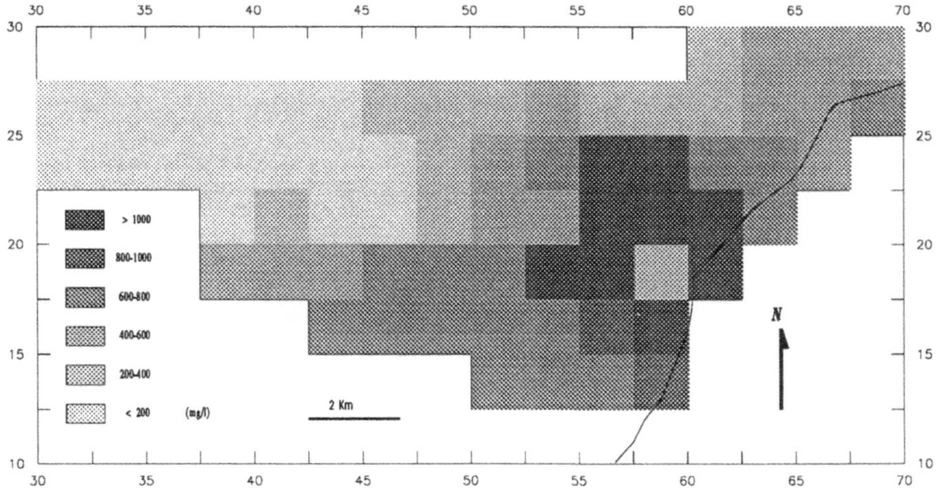

Fig. 5.- Estimated averaged block chloride values.

An important advantage of the method used is the possibility of the local conditional density function calcul (fig. 6) which makes it possible to obtain the conditional probability maps (fig. 7).

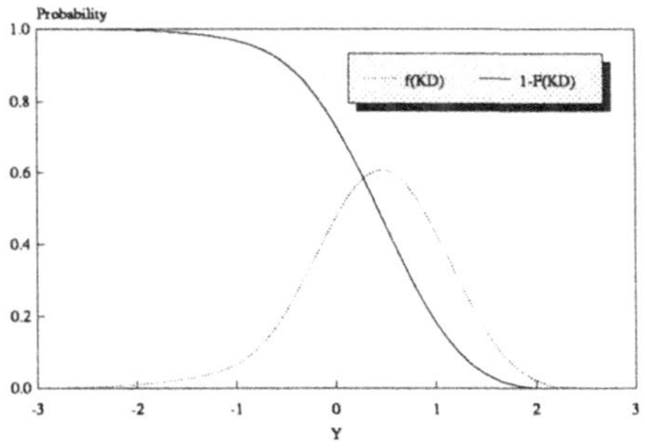

Fig.6.- Density and inverse distribution functions for a estimated block.

This approach, withouth a doubt, represents a new focus to the hydrogeochemical data analysis, making possible to establish the probability that a determinated point or sector exceeds a critical value marked by the plants tolerance to chloride. When the limit value increases, logically the probability decreases locating the maximum in the positive anomaly sector. The positives anomalies are present even in very high chloride contents (1000 mg/l).

Moreover, the inverse distribution function ($1\text{-}F_{KD}$) allows the obtention of maximum and minimum values between which the variable oscillates in a determinated block or panel when we establish a fixed probability interval. This calculation, direct consequence of the $1\text{-}F_{KD}$ analysis has, in my opinion, important applications in the

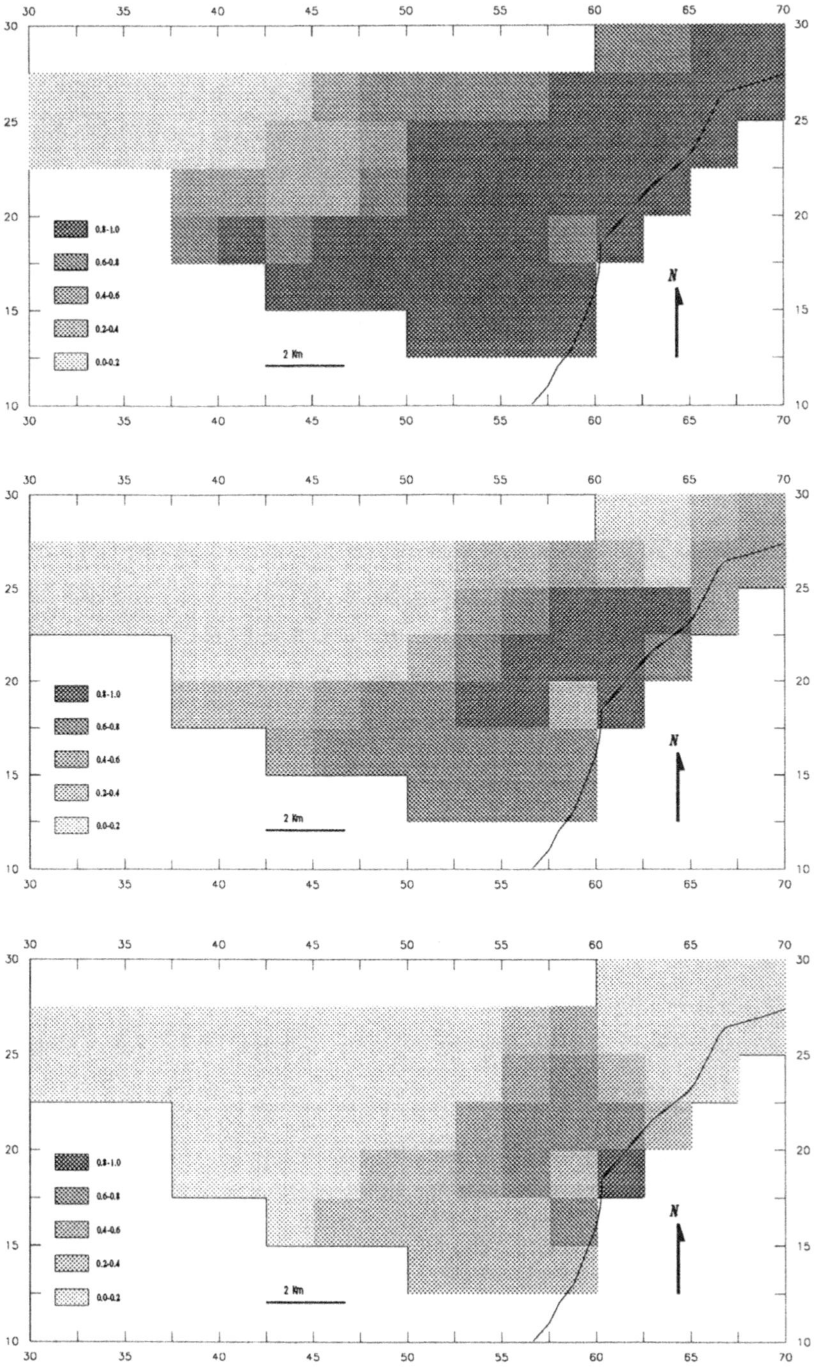

Fig.7 .- Conditional probability maps for 200, 600 and 1000 mg/l limits.

geological data treatment. It allows to add to the estimation information ("the most probable"), the one derived from the oscillation range in probability function. Figure 8 shows the obtained values for the 0.3 and 0.5 probability levels.

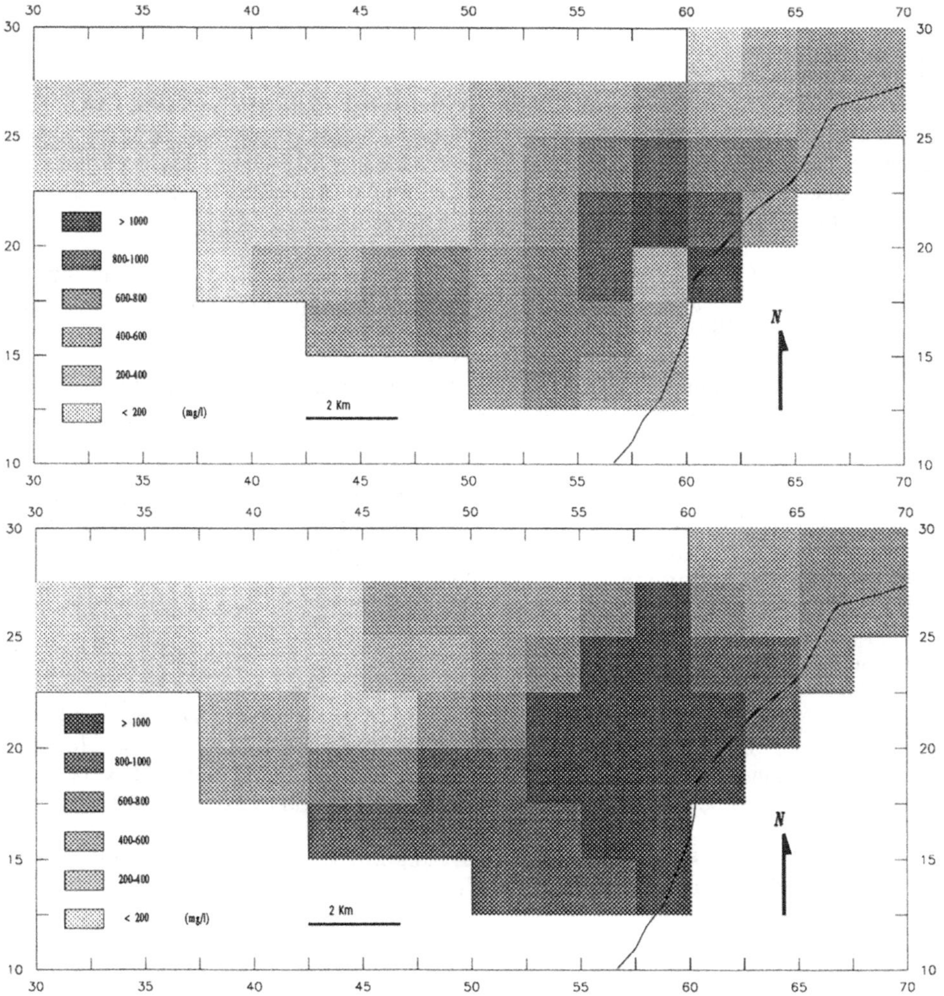

Fig. 8.- Obtained values maps for 0.3 and 0.5 probability levels.

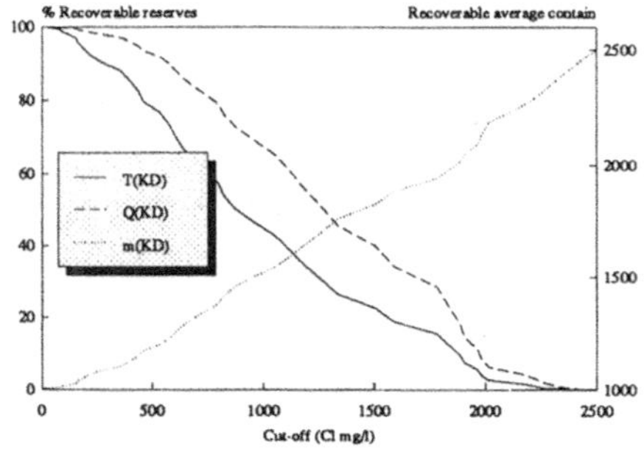

Fig. 9.- Transfer functions for a subsector of the aquifer system.

The seriousness of the problem has been shown with the transfer function estimation (fig.9) in a subsector of coordenates X: 52.5 to 60.0 and Y: 17.5 to 22.5. In this area, the most affected by the contamination, 30 % are above the 600 mg/l level and only 7 % are above the 200 mg/l one (human proper level).

4. Conclusions

The use of the non linear geostatistical methodology (DK) that has been extensively validated for solving estimation problems in several areas of the Earth Sciences, specially in the local recoverable reserves, it has been shown very efficient in the hydrogeochemical data treatment and the anomalies probabilistic analysis. This method presents many important advantages, among others:

1st) Characterization of the estimated value, both punctual and any desired support (area or block), in function of the experimental data and the regional variability structure of the phenomenon.

2nd) Obtaining the kriging error, an essential aspect to be able establish the confidence level of the estimated value. This error is minimum and even smaller that the one obtained by another linear geostatistical method (OK).

3rd) Calculation of the local probability distribution. This makes it possible to establish the probability that the estimated value exceeds a determinated limit value and inversely, we can know the upper and lower estimated values inside a probability range, that complete the information derived from the estimated values.

Acknowledgments

The data used in the practical example of this work have been taken of the results volume of the Collaboration Contract between the IARA and the University of Granada "Caracterización Hidrogeoquímica del Campo de Dalías (Almería)", we are grateful to Dr. Pulido-Bosch, responsable of it, for his collaboration and help.

The work in this paper was supported by the Dirección General de Investigación Científica y Técnica (DGICYT) under grant PB89-0015.

References

Alli,M.M.; Nowatzki,E.A. and Myers,D.E. (1990) "Probabilistic Analysis of Collapsing Soil by Indicator Kriging" Mathematical Geology, 22(1), 15-39.

Chica-Olmo,M. (1987) "Análisis Geoestadístico en el estudio de la explotación de los recursos minerales". Tesis Doctoral. Dpto. Geodinámica. Universidad de Granada. 387 p.

Journel,A. (1982). "Indicator approach to spatial distributions" Proc. 17th APCOM Simposium, Colorado, Golden, 793-806.

EVALUATING THE UNCERTAINTY OF LEAKAGE RATES OF TEJO AND SADO GROUNDWATER SYSTEM

LUÍS RIBEIRO

CVRM, IST - Av. Rovisco Pais, 1096 Lisboa Codex, Portugal

ABSTRACT : The Miopliocene groundwater system of Tejo and Sado, the most important of Portugal, consists of two main aquifers separated by an aquitard. The lower one is confined and contributes to urban and industrial water supply in a great percentage. In order to evaluate the role of these resources, both numerical and analytical models were implemented, since the 1980 Setúbal Project, to evaluate and predict the response of the groundwater system, in terms of variation of hydraulic heads or flow rates, subject to different pumped situations. Amongst the principal results of this project, is the importance of leakage mechanisms in the hydraulic characterization of the groundwater system. In order to evaluate the uncertainty of the leakage rates, a geostatistical approach based on conditional simulation techniques is used here to study the influence of the spatial variability of the transmissivity and the leakage parameter via the numerical model. A special emphasis is given to the uncertainty analisys of phenomena induced by leakage such as : contamination from the sea or from the estuary and capture from pumping, and to the analysis of dispersion of up and downward leakage rates in relation to the particular system boundary conditions. Both steady state and transient models are used and results are analysed in a global sense (probability distribution function outputs) and in a grid by grid sense (boxplots outputs) .

1. INTRODUCTION

From Freeze classical study (Freeze, 1975) to the first stochastic models with spatial correlation scale; from the spectral methods to the more sophisticated models to solve groundwater partial differential equations; from traditional Monte Carlo approach to Conditional Simulation theory, Stochastic Hydrogeology has seen the development of a lot of models and/or contributions for characterizing aquifer uncertainty.

Geostatistics plays an important role on the development of groundwater stochastic methods because it introduces relevant improvements on the definition of the spatial variability of the phenomenon.

Nowadays there is a growing need to look for methods which incorporate the maximum of available hydrogeological information in a stochastic-deterministic manner, in such a way that all the estimation and uncertainty calculations should be conditioned upon not only experimental data but also consistent to the physical laws that govern the groundwater flux. This is a natural consequence of the increased applications of such models on real groundwater planning projects .

Conditional Simulation (CS) geostatistical tool combined with groundwater numerical modelling presents all of these advantages for the analysis of hydraulic flow of the aquifer systems. This paper presents a study about the uncertainty of the leakage rates subject to the influence of the spatial variability of transmissivity (T) and leakage factor (TK) .

2. THE IMPORTANCE OF LEAKAGE MECHANISMS ON TEJO AND SADO GROUNDWATER SYSTEM

The groundwater system of Tejo and Sado covers an area of about 8,000 km^2 which includes, but extends beyond, the Setúbal Peninsula, as shown in fig. 1 .

The area corresponds to the Mio-Pliocene detrital sedimentary basin of the Lower Tejo and the Lower Sado. On the northwest, the basin is bounded by Mesozoic limestones and sandstones, and on the east and south by Paleozoic and Precambrian bedrock. On the west, the basin extends beyond the coastline under the Atlantic Ocean. The aquifer system is discussed in detail by Coma-Guillen,1978 . It consists of an unconfined and a confined aquifers separated by a leaky confining layer.

Most of the past and the current groundwater development is from the confined aquifer and is located in Setúbal Peninsula. Within the peninsula, the aquifer system forms a sub-basin. On the south, the confined aquifer outcrops along the northern edge of the Serra da Arrábida range and on the north, the confined subcrops the alluvial bottom sediments of the Tejo estuary. The aquifer increases in thickness towards the center of the peninsula where it is about 400 m thick. The overlying unconfined aquifer is about 50 m thick and is separated from the confined aquifer by a 20 m thick aquitard, providing a leaky hydraulic connection between those two. The unconfined aquifer is also hydraulically connected to the Tejo and Sado estuaries and to the Atlantic Ocean, receiving recharge from precipitation and it discharges into the local streams, the estuaries and the ocean; it exchanges water with the confined aquifer through up or downward leakage .

Preliminary computations using a analytical model, the Hantush and Jacob unsteady leaky aquifer equation, (cf. Papadopulos,1979) and applied to the 1972-78 period indicate that the confining layer is leaky with a relatively high vertical hydraulic conductivity of 5.6×10^{-9} m/s and a leakage factor of about 6,000 m. Furthermore, the analysis indicated that the system would reach a steady-state within a relatively short period of time, probably in the order of 1 to 2 years, if the pumping rates were to be maintained constant for an extended period of time.

For these reasons, the continued development of pumping rates from the deep aquifer will be balanced by an increase in downward (*drp*) leakage and a decrease, or even a reversal, in upward (*drn*) leakage .

From this hydrogeological context, two situations are feasible :

A - Capture : the downward leakage induced by pumpage .

B - Contamination : the saltwater intrusion from downward leakage (*drmar*) induced by pumping in areas where the cones of depression extend beneath the Tejo estuary or the ocean .

3. SOME PREVIOUS WORKS IN THE FIELD OF STOCHASTIC HYDROGEOLOGY

There are basically three different approaches in the field of stochastic hydrogeology: analytical, semianalytical and numerical .

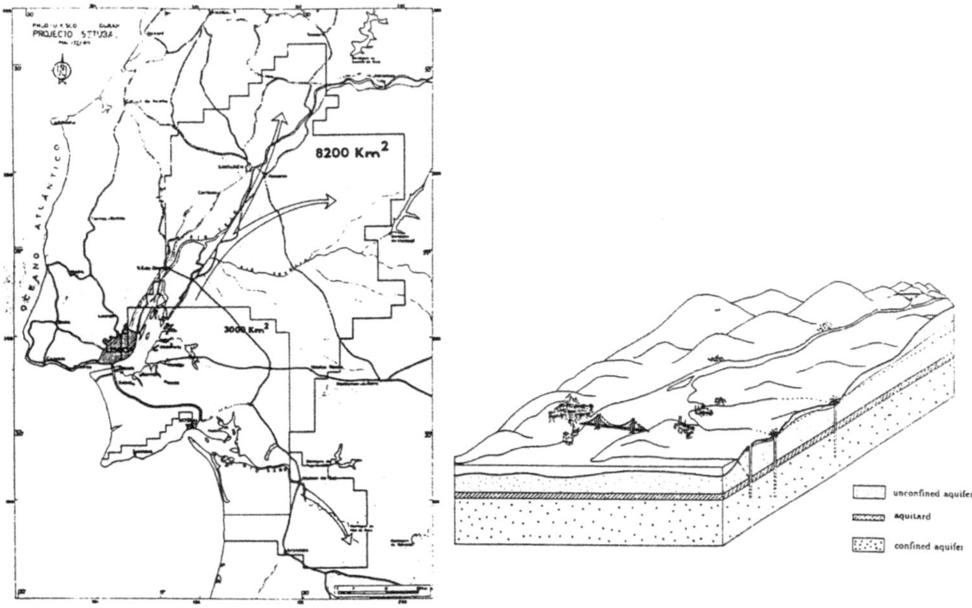

Fig. 1 - Plan view of the Tejo and Sado groundwater system focusing Setúbal Peninsula area and a 3D representation .

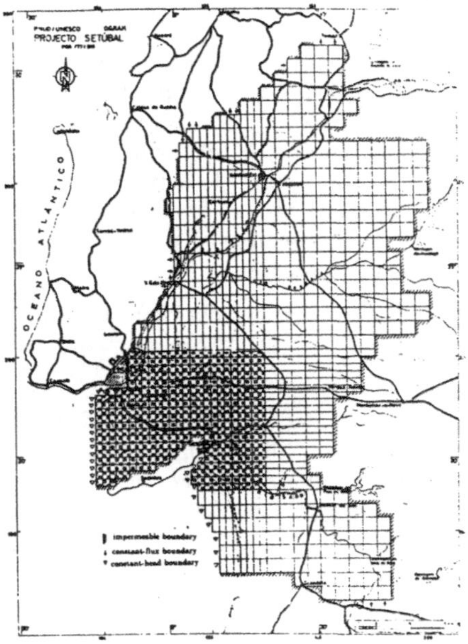

Fig. 2 - Finite difference grid used in flow simulations of the Tejo and Sado groundwater system, focusing the study area .

The analytical approach uses spectral or perturbation methods to solve the groundwater flow stochastic partial differential equation (PDE). Evaluation of uncertainty is restricted to hypothetical infinite aquifers or finite groundwater systems with very simple boundary conditions, treated as a continuum media, with a relative small variability of the parameters. The results are analytical expressions relating the two first moments (expected value and variance) of the variables such as heads (H) and T .

Gelhar(1976) gives the classic analytical result on head variances for steady flow. This moment is directly proportional to the mean hydraulic gradient, variance of the logarithm of transmissivity (LT) and its correlation scale .

Particularly important in this study, is also the expression derived by Gutjahr et al., 1978 , for flow through an aquitard. The equation relates the variance of the error of flow estimates with the coefficient of variation of the aquitard hydraulic resistivity, its correlation length and some geometric quantities .

The second approach uses perturbation methods, mixing analytical and numerical techniques. The moment estimation is carried out by a first order Taylor series approximations normally on a finite difference or on a finite element spatial domain. Although the computational effort is nearly always modest the truncation of higher order terms can generate large errors in the estimates.

Jones (1989) compared the solutions obtained by the semianalytical approach and the Monte Carlo one in a highly idealised confined aquifer, T homogeneous with an exponential spatial covariance structure and steady-state conditions. Pumping and no pumping scenarios were considered and the effects of change in the variance of LT, in its correlation scale and in the boundary heads on the head variance (σ^2_H) were analysed . The author concluded that the semianalytical method is not very robust because it underestimates σ^2_H in the vicinity of the well and it overestimates it outward.

Comparing to the last ones, Monte Carlo approach is the most powerful because the PDE is solved through groundwater numerical model runs by means of repeated simulations. This provides a complete relationship between the Probability Distribution Functions (PDFs) of variables and parameters. It is not restricted to specific expressions between the first moments. Therefore, the methodology can be applied to any complex aquifer system subject to any hydrogeological scenarios. The methodology can be improved by using the geostatistical approach in order to build a stochastic model conditioned upon measured values, creating images more consistent to the real data : The CS approach .

Delhomme,1979 was the first to use CS for characterizing the uncertainty of a T field and analysing its effect on predicted H values. Two case studies were studied: a highly idealised aquifer and a real one .

More recently Gómez-Hernández et al., 1989 used both non-CS and CS methods to inspect the influence of spatial variability of hydraulic condutivity, leakance and recharge properties of an hypothetical unconfined aquifer with steady-state conditions on effective groundwater model parameters, using 9 different case studies, exploring the magnitude of the correlation length and zonation .

4. EVALUATING THE UNCERTAINTY OF LEAKAGE RATES

4.1 DESCRIPTION OF THE DIFFERENT SCENARIOS

We intend to study the influence of the spatial variabilities of T of the confined aquifer and of TK of the aquitard on the up and downward leakage rates of the Tejo and Sado groundwater system, restricted to the Setúbal peninsula area and for specific pumping conditions. Two analyses are carried out : 1990 steady-state and 1961-78 transient .

Flow simulations employs the finite difference model developped by Trescott,1975, which was implemented and calibrated during the 1979 Setúbal Project (Projecto Setúbal, 1980). It consists of 1109 blocks with variable space gridding . Boundary conditions include impermeable, constant head and constant flow limits (see fig. 2) .

Groundwater model results were analysed in a global sense : total values of *drp, drn* and *drmar* obtained for the region are present in terms of PDFs and in a discretised sense; in terms of boxplots of the net block leakage rate .

4.2 GENERATION OF CONDITIONAL SIMULATIONS OF T

A preliminar geostatistical study of T (Ribeiro,1991) points out the existence of two independent subregions bounded by an important geological accident : The Setúbal-Alcochete graben, with higher T values in the E region than in the W region (see figs. 3) . This situation was taking account into the variogram calculations.

As we expect, all T values show a lognormal distribution. Therefore subsequent calculations were performed with LT (base *e*) values.

For West region the data consists of 37 T measured values and 351 specific capacity values. Statistics for LT in this area are m= -6.07 and σ^2=1.55, with original values in m^2/s . For East region the data consists of 34 T measured values and 166 specific capacity values. Statistics for LT in this area are m= -4.36 and σ^2=0.857, with original values in m^2/s .

The location of the points can be seen in figs 3a and 3b .

The LT experimental omnidireccional variograms were fitted with spherical models plus a nugget structure (see figs. 4 and 5) :

For each area 100 CS of LT were generated by the turning band method and conditioned upon all the available information. Two of these CS are shown in fig. 6 .

a)

b)

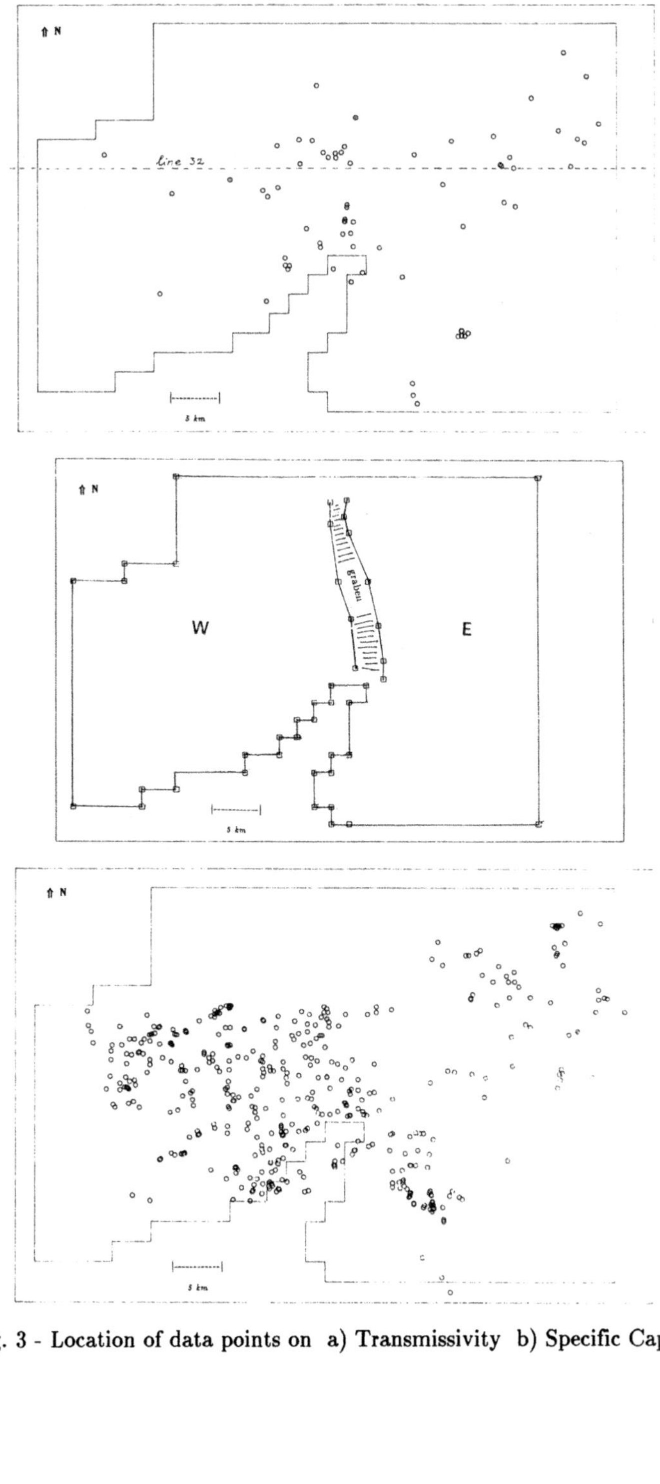

Fig. 3 - Location of data points on a) Transmissivity b) Specific Capacity .

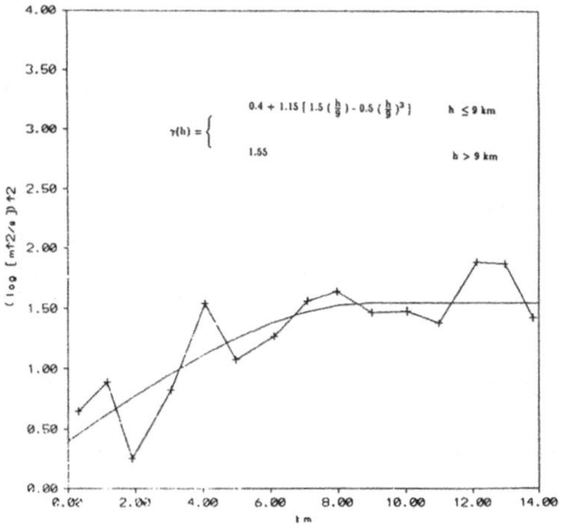

Fig. 4 - Experimental omnidireccional variogram for LT and spherical model for zone W.

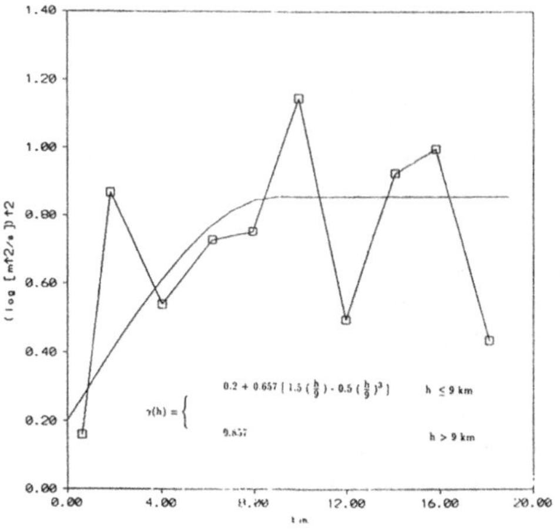

Fig. 5 - Experimental omnidireccional variogram for LT and spherical model for zone E.

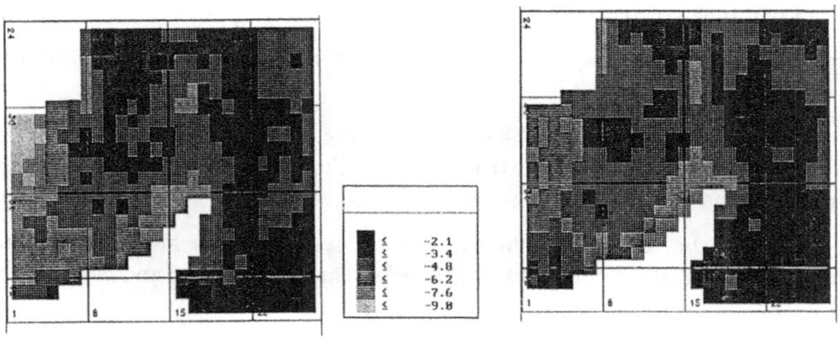

Fig. 6 - Two conditional simulations of LT (original values in m^2/s) .

4.3 GENERATION OF NON CONDITIONAL SIMULATIONS OF TK

Owing to the lack of available information about TK, this was assumed to be lognormally distributed (LTK) with spatial correlation described by a spherical variogram (sill=0.52 and range=9 km) plus a nugget effect (=0.17). The expected value of TK is 2.8 e-10 s^{-1} (from Papadopulos analysis) and its standard deviation 2.8 e-10 s^{-1}.

The Monte Carlo approach was used to generate another 100 non-CS of LTK .

5. ANALYSIS OF RESULTS

These CS and non-CS served as input for groundwater flow simulations. For assuring independence of results for both T and TK analysis, model runs were carried out separately. All the remaining parameters were left constant (results from the 1979 model calibration) .

No significant water-level declines were observed in the confined aquifer during 1990, in consequence the system was simulated as steady-state for that period of time. Boundary conditions were the ones specified in 4.1. Pumping wells in Setúbal Peninsula are mainly from industries and are located near estuaries and coastal areas and quantities pumped vary from 20 l/s to values upper 300 l/s (total amount by grid) .

Table 1 shows the statistics of global leakage results from 1990 steady-state groundwater flow simulations based on 100 CS of LT and 100 non-CS of LTK. Note that *md* is the median and *ch* is the relative dispersion coefficient and have the same signification on exploratory data analysis as the coefficient of variation.

We can draw the following conclusions :

i) - Values of *m* and *md* are very similar for all leakage terms, suggesting symmetric distributions .

ii) - There is a reduced dispersion for all the rates (cf. *ch* values) with the only exception of the upward terms on pumping conditions .

iii) - For downward leakage, dispersion decreases from a scenario without pumping to one with pumping and inversely for upward leakage .

The reason for the first result lies maybe on the fact that the volumes of water released from the upper aquifer contributed to an attenuation of the dispersion, because of their contribution to the modification of aquifers heads (self-correction effects) .

For the second result the reason is, that the upward leakage areas are mainly located outside the pumping areas, near the boundaries which affect head gradients and with no real measures to be conditioned .

iv) - As one expects, the process of conditioning decreases the uncertainty of groundwater model results (comparing the results of simulations using T or TK) . However for the upward leakage in pumping situation there is a slight increase. This is due to the lack of conditioning upon T data in the corresponding areas of upward leakage. For these reasons, high variances make some of the simulations go "wild" in this zone as well as go the dispersion of groundwater model results .

Note that the analysis provides a probabilistic quantitative evaluation of leakage rates: From fig. 7 we can calculate what is the probability that saltwater intrusion by leakage induced by pumping, exceeds some threshold value . For instance from that figure we conclude that there is a probability of 95%,with these pumping conditions, that total saltwater contamination exceeds 0.56 m^3/s.

Influence of the variability of on leakage rates

	[T] drpl	[T] drp2	[TK] drp3	[TK] drp4	[T] drn1	[T] drn2	[TK] drn3	[TK] drn4	[T] drmar2	[TK] drmar4
m	.448	4.026	.563	4.058	-.820	-.225	-1.027	-.327	.606	.586
σ	.026	.047	.048	.096	.0715	.0637	.095	.067	.026	.064
md	.448	4.029	.564	4.068	-.815	-.216	-1.03	-.328	.609	.590
ch(%)	7.59	1.56	12.4	3.46	10.6	31.2	14.2	29	5.75	10.9

** Values in m^2/s **

legend:

drpl and drp3 – downward leakage [without pumping]

drp2 and drp4 – downward leakage [with pumping]

drn1 and drn3 – upward leakage [without pumping]

drn2 and drn4 – upward leakage [with pumping]

drmar2 and drmar4 –downward leakage from sea and estuary [with pumping]

****

Table 1- Influence of the variability of T and TK on the leakage rates of Tejo and Sado groundwater system. Flow simulations results are based on CS of LT and non-CS of LTK .

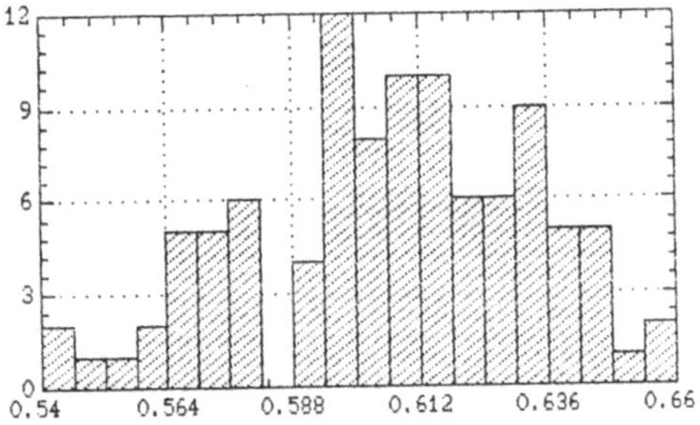

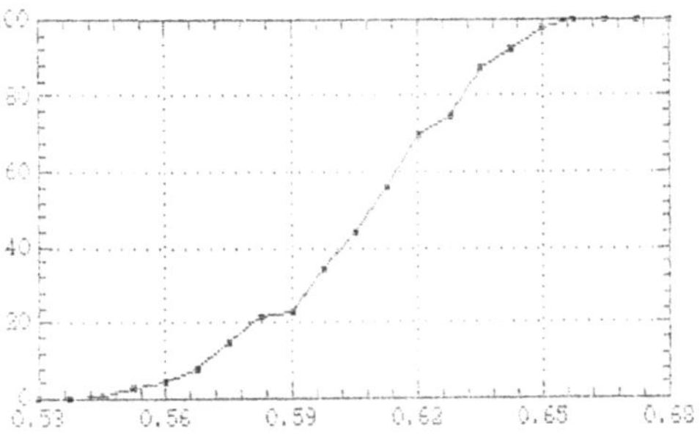

Fig. 7 - Histogram and Cumulative Probability Distribution Function of *drmar* resulted by 100 steady-state flow simulations based on 100 Conditional Simulations of LT (values in m^3/s) .

A grid by grid analysis provides cross sections of the variability along the columns or the lines of the numerical model. Figures 8 show the total pumping rates by grid and the uncertainty of the net leakage rates due to the spatial variability of T along the line 32 of model (see fig. 3a).

Besides the smoothing effect of the groundwater flow equations observed in this analysis we see that the uncertainty of the net leakage is bigger in areas of pumping wells and in zones near the boundary nodes (constant heads) .

This is due indirectly to the modification of the head gradients and of the head difference between upper and lower aquifers caused by the local hydrogeological situations which increases the variance of the piezometry (cf. the analytical expression derived by Gelhar) .

Finally, a transient state analysis was carried out based on the 1961-1978 historical pumping (see fig. 9c). The simulations use the same set of T and TK images as the ones used in steady state .

Figures 9a and 9b show the annual evolution of the dispersion of *drp* and *drn* due to the uncertainty of T . There are some conclusions to draw :

v) - The dispersion of *drp* and *drn* maintains more or less the same magnitude across the years.

vi) - Upward leakage dispersion is bigger than the other leakage components for the reasons pointed out in iv) .

6. CONCLUSIONS

Conditional Simulation approach coupled with groundwater numerical modelling was used to study the influence of the spatial variabilities of T and TK on the uncertainty of leakage flow rates across the aquitard of Tejo and Sado system .

The methodology shows considerable advantages over the classical methods based on a priori variances. This is due to the conditionning effect, that provides an uncertainty reduction of the model outputs being also consistent with the particular physical laws of groundwater motion .

The different behaviour of dispersion coefficients obtained for up and downward leakage rates is mainly related to hydrogeological factors that affect the head gradients, such as pumping and boundary conditions, the magnitude of the pumping rates and the amount of data needed for conditioning purposes .

This geostatistical-deterministic methodology is also a powerful tool for probabilistic resource evaluation and for the risk analysis of saltwater contamination in Tejo and Sado groundwater management programs .

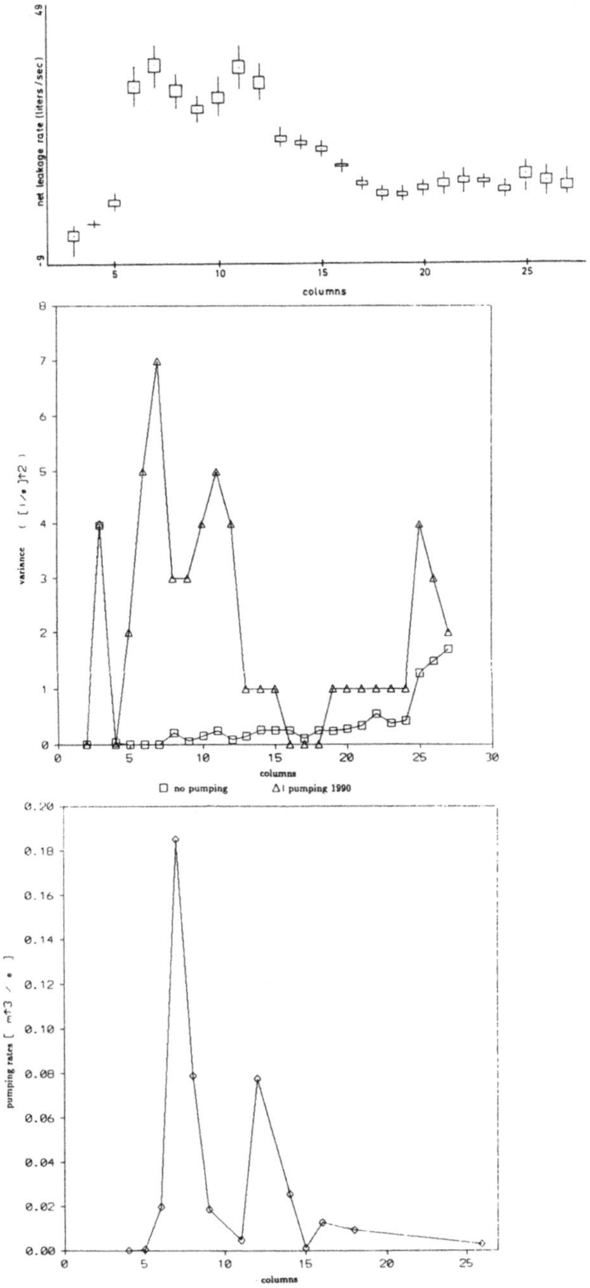

Fig. 8 - Cross-section along line 32 : a) Boxplots and b) Variance of net leakage rate result by 100 steady-state flow simulations based on 100 Conditional Simulations of LT; c) Pumping rates .

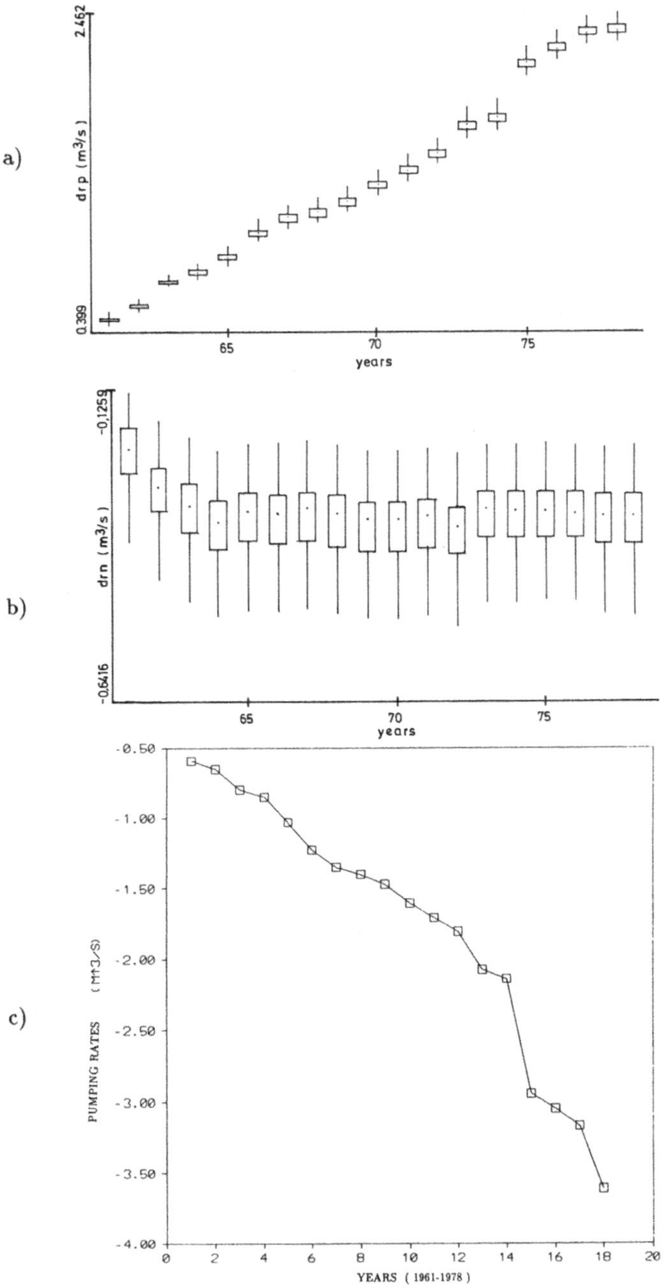

Fig. 9 - Analysis of the variability of a) *drp* and b) *drn,* resulted by 100 transient flow simulations based on 100 Conditional Simulations of LT, from 1961 to 1978 ; c) - Total pumping rates

7. AKNOWLEDGEMENT

The author is grateful to Jesús Carrera for his helpful comments and suggestions.

8. REFERENCES

Coma-Guillen, J.E. (1978) - "Estudio de los Recursos en Aguas Subterraneas de la Peninsula de Setúbal", *Projecto Setúbal, PNUD-UNESCO-DGRAH, internal report,* Lisboa .

Delhomme, J.P. (1979) - "Spatial Variability and Uncertainty in Groundwater Flow Parameters : A Geostatistical Approach", *Water Resour. Res., 15(2), 269-280 .*

Freeze, R.A. (1975) - "A Stochastic-Conceptual Analysis of One-Dimensional Groundwater Flow in Nonuniform Homogeneous Media", *Water Resour. Res., 11(5), 725-741 .*

Gelhar, L.W. (1976) - "Effects of Hydraulic Conductivity Variations on Groundwater Flows", *2nd International Symposium on Stochastic Hydraulics, Lund .*

Gutjahr, A. Gelhar, L.W., Bakr, A.L., MacMillan, J.R. (1978) - "Stochastic Analysis of Spatial Variability in Subsurface Flows - 2. Evaluation and Application", *Water Resour. Res., 14(5), 953-959 .*

Goméz-Hernández, J.J. e Gorelick, S.M. (1989) - "Effective Groundwater Model Parameter Values: Influence of Spatial Variability of Hydraulic Conductivity, Leakance and Recharge", *Water Resour. Res., 25(3), 405-419 .*

Jones L. (1989) - "Some Results Comparing Monte Carlo Simulation and First Order Taylor series approximation for Steady Groundwater Flow", *Stochastic Hydrol. Hydraul. 3, 179-190 .*

Papadopulos S. (1979) - "Preliminary Assessment of the Effects of Present and of Proposed Future Groundwater Development in the Setúbal Península", *Projecto Setúbal, PNUD-UNESCO-DGRAH,* Lisboa .

Projecto Setúbal (1980) - "Étude des Éaux Souterraines de la Peninsule de Setúbal", *PNUD-UNESCO-DGRAH, final report, 98p.,* Lisboa .

Ribeiro, L. (1991) - "Caracterização Geoestatística de Sistemas Hidrogeológicos", *Ph.D. thesis, Technical University of Lisbon-IST, 415p.,* Lisboa .

Trescott, P. (1975) - "Documentation of Finite-Difference Model for Simulation of Three-Dimensional Groundwater Flow", *U.S.G.S., Open File Report 75-438 .*

GEOSTATISTICS TO ASSIST HYDROGRAPHIC SURVEY DESIGN

Michel Dagbert[1],Peter Kielland[2],Manon L'Esperance[1],Andrew Cowan[3]

[1] : Geostat Systems International Inc. , Montreal , Canada
[2] : Canadian Hydrographic Survey , Ottawa , Canada
[3] : Dept. of Mathematics , Waterloo University , Canada

ABSTRACT : The Canadian Hydrographic Service has initiated a research project to investigate the application of geostatistics to the processing of bathymetric data . A promising field is the on-line variogram analysis of depth soundings along survey lines to characterize the roughness of bottom topography and determine areas where additional lines are required. More specifically , the hydrographer specifies a limit interpolation error which defines a confidence corridor around survey lines based on local variograms calculated in moving intervals along those lines. An extensive cross-validation of those confidence corridors has been conducted in various zones of mixed submarine topography. Results of that cross-validation have been used to adjust the methodology and parameters used : length of intervals along survey lines and overlap between intervals, trend model, local variogram model, relationship between limit error and distance from survey lines, search zone for depth interpolation in-between lines, pooling of local variograms, ordinary vs universal kriging. This methodology has been implemented in a PC based prototype program called HYDROSTAT.

INTRODUCTION

Hydrographers measure and chart the depth of inland or marine waters . Navigators use the hydrographic charts which must then be as precise or at least as safe as possible . Most hydrographic measurements are still being obtained by single-beam echo-sounding devices that record the depth below a vessel which follows a survey line. Survey lines are generally parallel with a fixed spacing of at least 50 or 100m . Along the survey lines , we have a continuous recording of depth hence the density of measurements is very high , typically 5 to 15 m depending of vessel speed and signal digitizing frequency .

With their mandate to produce safe nautical charts in mind , hydrographers have spent a great deal of efforts to control positioning and depth measurement errors along the survey lines (Kielland and Dagbert , 1991) . In the process , they may have forgotten another important aspect of navigation safety i.e. interpolation errors between lines . Those errors automatically creep in as soon as the hydrographer starts contouring depth from survey data . Even when an hydrographic map is produced with just the raw data and no contours , there is an implicit interpolation being made i.e. depth around a survey point is constant and equal to the depth of that survey point .

Depth interpolation errors are related to be spacing between the survey lines and the roughness of the bottom topography . In fact , the spacing should vary with the roughness: in zones of very flat topography , we can probably content ourselves with lines at 200 or even 400m whereas in a complex shoal area , we may need 50m or less between lines to keep interpolation errors within acceptable limits . Over the years, hydrographers have developed rules of thumb for deciding line spacing . It is generally tied directly to survey scale . For example , in Canada , survey line spacing in the field should result in 2 soundings/cm on the map .

Geostatistics seems to be the right tool to relate depth interpolation errors to survey line spacing and bottom roughness in an objective way . In fact , it has already been used with success in that discipline (Chiles and Chauvet (1974) ; David and Crozel (1986) ; Herzfeld (1989)) . In the work that we present here , we are trying to take advantage of the particular configuration of hydrographic survey data i.e. plenty of values along parallel and equally-spaced lines . Also , we are not that much interested in the interpolation and contouring from the survey lines but rather by the right prediction of interpolation errors and their variation with line spacing and local bottom topography .

A SIMPLE EXERCISE

The best way to introduce the subject is to look at some hydrographic data . On figure 1, we have a map of a 0.7 x 1.2 km survey area in the Pacific Ocean , off the coast of British Columbia . The zone is recognized by 4 N-S lines at 200m distance with an average spacing of 7m between survey points along lines .

Because of the high density of data along lines , the corresponding depth profiles are very well known . Then , it is possible to manually connect those profiles by contour lines which indicate a somewhat E-W hilly structure superimposed on a smooth regional slope to the west . As we mentioned above , as soon as we draw contours between the survey lines, we are doing some interpolation from known values which means that an error is attached to each interpolated value . The hydrographer who is supposed to provide a "safe" nautical chart from those data would obviously ask the question : how large can this error be? In more practical terms , could it be that a predicted depth of 35m turns out to be a mere 5m, hence a potential hazard for navigators .

In the particular case of figure 1, since we had in fact survey lines every 100m in that zone, we can actually see the magnitude of interpolation errors at about half distance from lines (i.e. where they are supposed to be the worst). Now if we confront the actual depths on those three lines to the interpolated depths from the contours, we can calculate and plot profiles of experimental errors (Figure 2) . Best line is the third one to the right (bottom one on figure 2): the profile of interpolated values closely follows that of real values except for an isolated peak which could not be inferred from the adjacent lines. The south part of the first line (top profile on figure 2), is fine but all depth values from the north part of that line have been underestimated. Worst line is probably the second one, right in the center and just going through the middle of the hill . At that place, the topography is more complex than anticipated

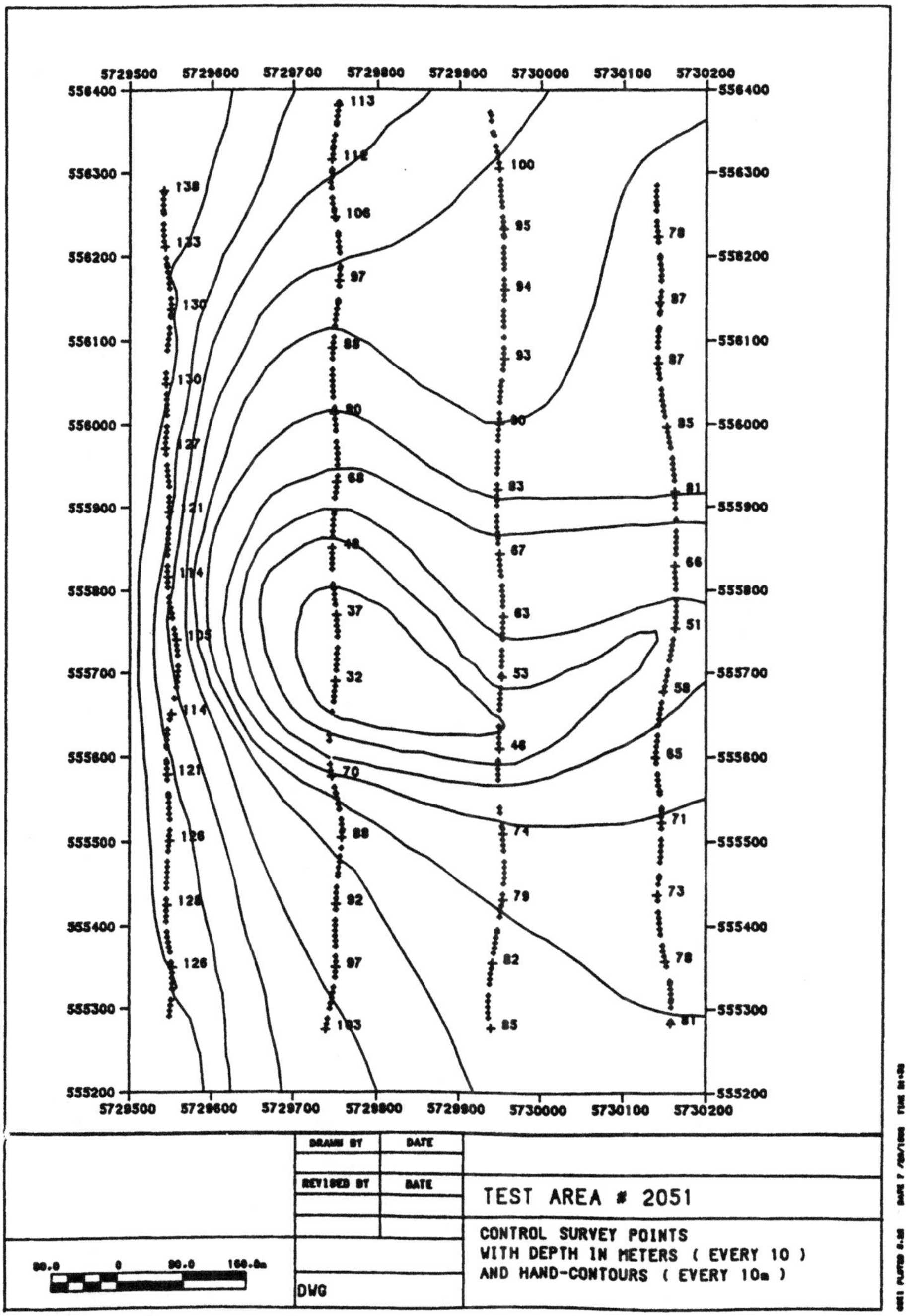

Figure 1 : Map with original survey lines , depth values and contours every 10 m

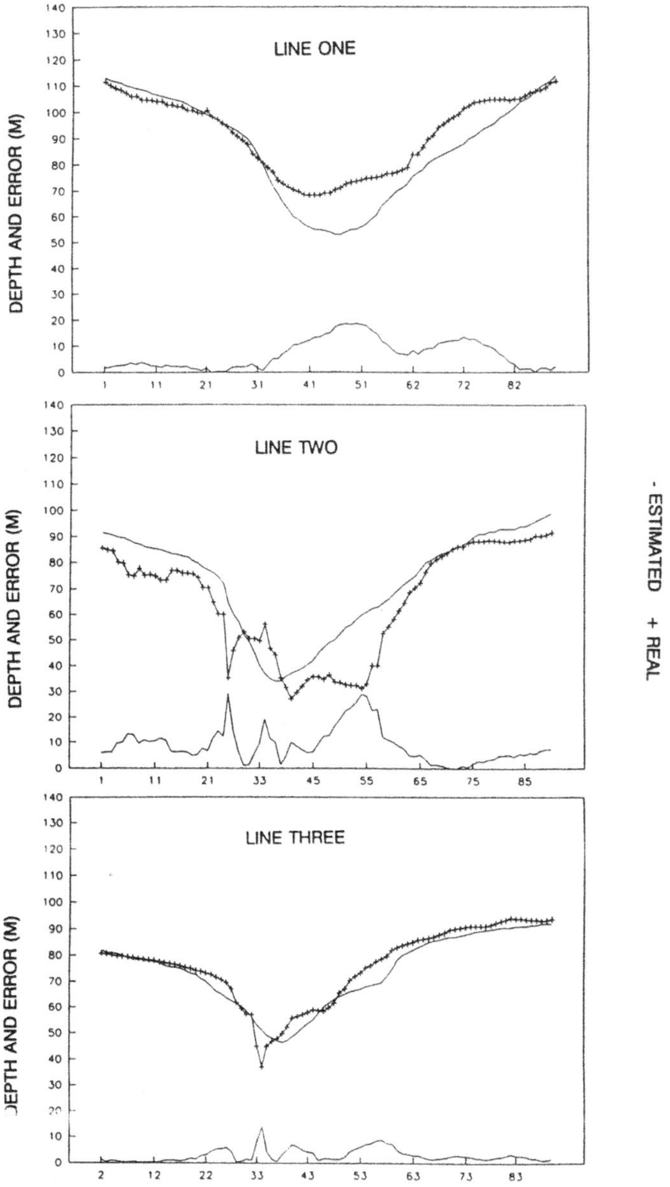

Figure 2 : Profiles of actual and interpolated depth and interpolation errors on the three new lines from the adjacent lines : instead of the single hill , we now see two of them with an intermediate valley and slopes are much steeper . As a result , we have many interpolation errors exceeding 10m . The histogram of experimental errors is rather symmetric with a mean close to zero but a standard deviation of 8.5 m . A common feature of all those diagrams is that we tend to have larger interpolation errors over the slopes and top of the hill than in the north and south flanks . In other words , interpolation is more difficult in a rough area than in a smooth area .

A STANDARD APPROACH TO ERROR PREDICTION

Evaluating the magnitude of interpolation errors from control points has been the *raison d'être* of geostatistics since its beginnings . Faced with that problem , the geostatistician would typically model the variation of depth values with distance using a variogram or a covariance and then use that variogram or covariance to krige the unknown depth at grid points in-between the survey lines . A mapping of kriging standard errors should provide the answer to the question of the hydrographer .

Back to our original example with 541 survey points on the 4 N-S lines , because of the high density of control points along the lines , we don't have any problem to model a N-S variogram . However , since we just have 4 lines , the recognition and modelling of a potential anisotropy along other directions is more difficult .

A second difficulty arising from the particular configuration of control points is the search for neighbors in the ordinary kriging of depth at grid points . If we are to keep all the control points but , at the same time , do the kriging with a reasonable number of neighbors (say less than 50) , we have to limit the search zone to an ellips with a long radius perpendicular to lines and equal to about the spacing between lines (200m) and a short radius parallel to lines and equal to only a fraction of the line spacing (30m). With that set-up , the kriged depth estimate halfway between lines is merely the average of depth values on the two adjacent lines and at the same latitude i.e. it is very close to the manual contouring estimate of Figure 1 .

If we contour the kriging standard error , we end up with a typical "finger" diagram (Figure 3) with a vitually zero error on the lines and a maximum error of 8m half-way between lines (contours have not been labelled for clarity purposes but the general trend is obvious) . That diagram is typical of error maps with control points on lines when standard geostatistics is used to infer the magnitude of interpolation errors between control points (see examples in David and Crozel (1986) and Oliver and Webster (1990)) . Overall , the average standard error on interlines, 7.9m, is in accordance with the standard deviation of experimental errors from manual contouring i.e. 8.5m.

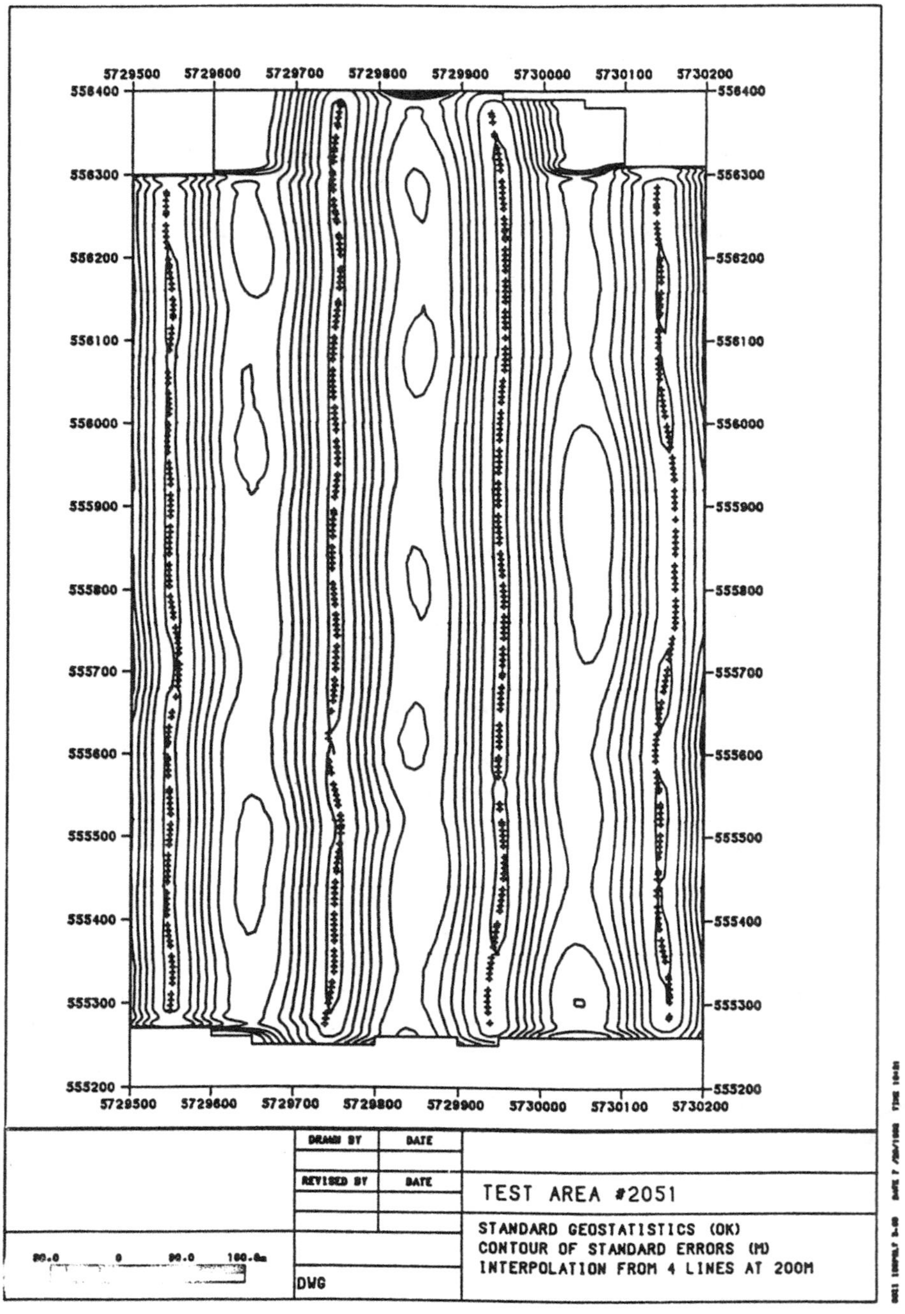

Figure 3 : Standard geostatistics : kriging standard error with contours every 1 m

LOCAL GEOSTATISTICS WITH DATA ON LINES

Even if the overall statistics of predicted errors that we get from the application of standard geostatistics are fine , what was worrying hydrographers at the C.H.S. was that we were predicting the same error in smooth and rough areas , in other words, the interpolation error was just a function of distance to survey lines . Then , they kept asking about some "local" geostatistics where the depth variogram or covariance would be redefined in each small area , just to reflect the local morphology of the sea bottom . Previous geostatistical studies in other fields (Isaaks and Srivastava , 1988) had indicated that , control data permitting , local variograms or covariances were doing a better job in error prediction than a single average model over the entire field of study . In fact , bathymetric survey data , with their high concentration along lines, seem particularly appropriate to some form of "local" analysis , although we just get insight into variations along a single direction , that of the lines .

Intervals for variogram computation

Variograms are computed in sliding intervals along the survey lines . In defining the length of those intervals , consideration must be given to 1) the average number of survey points in the interval : it is hard to derive a meaningful variogram with less than 20 or even 30 points 2) the spacing between lines : variograms will be used to derive interpolation errors between lines . Then , keeping in mind the old rule-of-thumb that variograms are useable only for distances of less than half the length of intervals, we normally should use intervals at least one line spacing long . If we combine those two rules , in our example with lines at 200m and survey points every 7m , we would use intervals at least 200m long since we would get about 30 points in each of them . Obviously, the longer the interval , the less "local" is the resulting variogram (but the better defined it is) . Now , to have a smooth transition from one interval to the next along the same line, it seems that using overlapping intervals is a good practice . Practically , we tend to use a degree of overlapping of 2/3 i.e., with intervals of 30 points , we move by only 10 points each time .

To trend or not to trend

As we restrict data to fairly small intervals , we tend to see fairly strong trends on the slopes of the hills and valleys of our survey area . Then the question arises as to whether we should not "de-trend" the depth values of our interval before we compute a variogram or a covariance for those values . Obviously , the degree of the local trend model that we fit on our interval data has a significant impact on the magnitude of the variogram values . Cross-validation exercises with decimated line data tend to show that any de-trending generates variograms which are too low i.e. predicted interpolation errors from those variograms are optimistic compared to the actual errors . In fact , the overall variogram which we calculated in the previous section without any "de-trending" was all right in the prediction of the average interpolation error .

1-D variogram model

Once we have calculated the experimental 1-D variogram from data in an interval , we have to model it in order to use it for prediction of interpolation errors or kriging between lines . Since , for a standard survey area with a dozen of lines , we end up with several hundreds of overlapping intervals , we can't afford to look at each of them and manually or manumatically model it . We use an automatic procedure with a model equation of fixed type . Again , based on experimental work with hydrographic data of various origin , we found that the power model : $V(h)=ah^b$ accomodates most of the experimental variograms of interval depth values . The two parameters of each model , a and b , are found by log-linear weighted least-squares on the experimental variogram values with distance values (weight is the number of pairs). This modelling is limited to the growing part of the experimental curve i.e. for short distances .

Error with distance from lines

Once a variogram has been modelled in an interval along a line , **assuming that depth variations around that interval are isotropic** , it is possible to calculate the standard error for the interpolated depth at any point between the lines . Obviously this standard error is maximum at the halfway distance between the two lines . Hence , we have simulated the kriging of a point halfway between two parallel lines and using a linear variogram .If there are continuous control data on the two lines , we have found that 1) there is no advantage in using control points on lines beyond a distance equal to half the distance between the lines (kriging weights of those points are almost nil) 2) the kriging variance is just 12% less than what it was with only two points i.e. 88% of the variogram value for a distance equal to the half spacing between the lines . This rule seems to stand for variograms which are not strictly linear .
We can use this relationship between interpolation error variance and spacing between lines the other way around : given a limit variance , what is the spacing between lines to achieve that variance where it is maximum , i.e. halfway between lines . More practically , instead of a variance , the hydrographer defines a limit interpolation error at a given confidence level (e.g. 95% , 90% or 80% confidence) which is then converted into a variance using a normality assumption . For example, if our limit interpolation error at a 95% confidence is 1m and our "local" variogram is $V(h) = 0.005h^{1.2}$, we need a spacing between lines d such that :

$$(1.0/2)^2 = 0.88 \times 0.005 \times (0.5d)^{1.2} \quad \text{or } d = 58\text{m}$$

Hence, given a limit interpolation error, we can relate it to a maximum spacing between lines for each local variogram along the lines . One way to map those maximum spacing values is by drawing corridors around lines with a width equal to the maximum spacing . Any point within the corridors would be interpolated with an error less than the limit error at the selected confidence level . Any gap between corridors around adjacent lines indicate a zone where the limit error is likely to be exceeded .

Figure 4 is a confidence corridor map for our example data with 4 survey lines at 200m distance . It corresponds to a limit error of 10m at a 95% confidence level (this number may seem pretty large but it is of the same order of magnitude as the experimental errors that we occasionally had on the interlines) . We clearly see that we get narrow corridors on the slopes and top of the center hill . Given the limit error, the 200m spacing is adequate in the north and south part of the area .

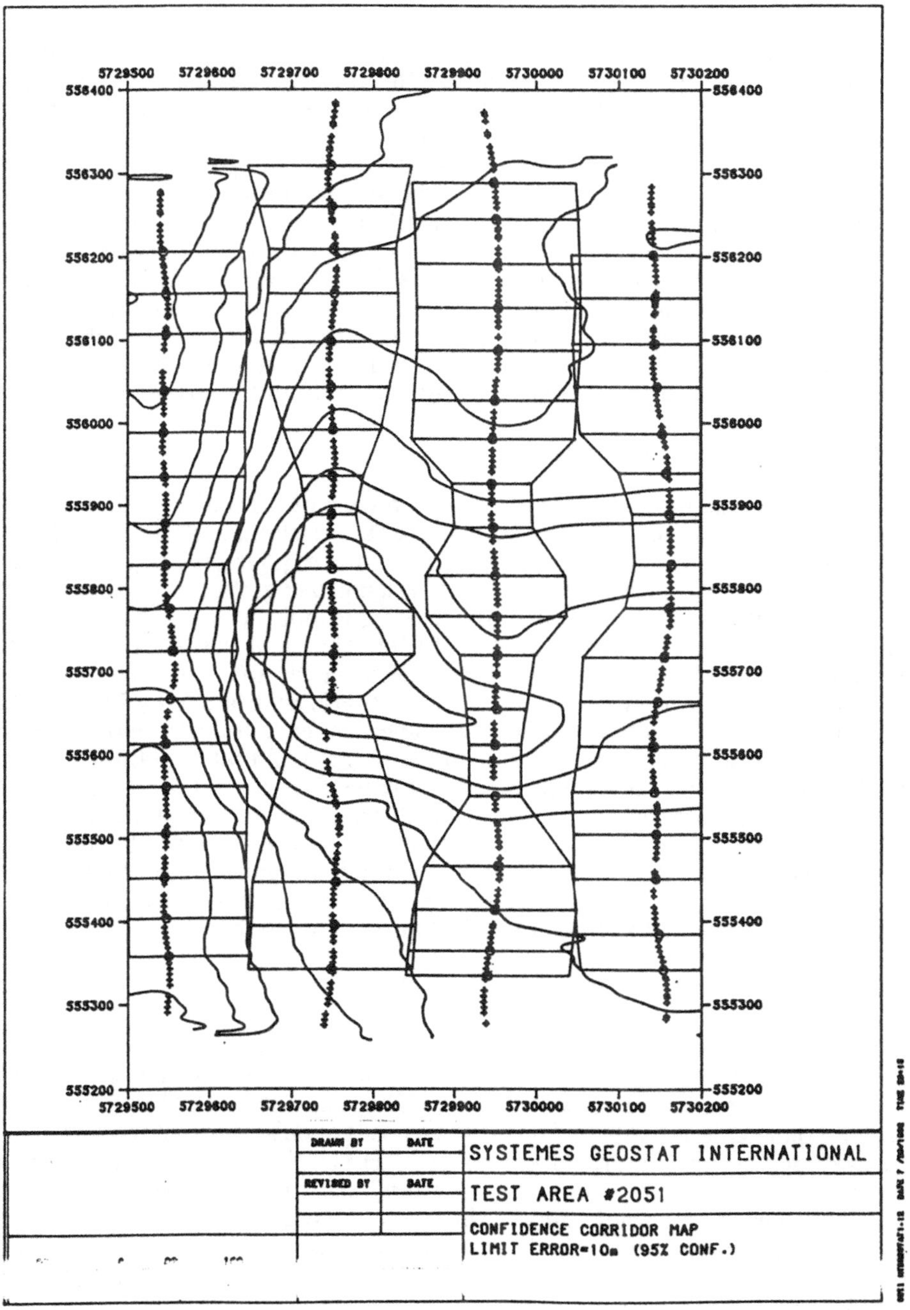

Figure 4 : Confidence corridor map for a 95% confidence limit interpolation error of 10m

2D variogram interpolation

So far , our analysis has been unidimensional i.e. we just have local variograms for specific intervals along the survey lines . In order to conduct a full 2D interpolation between survey lines , we need extend our set of 1D models into a mosaic of 2D models . As explained above , the 2D extension of the 1D variograms is simply done by assuming that local variations are isotropic . We can get intermediate variograms between lines by just interpolating our two variogram model parameters a and b on a regular grid covering the survey area. Right now , this interpolation is done by simple inverse-distance . Obviously this concept of smoothly varying variogram leads to theoretical inconsistencies but as long as the same variogram is used in any kriging system , kriging results are numerically consistent . A final check of the interpolated variograms is done by re-estimating selected control points on lines with control values on adjacent lines . The ratio of variance of actual errors to average predicted error variance is used as a correction factor for all local and interpolated variograms .

Final kriging and error mapping

Once variograms have been interpolated in cells of a regular grid , we can proceed with the last step of our analysis i.e. the interpolation of depth values and the derivation of interpolation standard errors on a regular grid over the survey area . This interpolation is done by ordinary kriging using the variogram power model previously interpolated in the cell where the grid point is located . At that stage , we still have to face the problem of how to find the right control points to use in the kriging when we have a high density of them on distant lines . Our solution to this problem has been to define a regular search grid with a survey line roughly at the center of each column of that grid . Whenever we have to krige depth at a grid point, we pick at random two survey points in the search cell that contains the grid point as well as in the eight surrounding cells . Figure 5 is a contour map of the kriging standard errors obtained by this procedure. Like in the standard geostatistical approach (Figure 3) , we see some fingering of the contours between the survey lines since, like before, the interpolation error is virtually nul on those survey lines . However, we clearly see an increase of standard errors in the central part of the area i.e. over the slopes and top of the hill that we have there .

The HYDROSTAT program

The methodology that we just described has been coded into an interactive graphics PC-based program called HYDROSTAT . This program is designed to process ASCII files with control values on survey lines . It shows the control points on the screen with a color scale for values. Once the user has specified an interval and skip length as well as a limit interpolation error , HYDROSTAT calculates and models variograms in the sliding intervals along the lines and then produces the confidence corridor map for the whole area . Alternatively , the user can pick any interval on a line and have the depth profile in that interval plus the variogram of that profile with fitted model on the screen . The same program also offers interactive cross-validation, gridding (with kriging) and contouring functions .

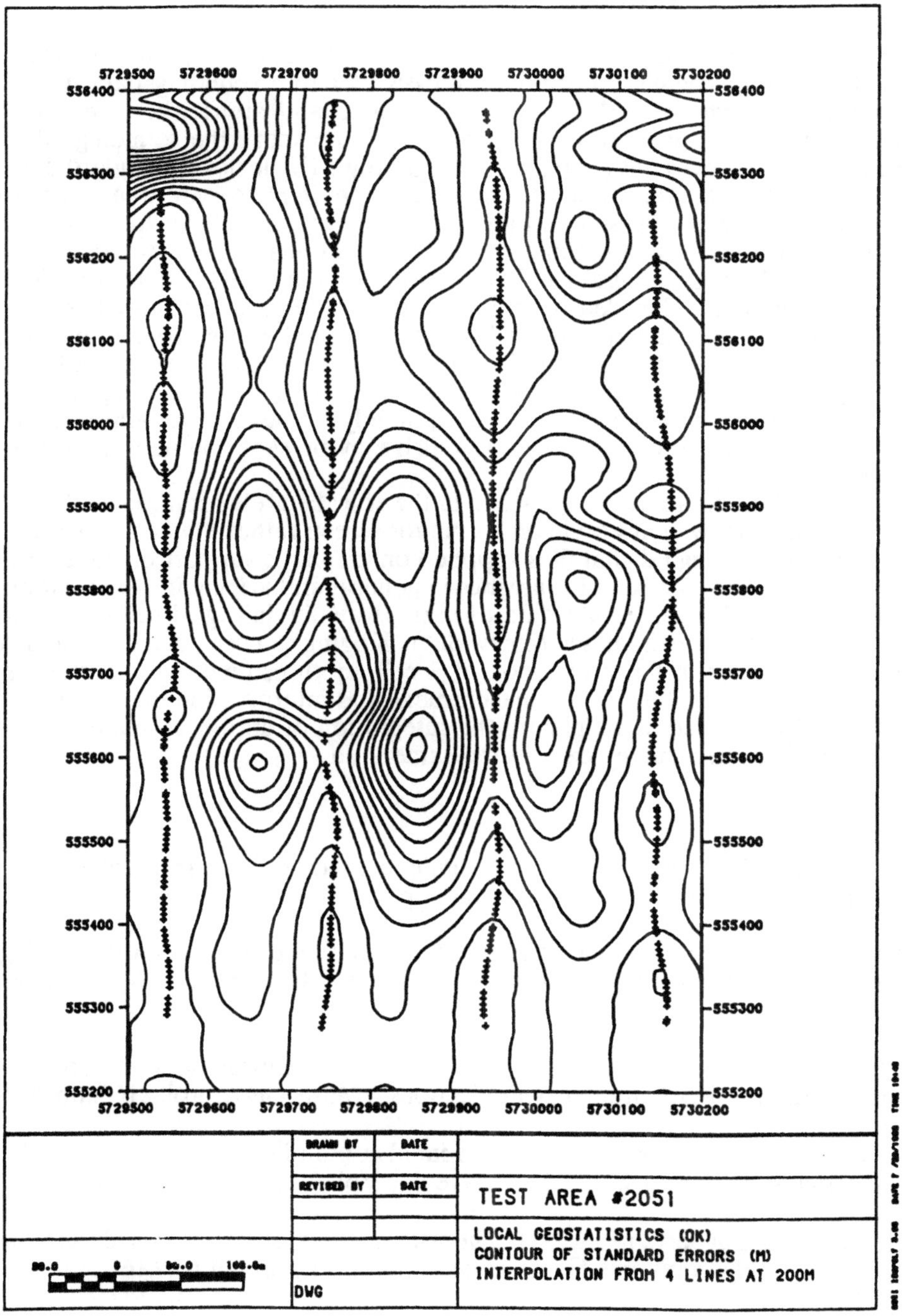

Figure 5 : Local geostatistics : kriging standard error with contours every 1 m

CONCLUSIONS

At this time , the main purpose of the HYDROSTAT program and underlying methodology is to conduct an on-line analysis of depth sounding data in order to determine areas where line coverage is adequate and areas where interlines are needed . Of course , the idea is to optimize , on the spot , the sounding effort and concentrate it in places with roughest topography . Right now , standard surveying procedures call for the same line spacing everywhere in the survey area . Once survey is completed , data is analysed and any dubious zone might have to be re-surveyed in more details which means time delays and additional costs . Another potential application of the same program and methodology is the analysis of historical data for project priorization . The idea is to detect zones of high uncertainty based on the spacing of the original survey lines and the roughness of bottom topography. Finally, it is conceivable to use the predicted standard interpolation errors derived from kriging to produce "safety-biased" nautical charts . In fact, cartographers who currently prepare those charts have a definitely conservative approach in the interpretation of shoals or rugged areas which may present potential hazards to the navigation .

The future of hydrographic mapping lies in the so-called "electronic charts" i.e. maps that show any window at any scale on a monitor screen rather than on paper and that can be updated through digital files supplied on magnetic support or transmitted on phone lines . In this context , geostatistics in general and kriging in particular appear like promising tools to integrate data of various quality into grid cell estimates and update those grid cell estimates with new data . Kriging can use positioning and depth measurement errors in the production of a cell depth estimate with corresponding error which would combine both measurement and interpolation errors . Those cell errors can also be used as weights in the integration of an old estimate with a new one derived from new survey data in the same area.

REFERENCES

Chiles, J.P. and Chauvet, P. (1974) " Kriging: a method for cartography of the sea floor", International Hydrographic Review, pp. 25-41

David, M. and Crozel, D. (1986) "Automated mapping of the ocean floor using the theory of intrinsic functions of the order k", Marine Geophysical Research, Vol. 8, pp. 49-74

Herzfeld, U.C. (1989) " Variability of submarine morphology : problems of deregularization and cartographical implications ", Math. Geol., Vol. 21(7),pp. 693-713

Isaaks E.H. and Srivastava , M. (1988) " Spatial continuity measures for probabilistic and deterministics geostatistics" , Math. Geol. , Vol. 20 , No 4 , pp. 313-342

Kielland, P. and Dagbert , M. (1991) "The use of spatial statistics in hydrography", in Proc. Canadian Hydrographic Conference , Rimouski , pp. 29.1-29.16

Oliver, M.A. and Webster, R. (1990) " Kriging : a method of interpolation for geographical information systems ", Int. Journal Geographical Information Systems, Vol. 4, No 3, pp. 313-332

ESTIMATION OF SPATIAL CROSS-COVARIANCES BY MAXIMUM LIKELIHOOD CROSS-COVALIDATION: APPLICATION TO HYDRAULIC HEADS AND TRANSMISSIVITIES IN A SANDY AQUIFER

J. SAMPER, J.C. CUCHI and R. PONCELA

Universidad Politécnica de Cataluña
C/ Gran Capitán s/n
08034, Barcelona, Spain

ABSTRACT

The proper characterization of the spatial cross-covariance of hydraulic head (h) and transmissity (T) is a critical issue in Stochastic Subsurface Hydrology. Existing methods to construct such cross-covariance often rely on the partial differential flow equation or on its numerical solution. Direct estimation from h and T raw data has received less attention because of the paucity of T data and the lack of a well established method for estimating cross-covariances. The common practice of using the same type of models for cross-covariances as for autocovariances (i.e. exponential, spherical, ...) does not guarantee positive definiteness. A way to overcome this problem consists of modeling the cross-covariance of Z_1 and Z_2 as a function of the covariances of Z_1, Z_2 and that of a linear combination of Z_1 and Z_2. These single covariances are then estimated using methods such as cross-validation. We present a maximum likelihood cross-covalidation (MLCC) method that overcomes most limitations of existing methods. MLCC has been applied to synthetic data as well as to hydraulic heads and log transmissivities in a sandy aquifer near the Doñana National Park in Southern Spain. Both variables exhibit spatial drifts which are filtered by a generalized least squares method.

1. Introduction

Many hydrogeological variables show spatial cross-correlation. Examples of such variables are hydraulic head and transmissivity, specific capacity and transmissivity, water level and topographic elevation, the concentration of dissolved chemical species, etc. The application of cokriging to these cases has shown to be of great interest (Kitanidis and Vomvoris, 1983; Ahmed and De Marsily, 1989; Hoeksema et al., 1989). In dealing with correlated variables, one is faced with the problem of estimating their cross-covariance. Existing methods for the estimation of cross-covariances suffer severe limitations. The common practice of using for cross-covariances the same type of models used for covariances (i.e. exponential, spherical, ...) does not guarantee positive definiteness. A way to overcome this problem consists of modeling the cross-covariance of two variables Z_1 and Z_2 as a function of the autocovariances of Z_1, Z_2, and that of a linear combination of Z_1 and Z_2. These autocovariances are then estimated using methods such as cross-validation. Cross-validation of the cross-covariance, however, is complicated by the fact that multiple measures of goodness can be defined. In this paper we present a maximum likelihood cross-covalidation (MLCC) method that overcomes most limitations of existing methods. MLCC is a natural extension of the maximum likelihood cross-validation concept of Samper and Neuman (1989 a) to the case of several correlated variables. MLCC allows one to use model identification criteria for the optimum selection of autocovariance and cross-covariance models. In addition, it can provide confidence regions for their parameters.

2. Cross-validation for a single variable

Cross-validation is a technique used for validating semivariogram models, although it has also been used in the context of automatic parameter estimation (Bastin and Gevers, 1985; and Samper, 1986). Let $Z(\mathbf{x})$ be a second order stationary random function with autocovariance $C(\mathbf{h})$ or semivariogram $\gamma(\mathbf{h})$ and let $Z_1, ..., Z_N$ be the values of $Z(\mathbf{x})$ at N measurement points $\mathbf{x}_1, ...\mathbf{x}_N$. Cross-validation consists of obtaining a kriged estimate, $\hat{Z}_i$, of Z_i on the basis of data at points other than $\mathbf{x}_i$, according to

$$\hat{Z}_i = \sum_{m \in N_i} \lambda_i^m Z_m \tag{1}$$

where N_i is the subset of points used for estimating Z_i and λ_i^m are the kriging coefficients obtained by solving the kriging equations.

Cross-validation can be performed using global or local neighbourhoods. The properties of each type are reviewed by Samper (1986). In local cross-validation (Bastin and Gevers, 1985; Samper and Neuman, 1989a) the subset of kriging points N_i changes from point to point and so does the kriging matrix. As shown by Dubrule (1983), with global cross-validation the kriging matrix is the same for all points and needs to be inverted only once. This may result in a substantial saving of computer time. Dubrule considered only cross-validation of a single variable. Later we will demonstrate how Dubrule's result can be extended to the multivariate case for which global cross-validation outperforms significantly its local equivalent. Dubrule's proof for a single variable is based on a duality property of the kriging equations. Our proof is based on a simpler and more intuitive property of the kriging equations that makes it easier to extend it to the multivariate case.

It is well known that in the absence of measurement errors, kriging is an exact interpolator. This means that $\hat{Z}_i = Z_i$ if point i is used for estimating Z_i. Therefore, if all the N points are used for the estimation of Z_i in (1), the trivial solution will be obtained. This, however, can be avoided by explicitly imposing that the weight λ_i^i given to Z_i be equal to zero. The corresponding kriging equations are obtained in the usual way by requiring unbiasedness and minimum variance of the estimate. By so doing one obtains

$$\sum_m C_{nm}\lambda_i^m + \mu_i + \varepsilon_i\delta_{in} = C_{in} \qquad \forall\, n \tag{2}$$

$$\sum_m \lambda_i^m = 1 \tag{3}$$

$$\lambda_i^i = 0 \tag{4}$$

where $C_{mn} = C(\boldsymbol{x}_m - \boldsymbol{x}_n)$, μ_i is the Lagrange multiplier corresponding to the unbiasedness condition (Eq. 3), δ_{in} is Kronecker's delta and ε_i is an additional multiplier needed to account for Eq. 4. Notice that Equations (2) - (4) differ from the usual kriging equations in two ways. First, there is an additional equation (Eq. 4). Secondly, there appears a new additional unknown (ε_i). As shown later this unknown has a well defined meaning.

Equations (2) - (4) can be rewriten in matrix form as

$$\boldsymbol{A}\,\boldsymbol{u}_i + \varepsilon_i\mathbf{1}_i = \boldsymbol{d}_i \tag{5}$$

$$\mathbf{1}_i^t\,\boldsymbol{u}_i = 0 \tag{6}$$

where $\boldsymbol{A}$ is a $(N+1)\times(N+1)$ square matrix defined as

$$\boldsymbol{A} = \begin{bmatrix} \boldsymbol{C} & \mathbf{1}_N \\ \mathbf{1}_N^t & 0 \end{bmatrix} \tag{7}$$

where again $\boldsymbol{C}$ is the $N\times N$ covariance matrix of the N data locations and $\mathbf{1}_N$ is a $(N\times 1)$ column vector with all entries equal to 1. Vectors $\boldsymbol{u}_i$, $\mathbf{1}_i$, and $\boldsymbol{d}_i$ are $(N+1)\times 1$ column vectors defined as

$$\boldsymbol{u}_i = (\lambda_i^1,\ \lambda_i^2,\ \ldots\lambda_i^i,\ \ldots,\ \lambda_i^N,\ \mu_i)^t$$

$$\boldsymbol{d}_i = (C_{i1},\ C_{i2},\ \ldots,\ C_{ii},\ \ldots,\ C_{iN},\ 1)^t$$

$$\mathbf{1}_i = (0,\ 0,\ \ldots,\ 1,\ \ldots,\ 0,\ 0)^t$$

Notice that all entries of vector $\mathbf{1}_i$ are equal to zero except the i-th which is equal to 1. Vector $\boldsymbol{d}_i$ coincides with the i-th column of $\boldsymbol{A}$. Let $\boldsymbol{B}$ be the inverse of $\boldsymbol{A}$ in Eq. (7). Premultiplying (5) by $\boldsymbol{B}$ one obtains

$$\boldsymbol{u}_i = \mathbf{1}_i - \varepsilon_i\,\boldsymbol{b}_i \tag{8}$$

where $\boldsymbol{b}_i$ is the i-th column of $\boldsymbol{B}$. Performing the scalar product of $\mathbf{1}_i$ and $\boldsymbol{u}_i$ in Eq. (8) one gets

$$\mathbf{1}_i^t\,(\mathbf{1}_i - \varepsilon_i\boldsymbol{b}_i) = 1 - \varepsilon_i B_{ii} = 0$$

which indicates that the multiplier ε_i is equal to the reciprocal of the i-th diagonal entry of $\boldsymbol{B}$, that is,

$$\varepsilon_i = \frac{1}{B_{ii}} \tag{9}$$

Substitution of (9) into (8) leads to:

$$\lambda_i^m = \delta_{im} - \frac{B_{mi}}{B_{ii}} \tag{10}$$

These results are identical to those obtained by Dubrule (1983) using the duality property of the kriging equations.

Cross-validation errors e_i are computed as

$$e_i = \hat{Z}_i - Z_i \tag{11}$$

and their covariance matrix $\boldsymbol{V}$ as

$$V_{ij} = C_{ij} + \sum_m \sum_n \lambda_i^m \lambda_j^n C_{mn} - \sum_m \lambda_i^m C_{mj} - \sum_n \lambda_j^n C_{ni} \tag{12}$$

This is the general expression of matrix $\boldsymbol{V}$, which for global cross-validation can be reduced to a much simpler form by noticing that:

(a) The second term can be expanded using the kriging equations for the i-th cross-validation point

$$\sum_n \lambda_j^n (\sum_m C_{mn} \lambda_i^m) = \sum_n \lambda_j^n (C_{ni} - \mu_i - \varepsilon_i \delta_{in}) = \sum_n \lambda_j^n C_{ni} - \mu_i - \varepsilon_i \lambda_j^i \tag{13}$$

(b) The third term can be expanded using the j-th equation of point i, according to which

$$\sum_m \lambda_i^m C_{mj} = C_{ji} - \mu_i - \varepsilon_i \delta_{ij} \tag{14}$$

Substituting (13) and (14) into (12) it follows that

$$V_{ij} = \varepsilon_i \left(\delta_{ij} - \lambda_j^i\right) \tag{15}$$

Notice that the diagonal terms of $\boldsymbol{V}$, i.e. the cross-validation variances, coincide with the multipliers ε_i. According to (10) and (11), this covariance can also be written as

$$V_{ij} = \frac{B_{ji}}{B_{ii}\ B_{jj}} \tag{16}$$

which coincides with the results of Samper and Neuman (1989a).

Summing up, global cross-validation requires computing once the inverse $\boldsymbol{B}$ of matrix $\boldsymbol{A}$. Both the errors e_i and their covariance matrix are simple functions of the entries of

B. Later we will demonstrate that the properties of global cross-validation for a single variable can be generalized to the general case of several correlated variables.

The aim of cross-validation is to meet the following properties at $M(M \leq N)$ locations: Mean error close to zero; mean square error as small as possible; dimensionless mean square error (DMSE) close to one; sample variance of cross-validation errors close to the average kriging variance; correlation coefficient of kriged and observed values close to one; and normalized errors e_i/σ_i plotting as a straight line with a unit slope on normal probability paper. Instead of relying on these criteria, it is better to use maximum likelihood approaches such as those of Bastin and Gevers (1985), Samper (1986) and Samper and Neuman (1989a). Let $\boldsymbol{e}$ be the vector of the M Gaussian cross-validation errors. They have zero mean and a covariance matrix $\mathbf{V}$ given by (12). Let $\boldsymbol{P}$ be the vector of the N_p covariance parameters (nugget, sill, range, etc..). The negative log-likelihood (NLL) of $\mathbf{P}$ given $\mathbf{e}$ is

$$NLL = M\ ln 2\pi + ln \mid \mathbf{V} \mid + \mathbf{e}^t\ \mathbf{V}^{-1}\ \mathbf{e} \tag{17}$$

where $\mid \mathbf{V} \mid$ indicates determinant of $\mathbf{V}$ and $\mathbf{e}^t$ means the transpose of vector $\mathbf{e}$.

Maximum likelihood cross-validation is well suited for the problem of selecting the most appropriate semivariogram model among a set of alternative functional forms. The literature contains a number of model identification criteria developed in the context of maximum likelihood estimation. Samper and Neuman (1989b) used these criteria for the selection of covariance models. Other properties of the method, such as its ability to provide confidence regions for parameter estimates and its applicability to noisy, regularized and nonstationary variables are described by Samper (1986) and Samper and Neuman (1989a, b).

3. Cross-covalidation

The maximum likelihood cross-validation method for estimating autocovariance functions can be extended to the case of several correlated random functions. In this case, both autocovariances and cross-covariances are validated and estimated. For this reason the method is referred as **Maximum Likelihood Cross-Covalidation** (MLCC).

Let $Z_1(\mathbf{x}),\ Z_2(\mathbf{x}), \cdots, Z_{N_v}(\mathbf{x})$ be N_v second order stationary functions which have been measured simultaneously at N locations $\mathbf{x}_1, \mathbf{x}_2, \cdots, \mathbf{x}_N$. Let $C_{\alpha\beta}(\mathbf{h})$ be the cross-covariance of $Z_\alpha(\mathbf{x})$ and $Z_\beta(\mathbf{x})$. Greek letters are used to denote variables ($Z_\alpha,\ Z_\beta, \cdots$) and Latin letters for locations ($\mathbf{x}_i, \mathbf{x}_m, \cdots$). Global cross-covalidation consists of obtaining a cokriging estimate, $\hat{Z}_\alpha(\mathbf{x}_i)$, of $Z_\alpha(\mathbf{x}_i)$ on the basis of all the data according to:

$$\hat{Z}_\alpha(\mathbf{x}_i) = \sum_{\beta=1}^{N_v} \sum_m \lambda^m_{i,\alpha\beta}\ Z_{m\beta} \tag{18}$$

where $Z_{m\beta} = Z_\beta(\mathbf{x}_m)$ and $\lambda^m_{i,\alpha\beta}$ is the weight given to $Z_{m\beta}$ to estimate $\hat{Z}_\alpha(\mathbf{x}_i)$. For convenience we will use the matrix notation of Myers (1982). Now $\hat{\mathbf{Z}}(\mathbf{x}_i)$ is given by:

$$\hat{\mathbf{Z}}(\mathbf{x}_i) = \sum_m \mathbf{L}^m_i\ \mathbf{Z}(\mathbf{x}_m) \tag{19}$$

where $\mathbf{L}_i^m$ is a $(N_v \times N_v)$ matrix with entries equal to $\lambda_{i,\alpha\beta}^m$, and $\mathbf{Z}(\mathbf{x}_m)$ is the vector of data values at point $\mathbf{x}_m$. For second order stationary variables, matrices $\mathbf{L}_i^m$ satisfy the following cokriging equations:

$$\sum_m \mathbf{C}(\mathbf{x}_n - \mathbf{x}_m)\, \mathbf{L}_i^m + \boldsymbol{\mu}_i + \boldsymbol{E}_i\, \delta_{in} = \mathbf{C}(\mathbf{x}_i - \mathbf{x}_n) \qquad \forall\, n \tag{20}$$

$$\sum_m \mathbf{L}_i^m = \mathbf{I} \tag{21}$$

$$\mathbf{i}_\alpha^t\, \boldsymbol{L}_i^i\, \boldsymbol{i}_\alpha = 0 \qquad \forall \alpha \tag{22}$$

where $\mathbf{C}(\mathbf{x}_m - \mathbf{x}_n)$ is a square $(N_v \times N_v)$ matrix having entries equal to $C_{\alpha\beta}\,(\mathbf{x}_m - \mathbf{x}_n)$, $\boldsymbol{\mu}_i$ is also a $(N_v \times N_v)$ matrix of Lagrange multipliers, $\mathbf{I}$ is the identity matrix of order N_v, and $\boldsymbol{i}_\alpha$ is the α-th column of $\boldsymbol{I}$. Matrix $\boldsymbol{E}_i$ is a diagonal $(N_v \times N_v)$ matrix whose entries $\varepsilon_{i\alpha\alpha}$ are the Lagrange multipliers corresponding to the conditions of Eq. 22 which state that $\lambda_{i,\ \alpha\alpha}^i$ must be zero for all α and i. These conditions are needed to avoid the trivial solution that one would obtain otherwise. Let $\boldsymbol{\Omega}$ be the $(N_v \times N_v)$ zero matrix. Defining the following matrices:

$$\mathbf{A} = \begin{bmatrix} \mathbf{C}(\mathbf{x}_1 - \mathbf{x}_1) & \mathbf{C}(\mathbf{x}_1 - \mathbf{x}_2) & \cdots & \mathbf{C}(\mathbf{x}_1 - \mathbf{x}_N) & \mathbf{I} \\ \mathbf{C}(\mathbf{x}_2 - \mathbf{x}_1) & \mathbf{C}(\mathbf{x}_2 - \mathbf{x}_2) & \cdots & \mathbf{C}(\mathbf{x}_2 - \mathbf{x}_N) & \mathbf{I} \\ \vdots & \vdots & & \vdots & \\ \mathbf{C}(\mathbf{x}_N - \mathbf{x}_1) & \mathbf{C}(\mathbf{x}_N - \mathbf{x}_2) & \cdots & \mathbf{C}(\mathbf{x}_N - \mathbf{x}_N) & \mathbf{I} \\ \mathbf{I} & \mathbf{I} & & \mathbf{I} & \boldsymbol{\Omega} \end{bmatrix}$$

$$\boldsymbol{U}_i = \left[\boldsymbol{L}_i^{1t},\ \boldsymbol{L}_i^{2t},\ \ldots,\ \boldsymbol{L}_i^{it},\ \ldots, \boldsymbol{L}_i^{Nt},\ \boldsymbol{\mu}_i\right]$$

$$\boldsymbol{D}_i = \left[\boldsymbol{C}(\boldsymbol{x}_i - \boldsymbol{x}_1),\ \boldsymbol{C}(\boldsymbol{x}_i - \boldsymbol{x}_2),\ \ldots,\ \boldsymbol{C}(\boldsymbol{x}_i - \boldsymbol{x}_i),\ \ldots,\ \boldsymbol{C}(\boldsymbol{x}_i - \boldsymbol{x}_N),\ \boldsymbol{I}\right]$$

$$\mathbf{1}_i = \left[\boldsymbol{\Omega},\ \boldsymbol{\Omega},\ \ldots,\ \boldsymbol{I},\ \ldots,\ \boldsymbol{\Omega},\ \boldsymbol{\Omega}\right]$$

Equations (20)-(22) become:

$$\boldsymbol{A}\,\boldsymbol{U}_i + \mathbf{1}_i\,\boldsymbol{E}_i = \boldsymbol{D}_i \tag{23}$$

$$\boldsymbol{i}_\alpha^t\, \mathbf{1}_i^t\, \boldsymbol{U}_i\, \boldsymbol{i}_\alpha = 0 \qquad \forall \alpha \tag{24}$$

The unknowns $\boldsymbol{U}_i$ and $\boldsymbol{E}_i$ are obtained by computing first the inverse $\boldsymbol{B}$ of matrix $\boldsymbol{A}$, then premultiplying (23) by $\boldsymbol{B}$ and substituting $\boldsymbol{U}_i$ into (24). Notice that $\boldsymbol{D}_i$ is the i-th column block of matrix $\boldsymbol{A}$ and therefore $\boldsymbol{B}\boldsymbol{D}_i = \mathbf{1}_i$. The entries $\varepsilon_{i\alpha\alpha}$ of matrix $\boldsymbol{E}_i$ are given by

$$\varepsilon_{i\alpha\alpha} = \frac{1}{B_{ii}^{\alpha\alpha}} \tag{25}$$

where $B_{ii}^{\alpha\alpha}$ is the α-th diagonal entry of the (i, i) block of matrix $\boldsymbol{B}$. The cokriging weights matrices $\boldsymbol{L}_i^m$ are given by

$$\boldsymbol{L}_i^m = \delta_{im}\boldsymbol{I} - \boldsymbol{B}_{mi}\boldsymbol{E}_i \tag{26}$$

where $\boldsymbol{B}_{mi}$ is the (m, i) block of $\boldsymbol{B}$.

The aim of maximum likelihood cross-covalidation is to estimate the set of autocovariance and cross-covariance parameters by minimizing the negative log-likelihood. In this case the vector $\mathbf{e}$ of errors is made up of the errors $e_{i\alpha}$ for each point i and for each variable Z_α. This vector has $M \times N_v$ components corresponding to each of the N_v variables and M locations at which validation is performed. These errors are computed as the difference between cokriged and measured values

$$e_{i\alpha} = \hat{Z}_\alpha(\mathbf{x}_i) - Z_\alpha(\mathbf{x}_i) \tag{27}$$

Their covariance at two locations $\mathbf{x}_i$ and $\mathbf{x}_j$ is given by

$$\mathbf{V}(\mathbf{x}_i - \mathbf{x}_j) = \mathbf{C}(\mathbf{x}_i - \mathbf{x}_j) - \sum_m \mathbf{L}_i^m \, \mathbf{C}\,(\mathbf{x}_m - \mathbf{x}_j) - \sum_n \mathbf{C}(\mathbf{x}_i - \mathbf{x}_n)\mathbf{L}_j^{nt} +$$

$$+\sum_m \sum_n \mathbf{L}_i^m \mathbf{C}(\mathbf{x}_m - \mathbf{x}_n)\mathbf{L}_j^{nt} \tag{28}$$

which using Equations (20)-(22) reduces to

$$\boldsymbol{V}_{ij} = (\delta_{ij}\boldsymbol{I} - \boldsymbol{L}_j^{it})\,\boldsymbol{E}_i \tag{29}$$

Substituting here the expression of the cokriging matrix $\boldsymbol{L}_j^i$ (given by Eq. 26), one finally obtains:

$$\boldsymbol{V}_{ij} = \boldsymbol{E}_j\,\boldsymbol{B}_{ij}^t\,\boldsymbol{E}_i \tag{30}$$

Notice that the cokriging variances, i.e. the diagonal entries of $\boldsymbol{V}_{ii}$, reduce to $\frac{1}{B_{ii}^{\alpha\alpha}}$ and therefore coincide with the multipliers $\varepsilon_{i\alpha\alpha}$ (see Eq. 25).

The optimum parameters $\hat{\mathbf{P}}$ are those that minimize the following objective function J:

$$J = MN_v \, ln\, 2\pi + ln|\mathbf{V}| + \mathbf{e}^t\mathbf{V}^{-1}\mathbf{e} + \phi \tag{31}$$

where

$$\phi = \sum_{\alpha=1}^{N_v} \sum_{\beta=\alpha+1}^{N_v} w_{\alpha\beta}^2 \left[\int_0^\infty [C_{\alpha\beta}^2(\mathbf{h}) - C_{\alpha\alpha}(\mathbf{h})C_{\beta\beta}(\mathbf{h})]^2 d\mathbf{h}\right]^{-1} \tag{32}$$

The first three terms in (20) correspond to the negative log-likelihood (NLL) of the errors $\mathbf{e}$ while ϕ is a penalty function introduced to ensure that the cross-covariances $C_{\alpha\beta}(\mathbf{h})$ satisfy the Cauchy-Schwarz inequality. $\mathbf{V}$ is a $(M\ N_v \times M\ N_v)$ block matrix having blocks $\mathbf{V}(\mathbf{x}_i - \mathbf{x}_j)$ given by (30), and $w_{\alpha\beta}$ are penalty coefficients.

Following Myers (1988), cross-covariances $C_{\alpha\beta}(\mathbf{h})$ can be written as

$$C_{\alpha\beta}(\mathbf{h}) = \frac{1}{2a_\alpha a_\beta}[C^+_{\alpha\beta}(\mathbf{h}) - a^2_\alpha C_{\alpha\alpha}(\mathbf{h}) - a^2_\beta C_{\beta\beta}(\mathbf{h})] \quad (33)$$

where $C^+_{\alpha\beta}(\mathbf{h})$ is the autocovariance of $Z^+_{\alpha\beta}(\mathbf{x})$ defined as $Z^+_{\alpha\beta}(\mathbf{x}) = a_\alpha Z_\alpha(\mathbf{x}) + a_\beta Z_\beta(\mathbf{x})$. The estimation of cross-covariances is transformed into a problem of estimating autocovariances. The vector of parameters $\mathbf{P}$ consists of the parameters that define the autocovariances $C_{\alpha\alpha}(\mathbf{h})$ of the original variables and those of the autocovariances, $C^+_{\alpha\beta}(\mathbf{h})$, for each couple of variables $Z_\alpha(\mathbf{x})$ and $Z_\beta(\mathbf{x})$. In other cases the analytic expression of the cross-covariance $C_{\alpha\beta}(\boldsymbol{h})$ can be derived from physical equations relating $Z_\alpha(\boldsymbol{x})$ and $Z_\beta(\boldsymbol{x})$. This is the case of log-transmissivity and hydraulic head which are related through the partial differential equation of groundwater flow. Approximate analytic expressions for their cross-covariance can be found in Dagan (1989).

3.1 MINIMIZATION ALGORITHM

The optimum parameters $\hat{\boldsymbol{P}}$ are found by minimizing J in (31). Assuming uncorrelated cross-validation errors and working with a single variable, Samper and Neuman (1989 a, b) found efficient the use of conjugate gradient and quasi Newton methods in conjunction with adjoint state theory. Samper et al. (1992) showed that the use of adjoint state methods for computing the gradient $\boldsymbol{g}$ of J looses its advantages when the spatial correlation of the errors is taken into account. In this case it is better to compute $\boldsymbol{g}$ using the analytic method described below. These authors also investigated the use of Gauss-Newton methods such as that of Marquard-Levenberg according to which the parameters at the (r+1)-th iteration are obtained as

$$\mathbf{P}^{(r+1)} = \mathbf{P}^{(r)} - (2\mathbf{F})^{-1}\mathbf{g}\Big|_{\mathbf{P}=\mathbf{P}^{(r)}} \quad (34)$$

where $\mathbf{F}$ is an approximation of the Hessian matrix of NLL which can be taken as the Fisher information matrix defined as

$$F_{pq} = \frac{1}{2} E\left(\frac{\partial^2 NLL}{\partial P_p \partial P_q}\right) \quad (35)$$

The inverse of the Fisher matrix at the optimum is the Cramer-Rao lower bound for the estimation error covariance matrix of the parameter vector

$$\mathbf{F}^{-1} \leq \mathbf{V}_p = E[(\mathbf{P} - \hat{\mathbf{P}})(\mathbf{P} - \hat{\mathbf{P}})^t] \quad (36)$$

The inverse of $\mathbf{F}$ is generally used as an approximation of $\mathbf{V}_p$.

According to our limited experience in minimizing J for multiple variables, we tend to conclude that, similar to the univariate case, quasi Newton methods with line search perform better than Gauss-Newton methods.

3.2 COMPUTATION OF THE GRADIENT

Contrary to the univariate case in which adjoint state methods are well suited for the computation of the gradient of NLL, for more than one variable it is more efficient to compute this gradient using the analytical method (Samper et al., 1992), which is based on computing directly the derivatives of NLL with respect to each parameter. The first order derivative with respect to a parameter P_k is given by

$$\frac{\partial NLL}{\partial P_k} = Tr\left(\mathbf{V}^{-1}\frac{\partial \mathbf{V}}{\partial P_k}\right) + 2\mathbf{e}^t\mathbf{V}^{-1}\frac{\partial \mathbf{e}}{\partial P_k} - \mathbf{e}^t\mathbf{V}^{-1}\frac{\partial \mathbf{V}}{\partial P_k}\mathbf{V}^{-1}\mathbf{e} \tag{37}$$

where $Tr(\mathbf{A})$ denotes the trace of matrix $\mathbf{A}$. The derivatives of $\mathbf{V}$ and $\mathbf{e}$ can be written in terms of the derivatives of the cokriging weights. The derivative of the vector $\boldsymbol{e}_i$ of cross-validation errors at point i is obtained from (19) and (27):

$$\frac{\partial \boldsymbol{e}_i}{\partial P_k} = \sum_m \frac{\partial \boldsymbol{L}_i^m}{\partial P_k} \boldsymbol{Z}_m \tag{38}$$

Similarly, from (29) it follows that the derivative of the block $\boldsymbol{V}(\boldsymbol{x}_i - \boldsymbol{x}_j)$ is given by

$$\frac{\partial \boldsymbol{V}(\boldsymbol{x}_i - \boldsymbol{x}_j)}{\partial P_k} = -\left(\frac{\partial \boldsymbol{L}_j^{it}}{\partial Pk} \boldsymbol{E}_i - \boldsymbol{L}_j^{it} \frac{\partial \boldsymbol{E}_i}{\partial P_k}\right) \tag{39}$$

According to (26), the derivatives of the cokriging weights can be written as a function of the derivatives of matrix $\boldsymbol{B}$

$$\frac{\partial \boldsymbol{L}_i^m}{\partial P_k} = -\left(\frac{\partial \boldsymbol{B}_{mi}}{\partial P_k} \boldsymbol{E}_i + \boldsymbol{B}_{mi} \frac{\partial \boldsymbol{E}_i}{\partial P_k}\right) \tag{40}$$

Finally the derivative of $\boldsymbol{B}$ is computed from:

$$\frac{\partial \boldsymbol{B}}{\partial P_k} = -\boldsymbol{B}\frac{\partial A}{\partial P_k}\boldsymbol{B} \tag{41}$$

4. Applications to synthetic and real data

The ability of the MLCC method to simultaneously estimate autocovariances and cross-covariances using synthetically generated variables has been illustrated by Samper et al. (1992). They considered a couple of two-dimensional stationary random functions having spherical covariances. Contrary to sample covariances which tend to underestimate the true values, MLCC estimates always agree well with the true covariances.

MLCC has also been applied to the estimation of covariance structures of hydraulic heads and log-transmissivities in a sandy unconfined aquifer located near the Doñana National Park in Southern Spain. Aquifer thickness ranges from 100 to 150 m and

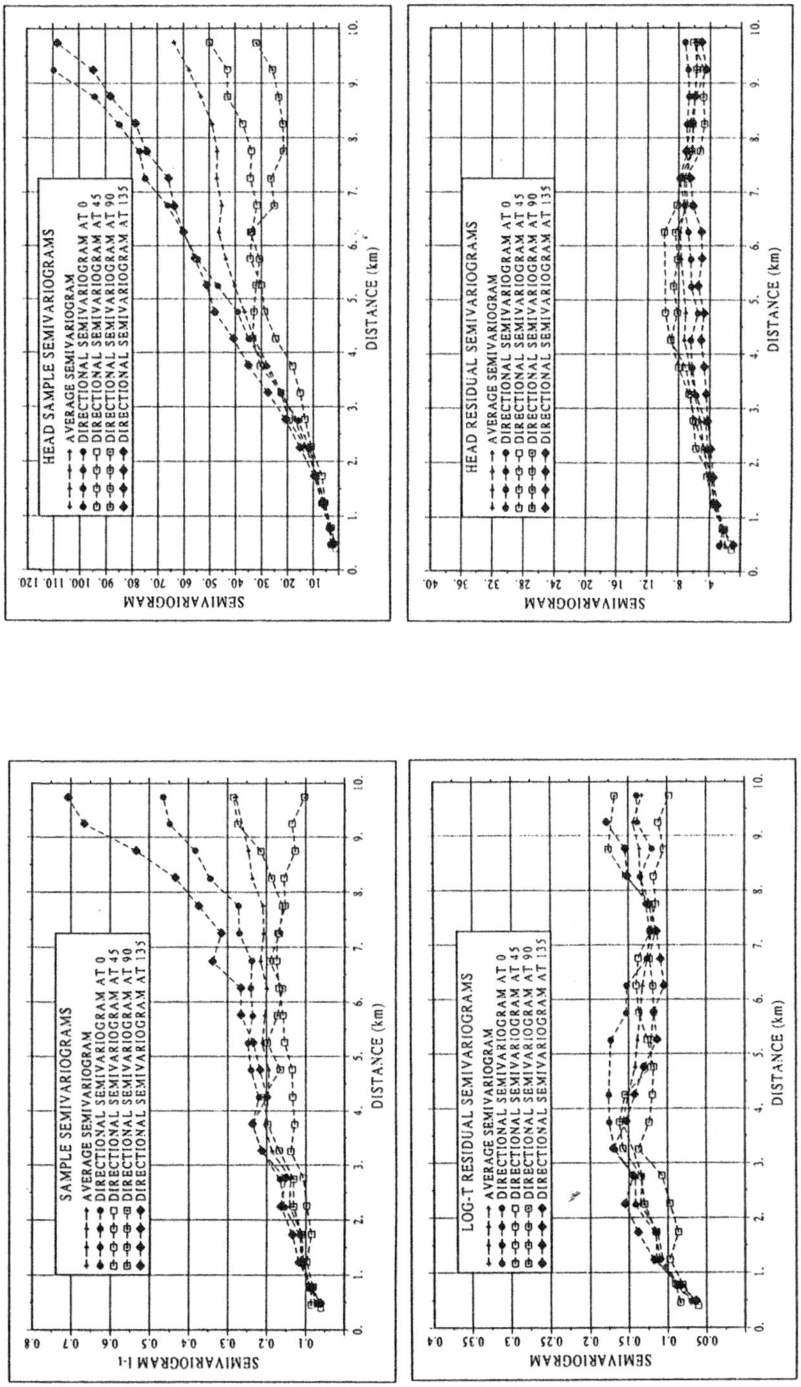

Figure 1. Sample semivariograms of raw hydraulic head and log10-T data. Also shown are the semivariograms of the second order residuals.

transmissivities vary between 10 and more than 1000 m^2/d. A detailed account of the hydrogeology of the area is given by Hollis et al. (1989) and Llamas (1990). Both hydraulic heads and log-transmissivities exhibit clear spatial drifts as shown by their sample semivariograms (Figure 1). Second order polynomial drifts were fitted using a generalized least squares methos (Samper, 1986). The resulting residuals have stable and almost isotropic semivariograms (see also Figure 1) Figure 2 shows the shapes of the deterministic trends of log-T and h. also shown in the location of the 242 data locations used for this study. The cross-covariance is non-symmetric, anisotropic and shows large fluctuations. This kind of cross-covariance cannot be modeled using Myers method wich works with the autocovariance of the sum (recall Eq. 33). The alternative consists of using theoretical cross-covariance models of h and log-T derived from various approximations of the governing partial differential flow equation (Dagan, 1989). These models, however, are only valid for unbounded domains, with parallel flow and assume no measurement errors. The samples cross-covariance of log-T and h data for the Doñana aquifer reflect some features of the theoretical models such as the zero cross-covariance at the origin. Other properties such as the antisymmetry, however, are not shown in the sample cross-covariances.

These discrepancies between the theoretical and the sample cross-covariances could be caused by: (1) large log-T interpretation errors (the nugget of log-T is almost one fourth of the total sill), (2) the fact that the wells are screened along selected intervals which causes the head measurements to represent depth-averaged values, and (3) measurement h errors and time variability of heads.

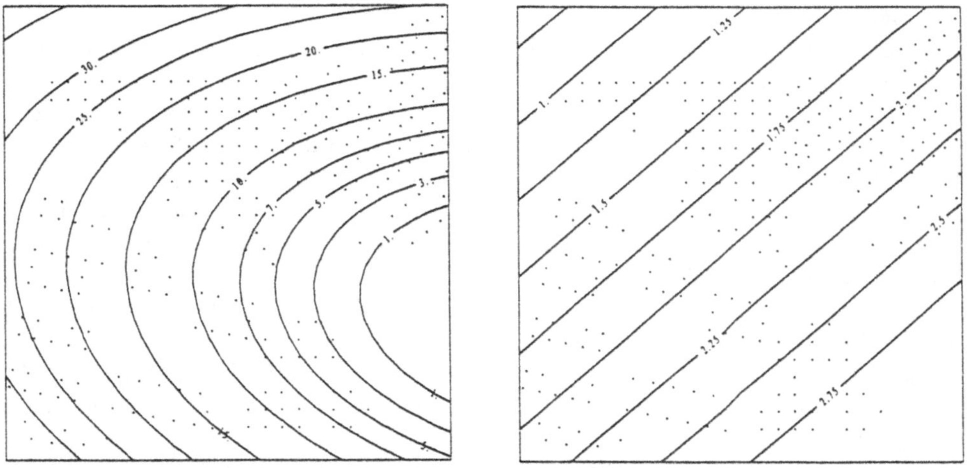

Figure 2. Second order polynomial drifts for hydraulic heads (left) and log10-Transmissivity data (right)

5. Conclusions

A maximum likelihood cross-validation method for estimating autocovariances and cross-covariances has been presented. The method is a natural extension of the maximum likelihood cross-validation concept, a method successfully used for estimating autocovariances. To overcome the lack of theoretical positive definite cross-covariance models, cross-covariances are modeled as combinations of autocovariances. It is shown that cross-covalidation using a global neighbourhood results in simpler expressions which require less computation time. The method has been tested with synthetic and hydrogeological variables.

6. Acknowledgements. This work has been funded by the CICYT Grant "NAT91-1001" from Spanish Government.

7. References

Ahmed, S. y G. De Marsily (1989) "Cokriged estimates of transmissivities using jointly water level data", In: Geostatistics, III Int. Geostatistics Congress, Avignon, France, M. Armstrong Ed., Kluwer Acad. Pub. Vol 2, 615-628.

Bastin, G. and M. Gevers (1985) "Identification and optimal estimation of random fields from scattered point-wise data", Automatica, 21(2): 139-155.

Dagan, G. (1989) "Flow and Transport in Porous Formations", Springer-Verlag, New York, 465 pp.

Dubrule, O. (1983) "Cross validation of kriging in a unique neighborhood", Mathematical Geology, 15(6): 687-699.

Hollis, T., P. Heurteaux and J. Mercer (1989) "The implication of Groundwater Extractions for the long Term Future of the Doñana National Park" Report of the WWF/IUCN/ADENA Mission to the Doñana National Park, May 1989, 60 pp.

Kitanidis, P.K. and E.G. Vomvoris (1983) "A geostatistical approach to the inverse problem in groundwater modelling (steady state) and one-dimensional simulations", Water Resour. Res., 19(3), 677-690.

Llamas, M.R. (1990) "Geohydrology of the Eolian Sands of the Doñana National Park (Spain)" Catena Supplement 18, 145-154.

Myers, D.E. (1982) "Matrix Formulation of Co-kriging", J. Math. Geol., 14(3), 249-257.

Myers, D.E. (1988) "Multivariate Geostatistical Analysis for Environmental Monitoring", Sciences de la Terre, 27: 411-427.

Samper, F.J., (1986) "Statistical Methods of Analyzing Hydrological, Hydrochemical and Isotopic Data from Aquifers", Ph.D. Diss., Dept. of Hydrology and Water Resour., Univ. of Arizona, Tucson.

Samper, F.J. and S.P. Neuman (1989a) "Estimation of Spatial Covariance Structures by Adjoint State Maximum Likelihood cross-validation: 1. Theory", Water Resour. Res., 3(25), 351-362.

Samper, F.J. and S.P. Neuman (1989b) "Estimation of Spatial Covariance Structures by Adjoint of State Maximum Likelihood cross-validation:2. Synthetic Experiments", Water Resour. Res., 3(25), 363-372.

Samper, F.J., J.C. Cuchí and J. Carrera (1992) "Estimation of spatial cross-covariances by maximum likelihood cross-covalidation" Sci. de la Terre, Ser. Inf., Nancy, 31:121-131.

Spatial structure of hydraulic conductivity in various porous media - problems and experiences

M.-Th. Schafmeister & A. Pekdeger
Institute for Geology, Geophysics
and Geoinformatics
Wichernstr. 16, D-1000 Berlin 33
GERMANY

ABSTRACT: Geostatistical simulation techniques, both Gaussian as well as indicator simulation are practical tools to generate spatial realizations of aquifer parameters such as hydraulic conductivity for further use in deterministic groundwater-flow and - transport models. The application of simulation methods require modeling of the spatial variability of the considered parameter, i.e. the variogram and pdf models must be derived a priori. In actual cases of groundwater contamination generally only sparse information on hydraulic parameters is available. Their spatial structure is often completely unknown. The main scope of our work is to describe the spatial structure of hydraulic conductivity for some typical aquifers and scales in order to use this information in geologically comparable cases.

Different test sites in Northern Germany have been chosen to measure hydraulic conductivity. The measurements were performed with different methods depending on first the scale and second the orientation of the sampling pattern (horizontal, vertical). Large scale aquifers are represented by conductivity values derived from pumping tests in a horizontal plane whereas the spatial distribution of conductivity in smaller cross-sections is described by values derived from grainsize-curves. The extension of the test fields range from a few meters up to kilometers.

The test sites are mainly situated in pleistocene glaciofluvial loose sands. Additionally the spatial structure of grain-size distribution parameters and hydraulic conductivity along a cross-section through a poorly consolidated Triassic Solling sandstone as well as fluviatil sands in Canada are analyzed.

The large amount of data for some of the fields (100 to 200 samples) allows an exhaustive geostatistical study of the spatial variability of hydraulic conductivity of typical aquifer materials. Effects of sampling pattern on the evaluation of spatial structures and small-scale variability (nugget-effect) are analyzed as well as the so called scale-effect.

Additionally a brief example of modeling tracer-transport in an heterogeneous aquifer using indicator-simulation for generating the conductivity fields is presented. This example demonstrates quite clearly the effects of spatial variability of aquifer properties on transport behaviour.

1. INTRODUCTION

During the last decade a lot of research in geosciences dealt with the consequences of spatial variability of aquifer properties and oil-reservoir characteristics, respectively. Only a few of many authors e.g. Gelhar & Axness, 1983; Schafmeister & Burger, 1989; De Fouquet et al. 1989; Dimitrakopoulos 1989; Schafmeister & Pekdeger, 1989; Gomez-Hernandez & Srivastava, 1990 can be mentioned here. For more complete background see Schafmeister, 1990.
The main two questions of their studies were:

> How can spatial variability of aquifer and reservoir parameters be identified?

and

> What can be done to reproduce "Real-World" heterogeneity for further use in deterministic models when only few measurements are available?

Whereas the first question could be answered by means of variogram analysis the second problem is nowadays treated by either stochastic modeling-techniques or by simulation of random fields.

The main advantage of simulating random-fields is that it is based <u>only</u> on the knowledge of the p.d.f., i.e. its first two moments (mean and variance) and - in the case of space-dependent simulation (geostatistical S.) - the variogram $\gamma(h)$. This information is sufficient to generate numerous equal probable realizations of the considered property.

Using these realizations as input in deterministic models provides - besides average results of flow and transport which could easily be gained by inserting interpolated (kriged) conductivity fields - extreme flow and transport behaviour. Thus consequences on groundwaterdynamics and -quality, respectively can be investigated.

Some advanced techniques, e.g. Indicator Simulation (Gomez-Hernandez & Srivastava, 1990) can handle discontinuous, non-linear problems. This was found to be important when dealing with pleistocene aquifers in Northern Europe where hydraulic conductivity varies more than several orders of magnitude within distances of only few meters (Schafmeister, 1990). This is due to specific processes with changing transport media (ice, water) and varying energies (subglacial, fluviatil, limnic) during sedimentation of till deposits. As a result grain-size distributions (and consequently hydraulic conductivities) are both wide-spreading with continuous changes within lithofaciel units as well as exhibiting sharp steps among them.

There is no doubt that simulation techniques can and should be applied in practice in order to reflect the "true" variability in flow and transport. Results of deterministic models can thus be improved by not only average hydraulic conditions but additionally a whole spectrum of different, equal likely possibilities.

Procedures to generate random fields either purely random in space (Monte-Carlo-S.) or preserving the autocovariance-structure ("Nearest-Neighbour", Smith & Freeze, 1979a,b, Smith & Schwartz, 1980, 1981 a,b; Cholesky's decomposition of covariance matrix, Binsariti, 1980, Neuman 1984; "Turning-Bands", Matheron 1973, Delhomme, 1979) do not need necessarily measured data in the aquifer of interest. In the case of sparsely investigated areas where no or only few measurements of hydraulic conductivity are available the information on the spatial variability, i.e. mean, variance and variogram can be transferred from fairly well-known aquifers of comparable lithological conditions.

During further investigation of the problem-site the simulated random fields can be conditioned to measured data (Conditional Simulation) and thus reflect more closely the actual conditions.

Transferring information on the spatial structure of hydraulic conductivity to another area raises at least two problems: first the scale-effect in the sense of Dagan (1986) which is related to different sampling support, i.e. K-values derived from grain-size-analysis ("point"-measurements) or from long-term pumping-tests (large volumes).

Secondly a scale-effect depending on the size of investigation domain in relation to the sampling distances can be observed. This latter one to which we will refer may result in a nested structure of variogram and can only be detected when samples at all scales (distances) are available (Gelhar, 1986). If this information is given small-scale variability can be simulated separately from the variability on a regional scale.

What ever method is chosen either simulation of random fields or the stochastic approach to macro-dispersion as suggested by Gelhar & Axness 1983, an intensive structural analysis of measured data must be done a priori. But these information can be derived from lithologically comparable aquifers with an higher level of information. In the following some typical aquifers were analyzed with respect to the spatial variability of hydraulic conductivity (K). Geostatistical parameters such as mean and variance (m, σ^2) and variogram with sill (C), nugget-effect (C_0) and range (a) are evaluated.

2. DESCRIPTION OF DATA

Nine data-sets of measurements of hydraulic conductivity from different test-fields have been analyzed. These data differ either in

a) the method of measurement (short-term hydraulic tests, grain-size analysis),
b) sampling pattern and orientation (horizontal/vertical), or
c) the type and geographical position of sediment.

Four major groups can be distinguished:

<u>Data-sets 1, 2 and 3</u> are from fields situated in Berlin/Germany. The measurements of hydraulic conductivity are derived from short-term pumping tests, which have been interpreted under consideration of the prevailing hydraulic conditions (i.e. confined, half confined or phreatic).

Even though different students interpreted the pumping tests, comparable results were provided. The aquifers analyzed are all of younger pleistocene age. They consist of glacio-fluvial loose sands with intercalated clays. The values of hydraulic conductivity are in the order of 10^{-3} m/s.

Data-set 1 is situated in the south of Berlin. With 300 km^2 it is much larger than area 2 and 3 which are subregions of area 1. Sampling locations (wells) are distributed quite regulary with distances of more or less constantly 1 km.

Data-set 2 (Berlin-Kreuzberg) is adjacent to data-set 3 (Berlin-Neukoelln). The random sampling pattern with average distances of 200 m to 400 m are quite similar. The sizes of investigated areas are about 15 km^2 both.

In all pumping tests the cone of depression was small compared to the distance between wells.

Data-sets 4 and 5 originate also from Berlin (Gatow) and are actually one data-set which after showing a very distinct trimodal p.d.f. was seperated into three sub-sets of which only the two sandy fractions could be analyzed further. The pleistocene aquifer was analyzed by grain-size analysis on a vertical cross-section (fig. 1). The three different lithofacial units each represented by one sub-population of K-values are intensively toothed. Generally a "layer" of fine sands (data-set 5) is underlain by up to 10 m of thick glacial clay which is often interrupted by glacial channels. The basal "layer" consists of medium sands (data-set 4). Hydraulic conductivity of data-set 4 ranges in the order of 10^{-4} to 10^{-3} m/s whereas data-set 5 shows values of 10^{-5} m/s. The clay was not analyzed, a value of $1 \cdot 10^{-6}$ m/s was assumed.

In vertical direction the sampling pattern depends strongly on the thickness of distinguishable sub-layers which varies between 0.3 and 5 m. Horizontally the sampling distance is about 50 m, covering a length of several hundred m.

Data-sets 4 and 5 and additional information on the location of clay layers built the basis for modeling tracer-transport in a highly heterogeneous medium using the Indicator-Simulation technique.

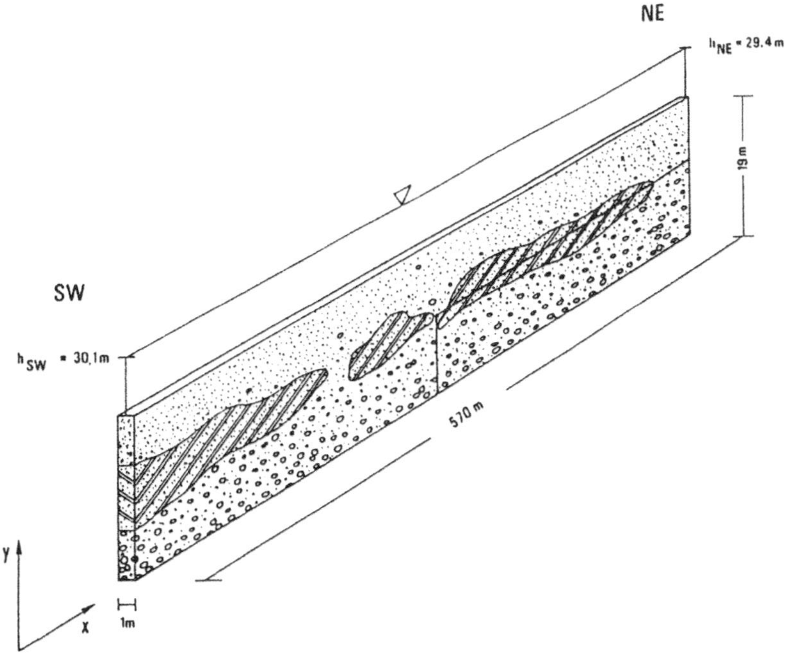

Fig 1: Schematic representation of the vertical cross-section Gatow (data-set 4 and 5).

Data-sets 6 and 9 contain hydraulic conductivity values derived from grain-size analysis of samples which were gathered in a sand-pit in North Germany (Bornhoeved/Kiel). The pleistocene sediments which on the first sight look quite homogeneous consist of alternating layers and lenses of fine and medium sands with values ranging in the order of 10^{-4} m/s. Data-set 6 comes from a cross-section of about

25 m long and 5 m deep with sampling distances of 2.5 m horizontally and 0.5 or 0.25 m vertically. Data-set 9 covers a much smaller cross-section of only 2 by 2 m^2 close to the sampling locations of the above described data-set 6. Cores were taken as dense as every 10 cm apart. Data-set 9 is the largest and most dense with 216 samples.

Data-sets 7 and 8 were chosen from different geological formations which nevertheless show similar grain-sizes (mainly fine sands). Though being of pleistocene age sediments in data-set 7 (Twin Lake/Ontario) represent loose fine sands with a high content of mica. The average K-value is $2.6 \cdot 10^{-4}$ m/s. Data-set 8 consists of samples along a cross-section of poorly consolidated fine sands which were sedimented in a terrestrial environment (triassic Solling-sequence, sm, Eschwege/Hessen). Both data-sets represent similar investigation scales with sampling distances of a few meters horizontally and less than half meter vertically.

Data-sets (No. 1 to 9) are classified in descending order of observation-scale.

3. RESULTS OF THE STRUCTURAL ANALYSIS

An often reported attribute of hydraulic properties e.g. transmissivity, hydraulic conductivity, storage coefficient being lognormally distributed (Hoeksema & Kitanidis, 1985) was also observed in our data. Therefore statistical and variogram analysis is performed on the ln-transforms of the original values which can generally be regarded as normally distributed. Results of the structural analysis are comprised in table 1.

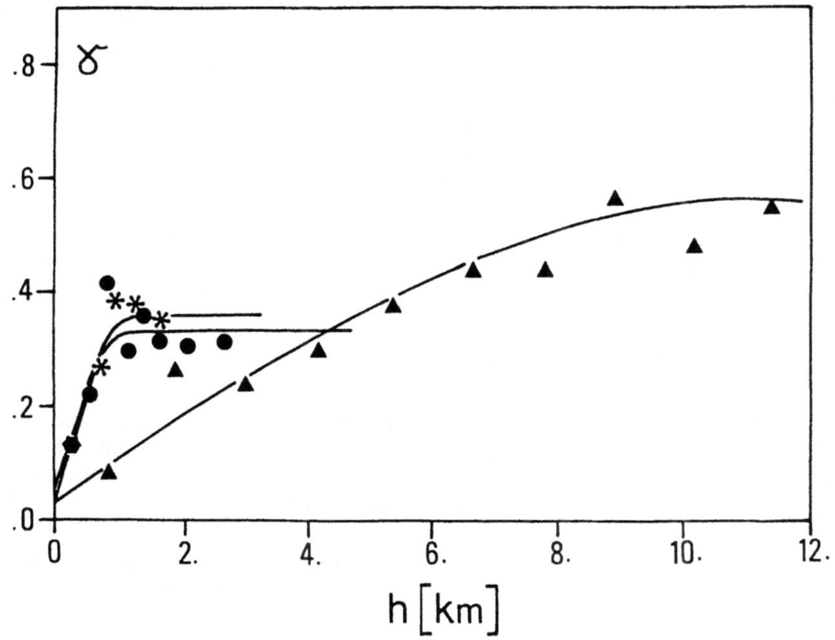

Fig 2: Variograms and adjusted spherical models for data-sets 2 (asterisks), 3 (circles) and 1 (triangles) respectively.

Considering data-set 2 and 3 which originate from two adjacent areas, a big similarity of p.d.f.'s is apparent. Statistical tests did not reject the hypothesis of one identical underlying population. Even though data are derived from pumping tests which as volume-integrating measurements result in more or less smoothed values a relatively high variability can still be observed. This variability is expressed by the - compared to the other investigated aquifers - high variation coefficient of 9 %.

The variograms of both data-sets could be fitted by a spherical model with ranges of about 1 km and nugget-variances of 9 and 14 % of total variance. These with respect to the rest of the aquifers analyzed high nugget-effects are due to the fact that no information on distances below 200 m is given.

Comparing the statistical parameters of data-set 1 to those of data-set 2 and 3 the most striking differences can be observed in the shape of the variogram. With given sampling distances of 1 km there is no evidence on the small-scale behaviour of hydraulic conductivity. Compared to those of data-set 2 and 3 the sill is almost doubled. The resulting spherical range is about 11 km (fig. 2). This example exhibits quite clearly the scale-effect according to Gelhar (1986); a nested-structure of variogram can be suggested here.

Tab. 1: Location, sampling geometry, lithofacial description of test sites; parameters of spatial structure of hydraulic conductivity.

No.	location	scale/orient.		sampling distance horiz.	vert.	main grain-size*	avg. K [m/s]	mean ln(K)	C.V. [%]	var. ln(K)	std. ln(K)	C_0+C **	N.E. [%]	range [m]	ratio of anisotr
1	Berlin	10-km	2D,h	1 km	-	mS	1.7E-03	-6.65	11	0.557	0.746	0.560	5	11000	-
2	Kreuzb.	km	2D,h	350 m	-	mS	1.8E-03	-6.49	9	0.350	0.592	0.350	14	1100	-
3	Neukoel.	km	2D,h	350 m	-	mS	1.1E-03	-6.94	8	0.313	0.559	0.330	9	1000	-
3a	Neukoel.	km	2D,h	350 m	-	mS	1.3E-03	-6.88	10	0.442	0.665	0.450	0	600	-
4	Gatow I	100-m	2D,v	50 m	0.3 - 5 m	mS	1.3E-03	-7.45	17	1.570	1.253	1.570	9	143	15
5	Gatow II	100-m	2D,v	50 m	0.3 - 5 m	fS	4.8E-05	-10.1	6	0.310	0.557	0.310	3	210	15
6	Bornh.II	m	2D,v	2.5 m	0.5(0.25)m	fS-mS	3.4E-04	-8.04	4	0.121	0.348	0.140	18	14	8
7	TwinL.	m	2D,v	5 m	0.3 m	fS	2.6E-04	-8.28	2	0.018	0.134	0.016	13	9.4	8
8	Eschw.	m	2D,v	2.5 m	0.5(0.25)m	fS,U'	8.9E-05	-9.51	6	0.362	0.602	0.420	36	7	11
9	Bornh.I	10-cm	2D,v	0.1 m	0.1 m	fS-mS	6.2E-04	-7.48	5	0.165	0.406	0.186	0	1	4

* mS: medium sand, fS: fine sand, U': clayey

** $C+C_0$: sill+nugget-effect, N.E.: nugget-effect as percent of total variance

Scale dependency of hydraulic conductivity can also be observed in data-sets 6 and 9. While the smaller sampling domain (Data-set 9) shows a horizontal range of 1 m a range of about 14 m could be observed when the total sampling area was expanded. Even the geometrical anisotropy in the vertical direction increased from 1/4 to 1/8 (fig. 3).

The increase of nugget-variance from 0 (data-set 9) to 18 % (data-set 6) is a clear evidence that in most cases the nugget-effect is only the sum of identified spatial structures.

The, at the first sight, surprising reduction of total variance can be explained by the fact that only in the case of a very dense sampling (data-set 9) with cores of 10 cm diameter could ensure that really the total variety of sedimentation textures is registered. Sampling with drill-cores often resulted in the loss of coarser parts of the sediments.

Beyond this the sampling distance of 5 m is not appropriate when the extension of the majority of structures e.g. sand lenses is smaller.

The coefficients of variation of 4 and 5 % respectively indicate that the considered fine sands of the Bornhoeved test-site exhibit less overall heterogeneity than the above described sediments from aquifers in Berlin. The observation of relatively high variability is supported by the fact that measurements in data-sets 6 and 9 are "point"-measurements (core samples).

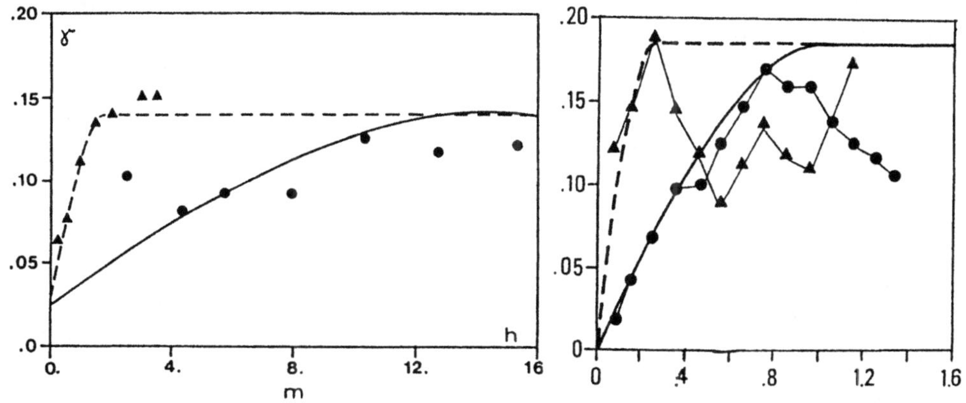

Fig. 3: Horizontal (circles) and vertical (triangles) variograms of data-set 6 (left) and 9 (right) and adjusted spherical models.

Compared to the other data, data-set 7 can almost be called homogeneous (C.V. < 2 %) which is not surprising considering the composition of these sediments. Nevertheless a spatial correlation structure with a range of about 9 m and a nugget-effect of 13 % of total variance can be evaluated. Tracer-experiments (Moltyaner & Killey, 1988a,b) give evidence on the heterogeneous shape of tracer-plumes which is due to the spatial variability of hydraulic conductivity.

Due to the much finer grains in the triassic, poorly consolidated sandstone probability density function of data-set 8 is shifted to the smaller values. The most striking difference to the other data-sets is the very high nugget-effect of 36 % of total variance which is observed in both main directions (parallel and orthogonal to the bedding). Though of different sedimentary background the spherical range of 7 m and a ratio of anisotropy of about one tenth, the spatial correlation corresponds well to parameters of data-sets 6, 7 and 8 which share the same investigation scale.

Data-set 4 and 5 figure two of three subpopulations of K-values in one single cross-section (the clayey parts of the cross-section are represented by one value). Contrary to the fine fraction (data-set 5) the coarser fraction (data-set 4) exhibits a rather high variability concerning the total variance as well as the nugget-effect. Despite of this the correlation structures show similar shapes with ratios of anisotropy of 1:15.

In order to perform conditional simulations for all three sub-units simultaneously additional indicator-variograms with two threshold-values (see below) were calculated. The spatial characteristics i.e. spherical ranges of 1 to 2 hundreds of meters and nugget-effects of up to 20 % of the sill resemble those of the variograms of the continuous K-values.

4. MODELING TRACER-TRANSPORT IN AN EXTREMELY HETEROGENEOUS AQUIFER BY MEANS OF INDICATOR SIMULATION

The aquifer in Berlin/Gatow was chosen to perform a study on macro-dispersion due to varying flow-vectors which result from the variabilty of hydraulic conductivity. A simple steady-state flow-field is built up by prescribed heads at the vertical boundaries of the section.

As described above the chosen cross-section 600 m long and 20 m deep exhibits a significant variability of conductivity with a range of more than four orders of magnitude representing aquitards as well as aquifers. This example gives a fairly good impression of a typical pleistocene aquifer in Northern Europe. The 165 data originate from grain-size-analysis of samples which were taken at 50 m apart wells with vertical spacing of a few decimeter up to a few meter.

Variograms of indicator-variables $I_0(x)$, $I_1(x)$ and $I_2(x)$, respectively, which represent either the presence or non-presence of a specific lithofacies were used to generate a number of equiprobable, conditional realizations of the considered cross-section (data-set 4 and 5, Berlin/Gatow). The subdivision of p.d.f. of hydraulic conductivity was performed with respect to the three distinguishable sub-populations which correspond well with lithofacies considerations. The definition of the indicator variable is:

$$I_0(x) = 1, \text{ if } K(x) < 1 \cdot 10^{-6} \text{ m/s, else } I_0(x) = 0$$
$$I_1(x) = 1, \text{ if } 1 \cdot 10^{-6} \text{ m/s} < K(x) < 1 \cdot 10^{-4} \text{ m/s, else } I_1(x) = 0$$
$$I_2(x) = 1, \text{ if } K(x) > 1 \cdot 10^{-4} \text{ m/s, else } I_2(x) = 0$$

After generation of the three different lithoface units by Indicator-Simulation, a spatial simulation of continuous K-values within these sub-areas was performed. The resulting random-fields of hydraulic conductivity were inserted into a 2-dimensional transport model with 38 by 19 finite-difference cells of 15 and 1 m cell-width, respectively (fig. 4). The cell-widths were chosen according to the observed ratio of anisotropy (1:15).

With a prescribed overall gradient of 0.12 % and an effective porosity of 0.25 flow-velocities within the transport-subgrid (lower right quarter of model) vary between 0.01 and 0.42 m/day.

The input of tracer was performed momentary into 10 cells at the lower part of the right boundary. The shape of the modeled tracer-plumes differ very significantly from the common Gaussian model. Each imaginable shape can be observed. Depending on the location of tracer-input, i.e. low- or high-velocity zone or zone where high conductivity values are immediately adjacent to low values, tracer plumes are subject either to an early splitting or to a transport as a more or less connected plume. Derived macro-dispersivity values for an observation distance of 200 m vary between 0.2 and 5 m. Compared to often used "rule-of-thumb" values for longitudinal dispersivity which range between one hundredth and one tenth of transport distance our values are very small. However the wide spectrum of dispersivity values is remarkable and gives a good insight in the range of uncertainty in modeling solute-transport in groundwater (for more detail see Schafmeister, 1990).

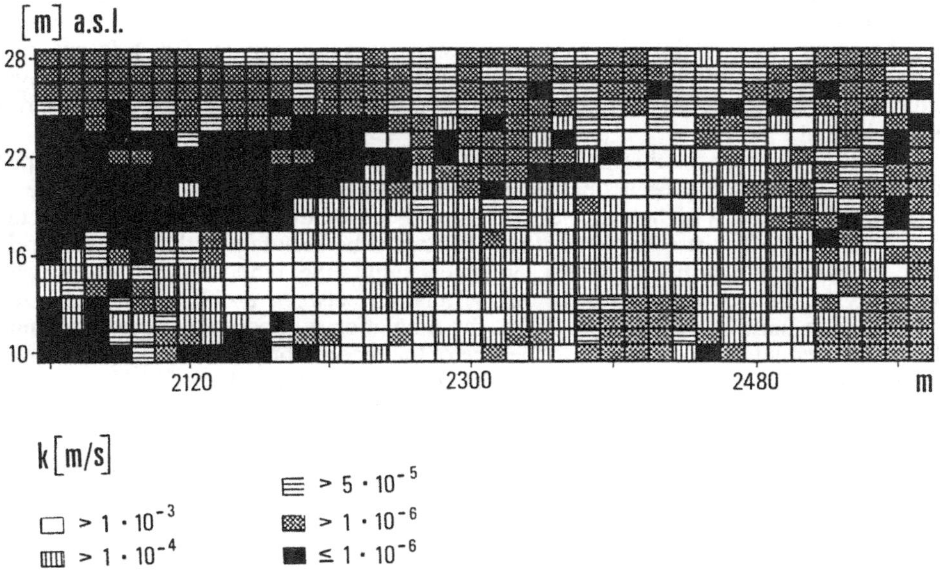

Fig. 4: A simulated realization of a cross-section in Gatow, data-sets 4 and 5 (10:1 vertical exaggeration).

5. CONCLUSIONS

The structural analysis of hydraulic conductivity in some typical porous aquifers shows that, depending on the sedimentary processes and environment, a big variety of simple statistical and spatial correlation characteristics can be observed. Some common features could be recognized although data are derived from different measurement methods.

There is - as some authors have already supposed before (Gelhar, 1986) - a significant dependency between scale of spatial correlation (i.e. variogram ranges) and scale of observation (investigation domain). Fig. 5 demonstrates the decrease of range (see solid line referring to the left axis) with decreasing scale of sample site. Data-sets 6 and 9 (Bornhoeved) and 1, 2 and 3 (Berlin), respectively give clear evidence that scale-effects can be observed within a unique aquifer, when an adequate sampling pattern is provided. Fitting a nested variogram-function would be an appropriate way of describing this phenomenon.

Often sampling distances do not allow evaluation of small-scale variabilities. Again data-set 1 is a good example for this. If data-set 2 and 3 which are a bit more detailed were not known variances beyond a distance of 1 km could not be considered or at least only be modeled by a rather high nugget-effect (here about 50 % of total variance! Fig. 2).

Bars in fig. 5 (referring to the right axis) represent the evaluated nugget-effect (N.E.) as percent of total variance. Not regarding data-set 8, where sediments are of different sedimentation and consolidation mode, no systematic dependency of investigation scale can be seen. Data-set 3 which could be fitted by two different but equally valid variogram-models (3: with N.E., 3a: without N.E.) demonstrates clearly that the nugget-effect often consists of a big portion of uncertainty about the short distance behaviour. It can be assumed that at any scale there will always be left at least a small amount of uncertainty due to measurement errors (i.e. nugget-effect). In general no clear separation between this uncertainty on the one hand and the portion of nugget-effect which is the sum of smaller, not identified spatial structures on the other hand can be carried out.

However, at this point it should be remarked that for modeling purposes where cell-sizes often exceed small-scale structures at several orders of magnitude it is sufficient to assume a high nugget-effect which incorporates both, small-scale structures and uncertainty. On the other hand it should always be ascertained that the appropriate scale of correlation-structure is chosen when simulating random-fields of hydraulic conductivity. This depends strongly on the relation between problem-size and size of discretization unit. Our suggestion is that for simulation purposes that variogram-model should be used which corresponds to the size of the model-domain in a way that its range is about 10 to 50 times bigger than the discretization unit but much smaller than total size of problem area.

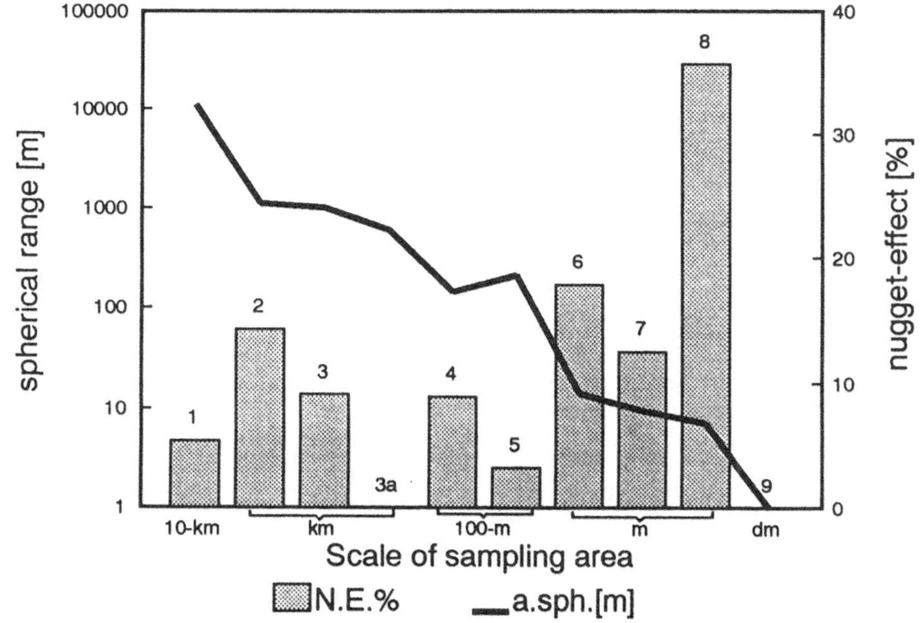

Fig. 5: Behaviour of spatial correlation characteristics versus scale of investigation-size

The assumption that the whole spectrum of variability can be observed when a sufficient number of samples, distributed more or less regulary in space are taken is not valid. The Bornhoeved data demonstrate quite clearly that even such narrow drilling-locations could not verify that all structures are recognized. Lenses of coarser sands which contribute highly to transport processes were simply not identified. Obviously a dense sampling pattern as the one we used for data-set 9 is not practical, a more irregular sampling pattern than the one of data-set 6 will at least help to evaluate total variation of data. Whether the small correlation-structures can be evaluated this way is still questionable.

The question whether information of one well-investigated, lithofaciel comparable aquifer can be transferred to an actual area of interest where no measured data are available can be answered positively. Test-sites 2 and 3 give clear evidence on this. Together with data-set 1 the data-sets in Berlin support the assumption that given the lithofacial characteristics of an aquifer and its scale it is possible to simulate random-fields of hydraulic conductivity on the basis of geostatistical parameters like those evaluated here. Therefore a catalogue for different porous aquifers and observation scales is desirable. This paper gives prelimenary results of present research at Free University Berlin which among its investigation aims at compiling this catalogue.

The brief case-study of modeling the evolution of tracerplumes on the basis of indicator-simulations demonstrated again the big importance of considering spatial heterogeneity of flow-field when modeling contaminant transport in groundwater.

6. REFERENCES

Binsariti, A.A.: Statistical analysis stochastic modeling of the Cortaro aquifer in Southern Arizona.- Ph.D. dissertation, Dept. of Hydrology and Water Resources, Univ of Arizona, Tucson, Arizona, 1980.

Dagan, G.: Statistical theory of groundwater flow and transport: Pore to laboratory, laboratory to formation and formation to regional scale.- Water Resources Research 22 (9), pp. 120S-134S, 1986.

De Fouquet, C., H. Beucher, A. Galli, C. Ravenne: Conditional simulation of random sets. Application to an argilaceous sandstone reservoir.- In: M. Armstrong (ed.): Geostatistics, 2, pp. 517 - 530, Kluwer Acad. Press, 1989.

Delhomme, J.P.: Spatial variability and uncertainty in groundwater flow parameters: a geostatistical approach.- Water Resour. Res., 15 (2), pp. 269 - 280, 1979.

Dimitrakopoulos, R.: Characterization of petroleum reservoirs using geostatistics.- In: M. Armstrong (ed.): Geostatistics, 2, p. 1010, Kluwer Acad. Press, 1989.

Gelhar, L.W.: Stochastic subsurface hydrology from theory to application. - Water Resources Research 22 (9), pp. 135S-145S, 1986.

Gelhar, L.W. & Axness, C.L.: Three-dimensional stochastic analysis of macrodispersion in aquifers.- Water- Resour. Res. 19 (1), pp. 161-180, 1983.

Gomez-Hernandez, J.J. & R.M. Srivastava: ISIM3D: An ANSI-C three-dimensional multiple indicator conditional simulation program.- Computer & Geosciences, Vol. 16, No. 4, S. 395-440, 1990.

Hoeksema, R.J. & P.K. Kitanidis: Analysis of the spatial structure of properties of selected aquifers.- Water Resources Research 21 (4), pp. 563-572, 1985.

Matheron, G.: The intrinsic random functions and their applications.- Adv. Appl. Prob. **5**, pp 438 - 468, 1973.

Moltyaner, G.L. & R.W.D. Killey: Twin Lake tracer tests: Longitudinal dispersion.- Water Resour. Res., 24 (10), pp. 1613 - 1627, 1988a.

Moltyaner, G.L. & R.W.D. Killey: Twin Lake tracer tests: Transverse dispersion.- Water Resour. Res., 24 (10), pp. 1628 - 1637, 1988b.

Neuman, S.P.: Role of geostatistics in subsurface hydrology.- In "Geostatistics for nature resources characterization." Proc. NATO-ASI, Verly, G., David, M., Journel, A.G.,Maréchal, A. (eds.), Part 1, pp. 787 - 816, Reidel, Dordrecht, The Netherlands, 1984.

Schafmeister, M.-Th. & A. Pekdeger: Influence of spatial variability of aquifer properties on groundwater flow and dispersion.- In: H.E. Kobus & W. Kinzelbach (eds.): Contaminant Transport in Groundwater, pp. 215 - 220, Balkema, Rotterdam 1989

Schafmeister, M.-Th. & H. Burger: Spatial simulation of hydraulic parameters for fluid flow and transport models.- In: M. Armstrong (ed.): Geostatistics, 2, pp. 629 - 638, Kluwer Acad. Press, 1989.

Schafmeister, M.-Th.: Geostatistische Simulationstechniken als Grundlage der Modellierung von Grundwasserströmung und Stofftransport in heterogenen Aquifersystemen.- Ph.D. dissertation, Schelzky & Jeep (Publisher), 143 p., Berlin 1990.

Smith, L. & F.W. Schwartz: Mass transport. 1. A stochastic analysis of macrodispersion.- Water Resour. Res. 16 (2), pp. 303 - 313, 1980.

Smith, L. & F.W. Schwartz: Mass transport. 2. Analysisi of uncertainty in prediction.- Water Resour. Res. 17 (2), pp. 351 - 369, 1981a.

Smith, L. & F.W. Schwartz: Mass transport. 3. Role of hydraulic conductivity data in prediction.- Water Resour. Res. 17 (5), pp. 1462 - 1479, 1981b.

Smith, L. & R.A. Freeze: Stochastic analysis of steady-state groundwater flow in a bounded domain. 1. One-dimensional simulations.- Water Resour. Res. 15 (3), pp. 521 - 528, 1979a.

Smith, L. & R.A. Freeze: Stochastic analysis of steady-state groundwater flow in a bounded domain. 2. Two-dimensional simulations.- Water Resour. Res. 15 (6), pp. 1543 - 1559, 1979b.

7. ACKNOWLEDGEMENTS

The authors like to thank G. Moltyaner and D. Killey for the kind submittance of information on the Twin-Lake test-site. Above this we wish to thank a number of students who made a tremendous effort on providing us with such a big amount of data.

STUDY OF SPATIAL AND TEMPORAL VARIATIONS OF HYDROGEOCHEMICAL VARIABLES USING FACTORIAL KRIGING ANALYSIS

P. GOOVAERTS* and Ph. SONNET**
Université Catholique de Louvain
*Unité de Biométrie, Place Croix du Sud, 2 Bte 16
**Unité de Géologie et Minéralogie, Place Pasteur, 3
1348 Louvain-la-Neuve, Belgium

ABSTRACT. This paper describes a multivariate geostatistical approach for analysing the space-time behavior of properties from data which are dense in space and scarce in time. The T sampled times are considered as T coregionalized variables which are decomposed, using Factorial Kriging Analysis, into spatial components associated with different spatial scales. The temporal behavior of the property of interest is then studied for each of the identified spatial scales. Linear combinations of the spatial components, called regionalized factors, are cokriged and mapped to display the permanent and temporary features for a given spatial scale. The method is illustrated with a study of the space-time behavior of calcium, chloride and nitrate contents in spring-waters originating from the Dyle watershed (Belgium).

1. Introduction

In order to arrive at a clear understanding of the behavior of many natural systems, variations, both in time and space, must be studied. Geostatistics provides tools well suited for analysing spatial data. To account for time effects, several authors have proposed to extend the usual geostatistical techniques into the space-time domain by treating time as an additional dimension (Rodriguez-Iturbe and Mejia, 1974; Rouhani and Hall, 1989; Rouhani and Myers, 1990; Séguret and Huchon, 1990). Egbert and Lettenmaier (1986) have described a multivariate hierarchical stochastic model, in which the measured variables are represented by the sum of three multivariate random fields, each associated with a given time scale. The method involves decomposing the original variables into time components and studying the spatial variation for each time scale. These approaches necessitate a considerable amount of data both in time and space. However, in many situations, there is a strong disparity between the information available in the spatial and temporal domains. In some cases (e.g. piezometric or rain gauges), multiple time points are sampled for a small number of locations. In other cases (e.g. water pollution investigations), samples are collected at numerous locations but on few occasions. Clearly, the density of sampling dictates whether it is the spatial or the temporal structure which can be estimated with reliability. For data dense in time and scarce in space, Rouhani and Wackernagel (1990) have proposed a multivariate approach, in which the time series at each of the N spatial locations are considered as N separate one-dimensional regionalized variables. After the decomposition of each time series into several temporal components, using Factorial Kriging Analysis (FKA), the spatial relationships between the different locations are studied for each time scale.

This paper focuses on situations where the sampling is denser in space than in time. By analogy to the Rouhani and Wackernagel's approach, we propose to perform a F.K.A. (Matheron, 1982) on the set of T times that is viewed as a realization of a set of T inter-correlated random functions. The method involves fitting a linear model of coregionalization to the T(T+1)/2 simple and cross-variograms, decomposing the T coregionalized variables into orthogonal components associated with different spatial scales, and studying the temporal behavior of the property for each detected spatial scale. In addition, the technique provides regionalized factors summarizing the permanent and temporary features of the property for each spatial scale.

The method is applied to a hydrogeochemical data set that includes the measurements at six years (1975,79,80,81,82,83) of the concentrations of three ions (Ca, Cl, NO_3) in 68 springs originating from the Dyle watershed (Belgium). The spatial and temporal behavior of each of the three ions is first investigated separately and a comparison of the three ions is then presented.

2. Theory

Let $\{z_t(\mathbf{x}_\alpha)\ ;\ t = 1,...,T\ \ \alpha = 1,...,N\}$ be a set of data resulting from the sampling of the regionalized variable Z at the same N spatial locations for each of the T times. The set of T regionalized variables $\{z_t(\mathbf{x})\ ;\ t = 1,...,T\}$ is interpreted as one particular realization of a set of T inter-correlated random functions $\{Z_t(\mathbf{x})\ ;\ t = 1,...,T\}$. Assuming second-order stationarity of the spatial increments $(Z_t(\mathbf{x}) - Z_t(\mathbf{x+h}))$, we have

$$E\left[Z_t(\mathbf{x}) - Z_t(\mathbf{x+h})\right] = 0$$

$$\frac{1}{2} E\left[\{Z_t(\mathbf{x}) - Z_t(\mathbf{x+h})\}\{Z_{t'}(\mathbf{x}) - Z_{t'}(\mathbf{x+h})\}\right] = \gamma_{tt'}(\mathbf{h})$$

For t'=t, $\gamma_{tt'}(\mathbf{h})$ is the simple variogram $\gamma_{tt}(\mathbf{h})$ and describes the spatial behavior of the variable Z at a given time t. For $t'\neq t$, $\gamma_{tt'}(\mathbf{h})$ is the cross-variogram which takes into account both the spatial and temporal variations. The cross-variogram is related to the simple variograms by (Myers, 1988)

$$2\gamma_{tt'}(\mathbf{h}) = \gamma_{tt}(\mathbf{h}) + \gamma_{t't'}(\mathbf{h}) - \gamma_{t-t'}(\mathbf{h}) \tag{1}$$

where $\gamma_{t-t'}(\mathbf{h})$ is the simple variogram of the difference between the values measured at times t and t'

$$\gamma_{t-t'}(\mathbf{h}) = \frac{1}{2} E\left[\{(Z_t(\mathbf{x}) - Z_{t'}(\mathbf{x})) - (Z_t(\mathbf{x+h}) - Z_{t'}(\mathbf{x+h}))\}^2\right] \tag{2}$$

In what follows, the difference $(Z_t(\mathbf{x}) - Z_{t'}(\mathbf{x}))$ will be referred to as the temporal change for a time lag $\Delta t=|t'-t|$. As the set of T variables only represents the same property being measured T times, one can consider averaging the quantities $\gamma_{tt'}(\mathbf{h})$ and $\gamma_{t-t'}(\mathbf{h})$ in the following way

$$\overline{\gamma}^{\Delta t}_{tt'}(\mathbf{h}) = \frac{1}{N_{\Delta t}} \sum_{t=1}^{T} \sum_{t'=t}^{T} \delta^{\Delta t}_{|t'-t|}\ \gamma_{tt'}(\mathbf{h}) \tag{3}$$

$$\overline{\gamma}^{\Delta t}_{t-t'}(\mathbf{h}) = \frac{1}{N_{\Delta t}} \sum_{t=1}^{T} \sum_{t'=t}^{T} \delta^{\Delta t}_{|t'-t|}\ \gamma_{t-t'}(\mathbf{h}) \tag{4}$$

where $N_{\Delta t}$ is the number of pairs of sampled times, a time lag Δt apart, and $\delta^{\Delta t}_{|t'-t|}$ is the Kronecker delta ($\delta^{\Delta t}_{|t'-t|}=1$ if $|t'-t|=\Delta t$ and $\delta^{\Delta t}_{|t'-t|}=0$ otherwise). $\bar{\gamma}^{\Delta t}_{tt'}(\mathbf{h})$ is the mean cross-variogram of the variables Z_t and $Z_{t'}$ repeatedly measured at a time lag Δt apart. It describes the spatial dependence between values of the variable Z observed at a time lag Δt apart. When $\Delta t=0$, $\bar{\gamma}^{\Delta t}_{tt'}(\mathbf{h})$ is simply the arithmetic mean of all the available simple variograms

$$\bar{\gamma}^{\Delta t=0}_{tt'}(\mathbf{h}) = \frac{1}{T}\sum_{t=1}^{T} \gamma_{tt}(\mathbf{h}) \tag{5}$$

The variogram $\bar{\gamma}^{\Delta t=0}_{tt'}(\mathbf{h})$ describes the average spatial structure of the variable Z. $\bar{\gamma}^{\Delta t}_{t-t'}(\mathbf{h})$ is the mean variogram of the temporal changes for the time lag Δt. It characterizes the regionalization of the temporal changes over a period Δt. From equations (1), (3) and (4), we obtain the following relation

$$\bar{\gamma}^{\Delta t}_{tt'}(\mathbf{h}) = \frac{1}{2\,N_{\Delta t}}\sum_{t=1}^{T}\sum_{t'=t}^{T} \delta^{\Delta t}_{|t'-t|}\left(\gamma_{tt}(\mathbf{h})+\gamma_{t't'}(\mathbf{h})\right) - \frac{1}{2}\,\bar{\gamma}^{\Delta t}_{t-t'}(\mathbf{h}) \tag{6}$$

The first right-hand side term is a linear combination of the T simple variograms $\gamma_{tt}(\mathbf{h})$ which is close to the arithmetic mean if the following conditions are met : the measurements have been made at regular time lags, Δt is a small multiple of that time lag and the variances of the T variables have the same order of magnitude. Hence, we take as approximation to this term the value of $\bar{\gamma}^{\Delta t}_{tt'}(\mathbf{h})$ for $\Delta t=0$, i.e. $\bar{\gamma}^{\Delta t=0}_{tt'}(\mathbf{h})$. Equation (6) can thus be rewritten as

$$\bar{\gamma}^{\Delta t=0}_{tt'}(\mathbf{h}) - \bar{\gamma}^{\Delta t}_{tt'}(\mathbf{h}) \approx \frac{1}{2}\,\bar{\gamma}^{\Delta t}_{t-t'}(\mathbf{h}) \tag{7}$$

Equation (7) states that if, for any time lag Δt, there is a difference between the mean cross-variograms $\bar{\gamma}^{\Delta t}_{tt'}(\mathbf{h})$ and $\bar{\gamma}^{\Delta t=0}_{tt'}(\mathbf{h})$, this difference is directly linked to the spatial variability of the temporal changes. Furthermore, if the temporal changes are spatially structured, this difference varies as a function of **h**.

While $\bar{\gamma}^{\Delta t}_{tt'}(\mathbf{h})$ and $\bar{\gamma}^{\Delta t}_{t-t'}(\mathbf{h})$ provide useful insight into the space-time behavior of the variable Z, a detailed study requires an exhaustive analysis of the T(T+1)/2 variograms. The variogram matrix $\Gamma(\mathbf{h})$ is a TxT symmetric matrix in which the diagonal and off-diagonal elements are respectively the simple and cross-variogram values for a lag **h**

$$\Gamma(\mathbf{h}) = \left[\gamma_{tt'}(\mathbf{h})\right] = \begin{bmatrix} \gamma_{11}(\mathbf{h}) & \gamma_{12}(\mathbf{h}) & \cdots & \gamma_{1T}(\mathbf{h}) \\ \gamma_{21}(\mathbf{h}) & \gamma_{22}(\mathbf{h}) & \cdots & \gamma_{2T}(\mathbf{h}) \\ \vdots & \vdots & & \vdots \\ \gamma_{T1}(\mathbf{h}) & \gamma_{T2}(\mathbf{h}) & \cdots & \gamma_{TT}(\mathbf{h}) \end{bmatrix} \tag{8}$$

If the measurements have been made at regular time lags and provided the rows of $\Gamma(\mathbf{h})$ are arranged in order according to time, the quantity $\bar{\gamma}^{\Delta t}_{tt'}(\mathbf{h})$ corresponds to the average of the principal ($\Delta t=0$) or secondary ($\Delta t>0$) diagonal elements. In the same manner, an approximate value for $\bar{\gamma}^{\Delta t}_{t-t'}(\mathbf{h})$ is twice the difference between the average of the principal diagonal elements and the average of the corresponding secondary diagonal elements.
Each variogram matrix $\Gamma(\mathbf{h})$ is a variance-covariance matrix of spatial increments which

describes the correlation structure associated with a given lag $\mathbf{h}$. Especially when the correlations among the variables change according to the spatial scale, Goovaerts (1992) has shown that the coregionalization matrices $\mathbf{B}^u$ provide a better description of the underlying correlation structures than the variogram matrices $\Gamma(\mathbf{h})$. The matrices $\mathbf{B}^u$ are estimated by modelling the variogram matrix using a nested variogram model composed of N_s basic variogram functions $g^u(\mathbf{h})$

$$\Gamma(\mathbf{h}) = \sum_{u=1}^{N_s} \mathbf{B}^u \, g^u(\mathbf{h}) \tag{9}$$

where $\mathbf{B}^u = \left[b^u_{tt'}\right]$ is a TxT positive semi-definite matrix which contains the coefficients $b^u_{tt'}$ that express the relative importance of the basic variogram function $g^u(\mathbf{h})$ in the modelling of $\gamma_{tt'}(\mathbf{h})$. Using the linear model of coregionalization involves decomposing the random functions $Z_t(\mathbf{x})$ into N_s orthogonal random functions $Z^u_t(\mathbf{x})$

$$Z_t(\mathbf{x}) = \sum_{u=1}^{N_s} Z^u_t(\mathbf{x}) \tag{10}$$

Each spatial component $Z^u_t(\mathbf{x})$ represents the behavior of the random function $Z_t(\mathbf{x})$ for the spatial scale u. Assuming second-order stationarity, the coregionalization matrix $\mathbf{B}^u$ is actually the variance-covariance matrix of the T components $Z^u_t(\mathbf{x})$. Therefore, each $\mathbf{B}^u$ describes the correlation structure associated with the spatial scale u.

By decomposing each $Z_t(\mathbf{x})$ into several spatial components $Z^u_t(\mathbf{x})$ one can study the temporal behavior of the variable Z for different spatial scales. For a given spatial scale u, the correlation between the different times is measured by the structural correlation coefficient $r^u_{tt'}$ (Wackernagel, 1988) computed from $\mathbf{B}^u$ as

$$r^u_{tt'} = \frac{b^u_{tt'}}{\sqrt{b^u_{tt}\, b^u_{t't'}}} \tag{11}$$

The T components $Z^u_t(\mathbf{x})$ associated with the spatial scale u are expected to share common features that can be interpreted as the time-independent - or permanent - spatial background. On the other hand, some spatial features are not found systematically every time but are observed a few times or even only once. These are here referred to as temporary features. To represent graphically these common or more specific spatial patterns, we map linear combinations of the spatial components, called regionalized factors. These factors are taken as the principal components extracted from the coregionalization matrices $\mathbf{B}^u$. In most cases, the regionalized factor summarizing the common features is the first principal component which accounts for most of the total variance. This first principal component corresponds to a size factor if all the elements of the matrix $\mathbf{B}^u$ are positive (Saporta, 1990) which is frequent when dealing with measurements of the same property at different times. All the samples are arranged along the size factor according to increasing values of all the variables (in average). Furthermore, when the correlations among the variables have the same order of magnitude, the size factor is proportional to the average of the original variables. The other regionalized factors outline opposition between several times or characterize one time only. If the T times are poorly correlated, the permanent spatial background is expected to contribute to a small extent to the spatial patterns of the $Z^u_t(\mathbf{x})$. In this case, the common factor could be associated with a principal component of higher order.

The values of the regionalized factors must be estimated by cokriging at the sampled points as well as at unknown points using the information provided by the T variables $Z_t(\mathbf{x})$. The estimator of the regionalized factor Y_v^u, i.e. the vth principal component (v=1,...,T) for the uth spatial scale (u=1,...,N_s), at a given point $\mathbf{x}_0$, is

$$Y_v^{u*}(\mathbf{x}_0) = \sum_{t=1}^{T} \sum_{\alpha=1}^{n} \lambda_{\alpha t}\, z_t(\mathbf{x}_\alpha) \tag{12}$$

where n is the number of data points used in the estimation and $\lambda_{\alpha t}$ is the weight assigned to the value of the variable $Z_t(\mathbf{x})$ at the αth location, $z_t(\mathbf{x}_\alpha)$. Minimizing the estimation variance under the constraint of unbiasedness leads to the following system of T(n+1) linear equations

$$\begin{cases} \sum_{t'=1}^{T} \sum_{\beta=1}^{n} \lambda_{\beta t'}\, \gamma_{tt'}(\mathbf{x}_\alpha,\mathbf{x}_\beta) - \mu_t = a_{tv}^u\, g^u(\mathbf{x}_\alpha,\mathbf{x}_0) \\ \sum_{\beta=1}^{n} \lambda_{\beta t} = 0 \end{cases} \tag{13}$$

for $t = 1,...,T \quad \alpha = 1,...,n$

with $a_{tv}^u = q_{tv}^u \sqrt{l_v^u}$ where l_v^u is the eigenvalue associated with the vth eigenvector $\mathbf{q}_v^u$ of the coregionalization matrix $\mathbf{B}^u$.

3. Case study

In more than 100 springs in the Dyle watershed (Belgium), concentrations of several ions were measured at irregular time intervals from 1975 to 1983. This case study includes the measurements at six years (1975,79,80,81,82,83) of the concentrations of 3 ions (Ca, Cl, NO_3) in 68 springs. All the springs take place in the same aquifer which is a layer of Eocene calcareous sand (50 m average thickness) overlaying a folded Paleozoic basement. Fig. 5 shows the location of the springs indicated by circles.

The mean and the variance of the 3 ion concentrations show no temporal trend (Table 1). In general, the correlation between the measurements made for different years, $r_{tt'}$, is high and varies between 0.70 and 0.97 (Table 2). For each year, the concentrations of the 3 ions are positively correlated, the weakest correlation being between Ca and NO_3 (Table 3).

Table 1. Mean and standard deviation of the three ions for each year.

	Ca (ppm)		Cl (ppm)		NO_3 (ppm)	
Year	$\overline{X}$	S	$\overline{X}$	S	$\overline{X}$	S
1975	100	36.6	45.7	15.1	26.2	10.4
1979	103	32.6	48.2	12.8	29.2	10.3
1980	106	33.8	51.0	14.1	30.7	11.4
1981	103	36.6	49.4	17.1	35.1	11.5
1982	108	34.7	54.3	16.3	32.2	11.6
1983	101	33.2	49.9	13.7	27.2	10.4

Table 2. Mean and range of the statistical correlation coefficient ($r_{tt'}$) and of the structural correlation coefficients ($r^2_{tt'}, r^3_{tt'}$) for each ion.

	Correlation coefficients		
Ion	$r_{tt'}$	$r^2_{tt'}$	$r^3_{tt'}$
Ca	0.93 (0.88-0.97)	0.83 (0.64-0.98)	0.98 (0.93-0.99)
Cl	0.80 (0.70-0.95)	0.87 (0.74-0.98)	0.82 (0.62-0.92)
NO3	0.83 (0.70-0.95)	0.93 (0.88-0.97)	0.84 (0.48-0.99)

Table 3. Correlation coefficients between the three ions for each year.

	Year					
Ion	1975	1979	1980	1981	1982	1983
Ca-Cl	0.59	0.42	0.44	0.48	0.52	0.52
Ca-NO_3	0.12	0.10	0.23	0.20	0.28	0.13
Cl-NO_3	0.51	0.57	0.69	0.68	0.75	0.64

For each ion, the 21 experimental isotropic simple and cross-variograms have been computed and averaged according to equations (3), (4) and (5). Due to the small number of sampled times, $\bar{\gamma}^{\Delta t}_{tt'}(h)$ and $\bar{\gamma}^{\Delta t}_{t-t'}(h)$ have been calculated only for Δt=0, 1 and 2 years. They are represented in Fig. 1. All the mean variograms $\bar{\gamma}^{\Delta t=0}_{tt'}(h)$ have a small nugget effect. The shape of the Ca variogram is dominated by a long-range structure whereas the two other variograms (Cl, NO_3) are characterized by the predominance of a short-range structure. A previous study based on a larger number of properties measured in 1975 in the same springs has provided a meaningful interpretation for these two structures (Goovaerts et al., 1992). The short-range structure can be linked to the presence of local sources of pollution such as farms, municipal solid waste dumps, or urbanized areas. On the other

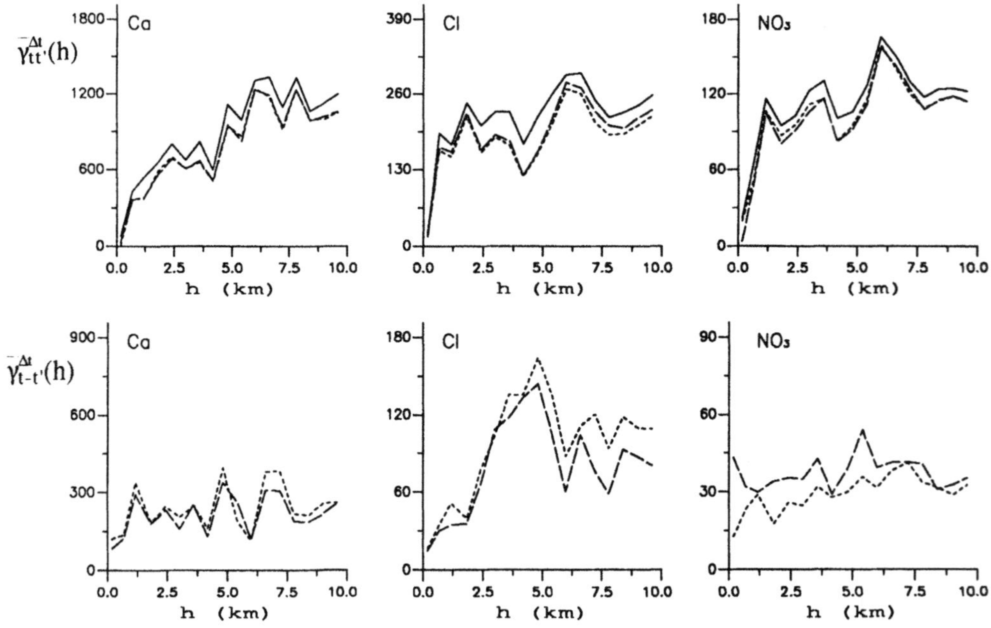

Fig. 1 Plots of $\bar{\gamma}^{\Delta t}_{tt'}(h)$ and $\bar{\gamma}^{\Delta t}_{t-t'}(h)$ for Δt=0 (—), 1 year (— —) and 2 years (- - -).

hand, the long-range structure is mainly due to changes of the geological characteristics of the aquifer occurring for large spatial scales. The relative contributions of the two structures to the variograms reflect the role played by the anthropic or the geological factors in the spatial pattern displayed by each ion. Indeed, the main source of NO_3 and Cl can be attributed to the human activity whereas the presence of calcite in the aquifer controls to a large extent the Ca content of the spring-water.

From $\Delta t=0$ to $\Delta t=1$, the decrease in the values of $\overline{\gamma}_{tt'}^{\Delta t}(h)$ is small because of the generally high correlation between the different years. For Cl, the difference between $\overline{\gamma}_{tt'}^{\Delta t=0}(h)$ and $\overline{\gamma}_{tt'}^{\Delta t=1}(h)$ varies as a function of the distance h, which indicates that changes occurring from one year to another depend on the spatial scale. As a matter of fact, $\overline{\gamma}_{t-t'}^{\Delta t}(h)$ show that the temporal changes are on the average spatially structured for Cl and NO_3 (for $\Delta t=2$ only) and unstructured for Ca. The most important and the best spatially structured temporal changes are observed for Cl and NO_3. These two ions are mainly influenced by the human activity which in turn is expected to vary in time and according to a spatial pattern.

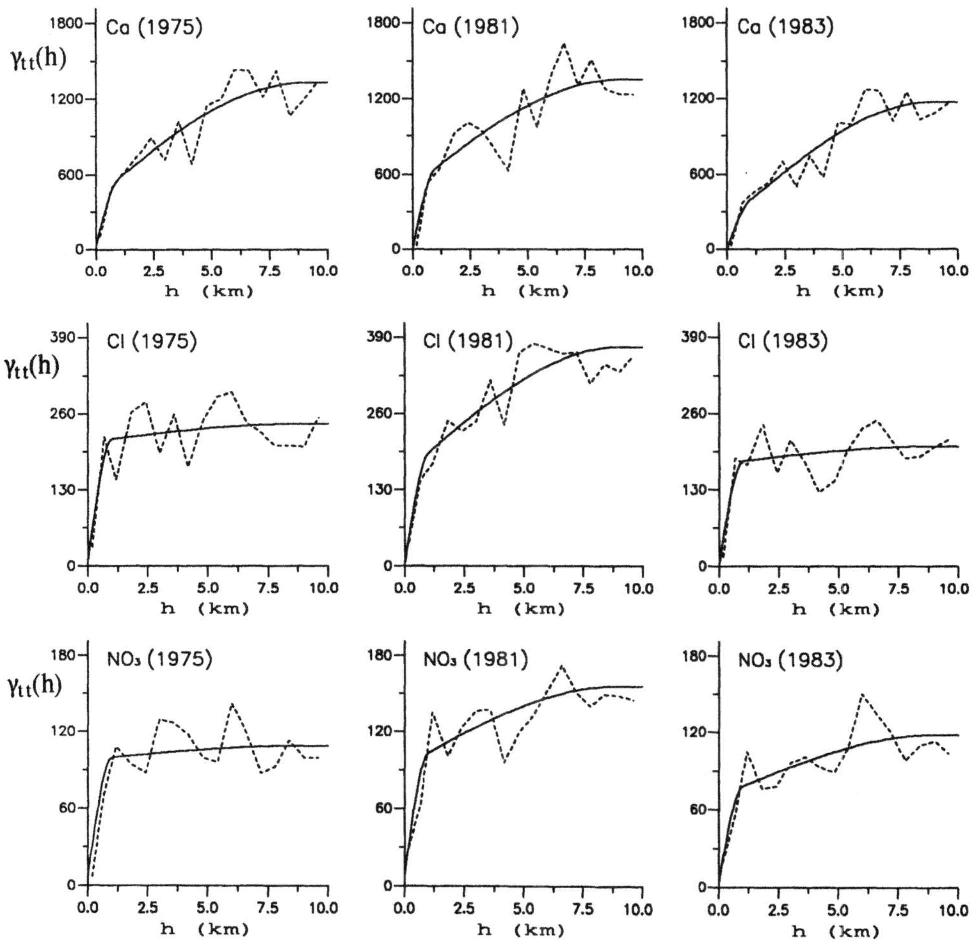

Fig. 2 Experimental simple variograms with the model fitted (solid line) for the concentrations measured in 1975, 1981 and 1983.

The experimental isotropic variograms were modelled using a nugget effect and two spherical schemes with a range of 1 km and 9 km (Fig. 2). The linear model of coregionalization was fitted using an iterative procedure developed by Goulard (1989). The structural correlation coefficients (Table 2) have been computed from the two coregionalization matrices $\mathbf{B}^2$ (range=1km) and $\mathbf{B}^3$ (range=9km) according to equation (11). For Ca, the correlation between the concentrations measured at different years is higher for large spatial scale. For Cl and NO_3, the correlation is higher for small scale. This indicates that the temporal persistence of the concentrations is the greatest for the spatial scale that contributes the most to the overall variation, i.e the large spatial scale for Ca and the small spatial scale for Cl and NO_3.

Fig. 3 gives the results of a principal component analysis performed on the variance-covariance matrix and the coregionalization matrices $\mathbf{B}^2$ and $\mathbf{B}^3$. The correlations between each variable and the first two principal components are laid down on a graph called the circle of correlations. In each case, the first two principal components account for at least 90% of the variance encountered at the corresponding spatial scale. The high positive correlation between the concentrations measured for the different years explains

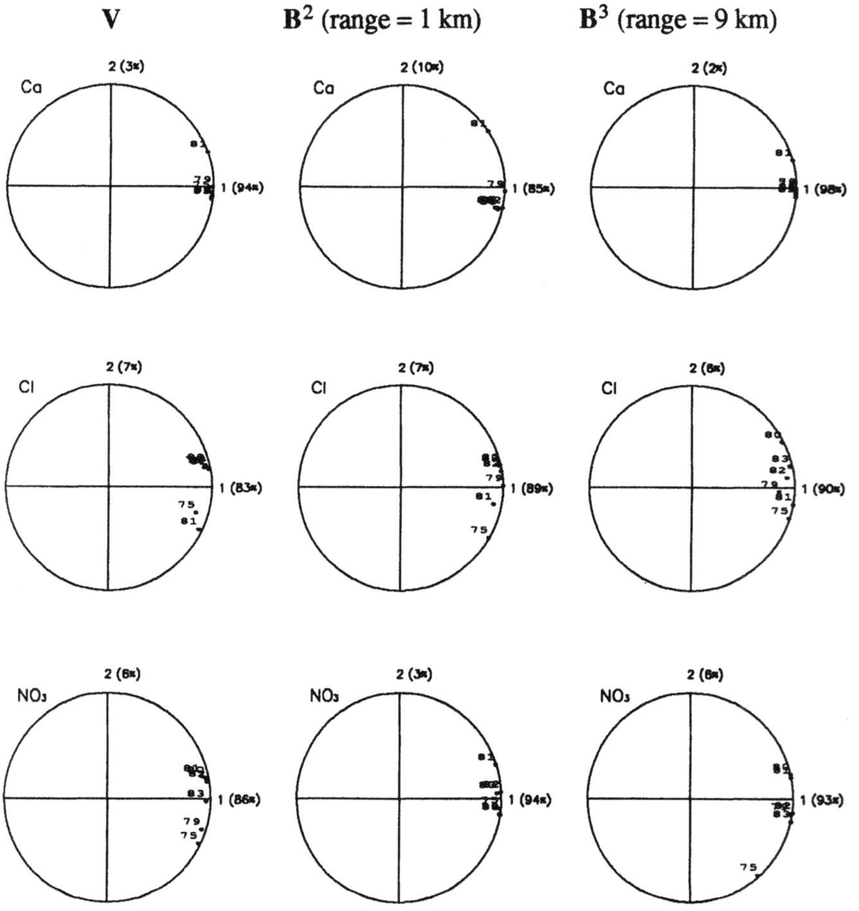

Fig. 3 Circles of correlations computed from the variance-covariance matrix (**V**) and the two coregionalization matrices ($\mathbf{B}^2$, $\mathbf{B}^3$) associated with the spherical models.

why there is a strong positive correlation between the years and the first principal component. Further, as the range of the classical or structural correlation coefficients is small in most cases (Table 2), each year contributes in a very similar way to the first principal component. As mentioned in the theoretical section, the first component can be identified with a size factor and, in this particular case, it is proportional to the average of the six measurements. The second component allows to differentiate or to regroup the years. However, in this case study, the proportion of variance explained by the second component is small (2-10%), which can make it sensitive to slight differences in the estimated values of $\mathbf{B}^u$. To assess this sensitivity, the linear model of coregionalization was fitted using two sets of weights for the lags, i.e. $w(h)=1$ and $w(h)=N(h)/(\gamma_{tt'}^{\Delta t=0}(h))^2$ where $N(h)$ is the number of pairs involved at the lag h. For each ion, large discrepancies were observed for the nugget component which is the least important structure (Figure 2) and is not interpreted here. For Cl, differences in the contribution of the years 1979 and 1982 to the long-range second component were found. Therefore, this component should be interpreted with caution. The other components are quite stable. The first column of circles of correlations shows that Cl and NO_3 behaved in a dissimilar way for the year 1975. The analysis of the coregionalization matrices enables us to distinguish between the two ions according to the spatial scale for which this distinct behavior is the most obvious : the small scale for Cl and the large scale for NO_3. This suggests that two genetic mechanisms have probably been operating at the same time but at two different spatial scales.

Figures 4 and 5 present the maps of the first two regionalized factors associated with each of the two spherical models. The long-range factors were cokriged at nodes of a regular grid (mesh dimensions : 1 x 1 km). In order to improve the map readability, the short-range factors were cokriged only at the locations of the springs. Regarding the long-range model (Figure 4), the first regionalized factor (F1) depicts features in the data that are somewhat continuous in space and permanent in time. A high value at a particular point indicates that, for each of the six sampled years, the ion content was high on the average due to a process acting at large spatial scale. The F1 maps show that Cl and NO_3 have more or less similar permanent spatial backgrounds. The north-south central area of low values corresponds to the valley of the Dyle River, where the thickness of the aquifer is smaller and, possibly, a closer hydraulic contact exists with surface waters. The central spot of high values is located where several springs drain the aquifer around a large sandpit which has been used for solid waste disposal for several decades. Other features which are observed on the F1 maps for Ca, Cl and NO_3 can be attributed to large-scale hydraulic or geological characteristics of the aquifer (thickness of the aquifer, depth of water travel path, presence of carbonate-rich layers) and to the geographical spread of polluting agents (large-scale use of nitrate-rich fertilizers). The second regionalized factor (F2) depicts large-scale features which are temporary. Such features could be related to the quantity of rainfall recharge or pumping which occurred before the springs were sampled. For Ca, the high correlation between the concentrations measured for the six sampled years explains the low variability of the second regionalized factor which is essentially due to year 1981. The small-range factors (Figure 5) summarize the main sources of variation that remain when the punctual and large-scale regional variations are filtered out. The first regionalized factor depicts the small-scale behavior of the ions that is observed on the average for each sampled year. A high value could be explained by the nearby presence of local permanent sources of pollutants (solid waste dumping site, septic tank for disposal of household wastes). A high or low value for the second factor indicates that the small-scale behavior of the ion content is variable in time. Possible causes could be the mixing with surface runoff water enriched in salt and nitrate leachate (salt used for road de-icing, nitrate from fields or cattle farms) or any sources of pollutants linked to human activity (leaks in sewage systems, for example).

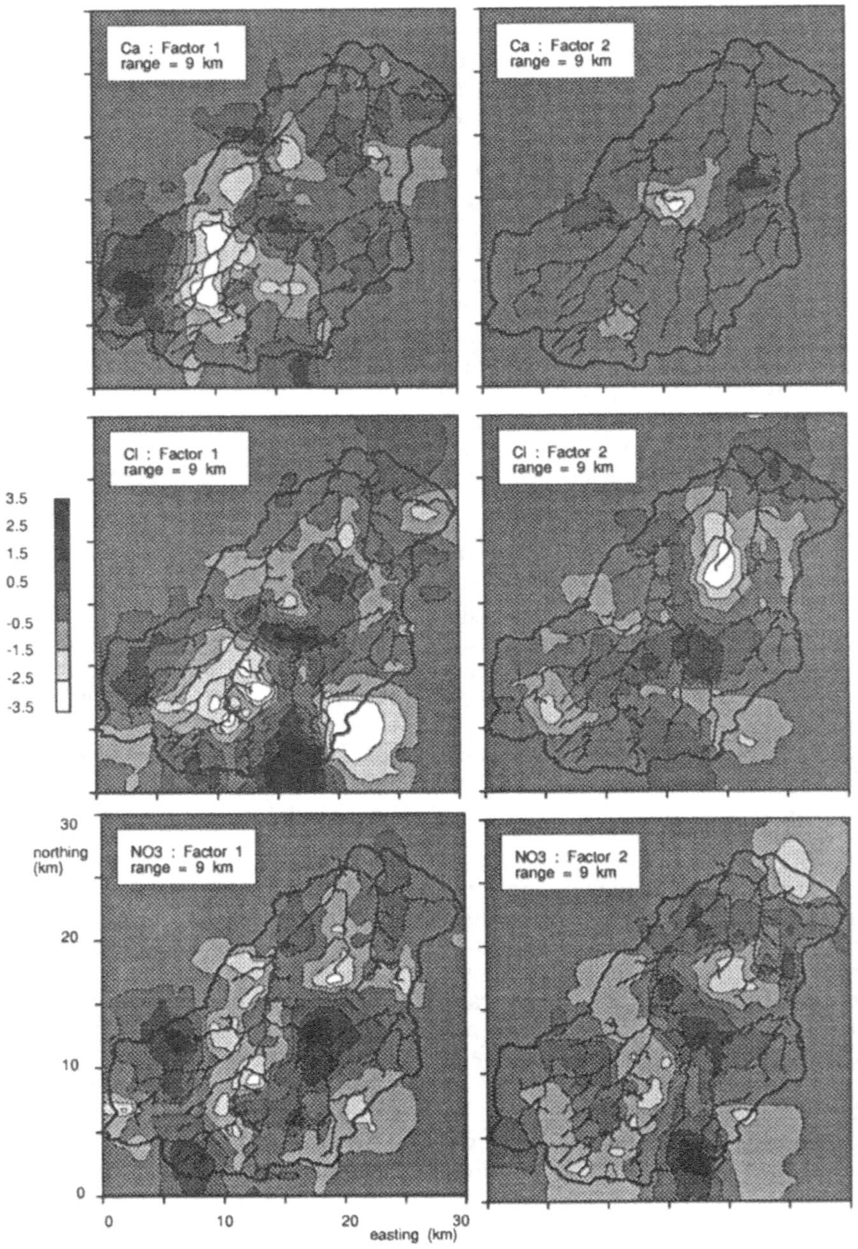

Fig. 4 Contour maps of the first two regionalized factors associated with the long-range spherical model. The factors were estimated by cokriging on a 1 km square grid. Lighter shading denotes lower values. The thick line marks the limit of the Dyle River catchment area ; the thin lines outline the river and its tributaries.

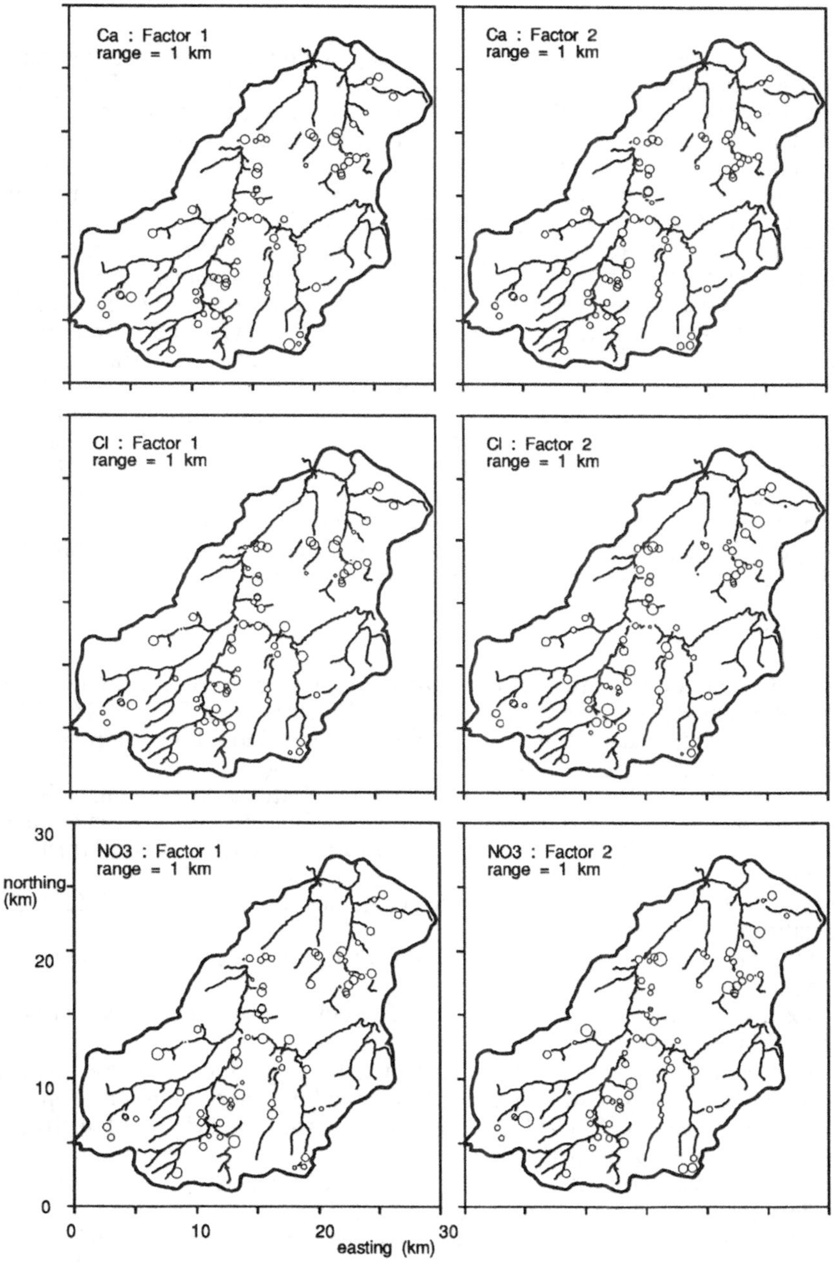

Fig. 5 Values of the first two regionalized factors associated with the short-range spherical model. The factors were estimated by cokriging at the locations of the springs. The radius of each circle is proportional to the value of the factor. The thick line marks the limit of the Dyle River catchment area ; the thin lines outline the river and its tributaries.

4. Conclusions

This paper demonstrates the suitability of a multivariate geostatistical approach for studying the space-time behavior of a property when only a few T times have been sampled. The first stage involves computing the mean variograms $\overline{\gamma}^{\Delta t}_{tt'}(\mathbf{h})$ and $\overline{\gamma}^{\Delta t}_{t-t'}(\mathbf{h})$ which describe the average spatial structure of the property and detect the regionalized nature of the temporal changes. The second stage requires decomposing each of the T times into spatial components, making it possible to characterize the time behavior for different spatial scales. The third stage entails performing a principal component analysis on the coregionalization matrices and cokriging the regionalized factors, thereby enabling the permanent and temporary features to be distinguished for each spatial scale.

The mean cross-variograms $\overline{\gamma}^{\Delta t}_{tt'}(\mathbf{h})$ clearly illustrate the evolution of the spatial relationships between Ca, Cl and NO_3 concentrations measured at different years as a function of the time lag Δt. The study of the mean variograms $\overline{\gamma}^{\Delta t}_{t-t'}(\mathbf{h})$ shows that temporal changes of Cl and NO_3 concentrations are spatially structured. The correlation between the different years is then expected to vary from one spatial scale to another. This is highlighted by the analysis of the coregionalization matrices which shows that, especially in 1975, the two ions behaved in a dissimilar way for each spatial scale. Based on the values of the regionalized factors, each spring can be classified according to temporal variation and spatial continuity. This makes it possible to ascertain whether or not a spring has abnormal behavior either in time or in space. Such information can prove valuable when deciding upon measures to protect underground water resources.

References

Egbert, G.D. and Lettenmaier, D.P. (1986) "Stochastic modeling of the space-time structure of atmospheric chemical deposition", Water Resour. Res. 22, 165-179.

Goovaerts, P. (1992) "Factorial Kriging Analysis : a useful tool for exploring the structure of multivariate spatial soil information", J. Soil Sci. 43(4), In press .

Goovaerts, P., Sonnet, Ph. and Navarre, A. (1992) "Factorial kriging analysis of spring-water contents in the Dyle river basin, Belgium", Water Resour. Res. (submitted).

Goulard, M. (1989) "Inference in a coregionalization model", in M. Armstrong (ed.), Geostatistics, Kluwer Academic Publishers, Dordrecht, pp. 397-408.

Matheron, G. (1982) "Pour une analyse krigeante des données régionalisées", Centre de Géostatistique, Report N-732, Fontainebleau.

Myers, D.E. (1988) "Multivariable geostatistical analysis for environmental monitoring", Sci. de la Terre, Sér. Inf. 27, 411-427.

Rodriguez-Iturbe, I. and Mejia, J.M. (1974) "The design of rainfall networks in time and space", Water Resour. Res. 10, 713-728.

Rouhani, S. and Hall, T.J. (1989) "Space-time kriging of groundwater data", in M. Armstrong (ed.), Geostatistics, Kluwer Academic Publishers, Dordrecht, pp. 639-651.

Rouhani, S. and Myers, D.E. (1990) "Problems in space-time kriging of geohydrological data", Math. Geol. 22, 611-623.

Rouhani, S. and Wackernagel, H. (1990) "Multivariate geostatistical approach to space-time data analysis", Water Resour. Res. 26, 585-591.

Saporta, A. (1990) Probabilités, analyse des données et statistique, Technip, Paris.

Séguret, S. and Huchon, P. (1990) "Trigonometric kriging : a new method for removing the diurnal variation from geomagnetic data", J. Geophys. Res. 95, 21,383-21,397.

Wackernagel, H. (1988) "Geostatistical techniques for interpreting multivariate spatial information", in C.F. Chungs et al. (eds.), Quantitative analysis of mineral and energy resources, D. Reidel Publishers, Dordrecht, pp. 393-409.

CHANNEL TUNNEL: GEOSTATISTICAL PREDICTION FACING THE ORDEAL OF REALITY

Raymonde Blanchin and Jean-Paul Chilès

BRGM, BP 6009, 45060 Orléans Cédex 2, France

ABSTRACT. Geometrical modelling of geological layers in the Channel Tunnel area used geostatistical techniques for the computing of predictions and confidence intervals. The results, especially those concerning the top of the Gault Clay, were used to design the route of the tunnel. During the tunnelling, subvertical boreholes were drilled downwards from inside the tunnels at various intervals, to check the real level and dip of the top of the Gault clay. This provided an outstanding opportunity to confront the geostatistical prediction with reality. A critical analysis of the discrepancies assessed the quality and reliability of the model, and validated the geostatistical methods.

KEYWORDS: Geostatistics; kriging; survey design; risk analysis; validation.

PROBLEMS AND METHODOLOGY

The presence of a favourable layer for boring a tunnel beneath the Channel between France and Great Britain, has been well known since the 1876-1882 surveys for such a project (see Fig. 1). This layer, Cenomanian Chalk Marl, is an ideal medium for tunnelling as it is composed of relatively soft and virtually impermeable chalk and clay. The Chalk Marl is overlain by the Grey Chalk, a highly porous layer, typically consisting of fractured and altered rocks, and is underlain by the Gault Clay that cannot be penetrated without serious civil-engineering problems.

The Channel Tunnel consists in fact of three tunnels two for transport and a smaller one for servicing. Although the optimization of the tunnel layout obviously had to take into consideration the slope and curvature constraints imposed by a high-speed railway, the primary constraints were geological and geotechnical: no geophysical borehole had to be intersected and the tunnels had to be sufficiently deep beneath the sea bed, and, above all, within the Chalk Marl formation. As this layer is inclined and only 30 metres thick, a precise knowledge of its top and bottom was essential. These horizons were determined by means of a geostatistical analysis of bathymetric and seismic-reflection surveys (1,500 km of seismic profiles and 19 boreholes during the 1986 and 1988 surveys), plus the data from 90 old boreholes drilled after 1958).Hereafter, we shall discuss primarily the top of the Gault clay (i.e. the bottom of the Chalk Marl), which is the most critical variable.

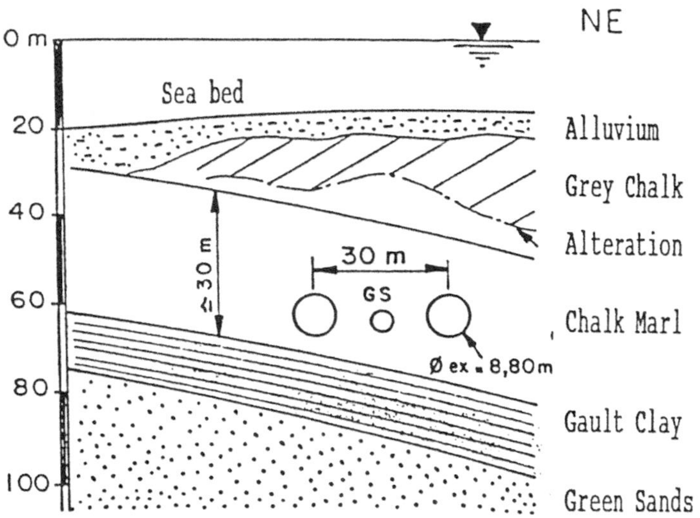

Fig. 1. Typical geological cross-section with the three tunnels.

STUDY OF THE TUNNEL AS A WHOLE

The principal objective was to prevent the risk of tunnelling within clay. Geostatistics, or more precisely kriging, was used to calculate a reliable digital model of the top of the Gault Clay and for producing representative maps and sections. The originality of kriging is to provide at each interpolated point both the estimate and its confidence level, characterized by a standard deviation. This enables the risk of a discrepancy between prediction and reality to be quantified. Blanchin *et al.* (1989) describes the geostatistical study of the overall survey. The interpretation was done in four phases:

1 - Estimation of the depth of the sea bed, S(x), from the bathymetric profile data.
2 - Estimation of the two-way time from sea-floor to top of the Gault Clay, T(x), from the seismic profile data.
3 - Estimation of the average velocity, V(x), from various geophysical data (retro-analysis at boreholes, sonic boreholes, fine-grain velocity analysis along profiles).
4 - Combination of the above results to provide the estimation of the depth to the top of the Gault Clay from sea level, G(x), by:

$$G(x) = S(x) + V(x)\,T(x)/2$$

The estimations take into account:

- the various sources of measurement error (location uncertainty, tide correction, migration of seismic reflectors, velocity calculations), their order of magnitude, and their degree of spatial correlation (for profile data);

- the local variability (degree of continuity, trend, anisotropy) of each variable, by means of a local variography by zones of about 2 km; this is essential as we want meaningful kriging standard deviations.

The kriging standard deviation was computed for each interpolated field. As the various estimates are independent, the precision σ_G concerning the top of the Gault is given by:

$$\sigma_G^2 = \sigma_S^2 + (\sigma_V^2 . T^2 + V^2 . \sigma_T^2 + \sigma_V^2 . \sigma_T^2)/4$$

In this expression, σ_S, σ_T, σ_V are the standard deviations attached to fields S, T, V, respectively, and all values depend on the location x.

GEOMETRICAL MODEL AND LAYOUT DESIGN

The investigation area was a corridor 1000 m wide centred on the main axis of the first planned layout. Figures 2 and 3 illustrate the results for part of the French side of the tunnel route. The general pattern of the top of the Gault Clay (Fig. 2) reflects the regional geological trend - an anticline whose axis is parallel to the layout. The boundary of the Gault Clay outcrop is shown on the map as a dotted line. The standard deviation (Fig. 3) displays well-marked minimum values near the profiles (good knowledge of seismic time) and boreholes (consistent knowledge of seismic time and average velocity); it increases from south to north, because the depth from sea floor to the top of the Gault increases, so that the uncertainties concerning the velocity have a greater impact in the north. Throughout the whole studied area, the standard deviation lies between 2 and 6 metres; nowhere along the submarine part of the tunnel route does it exceed 4 metres, and generally it falls between 2 and 3 metres.

When looking at the kriged model along the planned route, it appeared that in certain places the tunnel would intersect the Gault Clay or the Grey Chalk. This led the engineers to revise the layout in such a way as to maintain the tunnels nearly everywhere at least one standard deviation above the estimated value of the top of the Gault. Thus, the risk of penetrating the Gault Clay was reduced, but of course was not entirely excluded.

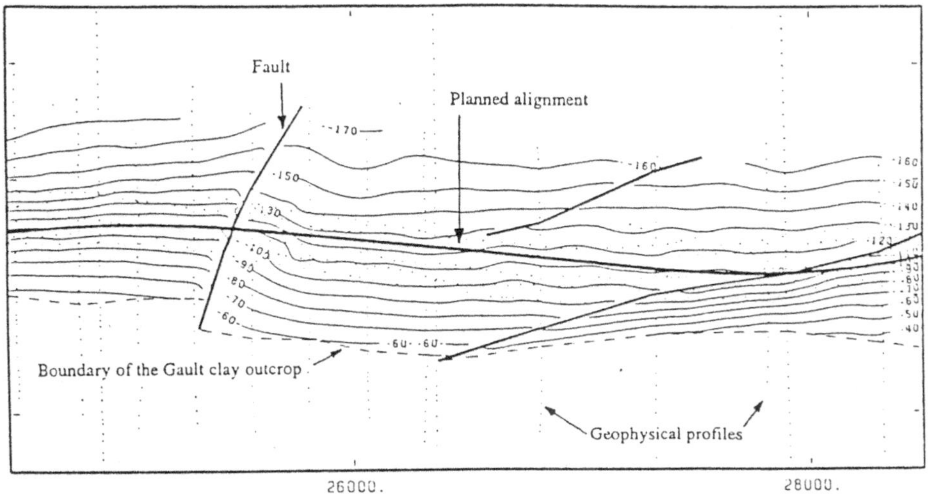

Fig. 2. Contour map of the elevation of the top of the Gault Clay (metres).

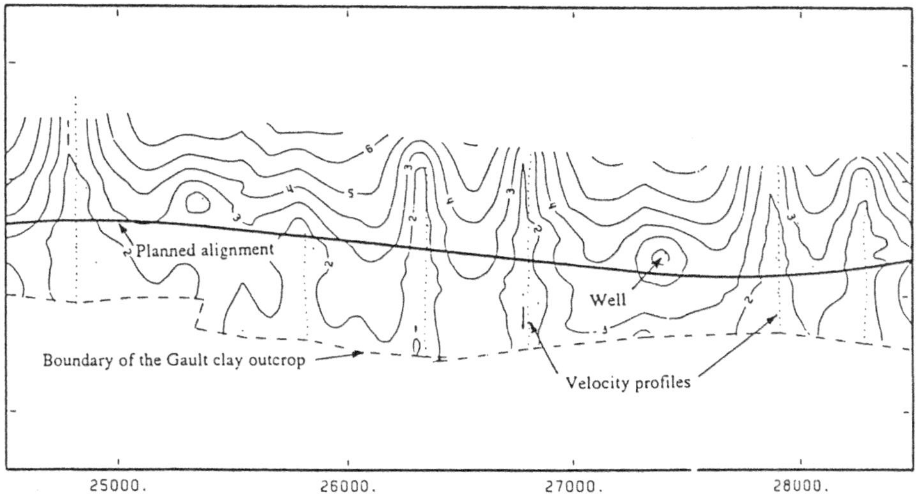

Fig. 3. Contour map of the kriging standard deviation of the Gault Clay (metres).

DESIGN OF COMPLEMENTARY SURVEYS

Cross-over excavations were planned at two locations, to enable the trains to pass from one tunnel to the other if necessary. As these cross-overs are 21 m in diameter and 200 m in length, their precise location and design had to be based on more accurate geological predictions than those used for the main tunnels. Complementary geophysical surveys were thus carried out which were designed from a geostatistical study that determined the data density and accuracy necessary to obtain the required precision (standard deviation = 1 m). It was found that the transverse seismic profiles had to be 25 m apart over the French cross-over and 100 m apart over the British cross-over, and that the seismic velocity had to be defined by at least four boreholes located on either side of the tunnel.(fig. 4)

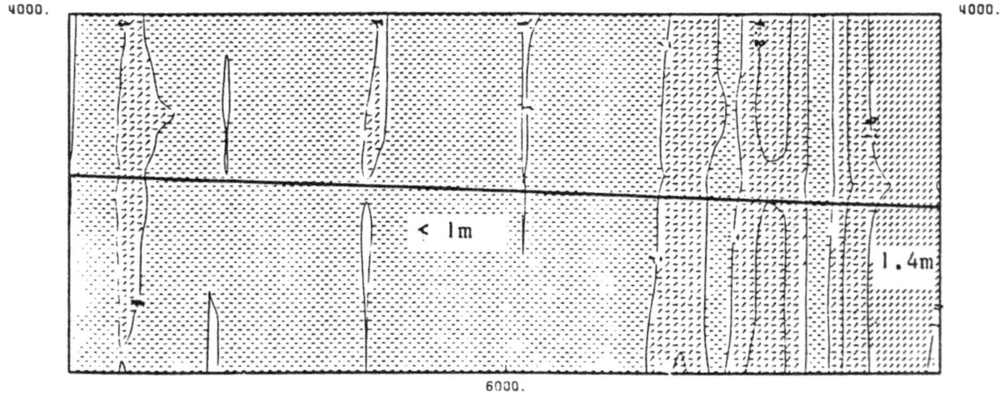

Fig. 4. Contour map of the kriging standard deviation of the Gaul Clay (metres) in the British cross-over area after complementary survey.

COMPARISON OF THE TWO MODELS

The 1988 PHASE II survey included a new variography and kriging excercise. The maps and cross-sections of the year before, were redrawn with a much greater precision (lower standard deviation) than for the original model. Figure 4 shows the map of the new kriging standard deviation of the top of the Gault for the British cross-over area. The objective of 1 m was effectively reached. In comparison with the first one, the new estimate can be considered as the reality. Figure 5 shows a comparison between the first estimate with its 68% confidence interval (± 1 standard deviation) and the second estimate for the British cross-over. It can be seen that the new estimate usually lies within the 68% confidence interval of the first estimate, and only exceptionally reaches the 95% confidence interval (± 2 standard deviations)

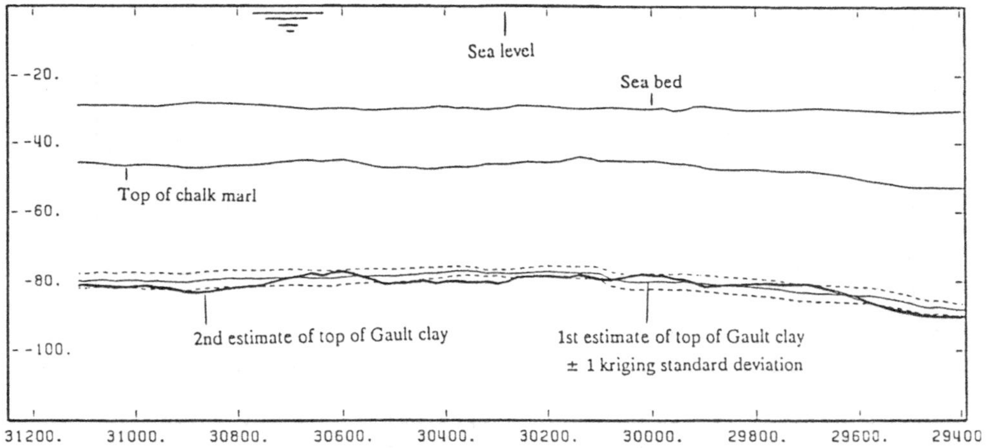

Fig. 5. Cross-section comparing the original interpretation with the new one along the British cross-over area

Since these estimations were carried out from different data sets, they can be considered as independent. The difference between the two estimations is then with mathematical expectation zero and with theoretical variance the sum of the two kriging variances. The corresponding theoretical standard deviation is on average 2.32 m. The observed differences at the 100 kriged points, with minimum and maximum values of –3.55 m and +2.61 m, respectively led to experimental values of 1.13 m for the average and 1.61 m for the standard deviation from zero. The errors seem to be lower than predicted, but perhaps have a small bias. In reality, the kriging errors of the successive points of the tunnel profile are highly correlated. As shown below, they display a structured behaviour corresponding to a 1-D integral range of about 700 m, so that the 100 kriged points are equivalent to only three independent data.

A similar study was done for the French cross-over (Blanchin and Chilès, 1992), but turned out to be meaningless: the tunnelling showed that the new estimate was not better than the first one, because of the presence of a 1965 geophysical borehole affected by a location error evaluated at 30 m in a zone with a dip of 22% which gave an error of 6 m in the depth values.

GEOSTATISTICAL PREDICTION AND REALITY

Pairs of boreholes were drilled downwards from the central service tunnel, which was bored ahead of the main tunnels, to determine the actual depth and dip of the Gault Clay horizon. On the French side, 54 borehole pairs were available over the first 13 km against 31 pairs over 15 km on the British side. The main objectives of this drilling were:

i) to compare the reality with predicted levels and identify the main components of inaccuracy leading to the discrepancies;
ii) to use the feedback information for reviewing the prediction for the boring of the main tunnels on either side of the service tunnel.

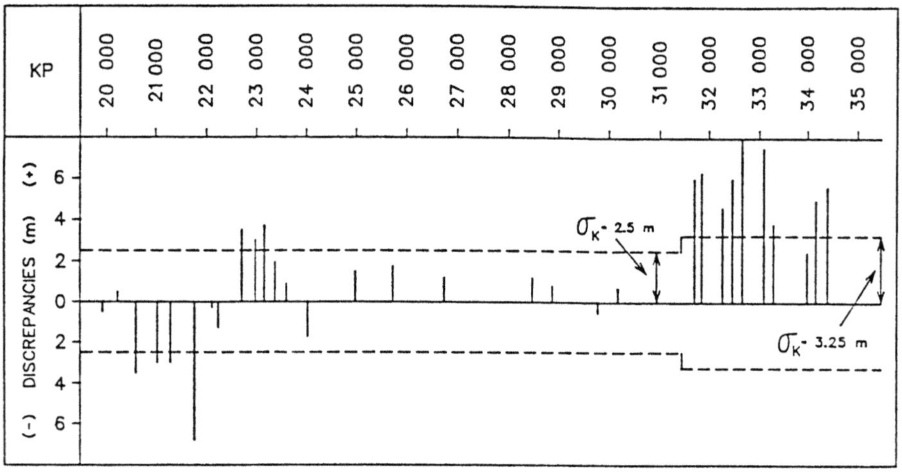

Fig. 6. Discrepancies observed on the British side of the service tunnel (reality minus prediction)

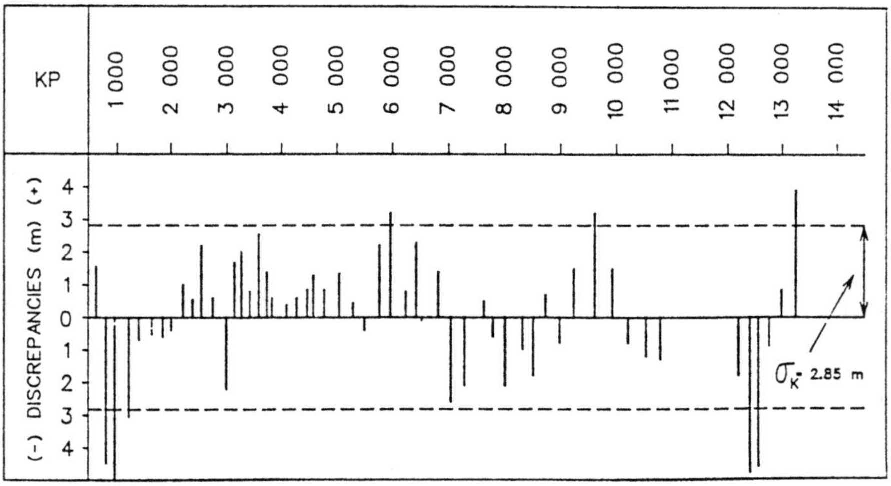

Fig. 7. Discrepancies observed on the French side of the service tunnel (reality minus prediction)

The differences "Reality-Prediction" are shown on figures 6 (British side) and 7 (French side). Their statistical characteristics are summarized in Table 1. The results concerning the British side lead to 1.7 m for the average difference, and 3.4 m for the standard deviation. The distribution of errors along the service gallery (Fig. 6) shows a good agreement of the predicted depth with reality: the discrepancies lie within the 95% confidence interval, $2\sigma_k$ ranging from 5 to 6.5 m, except for three boreholes. The trend that is apparent for the last ten boreholes (average difference: 5.6 m) has prompted a review of the earlier interpretation in the light of the new data.First, it is important to note that these ten discrepancies have been recorded in an area where the density of seismic profiles and borehole data is the lowest for the whole tunnel. This means that the reliability of the geostatistical mapping is low (σ_k = 3.25 m) compared with the average value for the British area ($\sigma_k \simeq$ 2.5 m). The re-interpretation showed that the major components of inaccuracy are:

- The lack of precision in the seismic-time pick: the re-interpretation over a 10-km-long area produces both positive and negative differences up to 3 ms for the two-way time, corresponding to 3.5 m in depth, but generally showing the re-interpreted horizon to be shallower.

- The errors in the calibration of the velocity data from the geophysical boreholes, due to the location uncertainty of the 1964-1965 boreholes: an uncertainty of 30 m on their location is considered as normal, which may introduce significant errors in the depth of the horizons in areas with a low dip; the revised velocity is some 5 to 10% less than the original one and accounts for 60% of the upward shift in level in the above-mentioned area.

Similar conclusions were drawn for the French part, even if the discrepancies are smaller than for the British part. Note, that the location uncertainty of the geophysical boreholes has been taken into account in the kriging process by means of a highly correlated error component in the velocity data.

	FRENCH SIDE (13 km)	BRITISH SIDE (15 km)
Number of borehole pairs	54	31
Minimum	- 5.00 m	- 7.00 m
Maximum	+ 3.90 m	+ 8.00 m
Average	+ 0.48 m	+ 1.70 m
Standard deviation from zero	2.02 m	3.40 m
kriging standard deviation	2.85 m	2.55 m

Table 1 -Statistical characteristics of the difference " Reality-Prediction"

In order to explain the discrepancies from a geostatistical view point, we carried out a variographics investigation, which is presented for the French part of the tunnel. As said earlier, the kriging errors on the seismic travel time as well as on the seismic velocity are spatially correlated. The main reasons for this are the following:

i) Even if the estimation is performed with a moving neighbourhood, successive interpolation points of the tunnel are estimated from almost the same data points.
ii) The distance from the tunnel to the longitudinal seismic profiles is not constant but varies.
iii) One part of the data uncertainty is correlated along the profiles, particularly concerning the velocity profiles (this is inherent in the fine grain-velocity analysis method).
iv) The variography of seismic time was carried out over successive homogeneous zones whose length ranges from 2 to 3 km in the French part of the tunnel.

The geometric part of the spatial correlations of the kriging errors is shown on the experimental variogram of the kriging standard deviation, which can be computed even if this variable is not a stationary random function. For the kriging standard deviation of seismic travel time (Fig. 8), two levels of correlation are visible. The first one (range $\simeq$ 300 m) is mainly due to the neighbourhood overlap and also to the uncertainty in time "picking", whereas the second one (range $\simeq$ 2000 m) is due to the geometry of the profiles and to the variations in the variogram parameters.

The variogram of the velocity kriging standard deviation (Fig. 9) is more irregular, for this kriging derives from a limited number of boreholes and not very accurate profiles. The short range component (300 m) corresponds to the distance between neighbouring velocity profiles, in addition to the fact that each profile is affected by a spatially correlated uncertainty. The most likely explanation for the second range (3000 m) is that it corresponds to the average distance between neighbouring sonic boreholes.

The kriging error (difference between prediction and reality) is also not stationary, for the kriging configuration (data points + kriged point) is not the same for two different kriged points. Its variogram is nevertheless instructive (Fig. 10). It resembles the previous ones as it has two ranges at 300 and 2000 m, which shows that the main part of its variability is of geometric origin. This variogram enables the 1-D integral range of the kriging error to be computed. The 1-D integral range A of a covariance C(h) is defined as:

$$A = \frac{1}{C(o)} \int_{-\infty}^{+\infty} C(h)\,dh$$

For example, for a spherical model with range "a", the 1-D integral range is $A = 3a/4$. For the model fitted here (Fig. 10), the 1-D integral range is equal to about 1000 m. The interest of the 1-D integral range is that, if the kriging error is known along a part of the tunnel of length L, all these correlated values are equivalent to $n = L/A$ independent values. (Lantuéjoul, 1991).

The 54 observations along the 13 km of the French side are thus equivalent to only $n = 13$ independent measurements. The average difference (+ 0.48 m) is then compatible with its theoretical standard deviation $\sigma_k/\sqrt{n} = 2.85 / \sqrt{13} = 0.8$ m.

Figure 11 shows the variogram of the relative kriging error (i.e., divided by its kriging standard deviation). Its sill is expected to be equal to 1. It is slight lower (0.80), thus indicating that the actual errors are smaller than expected. The difference is, however, not incompatible with the number of independent data.

Similar conclusions can be drawn for the British side, where the 1-D integral range is 700 m, and thus the number of equivalent independent data n = 21, with the difference that on this side the errors are higher than expected (but always compatible with n).

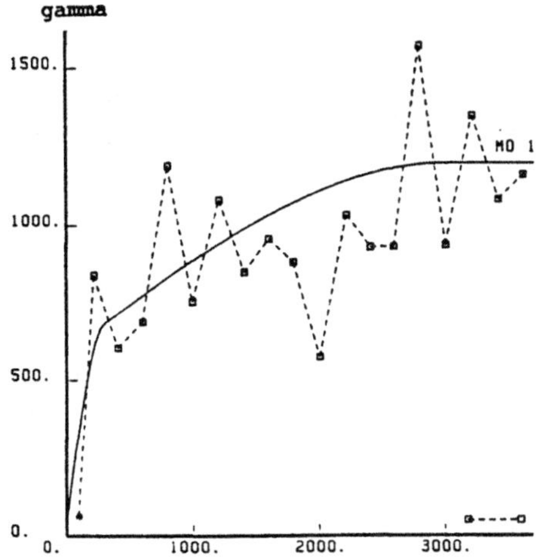

Fig. 8. Variogram of the kriging standard deviation of seismic time

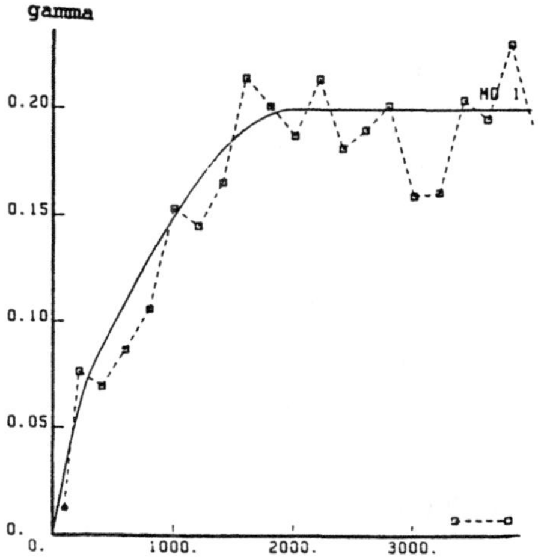

Fig. 9. Variogram of the kriging standard deviation of seismic velocity

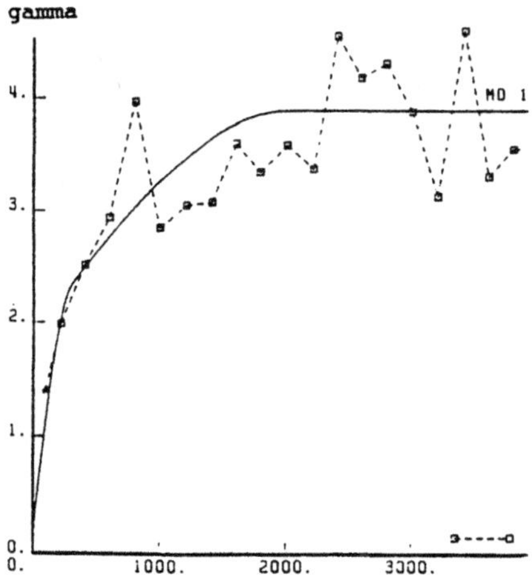

Fig. 10. Variogram of the kriging error (Reality-Prediction)

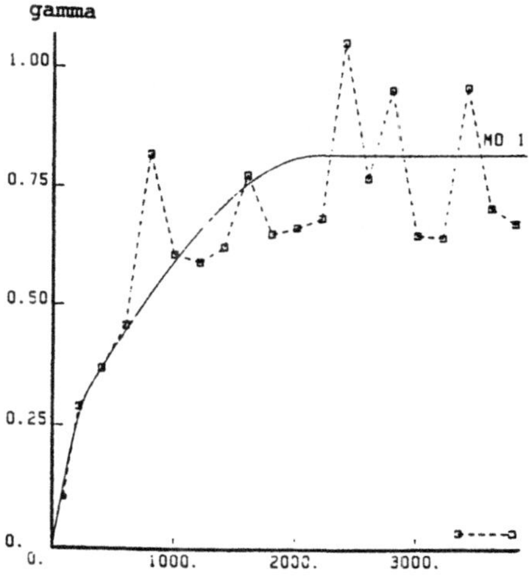

Fig. 11. Variogram of the relative kriging error

CONCLUSION

The use of geostatistical techniques led to accurate geometrical modelling of the geological layers involved in the Channel Tunnel project, and so optimized the design of the tunnel layout. The observations in the service tunnel are in a good agreement with the model and its predicted accuracy. An analysis of the differences between the geostatistical predictions and reality assessed the quality and the reliability of the model, and validated the efficiency of the geostatistical modelling.

Whereas most tunnelling machines started boring with a delay of several months, this careful geostatistical modelling, together with the combined, performances of machines and men, enabled the tunnelling to be completed on 28 June 1991, one month ahead of the initial schedule.

ACKNOWLEDGEMENTS

This is BRGM scientific contribution No.92025. The writing work was founded from an internal BRGM research grant.

REFERENCES

BLANCHIN, R. and J.-P. CHILÈS. (1992) - Geostatistical modelling of geological layers and optimization of survey design for the Channel tunnel.*Computer Graphics in Geology*, R.Pflug and J.-W. Harbaugh (Eds.), Springer Verlag, Berlin, pp. 251-256.

R.BLANCHIN, J.-P. CHILÈS and F. DEVERLY (1989) - Some applications of geostatistics to civil engineering. In *Geostatistics*, Armstrong M., (Ed.), Vol. 2, pp. 785-795. Kluwer Academic Publishers, Dordrecht, Netherlands.

C. LANTUÉJOUL (1991) - Ergodicity and integral range. *Journal of Microscopy*, Vol. 161, part 3, pp. 387-403.

Regularization of Hydraulic Conductivities: A Numerical Approach

J. Jaime Gómez-Hernández, Departamento de Ingeniería Hidráulica y Medio Ambiente, Universidad Politécnica, 46071 Valencia, Spain.

Abstract

Regularization techniques, i.e., techniques that allow the inference of the spatial statistics of block values from corresponding point statistics, are available for properties that average linearly. Hydraulic conductivities, which are known to average non-linearly in space, call for specific regularization algorithms. Realizations of point conductivity values are generated using a point covariance inferred from point (well) data. Then, corresponding realizations of block conductivity values are retrieved using limited flow simulations and spatially averaged Darcy's law. These joint realizations of point-block conductivity values provide training images for inference of corresponding statistics and cross-statistics used to generate further realizations of block values now conditional to point (well) data.

Introduction

In hydrogeology and petroleum engineering, modelers face the problem of assigning hydraulic conductivity values to the blocks of their numerical flow simulators. Unfortunately, block data are almost never available, and some kind of extrapolation exercise from either core data or pump test analyses is needed. The problem is aggravated if the modeler pursues a stochastic analysis of the response of the model, since then, generation of multiple realizations of block values is required. Such generation calls for knowledge of the spatial statistics of block values.

When the block value is a linear average of the points within the block, as is the case for porosity and metal grade, the inference of block statistics from point statistics is straightforward (Journel and Huijbregts, 1978; Vanmarcke, 1983). Let $Z(\mathbf{x})$ and $Z_V(\mathbf{y})$ be random functions representing point and block values related through the expression:

$$Z_V(\mathbf{y}) = \frac{1}{V} \int_{\mathbf{x} \in V(\mathbf{y})} Z(\mathbf{x}) d\mathbf{x} \tag{1}$$

It is straightforward to derive the expected value and covariance of $Z_V(\mathbf{y})$ given that $Z(\mathbf{x})$ is second order stationary with expected value m and covariance $C(\mathbf{h})$:

$$E\{Z_V(\mathbf{y})\} = E\{Z(\mathbf{x})\} = m$$

$$\mathrm{Cov}\{Z_V(\mathbf{y}_1), Z_V(\mathbf{y}_2)\} = C_V(\mathbf{y}_1 - \mathbf{y}_2) = \frac{1}{V^2} \int_{\mathbf{u} \in V(\mathbf{y}_1)} \int_{\mathbf{v} \in V(\mathbf{y}_2)} C(\mathbf{u} - \mathbf{v}) d\mathbf{u} d\mathbf{v}$$

If the point-block relation is not linear as in (1), regularization expressions are only available for very simple cases. For instance, if the block value is equal to the geometric mean of the point values within the block, a logarithmic transform of both block and point values results in a linear relationship that can be used to infer the block statistics.
However, block hydraulic conductivities do not, in general, relate linearly or geometrically to point conductivities[1]. This paper addresses the problems of how to determine block values from point data and of how to derive the block spatial statistics for the case of hydraulic conductivities.

Block hydraulic conductivities

Regularization of hydraulic conductivities is much more difficult than regularization of porosity or ore grade because there is no general explicit relation between point and block values. Such relation can be implicitly established through the flow equations by requiring that the flow behavior of a heterogeneous block be reproduced, in some average sense, by the behavior of a homogeneous block.
Unfortunately, there is no general agreement of what is meant by the flow behavior of a heterogeneous block, particularly because that behavior is a function not only of the point conductivities within it but also of the boundary conditions applied to it. This fact has two implications, first, one has to choose a set of boundary conditions to analyze the flow behavior of a block, and second, the block conductivity is also function of the points outside the block since it is the conductivities outside the block which determine the boundary conditions. As a consequence, the block conductivity is not intrinsic to the block.
An additional problem in determining block conductivities is that the resistance to flow in a heterogeneous block varies depending on the direction of the hydraulic gradient and so does the block conductivity. Therefore, the block conductivity is not, in general, a scalar value but rather a tensor that must be symmetric and positive definite.
A common approach to determine block conductivity tensors starts by assuming that the principal directions of the tensor are parallel to the cartesian axes *for all blocks*. The heterogeneous block is isolated and flow is solved using no-flow boundary conditions except for two opposing faces which have prescribed piezometric heads (Desbarats, 1987; Begg et al., 1989). In two dimensions, two flow problems are solved for each block with the boundary conditions displayed in Figure 1. From the solution of these flow problems, the total amounts of flow driven through the block are computed and their relation to the overall piezometric head gradient provides the components of the block conductivity tensor. Referring to Figure 1, let Q_x and Q_y be the total flows crossing the block for each of the two sets of boundary conditions, with L_x and L_y being the block side lengths, and H_1 and H_2 being the piezometric heads applied to the block sides. The block tensor components $K_{V,xx}$ and $K_{V,yy}$ are estimated by:

[1]By a slight abuse of language we shall speak of point conductivities to refer to conductivity values measured on a support much smaller than the block support.

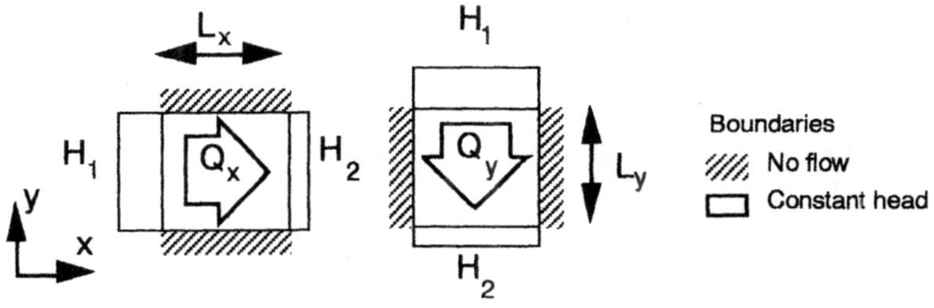

Figure 1: Determining block conductivity tensor components. A common approach in 2-D for determining block conductivities consists of solving the two flow problems displayed in the figure, computing the total flows Q_x and Q_y crossing the block, and relating them to the overall hydraulic gradient imposed by H_1 and H_2

$$
\begin{aligned}
K_{V,xx} &= -\frac{Q_x/L_y}{(H_2 - H_1)/L_x} \\
K_{V,yy} &= -\frac{Q_y/L_x}{(H_2 - H_1)/L_y}
\end{aligned}
\qquad (2)
$$

the off-diagonal component $K_{V,xy}$ is zero since it is assumed that the principal directions of the block tensor are parallel to the block sides.

This way of computing the hydraulic conductivity tensor is valid only if (1) the assumption about the principal directions is correct and (2) the values Q_x and Q_y are function of the hydraulic gradient but not of the boundary conditions. An aquifer that meets the first condition but all blocks fail the second one is displayed in Figure 2. In this figure, all blocks are intersected in their total horizontal extension by an impermeable shale. Because of the boundary conditions used to determine the tensor components, the resulting value for $K_{V,yy}$ would be incorrectly 0 for every block.

When the spatial distribution of hydraulic conductivities is not so anisotropic as that shown in Figure 2, the hydraulic conductivity variance is small and/or the block is much larger, or much smaller, than the conductivity spatial correlation there exist simpler methods to determine block conductivities such as power averaging (Deutsch, 1989) or streamtube methods (Begg and King, 1985).

An alternative approach to compute block conductivities

In the following an alternative approach is described that alleviates the problem. This approach is described in length in Gómez-Hernández (1991) borrowing some ideas from a similar approach proposed by White and Horne (1987).

There are three major differences with the traditional approach described above: (1) the principal directions of the block conductivity tensor need not be aligned with the cartesian axes, (2) the point conductivity values within an annular skin of given size around the block are included in the analysis, and (3) the flow response of the block to several alternative boundary conditions is considered.

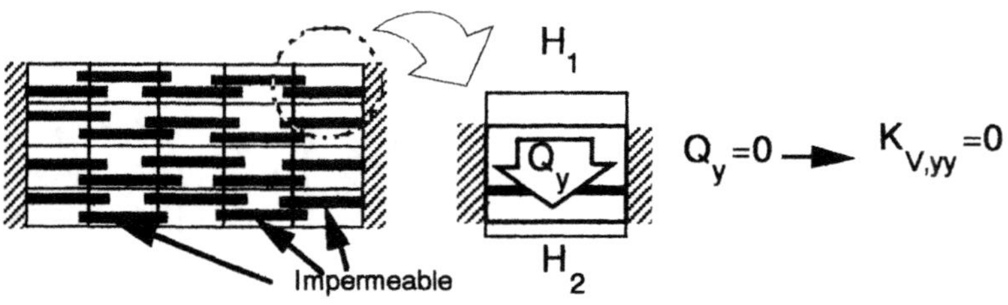

Figure 2: A pathologic example. Using the method described in Figure 1, all blocks in the figure would receive a zero vertical hydraulic conductivity $K_{V,yy}$.

Relaxing the assumption on the orientation of the principal directions implies that not only the principal components of the conductivity tensor must be computed but also their orientation with respect to the cartesian axes.
The analysis of flow including a skin around the block contributes to produce a flow pattern within the block closer to that had the block been embedded within the aquifer. That is, the influence of the neighboring conductivities is accounted for.
The use of several boundary conditions applied to the outer edges of the skin accounts for the behavior of the block under different hydraulic conditions. The block conductivity retained is that which best reproduces all flow patterns for the various boundary conditions considered, bearing in mind that no single set of block conductivities will match exactly the behavior of the block under all boundary conditions.

Define block conductivity as the symmetric positive definite tensor that relates average gradient to average specific discharge as given by a formulation of Darcy's law at the block scale:

$$\overline{\mathbf{q}} = -\mathbf{K}_V \cdot \overline{\nabla H} \ , \qquad \text{given boundary conditions,}$$

where the overbar indicates spatial average over the block V, $\mathbf{q}$ is the specific discharge vector and ∇H is the piezometric head gradient vector. In two dimensions, this expression is written in terms of its components as:

$$\left\{ \begin{array}{c} \overline{q_x} \\ \overline{q_y} \end{array} \right\} = - \left\{ \begin{array}{cc} K_{V,xx} & K_{V,xy} \\ K_{V,xy} & K_{V,yy} \end{array} \right\} \cdot \left\{ \begin{array}{c} \overline{\partial H/\partial x} \\ \overline{\partial H/\partial y} \end{array} \right\} \qquad \text{given b.c.'s} \tag{3}$$

It is important to stress that this equation cannot be solved for the components of the hydraulic conductivity tensor unless some boundary conditions are specified. In particular, this formulation yields the same block conductivities as in (2), if $K_{V,xy}$ is assumed zero and the boundary conditions of Figure 1 are used.
For the general case of $K_{V,xy} \neq 0$, we suggest solving the flow problem for a number of boundary conditions (a minimum of 2, from our experience 4 boundary conditions give good results). Each solution results in two linear equations with the three components of the block tensor as unknowns. This overdetermined system of linear equations is then solved using constrained least-squares optimization in order to determine the set of values that conform a symmetric positive definite tensor and give the best approximate solution in terms of minimum least square discrepancy between predicted and measured specific discharges.

For the example shown in the next section, 4 boundary conditions were selected, all with prescribed heads at the four sides of the skin. The values of the prescribed heads lie on a plane given by:

$$H(x, y) = Ax + By, \quad \forall x, y \text{ along the skin boundaries}$$

The parameters A and B are set to $(A, B) = \{(1,1), (1,0), (0,1), (1,-1)\}$. These 4 boundary conditions define an overall head gradient under which the block is analyzed in the directions parallel to the block sides and to the block diagonals.
This approach has been shown to give excellent results in providing block conductivity tensors that reproduce the small-scale flow behavior for large aquifers (Gómez-Hernández and Journel, 1990; Gómez-Hernández, 1991). However, being CPU-intensive this approach should be used only in those cases in which the assumptions underlying the simpler approaches mentioned at the end of the introduction are deemed inappropriate.

Regularization of block conductivities

Once a method for computing the block conductivities has been selected we suggest generating a synthetic image of block conductivities from which their expected value and covariances could be inferred. First, a realization of point conductivities is generated, then the block values corresponding to this realization are deduced following the procedure described in the previous section. This results in a synthetic image of block tensors. This synthetic training image has two important characteristics: it is based on a realization of point conductivities with the same spatial variability as displayed by the measurements; and, the computation of the block tensors is based on an averaging procedure that takes into account the physics of flow.
A two-dimensional example of a realization of point conductivities and the resulting components of the block conductivity tensors is shown in Figure 3. The point support is assimilated to a unit square pixel; the block support is a square of 10 by 10 pixels. The point conductivity realization was generated using a lognormal model: $Y = \ln K$ has a Gaussian distribution with mean $m_Y = 0$, variance $\sigma_Y^2 = 2.0$ and an anisotropic exponential covariance with principal axes of anisotropy aligned with the cartesian axes: range in the x-direction equal to 3 blocks (30 pixels) and range in the y direction equal to 0.3 blocks (3 pixels). A 400 by 400 pixel image displaying the spatial variability given by the previous model is shown at the top of Figure 3. Using the previously described approach to compute block conductivities, 40 by 40 block conductivity tensors were obtained from that point image. The components of the tensors are displayed separately at the bottom of Figure 3. It should be stressed that the two bottom images in Figure 3 correspond to a single set of square blocks. Each block is anisotropic to flow; therefore, the block conductivity tensor has different values in the x and y directions. Because the principal anisotropy directions are parallel to the block sides, the $K_{V,xy}$ components are negligible and thus were not displayed.
The three images of Figure 3 are used to compute the univariate distributions of both $K_{V,xx}$ and $K_{V,yy}$ (Figure 4). Both $K_{V,xx}$ and $K_{V,yy}$ can be modeled with a lognormal distribution, with $K_{V,xx}$ being about half an order of magnitude larger than $K_{V,yy}$. Similarly, the covariances of $Y_{V,xx} = \ln K_{V,xx}$ and $Y_{V,yy} = \ln K_{V,yy}$ and their cross-covariances with $Y = \ln K$ can be inferred numerically from these training images.

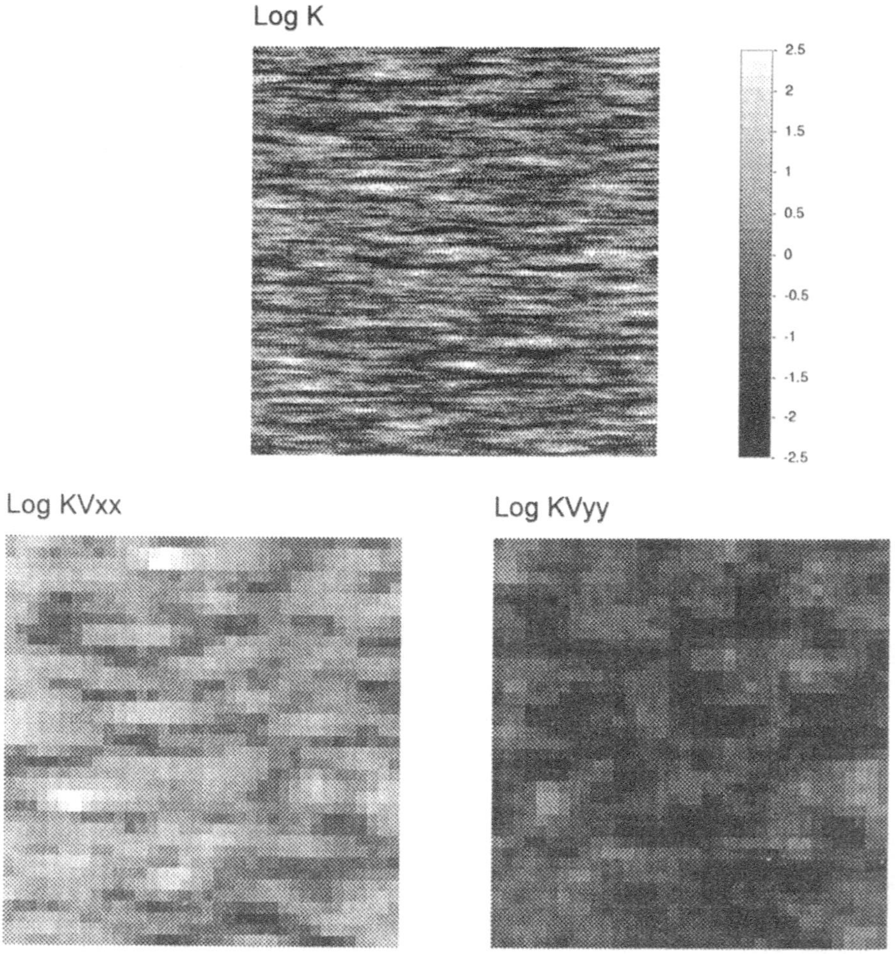

Figure 3: Synthetic training images. The top figure shows a map of point conductivities. The components of the corresponding block conductivities tensors are displayed in the bottom two figures.

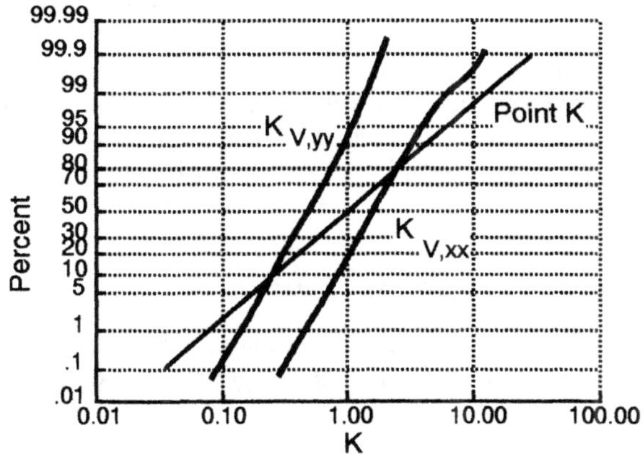

Figure 4: Lognormal probability plots for $K_{V,xx}$ and $K_{V,yy}$. The point conductivity probability plot is also shown for reference

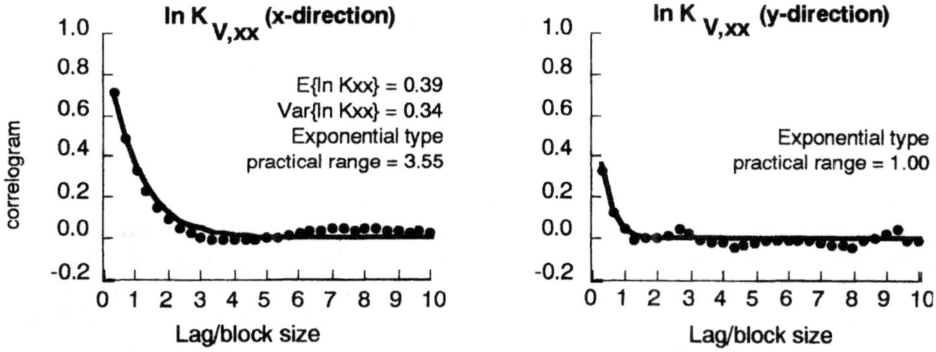

Figure 5: Correlation of $\ln K_{V,xx}$. This correlation function has been directly inferred from the values on the synthetic images of Figure 3. For reference the practical ranges of the point conductivity are 3 and 0.3 blocks in the x and y directions respectively.

As an example, the covariance for $\ln K_{V,xx}$ is shown in Figure 5 along with the corresponding fitted models.
Starting from an initial model of point hydraulic conductivity which is isotropic to flow (that is, the initial hydraulic conductivity is a scalar field) and with anisotropic spatial variability, a model is derived for block tensor hydraulic conductivities with each tensor component displaying spatial anisotropy and cross-correlation with the other component.

Validation of the regularization model

The next step consists of checking whether the statistics inferred from the training images are reliable enough to be used for the direct generation of block conductivity values.

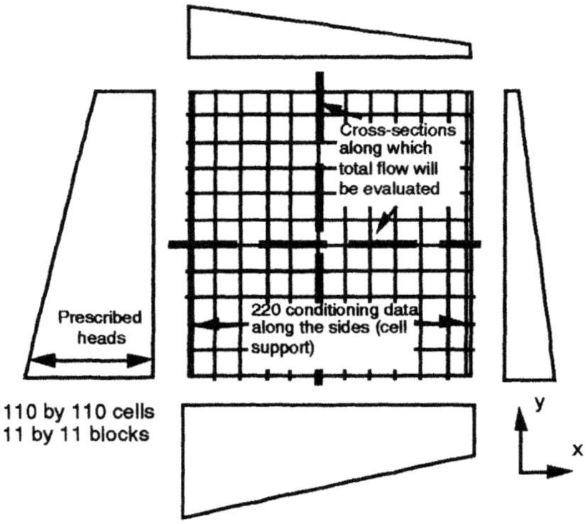

Figure 6: Validation exercise. An aquifer is discretized into 110 by 110 pixels taken to be at the scale of the measurements, then into 11 by 11 square blocks. A total of 200 data on the pixel support are available at the first and last column of the aquifer. Hydraulic heads are prescribed at the boundaries imposing an overall diagonal gradient through the aquifer. The cross-sections along which the total flow will be evaluated are located at the interfaces between the 5th and 6th block columns and the 5th and 6th block rows (50th and 51st pixel columns and 50th and 51st pixel rows, respectively)

First, a joint statistical model for the spatial variability of K, $K_{V,xx}$ and $K_{V,yy}$ is needed. Given the univariate lognormal character of the three variables, a multiGaussian model of the logarithms is deemed appropriate. Such multiGaussian model is fully characterized by the means, covariances and cross-covariances of the three variables involved; all these statistics were inferred from the synthetic training images in Figure 3.

A Monte-Carlo analysis is carried out for an aquifer for which 220 data are available along the two sides parallel to the y-axis (Figure 6). The objective of this analysis is to evaluate uncertainties on the total amount of flow passing through the two cross-sections shown in Figure 6.

Three Monte-Carlo analyses are carried out all conditional to the same 220 data.

The reference Monte-Carlo analysis consists of 200 conditional realizations of 110 by 110 $\ln K$ values. The flow problem is solved using a standard five-point finite difference scheme on each of these 110 by 110 cell fields and the total amount of flow through the cell interfaces between the 50th and 51st columns and through the 50th and 51st rows are computed. There is no regularization involved in the reference results. The resulting cumulative frequency distributions of flows are denoted Q_1 and shown in Figure 7.

The second Monte-Carlo analysis uses the previously generated 200 conditional realizations of point (unit pixel) $\ln K$ values. The block conductivity tensor is computed for each of the 11 by 11 blocks, as described in the previous to last section. Once the block conductivities are available, flow is solved over each of the 200 aquifers at the block scale. The total amount of flow through the block interfaces between the 5th and 6th columns and between the 5th and 6th rows are computed. Regularization is applied here to determine the block hydraulic conductivity tensors on exactly the same conductivity images used to compute the reference results. The resulting frequency distributions, denoted Q_2, are shown in Figure 7.

Finally, the third Monte-Carlo analysis considers the multiGaussian model for the joint spatial variability of $\ln K$, $\ln K_{V,xx}$ and $\ln K_{V,yy}$ inferred from the training images in Figure 3 to generate directly 200 realizations of the components of the block conductivity tensors conditional to the 220 (point) $\ln K$ data. Flow is solved at the block scale. The total flow going through the 5th and 6th columns and through the 5th and 6th rows are computed. The resulting frequency distributions denoted Q_3 are shown Figure 7. For the direct generation of the components of the block conductivity tensor conditioned to the $\ln K$ data, the public domain code GCOSIM3D that implements the algorithm of sequential simulation for jointly multiGaussian random variables was used (Gómez-Hernández, 1991; Gómez-Hernández and Journel, 1992). Regularization of hydraulic conductivity is implicitly taken into account through the use of the cross covariances between point and block values and between the components of the block tensors.

Figure 7 displays the frequency distributions corresponding to the 3 Monte-Carlo analyses for the total flows going through the cross-sections shown in Figure 6. The solid line corresponds to the reference frequency distribution Q_1. The dashed line corresponds to the frequency distribution resulting from the second Monte-Carlo analysis Q_2. The dotted line corresponds to the results from the third analysis Q_3. From these curves one can retrieve information such that 90% of the realizations result in total horizontal flows smaller than 20.

The choice, as a performance measure, of the reproduction of the total amount of flow through sections that cross the aquifer from side to side was made to detect possible bias on the reproduction of flow values in a block per block basis. If there is a bias, it will be more noticeable on an integrated measure such as the total flow. In the earlier stages of the development of the method it was found that incorrect approaches generally lead to biased results which sometimes were not so apparent on a block per block analysis of flow.

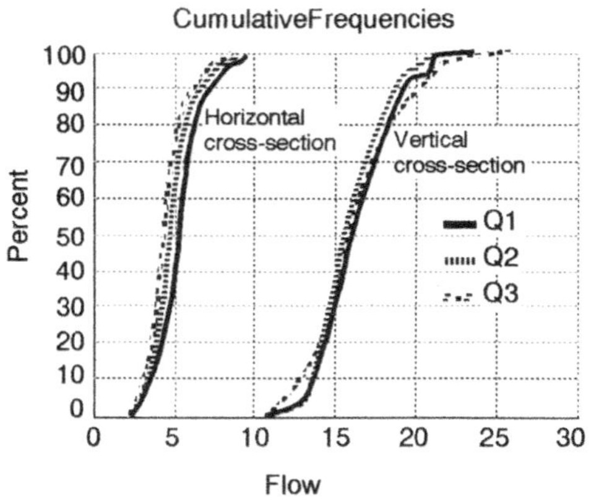

Figure 7: Uncertainty on flows. The uncertainty on the total flows crossing the two cross-sections shown in Figure 6 is measured by the cumulative frequency distributions resulting from 200 conditional realizations. The frequency distributions for the three Monte-Carlo analyses are shown. Q_1 corresponds to the reference analysis, Q_2 corresponds to the scale-up analysis, and Q_3 corresponds to the direct generation of block conductivities

The good reproduction of the reference frequency distributions Q_1 by both Q_2 and Q_3 indicates that, (1) the method for determining the block conductivity tensors is good and (2) that the inference of the statistics of the block conductivity tensor components from the synthetic training images is good enough to reproduce the uncertainty on the flow model response, at least, for the particular case analyzed here.

It must be stressed that the CPU time required to carry out the Monte-Carlo analysis Q_3 was more than 100 times smaller than the one required by either Q_1 or Q_2. This drastic reduction on execution time justifies the direct generation of block conductivities. Knowing the block statistics allows working directly at the block support with the consequent reduction in CPU time. It should be noted that the purpose of the methodology proposed is the direct generation of block conductivities for full analysis of flow at the block scale not to determine the overall flow going through the aquifer as measured by the validation exercise.

The major drawback of the proposed regularization technique for hydraulic conductivities is that it is specific to each case. If a different model for the spatial variability of point conductivities is used or the block size (or shape) is changed, a new set of training images must be generated and the covariance and cross-covariances inferred again. Another important drawback is the practical impossibility of mixing blocks of different sizes in the same realization. Although from a theoretical point of view, one could generate a synthetic training image for each of the block sizes being considered, modeling each block size as a different random function and inferring all corresponding cross-covariances would not be practical.

Conclusions

The statistics describing the spatial variability of block conductivities can be used for the direct generation of block conductivities used as input to numerical flow simulators. Unfortunately, no general explicit expression relating point and block values exist.

A numerical approach based on the use of a synthetic training image of block values is proposed. The building of such synthetic training image requires an algorithm to compute block conductivities from exhaustive knowledge of point values within the block. A new technique is proposed that would work in cases where traditional techniques fail. The spatial variability of point and block conductivities is then characterized through histograms and (cross)covariances from these (point and block) training images.

This approach to regularization and direct generation of block values has been tested with favorable results. The testing consisted of comparing the flow response of 200 realizations generated at the point scale with 200 realizations generated at the block scale conditional to the same point data. The generation of the block realizations was done directly using the joint spatial variability model of both block and point values inferred from synthetic images. For the particular case analyzed here, the proposed regularization technique performed well.

Acknowledgments

Financial support for this study was provided by the Stanford Center for Reservoir Forecasting. The author wishes to thank André G. Journel, Yoram Rubin and François Alabert for the brainstorming sessions at the origin of some of the ideas developed in this paper.

References

Begg, S. H., Carter, R. R., and Dranfield, P. (1989). Assigning effective values to simulator gridblock parameters for heterogeneous reservoirs. *SPE Reservoir Engineering*, pages 455–463.

Begg, S. H. and King, P. R. (1985). Modelling the effects of shales on reservoir performance: calculation of effective vertical permeability. *SPE 13529.*

Desbarats, A. J. (1987). Numerical estimation of effective permeability in sand-shale formations. *Water Resources Research*, 23(2):273–286.

Deutsch, C. V. (1989). Calculating effective absolute permeability in sand-shale sequences. *SPE Formation Evaluation*, 4(3):343–348.

Gómez-Hernández, J. J. (1991). *A Stochastic Approach to the Simulation of Block Conductivity Values Conditioned Upon Data Measured at a Smaller Scale.* PhD thesis, Stanford University.

Gómez-Hernández, J. J. and Journel, A. G. (1990). Stochastic characterization of grid-block permeabilities: from point values to block tensors. In *Proceedings of the 2nd European conference on the mathematics of oil recovery.*

Gómez-Hernández, J. J. and Journel, A. G. (1992). Joint simulation of multiGaussian random variables. In *Proc. of the 4th Int. Geostat. Congress.*

Journel, A. G. and Huijbregts, C. J. (1978). *Mining Geostatistics.* Academic Press, London.

Vanmarcke, E. (1983). *Random Fields, Analysis and Synthesis.* MIT Press, Cambridge, Mass.

White, C. D. and Horne, R. N. (1987). Computing absolute transmissibility in the presence of fine-scale heterogeneity. *SPE 16011.*

GEOSTATISTICAL FORMULATIONS OF GROUNDWATER COUPLED INVERSE PROBLEMS

J. CARRERA[1], A. MEDINA[2] and X. SÁNCHEZ-VILA[1]

[1] *Dep. Ingeniería del Terreno*
[2] *Dep. Matemática Aplicada III*
Universidad Politécnica de Cataluña
Barcelona, Spain

ABSTRACT

Geostatistical formulations of groundwater inverse problems are based on treating one or several model parameters as regionalized variables and have proven useful for groundwater flow. Extension to coupled flow and solute transport improves the estimation of both aquifer and geostatistical parameters. Specifically, log-transmissivity (Y) estimation errors are reduced when conditioning the estimation on not only head and Y measurements, but also on concentration measurements. Two alternative formulations for the estimation problem are presented. One is based on cokriging and the other one is based on maximizing the likelihood function. While results are encouraging, quasi-linear approximation used for computing errors covariance matrix appears to be somewhat optimistic (i.e., it underestimates actual uncertainties).

1. Introduction

The main difficulty in groundwater modeling arises from characterizing transmissivity(T) or hydraulic conductivity(K). Spatial variability of these parameters can be extremely marked, so that a proper representation of the T field may require many more measurements than usually available. Moreover, these are prone to error and, sometimes, local values are not representative of large scale conditions. As a result, when they are used indiscriminately in a groundwater model, predictions tend to be highly unrealible. This motivates a need to calibrate groundwater models before they can be used for predictions. State of the Art reviews of the inverse problem have been published (Yeh, 1986; Carrera, 1987; Ginn and Cushman, 1990).

Among the many approaches dealing with the inverse problem, we are particularly interested in the ones by Neuman (1980) and by Clifton and Neuman (1982) and by Kitanidis and Vomvoris (1983). The former proposed estimation of average transmissivities (actually, $log - T$'s) over the elements of the finite element grid that is used for solving the groundwater flow equation and minimized the differences between computed and measured heads at observation points with respect to $log - T$ values, while trying to keep these close to prior estimates, obtained by kriging. The latter formulated the problem as that of computing expected $log - T$ given $log - T$ and head point measurements. Their approach was later reformulated using cokriging (Hoeksema and Kitanidis, 1984). Even though the two approaches seem rather different, Carrera and Glorioso (1991) proved that if the method of Clifton and Neuman (1982) is posed in a Maximum Likelihood framework, the cokriging method can be viewed as a particular case of the iterative method.

The objective of this paper is to extend these methods to the case in which both head and concentration measurements are available. For this purpose, we will use the approach proposed by Medina et al (1990). Also relevant to the problem presented here is the work of Graham and MacLaughlin (1990), Keidser and Rosjberg (1991), Dagan (1985) and Ahmed and De Marsily (1987). The paper is organized as follows. First, the flow and solute transport equations are briefly described. The cokriging and iterative methods are then presented. Finally, applicability of the approach is illustrated with a synthetic example.

2. Flow and Solute Transport Equations

Groundwater flow is governed by:

$$\nabla(\boldsymbol{K}\ \nabla h) + w = S_s \frac{\partial h}{\partial t} \qquad on\ \Omega \tag{1}$$

where $\boldsymbol{K}$ is hydraulic conductivity tensor; h is head; w is a distributed sink/source term; ∇ is the "del" operator (divergence or gradient); S_s is storativity and t is time. Equation (1) is solved on the domain Ω, subject to appropriate boundary and initial conditions. Often, one works with vertically averaged values (Dupuit assumption). In such case, hydraulic conductivity is substituted by its vertical integral, transmissivity (T).

Derivation of the solute transport equation is immediate once the basic transport mechanisms, advection and dispersion, have been defined. Most difficulties in simulating solute transport arise from the manner in which dispersion is treated. The usual approach consists of postulating a dispersive mass flux, formally equal to Fick's first law:

$$\boldsymbol{J_D} = -\boldsymbol{D}\nabla c \tag{2}$$

where $\boldsymbol{J_D}$ is the dispersive mass flux and $\boldsymbol{D}$ is the dispersion tensor, whose principal component is:

$$D_L = \alpha_L|\boldsymbol{q}| + \phi D_m \tag{3}$$

which is parallel to the flux $\boldsymbol{q}$; α_L is called longitudinal dispersivity and has length dimensions; ϕ is porosity and D_m is molecular diffusion. A similar definition applies for transversal dispersion. With these definitions, the solute transport equation becomes

$$\nabla(\boldsymbol{D}\ \nabla c) - \nabla(\boldsymbol{q}c) + wc_e = \frac{\partial \phi c}{\partial t} \qquad on\ \Omega \tag{4}$$

where c_e is the concentration of water entering the aquifer ($c_e = c$ for the discharging portions of the aquifer). Equation (4) is solved subject to appropiate boundary and initial conditions (See Bear, 1979, for further details on these equations).

Equations (1) and (4) can be solved by means of standard numerical methods. When using backwards differences in time, (1) leads to a system of the form:

$$(\boldsymbol{A} + \frac{\boldsymbol{D}}{\Delta t})\boldsymbol{h}^{k+1} = \boldsymbol{b} + \frac{\boldsymbol{D}}{\Delta t}\boldsymbol{h}^{k} \tag{5}$$

where $\boldsymbol{A}$ is the conductance matrix, which is a linear function of $\boldsymbol{T}$, $\boldsymbol{h}$ is the vector of nodal heads and $\boldsymbol{b}$ is the vector of source terms. $\boldsymbol{D}$ is a storage matrix, k represents time step and Δt is the time increment. Equation 5 is solved sequentially in time.

A similar scheme is used for solving the solute transport equation. Later on, we will require computing the Jacobian matrices ($\partial \boldsymbol{h}/\partial \boldsymbol{p}$ and $\partial \boldsymbol{c}/\partial \boldsymbol{p}$), where $\boldsymbol{p}$ is the vector of unknown model parameters. These can be obtained by taking derivatives of (5) with respect to model parameters, as shown by Medina et al. (1990) and by Carrera et al. (1990).

FORMULATION OF THE ESTIMATION PROBLEM

As discussed earlier, the objective is to estimate element averaged values of $y = log-T$ on the basis of point measurements of T, h and c. Let $\boldsymbol{y}^*_m$ be the vector of $log - T$ measurements, $\boldsymbol{h}^*_m$ the vector of head measurements and $\boldsymbol{c}^*_m$ the vector of concentration measurements. Notice that, in transient problems, $\boldsymbol{h}^*_m$ and $\boldsymbol{c}^*_m$ may consist of many measurements at every observation well. Hence, their dimensions can be very large. We will denote $\boldsymbol{z}^*_m$ to the total vector of measurements, including $\boldsymbol{y}^*_m$, $\boldsymbol{h}^*_m$ and $\boldsymbol{c}^*_m$. We will assume that y is multinormal and that can be described by:

$$E[y(\boldsymbol{x})] \;=\; f(\boldsymbol{\theta}, \boldsymbol{x}) \;=\; \sum_{l=1}^{p} a_l f_l(\boldsymbol{x}) \tag{6}$$

$$Cov\;[y(\boldsymbol{x}_1)), y(\boldsymbol{x}_2)] \;=\; c_v\;(\boldsymbol{\theta}, \boldsymbol{x}_1, \boldsymbol{x}_2) \tag{7}$$

where $\boldsymbol{\theta}$ is the vector of statistical parameters, $f_l(\boldsymbol{x})$ are basis functions for defining the mean $log-T$, a_l are unknown coefficients (components of $\boldsymbol{\theta}$) and c_v is the covariance function. The latter is usually a fuction of $\mid \boldsymbol{x}_1 - \boldsymbol{x}_2 \mid$ rather than $\boldsymbol{x}_1$ and $\boldsymbol{x}_2$ separately.

Spatial variability of heads can be related to that of $log - T$ through the flow equation. First order expansion of heads leads to:

$$\boldsymbol{h}^*_m \;-\; \bar{\boldsymbol{h}}_m \;=\; (\boldsymbol{h}^*_m \;-\; \boldsymbol{h}_m(\boldsymbol{y})) \;+\; (\boldsymbol{h}_m(\boldsymbol{y}) \;-\; \bar{\boldsymbol{h}}_m) \simeq \boldsymbol{\varepsilon}_h \;+\; \boldsymbol{J}_h(\boldsymbol{y} \;-\; \bar{\boldsymbol{y}}) \tag{8}$$

where $\boldsymbol{\varepsilon}_h$ represents measurement and model errors, $\bar{\boldsymbol{y}}$ is the vector of block-averaged unconditional expectation of $log - T$ at the m blocks comprising the solution domain, $\bar{\boldsymbol{h}}_m$ is the vector of heads computed at observation points using $\bar{\boldsymbol{y}}$ and $\boldsymbol{J}_h$ is the jacobian matrix $[\partial \boldsymbol{h}_m/\partial \boldsymbol{y}]$. Equation (8) allows expressing the covariance of head deviates $(\boldsymbol{h}^*_m - \bar{\boldsymbol{h}}_m)$ and the cross-covariance between these and $log - T$ deviates as:

$$\boldsymbol{Q}_{hh} \;=\; E[(\boldsymbol{h}^*_m - \bar{\boldsymbol{h}}_m)(\boldsymbol{h}^*_m - \bar{\boldsymbol{h}}_m)^t] \;=\; \boldsymbol{J}_h \boldsymbol{Q}_{yy} \boldsymbol{J}^t_h \;+\; \boldsymbol{V}_h$$

$$\boldsymbol{Q}_{hy} = E[(\boldsymbol{h}_m^* - \bar{\boldsymbol{h}}_m)(\boldsymbol{y} - \bar{\boldsymbol{y}})^t] = \boldsymbol{J}_h \boldsymbol{Q}_{yy} \tag{9}$$

where $\boldsymbol{Q}_{yy}$ is the covariance matrix of block-averaged $log - T$, which can be easily obtained by averaging (7). $\boldsymbol{V}_h$ is the covariance matrix of ε_h and may be a function of $\boldsymbol{\theta}$.

Similarly, variability of concentrations can be related to that of y through the flow and solute transport equations. Following the same steps as for heads, one can write:

$$\boldsymbol{Q}_{cc} = E[(\boldsymbol{c}_m^* - \bar{\boldsymbol{c}}_m)(\boldsymbol{c}_m^* - \bar{\boldsymbol{c}}_m)^t] = \boldsymbol{J}_c \boldsymbol{Q}_{yy} \boldsymbol{J}_c^t + \boldsymbol{V}_c$$

$$\boldsymbol{Q}_{cy} = E[(\boldsymbol{c}_m^* - \bar{\boldsymbol{c}}_m)(\boldsymbol{y} - \bar{\boldsymbol{y}})^t] = \boldsymbol{J}_c \boldsymbol{Q}_{yy} \tag{10}$$

where, again, $\boldsymbol{V}_c$ is the covariance matrix of concentration measurement errors, which are assumed to be independent of $\boldsymbol{y}$, $\boldsymbol{h}$ and $\boldsymbol{c}$, and $\boldsymbol{J}_c$ is the concentrations jacobian.

The vector of statistical parameters $\boldsymbol{\theta}$ is obtained by maximizing its likelihood:

$$L(\boldsymbol{\theta}/\boldsymbol{z}_m^*) = \frac{1}{[(2\pi)^M |\boldsymbol{Q}_{zz}|]^{1/2}}.exp[-\frac{1}{2}(\boldsymbol{z}_m^* - \bar{\boldsymbol{z}}_m)^t\, \boldsymbol{Q}_{zz}^{-1}(\boldsymbol{z}_m^* - \bar{\boldsymbol{z}}_m)] \tag{11}$$

where $\boldsymbol{z}^t = (\boldsymbol{c}_m^t, \boldsymbol{h}_m^t, \boldsymbol{y}_m^t)$, M is the dimension of $\boldsymbol{z}_m (n_h + n_c + n_y)$, and $\boldsymbol{Q}_{zz}$ is the covariance matrix of $\boldsymbol{z}_m$, given by:

$$\boldsymbol{Q}_{zz} = \begin{bmatrix} \boldsymbol{Q}_{cc} & \boldsymbol{Q}_{ch} & \boldsymbol{Q}_{cy_m} \\ \boldsymbol{Q}_{hc} & \boldsymbol{Q}_{hh} & \boldsymbol{Q}_{hy_m} \\ \boldsymbol{Q}_{y_m h} & \boldsymbol{Q}_{y_m h} & \boldsymbol{Q}_{y_m y_m} \end{bmatrix} \tag{12}$$

where $\boldsymbol{Q}$ are covariance matrices. The cross-covariance between $\boldsymbol{h}_m$ and $\boldsymbol{y}_m$, $\boldsymbol{Q}_{hy_m}$ is given by $\boldsymbol{J}_h\, \boldsymbol{Q}_{yy_m}$, where $\boldsymbol{Q}_{yy_m}$ can be obtained from (7). $\boldsymbol{Q}_{cy_m}$ and $\boldsymbol{Q}_{hc}$ can be written in a similar fashion. $Log - T$ measurement errors are often very large and should be taken into account. They enter in $\boldsymbol{Q}_{y_m y_m}$, which is given by:

$$(\boldsymbol{Q}_{y_m\, y_m})_{ij} = c_v(\boldsymbol{\theta}, \boldsymbol{x}_i, \boldsymbol{x}_j) + \delta_{ij}\sigma_{y_m i}^2 \tag{13}$$

where i and j represent $log - T$ measurement points and $\sigma_{y_m i}^2$ is the variance of the measurement error at point i, which can also be a function of $\boldsymbol{\theta}$.

Estimation by Cokriging

Assuming that $\boldsymbol{\theta}$ is known, an estimate of the average of $log - T$ over the k-th block $\boldsymbol{y}$ can be obtained by minimizing the variance of a linear estimator:

$$\hat{y}_k = \sum_{l=1}^{n_c} \alpha_{lk}(c_{ml}^* - \bar{c}_{ml}) + \sum_{j=1}^{n_h} \mu_{jk}(h_{mj}^* - \bar{h}_{mj}) + \sum_{i=1}^{n_y} \beta_{ik} y_{mi}^* \tag{14}$$

Minimizing the variance of $\hat{y}_k$ with respect to α, β and μ, while imposing unbiasedness, leads to (see, e.g., Samper and Carrera, 1990, for similar derivations) a system of equations that can be written in matrix form as:

$$\hat{\boldsymbol{y}} - \bar{\boldsymbol{y}} = \boldsymbol{Q}_{yze} \boldsymbol{Q}_{zze}^{-1} (\boldsymbol{z}_{me}^* - \bar{\boldsymbol{z}}_{me}) \tag{15}$$

where the subindex e indicates extended matrices and vectors. Specifically z_m^* and $\bar{z}_m$ are extended by adding as many zeros at the end as coefficients are used in the expression of the mean $log-T(8)$, and $\boldsymbol{Q}_{yz}$ and $\boldsymbol{Q}_{zz}$ are extended as follows:

$$\boldsymbol{Q}_{zze} = \begin{bmatrix} \boldsymbol{Q}_{cc} & \boldsymbol{Q}_{ch} & \boldsymbol{Q}_{cy_m} & \boldsymbol{\omega}_{n_c p} \\ \boldsymbol{Q}_{hc} & \boldsymbol{Q}_{hh} & \boldsymbol{Q}_{hy_m} & \boldsymbol{\omega}_{n_h p} \\ \boldsymbol{Q}_{y_m c} & \boldsymbol{Q}_{y_m h} & \boldsymbol{Q}_{y_m y_m} & \boldsymbol{f}_{n_y p} \\ \boldsymbol{\omega}_{pn_c} & \boldsymbol{\omega}_{pn_h} & \boldsymbol{f}^t_{pn_y} & \boldsymbol{\omega}_{pp} \end{bmatrix} \tag{16}$$

$$\boldsymbol{Q}_{zye} = \boldsymbol{Q}^t_{yze} = \begin{bmatrix} \boldsymbol{Q}_{cy} \\ \boldsymbol{Q}_{hy} \\ \boldsymbol{Q}_{y_m y} \\ \boldsymbol{f}^t_p \end{bmatrix} \tag{17}$$

where $\boldsymbol{\omega}_{pn}$ is a $(p \times n)$-zero matrix, $\boldsymbol{f}_{pn_y}$ is the matrix of drift functions evaluated at the points where $log-T$ is measured (its $i-j$ component being $f_j(\boldsymbol{x}_i); j = 1, ..., p; i = 1, ..., n_y$) and $\boldsymbol{f}_p$ is a matrix of the averaged values of such functions at the finite element or finite differences blocks used for the solution of the flow equation.

Matrix $\boldsymbol{Q}_{zze}$ can be ill-posed. In fact, if head or concentration measurements are assumed to be error-free and their number exceeds the dimension of $\boldsymbol{y}$, then $\boldsymbol{Q}_{zze}$ is singular. Ways to overcome this difficulty will be discussed later. Substituting these weights into the expression for covariances allows us writing the estimation errors covariance matrix as:

$$\Sigma_e = E[(\hat{\boldsymbol{y}} - \boldsymbol{y})(\hat{\boldsymbol{y}} - \boldsymbol{y})^t] = \boldsymbol{Q}_{yy} - \boldsymbol{Q}^{-1}_{zze}\boldsymbol{Q}_{zye} \tag{18}$$

In examining the differences between the proposed formulation and standard cokriging, one may notice that cross covariances are not evaluated analytically. Rather, they are obtained from the flow and transport equations (recall equations (9) and (10)). In essence, this involves computing the sensitivity of heads and concentrations to $log-T$'s ($\boldsymbol{J}_h$ and $\boldsymbol{J}_c$). As mentioned earlier, $\boldsymbol{J}_h$ and $\boldsymbol{J}_c$ are computed by taking derivaties of the flow and transport equations for $log-T$ equal to $\bar{y}$. Actually, estimated $log-T$'s may be far from $\bar{\boldsymbol{y}}$, so that one might wonder whether $\boldsymbol{J}_h$ and $\boldsymbol{J}_c$ should be recomputed for the estimated $\boldsymbol{y}$. Obviously, this leads to an iterative process. In dealing with the case of using only head data, Carrera and Glorioso (1991) discuss extensively, and demonstrate empirically, the convenience of evaluating the jacobians at the estimated parameters. This, together with the potential ill-posedness of matrix $\boldsymbol{Q}_{zze}$ motivates the method proposed below.

Estimation by Iterative Maximum Likelihood

We discussed earlier the possibility of using maximum likelihood for estimating the vector of statistical parameters. Assuming that this is known (as it turns out, it suffices to leave it in terms of multiplicative unknown factors for the components defining variances), one may also formulate the estimation of block $log-T$ by maximizing the likelihood function:

$$L(\boldsymbol{\beta}/\boldsymbol{z}^*) = f(\boldsymbol{z}^*/\boldsymbol{\beta}) = \frac{1}{[(2\pi)^N \mid \boldsymbol{C}_z \mid]^{\frac{1}{2}}} exp[-1/2(\hat{\boldsymbol{z}} - \boldsymbol{z}^*)^t \boldsymbol{C}_z^{-1}(\hat{\boldsymbol{z}} - \boldsymbol{z}^*)] \tag{19}$$

where $\boldsymbol{C}_z$ is a block diagonal matrix with its three diagonal blocks being $\boldsymbol{V}_c, \boldsymbol{V}_h$ and $\boldsymbol{V}_y$, $(\boldsymbol{z}^*)^t = (\boldsymbol{c}_m, \boldsymbol{h}_m, \boldsymbol{y}^*)^t$ and N is the dimension of $\boldsymbol{z}$. L depends on model parameters through $\hat{\boldsymbol{z}}$, its first n_c components being $\hat{\boldsymbol{c}}$, computed concentrations, the next n_h components being computed heads, $\hat{\boldsymbol{h}}$, and the remaining m components being the computed parameters themselves. On the other hand, vector $\boldsymbol{\theta}$ of statistical parameters may be needed for especifying the values of $\boldsymbol{V}_h$, $\boldsymbol{V}_c$ and $\boldsymbol{V}_y$ and $\boldsymbol{y}^*$.

The main difference between (19) and (11), aside from the fact that the unknown in (11) is $\boldsymbol{\theta}$ and in (19) is $\hat{\boldsymbol{y}}$, arises from the differences between $\boldsymbol{z}^*$ and $\boldsymbol{z}_m^*$. While the last part of $\boldsymbol{z}_m$, $\boldsymbol{y}_m^*$, represents point measurements of $log - T$, $\boldsymbol{y}^*$ represents prior estimates. Glorioso (1989) shows that both formulations lead to identical estimates if $\boldsymbol{y}^*$ is obtained by kriging. This implies substituting $\boldsymbol{Q}_{zz}$ in (11) by $\boldsymbol{C}_z$ (i.e., using kriging for the covariance matrix $\boldsymbol{V}_y$).

Medina et al (1990) proposed the use of an automatic algorithm for maximizing L with respect to model parameters and some scaling statistical parameters, while estimating the remaining (hopefully few) statistical parameters by trial and error (they describe an aproach for doing so). This allows simplifying (19), whose maximization is equivalent to minimization of $-2lnL$. After dropping terms that only depend on statistical parameters, this leads to the "sum of squared residuals" objective function:

$$F = J_h + \lambda_c J_c + \lambda_y J_y \tag{20}$$

where $J_c = (\hat{\boldsymbol{c}} - \boldsymbol{c}_m^*)\boldsymbol{V}_c^{-1}(\hat{\boldsymbol{c}} - \boldsymbol{c}_m^*)$; $J_h = (\hat{\boldsymbol{h}} - \boldsymbol{h}_m^*)^t\boldsymbol{V}_h^{-1}(\hat{\boldsymbol{h}} - \boldsymbol{h}_m^*)$; $J_y = (\hat{\boldsymbol{y}} - \boldsymbol{y}^*)^t\boldsymbol{V}_y^{-1}(\hat{\boldsymbol{y}} - \boldsymbol{y}^*)$; $\hat{\boldsymbol{h}}$ is the vector of heads computed with parameters $\hat{\boldsymbol{y}}$ at observation points. Weighting factors λ_c and λ_y are used for acknowledging the fact that covariance matrices $\boldsymbol{V}_h$, $\boldsymbol{V}_c$ and $\boldsymbol{V}_y$ may not be accurately known. Rather, we assume that the true covariance matrices are $s_h^2\boldsymbol{V}_h$, $s_c^2\boldsymbol{V}_c$, $s_y^2\boldsymbol{V}_y$ where s_h^2, s_y^2 and s_c^2 are unknown scalars. These can be estimated as:

$$\hat{s}_h^2 = \frac{J_h}{n_h} \;;\; \hat{s}_y^2 = \frac{J_y}{n_y} \;;\; \hat{s}_c^2 = \frac{J_c}{n_c} \tag{21}$$

Theoretically, λ coefficients are defined as:

$$\lambda_y = \frac{s_h^2}{s_y^2} \qquad \lambda_c = \frac{s_h^2}{s_c^2} \tag{22}$$

However, in order to estimate s_h^2, s_c^2 and s_y^2 in (22) one must have first computed J_h, J_c and J_y. Hence, one should proceed iteratively. In practice, computer cost is greatly reduced and convergence facilitated by minimizing (20) for a sequence of decreasing λ_y's and starting the iterative process for a given λ_y from the parameters computed for the previous one. Ideally, this approach should lead to the λ values for which L is maximum. However, degenerate cases may occur for which L tends to infinite as λ_y goes to zero or infinite. In such cases, λ's are computed so as to ensure (22).

The objective function (20) can be minimized using different methods. We concentrate on Gauss-Newton methods, based on updating iteratively the parameters vector by a vector $\boldsymbol{d}$ given by:

$$\boldsymbol{H}\boldsymbol{d}^i = -\boldsymbol{g}^i \tag{23}$$

where $\boldsymbol{H}$ is a first-order approximation of the hessian matrix of F and $\boldsymbol{g}^i$ is its gradient, both computed using the parameters at iteration i, $\boldsymbol{y}^i$. They are obtained from direct derivation of F, neglecting second order derivatives for $\boldsymbol{H}$. Equation (23) may need modifications due to poor behavior of F. Following the suggestion of Cooley (1985), we have found Marquardt's method most useful.

Assuming that the basic Gauss-Newton method works, the parameters vector at iteration $i+1$ is derived from that at iteration i as:

$$\boldsymbol{y}^{i+1} = \boldsymbol{y}^i + \boldsymbol{H}^{-1}[\boldsymbol{J}_c^t \boldsymbol{V}_c^{-1}(\boldsymbol{c}_m - \boldsymbol{c}^i) + \boldsymbol{J}_h^t \boldsymbol{V}_h^{-1}(\boldsymbol{h}_m - \boldsymbol{h}^i) + \boldsymbol{V}_y^{-1}(\boldsymbol{y}^* - \boldsymbol{y}^i)] \qquad (24)$$

Equation (24) is applied iteratively; that is, $\boldsymbol{y}^1$ is computed from $\boldsymbol{y}^0$ and used for evaluating $\boldsymbol{J}$, $\boldsymbol{H}$, etc. This allows applying again equation (24) with the updated matrices for computing $\boldsymbol{y}^2$, etc. The process is repeated until convergence. Finally, a lower bound of the estimation errors covariance matrix is given by $\hat{s}_h^2 \boldsymbol{H}^{-1}$.

Example

Problem definition

In order to test the effectiveness of the approach, a synthetic example was generated. This section is devoted to summarize the results. The flow domain consists of a rectangular region of 200×90 m^2 (length and time units are somewhat arbitrary). The transmissivity field was generated using the sequential simulation code of Gómez (1991) with a linear variogram: $\gamma(h) = 0.017h$

This field is displayed in Figure 1a, which represents point values of transmissivity and is going to be assumed as the "true" $log - T$ field in what follows ($log - T$ increases with darkness). Because of instability problems associated with solute transport simulation, dispersivity cannot be significant smaller than element size. Yet, in order to avoid artificial smoothing, one is forced to use small dispersivities, comparable to the local scale. This problem was overcome by working with constant transmissivity over blocks of 8×5 m^2, while keeping the element size to 1×1 m^2. This allowed us to work with a dispersivity of 3 m. These values are smaller than the block size, hence suggesting physical plausibility, but comparable to element size, thus ensuring numerical stability.

In short, even though the "true" $log - T$ field is the one displayed in Figure 1a, solute transport was simulated using the block averages (Fig. 1b) which, as shown by Sanchez-Vila and Carrera (1992), does indeed provide a good representation of average flow rates. The flow and transport problem is outlined in Figure 1c. A prescribed flow rate of 0.4 m^3/day/m was prescribed at the left boundary, while head was prescribed to 0 m at the right boundary. The upper and lower boundaries were assumed impervious. A constant rate of 12 m^3/day is extracted from an internal node located where shown in Figure 1c. Flow takes place under steady-state conditions.

Simulation of transport was performed using the same 201×91 nodes grid, assuming that concentrations are initially zero and that a concentration of 100 g/m^3 (again, units are arbitrary) is injected continuously with the water that enters the domain across the left boundary. A thickness of 1m and a porosity of 0.1 were taken throughout the domain. Flow and transport over this refined grid (almost 20000 nodes) were simulated using FAITH

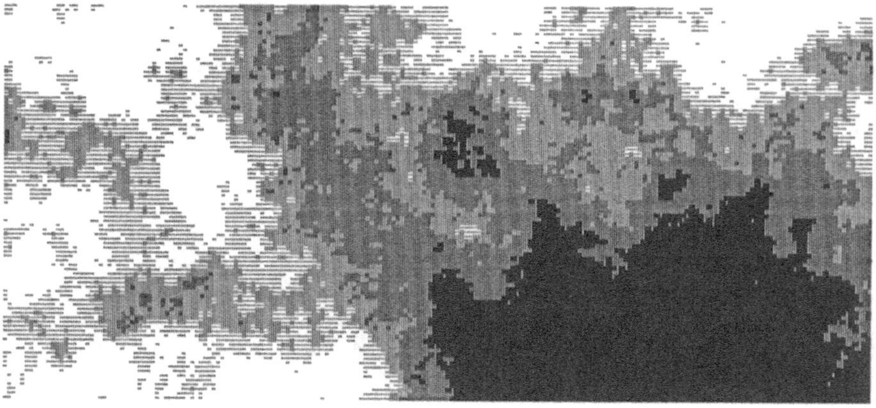

1a. True point y(=log-T) field

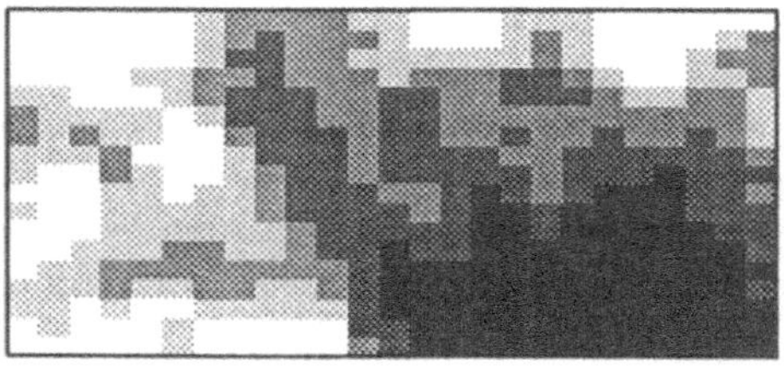

1b. True block y's for simulation

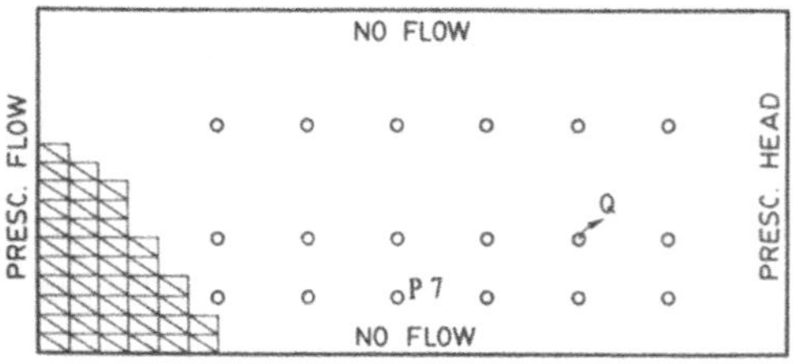

1c. Definition of flow problem

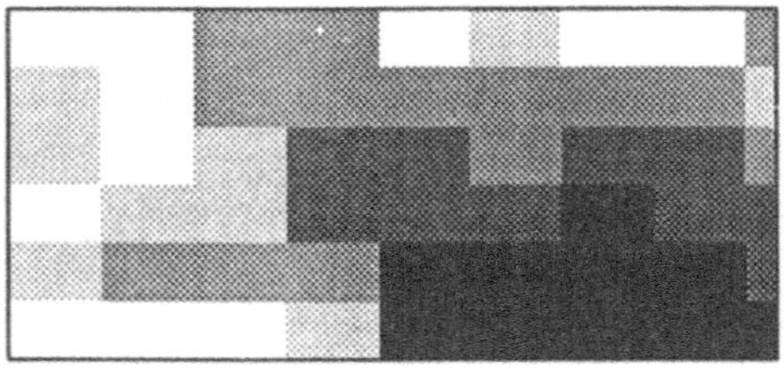

1d. True block y's for estimation

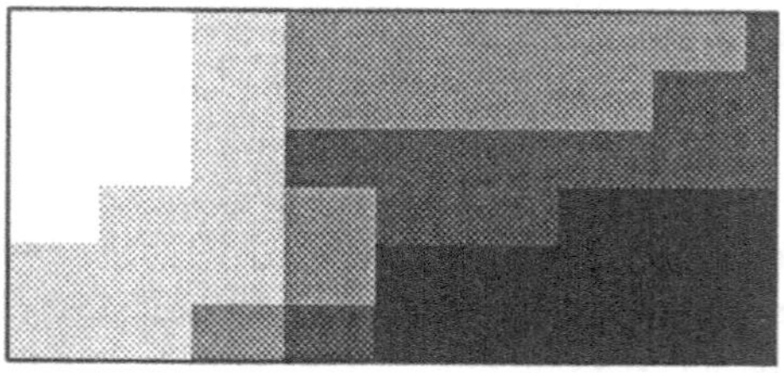

1e. Block y's from kriging

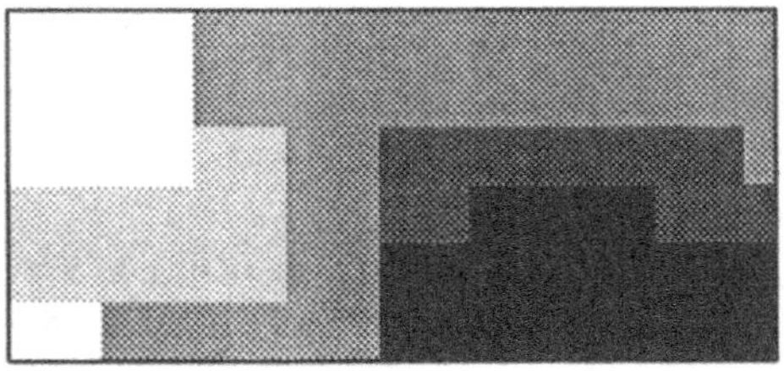

1f. Block y's from $y^* + h_m$

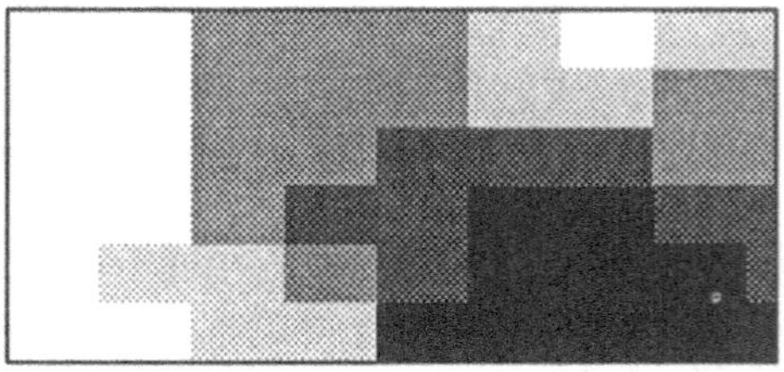

1g. Block y's from $y^* + h_m + c_m$

Figure 1: Description and results of the example

(Sánchez-Vila et al., 1992), a code specially designed to solve these equations over very fine grids on multiprocessor vectorial machines.

Concentrations computed at the 18 observation points shown in Figure 1c were recorded over time as actual measurements. Heads at 17 points (all but the pumping "well") were also recorded as measurements.

Formulation of the estimation

Using the proposed approach (iterative maximum likelihood), involves the following steps:

1- Define the model structure. This includes model geometry, boundary conditions, geostatistical structure of model parameters, blocks over which average $log-T$ is to be estimated, etc.

2- Use kriging for computing prior estimates of block averaged $log-T$ ($\boldsymbol{y}^*$) from point measurements ($\boldsymbol{y}_m$). Also, kriging errors covariance matrix is computed at this stage (Notice that one is not forced to use kriging, but any prior estimation method that produces $\boldsymbol{y}^*$ and $\boldsymbol{V}_y$ from any available data other than head and concentration).

3- Use the iterative approach described above to estimate model parameters (block-averaged $log-T$'s in our case).

Regarding step 1, model structure resembles the true one except that a much coarser grid was used, in order to simulate that in reality one will always be constrained by computer resources. Hence, the grid consists of only 494 nodes (A portion of the grid is shown in the lower left corner of Figure 1c). As for the blocks over which $log-T$ is going to be estimated, we used different discretizations. We will present the results obtained for 9×6 blocks (Figure 1d displays the arithmetic averages of true $log-T$'s over them).

Regarding step 2, block-kriging was performed using program KRINET (Olivella and Carrera, 1990). True $log-T$ at the 18 observation points were used as data (Notice that these are point values). In addition to block estimates, which are displayed in Figure 1e, KRINET also produces the covariance matrix $\boldsymbol{V}_y$.

Finally, $log-T$'s over the 54 blocks were estimated under several hypotheses. First, only head data were used, while its weight was varied. Then, concentration data were also included in the estimation. Results are discussed below.

Matrices $\boldsymbol{V}_h$ and $\boldsymbol{V}_c$, which quantify model and measurement errors, are unknown. In this case, the main components of the error can probably be attributed to truncation (coarse F. E. grid) or discretization of the $log-T$ field (large blocks). Lacking a good quantification of these, we treated them as diagonal matrices using standard deviations of 0.1 m for heads and of 10 g/m^3 for concentrations (one tenth of input concentration). We will recognize that these matrices are uncertain by letting them defined in terms of unknown factors s_h^2 and s_c^2 as discussed earlier.

Results

As mentioned above, estimation was first performed using only heads and prior estimates of $log-T$ obtained by kriging. The optimal λ_y was chosen as the one that is equal to the ratio s_h^2/s_y^2. As shown in Table 1, this leads to $\lambda_y = 0.1$, which results in the $log-T$ field displayed in Figure 1f. In order to quantify whether results are indeed reasonable we used

Table 2: Performance criteria of log-T estimated by kriging ($\boldsymbol{y}^*$), adding head data ($\boldsymbol{y}^*+\boldsymbol{h}_m^*$) or adding also concentration.

Data	MSE	CI_1	CI_2
$\boldsymbol{y}^*$	5.46 10^2	1.56	1.44
$\boldsymbol{y}^* + \boldsymbol{h}_m^*$	2.68 10^2	2.72	1.86
$\boldsymbol{y}^* + \boldsymbol{h}_m^* + \boldsymbol{c}_m^*$	2.01 10^2	4.18	3.31

$Log - T$'s computed after including concentrations, as well as heads and transmissivities, in the estimation process are displayed in Figure 1g. Visually, they lead to a notable improvement. While $log - T$ fields computed with heads and/or measured $log - T$'s are very smoothed versions of true $log - T$'s, addition of concentrations leads to a much better picture of $log - T$ variability. However, as shown in Table 2, performance criteria are rather disappointing. The MSE is only reduced slightly from the previous cases. This is at the cost of increasing consistency indices (i.e., reduction in errors is much smaller than anticipated by the covariance matrix). Overall, the model seems to properly reproduce observed heads and concentrations (Figure 2). Moreover, computed $log - T$'s are not far from block averages of true $log - T$'s (also shown in Figure 2).

It is interesting to notice that, even when only kriging was used, actual errors had been underestimated by the kriging covariance matrix (even though the ratio of actual to expected errors is of only 1.4, one would expect a much better behavior). In this context, underestimation of s_y^2, which was shown in Table 1, can only be attributed to the methodology, not to the particular realization under study (If anything, actual variability of this realization is larger than average!). This suggests that consistency of error estimates may be significantly improved by ignoring the s_y^2 correction to $\boldsymbol{V}_y$ (actually to the variogram). However, such a method would not address the basic issue. Carrera and Glorioso (1991) argue that underestimation of the covariance matrix is caused by the non-linear relationship between heads and $log - T$. This leads to non-linear dependence between $log - T$'s at different blocks. Such a dependence cannot be properly accounted for with the covariance matrix. If this is indeed the cause of problems, then the limitation is inherent to representing uncertainties by means of the covariance matrix. Alternative methods, such as non-linear confidence intervals (Vecchia and Cooley, 1985), should be used.

Summary and Conclusions

Two approaches have been presented for the estimation of groundwater flow and transport parameters from measurements of those parameters, heads and concentration. Emphasis has been given, both in the derivations and in the example, to the case of estimating log-transmissivities. The first approach is based on co-kriging and the second consists of iterative maximization of the likelihood function. It can be shown that the cokriging approach is a particular case of the iterative methods. Specifically, cokriging estimates and covariance matrix are identical to those obtained after the first iteration of the second approach, provided three criteria:

$$MSE = \frac{1}{q}(\hat{\boldsymbol{y}} - \boldsymbol{y}_t)^t(\hat{\boldsymbol{y}} - \boldsymbol{y}_t) \tag{25}$$

$$CI_1 = \left[\frac{1}{m}(\hat{\boldsymbol{y}} - \boldsymbol{y}_t)^t\boldsymbol{\Sigma}_y^{-1}(\hat{\boldsymbol{y}} - \boldsymbol{y}_t)\right]^{1/2} \tag{26}$$

$$CI_2 = \left[\frac{1}{m}\sum_{i=1}^{m}\frac{(\hat{y}_i - y_{ti})^2}{Var(\hat{y}_i)}\right]^{1/2} \tag{27}$$

where $\hat{\boldsymbol{y}}$ is the vector of estimated block $log - T$'s, $\boldsymbol{y}_t$ is the vector of "true" block $log - T$'s (average of point values; $\boldsymbol{y}_t$ are shown in Figure 1d), m is the number of blocks and $\boldsymbol{\Sigma}_y$ is the covariance matrix. the Mean Square Error (MSE) is an absolute measure of the overall error in estimated $log - T$'s. The other two indices measure the consistency between actual estimation errors and those predicted by the computed covariance matrix and variances. Ideally, these two should be close to one , although one would expect them to be above unity because computed $\boldsymbol{\Sigma}_y$ is a lower bound of the actual covariance matrix.

Table 1: Selection of optimal λ_y for estimation of block $log - T$'s from head and $log - T$ point measurements.

	Crit. for. λ_y Selection			Performance Criteria		
λ_y	s_h^2	s_y^2	s_h^2/s_y^2	MSE	CI_1	CI_2
10	1.50	0.008	176.	$4.93\ 10^{-2}$	4.01	3.91
1	0.56	0.12	4.66	$3.94\ 10^{-2}$	2.04	1.94
0.1	0.06	0.59	0.10	$2.68\ 10^{-2}$	2.72	1.86
0.05	0.023	0.76	0.03	$2.72\ 10^{-2}$	4.53	2.22
0.01	.0017	1.00	0.002	$2.91\ 10^{-2}$	15.82	3.99

Results shown in Table 1 allow us to verify that the optimal λ_y in the sense of minimum estimation errors is indeed the one for which s_h^2/s_y^2 equals λ_y. Moreover, the corresponding consistency indices are lower for $\lambda_y = 0.1$ than for most other λ_y values. It is interesting to notice that computed variances appear to be more consistent with actual errors than the full covariance matrix, as suggested by the fact that CI_1 is always larger than CI_2. This is consistent with the findings of Carrera and Glorioso (1991), who attributed it to the non-linear relationship between heads and concentrations. This leads to non-linear dependences between errors of estimated parameters. Such type of dependences cannot be properly accounted for by the covariance matrix.

Another interesting observation from Table 1 is that $s_y^2 = 0.59$. This suggests that prior estimates are less uncertain than implied by $\boldsymbol{V}_y$. Since this was computed using the true variogram, one should conclude that the studied realization of $log - T$ is slightly less variable than expected or that computed s_y^2 is in error (not a very large one, in any case). We shall come back to this issue later.

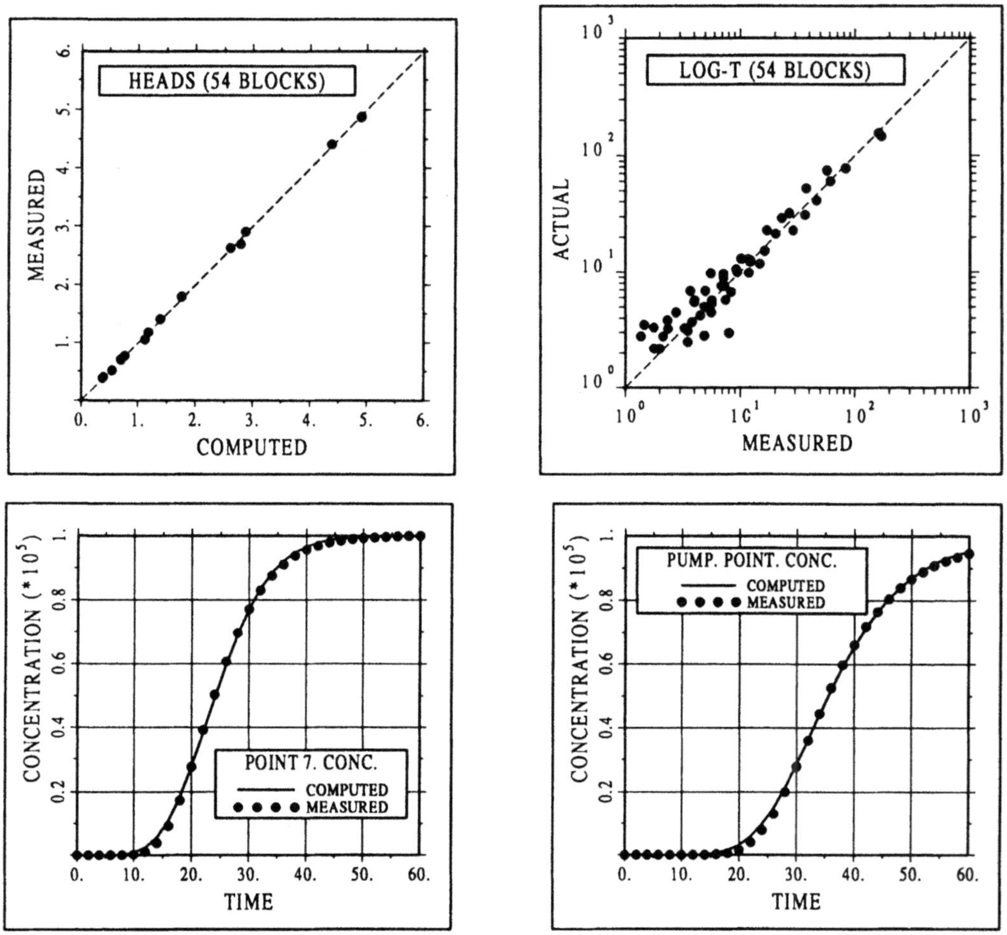

Figure 2: Summary of fitted heads, *log* – *T*'s and concentrations (only two observation points).

that the latter is started from the expected value of $log - T$.

The approach has been tested with an example, which has allowed us to show that considering concentration and head data improves the estimation and leads to identification of spatial variability trends which may not be apparent from $log - T$ point measurements. However, estimation improvements brought about by inclusion of head and concentration data are not well represented by the covariance matrix. In fact, the covariance matrix tends to be overly optimistic. In the example presented here, actual errors were about 1.4 times larger than expected when estimation was performed by kriging, they were 1.9 times larger than expected when including head data (That is estimation errors were reduced, but not as much as expected from the covariance matrix). The ratio between actual and expected errors grew to 3.3 when concentration data was also included. This optimistic evaluation of the estimation errors covariance matrix may be caused by the non-linear relationship of heads or concentrations with respect to $log - T$'s. This leads to non-linear dependence between estimation errors, which may not be well represented by the covariance matrix. Several alternatives have been mentioned to overcome these difficulties. However, the basic limitation inherent to representing uncertaintites by a covariance matrix will remain unles the uncertainties are greatly reduced.

Despite of this problems, it is clear that including concentration data in the estimation process leads to notable reductions in uncertainty and, what is more important, to improved estimation of spatial variability trends.

Acknowledgements

This work was funded by ENRESA (Spanish nuclear waste disposal company) and by the European Community through grant IT003 with CIMNE and by CICYT through grant PB1015 with UPC. Luis Ribeiro provided a very useful review.

References

Ahmed, S. and G. De Marsily (1987) "Comparison of geostatistical methods for estimating transmissivity using data on transmissivity and specific capacity", Water Resour. Res., 23(9), 1717-1737.

Bear, J. (1979) "Hydraulics of Groundwater", McGraw-Hill, New York.

Carrera, J. (1987) "State of the Art of the Inverse Problem Applied to the Flow and Solute Transport Equations", Groundwater Flow and Quality Modelling, Reidel, Dordrecht, 549-584.

Carrera J, F. Navarrina, L. Vives, J. Heredia and A. Medina (1990) "Computational aspects of the inverse problem" in Computational Methods in Subsurface Hydrology, CMP and Springer-Verlag.

Carrera, J. and L. Glorioso (1991) "On Geostatistical Formulations of the Groundwater Flow Inverse Problem". Advances in Water Resources, 14(5), 273-283.

Clifton, P.M. and S.P. Neuman (1982) "Effects of kriging and Inverse Modeling on Conditional Simulation of the Avra Valley Aquifer in Southern Arizona", Water Resour. Res., 18(4), 1215-1234.

Cooley, R.L. (1985) "A comparison of several methods of solving nonlinear regression groundwater flow problems", Water Resour. Res., 21(10), 1525-1538.

Dagan, G. (1985) "Stochastic modeling of groundwater flow by unconditional and conditional probabilities: The inverse problem", Water Resour. Res., 21(1), 65-72.

Glorioso, L.A. (1989) "Relación entre dos formulaciones para la solución del problema inverso de la hidrogeología subterránea", Tesina de Especialidad, Dep. de Ingeniería del Terreno, E.T.S.C.C.P., Barcelona, 53 pp.

Gomez, J.J. (1991) "A stochastic approach to the simulation of block conductivity fields conditioned upon data measured at a smaller scale". Ph. D. Dissertation, Department of Applied Sciences, Stanford University, California, USA.

Ginn, R. and J. Cushman (1990) "Inverse Methods for Subsurface Flow: A Critical Review of Stochastic Techniques", Stochastic Hydrol. Hydraul., 4, 1-26.

Graham, W. and D. McLaughlin (1990) "Stochastic analysis of nonstationary subsurface solute transport, 2, Conditional moments". Water Resour. Res. 25(11), 2331-2356.

Hoeksema, R.J. and P.K. Kitanidis (1984) "An application of the geostatistical approach to the inverse problem in two-dimensional groundwater modeling", Water Resour. Res., 20, 1003-1020.

Keidser, A. and D. Rosbjerg (1991) " A comparison of four inverse approaches to groundwater flow and transport parameter", Water Resour. Res., 27(9), 2219-2232.

Kitanidis, P.K. and E.G. Vomvoris (1983) "A geostatistical approach to the inverse problem in groundwater modeling (steady state) and one-dimensional simulations", Water Resour. Res., 19(3), 677-690.

Medina, A., J. Carrera and G. Galarza (1990) "Inverse modelling of coupled flow and solute transport problems", ModelCARE 90: Calibration and Reliability in Groundwater Modelling, IAHS Publ. no. 195, pp. 185-194.

Neuman, S.P. (1980) "A statistical approach to the inverse problem of aquifer hydrology, 3. Improved solution method and added perspective", Water Resour. Res., 16(2), 331-346.

Olivella, S. and J. Carrera (1990) "KRINET: Manual del usuario" E.T.S.I.C.C.P., Barcelona.

Sánchez-Vila, X. and J. Carrera (1992) "Comparación de diversas formulaciones estocásticas en el problema del escalado de transmisividades". Hidrogeología y recursos hidráulicos, Vol. XVII, pp. 515-533.

Sánchez-Vila, X., I. Colominas and J. Carrera (1992) "Manual de usuario del programa FAITH". E.T.S.I.C.C.P., Barcelona.

Samper, F.J. and J. Carrera (1990) "Geoestadística; Aplicaciones a la hidrología Subterránea", Ed. CIMNE, Barcelona.

Vecchia, A.V. and R.L. Cooley. (1987) "Simultaneous confidence and prediction intervals for nonlinear regression models, with application to a groundwater flow model", Wat. Resour. Res., 23(7), pp. 1237-1250.

Yeh, W.W.-G. (1986) "Review of parameter identification procedures in groundwater hydrology: The inverse problem", Water Resour. Res., 22(2), 95-108.

MULTIVARIATE GEOSTATISTICS FOR SOIL CLASSIFICATION

Giuseppe RASPA, Roberto BRUNO
Department of I.C.M.M.P.M.
Univ. of Rome "La Sapienza"
Via Eudossiana, 18
00184 Rome, Italy

Paola DOSI, Nicola FILIPPI
Regione EMILIA-ROMAGNA
Viale Silvani 4/3
40122 Bologna, Italy

Giuseppe PATRIZI
SINGEA S.r.l.
Via G.Mazzini, 15
48100 Ravenna, Italy

ABSTRACT. This paper deals with soil classification for semi-detailed mapping (scale 1:25000) in an area situated in the Emilia Romagna Region (Northern Italy). The available data consist of field and laboratory measurements of more than 750 soil samples spaced at various distances. For this study six variables were considered, namely three grain sizes, pH and 2 calcium contents. Principal Component Analysis was carried out in order to further reduce the number of variables. The three first factors extracted more than 80% of the global variance and seemed to synthesize adequately the soil characteristics and to be relevant to soil genesis. Geostatistical structural analysis of the factors showed nested structures at different scale. Thus it was possible to estimate their spatial components at the sample locations through a simplified version of factorial kriging. Then, a cluster analysis was applied sequentially over the groups of samples consecutively identified on the basis of specific combinations of the components of the factors. The procedure seems to help the paedologist to identify criteria for classifying and mapping soils at detailed scale.

1. INTRODUCTION

Soil is a complex, tridimensional covering that is usually considered to be a "continuum" since it is rarely possible to divide it into discrete units: a "soil-individual" does not have well-defined borders.

The purpose of a soils map is to illustrate the geographical distribution of the main soil features, describing the latter in terms both of genetic origin and of composition and properties (e.g. water and thermal regimes, chemical and physical characteristics). Reconnaissance or exploration maps at large scales (1:1,000,000-1:100,000) are useful for conveying general knowledge rapidly; semi-detailed maps (1:50-25,000) are mainly useful for feasibility or preliminary design purposes while detailed maps (1:10,000 and more detailed) are required for final designs.

Traditional soils mapping is based on the genetics of soils formation with the data being interpreted and extrapolated using "laws" that link specific soils characteristics with other, more easily identifiable features of the landscape. However, the variability of soils in space, which is defined by sampling the more significant parameters, does not always remain constant at all scales of mapping. Changes in scale often modify the terms of the problem. Whereas the borders between two units identified in a reconnaissance or semi-detailed map are normally based on inferences from landscape features, most of the borders in a detailed map must be verified by means of a very high number of observations on a regular grid.

The past decade has seen the development of computer-aided methodologies based on statistical and geostatistical techniques and designed both to study and quantitatively characterize soils variability (Iris, 1984; Webster and Oliver, 1985), and to permit the classification of soils by automatising the identification of suitable sized homogeneous units (Oliver and Webster, 1989; Wackernagel et al., 1988). This paper is a contribution towards this aim, aware that the potential of multivariate geostatistics in the earth sciences has yet to be fully exploited.

2. AIMS OF THE STUDY

The study aims at identifying and developing computer-aided soils-mapping tools that exploit the methods of multi-dimensional geostatistics. The traditional classification methodology is based on consolidated soils formation models and the maps produced are sufficiently reliable up to the semi-detailed scale. The present study focuses on semi-detailed up to detailed maps. The semi-detailed maps, when reliable,are usefull essentially for testing the inforrmatic methodology. The fact that the passage to the smaller scales is associated with increasingly more numerical and quantitative sorts of data is another reason why the study focuses on a scale where numerically based methodologies are more appropriate.

The envisaged procedure performs soils classifications by applying the techniques of multivariate statistical analysis and multivariate geostatistics (Wackernagel et al., 1989; Matheron, 1982).

3. THE METHODS USED

The most practical way to generate a map based on the point data collected by sampling in an area consists in
a) dividing the area into regular parts or cells,
b) estimating the studied parameters for each cell, and
c) clustering the cells on the basis of the estimated parameters.

It is not convenient to estimate all the original variables because the latter are correlated and co-kriging them would not be practical given the sizes of the matrices that would have to be inverted. The number of variables is therefore reduced beforehand, by applying factorial analysis. This procedure has the added avantage of generating independent factors, making it possible to limit the subsequent estimating operation to a kriging of the most representative factors.

3.1 Factorial Analysis

The type of factorial analysis to be carried out depends on what kind of data are available. The following are possible choices:

- Principal Components Analysis (PCA), which is adopted when all the original variables are numerical and continuous. This technique is well-known.
- Analysis of Multiple Correspondences (AMC), which is used when some qualitative variables are considered important for the classification. This technique consists in finding the factors of a classical correspondence analysis applied to a disjunctive table Z with n lines (number of individuals) and m columns (total number of modalities that the individual may assume) that satisfies the equation $m = \Sigma_{i=1,p}\, m(i)$, where p is the number of variables and m(i) is the number of modalities that the variable i may assume. If the variable is qualitative, m(i) is intrinsic to the variable. If m(i) is quantitative it depends on user choices and it may represent the number of intervals into which the field of values assumed by the variable is to be divided: e.g. if pH is the variable, the field might be divided into 3 classes: less than 6.5, between 6.5 and 7.5 and higher than 7.5. The individual disjunctive code would therefore assume the values 100,010,001 according to whether the measured value falls in the first, second or

third class.

3.2 Structural Analysis and Kriging of Factors

Having carried out a factorial analysis and interpreted the findings, the desired number of factors is selected and the spatial variability of each one is examined by calculating its variograms. The model variogram is then adjusted to highlight, for each factor, the structures that may be present. Thereafter, the spatial components of the factors are kriged. Briefly, the method can be described as follows. For more details of the method see Matheron, 1982 and Wackernagel, 1992.

Let $F_i(x_\alpha)$ be the value of the coordinate of factor i at the point x_α. If the variogram of factor i highlights p structures, the factor is divided into its p spatial components:

$$F_i(x_\alpha) = \Sigma_u F_{iu}(x_\alpha)$$

The estimate of the component u at the point x_0 is obtained using the linear combination:

$$F_{iu}(x_0) = \Sigma_\alpha \lambda_{u,\alpha} F_i(x_\alpha)$$

The following kriging system generates the values of the weights $\lambda_{\alpha,u}$:

$$\Sigma_\alpha \lambda_{\alpha,u} \gamma_i(x_\alpha,x_\beta) = \gamma_{iu}(x_0,x_\beta) \qquad \forall \ \beta = 1,n$$

where γ_i is the variogram function of factor i and γ_{iu} is the structural component u of the variogram γ_i.

The estimate of a factor's components provides a set of data that is structurally considerably more informative than the simple factors. The classification is therefore more effective, especially for subsequent, more detailed definitions.

3.3 Cell Clustering

The clustering on the cells for preparing the maps is the most delicate operation of the procedure. As the parameters used to define the cell clusters are the spatial components of the factors that intervene at different scales, there would be no sense in incorporating all the parameters simultaneously. However, this breakdown into spatial components can be exploited by generating a series of successive clusters, starting with one based on the components that are significant at the larger scale and passing gradually to the smaller scales, grouping and filtering the results as appropriate. The "path" of successive clusters thus created is clearly pregnant with potential and hence also with difficult choices.

The most effective path must be identified first, by trial and error and examining possible clusters at different degrees of detail. During the search for the most effective path, the clusters should be generated using sampling data rather than the "cells" constructed on the basis of spatial components. This saves time and streamlines the procedure without jeopardising its effectiveness. During experimental applications, the procedure has been applied to sample data, undertaking cell classification only after the path has been identified.

As the aim at this stage is to analyse the results and understand the phenomenon, no particular spatial constraints are included in the clustering algorithm as their inclusion would only complicate matters. Excessive discontinuity of the mapping units, if it should arise, can be addressed later by applying constrained clustering or by redefining the borders generated by a free cluster. It is, however, necessary to bear in mind that any discontinuity effect will be strongly mitigated by the fact that the classification will be based on estimates of the areal cells.

4. CASE STUDY

4.1 The Sample

The proposed procedure was applied using the data from a sampling campaign conducted in Piacenza plain in Emilia-Romagna Region, Northern Italy (Fig. 1). Observations and recordings had been made at over 800 stations for extremely

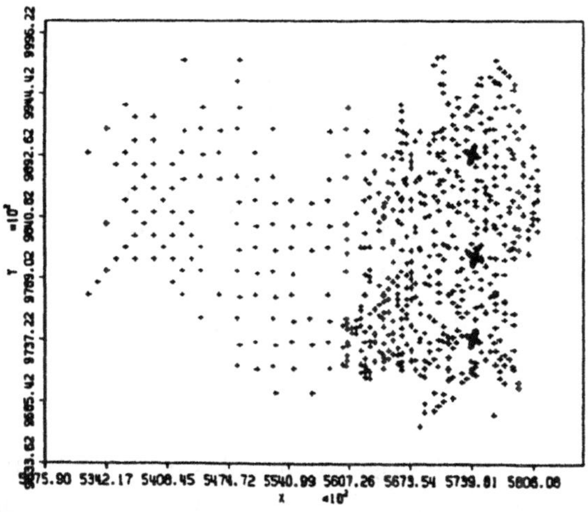

Fig.1 - Sample data location.

different magnitudes: quantitative and qualitative, characterised by nominal, ordinal or interval scales. Meaningful magnitudes for the purposes of the present study were considered.

The available samples could be divided into 3 groups on the basis of their positions (in plan and in depth) and the variables that had been recorded:

- (A) For about 200 stations positioned on a regular grid, 4 soil samples had been taken at different depths and 13 variables had been recorded: 4 grain sizes (coarse and fine sand, silt and clay), pH, active and total Ca, 3 wet-colour indices (hue, value and chrome) and 3 mottling indices (hue, value and chrome).
- (B) For about 180 stations located along 3 "crosses" formed by 2 transects involving about 30 stations each, 2 soil samples had been taken at different depths and 10 variables had been recorded: 4 grain sizes (sand, coarse and fine silt, and clay), pH, active and total Ca, organic matter, Cu and Zn ions.
- (C) For another 450 stations located randomly in the eastern half of the surveyed area, 1 surface soil sample had been taken and 6 variables had been considered: 3 grain sizes (sand, silt and clay), active and total Ca, and pH.

Given the lack of homogeneity of the above data, it was decided to use only the surface recordings (777 soil samples) and 6 variables: 3 grain sizes (sand, silt, clay), total and active Ca, and pH. This decision was motivated as follows:

- except for "hue", the data on colour were so incomplete as to signficantly reduce sample size; the variables "value" and "chrome" were therefore rejected a priori;
- a preliminary Principal Component Analysis (PCA) on the variables of group A at all depths revealed a low degree of significance for "hue"; hue extracts only 1-2% of the overall variance and generally behaves like an isolated factor. This variable was therefore also rejected;
- grain-size homogeneity could be enhanced by combining the 2 sand variables of group A (a preliminary study had also shown that no information would be lost) and the 2 silt categories of group B;
- the largest body of data is represented by the surface samples.

4.2 Principal Component Analysis

A PCA was carried out on the 6 original variables in order to obtain independent magnitudes (the factors). By examining the nature and importance of the correlations and the "extraction" capacity of each factor from total variance, it may be possible to reduce the number of variables to be considered.

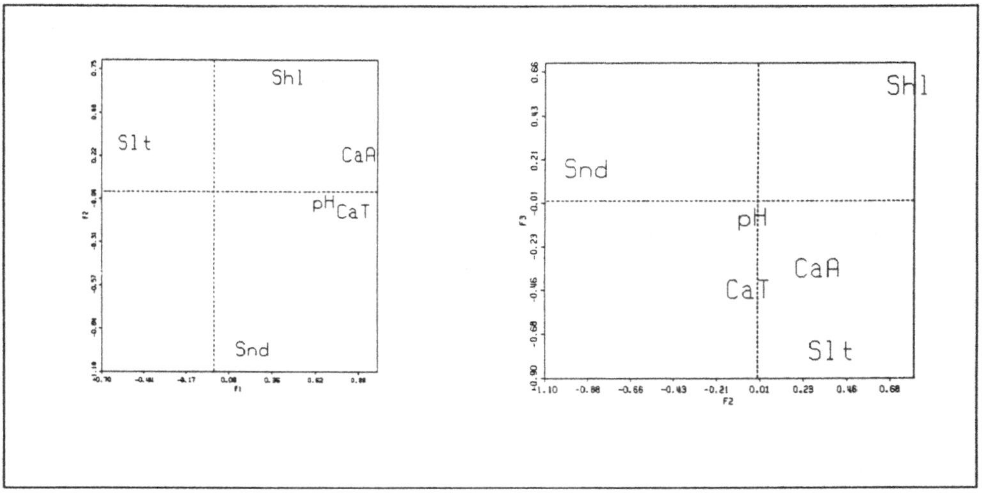

Fig.2 - PCA results: projections of the six variables on the two spaces F1/F2 and F2/F3.

Figure 2 illustrates a result of the PCA, produced by projecting the original variables into the space of the first 3 factors. The first factor is clearly linked mainly to the 2 variables for active and total Ca and to a lesser extent to the pH variable. The second factor discriminates between the large (sand) and fine grain sizes, whereas the third factor discriminates between silt and clay. As these first 3 factors account for over 80% of the total variance, factors 4 to 6 could be ignored.

4.3 Structural Analysis

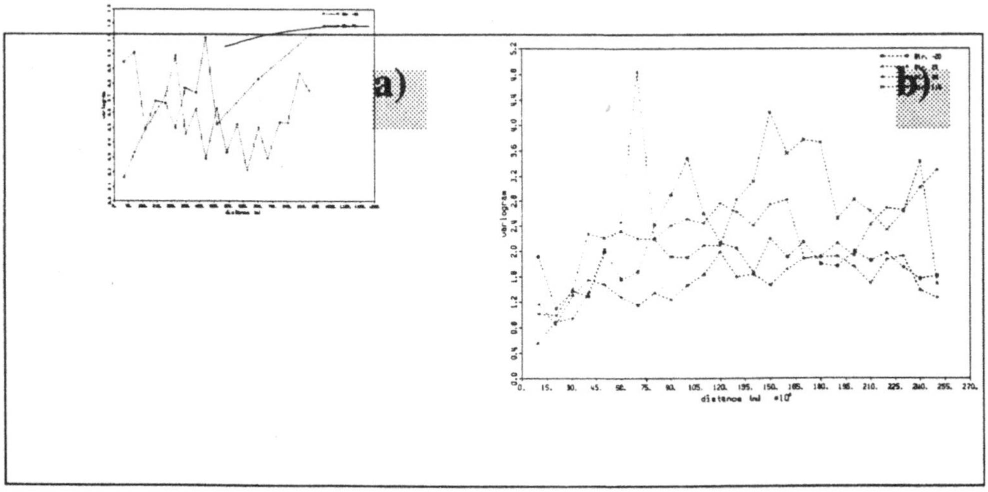

Fig.3 - Factor 1: experimental variograms at small (A) and large scale (B).

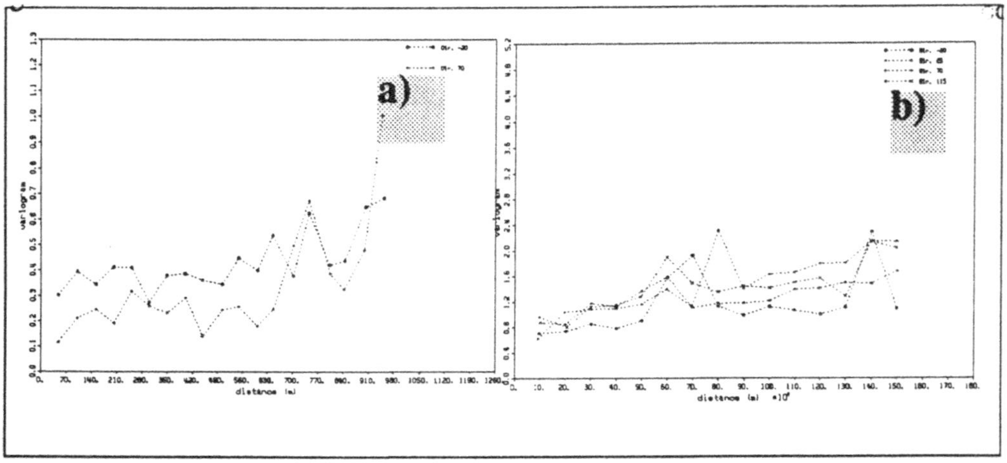

Fig.4 - Factor 2: experimental variograms at small (A) and large scale (B).

statistical structural analysis conducted on the first 3 factors allowed both to verify the stationarity of the variables and to model the 3 experimental variograms (Figs. 3-5).

The identified models are not simple ones. They reflect nested structures involving up to 4 components that can refer to different scales and that present anisotropies with respect to the conjugated directions 70°/-20°. Thanks to the 3 "crosses" of closely grouped samples from different zones, it was possible to examine the variability at small scale (0-1,000m). A three components variogram has been modeled for factor 1, with ranges of 200, 7,000 and 15,000m. A four components variogram has been

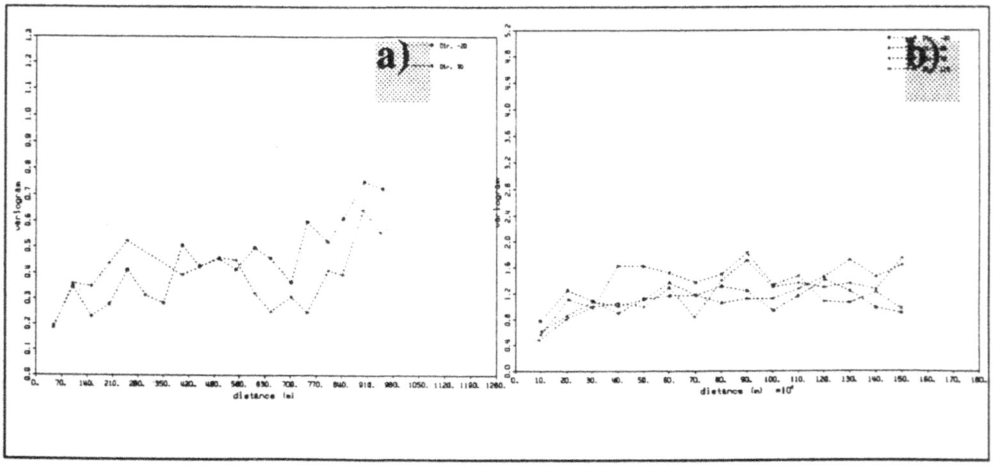

Fig.5 - Factor 3: experimental variograms at small (A) and large scale (B).

small scale (0-1,000m). A three components variogram has been modeled for factor 1, with ranges of 200, 7,000 and 15,000m. A four components variogram has been modeled for factor 2: a nugget effect and three structures with ranges of 100, 2,000 and 21,000m. A two component model has been modeled for factor 3 with ranges of 100 and 4000m.

The cross variograms calculated for all 3 factors confirmed the independence of their spatial variability.

4.4 Estimation of the Components

In view both of the objective of classifying the samples and the territory and the need to identify the appropriate clustering modalities by trial and error, the components at small, medium and large scales of all the considered factors were estimated for each station.

An estimate of the nugget component (present only for factor 2) makes sense for the sampling data. But, once the proper clustering modalities will have been identified, the procedure pass on to the estimation of the average values of the components for the elementary cells of the area and the nugget component will be virtually filtered. The 10 spatial components identified for the factors were thus computed for each station.

4.5 Cluster Analysis

Cluster analysis was then carried out using Ward's technique (Ward, 1963) and adopting different strategies.

For comparative purposes, a first reference cluster was computed using all 3 global factors of all the samples. The result brings out a significant irregularity of the spatial distribution of the identified groups.

Cluster analysis conducted individually on the single components of each factor, the groupings remaining unchanged, brought out the expected relationships between the regularity of the groups of samples and the scale of variability of the component. Consequently, and considering the average sampling distances (order of magnitude of about 1 km), the variability at small scale tends to produce effects that resemble a nugget effect. It was thus decided that the contribution of the smaller scale component should generally be filtered.

A procedure for sequential clustering was also experimented with success. This procedure makes it possible to make the following changes specifically for any subset:

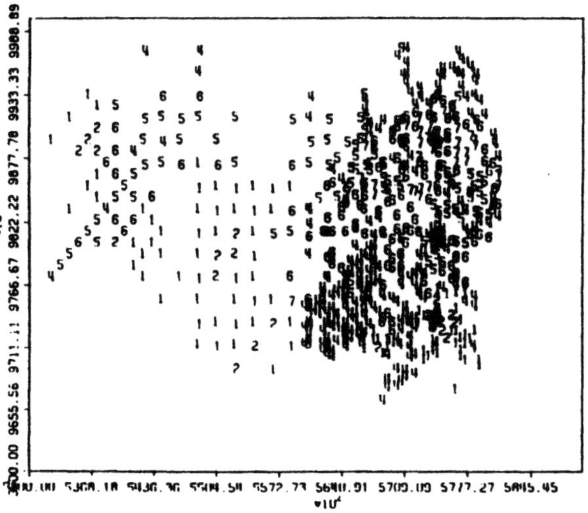

Fig.6 - Sequential clustering on samples.

- the nature of the variables to be considered for cluster analysis;
- the variability scale of the components for a same variable.

Fig. 6 shows the result of sequential (or branching) clustering carried out at two levels:

1) using the whole set of data, clustering was performed on the large-scale component of factor 1, thereby dividing the samples into 2 groups essentially by distinguishing between the regional components of Ca and pH;

2a) using the first subset of samples identified above, clustering was again performed to divide the subset into another 2 groups on the basis of the medium scale component of factor 1;

2b) using the second subset identified at the first level, clustering was performed to divide it into 4 groups on the basis of 2 variables made up of the sum of the medium and large scale components for factors 2 and 3.

The above case study clearly does not exhaust the potential and versatility of the application. The latter is the object of the ongoing experimentation whose immediate and specific aim is to calibrate the appropriate modality for clustering on the basis of the semi-detailed maps produced by traditional methods.

5. DISCUSSION OF THE RESULTS

Analysis of the spatial variability of soils, based on a small number of independent variables (factors) has been demonstrated to be a highly effective method, although the factors are not as immediately grasped as the original variables, whose use for the purposes of applicative interpretations of soil maps has been consolidated.

With regard to the use of factors to generate clusters characterised by internal homogeneity of behaviour, the following comments can be made.

Regarding the use of the original global factors:

- the geographical compactness of the classes and pattern congruence are unsatisfactory compared to the identifications of the different structures generated by the traditional approach;

Regarding the use of the spatial components of the factors:

- there is a greater robustness in the aggregation in an increasing number of clusters, to the extent that all the samples of each "cross" are located in the corresponding 3 groups;
- the number of classes and subclasses that can be satisfactoraly generated is significantly greater than the number obtainable using the global factors, under the condition described in the preceding point;
- the integrated use of the factors through their own spatial components at different scales permits a sequential classification at several stages that is reasonably satisfactory in terms of geographical compactness and pattern congruence compared to the structures obtained using the traditional procedure.

6. CONCLUSIONS

The fact that interpretative models of the associations between soils characteristics and landscape features can be identified, necessarily corroborates the opportunity of their use.

The use of "exploratory" methods like cluster analysis requires a decidedly smaller body of information. In fact these methods are limited to accessory variables whose presence is not a clear indicator of the mechanisms or causal processes that generate differences in soil properties and structures.

The ability to model the factors and/or the original variables in space becomes highly important in cases where no a priori hypotheses are available that describe the soils in a precise and well articulated manner irrespective of the scale. This is particularly true for the smaller scales, where it is extremely difficult to define a credible traditional models with an acceptable level of precision. The study proposes a tentative way for exploiting the spatial variability of the factors for classification purposes.

REFERENCES

1) Iris, J.M. (1984), "Analyse de la variabilité d'un paramètre pédologique", in Sciences de la terre, Sér.Inf.Géol. N.20, Part 1, J.J.Royer (ed.), CRPG, Nancy, pp.321-332.

2) Matheron, G. (1982), "Pour une analise krigeante des données réegionalisées", Centre de Géostatistique, ENSMP, Fontainebleau, N.732, 22p.

3) Oliver, M.A. and Webster, R. (1989), "Geostatistically constrained multivariate classification", in Geostatistics, Vol. I, Kluwer Academic Publisher, Dordrecht, pp.383-395.

4) Wackernagel, H. (1992), "Cours de Géostatistique multivariable, Centre de Géostatistique, ENSMP, Fontainebleau, C-146, 62p.

5) Wackernagel, H., Petitgas, P. and Touffait, Y. (1989), "Overview of methods for coregionalization analysis", in Geostatistics, Vol. I, Kluwer Academic Publisher, Dordrecht, pp.409-420.

6) Wackernagel, H., Webster, R and Oliver, M.A. (1988), "A geostatistical method for segmenting multivariate sequences of soil data", in Classification and Related Methods of Data Analysis, Ed.H.H.Bock, North-Holland, Amsterdam, pp.641-650.

7) Webster, R. and Oliver, M.A. (1985), "Utilisation exploratoire de la géostatistique pour la cartographie du sol dans la forêt de Wire (G.B.)", in Sciences de la terre, Sér.Inf.Géol., N.24, J.J.Royer (ed.), CRPG, Nancy, pp.161-173.

8) Ward Jr., J.H. (1963), "Hierarchical grouping to optimise an objective function", J.Amer.Stat.Assoc., Vol. 58, N. 301, pp.236-244.

GEOSTATISTICAL INTERPOLATION OF CURVES: A CASE STUDY IN SOIL SCIENCE.

M. GOULARD & M. VOLTZ *

Biometric Laboratory - I.N.R.A. Domaine St Paul - BP 91

84140 Montfavet, France

* *Soil Science Laboratory - I.N.R.A. - 2 Place Viala*

34060 Montpellier cedex 1, France

ABSTRACT. In various scientific fields properties are expressed by curves or functions varying over space. For predicting these curves or functions at unvisited sites with a spatial sample three geostatistical approaches can be chosen. Two of them use a multivariate approach with a cokriging step. The third uses a kriging-like method which predicts directly the curves in a geostatistical sense. The paper outlines the theoretical principles of these three approaches and then illustrates their application on a case study in soil science that concerns mapping of soil water retention curves.

1. Introduction

Geostatistics has been applied almost entirely to scalar variables. In some scientific fields, however, regionalizations are represented by curves, whose shape varies from place to place. This is the case in Soil Science where several soil properties, like pore size distribution, water retention characteristics are expressed by univariate functions. For estimating the shape of these functions at unvisited sites some kind of interpolation is needed.

Direct methods for interpolating curves are those that are suited to categorical data. The simplest methods are based on a tessellation of space into cells, and all points of a cell are assigned the value at the defining point of the cell. This is the approach that Wösten et al. (1985), and later Stein et al. (1991) adopted for predicting soil water retention curves; they used either pedological clustering or Dirichlet tessellation (Ripley, 1981) to construct their cells and choose their defining points. Both methods assume a very discontinuous model of spatial variation. This is likely to happen where the regionalization exhibits abrupt changes. Nevertheless, in many cases, this is only a very crude approximation and, thus, the use of interpolators predicting more continuous variations would be more realistic. The problem is therefore to define methods that can predict continuous changes of curves in space, and preferably with no bias and optimal in some sense (for example minimum variance of errors).

We have examined this problem in the common case where the curves are only known by a finite set of their points, and where a parametric model is fitted to them for reconstructing the whole curve.

In this paper, we present three geostatistically based methods for interpolating curves. The first method uses an appropriate framework for curves, in which is defined a kriging-like predictor, whereas both other methods transform the problem in a multivariate one. The performances of the methods were evaluated and compared on a case study concerning the mapping of soil water retention curves. Below we describe the theory underlying the methods, the study area, its sampling and our results.

2. Methods

2.1. A Kriging Like Approach

To interpolate curves we built an adequate framework for univariate real-valued functions. In this framework a random function field is a random field with values in a space of functions.

Let U be an open set in $\mathbb{R}$, and let us denote $\phi_{\mathbf{x}}$ the value of the random field at site $\mathbf{x}$ in $\mathbb{R}^2$, then $\phi_{\mathbf{x}}$ is a function from U to $\mathbb{R}$.

We shall assume that this random field is such that

$$E[\phi_{\mathbf{x}}] = \Phi$$

and

$$E[\| \phi_{\mathbf{x}} - \phi_{\mathbf{x}+\mathbf{h}} \|_f^2] = 2\Gamma(\mathbf{h})$$

where $\| \ \|_f$ is a norm defined on the space of functions and where Γ is a function that exists for all $\mathbf{x}$ and all $\mathbf{h}$ in $\mathbb{R}^2$ and depends only on $\mathbf{h}$. Γ will be called the functional variogram. As in the classical case of random field, if isotropy is assumed the functional variogram will depend only on $\| \mathbf{h} \|$.

In the following, $\| \ \|_f$ will be chosen such that it is defined from a scalar product. Hence, the functional variogram Γ will be a conditional negative function, and all authorized theoretical functions for traditional variograms may be used to model Γ. A first choice for $\| \ \|_f$ can be made by using the classical $\mathbb{L}^2$-norm:

$$\| \phi \|_f^2 = \int_U \psi(u)\phi(u)^2 du \tag{1}$$

where ψ is a positive function. If the function ϕ is smooth and exhibits k derivatives, the derivatives can be taken into account for the definition of the norm as follows

$$\| \phi \|_f^2 = \int_U \psi_0(u)\phi(u)^2 du + \sum_{l=1}^{k} \int_U \psi_l(u)\phi^{(l)}(u)^2 du \tag{2}$$

where $\psi_0, \ldots, \psi_k$ are positive functions.

Assuming that the random field is observed at some sites $\mathbf{x}_1, \ldots, \mathbf{x}_n$ (value at site $\mathbf{x}_i$ is denoted ϕ_i), we estimate $\Gamma(\mathbf{h})$ by :

$$\hat{\Gamma}(\mathbf{h}) = \frac{1}{W(\mathbf{h})} \sum_i \sum_{j \neq i} w(\mathbf{h}, \mathbf{x}_i - \mathbf{x}_j) \| \phi_i - \phi_j \|_f^2$$

where $w(\mathbf{h},\mathbf{x})$ is a function of the proximity between $\mathbf{h}$ and $\mathbf{x}$ and

$$W(\mathbf{h}) = 2\sum_{i}\sum_{j\neq i} w(\mathbf{h},\mathbf{x}_i-\mathbf{x}_j).$$

For example $w(\mathbf{h},\mathbf{x}) = 1$ if the distance between $\mathbf{h}$ and $\mathbf{x}$ is less than some ε and zero otherwise. For isotropic cases $\mathbf{h}$ is just replaced by $d = \| \mathbf{h} \|$ and w is a function both of d and the distance between the two points.

In this context, to predict the function ϕ at an unvisited site $\mathbf{x}_0$, we define the best linear unbiased predictor by :

$$\phi_0^* = \sum_{i=1}^{n} \lambda_i \phi_i$$

with $\lambda_i, i = 1,\ldots,n$ such that :

$$E(\phi_0^* - \phi_0) = 0 \text{ and } E(\| \phi_0^* - \phi_0 \|_f^2) \text{ minimum.}$$

If the norm is defined from a scalar product $\| \phi \|_f^2 = <\phi,\phi>_f$ then $\lambda_i, i = 1,\ldots,n$ are such that :

$$\begin{cases} \sum_{i=1}^{n} \lambda_i = 1 \\ -\sum_{i=1}^{n}\sum_{j=1}^{n} \lambda_i\lambda_j \Gamma(\mathbf{x}_i-\mathbf{x}_j) + \sum_{i=1}^{n} \lambda_i \Gamma(\mathbf{x}_0-\mathbf{x}_i) \text{ is minimum} \end{cases}$$

thus giving same formulas as the classical ordinary kriging. In the following we refer to this predictor as curve kriging predictor (CKP).

A more refined predictor could be defined by :

$$\phi_0^*(u) = \sum_{i=1}^{n} \lambda_i(u)\phi_i(u) \text{ for all } u \text{ in } U$$

where $\lambda_i(u)$ are now functions such that :

$$\sum_{i=1}^{n} \lambda_i(u) = 1 \text{ for all } u \text{ in } U$$

and

$$E(\| \phi_0^* - \phi_0 \|_f^2) \text{ is minimum.}$$

If we search for the values of the functions λ_i at only a finite set of values then this problem reduced to classical cokriging. If this is not the case the quantity under minimization can't be explained in terms of Γ and the minimization becomes rather complicated.

In practice function ϕ_i at an observed site $\mathbf{x}_i$ is only known for some values $u_1,\ldots,u_s$. This confronts us with the problem of estimating Γ, and more especially with the problem of estimating the norm $\| \phi_i - \phi_j \|_f$. A first answer can be given by taking $U = [Umin, Umax]$ and using for norm definition $\int_U \psi(u)\phi(u)^2 du$, then an estimator of $\| \phi_i - \phi_j \|_f^2$ will be $\sum_{l=1}^{s} \Delta_l \psi(u_l)((\phi_i - \phi_j)(u_l)^2$, where Δ_l are weights used for numeric integration. In this case, the framework reduces practically to that of Bourgault & Marcotte (1991), and Γ is nothing more than their multivariate variogram. A second answer is to use a set of parametric models to estimate the norm. We fit first a model $\phi(.,\theta)$ to the observed values of ϕ and estimate then $\| \phi_i - \phi_j \|_f$ by $\| \phi(.,\hat{\theta}_i) - \phi(.,\hat{\theta}_j) \|_f$. Note that this last approach can be used even if the u-values are different for the different observations sites, moreover we are not restricted to use one parametric model for all sites. Hereafter, we used the second solution.

2.2. Multivariate Approaches

In these approaches, the problem of interpolating curves is turned to the more classical problem of interpolating a vector of random variables. For this purpose, any curve at any position $\mathbf{x}$ in space is viewed as a set of values, in other words a vector of random variables. Two interpolation procedures may be distinguished. Both include a cokriging step.

In the first procedure, the vector of random variables is assigned the values of the function $\phi(u)$ at several prescribed values of u. Cokriging is then applied to predict, at unvisited sites, the values of $\phi(u)$ at the same prescribed u. Fit of the parametric model is done in a second stage for reconstructing the whole curve. The main assumption, here, is that cokriging individual values of $\phi(u)$ will honor the global form of $\phi(u)$. This procedure will be described later as "Cokrige first, Fit later " predictor (CFP).

In the second procedure, the vector is assigned the values of the coefficients θ of the parametric model $\phi(u,\theta)$ that has already been fitted to the observed curves. Consequently, cokriging is applied to predict the coefficients of the parametric model at the unvisited sites. This procedure will be named hereafter as "Fit first, Cokrige later" predictor (FCP).

Contrary to the curve approach both previous procedures exhibit some constraints on either the measurement of the curves or the spatial similarity of the curves over the mapping area. Procedure CFP requires the same number of values of the function $\phi(u)$ to be measured at all sampling points. Use of procedure FCP implies that all sampling and predicted points exhibit curves that can all be satisfactorily fitted with an identical parametric model.

2.3. Testing Procedure

To facilitate the comparison between predictors we divided the set of data of our study area in two separate samples; a prediction one and a validation one. The parametric model is predicted at the sites of the validation sample from the data available at the sites of the prediction sample. The measured values of $\phi(u)$ at the validation sites are then compared to the values computed by the parametric model $\phi^*(u)$ predicted at the same sites. For each predictor we computed the three following indices. The mean deviation, MD, is given by

$$MD = \frac{1}{m}\sum_{i=1}^{m}\frac{1}{k_i}\sum_{j=1}^{k_i}\left\{\frac{\phi_i^*(u_{ij}) - \phi_i(u_{ij})}{\sigma_j}\right\}$$

where m is the number of validation sites, k_i is the number of measured values of $\phi_i(u)$ at site i for $u_{ij}, j = 1, \ldots, k_i$ and σ_j is the empirical standard deviation for the corresponding u-value on the validation set. This quantity measures the bias of predicting the values of $\phi(u)$ for a range of values of u at unvisited sites. The mean square deviation, MSD, is

$$MSD = \frac{1}{m}\sum_{i=1}^{m}\frac{1}{k_i}\sum_{j=1}^{k_i}\left\{\frac{\phi_i^*(u_{ij}) - \phi_i(u_{ij})}{\sigma_j}\right\}^2$$

MSD measures the precision of predicting values of $\phi(u)$. The mean norm of deviation, MND, is defined as

$$MND = \frac{1}{mv_f}\sum_{i=1}^{m} \| \phi_i^*(.) - \phi(,\hat{\theta}_i) \|_f$$

where $\hat{\theta}_i$ is obtained by fitting the spline to the observed values at the test sites and v_0 is defined by :

$$v_f^2 = \frac{1}{m}\sum_{i=1}^{m} \| \phi(,\hat{\theta}_i) - \bar{\phi} \|_f^2$$

where $\bar{\phi}$ is the mean of the curves $\phi(,\hat{\theta}_i)$ for the validation set. MND is a measure of the global deviation between the interpolated and fitted parametric models of $\phi(u)$, which represent the observed curves. For the CFP procedure $\phi_i^*(.) = \phi(.,\tilde{\theta}_i)$ where $\tilde{\theta}_i$ is obtained by fitting to the cokriged values. For the FCP procedure $\phi_i^*(.) = \phi(.,\theta_i^*)$ where θ_i^* is obtained by cokriging the parameters, and for the CKP procedure $\phi_i^*(.)$ is the "kriging-like" predictor.

3. Case Study

3.1. Study area

The region for this study occupies part of the alluvial plain of the river Rhône near Yenne in Savoie, France. The sampled area is 92 ha in extent. Several types of alluvial soils were distinguished, but most of the soil's variation follows a lateral trend perpendicular to the direction of the alluvial bed. In a previous work concerning the mapping of clay content using different procedures, Voltz and Webster (1990) showed in the same area that kriging performed well because the soil varies gradually. In principle this should also be true for other soil properties and stresses the interest of using more continuous predictors than nearest neighbour methods.

3.2. Sampling Design

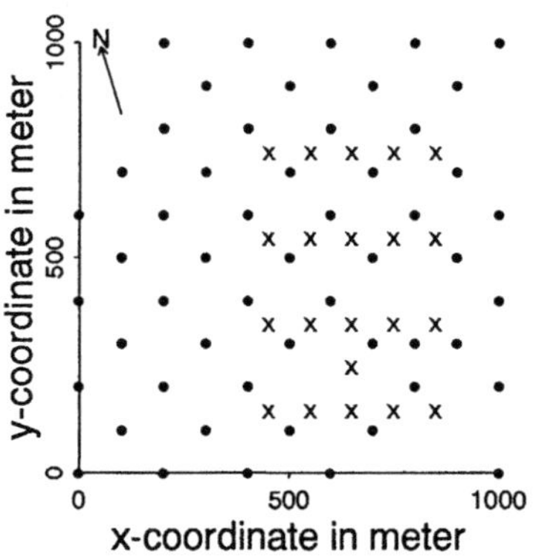

Figure 1: Disposition of the sample on the area. Dots are the prediction set of data and crosses are the validation one.

The sampling scheme consisted of two interlaced grids: one was a square grid with points spaced at 141 m intervals (prediction set), the other was a rectangular grid with points spaced at 200 m intervals in the y direction and at 100 m intervals in the x direction (validation set). Because of practical sampling difficulties, the initial sampling scheme could not be entirely carried out. We took therefore some additional samples to complete the number of observations.

Figure 1 shows the final sampling design. At each sampling location undisturbed soil aggregates were collected at a constant depth of 40 cm, characteristic of the topsoil. The water retention curves represented by the soil water content at 8 values of capillary pressure (-1 kPa, -5 kPa, -10 kPa, -20 kPa, -40 kPa, -90 kPa, -300 kPa, -1500 kPa) were determined on the soil aggregates by standard laboratory procedures. In the following, corresponding u-values for the curve will be (1,5,10,20,40,90,300,1500).

3.3. Fit of a Parametric Model

Although there is a range of parametric models that are usually fitted to soil water retention curves (Van Genuchten and Nielsen, 1985), we preferred to fit a more universal model: polynomial splines (Wegman & Wright 1983). This model consists of functions built up from polynomials joined end to end.

Here we assume that the interval U, $U =]Umin, Umax[$, is subdivided by choosing boundary points t_i referred to as knots, and the function in the interval $[t_i, t_{i+1}]$ is a polynomial of degree $k-1$ for a spline of order k. The crucial thing is that the polynomials are chosen such that the derivatives at the boundary are equal up to

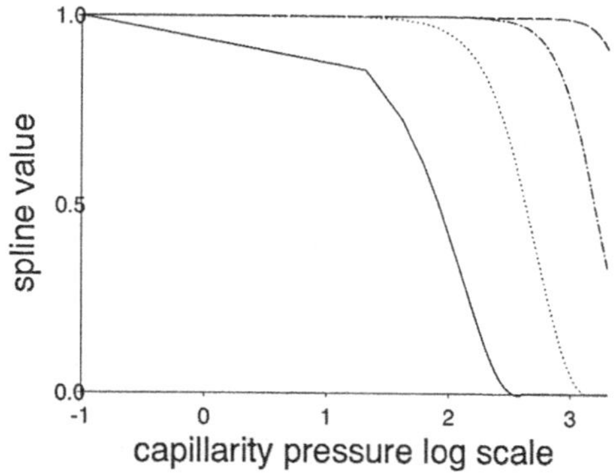

Figure 2: Basic polynomial splines $1 - I$ used to define the parametric model.

order $k - 2$. The knot sequence $\{t_1, \ldots, t_{n_1+k}\}$ is chosen such that :

$$\begin{cases} t_1 \leq \ldots \leq t_{n_1+k} \\ t_k \leq Umin \\ t_{n_1+1} \geq Umax \\ t_i < t_{i+k} \text{ for all } i \end{cases}$$

In fact for application a set of basic splines M_i is required such that the polynomial spline is $\sum_{i=1}^{n_1} a_i M_i$ (de Boor 1978). The M-spline family is often used, this set is defined by the fact that each M_i is a density on $[t_i, t_{i+k}]$ and is defined by the recursion :

$$\begin{cases} M_i(u \mid 1, t) = & \frac{1}{t_{i+1}-t_i} \text{ if } t_i \leq u \leq t_{i+1} \\ = & 0 \text{ otherwise} \\ M_i(u \mid k, t) = & \frac{k[(u-t_i)M_i(u|k-1,t)+(t_{i+k}-u)M_{i+1}(u|1,t)]}{(k-1)(t_{i+k}-t_i)} \end{cases}$$

With M-spline one can define I-spline (Ramsay 1988) by :

$$I_i(u \mid k, t) = \int_{Umin}^{u} M_i(v \mid k, t) dv$$

thus yielding monotone spline. Note here that $I_i(\mid k, t)$ is a polynomial of degree k with k-1 continuous derivatives.

In our study, to fit the soil water retention curves, we used monotone decreasing polynomial spline of degree 3:

$$f(u) = \sum_{i=1}^{n_1} a_i(1 - I_i) \text{ with } a_i \geq 0 \text{ for } i = 1, \ldots, n_1.$$

For the knots we chose the sequence (0, 0.5, 1, 400, 1500, 5000, 10000); so, the polynomial splines exhibited four parameters to estimate. Figure 2 shows the basic elements $1 - I_i, i = 1, \ldots, 4$ used to define the parametric model and Figure 3 shows an exemple of fit of an observed soil water retention curve.

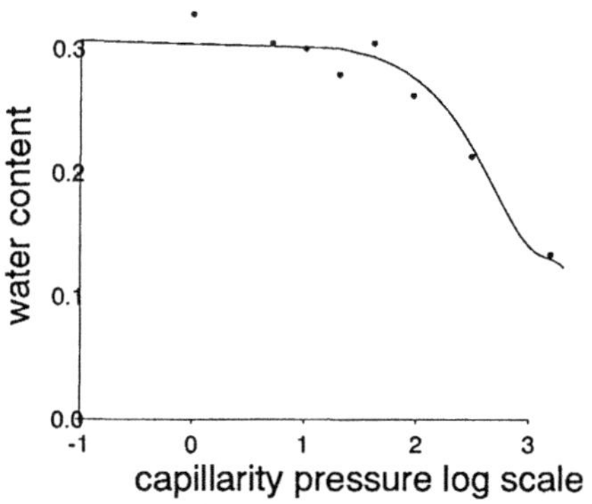

Figure 3: Example of fit of an observed soil water retention curve. Plot of water contents against log of the capillarity pressure values. Dots are the observed values, plain line is the fitted curve.

3.4. Structural Analysis

In the CKP procedure, we first computed $\hat{\Gamma}$ by using norm (1) and choosing $\psi(u) = 1$ for all u in the interval $U =]0, 2000[$ and $\psi(u) = 0$ otherwise.

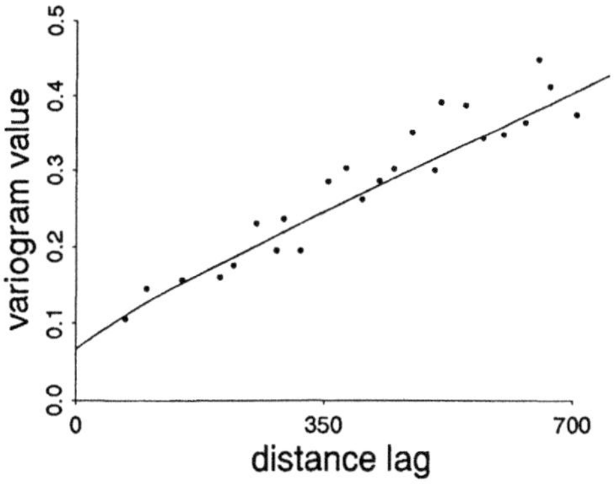

Figure 4: Functional variogram used in the curve kriging procedure. Dots are the estimated values $\hat{\Gamma}(h)$ of the functional variogram and plain line is the fitted nested model. Distance lags are in meter.

Note that, as we use polynomial splines, we have:

$$\| \phi(.,\hat{\theta}_i) - \phi(.,\hat{\theta}_j) \|_f^2 = \| \sum_{l=1}^{4} (\hat{\theta}_{il} - \hat{\theta}_{jl})(1 - I_l(. \mid 3, t) \|_f^2 = (\hat{\theta}_i - \hat{\theta}_j)^T Q (\hat{\theta}_i - \hat{\theta}_j)$$

$$\text{with } Q_{ll'} = \int_U (1 - I_l(u \mid k, t))(1 - I_{l'}(u \mid k, t)) du$$

As a remark, note here that the functional variogram is a multivariable variogram with metric Q (Bourgault & Marcotte 1991).

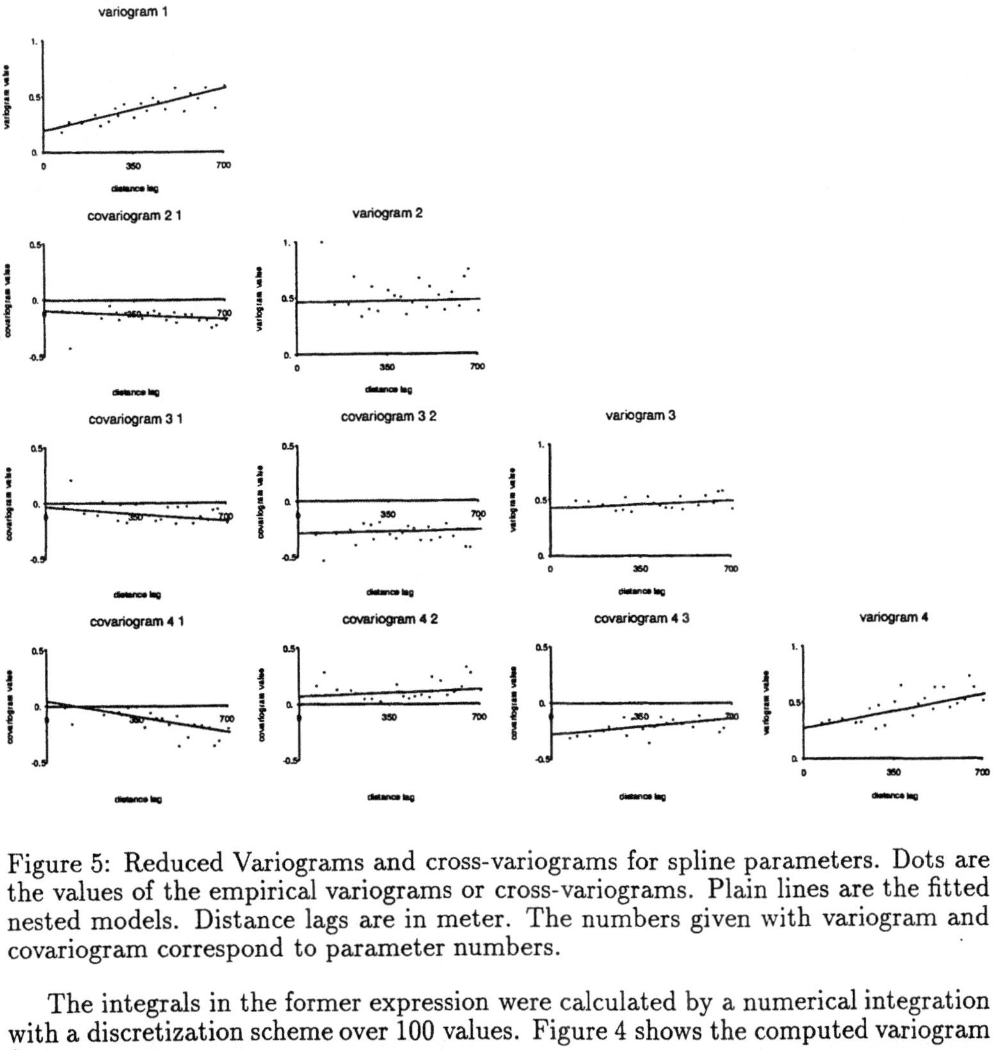

Figure 5: Reduced Variograms and cross-variograms for spline parameters. Dots are the values of the empirical variograms or cross-variograms. Plain lines are the fitted nested models. Distance lags are in meter. The numbers given with variogram and covariogram correspond to parameter numbers.

The integrals in the former expression were calculated by a numerical integration with a discretization scheme over 100 values. Figure 4 shows the computed variogram $\hat{\Gamma}$. The curve is that of a nested model fitted to the sample values, the plotted points, by weighted least-squares approximation. The nested model included three components: nugget, linear and spherical with a 150 m range. We also tried norm (2) for computing $\hat{\Gamma}$ taking into account first derivatives. Again, because we use polynomial splines, the norm may be transformed:

$$\| \phi(.,\hat{\theta}_i) - \phi(.,\hat{\theta}_j \|_f^2 = (\hat{\theta}_i - \hat{\theta}_j)^T Q (\hat{\theta}_i - \hat{\theta}_j) + (\hat{\theta}_i - \hat{\theta}_j)^T Q_1 (\hat{\theta}_i - \hat{\theta}_j)$$

$$\text{with } (Q_1)_{ll'} = \int_U M_l(u \mid k,t) M_{l'}(u \mid k,t) du$$

Using this second norm resulted in values of $\hat{\Gamma}$ virtually identical to those computed with norm (1).

In the CFP and FCP procedures, coregionalization analyses are necessary. In the context of an exploratory analysis of the variability of the soil water retention curves at the same study area, Voltz et al. (1991) already estimated the empirical cross-variogram matrix of the soil water contents at the eight capillary pressures. They fitted a linear coregionalization model using an automated procedure (Goulard and Voltz, 1992). The fitted model was a combination of a nugget effect and a linear model, and has also been adopted here in the CFP procedure for cokriging.

In the FCP procedure, the cross-variogram matrix of the parameters of the polynomial splines was estimated and fitted , as previously, with a linear coregionalization model. Figure 5 shows the matrix of observed and fitted variograms. A nested model combining nugget effect and linear components worked well again.

However it must be stressed that two parameters on four exhibit a very loose spatial structure. Moreover, if we examine the statistical distributions of all parameters (Figure 6), the second parameter appears as highly skewed. These both aspects of the variability of the parameters of the spline functions may affect the accuracy of their mapping.

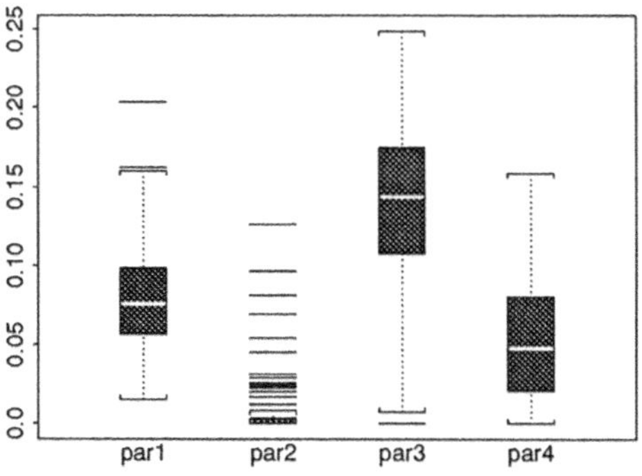

Figure 6: Distribution of spline parameters.

3.5. Comparison of the different procedures

Table 1 gives for each procedure the mean deviation and the mean square deviation between predicted and observed values of the soil moisture curves at the validation sites. The mean norm of deviation between the predicted spline models of the soil moisture curves and the fitted spline models is included as well.

Mean deviations of CFP and CKP procedures are low and indicate the unbiasedness of these methods. This is not the case of the FCP procedure which presents some bias. As suggested above, this may be explained by the skewed distributions of some parameters of the spline model, for which simple cokriging does not account. Concerning the mean square deviations, procedure CFP appears to be the most accurate, whereas both other procedures present similar performance (The greatest value for

Predictor	MD	MSD	MND
CKP	-0.008	0.471	0.663
CFP	0.006	0.454	0.660
FCP	-0.07	0.497	0.679

Table I: Mean deviation, mean square deviation and mean norm deviation for the testing procedure.

the FCP procedure may be a consequence of the bias). However this result must be balanced against the fact that our testing method favors the CFP procedure in terms of the MSD criterion. We measured the values of the soil moisture curves at the same set of abscissa for the prediction sites and the validation sites. In this context, cokriging first, fitting later always guarantees minimum variance of predicting the values of the curves, both during interpolation and during fit; whereas the fitting first procedure only guarantees minimum variance during the fit because the interpolation step concerns the parameters or the whole function and not the values. Examining the mean norm of deviation was therefore complementary, for it compares the whole shape of predicted and fitted curves. As expected, differences in MND between the procedures CKP and CFP were smaller, although the rank remained. From the MND point of view CKP and CFP were similar and better than FCP.

Nevertheless, even in terms of MND the testing procedure still favors slightly the CFP predictor. A better procedure would have been to perform a kind of cross-validation. It involves eliminating from the prediction sites the observed point at one pressure and predicting their values at the validation sites from the observed points at the other pressures. This must be done in turn for all measured pressures. However, in this case, the problem of the choice of the knots for the spline polynomial functions would have generated too much distorsion. So we could not use it here. This study will be pursued with another parametric model for curves in the future.

3.6. Conclusion

In this paper, we presented three procedures for predicting at unvisited sites regionalizations that are expressed by curves. The result of a comparison of these methods on a case study exhibited only small differences between the CFP and CKP methods. The cokrige first, fit later method worked slightly better. The most significant difference was the existence of a significant bias for the fit first, cokrige later predictor and the unbiasedness of both other methods.

Finally, we found that the kriging like predictor worked reasonably well and was in addition very attractive from a practical point. This method only requires the estimation and the fit of a unique variogram, allows to use different parametric models of curves within the mapping area and does not impose to know the points of the curves at the same abscissa for all observation sites.

4. Acknowledgements

We thank MM. J.L. Aznar, S. Conventi and O. Huttel for their technical help. We also thank PIREN (Paris) for funding the research at Yenne.

5. References

Bourgault G. and Marcotte D. (1991) " Multivariable Variogram and Its Application to the Linear Model of Coregionalization". Math. Geol., 23 (7), 899-928.

de Boor C., (1978) A Practical Guide to Splines. Springer-Verlag, New-York.

Goulard M. and Voltz M. (1992) "Linear Coregionalization Model: Tools fo Estimation and Choice of Cross-Variogram Matrix". Math. Geol., 24 (3), 269-286.

Ramsay J.O. (1988) "Monotone Regression Splines in Action", Statistical Sciences 3 (4), 425-461.

Ripley (1981) Spatial Statistics. Wiley, New-York.

Stein A., Staritsky I.G., Bouma J., Van Eijnsbergen A.C., Bergt A.K. (1991) "Simulation of moisture deficits and areal interpolation by universal cokriging". Water Resour. Res., 27 (8), 1963-1973.

Van Genuchten M.Th. and Nielsen D.R. (1985) "On describing and predicting the hydraulic properties of unsaturated soils". Ann. Geophys., 3 (5), 615-628.

Voltz M., Bornand M. and Goulard M. (1989) "Estimation spatiale des paramètres hydrodynamiques du sol a l'echelle d'une plaine alluviale". in G. Vachaud (ed.), Influence à l'échelle régionale des couvertures pédologiques sur les bilans hydriques et minéraux, Piren, Paris, pp. 81-109.

Voltz M. and Webster R. (1991) "A comparison of kriging, cubic splines and classification for predicting soil properties from sample information". Journal of Soil Science, 41, 473-490.

Wegman E.J. and Wright I.W. (1983) " Splines in Statistics". J.A.S.A., 78, 382, 351-365.

Wösten J.H.M., Bouma J. and Stoffelsen G.H. (1985) "Use of soil survey data for regional soil water simulation models". Soil Sci. Soc. Am. J., 49, 1238-1244.

QUANTITATIVE STUDY OF AN EVOLUTIONARY SOIL SEQUENCE

Michel GRZEBYK
Centre de Géostatistique
Ecole des Mines
35 rue Saint-Honoré
77305 Fontainebleau
France

ABSTRACT. The pedological cover of the basin of the Upper Rio Negro in Amazonia is composed mainly of several kinds of ferrallitic soils and podzol. Some soil scientists think there should be an evolutionary sequence among these morpho-pedological units. It would correspond both to a lowering and a flattening of the relief of the ferrallitic soils, with the podzol as its ultimate stage.

This study consists in looking for quantitative arguments which could confirm this hypothesis, using only the map of the morpho-pedological cover of the basin. In the first stage contacts between two units are analysed, using a coordination index. Then, we generalize this process to larger and larger neighbourhoods of each unit. The use of vicinity functions allows us to measure the preferential spatial association between one unit and the others.

1 Introduction.

The pedological cover of the Basin of the Upper Rio Negro (Amazonia) consists mainly of ferrallitic soils and podzols [2]. A map of this cover has been made by O.R.S.T.O.M.. Figure 1 gives a representation where the soil has been divided in 10 morpho-pedological units. This classification depends on the pedological type and the form of the cover. To explain the coexistence of these types of soil in this basin, Dubroeucq *et al.* [2] have proposed the existence of an evolutionary sequence among them, corresponding to the lowering and the flattening of the relief of the ferrallitic soils in the following order: ferrallitic mantle of polyconvex hills (1), ferrallitic mantle of flat hills (2), level ferrallitic mantle (3), ending with the level podzolic mantle (4).

This article aims at verifying this hypothesis quantitatively, using only the map which was digitized out to a 0.5 km x 0.5 km raster. No mention will be made of the resulting bias. The reasoning is based on the following idea: if an evolutionary sequence exists, it should influence the spatial organization of the units. Two successive units in the sequence should be contiguous and extend themselves close to each other in space. We can measure the proximity between units to verify the hypothesis. Two ways are proposed. The first one is based on the study of contacts between units: two successive units are expected to have preferential contacts. Some indexes, computed

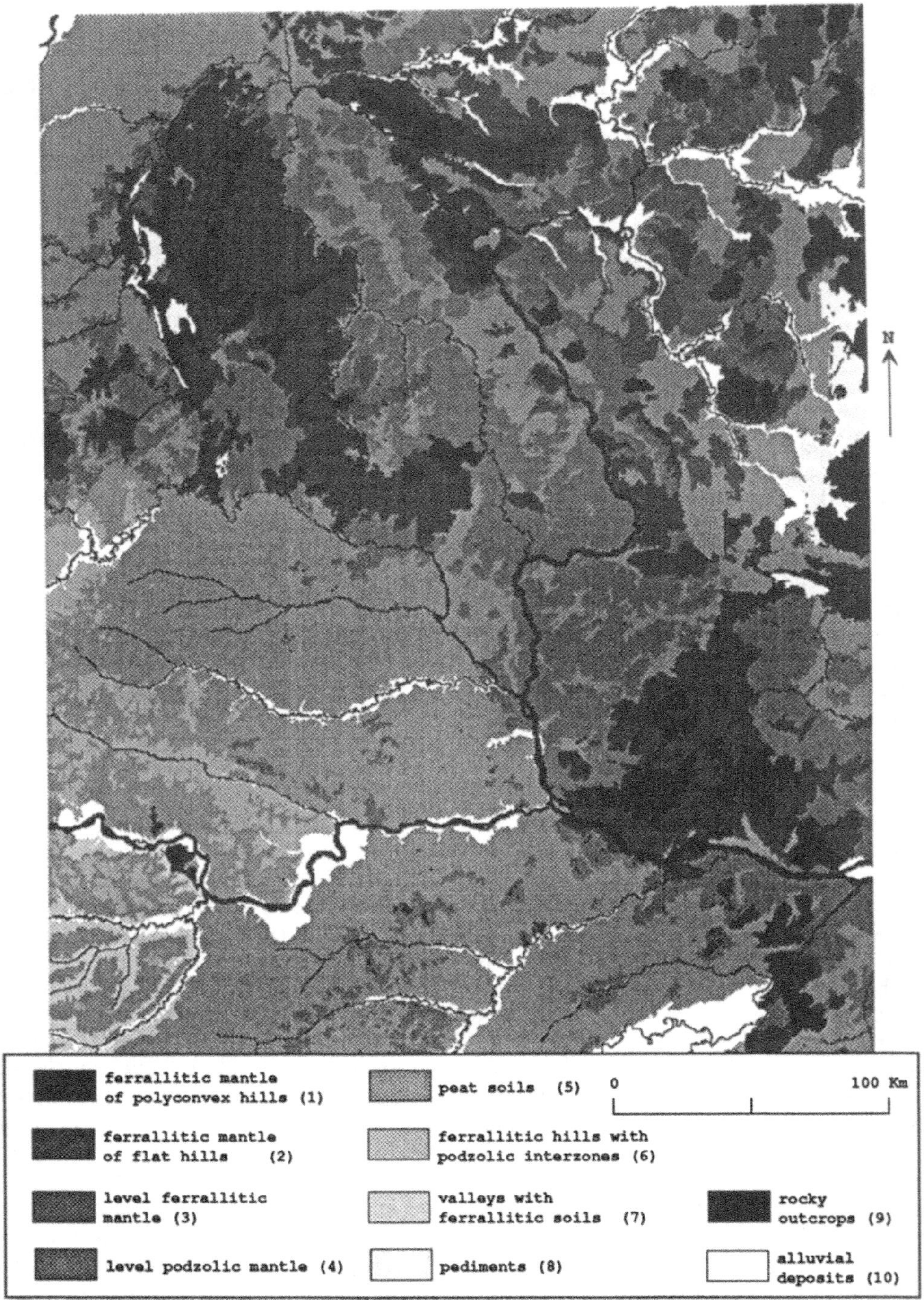

Figure 1: Map of the soil cover of the Basin of the Upper Rio Negro (Amazonia).

with length of interface provide simple tests to compare contacts between units but they are imperfect. The second approach attempts to improve them and enrich the processes. Instead of limiting the analysis to the interfaces, the view will be enlarged to larger and larger neighbourhoods of each unit. In this case, the neighbourhood of a unit of the sequence should contain a large proportion of its previous and following units in the sequence.

In this approach, we consider that the map is an exact and exhaustive representation of the phenomenon. So we do not need to define a probabilistic background to perform the analysis. However, the different indexes are computed from a discretized version of this map, introducing some error when computing surfaces and perimeters. These errors should be evaluated, but since they decrease with the size of the discretisation lag [6], we assume that the present lag is small enough to make them negligible in this study (which should be verified all the same).

2 Properties of contacts.

Let n be the number of units and let X_j denote the jth unit.

We can try to characterize the distribution of contacts between the units by comparing the length of the interface of two units, $L(i,j)$, with the perimeter of one of them, $L(i)$.

2.1 Proportions of contact.

The simplest way to compare these lengths is to form the ratio:

$$p_i^j = \frac{L(i,j)}{L(i)}$$

which is the proportion of contact of the unit X_i with X_j [3, 5]. This ratio can be used in the following way. If we want to characterize the contacts of the unit X_i:

- if all its contacts are distributed uniformly with the others, $p_i^j = 1/(n-1)$ for all j.
- if there exists a preferential contact with X_j, then $p_i^j > 1/(n-1)$
- in case of scarce contacts, $p_i^j < 1/(n-1)$

To simplify the use of these proportions, we write $\tilde{p}_i^j = (n-1)p_i^j$ and compare it to 1. The results are presented in Table 1. This index is not symmetrical with respect to i and j. This leads to three situations for every pair (X_i, X_j):

- both $\tilde{p}_i^j$ and $\tilde{p}_j^i$ are greater than 1 (values in bold digits in the table).
- $\tilde{p}_i^j$ and $\tilde{p}_j^i$ are less than 1
- only one value, say $\tilde{p}_i^j$, is greater than 1 (the values greater than 1 are underlined in the table).

$j \backslash i$	1	2	3	4	5	6	7	8	9	10
1		**5.60**	1.81	0.61	0.00	0.40	0.00	0.02	0.38	0.19
2	**1.96**		**2.56**	**3.16**	0.00	0.28	0.04	0.12	0.54	0.35
3	0.38	**1.55**		**5.88**	0.01	0.35	0.00	0.03	0.35	0.45
4	0.09	**1.32**	**4.04**		**1.17**	0.53	0.33	0.04	0.13	**1.34**
5	0.00	0.00	0.00	**7.97**		0.00	0.00	0.00	0.00	1.00
6	0.44	0.89	1.80	3.98	0.00		0.00	0.11	0.08	1.70
7	0.00	0.27	0.0	5.94	0.01	0.00		0.00	0.00	2.75
8	0.05	0.94	0.35	0.80	0.00	0.28	0.00		**4.63**	1.95
9	0.54	2.20	2.37	1.31	0.00	0.10	0.00	**2.33**		0.19
10	0.10	0.54	1.13	**4.88**	0.53	0.82	0.56	0.36	0.06	

Table 1: $\tilde{p}_i^j$: proportions of contact multiplied by $(n-1)$.

This litigious case may have several origins. In particular, the bigger units X_j (in regard to their size or their perimeter) are favoured: $\tilde{p}_i^j$ is greater than 1 for most of the units X_i. This is the case with the podzols (X_4) which cover 42% of the region. So we can state that two units have preferential contacts when both $\tilde{p}_i^j$ and $\tilde{p}_j^i$ are greater than 1 and build a minimum graph (Figure 2) representing the relations that are put forward. These relations can be interpreted as relations of successiveness in a sequence. In this sense, the expected sequence appears clearly in the graph.

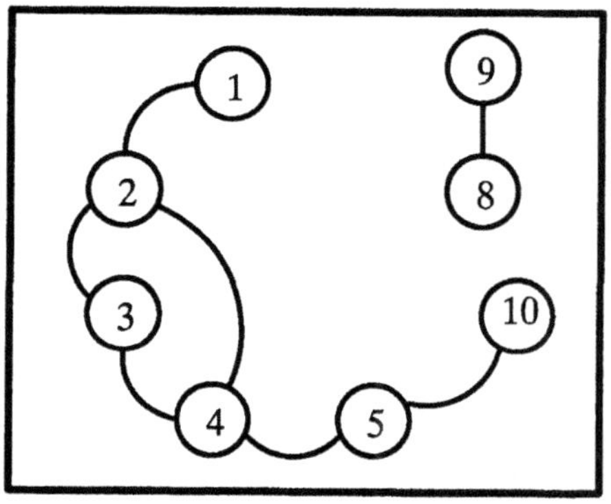

Figure 2: Relations of preferential contact given by p_i^j.

Nevertheless, there appears a relation between X_4 (the greatest unit) and X_2. Does it correspond to a real affinity or is it only due to the size of X_4 as mentioned above? Besides, since the proportion of contacts are not symmetrical, it is difficult to quantify the strength of a relation.

2.2 Index of coordination.

The following index, proposed by Jeulin [3, 5], may be a simple solution to overcome these drawbacks. It is defined as :

$$C(i,j) = \frac{L(i,j)L}{L(i)L(j)} \quad \text{where} \quad L = \sum_{k=1}^{n} L(k)$$

$j \backslash i$	1	2	3	4	5	6	7	8	9	10
1		**4.04**	0.79	0.18	0.00	0.92	0.00	0.10	**1.10**	0.21
2	**4.04**		**1.12**	0.95	0.00	0.64	0.19	0.68	**1.58**	0.39
3	0.79	**1.12**		**1.77**	0.01	0.79	0.00	0.15	**1.04**	0.50
4	0.18	0.95	**1.77**		**2.40**	**1.20**	**1.79**	0.24	0.39	**1.47**
5	0.00	0.00	0.01	**2.40**		0.00	0.09	0.00	0.00	**1.10**
6	0.92	0.64	0.79	**1.20**	0.00		0.00	0.64	0.23	**1.87**
7	0.00	0.19	0.0	**1.79**	0.09	0.00		0.00	0.00	**3.04**
8	0.10	0.68	0.15	0.24	0.00	0.64	0.00		**13.64**	**2.14**
9	**1.10**	**1.58**	**1.04**	0.39	0.00	0.23	0.00	**13.64**		0.18
10	0.21	0.39	0.50	**1.47**	**1.10**	**1.87**	**3.04**	**2.14**	0.18	

Table 2: index of coordination (multiplied by $n/(n-1)$).

This index is symmetrical in i and j. In fact, it compares p_i^j to $L(j)/L$, or equivalently p_i^j to $L(i)/L$. Furthermore, it has a more global meaning; if all the contacts of all the units are equally distributed, then:

$$\forall i, j, \ p_i^j = \frac{1}{n-1} \text{ and } \frac{L(j)}{L} = \frac{1}{n}$$

$$\text{so that } C(i,j) = \frac{n}{n-1}$$

On the contrary, if there is a preferential contact between X_i and X_j, then $C(i,j) > n/(n-1)$, whereas $C(i,j) < n/(n-1)$ reflects a weak contact. Table 2 presents the results of our case study in which the values have been multiplied by $n/(n-1)$. The values corresponding to a preferential contact are printed bold. As above, the relations put forward are represented in a graph (Figure 3). Compared to the proportion of contact, this graph has been enriched with new links, corresponding to ambiguous cases and some small units have more weight. Concerning the sequence, a chain links each of its units in the right order. Nevertheless, extensive units, such as the podzols (X_4) are still favoured (this unit has 5 links over 9 possible units).

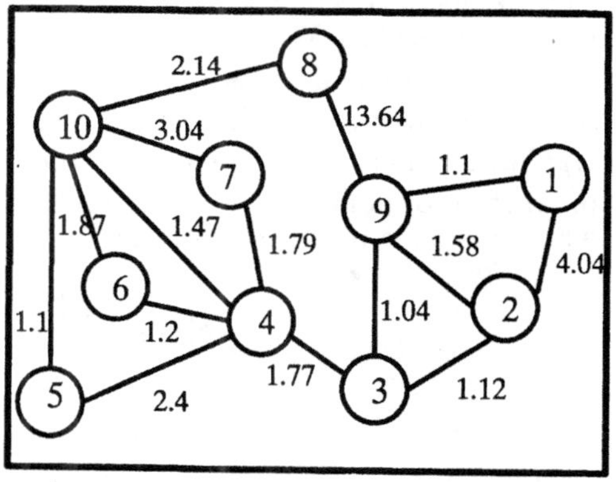

Figure 3: relations of preferential contact given by $C(i,j)$.

These approaches, only based on contacts, are obviously imperfect and restrictive. Imperfect because they take account neither of the irregularities of contact, that increase the length of interfaces, nor of the size of the units. Restrictive because the spatial association cannot be characterized by contacts alone (and further more by the length of interfaces alone!) Therefore, we are going to scan larger and larger neighbourhoods of each unit and see which other units surround it.

3 Vicinity properties.

This approach has been described by Jeulin [4, 5]. It consists of measuring and comparing the proportion of surface of the other units in neighbourhoods of a unit X_k.

3.1 Construction and definition of the vicinity function.

The neighbourhood of a unit, say X_k, is built by dilating it [1, 7] using a disk of radius r, to give $D_r(X_k)$ (Figure 4) , which is the set of points whose distance to X_k is smaller than the radius r:

$$D_r(X_k) = \{x | d(x, X_k) < r\}$$

Then, for every unit X_j ($j \neq k$), we measure the surface contained in this dilated unit: $S(D_r(X_k) \cap X_j)$, and compute the histogram of the distance of a point in X_j to (the boundary of) X_k:

$$F_k^j(r) = \frac{S(D_r(X_k) \cap X_j)}{S(X_j)}.$$

This is the proportion of points in X_j whose distance to X_k is less than r. This function could be used to characterize the proximity of X_j in relation to X_k, but it is better to compare it to the proportion of points outside of X_k whose distance to X_k is less than r:

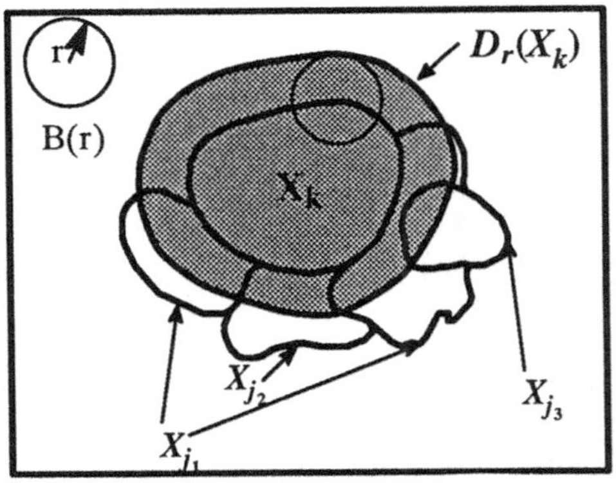

Figure 4: Neighbourhood of X_k defined by dilation with the disk $B(r)$.

$$F_k(r) = \frac{S(D_r(X_k) \cap (\bigcup_{j \neq k} X_j))}{S(\bigcup_{j \neq k} X_j)}$$

This leads to the definition of the vicinity function :

$$\rho_k^j(r) = \frac{F_k^j(r)}{F_k(r)}$$

3.2 Properties of the vicinity functions.

3.2.1 Why "vicinity function" ?

The first property is the basis of this analysis: these functions allow us to quantify the way in which each unit X_j surrounds another unit (of reference) X_k, for every radius r:

- if $\rho_k^j(r) = 1$ for some units X_j, none of them is preferentially present in the vicinity of size r of X_k.

- if $\rho_k^j(r) > 1$ for some radii r, we can expect a spatial affinity of X_j toward X_k, at the distance r.

- if $\rho_k^j(r) < 1$ for some r, then there is a kind of repulsion of X_j toward X_k at these distances.

So, concerning the sequence, for two consecutive units X_i and X_j we can expect that $\rho_i^j(r) > 1$ and $\rho_j^i(r) > 1$ for some r. Unfortunately, these functions are not symmetrical in j and k, making their use complex.

3.2.2 Range of extension.

As $F_k^j(r)$ is a proportion, $\forall r, \quad 0 \leq \rho_k^j(r) \leq 1/F_k(r)$ and $\rho_k^j(r)$ is bounded by a decreasing function from $+\infty$ to 1. So, the larger is r , the less meaningful is the notion of affinity; the small ranges are the most important.

For each unit X_j, the upper bound is reached when $X_j \subset D_r(X_k)$. Hence, if R_k^j is the smallest radius for which $X_j \subset D_r(X_k)$, any point of X_j is less than R_k^j from X_k: we can call this radius the range of extension of X_j around X_k. Then, if $r \geq R_k^j$, $\rho_k^j(r) = 1/F_k(r)$, $\rho_k^j(r) \geq 1$ and there is a spatial affinity of X_j toward X_k (which is even the most important at that distance).

Moreover, when $D_r(X_k)$ covers all the field, $F_k(r) = 1$ and for all j, $F_k^j(r) = 1$ so that $\rho_k^j(r) = 1$. The smallest radius R_k such that $F_k(r) = 1$ gives indications as to the size of the unit X_k as well as the dispersion of its components through the field.

The comparison of R_k^j and R_k may be instructive: if $R_k^j \ll R_k$ we can actually say that the components of X_j are preferentially located in an annulus of radius R_k^j around X_k. Otherwise, any interpretation is unwise. However, the range of extension may only be useful in special cases, when units are divided into rather small components, or when their sizes are similar.

3.2.3 Border effect.

In practice, the field is always bounded, which greatly influences the values of $\rho_k^j(r)$, R_k^j and R_k. $D_r(X_k)$ is only completely known on the whole field eroded by the disk of radius r (Figure 5) . To take this into account, we have to restrict the computation of $\rho_k^j(r)$ on this induced field that shortens when r increases. Then, a larger and larger part of the border of the map is ignored. To limit this effect, it is better to compute these functions for radii that are small with respect to the size of the field. In this study, the dimensions of the map (300 km x 400 km) are much higher than the distances for which the notion of vicinity and spatial affinity are useful. So, if we

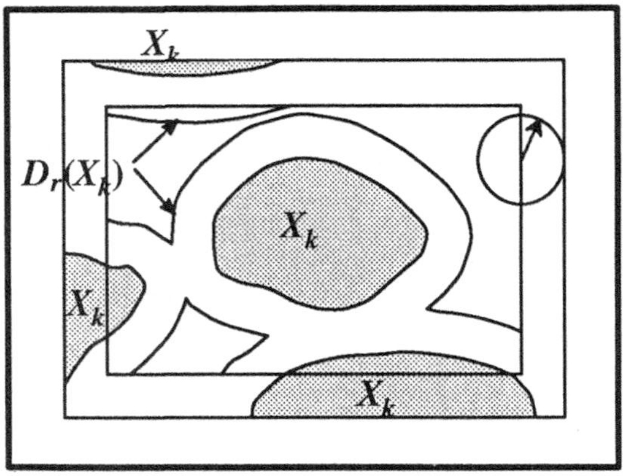

Figure 5: Reduction of the field: the dilated unit is only known on the eroded total field.

limit our analysis to radii up to 20 km, reducing the field up to 78%, we can expect to limit this border effect.

3.2.4 Value at the origin.

The value of $\rho_k^j(r)$ at the origin shows the link of this section with the previous one. As r decreases to 0, $\rho_k^j(r)$ tends to a (finite) limit if the length of the perimeter of X_k is finite (which is the case in this study since the map has been digitized):

$$\lim_{r\to 0}\rho_k^j(r) = \frac{L(j,k)}{L(k)}\frac{S(\bigcup_{j\neq k} X_j)}{S(X_j)}$$

This is the vicinity index of Chermant [1] which could be used to characterize the contacts as in the previous section. This shows that this analysis of vicinity is complementary to the analysis of contacts and richer than it.

3.2.5 Smoothing of interfaces.

The effect of the irregularities of interfaces (the main disadvantage of the properties of contact) will appear with the vicinity functions, because of a basic property of the dilation. The dilation of radius r absorbs inclusive irregularities (as holes and gulfs) of width less than r (Figure 6a).

The effect of such irregular contacts between X_j and X_k is revealed by a characteristic behaviour of $\rho_k^j(r)$ around the origin:

- $\rho_k^j(0)$ tends to be large, because of a long interface between X_j and X_k,

- $\rho_k^j(r)$ decreases (more or less) abruptly from this value as the small inclusions are absorbed.

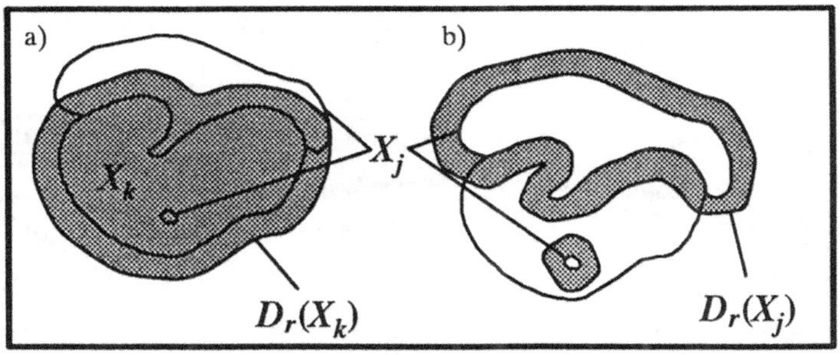

Figure 6: Effect of the dilation on inclusive irregularities (a) and exclusive irregularities (b).

The functions $\rho_1^9(r)$, $\rho_2^9(r)$, $\rho_3^9(r)$ give a good illustration of this phenomenon (Figures 7-a,b,c), because of numerous rocky mountains (unit 9) throughout the region, in the units of ferrallitic soils (units 1, 2, 3).

However, the reverse situation of protruding irregularities of X_k are not revealed systematically (Figure 6b). The functions $\rho_9^1(r)$, $\rho_9^2(r)$ and $\rho_9^3(r)$ do not have a characteristic behaviour around the origin.

3.3 Results.

The vicinity functions have been computed up to the radius $r = 20$ (40 lags) as mentionin in section 3.2.3. To facilitate the analysis of these 90 curves, they have been gathered according to the dilated unit (referring to the index k in this section) in Figure 7. With this representation, for each unit X_k, we can determine which unit X_j seems to be attracted into its vicinity when $\rho_k^j(r) > 1$

Concerning the expected sequence, for two of its consecutive units X_i and X_l, both $\rho_i^l(r)$ and $\rho_l^i(r)$ are greater than 1 for almost all radii in $[0, 20]$. Further more, their values are greater than those of $\rho_i^j(r)$ $\rho_l^j(r)$ of the other units. However, for the level ferrallitic mantel(unit 3), $\rho_3^4 \in]1, 0.95]$ for $r \in [3.5, 20]$ whereas $\rho_3^4(0)$ is the largest value (if we assume that the large value of $\rho_3^9(0)$ is due to the existence of numerous inselbergs throughout the level ferrallitic mantel) and $\rho_3^2 < 1$ for $r < 2$ then increases and reaches the largest value of the vicinity functions of unit 3. This may be explained by the fact that the level podzolic mantle (unit 4) consists of a giant podzolic plain in the south-west and other smaller components in the north-east. These small components tend to make $\rho_3^4(r)$ large for small r but the giant podzolic plain tends to diminish $\rho_3^4(r)$ for larger r. In fact, the larger are the components, the more difficult it is to evaluate their attraction by smaller ones. For small radii, the vicinity functions are more influenced by the contacts than by the extension of the components.

As for the association of the ferrallitic mantle of flat hills(unit 2) and the level ferrallitic mantel(unit 3), both ρ_3^2 and ρ_2^3 are increasing near the origin. This results from the occurence of some narrow valleys of alluvial deposits (unit 10) which separate

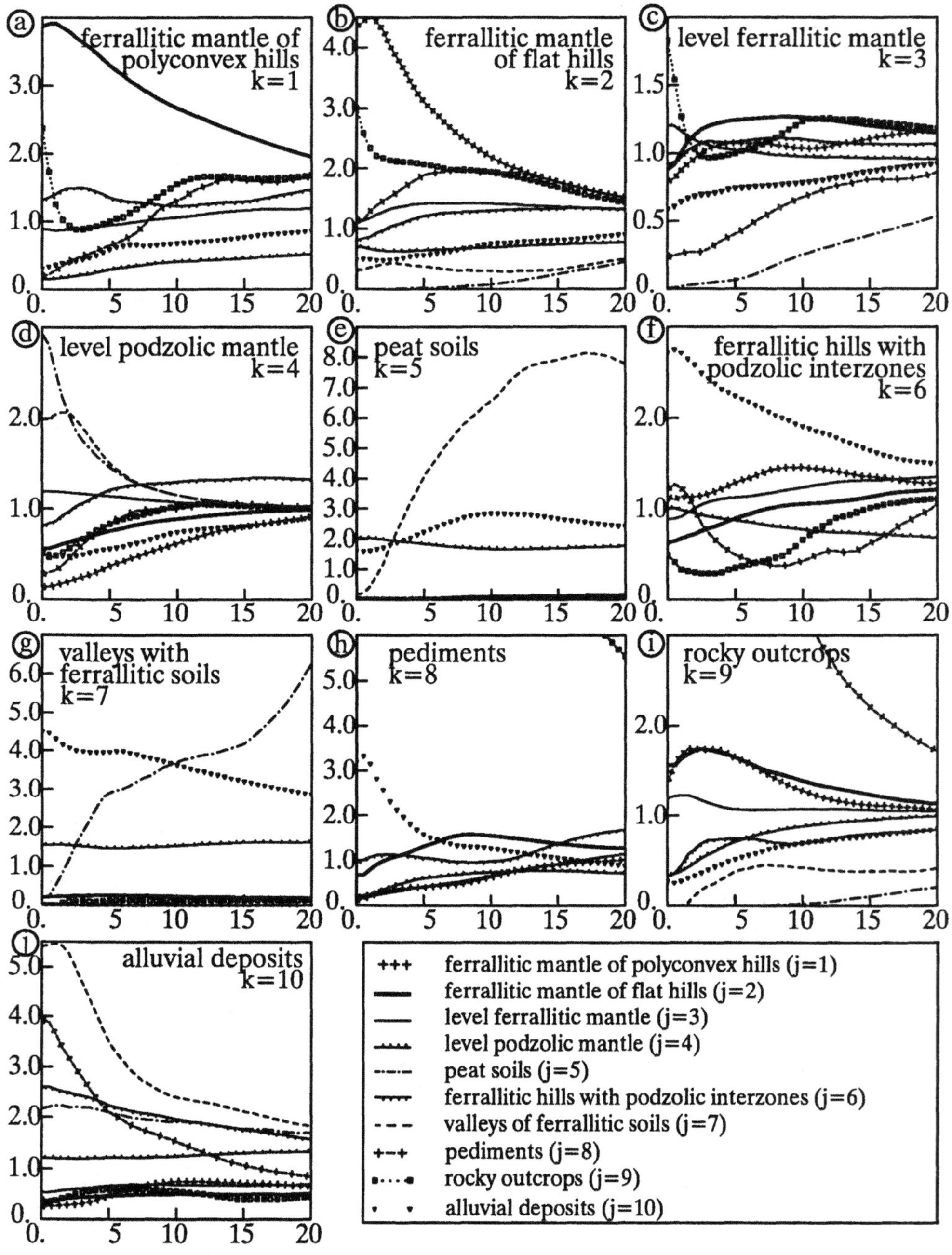

Figure 7: vicinity functions $\rho_k^j(r)$. Each graph gathers the vicinity functions of the dilated unit. (X-axis: r in km; Y-axis: $\rho_k^j(r)$)

$k \backslash j$	1	2	3	4	5	6	7	8	9	10
1										
2	17							16		
3										
4					9.5	13	5	15		12.5
5							19			
6										
7										
8										
9								17		
10							16.5			

Table 3: Ranges of extension R_k^j.

these two units in the north-east region of the map. This reduces the value of their vicinity functions for small radii, but makes them increase for larger radii, when the dilation has absorbed the small valleys. There is a kind of deferred contact between X_2 and X_3 which could not have been detected by the study of real contact in the previous section.

The ranges of extension (Table 3) are of little use in this case. R_k^j has only been reached for small units X_j covering less than 6% of the surface, and no R_k is less than 20 km. However, we can see that any point in the ferrallitic mantle of polyconvex hills (unit 1) is less than 17 km from the ferrallitic mantle of flat hills (unit 2). Other units of the sequence are larger and their ranges of extension are not computable.

From this, we can say that the hypothesis of the existence of the soil sequence is reinforced.

4 Conclusion.

This study, based on simple measures of perimeter and surface, tends to confirm the hypothesis of the existence of the soil sequence. These measures are used to try to characterize the spatial association of the units. The properties of contacts are the simplest to be computed but their interpretation is limited. The vicinity functions give a richer characterization because they enable us to analyse the spatial association beyond the interfaces of the units.

However, some difficulties arose through the study. The size of the components of the units seems to have an effect on the computed parameters and the analysis of neighbourhoods introduces border effects. It would be proper to determine to which extend this influences our results.

References

[1] Coster, M. and Chermant,J.L. (1985) Précis d'analyse d'image. C.N.R.S., Paris.

[2] Dubroeucq, D. Pédro, G. and Volkoff, B. (1991) "La couverture pédologique du bouclier du nord de l'amazonie (bassin du haut rio négro). séquence évolutive des sols et son rôle dans l'aplanissement généralisé des zones tropicales perhumides." *C. R. Acad. Sci. Paris, Série II*, 312:663–671.

[3] Jeulin, D. (1979) "Morphologie mathématique et propriétés physiques des agglomérés de minerais de fer et du coke métallurgique." thèse de docteur-ingénieur en sciences et techniques minières, Ecole des Mines, Paris, France.

[4] Jeulin, D. (1988) "Mathematical morphology and materials image analysis." *Scanning Microscopy*, Supplement No 2:165–183.

[5] Jeulin, D. (1990) "Analyse morphologique de la répartition spatiale." In *Traitement et Analyse d'Images en Microscopie à Balayage par Sonde Electronique.* ANRT (Groupe Microanalyse et MEB), Paris.

[6] Matheron, G. (1970) La théorie des variables régionalisées et ses applications, volume 5 of *Cahiers du Centre de Morphologie Mathématique.* E.N.S.M.P., Fontainebleau.

[7] Serra, J. (1982) Mathematical morphology and image analysis, volume 1. Academic Press, London.

CLASSIFICATION OF HIGH SPECTRAL RESOLUTION IMAGERY USING AN INDICATOR KRIGING BASED TECHNIQUE

FREEK D. VAN DER MEER
International Institute for Aerospace Survey and Earth Sciences (ITC)
Department of Earth Resources Surveys
350 Boulevard 1945, P.O. BOX 6, 7500 AA, Enschede, The Netherlands

ABSTRACT. High spectral resolution images (AVIRIS) providing detailed information on the surface mineralogy have been used to evaluate a new indicator kriging based classification technique. This technique directly uses spectral information derived from AVIRIS data instead of information from training areas studied in the field. A small corner of the image was selected for its known occurrences of both kaolinite and alunite. Three "conventional" classification methods were applied as well as the new indicator kriging based technique and results were evaluated using shape characteristics of the classes. Indicator kriging performs better than the conventional methods and gives information on the reliability of the classification.

INTRODUCTION

Remote sensing of earth's surface from aircraft and from spacecraft provides information not easily acquired by surface observations. Until recently, the main limitations of remote sensing were that no subsurface information could be obtained and that surface information lacked specification. Conventional scanners (e.g. Landsat MSS and TM, and SPOT) acquire information in a few separate spectral bands of various widths, thus filtering to a large extent the reflectance characteristics of the surface (Goetz & Rowan, 1981). Therefore, new scanner types were developed with high spectral resolution yielding new image processing techniques to cope with the increased amount of data. The use of indicator kriging as classification routine is discussed in this paper using high spectral resolution imagery although the technique is also valid for conventional scanner data.

IMAGING SPECTROMETRY

The use of high spectral resolution remotely sensed imagery for mineralogic mapping was first demonstrated in spectral laboratory studies (e.g. Hunt, 1977). In 1981, airborne spectrometer data were acquired using a sensor developed by the GER corporation. The first imaging device was the Airborne Imaging Spectrometer (AIS), developed by NASA, and in 1987 the improved version called the Airborne Visible/Infrared Imaging Spectrometer (AVIRIS; see Macenka

& Chrisp, 1987) from which data was used in this study. This scanner makes possible the simultaneous collection of images in 224 contiguous bands resulting in a complete reflectance spectrum for each 20*20 m. picture element (pixel) in the 400 to 2500nm region with a sampling interval of 10 nm (Porter & Enmark, 1987). Many surface materials have diagnostic absorption features that are 20-40nm in width (Hunt, 1977). Therefore, spectral imaging systems which have 10nm wide bands can produce data with sufficient resolution for resolving these features and subsequent direct identification of those materials (Goetz, 1991). Analysis of high spectral resolution imagery for mineral identification involves three steps: (1) the pre-processing of the data to convert raw spectra into reflectance spectra corrected for atmospheric influences, (2) extraction of absorption features characterizing surface materials of interest, and (3) evaluating for each pixel whether the absorption feature is present or absent at the wavelength (Okada et al., 1991).
This paper shows the potential use of indicator kriging based techniques for image classification (the third processing step mentioned above) of remotely sensed imagery in general and high spectral resolution data in particular.

GEOLOGY AND SPECTRAL CHARACTERISTICS OF SURFACE MATERIALS

The area of study, the Cuprite mining district (figure 1) is situated in western Nevada. Hydrothermally altered rocks about 12 km2 in extent including minor sulphur, silica, and precious metals are well exposed and nearly devoid of vegetation. The eastern half of the district, which is discussed here, is an area of extensive hydrothermal alteration within a sequence of rhyolitic welded ash flow and air fall tuffs. The altered rocks are divided into three mappable units: silicified rocks, opalized rocks, and argillized rocks. Silicified rocks, containing abundant hydrothermal quartz, form a large irregular patch extending from the middle to the south end of the area. Opalized rocks contain abundant opal and as much as 30% alunite and kaolinite. Locally, an interval of soft, poorly exposed material mapped as argillized rock separates fresh rock from opalized rock. In the argillized rocks, plagioclase is altered to kaolinite, and glass is altered to opal and varying amounts of montmorrillonite and kaolinite (Abrams et al. (1977).
The raw AVIRIS data were normalized to correct for atmospheric influences and solar irradiance drop off toward longer wavelengths by applying the flat-field correction. This was done by dividing the original pixel values by the DN value of an area which has little or no topographic relief and known spectral reflectance characteristics (in this study the Stonewall Playa area; noted with A in figure 1). Spectral features in the surface material become more apparent in the normalized spectrum because the removal of systematic effects makes it possible to display the data at their full radiometric resolution (see Rast et al., 1991 for a complete review).
The potential use of Imaging Spectrometry for mapping of hydrothermal alteration minerals in the Cuprite mining district has been shown by many workers (e.g. Vane, 1987) therefore the present analysis merely confirms their findings.
Figure 2 shows normalized spectra from areas B, C and D in figure 1. Area B is known as Kaolinite Hill for its extensive and well exposed deposit of the clay mineral kaolinite. The wavelength positions of the double OH absorption features at 2160 and 2200nm which is seen in laboratory spectra (VanderMarel & Beutelspacher, 1976) and resolved in the AVIRIS spectrum, was the main

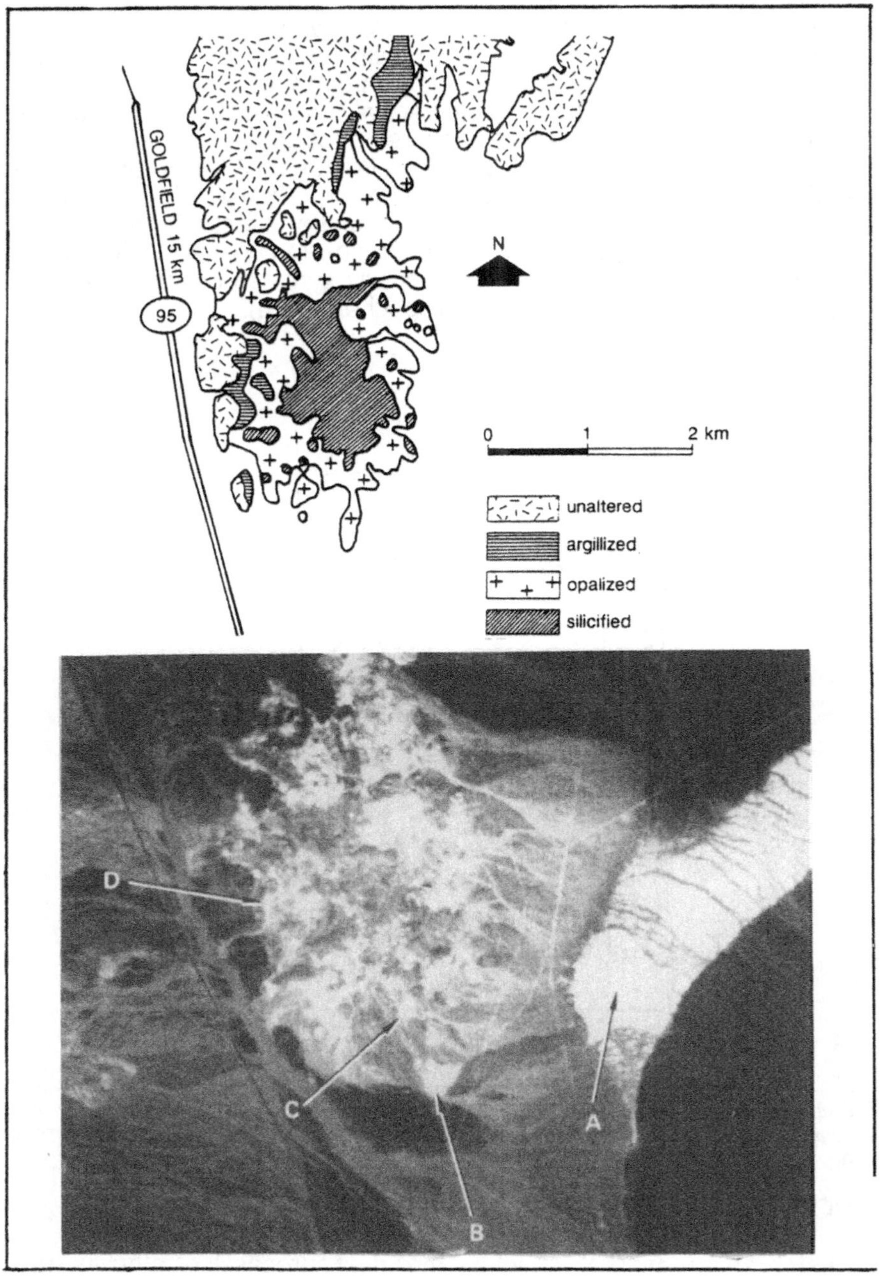

Figure 1: AVIRIS image of the Cuprite mining district (below) in band centered at 1014nm (A=Stonewall Playa; B=Kaolinite Hill; C=Alunite occurrence; D=Buddingtonite occurrence). The box indicates the area selected for classification. The upper map is a display of the alteration zones after Ashley (1977).

criterium used for positive identification. Furthermore, the characteristic asymmetric shape of the spectrum towards long wavelengths is well seen. Other clay minerals which are difficult to distinguish from kaolinite visually are alunite and buddingtonite. Alunite occurs widely throughout the Cuprite mining area while buddingtonite has been observed at few locations at the site. The occurrence of buddingtonite was discovered through the analysis of AIS spectra (Goetz & Srivastava, 1985). An absorption feature at 2020nm and a vibrational absorption feature due to NH4 (Krohn & Altaner, 1987) at 2110nm are the main diagnostic features distinguishing buddingtonite spectrally from other clay minerals. Alunite is characterized by absorption features at 2160 and 2200nm due to OH frequency stretching and a nearly symmetrical shape in the 2080-2280nm region (VanderMarel & Beutelspacher, 1976).

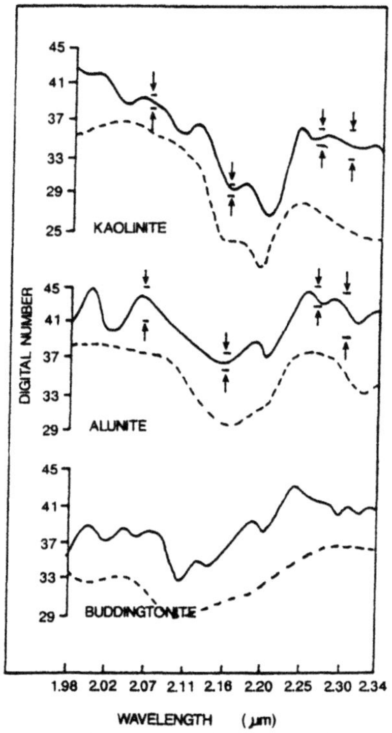

Figure 2: Mineral spectra for Kaolinite (top), Alunite (middle) and Buddingtonite (bottom) derived from AVIRIS imagery using a 5x5 pixel area at locations indicated by B, C, and D respectively in figure 1. Dotted curves represent laboratory spectra of the investigated minerals. Arrows enclose the ranges used for indicator classification in the four bands used.

METHODS FOR SUPERVISED CLASSIFICATION

After establishing mineral spectra, the next processing step is to apply classification rules to the image data to investigate spatial distribution of mineral occurrences. Conventional classification methods selecting bands on the shoulders and centre of absorption features will be used and compared to an indicator kriging based method selecting the same bands.

conventional classification methods

The processing strategy for information extraction from multiple datasets usually involves image classification. Classification techniques are based on computer assisted recognition of surface materials from their characteristic spectral properties in various wavelength bands. During investigation of imagery and during field check, areas known to represent the outcrop of a mineral or rock of interest are being selected as training areas. For each band which is used in the classification the spectral range for each class is estimated from the training pixels. Combining the information for various bands results in an N-dimensional histogram (the feature space) from which all classes can be separated according to their spectral response. The next step is to give the computer a set of rules to classify all pixels in an image by comparing their spectral response in various bands with those of the training pixels. The simplest method of doing so is box classification by which the two dimensional feature spaces are divided into rectangular boxes. The boundaries of the boxes representing the spectral ranges of DN values for the two bands within known area of the surface categories. Box classification compares all unknown pixels with the boxes; if they fall within one box they are assigned to the relevant class, if they fall in no box they remain unclassified (figure 3b).

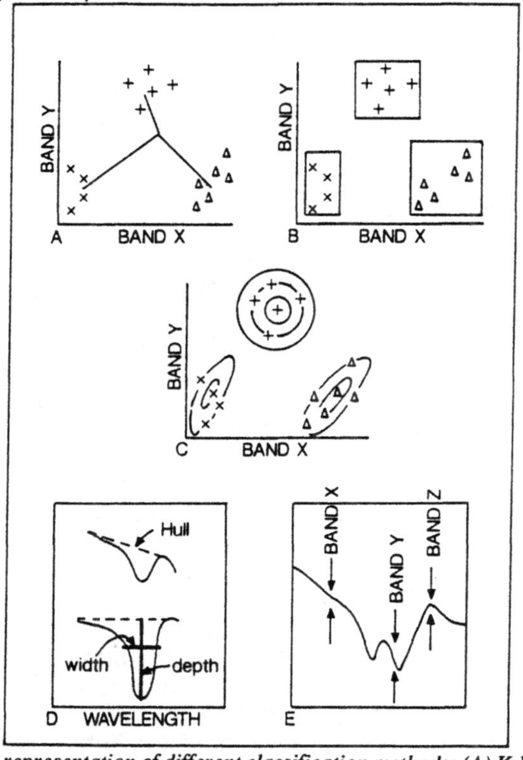

Figure 3: Schematic representation of different classification methods: (A) K-Nearest Neighbour analysis, (B) Box classification, (C) Maximum likelihood decision rules, (D) absorption parameters used for the analysis of AVIRIS data, and (E) indicator kriging definition of bands and upper and lower limits (see text for explanation).

A more sophisticated set of rules is provided by the K-nearest neighbour analysis which employs the statistics of the data from training areas. This method, first identifies the mean for each class used and secondly assigns the unknown pixel to the class with the closest mean.
A userdefined search radius is adopted; if this distance is exceeded the pixel will remain unclassified (figure 3a).
A refinement to this method is the maximum likelihood classification method which also uses the variance of the training samples. If DN values within the training areas are assumed to be normal distributed for each band, histograms can be considered to be bell-shaped. Dependent on the variance, the further a DN is from the mean, the less the probability is that it represents the class. In a two dimensional feature space, DN values from training pixels form a elliptical cloud which can be contoured to show the decrease of the probability from the mean value. The plot of DN for any unknown pixel can then be assessed in this probabilistic context by calculating the likelihood of the pixel belonging to each of the predefined classes. The unknown pixel is assigned to the class for which the likelihood (=probability) is maximized (figure 3c).
The main disadvantages of these conventional classification methods is that training pixels have to be selected which is often a tedious task. Furthermore, none of the methods discussed gives information on the reliability of the classification results and only maximum likelihood classification defines the probability of a pixel belonging to a class.
Analysis of high spectral resolution imagery uses spectral bands on the shoulders and centre of an absorption feature to deduce maps for the wavelength position, width, depth, symmetry, and Hull quotient of absorption features to investigate the spatial distribution of surface mineralogy (figure 3d).

THE INDICATOR KRIGING METHOD

The new classification method, the indicator kriging based method, allows for multispectral classification without prior knowledge of ground truth. Instead of using training pixels to define spectral ranges of classes of interest, this method directly employs the mineral spectra derived using AVIRIS or other type of scanner data.
Indicator kriging is one of the non-parametric geostatistical techniques which discretizes the histogram into several classes and carries out interpolation separately for every class (Fytas et al., 1989). The principal difference between ordinary kriging and indicator kriging is that indicator kriging works on (0,1) transformed data according to several threshold levels. Therefore the final result of indicator kriging is a set of probabilities that a block exceeds a specific set of cut-off values.
Classification using the indicator kriging approach starts by defining the spectral bands that contain "key information" on the spectral response of a certain ground class of interest. This step is done by extracting absorption features characterizing a mineral of interest from field spectral analysis or laboratory measurements. Bands on the shoulders of the absorption feature and bands on the centre of the feature are subsequently used for further analysis. This first step requires information on the spectral characteristics of certain minerals and not necessary on field geology which is obligatory for proper selection of training areas. The next step is to define the spectral range in each band by means of

setting upper and lower limits for the DN values in those spectral bands. Upper and lower limits in each bands are indicated by the area enclosed in between the arrows in figures 2 and 3. For all of these limits the data are being transformed into 0 and 1 values; 1 if they are below the threshold and 0 if they are above the threshold. This results in two binary (0,1) maps for each spectral band used in the classification; one for the upper limit and one for the lower limit. Subsequently, the binary maps are interpolated using the ordinary kriging procedure. By interpolating the 0,1 map for the upper limit we gain a map representing the probability that the value of a block lies below the indicated upper limit. The interpolation for the lower limit results in a map representing the probability that the block value lies above the lower limit. Combining these we calculate the probability that the block DN value is higher than the lower limit and smaller than the upper limit (e.g. the probability of a block having a spectral DN value in between the predefined range corresponding to the spectral response of the mineral of interest). The size of the blocks to produce output can be defined by the user and is not necessary of the same size as the input pixel size. This provides a means of extrapolating to areas smaller than the pixel size, although at very small block sizes one should consider support problems. Repeating this procedure for all key bands results in a set of probability maps (one for each band) which can be integrated by calculating the joint probability which is a measure for the likeliness that a pixel belongs to a certain class. This image is used as input for the classification. By setting tolerances on the minimum probability acquired for each class, pixels can be classified with a predefined accuracy.

Figure 4: Classification results for ground classes Kaolinite and Alunite using (A) Box classification, (B) K-Nearest neighbour analysis, (C) maximum likelihood decision rules, (D) indicator kriging using probability values greater than 0.5, and (E) indicator kriging using probability values greater than 0.

RESULTS

Different classification methods have been applied to a small (50x20 pixels) area located in the northwestern part of the image (figure 1). This are was chosen for its known occurrence of both kaolinite and alunite rich deposits. Therefore classification focused on these two minerals using bands centered at 2060, 2160,

2260 and 2300nm. From the mineral spectra derived from AVIRIS it can be shown that both alunite and kaolinite have an absorption feature at 2160nm which differs significantly in depth. Furthermore, the shape of the spectra over the spectral range 2060-2300nm is different (figure 2).
Classification results of conventional methods are shown in figure 4. Training samples were chosen at sites indicated by Kaolinite Hill and Alunite in figure 1. For the indicator classification method the threshold values used in all four bands are shown in figure 2. Indicator variograms (figure 5) are well defined at low cutoffs but tend to become conspicuous at high cutoffs.

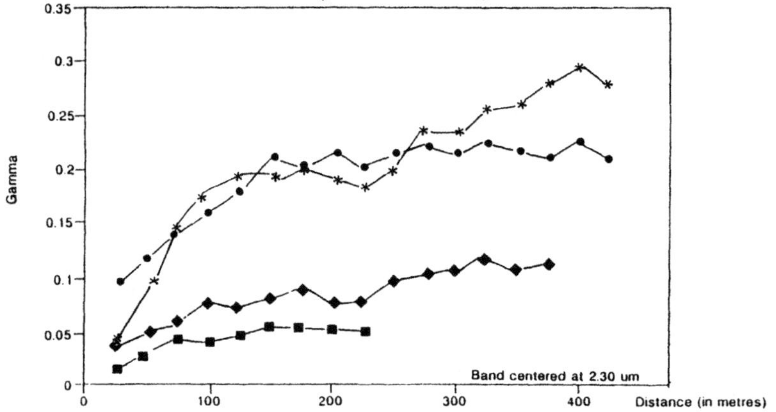

Figure 5: Example of an indicator variogram for binary images of the band centered at 2.3μm. Box symbol is the variogram for the upper limit used for class Alunite, diamond symbol is the variogram for the lower limit used for class Alunite, Star symbol is the variogram for the upper limit used for class Kaolinite, and circle symbol is the variogram for the lower limit used for class Kaolinite.

By carefully examining all the variograms we can notice that at higher cutoff levels the range is increasing and the behaviour is becoming more erratic. However, most of the variograms are well behaved and they have been fitted to spherical models. Moreover ot could be shown that the variograms used at cutoff levels for class alunite showed a anisotropy of the geometrical type which is more continuous in NW-SE directions.
Indicator kriging derived probability maps for alunite and kaolinite are shown in figure 6. These have been transformed into classified images by using all probabilities greater than zero (referred to as the indicator maximum result) and by using all probabilities greater than 0.5 (referred to as the indicator minimum results). The classified images for the indicator kriging based method are shown in figure 4.
Since the kriging error is dependent only on the semivariogram and therefore only on the data configuration in relation to the configuration of blocks to be estimated, a assessment of the reliability of the outcome can be made. If the variogram is known and the spacing and size of the input data (in this case 20x20 m pixels) for each output block geometry the maximum error can be calculated (McBratney et al., 1981). The maximum error for each upper and lower limits in each band has been evaluated for the class alunite (table I). Input data are on 20x20 m regular grid whereas output has been produced on grids of 1x1, 10x10, 20x20, 40x40, 80x80, and 100x100 m. An exponential decrease of error is clearly seen with increasing blocksize although the difference for block

sizes exceeding the pixel size is insignificant. For this study, output was produced on a 10x10 m. grid yielding a average error of 18%.

table I: Kriging standard deviation for estimation of class alunite in four bands used.

block size	band 2060nm		band 2160nm		band 2260nm		band 2300nm	
	lower	upper	lower	upper	lower	upper	lower	upper
1	0.15	0.10	0.24	0.25	0.22	0.16	0.23	0.24
10	0.09	0.06	0.15	0.20	0.10	0.10	0.20	0.17
20	0.07	0.06	0.10	0.20	0.10	0.10	0.10	0.10
40	0.06	0.05	0.10	0.10	0.09	0.08	0.09	0.10
80	0.06	0.04	0.08	0.09	0.06	0.06	0.09	0.09
100	0.06	0.04	0.01	0.02	0.05	0.01	0.08	0.07

EVALUATION OF RESULTS

Mathematical morphology (Serra, 1982) quantifies the geometrical structure and texture of an object. Basic transformations of binary patterns are erosion, opening and closing (Fabbri, 1984) from which synthetic measurements of shape are derived quantifying the visual aspects of a class (Fabbri et al., 1990). In this study, measurements used were (Durand & Flouzat, 1985): (1) the index of morphologic compactness (IC(c)), an indication for the roundness of particles in a class. The index varies from 0 to 1 where IC(c)=1 is a class composed of entities with rounded edges and IC(c)=0 is a class composed of entities with a blocky character, (2) the connectivity number (NC(c)), defining the number of particles in class c minus the number of holes in them. The higher the number of entities, the greater the disseminated aspect of the class, and (3) the boundary density (DF(c)) representing application to a given class of the shape coefficient for isolated objects, and provides information on the physiognomy of edges in that class.

Figure 6: Probability that a pixel belongs to class Alunite or Kaolinite. By setting tolerances on the probability values, pixels can be classified. From these images, the final classification was deduced for P > 0.5 and P > 0.

The morphological compactness of classes alunite and kaolinite in the box classified image and the K-nearest neighbour image are low giving them a blocky aspect (figure 7). Maximum likelihood classification yields better results, however, the indicator kriging classified images score best with this respect. The connectivity number of all methods is high, especially for the class kaolinite, although maximum likelihood classification and the indicator kriging based methods perform best with this respect. The boundary density for the indicator kriging based methods is significantly lower than for the conventional classification techniques, thus indicating less isolated objects.

The classified images were also compared with the outcrop patterns deduced from geological field data documented by Ashley (1977) and Albers & Stewart (1972), the results are shown in figure 8. For each mineral occurrence and for each classification method, pixels were counted for the four combinations possible: (1) mineral occurrence in both the field and in the classified image, (2) mineral occurrence in the field only, (3) mineral occurrence in the classified image only, and (4) mineral does not occur in the field and in the image.

From figure 8 it is apparent that classification results for alunite a better with respect to kaolinite. The amount of misclassification is less than 10% for maximum likelihood classification and indicator classification using probability values greater than zero.

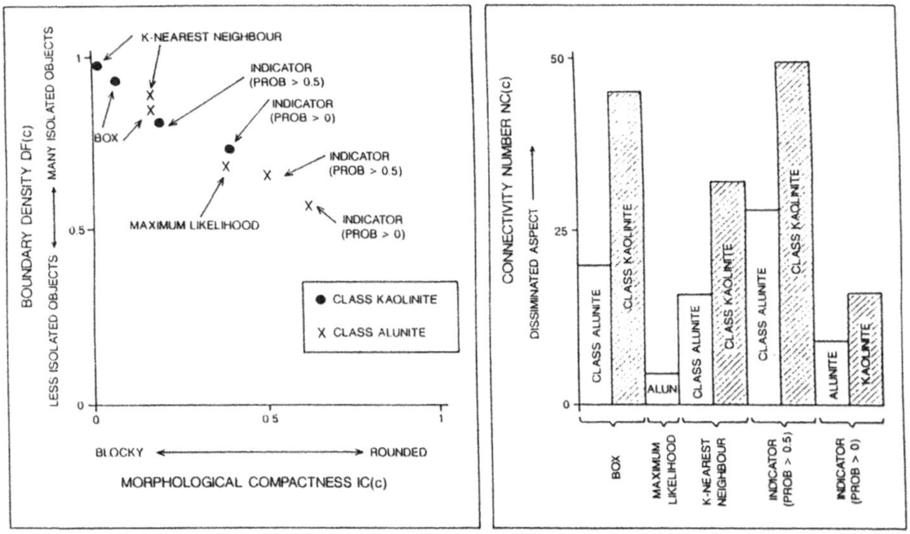

Figure 7: Synthetic measurements of shape of classes Kaolinite and Alunite in images classified using the methods described in this paper. Conventional classification methods yield significantly more blocky classes with more isolated objects. The disseminated aspects of classes classified using the conventional classification methods is generally higher than when using the indicator kriging based classification technique.

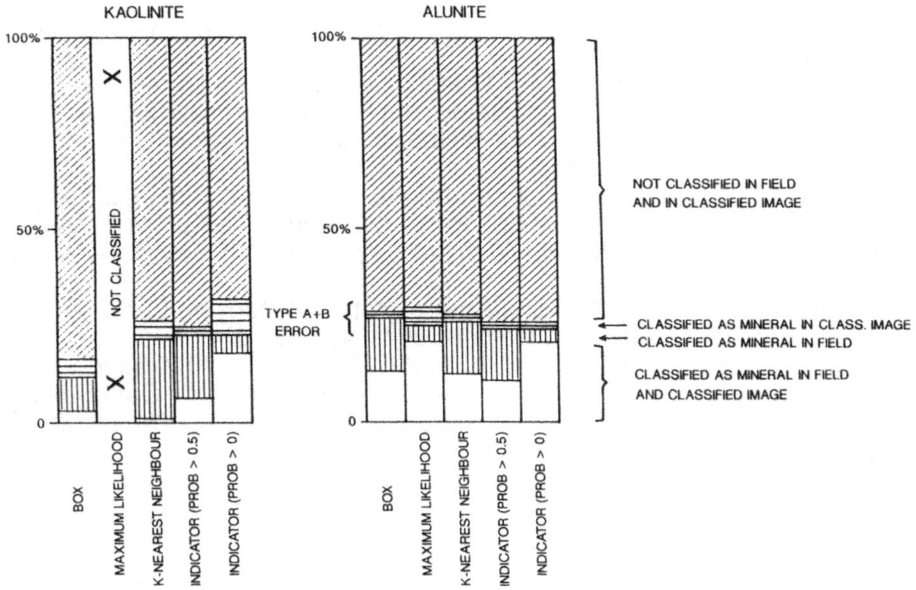

Figure 8: Comparison of classification results with known field geologic data from Albers & Stewart (1972) and Ashley (1977). The overall type A and B errors in classification is lower for the indicator kriging based method than for the conventional classification methods.

CONCLUSIONS

Direct identification of earth surface materials on the basis of their unique spectral reflectance features is now possible using imaging spectrometry (e.g. AVIRIS) in comparison with laboratory spectra. New image processing and analysis methods have to be developed in order to cope with the large amount of data available. Indicator kriging techniques as classification routines for image analysis allows for direct use of spectral information without prior knowledge of the geology of the area. A major advantage of this new technique is that it gives access to the reliability of the results and that areas smaller than the pixel size can be classified. In terms of visual aspects of classes, the new method performs better than the conventional methods.

REFERENCES

Abrams, M.J., Ashley, R.P., Rowan, L.C., Goetz, A.F.H. and Kahle, A.B. (1977) "Use of imaging in the 400-2360nm spectral region for alteration mapping in the Cuprite mining district, Nevada", US Geol. Survey Open file Report 77-585, pp. 1-19.

Albers, J.P. and Stewart, J.H. (1972) "Geology and mineral deposits of Esmeralda County, Nevada", Nevada Bur. Mines and Geology Bull. 78, pp. 1-80.

Ashley, R.P. (1977). "Geologic map and alteration map, Cuprite mining district, Nevada.

Durand, M.A. and Flouzat, G. (1985) "Quantification de l'aspect visuel des images classees", Photo interpretation 3, fascicule 5, 47-51.

Fabbri, A.G. (1984) Image processing of geologic data, Van Nostrand Reinhold publishers, New York.

Fabbri, A.G., Kushigbor, C.A., Valenzuela, C.R. and van der Meer, F.D. (1990) "Automated strategies for geometric characterization in geological remote sensing", 4th International Conference of Geoscience information (GEOINFO-4), 24-29 june 1990, Ottawa, Canada.

Fytas, K., Chaouai, N.E. and Lavigne, M. (1989) "Indicator kriging performance in gold deposit evaluation", Paper, 58, 91th Annual General meeting of the Canadian Institute of Mining and Metallurgy, 30 april, 3 may, Quebec City, Quebec, Canada.

Goetz, A.F.H. (1991) "Imaging Spectrometry for studying Earth, Air, Fire and Water", EARSel Advances in Remote Sensing 1, 3-15.

Goetz, A.F.H. and Srivastava, V. (1985) "Mineralogical mapping in the Cuprite mining district, Nevada", in: G. Vane and A.F.H. Goetz (eds.), Proceedings of the Airborne Imaging Spectrometer Data analysis Workshop, JPL Publication 85-41, Pasadena, California, pp. 22-31.

Goetz, A.F.H. and Rowan, L.C. (1981) "Geologic remote sensing", Science 211, 781-791.

Hunt, G.R. (1977) "Spectral signatures of particulate minerals in the visible and near infrared", Geophysics 42, 501-513.

Krohn, M.D. and Altaner, S.P. (1987) "Near Infrared detection of ammonium minerals", Geophysics 52, 924-930.

Macenka, S.A. and Chrisp, M.P. (1987) "Airborne Visible/Infrared Imaging Spectrometer, Design and Performance", in: G. Vane (ed.), Airborne Visible/Infrared Imaging Spectrometer AVIRIS (A description of the sensor, ground data processing facility, Laboratory Calibration, and first results, JPL Publication 87-38, Pasadena, California, pp. 89-97.

McBratney, A.B., Webster, R. and Burgess, T.M. (1981) "The design of optimal sampling schemes for local estimation and mapping of regionalized variables I: theory and method", Computer & Geosciences 7, 331-334.

Okada, K., Ohnuma, T. and Watanabe, H. (1991) "Geologic mapping of Keping uplift, Northwestern margin of Tarim basin based on absorption features using hyper-multispectral image data", Second annual JPL airborne Geoscience Workshop: AVIRIS, may 20-21, Pasadena, California.

Porter, W.M. and Enmark, H.T. (1987) "A system overview of the Airborne Visible/Infrared Imaging Spectrometer (AVIRIS)", proceedings of the 31th annual International Technical Symposium, 16-21 august 1987, Soc. of Photooptical Instrumentation Engineers V834, 22-31.

Rast, M., Hook, S.J., Elvidge, C.D. and Alley, R.E. (1991) "An evaluation of techniques for the extraction of mineral absorption features from High Spectral Resolution Remote Sensing data", Photogrammetric Engineering & Remote Sensing 57, 1303-1309.

Serra, J. (1982) Image analysis and Mathematical Morphology, Academic Press, New York.

VanderMarel, H.W. and Beutelspacher, H. (1976) Atlas of infrared spectroscopy of clay minerals and their admixtures, Elsevier, New York.

Vane, G. (1987) "First results from the Airborne Visible/Infrared Imaging Spectrometer (AVIRIS)", in: G. Vane (ed.), Airborne Visible/Infrared Imaging Spectrometer AVIRIS, JPL Publication 87-38, Pasadena, California, pp. 89-97.

SPATIAL INTERPOLATION OF SOIL MOISTURE DATA WITH UNIVERSAL COKRIGING.

A. STEIN
Dept. of Soil Science and Geology
Agricultural University
P.O. Box 37
6700 AA Wageningen
The Netherlands

ABSTRACT. Moisture deficits caused by groundwater extraction are land qualities that need to be interpolated for certain areas of land. Because these data are difficult to collect, and may be correlated with basic soil data, cokriging is used. Standard cokriging routines, however, do not allow the existence of a trend. Therefore, cokriging has been formulated for predicting multivariate and non-stationary data. Attention is focused on multivariate random functions. An important type of non-stationarity is defined in terms of multivariate increments that are stationary of order k. The covariance between increments of different variables is modelled by the pseudo-cross-covariance function. Cokriging equations and the induced cokriging equations are given. The procedures have been applied in a study of moisture deficits in the Netherlands. Parameters of the pseudo-cross-covariance function are estimated using Restricted Maximum Likelihood. Universal cokriging resulted in an improvement of the predictions in terms of the mean kriging error for a set consisting of 100 test data with 20-30%. No reduction, however, in the mean squared error between predictions and observations (MSE) was observed.

1. INTRODUCTION

In certain regional studies there is a growing need to measure and interpret characteristics that require relatively advanced procedures for measuring them. In this presentation a typical example is given on the land quality moisture deficit (Stein et al., 1989a), whereas other variables like the concentrations of heavy metals in the soil (Leenaers et al., 1989) and rain gauge data, (Krajewski, 1987) may be treated in a similar way. Moisture deficits are calculated with an agricultural simulation model, which typically needs data on many variables. Hence, calculation is usually costly. If moisture deficits are correlated with other more easily measured regionalized variables, like the mean highest groundwater level, then cokriging is an appropriate prediction technique (Journel and Huijbregts, 1978; Myers, 1982; Wackernagel et al., 1989; Stein and Corsten, 1991). The easily measured variable may serve as a co-variable to predict the value of the more difficult or more expensive property at an unvisited location. In this paper attention is focused on a description of universal cokriging and how it can be used for predicting moisture deficits.

In this presentation universal cokriging is defined using multivariate random functions and multivariate increments. Attention will be given to stationarity of increments, which is defined as pseudo-stationarity of random functions. The cokriging equations and the induced equations are derived without using Lagrange multipliers. The parameters of a polynomial pseudo-cross-covariance function, describing the covariances between increments, are estimated using Restricted Maximum Likelihood.

2. UNIVARIATE AND MULTIVARIATE RANDOM FUNCTIONS

In this section the main concepts of random functions and their increments will be given. First, the univariate and multivariate random functions will be considered, next increments of a certain order and the stationarity thereof will be defined. Finally the covariance structure between increments will be defined.

Data collected in space may be considered as realizations of a regionalized variable, generated by a random function (Cressie, 1992). A univariate random function (URF) in a d-dimensional region S will be denoted by $Y(\xi)$ where the location in S is given by ξ. For example, $Y(\xi)$ may be the moisture deficit for grassland and S an area in the eastern Netherlands. Since attention focuses on the quality of land related to crop yield, d is equal to 2. A random function can never be observed everywhere, only at a finite number of locations. Hence, a finite restriction of a URF $Y(\xi)$ is defined as that function restricted to a finite number n of observations at locations ξ_i for i=1,...,n. The vector of n values $Y(\xi_i)$ is denoted by Y. To be clear: $Y(\xi)$ is the URF and Y is the finite restriction.

In many practical studies several variables are observed simultaneously (Myers, 1988). Therefore, consider a collection of p univariate random functions. The variables will be denoted by $Y_1(\xi)$ through $Y_p(\xi)$, where, as before, the location in S is indicated by ξ. This presentation will focus on an example with p = 2, where $Y_1(\xi)$ is the moisture deficit for grassland and $Y_2(\xi)$ the mean highest groundwater level. It is the simultaneous behavior of the p variables in terms of their spatial structure and their interrelationship that is of interest. A multivariate random function (MRF) is defined for every $\xi \in S$ as a p-variate random variable $Y(\xi) = (Y_1(\xi),...,Y_p(\xi))'$. An MRF is used to describe simultaneously variables that may be mutually dependent. This distinguishes an MRF from p univariate functions (Dowd, 1989). A finite restriction of an MRF consists of finite restrictions of each component to a finite number n_u of observations at locations ξ_{ui} for u=1,...,p. The vector of $n=n_1+...+n_p$ values $Y_u(\xi_{ui})$, u=1,..,p, i=1,..,n_u is denoted by Y.

Next, finite linear combinations of an MRF are defined to allow linear predictions to be determined at unvisited locations. A finite linear combination of an MRF $Y(\xi)$ is the sum of finite linear combinations of the URFs $Y_u(\xi)$:

$$\lambda' Y = \sum_{u=1}^{p} \lambda_u' Y_u = \sum_{u=1}^{p} \sum_{i=1}^{n_u} \lambda_{ui} Y_u(\xi_{ui}) \tag{1}$$

where λ is a vector of n unknown weights, partitioned in p subvectors λ_u, for u=1,...,p, of n_u elements and the prime indicates the transpose of a matrix or a vector.

To deal with non-stationary data, it is necessary to consider increments, being a special class of linear combinations. For a URF $Y(\xi)$ it is well known that a (k+1)th-order increment defined for any non-negative k, is a finite linear combination $\lambda'Y$ with the property that it eliminates polynomials of degree ≤ k, i.e. $\lambda'\pi = 0$ for all finite restrictions of polynomials $\pi(\xi)$ of degree ≤ k (Matheron, 1973; Delfiner, 1976). For an MRF $Y(\xi)$ the finite linear combination $\lambda'Y$ is called an increment of order k+1 iff. each of its components is an increment of order k+1. Although different orders may be used for the different components, only the common case that they are all equal will be considered.

An MRF, just like a URF, should have a consistent family of distributions for finite restrictions of $Y(\xi)$. If first and second moments of the family of distributions exist, the p-variate expectation vector is defined as $\mu(\xi)=(\mu_1(\xi),...,\mu_p(\xi))'$, where $\mu_u(\xi) = E[Y_u(\xi)]$, u=1,...,p, and E[.] denotes the mathematical expectation. Also, for any two locations ξ and ζ the p-variate covariance function $G(\xi,\zeta)$ is defined as the p x p matrix $\{g_{uv}(\xi,\zeta)\}$ with the typical element $g_{uv}(\xi,\zeta) = \mathrm{cov}(Y_u(\xi),Y_v(\zeta))$. The expectation and the variance of an arbitrary finite linear combination $\lambda'Y$ are given by:

$$E[\lambda'Y]=\lambda'\mu=\sum_{u=1}^{p}\lambda_u'\mu_u=\sum_{u=1}^{p}\sum_{i=1}^{n_u}\lambda_{ui}\mu_u(\xi_{ui}) \tag{2}$$

$$Var[\lambda'Y]=\lambda'G\lambda=\sum_{u=1}^{p}\sum_{v=1}^{p}\lambda_u'g_{uv}\lambda_v \tag{3}$$

where λ and λ_u are defined before and μ_u, g_{uv}, μ and G are the finite restrictions of $\mu_u(\xi)$, $g_{uv}(\xi,\zeta)$, $\mu(\xi)$ and $G(\xi,\zeta)$, respectively.

Next, more general models to describe the randomness of a MRF will be considered. Assume that only expectations and covariances of increments of order k exist. Therefore, there exists a multivariate expectation function $\mu(\xi)$ such that the expectation of an increment $\lambda'Y$ can be calculated with eq. (2). The function $\mu_u(\xi)$ is not an expectation in the ordinary sense anymore, because it is not unique: if $\pi(\xi)$ is a polynomial of degree ≤ k, then $\mu(\xi)+\pi(\xi)$ may also serve as the expectation. It is therefore termed the *pseudo*-expectation. Also, there exists a multivariate matrix function $G(\xi,\zeta)$ such that the variance of an increment $\lambda'Y$ can be calculated with eq. (3). The function $g_{uv}(\xi,\zeta)$ is also not a covariance in the ordinary sense: if $\pi_i(\xi)$ are polynomials of degree ≤ k for i=1,...,4, and $f_1(\xi)$ and $f_2(\xi)$ are any functions, then using $G^*(\xi,\zeta) = G(\xi,\zeta)+\pi_1(\xi)f_1(\zeta)+f_2(\xi)\pi_2(\zeta)+\pi_3(\xi)\pi_4(\zeta)$ is equivalent to using $G(\xi,\zeta)$, as follows from the definition of increments. Therefore the functions $g_{uv}(\xi,\zeta)$ are termed *pseudo*-covariance functions if u = v, and *pseudo*-cross-covariance functions if u ≠ v. The multivariate matrix function $G(\xi,\zeta)$ is also called a pseudo-covariance function. Eq. (3) need to be positive only for increments; therefore the pseudo-covariance functions need to be pseudo-positive definite only. The expectation and the variance of a linear combination that is not an increment needs not exist and if its variance exists it needs not be positive.

The collection of λs yielding pseudo-positive definiteness is called the collection of permissible λs (Christakos, 1984). It forms a finite-dimensional linear space within the space of all λs. The elements of any basis of the space can be arranged as rows of the matrix C. Any linear combination of the elements of the vector Z = CY is an increment. Conversely, any increment of the observations can be expressed as a linear combination of the elements of the vector Z = CY (Kitanidis, 1983; Barendregt, 1987). For example, the matrix

$$C = \begin{pmatrix} C_1 & 0 & \dots & 0 \\ 0 & C_2 & \dots & 0 \\ . & . & . & . \\ 0 & 0 & \dots & C_p \end{pmatrix} = (I - X(X'X)^{-1}X') \tag{4}$$

can be used, where X is a partitioned matrix with blocks X_{uv}. For u ≠ v the blocks X_{uv} are equal to zero, whereas the blocks X_{uu} contain multivariate monomials up to the degree of non-stationarity k of the coordinates of the n_u observation locations ξ_{ui}, i=1,...,n_u of the uth variable. The number of such monomials is given by m. In a two-dimensional region, m = 1 if k = 0, m = 3 if k = 1, and m = 6 if k = 2. The last m rows of X_{uu} may be skipped, as the rank of C_u equals n_u-m, yielding a matrix C of n - p x m rows and n columns.

In order to treat stationarity of multivariate increments, attention will be paid first to stationary MRFs. An MRF Y(ξ) which has the first and second order moments is called stationary iff. both E[Y(ξ+**h**)] and Var[Y(ξ+**h**)] are independent of any translation vector **h**. Now consider a filtered URF, defined for any ξ ε S as $Z(\xi) = \Sigma\lambda_i Y(\xi_i+\xi)$; a filtered MRF is defined as an MRF consisting of p filtered URFs. A URF Y(ξ) is called *pseudo-stationary of order k*, if Z(ξ) is stationary for every increment λ'Y of order k. Likewise, an MRF is called pseudo-stationary of order k, if each of its components is pseudo-stationary of order k. Such a URF and MRF are said to have k^{th} order stationary increments (Ito, 1953). Stationarity of increments corresponds to pseudo-stationarity of random functions, and may be interpreted as non-changing expectation and variance with regard to translations. For example, the Wiener process Y(ξ) with E[Y(ξ+**h**)-Y(ξ)] = 0 and Var[Y(ξ+**h**)-Y(ξ)] = |**h**| is pseudo-stationary of order 1, whereas Var[Y(ξ)] does not exist. This agrees with (Gel'fand-Vilenkin, 1964, p. 264-265). Following (Matheron, 1973) one may call this process an intrinsic random function of order 0.

An MRF Y(ξ) pseudo-stationary of order 1 has a pseudo-expectation and a pseudo-covariance which are unique apart from polynomials of degree 0. If μ(ξ) is constant, but μ(ξ) ≠ 0 one may use increments of order 2 in order to obtain increments with expectation equal to zero. An example of an increment of order 1 of such an MRF is the difference Y(ξ) - Y(ζ). In this case, the opposite of the semivariogram H, defined by H(ξ,ζ) = (1/2)*Var[Y(ξ) - Y(ζ)], is equivalent to G(ξ,ζ). For first order increments the variances of the constituting MRFs, do not necessarily exist. If they do exist, however, and are equal to $G_0(\xi,\zeta)$, say, the well-known relations hold:

$$H(\xi,\zeta) = -\tfrac{1}{2}G_0(\xi,\xi) - \tfrac{1}{2}G_0(\zeta,\zeta) + G_0(\xi,\zeta)$$

(Journel and Huijbregts, 1978). For this case one may define H(**h**) = H(ξ,ξ+**h**) and use H(ξ-ζ) subsequently.

For a pseudo-stationary MRF Y(ξ) of order k+1 the pseudo-expectation function is a multivariate polynomial of degree ≤ k+1, which is unique apart from polynomials of degree k. If increments are stationary but their expectation is not equal to zero, the use of an increment of an order 1 step higher, will eliminate the pseudo-expectation function. By definition, polynomials of degree ≤ k are eliminated, because for $Y = Y_1 + \pi$, where

π is the finite restriction of a polynomial of degree $\leq$ k one has that $\lambda' Y = \lambda' Y_1$. Therefore, equivalence classes arise in MRFs, with respect to polynomials of degree k. For such increments, an extended class of pseudo-covariance functions is available.

Finally, an MRF $Y(\xi)$ is called rotation invariant (or: isotropic) iff. for any u and any v $g_{uv}(\xi,\zeta)$ depends on the length of the distance vector r between ξ and ζ only, and not on its direction. In this case one may use $g_{uv}(\mathbf{r})$ instead of $g_{uv}(\xi,\zeta)$.

3. ESTIMATING THE PARAMETERS OF A LINEAR PSEUDO-COVARIANCE FUNCTION WITH RESTRICTED MAXIMUM LIKELIHOOD.

We will now pay attention towards modelling the PCFs and estimating of its parameters. It will be assumed throughout this section that the PCF is the sum of the functions $g_{ii}(r)$ and $g_{ij}(r)$, each isotropic and linearly dependent on a vector of parameters, γ. The pseudo covariance function of the ith variable is denoted with $g_{ii}(r)$ and the pseudo-cross-covariance function describing the spatial interaction between the ith and the jth variable with $g_{ij}(r)$:

$$g_{ij}(r)=\gamma_{ij,0}*\delta_K(r)+\gamma_{ij,1}*r+\gamma_{ij,3}*r^3+\gamma_{ij,5}*r^5 \tag{5}$$

Here r is the distance between observation points and the Kronecker function, $\delta_K(r)$, equals 1 if r = 0 and equals 0 if r $\neq$ 0. The elements of the vector γ are denoted with $\gamma_{ii,\alpha}$, for i,j=1,..,p and α = 1,...,k+2, α depending on the degree k of the polynomial trends: $\gamma_{ii,2}r^3$ can be included only if k exceeds 0, and $\gamma_{ii,3}r^5$ only if k exceeds 1, due to requirement of positive-definiteness (Delfiner, 1976).

The variance of increments of the observations, $Q = \mathrm{Var}[Z] = E[ZZ'|\gamma] = E[CYY'C'|\gamma], = CE[YY'|\gamma]C'$ has a structure similar to (5), i.e.

$$Q=C\left(\sum_{i,j=1}^{p}\sum_{\alpha=1}^{k+2}\gamma_{ij,\alpha}G_{ij,\alpha}\right)C' = \sum_{i,j=1}^{p}\sum_{\alpha=1}^{k+2}\gamma_{ij,\alpha}Q_{ij,\alpha} \tag{6}$$

where the matrices $Q_{ij,\alpha} = CG_{ij,\alpha}C'$ are of size N x N. Elements of the matrix $G_{ij,\alpha}$ are obtained from each of the constituting parts of (5). The number of terms in eq. (6) is equal to $M = \frac{1}{2}p(p+1)*(k+2)$.

The vector Z is assumed to have a Gaussian distribution with zero mean and variance matrix Q, which is linear in the parameter vector γ. As shown by standard methods the likelihood equations are

$$Tr(Q^{-1}Q_iQ^{-1}[ZZ'-Q])=0 \tag{7}$$

for i=1,...,M, where Tr(.) denotes the trace of a square matrix. Eq. (7) is equivalent to the following system of linear equations:

$$a=B\gamma \tag{8}$$

with $\{a\}_i = Tr(Q^{-1}Q_iQ^{-1}Z.Z')$, $\{B\}_{ij} = Tr(Q^{-1}Q_iQ^{-1}Q_j)$ and $\{\gamma\}_i = \gamma_i$, i = 1,...,M, j = 1,...,M. After a prior estimate of Q, these equations can be solved by determining the γs and then calculating a new Q. This is repeated until convergence is achieved. The procedure, commonly known as restricted maximum likelihood (REML), maximizes the likelihood of the parameter γ for the observed increments.

A key statistic for distinguishing between various degrees of trends is Akaikes information criterion (Akaike, 1974), defined as:

$$AIC = l(Z|\gamma) + 2*M \tag{9}$$

where $l(Z|\gamma)$ is the log-likelihood of the increments Z as a function of the vector γ of parameters. The degree of the trend which minimizes (9) is regarded as the appropriate degree of non-stationarity.

4. COKRIGING

Multivariate random functions and their increments may be used to predict one of its components in unsampled locations. This is known as cokriging. An important aim of cokriging is to improve the prediction by using the information on the other components that have generally been better sampled.

Consider a set of locations, a part where $Y(\xi)$ has been observed and another part to predict the values one of its components. The finite restriction of the MRF $Y(\xi)$ is partitioned into $Z = (Y'\ Y_P')'$ and the weight vector λ is partitioned accordingly into $(\lambda'\ \lambda_P')'$. Throughout this section P is used to label the part where a prediction is to be made. The aim is to predict a linear combination $\lambda_P'Y_P$ by a linear predictor $\lambda'Y$. In most practical cases where the aim is to predict the value of the main component $Y_1(\xi)$ in one single location, Y_P contains one single element and $\lambda_P = 1$. If the aim is to predict the average value in an area of an arbitrary shape, Y_P consists of elements located in that area and λ_P will be a vector of equal weights summing to 1. If the aim is to predict the gradient (in one direction) of the main component along 6 equally spaced points Y_P contains elements located in those six points and $\lambda_P = 1/70(-5,-3,-1,1,3,5)'$.

Linear unbiased increment (LUI) predictors of $\lambda_P'Y_P$ are defined as predictors linear in the variables Y such that the prediction error $\varepsilon'Z = \lambda'Y - \lambda_P'Y_P$ has zero expectation, i.e. $E[\varepsilon'Z] = 0$ and is an increment with expectation equal to zero. The size of the prediction error is given by its mean squared error $E[\varepsilon'Z]^2 = var[\varepsilon'Z]$. The BLUI-predictor (best LUI-predictor) is defined as the LUI-predictor with minimum mean squared error.

A zero-expectation increment $\varepsilon'Z$ satisfies $\lambda'X - \lambda_P'X_P$ where X and X_p contain the monomials of the coordinates of the observations locations and the prediction locations, respectively, both up to the degree of the polynomial trend k. The covariance structure is represented by the finite restriction of the pseudo-covariance function, i.e. the matrix Γ, partitioned as

$$\Gamma = \begin{pmatrix} G & G_p \\ G_p & G_{pp} \end{pmatrix} \tag{10}$$

The sub-matrix G contains the pseudo-covariances and pseudo-cross-covariances between the elements of Y, i.e. between the observation part of Z, the sub-matrix G_P (= G_P') contains the pseudo-covariances and the pseudo-cross-covariances between the elements of Y and those of Y_P and the sub-matrix G_{PP} contains the pseudo-covariances between the elements of Y_P.

It can be proved that a prediction error ε'Z has minimum variance iff. it has zero covariance with all unbiased predictors of 0 (Stein et al., 1991b). This yields the following partitioned matrix equation:

$$\begin{pmatrix} G & X \\ X' & 0 \end{pmatrix}\begin{pmatrix} \lambda \\ t \end{pmatrix}=\begin{pmatrix} G_P\lambda_P \\ X_P'\lambda_P \end{pmatrix} \tag{11}$$

where t is any vector with m elements. If λ satisfies eq. (11) for some t, then $\varepsilon'Z = \lambda'Y - \lambda_P'Y_P$ has minimum variance and therefore $\lambda'Y$ is the BLUI-predictor of $\lambda_P'Y_P$. Therefore equation (11), the so-called 'cokriging'-equation, characterizes the BLUI-predictor of $\lambda_P'Y_P$. Solving λ with the first cokriging equation for given t and inserting it into the second equation gives as a solution $t = -(X'G^{-1}X)^{-1}(X_P-G_PG^{-1}X)\lambda_P$, which makes it possible to solve for λ. Thus the BLUI-predictor T of $\lambda_P'Y_P$ is obtained as

$$T=\lambda_P'(X_pB+G_pF) \tag{12}$$

where B is defined as $(X'G^{-1}X)^{-1}X'G^{-1}Y$ and F as $G^{-1}(Y-XB)$. Note that B and F are completely determined by the observed values of the MRF and their locations and are independent of Y_P.

The first term of (12), $\lambda_P'X_PB$, may be defined as the pseudo-trend XB. This is not an estimate of the pseudo-expectation vector μ, since μ is an artifact. However, $\lambda_P'X_PB$ itself is a LUI-predictor of $\lambda_P'Y_P$. It is of a polynomial form in the coordinates of the locations of Z and is obtained using generalized least squares for the estimating the regression coefficients B. The second term, $\lambda_P'G_PF$, is free from polynomial effects and from μ-effects. It improves $\lambda_P'X_PB$ as a predictor of $\lambda_P'Y_P$ by using a linear combination of the residuals Y-XB with respect to the estimated pseudo-expectation.

The mean square error of the BLUI-predictor is obtained as

$$\varepsilon'\Gamma\varepsilon=\lambda_P'(G_{PP}-G_P'G^{-1}G_P+X_a'VX_a)\lambda_P \tag{13}$$

with $V = (XG^{-1}X)^{-1}$ and $X_a = X_P-G_PG^{-1}X$. Eq. (13) defines the kriging error, equal to the variance of the prediction error.

Another form of the cokriging equations is obtained as follows. The best predictor for Y is considered to be Y itself, being an exact interpolator. Therefore from (12) we have XB + GF = Y, by substituting X for X_p and G for G_p. Moreover, unbiasedness assures that the columns of the matrix F are zero-expectation increments, i.e. X'F = 0.

The two requirements yield the following partitioned matrix equation:

$$\begin{pmatrix} G_{OO} & X_O \\ X'_O & 0 \end{pmatrix} \begin{pmatrix} F \\ B \end{pmatrix} = \begin{pmatrix} Y \\ 0 \end{pmatrix} \tag{14}$$

Solving F from the first cokriging equation for a given B and substituting it in the second equation gives expressions for B and F as already mentioned. Equation (14) is the so-called induced equation (Myers, 1988).

5. MOISTURE DEFICITS IN THE MANDER AREA

Attention will now be focused on the problem of moisture deficits that motivated this study. The 404 ha Mander area, located in the eastern part of the Netherlands has been used for groundwater extraction for the drinking-water supply during the past 30 years (Bregt et al., 1991; Stein et al., 1991a). Water tables in the area were lowered, causing water shortage. In dry periods crop yields, mainly grassland, of some of the local farmers decreased. By means of a detailed 1:10,000 soil survey in 1985 a total of 499 observations was obtained on basic soil properties, like rooting depth, organic matter content, clay percentage and former and present mean highest and mean lowest water table. Three different soil types are distinguished (Table 1). Current water tables were measured. Levels that occurred before the period of water extraction were estimated by the soil surveyors using soil mottling criteria. A general lowering of 50 cm of the groundwater table was obtained during the 30 year period of groundwater extraction.

Simulation models were used to calculate two different moisture deficits for grassland under the present hydrological situation. The yearly average moisture deficit for the period 1956 through 1985 has been determined as well as the moisture deficit for 1976 an extremely dry year. Calculations of this model are based on observations of soil variables as discussed by Bouma et al. (1980), on meteorological data and on data on groundwater extraction. The model distinguishes between the topsoil and the subsoil, composed of different layers, defined in terms of soil texture and organic matter content. Moisture retention curves for both top- and subsoil and hydraulic conductivity curves for the subsoil are based on measured data in well defined soil layers for all 499 points. The mean values of MD30 equals 11.0mm, and that of MD76 equals 59.5mm.

In this study universal cokriging has been used to predict moisture deficits, using the Mean Highest Groundwatertable (MHW) as a covariable. Correlation coefficients with MD30 and MD76 were equal to 0.636 and 0.587, respectively. A test set of 100 points was selected in advance to compare the quality of the predictions by means of the Mean Squared Error of the Predictions (MSE) as well as with the mean variances of the prediction error (MVP).

Since it was possible that the two moisture deficits exhibited a trend, the model was set up with k=2, allowing for a quadratic trend. For bivariate increments the spatial dependence is described by pseudo-covariance functions according to eq. (5) with p = 2. To estimate simultaneously the coefficients $\gamma_{ii,\alpha}$, and $\gamma_{ij,\alpha}$ by minimizing the negative log-likelihood function, the total data set was too large to be handled efficiently. Therefore, 9 sets, each comprising 25 observations, were selected at random from the original data. The coefficients were estimated for each of these sets. Non-permissible coefficients were

TABLE 1. Mean, standard deviation (SD) and coefficient of variation (CV) for calculated moisture deficits and current and the previous mean highest groundwater level (MHW and PMHW, respectively) for the different soil types.

		MD30 [mm]	MD76 [mm]	MHW [cm]	prev MHW [cm]	N_{obs}
All soil types	μ	11	59.5	88.8	42.9	399
	σ	15.3	52.1	74.1	55.3	
	C.V.	1.4	0.88	0.83	1.29	
Haplaquods	μ	10.7	61.9	79.0	32.5	226
	σ	15.5	46.6	49.6	28.6	
	C.V.	1.4	0.75	0.63	0.88	
Plaggepts	μ	27.2	120.6	210.1	140.3	51
	σ	15.0	51.1	103.7	90.3	
	C.V.	0.6	0.42	0.49	0.64	
Humaquepts	μ	4.7	29.6	56.5	21.5	122
	σ	9.1	36.6	40.0	22.6	
	C.V.	1.9	1.24	0.71	1.05	

assigned the value 0 (Delfiner, 1976). The final coefficients were obtained by averaging the coefficients obtained for each of the 9 sets.

On evaluation of Akaikes criterion the three variables MD30, MD76 and MHW show a second order trend. This is not surprising, because low changes in the water tables, are associated with humaquepts and haplaquods, mainly located in the western part of the area, and covering the largest part of the area (appr. 85%), whereas the values are high for plaggeptic soils which occur in a relatively small area (15%) in the eastern part (Table 1). Moisture deficits are correlated with these changes in water tables.

The vectors of estimated coefficients for the pseudo-covariance functions γ are given in Tables 2a and 2b. The coefficients which are estimated for the pseudo-covariance functions for bivariate increments (Table 2b) agree with the corresponding coefficients for the pseudo-covariance functions for univariate increments (Table 2a). For example, the coefficients $\gamma_{22,1}$ and $\gamma_{22,2}$ from Table 2b correspond with the coefficients $\gamma_{11,1}$ and $\gamma_{11,2}$ for MHW in Table 2a.

The predictions obtained with universal cokriging are compared with those from universal kriging by means of predicting the values for the set of test data. When universal kriging was applied a neighborhood of 18 observations was used. For universal cokriging the neighborhood was extended to include 19 observations for the covariable. To evaluate the precision of the estimates at the 100 test locations the MSE (the mean of the squared differences of predicted and observed values) and the MVP (the mean of the variances of the prediction errors) is calculated. These are given in Table 2. The MSE values are approximately the same for both universal cokriging and for universal kriging for the two moisture deficits considered, whereas the MVP values for universal cokriging are reduced with 20% (for MD76) and 30% (for MD30), respectively, as compared with

TABLE 2a. Estimated coefficients for pseudo-covariance functions for the different URFs.

Variable		$\gamma_{11,1}$	$\gamma_{11,2}$	$\gamma_{11,3}$	$\gamma_{11,4}$
MD76	[mm]	885	-82	0	-0.92
MD30	[mm]	48.3	-5.9	0	-0.04
MHW	[cm]	337	-114	0	-1.05

TABLE 2b. Simultaneously estimated coefficients for pseudo-covariance functions and pseudo-cross-covariance functions for MD30 and MD76 with MHW.

		MD30	MD76
MD	$\gamma_{11,1}$	33.4	774.4
	$\gamma_{11,2}$	-5.2	-36.0
	$\gamma_{11,3}$	0	0
	$\gamma_{11,4}$	0	0
MHW	$\gamma_{22,1}$	411	499
	$\gamma_{22,2}$	-77	-56
	$\gamma_{22,3}$	0	0
	$\gamma_{22,4}$	0	0
MD x MHW	$\gamma_{12,1}$	36.6	324.2
	$\gamma_{12,2}$	-15.8	-33.0
	$\gamma_{12,3}$	0	0
	$\gamma_{12,4}$	0	0

Table 3. The Mean squared error (MSE) and the mean variance of the prediction error (MVP) obtained by means of universal kriging (UK) and universal cokriging (UCK).

	MSE		MVP	
	UK	UCK	UK	UCK
MD30	66	66	65	42
MD76	1219	1229	1190	989

universal kriging. Based on the reduced MVP value it is tentatively concluded that universal cokriging increases the precision of the predictions. Finally, the pseudo-covariance function has been used to create an MD30 map (Fig. 1).

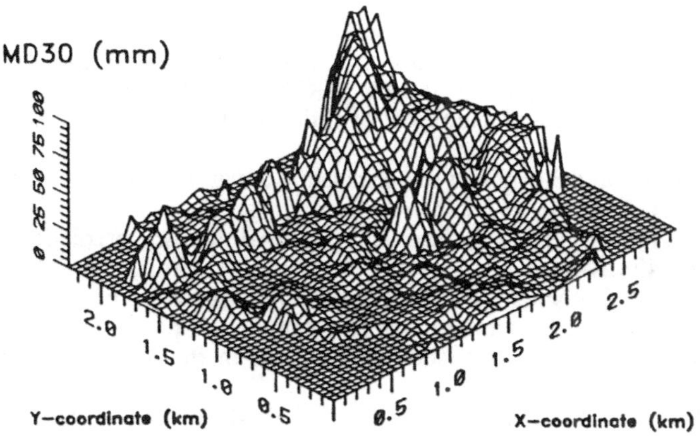

Fig. 1. 30-Year average Moisture Deficit for grassland obtained by means of universal cokriging with the mean highest Groundwater level as a co-variable.

6. CONCLUSIONS

1. Universal cokriging is an extension of current geostatistical prediction and interpolation procedures; it is now being used routinely in environmental studies such as soil science.
2. Universal cokriging allows one to predict values optimally at a single locations, at multiple points (and hence as an approximation of a block with an arbitrary shape) or on a gradient.
3. Restricted maximum likelihood estimation of the parameters of a pseudo-covariance function is straightforward. Only computer speed and memory size may be prohibitive for some practical studies.
4. More research is needed to fully explore the possibilities and limitations of universal cokriging.

ACKNOWLEDGEMENT

This study was carried out with a grant from the Netherlands Integrated Soil Research Programme, PCBB.

7. REFERENCES

Akaike, H. (1974) "A new look at the statistical model identification", IEEE Trans. Automat. Control, AC-19, 716-723.

Barendregt, L.G. (1987) "The estimation of generalized covariance when it is a linear combination of two known generalized covariances", Water Resour. Res. 23, 583-590.

Bouma, J., De Laat, P.J.M., Awater, R.H.C.M., Van Heesen, H.C., Van Holst, A.F., and Van de Nes, T.J. (1980) "Use of soil survey data in a model for simulating regional soil moisture regimes", Soil Sci. Soc. Am. J., 44, 808-814.

Bregt, A.K, McBratney, A.B., and Wopereis, M.C.S. (1991) "Construction of isolinear maps of soil attributes with empirical confidence limits", Soil Sci. Soc. Am. J., 55, 14-19.

Christakos, G. (1984) "On the problem of permissible covariance and variogram models", Water Resour. Res. 20, 251-265.

Corsten, L.C.A. (1989) "Interpolation and optimal linear prediction", Statistica Neerlandica 43, 69-84.

Cressie, N.A.C. (1992) Statistics for spatial data, Wiley, New York.

Delfiner, P. (1976) "Linear Estimation of Nonstationary Spatial Phenomena", in M. Guarascio et al. (eds.), Advanced Geostatistics in the Mining Industry, Reidel, Dordrecht, 49-68.

Dowd, P.A. (1989) "Generalized Cross-covariances", in M. Armstrong (ed.), Geostatistics, Reidel, Dordrecht, 151-162.

Gel'fand, I.M. and Vilenkin, N. (1964) Generalized Functions 4, Academic Press, New York.

Ito, K. (1953) "Stationary random distributions", Mem. Coll. Sci. Univ. Kyoto 28, 209-233.

Journel, A.G. and Huijbregts, C.J. (1978) Mining geostatistics, Academic Press, London.

Kitanidis, P.K. (1983) "Statistical Estimation of Polynomial Generalized Covariance Functions and Hydrologic Applications", Water Resour. Res. 19, 901-921.

Krajewski, W.F. (1987) "Cokriging Radar-rainfall and Rain Gage Data", J. Geoph. Res. 92, 9571-9580.

Leenaers, H., Okx, J.P. and Burrough, P.A. (1989) "Co-kriging: an accurate and inexpensive means of mapping flood plain soil pollution by using elevation data", in M. Armstrong (ed.), Geostatistics, Reidel, Dordrecht.

Matheron, G. (1973) "The intrinsic random functions and their applications", Adv. Appl. Prob. 5, 439-468.

Myers, D.E. (1982) "Matrix Formulation of Co-kriging", Math. Geol. 14, 249-257.

Myers, D.E. (1988) "Multivariable geostatistical analysis for environmental monitoring", Sci. de la Terre 27, 411-427.

Stein, A. and Corsten, L.C.A. (1991) "Universal Kriging and Cokriging as a Regression Procedure", Biometrics 47, 575-587.

Stein, A., Staritsky, I.G., Bouma, J., Van Eijnsbergen, A.C., and Bregt, A.K. (1991a) "Simulation of moisture deficits and interpolation by universal cokriging", Water Resources Research 27, 1963-1973.

Stein, A., Van Eijnsbergen, A.C., and Barendregt, L.G. (1991b) "Cokriging non-stationary data", Mathematical Geology 23, 703-719.

Wackernagel, H., Petitgas, P. and Touffait, Y. (1989). "Overview of methods for co-regionalization analysis". In: Geostatistics, Reidel, Dordrecht, 409-421.

APPLICATION OF CORRESPONDENCE ANALYSIS AND FACTORIAL KRIGING ANALYSIS: A CASE STUDY ON GEOCHEMICAL EXPLORATION

Jiménez-Espinosa, R[1], Sousa, A.J.[2] and Chica-Olmo, M.[1]

1)Dpt.Geodinámica-IAGM, Unv.Granada-CSIC. Av.Fuentenueva s/n. 18071 Granada, Spain.
2)Centro Valorizaçao Recursos Minerais, IST, Av.Rovisco Pais, 1096 Lisboa CODEX, Portugal.

ABSTRACT

The main objective of this paper is to present the results of joint application of Correspondence Analysis and Factorial Kriging Analysis in order to set up geochemical patterns and anomalies.

This study was carried out on 602 soil samples taken from a region of northwest Spain, where thirteen chemical elements were analyzed.

2 Correspondence Analysis procedure was applied to classes of variables, in order to identify and explain patterns of the elements, seeking for synthetic indices (factors) related to particular associations of variables.

The first factor represents an important association of elements linked to the N-S main lithological trend which is, probably, related to the mineralizations in this area. Experimental and theoretical variograms of this factor were calculated , afterwards Factorial Kriging Analysis was used to split the 1st factor into different spatial scales.

INTRODUCTION

Factorial Kriging Analysis is a geostatistical tool which has been used in geochemical exploration for filtering purposes (Sandjivy, 1984). This technique allows to estimate spatial components corresponding to different scales of variability, revealed by simple and cross variogram functions.

Nevertheless, as is often in the case of geochemical exploration, raw data can not be used directly to derive a numerical characterization of geochemical behavior. This is, mainly, due to geochemical variables strongly depend on accuracy of measurement performed in the field, sampling errors and presence of disturbing outliers. Moreover, variable values are often below detection limits and some of them are artificially trimmed, due to the lack of precision of measurement techniques. Also, each variable considered separately could not respond to a real geochemical anomaly; multivariate analysis is thus required.

For the given reasons, Correspondence Analysis (Benzécri, 1973; Greenacre, 1984) was chosen for summarizing geochemical information in quantitative indices based on categorical variables. Resultant factors do not take into account the cross spatial variability of variables, they can be viewed as non-linear synthesis of essential associations of elements. Then, Structural Analysis and later Factorial Kriging may be used to identify and filter spatial structures of some of these factors in order to estimate local and regional components, drawing main lithological features and identifying spatial patterns and meaningful anomalies.

The methodology proposed here to deal with specific characteristics of geochemical variables (presence of values below detection limits, gross outliers, sampling errors and artificial trimming) contains two main steps:

i) Correspondence Analysis (CA) of categorical variables obtained according to particularities of each element, producing quantitative indices that are geochemically interpretable.

ii) Factorial Kriging Analysis (FKA) applied to the indices resulting from CA in order to decompose these indices in their spatial components.

In the case study presented in this paper, the methodology was used with the aim of establishing geochemical patterns and anomalies in a region of Galicia (NW Spain) where metal sulphide ores mineralizations occur.

DESCRIPTION OF RESEARCHED AREA

Geology and occurrence of mineralizations

The area of study is a 1.2 by 2.8 Kms rectangle located near the Fervenza water reservoir in La Coruña province, NW Spain (Figure 1). Geologically it appears in the northwestern part of the Central Iberian Zone. A series of contacts between two main structural and petrological units, Malpica-Tuy unit (MTU) and Schist Domain of Galicia-Tras os Montes (SDGTM), occur in this area. Also, important amounts of Hercynian Granites (HG) intruding both units outcrop (Gil-Ibarguchi and Ortega-Gironés, 1985). The MTU is a narrow band of basement composed of mica schists, paragneisses and calc-alkaline to peralcaline orthogneisses with minor amounts of eclogites. The SDGTM consists of a thick sequence of metasediments, gneisses and basic rocks. These two units are juxtaposed by an important thrust, being the SDGTM the relative autochthonous and the MTU the alochthonous (Farias-Arquer, 1992). Just in this contact area a N-S ductile shear zone appears, along which cataclastic rocks are found. Later E-W fragile deformations are generalized in the area.

The mineralization is made up, mainly, of metal sulphides ores and quartz. It is disseminated in quartz layers and appears associated with orthogneisses and granites next to the cataclastic rocks, and even in these cataclastic rocks (Figure 1). Along cataclastic rocks and taking advantage of its fractured texture, hydrothermal circulations of metal-rich fluids took place during emplacement of Hercynian Granites and produced this sulphide mineralization (Agudo et al., 1989).

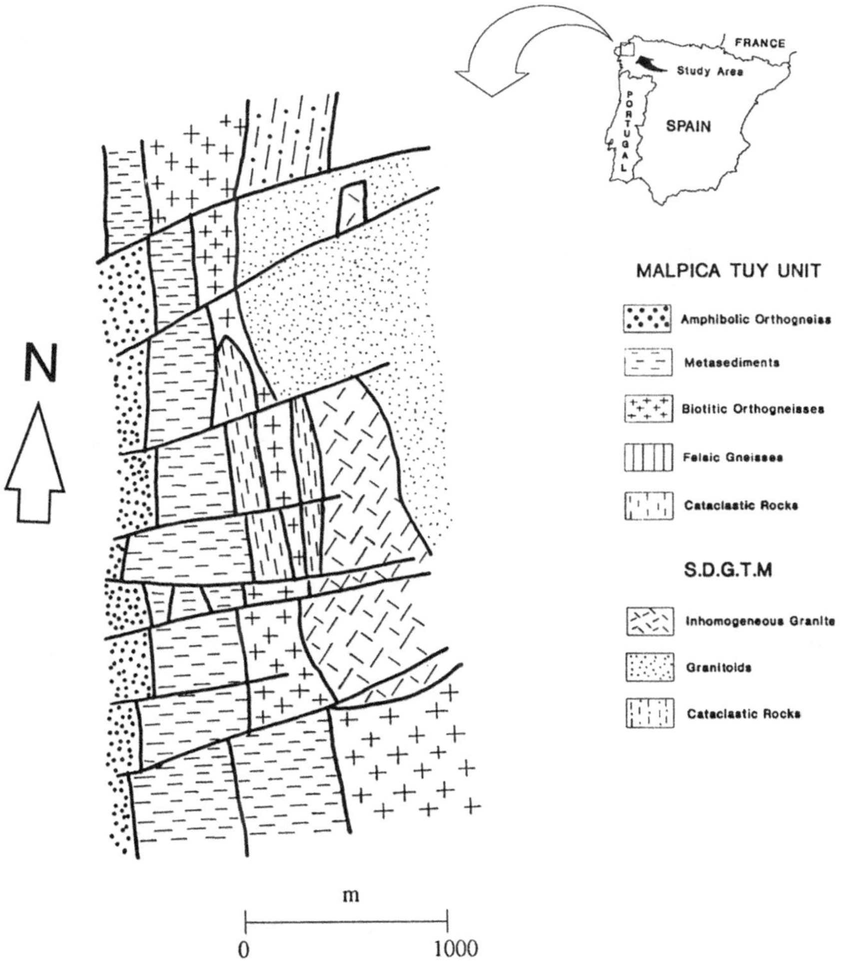

Figure 1.- Geologic map of the study area (adapted from Agudo et al., 1989).

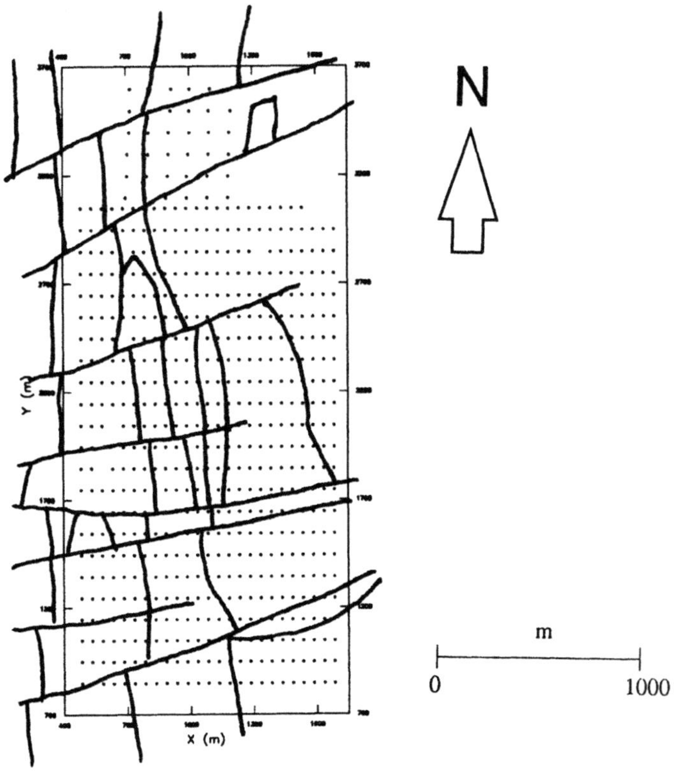

Figure 2.- Map of the area showing samples sites in a regular 50x100 m. grid.

Table 1.- Basic statistics of variables studied.

	Mean	St-Dev.	Min.	Q1	Median	Q3	Max.
As (ppm)	58.5	144.1	5.0	12.0	27.5	58.5	3000.0
Sn (ppm)	4.6	4.4	2.5	2.5	2.5	6.0	72.0
Zn (ppm)	81.3	44.3	14.0	53.0	74.0	102.0	560.0
Pb (ppm)	78.2	95.2	35.0	55.0	67.0	85.0	2300.0
Cd (ppm)	1.9	0.7	0.5	1.0	2.0	2.0	5.0
Co (ppm)	7.2	5.4	2.5	2.5	6.0	10.0	45.0
Ni (ppm)	27.2	15.0	2.5	16.0	26.0	36.0	110.0
Ba (ppm)	399.6	165.4	50.0	303.0	382.0	473.0	1600.0
Fe (%)	5.7	1.8	0.9	4.3	5.7	7.0	14.3
Cr (ppm)	56.2	35.0	5.0	32.0	51.0	73.0	296.0
Be (ppm)	2.9	1.7	0.5	2.0	3.0	3.0	25.0
Cu (ppm)	26.9	20.0	5.0	14.0	22.0	34.0	211.0
Au (ppb)	10.0	20.8	5.0	5.0	5.0	10.0	340.0

Collection of samples and data

Data set used for this study comprises 602 soil samples taken below a well developed argillaceous layer, collected at 50x100 meters grid (Figure 2). In each sample thirteen chemical elements were analyzed (As, Sn, Zn, Pb, Cd, Co, Ni, Ba, Fe, Cr, Be, Cu, Au).

Basic statistics of these elements are shown in Table 1. Some variables, like As, Pb and Ba, present very large maximum values. Over 75% values of Sn and Au are equal or less than detection limits. In the case of Cd, more than 50% of data are equal to 2. Many other variables have a small number of different values.

CORRESPONDENCE ANALYSIS RESULTS

In order to deal with those characteristics of chemical elements, whole variables were categorized. For each variable, a three class-histogram was calculated, one class for low values, another one for medium ones and a third class for high values. Therefore, we obtained three data groups per each chemical element in which values have the same weight: a set grouping the data that could be extreme values (sampling errors, outliers or true anomalous values); another set joining values below detection limit and those truly low; and a last data group having medium values. All sample values were recoded with three indicators according their class. Thus, the original 602x13 raw data matrix was then transformed in a 602x39 indicator matrix.

CA was performed on the 39x39 Burt matrix (Greenacre, 1984) obtained from that indicator matrix. Eigenvalues and percentage of explained variance for the first six axes are given in Table 2. These six factors account for 81.5% of total variance.

Table 2.- Eigenvalues, percentage and cumulated percentage of variance accounted for each axis.

	Eigenvalues	Explained Variance (%)	Cumulated Explained Variance (%)
F1	0.1484	51.81	51.81
F2	0.0396	13.83	65.62
F3	0.0186	6.52	72.17
F4	0.0119	4.17	76.34
F5	0.0076	2.68	79.02
F6	0.0072	2.51	81.53

An important group of elements, associated with mineralization of metal sulphide ores (Zn, Co, Ni, Cr, Cu, Fe), is strongly related with the first axis. It opposes on the first axis high classes (positive loadings) to low ones (negative loadings). Second axis separates the extreme values of these variables from intermediate ones (negative coordinates on F2). The projections of the variables on the plane defined by axes 1 and 2 (Figure 3) show a "horseshoe" effect (Greenacre, 1984) corresponding to a non-linear relation between these two axes. Accordingly, the first axis alone can provide a correct ordering of the variables strongly related with it.

Supplementary projections of the samples (Greenacre, 1984) on the principal axes produce one index (as a new variable), *I1*, which summarizes an important association of chemical elements (Zn, Co, Ni, Cr, Cu, Fe). These elements have interesting implications in the metal sulphide ore mineralization.

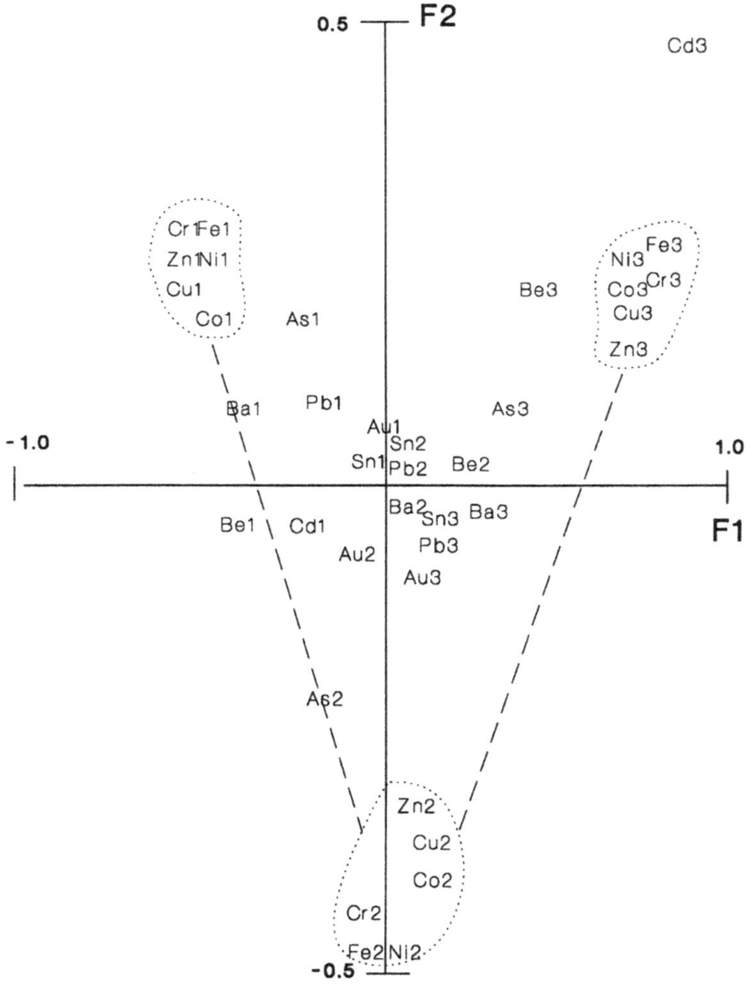

Figure 3.- Graphical display of the variables on the first vectorial plane. Each variable is represented by a code composed by the chemical symbol and the number 1, 2 or 3 corresponding to low, intermediate and high values, respectively.

STRUCTURAL ANALYSIS OF *I1*

To account for spatial structure, variograms of *I1* were computed in the main directions of grid sampling (Figure 4). As previously noted, the geological control of the mineralization yields a N-S trend along the band of cataclastic rocks, giving as a result a more continuous N-S variogram. This spatial pattern, with an anisotropy ellipse N-S clearly oriented, is perfectly in agreement with those of individual elements correlated with the factor 1 (see Jiménez-Espinosa, 1991).

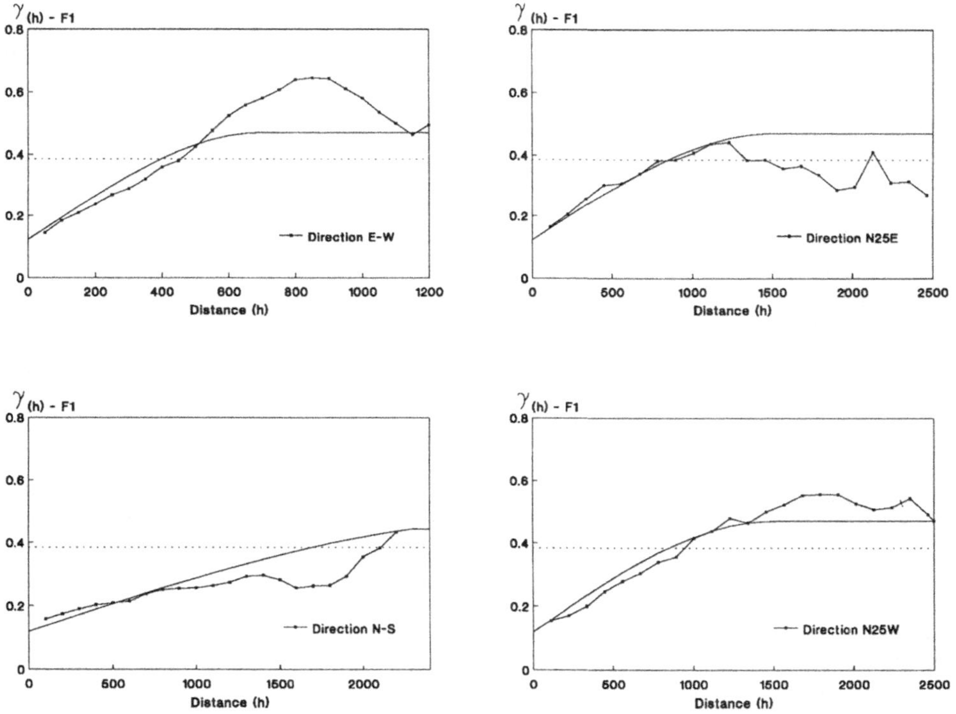

Figure 4.- Experimental and fitted variograms of index *I1*, in the main grid directions.

Table 3.- Fit Parameters: nugget, sill and ranges for index *I1*.

	C0	C1	A1 E-W	A1 N25E	A1 N-S	A1 N25W
I1	0.12	0.35	700	1500	3000	1500

A model composed of a nugget effect of around 25% of *a priori* variance and an anisotropic spherical structure was fitted. The model parameters are indicated in Table 3.

FACTORIAL KRIGING ANALYSIS OF INDEX *I1*

The previous structural analysis carried out led to consider the spatial variability of *I1* index as the sum of two elemental schemes, C0 and C1. These are the basis for estimating the corresponding spatial components (Y0 and Y1) for *I1* using univariate Factorial Kriging Analysis. Each spatial variation scale would be related to a fitted structure in the experimental variograms. They consist of a small scale component, Y0, represented by the nugget effect (C0) and another anisotropic one, Y1, representing the regional structure of the phenomenon. Sandjivy (1984) suggests that anomalies are associated with small-scale structures, in this case anomalous samples would be considered to be strongly correlated with the C0 component, and the first component to be related to the background.

Maps of raw data, Y0 and Y1 were made for variable *I1* (Figures 5, 6 and 7, respectively). These maps were made by means of Ordinary Kriging and using the results obtained from FKA. Analysis of the raw data revealed an area with high values in the western zone along lithological N-S trend. With this method we hoped to establish which of these high values samples belong to the background and which can be considered as anomalies. Thus, it carries out a filtering of the raw values of variable *I1*.

Results obtained by applying Factorial Kriging Analysis show that the diffused distribution of values in the raw data map is not preserved in the great scale component map (Y1). Thus, the values belonging to the geochemical background become defined in Y1 map. This effect can be clearly seen by comparing the corresponding figures for the *I1* raw data and the Y1 component (Figures 5 and 7). To determine which samples with high values are truly anomalies, it is enough to analyze the Y0 component map and to look for high correlated points to Y0 component, obviously, those points with elevated Y0 values could be considered anomalous.

In this way and viewing *I1* raw map, it is worth pointing out the existence of a N-S zone in the western part with concentration of high values. This area was characterized by the circulation of rich-metal fluids along the cataclastic rocks. This phenomenon was conditioned by the generalized N-S thrusting in this area, giving its characteristic trend. Apart from this, an intense E-W later hercynian faulting confuses the general trend.

There are no mineralizations in the eastern zone, occupied by a type of non-homogeneous granites, giving a non favorable area for metal sulphide precipitation.

In the Y1 map we could see how the general trend appears delimited, characterizing the geochemical background. The map of Y0 shows clearly a anomalous zone belonging to the richest area in the central region where the cataclastic rocks and surrounding materials appears. Also in the northern area where homogeneous granites outcrop it is possible distinguish another anomalous area. These could be interesting areas for future exploitation of related elements to index *I1*.

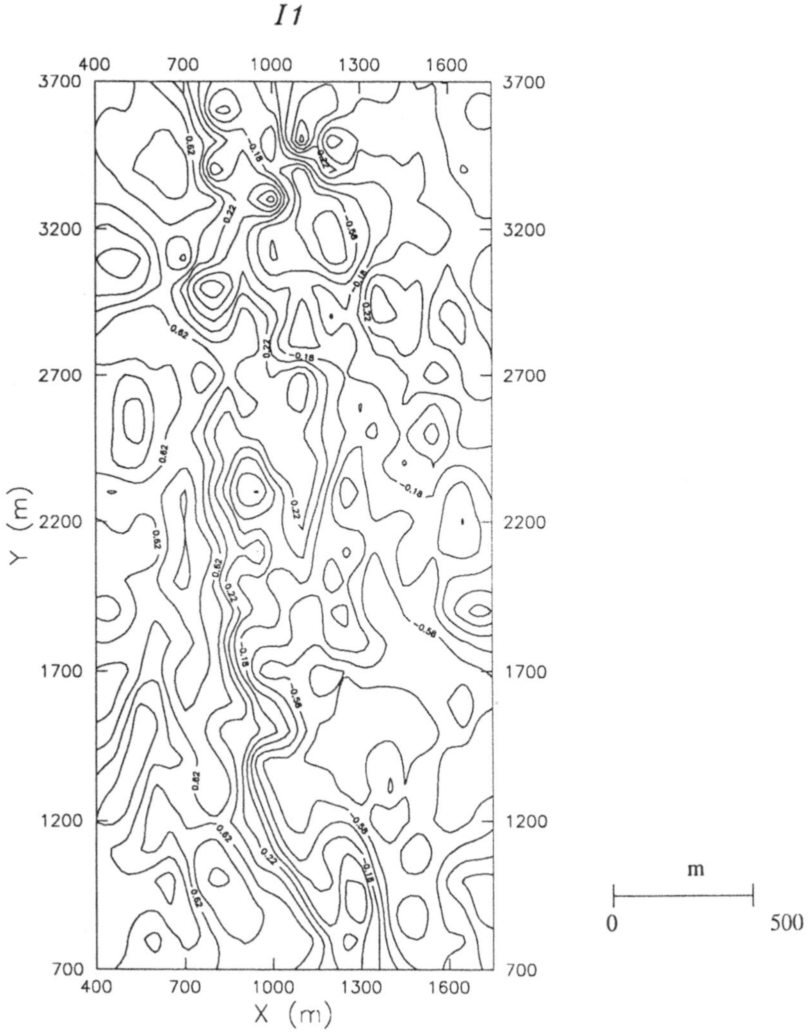

Figure 5.- Map of *I1* raw data.

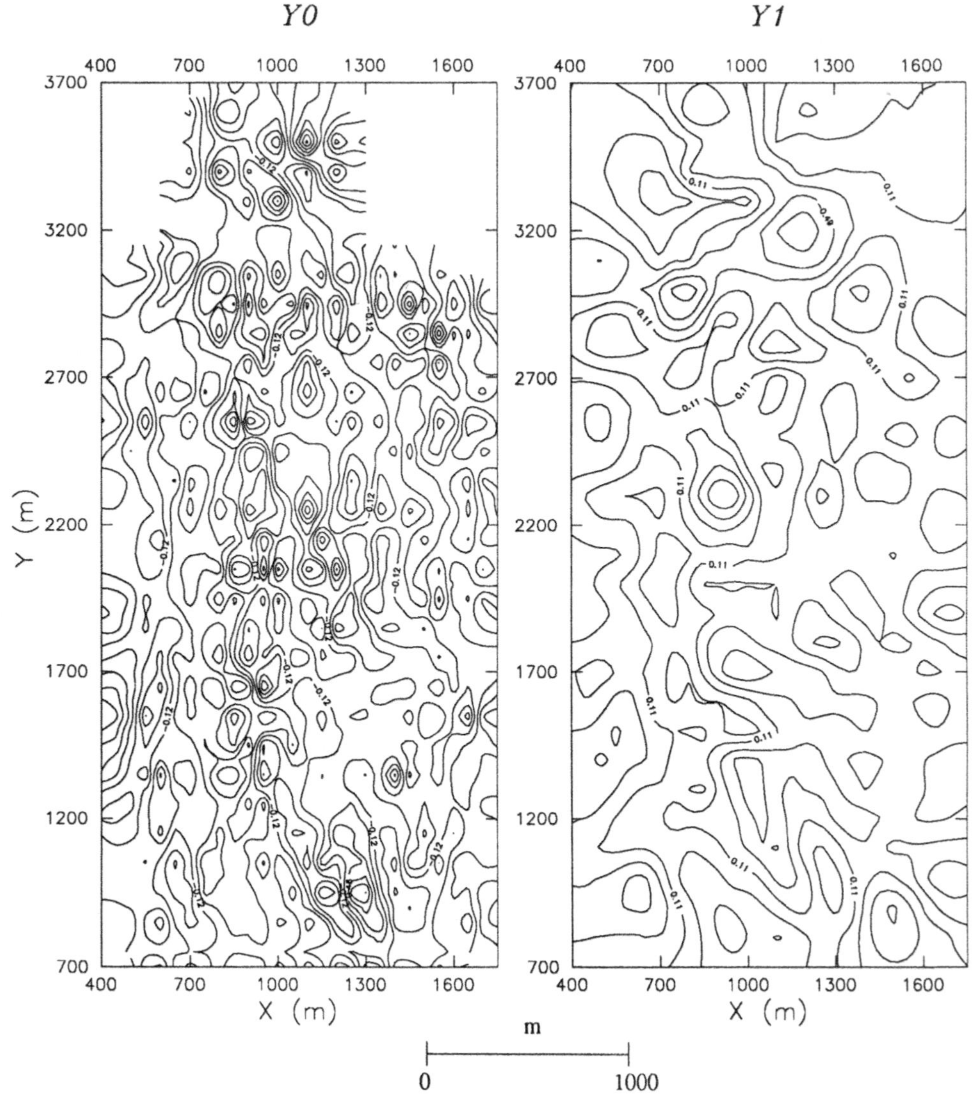

Figure 6.- Map of component Y0 from Factorial Kriging Analysis of *I1*.

Figure 7.- Map of component Y1 from Factorial Kriging Analysis of *I1*.

CONCLUSIONS

The case presented in this paper suggests a methodology suitable for the purpose of identifying geochemical patterns using categorical data. We applied it to a geochemical data set taken from a region of northwest Spain.

Correspondence Analysis was used to establish one index, *I1*, that gave results about an association of elements (Zn, Co, Ni, Cr, Cu, Fe) which were highly correlated among themselves and furthermore accounted strongly for factor 1 (F1).

The methodology used in this paper was based on the treatment of an association of elements, which characterizes metal sulphide mineralization, by means of a geostatistical approach of the index *I1*. Thus, Factorial kriging Analysis was applied to obtain a decomposition of the new regionalized variable, *I1*, and obtaining geochemical patterns and anomalies of this index but representative of the whole association.

Results of FKA showed how it was possible to split up raw variable to obtain two different maps: Y0 map representative of anomalous samples, located around cataclastic rocks, and Y1 map for the geochemical background.

The methodology proves to be as useful to treat the specific case of geochemical variables. Also, it become simpler than multivariate method of Factorial Kriging based on the whole spatial cross correlations between variables.

ACKNOWLEDGMENTS

A grateful acknowledgment is made to ENADIMSA (Spain) for providing of data. We would like to thank referees for their valuable comments and suggestions. This study was supported by the *Dirección General de Investigación Científica y Técnica* of Spain under the grant PB89-0015 and *Consejería de Educación y Ciencia* of *Junta de Andalucía* (Research Group 4020). We also thank the Research Group of *Centro de Valorizaçao de Recursos Minerais (Planeamiento Mineiro)* of *Instituto Superior Técnico* (University of Lisbon).

BIBLIOGRAPHY

Agudo-Fernández, M.L., Hernán-Reguera, P. and Maldonado-Beltrán, M. (1989) "Prospección Geoquímica de Detalle de Mineralizaciones Auríferas Ligada a Cizalla Dúctil. Galicia NW España", III Congreso de Geoquímica de España, 1, III 33-50.

Benzécri, J.P. (1973) L'Analyse de Données. Tome 1: La Taxonomie. Tome 2: L'Analyse des Correspondences. Dunod, 320 pp.

Farias-Arquer, P. (1992) "El Paleozoico Inferior de la Zona de Galicia-Tras os Montes (Cordillera Herciniana, NW de España)". In: Gutiérrez-Marco, J.G. et al. (Eds), Paleozoico Inferior de Ibero-América. Universidad de Extremadura.

Gil-Ibarguchi, J.L. and Ortega-Gironés, E. (1985) "Petrology, Structure and Geotectonic Implications of Glaucophane-bearing Eclogites and Related Rocks from the Malpica-Tuy (MT) Unit, Galicia, Northwest Spain", Chemical Geology, 50, 145-162.

Greenacre, M. (1984) Theory and Applications of Correspondence Analysis, Academic Press, 364 pp.

Jiménez-Espinosa, R. (1991) "Estudio Geoestadístico de Datos de Prospección Geoquímica" Tesis de Licenciatura. Universidad de Granada, 202 pp.

Sandjivy, L. (1984) "The Factorial Kriging Analysis of Regionalized Data. Its Application to Geochemical Prospecting" In G. Verly et al. (Eds), Geostatistics for Natural Resources Characterization, Reidel, pp. 559-572.

ASSESSING SIMILARITIES AND DIFFERENCES AMONG MAPS: A STUDY OF TEMPORAL CHANGES IN DISTRIBUTION OF NORTHERN SHRIMP (*PANDALUS BOREALIS*) IN THE GULF OF ST.LAWRENCE

YVAN SIMARD
Institut Maurice-Lamontagne
Ministère des Pêches et Océans
C.P. 1000, Mont-Joli,
Québec, Canada G5H 3Z4

DENIS MARCOTTE
Département de Génie Minéral
Ecole Polytechnique
C.P. 6079, Succ. A, Montréal,
Québec, Canada H3C-3A7

ABSTRACT.

A set of shrimp biomass random samples collected yearly in the Gulf of St.Lawrence is analyzed to evidence changes in spatial organization during the 1980s. A gaussian transformation, weighted for the Voronoi polygon of each sample, is first applied to standardize the data for interannual comparisons. Cross-validation is used to optimize the kriging parameters, which resulted in the choice of a global stationary model for mapping the biomass. Two dissimilarity indexes, taking the map error into account, are developed to compute interannual or global dissimilarity maps and to measure the total similarity between years. The similarity matrix is the object of cluster analysis to classify the maps by resemblance. Results are interpreted in terms of variability of shrimp spatial organization.

1. Introduction

The northern shrimp of the Gulf of St.Lawrence (Fig. 1) is an important resource exploited by a fleet of trawlers on eastern Canada (Savard and Hurtubise 1991). The shrimp live for 5-6 years (Simard et al. 1990). The recruitment of the population is annual and synchronized by larval hatching during spring (Ouellet et al. 1990). Fishing is selective for adult shrimp, older than 3 years. These latter are found on the deep grounds of the Gulf, mainly from 150 to 300 m. Even though northern shrimp is rather sedentary compared to fish which undergo large-scale migrations, there is a slow meso-scale migration of the shrimp to deeper grounds throughout their life cycle (Simard and Savard 1990). Biomass at a given location is however mostly determined by a balance of recruitment and import versus mortality and export. Therefore, there is a high probability of interannual spatial variation, according to fluctuations in (1) recruitment strength, (2) mortality and (3) the environment, which controls many aspects of population dynamics and distribution. Understanding these spatial changes is desirable to reveal the fundamental mechanisms driving the shrimp ecology in the Gulf and, ultimately, to precisely predict its spatial distribution and global abundance for the exploitation and management of the fishery. Spatial variations also strongly interfere with sampling strategies, generating uncontrolled variability and hindering optimal

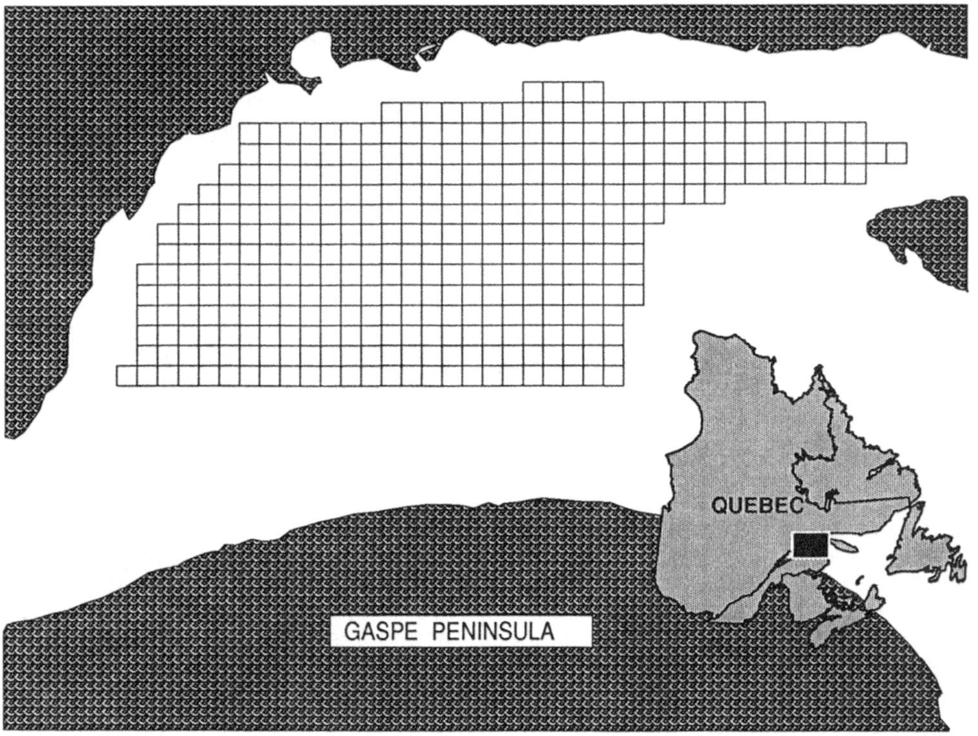

Figure 1. Map of the north-western Gulf of St.Lawrence showing the study area and the kriging grid used (5 x 5 km).

allocation of the samples with stratified plans for routine global biomass measurements (Cochran 1977).

Efficient methods for comparisons of distribution maps through time in numerical ecology are, however, deficient. There is a general need for such methods, for different purposes, including the measure of significant changes which occur in maps of abundance obtained for different times, often with different sample numbers and locations, or for measuring the correlation of abundance maps with maps of environmental variables. The objective of this paper is to exploit geostatistics in the development of such methods for maps comparisons. The aim is studying the interannual spatial variability of the shrimp biomass and objectively compare its annual maps, when the number of samples and their locations vary from year to year and when the global biomass also fluctuates yearly. Questions asked were: (1) how the relative spatial distribution of shrimp changes? (2) is shrimp randomly located in the studied region or show consistent patterns? (3) considering the local error of annual maps of biomass, how differ the maps locally and globally? (4) can zones of high variability and more stable zones be identified?

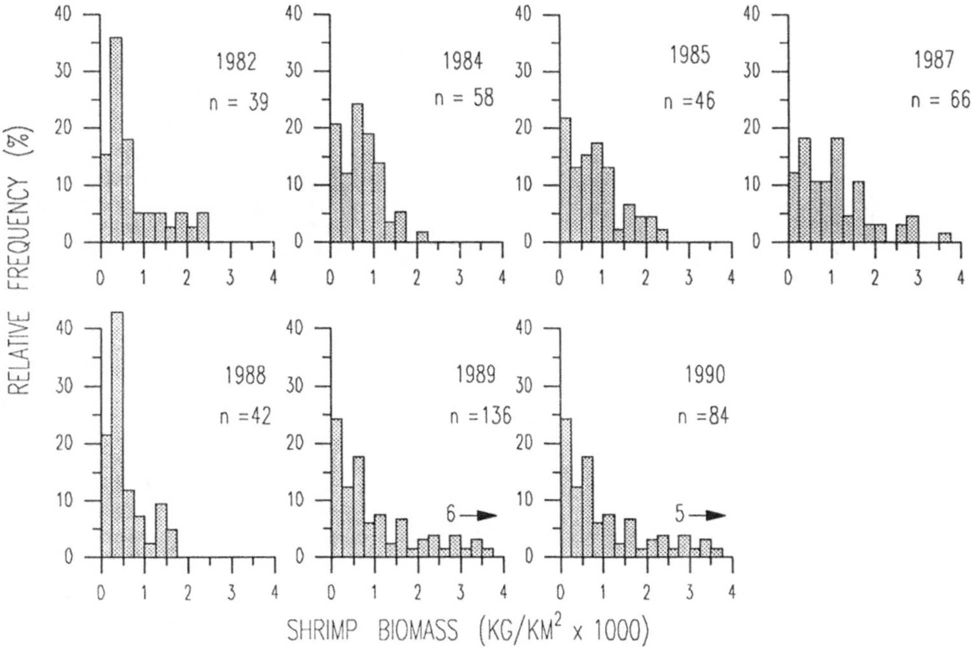

Figure 2. Annual histograms of the shrimp biomass samples.

2. Methods

Every year from 1982 to 1990, with the exception of 1983 and 1986, northern shrimp biomass (kg/km^2) was sampled with bottom trawls at 39-136 stations in north-western Gulf of St.Lawrence (Fig. 1), from August to October (details in Simard 1991). The mean biomass showed strong interannual fluctuations and its distribution was highly skewed (Fig. 2).

To allow interannual comparisons, the biomass data of each year were submitted to a weighted gaussian transformation. A cumulative probability curve was computed from the samples, weighted for their respective Voronoi polygon of influence, and was used to assign the gaussian value N(0,1) to each sample. The final result of this transformation is a gaussian value that is compensated for the respective influence of each sample in the study area. This way, two random surveys done at the same time would give about the same map of gaussian values, which would not be the case for a simple gaussian transformation. Since locations of samples varied from year to year, this weighting procedure makes uniform the gaussian transformation from year to year.

The omnidirectional variograms of the transformed data were computed separately for each year and globally for all years, the sill being held fixed at the expected value of 1.0. The estimation performance of the kriging method (neighbourhood size, ordinary or simple kriging, annual or global variograms) was explored with

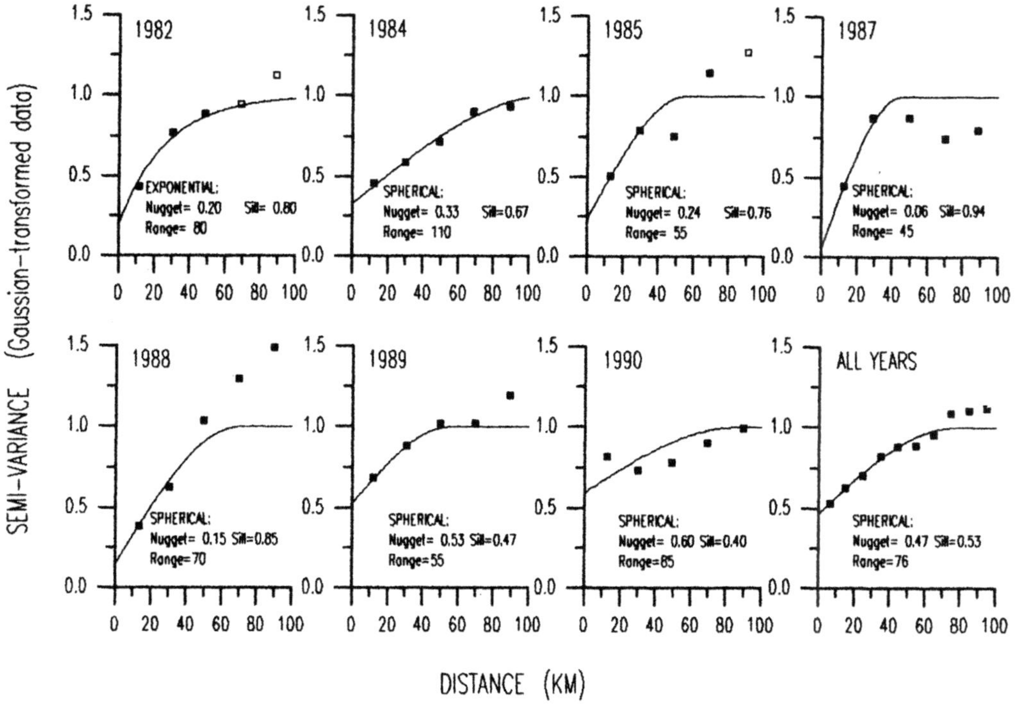

Figure 3. Omnidirectional variograms of the gaussian-transformed shrimp biomass computed separately for each year and globally for all years combined. All points include more than 50 pairs, except empty symbols (25-50 pairs).

cross-validation. The optimal method found was then used to compute the annual maps of the shrimp biomass. To compare the maps, the following relative dissimilarity index taking the map error into account was developed to compute a dissimilarity map.

$$d_{0_{(t_1,t_2)}} = \frac{\frac{1}{n_0} \sum_{i=1}^{n_0} \left| z^*_{(i,t_1)} - z^*_{(i,t_2)} \right| \cdot w_{(i,t_1)} \cdot w_{(i,t_2)}}{R_{max}} \qquad (1)$$

where:

t_1 *and* t_2 = *years* t_1 *and* t_2

$z_{(i,t)}$ = *kriging estimate at the grid point i for time t*

n_0 = *number of points in the neighbourhood*

$w_{(i,t)} = 1 + \sigma_{k_{(i,t)}}$

R_{max} = max[*numerator*]$_{(for\ all\ years)}$

This index is alike the similarity index proposed by Gower (1971) for multivariate sample comparisons in ecology. It varies from zero, when the local estimates of the maps are identical, to 1.0 when they present the maximum dissimilarity observed in the set of maps considered. The total dissimilarity between two maps can be obtained in the following two ways: (1) by taking the average of all local d_0 of the maps or (2) by using the above formula and replacing n_0 by N_k, the total number of kriged estimates. In this paper, we have used the first of these two alternatives. The similarity is given by the complement of the index ($S=1-d$).

Similarly, to map the global relative dissimilarity existing between the whole set of maps, taking into account the local error associated with each map, the following index was developed. Alike the other index it varies from 0, when all maps and their error are the same for the local estimate considered, to 1.0, the maximum local dissimilarity for the whole set of maps.

$$d_{0_{(all\ years)}} = \frac{\frac{1}{n_0}\sum_{i=1}^{n_0}\frac{1}{m}\sum_{t=1}^{m}\left|z^*_{(i,t)}-\bar{z}^*_{(i,*)}\right| \cdot w_{(i,t)}}{RG_{max}} \qquad (2)$$

where:

m = *number of years*

n_0 = *number of points in the neighbourhood*

$z_{(i,t)}$ = *kriging estimate at the grid point i for time t*

$$\bar{z}^*_{(i,*)} = \frac{\sum_{t=1}^{m} z^*_{(i,t)} \cdot \frac{1}{w_{(i,t)}}}{\sum_{t=1}^{m}\frac{1}{w_{(i,t)}}}$$

$w_{(i,t)} = 1 + \sigma_{k_{(i,t)}}$

RG_{max} = max[*numerator*]

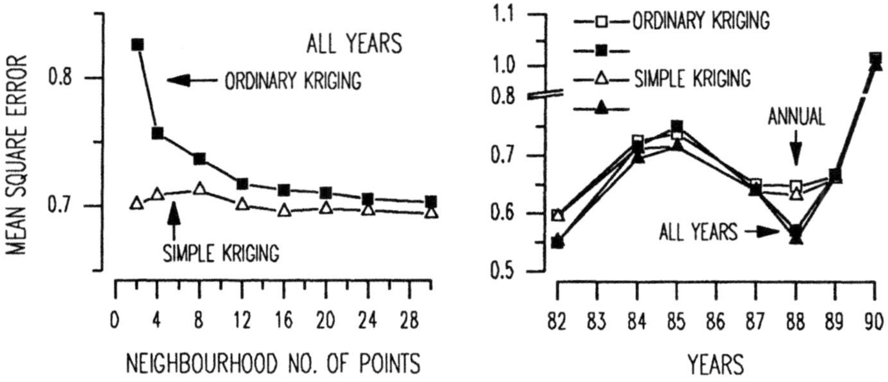

Figure 4. Cross-validations for the kriging method: (left) for the size of neighbourhood using the global variogram, and (right) for simple or ordinary kriging using the annual or global variograms with 24 neighbours.

3. Results and Discussion

Except for 1990, where random variations seem to prevail, all annual variograms and the global variogram were easily modeled (Fig. 3). The global variogram indicated a structural component representing 53% of the total variability, leaving 47% of the variability unresolved. The range of this structural component was about 75 km, indicating the presence of spatial patterns with a scale reaching 150 km. Cross validation for an optimal kriging neighbourhood was performed with the global variogram. For ordinary kriging, the mean square residuals between predicted and observed values decreased rapidly with the size of the neighbourhood and stabilized at 12 neighbours (Fig. 4). In contrast, simple kriging was not sensitive to the size of neighbourhood, as expected because of a greater screen effect. Simple kriging had systematically lower errors than ordinary kriging. This indicated a probable better fit of the stationary model (simple kriging) than the quasi-stationary model (ordinary kriging). Both models converged for large neighbourhood. A neighbourhood of 24 points was chosen for subsequent analyses.

Cross validation for comparing the relative performance of simple and ordinary kriging (Fig. 4) showed that simple kriging had always lower mean square error than ordinary kriging. For simple kriging, the global variogram was always better than the annual variograms. This was also true for ordinary kriging, except 1985. This good performance of the global variogram indicated that the probabilistic spatial model of shrimp biomass was relatively stable among years. The fluctuations of the annual variograms seemed therefore to represent variations around a more general model and may have resulted from the imprecision of the annual variograms, due to their low number of samples. The variance of the

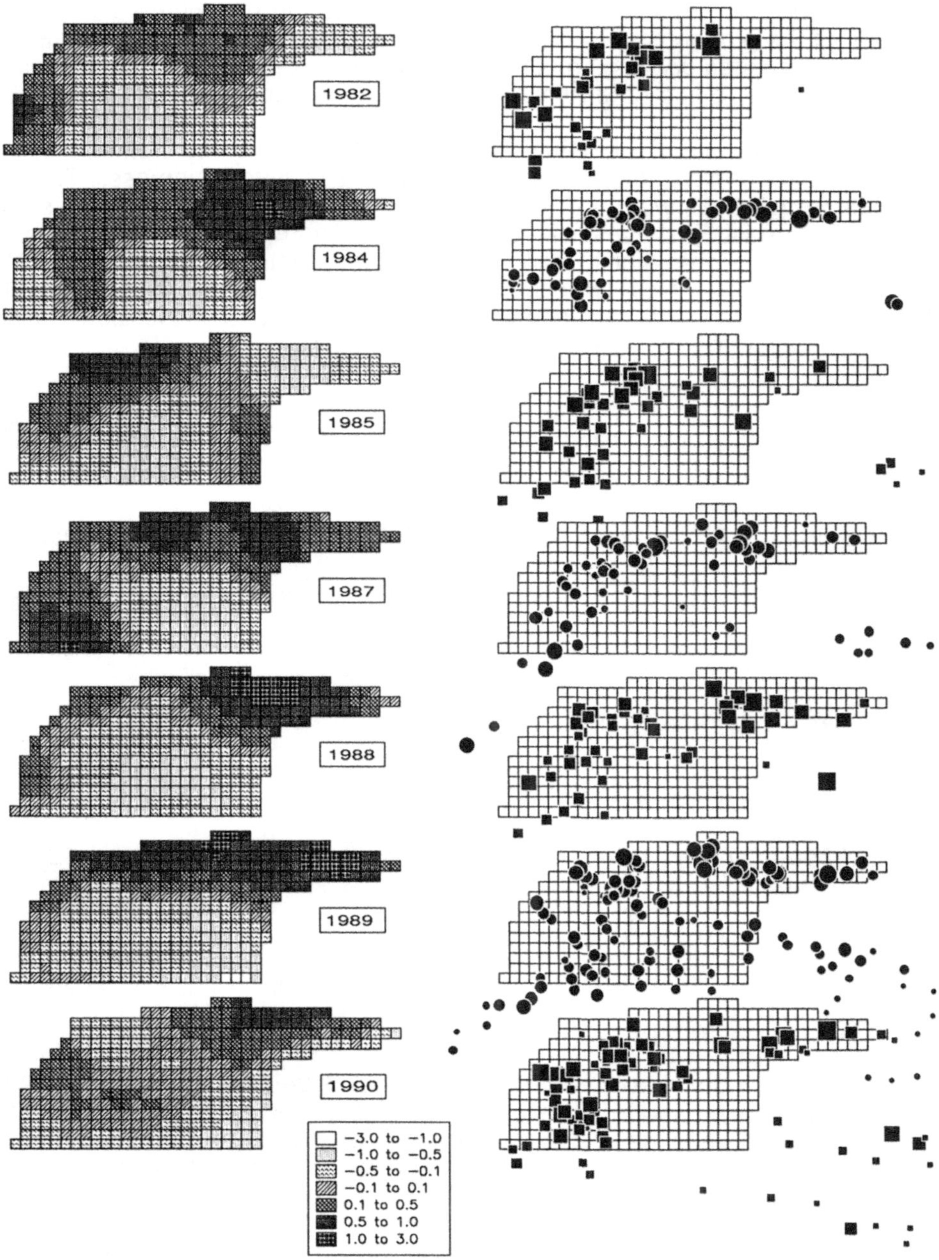

Figure 5. Annual hatch maps of the kriged estimates of the gaussian-transformed shrimp biomass and corresponding maps of samples locations with symbol size proportional to biomass.

standardized residuals was also generally closer to the expected value of 1.0 in the case of the global model (1.10)than in the case of annual models (1.32), for both simple or ordinary kriging. Simple kriging using the global variogram was therefore chosen to krige the annual maps of shrimp biomass.

Kriging on blocks of 5 x 5 km was performed for mapping the biomass (Fig. 5). The south-east portion of the region had systematically low biomass. The zone of maximum biomass was mainly located in the northern portion of the studied region. The east-west position of maximum biomass, however, changed considerably from one year to another, with high variations in the north-eastern end of the region. This later zone had minima in 1982 and 1985. This corresponded to years of poor sampling in this area (Fig. 5). It is therefore possible that these minima partly reflected this sampling bias.

Table 1. Average similarity (*S*, upper half) and Pearson correlation coefficient (lower half) between the kriged maps of shrimp biomass.

Year	1990	1989	1988	1987	1985	1984	1982
1982	0.76	0.73	0.77	0.75	0.81	0.74	1
1984	0.78	0.80	0.82	0.82	0.73	1	.45
1985	0.73	0.70	0.70	0.70	1	.20	.63
1987	0.78	0.81	0.79	1	.20	.64	.51
1988	0.79	0.86	1	.65	.00	.81	.52
1989	0.77	1	.85	.69	.16	.74	.46
1990	1	.59	.62	.47	-.18	.51	.26

Global similarity indexes and correlation coefficients between the maps indicated strong resemblances and differences among years (Table 1). These coefficients cannot be tested because of the spatial dependence of the data, but they are sufficiently high to invalidate the hypothesis of random localization of shrimp biomass in the study region from one year to another. The most similar spatial organizations of the shrimps were 1988 and 1989 and the most different years were 1985 with 1990 or 1988. The local dissimilarity for those years from equation (1) are illustrated in Fig. 6. The similarity index computed between maps gave general results similar to the correlation coefficient, with slight variations though (Fig 7). The resemblance between maps was therefore slightly different when the map error was taken into account, with the similarity index used here,compared to correlation coefficients which ignored the error. This resulted in differences in

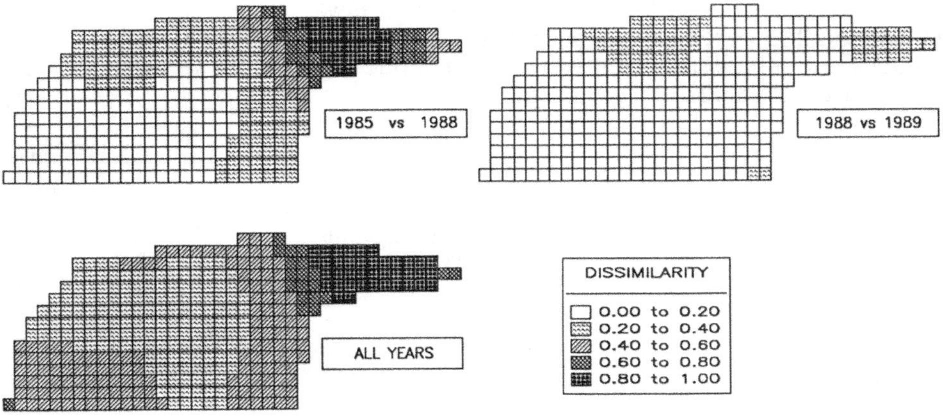

Figure 6. Dissimilarity maps comparing the kriged annual maps of shrimp biomass by pairs or altogether.

classifications of the maps from a cluster analysis performed on the matrices of table 1, using the complete linkage algorithm (c.f. Legendre & Legendre 1983) (Fig. 7). The general result of the classification was the same for both indexes, but the clusters sequences were different. The years 1982 and 1985 were clearly distinct from the other years, mainly due to the low biomass in the north-eastern end of the study region (Figs. 5 & 6). The map of global dissimilarity among all years reflected this latter fact (Fig. 6). It also pointed out that the more stable area was the centre and west of the study region, the rest of the region having intermediate levels of dissimilarity. A principal component analysis of the kriged maps, taking each annual kriged estimate as a variable, confirmed that the centre of the region was quite stable and that the first component of the total variance (65% of variance) was strongly correlated with latitude.

4. Conclusion

The results indicated that the northern shrimp biomass variability in the north-western Gulf of St.Lawrence is about equally represented by meso-scale spatial structures and unresolved small scale variability. A global stationary model was found to perform better than the annual models in estimating the biomass of each year. The spatial organization exhibited important changes from 1982 to 1990, which were discernible despite the facts that the global biomass changed considerably during this period and that the samples were located differently each year. The richest zones were located in the northern and western periphery of the region, varying yearly in location and size, and the south-eastern areas had systematically low biomass. The north-eastern end of the study region is the most variable zone, exhibiting alternation of minima and maxima, but variable density of sampling may have affected the comparisons. In contrast the centre and west of the region was relatively stable. This information could be considered advantageously for sample allocation in stratified sampling plans. The causes of these spatial changes of shrimp biomass must now be investigated carefully, by looking at recruitment, mortality and transport of biomass and at changes of habitat characteristics.

Acknowledgments

We are indebted to Louise Savard and to late Estelle Laberge for the collection of shrimp biomass samples.

References

Cochran, W. (1977) Sampling techniques, John Wiley and Sons, New York

Gower, J.C., (1971) "A general coefficient of similarity and some of its properties",Biometrics 27, 857-871.

Legendre, L. and Legendre, P. (1983) Numerical ecology, Elsevier, Amsterdam.

Ouellet, P., Lefaivre, D., and Koutitonski, V. (1991) "Distribution of shrimp (*Pandalus borealis*) larvae and hydrographic pattern in the northern Gulf of St.Lawrence", Can. J. Fish. Aquat. Sci. 47, 2068-2078.

Savard, L. and Hurtubise, S. (1991) "Evaluation des concentrations de crevettes nordiques (*Pandalus borealis*) du nord du golfe du Saint-Laurent (divisions 4rst de l'OPANO)", Can. Atl. Fish. Adv. Com. Res. Doc. no. 91/71, 1-35.

Simard, Y. (1991) "The northern shrimp biomass in the Sept-Iles region of the Gulf of St.Lawrence during the 1980's: a geostatistical estimation", Can. Atl. Fish. Adv.Com. Res. Doc. no. 91/79, 1-27.

Simard, Y., Brunel P., and Lacelle J. (1990)"Spatio-temporal variability and growth of pre-recruits of the northern shrimp (Pandalus borealis) in the western gulf of St.Lawrence as related to hydrographic conditions" Can. J. Fish. Aquat. Sci. 47, 1526-1533.

Simard, Y. and Savard, L. (1990) "Variability, spatial patterns and scales of similarity in size-frequency distributions of the northern shrimp (*Pandalus borealis*) and its migrations in the Gulf of St.Lawrence", Can. J. Fish. Aquat. Sci. 47, 794-804.

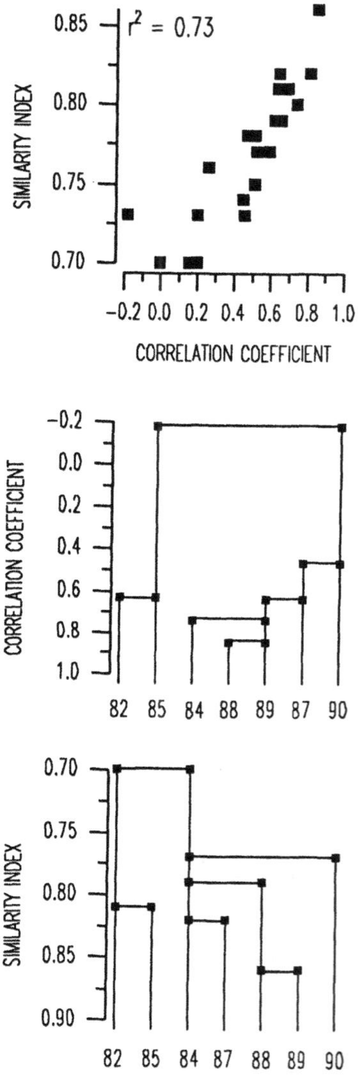

Figure 7. Scatter plot and cluster analysis from matrices of table 1 using complete linkage algorithm.

USING GEOSTATISTICS FOR FOREST INVENTORY WITH AIR COVER: AN EXAMPLE

C. de FOUQUET*, D. MANDALLAZ**
* Ecole des Mines, Centre de Géostatistique, Fontainebleau (France)
** Institut für Wald–und Holzforschung – E.T.H. – Zürich (Suisse)

ABSTRACT

The forest of Zürichberg was sampled in order to determine a method to be used for forest inventory. It was carried out in two ways:
– by an accurate but costly terrestrial survey on regularly spaced plots;
– by an aerial photographic campaign. The photos give, at low costs, large quantities of less accurate qualitative information.
As a control, a small area of the forest was sampled exhaustively.
Two variables are studied : the stem density and the surface area per hectare. Let X denote either one of these two variables.
The comparison between the aerial survey with the ground one on the same plots, allows to define an auxiliary variable Y calculated for each qualitative class, and whose value is equal to the average of X for the class. So fitted this variable Y takes the general sense of a regularised of X, with which it verifies the Cartier relation:

$$E(X|Y) = Y$$

The structural analysis shows that in the decomposition

$$X = Y + R$$

the residual R is without spatial correlation with Y. This is thus a case where cokriging of X and Y is factorized, and is reduced to the kriging of Y and R separately.
On the small test area this estimation is then compared with the exhaustive count.

I – INTRODUCTION

Forest inventory needs efficient estimation techniques combining terrestrial data – accurate but expensive – with qualitative auxiliary information provided by thematic maps based on the interpretation of aerial photographs – available in large quantities and at low costs, but generally only weakly correlated with the terrestrial measurments.
The Zurichberg forest (218 ha) in Switzerland was chosen for preliminary investigation allowing methodological studies before performing inventories over large areas.
Systematic cluster sampling on a 120m x 75m grid was used to sample auxiliary information, that was provided by a stand map based on aerial photos. The cluster (or satellites) consists of 5 points arranged in a cross whose branches are 40m. along Ox and 30m. along Oy. The first phase sample resulted in 298 clusters and 1203 points (the

effective number of points in a cluster varies for the cluster located at the bordure of the forest or near lakes or roads). The terrestrial inventory rests upon 300 m^2 circular plots, each plot corresponding approximatly to a circle with a 10m radius, following the same cluster structure on a 1:4 subgrid, i.e. 240m x 150m (73 clusters, 298 plots); see figure I−1. For validation purpose, a small area of 17 ha has been fully inventoried with determination of the tree coordinates (4784 trees).
The variable of interest are the stem density (number of stem per ha., variable tige) and the basal area density (total surface area of the trees at 1.3m above the ground, in m^2 per ha., variable surface). These quantities are attributed at the center of the terrestrial plots.
The qualitative auxiliary variables are :

– ES development stage; 4 levels from 3 for pole to 6 for old timber stage.
– MG degree of mixture; 2 levels : 1 for dense and 2 for sparse.
– DG crown closure index; 2 levels : 1 for predominantly coniferous, 2 for deciduous.

The aim of the study is to develop an estimation method of the quantitative variables of interest combining the purely qualitative information with the quantitative data of the terrestrial survey.
As the stems density and the basal area are known at the same place, these variables will be considered separately for the purpose of the linear estimation.

II – COREGIONALIZATION MODEL OF THE QUALITATIVE AND QUANTITATIVE VARIABLES

a) Conventional value of the qualitative variables

The numerical value given to the qualitative variables is purely conventional. As these variables indicate belonging to a class, several models can be considered:

– work with indicator families which implies having consistent models of cross covariances both for the indicators associated with the same variable, (for example the indicators of the 4 values of ES), for the indicators associated with the classes of 2 distinct variables, (for example the indicator of the class ES = 3 and that of the class MG = 2), for the indicators and the quantative variables and for the quantative variables. Being sure of the consistency of the set of covariances is not always easy.

– convert the qualitative variables into quantitative variables which makes it possible to use only a limited number of variables and facilitates the fitting of a coregionalization model.

The last solution was retained.

b) Factorization model

In order to prevent the qualitative variables from intervening by their conventional value, they have been treated as class indicators. The cross table of the 2 values of DG and MG and the 4 values of ES determines 16 classes representing the state of the plots.So let X be a quantitative variable. On condition that the variables are stationary on the field, the mean value of X can be associated with each one of the 16 indicator classes $(J_i)_{i=1,16}$ for the plots of the class. Even if it entails regrouping the classes where the mean of X would be identical, each class is then represented by this new variable Y.

More precisely, X and (J_i) being stationary R.F., we determine the following two Random Functions, where i denotes the class number

$$Y=E(X|(J_i)_{i=1,16}) \text{ and } R=X-Y \qquad (I-1)$$

These functions are such that:

$$E(Y) = E\,(E(X|(J_i)_{i=1,16})) = E(X) \text{ thus } E(R) = 0$$

and moreover the residual expectation being zero for each class:

$$E(R|Y) = 0$$

In all the x points, the random variables Y and R are uncorrelated. Assuming the absence of spatial correlation of the R.F. Y and R (zero cross covariance) the decomposition

$$X = Y + R \qquad (I-2)$$

performs the factorization of X. We know that in this model the cokriging of X by Y amounts to the separate kriging of each of the factors Y and R. As the errors of estimation of the factors are uncorrelated, the variance of estimation of X is the sum of the variances of estimation of Y and of R.

From (I−1) we have the well known Cartier relation

$$E(X|Y)=E(X|(J_i)_{i=1,16})=Y \qquad (I-3)$$

The R.F. Y can therefore be interpreted as a regularization of X. We know that for each convex function f

$$E(f(X)) \geq E(f(Y))$$

Here, the regularization takes place on each of the 16 respective indicator classes :

$$Y(x) = \sum_i Y_i \; J_i(x)$$

where

$$Y_i = \frac{1}{p_i} \int X(x)\; J_i(x)\; dx$$

with

$$p_i = \int J_i(x)\; dx$$

c) Checking the model

In the stationary case, the factorization (I−2) will be accepted if the experimental variograms are adjustable by a model such as:

– the covariance of X is the sum of the covariance of Y and that of R.
– the cross covariance of Y and R is zero.
– the cross covariance of X and Y (resp. R) is proportional to the simple covariance of Y (resp. R).

d) Applying the model to the Zurichberg data

On the 298 terrestrial plots, only 14 classes are represented. After checking the stationarity, we will calculate Y for the classes that are large enough (by simple arithmetic mean; we could equally carry out the kriging of the mean value on the area where the class points meet). In each point we then have the residual

$$R(x) = X(x) - Y(x).$$

Y therefore takes as many values as there are classes.

The qualitative data from the aerial survey are then converted by allocating to each point x of this grid the value of Y which fits the class of the cross table of ES, DG and MG. At the points where only the aerial survey was carried out, the residual R is unknown.
The experimental variograms of X,Y and R as well as the cross variograms are calculated on the ground data. A model of the variogram proving the properties of the §c is then adjusted. We shall check the similarity of the experimental variogram of Y, calculated with the ground data and with that calculated with the aerial data.
Kriging in any point is carried out in two stages:
– the kriging of Y, carried out from the dense grid of the aerial survey
– the kriging of R, carried out from the looser grid of the terrestrial plots.
At a data point x where Y only is known, X is estimated by adding to Y(x), the kriging of the residual of R in this point. In the model, the expectation of R being zero, we could carry out a kriging without universality conditions. In a point far from the terrestrial plots, X will fit Y (known by conversion of the qualitative variables, or estimated). In practice, in order to reduce the importance of assuming stationarity, the two krigings were carried out with universality conditions.

e) The question of empty classes

What should be done when there are few plots in the class or when the class is absent from the terrestrial survey and present on the aerial survey ?
One solution consists in regrouping the small classes with other classes, taking into account analogies (e.g. regrouping classes of neighboring stages of development, having the same coverage degree and the same mixture degree). If one of the classes is just very poorly represented, but is quite different from the neighboring class on the ground, we risk making errors which could be locally important. The safest way would be not to carry out an estimation of these classes.
Another solution consists in "extrapolating" the conversion of the qualitative variables of Y. Besides, a linear regression of X on ES, DG and MG provided a "predicted" variable pX. For the larger classes, pX is very close to Y (see §III). For the classes which are absent from the terrestrial survey and present on the aerial survey, we could put Y(x)=pX(x). However it would be better to validate the pX variable by analogy with other sites.

III VARIOGRAPHIC STUDY

a) Mean by class and residuals

Table III–1 summarizes the statistics by class. The variance of the residuals R is high (figure III–2). The coefficient of correlation of the residual R and of the variable X is equal to the ratio of their standard deviation; indeed

$$\varrho = \frac{\mathrm{Cov}\,(X,R)}{\sigma_X\sigma_R} = \frac{E(\,XR\,)}{\sigma_X\sigma_R} = \frac{E(\,R^2) + E(YR)}{\sigma_X\sigma_R} = \frac{\sigma_R}{\sigma_X}$$

The result is analogous for the X and Y variables.
For the variable X and its residual R, the numeric values are as follows:

– tige : $\varrho_{XR} = 0.67$ $\varrho^2_{XR} = 0.45$

– surface : $\varrho_{XR} = 0.88$ $\varrho^2_{XR} = 0.78$

The scatter diagram of the mean per class of the variables "tige" and "surface" gives a synthetic image of the different classes (figure III−2): for older stands the number of stems decreases and the basal area increases, except for the oldest stand (ES=6) where the surface decreases a little. The density of the stand plays an important role for deciduous trees and a lesser one for conifers. It becomes without importance for old stands with a majority of conifers (ES=5).

b) Variance analysis model : "predicted" values

As all the 16 classes were present on the aerial survey data, the decomposition I−2 is modified according to an analogous model

$$X=Y'+R' \qquad (III-1)$$

where Y' is the result of a variance analysis model with 6 parameters− general mean, 3 degrees of freedom for the development stage, 1 for the density, 1 for the mixture degree; the interactions are not significant. In our case, this model is numerically equivalent to the decomposition I−2, which would correspond to a model with complete stratification with 14 parameters, one for each class (figure III−3). The coefficients of correlation of the variables X and Y (respectively X and R) and X and Y' (respectively X and R') are very close:

− tige : coefficient of correlation	$\varrho_{XR} = 0.67$;	$\varrho_{XR'} = 0.69$
coefficient of determination :	$1 - \varrho^2_{XR} = 0.55$;	$1 - \varrho^2_{XR'} = 0.53$
− surface : coefficient of correlation :	$\varrho_{XR} = 0.88$;	$\varrho_{XR'} = 0.90$
coefficient of determination :	$1 - \varrho^2_{XR} = 0.22$;	$1 - \varrho^2_{XR'} = 0.19$

which corresponds to the following variances:

− tige : mean per class	Var X = 24229.;	Var X' = 23203.5
residuals	Var R = 19796.;	Var R' = 20822.
− surface : mean per class	Var X = 36.5 ;	Var X' = 31.2
residuals	Var R = 130. ;	Var R' = 135.

The approximation III−1 is more precise for the variable tige than for the variable surface.

c) Variography

The experimental variograms which were calculated for the decomposition III−1 with the terrestrial count data show that the residual model is adequate for the variable tige; it seems more approximative for the basal area (Figure III−4). This well known finding can be due to the fact that aeral photographs allow for a good estimation of the number of trees (what are easy to count), but not for their basal aera and timber volume, because these are not visible.

The variographic model fitted for the kriging is an anisotropic model for the mean values per class (experimental variogram calculated on all the plots) and isotropic for the residuals (experimental variogram calculated on the terrestrial plots − Figure III−5).

IV − COMPARING DOUBLE KRIGING WITH THE COMPREHENSIVE COUNT

For the kriging the forest was divided into 8 equal sectors. The neighborhood used covers all the plots inside the sector for the "predictions" and the 298 terrestrial plots for

the residuals (unique neighborhood).The numerical integrations were performed with the same discretization grid of 25mX25m .The kriging variance for the forest is obtained by combining the kriging variances of the 8 sectors.
The results are compared with other estimators commonly used:
– "classical estimator", used in nearly all forest inventories (W.G. Cochran, 1977;D. Mandallaz, 1991); this design–based regression estimate assumes that the centers of the clusters were drawn independently and uniformly in the forest, whereas they are actually on a randomly chosen systematic grid.
– kriging with measurement errors; in this well known technique (G. Matheron, 1970) the average of square residual calculated on the data points determines the variance of the additionnal nugget effect between these points.
For the kriging of the predictions on the small domain a unique neighborhood with 192 plots (out of which 92 inside) was used. As for the 8 sectors, the residuals are kriged with the data of the 298 terrestrial plots, 19 being inside the small domain. In order to make a precise comparison, this small domain is divided into squares of 0.25 ha (50m x 50m), for which the real values are known. For the squares located on the boundary, the kriging value is converted according to the intersection of the square with the small domain.
For comparison, the residuals were also kriged with a moving neighborhood on the small domain; this neighborhood always gives poorer results, because on this area the 19 terrestrial plots take significantly lower values. As expected the previous neighborhood using plots located inside and out of the area is preferable.
Table IV–I shows the results and in particular the comparison (made on the small domain) of the kriging carried out according to the decomposition III–I (double kriging) with the actual value. Among the estimators studied, this factorization provides the best results, not only for the estimation value – small surfaces included –, but also for the reliability of the variance. Figure IV–I shows that the empirical 95% confidence interval based on the kriging standard deviation is reliable, either for the estimation of 1 ha or 0.25 ha squares.
Table IV–I shows that the quality of the estimation remains, even when the factorization III–I is not perfectly suited, as is the case for the surface.
The robustness of the kriging estimator remains if the terrestrial inventory is reduced to 1:4 . The study was made again – regression, variography and kriging – using only 75 terrestrial plots instead of 298, while retaining the 1203 aeral plots ; the range of the variogram of residual is taken as the same as before whereas the sill changes. On the small area, the estimated values vary little (Table IV–2), whereas the "classical estimator" used for forest inventories diverge.
The reliability of the double kriging, even with a relatively small terrestrial sample allows to use it for global and local inventory at reasonable costs.

V – CONCLUSION

The factorization proposed makes it possible to use the information brought by the qualitative variables, avoiding the artefacts which result from a numerical coding of these variables. The validity of the factorization can be easily checked by considering the cross variograms of the variable and its factors. For this application the question of the empty classes was easily solved because of the analogy between the factorization and a model resulting from an analysis of the variance; this model could provide the numerical values for the unsampled classes. Unlike this model, the factorization can be applied in cases where the relations between variables are clearly not linear.

This method – regression on the classes and factorization of the qualitative variable according the regression and a residual – is potentially useful in other fields, especially in soil sciences.

In the application presented here, kriging the factors brings about an improvement vis–a–vis the usual methods of estimation in forest inventory.

ACKNOWLEDGMENTS

We should like to thank Messrs. J–P. BEHAXETEGUY, E. ILDEFONSO and Mrs. DONG Y. for taking part in this study.

BIBLIOGRAPHY

G. MATHERON, 1970. La théorie des variables régionalisées et ses applications. Les cahiers du Centre de Morphologie mathématique de Fontainebleau, Fascicule 5. E.N.S.M.P.,Fontainebleau (France).

G. MATHERON, 1981. La sélectivité des distributions. Note 686, E.N.S.M.P., Centre de Géostatistique, Fontainebleau (France).

D. MANDALLAZ, 1991. A unified approach to sampling theory for forest inventory based on infinite population and superposition models. Chair of forest inventory and planning, E.T.H., Zürich (Schweiz).

W.G. COCHRAN, 1977. Sampling techniques, Wiley, New–York.

J. RIVOIRARD, 1991. Introduction au krigeage disjonctif et à la géostatistique non linéaire. Cours de géostatistique non linéaire (2ème édition), E.N.S.M.P., Centre de géostatistique, Fontainebleau (France).

For the interested reader, further references for geostatistical methods in forestry :

J. MARBEAU, 1976. Géostatistique forestière – état actuel et développements nouveaux pour l'aménagement en forêts tropicales. Thèse ENSMP, Paris.

B. MATERN, 1960. Spatial variation. Reports of the Forets Research Institute of Sweden, 49, 5, 1–114.

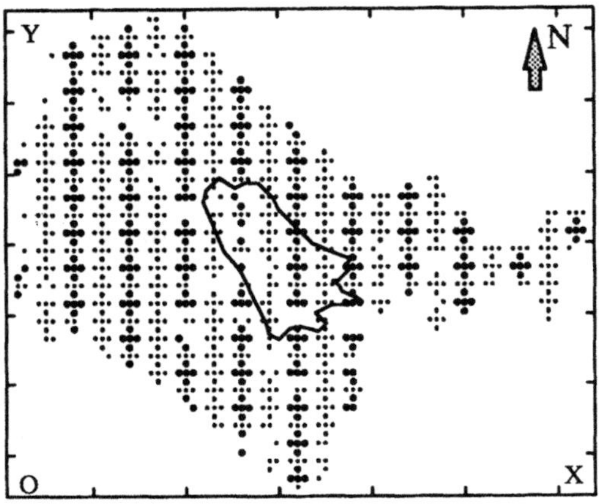

Figure I−1 Location of data

• terrestrial plots

· or • aerial survey

ES	DG 1 : dense 2 : sparse	MG 1 : conifer 2 : deciduous	Nbr of plots per class	Stem density : tige mean per class	Std. deviation of residual	Basal area : surface mean per class	Std deviation of residuals
3	1	1	16	790	275	21.6	8.2
	1	2	4	717	146	30.0	5.0
	2	1	7	733	289	19.7	7.8
	2	2	4	442	265	12.0	7.5
4	1	1	8	400	108	34.1	11.2
	1	2	absent	–	–	–	–
	2	1	13	477	202	27.6	7.6
	2	2	8	425	115	27.0	6.9
5	1	1	6	289	42	36.0	6.0
	1	2	29	261	85	41.9	11.7
	2	1	56	301	117	36.8	10.3
	2	2	109	247	125	30.5	11.6
6	1	1	3	300	56	48.4	7.2
	1	2	7	210	13.5	37.9	4.7
	2	1	absent	–	–	–	–
	2	2	28	180	90.6	28.0	17.5

Table III−1 : Terrestrial plots : statistics per classes

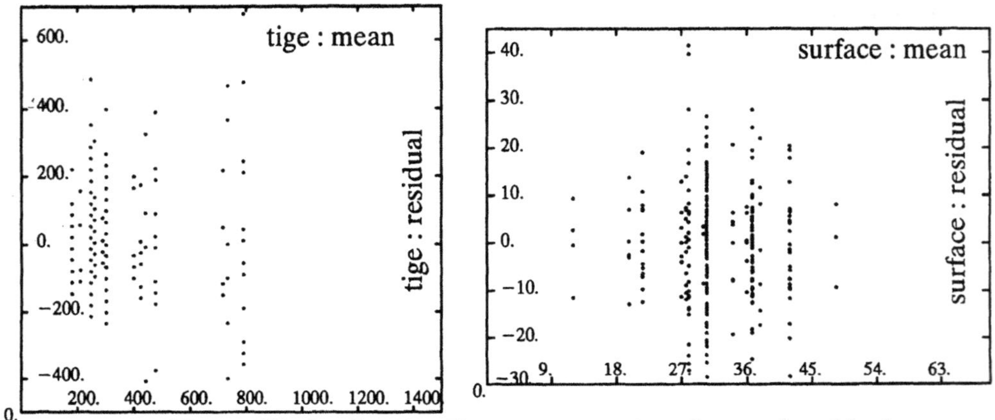

Figure III−1 : Scatter diagram : mean per class and residual

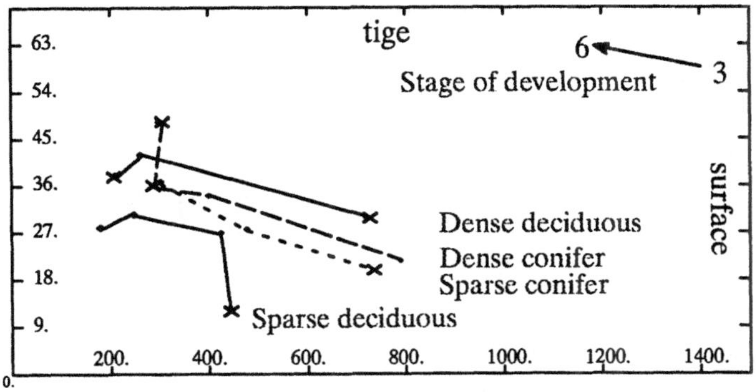

× *less than 8 plots in the class*

Figure III−2 : Terrestrial placettes : Scatter diagram − mean per class of the basal area versus the stem density

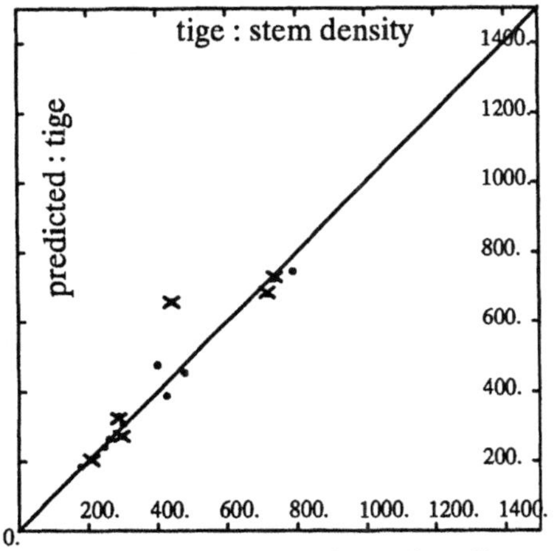

a) number of trunks − "tige"

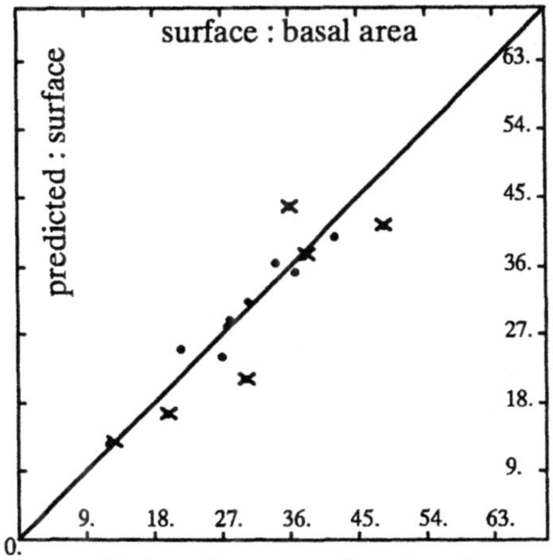

b) basal area − "surface"

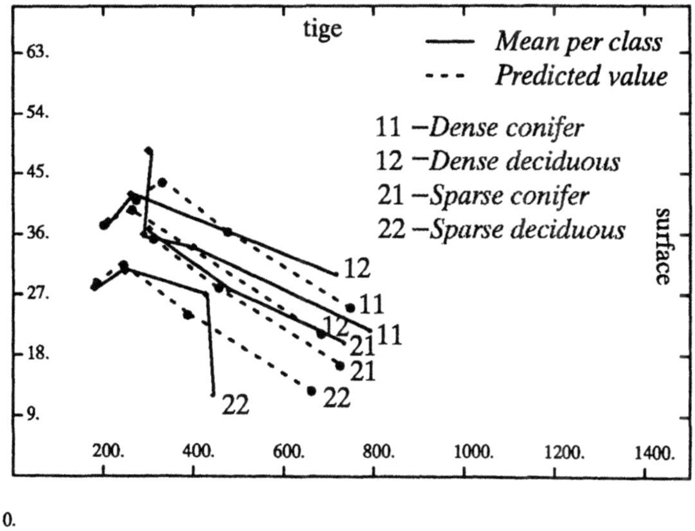

c) Scatter diagram : basal area versus stem density

Figure III−3 : Mean per class Y and predicted value Y'

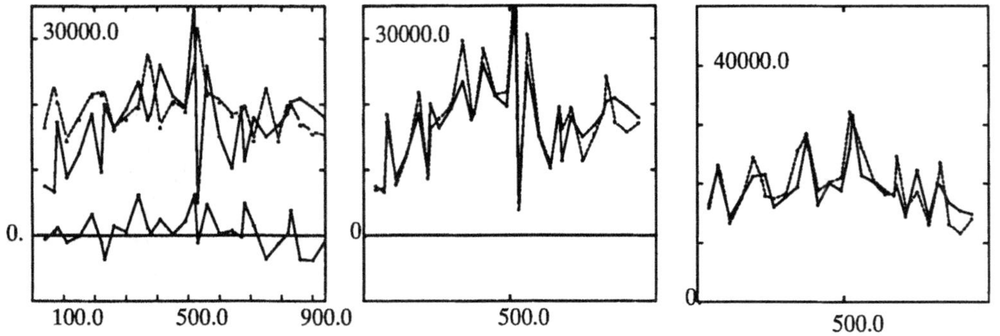

prediction Y', residual R' and cross variogram Y'R' | prediction Y' and cross variogram X, Y' | residual R' and cross variogram X,R'

a) stem density

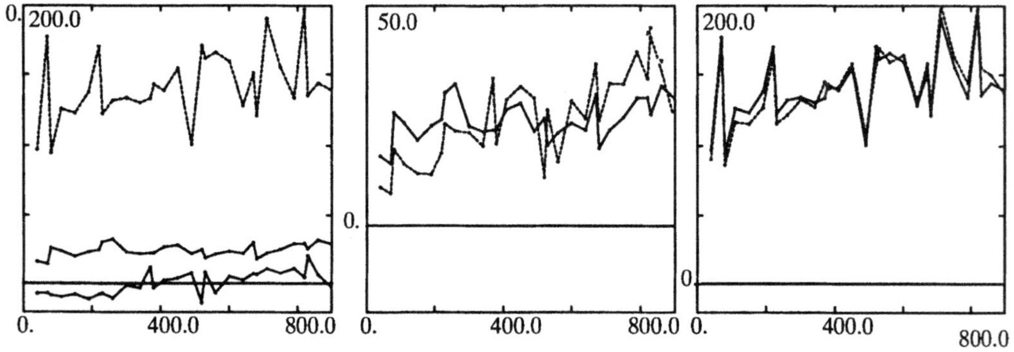

prediction Y', residual R' and cross variogram Y'R' | prediction Y' and cross variogram X, Y' | residual R' and cross variogram X,R'

b) basal area

Figure III–4 : Cross variograms – variable X, prediction Y' and residual R'

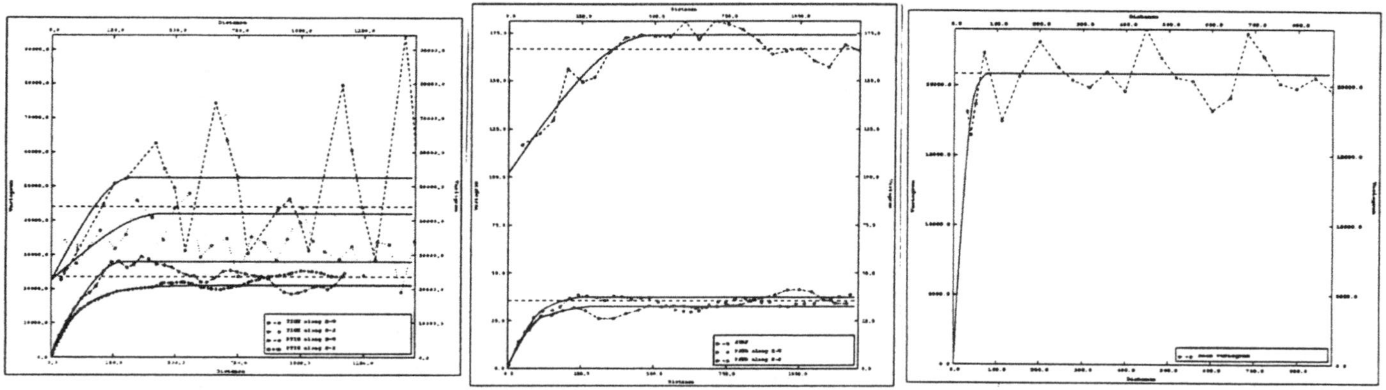

a) Stem density "tige" b) Basal area "surface" c) Stem densiry "tige"
terrestrial plots X_α and prediction Y'_α residual R'_α

Figure III–5 : Fitting of experimental variograms

LARGE DOMAIN						
	Actual value	Kriging from terrestrial survey	Double Kriging : factorised model predictions + residuals	Double Kriging **: reduced terrestrial inventory 1 : 4	Kriging with measurement errors on predictions	Classical Estimators*
Tige	-	330.52 (13.18)	325.84 (11.15)	313.36 (13.27)	327.33 (6.45)	325.08 (12.12)
Surface	-	31.92 (0.70)	31.17 (0.71)	31.05 (1.32)	31.25 (0.08)	31.23 (0.89)
SMALL DOMAIN						
Tige	280.23	294.35 (40.56)	281.67 (26.91)	294.97 (15.10)	288.31 (20.38)	257.14 (48.59)
Surface	29.6	27.68 (2.08)	29.46 (1.39)	28.88 (1.50)	29.00 (1.26)	24.00 (3.70)

* : *Classical sampling method with regression taking account of the random number of placettes per satellite (cf. MANDALLAZ, 1991).*
** : *The standard deviation is calculated from the experimental variance of the reduced inventory* *( ): Standard deviation*

Table IV–1 : Kriging results – Comparison with actual values in the small domain.

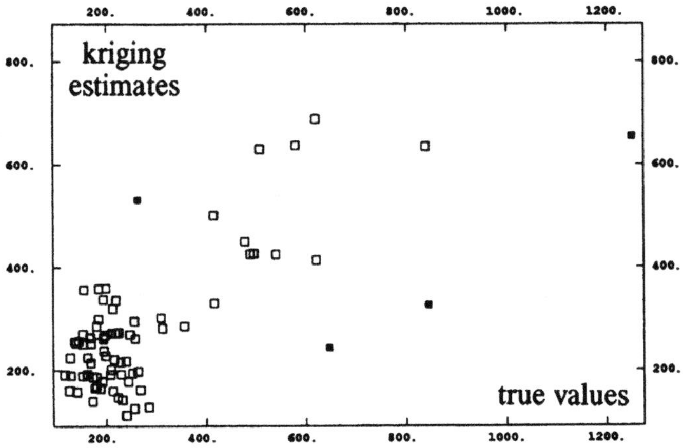

a) squares of 0.25 ha

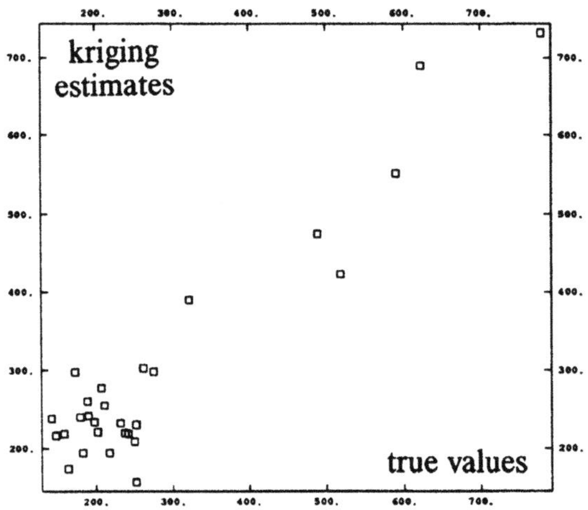

b) squares of 1 ha

Figure IV-1 : Comparison between the kriging with the factorized model and the actual values in the small domain; variable stem density "tige"

The white squares unlike the black ones denote that the true value is in the 95% confidence interval. There is good agreement.

GEOSTATISTICAL ESTIMATION OF ANIMAL ABUNDANCE IN THE KRUGER NATIONAL PARK, SOUTH AFRICA

FRANCOIS E. STEFFENS
Statistics Department
University of South Africa
P.O. Box 392
Pretoria
0001 South Africa

1. INTRODUCTION

The Kruger National Park comprises 19 485 square km in the north-eastern corner of the Transvaal, bordering Mozambique to the east and Zimbabwe to the north. This is an area with a rich biological diversity. The animal population includes a large number of herbivore species such as buffalo, giraffe, elephant, warthog, blue wildebeest, zebra and kudu, with impalas the most abundant species.

An aerial census is conducted annually during the months of May to August in order to count the number of animals of the larger herbivore species (Viljoen, 1989). Some 16 species form the main target of the survey, and in addition environmental factors, such as the availability of water and the condition of the plants are also noted. A fixed-winged aircraft (presently a twin-engined Partenavia P68 Observer) is used for this purpose. The Park has been divided into 8 census compartments and further subdivided into 66 census blocks. Each census block is surveyed in 800 metre strips, with teams of two observing the animals within 400 metres of the aircraft on each side. The aircraft flies 65-70 m above the ground at a speed of 95-100 knots.

The main purposes of the annual survey are to monitor the trends in the populations for management purposes, and to relate these trends to other components of the ecosystem. The survey has become rather expensive and manpower-intensive, and it has been decided to study the possibility of conducting a sample survey rather than a census, in case the full survey should become too expensive at some time in the future. Also, the survey is done entirely in the dry season, and there are some uncertainties as to the movements of the animals during summer. While it would be impossible to repeat the full survey during the summer months, it might be feasible to conduct a sample survey on a limited scale.

The theory of survey sampling (Cochran, 1977) is based entirely on the assumption of independence. Thus the individuals in the population are numbered and the numbers thrown into the proverbial hat. A random sample of the required sample size is then drawn and the population characteristics are estimated, regardless of the geographical location of each sample unit. The estimates obtained relate to the population as a whole, not to any subpopulation. In the case of stratified sampling the strata are often defined geographically, and thus the characteristics of each stratum may be estimated, but there is no attempt at local estimation of individuals or groups of individual members of the population. A geostatistical approach may help in solving the problem, and this is the subject of this paper.

Geostatistical techniques have been applied to ecological studies involving fish (Pereira and Soares, 1989) and birds (McNeill, 1991) but the techniques do not seem to have filtered through to the Statistical Ecology literature (see Ludwig and Reynolds, 1988, for example, where the term "spatial dependence" is used, but the geostatistical approach does not feature).

2. THE DATA

The data used in this study relate to the 1990 census of census block G, the southernmost census block. It was decided to concentrate in the present study on six species, namely giraffe, impala, kudu, warthog, blue wildebeest and zebra. These are the most abundant species in the area, and there are some indications of interactions between these species, e.g. several species may sometimes be seen grazing together.

The raw data contained coordinates for each herd seen, as well as the number of animals in the herd. Attempts at modelling the spatial properties of these data, using the techniques developed for mining, were not very successful. This brought home a very definite difference between mining data and ecological survey data. In the case of the survey data, the spaces between the observed herds should actually be filled with zeroes, because there were no animals in those spaces. In the case of mining data, the minerals may not have been observed except at certain locations, but the assumption is that the mineral deposit continues in the unobserved locations. This difference creates a complication when working with point data obtained from an ecological survey. This difficulty may be overcome by working with block data. In the present study the area was subdivided into 800 m × 800 m blocks, and the total number of animals of each species in each block was computed. The shape of the census block is shown in figure 1.

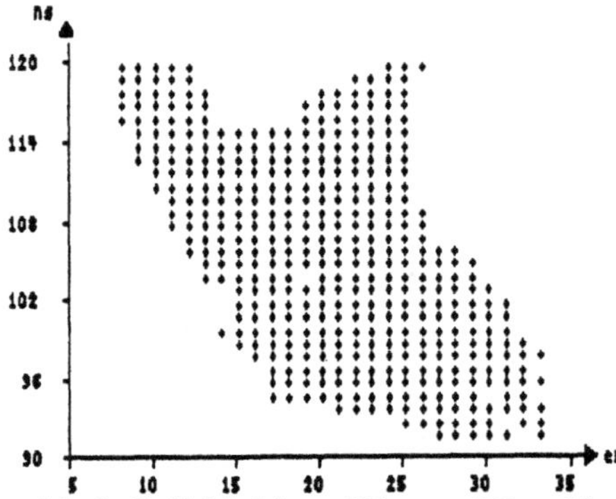

Figure 1. Census block G divided into 800 m × 800 m blocks

Histograms of the herd sizes of individual herds are shown in figure 2. Note that impala herds may exceed 150, while giraffe, kudu and warthog groups rarely exceed 10 in number. Histograms of the numbers of animals per block are shown in figure 3. In the case of impala there were more than 800 individuals in one block, in the case of zebra almost 180, while kudu, warthog and giraffe were much less abundant. Attempts at modelling the herd size distribution have not been very successful thus far, partly because the distributions have such a long tail to the right. In addition the social structure of these animals is very complex, with non-dominant males often found in separate herds, the composition of the herds changing during the mating season etc. In order to obtain a sensible distribution one will probably have to distinguish between the different types of herds, and such data are not available at present. An aerial census is probably not suitable to obtain such data anyway, since herds may spread out during grazing, and various smaller groups observed may actually belong to one herd.

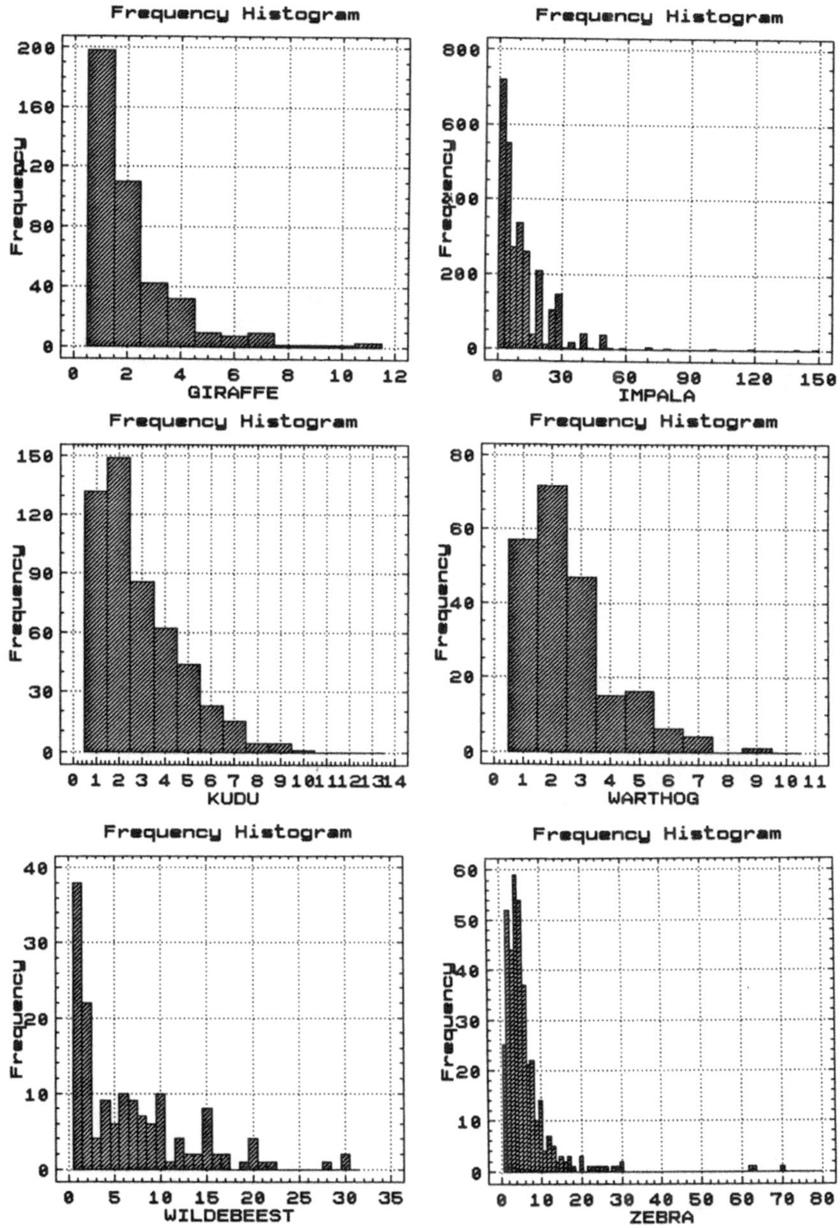

Figure 2. Histograms of individual herd sizes.

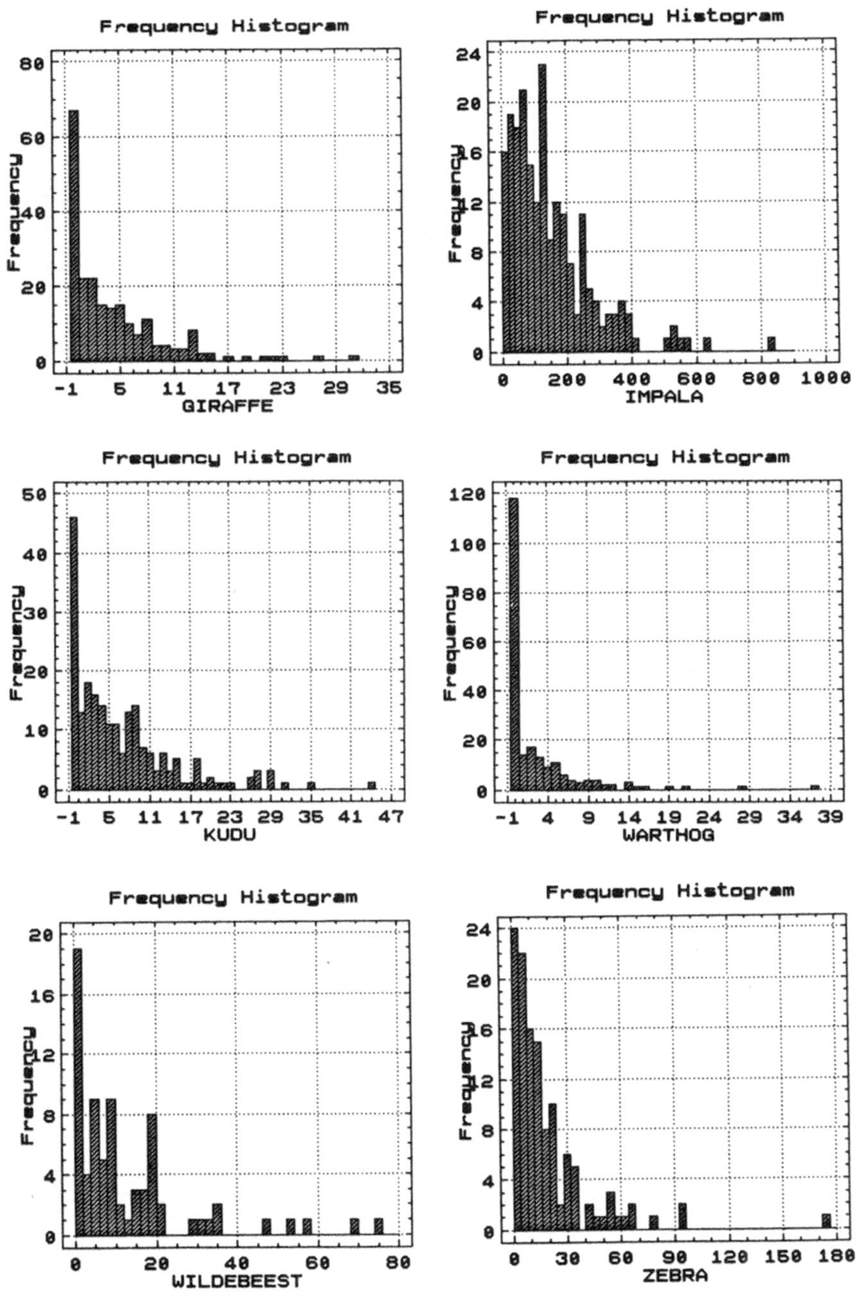

Figure 3. Histograms of number of animals per block

3. VARIOGRAM MODELS

Using the block values, experimental variograms were computed for the six species. A definite structure was observed, and in all cases a reasonable fit was obtained using a spherical variogram plus a nugget effect. Some of the fitted variograms are shown in figure 4. The unit of distance in figure 4 is 800 m. A range of 5 units (i.e. 4 km) was used throughout. In addition, cross-covariograms were computed for each pair of species, and in many cases these showed a strong structure as well, indicating that counts of one species may be helpful in estimating the numbers of another species. In this application it would of course not be reasonable to assume that one species was counted and the other not, but at least both species will have been counted in blocks adjacent to the one being estimated.

In order to ensure a positive definite system, the same type of variogram model (spherical with range 5 plus nugget) was fitted to the cross-covariograms. The cross-covariograms involving impalas, warthogs and blue wildebeest are shown in figure 5. The constants of the system of variograms and cross-covariograms are shown in table 1.

	Giraffe	Impala	Kudu	Warth	Wildeb	Zebra
Giraffe	**17.5**	30	5	0.5	0	8
	7.5	150	3	4.0	8	9
Impala	30	**8000**	130	0	100	550
	150	**13000**	50	300	500	150
Kudu	5	130	**27**	3	-2	0
	3	50	**35**	-5	2	28
Warthog	0.5	0	3	**12**	4	0
	4.0	300	-5	**20**	28	5
Wildebeest	0	100	-2	4	**50**	30
	8	500	2	28	**50**	35
Zebra	8	550	0	0	30	**250**
	9	150	28	5	35	**75**

Table 1 Constants of the system of variograms and cross-covariograms. The first figure is the nugget effect and the second figure is the multiplier for the spherical part. In all cases the range is 5.

The variograms and cross-covariograms involving impala, warthog and blue wildebeest looked particularly promising. In one case, namely the cross-covariogram between warthog and kudu, the model had a negative coefficient for the spherical term, indicating that the two species are negatively correlated, i.e. kudu is abundant where warthogs are rare and vice versa. Some of the variograms showed a lot less structure than those in figure 4, e.g. the variogram of zebra in which the nugget effect is 250 whereas the coefficient for the spherical part is only 75.

Isotropic models were used throughout, because no indication of anisotropy was found in the experimental variograms and cross-covariograms. No sign of severe non-stationarity was found in the limited study area, but it is possible that different variogram models may hold in different census blocks. The environmental variables, particularly the availability of water, should also prove to be important.

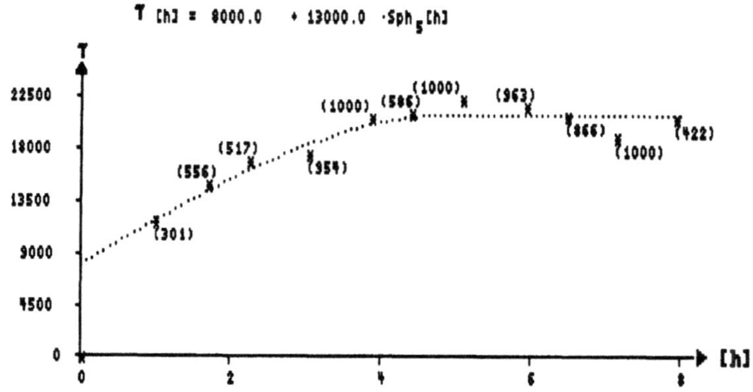

(a) Experimental variogram and fitted model for impala

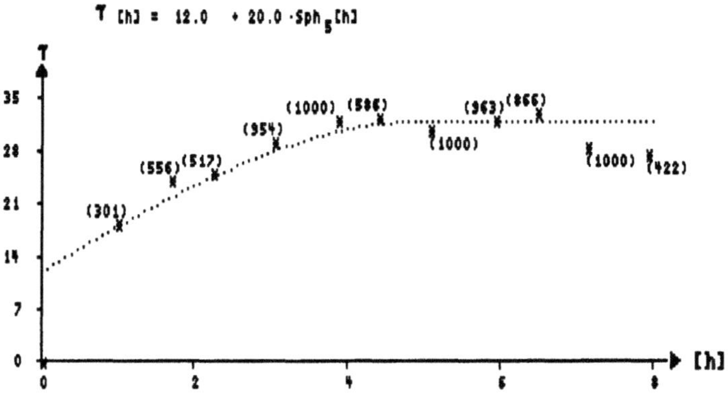

(b) Experimental variogram and fitted model for warthog

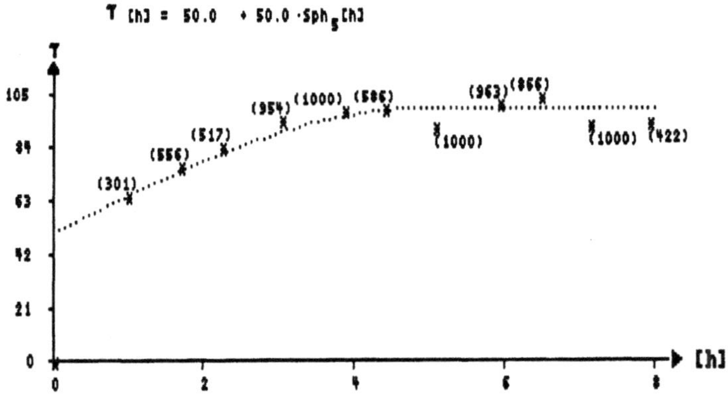

(c) Experimental variogram and fitted model for blue wildebeest

Figure 4 Variograms for (a) impala, (b) warthog and (c) blue wildebeest

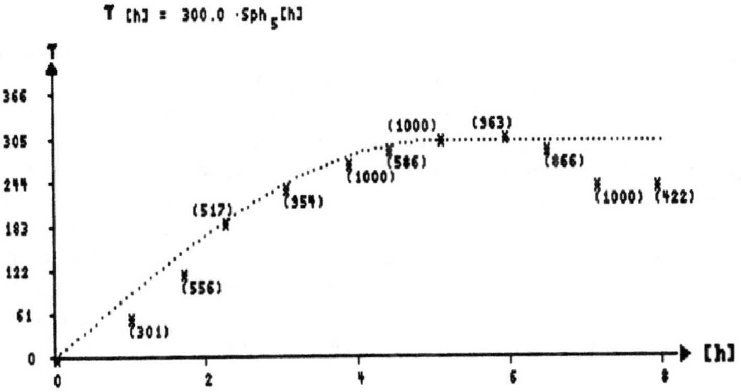

(a) Cross-covariogram between impala and warthog

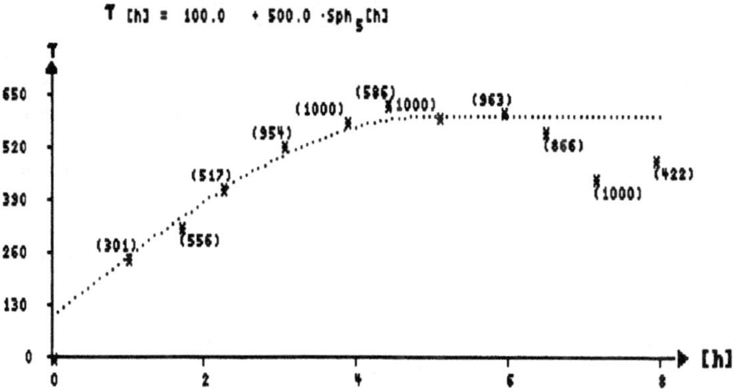

(b) Cross-covariogram between impala and blue wildebeest

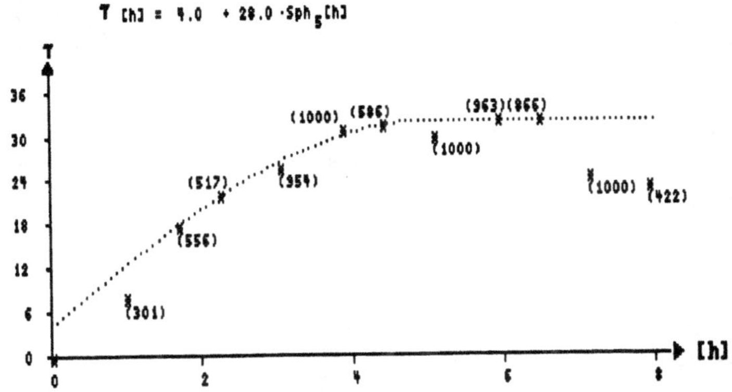

(c) Cross-covariogram between warthog and blue wildebeest

Figure 5 Cross-covariograms between impala, warthog and blue wildebeest

4. KRIGING AND COKRIGING

In order to simulate the flight paths of the actual aerial census, all the blocks with the same east-west coordinate (X) were assumed to form one flight path. Thus in this simulation all flight paths are assumed to run north-south. This is not the case, in fact, but for simulation purposes it should suffice. Next, the sampling plan is assumed to be a systematic sample in which every second flight path is observed. A completely random sample would entail gaps in the data, and the variogram models showed that the range of influence is only 4 km. For simulation purposes two runs were completed. In the first run the data for flight paths 1, 3, 5, ... were assumed unknown and these were estimated from the data for flight paths 2, 4, 6, ... while in the second run the roles were reversed. The variogram models were not re-estimated for each half of the data, although that is what would have to be done in practice, unless it turns out that historically estimated variogram models are appropriate for new data.

A further complication involves the irregular shape of the census block, resulting in different flight path lengths. In order to avoid such complications, a rectangular 13×9 set of blocks were selected for estimation. Ordinary point kriging and ordinary point cokriging were employed to estimate the "missing" values, and these could then be compared with the actual values. The "missing" values were actually block values, but since the data referred to exactly the same block sizes, point kriging was found to be more appropriate. The results of the ordinary kriging estimation are shown in tables 2 and 3. In order to try to improve on these estimates, the impala counts were re-estimated using the other species (one at a time) as covariates. The results of this exercise are shown in tables 4 and 5. The ordinary cokriging estimates were in fact an improvement.

As a measure of comparison of the various kriging and cokriging estimates, the following root mean squared error was used:

$$s = \left[\frac{1}{n} \sum_{i=1}^{n} (\text{Observed value} - \text{Estimated value})^2 \right]^{0.5}$$

The values of s for the ordinary kriging as well as the ordinary co-kriging are given below. For comparison s(K), the square root of the average kriging variance, is also given, as was computed from the output of the kriging program. (The kriging programs used in this study are contained in the Geostat Toolbox (Froidevaux, 1989)).

	Even flight paths		Odd flight paths	
Species	s	s(K)	s	s(K)
Giraffe	5.01	4.79	5.10	4.80
Impala	136.3	120.1	142.1	119.8
Kudu	7.32	6.73	7.22	6.71
Warthog	4.26	4.67	5.68	4.66
Wildebeest	7.84	8.83	8.76	8.81
Zebra	14.7	17.7	13.0	17.7

Table 2 Root mean squared error s and average kriging standard deviation s(K) for ordinary kriging

	Flight path no.	12	14	16	18	20	22	Total
Giraffe	Observed	32	15	34	33	42	50	206
	Estimated	37	23	34	41	37	38	210
	Flight path no.	13	15	17	19	21	23	Total
	Observed	12	33	40	40	24	44	193
	Estimated	25	26	35	35	43	52	216
	Flight path no.	12	14	16	18	20	22	Total
Impala	Observed	719	695	1691	2656	1020	861	7642
	Estimated	805	1157	1399	1788	1747	1310	8206
	Flight path no.	13	15	17	19	21	23	Total
	Observed	1042	1181	1544	1901	1472	939	8079
	Estimated	736	1402	2388	2007	1085	951	8569
	Flight path no.	12	14	16	18	20	22	Total
Kudu	Observed	183	83	40	57	45	28	436
	Estimated	104	100	71	70	55	41	441
	Flight path no.	13	15	17	19	21	23	Total
	Observed	121	74	59	67	26	43	390
	Estimated	123	65	54	57	37	27	363
	Flight path no.	12	14	16	18	20	22	Total
Warthog	Observed	18	26	47	34	6	17	148
	Estimated	12	30	58	46	15	15	176
	Flight path no.	13	15	17	19	21	23	Total
	Observed	9	36	62	20	9	17	153
	Estimated	27	48	49	23	11	17	175
	Flight path no.	12	14	16	18	20	22	Total
Wildebeest	Observed	18	39	36	45	9	17	164
	Estimated	45	67	85	60	19	13	289
	Flight path no.	13	15	17	19	21	23	Total
	Observed	52	58	87	20	15	9	241
	Estimated	32	41	43	29	16	18	179
	Flight path no.	12	14	16	18	20	22	Total
Zebra	Observed	101	89	65	78	15	91	439
	Estimated	161	133	91	58	74	85	602
	Flight path no.	13	15	17	19	21	23	Total
	Observed	106	145	24	49	94	73	491
	Estimated	105	83	68	54	69	84	463

Table 3 Observed values and ordinary kriging estimates of animal counts

	Even flightpaths		Odd flightpaths	
Co-species	s	s(K)	s	s(K)
Giraffe	131.4	119.7	135.9	127.2
Kudu	132.4	119.7	136.2	127.3
Warthog	130.0	119.4	133.3	127.0
Wildebeest	131.3	119.6	134.6	127.3
Zebra	132.7	119.5	134.5	127.2

Table 4 Root mean squared error s and average kriging standard deviation s(K) for ordinary cokriging for impala

From table 4 it would seem that warthog is the best covariate to estimate impala, using ordinary cokriging. In the case of the even-numbered flight paths the root mean squared error is 8.5% less, using cokriging with warthog as covariate, than with ordinary kriging. In the case of the odd-numbered flight paths the saving is smaller (4.4%). This same exercise was repeated for the other species, but the cokriging did not produce substantially lower root mean squared errors in these cases.

Flight path no.	12	14	16	18	20	22	Total
Observed Impala	719	695	1691	2656	1020	861	7642
Estimated (using Giraffe)	790	1134	1415	1801	1738	1310	8188
Estimated (using Kudu)	808	1149	1410	1788	1755	1308	8218
Estimated (using Warthog)	790	1177	1468	1789	1715	1316	8255
Estimated (using Wildeb.)	815	1182	1414	1779	1728	1311	8229
Estimated (using Zebra)	829	1148	1390	1827	1747	1305	8246
Flight path no.	13	15	17	19	21	23	Total
Observed impala	1042	1181	1544	1901	1472	939	8079
Estimated (using Giraffe)	718	1233	2150	1813	970	908	7792
Estimated (using Kudu)	707	1220	2169	1823	975	912	7806
Estimated (using Warthog)	708	1258	2175	1790	963	920	7814
Estimated (using Wildeb.)	699	1242	2185	1822	959	902	7809
Estimated (using Zebra)	703	1223	2179	1847	971	874	7797

Table 5 Observed values and ordinary cokriging estimates of impala counts

5. DISCUSSION

The root mean squared errors and kriging standard deviations in table 2 are in reasonable agreement. In a few cases s(K) underestimates the error (e.g. in the case of Impala) but in other cases it overestimates the error (e.g. in the case of Zebra).

The estimates of the number of animals in individual flight paths, given in table 3, are in some cases rather far out (notably in the case of Impala) but the totals are on the whole acceptable. It should be realised that the actual counts are not absolutely accurate in any case.

Co-kriging did not provide spectacular improvements in the estimates. Other environmental factors, such as the condition of the grazing and the availability of water, should perhaps provide better covariates when they are available (as they will be in practice). If a decrease in the estimation variance is translated into less intensive sampling, however, which in turn may mean a huge saving in the cost of the census, then such research is certainly worthwhile.

The present study is still a preliminary one, based only on one area in a very large park, and the results are not yet ready for implementation. The census is presently a complete census lasting about five months every year, but if at some time in the future it may become necessary to cut down on the cost of counting the animals, then the knowledge will be available. Also, sampling may enable the animals to be counted selectively at other times of the year in addition to the present winter count.

ACKNOWLEDGEMENT

The author wishes to thank the National Parks Board of South Africa for approval of the research project on sampling as an alternative to a census, and in particular Mr. Petri Viljoen for his cooperation and many helpful discussions.

REFERENCES

Cochran, W.G. (1977) Sampling Techniques (3rd Ed.) Wiley, New York.

Froidevaux, Roland (1989) Geostatistics Toolbox. Unpublished report, FSS International, Troinex, Switzerland.

Journel, A.G. and Huijbregts, C.J. (1978) Mining Geostatistics, Academic Press, London.

Ludwig, John A. and Reynolds, James F. (1988) Statistical Ecology. A Primer on Methods and Computing, Wiley, New York.

McNeill, Lindsay (1991) "Interpolation and smoothing of binomial data for the Southern African bird atlas project", South African Statistical Journal 25, 129-143.

Pereira, H.G. and Soares, A.O. (1989) "Application of Geostatistics to groundfish survey data", in M. Armstrong (ed.) Geostatistics, Vol. 1, Kluwer Academic Publishers, 459-467.

Viljoen, P.C. (1989) "Ecological aerial surveys in the Kruger National Park", Unpublished Report.

ESTIMATING THE RISK OF CHILDHOOD CANCER

M. A. OLIVER[1], Ch. LAJAUNIE[2], R. WEBSTER[3], K. R. MUIR[4] & J. R. MANN[5]

[1] *School of Geography, The University, Birmingham B15 2TT, England*
[2] *Centre de Géostatistique, 35 rue St-Honoré, 77305 Fontainebleau, France*
[3] *ETH Zürich, Grabenstrasse 3, 8952 Schlieren, Switzerland*
[4] *Queen's Medical School, The University, Nottingham NG7 2UH, England*
[5] *The Children's Hospital, Ladywood Middleway, Birmingham B16 8ET, England*

SUMMARY: The incidence of childhood cancer and the apparent risk of a child's getting the disease vary from place to place. We have analysed the incidence in the West Midlands of England from 1980 to 1984 treating local frequencies of the disease as estimates of the risk. The risk was first shown to be heterogeneous over the Region, and then analysed geostatistically. The variogram of the risk was calculated from the frequencies, taking account of the varying population of children, and modelled. Covariances of the frequency and cross covariances between the risk and the frequency were derived, and from them and the frequencies the risk was estimated over the Region by cokriging. The risk variogram is strongly structured, and the kriged estimates when mapped showed a very uneven distribution; there are patches of high risk, especially in the rural areas, and others, notably the more densely populated regions, where the risk is small.

INTRODUCTION

The incidences of human diseases vary from place to place. The patterns of some of them are distinct and easily explained. Many others are not, and this is especially true of rare diseases such as childhood cancer. Much effort has been devoted to analysing the spatial distribution of this disease because of its elusive aetiology and its putative association with nuclear installations. There have been two principal approaches. One is based on cell counts and includes the Potthoff and Whittinghill (1966) test and Poisson probability mapping (Muir *et al.*, 1990). Results in both depend on an arbitrary choice of cell size: large cells cause too much smoothing; small cells emphasize local fluctuation; and counts in neighbouring cells play no part. The other approach is based on the distances between cases, and it goes some way to avoiding these disadvantages. It includes the nearest neighbour analysis of Cuzick and Edwards (1990) and of Besag and Newell (1991), and the 'geographical analysis machine' of Openshaw *et al.* (1988). Each method has its shortcomings, and it is fair to say that none has yet provided conclusive results.

It seems to us that it is the underlying risk of developing cancer that is of prime concern. Bithell (1990) made the same observation, and attempted to estimate the risk by the kernel method. We have explored geostatistics for analysing the distribution of childhood cancer in the West Midlands of England. The disease is rare, and this adds to the difficulties for geostatistical analysis. Our aim was to analyse its distribution, taking the varying density of population into account, and to estimate the risk of a child's developing cancer over the Region.

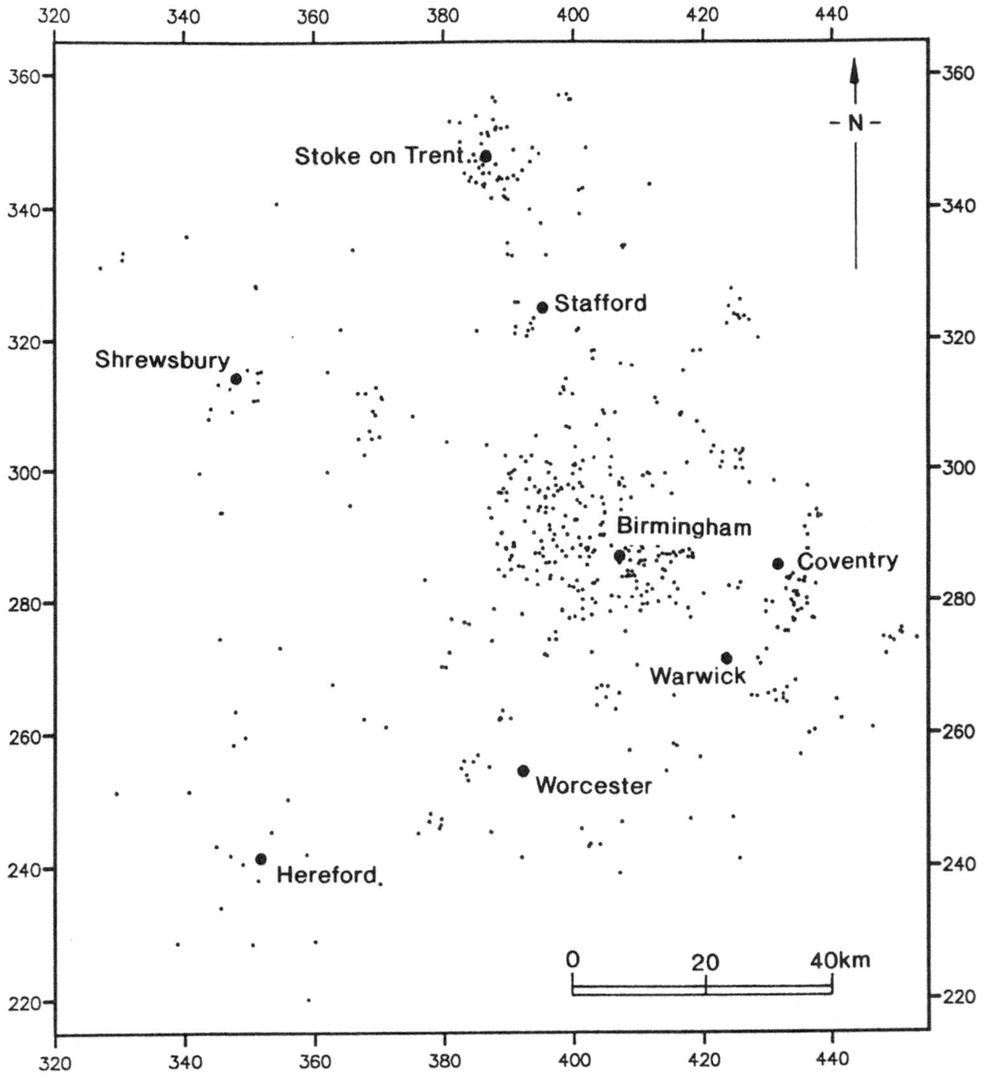

Figure 1. Cases of childhood cancer in the West Midlands Health Authority Region from 1980 to 1984 inclusive. Each dot represents a diagnosed case, and its position is that of the child's home.

THE DATA AND SUMMARY OF STATISTICS

The West Midlands Health Authority Region extends over about 25 000 km^2. It comprises rural, industrial-cum-urban, and suburban environments. One tenth of the children of England and Wales live in it, and this is the number of children at risk. We considered all cases of cancer in children under fifteen years of age during the five years from 1980 to 1984 inclusive. Fig. 1 shows where the children were living at the time of diagnosis. The density of the cases varies enormously: there are many more cases in towns than in rural areas, if only because more children live there.

As a first step to estimating the local risk we calculated a local frequency, i.e. the ratio of the number of cases of cancer within each electoral ward to the number of children living in that ward. The frequencies are referenced by the coordinates of the centroids of the electoral wards. The wards are not equal in area as we should like, but they are the main source of data on the numbers of children. The figures are from the National census of 1981.

Table 1 summarizes the statistics of the disease compiled ward by ward during this time. The mean frequency for the Region, i.e. number of cases divided by the number of children, is 0.000528. The average of the ward frequencies is somewhat different, Table 1, because the weight of an individual case in this average is inversely proportional to the number of children in the same ward. More than half of the wards, 494, had no childhood cancer diagnosed, and few wards had more than 1 case each. So the distribution is very skew, see Table 1.

Table 1. Summary statistics

Number of wards	838
Number of cases	591
Number of wards with cases	344
Number of wards with no cases	494
Mean of ward frequencies	0.000586
Maximum frequency	0.01307
Variance	1.3225×10^{-6}
Skew	4.10

TESTING THE CONSTANCY OF RISK

We then tested the frequencies for heterogeneity without regard to spatial location. We assumed that there is an underlying risk, $R(\mathbf{x})$, of a child's developing cancer to which all children are exposed. The average risk for a particular ward i of area v_i is :

$$R(v_i) = \frac{1}{|v_i|} \int_{v_i} R(\mathbf{x}) \mathrm{d}\mathbf{x}. \tag{1}$$

The risk is such that a child chosen at random and living in the ward i has a chance of developing cancer during a time interval δt equal to $\delta t \times R(v_i)$. We also assumed that two children living in the same or two different wards, i and j, behave independently of one another when the risks $R(v_i)$ and $R(v_j)$ are known. Therefore, the function R is the only possible source of correlation among the cases, and this seems reasonable for a non-contagious disease.

Our knowledge of the underlying risk can come only from the observed frequencies, $F(v_i)$, and the problem is one of using these to make statements about R. As a consequence the number of cases, $C(v_i)$, observed in the ith ward is a binomial variable:

$$C(v_i) = \mathrm{Bi}[P, n(v_i)], \tag{2}$$

where $P = \delta t.R(v_i)$ is the probability that a child develops the disease, and $n(v_i)$ is the number of children in the ward. Its value is independent of the number of cases in any other ward. Without any loss of generality we can assume that $\delta t = 1$, and this simplifies the notation. The frequency is then

$$F(v_i) = C(v_i)/n(v_i).$$

The observed $F(v_i)$ is drawn from a binomial distribution that depends on the risk, $R(v_i)$, and $n(v_i)$. For the ward i and for fixed risk R the expectation and variance are given by:

$$\mathrm{E}[F(v_i) \mid R] \;=\; R(v_i) \tag{3}$$

and

$$\mathrm{Var}[F(v_i) \mid R] \;=\; \frac{R(v_i).\{1 - R(v_i)\}}{n(v_i)}. \tag{4}$$

We tested the null hypothesis, H_0, that the risk is constant from ward to ward:

$$[H_0] \;:\; R(\mathbf{x}) \;=\; p \;=\; constant, \tag{5}$$

where p is estimated from the numbers of cases in the wards by:

$$\hat{p} = \sum_{i=1}^{N} C(v_i) / \sum_{i=1}^{N} n(v_i). \tag{6}$$

We used the statistic T as a measure of heterogeneity:

$$T \;=\; \frac{1}{N\hat{p}(1-\hat{p})} \sum_{i=1}^{N} n(v_i)\{F(v_i) - \hat{p}\}^2. \tag{7}$$

If the differences $F(v_i) - \hat{p}$ are larger than expected, and likewise T itself is, then we reject the null hypothesis.

The distribution of T was obtained for actual values of $n(v_i)$ and for p equal to the estimate $\hat{p} = 0.528 \times 10^{-3}$ by repeated simulations. Critical thresholds of T at 90% and at 95% confidence levels were respectively 1.091 and 1.121. Under the null hypothesis H_0 T has less than a 1 in 20 chance of exceeding 1.121. We therefore rejected H_0 if the observed value of T exceeded that threshold.

To determine the effect of the uncertainty in estimating p by equation (6) the distribution of T was obtained for $p = 0.4 \times 10^{-3}$ and $p = 0.6 \times 10^{-3}$, based on the standard deviation of $\hat{p}$, to give the critical values in Table 2. Values of T exceeding 1.098 and 1.132 led us to reject H_0 for p in the range $[0.4 \times 10^{-3}, 0.6 \times 10^{-3}]$ at levels 90% and 95% respectively. The value obtained for T was 1.19, and we felt safe in rejecting the null hypothesis: we could assume that the risk of a child's developing cancer in the West Midlands varied significantly from one ward to another during 1980 - 1984 and that we could proceed profitably with the spatial analysis.

Table 2. The distribution of T for different probabilities and constant risk, p.

	Probability	
p	90 %	95 %
0.4×10^{-3}	1.098	1.132
0.528×10^{-3}	1.091	1.121
0.6×10^{-3}	1.088	1.119

ESTIMATING THE RISK GEOSTATISTICALLY

To assess the local risk of a child's developing cancer we follow the usual rationale and assume that it is a particular realization of a random function. We also assume that the function is second order stationary, with expectation $\mathrm{E}[R(\mathbf{x})] = \mu$. With this model we can represent the average risk for a ward i of area v_i by the spatial average:

$$R(v_i) = \frac{1}{v_i} \int_{|v_i|} R(\mathbf{x})\, \mathrm{d}\mathbf{x}.$$

In the West Midlands the wards do not have equal area, and so the variances of the $R(v_i)$ are unequal. Nevertheless, the areas are not hugely different, and to simplify the analysis we shall take them as being equal: $v_i \approx v$. We shall use this approximation systematically in the rest of the paper, and all variables will have the support v; thus, $R(\mathbf{x})$ will denote the value for a ward centred at x, previously denoted by $R_v(\mathbf{x})$. Lajaunie (1991) has elaborated the solution for the more general problem.

With this simplification, the risk for a ward can be considered as the value at the ward centroid of the moving average process:

$$R_v(\mathbf{x}) = \frac{1}{|v|} \int_v R(\mathbf{x} + \mathbf{x}')\, \mathrm{d}\mathbf{x}'.$$

Our task now is to estimate this risk everywhere in the Region so that we can map it. We do not know the risk anywhere: we have only the frequencies of the disease, $F(\mathbf{x}_i)$, and it is from these that we must derive our local estimates of R at points $\mathbf{x}_0$, thus:

$$\hat{R}(\mathbf{x}_0) = \sum_{i=1} \lambda_i F(\mathbf{x}_i), \tag{8}$$

As $\mathrm{E}[F(\mathbf{x}_i)] = \mathrm{E}[R(\mathbf{x}_i)]$, the estimate is unbiased if the weights, λ_i, for n data sum to 1, and subject to this we want to minimize the estimation variance,

$$\sigma_R^2(\mathbf{x}_0) = \mathrm{E}[\{\sum_{i=1}^{n} \lambda_i F(\mathbf{x}_i) - R(\mathbf{x}_0)\}^2],$$

which we can do by solving the kriging equations:

$$\sum_{i=1}^{n} \lambda_i C_{ij}^F + \psi = C_{j0}^{RF} \quad \text{for all} \quad j, \tag{9}$$

$$\sum_{i=1}^{n} \lambda_i = 1. \tag{10}$$

Here C_{ij}^F is the covariance of F between wards centred on $\mathbf{x}_i$ and $\mathbf{x}_j$, and C_{j0}^{RF} is the cross covariance between the frequency in the jth ward and the risk at the point $\mathbf{x}_0$ for which we want the estimate. The quantity ψ is the familiar Lagrange multiplier in ordinary kriging.

The procedure is effectively cokriging for which we need not only the covariances of the frequencies but also the cross covariances between the frequencies and the risk. The exercise is not straightforward, however, because the numbers of children at risk in the wards are not constant, and so the variance is not stationary. Both sets of covariances can be deduced from the covariances of the risk, C_{ij}^R, as follows (Lajaunie, 1991):

$$C_{ij}^{RF} = C_{ij}^R \tag{11}$$

and

$$C_{ij}^F = \left\{1 - \frac{\delta_{ij}}{n_i}\right\} C_{ij}^R + \frac{\delta_{ij}}{n_i}\,\mu(1-\mu), \tag{12}$$

where μ is the mean risk, n_i is the number of children in the ith ward, and δ_{ij} is the Kronecker delta, taking the value 1 when $i = j$ and 0 otherwise. We still have to estimate the C_{ij}^R. To obtain them we compute first the empirical variogram of the frequencies, γ_F, and then estimate the variance and variogram of the risk, σ_R^2 and γ_R, respectively. From these we can deduce the covariances required for equation (9), as follows.

Variogram of frequency. We calculated semivariances of the frequencies from the data using the usual computing formula:

$$\hat{\gamma}_F(\mathbf{h}) = \frac{1}{2M(\mathbf{h})} \sum_{i=1}^{M(\mathbf{h})} \{F(\mathbf{x}_i) - F(\mathbf{x}_i + \mathbf{h})\}^2, \tag{13}$$

where $F(\mathbf{x}_i)$ and $F(\mathbf{x}_i + \mathbf{h})$ are the frequencies in the wards centred at $\mathbf{x}_i$ and $\mathbf{x}_i + \mathbf{h}$ separated by $\mathbf{h}$, and $M(\mathbf{h})$ is the number of pairs of places for which data are available at that lag. The variation seemed to be isotropic. Therefore, the lag could be treated as a scalar in distance only, $h = |\mathbf{h}|$, and it was incremented in steps of 8 km to 120 km to produce the experimental variogram, $\gamma_F(h)$, of the frequencies, Fig. 2a.

Variogram of risk. We can now derive the formula for the variogram of the risk. The conditional independence of pairs of frequencies and relations (3) and (4) lead to the conditional differences:

$$\begin{aligned} \mathrm{E}[\{F(\mathbf{x}_i) - F(\mathbf{x}_i + \mathbf{h})\}^2 \mid R] &= \{R(\mathbf{x}_i) - R(\mathbf{x}_i + \mathbf{h})\}^2 + \frac{1}{n(\mathbf{x}_i)} R(\mathbf{x}_i)\{1 - R(\mathbf{x}_i)\} \\ &\quad + \frac{1}{n(\mathbf{x}_i + \mathbf{h})} R(\mathbf{x}_i + \mathbf{h})\{1 - R(\mathbf{x}_i + \mathbf{h})\}, \end{aligned} \tag{14}$$

in which $n(\mathbf{x}_i)$ and $n(\mathbf{x}_i + \mathbf{h})$ are the numbers of individuals in the wards centred at $\mathbf{x}_i$ and $\mathbf{x}_i + \mathbf{h}$ respectively. This leads to an expectation

$$\mathrm{E}[\{F(\mathbf{x}_i) - F(\mathbf{x}_i + \mathbf{h})\}^2] = 2\gamma_R(\mathbf{h}) + \{\mu(1 - \mu) - \sigma_R^2\} \left\{ \frac{n(\mathbf{x}_i) + n(\mathbf{x}_i + \mathbf{h})}{n(\mathbf{x}_i).n(\mathbf{x}_i + \mathbf{h})} \right\}, \tag{15}$$

where σ_R^2 is the variance of the underlying risk and μ is the mean estimated without bias by $\hat{\mu}$, the average of the data. Clearly, the variance obtained depends on the spatial location, because the population of children, $n(\mathbf{x}_i)$, varies from ward to ward. Consequently, equations (13) and (14) do not define a variogram in the strict sense. However, the average of these variances over the whole area under study is estimated correctly by the empirical variogram of the frequencies obtained using equation (13). This suggests the following relation between the estimated variogram of the risk and the empirical variogram of the frequencies:

$$\hat{\gamma}_R(\mathbf{h}) = \hat{\gamma}_F(\mathbf{h}) - \frac{1}{2}\{\hat{\mu}(1 - \hat{\mu}) - \hat{\sigma}_R^2\} \overline{\left\{ \frac{n(\mathbf{x}_i) + n(\mathbf{x}_i + \mathbf{h})}{n(\mathbf{x}_i).n(\mathbf{x}_i + \mathbf{h})} \right\}}, \tag{16}$$

in which the quantity beneath the bar is the average over all pairs of wards involved in the calculation of $\hat{\gamma}_F(\mathbf{h})$. McNeill (1991) arrives at the same result by a similar approach.

We have no value for $\hat{\sigma}_R^2$, and direct estimation based on the variance relation between R and F would be very unreliable. But with large $n(\mathbf{x}_i)$ and $n(\mathbf{x}_i + \mathbf{h})$ the contribution of σ_R^2 to the estimate of $\gamma_R(\mathbf{h})$ in equation (16) is much smaller than γ_R

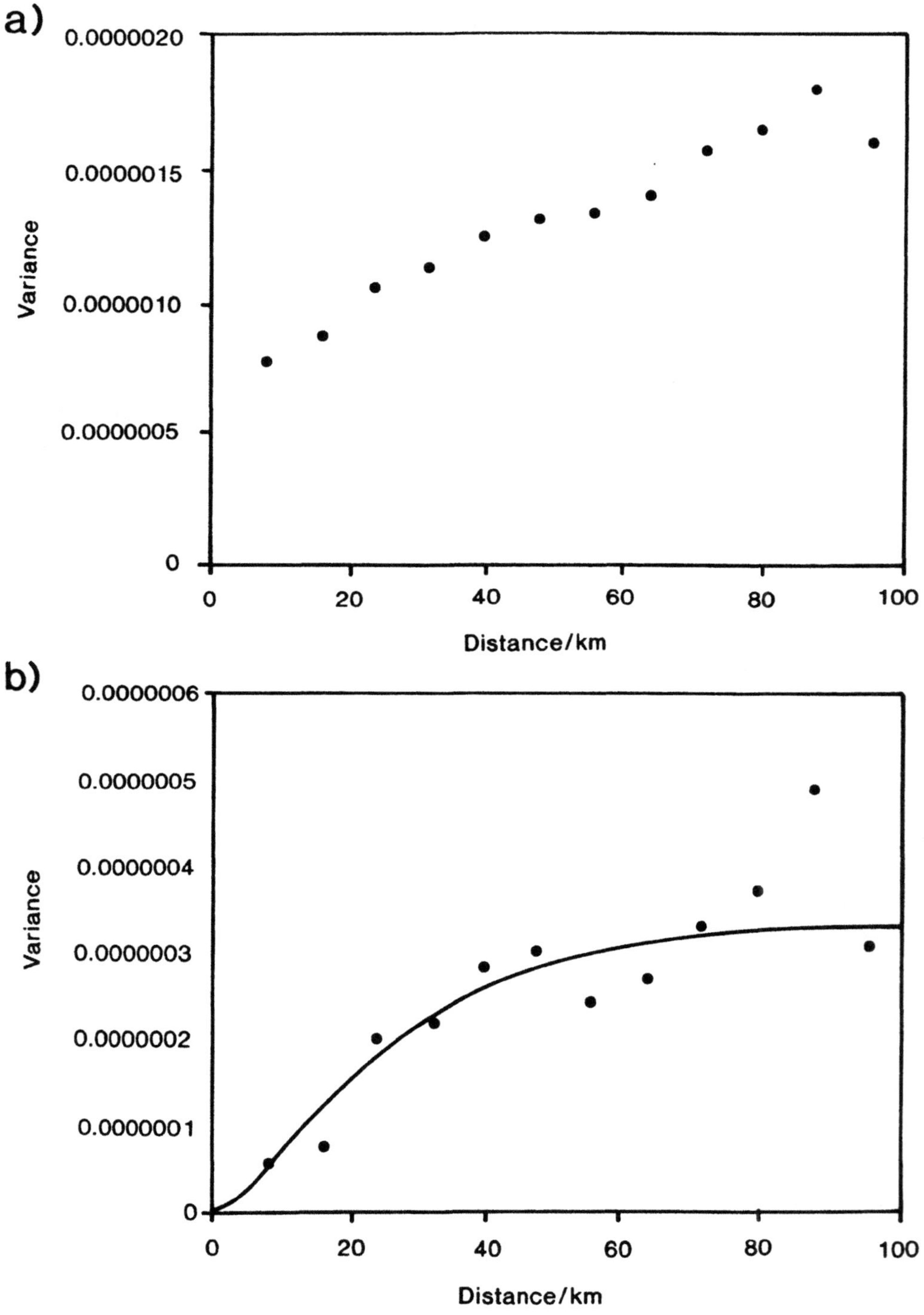

Figure 2. Variograms of childhood cancer in the West Midlands. (a) The variogram of frequency, $\gamma_F(h)$, computed using the usual formula. (b) The variogram of risk, $\gamma_R(h)$, with the fitted model shown as the solid line.

itself, so that we may disregard the variance initially and obtain a first approximation to $\gamma_R(\mathbf{h})$ without it. Once the experimental variogram has been obtained a model can be fitted to it (see below), and we can use its sill as an estimate of σ_R^2. We can then insert this value into equation (16), obtain a new and more accurate variogram, which in turn can be modelled to obtain a new value of $\hat{\sigma}_R^2$. The process can be repeated until the results have converged.

We calculated semivariances, $\hat{\gamma}_R(h)$, from the frequencies in this way, and the results are shown by the set of points in Fig. 2b. In the event they converged after only one iteration, see Table 3. Among the several plausible functions to describe the experimental variogram Whittle's elementary correlation (Whittle, 1954) fitted best in the least squares sense. In its isotropic form it is

$$\gamma(h) = c\left\{1 - \frac{h}{r}K_1\left(\frac{h}{r}\right)\right\}, \tag{17}$$

where c is the sill, estimating the *a priori* variance, σ_R^2, r is a distance parameter, and K_1 is the modified Bessel function of the second kind. This function is the solid line in Fig. 2b. Table 3 gives the coefficients of the final model. Whittle's function approaches its sill asymptotically, and so sets no definite limit to the spatial correlation. For practical purposes we can take the lag at which it reaches 95 % of the sill as the limit, and this is about 50 km. We note also that, being bounded, the variogram is that of a second order stationary process.

Table 3. Estimates of parameters for the risk variogram.

	Model Parameters	
	c	r/km
First estimates	0.3336×10^{-6}	18.23
Estimates after iteration	0.3340×10^{-6}	18.02

The final model of the risk variogram was then easily converted to the covariance function for kriging the risk by

$$C^R(\mathbf{h}) = \sigma_R^2 - \gamma_R(\mathbf{h}) \,. \tag{18}$$

RESULTS

Interpreting the variogram.

The risk variogram, Fig. 2b, suggests that the risk of a child's developing cancer has a coarse patchy distribution; i.e. the risk is large in some parts of the Region and small in others. Its nugget variance is zero as we would expect: the underlying risk is continuous, and the wards, which constitute the supports, are contiguous.

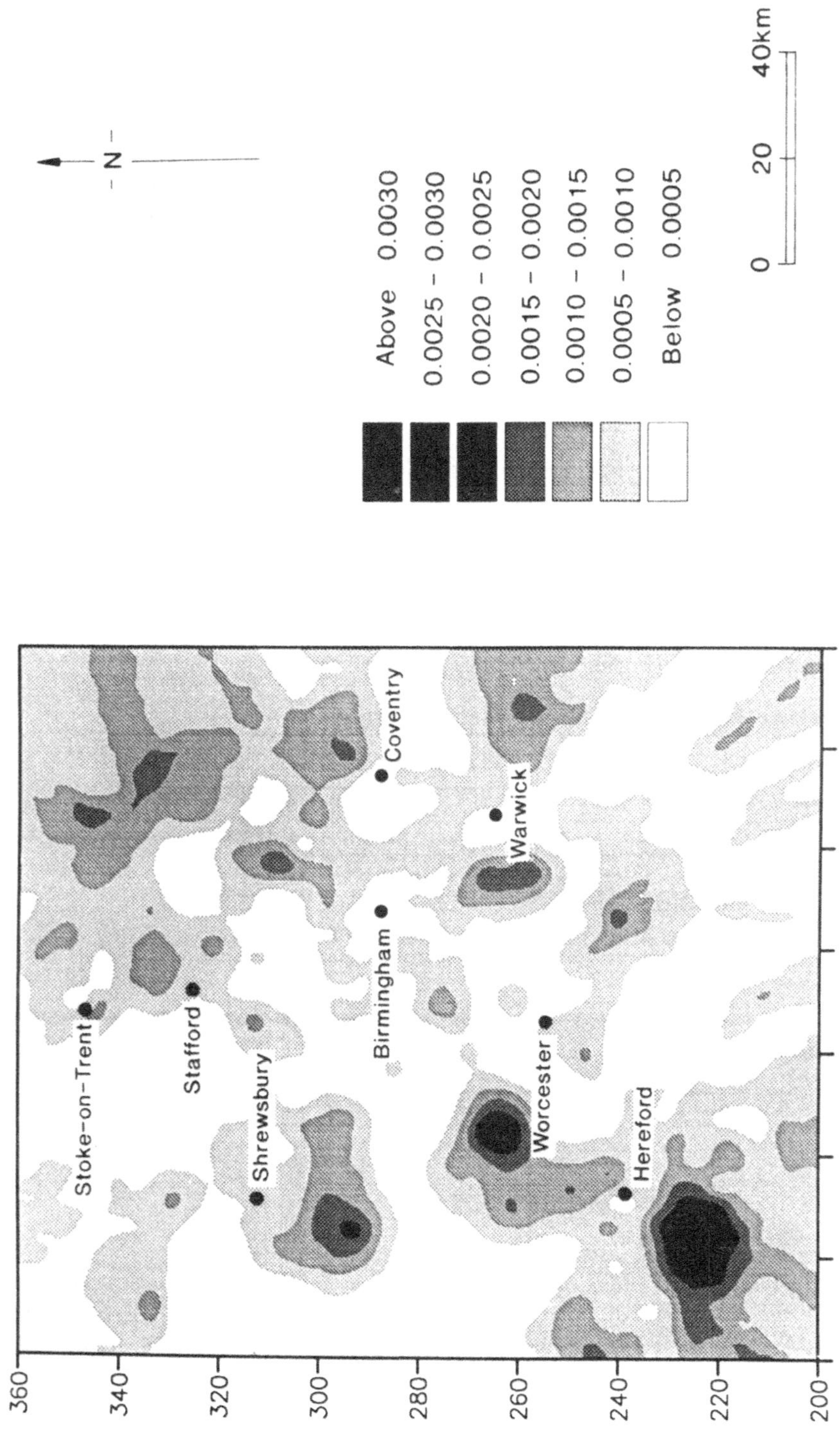

Figure 3. Map of the estimated risk of childhood cancer in the West Midlands made by quasi-punctual kriging.

The semivariances of frequency are much larger than those of the risk, Fig. 2a, and the projection of the empirical values to the ordinate has an appreciable nugget variance. This nugget is the principal difference between the two variograms. It represents the error in estimating the risk from the few cases among the finite populations. It may also be thought of as the chance of a child's developing the disease that is quite unrelated to the underlying risk.

Mapping the risk

We converted $\gamma_R(h)$ to the covariance function, $C^R(\mathbf{h})$, using equation (18), and from it we obtained the values for the kriging equations. The risk and the associated estimation variance were estimated at 2 km intervals on a square grid over the Region with n, the number of local data (wards) set to a maximum of 20. We then 'contoured' the estimates to produce an isarithmic map, and Figure 3 is the result. The risk has a patchy distribution, with areas where the risk is large and others where it is small. The rural areas and the outer suburbs have the largest underlying risk, especially in the southwestern and western parts of the Region, whereas the risk is small in the towns and cities in general. This contrasts with the general impression gained from the distribution of cases in Fig. 1.

The estimation variances from kriging the risk were mapped in the same way as the risk. The estimation variances are large at the edges of the region, and they mark the boundary of the West Midlands Health Authority Region clearly. The smallest values occur in the urban areas where the wards are smallest and where in consequence the data are most dense. However, the estimation variances are only a little larger in the rural areas where the data are more sparse.

DISCUSSION AND CONCLUSION

This analysis shows how the data for a rare disease such as childhood cancer can be analysed geostatistically with advantage. With a rare disease the rates of incidence are always going to be small. Local counts will be similarly small, and the errors associated with them will be large therefore. For this reason large sets of data will be needed to use geostatistical tools with confidence. We did not know where all the healthy children lived during the five years of the study, and so we had to work on data for electoral wards, which are not all of the same area. In the usual geostatistical setting this implies computing a variogram over a variable support, and the result is some kind of average variogram over the range of support. In our case, in which our data are incidences based on counts from known populations and of known error, this is probably less important because we already take into account the varying parent population. However, more research is necesary to improve estimation of the covariance, to compare different methods of estimating the risk, and to take account of the variation in ward areas, v_i.

The test for homogeneity in the risk together with the kriged map of the risk, Fig. 3, suggests strongly that the chance of a child's developing cancer is not purely random in the West Midlands.

ACKNOWLEDGEMENTS

We thank Dr C. Daly for his advice, Mrs J. Dowling for preparing the figures, the West Midlands Regional Health Authority and the Special Trustees of the Research Endowment Fund for financial support, the Birmingham and West Midlands Regional Cancer Registry and Dr G. J. Draper and Mr C. A. Stiller for access to data; medical records officers, consultants and general practitioners for access to their records and histopathology material; and the Office of Population Censuses and Surveys for the 1981 population data.

REFERENCES

BESAG, J. & NEWELL, J. 1990. The detection of clusters in rare diseases. *Journal of the Royal Statistical Society, Series A* **154**, 143-155.

BITHELL, J. F. 1990. An application of density estimation to geographical epidemiology. *Statistics in Medicine* **9**, 691-701.

CHOYNOWSKI, M. 1959. Maps based on probability. *Journal of the American Statistical Association* **54**, 385-388.

CUZICK, J. & EDWARDS, R. 1990. Spatial clustering for inhomogeneous populations. *Journal of the Royal Statistical Society, Series B* **52**, 73-104.

LAJAUNIE, C. 1991. *Local Risk Estimation for a Rare Noncontagious Disease based on Observed Frequencies.* Centre de Géostatistique de l'Ecole des Mines de Paris, Fontainebleau, Note N-36/91/G.

MCNEILL, L. 1991. Interpolation and smoothing of binomial data for the Southern African Bird Atlas project. *South African Statistical Journal* **25**, 129-146.

MUIR, K. R., PARKES, S. E., MANN, J. R., STEVENS, M. C. G., CAMERON, A. H., RAFAAT, F., DARBYSHIRE, P. J., INGRAM, D. R., DAVIS, A. & GASCOIGNE, D. 1990. 'Clustering' - Real or Apparent? Probability maps of childhood cancer in the West Midlands Health Authority Region. *International Journal of Epidemiology* **19**, 853-859.

OPENSHAW, S., CRAFT, A. W., CHARLTON, M. & BIRCH, J. M. 1988. Investigation of leukaemia clusters by use of geographical analysis machine. *The Lancet*, i; 272-273.

POTTHOFF, R. F. & WHITTINGHILL, M. 1966. Testing for homogeneity. I. The binomial and multinomial distributions. *Biometrika* **53**, 167-190.

WHITTLE, P. 1954. On stationary processes in the plane. *Biometrika* **41**, 434-449.

SOME EPIDEMIOLOGIC APPLICATIONS OF KRIGING

DANIEL WARTENBERG
Department of Environmental and Community Medicine
UMDNJ-Robert Wood Johnson Medical School
675 Hoes Lane
Piscataway, NJ 08854

Environmental epidemiology is the study of the association between exposure to environmental agents and the occurrence of disease. Exposure data typically are expensive and difficult to collect, and epidemiologists try to make efficient use of all available data. Nonetheless, quantification of exposure is often the weakest link in the investigation of the chain of causation, and exposure assessment is the area most subject to improvement. In this study, I try to improve the use of exposure data by exploring methods which may increase the statistical power of the epidemiologic assessment. Since many environmental exposures are coherent in space (e.g., a plume of polluted air downwind of a factory, a path of polluted water downstream from a discharge point), the application of geostatistics offers a wide range of opportunities for improving exposure models and providing more meaningful epidemiologic assessment. In this paper, I discuss two applications of kriging to epidemiology from a recent investigation of a cancer excess on Cape Cod, Massachusetts. I note at the outset that this study is mainly exploratory and, given the nature of the environmental monitoring data used, substantive results should be interpreted cautiously and viewed as suggestive at best.

Cape Cod is a peninsula on the southeast end of Massachusetts. An excess of cancer incidence on Upper Cape Cod was reported by the Massachusetts Department of Public Health based on data for 1982-1984. To determine possible causes of the observed excess, investigators proposed a case-control study examining all cases of nine types of newly diagnosed cancers among permanent Upper Cape residents reported to the Massachusetts Cancer Registry from 1983-1986 (Aschengrau and Ozonoff 1991). In a case control study, one compares the distribution of exposures among a sample of those with the disease of concern to the distribution of exposures among a sample of those without the disease. The current study was designed to assess the association between a set of environmental exposures and the distribution of the cancer among the permanent population of the region. This study looked at two of these environmental exposures that were believed to have characteristic spatial patterns: underground drinking water contamination by organic chemicals and microwave radiation from a radar facility known as PAVEPAWS.

In both situations, as is typical in many epidemiologic investigations, exposure measurements were made prior to the design of the case-control study and were not geographically coincident with subject residence locations. The fact that exposure measurements preceded disease detection is not necessarily a problem since cancer takes many years to develop after an individual is exposed. Ideally, however, in conducting an epidemiologic assessment, one would take a series of measurements over time prior to diagnosis, and would model the latency of cancer incidence after exposure. The lack of spatial coincidence between the exposure measurement locations and subject residence locations precluded direct epidemiologic evaluation and required geographic interpolation.

Another concern in conducting an epidemiologic study is limiting possible investigator bias. To do so it is necessary to evaluate or categorize each subject's exposure independent of that person's disease status. Thus, exposure evaluations and kriging were conducted without knowledge of the disease status of the study subjects, and only after the exposure values were determined were case and control locations compared with respect to exposures.

In the first investigation, groundwater measurements were interpolated to provide exposure estimates for each subject. I conducted a simulation study to compare the statistical power of a variety of interpolation methods, including kriging, for detecting a significant exposure-disease relationship. After verifying the suitability of kriging for this purpose, I applied this interpolation methods to the real data.

In the second example, the microwave radiation exposure data were sparse overall and the sampling distribution was highly clustered spatially. I investigated the interpolation of the environmental samples but found it problematic due to the data distribution. Instead, case-control status was interpolated for each exposure measurement location (without regard to the measured exposure value) and then compared these values with the observed exposures. To investigate the validity of the conclusions, I ran a set of simulations in which the case and control locations were permuted randomly.

The epidemiologic measure of association used for these data is the odds ratio, an estimate of the relative risk. Case and control subjects are selected without regard to exposure status. Exposure status is then determined for each subject. The odds ratio is calculated as the ratio of the odds of being exposed given that one is a case to the odds of being exposed given that one is a control. The notion underlying this approach is that since cases and controls are selected on the basis of disease status but without knowledge of exposure, comparisons of exposure status (evaluated retrospectively) can be used to infer association and possible causation (Rothman 1986). Since cancer incidence reporting is relatively complete for this database (Richard Clapp, personal communication), it is unlikely that there is bias in the disease status data. The exposure determinations are the critical data for assessing exposure-disease relationships.

Drinking Water Contamination

The Ashumet Valley area is the site of a contaminated groundwater plume approximately 11,000 feet long and 3,500 feet wide. The plume contains a variety of contaminants including a sampling of volatile organic compounds, VOCs, and substances characteristic of sewage (Aschengrau and Ozonoff 1991). The contaminants are believed to have come from a variety of sources not specifically defined for the purposes of this study. Residents of this area received the drinking water from private wells that drew directly from the groundwater and thus may have ingested these substances. Many of the substances are known or suspected carcinogens when ingested via drinking water. For our investigation, we wanted to know which wells were contaminated and the specific level of contamination in each well. Since the multiple source contamination problem was difficult to model directly, the measured levels of contaminants at a few arbitrary wells were used to estimate exposure for our epidemiologic assessment.

To test the statistical power of using kriging for interpolation of exposure data, with colleagues I undertook a series of simulation studies (Wartenberg et al. 1991). First, two hypothetical plumes were modeled: a step-function plume and a parabolic plume. We used 25 randomly placed measurement locations and included random measurement error. We compared the association of the interpolated measurement values with the status of 50 randomly located, hypothetical cases and 50 randomly located, hypothetical controls. To derive the interpolated values, we used three methods: assigning the value of the nearest neighbor to the subject location (i.e, a direct assignment method), using inverse distance squared weighting on all observed points (i.e., a weighted average method), and using kriging with a variogram based on the hypothetical plume model (i.e., a weighted average method tuned to the data). The resulting interpolation data were assessed using a t-test for direct comparison of exposure among hypothetical cases and controls and an odds ratio as the traditional epidemiologic index of effect.

For the kriging estimator, variogram parameters were needed. To derive estimates, linear variograms were fit to the model data. Figure 1 shows the model variogram data (labeled as Truth 144) and the fitted variogram for the step function plume (dashed line). The fitted variogram had a nugget value of 0, a sill of 0.27 and a range of 6. The fitted variogram for the parabolic plume had a nugget value of 0, a sill of 0.06 and a range of 5.5 (not shown). The results of the simulations (Table 1) showed that mean exposures between hypothetical case and control households were not statistically significantly different, and that mean odds ratios, while elevated, were not statistically significant. Surprisingly, inverse distance squared weighting interpolations outperformed kriging although the difference was not statistically significant. We suspected that this differential was due to our use of a variogram based on the hypothetical plume rather than the observed data.

To investigate this question further, I undertook a second, more detailed simulation study (Wartenberg 1992). Since there was no appreciable difference in the statistical results for the different plumes (although the step-function plume showed greater differences among hypothetical cases and controls), I used only the step-function plume in the second simulation study. Rather than picking arbitrary values for the other parameters of the model, I did a series of simulations for which I varied each parameter independently. The parameters considered were: number of samples (25 to 150), measurement error (0.75 to 5.0 units, where clean=0.0 and contaminated=1.0), magnitude of hypothesized relative risk of exposure to cases relative to that among controls (1=no association to 5=strong association), number of total subjects (for a 1:1 case to control ratio, 50 to 300 subjects), number of controls (for 50 cases, 25 to 150 controls), and the ratio of cases to controls (for 300 subjects, from 1:1 to 1:9 cases to controls). Finally, two kriging estimators were used, one with a variogram based on the hypothetical data (as in the first simulation study) and one with a variogram based on simulated measurement data (i.e., including measurement error). For each set of parameters, 500 simulations were run.

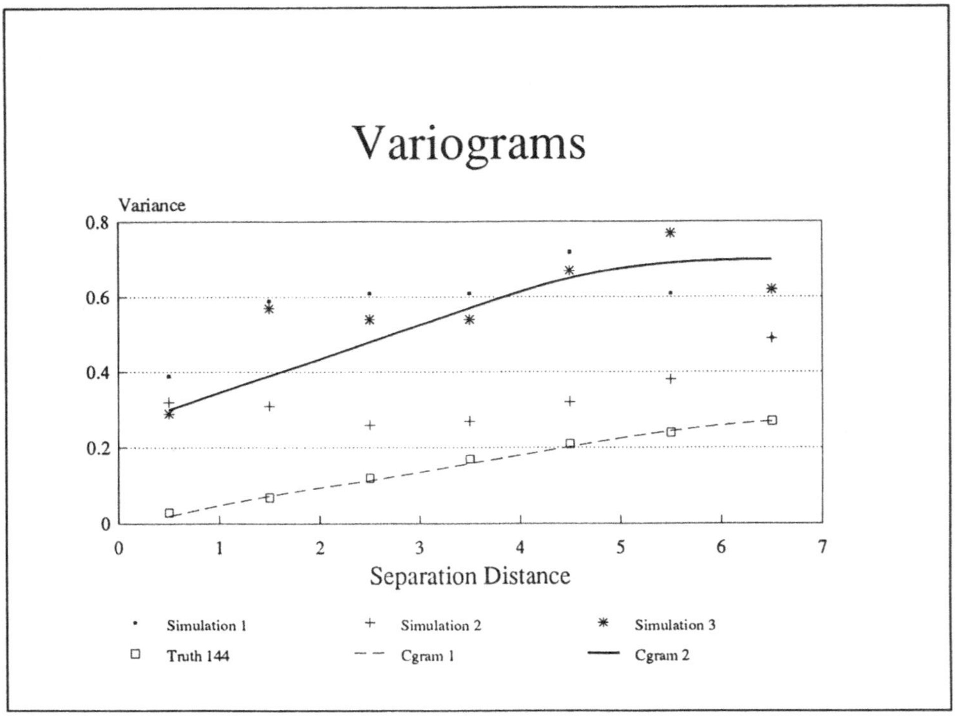

Figure 1: Variograms

There was great variability in the variogram data from simulation to simulation (Figure 1). To conduct our simulations, arbitrary simulations were fit by eye to a linear variogram model consistent with the general pattern of the points (Figure 1--solid line). This variogram was similar in shape to the variogram fit to the hypothetical data but unlike the hypothetical data included a substantial nugget effect. Selected results for the power calculations from this study are shown in Figure 2. (Results for data with no measure error are labeled as TRU; results for kriging with the variogram based on data with no measurement error are labeled as KRG1; results for kriging with the variogram based on data with measurement error are labeled as KRG2; results using nearest neighbor values are labeled as NN; results using the mean of the 8 nearest neighbors are labeled NNAVE; results using the means of the 8 nearest neighbor values weighted by the inverse square of their distance are labeled as ISD; results using the hypothetical value with measurement error at the subject location is labeled as MSR.) These plots show that kriging with the variogram estimated from the simulated measurement data outperformed all other methods evaluated. The statistical power of the methodologies varied most dramatically with the relative risk, the exposure measurement error (or noise) and the number of subjects. Yet, even for moderate risks with a relatively small study design, this approach has adequate power to detect the hypothesized effect. For example, with relative risks as low as 3, with random measurement noise equal to twice the true range of the data variation, and with sample sizes exceeding 150 subjects (75 cases and 75 controls), the best methods showed power exceeding 60%. The number of exposure samples, the number of control subjects and the case to control ratio had markedly smaller effects. The geometric mean odds ratios resulting from the simulations (not shown) showed a much smaller dependence of these parameters, although again the relative risk, the exposure measurement error (or noise) and the number of subjects had the largest effects. Complete results are available from the author.

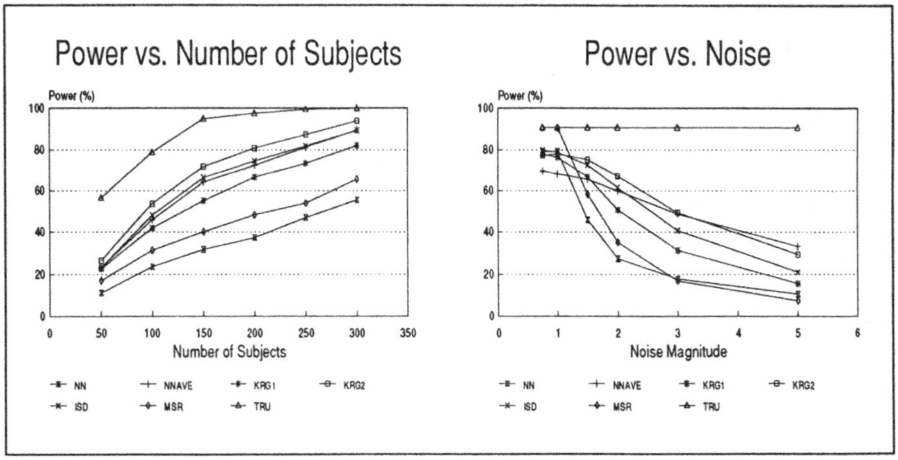

Figure 2: Simulation Results

From this study, I concluded that kriging is a useful tool for interpolating groundwater contamination data. Provided that the plume is relatively smooth, as often is suggested by groundwater hydrologists, epidemiologists should concentrate on improving the accuracy of the exposure measurements and on increasing the number of subjects in the study to maximize the statistical power. According to these simulations, fewer exposure samples, if measured more accurately, are better than more measurements with less accuracy. Kriging, while only slightly better than more arbitrary weighting schemes, is consistently better and would likely be even more so in situations of less coherent spatial patterns.

I finally applied kriging to the observed groundwater contamination data. Groundwater quality measurements were made at 59 locations in the region in which study subjects resided. Neither of the two principal contaminants showed statistically significant linear or quadratic trends (drift). Kriging results using presence or absence of either contaminant as an indicator variable show a relatively smooth, well-dispersed plume. Once the location of this plume was determined, exposure values were assigned to each subject living in this region. Two cases and five controls were found to live within the confines of the plume. These data were too sparse for further epidemiologic assessment. No association was detectable, although the data distribution preclude definitive assessment of risk from the contamination.

Microwave Radiation

PAVEPAWS is a Precision Acquisition of Vehicle Entry, Phased Array Warning System operated by the Cape Cod Air Station on Flat Rock Hill. It is designed to detect and track incoming, sea-launched ballistic missiles and to provide support for tracking objects in space orbit. It is a stationary, computer controlled radar transmitter that produces a flat blanket of electromagnetic radiation by sending out electronic pulses in different directions. Emission frequencies are equally spaced between 420 MHz and 450 MHz across a 240° sector pointing offshore, approximately from NNW to SW (Aschengrau and Ozonoff 1991). A number of study subjects live within the path of the emissions.

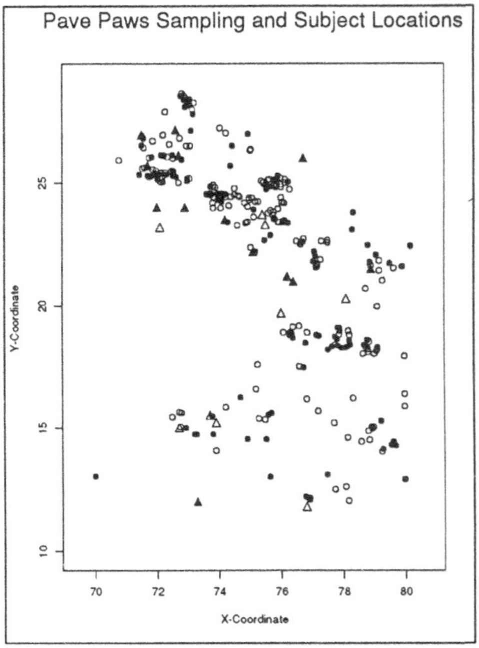

Figure 3: PAVEPAWS Data

Measurement of the electromagnetic field resulting from PAVEPAWS is very difficult technically and results in a reported margin of error of ±60% (Aschengrau and Ozonoff 1991). Data for the average field density were reported for 23 locations and for peak field density at 12 of those same locations. The sampling site locations were selected to reflect the areas of greatest population density (towns of Bourne, Sandwich, Mashpee and Falmouth). The biologically absorbed dose of this radiation resulting from these fields is mediated by a variety of factors including the source configuration, the resonant frequency of a person, the size and shape of the person, and the radar power output. There were 304 subject residences in this region, 132 cases and 172 controls.

Figure 3 shows a plot of the data in an arbitrary coordinate system. The open circles show the location of control subjects and the solid circles show the location of case subjects. The open triangles show the location of exposure measurements below the detection limit and the solid triangles show the location of the exposure measurements above the detection limit. There is no obvious pattern in the data.

The first analytic approach I used on these data was interpolation of the measured microwave average field densities to each subject location, and comparison of these values among cases and controls using traditional epidemiologic measures. Review of the spatial location of the measurement sites showed them to be highly clustered and nearby measurement values to be inconsistent. Interpolation, and even estimation of the variogram, was problematic. There was no linear or quadratic trend (drift) in the data. A kriging-derived contour map of the measured exposure data (using a linear variogram) superimposed on the location of study subjects did not reveal an obvious pattern (Figure 4). Given the lack of resolution of the exposure data, these data do not warrant further epidemiologic analysis.

As an alternative approach, I interpolated subject status (case=1; control=0) to each measurement location using indicator kriging (Journel and Huijbregts 1978; Journel 1983). First, the drift in the data was evaluated and there were no statistically significant linear or quadratic trends. Then, the binary data were fit to a variogram. These data, when analyzed in bands of ±22.5 degrees, showed substantial anisotropy, and fairly noisy variograms (Figure 5). The 45 degree direction showed the most predictable pattern, generally increasing with increasing separation distance. Other bands did not show consistently increasing variation, and the isotropic variogram was even less distinct. The data were fit to a variogram with a nugget effect of 0.2, a sill of 0.1 and a range of 4. Ordinary kriging was then used to interpolate probabilities of being a case for each exposure measurement location (Figure 6).

To evaluate the results of this interpolation epidemiologically, we regressed the measured exposure values on the probabilities. We also estimated case probabilities using inverse distance squared weighting, and regressed the exposure values on these probabilities. The results did not show a statistically significant association (Table 2). The exposure estimates derived using kriging showed a positive slope, indicating that increasing exposure is weakly associated with an increased probability of being a case. Further, when a large outlier was removed, the strength of the association increased as did the size of the regression coefficient. Similar analyses with inverse distance squared weighting interpolation showed substantially weaker associations and did not improve similarly with deletion of outliers. While not conclusive because of the questionable quality and sparsity of the data, this surprising result encouraged further exploration of the data although it did not seem appropriate to tune further the regression based on the kriging estimates.

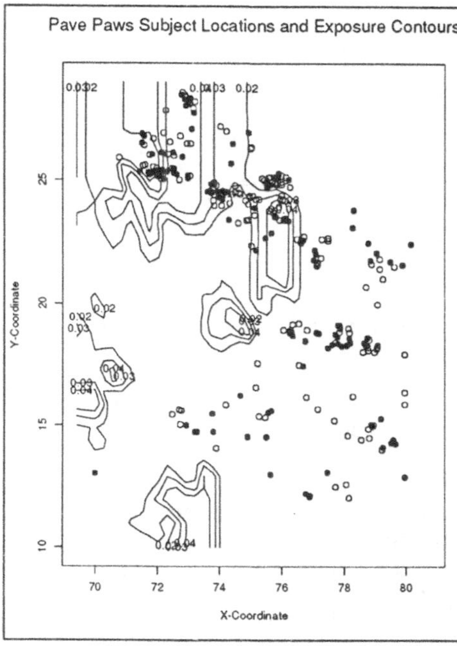

Figure 4: Exposure Interpolation

To further study this problem, we undertook a simulation of an alternative model using the observed data. If, indeed, exposure to high electromagnetic fields from microwaves is a risk factor for cancer, one would expect cases, on average, to be closer to the high exposure locations than the controls, on average. To investigate this, the mean distance for all observed cases to each observed exposure location was compared with the mean distance for all observed controls to each observed exposure location. The underlying hypothesis is that for high exposures mean case distance should be smaller than mean control distance, and that for non-detectable exposures there should be no difference. Since a test of this hypothesis is affected by the location of the subject residences, a simulation was used to determine the null distributions for the distances by randomly permuting all the cases and controls among subject locations (i.e., reallocated all subjects to one of the originally occupied locations without reference to their disease status) and calculating the mean case distance to each site to each site. This process was repeated 1,000 times and the results were used to calculate the distribution of mean case distances under the null hypothesis of no association. Then, I compared the observed mean case distances to the tails of these simulated distribution. This approach is strongly ad hoc and meant

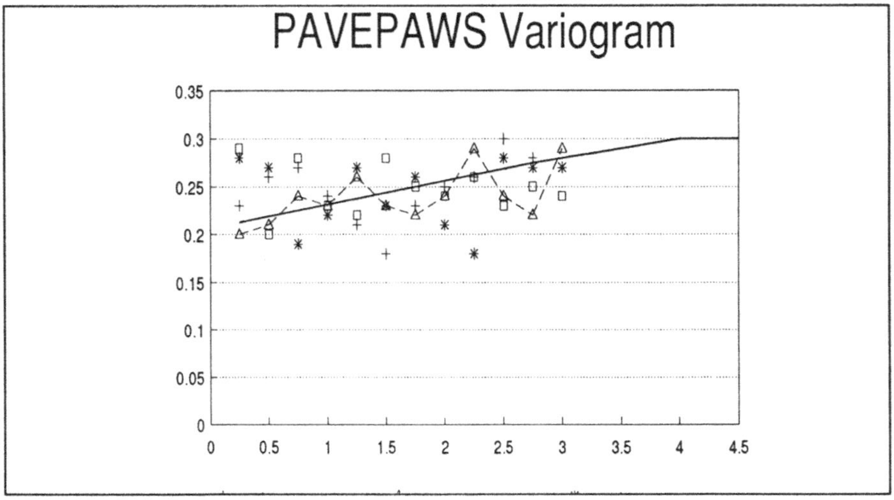

Figure 5: PAVEPAWS Variogram

only as another exploratory evaluation to help determine if further field investigation is warranted.

In this set of simulations the results were not suggestive of an association (Table 3). That is, the mean distance of controls was less than that for cases more often for non-detectable situations. However, this may in part be due to the greater number of control subjects in the study, the small number of measurement locations and the instability of the indices due to the small numbers of samples. The odds ratio is less than 1.0, suggesting a protective effect.

The simulations were run to correct for the number and location of subjects, as well as other potential distributional biases. These results of these simulations were similar but less extreme. For non-zero exposures, the mean distance of the cases was less than the average simulated mean distance 9 out of 14 times using a cutoff of the 30%. For zero exposures, the mean distance of cases was less the simulated mean distance 7 of 9 times. This resulted in an odds ratio of about 0.5, implying that controls are twice as likely to be closer to sites with measurements above background as they are for sites with no detectable radiation. This implies a protective effect.

These simulation results are not consistent with the observed cases being closer to the measurement sites than expected with exposures greater than 0, and thus are not suggestive of increased risk from microwave exposure. Once again, however, because of the severe limitations of the data, the interpretation of this analysis can be resolved only through further study. For example, I have not considered how the field falls off with distance from each site, I have assumed isotropy, I have not considered the time or date of the measurements, nor the ways in which the fields may vary over time. The cancer data were not adjusted for age, gender, birth cohort or years of residence (i.e., total exposure). And, perhaps most importantly, all cancers types have been lumped together rather than being analyzed by specific anatomical sites. (Preliminary analysis by cancer site did not reveal substantially different patterns. Further analysis is not worthwhile without more extensive exposure data.) Further, those types of cancer most likely to be associated with microwave radiation, such as leukemia and brain cancer, are to rare for meaningful interpretation. Given the limited size of the exposure data set and the reported imprecision in the data, I do not believe further analysis of these data is warranted. Rather, because of the equivocal results comparing exposure to microwave radiation from the PAVEPAWS site being associated with increased risk of cancer, additional data are needed. I recommend that a more thorough and systematic exposure assessment be conducted throughout this region and that the health outcome data be reanalyzed.

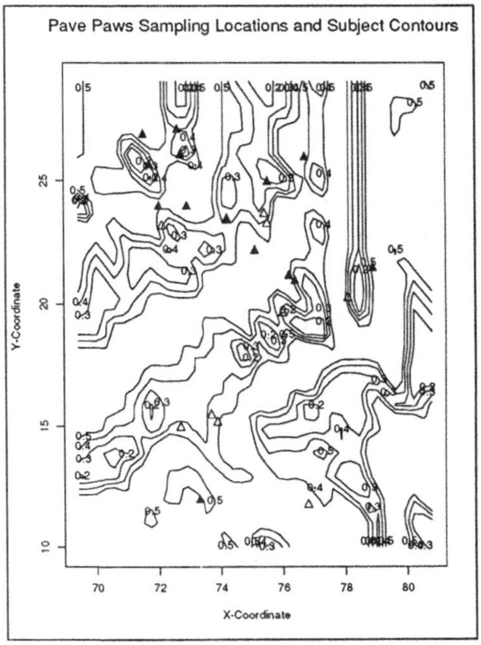

Figure 6: Subject Interpolation

Discussion

Kriging is a methodology little explored in epidemiology. While it is widely acknowledged that exposure assessment is often the weakest link in an epidemiologic study, most efforts at improvement focus on enhanced measurement technology (e.g., biochemical markers as measures of delivered dose) rather than more sophisticated statistical methodology. While this improvement in measurement technology is advantageous, one can also enhance one's epidemiologic efficiency by improving one's use of whatever data are available. This study documents two ways in which the utility of exposure measurements can be improved. First, statistical techniques that are more sensitive to particular aspects of the data, in the way that kriging is responsive to spatial dependence, can improve the power of a given evaluation. Second, computer simulations can be used to specify more accurately the distributions for a given statistical test and the power of an evaluation under specific null and alternative hypotheses.

In the first case study, geostatistics were used to interpolate sparse exposure data to study subject locations. We modeled a typical groundwater contamination event and compared a variety of interpolation algorithms for assessing exposure. Application of these interpolation models to observed drinking water contamination data and subsequent epidemiologic analysis showed no association between contamination and cancer incidence. However, not many subjects lived within the plume, precluding detailed epidemiologic assessment. The simulations showed that kriging, as compared to other interpolation methods, was more complex to implement but had substantially greater statistical power for detecting an association between contamination and disease.

In the second case study, geostatistics were used for interpolation of disease data when the exposure data were sparse and cluster sampled. Because interpolation of cluster sampled data resulted in large errors of estimation, we interpolated more evenly distributed disease incidence data to derive disease probabilities at each resulted in large errors of estimation, we interpolated more evenly distributed disease incidence data to derive disease probabilities at each exposure measurement location and compared these values with the observed exposure data. Again, kriging proved superior to other methods of interpolation although only a weak, statistically not significant association was found with measured microwave field strengths and cancer incidence. Simulation studies evaluating the distance of subjects from exposed sites gave different results which were not consistent with the hypothesized positive association. The lack of consistency of the results of these two methods of evaluation, in the face of sparse and idiosyncratically located data, is suggestive of no association although additional data collection and analysis would help clarify these issues..

These applications are particularly noteworthy in that they both utilize extant environmental data in the study of possible association with adverse health outcomes. Many have argued that such data, compiled for a particular purpose other than epidemiologic investigation, are not useful for the study of exposure-disease relationships. I believe that this broad brush dismissal of extant environmental contamination data is unwarranted and that specific issues can be addressed in a meaningful way. Certainly, collection of new data with a sampling design tailored to address a specific question of concern is preferable, but in times of shrinking budgets, we must learn to make use of existing resources. Further, these data could be helpful in designing a new sampling program. I believe I have shown that, in certain circumstances, one can use such data to prioritize and make decisions regarding follow up to apparent disease excesses.

The second application is specially interesting because it addressed data that were cluster sampled. It is not uncommon for field researchers to concentrate their sampling efforts in areas where contamination is greatest, trying to delineate boundaries of contamination and characterize the volume of contaminant. While understandable, such a strategy may overlook small pockets of contamination and inadequately characterize areas of low, but detectable, contamination. Such cluster sampling greatly complicates the statistical and spatial pattern analysis of the measurements because the data are collected in a biased manner. In this instance, it was fortunate to have an alternate data set, the cases and controls, that could be manipulated for eventual hypothesis testing.

In both case studies, kriging proved a useful and powerful tool for the assessment of epidemiologic data. By providing more accurate estimates of exposure and disease through incorporation of spatial dependence and directionality, kriging helped improve the statistical power of the analyses. These examples are just a couple of ways in which such tools can be used in epidemiology. I look forward to more innovative approaches in future studies.

Acknowledgements

This project was supported in part by NIEHS Center Grant ES05022. The author thanks Ann Aschengrau, Patricia Coogan, David Ozonoff, and Christopher Uchrin for helpful discussions and suggestions. The Massachusetts Department of Public Health and the Massachusetts Department of Environmental Protection provided funding for the Upper Cape Cancer Incidence Study (Aschengrau and Ozonoff 1991).

References

Aschengrau, Ann and Ozonoff, David (1991) "Upper Cape Cancer Incidence Study Final Report", submitted by the Boston University School of Public Health to the Massachusetts Department of Public Health.

Journel, Andre (1983) "Nonparametric estimation of spatial distributions," Mathematical Geology 15, 445-468.

Journel, Andre and Huijbregts, Ch. J. (1978) Mining Geostatistics, Academic Press, London.

Rothman, Kenneth (1986) Modern Epidemiology, Little, Brown and Co., Boston.

Wartenberg, Daniel (1992) "Geographic interpolation of exposure data for epidemiologic investigation", manuscript.

Wartenberg, Daniel, Uchrin, Christopher and Coogan, Patricia (1991) "Estimating exposure using kriging: A simulation study", Environmental Health Perspectives 94, 75-82.

Table 1: Plume Simulations

	Parabolic Plume			Step Function Plume		
	Mean Exposure		Odds Ratio	Mean Exposure		Odds Ratio
	Cases	Controls		Cases	Controls	
Truth	0.16	0.08	2.31	0.50	0.23	3.83
NN	0.15	0.08	1.33	0.46	0.25	1.72
ISD	0.14	0.10	1.76	0.42	0.28	2.26
Krige	0.14	0.08	1.72	0.43	0.24	1.21

No odds ratios were statistically significant at $p \leq 0.05$.

Table 2: PAVEPAWS Regression Results

All Data

	slope	intercept	p-value	adj. r^2
ISD	-0.53	0.48	0.73	<0%
Krige	0.80	0.48	0.47	<0%

Deleting Largest Outlier

	slope	intercept	p-value	adj. r^2
ISD	-1.02	0.49	0.73	<0%
Krige	3.36	0.46	0.10	8.5%

Table 3: Cluster Sampled Simulation Results

Observed Data

	$Dist_{cancer} < Dist_{control}$	$Dist_{cancer} > Dist_{control}$
Exposure > 0	1	13
Exposure = 0	3	6

Simulated Data

	$Dist_{cancer} < Dist_{p=0.05}$	$Dist_{cancer} > Dist_{p=0.05}$
Exposure > 0	9	5
Exposure = 0	7	2

GEOLOGICAL AND STRUCTURAL CONTROL IN KRIGING

P.A. Dowd
Department of Mining and Mineral Engineering
University of Leeds
Leeds LS2 9JT
England, U.K.

ABSTRACT

The paper describes an adaptation of the method of indicator conditioned estimation proposed by Zhu and Journel (1990) with the indicators defined as geological and/or structural categories. Two types of category classification are distinguished : mutually exclusive and non-mutually exclusive. In both cases reductions in the cokriging systems are suggested by means of simple orthogonal transformations.

The methods are illustrated by two case studies in which the effects of the controls on grade estimation and ore/waste discrimination are demonstrated.

INTRODUCTION

Many types of geological and structural information which are required for estimation can be specified in the form of categories. For example :

(a) The sequence of units comprising a stratigraphic orebody. For grade estimation in such orebodies it is usually essential to take account of the structural controls exerted on grades by the constituent units of the sequence.

(b) Rock types and lithological units. These often have a significant and controlling effect on grade distributions.

(c) The presence or absence of additional variables such as silica and sulphides in gold mineralisations.

Two types of categorical classification can be distinguished : those for which the categories are mutually exclusive and those for which they are not. In either case indicators can be used to classify the data into categories :

$$I_k(x) = \begin{cases} 1 & \text{if } x \in \text{category } k \\ 0 & \text{otherwise} \end{cases} \qquad \text{....... (1)}$$

This is slightly different to the usual coding used for indicator kriging in which a variable is coded according to excedance of specified threshold values. In these applications a different indicator variable must be defined for each category.

Let $\{x_i, i=1,n\}$ be the set of locations at which the grade variable is recorded and let $\{x'_j, j=1,m\}$ be the set of locations at which the category indicator variables are recorded. It is assumed that all category indicator variables are recorded at all locations $x'_j : \{x_i\} \subset \{x_j'\}$.

The estimated grade at a given location can then be found by cokriging using grade and category indicator variables in a manner similar to that suggested by Zhu and Journel (1990) :

$$Z^*(x) = \sum_{i=1}^{n} \lambda_i Z(x_i) + \sum_{k=1}^{K} \sum_{j=1}^{m} \nu_{kj} I_k(x_j') \quad \ldots. (2)$$

with $m \geq n$.

The weights λ_i and ν_{kj} are the solutions to the cokriging equations :

$$\sum_{i=1}^{n} \lambda_i C_Z(x_i\ x_\ell) + \sum_{k=1}^{K} \sum_{j=1}^{m} \nu_{kj} C_{ZI_k}(x'_j\ x_\ell) + \mu_0 = C_Z(x\ x_\ell) \qquad \ell = 1,n$$

$$\sum_{i=1}^{n} \lambda_i C_{ZI_{k'}}(x_i\ x'_r) + \sum_{k=1}^{K} \sum_{j=1}^{m} \nu_{kj} C_{I_k I_{k'}}(x'_j\ x_r) + \mu_{k'} = C_{ZI_{k'}}(x\ x'_r) \qquad r = 1,n \quad k' = 1,K$$

$$\sum_{i=1}^{n} \lambda_i = 1 \qquad \sum_{j=1}^{m} \nu_{kj} = 0 \qquad k = 1,K$$

or, in dual form (Galli et al., 1984, Royer and Vieira, 1984) :

$$Z^*(x) = \sum_{i=1}^{n} \alpha_i C_Z(x_i\ x) + \sum_{k=1}^{K} \sum_{j=1}^{m} \beta_{kj} C_{ZI_k}(x_j'x) + \omega_0$$

$$I^*_k(x) = \sum_{i=1}^{n} \alpha_i C_{ZI_k}(x_i\ x) + \sum_{k'=1}^{K} \sum_{j=1}^{m} \beta_{k'j} C_{I_k I_{k'}}(x_j'x) + \omega_k$$

subject to :

$$\sum_{i=1}^{n} \alpha_i C_Z(x_i\ x_\ell) + \sum_{k=1}^{K} \sum_{j=1}^{m} \beta_{kj} C_{ZI_k}(x'_j\ x_\ell) + \omega_0 = z(x_\ell) \qquad \ell = 1,n$$

$$\sum_{i=1}^{n} \alpha_i C_{ZI_k}(x_i\ x'_r) + \sum_{k=1}^{K} \sum_{j=1}^{m} \beta_{kj} C_{I_k I_k}(x'_j\ x_r) + \omega_{k'} = I_{k'}(x'_r) \qquad r = 1,n \quad k' = 1,K$$

$$\sum_{i=1}^{n} \alpha_i = 0 \qquad \sum_{j=1}^{m} \beta_{kj} = 0 \qquad k = 1,K$$

Or alternatively, these equations can be expressed in matrix form (see Myers, 1982).

In some applications the grade variable, Z(x), can be replaced by an additional indicator variable defined by an operating cut-off grade; the resulting estimator can then be regarded as a composite index for discriminating between ore and waste in detailed mine planning. This is particularly useful in many low grade gold orebodies where the erratic nature of the mineralisation makes ore/waste discrimination very difficult.

The main objection to using such formulations is the amount of modelling and inference required. Some simplifications, which avoid excessive modelling, are given below.

CATEGORY CLASSIFICATION

Mutually exclusive classification

If the categories are mutually exclusive then any correlations between the indicators are spurious and are generated by the fact that they form a closed array :

$$\sum_{k=1}^{K} I_k(x) = 1$$

In this case, if $E[I_k(x)] = p_k$ then $Var[I_k(x)] = p_k(1 - p_k)$, $E[I_k(x)I_{k'}(x)] = 0$ and the covariance between any two indicator variables is thus :

$$Cov[I_k(x)I_{k'}(x)] = - p_k p_{k'}$$

The covariance and cross-covariance functions of the indicators are related by :

$$\sum_{k'=1}^{K} C_{I_k I_{k'}}(h) = 0 \qquad \text{.... (3)}$$

In addition :

$$E[Z(x)] = \sum_{k=1}^{K} E[Z(x)I_k(x)]$$

and, as a consequence :

$$\sum_{k=1}^{K} C_{ZI_k}(h) = 0 \qquad \text{.... (4)}$$

Analagous results to (3) and (4) can be obtained for variograms and cross-variograms provided there are no offset lag correlations.

As pointed out by Pawlowsky (1985), such relationships result in singular kriging matrices. However, if there are K mutually exclusive categories then only K-1 category indicators need be defined and the singularity of the matrices is avoided. However, there is no guarantee of consistent results and there is a problem of bias towards negative values of the cross-covariances as pointed out by Pawlowsky (1985). For example, the implicitly coestimated indicators should be such that :

$$\sum_{k=1}^{K-1} I_k^*(x) \leq 1$$

Short of imposing constraints on the dual formulation of cokriging given above this condition will not, in general, be satisfied.

In the mutually exclusive case a single indicator variable can be defined by :

$$S(x) = \sum_{k=1}^{K} k\, I_k(x) \qquad \text{.... (5)}$$

This variable was used for simulation purposes by Matheron et. al. (1987). The covariance function of the new variable and those of the original category indicators are related by :

$$\text{Cov}[S(x)S(x+h)] = \sum_{j=1}^{K} \sum_{k=1}^{K} j\, k\, \text{Cov}[I_j(x)I_k(x+h)]$$

and similarly for variograms and cross-variograms provided that there are no offset lag correlations.

This approach does not allow for the estimation of the probability of occurrence of the individual categories, i.e. $I_k^*(x)$. However a method suggested by Matheron (1979) in another context can be used to obtain such estimates. It can be shown that S(x) is autokrigeable if and only if :

$$\sum_{k'=1}^{K} k' \text{cov}[I_k(x)I_{k'}(x+h)] = \theta_k \, \text{cov}[S(x)S(x+h)]$$

This provides a way of defining each individual indicator as the sum of two orthogonal components :

$$I_k(x) = Y_k(x) + \theta_k S(x)$$

with :

$$\sum_{k=1}^{K} k\, Y_k(x) = 0 \qquad \text{and} \qquad \sum_{k=1}^{K} k\, \theta_k = 1$$

which can be kriged independently. By assuming that the $Y_k(x)$ are in intrinsic coregionalisation the values of these variables can be kriged independently using the same covariance function. This approach is developed further in the following section.

Non-mutually exclusive classification

Non-mutually exclusive classification applies to the presence or absence of certain control variables such as the gold mineralisation example given above.

If the categories are not mutually exclusive then K category indicators must be defined and the simplifying definition given in (5) cannot be applied. There is similarly no implicit constraint on the co-estimated category indicators. In this case the system can be simplified by defining orthogonal transforms of the category indicators and the grade variable. One approach would be to use Principal Components as suggested by Suro-Perez and Journel (1990), another is to use a dummy variable defined as the sum of the category indicators at each location as follows.

Define :

$$Z(x) = \theta_0 S(x) + Y_0(x) \qquad \text{where} \qquad S(x) = \sum_{k=1}^{K} I_k(x)$$

Orthogonality of $Y_0(x)$ and $S(x)$ can be imposed by setting :

$$\theta_0 = \frac{\sum_{k=1}^{K} \text{cov}[Z(x)I_k(x+h)]}{\sum_{k=1}^{K}\sum_{k'=1}^{K} \text{cov}[I_k(x)I_{k'}(x+h)]} \qquad \text{..... (6)}$$

a value for θ_0 can be obtained by putting $h = 0$ in (6) :

$$\theta_0 = \frac{\text{cov}[Z(x)S(x)]}{\text{var}[S(x)]}$$

$Y_0(x)$ and $S(x)$ are orthogonal for $h = 0$. If it is assumed that they are orthogonal for all h then they can be estimated independently by separately kriging the value of each variable. Note that Y_0 is only defined at locations $\{x_i\}$ whereas S is defined at $\{x_j'\}$. The kriged estimates at a given location x_0 are :

$$Y_0^*(x_0) = \sum_{i=1}^{n} \lambda_i Y_0(x_i) \qquad\qquad S^*(x_0) = \sum_{j=1}^{m} \beta_j S(x_j')$$

where the weights λ_i and β_j are determined from the usual kriging systems using the covariances of the $Y_0(x)$ and the $S(x)$ respectively.

However, if estimates of $I_k(x)$ are required a compatibility condition must be imposed on the estimates of the sum and the individual components :

$$\sum_{k=1}^{K} I_k^*(x) = S^*(x)$$

This condition is automatically achieved by a complete cokriging using all K indicators (there is assumed to be no undersampling of the category indicator variables). But as this would defeat the purpose of the original simplification, the following approach, suggested by Matheron (1979), is used.

Define : $$I_k(x) = Y_k(x) + \theta_k S(x) \qquad \text{for } k = 1,K$$

If : $$\theta_k = \frac{\sum_{k'=1}^{K} \text{Cov}[I_k(x)I_{k'}(x)]}{\text{Var}[S(x)]} = \frac{\text{cov}[I_k(x)S(x)]}{\text{var}[S(x)]}$$

then : $$\sum_{k=1}^{K} \theta_k = 1 \qquad \text{and} \qquad \sum_{k'=1}^{K} \text{Cov}[Y_k(x)Y_{k'}(x+h)] = 0$$

The $Y_k(x)$ are orthogonal to $S(x)$ and the values of the two variables can thus be estimated independently by separately kriging them.

A further simplification is introduced by assuming that the Y_k are in intrinsic coregionalisation so that the complete cokriging of the $Y_k(x)$ from $\{Y_{k'}(x'_j), k'=1,K\}$ is the same as the individual kriging of $Y_k(x)$ from $\{Y_k(x'_j)\}$.

At a given location x_0 the Y_k value can be kriged from the surrounding Y_k data values:

$$Y_k^*(x_0) = \sum_{j=1}^{m} \alpha_j \, Y_k(x'_j)$$

where, because the Y_k are intrinsically coregionalised, the weights α_j are identical for each Y_k and are determined from the usual kriging systems using the intrinsic component of the covariances of the $Y_k(x)$. Matheron (1979) shows that, by minimising the weighted error variance $\Sigma_k W_k \, \mathrm{var}[I_k(x) - I_k^*(x)]$, with $W_k = 1/\theta_k$, the intrinsic component of the covariance function required for kriging is :

$$C(h) = \sum_{k=1}^{K} \frac{1}{\theta_k} \mathrm{cov}[I_k(x)I_k(x+h)] - \mathrm{cov}[S(x)S(x+h)]$$

and thus the only models necessary are the covariance functions of the individual $I_k(x)$ together with the covariance function of S; no cross-covariance functions are required.

The estimated indicator values are then :

$$I_k^*(x_0) = \theta_k \sum_{j=1}^{m} (\alpha_j - \beta_j) \, S(x_j') + \sum_{j=1}^{m} \beta_j \, Y_k(x_j')$$

These values give the estimated probability of occurrence of each category at x_0.

CASE STUDIES

These methods are illustrated by application to two case studies in which the results of kriging with and without the categorical controls are compared. In particular, the ability to account for stratigraphic controls is demonstrated.

Case study : low grade gold mineralisation

This study is based on an erratic gold mineralisation in which some 85% of the gold is associated with more or less vertically dipping quartz veins, the remainder, which is of little economic consequence, is associated with sulphides. The orebody is sampled by percussion drilling at an inclination of 45° in the plane perpendicular to the general strike of the quartz veins. There is no statistically significant correlation between the quartz and gold contents of samples. This may be as a result of the gold being associated with the contacts between granodiorite and quartz rather than directly with quartz in which case gold would be correlated with the number of such contacts in the sample rather than with the proportion of quartz; because of the nature of the sampling this information is not available. A study of the drillhole samples indicated that 96% of all non-sulphide samples with a significant gold grade and no quartz were within 3m of a sample containing quartz in the same drillhole. In addition, both quartz veining and gold occurrence appear to be associated with various intrusive units of porphyries and breccias.

In practice, the sample data are used together with meticulous and time consuming face and bench mapping to determine whether a block is ore or waste. Geostatistical and manual methods which use sample data alone result in the classification of many ore blocks as waste. After blasting of assumed waste blocks the broken rock is checked for significant quartz veining and this may result in reclassification of the block.

A quartz category indicator is defined as follows :

$$I_1(x) = \begin{cases} 0 & \text{if sample has no quartz and there is no quartz within 3m of the sample} \\ 1 & \text{if sample contains quartz or there is an occurrence of quartz within 3m of the sample} \end{cases}$$

Blasthole cuttings are not analysed for gold because it is believed that the orientation of the quartz veins would result in biased sampling from vertical holes. However, the presence or absence of quartz in the cuttings is noted and is used to assign indicator values to these samples. Quartz veins and fractures in a block can also be noted from exposure on the level above and from the bench face and these can be used in assigning quartz category indicator values to the block to be estimated.

For the purposes of this study the significant breccias and porphyries are treated as a single category and a second category indicator is defined by their presence/absence in drillhole and blasthole samples :

$$I_2(x) = \begin{cases} 0 & \text{if units absent} \\ 1 & \text{if units present} \end{cases}$$

The primary variable (Z in the original formulation) is an indicator defined by the operating cut-off grade :

$$I_0(x) = \begin{cases} 1 & \text{if grade} \geq \text{cut-off grade} \\ 0 & \text{otherwise} \end{cases}$$

Variograms of these three indicators are shown in figures 1, 2 and 3.

Ore blocks are defined as those blocks which are identified by mapping and sample data to be above the cut-off grade and after mining confirmed to be so. Approximately 15% of blocks identified as ore by this method ultimately prove to be below cut-off grade. A test area, comprising 87 blocks, was selected within which tight control could be exercised over the grade determination of mined blocks. The blocks were assigned an actual indicator value of either 1 or 0 depending on whether the mined block was above or below cut-off; all other blocks are assumed to be waste and were assigned an actual indicator of 0. A composite ore index was then cokriged from I_0, I_1 and I_2 using the methods given above and the regression of actual indicator on estimated index was calculated. Conditional histograms of estimated index given actual index were also

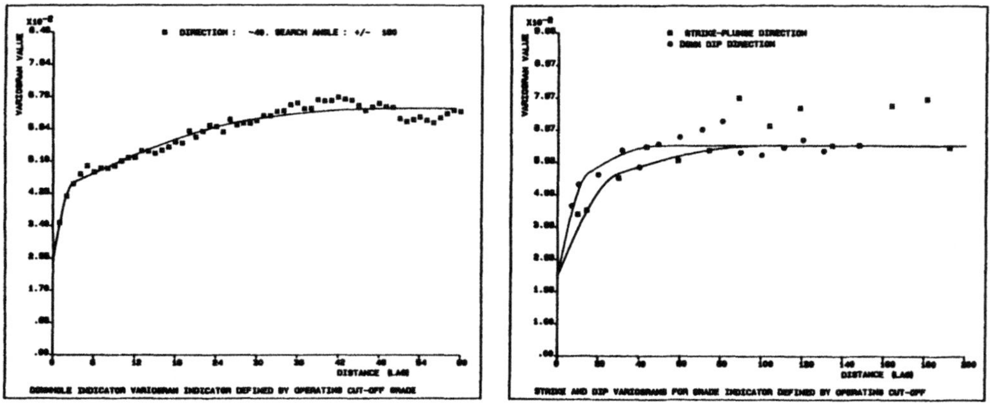

FIGURE 1 : grade indicator variograms

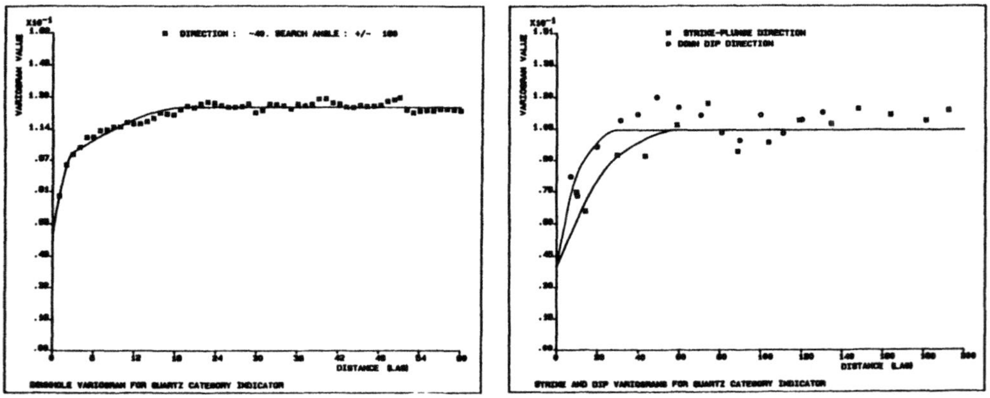

FIGURE 2 : quartz category indicator variograms

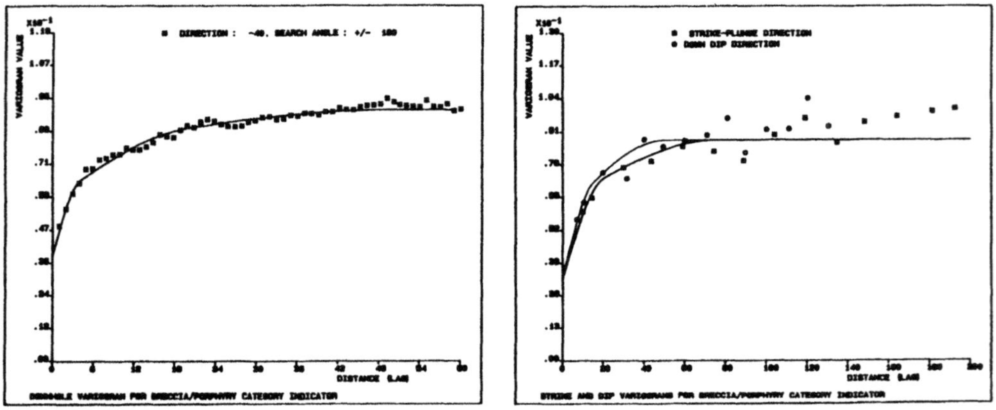

FIGURE 3 : breccia/porphyry category indicator variograms

calculated. The results are summarised in figures 4 and 5.

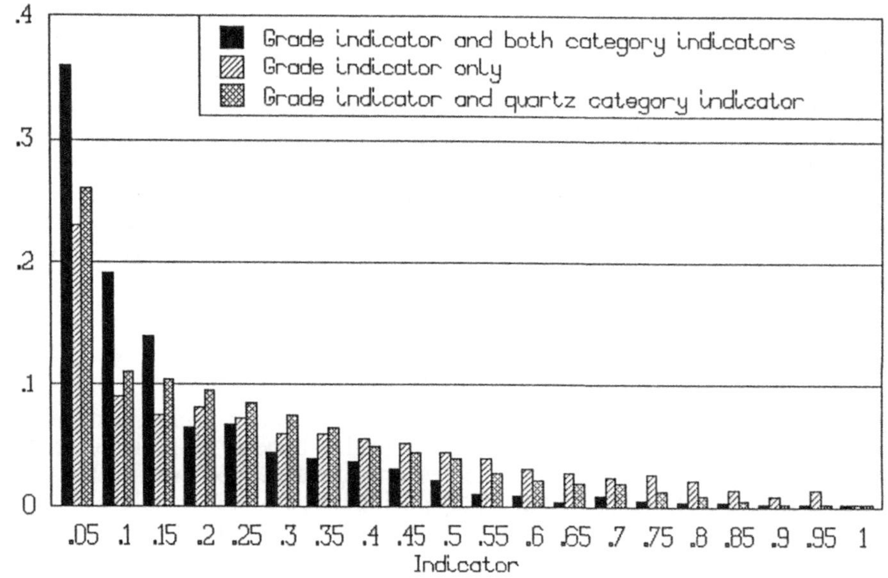

Figure 4 : histograms of $I^*(x)|I(x) = 0$

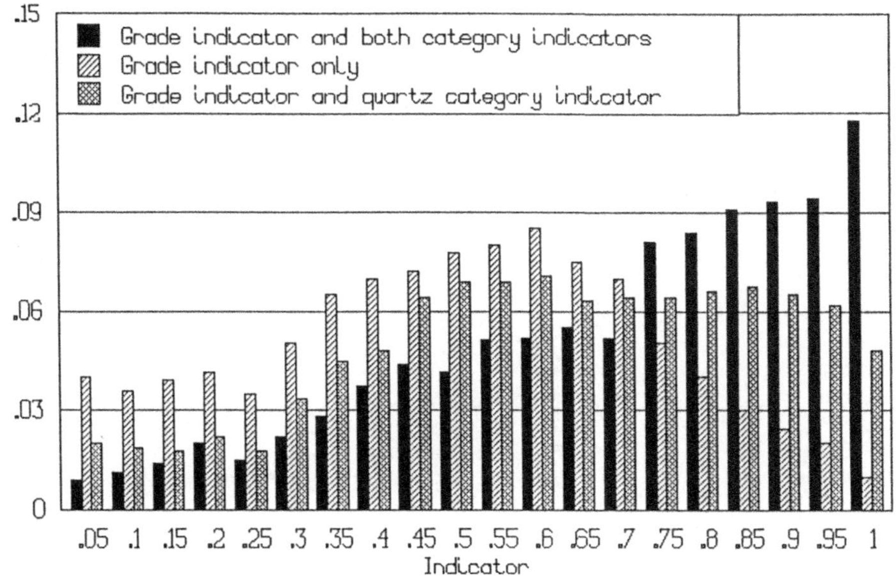

Figure 5 : histograms of $I^*(x)|I(x) = 1$

The use of the category indicators significantly improves the ability of geostatistical methods to discriminate between ore and waste blocks. From figures 4 and 5 the identification of waste blocks appears to be more reliable than the identification of ore blocks. This is almost entirely due to the fact that the cut-off grade is greater than the median grade. As the cut-off grade increases the histogram of $I^*(x)\,|\,I(x)=1$ becomes more uniform whilst the histogram of $I^*(x)\,|\,I(x)=0$ becomes more positively skewed; as the cut-off grade decreases the histogram of $I^*(x)\,|\,I(x)=0$ becomes more uniform whilst the histogram of $I^*(x)\,|\,I(x)=1$ becomes more negatively skewed. In this case study most of the improvement in the identification of the waste blocks is achieved by the incorporation of quartz category control in the estimation. In the test area estimated composite indices (based on all three indicators) less than 0.35 successfully identify 95% of all waste blocks.

Case study 2 : disseminated nickel deposit

This study concerns a disseminated nickel deposit which contains erratic occurrences of high grade, massive ore. The high grade mineralisation appears to be associated with adjacent veins which form separate orebodies and is associated with either of two lithologies; in some areas it can also be broadly distinguished by a threshold nickel value. Twenty-two lithological units are recorded although for the purposes of this study a number of classifications were combined giving a total of five lithological categories : massive sulphide, ultramafic, porphyry, quartzite and dunite/serpentenite.

Each unit is defined by a category indicator variable and the classification is mutually exclusive. Figure 6 shows the downhole indicator variograms for the five lithological units. Figure 7 shows the corresponding downhole variogram for the weighted sum indicator (Σ_k k $I_k(x)$). The results of grade estimation with and without the category indicators are shown in figure 8. Although there is still some bias in the category coestimated grades it has been reduced significantly.

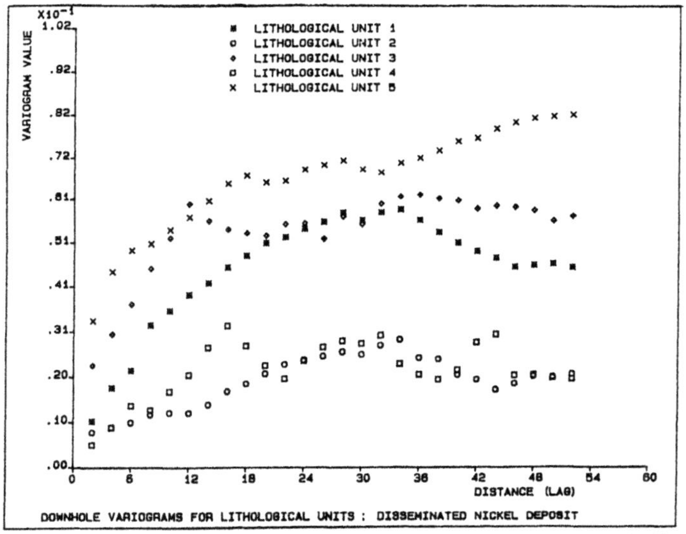

FIGURE 6 : Downhole variograms for individual category indicators

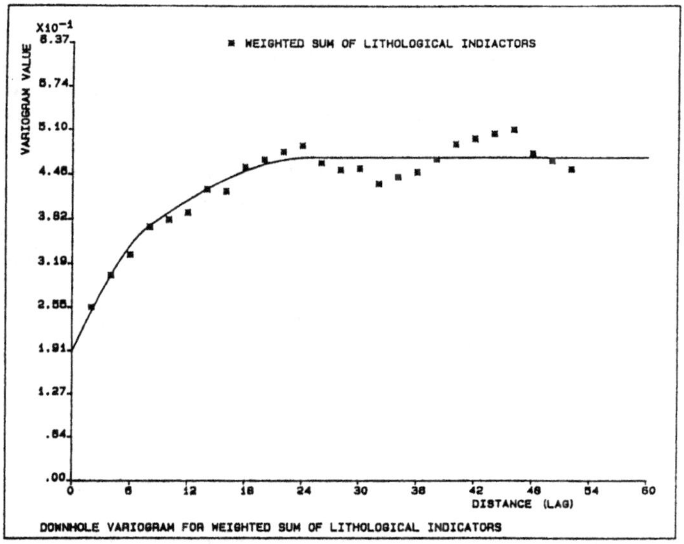

FIGURE 7 : downhole variogram of weighted category indicators

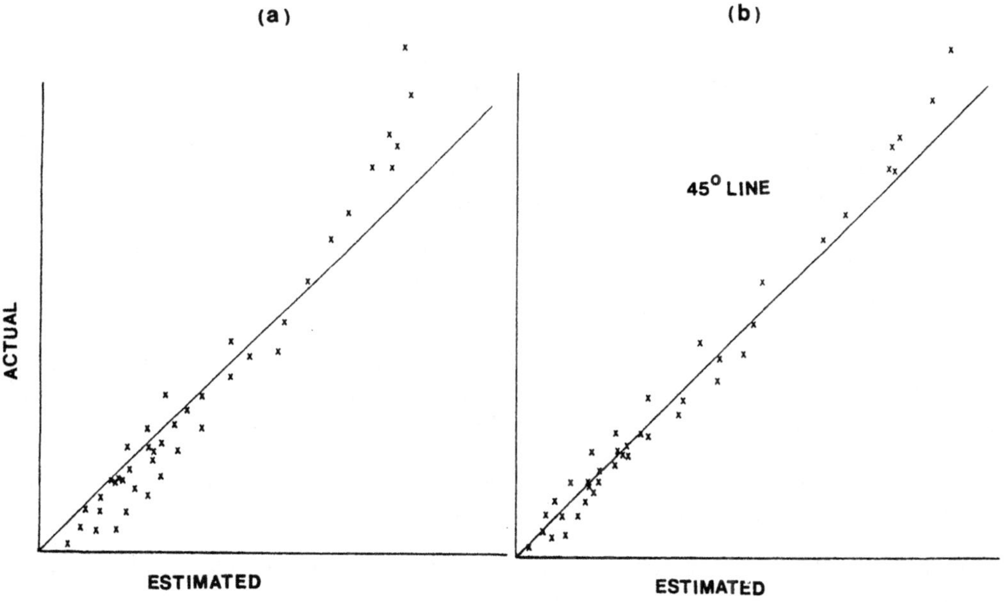

FIGURE 8 : comparison of actual and estimated block grades
(a) grades only (b) grade and category coestimation

CONCLUSIONS

The methods described here provide an effective means of introducing structural and/or geological control into the kriging process. It is possible to reduce the cokriging modelling, inference and calculation requirements by making simple transformations and assumptions and thereby dealing with a large number of category variables. The methods described here rest on the assumption of intrinsic coregionalisation of the auxiliary variables. In both case studies given here cross-variograms were calculated to check the assumption of intrinsic coregionalisation. In the gold case study this assumption was only approximately satisfied whilst for the nickel case study there was excellent agreement. In practice, however, with a large number of categories verification of this assumption would negate the time saving advantages of making it in the first place. From practical applications it appears that the methods are fairly insensitive to the assumption. In the two case studies described, significant improvement in estimates was achieved by incorporating structural and geological control by means of co-estimated category indicators.

Myers (1984) gives a method for reducing the size of a system of cokriging equations.

Another approach to these applications would be by way of the external drift methods (Marechal (1984), Galli and Meunier (1987), Delfiner et. al. (1983)

REFERENCES

Delfiner, P., Delhomme, J.P., Pelissier Combescure, J. (1983) Application of geostatistical analysis to the evaluation of petroleum reservoirs with well logs. Twenty-fourth Annual Logging Symposium, Calgary, Canada.

Galli, A. and Meunier, G. (1987) A study of a gas reservoir using the external drift method. in Matheron, G. and Armstrong, M. (eds.) Geostatistical case studies, D. Reidel publishing company, Dordrecht, Netherlands, pp 105-120.

Galli, A., Murillo, E. and Thomann, J. (1984) Dual kriging - its properties and its uses in direct contouring. *Geostatistics for natural resources characterization* NATO ASI Series C : Mathematical and Physical Sciences vol 122 part 2 pp 621 - 634.

Marechal, A. (1984) Kriging seismic data in the presence of faults. in Verly, G. (ed.) Proceedings of the NATO A.S.I. on Geostatistics : Geostatistics for natural resource characterisation. pub. Reidel, Dordrecht, Netherlands, pp 271-294.

Matheron, G. (1979) Recherche de simplification dans un probleme de cokrigeage. Centre de Geostatistique, Fontainebleau, Note N-698.

Matheron, G. (1982) La destructuration des hautes teneurs et le krigeage des indicatrices. Centre de Geostatistique, Fontainebleau Note N-761

Matheron, G., Beucher, H., de Fouquet, C., Galli,A., Guerillot, D. and Ravenne, C. (1987) SPE 16753, paper presented at 62nd Annual Technical Conference of the Society of Petroleum Engineers, Dallas, USA 27-30 September 1987.

Myers, D. (1982) Matrix formulation of cokriging. Mathematical Geology, vol. 14, No. 3, pp 249-257.

Myers, D. (1984) Cokriging - new developments. in Verly, G. (ed.) Proceedings of the NATO A.S.I. on Geostatistics : Geostatistics for natural resource characterisation. pub. Reidel, Dordrecht, Netherlands, pp 295 - 305.

Pawlowsky, V. (1985) On spurious spatial covariance between the variables of constant sum. *Sciences de la Terre, Series Informatiques*, No. 21, pp 107 - 113.

Royer, J.J. and Vieira, P.C. (1984) Dual formalism of kriging. *Geostatistics for natural resources characterization* NATO ASI Series C : Mathematical and Physical Sciences vol 122 part 2 pp 691 - 702.

Suro-Perez, V. and Journel, A.G. (1990) Stochastic simulation of lithofacies : an improved sequential indicator approach. *ECMOR*, Technip. Paris pp 3 - 10.

Zhu H. and Journel, A.G. (1990) Indicator conditioned estimator. *Transactions of the Society for Mining, Metallurgy and Exploration.* volume 286, pp 1880 - 1886.

THE DEVELOPMENT OF AN INTEGRATED EVALUATION SYSTEM FOR A WITWATERSRAND GOLD MINE

F A Camisani-Calzolari, G Knox, A Woodend and M Lorenzini

Anglovaal Limited, P O Box 62379, Marshalltown, Johannesburg

ABSTRACT

Anglovaal, in 1966, had already adopted a computerised ore reserve evaluation system for its gold mines. Though computers and programs were still at an early stage of development, the set of programs developed at that time were well advanced and met the needs of the mines for many years. In practice, the system was using an estimation method close to Simple Kriging as later developed in France.

A series of shortcomings emerged during the years. They were further aggravated by the necessity of mining more selectively to meet dropping grades, rising costs and disappointing metal prices.

This paper presents the recently completed new system and addresses the practical and theoretical inadequacies of the old system. Designed in-house, the new system eliminates these shortcomings and problems.

The upgrade aims to:

- Enable evaluation to be carried out at the mines by local personnel;
- Overcome the rigidity of the old system;
- Ensure data validation and security;
- Increase the accuracy of evaluation;
- Facilitate the communication between the various mining disciplines;
- Introduce graphics for the interpretation of the deposit and the presentation of results.

INTRODUCTION

The estimation of ore reserves in a South African gold mine has always been problematic. The complexity and erratic nature of gold distribution was to a great extent already appreciated at the beginning of the exploitation of the Witwatersrand. De Launay (1896) observed that, regardless of the reef in the Witwatersrand, variations in grade from 2 g/t to 100 g/t are regularly found in samples taken a few centimetres apart. This occurs without any apparent lithological change, with very accurate sampling procedures and repeated sampling. He also noted the "irregular, nearly paradoxical" fact that although the grade is extremely variable from point to point of a face, once averaged over a face of adequate length and compared to the grade of the successive face advances, the variation becomes greatly reduced in such a way that the average of a working section of a mine is close to the average grade of an adjoining section. In other words the grades of the different sections of a mine are relatively "constant". Further, variations in grade are confined to geologically and sedimentologically uniform units within the deposit

The major factors influencing the estimation of a gold deposit - namely nugget effect, variance, support and subdivision of a deposit into uniform areas - were already appreciated, though not fully quantified, one hundred years ago.

Recently, the application of mathematics and statistics and the use of computers have made the evaluation more precise and efficient. A more accurate evaluation and the quantification of risk or confidence in the estimation by using a versatile and flexible computerised valuation system accommodating the needs of the individual mines are now essential because mines are moving toward low-grade areas and working costs are escalating resulting in many operations becoming marginal.

HISTORY OF EVALUATION AT ANGLOVAAL

Krige and Ueckerman (1963) proposed the use of value contour maps showing regional trends to be used together with the average of block periphery samples to predict the content of blocks of ore. This proposal established an area of interest with workers in the U.S.A. who were actively developing techniques under the general heading of Trend Surface Analysis (Knox, 1990).

In 1964 the South African Chamber of Mines (COM) hosted and sponsored a research and development project to develop a computerised ore reserve system. A series of six programs (Pirow *et al.*, 1966) were written in CLEO. During 1966 the system was tested and became operational. Since then, Anglovaal has routinely carried out ore reserve estimations using this system and its upgrades.

Because computers were then relatively expensive to run, the system was designed to use the minimum amount of CPU time. In order to do so a considerable amount of the operations and data processing was manually performed. It was then much cheaper to do as much manual input as possible (to generate a complete run of ore reserves for an average gold mine with 40 000-50 000 values amounted to the equivalent of U.S. $ 40 000 in today's terms).

Anglovaal was the only group that had the weighting table construction expertise and soon became the only user of the system. With the advent of plotters, time-sharing, remote terminal access and greatly increased core size and speed, it was decided in early 1972 to redesign and extend the system so that the labour involved at the mine for the annual ore reserve estimation would be cut to the barest minimum. All programs were re-written in FORTRAN. Early in 1975 BASIC conversational front-end programs were developed to simplify the routine running of the suite of programs. Ore reserves were run on a bureau basis using mainframes at the University of the Witwatersrand or commercial computer companies. From 1979 the responsibility for the routine running of the ore reserves was passed to the newly established Computer Department of the Anglovaal Head Office. It was only in April 1983 that a PRIME 550-II computer was acquired and installed and ore reserve runs were done routinely at Anglovaal Head Office.

Though the package was modified through the years, it kept the original weighted moving average model based on a standard pattern defined by Krige (1966). This consisted of squares, the size of which increases with distance from the central square to be evaluated. Samples falling within each square were averaged. Tables of weights were developed for different squares based on covariances in the various directions. Sample-weight products were accumulated for informed squares, the weighted mean for the area was added and the result divided by the sum of weights to arrive at the estimate of the central square. No change of square size was allowed either in the regularisation or evaluation phase. The estimated values obtained on a grid of 100x100 foot squares were integrated into the irregular polygons (mining blocks) to get the final block values.

The printing facilities allowed for listing the ore reserves by block, reef, availability codes and grade categories. Graphics remained confined to plots of grade contours and mining block outlines.

THE PHILOSOPHY OF THE NEW SYSTEM - PRACTICAL AND THEORETICAL CHOICES.

In recent years, increasing mining costs have forced mine management to look for more flexible and precise methods of ore reserve estimation. This translates into the need for a system that can give rapid results, as data become available, and a better confidence in the estimated value.

Selective mining in a Witwatersrand gold mine is particularly difficult due to rock-mechanics implications and the logistics of scattered mining. Total evaluation and ad hoc statistical and geostatistical studies of critical areas are strategically important for financial mining policies and fundamental to production planning and scheduling.

In order to meet these needs, it was decided in 1990 to redesign the package.

The system has been designed as an integrated evaluation package, in which each part of the evaluation follows a logical path via chained menus. The following requisites had to be met:

a) The system was to be run at the mine with technical support from Head Office. This implied that the system had to be user-friendly and easy to understand, in order to facilitate the detection of any problem or errors in the evaluation.

b) The user, not necessarily a statistician or computer specialist, would have to be trained to run the system independently.

c) Head Office was to continue updating the theoretical and computerised aspects of the system with regard to further needs or requests from the mines. Head Office was to keep the control of any major input and output. For this reason it was to be designed in such a way as to use both PC's (low running cost) and Main Frame (large memory and easy access).

The package has been structured to handle the total evaluation of the ore reserves in a large two dimensional mine. It can easily be used for evaluating areas of any size within the mine or for special projects.

The major topics addressed are the following:

1) Data capture and validation;
2) Database and data storage;
3) Statistical tools and semivariogram calculation;
4) Interactive graphic modelling;
5) Estimation process;
6) Graphics facilities

1. DATA CAPTURE AND VALIDATION

Data capturing is done on site using digitisers. Once captured, the sample data is transferred to the Head Office mainframe via dedicated modem lines and stored in the master file. In addition to the coordinates the following variables are stored: channel width (cm), stope width (cm), gold value (cmg/t), uranium value (cmkg/t) and dip of reef in degrees from the horizontal. Two other variables have been added, namely footwall waste and a reliability code (to discard from the evaluation all values affected by incomplete information). These variables also help to highlight areas where mining is inefficient.

On the practical side, particular attention has been given to the training of the people involved to make them fully aware of the importance of the process. The amount of data captured in a single month (about 2000 values at each month-end) creates practical problems in terms of accuracy and reliability in the digitising. Graphics on screen or plotted are used to check the input data.

2. DATABASE AND DATA STORAGE

For security the main master files are kept in database format with password protection. In addition, access to the data for any purpose (e.g. update, deletion, etc.) is via programs which keep an audit log. The current generations of computers have addressed many of the problems associated with data storage and access. Data storage can now use database technology which allows fast access on very large files (approximately 500 000 samples), organised in any convenient structure or design. The database structure is based on geographical coordinates and the object is to allow the quick access to the data of any area. Because the system allows for regularisation over any rectangular or squared input block size, also regularised data have been organised into a data base. In addition other data bases have been created to contain the output grids of kriging programs, outlines and values of irregular mining blocks.
The quick extraction of data from a selected area combined to interactive graphic facilities to define the outlines of the area involved in the processing, allows the system for any declustering required. This procedure is not used for routine production purposes.

3. STATISTICAL TOOLS AND SEMIVARIOGRAM CALCULATION

Most of the commercially available systems do not allow the use of very large sets of data. Homogeneous areas in the Witwatersrand deposit are often very extensive with more than 100 000 sample values.

The uni- or bi-variate statistical tools, and semivariogram calculation allow for log transformation of the data and the determination of the additive constant (third parameter). The semivariogram calculation is very flexible. It calculates simultaneously the covariances for any variable up to eight in any selected area within the mine (currently this is confined to three variables, namely gold (cmg/t), uranium (cmkg/t) and channel width (cm)).

It can statistically analyse the lags in different directions in order to determine the "optimal" one for the calculation, in this way saving a large amount of C.P.U. time. This can be convenient when using irregularly spaced sample data.

In some cases when the spacing between the information is larger than the semivariogram range, only classical statistics can be used. For this reason, Sichel's "t" estimation has been included in the system.

4. INTERACTIVE GRAPHIC MODELLING

The interactive modelling of experimental semivariograms has been designed to be quick and easy to use. The program allows for any model (including lognormal de Wijsian) and can determine the geometric anisotropy and fit the anisotropy ellipse. At the end of the modelling process all the parameters are saved in the relevant master file. From here they can be retrieved to be checked against the experimental semivariograms calculated using the latest values as mining operations advance. Each homogeneous area has a model for each of the three variables, namely gold (cmg/t), uranium (cmkg/t) and channel width (cm). The stoping width-channel width relationship is peculiar to the mines in the Witwatersrand basin and was originally developed by Anglovaal. Through the years, it has proved to be a useful and effective tool for controlling stoping efficiency. Though the evaluation of ore reserves is carried out in cmg/t (for gold), the tonnage of ore expected to be exploited from a stope and its actual grade in g/t is fundamental to mine planning. Both these variables are directly related to the stoping width. The method consists of plotting the channel width, which is the thickness of the mineralised reef, against the stoping width which is measured 3m back from the face for each sample. The resulting experimental curve is interactively modelled using three linear functions:

- The first represents samples taken in areas of narrow reef. The stopes were mined at a minimum and constant width. This portion is usually fitted by a horizontal line that represents the average width at which narrow reefs are mined in a specific area of the mine;
- The second portion is a transition phase which occurs in stopes where the thickness of the channel starts to have some influence on the stoping width;
- The third segment of the graph represents stoped out areas where the channel width has reached thicknesses which directly and equally influence the stoping width. In other words, the scatter of sample points is spread along a 45° line.

The three equations of the fitted lines appear on the top of the graph (Fig. 1). The histogram is the percentage frequency distribution of the channel width. This graph allows the miners to set realistic stoping width targets, which in turn are used in the evaluation of ore reserves.

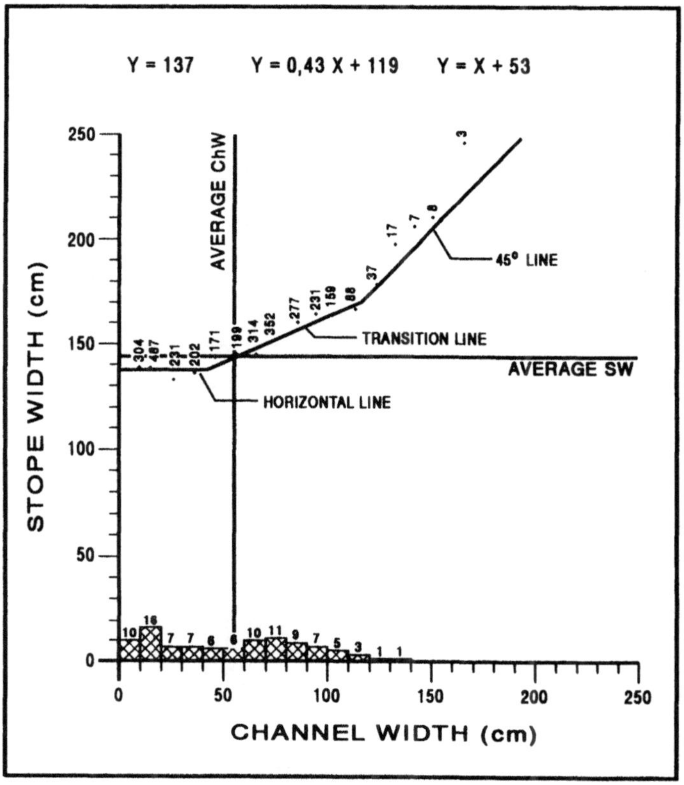

Fig. 1 Channel Width / Stoping Width relationship

5. ESTIMATION PROCESS

Although of extreme importance, the evaluation calculation in this system is but a portion of the entire package . The philosophy adopted in this system is to optimise the use of input data and geological knowledge, and to use the simplest appropriate evaluation method.

The evaluation, which consists mostly of extrapolation, does not improve if non-linear methods are used. Studies carried out in mined-out areas on Loraine Gold Mine using different methods showed no significant difference in the results (Lorenzini, 1989).

Method	Statistics		Corr.(Z/Z^*)	E(Z/Z^*)	$\Delta(Z-Z^*)$ cmg/t		% Profit (cut-off at 600 cmg/t)
TRUE	N	114	1	1			
	Min	327			Min	-	100%
	Max	1327			Max	-	
	Mean	751			Mean	-	
	Var.	49197			Var.	-	
SK	N	114	0.62	1.08			
	Min	487			Min	-327	94.7%
	Max	1190			Max	560	
	Mean	725			Mean	26	
	Var.	16334			Var.	30190	
OK	N	114	0.62	0.56			
	Min	229			Min	-683	90.3%
	Max	1516			Max	603	
	Mean	730			Mean	22	
	Var.	61118			Var.	42092	
SDK	N	114	0.64	1.13			
	Min	413			Min	-291	94.1%
	Max	1102			Max	547	
	Mean	723			Mean	28	
	Var.	15938			Var.	29183	
DOK	N	114	0.62	0.56			
	Min	222			Min	-695	89.7%
	Max	1473			Max	601	
	Mean	728			Mean	24	
	Var.	60404			Var.	42086	

Table 1 Simple and Ordinary Kriging (SK,OK) and their Disjunctive equivalents (DSK,DOK) are compared using different statistical tools. % Profit is calculated as {%[Ton(Revenue-Cost)]}. (After Lorenzini, 1989)

Table 1 gives the results of different techniques applied to 114 panels each with a minimum of 20 informed small blocks (8 m x 8 m in size) in the kriging neighbourhood. The kriging pattern is shown in figure 2. Because of no change of support the universality condition has been applied to the Disjunctive Kriging. Uniform Conditioning has been also taken into consideration giving similar results as Disjunctive Kriging. In order to obtain a more critical comparison between the four methods (SK,OK,DSK,DOK) and the "true" values, the following parameters have been used:

a) Correlation coefficient
b) Slope of regression line
c) Estimation error in cmg/t (Min,Max,Mean,Var.)
d) Relative profit at cut-off of 600 cmg/t

The results indicate that there is no significant difference between the various methods with SK and DSK giving slightly better relative profit and regression line slope

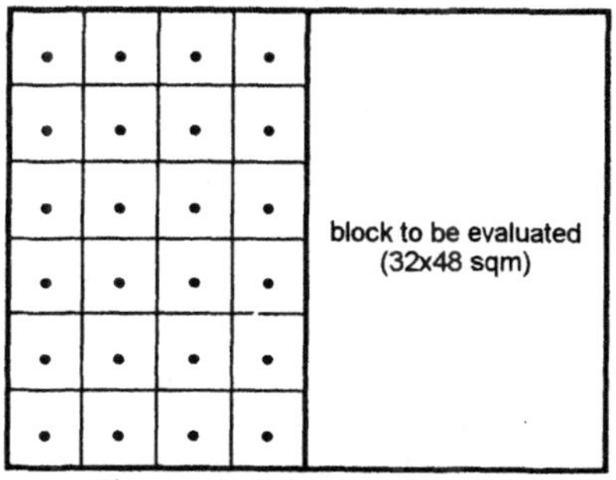

Fig. 2 Kriging pattern.

Similar results were found for other areas in the Witwatersrand basin (Krige *et al.* 1989). Thus, discarding simulation for practical reasons (time, cost etc.), linear (and log-normal) kriging techniques have been selected as the most convenient methods for the routine evaluation on the gold mine.

Tests at Hartebeestfontein Gold Mine showed that the old system gave a globally unbiased estimate (Fig. 3). This follows as the old system is very close to Simple Kriging (normal or log-normal).

In fact in the normal case, the estimator

$$Z^* (x) = (1 - I^{\alpha}) m + I^{\alpha} Z_{\alpha} (x)$$

contains the factor

$$(1 - I^{\alpha})$$

which, if the position and number of samples utilised in the kriging are not adequate in relation to the geostatistical structure (e.g. one-sided samples), approaches unity.

This signifies that under these condition the estimator would be greatly influenced by the mean and would tend to excessively smooth the block values toward the regional mean.

The evaluation is thus globally unbiased but the results are locally inaccurate and conditionally biased.

In fact, the mean used in SK is calculated from the available samples collected within large geologically homogeneous sections of the deposit and, due to the variability of the reefs, practically never respects the local grade. In addition the mines are moving toward low grade areas and the mean values are slightly biased towards high grades. In this case Ordinary Kriging (OK) can give a better indication of block grades than Simple Kriging (SK), but only if more values from a larger neighbourhood are used.

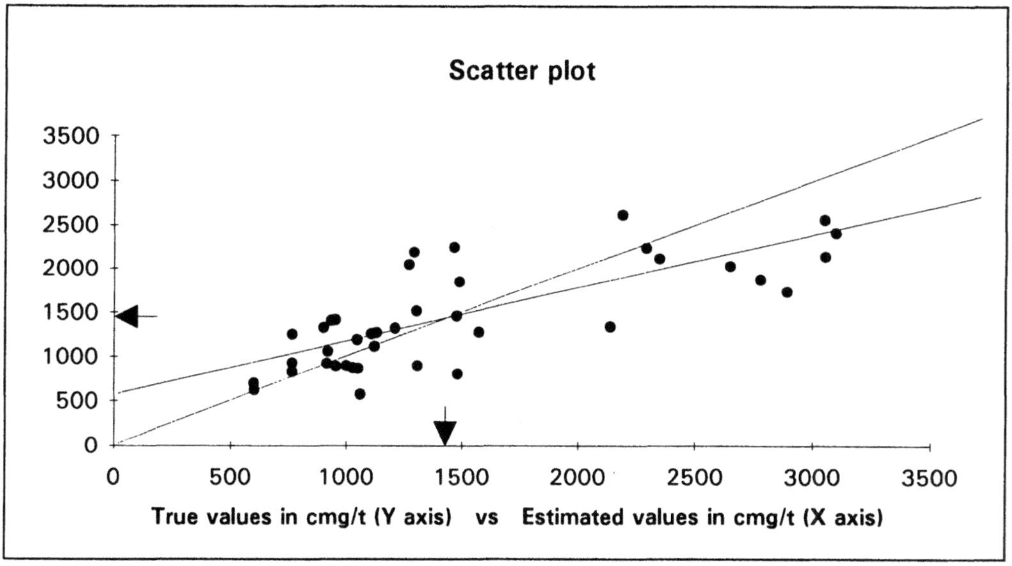

Fig. 3: Correlation between true values (1991) and estimated value (1990) in 43 blocks mined out at more of 90%. The evaluation is globally unbiased with true values mean = 1459 cmg/t and the estimated values mean = 1467 cmg/t . The slope is .587, the intercept = 560 cmg/t and the correlation coefficient is 0.76

Further tests have indicated that, in most instances, SK, using the mean calculated from the closest values to the centre of the block within an appropriate search radius (approximately 100-200 values) and using a smaller kriging neighbourhood than required by OK, gives better results than OK and has the advantage of saving computer time ("Local Simple Kriging").

Local Simple Kriging (LSK) also reduces the negative influence on the evaluation of the particular geographical distributions of the kriging weights found in a one-sided sample pattern (the most frequent and inconvenient is the "inverse screen effect").In this instance the model, the support and the neighbourhood play an important role and must be tested thoroughly.

Figures 4 and 5 (Lorenzini, 1989) show the distribution of weights in a fixed pattern evaluation for different methods (SK or LSK and OK) and give an example of "inverse screen effect" (Fig. 4).

For these evaluation (Loraine Gold Mine) the following parameters have been used:

a) semivariogram model:
73 100 + 55 500 sph(12) + 27 400 sph(33)+29 100 sph(130)

b) direction of maximum variability: -13 degrees from E-W axis

c) anisotropy coefficients: I structure (40), II structure (32), III structure (50)

d) dimension of input block: 8 x 8 square meters

e) dimension of output block (block to be evaluated); 32 x 48 square meters

f) neighbourhood: 24 regularised 8m x 8m blocks left-sided

5.93	5.73	4.17	7.31	block to be evaluated (32x48 sqm)
3.45	2.03	1.87	6.14	
2.67	1.72	1.43	6.50	
2.62	1.87	1.41	6.82	
2.75	2.43	1.80	8.12	
4.78	4.09	3.64	12.5	

Fig. 4 Spatial distribution of the kriging weights from OK. Kriging gives higher weights to far away samples than closer ones ("inverse screen effect"), that is, the fourth column from the block to be evaluated has higher weights than the second one. The kriging variance is 43 500.

1.31	1.11	1.61	4.04	block to be evaluated (32x48 sqm)
0.58	0.78	0.90	0.98	
0.47	0.62	0.79	4.47	
0.49	0.59	0.90	4.77	
0.59	0.59	1.14	5.55	
0.82	1.09	2.07	7.96	

Fig. 5 Spatial distribution of the kriging weights from SK (LSK). The weight of the mean is 52% (very high) and the kriging variance is 30 783 (S.D. 175.5). The regional mean value and the variance are 722 cmg/t and 184 967 respectively.

The degree of skewness of the distribution of gold and uranium is taken into account in deciding whether to use log-transformed or raw data. The system allows for both approaches and the former is often necessary. The semivariograms are more stable if calculated with log-transformed data which also prevent the evaluation from being overly influenced by the high variability of values.

While theoretically more sound in the evaluation of Witwatersrand deposits, log-normal kriging encounters some practical problems and requires considerable monitoring mainly because of the difference between the geometric mean and the population mean. In fact, the log of the mean of a population is equal to the log of the geometric mean plus half of the log variance of the population. Unfortunately, the population variance is not known *a priori*, even if it is positively related to the size of the area to which the blocks to be valued belong.

It follows, then, that the larger the area in which the block is contained, the higher, theoretically, the correction factor that must be applied to the geometric mean. Also, as the number of samples decreases and/or their location is asymmetric, the error of estimation increases. As this error is utilised in the back-transform procedure, there is a high risk of obtaining a log-normal correction factor which is either too high or too low for the actual set of samples used in the kriging, and hence getting overvaluation or under-valuation (Camisani-Calzolari, 1992).

South African gold mines use a parameter called Block Factor (B.F.). This parameter is the ratio between the gold accumulation calculated from current sampling in the mined out portion of a block and the gold accumulation of that block estimated before mining commenced. The weighted average of this parameter for all the mined blocks is calculated every month and is expected not to deviate significantly from 100%. Consistent one-sided deviations over extended periods will influence mine-planning and the profitability of the mine. Though this factor is not statistically valid, it has been inserted for the practical reasons mentioned above.

Gold mining becomes more risky every day as capital and operational costs increase and grades drop. Mines or parts thereof become marginal or unpaying. With the awareness of this, and improved acceptance of geostatistics by mine management, it has now become essential to give some measure of confidence in the estimated value for each block. The 90% central confidence limit has been accepted as a practical measurement and has proved to be adequate in making management aware of the varying reliability of the evaluation of block values.

Studies were undertaken to examine the possibility of including samples from underground boreholes in the routine valuation. As the ore reserves are calculated from peripheral values, the few borehole samples collected within or near the blocks to be evaluated become important and acquire considerable weight in the evaluation. However, tests carried out in some mined-out areas of one of the mines have indicated that values of underground boreholes are frequently biased toward low grades and no significant correlation was found between borehole and chip sample values. One reason for this is that the carbonaceous material which contains most of the mineralisation can easily be washed away in underground drilling. To overcome this problem a reliability code is now being applied to each borehole value.

6. GRAPHICS FACILITIES

A way of improving the evaluation is through better understanding of the geology and of the many mining constraints. It is essential to marry the statistically defined grade structure to the geological interpretation, particularly at the level of local evaluation. The reefs are variable and complex due to the many geogenetic factors influencing the mineralisation. Palaeo-highs, pebble size and localised grade trends are valuable in arriving at an improved evaluation. The use of graphics is the easiest way to define and study their influence in the evaluation. Moreover, graphics are often the only means of communicating with management and making the evaluation readily understandable.

CONCLUSION

The development of this system has emphasised the importance of the following points:

IN-HOUSE EXPERTISE

Each mine has specific evaluation problems, grade distributions in the reefs, geological and structural characteristics, reporting systems and policies that make it unique. The experience of Anglovaal is that an evaluation package has to be tailor-made for the mine and this is lacking in commercial packages. In addition the package is never finished and is frequently subject to change. In-house expertise has to be built in order to carry out the necessary improvements, to understand the evaluation process, to train the user and to maintain security. External expertise is nevertheless important but should be confined to specific well-defined tasks.

EVALUATION AT THE MINE

The evaluation has to be done at the mine, where the data are captured and where there is full understanding of the local conditions and needs. To remove the evaluation from the mine will inevitably result in incorrect data in the master-files, inadequate estimates and loss of credibility in the results by mine management. Vetting from Head Office is essential in order to maintain the required high standard and accuracy.

EVALUATION MUST BE A COOPERATIVE EFFORT

The evaluation cannot be done in isolation and be expected to be accurate. Geology, mining methods, planning, production policies, pay-limits, selectivity, blending, to name but a few are all relevant, and failure to appreciate them lessens the evaluation in the long run and eventually leads to rejection by mine management.

ADEQUATE CHECKS IN DATA CAPTURING AND VALIDATION

Correct data is the single most important prerequisite for an accurate evaluation. Particular care has to be given to this facet of the problem, thus the training and motivation of people involved is equally important.

EVALUATION METHODS

The evaluation system has to be flexible in order to apply the most reliable procedure. Such flexibility has to apply to all aspects of the evaluation down to the square size. Linear kriging has proved to be valid and practical for the evaluation of Wits-type deposits.

GRAPHICS

Graphic tools have become a necessity not only for presentation of results in a way that is unambiguous and readily understood but also for improving the interpretation of the deposit. The last aspect makes an important contribution to the accuracy of the estimation.

However, a geostatistically sound estimation procedure is a necessary but not sufficient condition for a successful ore reserve evaluation system. Without informed underground control, and incentive schemes which actively promote the total extraction of reef with minimum waste, the best estimation remains a back-room intellectual exercise. As the evaluation process moves from Head Office to the mine it provides a unique opportunity to open communication channels between the different departments (Sampling, Survey, Geology, Planning, Mining and Metallurgy) so that they can work as a team in extracting maximum profit from the evaluation.

ACKNOWLEDGEMENTS

The authors are grateful to Dr Isobel Clark, L A E Carter and S J C L Peeters for critically reviewing the manuscript, Mrs C M Maritz for drawing the figures and Mrs N Moodley and Mrs C Empey for typing the manuscript. The Management of Anglovaal is thanked for allowing this paper to be published.

REFERENCES :

Camisani-Calzolari, F.A. (1992) "The Evolution of Geostatistics in South Africa in the Evaluation of Witwatersrand Gold Deposits", In Press, Geological Society of South Africa, Johannesburg.

De Launay, L. (1896) Les Mines d'Or du Transvaal, Baudry et Cie, Paris.

Knox, G.M. (1990) Ore valuation at Anglovaal, Anglovaal Limited, Johannesburg, Internal Report.

Krige, D.G. and Ueckermann, H.J. (1963) " Value contours in improved regressions techniques for ore reserve valuations", J.S. Afr. Inst. Min. Metall., Johannesburg, May 1963, pp 429 - 452.

Krige, D.G. (1966) "Two dimensional weighted moving average trend surfaces for ore valuation", Proc. Symp. on Mathematical Statistics and Computer Applications in Ore Valuation, J.S. Afr. Inst. Min. Metall., Johannesburg, March 1966, pp 13-38.

Krige, D.G., Guarascio, M. and Camisani-Calzolari, F.A. (1989) "Early South African Geostatistical techniques in to-day's perspective", Geostatistics, M. Armstrong. (Editor), Kluwer Academic Publishers, Dordrecht, Vol. 1, pp 1-19.

Lorenzini, M. (1989) Algoritmi di Geostatistica non Lineare e Previsione di Selettivita` in Miniera, Unpublished MSc Thesis, University of Trieste.

Pirow, P.C., Siebert, J.F., Knox, G.M., Claassen, P. and Lawton, J.C. (1966) "Computer programs for the estimation of a two dimensional trend surface using a weighted moving average", Proc. Symp. on Mathematical Statistics and Computer Applications in Ore Valuation, J.S. Afr. Inst. Min. Metall., Johannesburg, March 1966, pp 80-105.

SAMPLING OF OREBODIES WITH A HIGHLY DISPERSED MINERALIZATION

Wynand Johannes KLEINGELD
Anglo American Corporation
P.O. Box 41848
2024 Craighall
Republic of South Africa

Christian LANTUEJOUL
Centre de Géostatistique
Ecole des Mines
35 rue Saint-Honoré
77305 Fontainebleau
France

ABSTRACT. This paper presents several results on how to sample highly dispersed type orebodies, e.g. alluvial gold or diamond deposits. Two major difficulties are encountered when sampling such orebodies. The first is of a geometric nature. Mineralization is not homogeneous, but is distributed in patches. The second difficulty is that the mineralization within the patches is highly variable. As a consequence, two different factors at two different scales account for the grade variability of the orebody. The first difficulty results from the spatial distribution of the patches and the second from the high dispersion of the mineralization within the patches. Both factors can be very important. In the case where the second factor is dominant, it is possible to sample so as to reduce its impact. This can be achieved by choosing a sample support appropriate to the size of the patches.

1 Introduction

One of the most important elements of any evaluation exercise is certainly the sampling. Before elaborating on problem areas it is helpful to review the areas where sampling is involved. For the purpose of this paper, we shall exclude the sampling and recovery problem in the laboratory.

In the evaluation of mineral deposits we use a phased approach to sampling and in financial assessment. During the *exploration phase* little sampling information is available. The main objective is to see if further work is warranted on a comparative basis with other deposits. Decisions are made on geological and specialised technical qualitative grounds. In the *prefeasability phase* more sampling results are available and global ore reserves as well as grade/tonnage relationships are estimated to assess the potential of the orebody. At the *feasibility phase* detailed sampling results are available to produce local ore reserve estimates so as to assess how well the potential of the orebody can be realized. During the *exploitation phase* detailed sampling results are available to produce local ore reserve estimates. The emphasis is on realizing the estimates made and a detailed reconciliation of depleted (i.e estimated) and actual production recovery results are necessary to manage and audit the operation.

In all of the above phases the sampling aspect is of paramount importance and if neglected it will give rise to unrepresentative estimates which could have a disastrous effect.

This paper deals with the sampling of deposits where the mineralization is not homogeneous, but is concentrated in trapsites (Sichel, 1973). Examples can be found among vein type and alluvial deposits of diamonds or gold (see Figure 1). In such deposits, two different factors account for the grade variability at two different scales, firstly the spatial distribution of the trapsites, and secondly the dispersion of the mineralization within the trapsites. Both factors can be very important.

Figure 1: Example of trapsites in an alluvial diamond deposit.

The existence of several scales of variability makes sampling a very complex operation. As a matter of fact, a set of samples of a given size may not account for all of the scales of variability. Using many small samples, traps are well detected, but their mineralized contents are poorly assessed. Using a limited number of larger samples, the richness of the traps is better known, but a significant number of traps are missed. These results will be demonstrated by plotting the estimation variance of a block grade versus the sample size (cf. section 3), and by comparing the distributions associated with samples of various sizes (cf. section 4). For didactic reasons, we have chosen to make a preliminary presentation based upon a probabilistic model which is discussed in section 2. Industrial experiments are being carried out and the results will be made public once sufficient information has been gathered.

2 A probabilistic model

The model considered here is the Cox point process (1955). To present it, we refer to the case of a two-dimensional precious stones deposit. All the parts of the deposit do not have the same chance of containing precious stones. Depending on various geographic, topographic and petrographic conditions, some parts may be potentially richer than some others (Oosterveld et al., 1987). This potential is conveniently modelled using a second order stationary random function, say Θ. The stones are then spatially distributed according a Poisson point process with the random intensity Θ. In other words, given the potential Θ, the number of stones in a region V is Poisson distributed with mean equal to the potential of V

$$\Theta(V) = \int_V \Theta(x)\, dx$$

Furthermore, under the same condition, the numbers of stones in two disjoint regions V_1 and V_2 are independent.

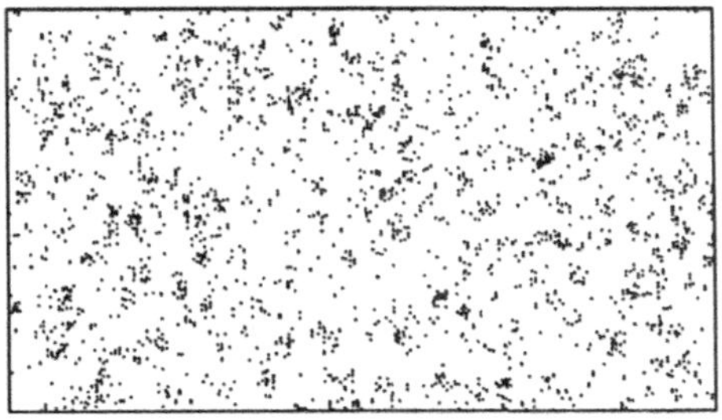

Figure 2: Example of a realization of a Cox process.

Since the potential has spatial correlations, the numbers of stones in two disjoint samples are not independent. The covariance between the number of stones in two arbitrary samples v and v' is

$$Cov\,\{N(v), N(v')\} = m|v \cap v'| + \int_v \int_{v'} C(x-y)dydx$$

where m denotes the mean value of the potential, and C stands for its covariance function. The right hand side of this formula is made up of two different terms. The first one depends only on the intersection of both samples. It is the covariance of a Poisson point process of intensity m. The second one stems from the spatial variability of the potential. It should be noted that this formula is quite similar to the one that holds in the Neyman-Scott or cluster process (Matérn, 1986; Deverly, 1984; Kleingeld, 1987). The similarity is not fortuitous as it can be shown under mild assumptions that the cluster process is a particular case of a Cox process. If the potential is large at a point x, then the potential is also large in the vicinity of x, and there are good chances for a cluster of stones to be observed. But in a Cox process, the clusters are not in general spatially distributed according a Poisson point process.

The variance of the number of stones in v can be written

$$Var\{N(v)\} = m|v| + \int_v \int_v C(x-y)dydx$$

or equivalently

$$Var\{N(v)\} = m|v| + \int_{\mathbf{R}^2} C(h)K_v(h)dh$$

where $K_v(h) = |v \cap v_h|$ is the geometric covariogram of v. In order to have a better understanding of the behaviour of $Var\{N(v)\}$ as $|v|$ increases, the following two cases are considered:

i) if $|v|$ is very small w.r.t. the range of C, then $C(h) \approx C(0) = \sigma^2$ wherever $K_v(h) \neq 0$, and since the integral of K_v is equal to $|v|^2$, we obtain

$$Var\{N(v)\} \approx |v|(m + \sigma^2|v|)$$

and by neglecting the second order terms in $|v|$

$$Var\{N(v)\} \approx |v|m \qquad (|v| \approx 0)$$

In other words, only the Poisson term acts for small samples.

ii) if $|v|$ is very large w.r.t. the range of C, then $K_v(h) \approx K_v(0) = |v|$ wherever $C(h) \neq 0$, and therefore

$$Var\{N(v)\} \approx |v|\left(m + \int_{\mathbf{R}^2} C(h)dh\right) \qquad (|v| \longrightarrow +\infty)$$

Let A be the integral range of the random intensity Θ

$$A = \frac{1}{\sigma^2}\int_{\mathbf{R}^2} C(h)dh$$

(Yaglom, 1986; Matheron, 1989; Lantuéjoul, 1991). A has the dimension of an area and is supposed here to be finite. Then the variance can be written as

$$Var\{N(v)\} \approx |v|(m + \sigma^2 A) \qquad (|v| \longrightarrow +\infty)$$

A physical interpretation of the coefficient $m+\sigma^2 A$ is as follows. Everything happens as if the sample v has been divided into a large number of independent subsamples of size A, and as if a random value has been assigned in each subsample, namely A times a Poisson variable with a mean value generated according the stationary distribution of Θ.

We have just seen that the variance of $N(v)$ has a linear behaviour for sample v of a small, as well as of a large support size. More generally, if $C \geq 0$ (the covariance of the intensity has no hole effect), we have

$$|v|m \leq Var\{N(v)\} \leq |v|(m+\sigma^2 A)$$

which shows that the curve $Var\{N(v)\}$ versus $|v|$ lies in a domain limited by two lines (see Figure 3).

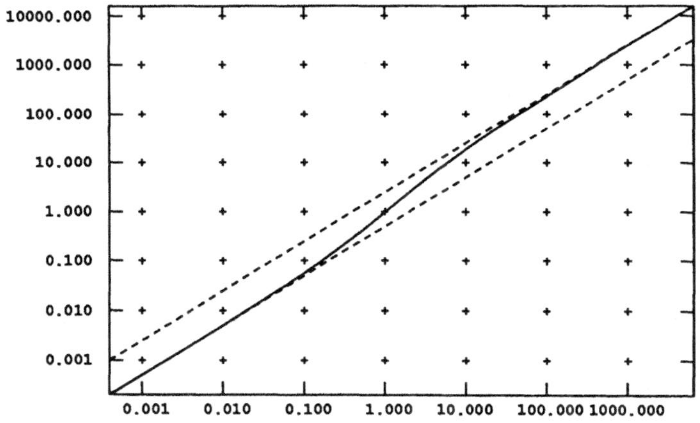

Figure 3: Plotting the variance of the number of stones in a sample versus the sample support size.

Since $N(v)$ is a Poisson variable with mean equal to the potential $\Theta(v)$ of v, the generating function of $N(v)$ is

$$E\left\{s^{N(v)}\right\} = E\left\{e^{(s-1)\Theta(v)}\right\}$$

In particular, the proportion of barren samples is

$$P\{N(v)=0\} = E\left\{e^{-\Theta(v)}\right\}$$

3 Estimation variance of a block grade

Now let v be a sample within a block V. To what extent is the sample v representative of the block V? To answer such a question, we need a criterion. The one that has been retained in this section is the estimation variance of the grades, that is $Var\{Z(v) - Z(V)\}$, where

$$Z(v) = \frac{N(v)}{|v|} \quad Z(V) = \frac{N(V)}{|V|}$$

Once again, the estimation variance can be be split into two different components. Explicitly, we have

$$Var\{Z(v) - Z(V)\} = V_1 + V_2$$

with

$$V_1 = m\left(\frac{1}{|v|} - \frac{1}{|V|}\right)$$

$$V_2 = Var\left\{\frac{\Theta(v)}{|v|} - \frac{\Theta(V)}{|V|}\right\}$$

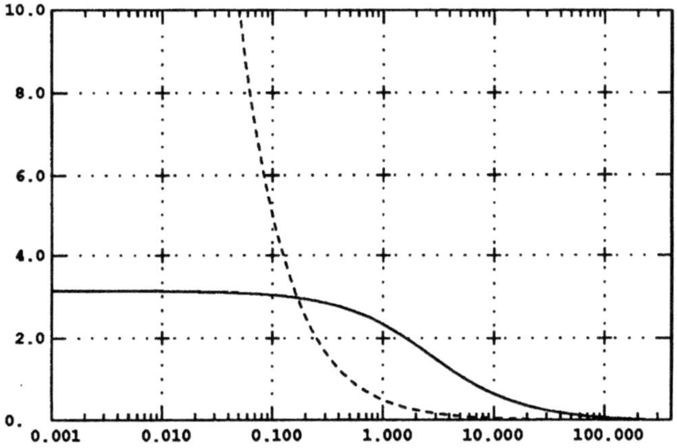

Figure 4: Plotting the two components of a block estimation variance versus the sample support size. In broken line the Poisson component. In solid line the potential component.

The first component is the estimation variance of a Poisson point process of intensity m. For this reason it is termed *Poisson component.* It is a hyperbolic function of the sample size. The second component, or *potential component*, stems from the spatial variability of the potential. It starts off from a finite value

$$V_2 \approx Var\left\{\Theta(x_0) - \frac{\Theta(V)}{|V|}\right\} \quad (|v| \longrightarrow \{x_0\})$$

and decreases as the sample support size becomes larger (cf. Figure 4).

The Poisson component depends only on the mean value of the potential. The second component involves the variance of the potential as well as its integral range. In the case where the potential has a large coefficient of variation or a large integral range, the contribution of the second component can be relatively large. Figure 5 shows the estimation variance for several variances of the potential.

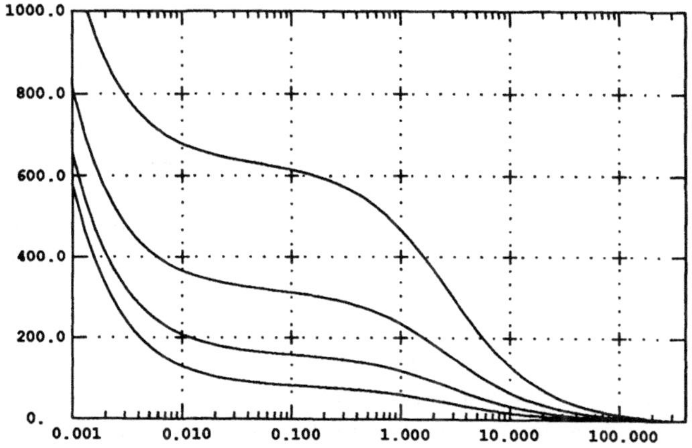

Figure 5: Plotting the block estimation variance versus the sample support size. Several curves are given depending on the variance of the potential.

On these curves, a slow decrease in the variance can be observed for moderately small sample support sizes. This is a visible consequence of the influence of the potential component. Physically speaking, the presence of such a kink is related to the size and type of trapping mechanism present. This effect is called the *trapsite effect* and was first identified by Kleingeld (1987) while studying diamond and gold alluvial type deposits. In Figure 6 the relative importance of the potential component is displayed. It turns out that the influence of the potential component is totally determinant provided that the sample size is large enough. Moreover, the larger the variance of the potential or the larger its integral range, the smaller the sample support size that makes the second component determinant.

What sample support size should be taken to ensure that sampling is representative? Of course, the larger the samples, the better their representativity. However, even a moderate increase in the sample support size may not significantly change the estimation variance, due to the presence of the kink (cf. Figure 5). In the case where

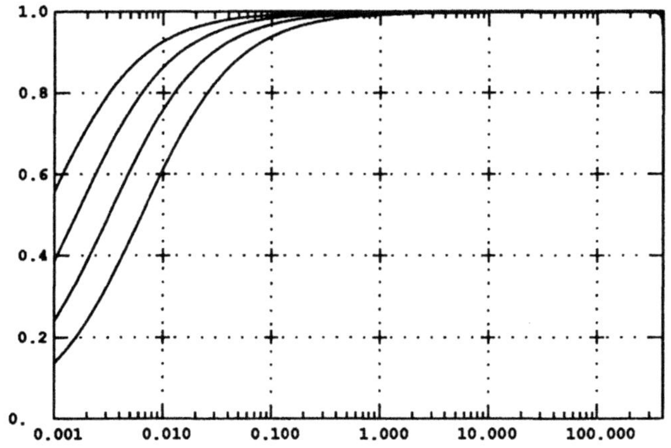

Figure 6: Plotting the relative importance of the potential component in the block estimation variance versus the sample support size. Several curves are given depending upon the variance of the potential.

the potential component is largely influential, it is clear that the sample support should be chosen so as to cope with the effects of this component. In what follows, the authors advise to take large enough samples so as to make the influence of the trapsite effect negligible. The practical implications of this are considerable as a large sample support is difficult to obtain in practice. For this reason it is necessary to replace each large sample by a cluster of smaller subsamples. The support of the subsamples is imposed by the available equipment. Their number is chosen so as to have the extension variance from the clusters of subsamples to the large samples as small as possible.

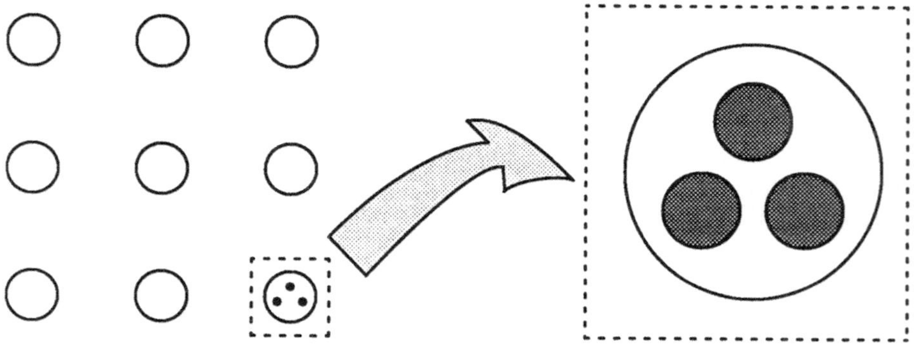

Figure 7: Replacing large samples by clusters of subsamples.

4 Comparison of two sampling procedures

Suppose that we want to estimate the number of stones in a block using the number of stones in some samples. Which is better? To use some large samples? Or to use many little samples? If a criterion such as the estimation variance was used, in most of the cases, the second solution would be preferable. It is not sure however that this criterion is really appropriate. As a matter of fact, the sample support should be chosen depending on the objectives at stake. To fix ideas, let us consider two samples v and v' such as in Figure 8. Both samples have the same area. v is a large, connected sample, whereas v' is a family of small isolated subsamples. In what follows, we are going to observe how the deposit under study is perceived through these two sampling procedures.

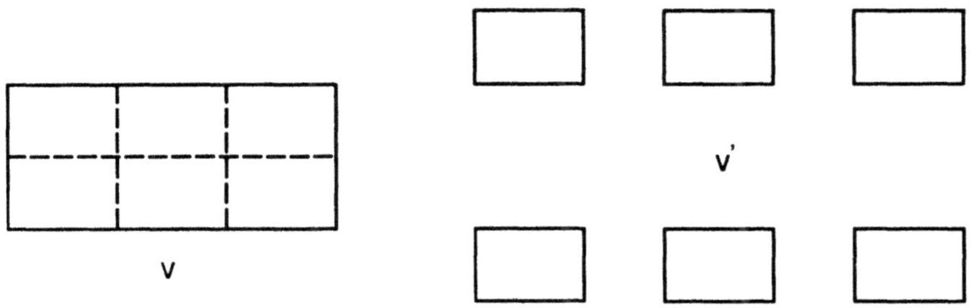

Figure 8: The two samples v and v' have the same area. Which one is more representative?

At first, let us consider the potential of the deposit within the two samples v and v'. Since both samples have the same area, they have the same average potential

$$E\{\Theta(v)\} = E\{\Theta(v')\}$$

The close "subsamples" of v are highly correlated, which is not the case for those of v'. As a consequence, the potential of v is more variable than the one of v', and in general we have

$$Var\{\Theta(v)\} \geq Var\{\Theta(v')\}$$

This suggests that the potential of v is *more selective* than the one of v', in the sense that

$$E\{\phi[\Theta(v)]\} \geq E\{\phi[\Theta(v')]\}$$

for any convex function ϕ (Matheron, 1984). In particular, if we take $\phi(t) = \exp(-t)$, we obtain

$$E\left\{e^{-\Theta(v)}\right\} \geq E\left\{e^{-\Theta(v')}\right\}$$

that is

$$P\{N(v) = 0\} \geq P\{N(v') = 0\}$$

which means that the traps are better detected by many small samples than by fewer large ones. One might think that such a result gives a definitive advantage to the small samples. Unfortunately things are not so simple as we are going to see now.

If the potential of v is more selective than the one of v', then the distribution of $N(v)$ is also more selective than the distribution of $N(v')$ (See Appendix for the proof). Practically speaking, this means that a campaign made up of large samples is not so capable of detecting the traps, but in contrast to this, can provide more information on the contents of the detected traps. In order to bring out the change in selectivity as the sample size increases, consider a square trap of $1m^2$ containing 20 independent uniform stones. This trap is hit by a sample, and we assume that the sample is not barren. Figure 9 shows the mean number of stones recovered in the sample versus the sample size. As we can see, about 6 stones are recovered when the sample has the size of the trap, that is slightly more than a quarter of the total number of stones in the trap. Even when the sample size is 1000 times the trap size, only 90% of the stones are recovered.

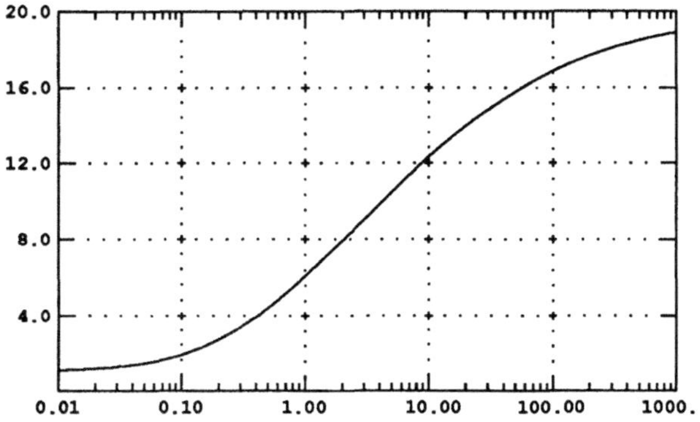

Figure 9: A square trap of $1m^2$ contains 20 stones. This Figure displays the average number of stones recovered in a non barren sample versus the sample support size.

In the case where the deposit is made up of isolated traps and where the richness of the traps is highly variable, the gain in selectivity by taking large samples is of special importance if selective mining is required.

5 Conclusions

The authors believe that the taking of representative samples in a highly dispersed type of mineralization is an important problem. In the case where the mineralization is concentrated in trapsites, the sampling campaign should integrate the aspects of trapsite location and richness. To that end, any sampling of such deposits requires a two-phased approach. The first phase is directed to estimating the location of the trapsites (using small samples). The second phase is directed to assessing the richness of the trapsites (using large samples) to cater for selective mining requirements. Although a definitive demonstration and a solution based on a representative industrial study is still outstanding, it is important that the implications of the problem be understood by ore evaluators in mining companies.

Acknowledgements: The authors wish to express their gratitude to G. Matheron and M.M. Oosterveld for their helpful advices and suggestions.

References

Cox D.R. (1955) "Some statistical models connected with series of events", Journ. Royal Stat. Soc., B17, pp. 129,164.

Deverly F. (1984) "Echantillonnage et Géostatistique", Geostatistics Doctoral Thesis, School of Mines of Paris.

Feller W. (1971) An introduction to Probability Theory and its Applications, Vol. 2, Wiley, New-York.

Kleingeld W.J. (1987) "La Géostatistique pour des variables discrètes", Geostatistics Doctoral Thesis, School of Mines of Paris.

Lantuéjoul Ch. (1991) "Ergodicity and Integral Range", Journal of Microscopy, Vol. 161-3, pp. 387-404.

Matérn B. (1986) Spatial Variations (second edition), Lecture Notes in Statistics, 36, Springer-Verlag, Berlin.

Matheron G. (1984) "The Selectivity of the distributions and the Second principle of Geostatistics", in Verly G. and al. (eds.) "Geostatistics for Natural Ressources Characterization", NATO ASI Series C, Vol. 122, Reidel, Dordrecht, pp. 421-433.

Matheron G. (1988) Estimating and Choosing, Springer-Verlag, Berlin.

Matheron G. (1989) "Two classes of isofactorial models", in Armstrong M. (ed.) "Geostatistics", Kluwer Academic Publishers, Dordrecht, pp. 309-322.

Oosterveld M.M., Campbell D. and Hazell K.R. (1987) "Geology related to statistical evaluation parameters for a diamondiferous beach deposit", in Lemmer I.C. and al. (eds.) "Application of Computer methods in the Mineral Industry", Vol. 3, The South African Institute of Mining and Metallurgy, Johannesburg, pp. 129-136.

Sichel H.S. (1973) "Statistical valuation of diamondiferous deposits", in Salamon M.D.G. and Lancaster F.H. (eds.) "Application of Computer methods in the Mineral Industry", The South African Institute of Mining and Metallurgy, Johannesburg, pp. 17-25.

Yaglom A. M. (1986) Correlation Theory of Stationary and related Random Functions, Springer Series in Statistics, New York.

Appendix. Let N and N' be two Poisson variables with random parameters X and X'. Suppose that X is more selective than X'. The aim of this appendix is to prove that N is more selective than N'.

According to Matheron (1984), one has to show that

$$B(z) \geq B'(z) \quad \forall z > 0$$

where B and B' are the conventional benefits of N and N', namely

$$B(z) = E\{(N - z)1_{N \geq z}\} \qquad B'(z) = E\{(N' - z)1_{N' \geq z}\}$$

Since N and N' take discrete values, it suffices to establish that

$$B(n) \geq B'(n) \quad \forall n \geq 1$$

By conditioning with respect to X

$$B(n) = E\{E\{(N - n)1_{N \geq n}|X\}\}$$

and expanding the conditional expectation into a series, the conventional benefit can be written as

$$B(n) = E\{\phi_n(X)\}$$

where

$$\phi_n(x) = \sum_{i=n}^{+\infty}(i - n)e^{-x}\frac{x^i}{i!}$$

Now, it is sufficient to show that ϕ_n is convex to complete the proof since X is more selective than X'. A simple calculation gives

$$\phi_n''(x) = e^{-x}\frac{x^{n-1}}{(n-1)!} \geq 0$$

INDICATOR APPROACH TO CHARACTERIZE FRACTURE SPATIAL DISTRIBUTION IN SHEAR ZONES

O. TAVCHANDJIAN and A. ROULEAU
Université du Québec à Chicoutimi, Dpt. Sci. Appl.
Chicoutimi (Qc),Canada
D. MARCOTTE,
Ecole Polytechnique de Montréal, Dpt. Génie Minéral
Montréal (Qc), Canada

ABSTRACT

Fracture density can be used as the basic parameter to analyze the spatial distribution of fractures in rock. In shear zones, fracture density can be defined as a function of the number, the length and the width of discontinuities. Fracture density values were obtained from fracture parameters measured on mine drift walls at three levels in each one of two mines of the Chibougamau area: Henderson and Copper Rand.

Frequency distributions of density show a strong positive skewness. Classical geostatistical procedures applied to the raw fracture density show inconsistencies in the variograms due to extreme values representing specific types of discontinuities (e.g. shear, tension vein). An indicator approach applied to field data shows a close relationship between variogram range and deformation intensity. Indicator kriging maps suggest spatial relations between fracture sets and are used to quantify and verify geomechanical models.

INTRODUCTION

Geostatistics have been developped for ore reserve estimation (Matheron, 1963; Journel and Hjuibregts, 1983; David, 1989). These statistical methods, based on the theory of "regionalized variables" have been applied to a variety of problems in the earth and environmental sciences. Chiles (1988) have listed a number of geostatistical applications in the study of fractures. Variogram and kriging procedures have been applied both to estimate and to model the geometry of fracture network (Brière and Razack, 1982; Long and Billaux, 1987) or the spatial distribution of fracture parameters (Girardi and Matar de Sarquis, 1984; Young, 1987).

All these studies were performed in rock masses affected only by tension fractures (mainly joints). Fracture data came from rock exposures, drill-holes or remote sensing maps. The particularities of shear zones, which have been poorly investigated include : the superposition of brittle and ductile structures in the rock mass, the presence of discontinuities with significant width in respect to their length, and the importance of the fracture network for water circulation, rock mass stability and mineralized fluids transport and deposition.

The purpose of this paper is to adapt existing geostatistical technics to the structural particularities of shear zones. In a first step we define the basic parameter used to characterize fracture spatial distribution. Then, the variogram procedure applied to the identified fracture sets is discussed. Finally an indicator approach is proposed to take into account the coexisting components of deformation. The results are used to confirm and quantify existing geomechanical models (Tchalenko, 1968).

SOURCE OF DATA

Fracture surveys were executed at the Henderson and Copper Rand mines in the Chibougamau area, Quebec. These deposits are located in major shear zones (Fig.1) hundred of meters in width and several kilometers in extent. The geomechanical behaviour of the rock mass ranged from brittle to ductile during the formation of these structures; this behaviour has been correlated with the evolution of mineralizing fluids by Archambault et al. (1984).

Geometric as well as genetic parameters were systematically recorded on mine drift walls at three levels in each mine. Geometric parameters included : the location of fracture trace center on the wall, its length, width, orientation, and the number of observable terminations. Genetic parameters were the nature of fracture (shear, tension vein, en-parallel, en-echelon, isolated) the nature of their infilling, and the apparent direction of movement when observable. This database was stored in computer files.

Based on stereographic projection, a total of nine different fracture sets have been defined at the Henderson mine and four fracture sets at the Copper Rand mine.

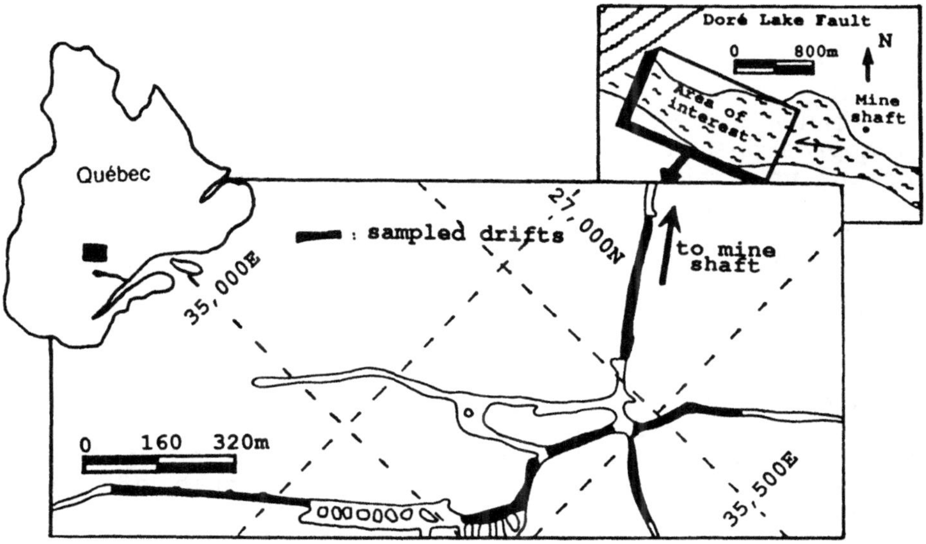

Figure 1 : Location of fracture sampling at the 1910' level of Copper Rand mine

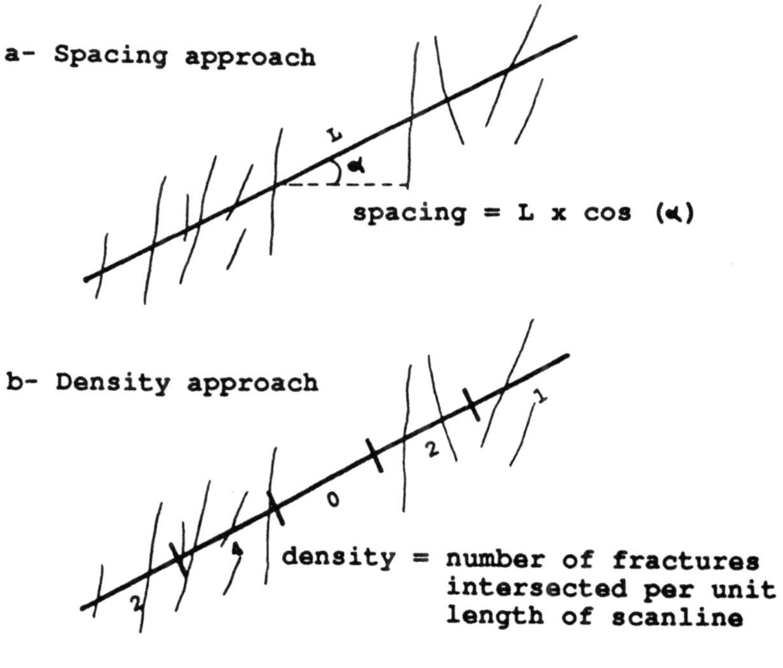

Figure 2 : Two approaches to characterize fracture spatial distribution

FRACTURE DENSITY ANALYSIS

Definition of fracture density

Fracture spatial distribution can be analyzed by at least two approaches (Fig.2). The calculation of spacing between consecutive fracture traces is widely used in geotechnics (Priest and Hudson, 1976). The other approach is based on the estimation of fracture density in a reference unit length, unit area or unit volume (Girardi and Matar de Sarquis, 1984). The selection of the appropriate approach depends on the source of data, the structural setting prevailing in the rock mass and the objectives of the study.

Shear zones are characterized by the presence of fractures with significant width with respect to their length. These deformation zones (e.g. mylonitized and mineralized zones, tension veins often filled with quartz and carbonates) can originate from the growth of a single fracture plane or from the junction of many planes. In this latter case, it is not realistic to applied the spacing approach between all the visible planes. Therefore, the density approach is selected as more appropriate for the tectonic environment typical of shear zones, encountered at the Copper Rand and Henderson mines.

For each fracture set identified on the basis of genetic and geometric parameters fracture density was estimated by the following relation (Fig.3) :

$$D_k = \sum_{i=1}^{N} (l_{i,k} \times e_i) \quad / \quad S_k$$

where D_k is the density for cell k, S_k is the total area of cell k ($4m^2$ in this case), N is the number of fractures intersecting cell k, $l_{i,k}$ is the length of fracture i in cell k and e_i is the width of fracture i. All the sampling planes do not have the same orientation (Fig.1). This constitutes a sampling bias that can be corrected by the following sinus correction :

$$L = l / \sin(b)$$

where b is the angle between the wall and the normal to the average orientation of the fracture set. L represents the distance along the drift required to estimate fracture density on a virtual cell with length l along the normal to the mean fracture orientation (Fig.3).

Statistical analysis of density

Frequency distributions of regularized fracture density data are characterized by a strong positive skewness, as illustrated on figure 4 for set I of the 1910′ level of the

a- Fracture density calculation

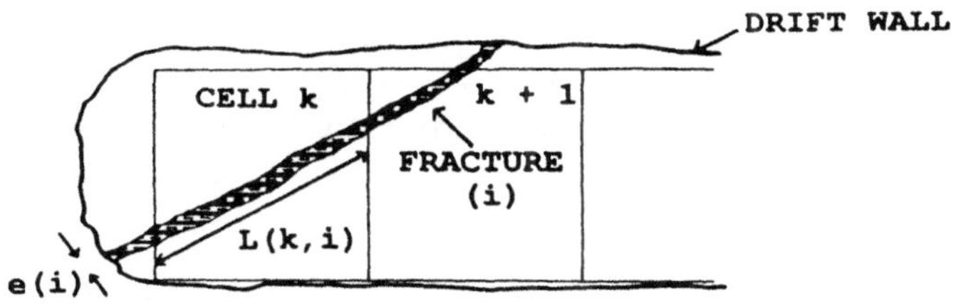

b- Angular bias correction

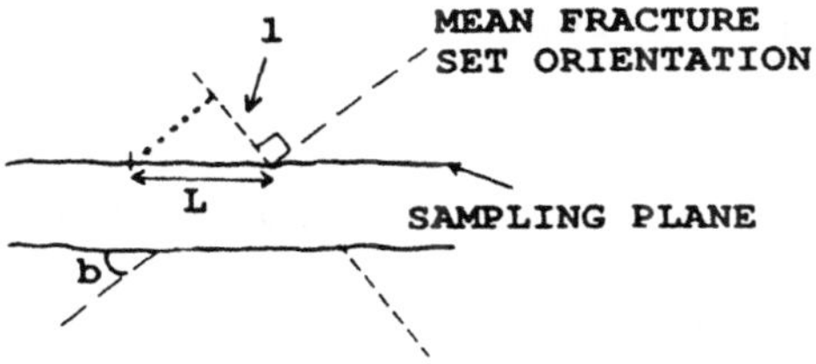

Figure 3 : Fracture density calculation and bias correction

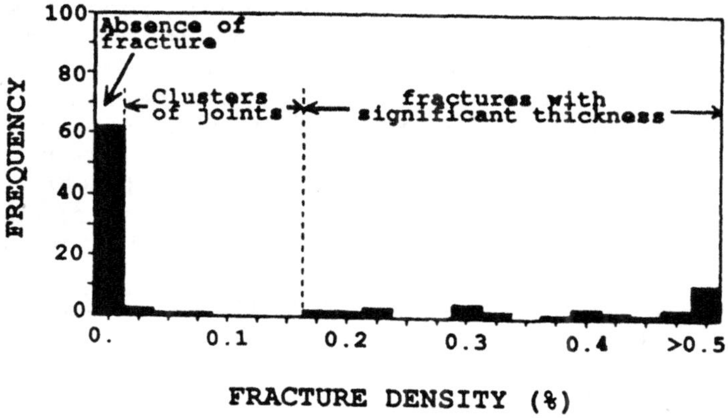

Figure 4 : Frequency distribution of fracture density for set I at the 1910' level of Copper Rand mine

Copper Rand mine. The multimodal aspect of these distributions suggests the superposition of three different types of discontinuity for each fracture set : the absence of the set (nil density), its presence as clusters of joints (density < 0.15%) and its presence as major fractures (shears or tension veins with density > 0.15%).

Variogram of fracture density

For every mine level investigated, variograms of raw fracture density present anisotropies and erratic values linked to the nature of the data. Extreme values contribute to nearly all the variability in the raw data. Because of the geometry of the drifts and the relative location of the extreme values, these values do contribute in the calculation of only some of the directional variograms causing the observed zonal anisotropy. In these directions, the variograms appear as series of bumps and holes, following their presence or absence at that lag.

INDICATOR APPROACH

Extreme values in geology often represent a distinct natural phenomenon from other observations. This is the case in certain precious metal deposits or for earthquake magnitude as examples. Journel and Arik (1988) showed that for a gold deposit, extreme values correspond to the presence of hydrothermal veins, while other observations were characterizing background dissemination from these veins. These two classes of mineralisation presented different spatial structurations. These observations can be applied to fracture density in shear zones, where high values are characteristic of shears or veins, and low values indicate clusters of small fractures and joints (Fig.4).

This particular association between density and type of deformation suggests the use of indicator coding of the density, in order to reflect the different types of deformation encountered. For each one of the identified fracture sets, three classes of intensity of deformation (nil, brittle and ductile) were defined and coded as :

$F_n I_k$ = 1, if the density of fracture set n is in class k
or = 0, otherwise

Indicator Variography

Results of the modeling of fracture indicators are illustrated with the example of the 1910' level in the Copper Rand mine (Table 1, Fig.5). All the models presented in table 1 are omnidirectionnal, since no significant

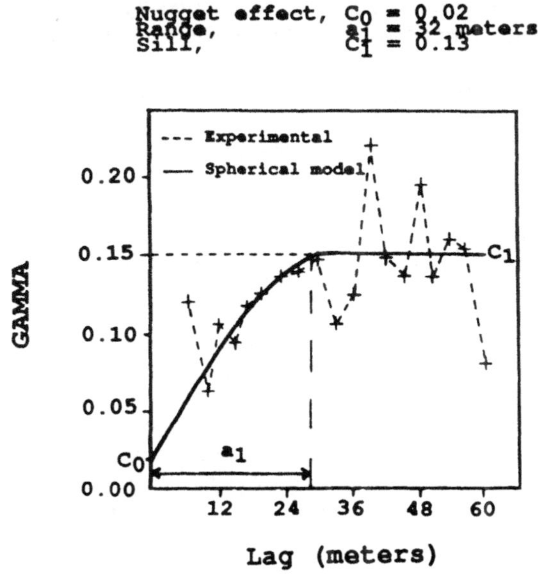

Figure 5 : Variogram modeling for indicator F2I2 at the 1910' level of Copper Rand mine

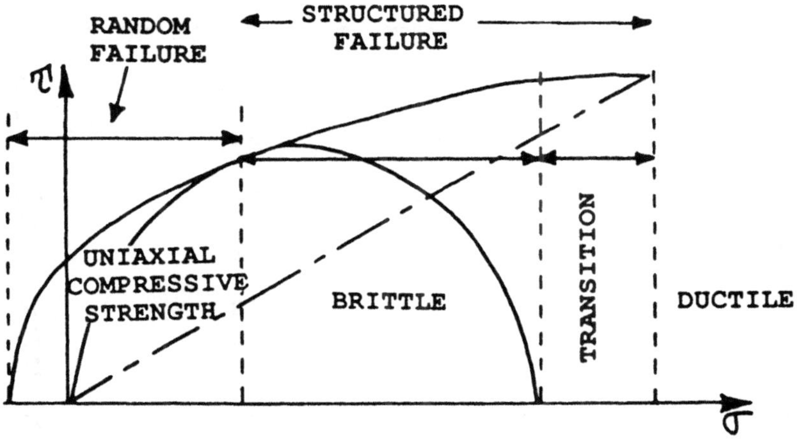

Figure 6 : Distribution of mechanical properties in rocks

anisotropy has been observed on variography. This absence of anisotropy on the indicator variables support the previous interpretation that the zonal anisotropy observed on the raw density was caused mainly by a few extreme values.

Direct and cross variograms for each fracture set are related by :

G (I1,I2) = (G (I3,I3) - G (I1,I1) - G (I2,I2)) / 2
G (I1,I3) = (G (I2,I2) - G (I1,I1) - G (I3,I3)) / 2
G (I2,I3) = (G (I1,I1) - G (I2,I2) - G (I3,I3)) / 2

It was verified that the experimental cross-variograms matched satisfactorily the above values after modelling of the direct variograms.

Two parameters can be considered to quantify the relative spatial structuration of each indicator. The range (a_1) of the spherical model defines the scale of spatial structuration, whereas the ratio $C_0/(C_0+C_1)$ indicates the part of the random component in the spatial structuration of the parameter.

Table 1 : Summary of indicator variogram modeling at the 1910' level of Copper Rand mine

	Nugget effect (C0)	Sill (C1)	range (a1)	ratio C0/(C0+C1)
F1I1	0.17	0.08	**20**	**0.68**
F1I2	0.07	0.14	**28**	**0.39**
F1I3	0.15	0.05	**33**	**0.75**
F2I1	0.23	nil	**nil**	**1.00**
F2I2	0.08	0.08	**28**	**0.50**
F2I3	0.10	0.10	**32**	**0.50**
F3I1	0.13	0.11	**38**	**0.54**
F3I2	0.09	0.03	**18**	**0.75**
F3I3	0.08	0.12	**32**	**0.40**
F4I1	0.04	0.11	**38**	**0.27**
F4I2	0.04	nil	**nil**	**1.00**
F4I3	0.04	0.10	**28**	**0.29**

Table 1 indicates that for every fracture set, the range is larger for I3, the ductile component, than for I2, the brittle component. The same observation holds at other levels of Copper Rand mine and .at the Henderson mine. However, the ratio $C_0/(C_0+C_1)$ does not show any such systematic behaviour between the various indicator variograms.

The fact that the range increases with the intensity of the deformation can be related to the greater influence of small scale heterogeneities when the deformation is brittle as compared to ductile deformation. The larger the confining stress, the weaker the influence of small scale heterogeneities on the resulting deformation. As the confining stress increases, the deformation changes from brittle to ductile and the structured component of fracturating overcomes gradually its random component (G.Archambault, pers.comm; Fig.6) and failure will develop according to the geomechanical model of Tchalenko (1968).

Indicator Kriging

Kriging estimations were also performed for fracture indicators. As no significant anisotropy has been observed, a circular searching window was used with a radius of 100 meters, and minimum and maximum numbers of observations fixed respectively to 1 and 8.

Kriging maps can be used to study the evolution of the spatial distribution of fracture indicator from one mine level to another. These comparisons of the relative distribution of indicators allow a correlation of their presence with regional structures. On the maps presented in figures 7 and 8, iso-contour values range from 0 (nil probability) to 1 (100% probability). The estimated probability of the three indicators was verified to equal approximatively 100% for each set.

Fracture set I at the Copper Rand mine represents shear fractures trending north-east, and cutting accross schistosity and mineral stretching lineation associated with the Copper Rand shear zone. These fractures can be associated with the Doré Lake fault, a regional structure trending north-east which cut across the Copper Rand deposit at its western edge (Fig.1). Figure 7 shows an increase in the presence of indicator F1I3 at the western edge of the 1910' level. This observation supports the hypothesis that fracture set I can be used as a structural signature of the regional Doré Lake fault.

Figure 8 shows the spatial distribution of indicator F1I3 at the 2200' level of the Copper Rand mine. This indicator tends to disappear with depth. Similar observations made for other structural indicators may be correlated with an increase in the deformation intensity,

contour values are expressed in percent

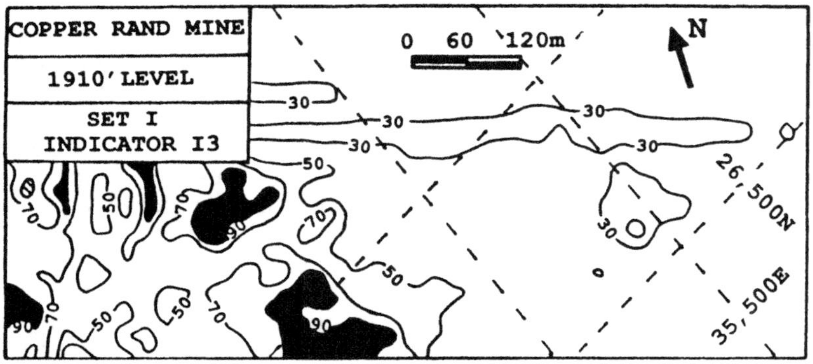

Figure 7 : Kriging map of indicator F1I3 at the 1910' level of Copper Rand mine.

contour values are expressed in percent

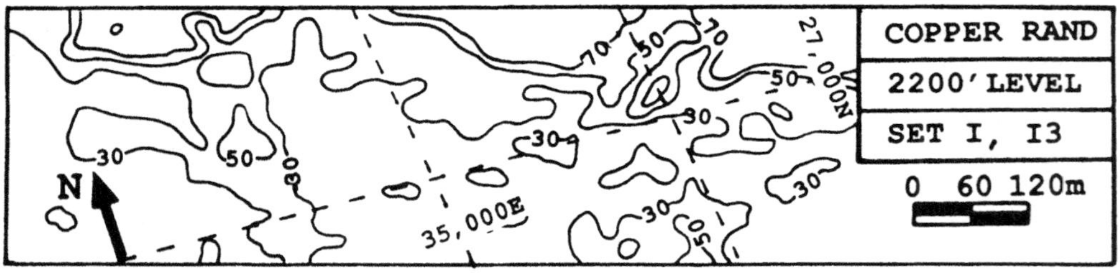

Figure 8 : Kriging map of indicator F1I3 at the 2200' level of Copper Rand mine.

and the evolution of rock foliation linked to the Copper Rand shear zone.

It seems paradoxical that the development of the Copper Rand shear zone has affected the spatial distribution of fractures formed subsequently to this large tectonic structure. A possible explanation is that the north-east trending fracture planes represent conjugate fault set created synchronously with the Copper Rand shear zone. Only the planes preserved by the first tectonic stage were used as preferential weakness planes during the Doré Lake fault development. This interpretation has been used by Archambault et al. (1984) to explain some structural features at a regional scale in the Chibougamau area.

CONCLUSIONS

Fracture density was used as the basic parameter to quantify the spatial distribution of fractures in shear zones. Frequency distributions of fracture density show a strong positive skewness due to the presence of extreme values. These high values represent fractures with significant thickness with respect to their length and they lead to chaotic, strongly anisotropic experimental variograms. An indicator coding, corresponding to three well defined classes of intensity of deformation, provided an easier approach to model the variography.

The kriging of these indicators for the various fracture sets provided an interpretable spatial distribution of fracture density in the shear zones of the Copper Rand and Henderson mines. This method takes into account the information on the structural geology in the statistical analysis of genetical and geometrical parameters of fractures, in order to discriminate their different geomechanical components. The same approach can be used in any underground excavations located in shear zones affected by brittle and ductile deformation.

The variograms of the indicators and the kriging maps provided tools for the geomechanical interpretation. A positive correlation was evidenced between the range of the variogram and the importance of the deformation. Also the kriged maps of fracture indicators support the interpretation by Archambault et al. (1984) of a preferential development of fractures associated to the Doré Lake fault, along the prior fractures related to the shear zones of the Copper Rand and the Henderson mines.

ACKNOWLEDGMENTS

We would like to thank Dr G. Archambault for participating in discussions. This research has been financially supported by the FCAR Funds and the UQAC Fundation.

REFERENCES

Archambault, G., Guha, J., Tremblay, A. and Kanwar, R. (1984) "Implication of the geomechanical interpretation of the Copper Rand Deposit on the Dore Lake Shear Belt", C.I.M. Special 34, 300-318

Chiles, J.P. (1988) "Fractal and Geostatistical Methods for Modeling of a fracture network", Math. Geol. 20, 631-654

David, M. (1989) Handbook of applied advanced geostatistical ore reserve estimation, Elsevier Co., Amsterdam.

Girardi, J.P. and Matar de Sarquis, M. (1984) "Selection of a minimum fracturation density area by geostatistics in crystalline rock", in Verly and al. (eds), Geostatistics for Natural Resources Characterization, pp. 759-773

Journel, A. and Hjuibregts, K. (1983) Mining Geostatistics, Elsevier Co., Amsterdam

Journel, A. and Arik, A. (1988) "Dealing with outliers high grade data in precious metals deposits", in Fytas and al. (eds), First Canadian Conference on Computer Application in the Mineral Industry, pp. 161-172

Long, J. and Billaux, D. (1987) "From Field Data to Fracture Network Modeling", Water Resources Research 23, 1201-1216

Matheron, G. (1987) "Le Krigeage", Mémoires du BRGM 24, 171p

Priest, S.D. and Hudson, J.A. (1976) "Discontinuity spacings in Rock", Int. J. Rock Mech. and Min. Sci. 13, 135-148

Razack, M. and Brière J. (1982) "Caractérisation géostatistique de champs de fractures - Implications pour la reconnaissance des systèmes fissurés discontinus", Doc. BRGM 45, 161-172

Tchalenko, J.S. (1968) "The evolution of kink bands and the development of compression textures in sheared clays", Tectonophysics 6, 159-174

CONDITIONAL SIMULATION: A MINE PLANNING TOOL FOR A SMALL GOLD DEPOSIT[1]

M. S. Nowak[1], R. M. Srivastava[2] and A. J. Sinclair[1]
[1]Mineral Deposit Research Unit (MDRU), Department of Geological Sciences, The University of British Columbia, 6339 Stores Road, Vancouver, B.C. Canada V6T 1Z4
[2]FSS International, 4106 Staulo Crescent, Vancouver, B.C. V6N 3S2
(MDRU Contribution 011)

ABSTRACT
The JM zone of the Shasta property in the Toodoggone camp of northern British Columbia, Canada, forms the basis for this study. The deposit is a fracture-controlled zone of gold-quartz-carbonate mineralization in volcanic rocks, elongate in a northerly direction but with no sharp, well-defined contacts to ore grade material. A population of high grade assays, irregularly located within the mineralized zone, seriously complicates reserve estimation and mine planning.

Conditional simulation is used to supplement more traditional reserve estimation techniques for the JM zone. With the erratic nature of the deposit expressed by high grade material having little lateral continuity, estimation procedures typically produce a biased view of the reserves/resources. Simulation, on the other hand, succeeds in portraying more realistically the complexity of the ore\waste contact. Using historical economic parameters from the recent mining operations, we show how ore\waste selection based on simulation can supplement any estimation study to improve production planning.

INTRODUCTION
The Shasta gold deposit is in the Toodoggone area of northern British Columbia, Canada (Figure 1), about 230k northeast of Stewart, B. C., a town at the south end of the Alaska panhandle. Several small epithermal gold deposits have been discovered in the area over the past decade, two of which have seen limited production to date. The deposits occur in Early Jurassic alkali-rich volcanic rocks as gold-bearing quartz veins with adularia, sericite and calcite. Extensive zones of alteration are associated with the veins.

Mineralization in the JM zone of the Shasta deposit is contained in a crudely tabular zone that strikes 150°, dips steeply to the east, with a thickness that ranges from 1m to 22m. Within this broad zone of mineralization/alteration are a series of much smaller tabular high grade (Au and Ag) zones that represent potential ore; some of the wider

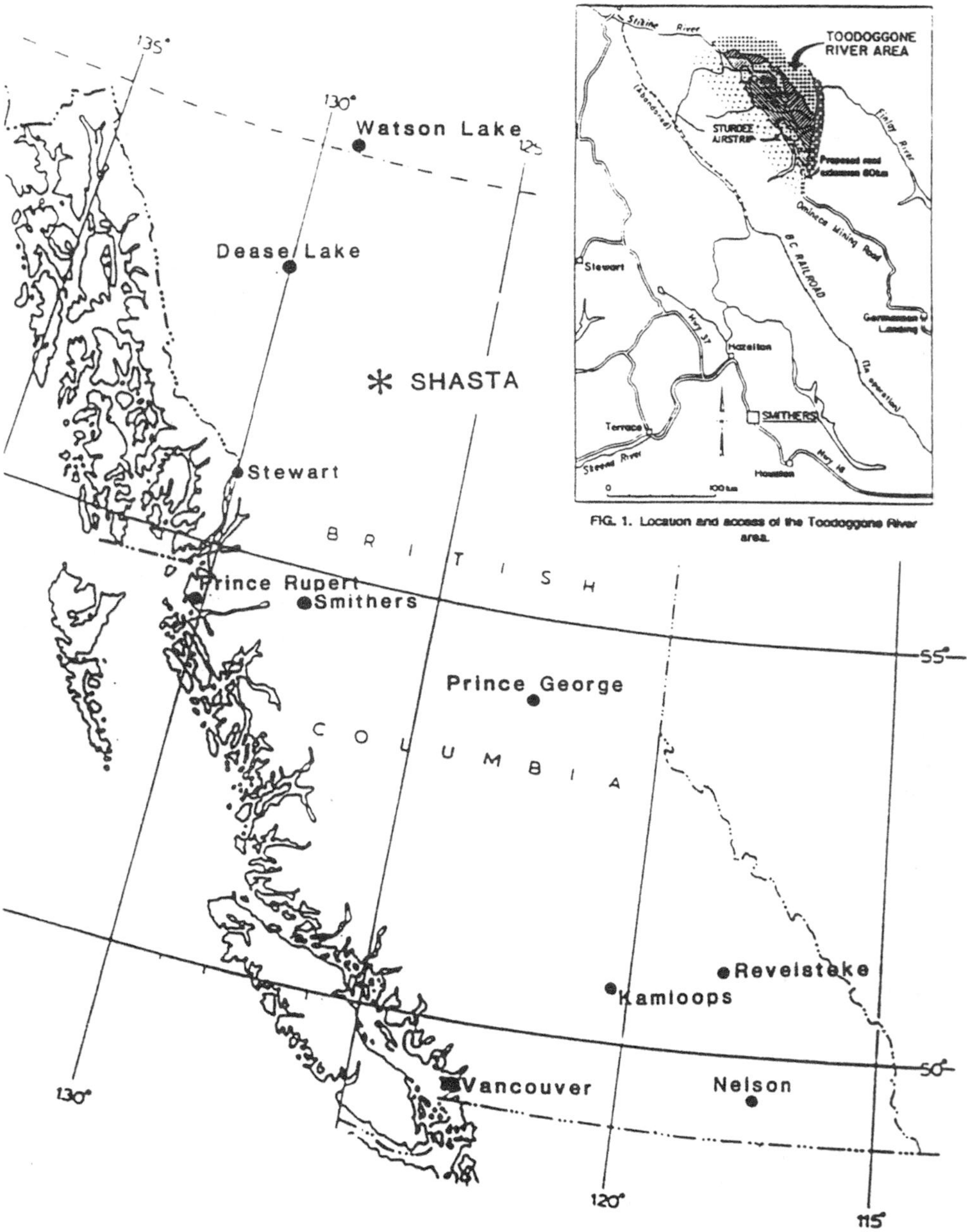

Figure 1: Location of the JM zone of the Shasta Deposit, northern British Columbia, Canada

of these higher grade shoots appear to be continuous between sections, perhaps for a total length of about 200 m. A limited duplicate sampling program demonstrated wide local variability in grades but insufficient close spaced data were obtained to quantify local continuity.

GEOSTATISTICAL ESTIMATION

For the JM zone we have demonstrated (Nowak et al, 1991), (i) the multimodal, lognormal character of both Au and Ag assays, (ii) the spatial limitation of the JM zone within which reserve estimates are warranted, and (iii) a sufficiently wide range of sample lengths to warrant treatment of metal accumulations rather than using individual samples. In familiarizing ourselves with the data and the geological characteristics of the JM zone we recognize two important data populations viz. relatively high grade assays and relatively low grade assays separated by cutoff grades of 50 g Ag/t and/or 1 g Au/t. The choice of cutoff grades was based on extracting a mixture of two lognormal populations from cumulative curve of declustered values (cf. Sinclair, 1976). These populations slightly overlap. Cutoff grades are 97.5 percentiles of low grade populations of Au and Ag. Our analysis indicates the presence of a proportional effect in the assay data but showed quasi-stationarity.

We approached reserve estimation at Shasta in two different ways: (i) The estimation of total high grade (i.e. >50 g Ag/t and/or >1 g Au/t) contained within the JM zone; and (ii) the estimation of resources represented by contiguous drill hole intersections which carry the largest quantity of silver, and are not diluted by grades lower than 50 g/t. Because of a very high correlation (r=0.95) between Ag and Au this intersection also insures the largest quantity of gold.

The first of these two estimations provides us with a total high grade resource, not all of which will be accessible for mining. The second estimation goal relies on the assumption that the "best undiluted high grade intersections" have sufficient continuity that they will be amenable to mining. This is a realistic assumption in view of the recognized continuity of their breccia/stockwork characteristics. The nature of physical continuity of the mineral-controlling structures is much better appreciated after some underground development. Consequently, the use of the "best undiluted high grade intersection" during the exploration phase would have been much more tenuous than in our retrospective analysis. These two grade/tonnage estimates should, therefore, bracket the eventual recoverable reserves if mining is based on a comparable cutoff grade.

Relative variogram models were determined in order to account for a proportional effect identified in the data. Despite the relatively small data base, experimental variograms for all variables considered (thickness and metal accumulations) could be fitted easily with spherical models although some assumptions were required in order to develop anisotropic, 3-dimensional models (particularly in a direction perpendicular to the tabular form of the JM zone).

Variogram models were tested using a cross validation procedure whereby individual data points were estimated from nearby data using a variety of methods (polygonal, inverse distance squared, kriging with isotropic model, kriging with anisotropic model). Various estimates were compared with known values. Kriging with an anisotropic variogram model was demonstrably the best cross validation technique and was the procedure selected for establishing mineral inventory.

We have defined a block size of 25 x 25 x t m3 for estimation purposes, where t is the average kriged thickness (perpendicular to the tabular form of JM zone) of the block to be estimated. In the broad mineralized zone we identified global reserve of 261,000 tons averaging 4.0 g Au/t and 221 g Ag/t. The tonnage of "best undiluted zones" totaled 128,000 tons averaging 6.4 g Au/t and 370 g Ag/t. Average thickness of the broad zone is 5.8 m and average thickness of the best undiluted zone is 3.5 m.

A comparison of estimates by kriging Ag and Au average grades of two small stopes with the production data from these stopes (Nowak, et al, 1991) shows that in one area mined out grades are comparable to estimated grades. In the second area, kriging results do suggest that a substantial proportion, perhaps up to 50%, of the ore grade material has not been recovered in some parts of the JM zone where recent mining has taken place.

The difficulty in making estimates that compare well with the production of the JM zone stems from the limited data available to us and the limited continuity of high gold grades within the broad mineralized zone. Consequently, we have been led to investigate the potential of conditional simulation in such cases.

PROBABILITY FIELD SIMULATION

Simulation methodology

The probability field simulation method (Froidevaux, 1992) is a conditional simulation i.e. it honours the data values. As implemented on this study the major steps of the method which produces a grid of simulated points are:

1. Define the conditioning data; original values will be transformed to indicators dependent on the cutoff value selected.

2. Calculate a variogram of the variable being simulated; this variogram will describe the pattern of spatial continuity for the probability field and it will be used in estimating the local cumulative distribution function, Fz, by indicator kriging.

3. Create a three dimensional grid of points.

4. Create a field of random numbers, U(x), in such a way that their global distribution is uniform between 0 and 1, and their variogram has the same shape as that of the variable being simulated.

5. At each location xi where a simulated value is required, estimate values of the local cumulative distribution function by ordinary indicator kriging.

6. At each location, xi, extract either 0 or 1 as the simulated value by drawing randomly, using the value U(xi) to draw from the local cdf at xi.

Conditioning data
Gold assay data from 48 exploration drill holes which intersect the JM zone of the Shasta deposit were used for our analysis. Drill core samples had, with few exceptions, a minimum length of 0.5m and a maximum length of 1.5m. The majority of samples had a length of 0.5m, 1.0m or 1.5m.

To convert those data of different support to point data within each drill hole, new points were established at intervals of 0.4m and each point was assigned a value of the nearest assay. This procedure transforms drill hole assays of variable support (length) into a regular 1-dimensional data array for each drill hole. This method will affect the estimates of the variogram for distances shorter than about 1.5m. The effect is to decrease variance as the lag approaches 0.4m for directions parallel to drill holes (i.e. across the JM structure). Variances at 0.4m were not used in developing variogram models across the structure. The separation distance insures that every original assay is represented at least once in the conditioning data, and avoids the problem of points coinciding with the contact between samples.

Choice of a Cutoff for Indicator Transform
Individual sample limits were selected to coincide with visual features such as breccia and stockwork boundaries or presence of argentite wherever possible. Thus, adjacent assay pairs can differ widely, ranging from background values to highly anomalous values. This fact may indicate the presence of two populations. It should be stressed that these populations are not the result of multi-stage mineralization processes, but relate to the style of mineralization. Probability plots (Sinclair, 1976) of declustered Au values show that 1.0 g Au/t is an effective threshold, hence, the indicator transform of the data value, zi, is defined as follows:

$$\text{indi} = \begin{cases} 0, & \text{if } zi < 1 \\ 1, & \text{if } zi >= 1 \end{cases}$$

Indicator variogram analysis
Three dimensional indicator variograms calculated on point data values show anisotropic continuity with the longest range plunging gently to

the north and parallel to the strike direction of the JM zone. This is supported by a structural model proposed by Marsden (1990). After having established the geometric anisotropy from a variogram analysis, the coordinate system was rotated parallel to the principal anisotropy axes. This rotation made the X-axis parallel to the direction of principal anisotropy and the Y-axis parallel to the direction of minimal anisotropy (across the structure). Figure 2 (a) (b) (c) shows experimental variograms calculated in the rotated coordinate system. Note, that the down dip indicator variogram (along the Z axis) is very noisy, hence, extremely difficult to model. The main reason is the sparsity of data values in this direction. Based on a structural model developed by Marsden (1990), and on the underground mapping in the stope areas by Holbek (personal communications, 1991), as well as on the apparent bottoming of mineralization at shallow depth, it is known that the down dip range of continuity is shorter than along the strike direction. Although the drill hole data are too sparse to define the down dip variogram, we are fortunate to have some indirect information on anisotropy from underground workings. The ratio of height to length of two of the stopes is approximately 1:3, and of the third is 1:2, suggesting that the maximum down dip range of continuity is about half of that of the principal direction of continuity. Hence, we assume anisotropy factors of 0.1 and 0.5 for models to be used for creating the spatially correlated probability field

Three dimensional grid

The rotation of the coordinate system enabled an efficient choice of the area of interest. The area covered by points to be simulated is 500m along the strike direction, 55m wide across the structure, and it extends 70m down dip. The separation distance of grid nodes is 5m along the strike direction and down dip, and 1m across the structure. The spacing in any direction should be fine enough to allow reproduction of the range of the variogram in that direction. The range may be difficult to verify for simulated data if the spacing is greater than one fourth of the range (Luster, 1985). For a grid node to be simulated a minimum of two data values had to be found within a search radius of 40m along the strike direction, 20m down dip, and 4m across the structure. The anisotropy of this search ellipsoid was determined from anisotropy evident in sample variograms shown in Figure 2. From a total of 84,840 grid nodes within the area, 61,674 were simulated.

Probability field of spatially correlated values

Luster (1985) shows that we can create a field Y(x) of random numbers that has a predefined circular model and spherical model variogram functions in 2-d and 3-d respectively by the following procedure:

1. Generate a grid of independent random numbers R(xi) uniform over the interval (0,1).

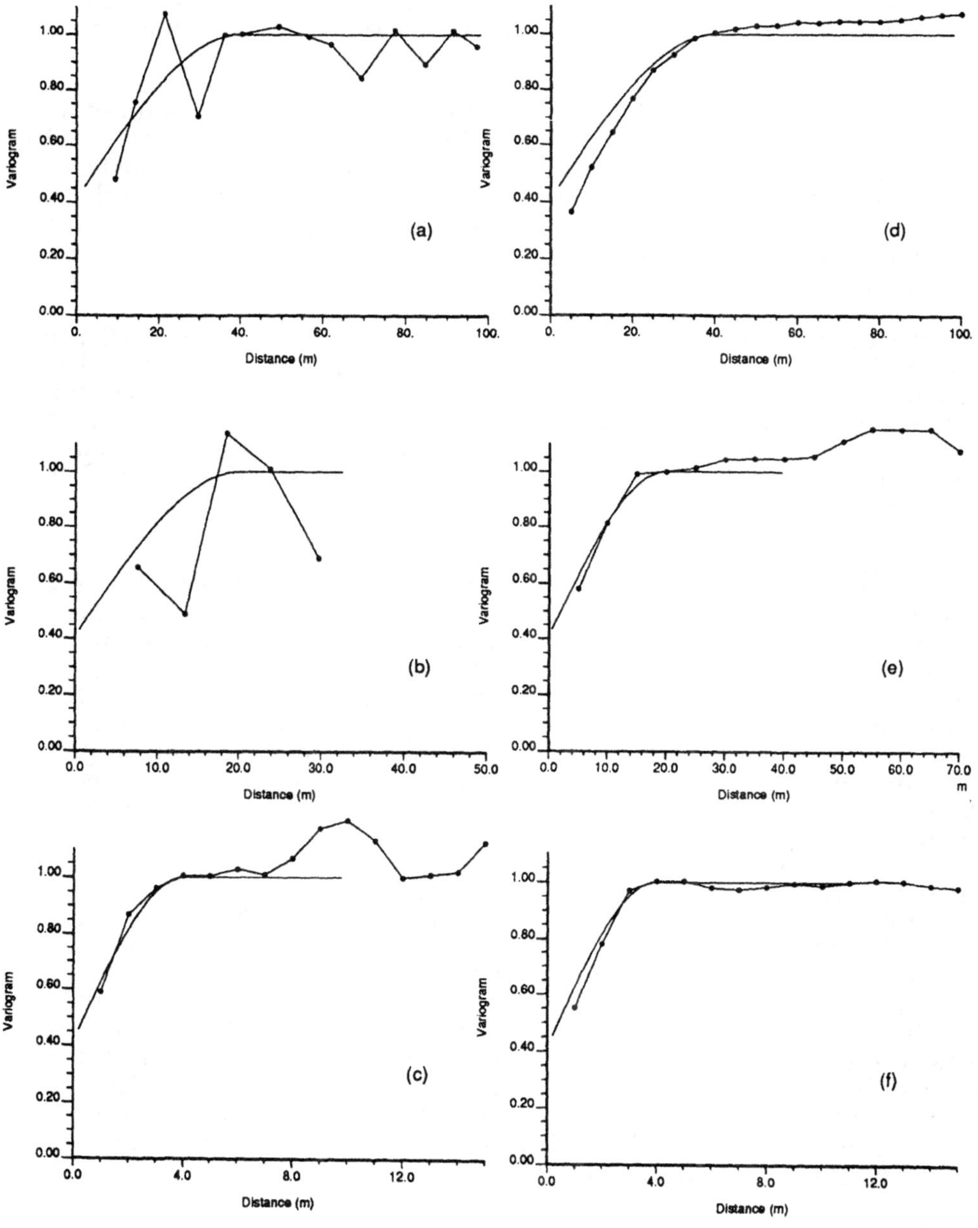

Figure 2: Experimental indicator variograms of JM zone conditioning data and their fitted models (a) strike direction, (b) down dip, (c) perpendicular to strike and dip; and variograms of simulated data (dots) compared with models (smooth) for experimental data (d) strike direction, (e) down dip, and (f) perpendicular to strike and dip.

Random numbers, generated by the linear congruential method, were shown to be random by Chi-square and k-dimensional uniformity tests (Bratley, 1983).

2. Attribute to each grid node xi a value Y(xi) equal to the sum of all R(xi) numbers located inside the ellipsoid. The size of the ellipsoid is determined by the ranges of continuity in three dimensional space, known from the variogram analysis of conditioning data. In our case the axes of the ellipsoid are 40m, 20m and 4m. This ellipsoid is positioned in space in such a way that its major axis is parallel to the major axis of anisotropy.

Because of RAM memory constraints the random numbers R(xi) are generated progressively as we move the ellipsoid in space. The realization Y(xi) has an anisotropic 3-dimensional spherical variogram. Next, the Y(xi) was transformed to a standard normal random variable YN(xi). To create a field of uniform random numbers U(xi) of the same spatial continuity, the formula U(xi) = G(YN(xi)) was used, where G is the standard normal cumulative distribution function.

The variogram of the probability field generated by this procedure has a zero nugget effect. Based on variograms calculated from conditioning data the relative nugget effect is 0.45. If this nugget effect is the result of errors committed during sampling and analytical procedure it should not be modeled into the probability field. After all, we are trying to simulate the pattern of spatial continuity which exists in reality not the errors committed due to sampling. This nugget effect was retained for estimating of local cdf by kriging. Our dual approach to the nugget effect appears to be an acceptable methodology because variograms of simulated data compare well with experimental variograms (Figures 2 d, e, f).

Estimate of a local cumulative distribution function
The estimate of the local cdf(xi) may be calculated by simple or ordinary kriging. Simple kriging creates estimates which are less conditioned on the data in the neighbourhood, and there is a maximum spatial mixing for a given variogram function. Because of the relatively few data available, and the concentration of high grade mineralization in two areas, the estimate of the global mean within the area covered by the grid may be unreliable. Hence, we used the ordinary kriging method to estimate each point for which a minimum of two data values were found within a search ellipsoid with axes equal to the ranges of continuity in the principal directions.

Creation of a Grid of Simulated Values
For each estimated point xi in the grid, a value either 0 or 1 was assigned according to the formula:

$$pfs(xi) = \begin{cases} 1, \text{ if } U(x) < cdf^*(x) \\ 0, \text{ if } U(x) >= cdf^*(x) \end{cases}$$

Figure 2 (d) (e) (f) shows the variograms of simulated data calculated along the three major anisotropy axes.

RESULTS

A total of six simulations were completed for the central part of the JM zone. The geometry of high grade mineralization can be appreciated by viewing simulated values on either grey scale sections and plans or on isometric diagrams. An example of the output from two different simulations for level 1280 is given in Figure 3. The long section includes the richest ore shoot known on the property. The distribution of grades is erratic within the simulated high grade zones. Very small areas of high grade appear in the footwall and hangingwall of the main high grade zone. The central part of the JM zone consists of a core formed by a vein stockwork with less intense stockwork development in the footwall and hangingwall. Typically the core stockwork contains one or more breccia zones which commonly contain the highest grades. However, small high grade sections are present in relatively weakly veined rock peripheral to breccia zones. All of these small scale features would be impossible to demonstrate with any kriging estimation procedure. There are substantial differences between the two simulations shown in Figure 3. This demonstrates uncertainty in the location of high grade structures in the JM zone based on the available conditioning data (48 drill holes).

The simulation output could be viewed as isometric diagrams to improve our understanding of the irregular geometry and the complexity of the deposit and is therefore a valuable aid for mine planners. It may be viewed as the helping tool for mine planners. In our case it is immediately apparent that the extent of dilution regardless of the mining method used may be quite substantial. Although simulation is not an efficient estimation procedure it reflects the underlying structure of the mineralized zone and permits estimation of dilution both internal and external to high grade zones. To check for the extent of dilution in the already mined out areas, two large sections of size 150m x 20m x 20m (B and C stope) and 60m x 30m x 15m (D stope) were chosen. Within each one of them the proportion of waste material was calculated. In the B and C stope area dilution, based on six simulations, may be as high as 75% and as low as 65%. In the D stope area dilution ranges from 38% to 46%. Thus, if a variable used in a conditional simulation is an indicator of the absence or presence of mineralization, we are able to calculate for each simulation run the extent of dilution for a selective mining unit (SMU) of any size. A large number of simulation runs may produce a distribution of a proportion of waste within SMU. It would be extremely difficult to calculate dilution from any estimation methods because they produce smooth output.

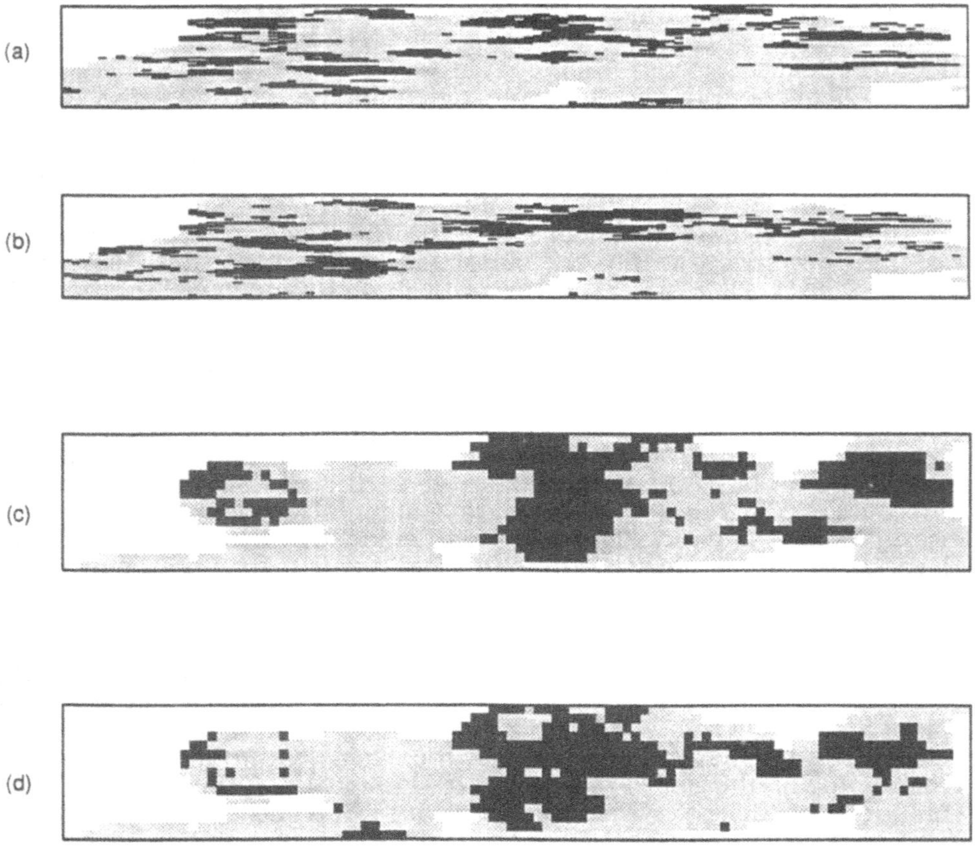

Figure 3: Gray scale maps of JM zone simulations; dark is high grade, stippled is low grade; (a) and (b) are plan views of two simulations of 1280 level ; (c) and (d) are vertical long sections of the same two simulations shown in the area of known highest grade

CONCLUSIONS

A geostatistical study (Nowak et al, 1991) of the JM zone clearly demonstrates the usefulness of geostatistical methods for resource estimation of small precious metal deposits. The limited data available for estimation purposes, however, are inadequate as a base for mine planning because there is insufficient information on the detailed distribution of recoverable reserves. This arises because of the contrasting effects of smoothing due to kriging, and the abrupt changes in grade continuity that characterize the deposit.

Conditional simulation is recommended as a planning tool in the development of small deposits such as the JM zone. Because simulated results are conditioned to the available data, simulation is a crude estimation method that has the added advantage of reproducing the continuity that characterizes the deposit. Thus, conditional simulation appears, as a result of our work, to be useful basis for production planning.

ACKNOWLEDGMENTS

This study is part of research into mineral inventory of gold deposits in British Columbia undertaken through the Mineral Deposit Research Unit, Dept. of Geological Sciences, The University of British Columbia. Funding is from the Science Council of B.C. and the mineral industry. We are particularly indebted to Dr. R.M. Britten and Mr. P. Holbek of Homestake Mineral Development Co. for their assistance and financial support for our study of the JM deposit. R. Froidevaux contributed substantially to the development of the simulation procedure. This contribution is Paper 011 of the Mineral Deposit Research Unit.

REFERENCES

Bratley, P., B.L. Fox and L.E. Schrage, 1983, A guide to simulation; Springer-Verlag, New York, 383 p.

Froidevaux, R. 1992, Proceedings of the 4th International Geostatistics Conference, Troia, Portugal

Luster, G.R., 1985, Raw materials for Portland cement: applications of conditional simulation of coregionalization; Ph.D. thesis, Stanford University, 531 p.

Marsden, H.M, J.M. Moore, 1990, Stratigraphic and structural setting of the Shasta silver-gold deposit, north-central B.C.; B.C.M.E.M.P.R. Geological Fieldwork 1988, Paper 1990-1, p. 305-314

Nowak, M.S., A.J. Sinclair, R.M. Srivastava and P. Holbek, 1991, Mineral inventory studies of precious metal deposits in British Columbia: Shasta gold deposit; Tech. Rept. No. 1, Mineral Deposit Research Unit, Dept. of Geological Sciences, The University of British Columbia, 93 p. plus appendices.

Sinclair, A.J., 1976, Applications of probability graphs in mineral exploration; Spec. Vol. 4, Assoc. Explor Geochem., 95 p.

GEOSTATISTICAL TOOLS FOR THE DETERMINATION OF FUNDAMENTAL SAMPLING VARIANCES AND MINIMUM SAMPLE MASSES.

D. FRANCOIS-BONGARCON
Principal Geostatistician
Inco Exploration and Technical Services Inc.
Copper Cliff, Ontario, Canada

ABSTRACT

The theory and practice of sampling of particulate material have a critical economic importance at all phases of a mining project, from exploration to production and marketing. A powerful, complete theory was put together by P.Gy (1982), but its applications have met with limited success to date, due to the inadequacy of its sampling variance and minimum sample mass determination formulae to certain types of ore, such as precious metal ores. Surprisingly, they have also met with little interest on the part of the geostatistical community. This paper shows how a geostatistical, albeit unsophisticated approach made it possible to shed light on the results of P. Gy's theory, make them more readily understandable to the non-specialist, and generalize them so that they could apply to any type of ore. Comments are offered on the main geostatistical concepts involved and the validity of the model, along with areas in need of more theoretical research.

INTRODUCTION

At all the stages of a mining project, including marketing of the commodity, critical decisions, often involving huge investments or expenses, are made on the basis of estimated quantities derived from sample values. Whether one needs to know the quality of a concentrate shipment, determine the grade of a piece of diamond drill core, or make a judgement about the adequacy of sampling a mining face for grade control purposes, cascading sampling errors combine to make the decision process less than optimal, and the question of how representative each of these samples is remains a central one.

To fully understand this question, it is important to realize the economic importance of good control on sample reliability in mining projects. This, of course goes along with the proper use of the sample data, that is to say, proper use of general statistical and geostatistical tools, computerized or not. Here are a few examples to put the issue into context:

- A recent survey by a major investment counsel firm (Lassonde, 1990) shows that out of 49 mining projects identified as disappointments or failures, 31 were to be blamed on improper predictions of grades and reserves, which is directly linked with the control of sample data reliability and proper use.

- Based on its past experience, a major, progressive American gold mining company has calculated that the cost of not using the proper approach to sampling and grade control would amount to an average $6.70 per ounce of gold production, or in their case, a total of $ 10 million per year.

- It is now well established that bulk sampling programs can make the difference between success and failure, and sample towers are increasingly used for reliable determination of the grade of pilot workings. A sample tower based bulk sampling operation costs in excess of $ 1 million, but is universally viewed in the industry as an insurance policy against project failure. The usability of the results of this expensive precaution, however, largely rely on the quality of the sample data produced, and therefore on proper sample tower tuning and sample preparation procedures, which both imply resorting to sampling theory.

Concerns about the reliability of mining samples can actually be traced all the way back to antiquity, but general theories of the sampling variance with provisions for practical applications are fairly recent. Only one of them, Gy's theory, is both general and practical (Gy, 1982). Gy's theory of sampling variances, however, is not particularly appealing, in spite of its considerable value to the industry. The theory is difficult to understand, probably because it is exposed in very cryptic terms; this lack of clarity makes it close to impossible for the non-specialist to adapt it properly to particular cases, and the potential for misuse is large; finally, as we found, it does not seem to be general enough to apply well to precious or low grade metals, and therefore does need some generalization. On the other hand, it has been often felt that geostatistics should be able to help clarify the concepts and parameters involved in Gy's theory. However, except for a theoretical statistical demonstration by Matheron (1966) to help justify Gy's underlying statistical model, and developments by Deverly (1984) to simulate the nugget effect generated by microstructures, apparently left with nó follow-up, the geostatistical community has not been active in the research in mining sampling theory.

Beside sensitizing the geostatistical research community to a field which has economic importance to the mining industry, this paper, along with two others already published by the author (Francois-Bongarcon, 1991 and 1992), has several other underlying objectives:

- emphasise the difference of approach between modelling spatial correlations for macroscopic estimation exercises, and building explicative models of mineralization and propagation of variability;

- point out the main geostatistical concepts involved, which are related to matters of scale, ranges, and fractal properties of self-similarity of geological controls at various scales, and the areas where more rigorous theoretical models are needed to conduct more formal and more detailed developments;

- give full credit to an exemplary geostatistical approach that has allowed us to design major improvements to Gy's variance formulae with little more than a mere analysis of the concepts and quantities involved.

Some of the developments in this paper do not come with theoretical rigorous demonstrations, but rather rely on an in-depth analysis of the physical quantities and properties to be modelled. Although the research was successful in reaching its objectives, this paper probably brings more theoretical questions than answers, and it is hoped that it will help set the ground for more theoretical research and trigger more interest from geostatisticians.

THE FUNDAMENTAL ERROR IN THE SAMPLING OF BROKEN ORES

In the sequel, we will concern ourselves only with the problem of sampling broken ores as opposed to compact ore (the ore in place). Depending upon their nature and/or comminution stage, broken ores come in two categories: non-liberated and liberated. The ore is said to be liberated when the constituents of interest are completely separated, so that for all purposes, the lot to be sampled is known to be a collection of rock fragments, or particles, that are either pure mineral or pure gangue. We will assume here the general case of non-liberated ore, the other case being a simpler, particular case, addressed in a note at the end of this section. We will also only address Gy's sample variance formulae, which relate the sample mass to the sample precision variance in order to allow us to either predict the sampling precision, or to determine the minimum sample mass required to achieve a given precision. This sample variance is the residual, unavoidable variance, when all errors that can be controlled have been eliminated or minimized and the lot of ore successfully homogenized. It is referred to by Gy as *the variance of the Fundamental Sampling Error (FSE).* Other quantifiable errors such as the Analytical and Segregation errors have been addressed in other publications (Francois-Bongarcon, 1992) but will not be described here.

The FSE was addressed numerically, albeit empirically at the end of the nineteenth century, following works by Vezin, and since then, the relative sampling variance has been represented by the well known generic form:

$$\sigma_R^2(M_S,d) = K.\ d^\alpha \ / \ M_S \qquad \text{(I)}$$

This formula relates the relative precision variance σ_R^2 to either the maximum or nominal rock fragment size d, and to the sample mass M_S. It uses a sampling constant K and an exponent α that has had a tendency to fluctuate through history, mostly between values 2 and 3, often for intuitive, empirical, and sometimes wrong reasons, far removed from scientific considerations. The work presented here placed particular emphasis on giving that exponent a more rational status, closely related to De Wijs' absolute intrinsic dispersion of the grade, thus recognizing that the exponent α will, in general, vary with the nature of the ore.

Before making such claim, however, let us put the problem into context. In Gy's theory, formula (I) is actually exposed and verified in a particular case where $\alpha=2.5$ (at first glance, Gy's version of formula (I) seems to correspond to $\alpha=3$, but his sampling constant, K, as we will see, is not constant (!) but actually proportional to a "liberation factor" in $d^{0.5}$). In the course of our research, we found that for all the types of gold and nickel ores that we processed, a more correct value was closer to $\alpha=1.5$. This observation was in fact the starting point of this research. Similarly, it is particularly interesting to note that as soon as Gy's initial work was published by BRGM in 1966, an expert discussion panel presided by Mr. Testut was also subsequently published in which the only negative comments regarded the non-suitability of Gy's value of exponent α. Finally, we have been able to verify in a number of cases that once a correct value has indeed been determined for α, another cause for concern many people have expressed in the past regarding "unrealistic minimum sample masses" obtained from Gy's formula, seemed to vanish in practical applications.

GEOSTATISTICAL TOOLS FOR THE MODELLING OF THE FSE

Obviously, the appropriate macroscopic geostatistical model to describe a lot of broken ore after homogenization, is the pure nugget effect model. Here may be the reason why there has not been more focus on sampling theory by geostatisticians: at first glance at least, sampling broken ores is in the domain of classical statistical models and does not seem to call for a geostatistical approach.

In this usual geostatistical macroscopic model of the ore, microscopic scale variations of specific gravities can usually be ignored at the scale of the study, and the mean grades of interest are straight spatial averages over known, fairly large volumes. The corresponding regionalized variable (e.g. the grade) has no directly observable spatial correlations (i.e. no detectable variogram range at the scale of the study, suggesting a pure nugget effect covariance or variogram model). The variance of the grade of a sample of volume v out of a larger lot of volume V is therefore of the well known, classical form:

$$D^2(v|V) = (1/v - 1/V).A \quad \text{(II)}$$

where, for the record, $A = \int C(h).dh$ is the integral, over distance h, of the covariance function C(h) of the undetected "microstructures", extended to the entire space. Integral A is a constant parameter of the specified regionalized variable model, but will change throughout the successive grinding stages, to reflect the decrease in range of the microstructures still present within the rock particles.

In this model, for any physical support (e.g. data sample size) of measure p, we also have:

$$A = p.C_o(p) \qquad \text{(III)}$$

where $C_o(p)$ is the nugget variance of (i.e. that could be observed on the variogram of) support p. In the case of a single, well defined support, integral A therefore appears as a support-variance integral product. If q is another support, we also know that:

$$C_o(q) = (p/q).C_o(p) \qquad (IV)$$

It is useful to recall here that, as a consequence, regularization over a support q much larger than the original support p has a drastic reduction effect on the variance, provided it is entirely attributable to a pure nugget effect. Conversely, the regularization of transition structure, i.e. of a variogram with no nugget effect component, and with a range larger than the supports considered, can be shown to have a much weaker reduction effect on the overall variance of the variable. These trivial concepts are very helpful in sampling theory.

Formula (IV) is also a particular case of an empirical volume-variance relationship initially established on agricultural data (Smith, 1928). This relation has been experimentally verified to hold true in good approximation for a variety of macroscopic variogram models including mixes of nugget effects and transition structures (Zhang et al., 1990) and, put in modern geostatistical terms, expresses the relation between the variances of two volumes v and V in a field F as:

$$D^2(V|F) = (v/V)^b.D^2(v|F) \qquad (V)$$

where parameter **b** obviously varies from 0 (extreme continuity - relation exactly verified) to 1 (pure nugget effect - relation exactly verified again). In practice, **b** is generally comprised between 0.17 and 0.45 for the ore in place. We will later elaborate on this particularly important relation, its applications at very small scale, and its relation to the De Wijs model.

VARIANCE OF THE FSE IN THE CASE OF NON-LIBERATED ORES

In this section, the sample itself is assumed to contain fragments drawn at random (equiprobable model) from a lot of non-liberated ore, or equivalently, drawn in one single increment from the lot once it has been perfectly homogenized. The rigorous statistical calculation of the variance of the mean grade of NF fragments of varying masses m_i is heavy and lengthy, even under simplifying assumptions (Matheron, 1966), and even then, the result is only a first order approximation. We have instead derived the same result by analogy using simple geostatistical concepts presented above, in a way that gives the resulting formulae a readily interpretable form. The reasoning has been presented in more detail in a previous publication (Francois-Bongarcon, 1992), and only its general articulation will be summarized here.

Starting with the equal mass case where all the fragments have same volume, mass, and therefore specific gravity, equations (II) and (III) can actually be used to express the relative variance of the grade of sample S in lot L of mass M_L, i.e. here the FSE variance, as:

$$D_R^{\,2}(S|L) = \sigma_{FSE}^{\,2} = (1/M_S - 1/M_L).m.D_R^{\,2}(t|L) \qquad (VI)$$

In this equation, $D_R^{\,2}(t|L)$ is the relative variance of the fragment grade t within the lot and m the mass of each fragment.

By means of simplifications equivalent to those proposed in Gy's developments to account for the fact that we are in fact dealing here with entire distributions of values for the volumes, specific gravities and masses of the particles, the general unequal particle case is approximated by simpler successive formulae as follows:

First, the support-variance integral product $m.D_R^2(t|L)$ in (VI) is developed as a sum of elementary products:

$$\sigma_{FSE}^2 = (1/M_S - 1/M_L).\Sigma_i(m_i.\delta_i) \quad (i=1,...NL) \qquad (VII)$$

where M_L and NL are the mass and number of fragments of the lot L, and:

$$\delta_i = (m_i/M_L).(t_i - t_L)^2/t_L^2 \qquad (VIII)$$

is the contribution of fragment m_i to the total variance of the fragment grades t_i in the lot. Note that because of the varying fragment mass and volume, each fragment contribution to their total variance is now weighted by the fragment proportion in mass, m_i/M_L, rather than by 1/NL.

Then, the general case is approximated by the equal volume case, with the distribution of fragment volumes represented by its mean value $\bar{v}$ (in mass), but with the specific gravities ρ_i, which may be strongly correlated to the grades, still variable from fragment to fragment:

$$\sigma_{FSE}^2 = (1/M_S - 1/M_L).\bar{v}.\Sigma_i(\rho_i.\delta_i) \quad (i=1,...NL) \qquad (IX)$$

The variance term is now represented by a specific-gravity-variance integral product.

Applying Smith's relation (V) to relate the variance-specific-gravity product $\Sigma_i(\rho_i.\delta_i)$ in (IX) to a smaller fragment size, and reasoning to the limit where this smaller size reaches liberation size, one now obtains:

$$\sigma_{FSE}^2 = (1/M_S - 1/M_L).\bar{v}.(\bar{v}_\lambda/\bar{v})^b.\Sigma_\lambda(\rho_\lambda.\delta_\lambda) \quad (\lambda=1,...N_\lambda) \qquad (X)$$

the sub-index λ indicating the summation is performed at liberation size.

Finally, following Gy, a fragment shape factor, f, and a size distribution factor, g, can be introduced to express the volumes and mean volumes as $v_i=f.d_i^3$ and $\bar{v}=f.g.d^3$ [a justification is given by Francois-Bongarcon (1992)], so that in (X):

$$\bar{v}.(\bar{v}_\lambda/\bar{v})^b = f.g.d^3.(d_\lambda/d)^{3b}$$

For all purposes, f is equal to 0.5 for approximately spherical fragments, and g=0.25 has been shown by Gy to be the best value to use, assuming that the fragment nominal size d is defined by the size of the screen producing a 5% reject.

The sum of the $\rho.\delta$ products in (X) (Gy's mineralogical composition factor **c**) can easily be elicited at liberation size, using the grades 0 and 1 of the fragments of pure gangue and mineral, their total cumulated masses M_G and M_M, their respective specific gravities ρ_G and ρ_M, and the lot grade t_L in %:

$$c = \Sigma_\lambda(\rho_\lambda.\delta_\lambda) = \Sigma_{iM}(\rho_{iM}.\delta_{iM}) + \Sigma_{iG}(\rho_{iG}.\delta_{iG})$$

$$c = \Sigma_\lambda(\rho_\lambda.\delta_\lambda) = [(1 - t_L)/t_L].[\rho_M.(1 - t_L) + \rho_G.t_L] \qquad \text{(XI)}$$

$$(\, c \approx \rho_M/t_L \quad \text{for low grade ores})$$

So that, at last, we obtain the final practical formula:

$$\sigma_{FSE}^2 = (1/M_S - 1/M_L).f.g.c.d^{3.(1-b)}.d_\lambda^{3b} \qquad \text{(XII)}$$

with c defined in (XI), and in general, g = 0.25 and f = 0.5 (Gy, 1982), and b ϵ [0,1]. When the fragments are approximately equal in mass, then g = 1. In (XI), when the pure mineral is not pure metal, the grades to be used are grades in the mineral of interest, possibly converted from pure metal grades. In practice, when the liberation size d_λ is known or can be readily determined or safely estimated, (XII) can be used directly. In the other cases, (I) will be used, with parameters K and α fitted experimentally (Francois-Bongarcon, 1991).

Formula (XII) is of the same form as the historical formula (I) with α = 3.(1-b) (the term $1/M_L$ is usually negligible). It can be shown that (XII) is also identical to Gy's well known classical formula (Gy, 1982) when b takes the particular value b = 0.17 . It is important to stress that formula (XII) intimately relies on the validity of Smith's relation at small scale. However, not only does the formula contain Gy's formula as a simple particular case, but it has been confirmed experimentally in other cases in several ways, including successful determination of the gold particle maximum size in a gold deposit where α had been estimated at 1.5.

Note regarding liberated ores: formula (XII) is valid only when the liberation size has not been reached. When comminution reaches the nominal liberation size, the lot is deemed to be a collection of fragments of pure gangue and pure mineral. Further splitting of the fragments does not increase their grade variance any more. The volume ratio in Formula (V) and (X) then remains conventionally equal to 1 as $\bar{v}$ continues to decrease, and (XII) takes particular simplified forms also obtainable using the binomial law (Francois-Bongarcon, 1991).

DISCUSSION

As a starting point, let us outline the overall reasoning presented above, the scales of study, and the main model features.

First, at the scale of the lot to be sampled, after crushing and mixing, a macroscopic nugget effect model was used, while it was recognized that this model mostly reflected "invisible" but real, existing microstructures. The advantage of this nugget effect model over, say, classical statistics as per Gy's work, was the introduction of integral A. The interpretation of the one particular quantity that is inversely proportional to the measure (mass) of the sample in the variance as an integral product is what led us to find the proper calculation mechanism for the unequal particle case. The price to pay was a lack of rigor. What we got in exchange is a clear variance formula one can easily modify and adapt to new situations (for instance in a case where Smith's relation would not be appropriate), that has the only known, more rigorous formula as a particular case, and has been so far very well verified in practice.

In a second step, that integral support-variance product, which actually geostatistically characterizes the stage of comminution of the lot, was elicited and developed. At the scale of the rock particles in a given stage of comminution, we are not facing a nugget effect situation any more. On the contrary, at that scale, we must deal with whatever is left of the non-nugget part of the mineralization structure. We therefore reasoned about the ore "in-place", but within the particle boundaries. It is important to realize this shift in viewpoint to understand the tools used. Smith's relation was then conveniently used to bring the scale of calculation down to the smallest scale relevant to the problem: the liberation size. Smith's relation achieved more than just allowing us to spell out the ultimate parameters involved. In particular, at least at the scale of the rock particles, the scale of study is orders of magnitude smaller than the variogram macroscopic range (if it exists at all), so that only a model of autocorrelations with no finite variance is appropriate. Such a model automatically entails that the variance of a support within a larger support is a function of the larger support as well. Smith relation expresses the ratio of variances of two different supports within a common larger field in a way that eliminates the influence of the common field size . This is a necessary feature of the model that we should use, and Smith's relation ensures that consistency, allowing us to reason to the limit and go back to non-spatial statistics, in order to finally relate to the constitutive variance at liberation size (the true nugget effect of the mineralization) which should not depend on any particular field size.

As already mentioned, the validity of our final model (formula (XII)) still ultimately relies on the validity of Smith's relation as applied in (X). If its application is valid, then the reasoning to the limit should pose no problem. Assessing this validity, however, is not easy. Although the work by Zhang et al. (1990) is interesting and useful in macroscopic applications (and in fact sets the ground, in our opinion, for more implications than appears at first glance), it does not directly justify our use of Smith's relation to relate the variance of the rock particles at a given stage of the crushing/grinding process to their (constitutive) variance at liberation size, which can easily be several orders of magnitude smaller. The experimental conclusions derived by Zhang et al. just cannot be used across several orders of magnitude. As a matter of fact, the entire research on the subject is haunted with issues related to changes in scales, and what we really needed was a model that would *explain* rather than "fit" the propagation of the variability at various scales, or, in other words, *a genetic model of the ore grade*. Here is again a significant shift in viewpoint. We are used to variogram models chosen for their ability to reasonably fit the autocorrelations at the usual, macroscopic scale of the study. That fit needs only be valid for the distances that would be involved in, say, a kriging of the data, or even a simulation of grades at macroscopic scale. For instance, if the experimental variogram "seems" to have a range, a model with a range will, in principle, be appropriate for estimation purposes, and especially so if the distances involved are mostly within or close to such a range. Whether the range actually exists, i.e. is a true feature of the mineralization, would not usually be considered critically relevant. By the same token, any microstructure of a scale smaller than the support of the data will be lumped into a nugget effect model, and what that model actually summarizes is not considered, in the end, a very important issue.

In our particular endeavour, obviously, beside being probably impractical, it would just not be enough to "reasonably" fit a model at whatever scale is accessible. It is quite interesting to see that, at the onset of geostatistics, works by De Wijs and Krige actually put the emphasis on genetic models and the corresponding models of variograms. After conducting the present research, and in view of experimental results obtained by Gy as well as by the author, De Wijs' model may well be more fundamental than we actually thought. The next few remarks are hoped to trigger more discussion on this particular aspect of our work.

Let us start, for instance, with the concept of variogram range, and raise an apparently naive question: in view of the incredible natural variety of mineral deposits, geological settings, chemical environments and orebody sizes on one hand, and, on the other hand, of the numerous orders of magnitudes between the constitutive, often micron-size mineral grains and the metric or kilometric scale of exploration observations, is it not *quite a coincidence* that the experimental variogram outer ranges found in practice are basically almost always all of the same order of magnitude, i.e. from 30 to 300 metres ? Or should we conclude these ranges usually are nothing but artifacts, possibly due to the limited fields of observation ?

A model of variogram that takes a particular meaning in this work is the De Wijs model (Krige, 1978). This model has the advantage of allowing us to describe spatial correlations across many orders of magnitude of distances. This model is a model without sill. When plotted on an arithmetic distance scale, however, the shape of logarithmic De Wijsian variogram clearly appears, at the usual scale of the study (i.e. about 10-150 times the size of the linear equivalent of the data support), as being amenable to one or two models with sill (figure 1). This observation can be combined with the fact that periodicities, which are considered to almost always exist even at large scale, could easily alter the shape of the variogram in a way conducive to the interpretation of a sill which does not really exists, or, in the opposite case, of a drift. The finite and limited nature of the data field could also because for artifact-based ranges. Finally, the shape of the De Wijs variogram on figure 1 is reminiscent of the two nested structures often observed on variograms of the logarithms of gold grades, for which the raw data often clearly indicate a single structure with sill. Finally, maybe should we ask ourselves the question: Are variogram models with ranges actually the most appropriate in the study of natural mineral deposits ?

At the other end now, let us think of the observable nugget effect of the experimental variogram of, say, a gold deposit. We usually think of it as representing a mixing of under-scale microstructures and "true" nugget effect. Is it realistic to think that the variability can propagate from pure metal grain size (e.g. 100 microns) to the comparatively enormous size of a drill hole core, in a way that would preserve a significant nugget effect at large scale ? Or is the "true" nugget effect, if any, only due to the a lack of spatial correlation between the metal grains themselves? In the latter case, it is easy to verify that, in spite of a gigantic constitutive grain variance, its regularization over the core support will result in a practically negligible contribution to the total population variance, so that almost no "true" nugget effect can be accounted for at the scale of the macroscopic observation. We do not pretend to bring definitive answers to these questions, but they must be recognized as somewhat fundamental.

With these considerations in mind, reading again about De Wijs' works takes a resolutely new and refreshing view. Let us recall that his model of the mineralization started with two (fairly ambitious) concerns:

1.- explain the apparently universal lognormal look of gold grade frequency distributions, and why the lognormal law seems so omnipresent in ore deposits, and

2.- explain how the propagation of the ore grade variability in a mineral deposit can be universally explained at various scales covering numerous orders of magnitude.

In order to do so (Harris, 1984 - Krige, 1978), De Wijs proposed a curdling scheme in which the average grade m is distributed unequally between two halves of the deposit as (1-d).m and (1+d).m for a certain value of the curdling factor d between 0 and 1. Each of the halves is now split in two and subjected to the same metal redistribution with the same value of d, and the process repeated ad libitum. Parameter d can be eventually be replaced at each iteration by the realization of a random variable of mean d. After k such redistributions, the grades of the obtained units follows a *logbinomial* frequency distribution, with the following consequences: for k large, the distribution is very close to a lognormal distribution; the grades of larger units, assimilated to redistribution units of a lower order than the final ones, are also logbinomially distributed, hence explaining the apparent well known permanency of the lognormality; finally, the resulting spatial structure of the ore is "De Wijsian", i.e. the logarithm of the grade has a De Wijsian variogram.

Whether De Wijs' particular scheme must be viewed or not as naive for a mathematical model of the ore-generating process of the Earth, the fact is that it provides for a self-similarity at nested scales that fits well what geologists can and do observe in nature. It is not surprising then that Serra (1967, 1968) found the model useful to describe spatial correlations down to microscopic scales, and his interpretation of the De Wijs variogram as the sum of an infinite number of spherical variograms with ranges in geometric progression gives De Wijs' model an increased credibility.

The model appealed to the author when he realized that, to conceive the propagation of variability at increasing scales starting at liberation size, it is infinitely easier and more logical to imagine a series of similar, homothetic geochemical processes at the various scales, than it would be to avoid self-similarity. Mandelbrot's "fractal geometry of nature" could well be, as mentioned somewhere else, unescapable. Interestingly, if enough self-similarity is required from a model so that the variance ratio of two supports in a common field depend only upon the support ratio, then Smith's relation, already exact for a pure nugget effect and extreme continuity, provides the most obvious form for that ratio.

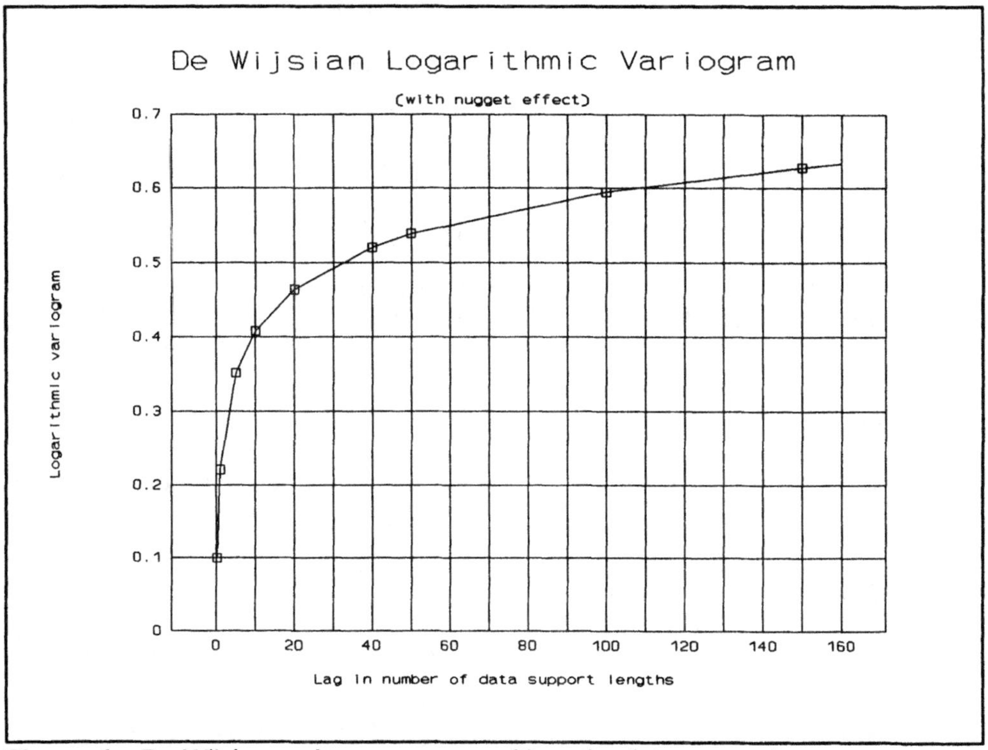

Figure 1 : De Wijsian variogram on an arithmetic distance scale

CONCLUDING REMARKS

While waiting for feedback from the geostatistical community, let us now throw in two observations to cement these issues together and relate them to our main subject:

1.- If, as is thought here, self-similarity in the variability propagation process at nested scales is an unescapable fact of life due to mother Nature, then Smith's relation is certainly a good translation of it. Reciprocally, if a model relying on the validity of Smith's relation is found efficient in practice for a large variety of ores, as is the case, then some indication has been gained that this self-similarity is of essence.

2.- The De Wijs model was used first to derive formula (XII), before the existence and possible universality of Smith's relation was suspected by the author. It is not difficult to verify that the De Wijsian model implies Smith's relation in good approximation, making Smith's relation a more general statement than the De Wijs model itself. The reasoning goes like this:

De Wijs variogram model on log-transforms:

$\gamma(h) \approx C_o + 3.a.Ln(h/u + 3/2)$ (u linear equivalent of data support)

logarithmic variance: $\beta^2(v|F) = a.Ln(F/v)$ (F = large field)

relative variance: $\sigma^2(v|F) = Exp(\beta^2) - 1$ (lognormal back-transform)

$\sigma^2(v|F)/\sigma^2(V|F) = [(F/v)^a - 1] / [(F/V)^a - 1] \approx (V/v)^a$ (Q.E.D.)

We therefore obtain Smith's relation with an exponent equal to De Wijs' logarithmic absolute intrinsic dispersion coefficient a. It would be interesting to see if a logbinomial back-transform would not conveniently give an exact result.

ACKNOWLEDGEMENTS

The management of I.E.T.S. is gratefully acknowledged for supporting this research and allowing the publication of this paper.

BIBLIOGRAPHICAL REFERENCES

Deverly, F. 1984. Doctoral thesis. Ecole Nationale Superieure des Mines de Paris. 158.

Francois-Bongarcon, D. 1991. CIM Bulletin. 84 (950): 46-57.

Francois-Bongarcon, D. 1992. in Proceedings of The XVth World Mining Congress. Madrid, May 1992.

Gy, P.M. 1982. Sampling of Particulate Materials, Theory and Practice. Elsevier. Amsterdam. 431.

Harris, D.P. 1984. Mineral Resources Appraisal. Clarendon Press. Oxford. 184-205.

Krige, D.G. 1978. Lognormal De Wijsian Geostatistics for Ore Evaluation. South African Institute of Mining and Metallurgy. 50.

Journel, A.G., Huijbregts, Ch.J. 1978. Mining Geostatistics. Academic Press. London. 600.

Lassonde, P. 1990. The Gold Book - The Complete Investment Guide to Precious Metals . Financial Times of Canada series. Penguin Books.

Matheron, G. 1966. Comparaison Entre les Echantillonnages a Poids et a Effectifs Constants. Revue de l'Industrie Minerale, August 1966. 609-621.

Matheron, G. 1971. The Theory of Regionalized Variables and its Applications, Les Cahiers du Centre de Morphologie Mathematique, Fascicule 5. Ecole National Superieure des Mines de Paris. 212.

Serra, J. 1967. Un Critere Nouveau de Decouvertes de Structures: le Variogramme. Sciences de la Terre. 12: 275-299.

Serra, J. 1968. Les Structures Gigognes: Morphologie Mathematique et Interpretation Metallogenique. Mineralium Deposita. 3: 135-154.

Smith, H.F. 1938. An Empirical Law Describing Heterogeneity in the Yields of Agricultural Crops. Agricultural Sciences. 28: 1-23.

Zhang, R., Warrick, A.W., Myers, D.E. 1990. Variance as a Function of Sample Support Size. Mathematical Geology. 22 (1): 107-121.

HOLE EFFECTS IN DIAMOND CORES FROM PALABORA

J.D.S.VIELER **I.CLARK**

Rio Tinto Management Services, Sandton and
Department of Mining Engineering,
University of the Witwatersrand, Johannesburg

1 ABSTRACT.

Palabora Mining Company has mined a large, low grade copper open pit in the Transvaal Province RSA for more than 25 years. The ore-body lies within an alkaline complex of intrusive pegmatoids. Residual fluids from more than one injection of carbonatite magmas exploited fractures and veins in the host rocks to deposit minerals of economic interest.

Initial geostatistical work on diamond cores indicated an interesting potential application of the "Hole-Effect" model. This is most significant in the horizontal direction and appears "damped" towards the vertical. The hole-effects have been highlighted by the application of multiple indicator cut-off grades; when samples from each of the major host rocks were isolated, hole-effects with different cycle lengths were apparent on application of different indicators; these may be signatures from the individual copper mineral deposition "regimes".

Subsequent comprehensive analysis of historical blastholes has led to the zonation of the ore-body by anistropy and grade. There is confidence that these models will be applicable at depth because of the remarkable vertical continuity of the lithology and mineralisation.

2 INTRODUCTION.

Palabora Mine is sited 500 Km north-east of Johannesburg, RSA (see Figure 1) and is amongst the largest copper open pits and processing facilities in the world. Palabora produces a host of other bi-products and is the site of the world's largest vermiculite reserves. Copper pit reserves are projected to the year 2001 and underground feasibility studies to extend mine life are in progress. Prior to the current study, the application of geostatistics to Palabora was very limited.

FIGURE 1.

LOCATION OF PALABORA MINE

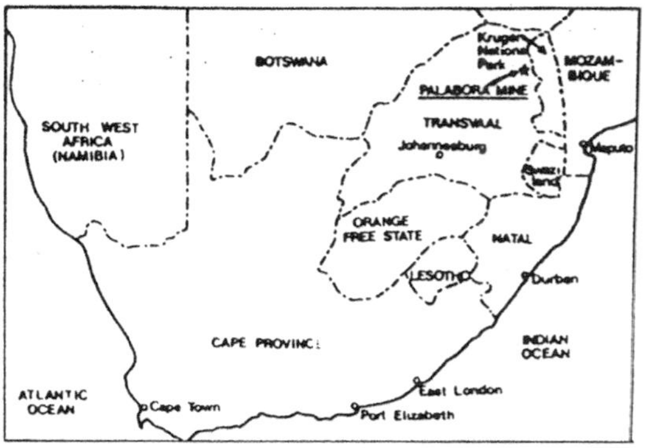

3 THE DEPOSIT AND THE SAMPLING DATA

3.1 Geological Summary.

The Palabora Igneous Complex shown in Figure 2, is the result of multiple alkaline intrusions (in order pyroxenite, syenite and ultrabasic pegmatoids) into Archean gneiss country rock. The age of the complex is generally assumed to be greater than 2060 million years.

Pyroxenite intruded first in a kidney-shaped stock 6,4 Km N-S by 2,6 Km E-W. A peripheral corona of feldspathic pyroxenite was formed by interaction with the gneiss. Numerous syenite plugs were forcibly injected into the gneiss surrounding the main pyroxenite. This was followed by an extended period of non-violent and partly metasomatic activity, forming vertically disposed pegmatoid pipes at three sites within the pyroxenite and causing fenitization of the gneiss at the external contacts.

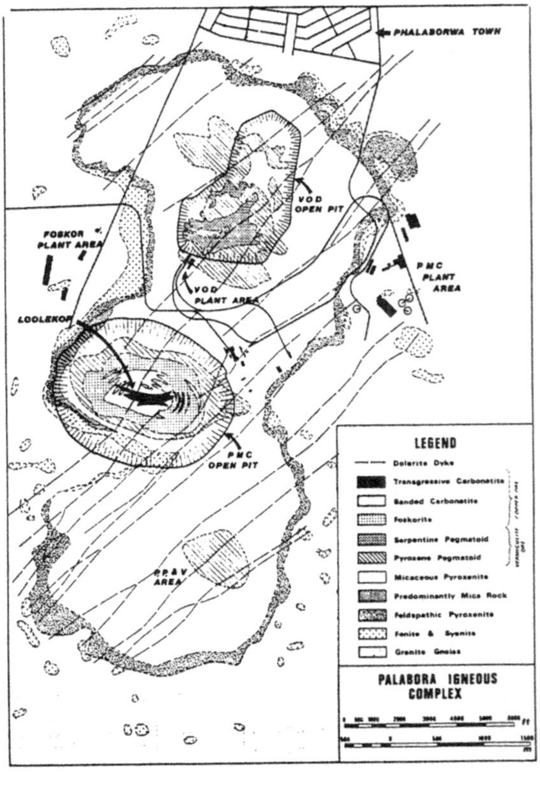

FIGURE 2.
GENERAL GEOLOGY OF THE PALABORA IGNEOUS COMPLEX

At the central site only ("Loolekop" see Figure 3), a dunite plug with some interstitial carbonatite was emplaced and then altered to foskorite. Banded carbonatite was then intruded at the centre and in concentric bands, bearing some copper sulphides. Subsequent fracturing of the entire infilling of the pipe and renewed igneous activity led to the intrusion of a dyke-like body of transgressive carbonatite at the intersection of two prominent zones of weakness. A divergent stockwork of transgressive veins cutting across all the older rocks was also developed. The bulk of the copper mineralisation accompanied this phase. Intensive post-carbonatite fracturing provided further channels for residual sulphide-rich fluids to permeate the transgressive carbonatite and, to a lesser extent, the older rocks. Shearing, brecciation, plastic flow and recrystallization occurred as the mass cooled. Some sulphides were remobilised and redeposited as valleriite at low temperatures.

Much later, the entire complex was invaded by barren dolerite dykes.

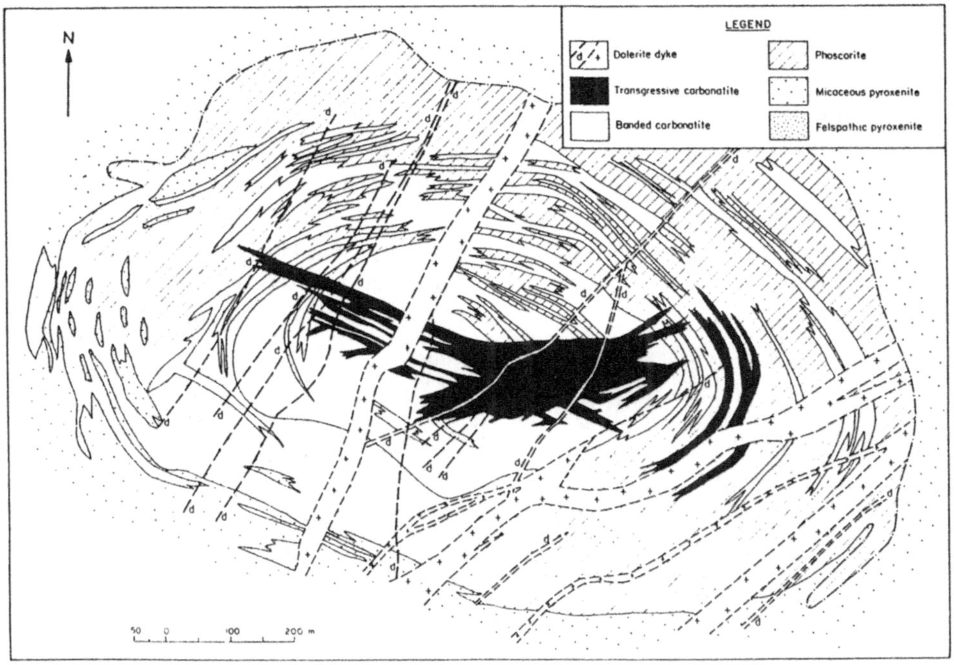

FIGURE 3. LITHOLOGY OF THE LOOLEKOP DEPOSIT

3.2 Lithologies and Ore Genesis.

Foskorite is composed of olivine, magnetite, apatite and phlogopite in variable proportions. Patches of calcite occur in addition to the carbonatite veining from later intrusives. Apatite and baddeleyite are abundant and economic but the magnetite is too high in titanium to be saleable.

Banded carbonatite is a magnetite-rich (20%) dolomitic sovite characterised by crude allignment of the magnetite in rudimentary layers concordant with the general concentric structure. Early, low-grade copper mineralisation accompanied this rock with bornite as the dominant phase. Transgressive carbonatite is similar without the crude banding. It is clearly cross-cutting and intrusive to the other rocks. It contains low-Ti magnetite, the bulk of the copper minerals (chalcopyrite, bornite, chalcocite and cubanite are dominant) and urano-thorianite.

The Palabora carbonatites are unusual and geochemically different from others, especially the high copper content. The sulphide textures are indicative of exsolution during cooling combined with reaction with late-stage fluids. The magnetite precedes the sulphides and its formation probably reduced the oxygen fugasity of the system, triggering sulphide precipitation. The sulphide deposition is estimated to have started at a temperature of 600 C, falling to 200 C for the valleriite.

Copper grades average 0.5% in the open pit and minimum mining units can exceed 1% Cu. Individual assays of diamond cores or blast-holes can record up to 10% Cu.

3.3. Sampling and Ore Reserves.

A total of 137 surface diamond drill holes were drilled between 1956 and 1976 intersecting the ore-body at depths up to 1.2 Km below surface. These holes indicate a remarkable vertical continuity of lithology, mineralogy and grade as shown in Figure 4. These cores were logged for structure, lithology and assayed in 5 feet (later 1.5 metres) sections for copper. Composites were assayed for subsidiary minerals.

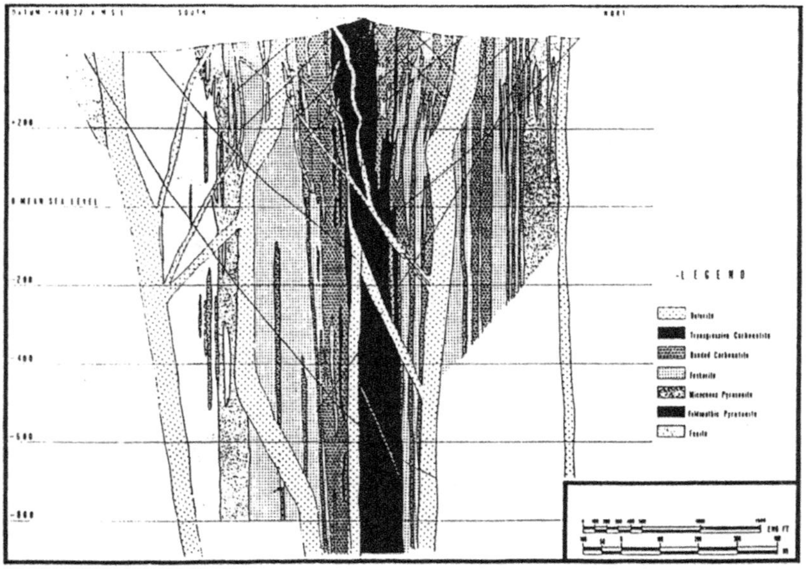

FIGURE 4. TYPICAL SECTION THROUGH LOOLEKOP DEPOSIT

The long term ore reserves were constructed by traditional methods from the diamond cores. Annual tonnage and grade experience factors are applied for long term planning. The chippings from production blast-holes in the open pit are sampled and assayed. The position and grade of every hole (250 000+) were recently captured on computer and this is now routine. The blast-hole assays are utilised for grade control and short-term mine planning. Blast-hole geological maps are produced of each bench, but carbonatites cannot be distinguished in chippings. A further 36 diamond holes (20 Km) are currently being drilled from an exploration shaft within the open pit as part of the underground feasibility studies.

All this information has recently been utilised for the development of experimental geostatistical models. These models, in conjunction with a comprehensive geological block model, will be used for the construction of future ore reserves.

4 GEOSTATISTICAL STUDIES OF DIAMOND CORES.

4.1 Down-the-Hole Semi-Variograms.

The diamond cores were initially studied by constructing down-the-hole semi-variograms. It was apparent from an early stage that the copper is closely associated with the carbonatite lithologies and that segmentation of the core by logged rock-type would be necessary. The rocks are interfingered and mixed throughout the deposit and therefore only continuous cores of 50m or longer were selected. This selection yielded about one hundred relatively uncontaminated lengths of the three main copper-bearing rocks, foskorite (FK), banded carbonatite (BC) and transgressive carbonatite(TC).

A semi-variogram of a length of trangressive carbonatite (LK016TC1) is shown in Figure 5a. The Nugget Effect is typically high, 0.6 of the total variance. The experimental model fitted is a Hole-Effect with a cycle period of 32 metres, with evidence of a second hole at 64 metres. As the holes are not as deep as the N.E., the model has been decayed to fit the data up to 40 metres.

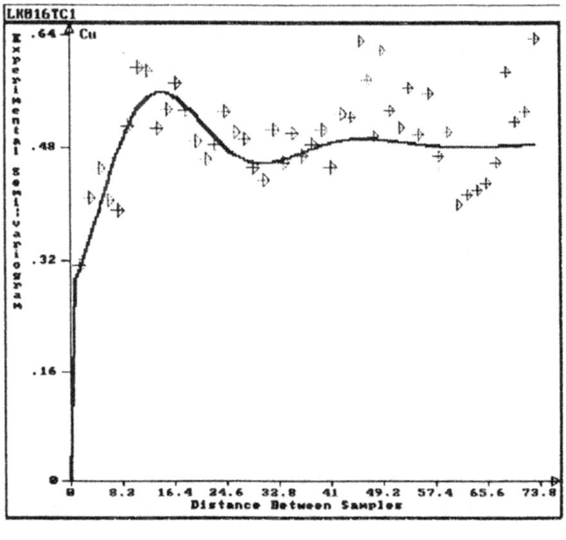

FIGURE 5a.
LK016TC1 DOWN-THE-HOLE SEMI-VARIOGRAM

Figures 5b to 5f are Indicator Semi-Variograms of the same length of core.

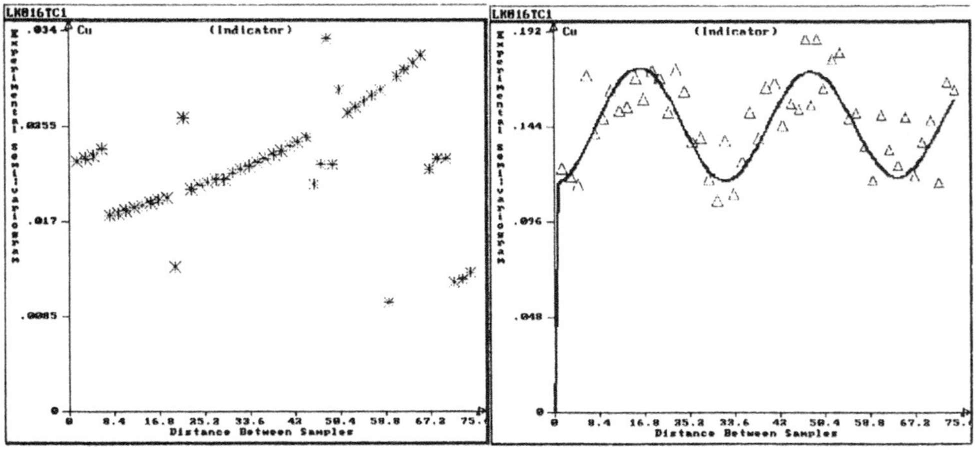

FIGURE 5b & c.LK016TC1 INDICATOR SEMI-VARIOGRAMS 0.2% & 0.6% Cu.

Figure 5b (0.2% Cu) shows a smoothly rising semi-variogram broken sharply at lags of 7, 20 and 45-50 metres. This type of semi-variogram is common for low and high indicator values. Figure 5c (0.6% Cu) shows well-developed holes at 32 and 64 metres (this cycle length corresponds to that of Figure 5a).

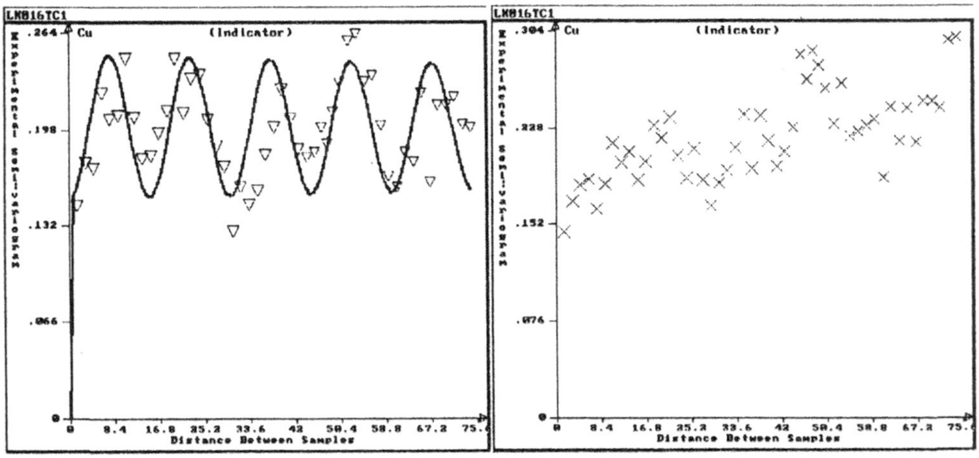

FIGURE 5d & e.LK016TC1 INDICATOR SEMI-VARIOGRAMS 0.8% & 1.4% Cu.

Figure 5d (0.8% Cu) clearly exhibits a different cycle period of 15 metres. Figure 5e (1.4% Cu) shows a generally rising but rather confused picture.

Figure 5f (2.7% Cu) shows remnants of a smoothly rising graph but is broken more frequently than in 5b.

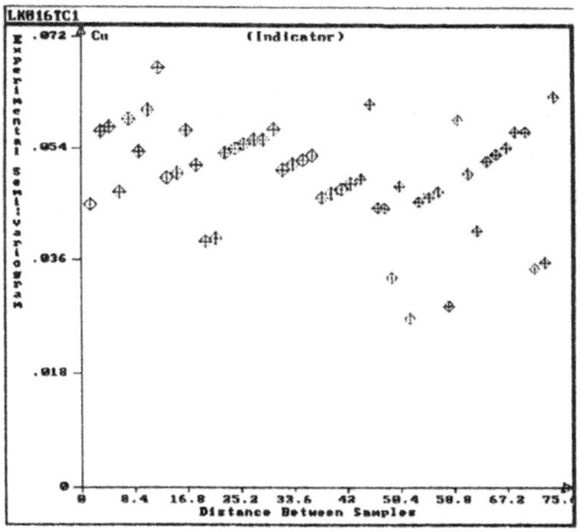

FIGURE 5f.

LK016TC1 INDICATOR SEMI-VARIOGRAM 2.7% Cu.

An example of banded carbonatite (LK122BC1) is shown in Figure 6a. The experimental model fitted is a Hole-Effect with a cycle period of 15 metres. In this case the holes are deep and the model has not been decayed. This graph is somewhat confused between lags of 20 and 40 metres.

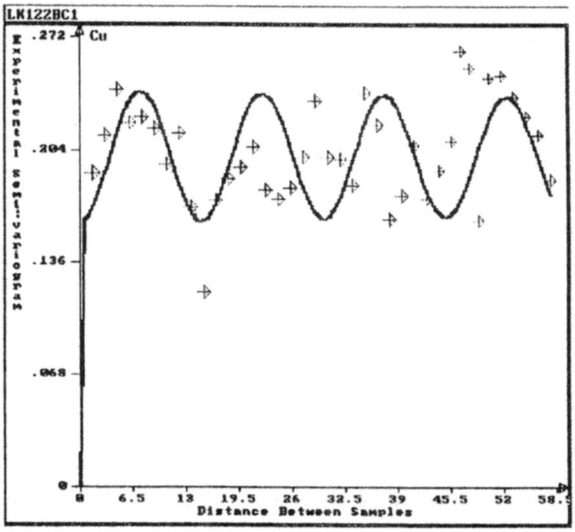

FIGURE 6a.

LK122BC1 DOWN-THE-HOLE SEMI-VARIOGRAM

Figures 6b to 6e are Indicator Semi-Variograms of the same length of core.

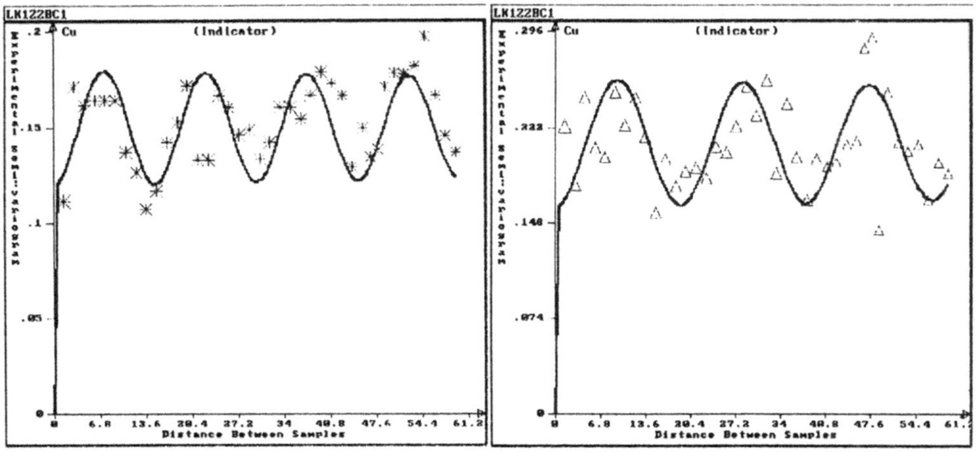

FIGURE 6b & c.LK122BC1 INDICATOR SEMI-VARIOGRAMS 0.2% & 0.8% Cu.

Figure 6b (0.2% Cu) exhibits a cycle period of 15 metres whereas Figure 6c (0.8% Cu) clearly shows a cycle period of 19-20 metres.

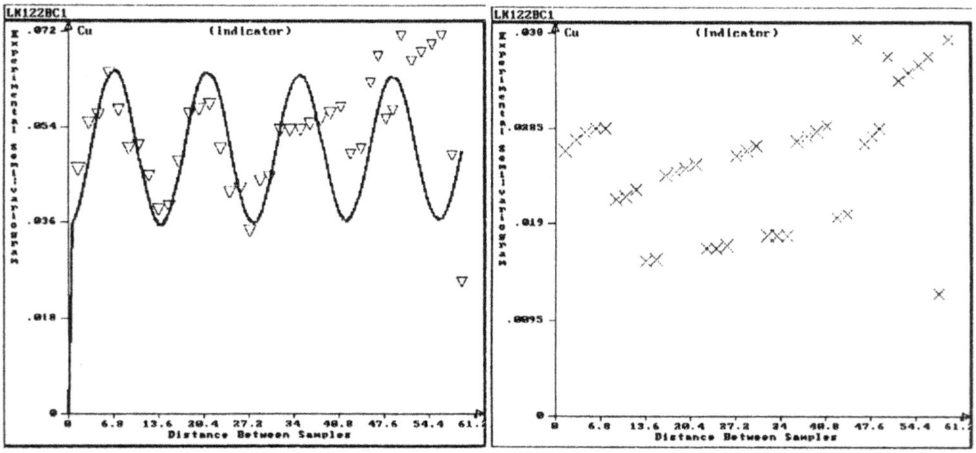

FIGURE 6d & e.LK122BC1 INDICATOR SEMI-VARIOGRAMS 1.4% & 1.6% Cu.

Figure 6d (1.4% Cu) clearly exhibits a cycle period of 14 metres. Figure 6e (1.6% Cu) shows evidence of a smoothly rising graph broken into "segments" of roughly 10 metres.

Similar examples can be found in the foskorite cores.

All the above cycle distances are along apparent dips and should be converted to the horizontal using the dips of the individual holes.

For each of the major rock types, three and sometimes more indicators have been found which yield cyclic indicator semi-variograms of different periodic lengths. The interpretation is that these spacings are signatures of the different phases of mineralisation. The infilling of the pipe was fractured by intermittent activity of varying intensity. The major joints so-formed were exploited by mineralised residual fluids, depositing copper and other minerals in the planes of weakness. Where later joints encountered copper deposited by a previous phase, remobilisation may have occurred, giving rise to some of the small-scale textures described by previous authors. The sharp breaks seen especially in the low and high grade indicator semi-variograms may be a measure of late stage fracture spacing.

4.2 Global Diamond Core Semi-Variograms.

Nearly all the diamond drilling was alligned perpendicular to strike. Figure 7 is a plan view of the drilling pattern over the central high grade core and Figure 8 is a section viewed from the East.

FIGURE 7.

PLAN VIEW OF DRILLING PATTERN

(not to scale - local coordinate system)

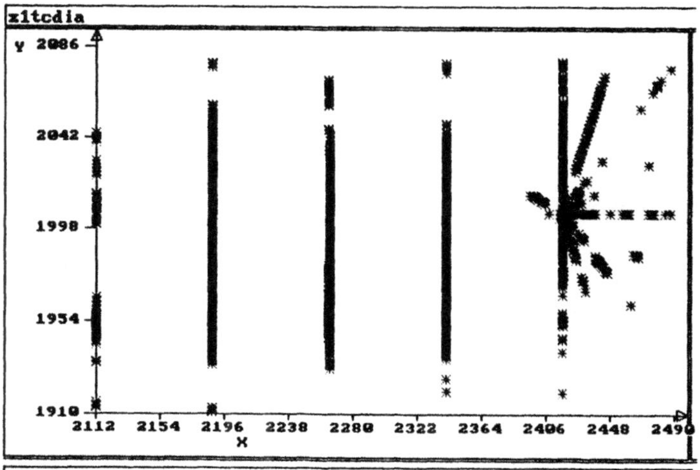

FIGURE 8.

SECTION VIEW OF DRILLING PATTERN

(not to scale - local coordinate system)

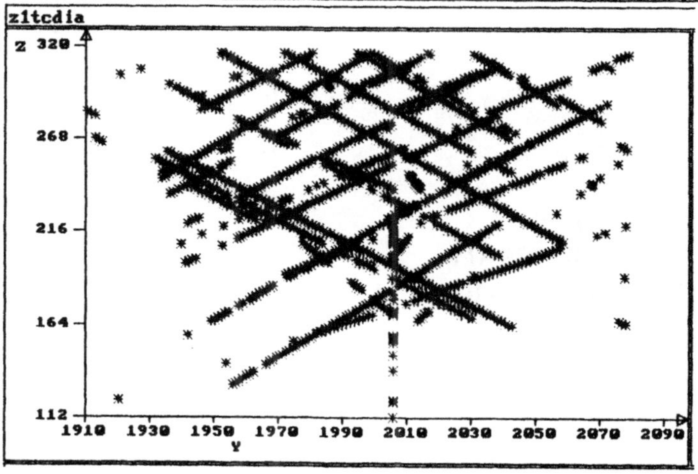

After classification of the entire suite by rock type, global semi-variograms were run (Figures 9,10 and 11 - the different symbols represent tolerances of 5, 10 and 15 degrees). These gave reasonable semi-variograms in the North-South (Figure 9) and Vertical (Figure 10) directions but there was an obvious lack of "infilling" in the East-West direction (Figure 11). Holes are evident at 80 & 160m North-South and the Vertical range is still to be established but exceeds 200m.

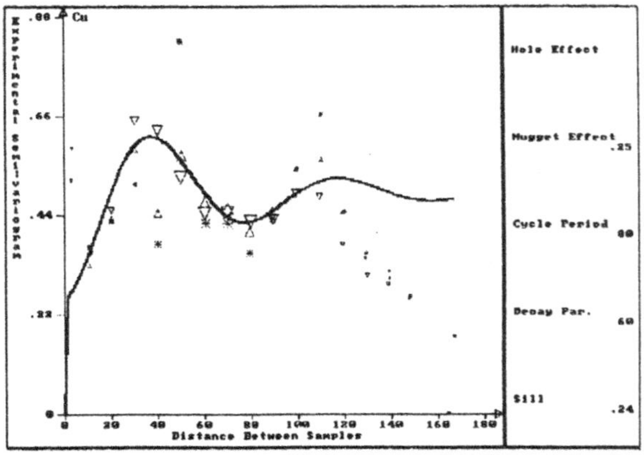

FIGURE 9.

NORTH-SOUTH
SEMI-VARIOGRAM
(TRANSGRESSIVE)

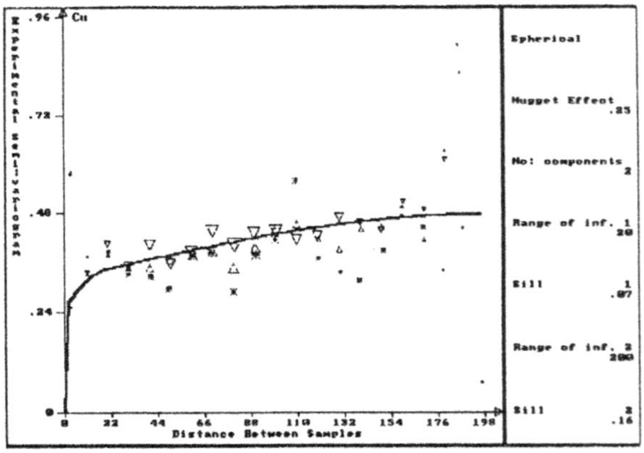

FIGURE 10.

VERTICAL
SEMI-VARIOGRAM
(TRANSGRESSIVE)

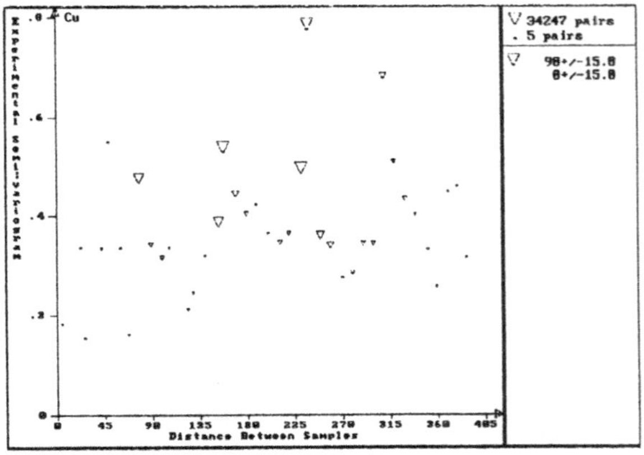

FIGURE 11.

EAST -WEST
SEMI-VARIOGRAM
(TRANSGRESSIVE)

5 GEOSTATISTICAL STUDIES OF BLAST HOLES.

A comprehensive study of the blast-holes from the open pit has been done. The data was split into 100m by 100m cells containing about 3 000 holes each. It was found that local anisotropies exist which can be related to the geology. There is a generally concentric pattern following the banded carbonatites (see Figure 3), sharply broken by very strong radial anisotropies in the transgressive carbonatites. On the basis of anisotropy and grade it was possible to divide the deposit into geostatistical provinces.

Semi-variograms from the main E-W trending transgressive carbonatite area are shown in Figure 12. These semi-variograms do not display the Hole Effects seen previously in the cores as when large quantities of data are analysed, the Hole Effects "counteract" each other and are smoothed out towards Two-Component Sphericals.

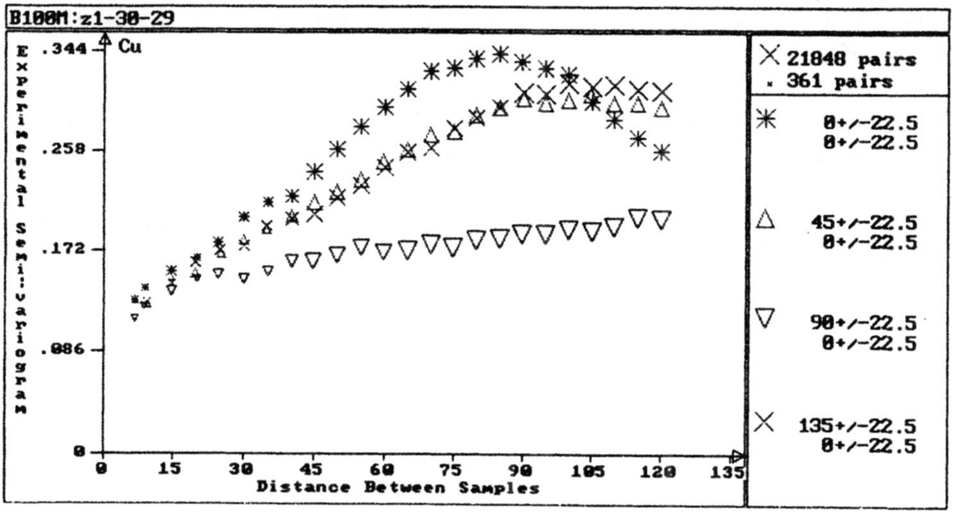

FIGURE 12. TYPICAL SEMI-VARIOGRAM OF BLAST-HOLES

6 DEFINITIVE MODELS FOR KRIGING.

The intention is to utilise both the blast-hole models (corrected for the difference in support) and the global diamond-core models to construct kriging systems. The blast-hole models will be particularly useful in deriving the E-W parameters because of the lack of diamond drilling in this direction (see Figure 11).

The underground reserves will be developed from the current exploration drilling, and several other deep holes which intersect the area below the projected bottom of the open pit.

7 CONCLUSIONS.

On a small scale, this copper deposit exhibits some extraordinary geostatistical behaviour. Almost every section of diamond core shows a Hole-Effect, sometimes dominant but often "recessive" within a Spherical- or Linear-type model.

As with most Hole-Effects, the behaviour disappears when large quantities of data are analysed simultaneously and Spherical Models have been fitted to the extensive blast-hole records for the purposes of developing kriging systems.

Nevertheless, it is interesting that the rocks contain evidence of individual phases within a complex mineralisation process.

8 SELECTED REFERENCES.

Aldous, R.T.H. Ore Genesis in Copper Bearing Carbonatites, Ph.D. Thesis, Imperial College of Science and Technology, University of London, 1980.

Journel, A.G and Froidevaux, R. Anisotropic Hole-Effect Modelling, Proceedings of the 17th APCOM Symposium, 1982 pp 572-585.

Leroy, A.J. and Lill, J.W. Case Study of Palabora Mining Company Limited, Surface Mining Symposium 1989.

SIMULATION OF MEDIUM AND SHORT TERM MINE PRODUCTION SCHEDULING USING GEOSTATISTICAL MODELLING

Vieira J.L., Travassos J.F. Gonçalves A. Santos N. and Muge F.

Centro de Valorização de Recursos Minerais
Instituto Superior Técnico
Lisbon, Portugal

ABSTRACT:

This paper presents results of a research project carried on with the aim of developing a software package dealing with the main phases of a mining project: The ultimate project of the mining method and the production planning. The results presented concern the simulation of the medium term and short term production schedulling of an underground mine (Neves Corvo-Mine). The methodology proposed in this paper is based on simulations of the integrated process operations, taking as input numerical geostatistical orebody models.

The simulation aims to answer relevant problems that arise at project level, the most important of which is to predict what will be the consequences in the ore-grade variability when adopting a certain mining method with its fixed geometric constraints.

1-INTRODUCTION

When choosing medium term production zones, the usual criteria is to maintain certain "in situ" characteristics of the mined ore constant, like its tonnage and average grade, as well as stabilizing the mining equipment needed to keep the production rate within preestablished limits.

Is it possible, in the short term level, to maintain these target parameters within an acceptable range, programming the production schedulling to minimize the grade variability of mined ore, providing a maximum of "internal" blending or is it necessary to perform blending operations (stockpiling, etc...) before the concentrator? In this case what should be the dimensions of the stockpiling?

Another significant question, of important account in mine planning, concerns the so called "information effect". In fact, the orebody boundaries definition and the selection of production zones, panels and stopes to be mined are based on ore grade estimates that depend strongly on the existing information.

In medium term planning the decisions concerning the selection of production zones, are taken on average kriging estimates of tonnage and grades. At this planning stage, the information usually consists of data coming from a regular mesh drilling campaign covering the entire orebody. This information is usually sufficient to estimate accurately the average value of ore grades within a defined medium term production zone. However due to the smoothing effect of the estimation procedure, it does not give an unbiased picture of the ore grade variability based on a smaller support like the selection block or exploitation unit. This means that the block recovery both in tonnage and in average grade can be strongly biased at this planning stage.

In short term planning, the decisions are made using the higher density of information coming from the sampling of the working faces.In this planning stage the miner usually has, in advance, a much better knowledge of the ore grade distribution in the working faces, planning the production schedulling accordingly to the fixed production targets.

2-SIMULATION OF THE INTEGRATED PROCESS OPERATIONS

The operations covered in the simulation procedure are the following:
- Numerical oreboby models (morphology, estimated and simulated block models);
- Definition of the exploitation layout;
- Medium-term planning;
- Short-term planning and production scheduling;
- Mining simulations and ore flows.

The first phase concerns the construction of numerical orebody models. These models are the input for the integrated process simulation. The geostatistical numerical models are of two types: kriging estimation models of block grades and conditional simulation models of blocks grades. In order to construct such models, a three dimensional definition of the orebody morphology is needed to confine the three-dimensional kriging estimates to the orebody boundaries. The estimated block model is used as input to define and estimate the exploitation modules. Based on the block grid model, the software is able to simulate the medium term planning decisions.

To settle some target parameters like ore tonnage and average grade, a production scheduling based on the interactive choice of medium term production zones was defined. Here the estimated block grid model was used.

The next step simulates, in the selected medium-term production zones, the short-term planning decisions. As explained before, this selection is based on a higher degree of information. So, to obtain this information for the short-term selection procedure, the simulated block model was used. The usual goal of the production scheduling is to maximize the internal blending. The software package simulates this by applying dynamic programming algorithms to define short-term mining sequences. An entire mining simulation was performed and the ore flows obtained were used to compare project alternatives.

2.1. - Numerical Orebody Models

The methods used to build the input models were: structural analysis, estimation by kriging and conditional simulation. Using octree encoding models with blocks of variable dimensions were developed (Ref. Travassos J.F., 1991). This encoding allows compression, union and intersection of block models created at different stages of the mine planning. The advantage of this is that the block models can be updated by taking into account more precise information coming from exploration survey campaigns. The construction of the model for the developed example concerning Neves-Corvo Copper Mine followed the steps described bellow:

i) - Based on indicator ore-type variograms at a cut-off of 2% Cu and a 4x4x4 m block size (Fig.1, Fig.2 and Fig.3) the orebody boundary was estimated by indicator kriging (Fig.4).

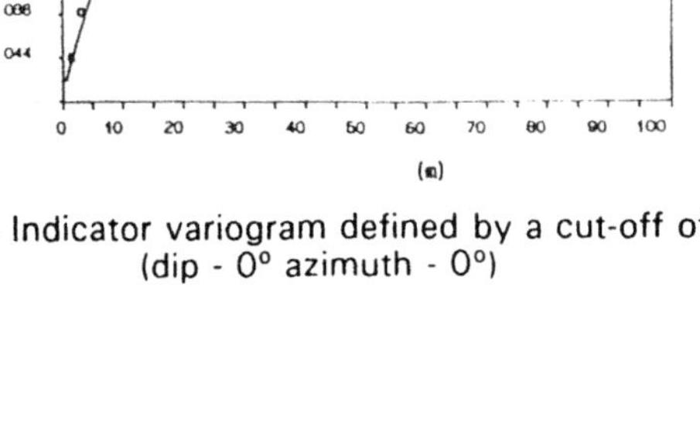

Fig. 1 - Indicator variogram defined by a cut-off of 2% Cu (dip - 0° azimuth - 0°)

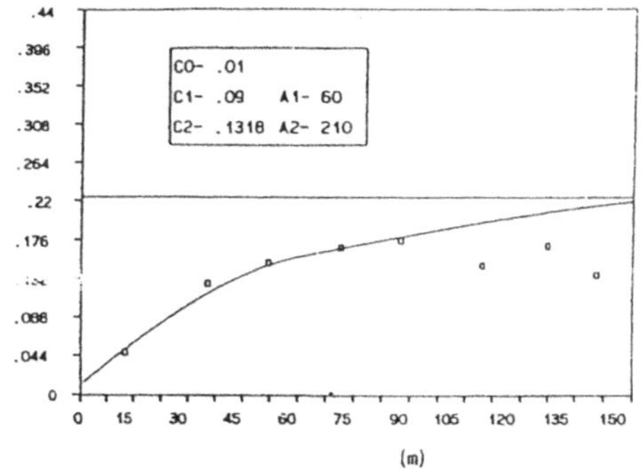

Fig. 2 - Indicator variograms defined by a cut-off of 2% Cu (dip: -33°, azimuth: 90°)

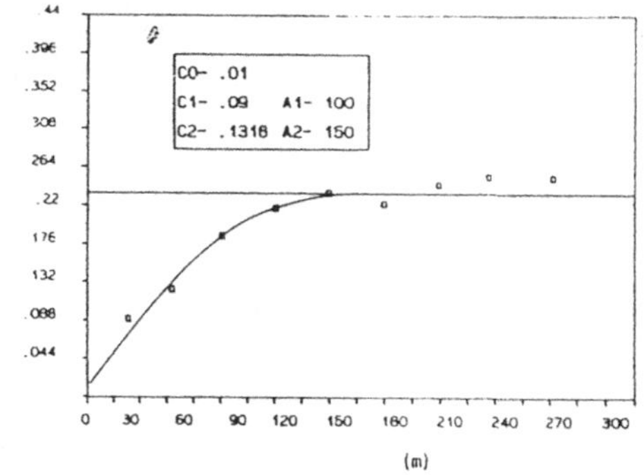

Fig. 3 - Indicator variograms defined by a cut-off of 2% Cu (dip: 57°, azimuth: 90°)

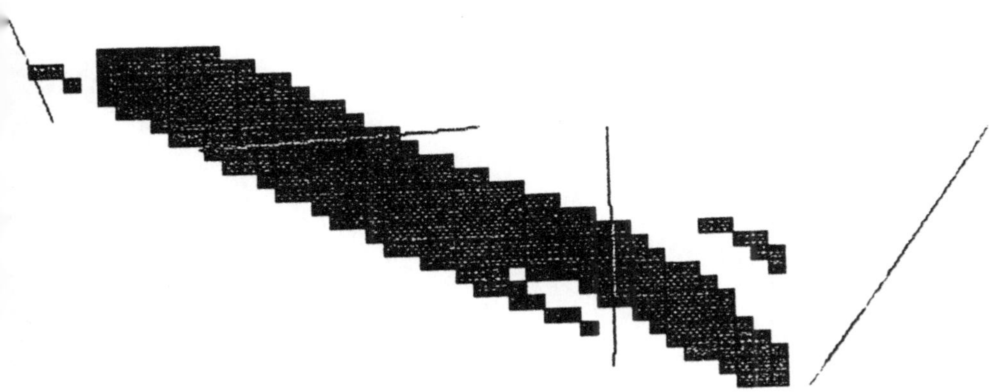

Fig. 4 - 4x4x4m block model (morphology of the orebody - cross-section 3)

ii) - Using new and more detailed information coming from exploration survey campaigns of a local zone, indicator kriging was also used to build a local model with blocks sized 1x1x1m defining the morphology zone between levels 812m and 832m. This model, constructed with the methodological aproach of i), was integrated in to the global model already developed (Fig 5).

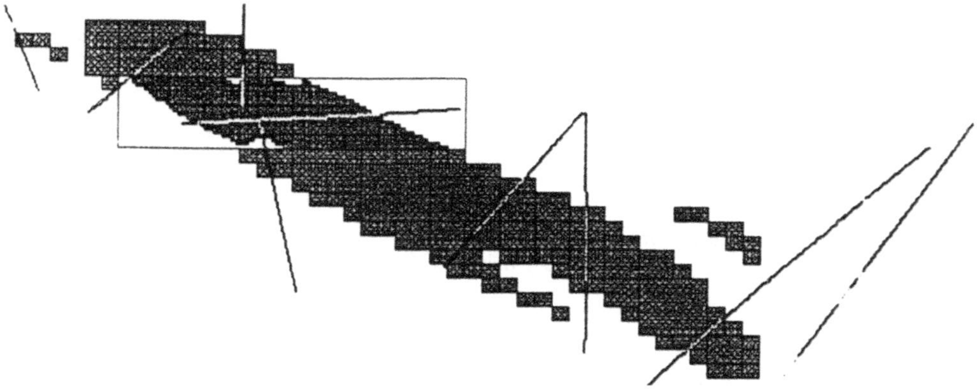

Fig. 5 - 1x1x1m Block model between levels 812m - 832m integrated in the 4x4x4m block model (updated morphology of the orebody (cross-section 3)

iii) - At this stage, two models were developed. An estimated block model and a simulated one. Variograms of copper grades were calculated (Fig. 6) and ordinary kriging was performed to build the estimated block model (Fig. 7). The second one was constructed using conditional simulation to simulate the copper grade variability. Both models have blocks of 1x1x1m in the space defined by the boundaries of the morphological model between levels 812m and 832 m.

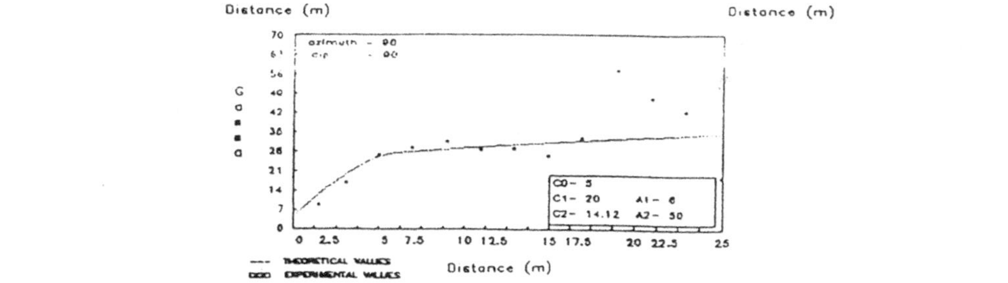

Fig. 6 - Structural analysis of the copper ore lens

iv) - Construction of two exploitation models (estimated and simulated models) supported by the models developed in iii) (Fig.7 and Fig. 8). These 8x4x4m block models were calculated through the mean of 1x1x1m blocks that are inside each exploitation block.

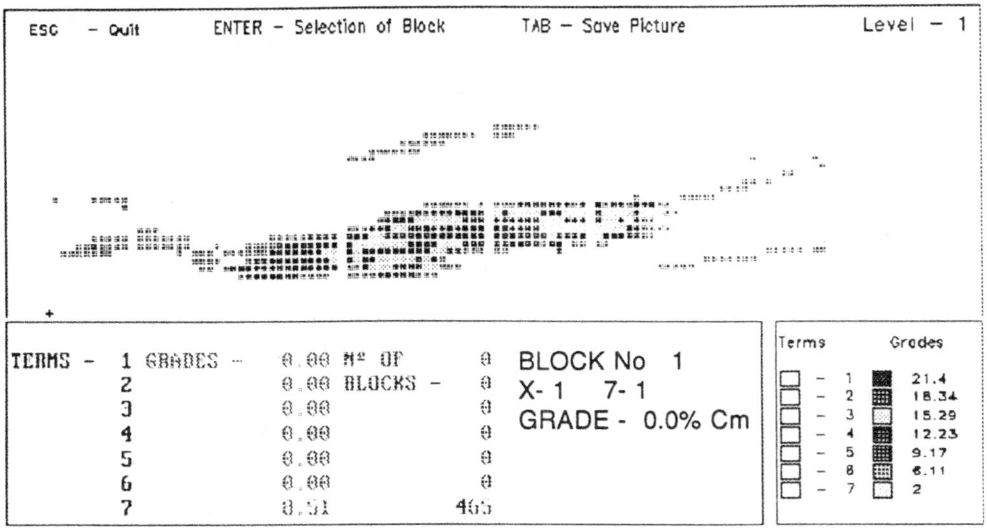

Fig. 7 - 8x4x4m estimated copper grade block model of level 812m - 816m

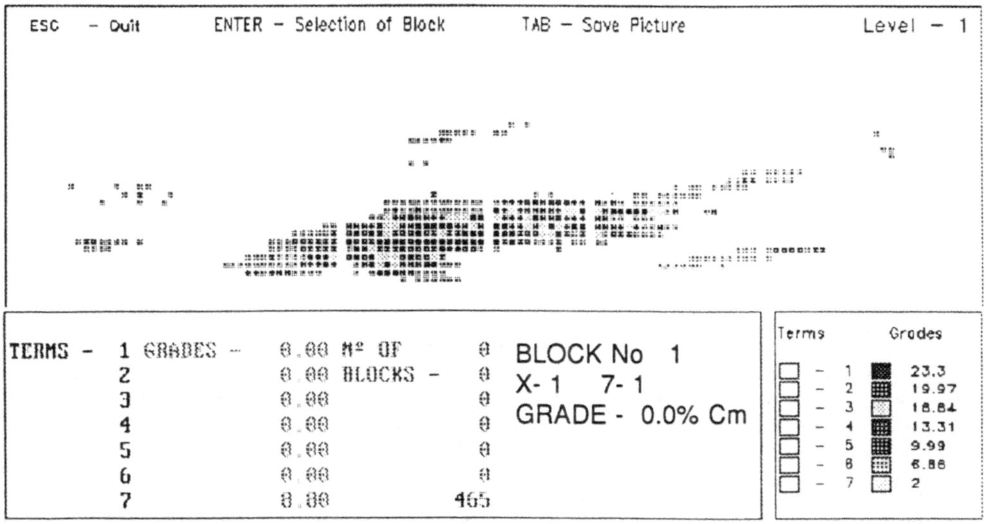

Fig. 8 - 8x4x4m simulated copper grade block model of level 812m - 816m

2.2. Definition of the Exploitation Layout

The cut-and-fill mining method was choosen for the exploitation of complex sulphide orebody located in the south of Portugal (Neves Corvo orebody).

Access to the ore will be assured from footwall drifts excavating an inclined drift to each mining level. From the bottom of a level, horizontal slices will be cut out (Fig. 9). Each slice will be divided into several mining panels by cross drifts. The different stopes will be worked from the cross drifts, adopting a retreat mining method with the backfilling from the hanging wall of the orebody's masses (Fig. 10).

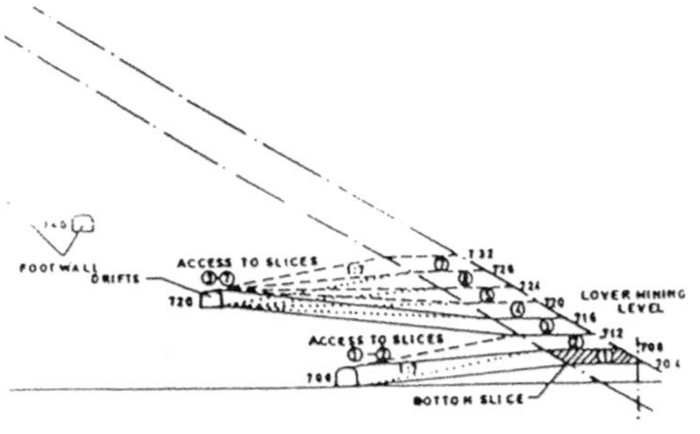

Fig. 9 - Transverse cross section

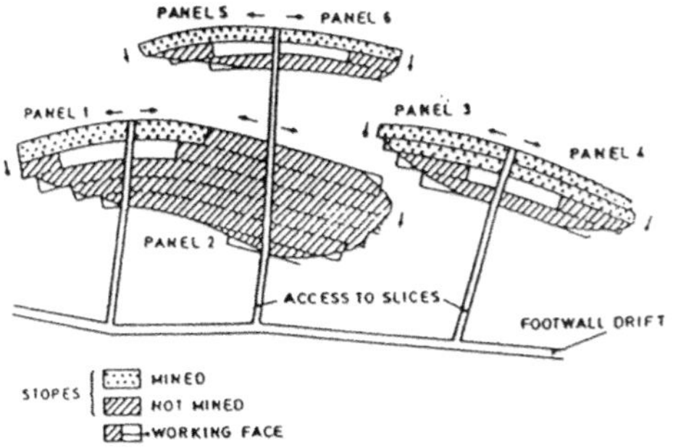

Fig. 10 - Horizontal cross section

2.3. Medium-Term Planning

The medium-term planning modules of the software package perform the interactive selection of medium-term selection zones. Initialy the block model is presented in the screen together with the exploitation layout. The selection procedure is made block by block and in each stage, the number of selected blocks is presented (ore tonnage) together with the respective average grade. The user can modify interactively the blocks assigned to a certain medium-term zone and perform the medium-term planning in order to meet defined targets such as the production rate, the average grade and the stabilization of mining equipment. Figure 11 shows an example of medium-term zone definition on a mining level. In this example, the annual rate of copper ore production was taken to be 1,300,000 tons according to the Neves-Corvo project.

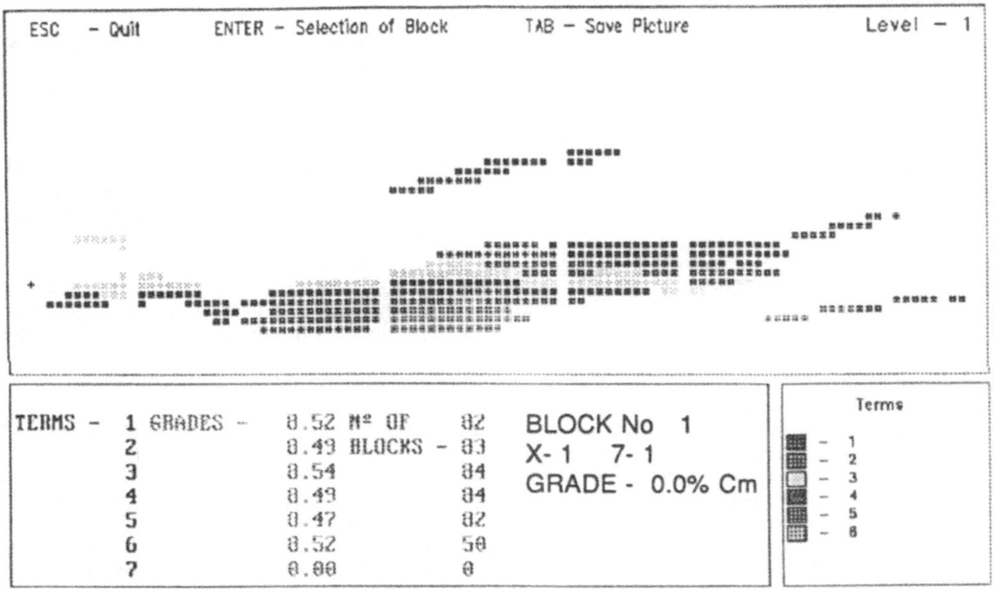

Fig. 11 - Interactive selection of all production zones (Neves-Corvo orebody).

The mining level (812 m to 816m) of the Corvo orebody is planned to supply half of the annual production, that is 650,000 tons of ore per year, which corresponds to a production of 2.500 tons/day or 4.0 x 8.0 x 4.0 of cubic meters, 4 blocks per day. Using these parameters into account, each medium-term production zone, corresponding to one month of production, was interactively defined to meet the following targets:

- constant rate of copper ore production at 54.000 tons, which correspods to mine 84 blocks of 4.0 x 8.0 x 4.0 cubic meters.
- constant average copper grades from zone to zone.

2.4 - Short-Term Planning and Production Scheduling

Once medium-term production zones were defined, the next step is to perform the short-term planning and production schedulling. The short-term planning modules of the software package were developed with the main objective of simulating the usual decisions made by the miner at this stage.

This was accomplished by using dynamic programming algorithms in order to minimize the ore variability, providing the maximum "internal" blending. The only interactive operations to be done by the user consist on the identification of mining stopes and the respective schedulling. The first aspect is performed by selecting the initial and final blocks belonging to each row. Doing this, the direction of the mining progression, in each stope is fixed. The schedulling of the stopes to be mined is also performed interactively by taking into account the already exposed geometrical constraints of the mining methods (Ref. Vieira J.L., 1992).

2.5 - Mining Simulations

Simulations of the integrated process operations are performed varying the degree of the medium-term updating procedure of ore grades estimates and the rate of ore production.

Assuming the rate production of 4 blocks/day, the following cases are considered:

CASE 1: The selection procedure in medium-term planning is based on estimated block grade values. No updating of ore-grade estimates was performed. Table I shows the main parameters of the schedule. The average and variance of ore-grades and the number of ore and waste blocks for each medium-term selected zone of level are shown. In the short-term planning, the definition of the mining sequences is based on simulated block grade values, assuming that the miner have access, in advance, to the ultimate information coming from the sampling of the working faces, updating the ore grade estimation block by block.

CASE 2: The selection procedure in the medium-term planning is also based on estimated block grade values, but, in this case, an updating procedure of the average of each medium-term production zone is considered. In order to take into account this situation, the simulated block-grid model was used to calculate the main parameters of medium-term scheduling for the same selected zones of case 1. In this case, the medium-term production zones are the same as for case 1 but the average grades are different and therefore also the mining sequence of each production zone. The results are presented in Table I.

Terms	Ore Blocks						Waste Blocks	
	Average (%)		Variance $(\%)^2$		Number			
	Case 1	Case 2	Case 1	Case 2	Case 1	Case 2	Case 1	Case 2
1	8.52	8.72	22.7	24.7	82	74	2	10
2	8.49	10.09	35.5	42.7	83	69	1	15
3	8.54	12.71	42.2	62.5	84	58	0	26
4	8.49	12.71	33.9	61.3	84	54	0	30
5	8.47	9.03	12.6	28.8	82	79	2	5
6	8.52	10.60	24.6	34.8	50	47	2	5
Total					465	381	7	91

Table I - Medium term scheduling parameters (Cases 1, 2)

Comparing the two cases one can easily see the consequences of the "information effect" in the medium-term planning:

i) The stabilization of ore-grades from zone to zone is apparently met in case 1. The situation exposed in case 2 gives a much more "realistic" picture of what occurs if the medium-term selection is based on estimated ore-grades without any updating procedure.

ii) The ore recovery in tonnage is systematically over estimated in case 1.

iii) The ore-grade variability that is kept approximately constant in case 1, varies from zone to zone and is higher than forecasted.

In the short-term planning, the simulated block grid model is again used and the minimization of the ore-grade variability is performed taking as target parameter the updated average of each medium-term production zone.

<u>CASE 3</u>: In this case the medium-term updating procedure of ore grades estimates is supposed to be complete. Here, the selection procedure in the medium-term planning is based on the block grid simulated values, assuming a complete redefinition of the medium-term production zones in function of the ultimate information coming from the sampling of working faces. Results are presented in Table II.

Terms	Ore Blocks			Waste Blocks	Total Blocks
	Average %	Variance $\%^2$	Number		
1	10.45	41.7	82	2	84
2	10.43	38.2	84	0	84
3	10.45	30.6	80	4	84
4	10.49	42.2	81	3	84
5	10.59	31.3	50	0	50
Total			377	9	386

Table II - Medium term scheduling parameters (case 3)

In this case the medium-term production zones are not the same as for cases 1 and 2. The target parameters are well stabilized from zone to zone namely the average of ore grades and ore tonnage recovery. In the short-term planning, the simulated block grid model is also used.

Assuming an increased rate of production of 6 blocks/day, two simulations of the integrated process operations are performed: case 4 with the same hypothesis as case 1 and case 5 equivalent to case 2.

2.6 - Ore Flows

Figures 12 to 14 present the daily ore flows, output of the mining simulation (case 1 to 5) for the mining level (812 m to 816 m) of Neves-Corvo Mine. This mining level corresponds to approximately 6 months of production for cases 1 to 3 (production rate of 4 blocks/day) and to about 4 months of production for cases 4 and 5 (production rate of 6 blocks/day).

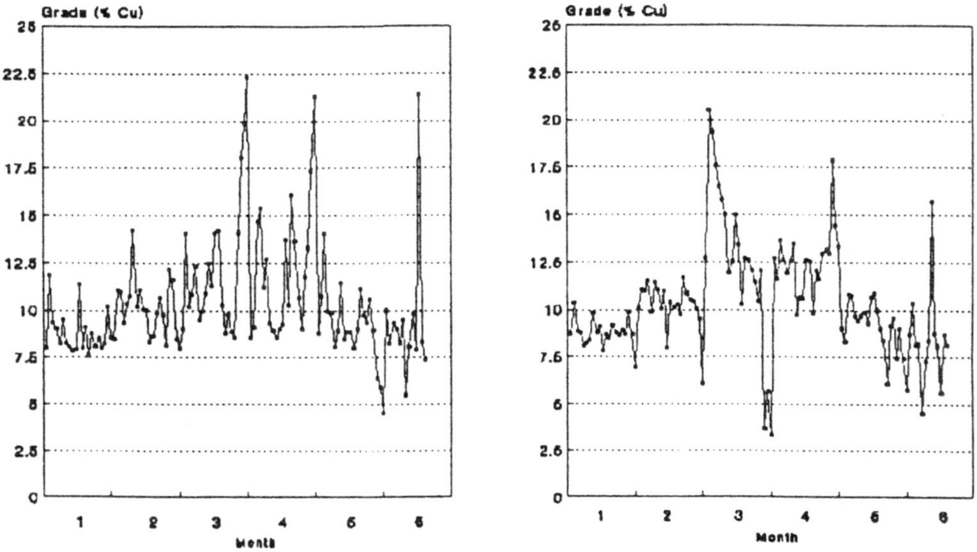

Fig. 12 - Daily ore flows (case 1 and case 2)

The ore grade variability decreases from case 1 to case 2 and also from case 4 to case 5, owing to the updating procedure of average grade for each medium-term production zone.

This decrease of variance is in about $1.5\%^2$ in both cases. The effect of increasing the rate of production is a decrease of about $2\%^2$ in the variance of the ore grades when comparing case 1 with case 4 and case 2 with case 5. The lower ore grade variability is attained in case 3, which corresponds to the complete medium-term updating of ore-grades estimates, showing a decrease in variance of about $7\%^2$ from case 1 and of about $6\%^2$ from case 2.

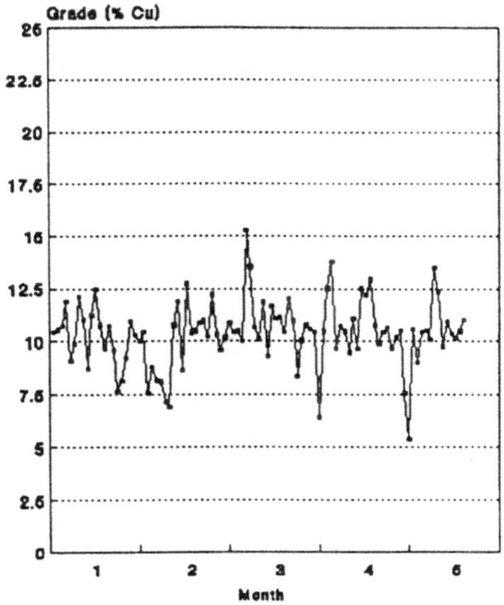

Fig. 13 - Daily ore flows (case 3)

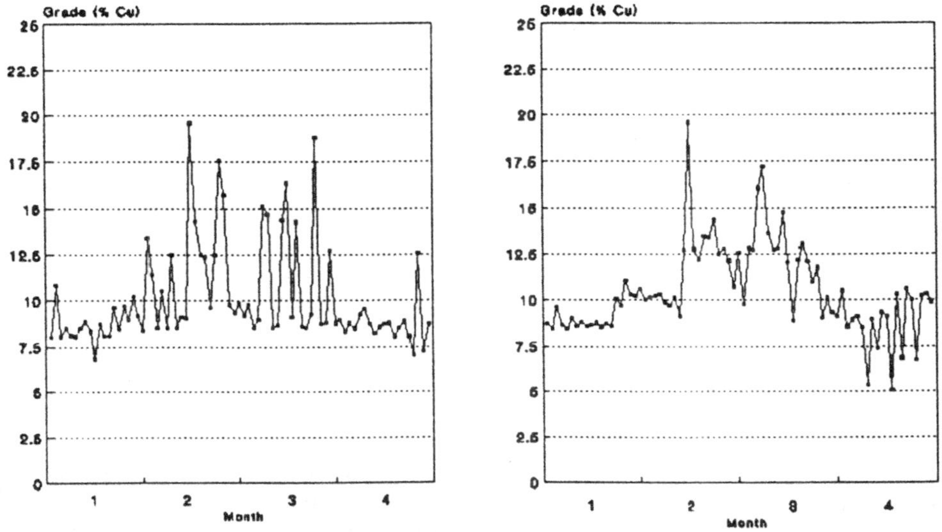

Fig. 14 - Daily ore flows (case 4 and case 5)

The real results coming from the mine and corresponding to the same period of production are in accordance with those obtained by the simulation of case 2, showing periods of high ore grade variability (daily fluctuations of copper grade, within one month of production, between 20% and 4 %).

The results shown by this simulation procedure suggest that stronger effort should be done in medium-term updating procedure of ore grade estimates. This is, the most important controlling factor of the ore grade variability.

Other output of mining simulation consists on studying the influence of blending operations (stockpiling) in the ore-grade variability. Figure 15 show the results of such study in terms of ore-grade variance evolution when performing an ore grade flow average within 2, 3, 4, and sucessive days.

In the Neves-Corvo mining project, the capacity of stockpiling is limited to 2-3 days owing to mined ore oxidation.

The results show, in general, clear decrease of about $2\%^2$ when comparing a 2days average variance with the initial daily variance. The lowest variance profile is again obtained by case 3.

It is important to note that in order to obtain the same initial variance (1 day) of case 3, it is necessary to increase 50 % the production rate and make an average of 5 days in cases 4 or 5. A 2 or 3 day stockpiling seems to be enough to absorb the great part of the initial daily ore-grade variability, but the most important controlling factor still remains to be the medium-term updating procedure of ore grade estimates.

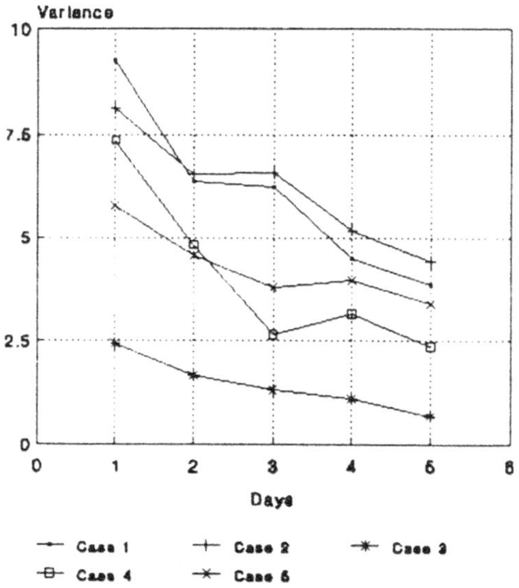

Fig. 15 - Variance evolution of ore grade averages

Another interesting feature is that, comparing cases 1 with 2 and 4 with 5, the values of the respective variances become similar, starting from the 2 days average variance. It is important to note that these variance are calculated within the period corresponding to an entire mining level.

The same variance calculated within a medium-term period (1 month) show quite a different picture. Apart from the third zone (term 3), where the situation is inverted, in all other zones the ore grade variance decrease is quite clear comparing case 1 with case 2. Figure 16 show the evolution of the ore grade variances averaged for all medium term except the third zone.

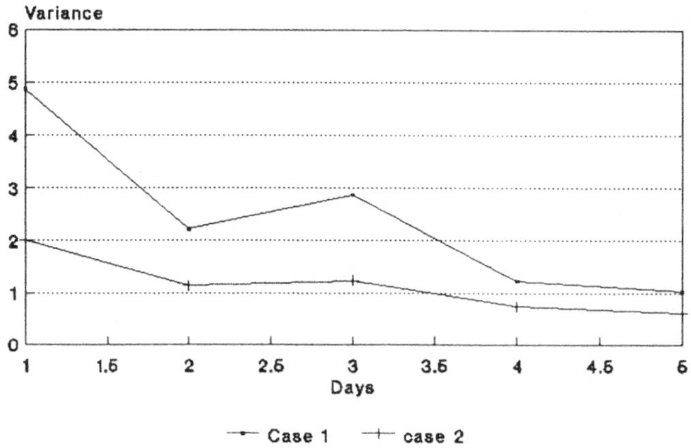

Fig. 16 - Variance evolution of ore grade averages in a medium term period

Indeed, the updating procedure simulated in case 2 is relative to the average grade of each medium-term production zone and so, it is expectable that, when calculating the variances within this one month period, a net decrease of variance is attained.

3. - CONCLUSIONS

The results obtained were quite encouraging. In fact,it is possible, by applying the developed software modules, to perform simulations of the mining process operations. The ore-grade variability found during the exploitation of the mine, can be forescasted by applying adequate modelling and integration of the various mining processes. The several developed software modules are now to be complemented and integrated into a complete system for underground mine planning purposes.

4. - ACKNOWLEDGEMENTS

The authors are grateful to the SOMINCOR managment the facilities given for this work.

5. - REFERENCES

Bellman, R. E., 1957, "Dynamic Programming", Princeton University Press, New Jersey, 342 p..

Bellman R., Dreyfus, 1962,"Applied Dynamic Programming", Princeton, Univ. Press.

Bruno R., Raspa G., Guarascio M., Rogado J.Q., Muge F.H., Santos N., 1989, "Computer Aided Planning of Underground Mining Exploitation", 22th Application of Computers and Operations Research in the Mineral Industry, Berlin, p.27 to 38.

Dowd P.A., 1987, "Dynamic Programming Applied To Grade Control in Sub-level Stoping", TIMM, 8 p.

Caupers D., Guerreiro L., Lobato, 1991 "O Planeamento Mineiro na Mina de Neves-Corvo. Sistemas e Métodos Utilizados", Departamento de Planeamento Mineiro , SOMINCOR.

Durão F.O., 1991, "Computer-Aided Design in Mineral Processing (SIMUL package)",23th International Symposium on the Application of Computers and O.R. in the Mineral Industry, Tucson, AZ, USA.

Muge F.H., Santos N., Vieira J.L. and Cortez L., 1991, "Dynamic Programming in Mine Planning and Production Scheduling", 23th Application of Computers and Operations Research in the Mineral Industry, Tucson, AZ, USA.

Muge F.H., Santos N., 1989, "Application of Dynamic Programming to the Grade Control and Scheduling in a Cut-and-Fill Mining Method", 21th Application of Computers and Operations Research in the Mineral Industry, Las Vegas, USA, p. 494 a 499.

Muge F.H., Pereira H.G., 1979, "Short-Term Planning in Sublevel Stoping Methods", 16th International Symposium on the Application of Computers and O.R. in the Mineral Industry. Ed T.J.O'Neil, p. 323 a 334.

Ribeiro L.T., 1982, "Dynamic Programming Applied to the Mining Sequence Optimization in a Sublevel Stoping Exploitation", 17th Application of Computers and Operations Research in the Mineral Industry, p. 494 a 499.

Rogado J.Q., Muge F.H., Santos N., Vieira J.L., Travassos J.F., 1990, "CADMIN - Computer Aided Design and Planning for Mine Exploitations", EPMESC III- International Conference on Education, Practice and Promotion of Computational Methods in Engineering using Small Computers, Macau.

Travassos, JF., 1991, "Sistema de Codificação por "Octrees". Aplicação à Modelização das Características "in situ" de um Jazigo Mineral", Ms. C. Thesis, IST, Lisboa, Portugal.

Vieira, J.L.,1991, "Aplicação da Programação Dinâmica à determinação de Sequências de Desmonte", Ms. C. Thesis, IST, Lisboa, Portugal.

GEOSTATISTICAL ESTIMATION OF A SUMMARY RECOVERY INDEX FOR MARBLE QUARRIES

Pereira H.G., Brito M.G., Albuquerque T., Ribeiro J.
CVRM - Mining Department - I.S.T.
Lisbon, Portugal

ABSTRACT

A summary index reflecting marble recovery is calculated by applying Correspondence Analysis to a set of non-homogeneous attributes observed in the working faces of quarries. The index is taken as a Regionalized Variable and estimated by kriging in the blocks to be extracted. Once validated by real production data, which matches the kriged values of the index, it is possible to use it for the purpose of guiding the exploitation planning procedures. A detailed case study regarding two Portuguese marble quarries is presented, to illustrate the proposed methodology.

INTRODUCTION

A major problem that arises in the exploitation planning of marble quarries is the lack of an objective criterion for optimizing the economical recovery of the material to be produced. In fact, current decisions regarding the exploitation policy depend strongly on casuistical "rules of thumb", which are subjectively derived from a set of factors, more or less linked to the fracture network prevailing in the quarry and to the expected rock fabric.
The point raised in this paper is that a quantitative summary index, encompassing *jointly* the set of attributes resulting from observations in the working faces of a quarry, can be used as a recovery measure of the material to be extracted, on which further exploitation planning relies, provided that an estimation procedure, based on Geostatistics, gives an accurate value of the blocks to be produced.

Apart from geotechnical applications (Soulié, 1983) and fracture network modeling (Miller,1979, Chilés, 1989), there is a gap in geostatistical literature regarding the estimation of small scale fracture characteristics in dimensional stone quarries, for planning purposes. One reason for this stems directly from the difficulty of defining a Regionalized Variable accounting for marble recovery, the role of which should be analogous to a variable such as the grade, in mine planning. Fracture spacing, length and orientation, rock type, etc., are features governing recovery, that must be balanced and summarized when exploitation planning algorithms are to be applied.

Hence, a new methodology, combining Correspondence Analysis, Geostatistics and Operations Research, was devised in the course of a research project aiming at the development of a dynamic

----

This work was supported by JNICT (Junta Nacional de Investigação Científica e Tecnológica), in the framework of a Research Project for optimizing the recovery of Portuguese marble quarries.

and interactive system for production planning in marble quarries. The first point of this methodology, tested in the case study presented in this paper, consists of the establishment of a recovery index by Correspondence Analysis (Benzécry, 1980, Greenacre, 1984), and on its estimation by kriging, being the entire procedure validated a posteriori by real production data and expert geological / engineering knowledge.

METHODOLOGY

Given *q* attributes related to marble recovery and a support *s* (the representative area of the observations), a set of *n* vectors containing the attribute values in the support is obtained by sweeping the available working faces of the quarry, using the support as a non-overlapping moving window. In most cases, the attributes are expressed by different data types - nominal, ordinal, numerical, Boolean variables. In order to keep track of soft information (which is, in general, relevant for the recovery index), numerical data were split into classes, and all attributes were encoded into approximately equal frequency disjunctive categories (Nanache, 1973). Moreover, in some cases, it is more feasible to assign a fuzzy grade of membership of a certain support to a given category, rather than a sharp Boolean indicator.
Hence, the general encoding scheme that includes all data types under a homogeneous format is outlined in Fig. 1. The basic experimental matrix X contains the values x(i,j), calculated by the fraction of the total observations in a given support *s* (row *i* in Fig. 1) that falls into the category *h* of attribute *k* (column j of Fig. 1). These values, defined in the interval [0,1], can be interpreted as a fuzzy grade of membership.

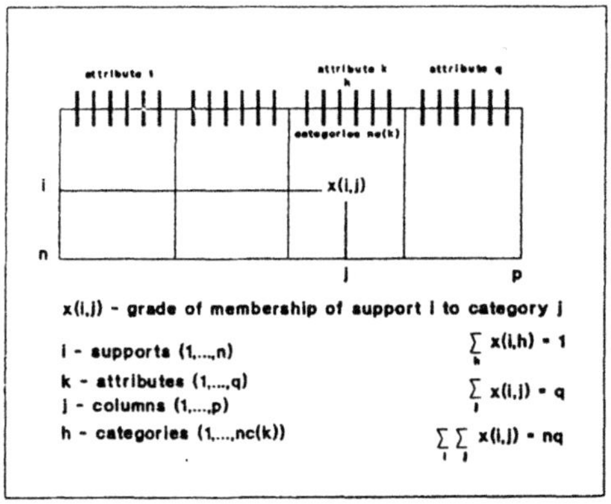

Fig.1 - The basic experimental matrix of supports vs. attributes encoded by a fuzzy disjunctive scheme

Now, the problem is to assign an index to each position of the support, summarizing the categorical data contained in the rows of the matrix of Fig. 1. This index must reflect the recovery of the material swept by the support, providing a set of numerical values in a pre-defined scale, the poles of which are given by the archetypes of GOOD and BAD recovery marble. These archetypes are constructed by analyzing the semantics of each attribute, from the viewpoint of the recovery, and choosing the more favorable classes of each attribute to define the GOOD pole, and the less favorable, to create the BAD pole. Once created the two poles of the scale,

Correspondence Analysis is used as a discriminant algorithm (Benzécri, 1980, Pereira, 1988), in order to produce the supplementary projections of the basic experimental matrix X (Fig. 1) onto the axis provided by the eigenvector of the inertia matrix derived from the two discriminant poles.

The supplementary projection of each support onto the discriminant axis is given by :

$$f(i) = \frac{1}{\sqrt{\lambda} \cdot q} \cdot \sum_{k=1}^{q} W(k) \cdot \sum_{j=nc(k-1)+1}^{nc(k)} x(i,j) \cdot p(j) \qquad [1]$$

where

- f(i) is the recovery index of support i in the scale limited by the two poles
- λ is the eigenvalue associated with the discriminant axis
- q is the number of attributes
- W(k) is the weight given to attribute k
- nc(k) is the number of categories of attribute k
- x(i,j) is the element of the basic experimental matrix of Fig. 1
- p(j) is the projection of category j onto the discriminant axis

It is worth noting that the index depends on three adjustable parameters that can be modified according to real data : the number and classe limits for the attributes, the support dimension, and the set of weights W introduced in equation [1].

Once constructed a first version of the index for a given quarry, variograms of this Regionalized Variable can be calculated, and kriging can be performed, aiming at the estimation of the recovery index in each block to be produced. Now intervenes the external information on real data concerning the commercial value of the blocks extracted after the establishment of the index and on expert knowledge about the behavior of the material to which the kriged index refers. The interactive process of calibration is achieved when the predicted kriged value of the index in the blocks used for validation matches the commercial range of values reported a posteriori for those blocks. In this process, the above mentioned adjustable parameters are modified interactively, according to expert geological / engineering knowledge, which is introduced into the global model by this bias.

CASE STUDY

In order to test the above described methodology with real data and provide a detailed illustration of the steps to be accomplished, two operating marble quarries, located in the map of Fig. 2, were selected for this case study.

For the purpose of the preparation of the basic experimental matrix X, the format of which was defined in Fig. 1, the following data capture procedure applied:

i) Production of an image of each working face by standard photography;

ii) Interpretation of the discontinuity sets by an expert geologist, who produces a transparency outlining the major features of the face;

iii) Digitalization of the relevant attributes using as a basis the interpreted document provided by the geologist;

iv) Sweeping of the working faces by the support, recording the observations taken in each position of the non-overlapping moving window.

The Estremoz Calcareous massif, is incorporated within a vast asymmetric anticlinorium (NW - SE direction), spreading out over 40 Km, from Sousel to Alandroal villages, in the Portuguese Ossa Morena zone.

The studied area was affected by two main folding phases followed by rupture processes, and regional metamorphism.

In some formations which integrate this area, three calcareous rock type occur:

- Crystalline limestones;
- Dolomite and dolomitic limestone;
- Calcitic marble.

The two quarries (Encostinha **E** and Rosal **R**), are located in the limit between the crystalline zone and the calcitic marbles, in the NE side of the anticlinorium near Borba.

The quarries are very close, producing **1.5 x 1.5 x 1.0 m commercial blocks** of "Mármore Rosa" (Pink marble of Borba), exploited in superficial levels.

One of the quarries, called **Encostinha**, with an area of 300 m^2 and 17 working faces distributed by two levels, presents an average **recovery** about **40%**.

The other, named **Rosal**, embraces an area of about 1500 m^2 with 29 working faces, showing a **recovery** of about **20%**.

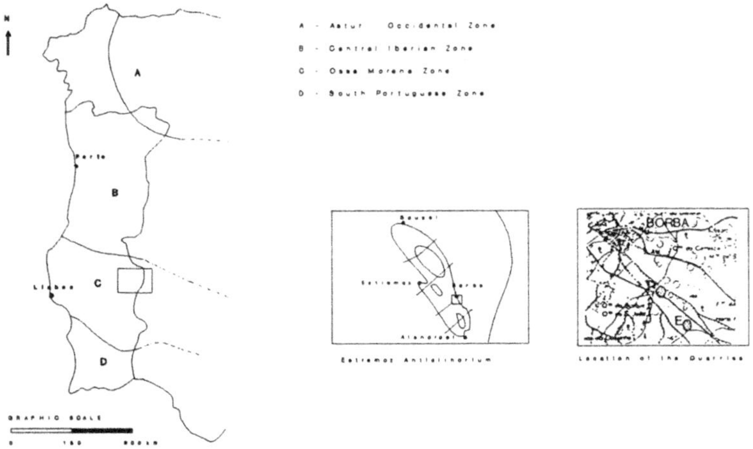

Fig. 2 - Location and characterization of the quarries. General geological settings

A schematic plan of the quarries, sketching the data capture procedure in a working face is given in Fig. 3. Each fracture is assigned to the support that contains its middle point.

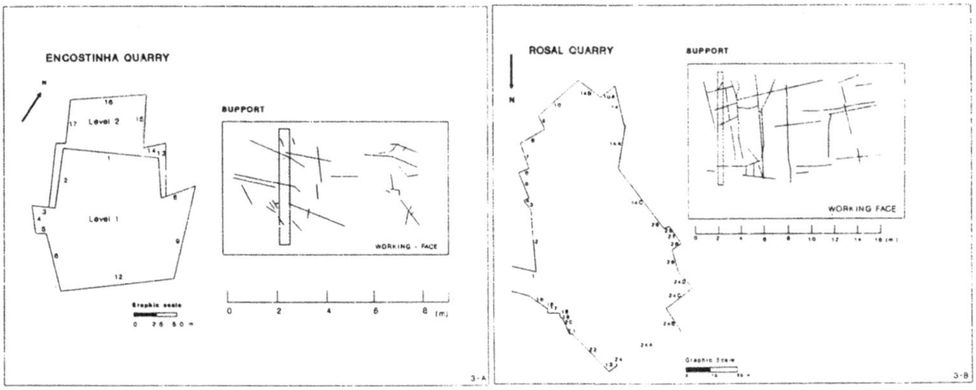

Fig. 3 - Example of data capture for assigning observations to supports

The attributes chosen for characterizing each fracture - Length, Curvature, Number of Intersections, Presence/Absence of Veinlets, Dip - are categorized into the classes given in Fig. 4, where the procedure for assigning each fracture to a certain attribute category can also be visualized. As previously mentioned, the relative importance of the retained attributes for the purpose of the study should be taken into account by the set of weights given in eq. [1].

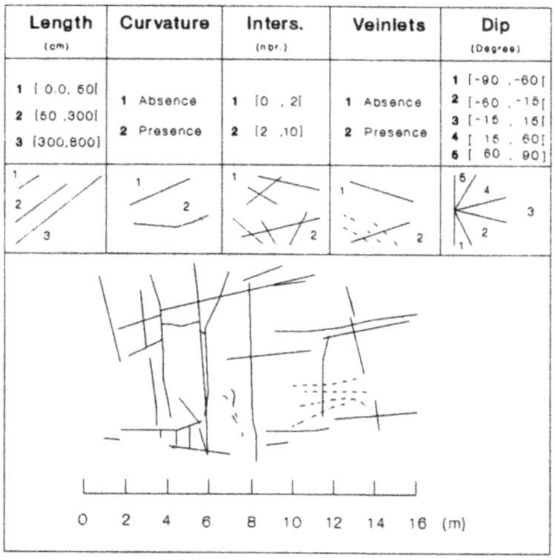

Fig. 4 - Example of the assignment of fractures to attribute categories

The relative frequency of each attribute category was calculated for each position of the support, being then merged with the variable density of fractures in a given working face, defined by the fraction of the total number of fractures in the working face contained in a given support. This variable was also split into classes, the limits of which are 0/.05, .05/.15, .15/1.
Once calculated the basic experimental matrix of Fig. 1, the archetypes of GOOD and BAD recovery index were established. In Fig. 5, it is shown an example of those two poles, compared with a set of experimental vectors of observed attributes.

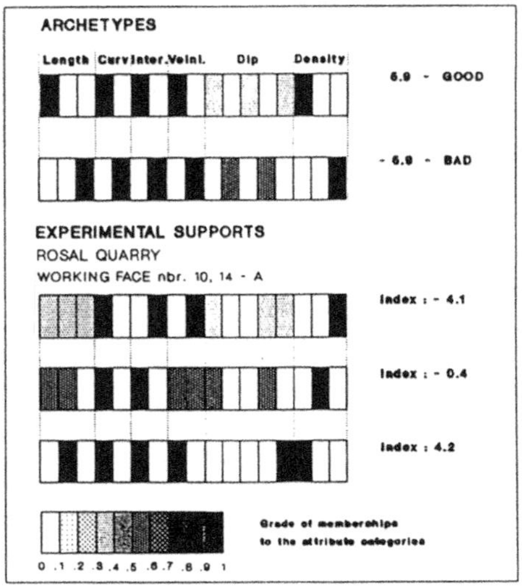

Fig. 5 - Example of GOOD and BAD archetypes of marble recovery, compared with experimental supports

Applying Correspondence Analysis as a discriminant technique for the above defined two poles, the values of the recovery index were obtained by the supplementary projection of the experimental vector characterizing each support onto the discriminant axis. The histograms of the recovery index for both quarries in 0.5 m supports, given in Fig. 6, show an important class of supports (index > 3) which are almost free of fractures (specially in Encostinha). On the other hand, the negative extreme class (index < -3) is more frequent in Rosal quarry, being these conclusions in agreement with the general characterization of the quarries, given in Fig. 2.

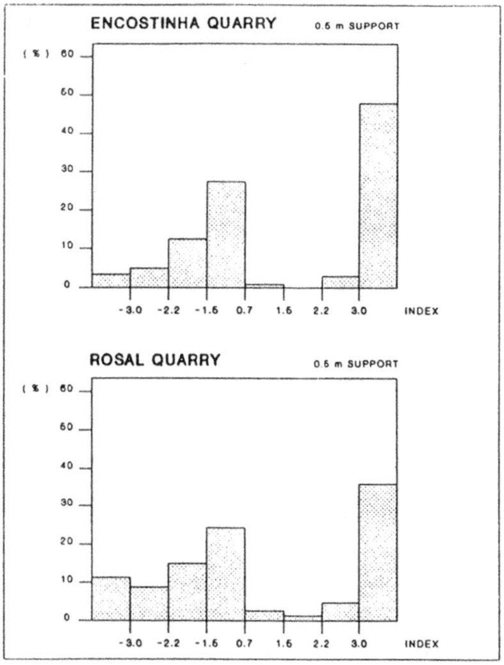

Fig. 6 - Histogram of the recovery index for the 0.5 m support

Variograms of the quality index were calculated for a set of different supports - 0.5, 1.0, 1.5 m, being the vertical dimensions constant and equal to 6 m. Under the assumption of vertical homogeneity, which is acceptable for the above given dimension, the location in space of the Regionalized Variable depends only on the x,y coordinates of the support barycentre, in a reference horizontal level. The omni-directional variograms are given in Fig. 7.

In the Encostinha quarry (A), a spherical structure is clear for the 0.5 m support, giving rise to a range of 6 m.

In the Rosal quarry (B), variograms for the three supports can be fitted by spherical models, being the range approximately equal to 12 m, in all cases.

Once studied the influence of the support change, it was decided to retain the spherical scheme fitted to the 0.5 m support as a good approximation of the theoretical model to be used in kriging (1.5 x 1.5 x 1.0 block).

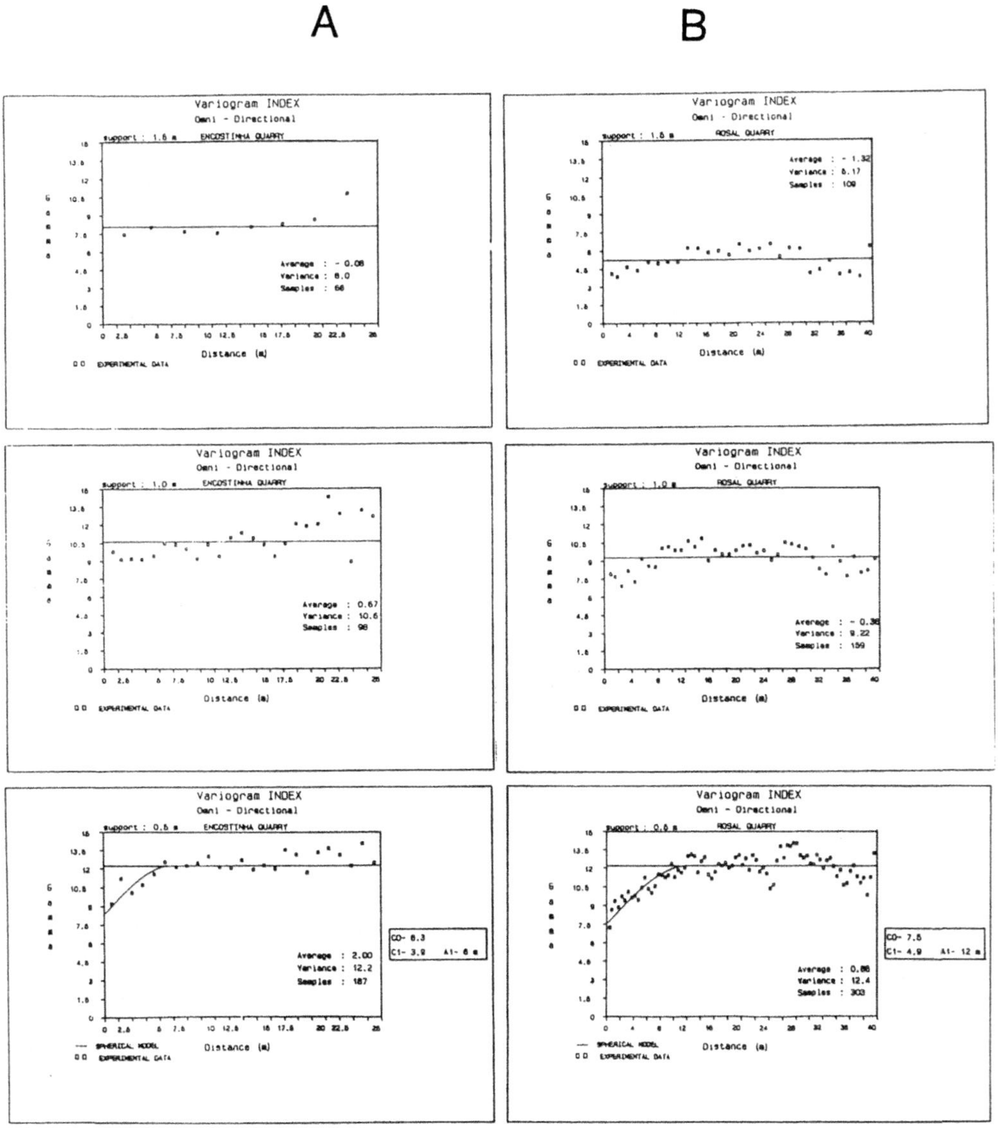

Fig. 7 - Omni-directional Variograms for 0.5, 1.0, 1.5 m supports

In Fig. 8, it is shown the map of the "raw" kriged index for the Encostinha quarry, derived directly from the Correspondence Analysis attribute categories scores, without using the set of weights W of equation [1].

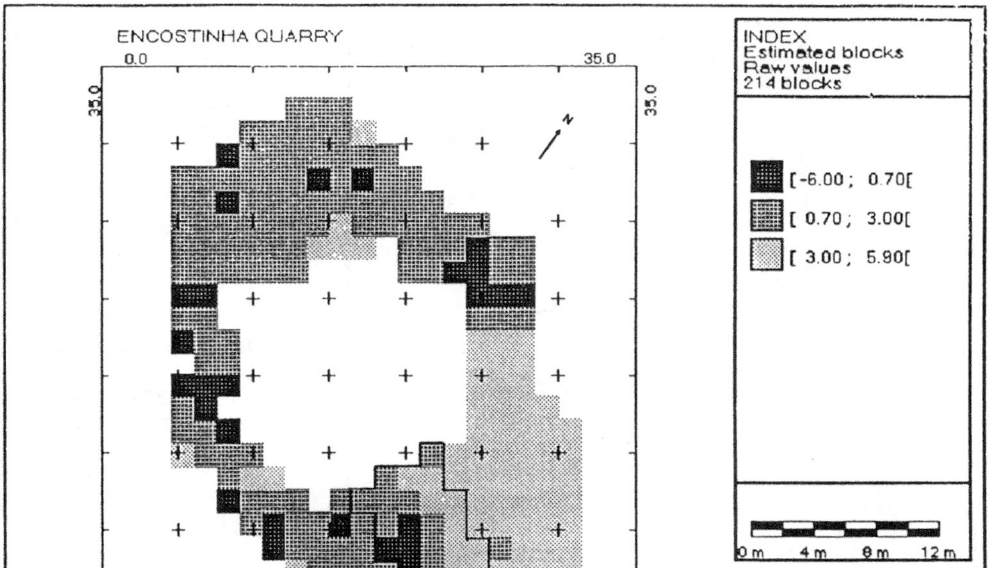

INDEX	Validation Data (%)	Estimated values	
		Validated Data (%)	Global Data (%)
[-6.0; 0.7[	23	19	13
[0.7; 3.0[	46	42	53
[3.0; 6.9[	31	38	37

Fig. 8 - "raw" kriged index for the Encostinha Quarry

In Fig. 8, it also depicted a validation zone, where the commercial value of a set of 26 blocks is known a posteriori. For this first version of the index, there is a poor fitting to real production data (15 % of mismatch in the validation zone). Hence, modifying interactively the set of weights in equation [1] until a perfect match in the validation zone is obtained, and repeating the entire procedure, a final kriged map, given in Fig. 9, is calculated.

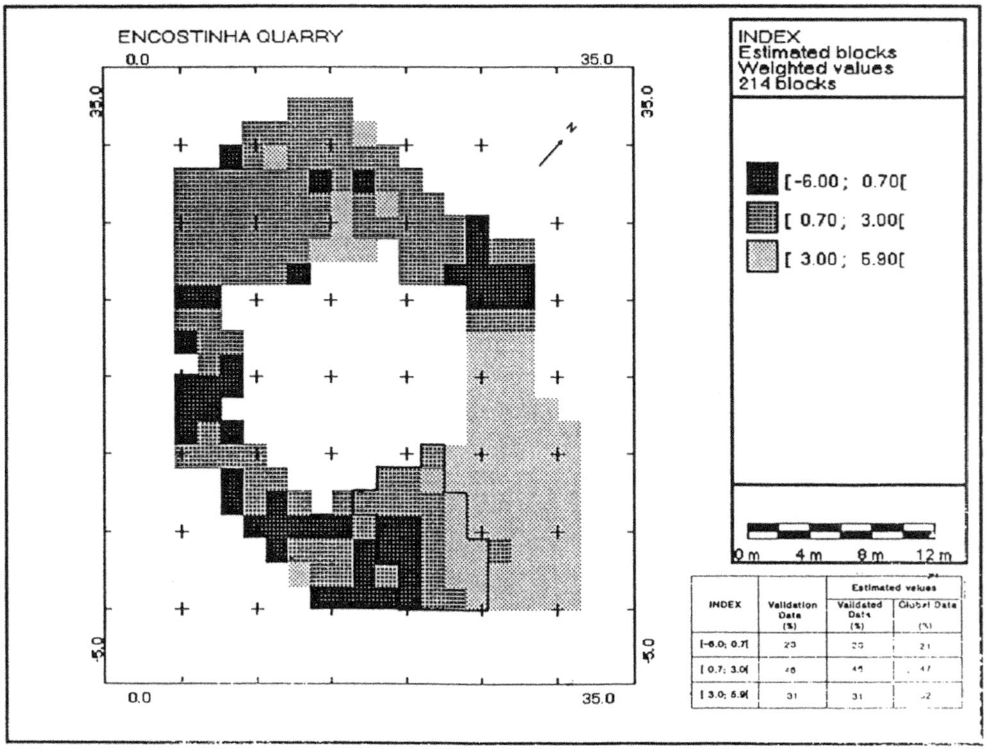

Fig. 9 - Final kriged recovery index map for the Encostinha Quarry

The set of weights to be applied to the attributes in order to get a perfect match in the validation zone is given in TABLE I.

Attributes	Lenght	Curvat.	Inters.	Veins	Dip	Density
Weights	0.15	0.03	0.20	0.025	0.175	0.42

TABLE I - Weights resulting from the validation procedure

In the case of the Rosal quarry, the validation criterion can not be the same as the one previously applied, since there is no a posteriori production data. The available external information concerns the global recovery of the marble to be produced, being this quarry considered, as a whole, significantly "worse" than Encostinha. This fact is apparent in Fig. 10 (B), where the proportion of high quality blocks is 18 %, compared with the value of 44 % found in Encostinha for the same set of weights (Fig. 10 (A)). These proportions agree approximately with the recoveries of 20 and 40 %, respectively, reported in Fig. 2. Moreover, small scale expert geological/engineering knowledge about Rosal is also reflected in Fig 10 (B), since there is field evidence showing that the material is strongly disturbed, in the Eastern faces of the quarry, by the contact with the surrounding schists. Furthermore, in the Northwest faces, the poor quality of the index shown in Fig. 10 (B) is partially explained by the negative influence of a dolomitic vein, arising locally (N30W).

Global comparison of the kriged index in the two quarries

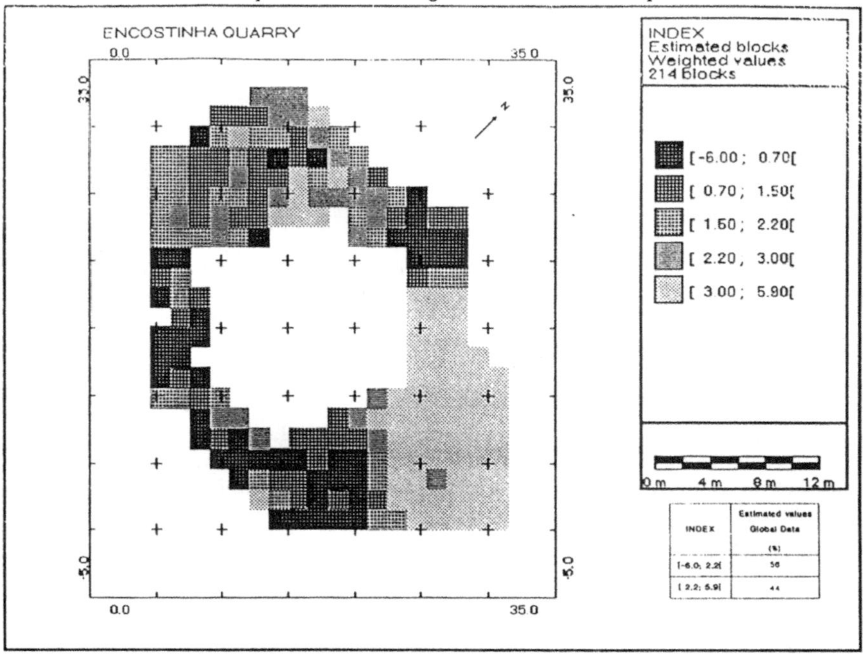

INDEX	Estimated values Global Data (%)
[-6.0; 2.2[	56
[2.2; 5.9]	44

Fig. 10 (A) - Encostinha Kriged Blocks

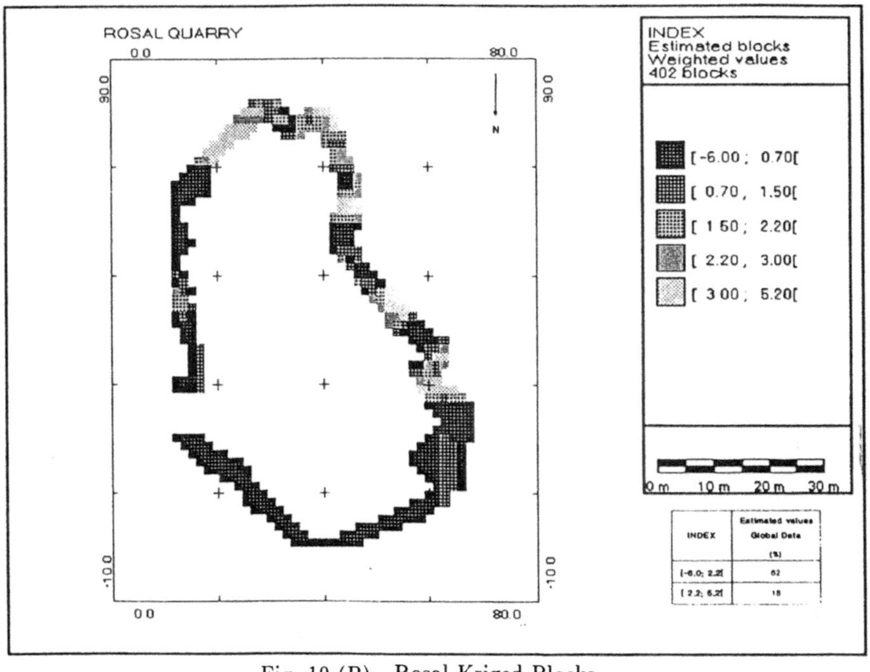

INDEX	Estimated values Global Data (%)
[-6.0; 2.2[	82
[2.2; 5.2[	18

Fig. 10 (B) - Rosal Kriged Blocks

CONCLUSIONS

The case study presented in this paper illustrates how to articulate Geostatistics with real data and expert geological/engineering knowledge, in order to validate a methodology for producing a summary recovery index in marble quarries. An interactive system, integrating the information available in a given quarry, is developed, aiming at the establishment of a reliable quantitative basis for further exploitation planning.

ACKNOWLEDGEMENTS

We are thankful to JNICT for granting the research conducted in the quarries of MARTOTI, to which a special acknowledgement is due, and to Dr. Rui Fernandes, who provided the validation data.

REFERENCES

Benzécri, J-P., 1980, Pratique de l' analyse des données, Dunod, 424 p.

Chilés, J-P., 1989, Three-Dimensional geometric modelling of a fracture network, in B. E. Buxton (Ed.) *Geostatistical, Sensitivity, and uncertainty methods for Ground-water flow and Radionuclide Transport Modeling*, Battelle Press, p. 361-385

Greenacre, M., 1984, Theory and Applications of Correspondence Analysis, Academic Press, 364 p.

Miller, S., 1989, Geostatistical analysis for evaluating spatial dependence in fracture set characteristics, Proceedings 16 APCOM, AIME, p. 537-545

Nanache, J-P., 1973, Influence do codage des données en analyse factorielle des correspondances, *Revue de Statistique Appliquée*, 1973, XXI, n. 2, p. 57-70

Pereira, 1988, Case study on application of qualitative data analysis techniques to an uranium mineralization, in C. F. Chung et al. (eds.) *Quantitative Analysis of Mineral and Energy Resources*. Reidel, p. 617-624

Soulié, M., 1983, Geostatistical applications in geotechnics, in G. Verly et al. (eds.) *Geostatistics for Natural Resources Characterization*, Reidel, p. 703-730

COAL PREPARATION AND GEOSTATISTICS - A BLEND OF TECHNOLOGIES TO FORECAST CLEAN COAL QUALITIES

SUKHENDU BARUA

ISLAND CREEK COAL COMPANY
250 WEST MAIN STREET
LEXINGTON, KENTUCKY 40507, USA

Abstract

Most of the geostatistical studies and applications in coal to date have focused on raw coal and clean coal at a single float gravity (generally 1.50 s.g.). In addition, most of the academic research has been done on ash/recovery prediction from core hole washabilities using geostatistics; sulfur prediction is still a problem.

Recently interest has developed in combining geostatistics and coal preparation plant simulation to predict clean coal qualities (Armstrong et al. 1989). The basic concept is that geostatistics would provide statistically realistic models of recovery, ash content, sulfur content, etc. of the coal deposit for each washing (separation) density. These washing characteristics could then be fed into a preparation plant simulator to forecast the clean coal quality. The combining of geostatistics with preparation simulation shows great promise, but the procedure described above in mixing these two technologies would be very difficult, if not impossible, to apply on a regular basis to predict clean coal qualities. The difficulty arises from the fact that the coal is cleaned by size and by gravity. The geostatistical models must consider all the cleaning gravities of the plant, as well as the size fractions of the run-of-mine coal, which could be different than the size consist of the core hole data.

This paper proposes a modification to the above approach that would make the clean coal prediction more practical and applicable. Using this method, the preparation plant simulation would first be run using the core hole washabilities data to estimate clean coal parameters at each core hole location; geostatistics would then be applied on the estimated clean coal parameters to predict qualities of a small coal block. This proposed technique is being applied in an underground coal mine operated by the Island Creek Coal Company and the results are discussed here.

Introduction

In the United States, geostatistics has proved its worth in ore reserve estimation for base and precious metals. The coal industry, however, has been slow in embracing geostatistics. The central reason behind this delayed response is that coal mining is primarily a production (tonnage)-driven, bulk commodity industry; production factors such as coal seam thickness and density are generally uniform and the limited quality variation can be easily adjusted for in the preparation process. With the passage of acid rain legislation included in the Clean Air Act Amendments of 1990, a great deal of uncertainty has been cast into the coal market with but one inescapable conclusion: coal buyers will become more selective in specifications for the purchase of coal. As a result, the attitude of the coal mining industry must start to shift towards an emphasis on coal quality to survive in the competitive market. Prudent coal cleaning measures and geostatistics can help in meeting some of these market pressures by forecasting the clean coal qualities of a block of coal, including an assessment of variability, to assure compliance with customer specifications before it is mined.

Mining Background

The coal deposit under study is the Upper Freeport seam, an Allegheny formation of Middle Pennsylvanian age, located in northern West Virginia. The seam thickness is fairly uniform with coefficient of variation (standard deviation / mean) 0.088. The average seam thickness is 2.29m (7.5ft). There are no major geological discontinuities within the boundary of present mining of the Upper Freeport Seam.

This seam is mined by a retreating longwall (229m wide X 1829m long; 750ft X 6000ft) and two three entry panel development sections. The development pillars are on 24.4m (80ft) centers and the cross cuts are 35.1m (115ft) apart. The annual production is 2.7 million metric tons raw coal. The longwall contributes about 85% of the mine production. The coal is deep cleaned in a washing plant located adjacent to the mine. The overall plant recovery (clean ton / raw ton) in 1991 was about 65%.

The coal is transported by rail to two major electric utility companies, each having a different set of specifications in terms of ash, sulfur, moisture, heating value, fixed carbon, and ash chemistry. This requires the plant to produce two different specification coals by adjusting plant operating parameters. More on this parameter adjustment will follow shortly.

Objective

The main objective of this paper is to layout a new approach, which combines the coal preparation and geostatistical techniques for predicting the monthly average clean coal qualities from a mine. Generally, the major quality concerns for both the customer and the producer are ash, sulfur, moisture, and heating value of the shipped coal. The penalty/premium criteria in a customer contract are normally based on these four parameters. Since the coal gains/looses moisture in transit due to weather, it is difficult to estimate the moisture content from core/channel data. The heating value of coal is linearly related to the ash content and therefore can be directly calculated from the ash estimate. This paper only deals with the ash and the sulfur content of the clean coal.

Approach

The underpinnings of this new approach for estimating clean coal parameters from core/channel data is to first process the core/channel data through the plant circuits. And then use any suitable estimating technique, kriging in this case, to estimate the monthly clean coal qualities. The core/channel data were processed in a computer program that attempts to mimic the cleaning process of the plant for a specific customer. The processed data were then used in a kriging program to estimate the average monthly clean coal qualities for that customer.

Core/Channel Data Processing

The core and channel data available at this mine were analyzed at size fractions similar to the plant sizes with washabilities at various specific gravities. The washability data give a theoretical picture of the coal separation by gravity. To induce the reality in the theoretical data, they are processed to imitate the plant operating state.

In a preparation plant, there are multiple circuits to handle different size coal. The plant in this study has four main circuits as shown in Figure 1. Plus 9.525mm (3/8") coal goes to the dense medium vessel, 9.525mm X 0.595mm (3/8" X 28M) coal is separated at the dense medium cyclone, 0.595mm X 0.149mm (28M X 100M) fine coal is sent to the hydrocyclone, and the -0.149mm (-100M) ultra fine coal is cleaned by froth flotation. The clean coal from each of the circuits is put on a clean coal belt and conveyed to the load out facility.

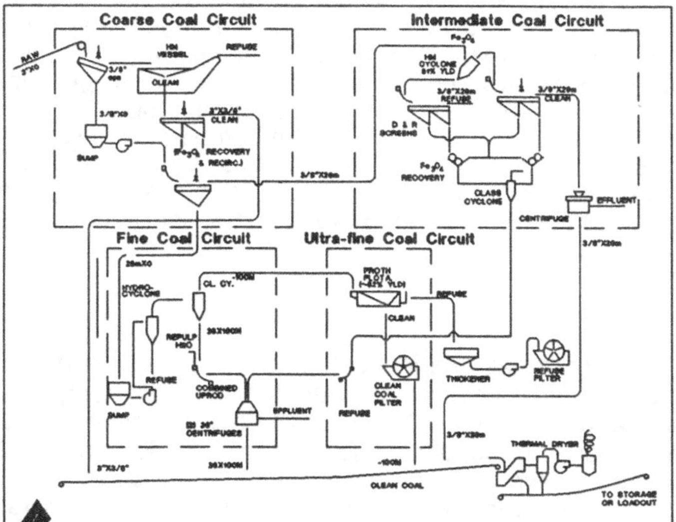

Figure 1: Coal Cleaning Plant Flowsheet

To get the plant operating parameters, a performance study was conducted on each plant circuit. The distribution curves, which define the performance of a cleaning equipment, of all the circuits except the flotation, were generated from the plant study. For the flotation circuit, various samples were collected to estimate the average values of recovery, ash, and sulfur contents. The theoretical washability data of a core sample when adjusted by the distribution curve and the plant feed size consist give a likely plant operating scenario for that core sample. The calculations are illustrated on a spreadsheet in Figure 2. The calculation logic of this spreadsheet was coded in a fortran program (WASHSIM) to automate the processing of all the core and channel samples.

The operating points on Figure 2 are 1.42, 1.50, and 1.90, which corresponds to the separating gravities at the heavy medium vessel, the heavy medium cyclone, and the hydrocyclone circuits respectively. When a customer's coal specification changes, one or all the above operating points are changed to produce that coal. The entire core/channel data set is run on the WASHSIM program to generate a clean coal data set. This data set becomes the main input to the kriging program. The WASHSIM program is designed in such a way that the plant operating points (i.e., separating gravities) can be changed as needed to produce a particular customer's coal. When the operating point of a cleaning equipment changes, the equipment's distribution curve also changes; the new curve values can be calculated from the old curve by employing the generalized distribution curve principle given by Gottfried and Jacobsen, 1977.

Geostatistics

Support

The top size of core and channel sample and the plant feed is 3 inches. The size consist of the feed, however, is not the same as the size consist of the core samples and all cores do not have the same size consists. The channel samples are generally pre-processed in a tumbler to produce a similar size consist as the plant feed. To establish the same support in size distribution, the theoretical washabilities of core samples were adjusted across the different sizes in such a way that the total raw ash (head ash) and raw sulfur (head sulfur) contents remain same before and after the adjustment. The wash plant simulation mentioned in the previous section was run on the adjusted washability data.

Variogram

The location of the drill/channel data and the mine workings are shown in Figure 3. The basic statistics of these data for raw, 1.50 float, and at the simulated plant float are shown on Table 1. The histograms of these three data sets show a nearly normal distribution in Figure 4. Experimental and the fitted variograms are shown in Figure 5. Verification of the model parameters was done by point kriging method. Choice of the model parameters was based on the least amount of mean error, smallest difference between weighted square error and average krig variance, and best visual fit.

Hole Id = AAA

Out-of-seam dilution = 0

		<----------Direct--------->				Dist.	<--------Fraction -------->		
Sink	Float	%Wt	Ashfd	Sulfd	%Wtfd	Curve	%Wtfr	Ashfr	Sulfr
A	B	C	D	E	F=(C/(OSD+100))	G	H=F*G	I=D*H	J=E*H
	1.30	0.00	0.00	0.00	0.00	100.00	0.00	0.00	0.00
1.30	1.35	20.50	7.40	1.25	20.50	99.80	20.46	1.51	0.26
1.35	1.40	22.37	15.76	2.45	22.37	84.08	18.81	2.96	0.46
1.40	1.45	0.00	0.00	0.00	0.00	20.00	0.00	0.00	0.00
1.45	1.50	2.81	20.69	3.16	2.81	3.67	0.10	0.02	0.00
1.50	1.55	2.34	23.40	4.48	2.34	0.82	0.02	0.00	0.00
1.55	1.60	1.44	27.00	4.62	1.44	0.00	0.00	0.00	0.00
1.60	1.70	0.00	0.00	0.00	0.00	0.00	0.00	0.00	0.00
1.70	1.80	0.00	0.00	0.00	0.00	0.00	0.00	0.00	0.00
1.80		50.54	66.83	4.13	50.54	0.00	0.00	0.00	0.00
		100.00			100.00		39.39	4.50	0.72
					Heavy Medium Vessel---->		39.39	11.43	1.83
	1.30	0.00	0.00	0.00	0.00	100.00	0.00	0.00	0.00
1.30	1.35	57.29	5.94	1.22	57.29	99.69	57.11	3.39	0.70
1.35	1.40	12.68	12.21	1.97	12.68	97.55	12.37	1.51	0.24
1.40	1.45	5.78	16.04	2.35	5.78	91.63	5.30	0.85	0.12
1.45	1.50	1.85	19.69	3.06	1.85	67.96	1.26	0.25	0.04
1.50	1.55	1.21	22.67	3.88	1.21	34.29	0.41	0.09	0.02
1.55	1.60	0.83	25.11	4.78	0.83	19.82	0.16	0.04	0.01
1.60	1.70	0.00	0.00	0.00	0.00	1.12	0.00	0.00	0.00
1.70	1.80	0.00	0.00	0.00	0.00	0.00	0.00	0.00	0.00
1.80		20.36	74.22	4.38	20.36	0.00	0.00	0.00	0.00
		100.00			100.00		76.61	6.14	1.13
					Heavy Medium Cyclone--->		76.61	8.01	1.47
	1.30	0.00	0.00	0.00	0.00	96.80	0.00	0.00	0.00
1.30	1.35	62.59	5.90	1.03	62.59	96.20	60.21	3.55	0.62
1.35	1.40	10.82	12.21	1.56	10.82	94.20	10.19	1.24	0.16
1.40	1.45	6.86	14.66	1.99	6.86	91.40	6.27	0.92	0.12
1.45	1.50	2.54	19.56	2.37	2.54	86.90	2.21	0.43	0.05
1.50	1.55	2.03	20.43	2.84	2.03	82.20	1.67	0.34	0.05
1.55	1.60	1.42	20.70	3.06	1.42	77.50	1.10	0.23	0.03
1.60	1.70	3.06	25.52	4.18	3.06	69.50	2.13	0.54	0.09
1.70	1.80	0.00	0.00	0.00	0.00	56.50	0.00	0.00	0.00
1.80		10.68	66.58	11.70	10.68	15.22	1.63	1.08	0.19
		100.00			100.00		85.40	8.34	1.32
					Hydro Cyclone------->		85.40	9.77	1.54

Plant Feed		<-----------------Circuits			---------------------->		<------- Adjusted	-------->	
Size	% Feed	Yield	Ash	Sulfur	Type	Sep. Grav	Yield	Ash	Sulfur
+3/8"	27.50	39.39	11.43	1.83	HM-Vessel	1.42	10.83	123.86	19.82
3/8" x 28M	50.50	76.61	8.01	1.47	HM-Cyclone	1.50	38.69	309.83	56.93
28M x 100M	13.00	85.40	9.77	1.54	Hydro Cyclone	1.80	11.10	108.44	17.11
100M x 0	9.00	70.00	8.26	1.83	Flotation		6.30	52.04	11.53
							66.92	594.17	105.39
			Final Simulated Values for Hole AAA:----->				66.92	8.88	1.57

Note: Distibution Curve and Flotation values were obtained from plant performance study.

Figure 2: A Spreadsheet Showing Plant Operating Adjustments of Washability Data

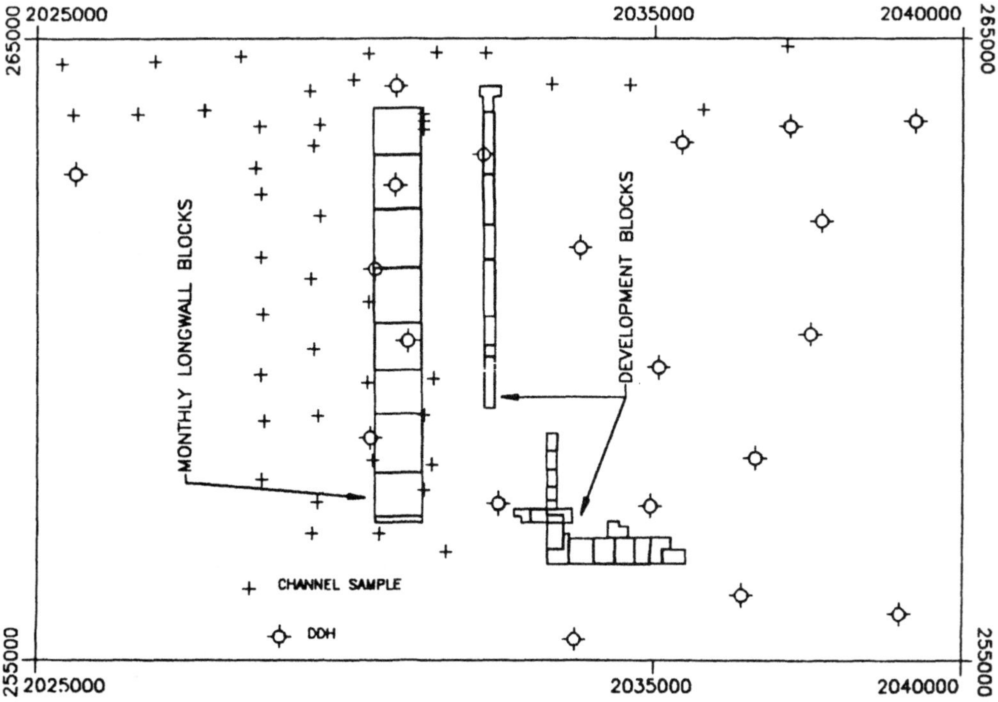

Figure 3: DDH and Channel Samples Around the Monthly Working Blocks

Table 1: Simple Statistics and Varigram Parameters of Raw, 1.50 Float, and Simulated Data

Coal Quality	N	Mean	Variance	Variogram	Nugget CO	Part. Sill C1	Range	Range
		%	%*%		%*%	%*%	Ft	Meter
Raw Ash	203	26.10	45.690	Spherical	21.00	34.00	16000	4877
1.50 Float Ash	203	10.33	1.066	Spherical	0.50	0.70	8000	2438
Simulated Ash	203	9.78	0.859	Spherical	0.45	0.50	7000	2134
Raw Sulfur	203	2.75	0.697	Spherical	0.30	0.50	8000	2438
1.50 Float Sulfur	203	1.72	0.187	Spherical	0.05	0.16	7250	2210
Simulated Sulfur	203	1.63	0.124	Spherical	0.05	0.09	9500	2896

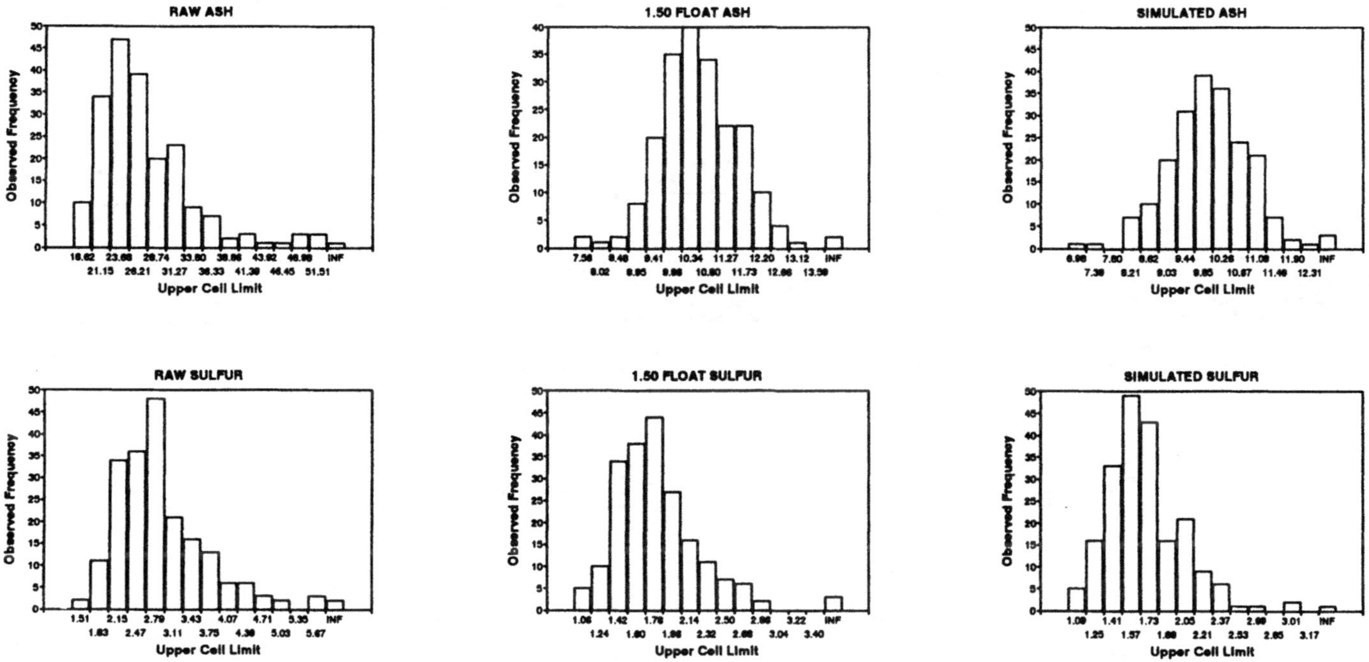

Figure 4: Histograms of Raw, 1.50 Float, and Simulated Data

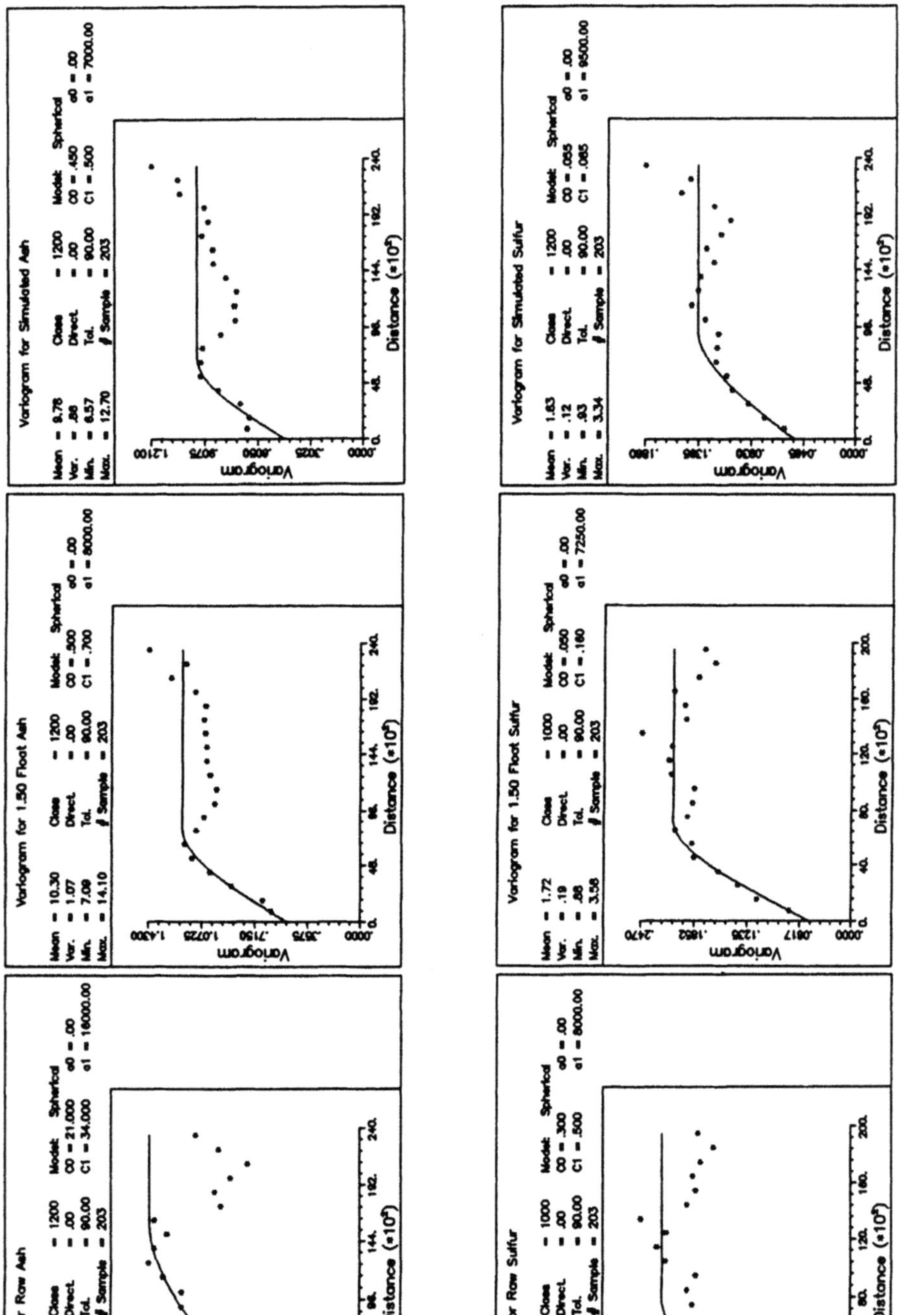

Figure 5: Experimental Variograms of Raw, 1.50 Float, and Simulated Data

It is interesting to note from Table 1 that the mean and variance of ash and sulfur have dropped drastically from the raw data to the clean data. The large reduction in the mean and variance from raw to clean is mainly due to the presence of highly variable in-seam partings, which drop out in the cleaning process. The simulated data show lower values than the 1.50 float data because the coarse material (about 27% of the total mass, see Figure 2) in the simulated case was cleaned at a lower specific gravity (1.42) than 1.50.

Although the mean and variance of ash and sulfur have changed after the coal samples were cleaned, the pattern of the experimental variograms in Figure 5 show a remarkable similarity among raw, 1.50 float, and simulated data. This suggests that the spatial relationship that existed in the raw data still remains even after the coal is cleaned either at 1.50 gravity or at the plant operating gravities with multiple circuits.

Kriging

Ordinary block kriging was performed on the actual 1991 monthly advances, (Figure 3), using the University of Arizona's kriging program STPM. The search radius was 1829m (6000ft) and the neighborhood sample size was 10 in all direction. The goal was to estimate the average and the range of ash and sulfur contents of the coal produced over a short time period (monthly) for each customer. The accuracy of any forecast is constrained by the accuracy of the variogram model and by the ability to predict future operational factors such as exact location of the production units, the relative productivity of different units, and the efficiency of the preparation plant.

The next step in the forecast process was to combine the monthly mean and variance estimated for each working area into a single mean and variance for the entire mine. Although the single mean value can be calculated by weight-averaging with tonnage, the average variance can not be calculated in the similar fashion due to sample dependency (spatial relation). The combined kriging variance technique suggested by Kim and Baafi (1983) can be used to calculate monthly variance. The premise of this technique is that the same data should be used in kriging all working areas no matter how far apart the working areas may be. This approach will face limitation if two working areas that contribute to the monthly tonnage fall into two different populations. An alternative method is to use Monte-Carlo simulation and to assume a normal distribution of sulfur values within each working area.

The simulated values (ash or sulfur) for the monthly production were generated as follows:

$$x_j = \Sigma_i \, w_i * [\, RN(0,1) * sk_i + zk_i \,]$$

where,

x_j	=	simulated monthly value on jth experiment
w_i	=	tonnage ratio i (area i tonnage / total monthly tonnage)
$RN(0,1)$	=	random number from a normal distribution with zero mean and one standard deviation

sk_i	=	krig standard deviation from area i
zk_i	=	krig estimate from area i
i	=	1 to number of working areas
j	=	1 to 500 experiments

The process was repeated 500 times to generate a population of possible mean values, from which the mean and the variance of clean coal ash and sulfur contents were calculated.

Results

The estimates generated using the process described above were compared with the actual customer shipments during the first nine months of 1991. During the fourth quarter of 1991, the mine had a longwall move and had faced some roof control and other production problems. The comparisons are listed on Table 2 and shown in Figure 6. Customer B did not receive any coal from the mine during May, June, July, and September of 1991. The customer B contract specification allows higher ash. Adjustments were made in the plant by raising the circulating gravities while producing the coal for customer B.

Table 2: Comparison Between Actual Shipments and Estimated Values, Year: 1991

Month	Diff: Actual Shipment - Estimated			
	Simulated Data		1.50 Float Data	
	Dry_Ash	Dry_Sul	Dry_Ash	Dry_Sul
Customer A				
January	0.40	-0.21	-0.10	-0.30
February	0.42	-0.04	-0.05	-0.13
March	0.66	0.01	0.33	-0.02
April	0.53	-0.03	0.14	-0.06
May	0.50	0.00	0.10	-0.05
June	0.66	0.12	0.23	0.05
July	0.98	0.12	0.50	0.05
August	0.62	0.08	0.19	0.01
September	0.54	-0.04	0.15	-0.09
Average:-->	0.59	0.00	0.17	-0.06
Customer B				
January	0.17	-0.13	0.21	-0.17
February	0.08	-0.01	0.13	-0.07
March	0.34	0.02	0.87	0.07
April	0.11	0.02	0.46	0.01
August	0.81	0.02	1.83	0.07
Average:-->	0.25	-0.01	0.59	-0.01

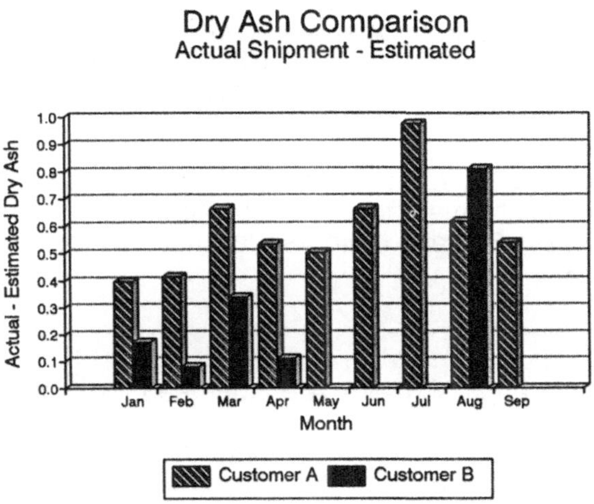

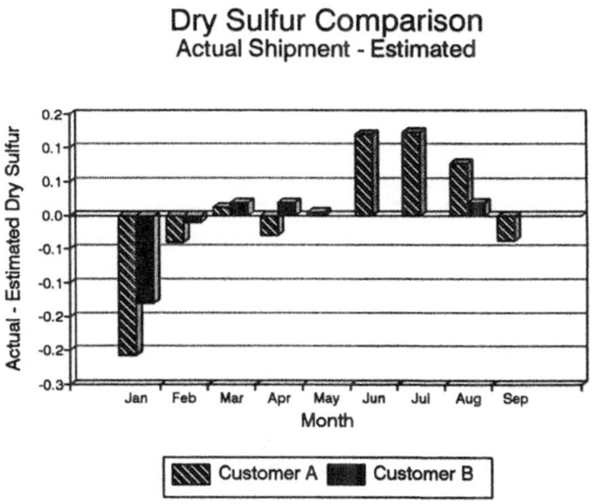

Figure 6: Comparison Between Actual Shipment and Estimated Values Using Simulated Data

The average differences between the actual shipment to Customer A and the estimated values (from the simulated data) were 0.59 for ash and 0.00 for sulfur. Similarly, for customer B, the differences were 0.25 and -0.01 respectively. All the differences were well within the 95% confidence limits for the estimates.

It appears from Table 2 that the simulated data do a better job of estimating the actual shipment than the 1.50 float data except the ash content of customer A. The 1.50 data, of course, do not distinguish between customer A and customer B.

Conclusion

This modified approach eliminates the need for statistical models (variograms) of coal qualities at differing washing (separation) densities as well as at various size fractions for each washing density. In a coal preparation plant, there are multiple cleaning circuits to handle different size coal; the coal separation takes place at different densities for each circuit and may even vary within an individual circuit for differing size fractions. If the washing density of a particular circuit changes, one can quickly rerun the WASHSIM program to generate a new set of clean coal parameters for each core hole and repeat the geostatistical estimation process.

If market requires a specific coal at a premium price then one can quickly analyze the potential of producing that coal by changing the operating procedures within the practical limits of the plant, lets say discarding the ultra fine coal, when the coal is insitu. Therefore, this approach offers a very valuable marketing tool for the sales staff. It is conceivable that a longwall panel can be sliced into different customer's specifications before the panel is mined to maximize the recovery of both the mine and the plant.

Acknowledgements

The general manager, plant manager, geologists, engineering vice-president, and the director of mergers and acquisition are gratefully acknowledged for their support in this study. Special thanks to the preparation engineer for his help in explaining the plant operation and for conducting the plant performance study.

References

Armstrong, M., Gillies, A. D. S., Wu, Y. Y., Just, G. D., and Lyman, G.D. (1989) "Review of the applications of geostatistics in the coal Industry", Geostatistics, Vol. 2, Kluwer Academic Publishers, Dordrecht, The Netherlands, 995-1005.

Gottfried, B. S. and Jacobsen, P. S. (1977) "Generalized Distribution Curve for Characterizing the Performance of Coal-Cleaning Equipment", United States Department of the Interior, Bureau of Mines Report of Investigations 8238.

Kim, Y. C. and Baafi, E. Y. (1983) "Combining Local Kriging Variances for Short Term Mine Planning", in Geostatistics For Natural Resources Characterization, 2nd NATO Advanced Study Institute, Edited by G. Verly, M. David, A. G. Journel, and A. Marechal; D. Reidel Publishing Company, 185-192.

ESTIMATING THE RECOVERABLE RESERVES OF GOLD DEPOSITS: COMPARISON BETWEEN DISJUNCTIVE KRIGING AND INDICATOR KRIGING

Jean-Paul CHILÈS [1] and Hong-Tao LIAO [2]

[1] BRGM, BP 6009, 45060 Orléans Cedex 2, France
[2] PETROSYSTEMS, 1 rue Léon Migaux, 91341 Massy Cedex, France

ABSTRACT. The practice and performance of disjunctive kriging (DK) and indicator kriging (IK) for estimating recoverable reserves are compared for two different gold deposits. Both estimates lead to similar results when there is no support change, but in the case of a support change, the choice of a consistent model is important. IK may give better results than a misspecified DK model, but Laguerre isofactorial models present more flexibility than the usual bigaussian model and improve the estimates.

INTRODUCTION

Many gold deposits are exploited from an open pit, making it possible to select ore at the level of small exploitation units. This ore is then sent to the mill or is disposed of as waste, according to whether its expected grade is higher or lower than a set cutoff value. The recoverable reserves depend on this cutoff grade. Its choice is based on economical considerations, and on the prediction of the proportions of ore and metal that will be selected. Any prediction of the recoverable reserves must take into account :

- the support effect : the selection is made at the level of 3D blocks but is based on data from drillholes or core samples ;
- the information effect : the final selection will be made according to an estimation of the grade based on the future preexploitation boreholes ;
- the local data : a feasibility study can be based on a global estimate of the recoverable reserves, whereas mine planning requires an estimation of the local recoverable reserves at the level of a panel or group of panels.

Two methods are commonly used for estimating recoverable reserves :

- disjunctive kriging (DK) (Matheron, 1973, 1974, 1976a, 1976b), which proposes a solution that is consistent from a theoretical viewpoint, but whose application has seemed complex for a long time ;
- indicator kriging (IK) (Journel, 1982) or indicator cokriging (ICK) (Journel, 1983), which are two facets of an attractive method because of its apparent simplicity ; however, this imposes the introduction of simplifications and approximations as soon as there is a support change.

In both cases, the grade is considered as (a realization of) a stationary random function (SRF) Z(x) where x is a point in $\mathbf{R}^3$. From a theoretical viewpoint, ICK and DK are equivalent when there is no support change : the former is based on the modelling of the direct and cross-covariances of all possible indicator functions, whereas the latter is based on the modelling of all possible bivariate distributions, and we have :

$$F_h(z,z') = P\{Z(x) < z \ \& \ Z(x+h) < z'\}$$
$$= E[1_{Z(x)<z} \, 1_{Z(x+h)<z'}] = C_{zz'}(h)$$

where F_h is the bivariate distribution of the pair (Z(x), Z(x+h)) and $C_{zz'}$ is the non-centred cross-covariance of the indicator functions corresponding to the cutoff values z and z' (here the indicator equals 1 when Z(x) is lower than the cutoff value and is equal to 0 otherwise). In practice, however, large differences can be observed because of the differences in the approach and because of the approximations that are made for IK or ICK with support change. This article presents the application of the two approaches to gold deposits from a practical viewpoint. DK will not be limited to the bigaussian model, which for a long time was the only one used (cf. Maréchal, 1984), though it is only one of several models. On the other hand, variants of IK and ICK, such as indicator-rank cokriging (also known as probability kriging or PK ; Sullivan, 1984), and other methods such as the multigaussian approach (Verly, 1983) will not be considered. All details concerning the methods used here and their application can be found in Liao (1990).

The ideal situation is that the results from preexploitation boreholes and exploited blocks are simultaneously available. However, the latter are seldom available as during ore processing the ore from several blocks is mixed. For this reason we can only mimic the real-life case by considering "blocks" defined from a combination of several sample units (ten 1-m-long cores ; 3x3 boreholes) and estimating them from a subset of the data (the four central cores ; one borehole among the 3x3 of the "block", plus neighbouring boreholes). We will only consider the support effect and ignore the information effect for the time being. This implies the assumption that the future selection is made on the knowledge of the true block grade. The type of estimate considered in this paper can be used for example for defining the next zone to be mined, but not for immediately deciding which blocks will be sent to the mill and which ones are the waste. The aim is thus to arrive for each block at an estimate of the recovery functions (of the cutoff value z), which are :

- $T(z;v) = 1_{Z(v)>z}$ for ore and
- $Q(z;v) = Z(v) \, 1_{Z(v)>z}$ for metal.

In both definitions, Z(v) denotes the average value of Z(x) within block v. As a next step, we will examine the local estimation of these recovery functions by DK and by IK or ICK, with or without support change.

THE DATA

The first data-set concerns a vein of the Salsigne orebody in France, exploited by open pit. Preexploitation boreholes were drilled on two 10- m-high benches on a 5m x 5m grid, and the Au grade was analysed for each of the ten 1-m-long cores for delimiting exploitation panels. The histogram of the grades presents 30 % of zero values and a long tail (0 up to 400 g/t ; average 10 g/t ; standard deviation 22 g/t). The variogram of the core grades seems to be isotropic ; it shows a relative nugget effect equal to 34 % and two nested spherical structures with ranges of only 3 m and 20 m (the variogram already reaches 90 % of its sill at a distance of 5 m).

The second data-set concerns a mining test of the Jiayu deposit. The ore lies within a 10 to 40m thick vein; 2-m-long vertical blastholes were drilled every 3.6m x 3.6m. Results from 865 blastholes belonging to the same 2-m-high bench are available. The histogram presents no zero effect: the Au grade ranges from 0.2 to 135 g/t (average 6.7 g/t; standard deviation 11.4 g/t). The variogram shows a moderate nugget effect (17 % of the sill), and ranges of 10 m and 50 m, thus displaying a better continuity than the Salsigne data-set.

ISOFACTORIAL MODELS

The calculation of recoverable reserves is based (explicitely for DK, implicitely for IK and ICK) on the modelling of the bivariate distributions concerning any pair of points or blocks, or of a point and a block. It is not obvious to derive bivariate models that are consistent for scattered points and/or blocks. Isofactorial models insure this internal consistency.

Isofactorial point models

If Y(x) is a SRF, G(dy) is its univariate distribution, and $F_h(dy,dy')$ is the bivariate distribution of the pair (Y(x), Y(x+h)), then the bivariate point distributions follow an isofactorial model if they have the form:

$$F_h(dy, dy') = \sum_{n=0}^{\infty} T_n(h)\, \chi_n(y)\, \chi_n(y')\, G(dy)\, G(dy') \qquad (1)$$

where $X_n(y)$ are normalized polynomials that are orthogonal relative to the univariate distribution G (the condition that X_n is a polynomial is not necessary, but we consider only this case):

$$\int X_m(y)\, X_n(y)\, G(dy) = \delta_{mn}$$

Anamorphosis

A basis of normalized and orthogonal polynomials can be associated with any univariate distribution (provided that it admits finite moments of any order). On the contrary, the set of all possible bivariate distributions seldom admits an isofactorial representation. We will only discuss two continuous univariate distributions for which isofactorial models exist: the normal and gamma distributions. If the studied variable is not normal or gamma, a transformation is performed in order to obtain this type of distribution. More precisely, the studied variable is considered to have the form:

$$Z(x) = \Phi(Y(x)) = \sum \phi_n\, \chi_n(Y(x)) \qquad (2)$$

where $\Phi(y)$ is the anamorphosis function, and ϕ_n are the coefficients of its polynomial expansion with χ_n. These polynomials are, up to a norming factor, the Hermite polynomials in the Gaussian case, and the Laguerre polynomials in the gamma case. The corresponding isofactorial models (1) are called Hermitian and Laguerre models.

Choice of the anamorphosis variable

Many past studies have used Gaussian anamorphosis (or transform). However, this transformation is not suited for distributions presenting a large zero effect, i.e. a significant proportion of grades equal to zero or a very low value, for these zero values can be represented by very different values of the Gaussian variable (for example from $-\infty$ up to 0 if there are 50 % of zero values), which creates artifacts. From the very beginning, Matheron (1973) proposed other models than the Hermitian one. In particular, the gamma distributions, thanks to their shape parameter α, enable the representation of histograms close to the Gaussian case (high value for α) as well as of histograms with a zero effect or a long tail (low value for α). The use of the corresponding isofactorial models was developed by Lantuéjoul (1985) and more widely by Hu (1988).

Since gold grades have highly dissymmetrical distributions, gamma transforms were used here. The study was also made with Gaussian transforms for comparison. In the gamma case, the shape parameter α has to be chosen. Four criteria were proposed by Hu (1988) : if the second parameter of the gamma distribution, the scale parameter, is chosen so as to reproduce the mean (this choice is arbitrary and of no consequence), the four criteria respectively preserve the variance, the median, the selectivity, or the normality index. A fifth criterion described by Demange et al. (1987) was used here, which preserves the percentage of waste below a given cutoff grade. Table 1 gives a comparison of the two anamorphosis types, showing that for both orebodies the first factor expresses a bigger part of the total variance in the gamma case than in the Gaussian one. This means that the actual gold distribution is closer to the chosen gamma distribution than to a normal distribution (the maximization of this part of the total variance could be a sixth criterion). The extra advantage of the gamma model is that it requires fewer factors than the Gaussian one, for a given accuracy of the representation of the anamorphosis function.

Orebody / Data	Anamorphosis	α	Factors	ϕ_1^2/σ^2
Salsigne / 1-m cores	Gauss Gamma	 0.34	27 16	0.42 0.89
Jiayu / 2-m holes	Gauss Gamma	 0.70	22 22	0.30 0.54

Table 1. Characteristics of the anamorphoses used (anamorphosis type, shape parameter α of the gamma distribution, number of factors, proportion of the total variance explained by the first factor).

Variography of the anamorphosis variable

Disjunctive kriging is based on the variography of Y(x). The data $Z(x_i)$ at the points x_i are transformed back to $Y(x_i)$ by inverse anamorphosis, either by inversion of relation (2), or from a ranking of the data. Figure 1a presents the variogram of the gamma variable Y(x) for the Salsigne orebody.

Fig. 1. Variography of Au grade for Salsigne (1m-core support) :
a) variogram of Y(x) ;
b) graph (γ_1, γ_2) of Y(x) ;
c) variogram of Z(x).

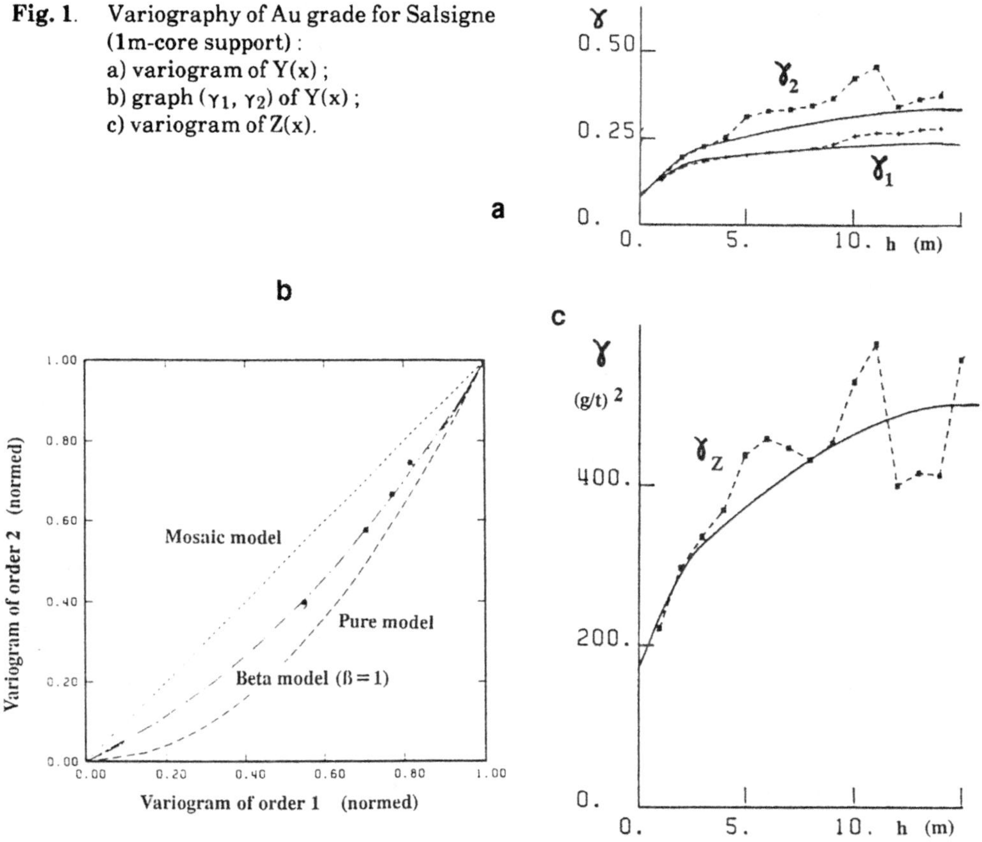

Types of bivariate distributions

The second problem is the characterization of the type of bivariate distribution, i.e. of the T_n in expression (1), which are the correlograms of the factors $X_n(Y(x))$. Most applications use a pure model (bigaussian or bigamma) for which $T_n(h) = \rho(h)^n$, where $\rho(h)$ is the correlogram of Y(x). This model corresponds to diffusion-type random functions. The other extreme is the mosaic model, with $T_n(h) = \rho(h)$, which corresponds to mosaic-type random functions. An intermediate possibility is the beta model, which provides a choice ranging from the mosaic model to the pure model according to the value of its parameter β, with :

$$T_n(h) = [\Gamma(\beta) / \Gamma(\beta + n)]\,[\Gamma(\beta\,\rho(h) + n) / \Gamma(\beta\,\rho(h))]$$

Choice of the bivariate distribution - the variogram of order 1

The choice of the form of $T_n(h)$ reacts on the variogram of order 1 of Y(x). Denoting the variogram of order 1 of Y(x) by $\gamma_1(h)$ and the ordinary variogram (of order 2) by $\gamma_2(h)$, the graph formed by the points ($\gamma_1(h)$, $\gamma_2(h)$) in a coordinate system (γ_1, γ_2) varies from a line (mosaic model) to a parabola (pure model). Figure 1b shows the example of choosing a beta model for the Salsigne data.

Consistency of the model with the variogram of Z

The variogram of Z can be expressed from the above parameters by :

$$\gamma_Z(h) = \Sigma\, \phi_n^2 [1 - T_n(h)]$$

This provides a way of checking the validity of the model fit. Figure 1c shows the variogram fit for Salsigne.

VARIOGRAPHY FOR IK

The IK of an indicator function only requires its variogram to be known. Similarly, the metal recovered above a given cutoff value z can be estimated by studying the service variable $Z(x)\, 1_{Z(x)>z}$. If we are interested in several cutoff values, as is the case when estimating recovery functions, many variograms have to be modelled. As will be shown below, this is also the case if there is a support change, even if only one cutoff value is studied at the level of the blocks. The structural analysis is even more complex in the case of ICK, as it requires knowledge of the direct and cross variograms at the various cutoff values. As for any cokriging, ICK is efficient insofar as it is based on the careful modelling of covariances or variograms. The problem is, that any covariance cannot be the covariance of an indicator function, and that the conditions required for this cannot be expressed in a simple manner (Matheron, 1987, 1989a). The complexity of the modelling (45 covariances for only nine cutoff values) means that ICK is practically never used. IK is usually performed with a single variogram model for all cutoff values, or sometimes with a model with one parameter that varies according to the cutoff grade, for example the relative nugget effect.

DK AND IK WITHOUT SUPPORT CHANGE

Once the parameters are known, DK, IK and ICK can be easily put into practice. This has been done for the Salsigne orebody, where the indicator of the average grade of a 10-m-long borehole was estimated from the average grades of its eight adjacent boreholes. Several methods were compared, i.e. Hermitian DK, Laguerre DK, IK, ICK. The conclusions are :

- The very low or very high cutoff values cannot be estimated locally. The reason is that the experimental error variance, obtained by comparing the estimated value with the true one, is hardly lower than the a-priori variance of the indicator function ; when the cutoff is equal to the median, DK and IK bring a gain of about 35 % over the a- priori variance ; this illustrates the well-known destructuration of high grades (Matheron, 1982).
- All the methods lead practically to the same precision : Hermitian DK and Laguerre DK because the grade distribution is more regular for boreholes than for cores, DK and ICK because they are theoretically equivalent, and IK and ICK because all variograms are very similar and reach already 90 % of their sill for the borehole separation.

Moreover, IK has been found to be not very sensitive to a modification in the variogram, and particularly to the adoption of a single variogram for all cutoff values (variogram of the median cutoff, or of Z). But we are mainly interested in the recoverable reserves with support change.

ISOFACTORIAL MODELS FOR A SUPPORT CHANGE

Continuous model and discrete model

The models for a support change are based on point-block models. Two types of model exist (Matheron, 1974, 1976b) that correspond to very different data configurations :

1. The continuous model takes into account the precise location of each datum with respect to the block ; this model must be used for example when the kriging configuration remains the same.
2. The discrete model takes no account of the precise location of each datum ; the studied domain is subdivided into identical blocks v_i, and each datum point is considered as randomly located within the block it belongs to. This model is the most commonly used.

In both cases, the regularized grade $Z_v(x) = 1/v \int_v Z(x+u)\,du$ is considered as transformed from a SRF $Y_v(x)$ by an anamorphosis function Φ_v, which can be expanded with normalized orthogonal polynomials $X^v{}_n(y)$ by :

$$Z_v = \Phi_v(Y_v) = \Sigma\, \phi_n^v X_n^v(Y_v)$$

Y_v is not equal to $1/v \int Y(x+u)\,du$, and the bivariate distribution takes the form :

$$F_v(dy_x, dy_v) = \Sigma\, T_n^V(x,v)\, \chi_n(y_x)\, \chi_n^V(y_v)\, G(dy_x)\, G_v(dy_v)$$

Choice of the anamorphosis variable of the blocks

The $\phi^v{}_n$ coefficients are derived from the ϕ_n values of the point anamorphosis and from the variogram of Y(x), whereby the user has to specify the distribution of Y_v. When a Gaussian transform is used for point grades, the same is done for the blocks. In the gamma case, a gamma transform is also used for the block, but with a parameter $\alpha' > \alpha$ closer to a normal distribution, because the distribution of Z_v is closer to normal than that of Z(x). Additional constraints limit the choice that is left to the user (see, for example, Hu, 1988 ; Liao, 1990).

Both models are, by construction, consistent with all variances and covariances between points and/or blocks. The discrete model always leads to a positive bivariate density, which is not guaranteed by the continuous model in case of an important support change. This may explain why the continuous model was not used before this study, even though it is the model to be used in case of a fixed configuration. No major problems were encountered during the present applications.

STRUCTURAL MODEL FOR IK WITH SUPPORT CHANGE

IK enables estimation of the indicator $1_{Z(x)>z}$ or its average value $1/v \int_v 1_{Z(x)>z}\,dx$. But the objective here is $1_{Z_v>z}$, which is very different. IK cannot directly solve this problem. An approximate solution consists of estimating the average point distribution within v, and then applying a support change algorithm to the result. In the present study, a local affine correction was assumed, which is based on the assumption that the reduced (i. e. centred and normed) point grade $(Z(x)-m)/\sigma$ and the reduced block grade $(Z_v - m)/\sigma_v$ have the same distribution. The histogram of Z_v can then be deduced from that of Z (x), by narrowing the latter around the mean by a factor σ_v/σ.

For a global estimation the global histogram of Z (x) is used, m is the mean of Z (x) and Z_v, σ^2 is the variance of Z (x), and σ_v^2 is the variance of Z_v; σ_v^2 can be computed from the variogram of Z (x). In case of a local estimation, the affine correction is made at the level of estimates of the local distribution, i.e. conditioned to the data included in the block and its neighbourhood. Journel (1983) suggested taking for m* and σ* the mean and standard deviation of the IK estimation F* (z) of the local point distribution of Z (x) within v, i.e.:

$$m^* = \int z\, dF^*(z) \qquad \sigma^{*2} = \int (z-m^*)^2\, dF^*(z)$$

This requires the calculation of F* (z) for all cutoff values. However, if all the IK are performed with the same variogram, m* coincides with the kriging Z_v^* of Z_v. Even if each IK is performed with its appropriate variogram it is preferable to take Z_v^* for m* rather than to compute m* by discretizing the above integral. Concerning σ* and σ_v^*, they intervene only as their ratio σ^*/σ_v^*, which can be simply taken to equal the a priori ratio σ/σ_v, as suggested by Journel (1983). This amound to making the assumption of a proportional effect, which is commonly found to be the case for gold deposits. Under this assumption,$(Z_v - Z_v^*)/\sigma_v$ and $Z(x) - Z_v^*)/\sigma$ have the same local distribution: the probability that Z_v exceeds the grade z is equal to the average probability that, within v, Z (x) exceeds the grade z' defined by:

$$z' = Z_v^* + \sigma/\sigma_v\,(z - Z_v^*) \qquad (3)$$

Note that, even if only one block cutoff z is being studied, the corresponding point cutoff z' varies from one block to another according to Z_v^*.

DK AND IK WITH SUPPORT CHANGE

The practice of DK for estimating the recovered ore and metal quantities at the level of blocks presents no particular problems, but nevertheless requires software with the corresponding algorithms (Liao, 1990).

Concerning IK, in the framework of the local affine correction described above , the recovery fonctions can be expressed by:

$$T_v^*(z) = [1/v \int_v 1_{Z(x) > z'}\, dx]^*$$

$$Q_v^*(z) = Z_v^* \,(1 - \sigma_v/\sigma\; T_v^*(z) + \sigma_v/\sigma\; [1/v \int_v 1_{Z(x) > z'}\, dx]^*$$

They are obtained by kriging the point indicator and the point service variable $Z(x)\, 1_{Z(x) > z'}$, with the cutoff z' related to z by (3). The practice is simple if the same variogram is used for all estimates, which was done here, since preliminary tests showed that the results are not very sensitive to a modification of the variogram, because of the high variability at the scale of borehole spacing. The variogram of Z was used, but it might be advisable to proceed differently for a better structured variable. For example, the kriging of the indicator functions could be performed with the variogram of order 1 of Z, which is equal to the integral over z of the indicator variogram at all cutoff values (Matheron, 1992).

COMPARISON OF THE RESULTS OBTAINED BY DK AND IK

The results obtained by DK and IK were compared on the two data sets:

- For Salsigne, the recovery functions of each 10-m-long borehole were estimated from the grades of the four 1-m-long cores (Fig. 2a; due to the symmetry of the configuration, four cores leave only two degrees of freedom).
- For Jiayu, the blastholes were grouped by "blocks" of 3x3 blastholes, and each block was estimated from one blasthole of the block and one blasthole in each of the eight adjacent blocks; two calculations were made, the one for a fixed configuration (Fig. 2b; the blastholes considered as "data" are the central blastholes of the blocks), the other for a variable configuration (Fig. 2c; a blasthole is randomly selected as "datum" among the nine holes making up each block).

Fig. 2. Kriging configuration of a " block " by " point " data:
a) borehole and central cores;
b) group of 3 x 3 blastholes and central blasthole of each neighbouring group;
c) group of 3 x 3 blastholes and random blasthole of each neighbouring group.

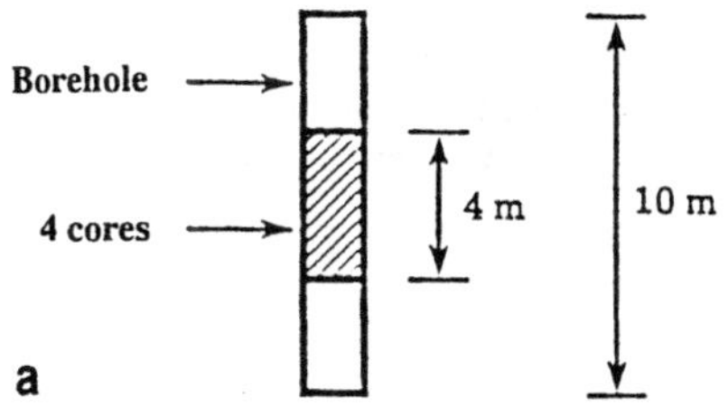

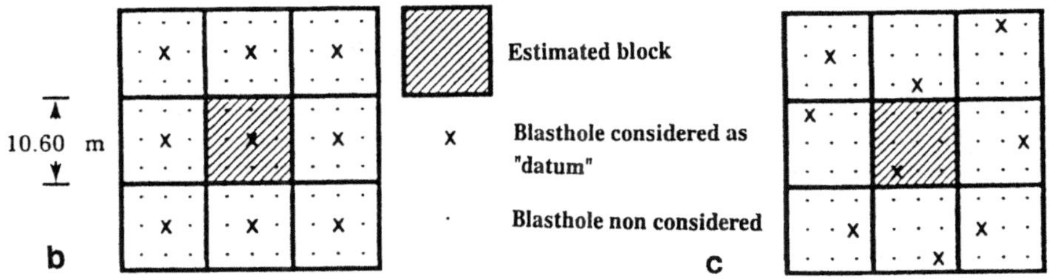

The recovery functions were estimated for each block:

- by DK with a continuous (fixed configuration) or discrete (variable configuration) isofactorial model; in each case, the calculation was made for a gamma transform, and also for a Gaussian one for comparison;
- by IK with the above model.

Switzer and Xiao (1988) have shown that a local estimation of recoverable reserves is usually not accurate. This is particularly the case here, as the studied phenomena present a nugget effect and a large spatial variability at the scale of the kriging neighbourhoods. The most important property required from these estimations is to provide unbiased estimates of the recovered ore and metal. Figure 3 presents the average error on the metal recovery functions, which were computed for the nine cutoff values that regularly discretize the block-grade histogram.

The main conclusions concerning Salsigne (Fig. 3 a), which are also apparent from the graphs of the average quadratic errors (see Liao, 1990), are:

- Laguerre DK (gamma anamorphosis and beta isofactorial model) leads to the best results.
- Bigaussian DK gives a bad estimate of the ore at high cutoff values and of the metal at low cutoff values.
- IK produces results with an overall quality that is similar to that of bigaussian DK ; in some cases they are better, but in other cases they are not as good.
- The beta isofactorial model, even though it leads to a better estimate of the bivariate distributions, is not much better than the bigamma model.

Fig. 3. Average error in the estimation of the recovered metal by DK or IK as a function of the cutoff grade:
a) Salsigne;
b) Jiayu – continuous case;
c) Jiayu – discrete case.

a

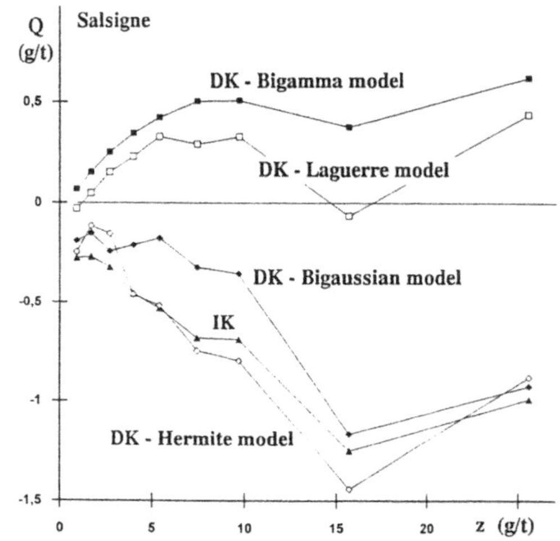

b c

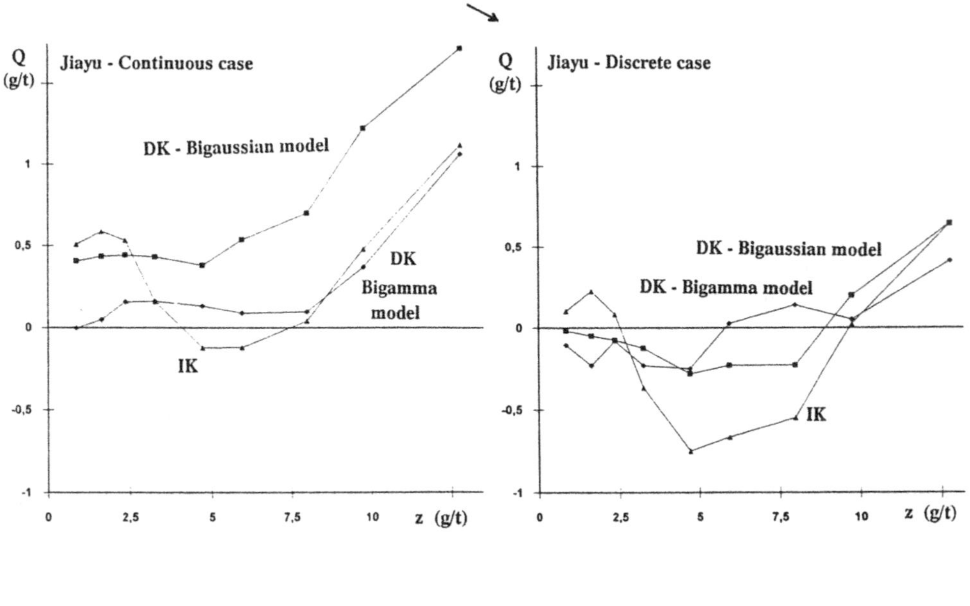

Concerning Jiayu (Fig. 3 b for the continuous case, Fig. 3 c for the discrete case), where pure isofactorial models have been fitted, the conclusions are:

- Bigamma DK gives good results.
- Bigaussian DK is not as good and may lead to large errors (estimate of recovered metal in the continuous case).
- IK commonly leads to major errors.
- The estimate usually is better in the continuous case than in the discrete one, which is normal as the former is obtained for a more regular configuration than the latter.

CONCLUSIONS

DK and IK or even ICK, have been applied to the estimation of local recoverable reserves of two different gold deposits, as one presents a zero effect and the other does not. The "point" and "block" supports were chosen in such a way that the estimate could be experimentally compared with reality. The size of the neighbourhoods (10 m in the first case, 30 m × 30 m in the second one) approximately corresponds to the variogram range. The main conclusion are:

- When there is no support change, the high variability hinders an accurate local estimation, so that IK, ICK and DK lead to results that are similar on average.
- When there is a support change, the practice of IK is not simple, except if a single variogram is used for all cutoff grades. This simplification can be made here without significant affect on the results, because the variogram in all cases presents a high variability.
- DK with a suitable isofactorial model, especially a Laguerre model with a gamma anamorphosis that is more suitable for representing a possible zero effect or a long tail, invariably leads to the best results.
- IK can lead to better results than a "blind" DK using an a-priori model, usually the bigaussian one.

In short, it is vital to base the estimates of local recoverable reserves on a model suited to the data. The gamma/Laguerre model offers more flexibility than the bigaussian one. Its practice is now well defined and presents no major difficulties (Hu, 1988; Liao, 1990). These results have to be confirmed by the study of other deposits. However that may be, the Laguerre model should not be taken as a new panacea, and other useful models exist (see Matheron, 1989 b).

ACKNOWLEDGEMENTS

This is the BRGM scientific contribution No. 92024. The work was founded from an internal BRGM research grant and from a scholarship from the People's Republic of China.

REFERENCES

Demange, C., C. Lajaunie, C. Lantuéjoul,and J. Rivoirard (1987): Global recoverable reserves : testing various changes of support models on uranium data. In *Geostatistical case studies,* G. Matheron and M. Armstrong (Eds.), Reidel, Dordrecht Holland.

Hu, L.Y. (1988) : Mise en oeuvre du modèle gamma pour l'estimation des distributions spatiales. Doctoral thesis, ENS des Mines de Paris, 142 p.

Journel, A.G. (1982) : The indicator approach to estimation of spatial distributions. Proceedings of the 17th APCOM International Symposium, Golden, Colorado, pp. 793-806.

Journel, A.G. (1983) : Non parametric estimation of spatial distributions. *Mathematical Geology*, Vol. 15, No. 3, pp. 445-468.

Lantuéjoul, C. (1985) : Trois modèles de changement de support à partir d'une distribution ponctuelle gamma. *Sciences de la Terre*, Série Informatique Géologique, No. 21, pp. 139-174.

Liao, H.T. (1990) : Estimation des réserves récupérables de gisements d'or - Comparaison entre krigeage disjonctif et krigeage des indicatrices. Doctoral thesis, Document du BRGM No. 202, BRGM, Orléans, France, 198 p.

Maréchal, A. (1984) : Recovery estimation : a review of models and methods. In *Geostatistics for natural resources characterization*, G. Verly et al. (Eds.), Reidel, Dordrecht, Holland, Part 1, pp. 385- 420.

Matheron, G. (1973) : Le krigeage disjonctif. Internal report N-360, Centre de Géostatistique, Fontainebleau.

Matheron, G. (1974) : Les fonctions de transfert des petits panneaux. Internal report N-395, Centre de Géostatistique, Fontainebleau.

Matheron, G. (1976a) : A simple substitute for conditional expectation : the disjunctive kriging. In *Advanced geostatistics in the mining industry*, M. Guarascio et al. (Eds.), Reidel, Dordrecht, Holland, pp. 221 - 236.

Matheron, G. (1976b) : Forecasting block grade distributions : the transfer functions. In *Advanced geostatistics in the mining industry*,M. Guarascio et al. (Eds.), Reidel, Dordrecht, Holland, pp. 237-251.

Matheron, G. (1982) : La destructuration des hautes teneurs et le krigeage des indicatrices. Internal report N-761, Centre de Géostatistique, Fontainebleau.

Matheron, G. (1987) : Suffit-il, pour une covariance, d'être de type positif? *Sciences de la Terre*, Série Informatique Géologique, No. 26, pp. 51-66.

Matheron, G. (1989a) : The internal consistency of models in geostatistics. In *Geostatistics*, M. Armstrong (Ed.), Kluwer Academic Publishers, Dordrecht, The Netherlands, Vol. 1, pp. 21-38.

Matheron, G. (1989b) : Two classes of isofactorial models. In *Geostatistics*, M. Armstrong (Ed.), Kluwer Academic Publishers, Dordrecht, The Netherlands, Vol. 1, pp. 309-322.

Sullivan, J. (1984) : Conditional recovery estimation through probability kriging - Theory and practice. In *Geostatistics for natural resources characterization*, G. Verly et al. (Eds.), Reidel, Dordrecht, Holland, Part 1, pp. 365-384.

Switzer, P., and H. Xiao (1988) : Approximate prediction intervals for spatial interpolation using bivariate information : an experiment. Technical report No. 118, Dept.of Statistics and Dept. of Applied Earth Sciences, Stanford University, California.

Verly, G. (1983) : The multigaussian approach and its applications to the estimation of local reserves. *Mathematical Geology*, Vol. 15, No. 2, pp. 259-286.

A CASE STUDY ON THE APPLICATION OF GEOSTATISTICS AND IMAGE ANALYSIS TO MINE PLANNING

Fernando MUGE, Pedro PINA, Gabriel LUÍS, João LUÍS,
Pedro PATINHA and Carlos GUIMARÃES

CVRM / IST
Av. Rovisco Pais
1096 Lisboa Codex - PORTUGAL

ABSTRACT : Geostatistical and image analysis techniques have been used in the mine planning of an underground wolframite mine (Panasqueira Mine, Portugal). The orebody is composed of a complex series of a near-horizontal quartz veins, associated with Hercynian granitic intrusion into the schist-greywake complex country rocks. Within the vein the ore minerals occur very erratically as individual large crystals which has made conventional chemical assay impractical. A technique based on the measurements of the areas of wolframite crystals exposed in the sidewalls of the mining faces has been used at the mine to estimate the ore reserves. It is based on the assumption that the area of wolframite per unit length can be considered an unbiased estimator of the product (accumulation) of the wolframite tenor and vein width. The purpose of this work is to study the cross-correlation between the spatial distribution of thickness and wolframite grade as a function of the measurement support. To characterize the spatial distribution of the wolframite crystals in the mining faces, textural parameters were obtained through mathematical morphology based image analysis. Results of this study could be applied to the estimation of the wolframite average grades of large panels for long term planning.

INTRODUCTION

The case study is related to mine planning in the wolframite underground mine of Panasqueira. The mine is in central Portugal (Beira Baixa province) on the southern flank of the Serra da Estrela mountains and about 30 km West of Fundão (300 km northeast of Lisbon) (see figure 1). It is a major tungsten producer in Europe and a substantial contributor to the world supply of this important metal. The mine concession covers about 35 km^2 in an irregular strip oriented NW-SE from the Cebola river in the North to the Zezere river in the South.

GEOLOGY

The oreboby is composed of a complex series of near horizontal hydrothermal quartz veins, associated with Hercynian granitic intrusion into the schist-greywacke complex country rocks. As Thadeu (1979) points out, the schist-greywake has been subjected to two phases of deformation. The first one has caused large folds on axes orientated between NNE-ENE. The second was the most important one and produced a near vertical axial-plane schistosity in a NW-SE direction. The ore deposit consists of many sub-horizontal hydrothermal quartz veins. Within the veins are wolframite, cassiterite and chalcopyrite in addition to important quantities of sulphides such as pyrite, sphalerite and pyrrhotite.

The veins cut across the bedding and foliation of the schist-greywacke rocks and are accepted to have been emplaced by the vertical dilation along openings created on the near-horizontal pre-existing joint sets.

As Williams (1985) points out : "... there has been considerable debate about the origin of the flat lying joint sets and the unusual orientation of the vertical dilation that caused the emplacement of the vein system within them ...".

Nevertheless it is commonly accepted that the vein material was formed by open space filling rather than by any replacement process (Kelly and Rye, 1979). Thadeu (1979) describes the fracture system along which the veins are emplaced as having dips of 10 degrees to the Northwest and of about 30 degrees to the Southeast.

Figure 1 - Location of the Panasqueira mine (from Williams, 1985).

SAMPLING METHOD

The ore minerals, and in particular the wolframite crystals, occur very erratically as individual large crystals in the mining faces, and so chemical assay based on small samples would produce very many nil values and a few very high large ones. Taking this in consideration, a technique proposed and developed by Mendes (1958) of measuring the areas of wolframite crystals exposed in the sidewalls of mining faces has been used at the mine for several years to estimate the ore reserves.

The principle of the method is that the area of wolframite exposed represents a measure of the horizontally projected grade (i. e. weight of wolframite per horizontal unit area):

$$Grade = 10 * S\, d_w / l$$

where l is the length of the vein run, in metres, on a vein having S (cm^2) of wolframite measured and d_w (g/cm^3) is the specific gravity of wolframite. The formula is valid under the Delesse principle of stereology stating that the areal density of profiles on sections is a unbiased estimator of the volume density of structures assuming that the sections are taken on isotropic uniform random planes.

As Williams (1985) points out, the sections are not pre-selected in the Panasqueira mine on the basis of the expected value of the areal density - the exact location of the section is determined by the point of break of a fired round, and there is no evidence to suggest that the breaks occur preferentially at points of abnormal wolframite content. The sections can therefore be considered as randomly located with respect to the areal density of wolframite. The area of wolframite per unit of length can therefore be considered an unbiased measure of the product (accumulation) of the wolframite tenor and the vein width. This regionalized variable is the basis of the ore reserve estimation at the mine and geostatistical methods have been applied by Sousa and Soares (1982) and Williams (1985) for the local estimation of the wolframite grade in a block.

In order to measure this variable, the following method of evaluation is in use at the mine:

. Divide the face into standard measurement lengths (usually 2 m)
. Measure directly the dimensions of each wolframite crystal exposed at the face
. Calculate the total area of wolframite crystals exposed (cm^2)
. Divide by the overall face width in metres
. The result is the face grade in Kg/m^2 of stope.

DATA AND BASIC STATISTICS

Data are collected every 2 m along North-South cross-cuts "Panels" and East-West "Drives" in each level of the mine.

Figure 2 shows the sample disposition in level 2. In this level 2226 values were obtained for both variables: average vein width of the quartz and the total area of the wolframite crystals exposed every 2 meters ("pintas").

Histograms of both variables and the respective scatter-plot are shown in figures 3 to 5.

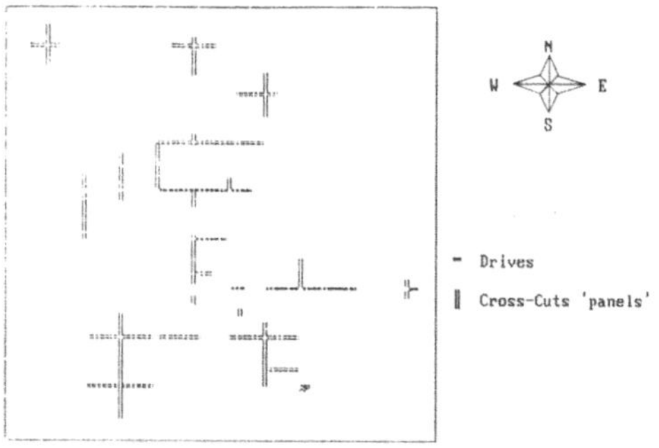

Figure 2 - Sample disposition in level 2.

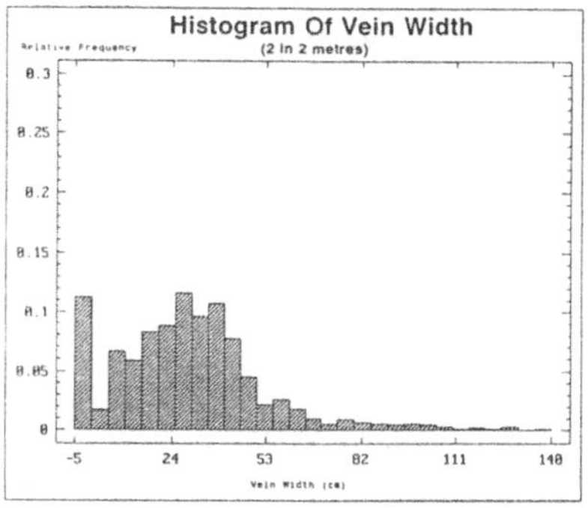

Figure 3 - Histogram of vein width

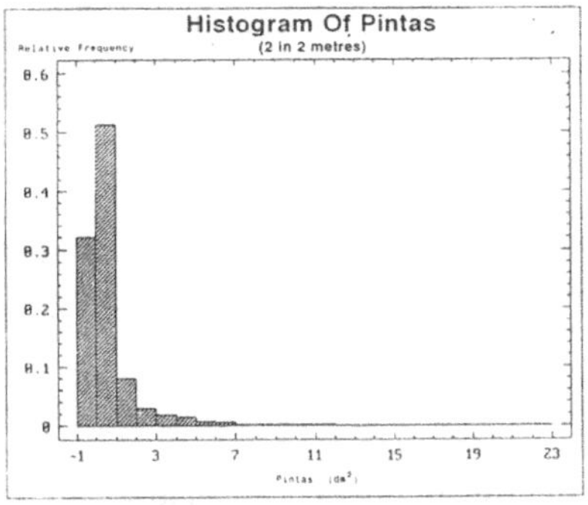

Figure 4 - Histogram of "pintas"

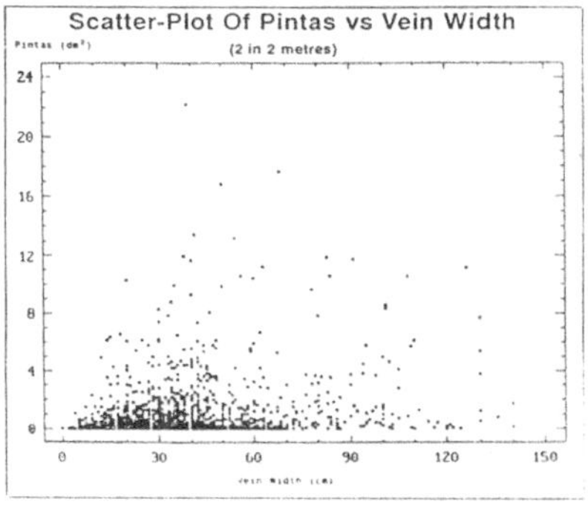

Figure 5 - Scatter-Plot of "pintas" vs. vein width

As it can be seen (figure 4) the variable "pintas" presents almost 30% of zeros and a very skewed distribution. The correlation coefficient between vein width and "pintas" is small (0.296), see also figure 5.

VARIOGRAPHY

Figures 6 to 8 show the variograms of both variables and the respective cross-variograms calculated along drives (direction 0°) and cross-cuts (direction 90°).

The variograms of the vein width have a small nugget, as expected, and two correlated structures. The first structure is isotropic with a range of 20 m and the second one somewhat anisotropic and oriented North-South (direction 90° of the cross-cuts).

The variograms of "pintas" have large nugget variances and also two structures: an isotropic one with a range of 5 m and an anisotropic one oriented East-West (azimuth 0° of the drivers). This second structure of the "pintas" variogram is not clearly defined in the North-South direction (azimuth 90°) (see figure 7b).

In order to have a better picture of the 2nd structure mainly for the "pintas" variogram, samples were grouped 2 in 2 thus constituting 4 m support samples.

Variograms and cross-variograms of these regularized variables are shown in figures 9 to 11. They exhibit the same behaviour as those on the 2 m support mainly in the second spatial structure.

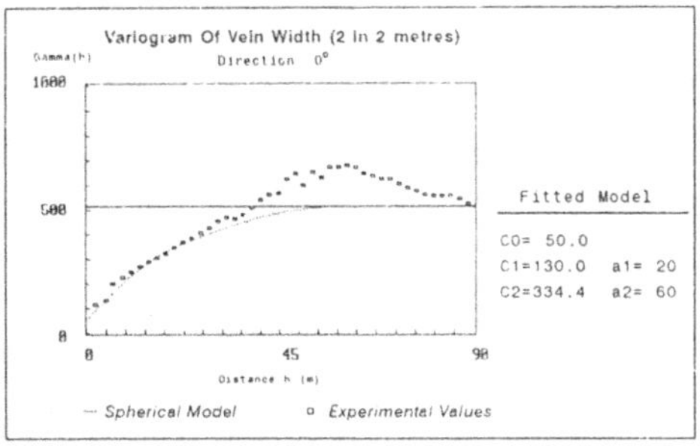

Figure 6a - Variograms of vein width (2 in 2 metres), direction 0°

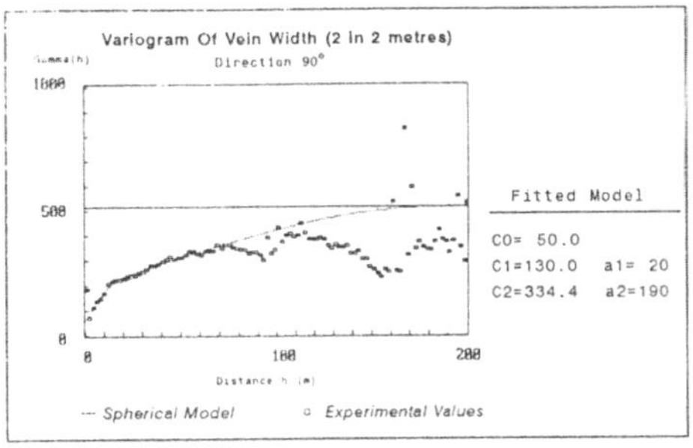

Figure 6b - Variograms of vein width (2 in 2 metres), direction 90°

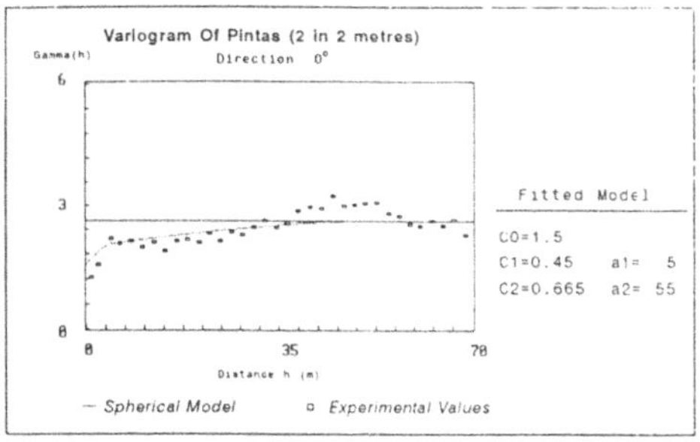

Figure 7a - Variograms of "Pintas" (2 in 2 metres), direction 0°

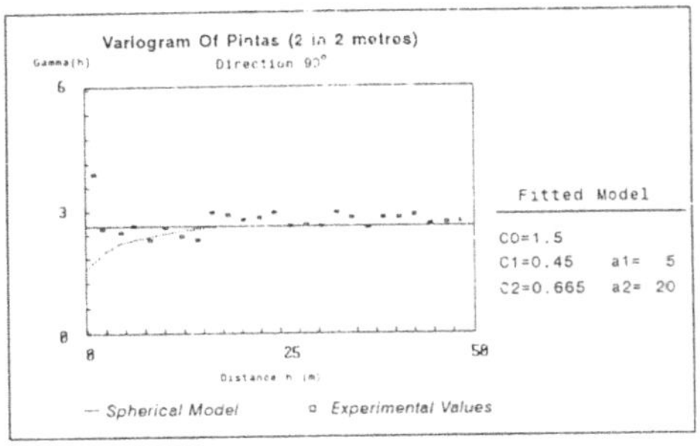

Figure 7b - Variograms of "Pintas" (2 in 2 metres), direction 90°

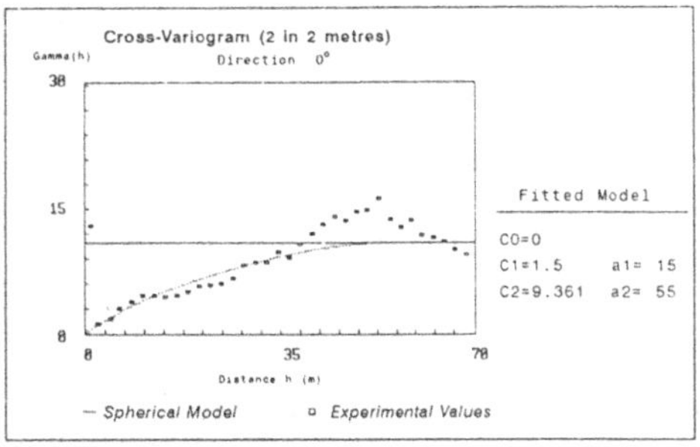

Figure 8a - Cross-variograms (2 in 2 metres) direction 0°

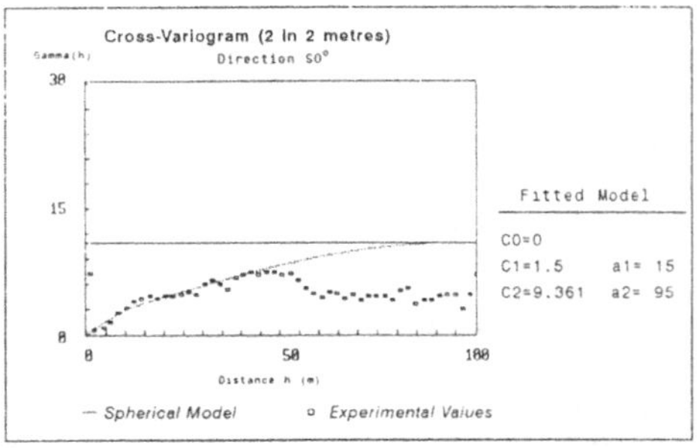

Figure 8b - Cross-variograms (2 in 2 metres), direction 90°

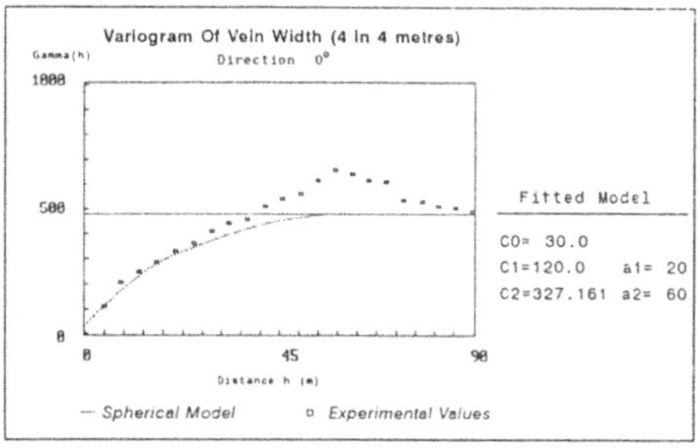

Figure 9a - Variogram of vein width (4 in 4 metres), direction 0°

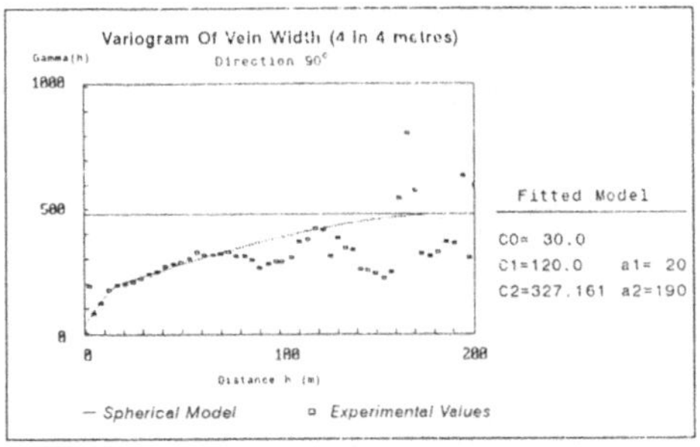

Figure 9b - Variogram of vein width (4 in 4 metres), direction 90°

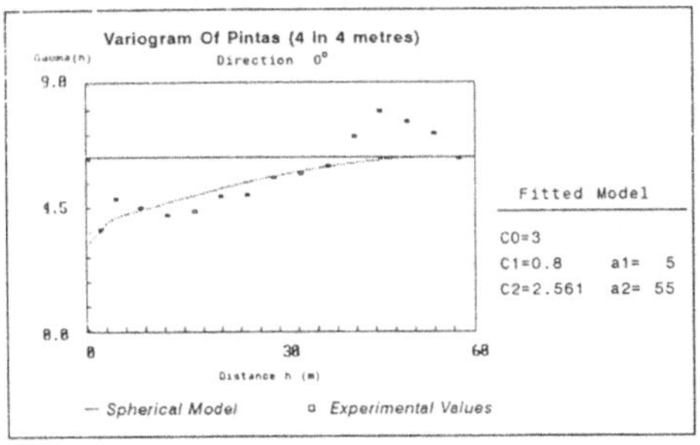

Figure 10a - Variogram of "Pintas" (4 in 4 metres), direction 0°

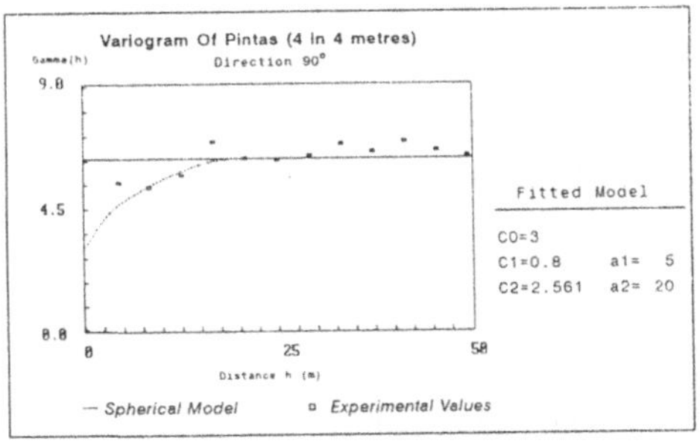

Figure 10b - Variogram of "Pintas" (4 in 4 metres), direction 90°

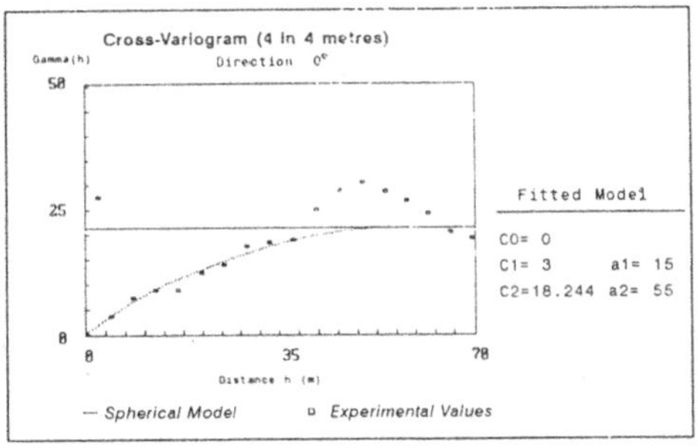

Figure 11a - Cross-variogram (4 in 4 metres) direction 0°

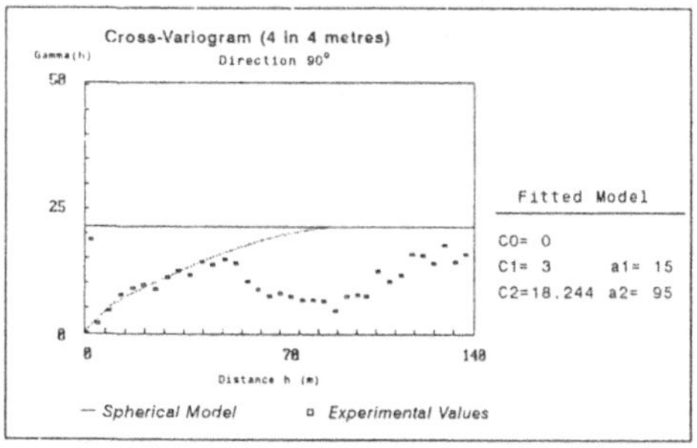

Figure 11b - Cross-variogram (4 in 4 metres), direction 90°

IMAGE ANALYSIS

To study the spatial distribution of the wolframite crystals in the mining faces more sensitively, images were taken and processed according as follows (Muge et al, 1990):

(i) - Image acquisition of the quartz vein (detection of the white "plateaus" in the image) and of the wolframite crystals (detection of dark zones with high contrast).

(ii) - Image treatment with identification of the quartz veins and wolframite crystals.

(iii) - Once the wolframite crystals were identified their number and area were easily measured. In particular, the weight of wolframite per horizontal unit area was calculated by measuring the total area occupied by the wolframite crystals. The average width of the quartz vein was also calculated by measuring the area occupied by it and dividing by the respective length.

Figure 12 presents an image of a vein showing wolframite crystals and other minerals. This image was processed and treated according to phases (i) and (ii) and the final image is presented in figure 13.

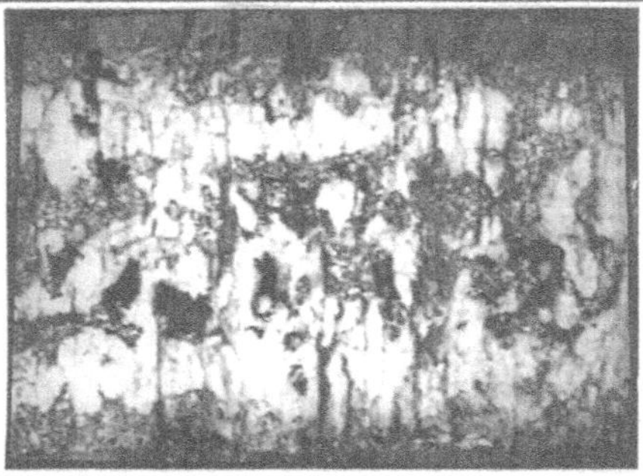

Figure 12 - Initial image of a vein

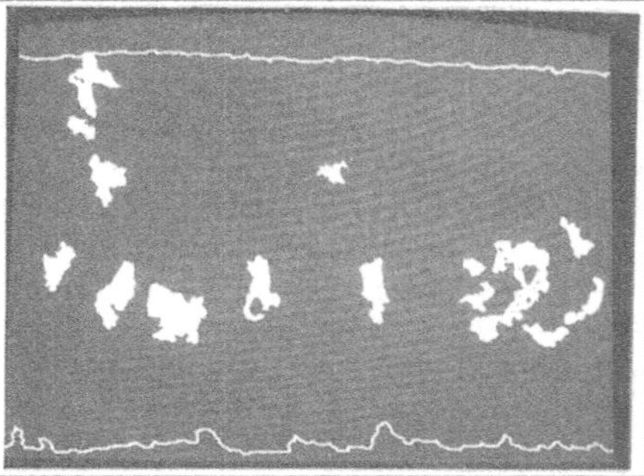

Figure 13 - Final image showing the quartz vein and wolframite crystals

Testing a Boolean model

Using the treated images, a Boolean model with Poisson polyhedron primary grains was tested for the distribution of the wolframite crystals (set X).In order to test this model two functions of X were calculated.

- The result of eroding the complementary set X^C (the pores) by a line segment:

$$Q(l) = Pr \{ l \subset X^c \}$$

- The covariance function of X^C :

$$C(h) = Pr \{ x \in X^c \wedge x+h \in X^c \}$$

In the case of a Boolean model (Serra, 1982), $Q(l)$ exhibits a negative exponential behaviour, which, in the specific case of Poisson polyhedron primary grains, has the following expression:

$$Q(l) = q \exp(-\pi \rho \theta' l) \tag{1}$$

Where: $q = \exp(-\theta')$ $\theta' = \frac{6\theta}{\pi^4 \rho^3}$

The two parameters θ and ρ are respectively the density of the Poisson process and the linear density of the Poisson polyhedron taken as a random set.
The covariogram of the Poisson polyhedron primary grains is the negative exponential:

$$K(h) = \left(\frac{6}{\pi^4 \rho^3}\right) \exp(-\pi \rho h) \tag{2}$$

and can be directly obtained through the covariance function of the set X (experimentally accessible):

$$C(h) = q^2 \exp[\theta K(h)] \tag{3}$$

Figure 14 shows that the log $Q(l)$ of the quartz vein (without the wolframite crystals) has a linear behaviour. Figure 15 presents the log of the geometric covariogram estimate obtained through expression (3) and shows that this function has a negative exponential behaviour thus corroborating the hypothesis that the wolframite crystals distribution can be modelled through a Boolean model with Poisson polyhedron primary grains.

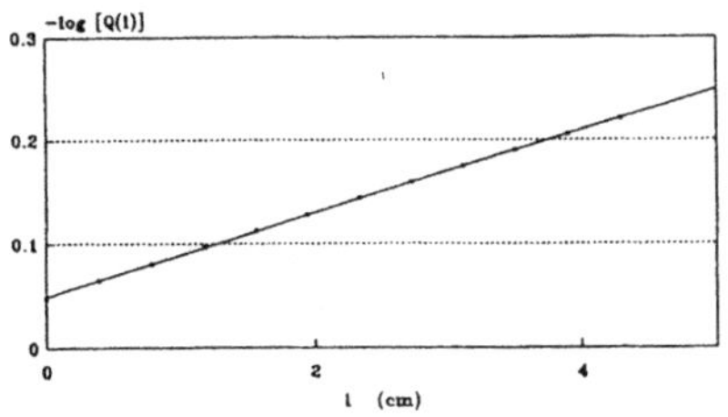

Figure 14 - Log of the linear erosion estimates of the pores (quartz vein without the wolframite crystals)

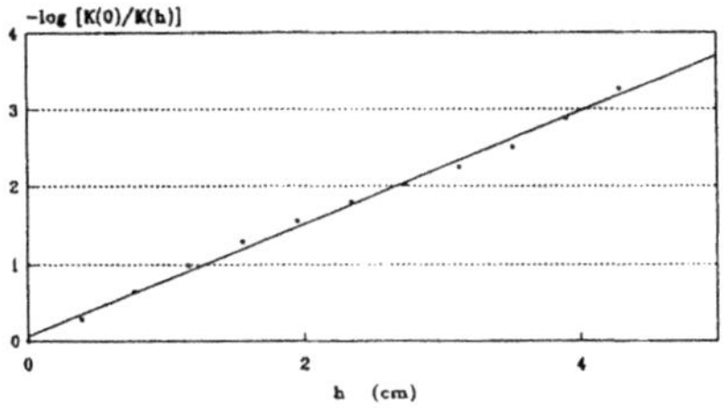

Figure 15 - Log of the geometric covariogram estimates of the wolframite crystals

Variography of the images

Using the above methods of image analysis , wolframite crystals were identified along 40 m of a mining face, as well as the hanging wall and footwall of the quartz veins. The surface of the crystals and the thickness of the quartz veins were measured each 7.8 cm and the respective variograms (figure 16 and figure 17), covariograms (figure 18) and correlation coefficients (figure 19) were calculated.

- The variogram of the vein width (figure 16) is very regular (null nugget effect) and is compatible with the model fitted for the 2 m support (see figure 6a)
- The variogram of "pintas" shows a smaller structure with a range of 50 cm (figure 17) responsible for the large nugget in the variogram for the 2 m support (figure 7a), and continues to reveal a large nugget (25%).
- The cross-variogram (figure 18) shows the same behaviour as the one calculated for 2 m support (see figure 8a)
- The correlation coefficient between these two variables is moderate (0.54)

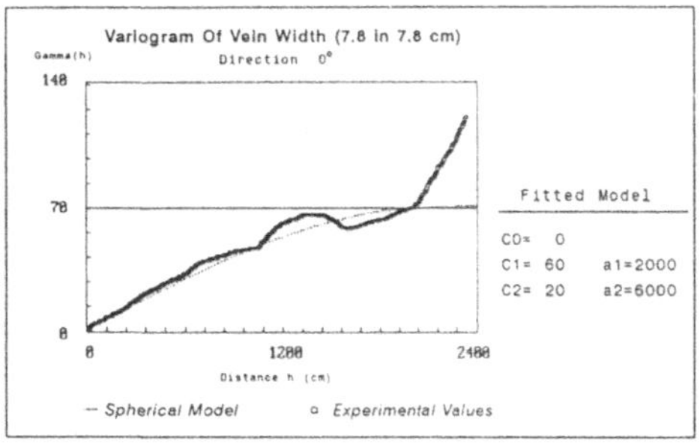

Figure 16 - Variogram of vein width (7.8 cm)

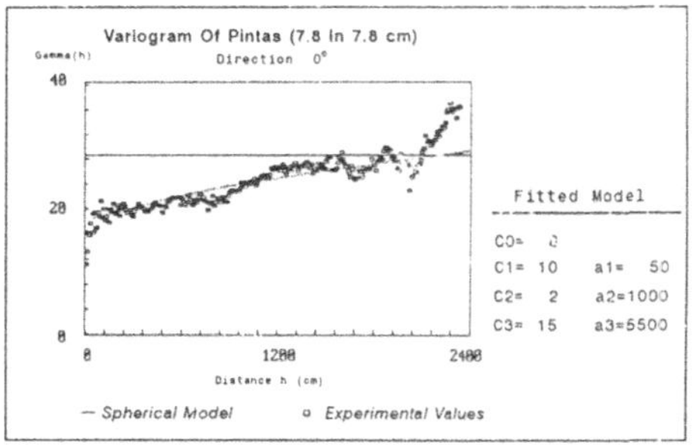

Figure 17 - Variogram of "pintas" (7.8 cm)

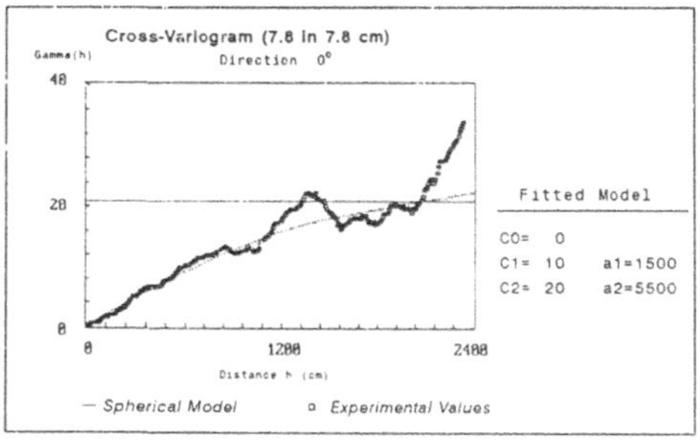

Figure 18 - Cross-Variogram (7.8 cm)

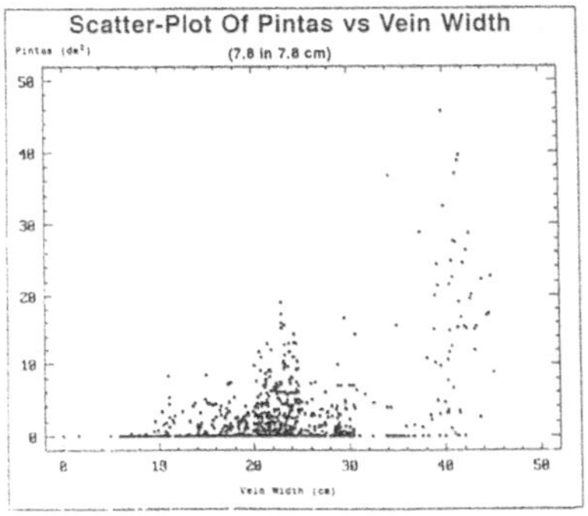

Figure 19 - Scatter-plot of "pintas" vs vein width

Cross-Correlation as a function of the support

The wolframite content and the vein width are weakly correlated when measured on the same sampling support. Nevertheless, taking into account that the distribution of the wolframite crystals can be modelled by a Boolean process, a strong correlation should be expected between the surface occupied by the quartz veins (proportional to its thickness) and the number of Poisson germs or primary grains (wolframite crystals) of the Boolean set.

To estimate the wolframite grade of large panels (100m x 100m) for long term planning, most of the available information consists of values of the thickness of the quartz veins cut by a D.D.H. campaign (owing to the erratic distribution of the wolframite crystals the drilling holes cut only the quartz veins). The study of cross-correlation of these two variables is therefore very important for planning.

Let us assume two regionalized variables X and Y whose correlation coefficient is:

$$\rho_{XY} = \frac{C_{XY}}{\sigma_X \sigma_Y} \tag{4}$$

where C_{XY} is the covariance between the two variables and σ_X^2 and σ_Y^2 are respective variances.

Let us now change the measurement support (l) of X and Y and study it influence on the correlation coefficient:

$$\rho_{X_lY_l} = \frac{\bar{C}_{X_lY_l}}{\sigma_{X_l}\sigma_{Y_l}} \tag{5}$$

Expression (5) can be rewritten in terms of the variogram:

$$\rho_{X_lY_l} = \frac{C_{XY}(0) - \bar{\gamma}_{X_lY_l}}{\sqrt{(C_X(0) - \bar{\gamma}_{X_l})(C_Y(0) - \bar{\gamma}_{Y_l})}} \tag{6}$$

where $C_{XY}(0)$, $C_X(0)$ and $C_Y(0)$ are respectively the covariance and the variances of X and Y.

Let us study the behaviour of $\rho_{X_lY_l}$ as a function of the linear size of the segment l in a domain where the variogram can be considered linear: $\gamma(l) = C_0 + \omega l$.
Expression (6) becomes:

$$\rho_{X_lY_l} = \frac{C_{XY}(0) - C_{0XY} - \omega_{XY} l/3}{\sqrt{(C_X(0) - C_{0X} - \omega_X l/3)(C_Y(0) - C_{0Y} - \omega_Y l/3)}} \tag{7}$$

And putting $K_{XY} = C_{XY}(0) - C_{0XY}$; $K_X = C_X(0) - C_{0X}$; $K_Y = C_Y(0) - C_{0Y}$

$$\rho_{X_lY_l} = \frac{K_{XY} - \omega_{XY} l/3}{\sqrt{(K_X - \omega_X l/3)(K_Y - \omega_Y l/3)}} \tag{8}$$

In order to study the sign of the derivative let us take only the first term of the quotient (the second term is by definition positive):

$$\frac{d\rho}{dl} = [\frac{1}{18}(\omega_X\omega_{XY}K_Y + \omega_Y\omega_{XY}K_X) - \frac{1}{9}\omega_x\omega_y K_{XY}]\, l - \\ - \frac{\omega_{XY}}{3}K_XK_Y + [(\frac{\omega_X}{6}K_Y + \frac{\omega_Y}{6}K_X)K_{XY}] \tag{9}$$

Dividing by $K = K_X K_Y K_{XY}$:

$$\frac{1}{K}\frac{d\rho}{dl} = [\frac{1}{18}(\frac{\omega_X}{K_X}\frac{\omega_{XY}}{K_{XY}} + \frac{\omega_Y}{K_Y}\frac{\omega_{XY}}{K_{XY}}) - \frac{1}{9}\frac{\omega_X}{K_X}\frac{\omega_Y}{K_Y}]\, l + \\ + \frac{1}{6}(\frac{\omega_X}{K_X} + \frac{\omega_Y}{K_Y}) - \frac{1}{3}\frac{\omega_{XY}}{K_{XY}} \tag{10}$$

The behaviour of the correlation coefficient depends on the relative contribution of the variogram and cross-variogram functions:

(i) If the corregionalization model is of intrinsic type:
$\frac{\omega_X}{K_X}=\frac{\omega_Y}{K_Y}=\frac{\omega_{XY}}{K_{XY}}$ then $\frac{d\rho}{dl}=0$ and so the correlation coefficient does not depend on the size of the support

(ii) If one of the variables is more irregular than the other $(\frac{\omega_Y}{K_Y}>\frac{\omega_X}{K_X}\approx\frac{\omega_{XY}}{K_{XY}})$, expression (10) can be simplified:

$$\frac{1}{K}\frac{d\rho}{dl}\approx\frac{1}{6}[\frac{\omega_Y}{K_Y}-\frac{\omega_X}{K_X}]-\frac{1}{18}\frac{\omega_x}{K_X}(\frac{\omega_Y}{K_Y}-\frac{\omega_X}{K_X})l \tag{11}$$

and dividing by $W=\frac{\omega_Y}{K_Y}-\frac{\omega_X}{K_X}>0$:

$$\frac{1}{K\,W}\frac{d\rho}{dl}\approx\frac{1}{6}-\frac{1}{18}\frac{\omega_X}{K_X}l \tag{12}$$

Expression (12) means that $\frac{d\rho}{dl}>0$ if $\omega_X l<3K_X$ which is always the case in the domain under study (linear behaviour of the variogram). Increasing the correlation coefficient under change of support can be observed in Figs 20 to 22. Figure 20 shows the values obtained from image analysis of 40 m of a mining face in which the support of the two variables has changed from 3.9 cm to several metres. Figures 21 and 22 were calculated using multiples of the 2 m support samples of level 2 respectively in the N-S and E-W directions.

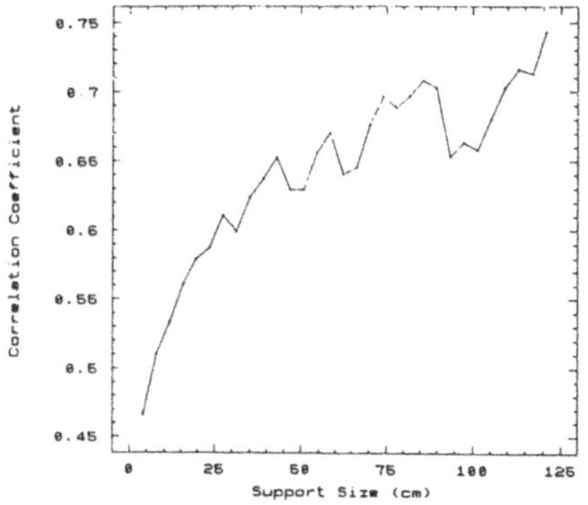

Figure 20 - Correlation coefficients obtained through image analysis

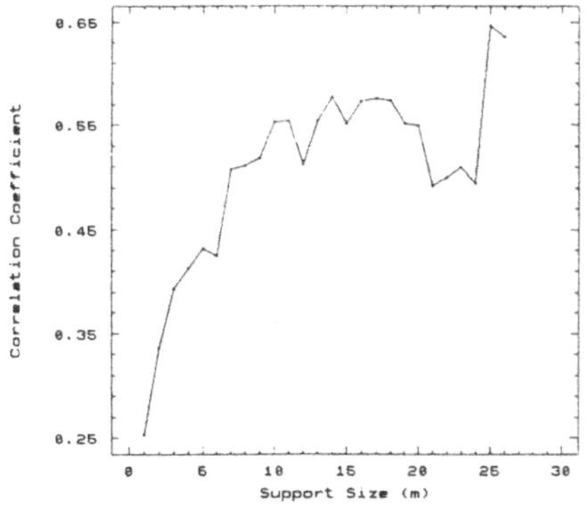

Figure 21 - Correlation coefficients for N-S direction

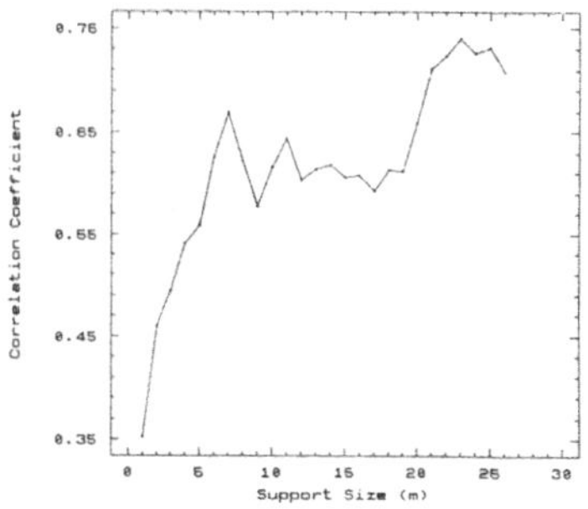

Figure 22 - Correlation coefficients
for E-W direction

Let us now study the behaviour of ρ when the measurement support (l) of only one variable (Y) changes. In this case expression (7) becomes:

$$\rho_{XY_l} = \frac{C_{XY}(0) - C_{0XY} - \omega_{XY}\, l/4}{\sqrt{C_X(0)\,(C_Y(0) - C_{0Y} - \omega_Y\, l/3)}} \tag{13}$$

putting $K_{XY} = C_{XY}(0) - C_{0XY}$; $K_X = C_X(0)$; $K_Y = C_Y(0) - C_{0Y}$:

$$\rho_{XY_l} = \frac{K_{XY} - \omega_{XY}\, l/4}{\sqrt{K_X(K_Y - \omega_Y\, l/3)}} \tag{14}$$

Note that the case under study supposes that the variable X is measured in the middle of the linear segment l where Y is calculated. The study of the derivative gives:

$$\frac{d\rho}{dl} = \frac{1}{6}\frac{\omega_Y}{K_Y} - \frac{1}{4}\frac{\omega_{XY}}{K_{XY}} + \frac{1}{24}\left(\frac{\omega_{XY}}{K_{XY}}\frac{\omega_Y}{K_Y}\right) l \tag{15}$$

The linear term is always positive, and the constant one depends on the relative contributions of the variogram of Y and the cross-variogram:

(i) If the model is of intrinsic type: $\left(\frac{d\rho}{dl}\right)_{l=0} = -\frac{1}{12}\frac{\omega_Y}{K_Y} < 0$
and, in general:

$$\frac{d\rho}{dl} = -\frac{1}{12}\frac{\omega_Y}{K_Y} + \frac{1}{24}\left(\frac{\omega_Y}{K_Y}\right)^2 l \tag{16}$$

Expression (16) means that $\frac{d\rho}{dl} < 0$ if $\omega_Y l < 2K_Y$ which is always the case in the domain under study. This means that in the intrinsic model the correlation coefficient always decreases with the increasing of the support l.

(ii) If the variable Y is more irregular than X: $\frac{\omega_Y}{K_Y} > \frac{\omega_{XY}}{K_{XY}}$, expression (15) gives:

$$\left(\frac{d\rho}{dl}\right)_{l=0} = \frac{1}{6}\frac{\omega_Y}{K_Y} - \frac{1}{4}\frac{\omega_{XY}}{K_{XY}} \quad (17)$$

Expression (17) means that $\left(\frac{d\rho}{dl}\right)_{l=0} > 0$ if $\frac{\omega_Y}{K_Y} > \frac{3}{2}\frac{\omega_{XY}}{K_{XY}}$.
For $l \neq 0$ the positive linear term of expression (15) will contribute to the increasing of ρ.
As l increases it is expectable that other imbricated structures will change the relation between $\frac{\omega_Y}{K_Y}$ and $\frac{\omega_{XY}}{K_{XY}}$. This is the case when there is a second more regular structure with: $\frac{\omega_Y}{K_Y} < \frac{3}{2}\frac{\omega_{XY}}{K_{XY}}$. In this case, it is expectable a decrease in the coefficient of correlation and, therefore, the existence of a maximum value for ρ. Figure 23 shows the values obtained from image analysis of 40 m of a mining face in which the support of the less regular variable ("pintas") has changed from 3.9 cm to several metres. Figures 24 and 25 were obtained using multiples of the 2 m support samples of level 2 for "pintas", maintaining the other variable (vein width) centred in the initial 2 m support.

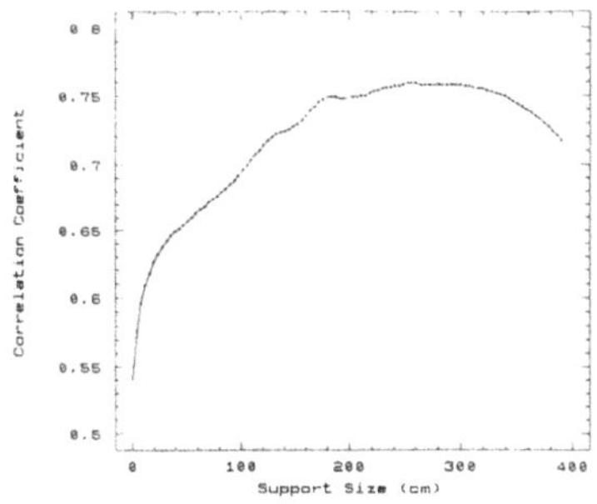

Figure 23 - Correlation coefficients obtained through image analysis

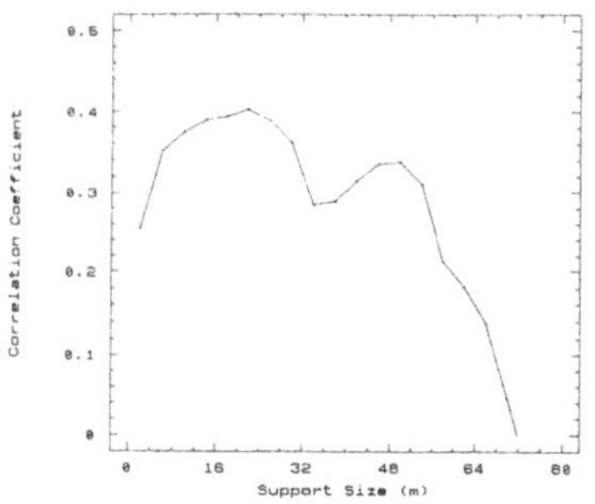

Figure 24 - Correlation coefficients for N-S direction

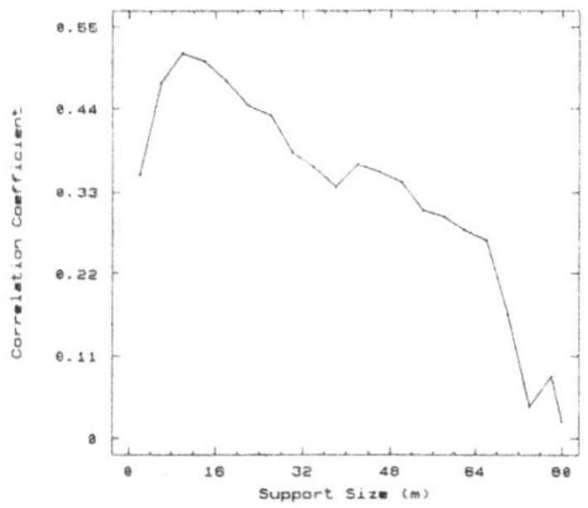

Figure 25 - Correlation coefficients for the E-W direction

CONCLUSIONS

The main conclusion of this research work refers to the possibility of using the vein width for the estimation of the wolframite grade of large panels for long-term planning. The fact that the correlation between the two variables increases when the most irregular one ("pintas") is taken in larger supports can be taken into consideration in a cokriging system of equations with variables measured in different supports. This fact opens the possibility of estimating wolframite grade in large panels using all information available: vein width cut by DDH and "pintas" measured along mining faces.

REFERENCES

Kelly W.C., Rye R.O. (1979) - Geologic, fluid inclusion and stable isotopic studies of the tin-tungesten deposits of Panasqueira (Portugal), *Economic Geology*, Vol.74, n° 8, pp 1721-1822.
Mendes F. (1958) - Determinação do valor dos filões - Processos usados nas minas da Panasqueira, *Técnica*, AEIST, Lisboa.
Muge F., Pina P., Guimarães C. (1990) - Application of Mathematical Morphology Based Image Analysis Techniques to Mine Planning, *2nd Codata Conference on Geomathematics*, Leeds, Sept. 1990 (to be published in *Sciences de la Terre*).
Serra J. (1982) - *Image Analysis and Mathematical Morphology*, Academic Press, London.
Sousa A.J., Soares A. (1982) - A Estimação Local de Reservas. Ensaio Metodológico de Aplicação à Volframite do Jazigo da Panasqueira, *Proceedings of the 3rd Annual Meeting of the Society of Mining Engineers of Portugal*.
Thadeu D. (1979) - Le gisement stano-wolframifère de Panasqueira (Portugal), *Chronique de la Recherche Minière*, n° 450.
Williams C. (1985) - *Ore Estimation and Mine Planning at Panasqueira, Portugal*, Ph.D. Thesis, Department of Mineral Resources Engineering, Royal School of Mines, London.

INDEX

Quantitative Geology and Geostatistics

Series Editor: F.M. Gradstein (*Canada*)

1. F. M. Gradstein, F. P. Agterberg, J. C. Brower and W. S. Schwarzacher: *Quantitative Stratigraphy*. 1985 ISBN 90-277-2116-5
2. G. Matheron and M. Armstrong (Eds.): *Geostatistical Case Studies*. 1987 ISBN 1-55608-019-0
3. *Cancelled*
4. M. Armstrong (Ed.): *Geostatistics*. Proceedings of the 3rd International Geostatistics Congress, held in Avignon, France (1988), 2 volumes. 1989 Set ISBN 0-7923-0204-4
5. A. Soares (Ed.): *Geostatistics Tróia '92*, 2 volumes. 1993 Set ISBN 0-7923-2157-X